Encyclopedia of Biotechnology in Agriculture and Food

CRC Press
Taylor & Francis Group
6000 Broken Sound Parkway NW, Suite 300
Boca Raton, FL 33487-2742

© 2011 by Taylor and Francis Group, LLC
CRC Press is an imprint of Taylor & Francis Group, an Informa business

No claim to original U.S. Government works

Printed in the United States of America on acid-free paper
10 9 8 7 6 5 4 3 2 1

International Standard Book Number: 978-0-8493-5027-6 (Hardback)

This book contains information obtained from authentic and highly regarded sources. Reasonable efforts have been made to publish reliable data and information, but the author and publisher cannot assume responsibility for the validity of all materials or the consequences of their use. The authors and publishers have attempted to trace the copyright holders of all material reproduced in this publication and apologize to copyright holders if permission to publish in this form has not been obtained. If any copyright material has not been acknowledged please write and let us know so we may rectify in any future reprint.

Except as permitted under U.S. Copyright Law, no part of this book may be reprinted, reproduced, transmitted, or utilized in any form by any electronic, mechanical, or other means, now known or hereafter invented, including photocopying, microfilming, and recording, or in any information storage or retrieval system, without written permission from the publishers.

For permission to photocopy or use material electronically from this work, please access www.copyright.com (http://www.copyright.com/) or contact the Copyright Clearance Center, Inc. (CCC), 222 Rosewood Drive, Danvers, MA 01923, 978-750-8400. CCC is a not-for-profit organization that provides licenses and registration for a variety of users. For organizations that have been granted a photocopy license by the CCC, a separate system of payment has been arranged.

Trademark Notice: Product or corporate names may be trademarks or registered trademarks, and are used only for identification and explanation without intent to infringe.

Library of Congress Cataloging-in-Publication Data

Encyclopedia of biotechnology in agriculture and food / editors, Dennis R. Heldman, Matthew B. Wheeler, Dallas G. Hoover.
  p. ; cm.
  Includes bibliographical references and index.
  ISBN 978-0-8493-5027-6 (hardcover : alk. paper)
  1. Agricultural biotechnology--Encyclopedias. 2. Food--Biotechnology--Encyclopedias. I. Heldman, Dennis R. II. Wheeler, Matthew B. III. Hoover, Dallas G. IV. Title.
  [DNLM: 1. Agriculture--Encyclopedias--English. 2. Biotechnology--Encyclopedias--English. 3. Food Technology--Encyclopedias--English. S 494.5.B563 E56 2011]

S494.5.B563E53 2011
630--dc22
                                                                                        2010019974

Visit the Taylor & Francis Web site at
http://www.taylorandfrancis.com

and the CRC Press Web site at
http://www.crcpress.com

Encyclopedia of Biotechnology in Agriculture and Food

Edited by
Dennis R. Heldman
Matthew B. Wheeler
Dallas G. Hoover

CRC Press
Taylor & Francis Group
Boca Raton London New York

CRC Press is an imprint of the
Taylor & Francis Group, an **informa** business

Encyclopedia of Biotechnology in Agriculture and Food

Editor-in-Chief

Dennis R. Heldman, Ph.D.
Heldman Associates, Mason, Ohio, U.S.A.

Topical Editors

Dallas G. Hoover, Ph.D.
University of Delaware, Newark, Delaware, U.S.A.

Matthew B. Wheeler, Ph.D.
University of Illinois, Urbana, Illinois, U.S.A.

Editorial Advisory Board

Carl A. Adams
Diversified Laboratory Testing, LLC, Mounds View, Minnesota, U.S.A.

Anne Bridges
Medallion Laboratories, Minneapolis, Minnesota, U.S.A.

Yves Bertheau
National Institute for Agronomic Research (INRA), Versailles, France

Geoffrey Dahl
University of Florida, Gainesville, Florida, U.S.A.

Adi B. Damania
University of California–Davis, Davis, California, U.S.A.

Nicki Engeseth
University of Illinois, Urbana, Illinois, U.S.A.

David W. Everett
University of Otago, Dunedin, New Zealand

G. David Grothaus
Pioneer Hi-Bred, Johnson, Iowa, U.S.A.

Peter J. Hansen
University of Florida, Gainesville, Florida, U.S.A.

Bob Hutkins
University of Nebraska, Lincoln, Nebraska, U.S.A.

Rolf D. Joerger
University of Delaware, Newark, Delaware, U.S.A.

Lokesh Joshi
University of Ireland, Galway, Ireland

Shawn Kaeppler
University of Wisconsin, Madison, Wisconsin, U.S.A.

Dietrich Knorr
Berlin University of Technology, Berlin, Germany

Huub L.M. Lelieveld
Unilever Research (Retired), Bilthoven, The Netherlands

Charles R. Long
Texas A&M University, College Station, Texas, U.S.A.

Calvin Qualset
University of California–Davis, Davis, California, U.S.A.

Viswambharan Sarasan
Royal Botanical Gardens, London, U.K.

George E. Seidel, Jr.
Colorado State University, Fort Collins, Colorado, U.S.A.

Raymond D. Shillito
BioScience Division, Bayer CropScience, Research Triangle Park, North Carolina, U.S.A.

Kimberly Magin Sutter
Monsanto Company, St. Louis, Missouri, U.S.A.

Bart Weimer
University of California-Davis, Davis, California, U.S.A.

Martin Wiedmann
Cornell University, Ithaca, New York, U.S.A.

Garrison Wilkes
University of Massachusetts, Boston, Massachusetts, U.S.A.

Curtis R. Youngs
Iowa State University, Ames, Iowa, U.S.A.

Contributors

Emmanuel Ohene Afoakwa / *Nutrition and Food Science, University of Ghana, Legon-Accra, Ghana*
Samuel E. Aggrey / *Department of Poultry Science, University of Georgia, Athens, Georgia, U.S.A.*
Sanjeev Anand / *Department of Dairy Science, South Dakota State University, Brookings, South Dakota, U.S.A.*
G.S. Anisha / *Department of Zoology, Government College, Kerala, India*
P. Anthony / *School of Biosciences, Plant Science Division, University of Nottingham, Loughborough, U.K.*
Gavin Ash / *School of Agriculture, Charles Sturt University, Wagga Wagga, New South Wales, Australia*
Neela Badrie / *Department of Food Production, University of the West Indies, St. Augustine, Trinidad and Tobago*
Ana Balasa / *Department of Food Biotechnology and Food Process Engineering, Berlin University of Technology, Berlin, Germany*
Charles Bamforth / *Department of Food Science and Technology, University of California–Davis, Davis, California, U.S.A.*
Manas R. Banerjee / *Research and Development Division, Brett-Young Seeds Limited, Winnipeg, Manitoba, Canada*
N. Bansal / *Department of Food and Nutritional Sciences, University College Cork, Cork, Ireland*
Pietro S. Baruselli / *Department of Animal Reproduction, University of Sao Paolo, Sao Paolo, Brazil*
D.K. Becker / *School of Life Sciences, Queensland University of Technology, Brisbane, Queensland, Australia*
Alaa El-Din A. Bekhit / *Department of Food Science, University of Otago, Dunedin, New Zealand*
Martin Bell / *Department of Food Science, University of Otago, Dunedin, New Zealand*
Alan Andrew Berryman / *Department of Entomology, Washington State University, Pullman, Washington, U.S.A.*
Yves Bertheau / *National Institute for Agronomic Research (INRA), Versailles, France*
Luciana Relly Bertolini / *Health Sciences Center, School of Medicine, University of Fortaleza, Fortaleza, Brazil*
Marcelo Bertolini / *Health Sciences Center, School of Medicine, University of Fortaleza, Fortaleza, Brazil*
Marcelo Bertolini / *Department of Food and Animal Sciences, School of Veterinary Medicine, Santa Catarina State University, Fortaleza, Brazil*
Kaustubh Bhalerao / *Department of Agricultural and Biological Engineering, University of Illinois, Urbana, Illinois, U.S.A.*
John Birch / *Department of Food Science, University of Otago, Dunedin, New Zealand*

James A. Birchler / Division of Biological Sciences, University of Missouri, Columbia, Missouri, U.S.A.

P.S. Bisen / Institute of Microbiology and Biotechnology, Barkatullah University, Bhopal, India

Elaine P. Black / Department of Food and Nutritional Sciences, University College Cork, Cork, Ireland

Gabriel A. Bó / Institute of Animal Reproduction, Cordoba (IRAC), Córdoba, Argentina

Jordi Giné Bordonaba / Department of Food Security and Environmental Health, Cranfield University, Bedfordshire, U.K.

Joseph J. Bozell / Biomass Chemistry Laboratories, Forest Products Center, University of Tennessee, Knoxville, Tennessee, U.S.A.

Debora Brand / Bioprocess Engineering and Biotechnology Division, Federal University of Paraná, Curitiba, Brazil

Jeff Broadbent / Department of Nutrition and Food Sciences, Utah State University, Logan, Utah, U.S.A.

Rodney J. Brown / College of Life Sciences, Brigham Young University, Provo, Utah, U.S.A.

Christine M. Bruhn / Center for Consumer Research, Department of Food Science and Technology, University of California–Davis, Davis, California, U.S.A.

Hikmet Budak / Faculty of Engineering and Natural Science, Biological Science and Bioengineering Program, Sabanci University, Istanbul, Turkey

Jennifer Lee Busto / Department of Plant and Environmental Protection Sciences, University of Hawaii at Manoa, Honolulu, Hawaii, U.S.A.

Javier Carballo / Food Technology Area, University of Vigo, Ourense, Spain

Kitty Cardwell / International Institute of Tropical Agriculture (IITA), Cotonou, Benin

Elaine Carnevale / College of Veterinary Medicine and Biomedical Sciences, Colorado State University, Fort Collins, Colorado, U.S.A.

M.G. Carter / Center for Regenerative Biology, University of Connecticut, Storrs, Connecticut, U.S.A.

D.E. Casanova / Faculty of Veterinary Science, National University of Buenos Aires, Buenos Aires, Argentina

Salvatore Ceccarelli / Biodiversity and Integrated Gene Management, International Center for Agricultural Research in the Dry Areas (ICARDA), Aleppo, Syria

Paul Chambers / Australian Wine Research Institute, Adelaide, South Australia, Australia

Claude P. Champagne / Food Research and Development Centre, Agriculture and Agri-Food Canada, Saint-Hyacinthe, Quebec, Canada

Haiqiang Chen / Department of Animal and Food Sciences, University of Delaware, Newark, Delaware, U.S.A.

Naveen Chikthimmah / Department of Food Science, Pennsylvania State University, University Park, Pennsylvania, U.S.A.

Paul Christakopoulos / BIOtechMASS Unit, Biotechnology Laboratory, School of Chemical Engineering, National Technical University of Athens, Athens, Greece

Jae-Kun Chun / Department of Food and Animal Biotechnology, College of Agriculture and Life Sciences, Seoul National University, Seoul, Korea

Sherrie Clark / Department of Veterinary Clinical Medicine, University of Illinois, Urbana, Illinois, U.S.A.

A. Conte / Institute for Biotechnology Research and Applications for Security and Promotion of Local Products and Quality (BIOAGROMED), University of Foggia, Foggia, Italy

Contributors

Tim Coolbear / *Fonterra Research Centre, Fonterra Cooperative Group Limited, Palmerston North, New Zealand*

Christopher Curtin / *Australian Wine Research Institute, Adelaide, South Australia, Australia*

Abhaya Dandekar / *Department of Plant Sciences, University of California–Davis, Davis, California, U.S.A.*

Michael R. Davey / *Department of Life Science, University of Nottingham, Nottingham, U.K.*

Howard V. Davies / *Department of Science Co-ordination, Scottish Crop Research Institute, Invergowrie, U.K.*

John Davison / *National Institute for Agronomic Research (INRA), Versailles, France*

Marzia De Giacomo / *Department of Veterinary Public Health and Food Safety, GMOs and Mycotoxins Unit, National Institute of Health (ISS), Rome, Italy*

Graciela Font de Valdez / *Lactobacillus Reference Center, National Council of Scientific and Technical Research (CERELA-CONICET), San Miguel de Tucumán, Argentina*

M.A. Del Nobile / *Department of Food Science, University of Foggia, Foggia, Italy*

Marie-Laure Delabre / *Fonterra Research Centre, Fonterra Cooperative Group Limited, Palmerston North, New Zealand*

F.A. di Croce / *Department of Animal Science, Institute of Agriculture, University of Tennessee, Knoxville, Tennessee, U.S.A.*

Andras Dinnyes / *BioTalentum Ltd., Godollo, Hungary*

John Dodds / *Dodds & Associates, Washington, District of Columbia, U.S.A.*

Hortense Dodo / *Department of Food and Animal Science, Alabama A&M University, Huntsville, Alabama, U.S.A.*

Doris H. D'Souza / *Department of Food Science and Technology, University of Tennessee, Knoxville, Tennessee, U.S.A.*

F. Du / *Center for Regenerative Biology, University of Connecticut, Storrs, Connecticut, U.S.A.*

Maret du Toit / *Department of Viticulture and Oenology, Institute for Wine Biotechnology, Stellenbosch University, Matieland, South Africa*

Stephen O. Duke / *Natural Products Utilization Research Unit (NPURU), U.S. Department of Agriculture/Agricultural Research Service (USDA/ARS), University, Mississippi, U.S.A.*

David A. Dunn / *Department of Pathobiology, College of Veterinary Medicine, Auburn University, Auburn, Alabama, U.S.A.*

Tri Duong / *Department of Poultry Science, Texas A&M University, College Station, Texas, U.S.A.*

J.L. Edwards / *Department of Animal Science, Institute of Agriculture, University of Tennessee, Knoxville, Tennessee, U.S.A.*

Matthew Escobar / *Department of Biological Sciences, California State University San Marcos, San Marcos, California, U.S.A.*

Edward R. Farnworth / *Food Research and Development Centre, Agriculture and Agri-Food Canada, Saint-Hyacinthe, Quebec, Canada*

Harry J. Flint / *Rowett Research Instutute, Aberdeen, U.K.*

William L. Flowers / *Department of Animal Science, North Carolina State University, Raleigh, North Carolina, U.S.A.*

C. Robert Flynn / *Arizona State University, Tempe, Arizona, U.S.A.*

Sonia Fonseca / *Food Technology Area, University of Vigo, Ourense, Spain*

Maria C. Garcia Fontán / *Food Technology Area, University of Vigo, Ourense, Spain*

Immaculada Franco / *Food Technology Area, University of Vigo, Ourense, Spain*
Brenda Frick / *Department of Plant Sciences, University of Saskatchewan, Saskatoon, Saskatchewan, Canada*
José Antonio Gabaldón-Hernández / *National Technology Center for Preservatives and Food, Molina de Segura, Spain*
S.R. Gao / *Center for Regenerative Biology, University of Connecticut, Storrs, Connecticut, U.S.A.*
Hugo S. Garcia / *Food Investigation Unit (UNIDA), Technological Institute of Veracuz, Veracruz, Mexico*
Duane L. Garner / *GametoBiology Consulting, Graeagle, California, U.S.A.*
Pierre Gélinas / *Food Research and Development Centre, Agriculture and Agri-Food Canada, Saint-Hyacinthe, Quebec, Canada*
Robert L. Gilbertson / *Department of Plant Pathology, University of California–Davis, Davis, California, U.S.A.*
Giorgio Giraffa / *Agricultural Research Council, Research Center for Forage and Dairy Productions (CRA-FLC), Lodi, Italy*
Robert A. Godke / *Department of Animal Sciences, Louisiana State University Agricultural Center, Baton Rouge, Louisiana, U.S.A.*
Janet R. Gorst / *Centre for Amenity and Environmental Horticulture, Benson Micropropagation, Brisbane, Queensland, Australia*
Ragini Gothalwal / *Institute of Microbiology and Biotechnology, Barkatullah University, Bhopal, India*
Simon Gowen / *Department of Agriculture, University of Reading, Reading, U.K.*
Jonathan Gressel / *Department of Plant Sciences, Weizmann Institute of Science, Rehovot, Israel*
M.N. Nierop Groot / *Wageningen UR Food and Biobased Research, Wageningen, The Netherlands*
Ann E. Hajek / *Department of Entomology, Cornell University, Ithaca, New York, U.S.A.*
Mee-Jung Han / *Department of Chemical and Biomolecular Engineering, Dongyang University, Gyeongbuk, South Korea*
Jaap C. Hanekamp / *Roosevelt Academy, Middelburg, and HAN-Research, Zoetermeer, The Netherlands*
John F. Hasler / *Bonner Peak Ranch, Laporte, Colorado, U.S.A.*
Ashraf N. Hassan / *Dairy Science, South Dakota State University, Brookings, South Dakota, U.S.A.*
Clair Hicks / *Department of Animal and Food Sciences, University of Kentucky, Lexington, Kentucky, U.S.A.*
Charles G. Hill, Jr. / *Department of Chemical Engineering, University of Wisconsin-Madison, Madison, Wisconsin, U.S.A.*
Kirsten A. Hirneisen / *Department of Animal and Food Sciences, University of Delaware, Newark, Delaware, U.S.A.*
Heikki Hokkanen / *Department of Applied Biology, University of Helsinki, Helsinki, Finland*
David Honys / *Institute of Experimental Botany, Academy of Sciences of the Czech Republic, Prague, Czech Republic*
Dallas Hoover / *Department of Animal and Food Sciences, University of Delaware, Newark, Delaware, U.S.A.*

Louis-Marie Houdebine / *Development and Reproduction Biology, National Institute for Agronomic Research, Jouy en Josas, France*

Marjorie A. Hoy / *Department of Entomology and Nematology, University of Florida, Gainesville, Florida, U.S.A.*

Hei-Ti Hsu / *Floral and Nursery Plants Research, U.S. Department of Agriculture/Agricultural Research Service (USDA/ARS), Beltsville, Maryland, U.S.A.*

Thom Huppertz / *NIZO Food Research, Ede, The Netherlands*

Charles Hurburgh / *Department of Agricultural and Biosystems Engineering, Iowa State University, Ames, Iowa, U.S.A.*

Robert W. Hutkins / *Department of Food Science and Technology, University of Nebraska, Lincoln, Nebraska, U.S.A.*

W. M. Ingledew / *Lallemand Ethanol Technology, Parksville, British Columbia, Canada*

Anna Janositz / *Department of Food Biotechnology and Food Process Engineering, Berlin University of Technology, Berlin, Germany*

Lee-Ann Jaykus / *Department of Food, Bioprocessing, and Nutrition Sciences, North Carolina State University, Raleigh, North Carolina, U.S.A.*

Jesse M. Jaynes / *Department of Biological and Physical Sciences, Kennesaw State University, Kennesaw, Georgia, U.S.A.*

Rolf D. Joerger / *Department of Animal and Food Sciences, University of Delaware, Newark, Delaware, U.S.A.*

Rojan P. John / *National Institute for the Scientific Research of the Water Ground Environment, Quebec City, Quebec, Canada*

Peter J.H. Jones / *Richardson Centre for Functional Foods and Nutraceuticals, University of Manitoba, Winnipeg, Manitoba, Canada*

Lokesh Joshi / *National Centre for Biomedical Engineering Science, National University of Ireland, Galway, Ireland*

Nicholas Kalaitzandonakes / *Department of Agricultural Economics, University of Missouri, Columbia, Missouri, U.S.A.*

Jaison Karavally / *Department of Food Science, Pennsylvania State University, University Park, Pennsylvania, U.S.A.*

Renuka Karuppuswamy / *Department of Food Science and Technology, University of New South Wales, Sydney, New South Wales, Australia*

James D. Kaufman / *University of Missouri, Columbia, Missouri, U.S.A.*

Alan Kelly / *Department of Food and Nutritional Sciences, University College Cork, Cork, Ireland*

George Kennedy / *Department of Entomology, North Carolina State University, Raleigh, North Carolina, U.S.A.*

Peter Kerr / *Vertebrate BioControl Cooperative Research Center, Wildlife and Ecology, Commonwealth Scientific and Industrial Research Organisation (CSIRO), Canberra, Australian Capital Territory, Australia*

Jay P. Kesan / *College of Law, University of Illinois, Urbana, Illinois, U.S.A.*

Kalmia E. Kniel / *Department of Animal and Food Sciences, University of Delaware, Newark, Delaware, U.S.A.*

Dietrich Knorr / *Department of Food Biotechnology and Food Process Engineering, Berlin University of Technology, Berlin, Germany*

Koffi Konan / *Department of Food and Animal Science, Alabama A&M University, Huntsville, Alabama, U.S.A.*

Duane Kraemer / *Department of Veterinary Physiology and Pharmacology, Texas A&M University, College Station, Texas, U.S.A.*

David A. Kudrna / *Department of Plant Sciences, University of Arizona, Tucson, Arizona, U.S.A.*

Monto H. Kumagai / *XTremeSignPost, Inc., Davis, California, U.S.A.*

Steve Labrie / *Department of Food Science and Nutrition, University of Laval, Quebec City, Quebec, Canada*

Kathryn D. Lardizabal / *Monsanto Company, Davis, California, U.S.A.*

Jean Guy LeBlanc / *Lactobacillus Reference Center, National Council of Scientific and Technical Research (CERELA-CONICET), San Miguel de Tucumán, Argentina*

Chang Yong Lee / *Department of Food Science and Technology, Cornell University, Geneva, New York, U.S.A.*

Sang Yup Lee / *Department of Chemical and Biomolecular Engineering, Korea Institute of Science and Technology (KAIST), Daejeon, South Korea*

Xingen Lei / *Department of Animal Science, Cornell University, Ithaca, New York, U.S.A.*

Huub L.M. Lelieveld / *Unilever Research (Retired), Bilthoven, The Netherlands*

Robert Levin / *Department of Food Science, Massachusetts Agricultural Experiment Station, University of Massachusetts, Amherst, Amherst, Massachusetts, U.S.A.*

Matthias Liebergesell / *Pioneer Hi-Bred International, Inc., Johnston, Iowa, U.S.A.*

Jun Liu / *Molecular Animal Biotechnology Laboratory, Szent Istvan University, Godollo, Hungary*

Y. Martin Lo / *Department of Nutrition and Food Science, University of Maryland, College Park, Maryland, U.S.A.*

Chris J. Lomer / *International Institute of Tropical Agriculture (IITA), Cotonou, Benin*

Charles R. Long / *Department of Veterinary Physiology and Pharmacology, Texas A&M University, College Station, Texas, U.S.A.*

Juan Loor / *Department of Animal Sciences, University of Illinois, Urbana, Illinois, U.S.A.*

K.C. Lowe / *School of Life and Environmental Sciences, University of Nottingham, Nottingham, U.K.*

Zivile Luksiene / *Institute of Materials Science and Applied Research, Vilnius University, Vilnius, Lithuania*

Susan MacIntosh / *Macintosh & Associates, Inc., Saint Paul, Minnesota, U.S.A.*

Kathrine Hauge Madsen / *Centre for Bioethics and Risk Assessment, Royal Veterinary and Agricultural University, Frederiksberg, Denmark*

Samantha A. Malusky / *Department of Animal Sciences, University of Illinois at Urbana-Champaign, Urbana, Illinois, U.S.A.*

Maria Marco / *Food Science and Technology, University of California–Davis, Davis, California, U.S.A.*

Michael J. Martin / *Department of Pathobiology, College of Veterinary Medicine, Auburn University, Auburn, Alabama, U.S.A.*

Hugh S. Mason / *Biodesign Institute, Arizona State University, Tempe, Arizona, U.S.A.*

H.C. Mastwijk / *Wageningen UR Food and Biobased Research, Wageningen, The Netherlands*

Baltasar Mayo / *Department of Microbiology and Biochemistry, Dairy Institute of Asturias (CSIC), Villaviciosa, Spain*

P.L.H. McSweeney / *Department of Food and Nutritional Sciences, University College Cork, Cork, Ireland*

Adriane Medeiros / *Bioprocess Engineering and Biotechnology Division, Federal University of Paraná, Curitiba, Brazil*

Karen Moore / *Department of Animal Sciences, University of Florida, Gainesville, Florida, U.S.A.*

Gretchen Mosher / *Department of Agricultural and Biosystems Engineering, Iowa State University, Ames, Iowa, U.S.A.*

Yasmine Motarjemi / *Quality Management, Nestlé, Vevey, Switzerland*

Fernanda Mozzi / *Lactobacillus Reference Center, National Council of Scientific and Technical Research (CERELA-CONICET), San Miguel de Tucumán, Argentina*

Charles Muangirwa / *Tropical Pesticides Research Institute, Arusha, Tanzania*

Denis J. Murphy / *School of Applied Sciences, University of Glamorgan, Cardiff, U.K.*

K. Madhavan Nampoothiri / *Biotechnology Division, National Institute for Interdisciplinary Science and Technology (CSIR), Kerala, India*

Syed Naqvi / *Food Safety Science Unit, Ontario Ministry of Food and Rural Affairs, Guelph, Ontario, Canada*

Luiz Nasser / *Department of Animal Reproduction, University of Sao Paolo, Sao Paulo, Brazil*

T.L. Nedambale / *Germplasm Conservation and Reproductive Biotechnologies, Agricultural Reserve Council, Animal Production Institute, Irene, South Africa*

Hudaa Neetoo / *Department of Animal and Food Sciences, University of Delaware, Newark, Delaware, U.S.A.*

Martina Newell-McGloughlin / *UC Systemwide Biotechnology Research and Education Program (UCBREP), University of California–Davis, Davis, California, U.S.A.*

Zivko Nikolov / *Biological and Agricultural Engineering, Texas A&M University, College Station, Texas, U.S.A.*

Goutam Nistala / *Department of Agricultural and Biological Engineering, University of Illinois, Urbana, Illinois, U.S.A.*

Estrella Nuñez-Delicado / *Department of Science and Food Technology, San Antonio Catholic University of Murcia, Guadalupe, Spain*

Björn Oback / *Ruakura Research Centre, AgResearch, Hamilton, New Zealand*

Moustapha Oke / *Food Safety Science Unit, Ontario Ministry of Food and Rural Affairs, Guelph, Ontario, Canada*

Roberto Olivares / *Department of Chemical and Biological Engineering, Chalmers University of Technology, Gothenburg, Denmark*

Lisbeth Olsson / *Center for Microbial Biotechnology, BioCentrum, Technical University of Denmark, Kongens Lyngby, Denmark*

Roberta Onori / *Department of Veterinary Public Health and Food Safety, GMOs and Mycotoxins Unit, National Institute of Health (ISS), Rome, Italy*

Rodomiro Ortiz / *Intensive Agroecosystems Program, International Center for Maize and Wheat Improvement (CIMMYT), Juarez, Mexico*

Fred Owens / *Pioneer Hi-Bred International, Inc., Johnston, Iowa, U.S.A.*

Gopinadhan Paliyath / *Department of Plant Agriculture, University of Guelph, Guelph, Ontario, Canada*

Gianni Panagiotou / *Center for Microbial Biotechnology, BioCentrum, Technical University of Denmark, Kongens Lyngby, Denmark*

Ashok Pandey / *Biotechnology Division, National Institute for Interdisciplinary Science and Technology (CSIR), Kerala, India*

Alyssa Panitch / *Arizona State University, Tempe, Arizona, U.S.A.*
Alistair Paterson / *Centre for Food Quality (SIBPS), University of Strathclyde, Glasgow, U.K.*
Carl A. Pinkert / *Department of Pathobiology, College of Veterinary Medicine, Auburn University, Auburn, Alabama, U.S.A.*
Jeffrey L. Platt / *Transplantation Biology Program, University of Michigan, Ann Arbor, Michigan, U.S.A.*
Bert Popping / *Molecular Biology and Immunology, Eurofins Scientific Group, Pocklington, U.K.*
Jesus M. Porres / *Department of Physiology, University of Granada, Grenada, Spain*
J.B. Power / *School of Biosciences, Plant Science Division, University of Nottingham, Loughborough, U.K.*
Sakkie Pretorius / *Department of Plant Science, University of the Free State, Bloomfontein, South Africa*
Csaba Pribenszky / *Department of Animal Breeding and Genetics, St. Istvan University, Budapest, Hungary*
Vernon G. Pursel / *Agricultural Research Service, U.S. Department of Agriculture (USDA-ARS), Beltsville, Maryland, U.S.A.*
Vanu R. Ramprasath / *Richardson Centre for Functional Foods and Nutraceuticals, University of Manitoba, Winnipeg, Manitoba, Canada*
Barbara Rasco / *Department of Food Science and Human Nutrition, Washington State University, Pullman, Washington, U.S.A.*
Alan Frank Raybould / *Product Safety, Syngenta, Bracknell, U.K.*
David F. Ritchie / *Department of Plant Pathology, North Carolina State University, Raleigh, North Carolina, U.S.A.*
Alfred L. Roca / *Department of Animal Sciences, University of Illinois, Urbana, Illinois, U.S.A.*
José Luiz Rodrigues / *Laboratory of Embryology and Reproductive Biotechnology, Federal University of Rio Grande do Sul, Porto Alegre, Brazil*
Maria R. Rojas / *Department of Plant Pathology, University of California–Davis, Davis, California, U.S.A.*
Peter Roupas / *Food and Nutritional Sciences, Commonwealth Scientific and Industrial Research Organisation (CSIRO), Werribee, Victoria, Australia*
Peter Sandoe / *Department of Agricultural Sciences (Weed Science), Royal Veterinary and Agricultural University, Frederiksberg, Denmark*
Luca Santi / *Biodesign Institute, Arizona State University, Tempe, Arizona, U.S.A.*
Tony Savard / *Food Research and Development Center, Agriculture and Agri-Food Canada, St-Hyacinthe, Quebec, Canada*
A.M. Saxton / *Department of Animal Science, Institute of Agriculture, University of Tennessee, Knoxville, Tennessee, U.S.A.*
Christine Scaman / *Food, Nutrition, and Health, University of British Columbia, Vancouver, British Columbia, Canada*
Katharina Schoessler / *Department of Food Biotechnology and Food Processing Engineering, Berling University of Technology, Berlin, Germany*
Lawrence B. Schook / *Department of Animal Sciences, University of Illinois, Urbana, Illinois, U.S.A.*
F.N. Schrick / *Department of Animal Science, University of Tennessee, Knoxville, Tennessee, U.S.A.*

Antje Schulz / *Department of Food Biotechnology and Food Process Engineering, Berlin University of Technology, Berlin, Germany*
Michael Schweizer / *School of Life Sciences, Heriot-Watt University, Edinburgh, U.K.*
George E. Seidel, Jr. / *Department of Animal Reproduction and Biotechnology Laboratory, Colorado State University, Fort Collins, Colorado, U.S.A.*
Miti M. Shah / *Arizona State University, Tempe, Arizona, U.S.A.*
Mark Shamtsyan / *Department of Technology of Microbiological Synthesis, St. Petersburg State Institute of Technology, St. Petersburg, Russia*
Thomas H. Shellhammer / *Department of Food Science and Technology, Oregon State University, Corvallis, Oregon, U.S.A.*
Louise VT Shepherd / *Plant Products and Food Quality Programme, Scottish Crop Research Institute, Invergowrie, U.K.*
M. Sinigaglia / *Institute for Biotechnology Research and Applications for Security and Promotion of Local Products and Quality (BIOAGROMED), University of Foggia, Foggia, Italy*
Iryna Smetanska / *Department of Methods in Food Technology, Berlin University of Technology, Berlin, Germany*
Albert L. Smith / *Fertility Lab Consulting, Deming, New Mexico, U.S.A.*
Edward H. Smith / *Cornell University, Ithaca, New York, U.S.A.*
Lawrence C. Smith / *University of Montreal, Montreal, Quebec, Canada*
M.K. Smith / *Queensland Department of Primary Industries, Brisbane, Queensland, Australia*
Geoffrey Smithers / *Food and Nutritional Sciences, Commonwealth Scientific and Industrial Research Organisation (CSIRO), Werribee, Victoria, Australia*
Carlos R. Soccol / *Bioprocess Engineering and Biotechnology Division, Federal University of Paraná, Curitiba, Brazil*
Monika Sodhi / *Department of Animal Sciences, University of Illinois, Urbana, Illinois, U.S.A.*
Pavan Kumar Soma / *Department of Nutrition and Food Science, University of Maryland, College Park, Maryland, U.S.A.*
Susanne Somersalo / *Dodds & Associates, Washington, District of Columbia, U.S.A.*
B. Speranza / *Institute for Biotechnology Research and Applications for Security and Promotion of Local Products and Quality (BIOAGROMED), University of Foggia, Foggia, Italy*
Leon J. Spicer / *Department of Animal Science, Oklahoma State University, Stillwater, Oklahoma, U.S.A.*
James L. Steele / *Department of Food Science, University of Wisconsin-Madison, Madison, Wisconsin, U.S.A.*
Duska Stojsin / *Trail Development Group, Monsanto Company, St. Louis, Missouri, U.S.A.*
L.Y. Sung / *Center for Regenerative Biology, University of Connecticut, Storrs, Connecticut, U.S.A.*
Ramanjulu Sunkar / *Deptartment of Biochemistry and Molecular Biology, Oklahoma State University, Stillwater, Oklahoma, U.S.A.*
Young-Joon Surh / *College of Pharmacy, Seoul National University, Seoul, South Korea*
Tetsuya S. Tanaka / *Department of Animal Sciences, Institute for Genomic Biology, University of Illinois, Urbana, Illinois, U.S.A.*
Leon Terry / *Department of Food Security and Environmental Health, Cranfield University, Bedfordshire, U.K.*

Wendu Tesfaye / *Department of Food Technology, Polytechnical University of Madrid, Madrid, Spain*

Evangelos Topakas / *BIOtechMASS Unit, Biotechnology Laboratory, School of Chemical Engineering, National Technical University of Athens, Athens, Greece*

María Inés Torino / *Lactobacillus Reference Center, National Council of Scientific and Technical Research (CERELA-CONICET), San Miguel de Tucumán, Argentina*

Patrick J. Tranel / *University of Illinois, Urbana, Illinois, U.S.A.*

A.M. Troncosco / *Department of Nutrition and Bromatology, University of Seville, Seville, Spain*

Ian A. Trounce / *University of Melbourne, Melbourne, Victoria, Australia*

Federico Trucco / *Department of Crop Sciences, University of Illinois at Urbana-Champaign, Urbana, Illinois, U.S.A.*

David Twell / *Department of Biology, University of Leicester, Leicester, U.K.*

Marcus Volkert / *Department of Food Biotechnology and Food Process Engineering, Berlin University of Technology, Berlin, Germany*

Eric Walters / *National Swine Research and Resource Center, Univerity of Missouri, Columbia, Missouri, U.S.A.*

Hua Wang / *Department of Food Science and Technology, Ohio State University, Columbus, Ohio, U.S.A.*

Yuan Wen / *William G. Lowrie Department of Chemical and Biomolecular Engineering, Ohio State University, Columbus, Ohio, U.S.A.*

Matthew B. Wheeler / *Department of Animal Sciences, Beckman Institute for Advanced Science and Technology, Institute for Genomic Biology, University of Illinois, Urbana, Illinois, U.S.A.*

Macdonald Wick / *Department of Animal Sciences, Ohio State University, Columbus, Ohio, U.S.A.*

Chakra Wijesundera / *Food and Nutritional Sciences, Commonwealth Scientific and Industrial Research Organisation (CSIRO), Werribee, Victoria, Australia*

Gerald E. Wilde / *Department of Entomology, Kansas State University, Manhattan, Kansas, U.S.A.*

Thomas P. Wilson / *Department of Food and Animal Sciences, Alabama A&M University, Normal, Alabama, U.S.A.*

Rod A. Wing / *Department of Plant Sciences, University of Arizona, Tucson, Arizona, U.S.A.*

Randall Wisser / *Department of Plant and Soil Science, University of Delaware, Newark, Delaware, U.S.A.*

Susan L. Woodard / *Biological and Agricultural Engineering, Texas A&M University, College Station, Texas, U.S.A.*

Brad Woonton / *Food and Nutritional Sciences, Commonwealth Scientific and Industrial Research Organisation (CSIRO), Werribee, Victoria, Australia*

Christine Wrenzycki / *Clinic for Cattle, Reproductive Medicine Unit, University of Veterinary Medicine, Hannover, Germany*

J. Xu / *Center for Regenerative Biology, University of Connecticut, Storrs, Connecticut, U.S.A.*

Shang-Tian Yang / *William G. Lowrie Department of Chemical and Biomolecular Engineering, Ohio State University, Columbus, Ohio, U.S.A.*

X. Yang / *Center for Regenerative Biology, University of Connecticut, Storrs, Connecticut, U.S.A.*

Laila Yesmin / *Research and Development Division, Brett-Young Seeds Limited, Winnipeg, Manitoba, Canada*

Curtis R. Youngs / *Department of Animal Science, Iowa State University, Ames, Iowa, U.S.A.*

Lucy Zakharova / *University of Missouri, Columbia, Missouri, U.S.A.*

Nicholas Zavazava / *Internal Medicine, University of Iowa, Iowa City, Iowa, U.S.A.*

Adel Zayed / *Paradigm Genetics, Inc., Research Triangle Park, North Carolina, U.S.A.*

Henry Zeringue / *Department of Bioengineering, University of Pittsburgh, Pittsburgh, Pennsylvania, U.S.A.*

Qifa Zhang / *National Key Lab of Crop Genetic Improvement, Huazhong Agricultural University, Wuhan, China*

Xudong Zhang / *William G. Lowrie Department of Chemical and Biomolecular Engineering, Ohio State University, Columbus, Ohio, U.S.A.*

Contents

Editorial Advisory Board .... *v*
Contributors .... *vii*
Topical Table of Contents .... *xxv*
Preface .... *xxxi*
Aims and Scope .... *xxxiii*
About the Editors .... *xxxv*

Alcoholic Beverages: Exogenous Enzyme Use / *Charles Bamforth* 1
Alcoholic Beverages: Starter Cultures / *Moustapha Oke, Syed Naqvi, and Gopinadhan Paliyath* 6
Allergies: Food and Peanut Risk Reduction / *Hortense Dodo and Koffi Konan* 11
Animal Cell Culture / *Shang-Tian Yang, Xudong Zhang, and Yuan Wen* 18
Animal Medicines / *Sherrie Clark* 25
Animals and Plants: Genetic Modification (GM) / *Louis-Marie Houdebine* 28
Animals: RNA Interference (RNAi) / *Luciana Relly Bertolini and Marcelo Bertolini* 33
Antibiotics Use in Food-Producing Animals / *Jaap C. Hanekamp* 39
Antimicrobial Packaging / *Haiqiang Chen and Hudaa Neetoo* 43
Aroma Compounds / *Adriane Medeiros, Debora Brand, and Carlos R. Soccol* 47
Arthropod Host–Plant Resistant Crops / *Gerald E. Wilde* 50
Artificial Chromosomes in Plants / *James A. Birchler* 53
Artificial Insemination / *William L. Flowers* 56
Bacteria: Gram-Positive / *Hua Wang* 60
Bacteriophages: Pathogen Control / *Clair Hicks* 64
Bananas: Somatic Cell Genetics / *D.K. Becker and M.K. Smith* 68
Beer Fermentation / *Thomas H. Shellhammer* 71
Bioactive Compounds in Mushrooms / *Mark Shamtsyan* 76
Bio-Based Chemicals: Renewable Carbon for Biorefinery Production / *Joseph J. Bozell* 82
Biocontrol Agents: Genetic Improvement / *Marjorie A. Hoy* 87
Bioconversions: Fuel Ethanol / *W.M. Ingledew* 91
Biodiesel: Enzymatic Production / *John Birch and Martin Bell* 95
Biological Control: Successes and Failures / *Heikki Hokkanen* 99
Biopesticides / *Gavin Ash* 103
Bioremediation / *Ragini Gothalwal and P.S. Bisen* 107
Biosensors / *Leon Terry and Jordi Giné Bordonaba* 112
Biotechnology / *Manas R. Banerjee and Laila Yesmin* 116
Biotechnology: Ethical Aspects / *Kathrine Hauge Madsen and Peter Sandoe* 120
Bovine Embryos: In Vitro Culture / *Karen Moore* 123
Cattle Embryo Transfer / *Curtis R. Youngs and Robert A. Godke* 129
Cereal Foods: Starter Cultures / *Claude P. Champagne, Pierre Gélinas, and Edward R. Farnworth* ... 133
Cereal-Based Grain Products: Fermented Indigenous Grains / *Neela Badrie* 138

Cheese Production: Nonstarter Culture Bacteria / *Giorgio Giraffa* 143
Cheese: Yeasts and Molds / *Steve Labrie* 147
Chemoprevention with Dietary Phytopharmaceuticals / *Young-Joon Surh and Chang Yong Lee* 151
Chitosan: Produced by Microorganisms / *Renuka Karuppuswamy* 156
Cloning: Breeding / *Rodomiro Ortiz* 159
Cloning: Nuclear Transfer / *X. Yang, M.G. Carter, S.R. Gao, F. Du, J. Xu, and L.Y. Sung* 163
Cloning: Stem Cells of Different Developmental Potency / *Björn Oback* 167
Cocoa Fermentation: Chocolate Flavor Quality / *Emmanuel Ohene Afoakwa and Alistair Paterson* ... 171
Cold Plasmas Used for Food Processing / *H.C. Mastwijk and M.N. Nierop Groot* 174
Corn Sweeteners: Enzyme Use / *Christine Scaman* 178
Crops: Feral De-Domestication / *Jonathan Gressel* 183
Cyclodextrins / *Estrella Nuñez-Delicado and José Antonio Gabaldón-Hernández* 187
Dairy Lactococci / *Baltasar Mayo* 191
Dairy: Fermented Products / *Elaine P. Black* 195
Dairy: Starter Cultures / *Robert W. Hutkins* 200
Drought and Drought Resistance / *Salvatore Ceccarelli* 205
Electric Field Stress on Plant Systems / *Ana Balasa, Anna Janositz, and Dietrich Knorr* 208
ELISA Assays: Microorganisms and Toxins in Foods / *Robert Levin* 212
Embryo Transfer Traits: Genetic Parameter Estimation / *F.A. di Croce, A.M. Saxton, J.L. Edwards, D.E. Casanova, and F.N. Schrick* 216
Enzymes / *Geoffrey Smithers, Peter Roupas, Brad Woonton, and Chakra Wijesundera* 220
Enzymes: Amylases / *Geoffrey Smithers, Peter Roupas, Brad Woonton, and Chakra Wijesundera* 223
Enzymes: Chymosin and Other Milk Coagulants / *P.L.H. McSweeney and N. Bansal* 227
Enzymes: Lipases / *Geoffrey Smithers, Peter Roupas, Brad Woonton, and Chakra Wijesundera* 231
Enzymes: Molecular Aspects of Chymosin / *Rodney J. Brown* 234
Enzymes: Proteases / *Geoffrey Smithers, Chakra Wijesundera, Brad Woonton, and Peter Roupas* 237
Estrus Synchronization / *Pietro S. Baruselli and Gabriel A. Bó* 240
Farm Animals: Embryo Transfer / *John F. Hasler* 243
Feeds: Genetically Modified / *Fred Owens and Matthias Liebergesell* 247
Fish Roe: Fermentation / *Alaa El-Din A. Bekhit* 251
Food Labeling / *Thomas P. Wilson and Barbara Rasco* 257
Food Regulations: Global Harmonization / *Yasmine Motarjemi* 262
Functional Foods / *Michael Schweizer* 269
Gametes and Embryos: Sublethal Hydrostatic Pressure Treatment / *Csaba Pribenszky* 272
Gene Expression Patterns: In Vitro Techniques in Oocytes and Preimplantation Embryos / *Christine Wrenzycki* 276
Genetic Engineering and Biotechnology: Biosafety and Environmental Impact / *Martina Newell-McGloughlin* 279
Genetic Engineering: Evolution / *John Davison and Yves Bertheau* 292
Genetically Modified Foods: Consumer Attitudes / *Christine M. Bruhn* 297
Genetically Modified Organisms (GMOs): Authorized, Unauthorized and Unknown / *John Davison and Yves Bertheau* 301
Genomic Resources: Genetic Conservation / *David A. Kudrna and Rod A. Wing* 306
Genomics Research: Livestock Production / *Monika Sodhi and Lawrence B. Schook* 310
Genomics: Animal Agriculture / *Juan Loor* 316
Genomics: Captive Breeding and Wildlife Conservation / *Alfred L. Roca and Lawrence B. Schook* ... 320
Herbicide-Resistant Crops / *Stephen O. Duke* 326
High Pressure Processing and Enzymatic Reactions in Food / *Alan Kelly and Thom Huppertz* 329

Contents

Horses: Commercial Oocyte Technologies / *Elaine Carnevale* 332
Insects and Mites: Biological Control / *Ann E. Hajek* 335
Intellectual Property and Plant Science / *Susanne Somersalo and John Dodds* 339
Interspecies Embryo Transfer / *Duane Kraemer* 342
Intracytoplasmic Sperm Injection (ICSI) / *Albert L. Smith* 345
Kimchi Fermentation / *Jae-Kun Chun* 348
Laboratory Animals: Cryopreservation of Oocytes and Sperm / *Eric Walters* 352
Lactic Acid Fermentation: Direct / *Rojan P. John, G.S. Anisha, K. Madhavan Nampoothiri, and Ashok Pandey* 355
***Lactobacillus plantarum* in Foods** / *Maria Marco* 360
Leavened Breads / *Neela Badrie* 363
Lemon Essential Oils: Bioactive Properties in Fermented Foods and Beverages / *M.A. Del Nobile, B. Speranza, M. Sinigaglia, and A. Conte* 367
Lipases in Foods / *Hugo S. Garcia and Charles G. Hill, Jr.* 370
Maize: Durable Resistance Breeding / *Randall Wisser* 375
Male Gametogenesis / *David Honys and David Twell* 381
Malolactic Fermentation in Wine / *Maret du Toit* 387
Mammalian Sperm Sexing / *Duane L. Garner* 390
Manipulated Embryos: Cryopreservation / *Andras Dinnyes, T.L. Nedambale, and Jun Liu* 394
Meat Fermentation / *Immaculada Franco, Sonia Fonseca, Maria C. Garcia Fontán, and Javier Carballo* 399
Meats: Proteomics / *Macdonald Wick* 402
Metabolite Extraction from Plant Tissues / *Iryna Smetanska* 407
Metabolomics and Genetically Modified Organisms (GMOs) / *Howard V. Davies and Louise VT Shepherd* 410
Microbial Molecular Biology / *Harry J. Flint* 414
Microbial Polysaccharides / *Graciela Font de Valdez, María Inés Torino, and Fernanda Mozzi* 418
Microbial Small Heat-Shock Proteins / *Sang Yup Lee and Mee-Jung Han* 422
Microfluids for Assisted Reproduction / *Henry Zeringue* 426
Miso Fermentation / *Naveen Chikthimmah and Jaison Karavally* 431
Nanoscale Biology: Engineering Applications / *Kaustubh Bhalerao and Goutam Nistala* 434
Nematodes: Biological Control / *Simon Gowen* 440
***Nicotiana benthamiana*: Tobamoviral Vectors Redirect Carotenogenesis** / *Monto H. Kumagai and Jennifer Lee Busto* 443
Nutraceutical Compounds: Feruloyl Esterases as Biosynthetic Tools / *Lisbeth Olsson, Gianni Panagiotou, Paul Christakopoulos, Evangelos Topakas, and Roberto Olivares* 448
Nutraceuticals and Functional Foods / *Nicholas Kalaitzandonakes, James D. Kaufman, and Lucy Zakharova* 454
Oil Crops: Genetically Modified / *Denis J. Murphy* 457
Oocytes and Embryos: Vitrification / *José Luiz Rodrigues* 460
Organic Farming / *Brenda Frick* 462
Organogenesis: In Vitro Plant Regeneration / *Janet R. Gorst* 466
Oxygenases in Food / *Estrella Nuñez-Delicado and José Antonio Gabaldón-Hernández* 469
Pediococcus / *Jeff Broadbent and James L. Steele* 474
Pest Management: Population Theory / *Alan Andrew Berryman* 479
Pesticides: History / *Edward H. Smith and George Kennedy* 482
Photosensitization and Food Safety / *Zivile Luksiene* 487
Phytases / *Xingen Lei and Jesus M. Porres* 491

Phytoremediation / *Adel Zayed* 495
Plant DNA Virus Diseases / *Robert L. Gilbertson and Maria R. Rojas* 499
Plant Genetic Resources: Effective Utilization / *Hikmet Budak* 504
Plant Pathogens (Viruses): Biological Control / *Hei-Ti Hsu* 509
Plant Pathogens: Pest Management / *Kitty Cardwell and Chris J. Lomer* 512
Plant Sterols / *Peter J.H. Jones and Vanu R. Ramprasath* 516
Plant Tissue Culture: Industrial Uses / *Kirsten A. Hirneisen and Kalmia E. Kniel* ... 519
Plant-Produced Recombinant Therapeutics / *Lokesh Joshi, Miti M. Shah, C. Robert Flynn, and Alyssa Panitch* 522
Policy and Regulation: U.S. Regulations / *Susan MacIntosh* 526
Polymerase Chain Reaction (PCR) / *Rolf D. Joerger* 531
Poultry Genetics and Breeding / *Samuel E. Aggrey* 536
Probiotic Bacteria Preservation / *Marcus Volkert, Antje Schulz, Katharina Schoessler, and Dietrich Knorr* 540
Probiotics / *Dallas Hoover* 545
Processing and Preservative Aids: Nisin and Other Bacteriocins / *Dallas Hoover and Haiqiang Chen* 549
Product Labeling: Policy and Regulation / *Bert Popping* 553
Propionibacteria / *Leon J. Spicer* 557
Protein Bioseparation: Plant and Animal Products / *Susan L. Woodard and Zivko Nikolov* ... 561
Protoplasts / *Michael R. Davey, P. Anthony, J.B. Power, and K.C. Lowe* 566
Protoplasts: Culture and Regeneration / *J.B. Power, Michael R. Davey, P. Anthony, and K.C. Lowe* ... 569
Protozoa and Parasites: Food Safety / *Kalmia E. Kniel* 572
Pulsed Electric Field Processing: Food Preservation / *Huub L.M. Lelieveld* 575
RNAs: Small Inhibitory / *Ramanjulu Sunkar* 579
Secondary Metabolites: Plant Cell and Hairy Root Cultures / *Iryna Smetanska* ... 583
Seeds: Transgenes and Genetic Modification (GM) / *Roberta Onori and Marzia De Giacomo* ... 587
Soy Sauce Fermentation / *Jean Guy LeBlanc* 591
Stem Cell and Germ Cell Technology / *Matthew B. Wheeler and Samantha A. Malusky* ... 594
Stem Cell Research / *Tetsuya S. Tanaka* 597
Superovulation in Mammals / *George E. Seidel, Jr.* 604
Temperate Climate Fruit Crop Pest Management (Plant Pathogens) / *David F. Ritchie* ... 608
Transgenes: Plant Breeding / *Duska Stojsin* 613
Transgenic Animals / *Vernon G. Pursel* 616
Transgenic Animals: Mitochondrial Genome Modification / *Carl A. Pinkert, Lawrence C. Smith, and Ian A. Trounce* 619
Transgenic Animals: Secreted Products / *Michael J. Martin, David A. Dunn, and Carl A. Pinkert* ... 622
Transgenic Crops: Environmental Concerns / *Alan Frank Raybould* 625
Transgenic Crops: Perennials / *Abhaya Dandekar and Matthew Escobar* 629
Transgenic Crops: Regulatory Standards and Procedures of Research and Commercialization / *Qifa Zhang* 633
Transgenic Livestock: RNA Interference (RNAi) / *Charles R. Long* 636
Transgenic Plant Risk: Coexistence and Economy / *Gretchen Mosher and Charles Hurburgh* ... 639
Transgenic Plants: Economic and Environmental Risks and Gene Flow / *Alan Frank Raybould* ... 643
Transgenic Plants: Protein Quality Improvements / *Jesse M. Jaynes* 647
Transgenic Plants: Wax Esters from / *Kathryn D. Lardizabal* 650
Tropical Agriculture: Pest Management / *Charles Muangirwa* 653
Utility Patents and Plant Innovation / *Jay P. Kesan* 657

Vaccine Production: Plants as Biofactories / *Luca Santi and Hugh S. Mason* 661
Vaccines and Anti-Infective Agents: Delivery via Lactic Acid Bacteria / *Tri Duong* 664
Vegetables: Fermentation Applications / *Tony Savard* 668
Vertebrates: Biological Control / *Peter Kerr* 672
Vinegar / *Wendu Tesfaye and A.M. Troncosco* 675
Viruses: Food Safety / *Lee-Ann Jaykus and Doris H. D'Souza* 680
Volatile Flavor Generation: Genetic Methods / *Tim Coolbear and Marie-Laure Delabre* 683
Weed Science / *Patrick J. Tranel and Federico Trucco* 686
Wine Fermentation / *Christopher Curtin, Paul Chambers, and Sakkie Pretorius* 689
Xanthan Gum: Bioreactors in Production / *Y. Martin Lo and Pavan Kumar Soma* 695
Xenogenic Tissue Use in Clinical Medicine / *Nicholas Zavazava* 700
Xenotransplantation / *Jeffrey L. Platt* 703
Yogurt Microbiology / *Ashraf N. Hassan and Sanjeev Anand* 707
Zebu Cattle: Timed Artificial Insemination / *Luiz Nasser* 711
Zygote Intrafallopian Transfer (ZIFT) / *Albert L. Smith* 714

Index 717

Topical Table of Contents

Animals

Animal Products

Animal Medicines / *Sherrie Clark*	25
Feeds: Genetically Modified / *Fred Owens and Matthias Liebergesell*	247
Vaccines and Anti-Infective Agents: Delivery via Lactic Acid Bacteria / *Tri Duong*	664
Xenogenic Tissue Use in Clinical Medicine / *Nicholas Zavazava*	700
Xenotransplantation / *Jeffrey L. Platt*	703

Genetic Modification

Animal Cell Culture / *Shang-Tian Yang, Xudong Zhang, and Yuan Wen*	18
Animals and Plants: Genetic Modification (GM) / *Louis-Marie Houdebine*	28
Animals: RNA Interference (RNAi) / *Luciana Relly Bertolini and Marcelo Bertolini*	33
Genomics Research: Livestock Production / *Monika Sodhi and Lawrence B. Schook*	310
Genomics: Animal Agriculture / *Juan Loor*	316
Genomics: Captive Breeding and Wildlife Conservation / *Alfred L. Roca and Lawrence B. Schook*	320
Poultry Genetics and Breeding / *Samuel E. Aggrey*	536

Cloning

Cloning: Breeding / *Rodomiro Ortiz*	159
Cloning: Nuclear Transfer / *X. Yang, M.G. Carter, S.R. Gao, F. Du, J. Xu, and L.Y. Sung*	163
Cloning: Stem Cells of Different Developmental Potency / *Björn Oback*	167

Fertility and Pregnancy

Artificial Insemination / *William L. Flowers*	56
Bovine Embryos: In Vitro Culture / *Karen Moore*	123
Cattle Embryo Transfer / *Curtis R. Youngs and Robert A. Godke*	129
Embryo Transfer Traits: Genetic Parameter Estimation / *F.A. di Croce, A.M. Saxton, J.L. Edwards, D.E. Casanova, and F.N. Schrick*	216
Estrus Synchronization / *Pietro S. Baruselli and Gabriel A. Bó*	240
Farm Animals: Embryo Transfer / *John F. Hasler*	243
Gametes and Embryos: Sublethal Hydrostatic Pressure Treatment / *Csaba Pribenszky*	272
Gene Expression Patterns: In Vitro Techniques in Oocytes and Preimplantation Embryos / *Christine Wrenzycki*	276
Horses: Commercial Oocyte Technologies / *Elaine Carnevale*	332
Interspecies Embryo Transfer / *Duane Kraemer*	342
Intracytoplasmic Sperm Injection (ICSI) / *Albert L. Smith*	345
Laboratory Animals: Cryopreservation of Oocytes and Sperm / *Eric Walters*	352

Fertility and Pregnancy (cont'd)

Male Gametogenesis / *David Honys and David Twell* 381
Mammalian Sperm Sexing / *Duane L. Garner* 390
Manipulated Embryos: Cryopreservation / *Andras Dinnyes, T.L. Nedambale, and Jun Liu* 394
Microfluids for Assisted Reproduction / *Henry Zeringue* 426
Oocytes and Embryos: Vitrification / *José Luiz Rodrigues* 460
Superovulation in Mammals / *George E. Seidel Jr.* 604
Zebu Cattle: Timed Artificial Insemination / *Luiz Nasser* 711
Zygote Intrafallopian Transfer (ZIFT) / *Albert L. Smith* 714

Transgenic Animals

Transgenic Animals / *Vernon G. Pursel* 616
Transgenic Animals: Mitochondrial Genome Modification / *Carl A. Pinkert, Lawrence C. Smith, and Ian A. Trounce* 619
Transgenic Animals: Secreted Products / *Michael J. Martin, David A. Dunn, and Carl A. Pinkert* 622
Transgenic Livestock: RNA Interference (RNAi) / *Charles R. Long* 636

Biotechnology (General)

Biotechnology / *Manas R. Banerjee and Laila Yesmin* 116
Biotechnology: Ethical Aspects / *Kathrine Hauge Madsen and Peter Sandoe* 120
Genetic Engineering: Evolution / *John Davison and Yves Bertheau* 292
Genetically Modified Foods: Consumer Attitudes / *Christine M. Bruhn* 297
Genetically Modified Organisms (GMOs): Authorized, Unauthorized and Unknown / *John Davison and Yves Bertheau* 301
Genomic Resources: Genetic Conservation / *David A. Kudrna and Rod A. Wing* 306
Metabolomics and Genetically Modified Organisms (GMOs) / *Howard V. Davies and Louise VT Shepherd* 410
Microbial Molecular Biology / *Harry J. Flint* 414
Nanoscale Biology: Engineering Applications / *Kaustubh Bhalerao and Goutam Nistala* 434

Bacteria

Bacteria: Gram-Positive / *Hua Wang* 60
Microbial Small Heat-Shock Proteins / *Sang Yup Lee and Mee-Jung Han* 422
Pediococcus / *Jeff Broadbent and James L. Steele* 474
Probiotic Bacteria Preservation / *Marcus Volkert, Antje Schulz, Katharina Schoessler, and Dietrich Knorr* 540
Probiotics / *Dallas Hoover* 545
Propionibacteria / *Leon J. Spicer* 557
Vaccines and Anti-Infective Agents: Delivery via Lactic Acid Bacteria / *Tri Duong* 664

Enzyme Systems

Biodiesel: Enzymatic Production / *John Birch and Martin Bell* 95
Enzymes / *Geoffrey Smithers, Peter Roupas, Brad Woonton, and Chakra Wijesundera* 220
Enzymes: Amylases / *Geoffrey Smithers, Peter Roupas, Brad Woonton, and Chakra Wijesundera* 223
Enzymes: Chymosin and Other Milk Coagulants / *P.L.H. McSweeney and N. Bansal* 227
Enzymes: Lipases / *Geoffrey Smithers, Peter Roupas, Brad Woonton, and Chakra Wijesundera* 231

Topical Table of Contents

    Enzymes: Molecular Aspects of Chymosin / *Rodney J. Brown* 234
    Enzymes: Proteases / *Geoffrey Smithers, Chakra Wijesundera, Brad Woonton, and Peter Roupas* 237
    Lipases in Foods / *Hugo S. Garcia and Charles G. Hill Jr.* 370
    Nutraceutical Compounds: Feruloyl Esterases as Biosynthetic Tools / *Lisbeth Olsson, Gianni Panagiotou, Paul Christakopoulos, Evangelos Topakas, and Roberto Olivares* 448
    Oxygenases in Food / *Estrella Nuñez-Delicado and José Antonio Gabaldón-Hernández* 469
    Phytases / *Xingen Lei and Jesus M. Porres* 491

Regulations

    Antibiotics Use in Food-Producing Animals / *Jaap C. Hanekamp* 39
    Biological Control: Successes and Failures / *Heikki Hokkanen* 99
    Food Labeling / *Thomas P. Wilson, Barbara Rasco* 257
    Food Regulations: Global Harmonization / *Yasmine Motarjemi* 262
    Intellectual Property and Plant Science / *Susanne Somersalo, John Dodds* 339
    Organic Farming / *Brenda Frick* 462
    Policy and Regulation: U.S. Regulations / *Susan MacIntosh* 526
    Product Labeling: Policy and Regulation / *Bert Popping* 553
    Transgenic Plant Risk: Coexistence and Economy / *Gretchen Mosher, Charles Hurburgh* 639
    Transgenic Plants: Economic and Environmental Risks and Gene Flow / *Alan Frank Raybould* 643
    Utility Patents and Plant Innovation / *Jay P. Kesan* 657

Foods and Beverages

Alcoholic Products

    Alcoholic Beverages: Exogenous Enzyme Use / *Charles Bamforth* 1
    Alcoholic Beverages: Starter Cultures / *Moustapha Oke, Syed Naqvi, and Gopinadhan Paliyath* 6
    Beer Fermentation / *Thomas H. Shellhammer* 71
    Bioconversions: Fuel Ethanol / *W.M. (Mike) Ingledew* 91
    Malolactic Fermentation in Wine / *Maret du Toit* 387
    Wine Fermentation / *Christopher Curtin, Paul Chambers, and Sakkie Pretorius* 689

Cereal and Baked Products

    Cereal-Based Grain Products: Fermented Indigenous Grains / *Neela Badrie* 138
    Leavened Breads / *Neela Badrie* 363
    Miso Fermentation / *Naveen Chikthimmah and Jaison Karavally* 431

Dairy Fermentation Products

    Cheese Production: Nonstarter Culture Bacteria / *Giorgio Giraffa* 143
    Cheese: Yeasts and Molds / *Steve Labrie* 147
    Dairy Lactococci / *Baltasar Mayo* 191
    Dairy: Fermented Products / *Elaine P. Black* 195
    Dairy: Starter Cultures / *Robert W. Hutkins* 200

Dairy Fermentation Products (cont'd)

Lactic Acid Fermentation: Direct / *Rojan P. John, G.S. Anisha, K. Madhavan Nampoothiri, and Ashok Pandey* 355

Yogurt Microbiology / *Ashraf N. Hassan and Sanjeev Anand* 707

Fermentation Products (Nondairy)

Chitosan: Produced by Microorganisms / *Renuka Karuppuswamy* 156

Cocoa Fermentation: Chocolate Flavor Quality / *Emmanuel Ohene Afoakwa and Alistair Paterson* 171

Fish Roe: Fermentation / *Alaa El-Din A. Bekhit* 251

Kimchi Fermentation / *Jae-Kun Chun* 348

Lactobacillus plantarum **in Foods** / *Maria Marco* 360

Lemon Essential Oils: Bioactive Properties in Fermented Foods and Beverages / *M.A. Del Nobile, B. Speranza, M. Sinigaglia, and A. Conte* 367

Meat Fermentation / *Immaculada Franco, Sonia Fonseca, Maria C. Garcia Fontán, and Javier Carballo* 399

Vegetables: Fermentation Applications / *Tony Savard* 668

Xanthan Gum: Bioreactors in Production / *Y. Martin Lo and Pavan Kumar* 695

Food Safety

Allergies: Food and Peanut Risk Reduction / *Hortense Dodo and Koffi Konan* 11

Bacteriophages: Pathogen Control / *Clair Hicks* 64

Biosensors / *Leon Terry and Jordi Giné Bordonaba* 112

ELISA Assays: Microorganisms and Toxins in Foods / *Robert Levin* 212

Photosensitization and Food Safety / *Zivile Luksiene* 487

Polymerase Chain Reaction (PCR) / *Rolf D. Joerger* 531

Protozoa and Parasites: Food Safety / *Kalmia E. Kniel* 572

Viruses: Food Safety / *Lee-Ann Jaykus and Doris H. D'Souza* 680

Ingredients

Aroma Compounds / *Adriane Medeiros, Debora Brand, and Carlos R. Soccol* 47

Bioactive Compounds in Mushrooms / *Mark Shamtsyan* 76

Corn Sweeteners: Enzyme Use / *Christine Scaman* 178

Cyclodextrins / *Estrella Nuñez-Delicado and José Antonio Gabaldón-Hernández* 187

Microbial Polysaccharides / *Graciela Font de Valdez, María Inés Torino, and Fernanda Mozzi* 418

New Products

Chemoprevention with Dietary Phytopharmaceuticals / *Young-Joon Surh and Chang Yong Lee* 151

Functional Foods / *Michael Schweizer* 269

Lactic Acid Fermentation: Direct / *Rojan P. John, G.S. Anisha, K. Madhavan Nampoothiri, and Ashok Pandey* 355

Meats: Proteomics / *Macdonald Wick* 402

Nutraceuticals and Functional Foods / *Nicholas Kalaitzandonakes, James D. Kaufman, and Lucy Zakharova* 454

Topical Table of Contents   xxix

Soy Sauce Fermentation / *Jean Guy LeBlanc* 591
Vinegar / *Wendu Tesfaye and A.M. Troncosco* 675
Volatile Flavor Generation: Genetic Methods / *Tim Coolbear and Marie-Laure Delabre* 683

Processes

Antimicrobial Packaging / *Haiqiang Chen and Hudaa Neetoo* 43
Cereal Foods: Starter Cultures / *Claude P. Champagne, Pierre Gélinas, and Edward R. Farnworth* 133
Cold Plasmas Used for Food Processing / *H.C. Mastwijk and M.N. Nierop Groot* 174
High Pressure Processing and Enzymatic Reactions in Food / *Alan Kelly and Thom Huppertz* 329
Probiotic Bacteria Preservation / *Marcus Volkert, Antje Schulz, Katharina Schoessler, and Dietrich Knorr* 540
Processing and Preservative Aids: Nisin and Other Bacteriocins / *Dallas Hoover and Haiqiang Chen* 549
Protein Bioseparation: Plant and Animal Products / *Susan L. Woodard and Zivko Nikolov* 561
Pulsed Electric Field Processing: Food Preservation / *Huub L.M. Lelieveld* 575

Plants

Byproducts

Bio-Based Chemicals: Renewable Carbon for Biorefinery Production / *Joseph J. Bozell* 82
Bioconversions: Fuel Ethanol / *W.M. (Mike) Ingledew* 91
Biodiesel: Enzymatic Production / *John Birch and Martin Bell* 95
Metabolite Extraction from Plant Tissues / *Iryna Smetanska* 407
Plant Sterols / *Peter J.H. Jones and Vanu R. Ramprasath* 516
Plant Tissue Culture: Industrial Uses / *Kirsten A. Hirneisen and Kalmia E. Kniel* 519
Plant-Produced Recombinant Therapeutics / *Lokesh Joshi, Miti M. Shah, C. Robert Flynn, and Alyssa Panitch* 522
Vaccine Production: Plants as Biofactories / *Luca Santi, Hugh S. Mason* 661

Genetic Modification

Animals and Plants: Genetic Modification (GM) / *Louis-Marie Houdebine* 28
Artificial Chromosomes in Plants / *James A. Birchler* 53
Bananas: Somatic Cell Genetics / *D.K. Becker and M.K. Smith* 68
Biocontrol Agents: Genetic Improvement / *Marjorie A. Hoy* 87
Crops: Feral De-Domestication / *Jonathan Gressel* 183
Genetic Engineering and Biotechnology: Biosafety and Environmental Impact / *Martina Newell-McGloughlin* 279
Nicotiana benthamiana: Tobamoviral Vectors Redirect Carotenogenesis / *Monto H. Kumagai; Jennifer Lee Busto* 443
Oil Crops: Genetically Modified / *Denis J. Murphy* 457
Organogenesis: In Vitro Plant Regeneration / *Janet R. Gorst* 466
Plant Genetic Resources: Effective Utilization / *Hikmet Budak* 504
Protoplasts / *Michael R. Davey, P. Anthony, J.B. Power, and K.C. Lowe* 566

Genetic Modification (cont'd)

Protoplasts: Culture and Regeneration / *J.B. Power, Michael R. Davey, P. Anthony, and K.C. Lowe* 569
RNAs: Small Inhibitory / *Ramanjulu Sunkar* 579
Secondary Metabolites: Plant Cell and Hairy Root Cultures / *Iryna Smetanska* 583
Seeds: Transgenes and Genetic Modification (GM) / *Roberta Onori and Marzia De Giacomo* 587
Stem Cell and Germ Cell Technology / *Matthew B. Wheeler and Samantha A. Malusky* 594
Stem Cell Research / *Tetsuya S. Tanaka* 597
Transgenes: Plant Breeding / *Duska Stojsin* 613
Transgenic Crops: Environmental Concerns / *Alan Frank Raybould* 625
Transgenic Crops: Perennials / *Abhaya Dandekar and Matthew Escobar* 629
Transgenic Crops: Regulatory Standards and Procedures of Research and Commercialization / *Qifa Zhang* 633
Transgenic Plants: Protein Quality Improvements / *Jesse M. Jaynes* 647
Transgenic Plants: Wax Esters from / *Kathryn D. Lardizabal* 650

Pest Control and the Environment

Arthropod Host–Plant Resistant Crops / *Gerald E. Wilde* 50
Biopesticides / *Gavin Ash* 103
Bioremediation / *Ragini Gothalwal, P.S. Bisen* 107
Drought and Drought Resistance / *Salvatore Ceccarelli* 205
Electric Field Stress on Plant Systems / *Ana Balasa, Anna Janositz, and Dietrich Knorr* 208
Herbicide-Resistant Crops / *Stephen O. Duke* 326
Insects and Mites: Biological Control / *Ann E. Hajek* 335
Maize: Durable Resistance Breeding / *Randall Wisser* 375
Nematodes: Biological Control / *Simon Gowen* 440
Pest Management: Population Theory / *Alan Andrew Berryman* 479
Pesticides: History / *Edward H. Smith and George Kennedy* 482
Phytoremediation / *Adel Zayed* 495
Plant DNA Virus Diseases / *Robert L. Gilbertson and Maria R. Rojas* 499
Plant Pathogens (Viruses): Biological Control / *Hei-Ti Hsu* 509
Plant Pathogens: Pest Management / *Kitty Cardwell and Chris J. Lomer* 512
Temperate Climate Fruit Crop Pest Management (Plant Pathogens) / *David F. Ritchie* 608
Tropical Agriculture: Pest Management / *Charles Muangirwa* 653
Vertebrates: Biological Control / *Peter Kerr* 672
Weed Science / *Patrick J. Tranel and Federico Trucco* 686

Preface

The discoveries in molecular biology, over the past 40 years, have stimulated an array of applications in every field with a relationship to biology. These applications have been referred to as biotechnology; a term with both positive and negative connotations. The origin of many of these applications has been in agriculture. In particular, the purification and amplification of DNA sequences in the 1970s acted as a stimulus for development of methods for transfer of DNA in plants and animals, as well as other organisms, in the 1980s. The applications continue to have significant impact on agricultural production and food manufacturing.

Current and potentially beneficial impacts of molecular biology on agriculture and food continue to be documented. Higher yields from both plants and animals have positive economic benefits for both producers and consumers. The benefits associated with reduced use of pesticides, herbicides and antibiotics should lead to positive impacts on the environment, as well as food safety. Ultimately, the foods available for the consumer will have improved quality and safety.

As occurs during the evolution and application of new science, concerns and questions about the potential negative impacts of the applications emerge. The economic benefits of a new technology may not be immediate and influence the acceptance of the technology. Biotechnology has introduced an array of social concerns associated with the potential for irreversible changes to plants, animals and the foods to be consumed by humans. These concerns are being addressed in the context of risk and risk analysis, but these are relatively new concepts for society and consumers.

The good and the potentially negative impacts of biotechnology in the food chain demand an authoritative reference, peer-reviewed by experts that offer timely information to all the stakeholders in this debate. Ultimately, the acceptance of biotechnology, as an application of molecular biology, depends on society's understanding of the science of molecular biology and the applications of risk assessment.

The primary focus of the *Encyclopedia of Agricultural and Food Biotechnology* is on the science and applications of molecular biology for agriculture and foods. The entries describe the concepts and processes used as new tools for the production of raw agricultural materials and the manufacturing of foods products, as well as concerns associated with these applications.

This reference contains over nearly 200 entries in areas related to biotechnology in agriculture and food, within the following general areas:

1. Descriptions and interpretations of research in molecular biology, with specific attention to the science associated with cloning of animals, genetic modification of plants and the enhancement of food quality.
2. Current and future applications of molecular biology, including entries on disease resistance in animals, drought resistant plants and improved health of the consumers through nutritionally enhanced foods.
3. Regulations associated with biotechnology, with specific attention to regulations on genetically modified organisms, the differences among regulations in different counties and impacts on the evolution of new applications.
4. Communication of the concepts of molecular biology and biotechnology, including contributions on interpretation of biotechnology, the biotechnology/environment interface and consumer acceptance of the products of biotechnology.
5. The future of biotechnology, with entries on risk assessment, food security, and genetic diversity.

The editors would like to acknowledge the contributions of the Editorial Advisory Board (EAB). The advisors have contributed to the content of this reference in a significant manner. Many of the topics and authors have been identified by members of the EAB. In addition, EAB members have been involved in the peer review of the contributions, either through review of the contributions, or by identification of peer reviewers. Finally, these colleagues have provided consistent encouragement as we progressed steadily toward the final goal.

In summary, we hope the *Encyclopedia of Agricultural and Food Biotechnology* will become an important reference in an evolving area of science and technology. References of this type should be beneficial in diffusing controversies about genetically modified foods and similar applications of molecular biology. By incorporating an array of entries, from basic science to applications, into a single reference, readers will be able to find information normally requiring many different resources. Hopefully, the *Encyclopedia of Biotechnology in Agriculture and Food* will become a reference for a broad array of readers with interest in the subject and will provide valuable insights into the important aspects of biotechnology and the potential applications.

Dennis R. Heldman; Editor
Matthew B. Wheeler; Topic Editor
Dallas G. Hoover; Topic Editor

Aims and Scope

BACKGROUND

The applications of molecular biology throughout the food chain will continue to evolve and future applications will have many positive impacts of the quantities and quality of foods. This reference should provide readers from many different backgrounds with the opportunity to gain a better understanding of current applications, and the potential applications of the future.

MISSION AND SCOPE

The *Encyclopedia of Agricultural and Food Biotechnology* focuses primarily on the science and applications of molecular biology for agriculture and foods, by offering short accessible overviews. These topics describe the concepts and processes being used as new tools for the production of raw agricultural materials and the manufacturing of foods products, as well as the fundamental concerns from a range of perspectives.

A total of 175 alphabetically-arranged entries cover the following areas related to biotechnology as applied to agriculture and food:

1. Descriptions and interpretations of molecular biology research, including topics on the science associated with cloning of animals, the genetic modification of plants and the enhanced quality of foods.
2. Current and future applications of molecular biology, with contributions on disease resistance in animals, drought resistant plants and improved health of the consumer through nutritionally enhanced foods.
3. Regulations associated with applications of biotechnology, with specific attention to regulations on genetically modified organisms, the differences among regulations in different counties and impacts on the evolution of new applications of biotechnology.
4. Communication of the concepts of molecular biology and biotechnology, including contributions on interpretation of biotechnology in the news and media, the biotechnology/environment interface and consumer acceptance of the products of biotechnology.
5. Benefits and concerns about biotechnology with overviews on topics such as risk assessment, food security, and genetic diversity.

The readers of the *Encyclopedia of Agricultural and Food Biotechnology* will be integral to future research in the many different disciplines related to biotechnology. The biologists and clinical nutritionists, as well as those researching genomics, animal physiology, and crop yields, all have a mutual interest in the topics being discussed. In addition, this reference will be valuable to the consumers of food products; researchers in policy positions and centers; and professionals associated with agribusinesses.

Each contribution is a brief overview of approximately 2,000 words and introduces basic background and concepts. Illustrations are used to communicate essential concepts and comprehensive reference lists allow readers to explore topics in greater depth. Each entry focuses on the interpretation of the science and the applications of molecular biology to animal agriculture, production of agricultural crops or manufacturing of foods. The *Encyclopedia* also addresses regulations related to applications of biotechnology in agricultural production and food manufacturing, and emphasizes the role of the regulatory system in ensuring the safety of the products evolving from biotechnology. Finally, topics related to communication of the scientific concepts of biotechnology to consumers are included, as are the major points of view.

In summary, the *Encyclopedia of Agricultural and Food Biotechnology* assumes an important educational role for an evolving science and technology. Controversies about genetically modified foods, with direct impact on broad segments of the population, are addressed in a manner that leads to improved understanding and communication. By integrating contributions from several dimensions, students, researchers, analysts, and the interested public who use the Encyclopedia will be able gain valuable insights about important aspects of biotechnology and the potential applications.

About the Editors

Dennis R. Heldman holds B.S. and M.S. degrees from The Ohio State University, and Ph.D. from Michigan State University. He has been Professor at Michigan State University, the University of Missouri and Rutgers, The State University of New Jersey. Dr Heldman has held positions at the Campbell Soup Company, the National Food Processors Association, The National Food Laboratory, and the Weinberg Consulting Group Inc. Currently, he is Principal in Heldman Associates, and served as President of IFT, the Society for Food Science and Technology from 2006–07. Dr. Heldman was elected Fellow in the International Academy of Food Science & Technology in 2006.

Dallas G. Hoover is Professor in the Department of Animal & Food Sciences, University of Delaware, Newark, joining the faculty in 1984. His B.S. degree is from Elizabethtown College (PA), M.S. from the University of Delaware, and Ph.D. from the University of Minnesota. His research interests include high pressure processing, food safety, use of probiotic cultures and sporeformer biology, and his are of teaching includes introductory food science and food microbiology.

Matthew B. Wheeler holds a B.S. and M.S. from the University of California-Davis and a Ph.D. from Colorado State University. He was a NIH post-doctoral fellow at the University of Virginia and the University of Wisconsin. Dr. Wheeler has been a Professor at the University of Illinois at Urbana-Champaign since 1989 and served as the President of the International Embryo Transfer Society from 2005–2006. Dr. Wheeler's responsibilities at the University of Illinois include teaching and research in biotechnology, reproductive biology and regenerative/stem cell biology in the Departments of Animal Sciences, Veterinary Clinical Medicine and Bioengineering. He is also a member of the Beckman Institute for Advanced Science and Technology and the Institute for Genomic Biology.

Encyclopedia of Biotechnology in Agriculture and Food

Alcoholic Beverages: Exogenous Enzyme Use

Charles Bamforth
Department of Food Science and Technology, University of California–Davis, Davis, California, U.S.A.

Abstract
Exogenous enzymes offer considerable opportunities to enhance process performance and product quality in the production of alcoholic beverages, notably beer, hard cider, and wine. These enzymes are for the most part hemicellulases, amylases, pectinases, and proteinases with a limited use of oxidoreductases. There is also a valuable application for use of decarboxylase in accelerating brewery fermentations.

INTRODUCTION

Worldwide, there is a diversity of alcoholic beverages, both those whose provenance is directly linked to alcoholic fermentation and those of much greater alcoholic strength achieved using a distillation phase subsequent to fermentation. The principle organisms involved in the formation of ethanol are from the genus *Saccharomyces*. Alcohol and carbon dioxide are produced as end products from a sequence of enzymes functioning within the intermediary metabolism of the yeast. The primary focus of this entry will be on the exogenous enzymes that have been advocated for addition to boost the process efficiency of alcoholic beverage production.

ENZYMES USED IN BREWING

There is a greater diversity of exogenous enzymes considered for use in brewing than any other alcoholic beverage.[1] This reflects both the substantially greater complexity of beer production than that of other beverages, as well as the greater technical advancement of the process.[2] Having said this, there remain large swathes within the brewing industry, which refuse to use process aids, including commercial enzymes. Table 1 summarizes the malting and brewing processes from a perspective of the impact of both endogenous and exogenous enzymes.

β-Glucanases

The endogenous endo-β-glucanases of malted barley are very heat-sensitive. If the β-glucan-rich walls of the starchy endosperm are not effectively degraded during germination, highly viscous glucan will be extracted into the wort during mashing. One approach to solve this problem is to achieve the hydrolysis of residual glucan using endogenous enzymes at a reduced mashing temperature;[3] however, it has been long advocated that heat-tolerant β-glucanases from microorganisms be used to allow degradation of the glucans at the higher temperature (circa 65°C) needed for the gelatinization and hydrolysis of starch, which would avoid the need for a low temperature mashing-in. These enzymes are particularly valuable if the brewer employs adjuncts or specialty malts that are rich in β-glucan, e.g., raw barley, flaked barley, *Torrefied* barley (barley subjected to intense preheating in a micronization regime), roast barley, and oats.

Thermotolerant β-glucanases may be derived from either bacteria or fungi. The most prominent example of the former is the enzyme from *Bacillus subtilis*; *Trichoderma viride* and *Penicillium funiculosum* are popular sources from fungi. The enzyme from *B. subtilis* has a mode of action closely similar to that from malted barley, insofar as it catalyzes the hydrolysis of β(1→4) linkages adjacent to β(1→3) bonds on the reducing side. In contrast, there is a diversity of enzymes within preparations derived from the fungi, including a range of endo- and exo-acting enzymes that attack both β(1→4) and β(1→3) linkages in endo- and exo-mode. As a consequence, there is a range of products, notably lower-molecular-weight oligosaccharides and some preparations are capable of hydrolyzing the β-glucan as far as the monomer, glucose.

Endoxylanases

Barley cell walls also contain approximately 20% arabinoxylan, and indeed, the walls of wheat, which is widely used as a brewing raw material, are primarily pentosan. Thus endoxylanases, which disrupt the β(1→4) linkages in the xylan backbone, have also been shown to be of benefit in enhancing brewhouse performance and for removal of downstream problems such as slow beer filtration. These

Table 1 Enzyme action in malting and brewing.

Process stage	Treatments	Events	Endogenous enzymology	Exogenous enzymology
Raw barley	Storage—perhaps to break dormancy	Hormonal changes—ill-defined	Few enzymes in raw barley: main ones carboxypeptidase and bound, inactive β-amylase	
Steeping	Water added, interspersed by air rests, to raise water content of embryo and endosperm; up to 48 hr at 14–18°C	Synthesis of hormones by embryo, hydration of "substrate" (starchy endosperm)	No apparent increases reported	
Germination	Controlled sprouting ("modification") of grain—typically 4–5 days at 16–20°C	Synthesis of enzymes by aleurone and migration into starchy endosperm; sequential degradation of cell walls, some protein, small starch granules and pitting of large granules	Solubilization of β-glucan by solubilases and endo-β-glucanase; degradation of arabinoxylans by arabinofuranosidase and endoxylanase; partial hydrolysis of proteins by endopeptidases and carboxypeptidase; development and limited action of α-amylase; splitting of β-amylase from protein Z; synthesis of bound and free limit dextrinase, and activation of latter	Microbial flora may contribute enzymes; opportunities for use of selected starter cultures
Kilning	Heating of grain through increasing temperature regime (50–220°C) for desired properties: enzyme survival, removal of moisture for stabilization, removal of "raw" flavors, development of "malty" flavors and color	Enzyme survival greater with low temperature start to kilning and lower final "curing" temperature. Increased heating of malts of increased modification (i.e., higher sugar and amino acid levels) gives increasingly complex flavors and colors via Maillard reactions	Some continued action of all enzymes at lower onset temperatures; but then solely an enzyme inactivation issue. Lability of enzymes endo-β-glucanase, limit dextrinase, lipoxygenase > endopeptidase > β-amylase, lipase > solubilase > α-amylase, peroxidase	
Malt storage	3–4 weeks ambient storage, otherwise wort separation problems later	Unknown, but may relate to development of cross-links between proteins through oxidation in mashed unstored malt	Lipoxygenase may catalyze this reaction (c.f. enhancement of bread-making by analogous reaction in wheat protein); lipoxygenase decays during malt storage	
Mashing	Extraction of milled malt at temperatures between 40 and 75°C	Enzymolysis continued; gelatinization of starch at >62°C	Continued β-glucanolysis favored at low temperatures—also possibly further proteolysis; starch degradation greatly facilitated by gelatinization; balance of enzymes acting faster at higher temperatures with increased destruction of more sensitive ones	Use of heat-stable β-glucanase from Bacillus or fungi comprises main use of exogenous enzymes in high malt mashes; use of glucoamylase to promote fermentability (light beers)

(*Continued*)

Table 1 Enzyme action in malting and brewing. (*Continued*)

Process stage	Treatments	Events	Endogenous enzymology	Exogenous enzymology
Use of adjuncts	Solid adjuncts used in brewhouse, taking advantage of malt enzymes (liquid sugars are products of acid and enzyme action in sugar factory and added at boiling stage)	Cereals with higher starch gelatinization temperatures than for barley need precooking before combining with main mash	Ditto—also a degree of dilution of malt enzymes, especially with high adjunct use	Use of highly heat-resistant α-amylase to promote gelatinization in cooker. Use of amylase, protease, β-glucanase mixtures in main mash
Boiling	1–2 hr at 100°C, before cooling	To sterilize, extract hops, concentrate, and kill all residual enzymes	No enzymology	
Fermentation	Wort pitched with yeast and fermented for 3–14 days at 6–25°C	Fermentation of glucose, maltose, sucrose, maltotriose to alcohol; enzymic production of various flavorsome compounds (alcohols, esters, fatty acids, sulfur-containing compounds etc.). Synthesis and removal of diacetyl as an offshoot of amino acid production	Embden–Meyerhof–Parnas pathway and offshoots	Addition of acetolactate decarboxylase to convert acetolactate precursor to acetoin, thereby circumventing "natural" route which is non-enzymic breakdown of acetolactate to diacetyl (butterscotch), which is slowly reduced by yeast enzymes to less flavor-active acetoin
Cold conditioning and filtration	−1°C for ≥3 days; then filtration	Precipitation, sedimentation, and removal of solids	Slow action by any enzymes "leaked" from yeast, e.g., proteinases: detrimental to foam	Filtration can be limited by viscous polysaccharides, ergo advantage of using β-glucanase in brewhouse (or fermenter). Some use papain as a haze-preventative—but risk of removing foam polypeptides
Package	Market-driven	Progressive deterioration by chemical reactions, including oxidation	Unpasteurized, "sterile-filtered" beer, may retain some of these enzymes	Use of glucose oxidase/catalase as an oxygen scavenger has been suggested

enzymes are from various fungi such as *Humicola insolens* and *Thermomyces lanuginosus*.

It has been shown that a meld of β-glucanase and xylanase is especially beneficial in mashes with less well-modified malts and high barley grists. The xylanase strips away the pentosan that encloses the β-glucan, making the latter more accessible to enzymatic degradation by glucanases.[4] It is unclear whether additional enzymes also enhance the process of removing entities from the pentosan, such as esterases stripping-off acetyl and feruloyl residues.

Proteinases

Endopeptidase enzymes, e.g., *from Bacillus subtilis* and *Aspergillus niger*, are sometimes advocated, especially for high adjunct mashes; however, the endogenous endopeptidases of malt should have completed their job in the germination of well-modified malt. Although residual polypeptides in beer are important for foam stability, some residual protein is disadvantageous as it potentiates haze formation. It seems that polypeptides originating in the hordein storage proteins of barley are especially problematic. (Hordein is a complex mixture of proteins contained in barley.) For many years, some brewers have used papain from the papaya latex as a colloidal stabilizer; however, papain is relatively non-specific and tends to hydrolyze desirable foaming molecules as well as the unwanted haze proteins. Recently, prolyl endopeptidase (produced by *Aspergillus* and *Myxococcus*) has been advocated for use, as it is specific to peptide bonds involving proline, an imino acid that is prevalent in hordein.[5] The enzyme may have added benefit in the production of beers free from the hordein-derived peptide sequences that are believed to make beers unsuitable for celiac sufferers.

α-Glucanases

For all but the most poorly modified malts and/or very high adjunct mashes, there is no shortage of the highly stable endogenous enzyme, α-amylase, the dextrinizing enzyme; however, β-amylase is more heat-sensitive (approximately 50% of the activity will be lost in 30 min of mashing at 65°C), and it may be limiting in mashes of less well-modified malt or when large amounts of adjunct form part of the grist, thereby diluting the level of endogenous enzyme. This leads to a reduced fermentability, as the main product of the action of this enzyme is maltose. β-Amylase from *Bacillus cereus* is available commercially for brewers who wish to boost levels of this enzyme in the mash.

The other enzyme even more likely to be limiting in a mash is limit dextrinase, which is bound to an inhibitor specific to it in malt, requiring unrealistically low mashing pHs to release it in an active form at sufficient levels. To boost fermentability, some brewers add pullulanase (e.g., from *Klebsiella aerogenes*, *Bacillus acidipullulyticus*, or *Bacillus subtilis*), or an $\alpha(1\rightarrow6)$ glucanase or glucoamylase (e.g., from *Aspergillus oryzae* or *Rhizopus niceus*), which is capable of hydrolyzing both $\alpha(1\rightarrow6)$ and $\alpha(1\rightarrow4)$ linkages. The sugar spectrum of wort when pullulanase is used is closer to that from a conventional wort (viz. maltose is the principle sugar), although for a wort produced using glucoamylase there is an overwhelming preponderance of glucose. This affects the action of the yeast and the balance of fruit-flavored esters that are produced. Glucoamylase is the enzyme that tends to be used in the production of light and "low-carb" beers. Whereas most exogenous enzymes tend to be added to the mash tun, glucoamylase is either added there or in the fermenter. The advantage of adding enzymes to the mash is that the activity is subsequently destroyed in the kettle boil. If an enzyme such as glucoamylase remains post-fermentation, then it may cause problems, e.g., a beer intended to contain the enzymes so as to produce a light beer may get inadvertently blended with a beer that is not destined to be fully attenuated.

Acetolactate Decarboxylase

The rate-limiting step of most brewery fermentations is not the achievement of the final specific gravity/alcohol content (attenuation) but rather the time taken to eliminate vicinal diketones, notably diacetyl, which impart undesirable flavors to most if not all beers. Diacetyl is produced from spontaneous oxidative decarboxylation of a precursor, acetolactate, which is an intermediate in the synthesis of the amino acid, valine. Acetolactate leaks out of the yeast cell during fermentation. The yeast, if sufficiently vigorous, will consume the diacetyl, reducing it to the much less flavor-active acetoin and then to butanediol. By adding acetolactate decarboxylase (e.g., from *Klebsiella aerogenes*), the brewer succeeds in converting acetolactate directly to acetoin, thereby circumventing the production of diacetyl.[6]

Oxidoreductases

Several such enzymes have been suggested for use in brewing but with no widespread use. An early suggestion was the combined use of glucose oxidase from *Aspergillus* and catalase to eliminate oxygen from beer while avoiding the accumulation of hydrogen peroxide. Similarly, it has been proposed that a combination of superoxide dismutase and catalase would destroy or avoid the development of the reactive oxygen species that promotes beer deterioration. Another proposal has been the use of laccase to simultaneously eliminate oxygen and also the polyphenols that promote haze formation. It is likely that the failure of these enzymes to achieve widespread application is because physical procedures for eliminating oxygen (e.g., use of de-aeration strategies based on purging or membrane technology) are increasingly efficient.

ENZYMES USED IN WINE-MAKING

The process of wine-making is simpler than that of beer-making, creating far fewer opportunities or needs for commercial enzymes. The most widespread application of enzymes in wine-making is use of pectinases (e.g., from *Aspergillus niger*) to reduce problems that pectins cause in the recovery of juice from the must. There is a demand for endo-β-glucanase to deal with problematic glucans produced in grapes that are infected with the noble rot mold, *Botrytis cinerea*, polysaccharides that otherwise jeopardize clarification and filtration. There is also occasional demand for proteinases to eliminate proteins that can interfere with wine clarity.

ENZYMES USED IN CIDER-MAKING

So, too, for the production of hard cider, pectinases are employed to facilitate the running of juice from the "cheese" (i.e., the crushed apple pulp). Blends of polygalacturonase and pectin methyl esterase may be used to counter haze problems.

CONCLUSION

There is a greater potential for use of enzymes in the brewing industry, reflecting a greater complexity in producing the product than is the case for other alcoholic beverages; however, there is an inherent reluctance by many brewers to use exogenous enzymes in beer-making such that enzymes are only employed if an opportunity exists that cannot be achieved in other ways (e.g., the use of glucoamylase in the production of light beers or, in the future, the use of prolyl endopeptidase in the development of traditional styles of beer suitable for those with celiac disease).

REFERENCES

1. Ryder, D.S.; Power, J. Miscellaneous ingredients in aid of the process. In: *Handbook of Brewing*; Priest, F.G., Stewart, G.G., Eds.; Taylor & Francis: Boca Raton, FL, 2006; 333–381.
2. Bamforth, C., Ed. *Brewing: New Technologies*; Woodhead Publishing: Cambridge, 2006.
3. Bamforth, C.W. β-Glucan and β-glucanases in malting and brewing: practical aspects. Brew. Dig. **1994**, *69* (5), 12–16, 21.
4. Scheffler, A.; Bamforth, C.W. Exogenous β-glucanases and pentosanases and their impact on mashing. Enzyme Microb. Technol. **2005**, *36* (5–6), 813–817.
5. Kabashima, T.; Fujii, M.; Meng, Y.; Ito, K.; Yoshimoto, T. Prolyl endopeptidase from *Sphingomonas capsulata*: isolation and characterization of the enzyme and nucleotide sequence of the gene. Arch. Biochem. Biophys. **1998**, *358* (1), 141–148.
6. Godtfredsen, S.E.; Rasmussen, A.M.; Ottesen, M.; Rafn, P.; Peitersen, N. Occurrence of α-acetolactate decarboxylases among lactic acid bacteria and their utilization for maturation of beer. Appl. Microbiol. Biotechnol. **1984**, *20* (1), 23–28.

Alcoholic Beverages: Starter Cultures

Moustapha Oke
Syed Naqvi
Food Safety Science Unit, Ontario Ministry of Food and Rural Affairs, Guelph, Ontario, Canada
Gopinadhan Paliyath
Department of Plant Agriculture, University of Guelph, Guelph, Ontario, Canada

Abstract

The alcoholic fermentation is a process of converting sugars into alcohol and carbon dioxide primarily by yeast. Alcoholic beverages include beer, wine, fermented cider, and distilled spirits. Traditionally, alcoholic beverages were obtained through spontaneous fermentation of musts or juices by wild yeasts found in the raw material or environment. The disadvantage of spontaneous fermentation is the inability to control the fermentative process and generate quality products in a consistent manner. Products from spontaneous fermentation are suitable for only small producers and specific beverages, which cannot be obtained otherwise. Industrial production of alcoholic beverages using starter cultures guarantees the production of high quality products in a consistent and timely manner. Apart from general demands for starter cultures from the view of quality, safety, technological effectiveness, and economics, many other attributes are considered when selecting strains for different alcoholic beverages. This entry describes the history, production, selection criteria, and the characteristics of ideal starter cultures.

INTRODUCTION

The successful manufacture of alcoholic beverages such as beer, wines, fermented cider, and distilled spirits relies on the presence, growth, and metabolism of specific microorganisms. These microorganisms are responsible for converting raw materials into fermented products with desired properties; therefore, it is important to ensure that fermentation always starts and proceeds in the same manner to generate a consistent product. Fermentation can be initiated using three different approaches: 1) the traditional method, which relies on indigenous microorganisms present in the raw materials (e.g., wine grapes) and equipment; 2) the back-slopping method, which uses a portion of a successful fermentation to initiate the fermentation of fresh raw material; and 3) use of starter cultures, which relies on the addition of characterized strains for the fermentation. Back-slopping is still used in beer processing, usually in small-scale production facilities or home brewing. Louis Pasteur demonstrated that fermentation and food spoilage are caused by microorganisms; this discovery together with the work of others scientists led to the isolation and identification of the organisms responsible for fermentation, as well as the development of product-specific starter cultures. The objective of this entry is to briefly review the history of starter cultures used for alcoholic beverages and their importance and use in the production of alcoholic beverages.

HISTORY OF STARTER CULTURES: CONCEPTS AND DEFINITION

Tremendous contributions to the development of the modern food microbiology were made by Pasteur, Lister, Koch, and Ehrlich in the 1860s.[1] While early scientific efforts in microbiology were directed to the study of infectious diseases, research focus broadened to include food microbiology in the 20th century. The use of starter cultures by the brewing industry first arose in Denmark in 1883 by Christian Hansen.[1] Using the dilution method, he isolated pure cultures of the brewing yeasts, *Saccharomyces cerevisiae* and *Saccharomyces carlsbergensis*, used for ale and lager types of beer, respectively. *S. cerevisiae* is also the common starter culture for distilled spirits, but more hybrids with the ability of *S. diastaticus* to ferment di- and tri-saccharides are available. Even though Pasteur and other microbiologists, such as Hermann Muller-Thurgau in 1890, demonstrated that wine fermentations were performed by yeast, wine yeast did not become commercially available as starter cultures for wine production until the 1960s.

IMPORTANCE OF STARTER CULTURES IN THE FOOD INDUSTRY

Starter cultures consist of microorganisms that are inoculated directly into unfermented food materials to prevail

Table 1 Comparison of the traditional[a] and modern methods of production of alcoholic beverages (e.g., beers, wines).

Traditional method	Modern method
Small scale (craft industry, homemade)	Large scale (in processing plant)
Non-sterile medium	Pasteurized or heat-treated medium
Varying quality	Consistent quality
Open	Contained
Manual	Automated
Significant exposure to contaminants	Minimal exposure to contaminants
Safety is a less of a concern	Safety is a major concern
Septic	Aseptic
Insensitive to time (length of fermentation difficult to predict)	Time-sensitive (optimized to a consistently short time period)

[a]A traditional method includes back-slopping or fermentation initiated by endogenous microbiota, whereas a modern method directly inoculates starter cultures into the unfermented materials.

over the existing microbiota and result in desired characteristics in the finished product. These characteristics include, but are not limited to, enhanced preservation, increased food safety, enhanced sensory attributes, novel functionality, increased economic value, and improved nutritional or health value. Advocates of the traditional manufacturing methods often claim that natural fermentation, whether by back-slopping or initiated by endogenous microbiota, yield products with better quality attributes. For instance, it is argued that naturally fermented wines are superior to wine made using starter cultures. For small-scale processing facilities, where time and quality are flexible, this attitude can be justified; however, for modern large-scale industrial production of alcoholic beverages, consistent product quality, predictable production schedules, and stringent quality control measures are critically important. The addition of concentrated microorganisms with defined attributes in the form of a starter cultures ensures that products are manufactured in a timely manner with consistent and predictable quality. A comparison of traditional and modern methods is shown in Table 1.

PRODUCTION OF ALCOHOLIC BEVERAGES

Alcoholic Fermentation and Yeast Starter Cultures

Although there are considerable differences among beer, wine, distilled spirits, and other alcoholic beverages, these products all require the metabolic action of yeasts, mostly *S. cerevisiae* or in the case of beers, usually *S. carlsbergensis*. Yeasts are unicellular fungi that reproduce asexually by budding or fission and are responsible for alcoholic fermentation. The fermentation of sugars results in the production of ethanol and carbon dioxide and can be summarized as:

$$C_6H_{12}O_6 \text{ (glucose)} + \text{yeast} \rightarrow 2CH_3CH_2OH \text{ (ethanol)} + 2CO_2 \text{ (carbon dioxide)}$$

The alcoholic fermentation pathway is shown in Fig. 1.

S. cerevisiae is undoubtedly the most important yeast species. The original habitat of *S. cerevisiae* or its progenitors is uncertain.[2] Strains, such as those used in winemaking, appear to occur in significant numbers only in human habitats.[3,4] Cavalieri et al.[5] showed the molecular presence of *S. cerevisiae* in wine from pottery jars dated 3150 B.C., buried in the tomb of King Scorpion I, one of the first Egyptian kings. *Saccharomyces paradoxus* (*S. cerevisiae* var. *tetrasporus*), often isolated from oak tree exudates, is believed to be the ancestral form. This species is occasionally isolated in nature from the intestinal tract of fruit flies (*Drosophila* spp.) and may be transmitted by bees and wasps, but the importance of the insects in the dispersal of *S. cerevisiae* is unresolved.[2] *S. cerevisiae* is usually absent on grapes. The presence of the wine yeast becomes noticeable

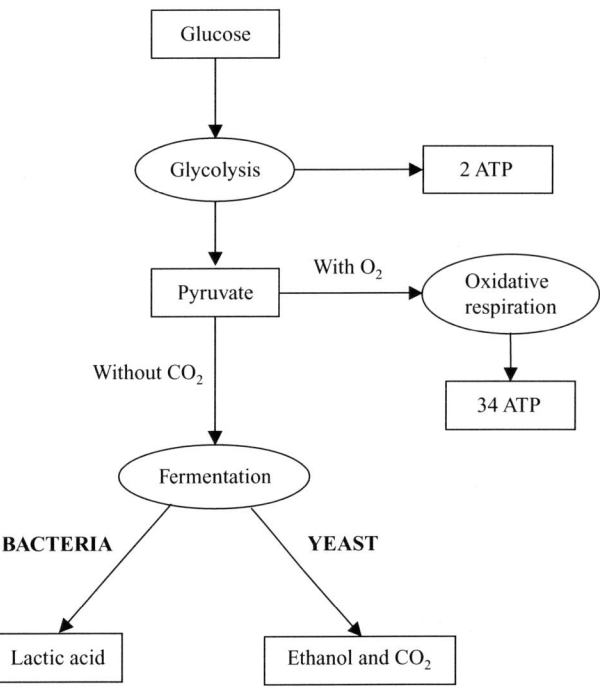

Fig. 1 Alcoholic and lactic acid fermentations: the Embden–Meyerhof pathway.

on grapes only near the end of fruit ripening. It is estimated that one healthy berry out of 1000 carries wine yeasts, although on the surface of damaged fruit the frequency may increase to approximately one in four.[6] Other wine yeasts, such as *Saccharomyces uvarum* and *Saccharomyces bayanus*, can also conduct effective alcoholic fermentations even though these varieties are less commonly used except for some specific winemaking situations. For example, *S. uvarum* is often preferred when the fermentation temperature is below 15°C; *S. bayanus* is mostly used in the production of sparkling wines and fino sherries. The other group of wine-related yeasts is the so-called "spoilage" yeasts, which include, but are not limited to, *Hanseniospora* sp., *Kloeckera apiculata*, *Candida stellata*, *Pichia membranefaciens*, *Hansenula* sp., *Metschnikowia* sp., *Sporobolomyces* sp., *Cryptococcus* sp., *Rhodotorula* sp., *Torulaspora delbrueckii*, and *Aureobasidium* sp. In the distilled spirit industry, the use of excess yeast obtained from the brewing industry is being replaced by the use of distilling hybrids with the ability to ferment some unfermentable starch-derived polysaccharides. Starter culture strain selection is based primarily on the desired flavor and sensory attributes of the final product. Other desired traits specific to alcoholic beverages are summarized in Table 2.

Table 2 Desirable properties of alcoholic starter cultures: yeasts and lactic acid bacteria (LAB).

Products	Purpose of use
Beer (yeast)	Ensure rapid fermentation
	Stable and tolerant to preservatives
	Produce desired flavor
	Lack off-flavors
	Flocculate (very flocculent are "gravelly;" non-flocculent are "powdery;" the ideal yeast is somewhere near to the middle of the range)
	Have proper attenuation (to achieve efficient conversion of sugars to ethanol)
	Tolerant to osmotic pressure, temperature, and handling stresses
	Growth at wide temperature range (5–18°C for lager and 15–25°C for ale beers)
Wine (yeast)	Able to complete fermentation (alcohol tolerance)
	Able to grow under high concentration of musts
	Able to metabolize large portion of malic acid in must
	Be tolerant to sulfur dioxide (SO_2)
	Tolerant to extreme fermentation temperatures
	Able to produce small or no amount of acetic acid, acetaldehyde, H_2S, and mercaptans, diacetyl, SO_2, higher alcohols
	Tolerant to high pressure
	Able to produce a desirable varietal flavor
	Able to produce a desirable amount of glycerol
	Able to produce inhibitory substances against spoilage microorganisms
	Give good flocculation, sedimentation, or agglomeration following fermentation and be able to help clarification
	Have low foaming
	Suitable for drying
	Have a low β-glucosidase activity
	Have low urea excretion
	Resistant to killer yeast
	Have low efficiency in converting sugar to alcohol
	Able to enable ester formation and to produce protease
Wine (LAB)	Be safe, free of any pathogen or toxic activity
	Free of any hygienic precarious infections or substances
	Dominate over the spontaneous microbiota
	Perform the required metabolic activity
	Have feasible propagation from the economical point of view
	Able to be preserved by freezing or freeze-drying with little practical loss of activity
	Stable under defined storage conditions for several months
	Easy handling
Spirits	Provide complete and rapid fermentation of wort sugars
	Able to complete fermentation in final 8–10% ethanol content of wash
	Able to grow well above 30°C
	Able to produce good flavor
	Tolerant to the osmotic stress of the initial sugar concentration in the wort (16–20° Brix)
	Lack flocculence
	Have minimum frothing

Malolactic Fermentation and Bacterial Starter Cultures

The malolactic fermentation (MLF) is essential for the deacidification of high-acid grape must. MLF involves the transformation of L-malic acid to L-lactic acid and carbon dioxide and is conducted by lactic acid bacteria (LAB; Fig. 1). MLF occurs naturally toward the end or after alcoholic fermentation. Many factors influence the process, including the pH of the medium, temperature, SO_2 content, alcohol content, and the presence or absence of phage.[7] Commercial starter cultures for inducing MLF became available in the 1980s and consist of strains of *Leuconostoc oenos* as single or multiple-strain preparations.[8] The advantage of initiating MLF by inoculation with starter culture is to have better control over the time of onset and rate of completion of MLF.[7] The general requirements for the LAB starter cultures with regards to safety, technological effectiveness, and economics are summarized in Table 2.

Some malolactic bacteria cannot effectively complete the MLF, and consequently winemakers must age wines for longer periods resulting in the possible formation of bioamines (biogenic amines). Spoilage bacteria and some malolactic strains can convert naturally occurring amino acids in wine into bioamines. The presence of bioamines can cause adverse reactions to some drinkers of red wine. An example of a bioamine found in wine is histamine, which can cause allergic reactions, headaches, asthma attacks, and skin rashes.[9] Efforts are currently underway to design malolactic bacteria with reduced ability to synthesize bioamines.

Starter Culture Technology

The most commonly used starter cultures in alcoholic beverage production is summarized in Table 3. The general flow chart for industrial production of starter cultures is shown in Fig. 2.[1,10]

APPLICATION OF MOLECULAR BIOLOGY TO THE DEVELOPMENT OF STARTER CULTURES

One of the important developments in the starter culture industry has been the application of molecular biology to improve starter cultures. It is now possible to modify or manipulate the phenotype of starter yeasts or bacteria, either by increased expression or inactivation of particular genes or by introduction of foreign (heterologous) DNA into selected organisms. The resulting genetically modified organisms have the capacity to generate products with better quality and safety attributes.

Several other genetic methods are available for improving yeasts used in alcoholic fermentation including clonal selection of variants, mutation and selection, hybridization, rare-mating, and spheroplast fusion.[11] Some of these methods alter defined regions of the genome, while others can be used to recombine or rearrange the entire genome.

Another trend in the science of starter cultures is the development of genetically engineered malolactic wine yeast. An example is the malolactic yeast, *S. cerevisiae* ML01, recently approved by the U.S. FDA and Health Canada for use in commercial wine production.[9] ML01, a genetically stable industrial strain, was constructed by integrating a linear cassette into the URA3 locus of wine yeast. The linear cassette contains the *Schizosaccharomyces pombe* malate permease gene (mae1) and the *Oenococcus oeni* malolactic gene (mleA) under control of the *S. cerevisiae* PGK1 promoter and terminator sequences. This strain has the potential to help solve issues related to bioamine formation as well as increase the organoleptic characteristics of wines.[9]

Table 3 Bacteria, yeasts, and molds used as starter cultures in alcoholic beverages.

Organisms	Products
Oenococcus oeni	Wine, both white and red. Recommended for chardonnay
Saccharomyces cerevisiae	Top-fermenting yeast. Used in ales. Ales are fruitier flavored than lager. Also used in stout beers (e.g., Guinness), specialty beers (e.g., low-calorie, low-alcohol beers)
Saccharomyces pastorianus	Bottom-fermenting yeast used in lager beer production
S. cerevisiae	Wine, both white and red
Hybrid (*S. cerevisiae/S. diastaticus*)	Distilled spirits, e.g., whisky
Malolactic yeast ML01[a]	Wine
Aspergillus oryzae	Saké, Chinese spirits
Candida spp., *Geotrichum candidum* and *Lactobacillus* spp.	Yeasts and bacteria used in the production of Pito, a Nigerian fermented drink (~3% alcohol) made of maize, sorghum, or both
S. cerevisiae, *Candida* spp.	Kaffir beer, palm wine
S. cerevisiae, *Schizosaccharomyces pombe*	Rum, aquavit, palm wine

[a] Genetically modified strain, called ML01 approved by the U.S. FDA to be used in commercial wine. ML01 malolactic yeast was created using the gene that catalyzes the conversion of malic acid to lactic acid from the malolactic bacterium *Oenococcus oeni* and added it to the common wine yeast strain, *S. cerevisiae*. Also added was the malic acid transport gene from *S. pombe*.

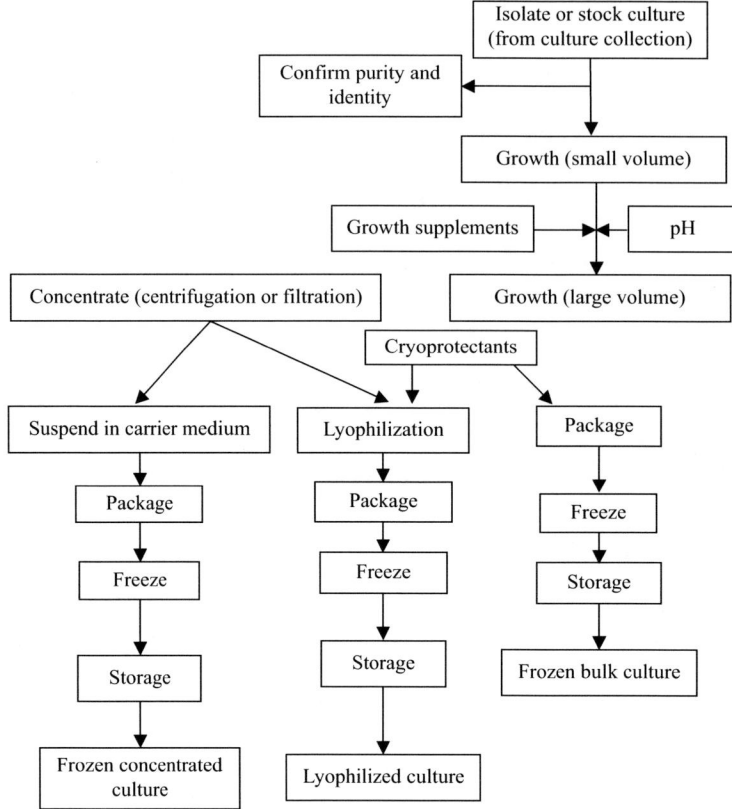

Fig. 2 Flow chart for starter cultures production. **Source:** Adapted from Buckenhuskes[10] and Hutkins.[1]

CONCLUSIONS

Alcoholic beverages heavily rely on starter cultures to ensure high volume production of products with consistent quality and safety in a timely manner. In addition to the general requirements for starter cultures that includes safety, technological effectiveness, and economics, commercial strains are selected primarily on the desired flavor and sensory attributes found in the final product. Other important attributes include the ability of strains to grow at high sugar concentrations, to produce adequate ethanol levels, and to flocculate under appropriate conditions. There is a potential for using molecular biology in the development of starter cultures with improved quality and safety attributes. A recent success is the introduction of the starter yeast *S. cerevisiae* ML01 in Canada and the United States; however, concerns still remain in the acceptance by consumers of alcoholic beverages produced using genetically modified organisms.

REFERENCES

1. Hutkins, R.W. Starter cultures. In *Microbiology and technology of fermented foods*; 1st Ed.; Hutkins, R.W., Ed.; IFT Press; Blackwell Publishing: Ames: Iowa, 2006; 69.
2. Phaff, H.J. Ecology of yeasts with actual and potential value in biotechnology. Microb. Ecol. **1986**, *12*, 31–42.
3. Jackson, R.S. *Wine Science: Principles, Practice, Perception*; Elsevier: United States, 2000; 307.
4. Ciani, M.; Mannazzu, I.; Marinangeli, P.; Clementi, F.; Martini, A. Contribution of winery resident *Saccharomyces cerevisiae* strains to spontaneous grape must fermentation. Antonie Van Leeuwenhoek **2004**, *85*, 159–164.
5. Cavalieri, D.; McGovern, P.E.; Hartl, D.L.; Mortimer, R.; Polsinelli, M. Evidence for *S. cerevisiae* fermentation in ancient wine. J. Mol. Evol. **2003**, *57* (Suppl. 1), S226–S232.
6. Mortimer, R.; Polsinelli, M. On the origins of wine yeast. Res. Microbiol. **1999**, *150*, 199–204
7. Lonvaud-Funel, A. La fermentation malolactique. In *Les Bacteries Lactiques. Actes du Colloque LACTIC 91*, Adria Normandie, Centre de Publication de l'Universite de Caen: F-14032 CAEN CEDEX France, 1992.
8. Hammes, W.P. Bacterial starter cultures in food production. Food Biotechnol. **1990**, *4*, 383–397.
9. Husnik, J.I.; Volschenk, H.; Bauer, J.; Colavizza, D.; Luoa, Z.; van Vuurena, H.J.J. Metabolic engineering of malolactic wine yeast. Metab. Eng. **2006**, *8*, 315–323.
10. Buckenhuskes, H.J. Selection criteria for lactic acid bacteria to be used as starter cultures for various food commodities. FEMS Microbiol. Rev. **1993**, *12* (1/3), 253–271.
11. Pretorius, I.S. Grape and wine biotechnology: setting new goals for the design of improved grapevines, wine yeast and malolactic bacteria. In *Handbook of Fruits and Fruit Processing*; Hui, Y.H., Barta, J., Cano, M.P., Gusek, T., Sidhu, J.S., Sinha, N., Eds.; Blackwell Publishing, Ames: Iowa, 2006; 453–489.

Allergies: Food and Peanut Risk Reduction

Hortense Dodo
Koffi Konan
Department of Food and Animal Science, Alabama A&M University, Huntsville, Alabama, U.S.A.

Abstract

Type I food allergy is increasing worldwide. It is an IgE-mediated hypersensitivity reaction to naturally occurring proteins. Many foods of animal and plant sources provoke allergic reactions. Several approaches have been proposed as a solution to food allergy; however, to date there is no cure. Strict dietary vigilance and total avoidance of the suspected food are the best management practices. Genetic modification of foods is a very promising approach to reduce and eliminate targeted allergens and to reduce the allergenic potency of food without collateral consequences on its nutritional quality. Success in the application of the tools of genetic modification to agricultural crops has been recorded with the silencing of allergen genes in many plants, such as apple, soybean, tomato, and rice. This approach is being applied to peanut in order to reduce peanut allergy risks.

INTRODUCTION

Allergy to food is usually classified as type I food allergy. It is an abnormal response of the immune system to a particular food component, usually naturally occurring proteins that are normally not harmful to the body. Sensitization is caused by the development of protein-specific IgE antibodies. Subsequent exposure to the allergenic proteins causes the IgE to trigger allergic reactions that may involve the skin (rash, hives, and eczema), swelling of the mouth, tongue, and throat, gastrointestinal tract (nausea, abdominal cramping and pain, diarrhea, and vomiting), respiratory system (sneezing, wheezing, nasal congestion, coughing, and difficulty in breathing), and cardiovascular system (drop in blood pressure, irregular heart beat, and cardiac arrest).

Food allergy affects millions of people around the world. Its prevalence is not precisely known, but varies by country and by food. The phenomenon is rapidly increasing in industrialized nations with an estimate of a 25% increase every 10 years in children and in the general population. Food allergy affects 6–8% of children younger than 3 years of age and approximately 4% of adults. In the United States, it is estimated that 1.1–5.3% of the population has a food allergy, which accounts for 35–50% of emergency room visits. Each year, about 30,000 food-induced anaphylactic events are recorded in the United States, of which 200 are fatal. In Canada, 1–2% of the general population is considered to be at risk of food anaphylaxis. In European countries, the prevalence of food allergy is estimated at 4.7 and 3.2% in children and adults, respectively. In those countries, life-threatening anaphylaxis occurs in 1 in 10,000 inhabitants and fatal anaphylaxis in 1 in 1 million inhabitants. A drastic increase in life-threatening and lethal anaphylaxis has been noted in the U.K. and Australia over the past 10 years. In France, there has been an increase of 28% in prevalence of Food Allergy between 2001 and 2006.

Despite the enormous diversity of food antigen exposure, only a few foods account for most food allergies. Bovine milk, egg, peanut, tree nuts, soy, wheat, fish, and shellfish account for more than 85% of documented food allergies. In infants from newborn to 3 years of age, the most frequent causes of food allergy are egg (77.5%) and milk (29.2%) while in children aged 3–15 years and adults, peanut is the most allergenic food affecting 49.3 and 10.1% of the population, respectively. Peanut is the most common cause of severe or fatal food-associated anaphylaxis in the United States, where it is implicated in 59% of the 63 deaths caused by food anaphylaxis. Unlike milk and egg allergies, which typically resolve during early childhood, peanut allergy tends to persist throughout the life. Although Peanut allergy can be resolved, its recurrence was observed in about 8% of children.

Yet to date, there is no cure for food allergy. Application of genetic engineering strategies to reduce the risk of food allergy by elimination of specific allergenic proteins is very promising. In this entry, we will provide an overview of food allergy, list some of the proteins identified as food allergens, and describe some of the current applications of genetic engineering approaches on agricultural crops to alleviate food allergy. An emphasis will be made on peanut, which is the source of the most severe form of food allergy.

MECHANISM OF FOOD ALLERGY

The human immune system produces five types of antibodies (IgG, IgM, IgA, IgD, and IgE) to defend the body against attacks by foreign invaders such as infectious organisms (bacteria, fungi, parasites, and viruses). A food allergy is a condition in which the immune system incorrectly identifies a food protein as a threat and attempts to protect the body against it. What makes a food protein an allergen is unknown; however, a few characteristics appear to be important in establishing a food protein as an allergen. These characteristics include their increased abundance in the food supply, their durability during food processing, and their resistance to digestion in the gastrointestinal tract.

Food allergies develop in genetically predisposed individuals when the immune system fails to induce oral tolerance to invading foreign food proteins, or when there is a breakdown in existing tolerance. The human gastrointestinal tract is the largest immunologic organ in the body. Lining this tract is a layer of epithelium under which is a mucosal immune system made of stroma of loose connective tissue populated by lymphocytes. The immune system is considered as a line of defense, which attempts to prevent foreign antigens to cross the intestinal epithelial barrier. In the process of digestion, dietary proteins are degraded in the gastrointestinal track and their epitopes are destroyed by gastric acidity and digestive enzymes, often resulting in the destruction of immunogenic epitopes. Proteins that are not digested and processed in the lumen of the gut contact the epithelium and the mucosal immune system beneath it. Then, they provoke protective local and systemic immunologic responses of which secretion of the local non-inflammatory IgA antibody is the initial response to take place. Escape proteins are presented to T-lymphocytes immune competent cells such as CD4+ and CD8+ T cells. This presentation is done by cells (macrophages, dendritic cells, and B-lymphocytes) that are called professional antigen-presenting cells (APCs) and which are able to take up the antigen, process it within the cell, and present it on the cell surface bound to a receptor known as a major histocompatibility complex (MHC) type II. Activated antigen-specific CD4+ and CD8+ T cells migrate to lymphoid organs as well as to target organs to suppress immune responses by producing the down-regulatory cytokines IL-10 and TGF-B. Consequently, the usual response to the antigens that have crossed the barrier is a systemic immunological unresponsiveness known as oral tolerance, which protects against the invading antigens on subsequent exposures.

Abrogation of tolerance or failure to induce tolerance due to defects in regulatory T cell activity leads to the induction and perpetuation of food allergy. Activated antigen-specific CD4+ T cells produce different kinds of cytokines including IL-4, IL-5, IL-9, and IL-13 that stimulate the production of IgE antibodies by B cells. Allergen-specific IgE antibodies have the capacity to bind with the high affinity FcεRI on mast cells and basophils to induce sensitization. Upon subsequent exposure to the allergen, membrane-bound IgE molecules get cross-linked and provoke degranulation of mast cells and basophil release of mediators. These mediators lead to a variety of cutaneous, gastrointestinal, respiratory, and systemic symptoms.

Allergic reactions to food proteins also occur independently from IgE. In contrast to IgE-mediated reactions that are rapid in onset, non-IgE-mediated disorders become evident hours to days after allergen ingestion. To date, the mechanisms underlying non-IgE-mediated food hypersensitivity reactions are not clear. It has been suggested that T cells play a role in the immunopathogenesis, also referred to as type IV- or cell-mediated hypersensitivity reactions.

FOOD PROTEINS IDENTIFIED AS ALLERGENS

Type I allergic reactions are caused by proteins that elicit specific IgE antibodies. Most food allergens are soluble proteins, and therefore, they are more tolerogenic than particulate allergens; however, solubility can change during food preparation, and peanut allergens, e.g., can become less soluble with progressive roasting, increasing the capacity of peanut-specific IgE binding to the protein.

Allergens of Animal Sources

The Pfam protein family database, which is a collection of protein families and domains, has classified protein sequences into families on the basis of sequence homology, 3-D structures, and function. Information from this database reveals that almost all the animal food allergens have homologs in the human proteome. Thus, animal food allergens lie at the limits of the capability of the human immune system to discriminate between foreign and self-proteins, and animal food proteins with a high degree of similarity to human homologs would be poorly immunogenic in human and hence less likely to become allergens.

Animal food allergens are classified into three main families, tropomyosins, EF-hand proteins, and caseins, along with 14 minor families. Tropomyosins play one of the key roles in muscle contraction. Vertebrate tropomyosins from mammals, birds, and fish have never been reported to be allergenic; however, allergenic tropomyosins are confined to the invertebrate groups, such as insects, crustaceans, and mollusks. The EF-hand family is composed almost of parvalbumins, which are subdivided into alpha- and beta-parvalbumins. Alpha-parvalbumins are not generally allergenic. They are abundant in the muscles of fish and amphibians and to a minor extent in the fast twitch muscle of birds and mammals. Many beta-parvalbumin allergens are found in a variety of fish species. Caseins are exclusively mammalian proteins found in milk. The IgE reactivity of the various types of caseins seems to be linked to their sequence similarities with human homologs; caseins that are the least like, human are more reactive.

Allergens from Plants Sources

Plant food allergens are identified in three families/superfamilies: 1) the prolamin superfamily; 2) the cupin superfamily; and 3) the Bet v 1 family. The prolamin superfamily comprises three major groups of plant food allergens, 2S albumins, non-specific lipid transfer proteins (nsLTPs), and cereal alpha-amylase/trypsin inhibitors. They have low molecular weight, are rich in cysteine, and are stable to thermal processing and proteolysis. The 2S albumins are a major group of storage proteins present in many dicotyledonous plants. They include major allergens from tree nuts and seeds, such as Brazil nut (Ber e 1 and Ber e 12), walnut (Jug r 1), sesame (Ses i 1), peanut (Ara h 2, Ara h 6, Ara h 7), and mustard (Bra j 1, Sin a 1). The nsLTPs play an important role in plant defense against fungi and bacteria. They have a wider distribution than 2S albumins, with over 100 sequences being available from seeds, fruits, and vegetative tissues, and major allergenic forms in peach (Pru p 1), apple (Mal d 3), and maize (Zea m 14). The family of cereal alpha-amylase and protease inhibitors mediates a certain degree of resistance to insect pests that feed on plant tissues. This protein family comprises allergenic members that are produced in wheat (Tri a 14), barley (Hor v 1), rice (Ory s 12), and rye (Sec c 1).

The cupins share two short conserved sequence motifs and have in common a double-stranded α-helix domain, which led to the name "cupin" based on the latin word for a small barrel or cask. Identified allergens in the cupin family include the 7S globulins of soybean (β-conglycinin), peanut (Ara h 1), walnut (Jug r 2), and 11S globulins of peanut (Ara h 3), and soybean (glycinin).

Two major allergens (Bet v 1 and Bet v 2) are identified in birch pollen. Both allergens show a high degree of sequence homology with several food allergens. Allergens sharing sequence homology with Bet v 1 include Mal d 1 of apple, Api g 1 of celery, Pru av 1 of cherry, Cor a of hazel nut, Dau c 1 of carrot, and Ara h 8 of peanut. Bet v 2 is a profilin and shares sequence homology with allergens such as Ara h 5 of peanut, Gly m 3 of soybean, Api g 4 of celery, and Pyr c 4 of pear. Individuals with pollen allergy frequently suffer from allergic symptoms after eating apple, cherry, apricot, pear, celery, carrot, peanut, and hazel nut because of cross-reactivity between the food allergen and Bet v 1 and/or Bet v 2

Food allergens are also identified in the cysteine protease family, a group of proteins with conserved Gln, Cys, His, and Asn residues at the active site. Food allergens belonging to this group include Act c 1 of kiwi, and the soybean Gly m Bd 30 K.

TREATMENT OF FOOD ALLERGY

To date, there is no cure for food allergy. Emergency treatments include epinephrine and antihistamines, which are administered only to alleviate the symptoms. The best management practice is to encourage strict dietary vigilance and total avoidance of the allergic food. Although this could reduce the likelihood of an accidental ingestion of the suspected food, it is difficult and stressful, as it could result in nutritional deficiency and negatively impact the health, development, and lifestyle of sensitive individuals.

Many efforts have been deployed to combat food allergy, including food processing, anti IgE vaccines or immunotherapy, peptide immunotherapy, and Chinese medicine. Alternatively, with the advent of genetic engineering, new tools and solutions such as the RNA interference (RNAi) technology are being proposed to alleviate food allergy by eliminating the culprits, the allergenic proteins.

GENETIC MODIFICATION TO ALLEVIATE THE RISKS OF FOOD ALLERGY

Progress in molecular biology and genetic engineering techniques has led to the isolation and cloning of genes encoding food allergens. Furthermore, now, there is the possibility to reduce the quantity of allergenic proteins in food by silencing the genes encoding them. Most of the allergenic proteins are present as isoforms, and genes encoding isoforms belong to multigene families with conserved stretches of identical sequences. Therefore, posttranscriptional gene-silencing mechanisms such as those induced by RNAi-based constructs are the most efficient approaches to create hypoallergenic foods by silencing all the isoforms of a target allergen.

RNAi was discovered in the nematode *Caenorhabilitis elegans* in which double-stranded RNA (dsRNA) was found to trigger the silencing of genes containing sequences identical to the dsRNA. It is a conserved mechanism among eukaryotes and triggers gene silencing through small interfering RNAs (siRNAs). siRNAs are first assembled into a multiprotein complex and then in an RNA-induced silencing complex (RISC), which seeks out and cleaves target mRNAs that are complementary to the siRNA. In plants, cosuppression and antisense technology were both found to trigger gene silencing through the cellular production of dsRNAs which activate the RNAi mechanism.

Success has been recorded with the reduction and/or elimination of specific allergens to reduce the risks of food allergy. Antisense gene constructs were used to downregulate a major 14–16 kDa allergen RA17 in rice. In soybean seeds, antisense and/or cosuppression-mediated gene-silencing approach was used to prevent the accumulation of the immunodominant Gly m Bd, a 30-kDa allergenic protein with no collateral alteration of any other proteins. In apple and tomato, RNAi constructs were used to silence the major allergens Mal d 1 and Lyc e 3, respectively. The silencing of Lyc e 3 led to a significant reduction of the allergenic potency of transgenic tomatoes, and the silenced phenotype passed on to the next generation of fruits. In in vitro assay, 10- to 100-fold higher amounts of extracts of transgenic fruits were required to induce the same amount of histamine release compared with extracts from wild-type fruits.

Peanut deserves particular attention as peanut allergy is relatively common, typically permanent, and often severe, resulting in most anaphylaxis shocks and occasional deaths. The incidence of peanut allergy doubled in children born between 1998 and 2003, and now affects over 4 millions Americans, mostly children.

A SUCCESSFUL STRATEGY TO SILENCING PEANUT ALLERGEN GENES

A genetic engineering strategy was established in our laboratory to eliminate major peanut allergens via RNAi and reduce the increasing risks of peanut food allergies. Steps taken include: 1) Isolation of the gene of interest; 2) Construction of the RNAi transformation cassettes using inverted repeats DNA fragments of targeted allergen genes; 3) Transformation of peanut hypocotyl via *Agrobacterium tumefaciens*; 4) Regeneration of kanamycin-resistant plantlets; 5) Molecular analyses to identify transgenic peanut lines; and 6) Immunological analyses of transgenic seeds to determine the level of reduction of targeted allergens.

Application of RNAi to Silence the Major Peanut Allergen Ara h 2

Isolation of genomic DNA encoding the major peanut allergens

Allergens are classified as "major" when they trigger a reaction in over 50% of the hypersensitive population. The three major allergens, Ara h 1, Ara h 2, and Ara h 3/4 are recognized by 65, 85, and 53% of the peanut hypersensitive population, respectively. A prerequisite to silencing peanut allergen genes is the acquisition of DNA fragments encoding the target proteins and their use as transgenes in the construction of RNAi-transformation cassettes. Although cDNA clones are known for all identified peanut allergens, isolation of the genomic DNA may provide additional information on the molecular structure of the genes encoding the major peanut allergens, Ara h 1, Ara h 2, and Ara h 3/4. Ara h 1 is a 63.5-kDa vicilin composed of 626 amino acid residues. Its genomic DNA of 1878 nucleotides is intercepted with 3 introns. Ara h 2, a 207-amino acid residue, is a 17.5-kDa glycoprotein with a full-length genomic clone of 622 nucleotides and no introns. In addition to being the most immunodominant allergen in peanut, Ara h 2 is over 52.5-fold more potent than Ara h 1. It has 76–83% and 76–85% nucleotide sequence homology with Ara h 6 and Ara h 7, respectively. Ara h 6 and Ara h 7 have no significant similarity; they are considered minor allergens with 38 and 43% reaction in a peanut-sensitive population, respectively. Ara h 3 and Ara h 4 are two variants of the same gene. Ara h 3/4 is a 60/61-kDa glycinin of 538 amino acid residues. The genomic DNA is 1614 nucleotides long, intercepted with 3 introns. Because Ara h 2 is the most immunodominant allergen in peanut, it was selected to demonstrate proof of concept in silencing peanut allergen genes via RNAi.

Construction of the RNAi transformation vector and production of fertile transgenic peanut

The vector pHANNIBAL was specifically designed to convert PCR products from a gene of interest into a highly effective intron hairpin RNA (ihpRNA) vector for efficient gene silencing in plants. For *Agrobacterium*-mediated transformation, the silencing cassette is spliced out from pHANNIBAL and subcloned into the unique *Not* I site of a binary plasmid pART27. Ara h 2 specific-silencing vector (pDK28) was constructed using a 265-bp genomic DNA. Plasmid pDK28 was mobilized into *A. tumefaciens* EHA105, and hypocotyl explants from peanut runner Georgia green were infected with *A. tumefaciens*. Fertile transgenic peanut plants were produced within 5–7 months with phenotype, plant growth rate, and reproduction similar to wild type.

The presence and integration of the transgene in the peanut genome was determined by molecular analyses

PCR and Southern hybridization were performed using genomic DNA from peanut leaves to identify true transgenic peanut plants. PCR primers were designed to target the Cauliflower Mosaic Virus 35S promoter (CaMV 35S) fragment not naturally present in peanut. About 44% of kanamycin-resistant plants screened were PCR-positive.

Southern hybridization was performed on DNA of 10 of the PCR positive plants using the CaMV 35S promoter fragment as probe. Selected PCR-positive lines were also Southern positive, confirming the stable transformation of regenerated peanuts. One single copy transgene integration was observed for all ten transgenic plants analyzed.

The level of reduction of the Ara h 2 protein in transgenic peanut seeds was determined by immunological analyses

ELISA and Western blots were performed using crude peanut extract (CPE) of seeds from Southern-positive plants

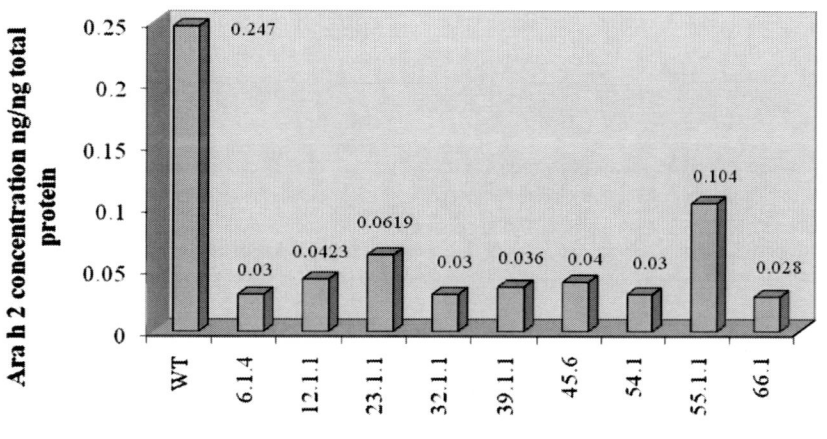

Fig. 1 Sandwich ELISA in crude peanut extracts (CPE) to quantify Ara h 2 content. WT, wild-type non-transgenic.

to determine the level of reduction and/or elimination of Ara h 2 in the seeds. Sandwich ELISA performed using CPE and Ara h 2 monoclonal antibody (mAb) revealed Ara h 2 content in wild-type peanut at 0.247 ng/ng total protein, and 0.028–0.104 ng/ng total protein in transgenic peanuts (Fig. 1). This represents a significant reduction ($p < 0.05$) ranging from 62.5 to 89.8% in the Ara h 2 content of transgenic peanut seeds compared to wild type.

SDS-PAGE and Western blots showed protein profiles and Ara h 2 detection in several transgenic peanut seeds (Fig. 2). The protein profiles of most of the transgenic seeds were similar to that of the wild type. The characteristic Ara

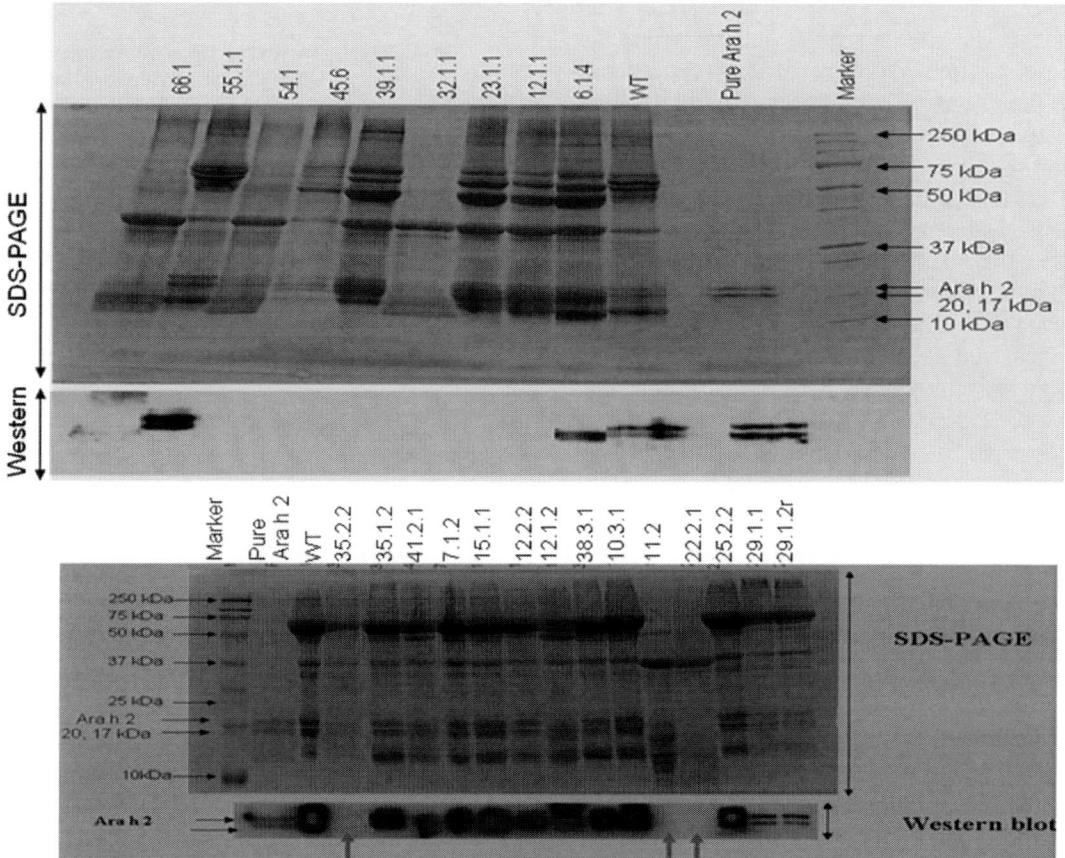

Fig. 2 Analysis of peanut seeds showing protein profiles (SDS-PAGE) in several transgenic seeds, and no detection of Ara h 2 protein (Western blots) in peanut extracts using monoclonal antibody specific to Ara h 2. Arrows indicate Ara h 2 not detected. WT, wild type.

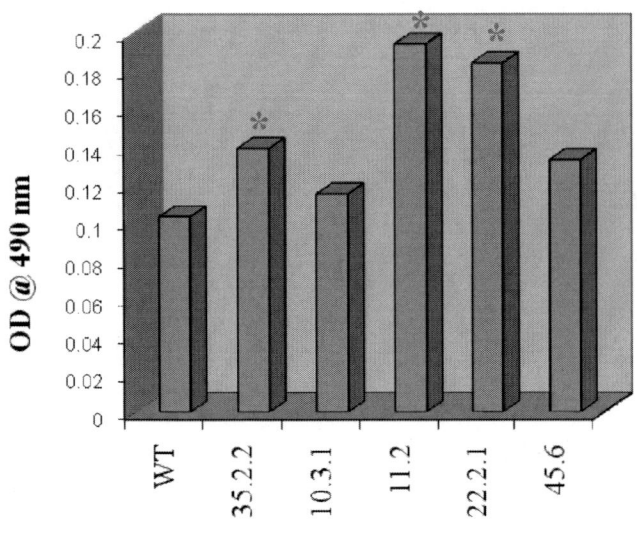

Fig. 3 Quantification of IgE-binding for serum IgE of selected CPE from four transgenic peanut seeds using competitive inhibition assays. *Indicates significant ($p > 0.05$) difference compared to WT (Duncan mean separation tests).

h 2 doublet migrated on SDS-PAGE as expected around 17–20 kDa in the purified Ara h 2 protein samples. This doublet was also visible in the protein profiles in wild-type control CPE as well as in CPE of some seeds from transgenic plants. The protein profiles of some transgenic seeds were different from that of the wild-type control, particularly with the disappearance of high molecular weight bands including the band corresponding to Ara h 1 (63.5 kDa). Total protein concentration of some seeds was the lowest (from less than $1/2$ to $1/4$) although the morphology of these seeds looked similar and their weight (0.62 and 0.6 g, respectively) was somewhat higher compared to the other seeds. In Fig. 2, the protein profiles of these two seeds were the most affected by the transformation. Hybridization with Ara h 2 mAb detected Ara h 2 in both the purified as well as the wild-type protein samples. Ara h 2 was not detected in 48 out of total 214 (22.4%) transgenic seeds screened (Fig. 2).

Competitive inhibition experiments were performed by preincubating CPE (250 μg/mL total protein) from wild-type and transgenic seeds with serum IgE prior to adding 100 μL of the resulting solution to microplate wells coated with 10-μg total protein from wild-type seed. The serum IgE was a pool serum from ten patients severely allergenic to peanut (serum IgE > 100 kU/L). Data in Fig. 3 show a significant ($p < 0.05$) reduction in IgE-binding capacity of transgenic peanut seeds 11.2, 22.2.1, 35.2.2, and 45.6 compared to wild type.

Together, the data obtained with PCR, Southern hybridization, Western Blot, indirect, and sandwich ELISA of transgenic peanut produced with the plasmid pDK28, corroborate and suggest that application of RNAi technology was a successful approach to decrease Ara h 2 content in peanut. This was translated in a significant decrease in the allergenic potency of two out of the three selected transgenic peanut T1 seeds.

CONCLUSION

Genetic engineering strategies applied to food crops to reduce the risks of food allergy are proving to be successful in rice, soybean, apple, tomato, and peanut. In soybean, no collateral alteration of non-targeted proteins was reported. In peanut, experiments are ongoing to study the progeny of seeds with reduced levels of Ara h 2 and silence in concert many allergens, which could result in a stable, more sustainable, and an allergy risk-free food.

BIBLIOGRAPHY

1. Bhalla, P.L.; Sing, M.B. Knocking out expression of plant allergen genes. Methods **2004**, *32*, 340–345.
2. Bock, S.A.; Munoz-Furlong, A.; Sampson, H.A. Fatalities due to anaphylactic reactions to foods. J. Allergy Clin. Immunol. **2001**, *107*, 191–193.
3. Bredehorst, R.; David, K. What establishes a protein as an allergen? J. Chromatogr. B **2001**, *756*, 33–40.
4. Breiteneder, H.; Radauer, C.A. Classification of plant food allergens. J. Allergy Clin. Immunol. **2004**, *113*, 821–830.
5. Burks, W.A.; Laubach, S.; Jones, S.M. Oral tolerance, food allergy, and immunotherapy: implications for future treatment. J. Allergy Clin. Immunol. **2008**, *121*, 1344–1350.
6. Dodo, H.W.; Konan, K.N.; Chen, F.C.; Egnin, M.; Viquez, O.M. Alleviating peanut allergy using genetic engineering: the silencing of the immunodominant allergen Ara h 2 leads to its significant reduction and a decrease in peanut allergenicity. Plant Biotechnol. J. **2007**, *6* (2), 135–145.
7. Fleischer, D.M.; Conover-Walker, M.K.; Christie, L.; Burks, A.W.; Wood, R.A. Peanut allergy: recurrence and its management. J. Allergy Clin. Immunol. **2004**, *114*, 1195–1201.
8. Gilissen, L.J.; Bolhaar, S.T.; Matos, C.I.; Rouwendal, G.J.; Boone, M.J.; Krens, F.A.; Zuidmeer, L.; Van Leeuwen, A.; Akkerdaas, J.; Hoffmann-Sommergruber, K.; Knulst, A.C.; Bosch, D. Van de Weg, W.E.; Van Ree, R. Silencing the major apple allergen Mal d 1 by using the RNA interference approach. J. Allergy Clin. Immunol. **2005**, *115*, 364–369.
9. Gleave, A.P. A versatile binary vector system with a T-DNA organizational structure conducive to efficient integration of cloned DNA into plants genome. Plant Mol. Biol. **1992**, *20*, 1203–1207.
10. Herman, E.M.; Helm, R.M.; Jung, R.; Kinney, A.J. Genetic modification removes an immunodominant allergen from soybean. Plant Physiol. **2003**, *132*, 36–43.
11. Jenkins, J.A.; Breiteneder, H.; Mills, E.N.C. Evolutionary distance from human homologs reflects allergenicity of

animal food proteins. J. Allergy Clin. Immunol. **2007**, *120*, 1399–1405.
12. Konan, N.K.; Viquez, O.M.; Dodo, H. Towards the development of a hypoallergenic peanut through genetic transformation. Appl. Biotechnol. Food Sci. Policy **2004**, 1(3) 159–168.
13. Koppelman, S.J.; Wensing, M.; Ertmann, M.; Knulst, A.C.; Knol, E.F. Relevance of *Ara* h 1, *Ara* h 2, and *Ara* h 3, in peanut-allergic patients, as determined by immunoglobulin E western blotting, basophile-histamine release and intracutaneous testing: *Ara* h 2 is the most important peanut allergen. Clin. Exp. Allergy **2004**, *34*, 583–590.
14. Sicherer, S.H.; Sampson, H.A. Peanut allergy: emerging concepts and approaches for an apparent epidemic. J. Allergy Clin. Immunol. **2007**, *120*, 491–503.
15. Tada, Y.; Nakase, M.; Adachi, T.; Nakamura, R.; Shimada, H.; Takahashi, M.; Fujimura, T.; Matsuda, T. Reduction of 14–16 kDa allergenic proteins in transgenic rice plants by antisense gene FEBS Lett. **1996**, *391*, 341–345.
16. Tang, G.; Reinhart, B.J.; Bartel, D.P.; Zamore, P.D. A biochemical framework for RNA silencing in plants. Genes Dev. **2003**, *17*, 49–63.
17. Viquez, O.M.; Summer, C.; Dodo, H.W. Isolation and partial characterization of the genomic clone of peanut allergen Ara h 2. J. Allergy Clin. Immunol. **2001**, *107*, 713–717.
18. Viquez, O.M.; Konan, N.K.; Dodo, H.W. Structure and organization of the genomic clone of a major peanut allergen gene, Ara h 1. Mol. Immunol. **2003**, *40*, 565–571.
19. Viquez, O.M.; Konan, K.; Dodo, H. Genomic organization of peanut allergen gene, Ara h 3. Mol. Immunol. **2004**, *41*, 1253–1240.
20. Wesley, S.V.; Helliwell, C.A.; Smith, N.A.; Wang, M.B.; Rouse, D.T.; Liu, Q.; Gooding, P.S.; Singh, S.P.; Abbott, D.; Stoutjesdijk, P.A.; Robinson, S.P.; Gleave, A.P.; Green, A.G.; Waterhouse, P.M. Construct design for efficient, effective high-throughput gene silencing in plants. Plant J. **2001**, *27*, 581–590.

Animal Cell Culture

Shang-Tian Yang
Xudong Zhang
Yuan Wen
William G. Lowrie Department of Chemical and Biomolecular Engineering, Ohio State University, Columbus, Ohio, U.S.A.

Abstract
Animal cell culture is widely used in the production of recombinant protein therapeutics. It also plays a key role in the fields of tissue engineering and cell-based assays. The key components in developing a successful animal cell culture process include cell source, media formulation, and bioreactor design, operation, and scale up. This entry provides an overview on the development of animal cell culture and its industrial applications.

INTRODUCTION

Mammalian cells are widely used in the biotechnology industry for recombinant therapeutic protein production.[1] The biopharmaceutical market was about 60–70 billion U.S. dollars in 2007 with a 13% annual growth rate. About 60–70% of recombinant biopharmaceutics on the market are currently produced by mammalian cells because they can provide proper protein folding, authentic post-translational modifications, and efficient protein secretion, which make their final products superior to those produced by bacteria, yeasts, and other organisms.[2] In addition to the application in the biopharmaceutical industry, animal cell cultures also provide a platform to investigate the physiology and biochemistry of cells, and can be used as models in studying diseases. They are also widely used in drug screening and discovery. The ability to continuously grow animal cells in vitro after removing them from animal tissues also opens up a plethora of opportunities for their applications in regenerative medicine and tissue engineering, mainly producing cells for therapeutic purposes and artificial tissues and organs for implantation.[3]

In general, animal cells have an average size of 10–30 μm in diameter and they all share the same cellular structures. Unlike bacteria, yeasts, filamentous fungi, and plant cells, animal cells do not have a cell wall and are much more vulnerable to shear stress when cultured in vitro. Also, animal cells are heterotrophic and rely on external energy sources such as glucose and amino acids for their growth and maintenance, which also depend on the presence of many growth factors and require delicate control on environmental factors. Changes in media composition and environmental factors can affect cell proliferation, gene expression, and protein production; stimulate differentiation; or even induce growth arrest or apoptosis. In this entry, we discuss the key factors affecting animal cell cultures, bioreactor design and operation for cell culture processes, and applications of animal cell cultures in industrial production of recombinant protein therapeutics, tissue engineering, and cell-based assays.

CELL SOURCE AND CHARACTERISTICS

Cells directly explanted from humans or animals are called primary cells, which have a limited life span and can not be maintained for long-term cultures in vitro. Treatment of chemical carcinogens, cancer-inducing viruses, DNA integration, and cell fusion are applied to alter cell growth property in order to establish immortalized cell lines, also known as transformed cell lines. Immortalized cell lines, which can grow and divide indefinitely under appropriate culture conditions, are the ones usually used in industrial production of recombinant protein therapeutics. Table 1 lists some industrial cell lines, including Chinese hamster ovary (CHO), murine lymphoid (NS0, Sp2/0), baby hamster kidney (BHK21), human embryo kidney (HEK-293), and human retina-derived (PER.C6) cells. Industrial cell lines usually have been genetically engineered to improve their growth and ability to produce the protein product at a high expression level with desirable biological properties.[4] Cells can be cryopreserved for months or years in liquid nitrogen at $-196°C$, which is an important technique for cell banking and delivery.

Most of animal cells are anchorage dependent and require the attachment to a solid surface for growth and maintaining their normal functions, while a few exceptions such as blood, cancer, and hybridoma cells can grow in free suspension. However, many anchorage-dependent cells can be adapted to grow in suspension, which is easier to use in stirred-tank bioreactors and to scale up in industrial cell culture processes.

Table 1 Common cell lines used in industrial production of recombinant protein therapeutics.

Cell line	Origin	Cell type	Product
CHO	Chinese hamster ovary	Epithelial	Recombinant glycoproteins: tPA, EPO, antibodies, Factor VIII
BHK21	Baby hamster kidney	Fibroblast	Vaccine, antibody, CD molecules, Factor VIII
Hybridoma	Fusion of B-cells and myeloma	Hybrid	Antibodies
NS0	Mouse myeloma	Lymphoblast	Antibodies
Sp2/0-Ag14	Fusion of mouse spleen cells and mouse myeloma	Hybrid	Antibodies
Vero	African green monkey kidney	Fibroblast	Human vaccine production
HEK 293	Human embryonic kidney	Fibroblast	Therapeutic protein and virus
PER C6	Human embryonic retinal cells	Retinal cells	Human proteins, including human monoclonal antibodies

CULTURING CONDITIONS

Solid Substrate

Except for blood and lymphatic cells, normal somatic cells natively reside in a relatively confined space and attach to neighboring cells or extracellular matrix (ECM). To culture animal cells in vitro usually requires a solid substrate for cell attachment. Cells are often grown as a monolayer culture on the surface of a T-flask, which is usually made of polystyrene with oxygen plasma surface treatment. To culture anchorage-dependent cells in suspension, microcarriers ranging from 90 to 330 μm in diameter can be used to provide large surface areas for cell attachment. Solid microcarriers allow cells to attach and grow only on the external surface (Fig. 1A), whereas macroporous microcarriers with 30–100 μm internal pores allow cells to have high-density in-growth (Fig. 1B). Cells grown in porous microcarriers are protected from shear stress and microcarrier collision in a stirred tank. The materials for making microcarriers can be cross-linked dextran, cellulose, polystyrene or polyethylene, often substituted with positively charged diethylaminoethyl (DEAE) or coated with gelatin to facilitate cell adhesion.

Polymeric materials such as polycaprolactone (PCL), poly(lactic-co-glycolic acid) (PLGA), poly(ethylene terephthalate) (PET), and poly(urethane) in the form of a microporous matrix or sponge and non-woven fibers have also been widely used to culture mammalian cells. Fig. 2 shows some polymeric scaffolds with cells. In tissue engineering, biocompatible materials such as PLGA, PCL, PET, collagen, and agarose form support matrices as tissue scaffolds to offer mechanical structures for cell attachment and templates for three-dimensional (3-D) tissue organization.[5] They promote cell adhesion and growth, and prevent inflammation and toxicity. One of the challenges in tissue engineering is to maintain the native characteristics and functions of cells and tissues similar to in vivo counterparts. Cells cultured on two-dimensional (2-D) surfaces could only grow with a thickness less than a millimeter due to the absence of in vivo capillary vessels. Foams, sponges, gels, and fibrous scaffolds with 3-D structures are thus widely used to increase the surface to volume ratio and to improve mass transfer especially oxygen delivery. These polymeric materials can also be made into nanofibers such as by electrospinning. These nanofibers can mimic the native environment provided by naturally occurring ECMs secreted by cells, as seen in Fig. 3.

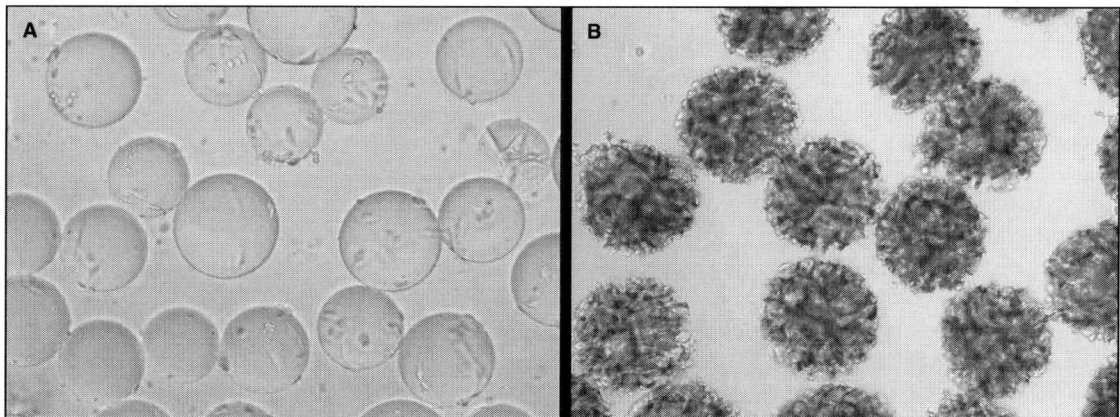

Fig. 1 Microscopic images of microcarriers with CHO cells. (A) Cells attached on the surface of solid microcarriers; (B) cells grown on the surface and in the pores of macroporous microcarriers.

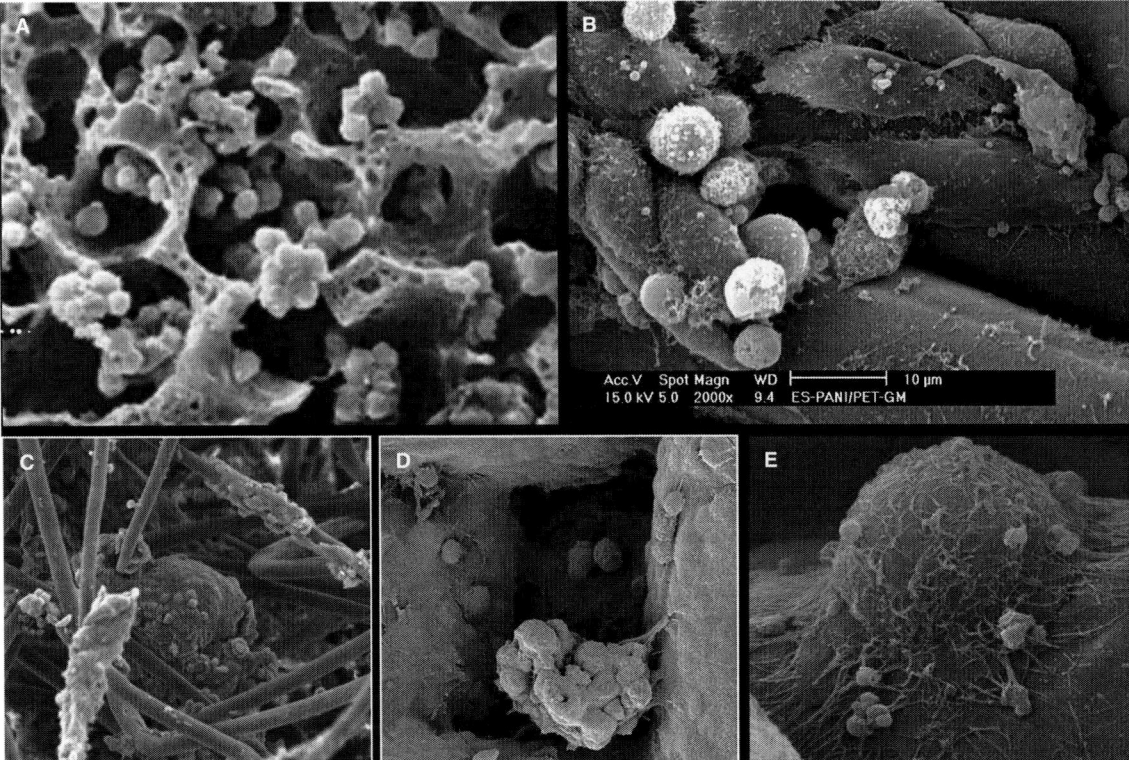

Fig. 2 SEM images of cells grown in different types of polymeric scaffolds. (A) Hepatocyte in polyurethane membrane; (B) astrocytes on PET fiber; (C) colon cancer HT29 cells in PET matrix; (D) breast cancer cells in PLGA matrix; and (E) embryonic stem cell on PET fiber.

Nanostructured polymers with tunable surface properties similar to the natural extracellular matrix can also be incorporated into 3-D scaffolds.[6] The nanoscale surface roughness may positively affect cell adhesion, proliferation, and function. Therefore, 3-D scaffolds with nanoscale features offer great promise for enhancing the biological performance of cell cultures, but they remain to be fully exploited.

Culture Media

Media provide nutrients for cell growth and usually contain balanced salt solutions (BSS), essential amino acids, glucose, vitamins, buffers, and antibiotics. The BSS provides inorganic salts required by the cells and usually have an osmolality between 280 and 350 mOsm/kg. A higher osmolality may increase cell-specific productivity but inhibit cell growth. Sodium bicarbonate and phosphates present in the BSS also act in a buffering capacity. Glucose is usually provided at several grams per liter as the main carbon and energy source for cell growth. High glucose concentrations may lead to high lactate and alanine production and inhibit cell growth, while glucose limitation can cause apoptosis or programmed cell death. Glutamine is also added as a carbon and energy source for cell growth, but it is also the main source of ammonium, a potentially toxic compound produced after glutamine utilization and decomposition. Media are often supplemented with different types of serum containing various growth factors and hormones necessary for cell growth. Serum also contains various adhesion factors that promote cell attachment and anti-trypsin activity. Serum components can also act as buffers and as chelators for labile or water-insoluble nutrients, bind and neutralize toxins, and provide protease inhibitors. Serum can also reduce oxidative injury to cells caused by ferric ions. Reduced serum conditions increase the susceptibility of cells to apoptosis.

However, serum may be contaminated with adventitious agents and cause costly downstream burdens. Serum-free media are thus more favorable for industrial processes. They were prepared by supplementing a basal medium with cell-line-specific growth factors, carrier proteins, additional lipids and amino acids, among others. These nutrient substitutes can be derived from, for instance, plant hydrolysates. However, chemically defined substitutes are more favorable for consistent performance and quality control. In addition, animal-component-free substitutes are required for the purpose of eliminating adventitious hazards.

pH

The medium pH may affect cell growth, lactate production, and product quality. The optimal pH value for animal cells is around 7.2. A pH value outside the range of 6.5–8.0 leads to apoptosis and decreased cell viability and productivity. Most of commonly used culture media contain

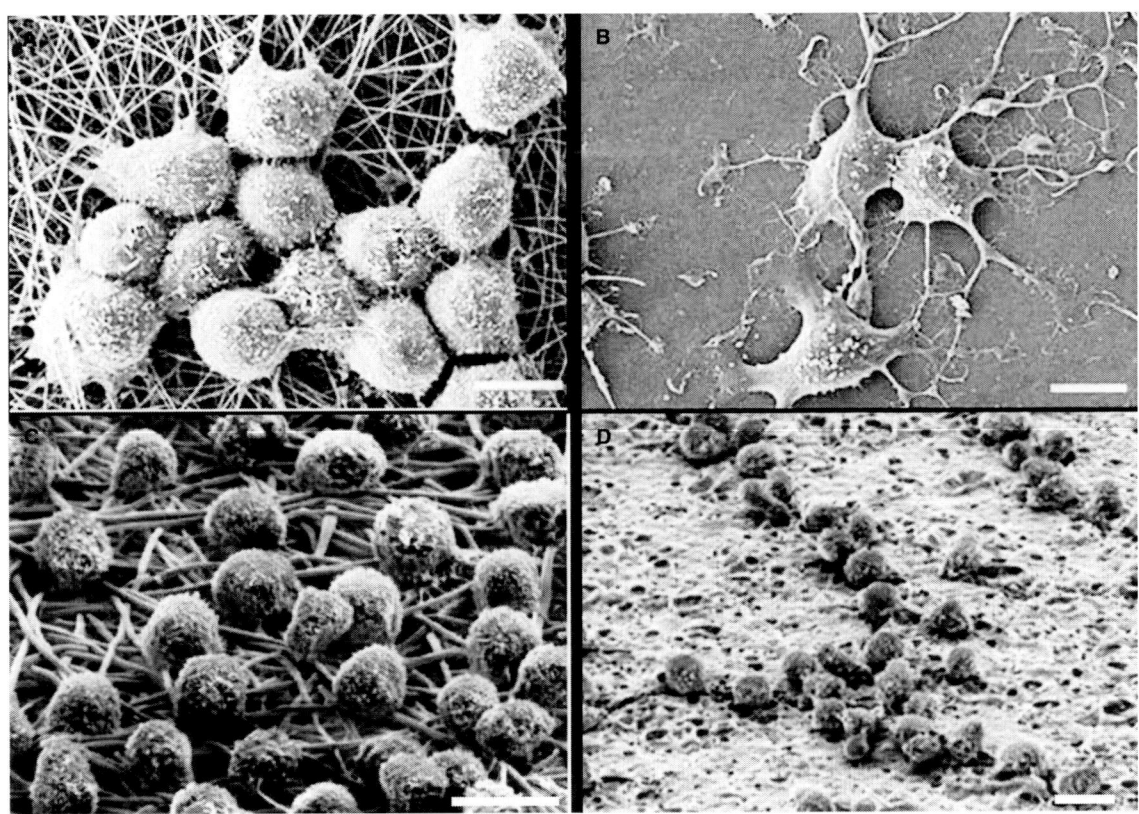

Fig. 3 SEM images of rat neural stem cells cultured on poly(ethersulfone) (PES) nanofibers (A) and film (B) and human cord blood hematopoietic stem/progenitor cells cultured on aminated PES nanofiber mesh (C) and film (D). All scale bars are 10 μm. Nanofibers exhibiting in vivo ECM structure can better support stem cells for their native morphology and faster proliferation. **Source:** (A) and (B) from Biomaterials.[7] (C) and (D) from Biomaterials.[8]

sodium bicarbonate, which in the presence of 5% CO_2 provides good buffering at pH around 7.4. The culture pH can be more tightly controlled, typically at 7.0, by base addition (NaOH or Na_2CO_3) and CO_2 sparging in a bioreactor.

Dissolved Oxygen

Cellular respiratory activities require oxygen as the ultimate electron receptor for generating energy. Since oxygen has a low solubility in water, continuous aeration is necessary to provide sufficient oxygen to cells in in vitro cultures. Aeration can be done through surface aeration, membrane aeration, or air/oxygen sparging, with sparging being the most convenient one for large-scale processes. The oxygen uptake rate (OUR) of animal cells ranges from 3×10^{-10} to 2×10^{-8} mg/cell/hr, and typically the culture medium is maintained at 20–50% air saturation. However, the optimal dissolved oxygen (DO) level for a cell culture process is cell line and process specific. In general, low DO levels limit cell growth and protein production, while excessive oxygen can be toxic to cells. For some stem cells, low oxygen tensions provide a favorable niche for cell proliferation or differentiation.

Temperature

In general, no animal cells can survive a temperature higher than 42–49°C. While the optimal temperature for growing insect cells is around 28°C, all mammalian cells are usually cultured at ~37°C. Temperatures lower than 37°C may lead to reduced metabolic activity and proliferation. However, increased cellular productivity and product quality were observed for some CHO cell cultures at temperatures between 30 and 37°C.

ANIMAL CELL BIOREACTORS

Depending on the scale, various types of bioreactors can be used for animal cell cultures. At the laboratory scale, multiwells, T-flasks, and spinner flasks are commonly used. For multiwells and T-flasks, cells grow to a monolayer on the surface of the culture vessel. Spinner flasks are used for growing cells in suspension, often with microcarriers to support anchorage-dependent cells. These bioreactors are simple, low-cost, and easy to use, but they generally lack the ability to control the pH, DO, etc. For larger industrial processes, stainless-steel stirred-tank bioreactors with mixing, temperature, pH, and DO controls are usually the

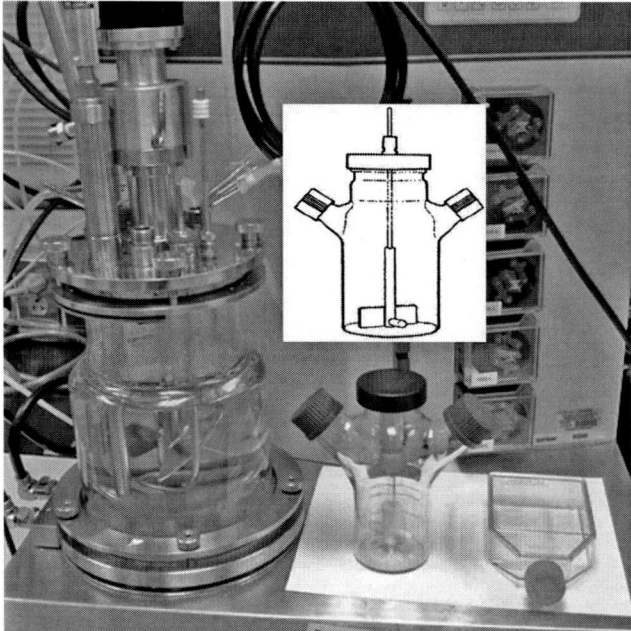

Fig. 4 Animal cell culture bioreactors commonly used in growing mammalian cells in the laboratory (from left to right: stirred tank bioreactor, spinner flask, T-flask).

choice because they are easier to sterilize, operate, and scale up. However, air-lift, fluidized bed, hollow-fiber, and fibrous-bed bioreactors have also been used. More complicated multi-tray and rotating-wall reactors have also been used in some tissue engineering applications. More recently, stirred and wave bioreactors with disposable plastic bags at various scales have also been used in industry at the seed preparation, preproduction, and even production stages. In addition, microfluidic and microscale bioreactor arrays have also been developed as scale-down models for high-throughput experimentation useful in cell-based assays as well as clone, medium, and process development. Fig. 4 shows a set of typically used bioreactors or vessels for small-scale animal cell cultures.

Agitation is commonly used to provide mixing and enhance mass transfer. Gentle shaking can provide effective mixing in small-scale cultures in microwell plates, shake-flasks, and disposable plastic bags. For larger-scale stirred-tank bioreactors, more vigorous agitation with impellers and gas sparging is used. However, animal cells are highly susceptible to shear damage. Careful design and operation of impellers for mixing and air sparging are thus important in a cell culture bioreactor. In operating a large-scale bioreactor, considerations are given to three critical issues: shear stress, mass transfer of oxygen and CO_2, and mixing time. Shear and bubbles from sparging and agitation are considered as the main causes of cell damage and death during culture. This problem can be partially solved by the addition of shear-protective surfactants such as Pluronic F68 and using microspargers to increase oxygen transfer rate and reduce sparging rate. However, CO_2 removal becomes difficult at low sparging rates. Large-bubble sparging can be used to adjust the balance between oxygen transfer and CO_2 removal. As the reactor scale becomes larger, the mixing time also increases, which may cause large local departures from the optimal pH during base feeding for pH control. Multipoint addition and well-mixed zone addition can be used to avoid the pH departure problem.[9]

In terms of feeding nutrients and metabolites control, there are mainly three modes of cell culture processes: batch, fed-batch, and perfusion. For a batch culture, there is neither addition nor removal of culture media, and the culturing time and final product titer are limited due to depletion of nutrients, such as glucose and glutamine, and accumulation of toxic metabolites, such as lactate and ammonia. Fed-batch cultures, which can prevent catabolite inhibition and undesirable metabolism caused by excessive carbon source and other nutrients by feeding them at a predetermined rate, can extend culturing period, increase cell density, and result in a high productivity that is up to 10 times of that in a batch culture. Because of these advantages and relatively easy operation, fed-batch cultures occupy more than 90% of current manufacturing processes in the biopharmaceutical industry.

Perfusion culture is the most sophisticated process for biologics production and can achieve the highest productivity. Feed medium is continuously supplied and spent medium without cells is continuously withdrawn at the same rate. To separate cells from the released spent medium, different cell retention devices such as centrifuges, spin filters, hollow fibers, and ultrasonic and gravity-based cell separators can be used. Perfusion culture can feed nutrients and remove toxic metabolites at the same time, so a high cell density (up to 10^8 cell/mL) over an extended period (up to several months) can be achieved. Generally, the volumetric productivity of a perfusion culture is several times higher than that in a fed-batch culture. Furthermore, perfusion cultures are better for producing fragile proteins prone to degradation, aggregation, and desialylation because of the relatively short residence time. Downstream processing can also be simplified with the cell-free broth produced in a perfusion culture.

Hollow-fiber bioreactor (HFB) consisting of a large number of parallel hollow fibers housed in a cylinder provides a high surface to volume ratio for cell attachment. HFB can support perfusion cultures with continuous feeding of nutrients and removal of toxic metabolites, and attain a high cell density up to 10^9 cells/mL fiber volume, comparable to the fresh tissue density in human body, so it can be used to produce artificial organ directly. Hepatocytes cultured in HFB have been studied as an artificial liver to treat patients with acute hepatic failure. As both a culture vessel and a retention device, HFB can also be used to produce biologics in perfusion cultures. However, its application is limited to small-scale preclinical production. Similar to hollow fiber bioreactors but with a better scale-up potential, a perfusion fibrous bed bioreactor (FBB) could support high-density culture (4×10^8 cells/mL) with high

Table 2 Some important biopharmaceutical products from animal cell cultures.

Type	Examples
Vaccines	Polio, Hepatitis A, Measles, Mumps, Rubella, Yellow Fever, Rabies, Influenza
Glycoproteins	Interferons: α-, β-, γ- Blood clotting factors: Factors VIII, IX Glycoprotein hormones: Erythropoietin (EPO) Plasminogen activators: t-PA
Monoclonal antibodies	Diagnostics Therapeutics—arthritis, psoriasis, ankylosing spondylitis (Enbrel, Remicade, Humira), breast cancer (Herceptin), colon cancer (Avastin), non-Hodgkin's lymphoma (Rituxan/MabThera), respiratory syncytial virus (RSV) infection (Synagis)
Hormones	Human growth hormone, Insulin, Interleukins (IL-1, IL-2, IL-3), Calcitonin, Parathyroid hormone
Growth factors	Colony-stimulating factors (CSFs)—immune system growth factors controlling the differentiation, growth, and activity of white blood cells (granulocytes and macrophages) Epidermal growth factor (EGF) for wound healing Insulin-like growth factor (IGF) for promoting tissue growth Platelet-derived growth factor (PDGF) for collagen deposition in tissue repair Nerve growth factor
Proteases	Urokinase

viability (>85%) and antibody productivity.[10] Cells cultured in the FBB can maintain normal cell morphology and functions for an extended period of more than several months, which is important to applications in tissue engineering and cell-based assays.

INDUSTRIAL APPLICATIONS

Animal cell cultures are increasingly used for production of recombinant glycoproteins, viral vaccines, and monoclonal antibodies in the biotechnology industry.[11] Table 2 lists various types of biopharmaceutical products from animal cell cultures. An emerging application area for animal cell cultures is regenerative medicine, in which how to mass produce cells and functional tissues for cell therapy and transplantation is a critical challenge that requires the use of advanced animal cell culture and tissue engineering techniques. With the discovery of various types of stem cells, this field has grown rapidly and is expected to make major impact on regenerative medicine.[12] Cells either inside the human body or *ex vivo* (e.g., Petri dish) can generate new tissues or organs with proper mechanical support of scaffolding materials and physiological support of growth factors, extracellular matrix proteins, and cytokines. Several tissue engineering products are now available for clinical applications, with many more at the research and development stages (see Table 3).

Table 3 Examples of cell and tissue types that have been successfully developed for clinical applications.

Product	Cell source/type	Targeted disease
Dermagraft® (skin)	Human fibroblast cells derived from newborn foreskin tissue cultured on a bioabsorbable polyglactin mesh	Diabetic foot ulcers
TransCyte™ (skin)	Human fibroblast cells derived from newborn foreskin tissue cultured on a nylon mesh coated with porcine dermal collagen	Burns
EpiCel® (skin)	Autologous cultured keratinocyte	Burns
CartiCel® (cartilage)	Autologous cultured chondrocyte	Articular cartilage injuries
BioSeed®—oral bone	Autologous cultured bone cells	Receding jaw bone
Islet sheet (islet)	Cultured donor pancreatic islets encapsulated in alginate	Diabetes
HepaMate™ (liver)	Extracorporeal cell-based bioartificial liver system	Severe liver failure
Prochymal™ (mesenchymal stem cell)	Cultured mesenchymal stem cells from healthy human donors	Craft versus host disease, Crohn's disease, cardiac disease, chronic obstructive pulmonary disease, and diabetes
MultiStem® (adult stem cell)	Cultured adult stem cells from healthy human donors	Bone marrow transplant support, acute myocardial infarction, and stroke

Table 4 Some examples of cell-based assays used in high throughput drug screening.

Applications	Cell type	Property and specialty
Neurotoxicity test	Neurons	Highly sensitive to external stimuli and provide electrical signals
Cardiotoxicity test	Cardiomyocytes	
Embryotoxicity test	Embryonic stem cells	The only truly immortal stem cells with a normal diploid karyotype. Pluripotent and can be differentiated into most somatic cell types
Detection of pathogens and superantigens	Immune cells	Recognition of external stimulus
Drug absorption test	Intestinal epithelial cell, e.g., Caco-2	One of the main barriers for oral drug
Metabolism and toxicity test	Hepatocytes	Detoxification and metabolism

Another important application of animal cell culture is cell-based assays.[13] Screening a huge number of bioactive compounds generated from combinatorial synthesis, and target identification and validation are the limiting steps in drug discovery. Conventional biochemical assays often generate many poorly qualified hits that are to fail but only after going through expensive and time-consuming animal experiments. Cell-based assays in a high-throughput fashion with relevant in vivo biological information can alleviate the huge early-phase drug screening burden and are thus becoming increasingly important in drug discovery. Various types of cells cultured in microwells or biochips have been used for this purpose.[14] The most critical challenge in using in vitro cell-based assays is how to maintain the native characteristics and functions of cells as an in vivo tissue representative. In order to create a more realistic representation of real human tissues, which include morphogenesis, cell metabolism, gene expression, differentiation, and cell–cell interactions, 3-D culture models that can mimic in vivo tissue environment have been established. Table 4 lists some of the commonly used cellular assays for high throughput drug screening.

CONCLUSION

Various types of animal cells have been cultured in vitro and used in various applications, including studying the physiology and biochemistry of cells, diseases, and drug effects, to produce recombinant protein therapeutics and viral vaccines for fighting diseases, and to engineer cells and tissues for use in regenerative medicine. The future of developing more products using animal cell cultures is very bright. However, commercial development of new cell culture processes remains a challenge in several areas, mainly cell line development, media optimization, and bioreactor design in both scale up and scale down.

REFERENCES

1. Yang, S.T.; Basu, S. Animal cell culture. In *Encyclopedia of Chemical Processing*; Lee, S., Ed.; Marcel Dekker: New York, 2005; 67–79.
2. Wurm, F.M. Production of recombinant protein therapeutics in cultivated mammalian cells. Nat. Biotechnol. **2004**, *22*, 1393–1398.
3. Yang, S.T.; Robinson, C. Tissue engineering. In *Encyclopedia of Chemical Processing*; Lee, S., Ed.; Marcel Dekker: New York, 2005; 3115–3128.
4. Birch, J.R.; Racher, A.J. Antibody production. Adv. Drug Deliv. Rev. **2006**, *58*, 671–685.
5. Li, Y.; Yang, S.T. Effects of three-dimensional scaffolds on cell organization and tissue development. Biotechnol. Bioprocess Eng. **2001**, *6*, 311–325.
6. Yang, S.T.; Ng, R. 3D cultures for growing and studying cells. Mater. Today **2007**, *10* (3), 56.
7. Christopherson, G.T.; Song, H.; Mao, H-Q. The influence of fiber diameter of electrospun substrates on neural stem cell differentiation and proliferation. Biomaterials **2009**, *30*, 556–564.
8. Chua, K-N.; Chai, C.; Lee, P-C.; Ramakrishna, S.; Leong, K.W.; Mao, H.-Q. Surface aminated electrospun nanofibers enhance adhesion and expansion of human umbilical cord blood hematopoiefic stem/progenitor cells. Biomaterials **2006**, *27*, 6043–6051.
9. Rose, S.; Black, T.; Ramakrishnan, D. Mammalian cell culture: process development considerations. In *Handbook of Industrial Cell Culture: Mammalian, Microbial, and Plant Cells*; Humana Press: Totowa, NJ, 2003.
10. Yang, S.T.; Luo, J.; Chen, C. A fibrous-bed bioreactor for continuous production of monoclonal antibody by hybridoma. In *Advances in Biochemical Engineering/Biotechnology, Biomanufacturing*; Zhong, J.J., Ed.; 2004; Vol. 87, 61–96.
11. Ozturk, S.S.; Hu, W.-S. *Cell Culture Technology for Pharmaceutical and Cell-based Therapies*; CRC Press: Boca Raton, FL, 2006.
12. Li, Y.; Yang, S.T. Stem cell-based tissue engineering. In *Encyclopedia of Agricultural, Food, and Biological Engineering*; Heldman, D.R., Ed.; Marcel Dekker: New York, 2004.
13. Yang, S.-T.; Zhang, X.; Wen, Y. Microbioreactors for high-throughput cytotoxicity assays. Curr. Opin. Drug Discov. Dev. **2008**, *11* (1), 111–127.
14. Wen, Y.; Yang, S.-T. The future of microfluidic assays in drug development. Expert Opin. Drug Discov. **2008**, *3* (10), 1237–1253.

Animal Medicines

Sherrie Clark
Department of Veterinary Clinical Medicine, University of Illinois, Urbana, Illinois, U.S.A.

Abstract

Discovery of medicines for use in animals that will help lower production costs and improve animal well-being by preventing and treating diseases is an ongoing process. The use of biotechnology and other scientific advances are extremely beneficial in the discovery and development of these new medicines. There are many new pharmaceuticals for use in agricultural and companion animals that have been the result of biotechnology techniques. Some of these pharmaceuticals may serve as substitutes for the use of antibiotics in animals, which would be beneficial in the recent concern for antibiotic-resistant bacteria in human infections. New or enhanced animal and plant-based vaccines have been developed through modern biotechnology, and many more are in the testing phase. Additionally, the development of rapid test kits for diagnosis of various disease conditions in livestock and companion animals has been accomplished with the use of biotechnology. All of these advances will hopefully improve our ability to protect humans and animals from impending threats to their environment and health.

INTRODUCTION

Advances in biotechnology have allowed scientists to develop new strategies and technologies to combat animal diseases and conditions that not only affect animals but also could be a human health risk. This entry will review new and improved pharmaceuticals that have been and are currently being developed for use in both agriculture and companion animal medicine. Also, the advancement of diagnostic medicine through biotechnology techniques will be discussed. In addition, many of the advances in animal medicines, vaccines and diagnostics will be discussed in relation to their effect on human medicine and public health.

BIOTECHNOLOGY IN DISCOVERY AND DEVELOPMENT OF MEDICINES

Numerous drugs and medicines have been discovered and developed through biotechnology. It is a technology based on biology and is defined as "any technological application that uses biological systems, living organisms, or derivatives thereof, to make or modify products or processes for specific use,"[1] according to the United Nations Convention on Biological Diversity.

In the last 10 years, the development of new pharmaceuticals through biotechnology techniques has grown immensely. With the threat of bioterrorism and the need for costly treatment procedures for various diseases, biotechnology has assisted modern medicine to offer new and improved drugs, vaccines, and diagnostic testing. The use of genetic-based marker systems has allowed scientists and clinicians the ability to detect various organisms. Fan and Bird[2] used an N^{pro}-disrupted marker bovine diarrhea virus to study the protein expression and establishment of persistence in the host. It is vital that these technologies be able to detect pathogenic strains of organisms in the host, such as avian pathogenic *Escherichia coli* (APEC), which have zoonotic potential. Ewers et al.[3] examined 46 different virulence-associated genes to distinguish outbreak and non-outbreak strains of APEC. They also revealed that most strains that were virulent in chickens belonged to sequence types that were associated with septicemia in humans. The use of these technologies enable those in the medical field to identify crucial genes and specific antigens, such as the recombinant forms of the *Sarcocystis neurona* surface antigens,[4] which can be targeted for drug development or diagnostic techniques.

DIAGNOSTIC DETECTION AND TECHNIQUES USING BIOTECHNOLOGY

The use of molecular genetics for the detection of various microorganisms and antimicrobial agents is invaluable in today's modern medicine. Several surveillance programs (DANMAP: Danish Integrated Antimicrobial Resistance Monitoring and Research Programme; NARMS: National Antimicrobial Resistance Monitoring System; and CAHFSE: Collaboration in Animal Health and Food Safety Epidemiology) have been established to monitor the incidence and quantity of antibiotic resistant organisms that are isolated from food animals and their food products.[5] These programs accomplish an enormous task that utilizes various microbiological and immunochemical techniques to detect these organisms. Although these methods may be limited by their sensitivity and low screening efficiency, they provide the ability to gather data on organisms that may be developing antibiotic resistance.[6]

Furthermore, to ensure a safe food supply and public health status, detection of drug residues by biotechnological methods is imperative. Researchers have developed miniaturized, high-sensitivity, integrated, high-throughput analysis systems that use small molecule microarray methods for detection of drug residues.[6] This nanotechnology utilizes small molecules as probes to determine the presence of certain genes or gene products in foodstuffs used for human as well as animal consumption as well as those associated with diseases in animals.[7] These sophisticated sensors can aid in the detection of a disease-causing particle, changes in body temperature, and other early biological changes that occur prior to the display of overt clinical symptoms of disease.[7] This would be an invaluable tool for the prevention rather than treatment of many animal conditions.

Detection of the presence of animal diseases as well as extraneous organisms in vaccines is extremely important.[8,9] With the introduction of molecular techniques, such as real-time polymerase chain reaction (PCR) and Western blot analysis, for characterization of various organisms, the detection of a veterinary pathogen is performed rapidly and with high accuracy.[8] In particular, PCR has enabled vaccine producers to analyze their products for extraneous agents and ensure that their vaccines are free of these organisms and safe for use in medical applications. This has become extremely important in the analysis of live and attenuated poultry and swine vaccines as a large population of animals will be infected and can have severe consequences.[9] The ability to detect vaccinated animals from naturally infected animals has been improved through the use of gene-deleted vaccines and addition of gene markers.[8,10]

VACCINE DEVELOPMENT AND APPLICATIONS

The use of targeted vaccination programs is one of the most cost-effective methods of preventing animal disease and the resulting loss of productivity. Recently, the development of vaccines derived from recombinant protein subunit and deoxyribonucleic acid (DNA) recombination technologies have been utilized in human and animal medicine.[10,11] Subunit vaccines are manufactured by identifying and purifying antigens produced by pathogenic bacteria and recombining the gene encoding the immunogenic antigen into a vector that is non-pathogenic and is capable of expressing the gene.[10,12] There are a variety of "protein expression systems" that can be used to express the antigenic glycoprotein. These systems include bacteria, yeast, insect cells, viruses, and plants.[11,12] DNA vaccines do not rely on protein expression by a vector, but "direct injection of the gene that codes for the immunogenic antigen."[10] The uptake of the DNA by the host animal allows for direct transcription and translation of the protective proteins by the animal itself without compromising the host's immune system.[12] Several animal vaccines against viruses, bacteria, rickettsia, protozoa, helminth, and ectoparasites have been produced by recombinant DNA technology in a relatively inexpensive and effective manner.[10,12,13] Additionally, there have been immunocontraceptive vaccines engineered and designed to control reproduction in livestock and companion animals.[10] These vaccines target specific hormones or antigens on the surface of sperm or oocytes and inhibit normal reproductive function of the host animal.[13]

Although DNA vaccines have been quite effective in controlling many viral diseases, it is important to distinguish between natural infection with the wild-type virus and immunization. With the ability to sequence the genome of pathogens and understand the function of the genes that make them pathogenic, specific modifications or deletions can be made to the viral genome with the intention to produce a stable vaccine. Through various biotechnology methods, several internationally important viral diseases such as classical swine fever and pseudorabies have been controlled by the use of gene-deleted or marker vaccine.[13]

DELIVERY SYSTEMS

With the development of new vaccines and pharmaceuticals from modern molecular biology, the major limiting factor in their application is an effective delivery system.[14] Many of the DNA and recombinant DNA vaccines require administration via injections and may not have a high immunogenic response without the addition of immunostimulants or adjuvants.[10–14] The coadministration of immunostimulating molecules or multigenic vaccines greatly improves the immune response of the host animal and clinical protection against the disease.[15] Also, some of these vaccines have some safety concerns related to incomplete inactivation and injection site reactions.[10–13] Vaccine manufacturers are aware of these challenges and have developed other delivery methods that include intradermal, intranasal, and gene-gun administration of DNA to mucosal sites of animals.[15]

Transgenic plant-based vaccines may be the most promising system for use in livestock production as the vaccines can be delivered orally to a large number of animals.[11] Some plants (corn) can be transformed into bioreactors for expression of antigenic glycoproteins (e.g., transmissible gastroenteritis) and then harvested, stored as it is, and directly fed to piglets, the target animal species.[11,13]

This delivery system would eliminate public health concerns of vaccine site reactions in meat, but still remains in the development and testing phases of production.[11,14]

The transport of DNA via vaccination is only one method of delivery of pharmaceuticals in an animal system. There are numerous other methodologies developed via biotechnology that have many applications in veterinary medicine for the delivery of drugs. These include the use of liposomes and other synthetic microspheres, virosomes, drug-targeting erythrocytes, and bacterial ghosts, "non-denatured envelopes derived from Gram-negative bacteria by protein-E-mediated lysis."[14] Many of these are still in their infancy in their uses in animal medicine, but great advances are being made through biotechnology to provide these drug delivery systems in an efficient and cost-effective manner to companion animal and livestock production systems.

CONCLUSION

With the advancement of molecular biology methods and other biotechnology, new and improved animal medicines and diagnostics have been developed that will greatly improve animal health and well-being. There are numerous hurdles that still need to be overcome, but the costs to perform certain assays and tests continue to decrease and become more efficient. The future of animal medicine development through biotechnological methods will continue to flourish through research for human medicine advancements that can economically be applied to animal systems. In turn, this should enhance the overall agricultural industry and provide a safe means to produce animal products.

REFERENCES

1. *The Convention on Biological Diversity* (Article 2: Use of Terms). *United Nations* 1992. Retrieved on October 5, 2008.
2. Fan, Z.; Bird, R.C. Generation and characterization of an N^{pro}-disrupted marker bovine viral diarrhea virus derived from a BAC cDNA. J. Virol. Methods **2008**, *151* (2), 257–263.
3. Ewers, C.; Antão, E.; Diehl, I.; Philipp, H.; Wieler L. Intestine and environment of the chicken as reservoirs for extraintestinal pathogenic *Escherichia coli* strains with zoonotic potential. Appl. Environ. Microbiol. **2009**, *75* (1), 184–192.
4. Hoane, J.S.; Gennari, S.M.; Dubey, J.P.; Ribiero, M.G.; Borges, A.S.; Yai, L.E.O.; Aguiar, D.M.; Cavalcante, G.T.; Bonesi, G.L.; Howe, D.K. Prevalence of *Sarcocystis neurona* and *Neospora* spp. infection in horses from Brazil based on presence of serum antibodies to parasite surface antigen. Vet. Parasitol. **2006**, *136* (2), 155–159.
5. Norby, B. Antimicrobial resistance in human pathogens and the use of antimicrobials in food animals: challenges in food animal veterinary practice. In *Current Veterinary Therapy: Food Animal Practice*, 5th Ed.; Anderson, D.E., Rings, D.M., Eds.; Saunders: St. Louis, MO, 2009; 479–488.
6. Peng, Z.; Bang-Ce, Y. Small molecule microarrays for drug residue detection in foodstuffs. J. Agric. Food Chem. **2006**, *54*, 6978–6983.
7. Scott, N.R. Nanoscience in veterinary medicine. Vet. Res. Commun. **2007**, *31* (Suppl 1), 139–144.
8. Schmitt, B.; Henderson, L. Diagnostic tools for animal diseases. Rev. Sci. Tech. Off. Int. Epizoot. **2005**, *24* (1), 246–250.
9. Ottinger, H.P. Monitoring veterinary vaccines for contaminating viruses. Dev. Biol. (Basel) **2006**, *126*, 309–319.
10. Vannier, P.; Martignat, L. New vaccines and new veterinary therapies derived from biotechnologies: examples of applications. Rev. Sci. Tech. Off. Int. Epizoot. **2005**, *24* (1), 215–299.
11. Streatfield, S.J. Plant-based vaccines for animal health. Rev. Sci. Tech. Off. Int. Epizoot. **2005**, *24* (1), 189–199.
12. Rogan, D.; Babiuk L.A. Novel vaccines from biotechnology. Rev. Sci. Tech. Off. Int. Epizoot. **2005**, *24* (1), 159–174.
13. Meeusen, E.N.T.; Walker, J.; Peters, A.; Pastoret, P.; Jungersen, G. Current status of veterinary vaccines. Clin. Microbiol. Rev. **2007**, *20* (3), 489–510.
14. Tabrizi, C.A.; Walcher, P.; Mayr, U.B.; Stiedl, T.; Binder, M.; McGrath, J.; Lubitz, W. Bacterial ghosts-biological particles as delivery systems for antigens, nucleic acids and drugs. Curr. Opin. Biotechnol. **2004**, *15*, 530–537.
15. Dufour, V. DNA vaccines: new applications for veterinary medicine. Vet. Sci. Tomorrow **2001**, *2*, 1–26.

Animals and Plants: Genetic Modification (GM)

Louis-Marie Houdebine
Development and Reproduction Biology, National Institute for Agronomic Research, Jouy en Josas, France

Abstract

Living organisms are genetically modified naturally at each generation. Ten thousand years ago, our ancestors used this observation to select plants and animals for their use. This selection is being pursued with or without artificial mutations and this led to create a number of organisms more or less capable of surviving autonomously. This was achieved by genetic modifications performed in a blind manner. Genetic engineering makes it possible the introduction of genes having a known function into plants or animals. The resulting genetically modified organisms (GMOs) may thus gain known and precise new genetic traits in one generation. This technique may thus shorten the selection process considerably and generate organisms with a much broader genetic diversity. GMOs started being used to improve agriculture, to study biological functions as well as human diseases, and to prepare pharmaceutical proteins. The success of GMOs is unprecedented in the agriculture story and poor as well as rich countries started using them. The use of GMOs raises problems that are not new but are amplified by the fact that genetic modifications, including via gene transfer, may generate toxic or allergenic food, induce uncontrolled gene dissemination, and makes non-equitable appropriation of GM varieties or breeds by agrofood companies easier.

INTRODUCTION

The fact that living organisms are genetically modified at each generation was perceived long before the proposal of the theory of evolution. When our ancestors domesticated some plants and animals, about 10,000 years ago, they could control their reproduction and thus proceed to their selection. This empirical approach has been extremely beneficial for human communities and still is, as essentially all our food as well as our pets and ornamental plants have been obtained by artificial selection. Indeed, this selection aims at "improving" the small number of species being used by human beings. This improvement is clearly beneficial for the users but not necessarily for the species in question. Carrots, silk worms, and other domesticated species would disappear rapidly if human beings did not assist them any more. Some species have lost the ability to cross with their wild relatives. Carrots might thus be considered as monsters if we refer to their wild relatives. It would certainly appear so if experimenters were capable of transferring genes into wild carrot to generate varieties similar to those we eat. The classical selection method is based on the fact that mutations occur spontaneously at each generation. The efficiency of selection, therefore, relies essentially on the capacity of species to identify the best individuals.

The heredity laws, discovered by G. Mendel, allowed a more rational selection. During the 20th century, the biologists discovered that mutation frequency can be increased considerably by using chemical mutagens or irradiations. Artificially mutated microorganisms are currently used for the preparation of our food. Similarly, more than 2500 lines of plants obtained by artificial mutations participate in the generation of varieties we currently eat. This approach is efficient but obviously imprecise. Indeed, the mutagens modified not only genes of interest that are generally unknown but also, inevitably, a number of other unknown genes in a random manner.

The generation of interspecies hybrid is also a long-term practice. The existence of mules exemplifies this point. During the 20th century, human beings created two new plant species. One is triticale resulting from an artificial crossing between wheat and rye. Triticale shows high yield of wheat and robustness of rye. Yet, this primary hybrid was unstable and could be stabilized by dislocating chromosomes using irradiation. This new species, which is currently used to produce feed, thus results from the blind transfer of the 25,000 genes from one species into the other. This kind of transfer occurs spontaneously in nature. Wheat thus appeared after the crossing of three species.

All these observations reveal that living organisms show remarkable plasticity and that they can be genetically modified, providing human beings with more and better food without generating strong risk. The discovery of DNA and genes offered very promising possibility to genetically modify living organisms in a more diverse and more precise

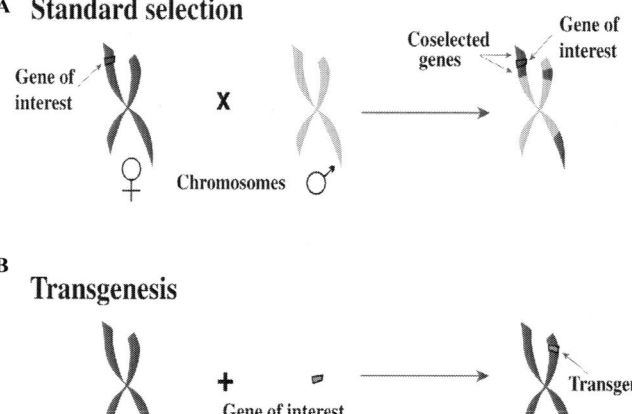

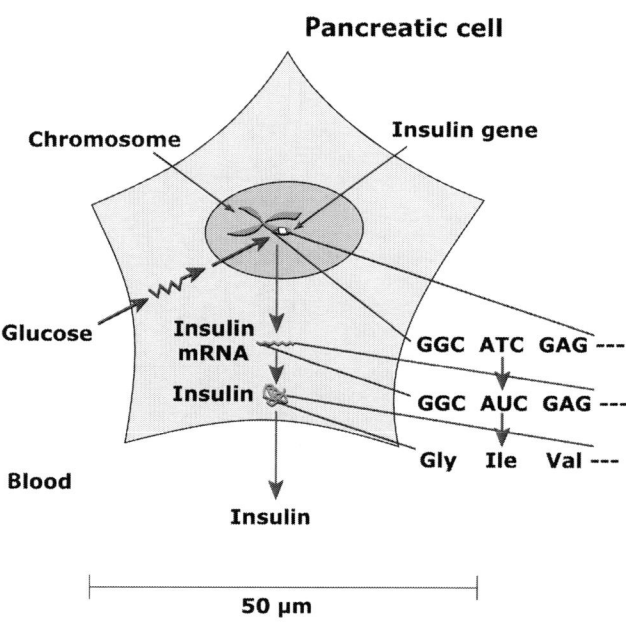

Fig. 1 (A) Conventional genetic selection relies on the random chromosome rearrangement during sexual reproduction and thus to the random distribution of the different gene versions in offspring. The gene of interest responsible for the expression of a valuable genetic trait, which is selected, is then unknown and the process implies the coselection of a number of unknown potentially deleterious genes surrounding the gene of interest. (B) Selection by gene transfer offers the possibility to add (or delete) a single gene having a known function into a living organism. This gene may be of various origins and may be optimized before being transferred.

Fig. 2 After a meal, glucose level in blood rises owing to digestion. The glucose must not stably exceed 1 g/L in human blood to avoid diabetes. On the contrary, glucose must be stored mainly in the form of lipids and retransformed progressively into glucose to provide cells with energy until the next meal. This is achieved by insulin. This hormone is a protein which is synthesized after the reversible activation of the corresponding gene by glucose. All the proteins are produced whenever needed by the same type of mechanisms.

manner than conventional selection (Fig. 1). The success of gene transfer and GMOs (Fig. 1B) relies on the identification of genes responsible for a major improvement of genetic traits.

Complex biological functions, such as lactation, depend on a number of genes, most of them having a limited impact on milk production. A coselection of these genes by the conventional method is therefore necessary (Fig. 1A). This selection method can also lead to the coselection of a number of unknown genes having potentially deleterious effects. Two varieties of potatoes obtained by conventional selection during the 20th century intoxicated consumers. These varieties had unexpected higher capacity to synthesize natural toxins, the solanines. The two selection methods have thus their advantages and drawbacks being more complementary than in competition. These two approaches cannot be fully controlled, as are all the manipulations of living organisms. Yet, it can be considered that the risk level is not fundamentally higher for the GMO approach than for the conventional method and that both methods are essentially of low intrinsic risk.

The large-scale use of GM plants to produce feed and food started in 1996 and some GM animals should be available in the future. The GMO approach raises some important problems that already exist in the conventional genetic selection. These concerns relate to food biosafety, gene dissemination, industrial property, and animal welfare. The state of the art in the GMO use is discussed below.

GENE TRANSFER METHODS

A gene must be considered as a coded information having DNA as physical support, rather than a piece of a chromosome, of a plant or of an animal. Genes are located essentially in cell nucleus and the translation of the encoded messages generates proteins which are the major biochemical actors in all the living organisms (Fig. 2). The genetic code which defines the correspondence between the order of the bases in DNA and the order of the amino acids in proteins, and thus of their biological activity, is the same for all the living organisms. A gene from an organism can thus work in another organism after some adaptations.

Genes can be isolated from any living organism or chemically synthesized giving the wanted messages needed to improve genetic traits. All our genes (about 25,000) are present in all our cells, yet, no more than 2000 are active in each cell type. The expression of the genes leading to the synthesis of the corresponding proteins is controlled by signals, known as promoters, located upstream of the genes. An isolated gene can be associated with different promoters and a promoter can be associated with different genes giving gene constructs which can be artificially transferred into different living organisms (Fig. 3).

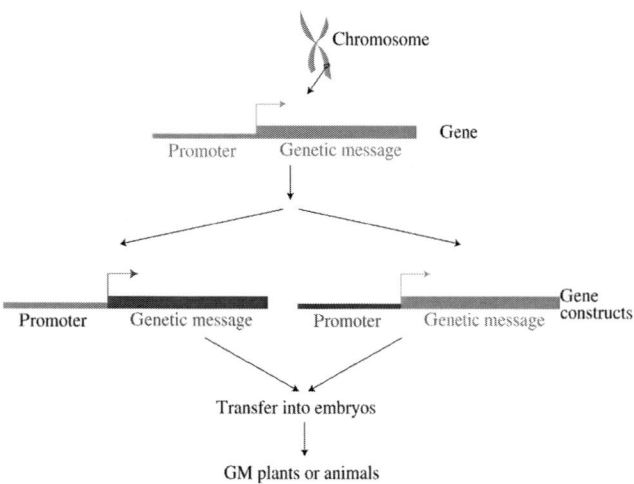

Fig. 3 Schematic representation of gene construction. The gene region containing a genetic message directing the synthesis of a protein can be associated with different promoters capable of driving the expression of the genetic message in different cell types and species.

A DNA fragment cannot enter easily into a cell; therefore, specific techniques must be implemented to reach this goal. In plants, the gene constructs are first introduced into the bacterium *Agrobacterium tumefaciens* that naturally can transfer some genes to plants. Alternatively, small metal bullets covered by the genes are projected into cells where they deliver their DNA. These methods are of sufficient efficiency to generate hundreds of lines from which the few best are kept to develop new varieties. In animals, genes must be injected into one-cell embryos or transferred via vectors (transposons and lentiviral vectors) having a high capacity to transfer and integrate foreign genes. Alternatively, the genes can be transferred to cells (sperm, embryonic stem cells, or somatic cells) that are used further to generate living organisms.[1] These methods have limited efficiency and offer lower possibility to generate a large number of GMOs than methods used in plants. These techniques do not allow the targeting of the genes into the recipient chromosomes. Other sophisticated techniques are used successfully for integrating genes into precise targeted sites in several animal species but presently not in plants.

THE GMOs USED AS FEED OR FOOD

Essentially four types of GM plants are being cultured at a large scale. These plants are soybean and rapeseed resistant to an herbicide, maize, and cotton resistant to insects. These plants are cultured in 21 rich countries and 11 poor countries representing 62–38% of the GMOs, respectively. More than 90% of the farmers using GMOs are in poor countries. The GMOs are cultured on 15% of the field area and 15% more GMOs are cultured each year. These GMOs made it possible the use of 14% less pesticides and provided farmers with more profit (6.4 billion dollars).[2]

These GMOs were generated to improve farmer task and income and to prepare feed. The GMOs presently available are only marginally used as food. It was indeed easy to improve culture techniques without modifying plant properties in a single step than generating plants with modified biological properties. A number of such projects are in progress not only in plants but also in animals. Milk of farm or experimental animals containing substances having antiviral or antibacterial activity are under study. Cows in which the prion gene have been deleted, pigs releasing less polluting phosphate, and cows secreting milk devoid of β-lactoglobulin (the main milk allergen) are also under study. Another example is fish, and mainly salmons, having accelerated growth.[3]

GMOs of second generation are being studied. Most of them have improved nutritional values as those fortified in vitamin A, iron, or omega-3 lipids. The perspectives offered by GMOs to provide human beings with more and better food appear very attractive.[4]

GM FOOD BIOSAFETY

Some genetic modifications, including via gene transfer, may give birth to organisms containing toxic or allergenic substances. Chances of producing such organisms are higher in plants as many of them contain natural toxins, and particularly pesticides, which have deleterious effects on predators, including human beings. This is the case for cotton, potatoes, rapeseed, and many others but not for maize, soybean, wheat, and many others. The chance of generating organisms containing high toxin levels is greater if the plant contains toxins before the genetic modification. The integration of the foreign gene at a non-controlled site may alter the physiology of the GMO. More importantly, the protein encoded by the foreign gene may interfere with the physiology of the organism. It must be noted that, as expected, no traces of GM plants were found in the milk and meat of animals fed with GMOs.[5]

Before being authorized for animal or human consumption, a GMO must fulfill all the criteria depicted in Fig. 4. The tests depicted in the legend of Fig. 4 are more or less strictly implemented according to the regulations of various countries. In the European Union (EU), a product containing less than 0.9% of an authorized GMO has not to be labeled. The mention "contains a GMO" must be indicated if the content is more than 0.9% of a GMO. The label "this is a GMO" must be added if the product is a GMO.

It is important to note that the evaluation of the varieties and breeds obtained by conventional selection are subjected only to an evaluation of their nutritional impact in some farm animals, at the most.

Fig. 4 The different control steps to which a GMO must be subjected before being accepted for animal or human consumption. Research using GMOs must be performed in different confined areas according to the risk. The confinements are divided into four classes. Only the GMOs of class 1 confinement, which corresponds to normal laboratories, greenhouses, or animal facilities, can be authorized for intended releases including field trials and large-scale culture. The safety evaluation of GM feed and food includes the following: 1) description of the original variety or breed and of the corresponding GMO; 2) structure of the gene transferred to the organism and of the gene found in the GMO; 3) comparison of the biochemical composition of the GMO and of the control organism; 4) evaluation of the acute toxicity of the protein encoded by the foreign gene; 5) evaluation of the general toxicity of the GMO by feeding rats for 90 days with the GMO or with the control organism; 6) evaluation of the nutritional value of the GMO by feeding farm animals with the GMO for weeks or months as they will be if the GMO is authorized; and 7) research of allergenic effects of the GMO. The international marketing of GMOs follows the rules defined by Codex Alimentarius and the World Trade Organization.

UNINTENDED GMO RELEASE

In open fields, the GMO pollen or seeds may disseminate in an uncontrolled manner. In some cases, this may be an agronomical problem. Maize is a hybrid seed that must be purchased every year, and it is not a wild plant in most countries. Soybean and cotton have very little chance to spread in wild space. Rapeseed seeds can stay for years in earth and keep their capacity to germinate. It can also crossbreed with wild plants in northern Europe. For this reason, available GM rapeseed resistant to herbicides is not authorized for culture in Europe. Another example is GM salmons having accelerated growth. It is known that the present aquaculture techniques do not prevent escape of some fish into the ocean. The best study cannot predict the impact of the escaped salmon on their wild relatives and other wild species satisfactorily.[6] The FDA thus will not authorize the large-scale production of these fish until appropriate physical or physiological confinements are found.

SIDE EFFECTS OF RELEASED GMOs

The possible side effects of the pesticides present in GM cotton and maize on local flora have been questioned. The Bt toxins used in GM plants are unstable proteins that target only a few categories of insects and show significantly lower deleterious effects on field flora than chemical pesticides.[7] The emergence of insects, particularly pyrale, in regions in which maize or cotton are cultured has not been observed so far. A protocol including the culture of non-GM maize in some fields followed by farmers appears sufficient to prevent emergence of pyrales resistant to Bt toxins. The simultaneous use of several toxins in the same GM plant is logically able to reduce considerably the emergence of resistant predators. Weeds resistant to the herbicide Roundup® have been shown to disseminate in wild space and the GM alfalfa approval has been recently revoked.[8] Wild plants resistant to Roundup are emerging in some regions in which this herbicide is extensively used. Moreover, a number of studies concluded that Roundup toxicity is low.[9]

Genetic modifications of farm animals have no intrinsic reasons to decrease their welfare. It may even increase the welfare of GM animals if they were generated to be resistant to diseases. This point must be examined on a case-by-case basis.

INTELLECTUAL PROPERTY OF GMOs

The plant varieties and animal breeds obtained by conventional selection are not patented and not easily patentable as they are not well-defined objects. The same is not true for GMOs. Logically, a GMO per se should not be patentable since this may unduly include the patenting of the GMO genetic background that belongs to all human communities. On the contrary, the methods to generate GMOs and the gene constructs can be patented.

The price of GM seeds is significantly higher than the seeds of normal plants. This may dissuade poor farmers from buying them. In several countries, the opposite is observed. The farmers still have a higher profit when they use GM seeds that give better yield, reduce the cost of pesticides, and diminish farmer intoxication.

CONCLUSION

Genetic selection is one of the major techniques that can provide human beings with abundant and healthy food. It is hardly conceivable that the GMO approach cannot bring an additional progress to conventional selection. This is already observed and will logically increase. Acceptability of GMOs by consumers remains limited in some countries and particularly in the EU. Problems raised by GMO use are real but unduly exaggerated. This is particularly the case for the GM food safety.[10,11] The environmental impact of GMOs is inevitable as it is the case for conventional plants. These problems do not appear insurmountable as soon as they are addressed on a case-by-case basis. The existing official guidelines can and must be improved and generalized, but their implementation represents a real application of the precautionary principle. The application of these guidelines is a limiting point for the use of GMOs in some poor countries and all should adopt them.

GMO patenting provides some companies with great power on food. Equitable sharing of profit and liberty to operate must be found.[12] GMO use is expected to be particularly profitable for poor countries. It cannot be reasonably expected that companies preparing GM varieties and breeds can develop projects generating no profit for them. Financial support from governments, international agencies, or NGOs might urge academic laboratories or even companies to develop specific and well-targeted non-profitable projects, as it is the case for some orphan diseases.

The present evolution of demography tends to generate an increasing food shortage that could be enhanced if a proportion of some plants are being used to generate biofuel. Use of GMOs cannot alone solve the complex problem of food shortage, but they may very significantly contribute to it. The well-controlled use of GMOs thus appears to fit more with the application of the precautionary principle than their rejection.

Ideally, the consumers should participate directly in the choice of the GMO projects to be developed. Experience has shown repeatedly that this approach is not realistic. Instead, the major operators, researchers, companies, and investors should take more into account the expectation of citizens, who on their own, should make more effort to become better informed on the real challenge of GMOs.

REFERENCES

1. Houdebine, L.M. *Animal Transgenesis and Cloning*; Wiley and Sons: London, 2003; 250.
2. Brief No. 34, 2005. Global Status of Commercialized Biotech/GM Crops: 2005, http://www.isaaa.org/.
3. Houdebine, L.M. Use of transgenic animals to improve human health and animal production. Reprod. Domest. Anim. **2005**, *40*, 269–281.
4. FAO. *Scientific Facts on Genetically Modified Crop*, 2004, Food & Agriculture Organization.
5. EFSA statement of the fate of recombinant DNA or proteins in meat, milk and eggs from animals fed with GM feed, 2007, European Food Safety Autority. http//:www.efsa.europa.eu.
6. Sundström, L.F.; Lõhmus, M.; Tymchuk, W.E.; Devlin, R.H. Gene–environment interactions influence ecological consequences of transgenic animals. Proc. Natl. Acad. Sci. U. S. A. **2007**, *104*, 3889–3894.
7. Marvier, M.; Mc Creedy, C.; Regetz, J.; Kareiva, P. A meta-analysis of effects of Bt cotton and maize on nontarget invertebrates. Science **2007**, *316*, 1475–1477.
8. Fox, J.L. US courts thwart GM alfalfa and turf grass. Nat. Biotechnol. **2007**, *25*, 367–368.
9. Williams, G.M.; Froes, R.; Munro, C. Safety evaluation and risk assessment of the herbicide Roundup and its active ingredient, glyphosate, for humans. Regul. Toxicol. Pharmacol. **2000**, *31*, 117–165.
10. Mc Hughen, A. Fatal flaws in agbiotech regulatory policies. Nat. Biotechnol. **2007**, *25*, 725–727.
11. EFSA. Statement of the scientific panel on genetically modified organisms on the analysis of data from 90-days rat feeding study with MON 863 maize, 2007, European Food Safety Autority. http//:www.efsa.europa.eu.
12. Hopkins, M.M.; Mahdi, S.; Patel, P.; Thomas, S.M. DNA patenting: the end of an era? Nat. Biotechnol. **2007**, *25*, 185–187.

Animals: RNA Interference (RNAi)

Luciana Relly Bertolini
Health Sciences Center, School of Medicine, University of Fortaleza, Fortaleza, Brazil

Marcelo Bertolini
Health Sciences Center, School of Medicine, University of Fortaleza, and Department of Food and Animal Sciences, School of Veterinary Medicine, Santa Catarina State University, Fortaleza, Brazil

Abstract

RNA interference (RNAi) is a highly conserved, highly regulated gene-silencing mechanism of natural occurrence in multicellular organisms. The gene silencing mechanism through RNAi can be induced either by small interfering RNAs (siRNA) or short hairpin RNA (shRNA). Such processes represent a powerful tool for studies in functional genomics, as RNAi can be manipulated in vitro or in vivo, in a transient or constitutive fashion, inducing sequence-specific post-transcriptional silencing of gene expression in cells, tissues, organs and systems, or even in a whole organism. In this entry, we briefly discuss the natural gene-silencing mechanism, or micro RNA (miRNAs), and the in vitro and in vivo applications of siRNAs and shRNAs to enhance animal health or animal production by devising antiviral therapies, for disease prevention, or to boost production traits, among other uses, as conditional RNA interference. Hence, the possibility of genetic engineering animals has become a reality beyond the classic methods of genetic modification, with the application of the RNAi technology promising to create safe and effective models that may in fact reach the herds in the future.

INTRODUCTION

RNA interference (RNAi) is a naturally occurring process in cells that has been conserved in multicellular organisms as diverse as plants, nematodes, insects, and mammals. The RNAi pathway promotes gene silencing by preventing protein formation. Hence, gene transcription occurs at the DNA level, but the transcribed messenger RNA (mRNA) is either blocked or destroyed before its translation, resulting in a protein downregulation (or a knockdown effect) in the cell.

Since its discovery over a decade ago, RNAi has been revolutionizing research in the life sciences. Albeit, the RNAi mechanism, was first described in the worm *Caenorhabditis elegans*,[1] the phenomenon per se was already known to occur in plants as a posttranscriptional gene silencing (PTGS) process.[2] As many other great discoveries in science, RNAi too was observed by accident, as an unintended consequence of experiments in genetic engineering. By adding additional copies of a gene to boost pigment expression in petunias, the engineered plants surprisingly turned white or exhibited a white/purple pattern, caused by the inhibition of the plant's endogenous gene. The authors named this process "gene silencing."[2] Soon after, related events started to be described in several eukaryotic organisms, including nematodes, fungi, insects, and mammals.

Most RNAs within a cell are involved with protein synthesis. These include mRNA, the intermediate transcript molecule between DNA and a functional protein and ribosomal RNA (rRNA), and transfer RNA (tRNA), the ribosomal structural and amino acid transfer molecules required to build a peptide chain in ribosomes during translation. However, a diverse repertoire of small RNAs that execute important roles in the control of gene expression, but that do not seem to encode for any functional protein, are also important components of the ribonucleic acid pool within cells. These molecules start as stem and loop structures, forming short hairpin, double-stranded RNAs (dsRNA) that are processed to produce small dsRNAs containing 20–25 nucleotides (nt) that subsequently mediate the silencing of specific genes at the posttranscriptional level. Such small RNAs are believed to be a primitive form of immune response to protect against the invasion of mobile genetic elements such as viruses and transposons (DNA sequences that can move around the genome of a given cell) into the genome. This sequence-specific gene expression inhibition by small RNAs has been designated small interfering or short interfering RNA (siRNA) and their activity per se is called RNAi.

In a practical view, the term siRNA has been preferably applied for the class of exogenous synthesized or in vitro-produced molecules that are used to obtain RNAi response in cells. However, cells can also express a class of endogenous short single-stranded RNA molecules called microRNAs (miRNAs). Specifically, miRNAs belong to a family of untranslated long RNA transcripts encoded by endogenous genes that are processed to produce mature miRNA, a phenomenon that also involves the RNAi machinery, as described below. Both miRNAs and siRNAs have a regulatory function by interfering with protein production. Once processed, miRNAs and siRNAs modulate translation by binding to their target mRNA, knocking down the complementary RNA preventing translation, or degrading the message using the RNAi machinery. Paradoxically, small dsRNAs molecules, either of endogenous (miRNAs) or exogenous (siRNAs) origin, can activate gene expression in mammalian cells in a sequence-specific, long-lasting manner, instead of silencing it, a phenomenon that has been recently termed as RNA activation or small RNA-induced gene activation (RNAa).[3] It is estimated that miRNAs are responsible for regulating a third of the human gene pool. The full extent of such gene silencing or activating pathways is yet to be uncovered.

RNAi can be used as a reliable and robust methodology that allows for the study of links between the expression of a given gene to its biological effect or phenotypical changes in an organism. This entry will consider the production and application of the siRNA-mediated RNAi phenomenon and the potential impact of such technology on animal biotechnology.

Gene Silencing Mechanism

The molecular mechanisms of the gene silencing pathway mediated by RNAi that results in the degradation of specific mRNA targets are starting to emerge. The RNAi pathway and siRNA functions are summarized in a two-step simplified diagram in Fig. 1.

The classical RNAi pathway is induced by long dsRNA molecules, also referred to as stem and loop double-stranded short hairpin RNA (shRNA). In this first step, known as the RNAi initiating step, large dsRNA are recognized, bound, and cleaved by an RNaseIII-like nuclease protein, named Dicer, resulting in active 21- to 23-nt double-stranded siRNA fragments, with 1- to 2-nt long overhangs on each side. On processing, ds-siRNA fragments are incorporated into a multiprotein complex known as RNA-Inducing Silencing Complex (RISC). Then, RISC unwinds the duplex siRNA into single-stranded siRNA. In this second step, the antisense strand of the siRNA guides RISC to its homologous mRNA. RISC-integrated endonucleases will then cleave and degrade the target mRNA.[4] Intermediate molecules and protein complexes in this mechanism remain largely unknown.

in vivo, miRNAs are derived from long single-stranded primary RNA molecules expressed by the cell, named pri-miRNA, which are enzymatically processed by the dsRNA-specific ribonuclease *Drosha* into 70–100 nt short

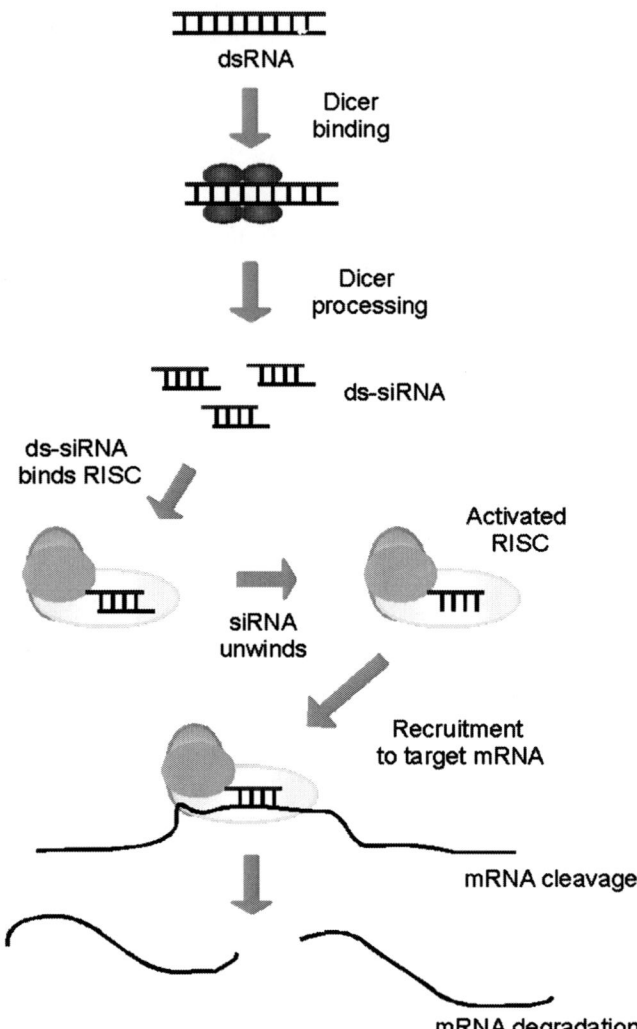

Fig. 1 Simplified schematic diagram of the RNAi mechanism. RNA interference is a posttranscription gene silencing mechanism through the degradation of homologous messenger RNA (mRNA) induced by short interfering RNA (siRNA). Once transcribed, precursor dsRNAs are recognized and processed by the enzyme Dicer. The double-stranded processed siRNA enters the RNA-induced silencing complex (RISC) that unwinds the siRNA and finds the homologous target mRNA for degradation.

stem-loops, called pre-miRNAs. The pre-miRNAs are transported to the cytoplasm to be digested by *Dicer*, a second double-strand-specific ribonuclease, resulting in a sequence-specific, single-stranded mature miRNA molecule with 21–23 nt. The mature miRNA is bound by a complex similar to RISC that participates in RNAi.

The understanding of how RNAi works in cells has allowed the exploitation of this endogenous pathway to silence genes of interest and, consequently, to study gene function and changes in phenotype in a number of different species. The introduction of specific, long dsRNA into cells has been used to investigate gene function in several organisms including trypanosomes, *Drosophila* sp., *C. elegans*, and mosquitoes. However, a natural antiviral defense is often activated after the introduction of long dsRNAs in mammalian cells, causing non-specific gene silencing (an interferon response) leading to apoptosis. In addition, unintended off-targeting (transcripts that are unintentionally downregulated by the siRNA) and the RNAa phenomenon can also affect results and cell survival. Consequently, in higher eukaryotes, synthetic or in vitro-produced specific siRNAs have been carefully designed and used to circumvent or minimize non-specific effects.

siRNA as a Tool for Functional Genetics

Gene silencing by RNAi has become a standard laboratory technique for investigating gene function by knocking down corresponding mRNAs. At least three critical steps are involved in the application of siRNA for gene silencing studies in higher eukaryotes, as outlined below.

1. The first key step involves the selection and designing of functional siRNA sequence(s) specific for the gene(s) of interest. Poorly designed siRNA sequences may cause non-specific effects, leading to problems in data interpretation and potential cell toxicity. As the genome of an increased number of animal species is unraveled, and as gene sequences, gene products, and their interactions are deciphered, the potential targets for the use of RNAi in any given species for any given application become immeasurable. Specialized software packages using specific and complex siRNA design algorithms are under a continuous process of development by companies and researchers for the generation of siRNAs molecules that may confer higher RNAi effects and lower unintended off-targeting or antiviral responses. Still, the ultimate confirmation of the effectiveness of a designed siRNA sequence is the evaluation of the effects upon testing in living cells.
2. A second step involves the generation and/or delivery of ds-siRNAs into the chosen cell or organism. With the identification of key components of the RNAi pathway, siRNA fragments can now be effectively produced either in vitro or in vivo. The choice for either system depends on the purpose or final application.

The in vitro synthesis is used for short-term studies to attain transient knockdown effects, for which two types of siRNA fragments can be produced: 1) multiple, distinct 21- to 23-nt long ds-siRNA fragments from the random digestion of a long dsRNA that is homologous to a segment of the gene of interest and transcript. A DNA sequence representing a target region of the gene of interest is linked to a bacteriophage gene promoter, which is subjected to in vitro transcription, producing single-stranded RNA molecules. Homologous RNA strands anneal producing dsRNA. Then, multiple and distinct siRNA fragments are created after exposure to a Dicer-like enzyme. This pool of multiple siRNAs specific to the target is then introduced into cells; and 2) identical ds-siRNA fragments highly specific to a target sequence of the gene of interest are chemically synthesized. Both systems have very effective knockdown responses, with the latter requiring more engineering and testing, but potentially causing minimized interferon response and lower side effects owing to the uniformity of the product.

The in vivo synthesis approach exploits the cell's own machinery by the use of expression vectors. These vectors mediate the production of transcripts containing shRNA, which will be endogenously digested to ds-siRNA fragments. Vectors can either be designed for a transient knockdown effect via episomal expression, or be integrated into the cell's genome for a constitutive expression of siRNAs with a persistent and specific knockdown of the target genes. In the near future, once technically mastered, such a transgenic approach may allow a stable, long-term phenotype in animals, without non-specific unintended effects, eliminating restrictions caused by the transient state of RNAi experiments.

Regardless of the siRNA production system, nucleic acids designed for an RNAi response need to be introduced or delivered into the cell as DNA (e.g., viral vector) or RNA (e.g., ds-siRNA fragments), as illustrated in Fig. 2. Because of its inefficiency, the cell delivery step has been one of the greatest hurdles for the widespread use of siRNAs in therapeutics or for in vivo studies in animals. in vitro-produced ds-siRNA fragments can be delivered directly into cells in culture via liposome-based transfection (lipofection) or by electroporation. Fragments are then expected to carry out the knockdown effect after binding to RISC. However, efficiency of lipofection or other delivery methods varies according to the cell type and culture conditions. This problem is being resolved by the continuous development of more efficient cell transfection agents or by the use of viral vectors for cell transduction.

The transfections of cells with plasmids or viral vector infection (transduction) are both designed to express dsRNA molecules in the cell that will then be endogenously generated into siRNA fragments. Viral vectors carrying the DNA sequence of interest have been successful delivery vehicles for studies in transgenesis and RNAi. Modified, replication-defective retroviruses, mainly from the lentivirus subclass, take advantage of the viral ability to infect cells with a high efficiency and randomly integrate them into the genome, resulting in constitutive, stable dsRNA expression. However, the random integration into the host genome may cause insertional mutagenesis, disturbing gene function in cells.

3. The last step in RNAi studies involves the monitoring of the efficiency of gene silencing. The gene knockdown effect can be evaluated by quantifying the levels of the transcript and/or protein of interest, or by the observation of the modified phenotype in cells or

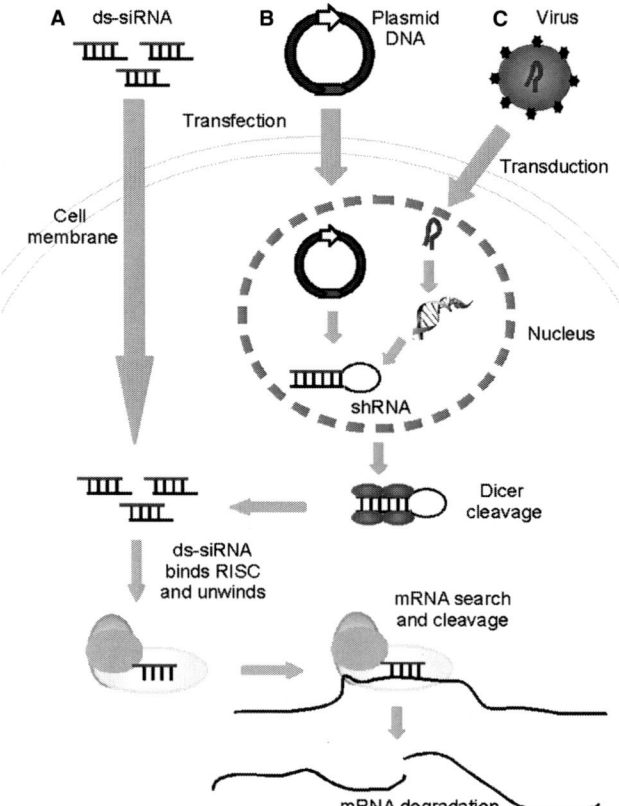

Fig. 2 Simplified view of the siRNA- and shRNA-mediated gene silencing in mammalian cells. Different delivery strategies and the processing of stem and loop structure of short hairpin RNA (shRNA) in the cell are shown. Once delivered into the cell's cytoplasm, in vitro-produced siRNA fragments (A) can be recruited to initiate RNA interference. Plasmids (B) or viral vectors (C) can also be used to in vivo transcribe shRNAs in the cell, which will endogenously enter the RNAi pathway. Plasmid DNA in (B) is shown as episomal expression, whereas viral vector in (C) is integrated into the host genome.

in the organism. The relative abundance of mRNA molecules, after reverse transcription to complementary DNA (cDNA), is usually measured by quantitative PCR (polymerase chain reaction), whereas the protein abundance is done by western blotting with specific antibodies to the target protein. The RNAi responses depend widely on the cell system one wishes to downregulate, but in general, a knockdown of 70% or more in protein abundance results in successful or expected phenotypes.

RNAi Applications in Animal Biotechnology

Enhanced productivity traits and resistance to infectious diseases have long been the goal of most genetic engineering programs tailored for livestock species. Since the production of the first transgenic livestock in 1985, a large number of transgenic animals have been born carrying a wide range of transgenes. Despite all efforts, the practical use of this technology has been restricted to medical applications, such as animal models for disease or production of pharmaceutical products in milk, with agricultural applications remaining largely a promise.[5]

The vector-mediated siRNA approach allows for the creation of transgenic animals with stable gene silencing using standard transgene technology (pronuclear microinjection) or by transduction of cultured cells for use in cloning procedures by nuclear transfer. By using RNAi, the silencing of specific genes of interest may not require embryonic stem cells or homologous recombination procedures. Consequently, the use of RNAi for the creation of transgenic animals to boost production traits or to enhance health fitness will likely be limited only by genetic, molecular, biochemical, and immunological understanding of targeted diseases, as well as by animal welfare issues and ethical considerations associated with trangenesis per se.

A fair number of approaches have been developed to introduce a transgene into an animal's genome with the potential to improve immune response to infectious diseases or to confer resistance to specific microorganisms, improving animal health. Many classical transgenic studies have demonstrated the potential of modifying food animals for production or health. The knockout of the prion gene, coding for the prion protein that causes spongiform encephalopathy in ruminants,[6] the development of cattle resistant to mastitis infection,[7] or even the engineering of livestock species to improve production or products (e.g., growth rate, wool, and milk), or to address the environmental issues are important examples of the benefits that can be attained through transgenesis. As the RNAi mechanism does not involve the production of any known protein, operating only at the RNA level, its use for genetic engineering may become a safer approach when the goal is the use in livestock species for commercial purposes. Thus, the use of RNAi technology, when combined with transgenic approaches and cloning procedures promises to encourage the resurgence of livestock transgenesis for agricultural applications.

Most potential RNAi experiments in livestock will ultimately deal with one of the major problems in agriculture, i.e., infectious diseases. The possibility to engineer RNAi transgenes into animal genomes to produce more resistant animals to infections not only will be beneficial to both animals and producers, but also to the market, as safer products with reduced human-transmissible diseases, such as avian influenza, can be produced.

Antiviral therapies

The sequence-specific nature of RNAi in combination with improvements in delivery systems (lentiviral transfer), are quickly becoming more attractive and safer for therapeutic applications. Some reports have described the efficient replication inhibition of viruses implicated in infectious diseases that affect livestock species and cause major social and economical problems throughout the world. Examples range from successful knockdown effects against human viruses in cells in culture to a whole animal approach (mouse model) using lentiviral vectors expressing shRNAs targeting the Hepatitis B virus.[8] Another important development involving RNAi technology is the demonstration that lentiviral vectors can be successfully used to deliver DNA constructs coding shRNAs into early stage embryos, producing live transgenic mice having the desired targeted silencing effect of the RNAi. This methodology should be easily adapted to the production of other animals, including livestock species. In addition, transient RNAi effects can also be used in preimplantation stage embryos to attain temporally- and/or developmentally-specific effects, for studies in functional genomics, epigenetics, and developmental and cell biology.

Disease prevention

The use of RNAi to prevent diseases in livestock is on the rise. RNAi has been used effectively in transgenic goats cloned by nuclear transfer using fibroblasts transduced by a lentiviral vector carrying a transgene expressing shRNA targeting the prion protein.[9] Scientists are also working on the possibility of using RNAi to protect poultry from the lethal influenza virus, highly contagious to humans, as seen in the Hong Kong outbreak in 1997, in the Netherlands in 2003, and in Southeast Asia in 2004. Researchers have previously shown that siRNA can be used to decrease lung virus titers and also protect mice from lethal challenge with a variety of Influenza A viruses.[10] Also, the effectiveness of stable RNAi expression in cultured cells to suppress the foot-and-mouth disease virus and porcine endogenous retroviruses has been clearly demonstrated.[11] The successes of such experiments have consolidated RNAi as a tool to block or virtually eliminate viral infections in animals, leading the way for the creation of livestock species more resistant to infections as diverse as, e.g., fowl plague, swine fever, coronavirus, avian flu, and brucellosis. Effective RNAi responses to diseases other than by viral particles, such as those caused by bacteria, fungi, protozoa, or parasites, are also a common goal. Once the molecular, biochemical, and immunological pathways of host–agent interaction and the establishment and progression of diseases are fully understood, RNAi systems can be devised to tackle one or more steps within the disease circuit, thereby promoting health.

Production traits

Any experiment in which endogenous negative feedback systems can be downregulated may also be of use in animals. For instance, mutations in the myostatin gene, a member of the TGF beta protein superfamily, or in its receptor lead to a genetic condition that enhances muscle mass in humans and animals, named double-muscle mutation. By knocking down myostatin via RNAi during development in livestock species, an increase in muscle mass is expected, which is economically desirable. In addition, RNAi-induced myostatin downregulation may be useful for the study of muscular dystrophy in animal models.

Other uses

RNAi technology can be used as a tool to allow achievements in other applications. A proof-of-principle example is the manipulation of the recombination machinery in cells in culture to increase site-specific transgene insertion.[12] By the downregulation of the non-homologous end joining pathway by RNAi, a significant increase in gene targeting frequency by homologous recombination was observed, with a concomitant decrease in random transgene integration. These results are the first step for more successful use of gene targeting for gain- or loss-of-function studies (functional genomics) in animals other than the mouse.

Conditional RNAi is another area in development, mostly in mice, that promises to be of practical application in the near future. In such procedures, dsRNA are expressed in a stage- or tissue-specific manner, by the use of promoters of genes that are only expressed at a given period in development or only in a desired tissue or organ. In addition, by using inducible promoters, i.e., gene promoters that can be activated or inactivated under certain circumstances (e.g., tetracycline- or zinc-activated promoters that will activate only when tetracycline or zinc are supplied to the animals) may allow an exogenous control of the RNAi-induced knockdown effects. Such technologies may prove valuable for the manipulation of an animal's metabolism, for instance, driving metabolic pathways to commensurate changes in the environment, improving productivity. Collectively, examples described above clearly demonstrate the nearly unlimited potential of the RNAi technology when applied to animal biotechnology.

CONCLUSIONS AND FUTURE IMPLICATIONS

The potential of RNAi, and perhaps RNAa, in animal biotechnology is as extraordinary as in any other area in health sciences. Advancements in RNAi evolved and matured rapidly since its discovery, becoming a very effective and powerful method to study gene function in model systems. The biggest advantage to the field of animal biotechnology came with the possibility to express dsRNA in a vector that may silence gene products in a stable fashion. Once the technology is fully understood and dominated, such transgenic animals can be produced by standard transgene technology or by transduction of a cell line for use in cloning by nuclear transfer.[4] As an emerging tool to generate stably targeted gene knockdowns, this technology is simpler than the traditional knockout experiments that involve physical removal/inactivation of a gene. RNAi is still at its early stages for use in livestock species, but promises to be an effective tool to improve: 1) animal production by boosting valuable traits in livestock animals; and 2) animal health by preventing, attenuating, or possibly eliminating infectious and non-infectious diseases in animals, having both practical applications of significant economical and scientific implications.

REFERENCES

1. Fire, A.; Xu, S.; Montgomery, M.K.; Kostas, S.A.; Driver, S.E.; Mello, C.C. Potent and specific genetic interference by double-stranded RNA in *Caenorhabditis elegans*. Nature **1998**, *391*, 806–811.
2. Napoli, C.; Lemieux, C.; Jorgensen, R. Introduction of a chimeric chalcone synthase gene into petunia results in reversible co-suppression of homologous genes in trans. Plant Cell **1990**, *4*, 279–289.
3. Li, L.C.; Okino, S.T.; Zhao, H.; Pookot, D.; Place, R.F.; Urakami, S.; Enokida, H.; Dahiya, R. Small dsRNAs induce transcriptional activation in human cells. Proc. Natl. Acad. Sci. U.S.A. **2006**, *103*, 17337–17342.
4. Elbashir, S.M.; Lendeckel, W.; Tuschl, T. RNA interference is mediated by 21- and 22-nucleotide RNAs. Genes Dev. **2001**, *15*, 188–200.
5. Clark, J.; Whitelaw, B. A future for transgenic livestock. Nat. Rev. Genet. **2003**, *4*, 825–833.
6. Richt, J.A.; Kasinathan, P.; Hamir, A.N.; Castilla, J.; Sathiyaseelan, T.; Vargas, F.; Sathiyaseelan, J.; Wu, H.; Matsushita, H.; Koster, J.; Kato, S.; Ishida, I.; Soto, C.; Robl, J.M.; Kuroiwa, Y. Production of cattle lacking prion protein. Nat. Biotechnol. **2007**, *25*, 132–138.
7. Wall, R.J.; Powell, A.M.; Paape, M.J.; Kerr, D.E.; Bannerman, D.D.; Pursel, V.G.; Wells, K.D.; Talbot, N.; Hawk, H.W. Genetically enhanced cows resist intramammary *Staphylococcus aureus* infection. Nat. Biotechnol. **2005**, *23*, 445–451.
8. Morrissey, D.V.; Blanchard, K.; Shaw, L.; Jensen, K.; Lockridge, J.A.; Dickinson, B.; McSwiggen, J.A.; Vargeese, C.; Bowman, K.; Shaffer, C.S.; Polisky, B.A.; Zinnen, S. Activity of stabilized short interfering RNA in a mouse model of hepatitis B virus replication. Hepatology **2005**, *41*, 1220–1222.
9. Golding, M.C.; Long, C.R.; Carmel, M.A.; Hannon, G.J.; Westhusin, M.E. Suppression of prion protein in livestock by RNA interference. Proc. Natl. Acad. Sci U.S.A. **2006**, *103*, 5285–5290.
10. Tompkins, S.M.; Lo, C.Y.; Tumpey, T.M.; Epstein, S.L. Protection against lethal influenza virus challenge by RNA interference in vivo. Proc. Natl. Acad. Sci. U.S.A. **2004**, *101*, 8682–8686.
11. Santos, T.; Wu, Q.; de Avilla Botton, S.; Grubman, M.J. Short hairpin RNA targeted to the highly conserved 2B nonstructural protein coding region inhibits replication of multiple serotypes of foot-and-mouth disease virus. Virology **2005**, *335*, 222–231.
12. Bertolini, L.R.; Bertolini, M.; Maga, E.A.; Madden, K.R.; Murray, J.D. Increased gene targeting in Ku70 and Xrcc4 transiently deficient human somatic cells. Mol. Biotechnol. **2009**, *41*, 106–114.

Antibiotics Use in Food-Producing Animals

Jaap C. Hanekamp
Roosevelt Academy, Middelburg, and HAN-Research, Zoetermeer, The Netherlands

Abstract
Antibiotics as used for animal medication in relation to food safety and toxicology are addressed and related primary food regulations are briefly discussed. The regulatory incongruity of zero tolerance is elucidated and the potential for hormetic amelioration is outlined.

INTRODUCTION

Setting scientific and policy standards that benchmark the benefits and risks of foods is of great consequence for industry, policymakers, and consumers. Food safety as a whole is more often than not defined as *chemical* food safety, meaning that food from whatever source is regarded as "safe" when man-made chemicals such as antibiotics and pesticides are absent or only present at very low concentrations. In this entry, antibiotics as a tool in animal rearing is discussed from the perspective of the consumer end-product. As antibiotics create benefits for the producers, the risks to consumers that similarly benefit from antibiotics in terms of food abundance and concomitant lower prices need to be balanced effectively.

On the one hand, the question has been raised whether the use of antibiotics, especially as growth promoters in animals, can result in resistance within human bacteria. Transfer of antibiotic resistance from bacteria in livestock to bacteria in humans to lower the effectiveness of human therapy is the point of concern here.[1–3] In 1969, the Swann Committee reported that there was a significant problem with regard to antimicrobial (mis)use in both human and veterinary practice. Conversely, potential toxicity as a result of exposure to antibiotic residues in food products is another point of regulatory concern. In this entry, we will focus on the latter by addressing the following issues: 1) the "nature" and use of antibiotics; 2) legislation; and 3) toxicological models of assessment. The focus of this entry with regard to regulatory requirements is somewhat European, although antibiotics regulation in relation to food safety has universal characteristics.

THE "NATURE" OF ANTIBIOTICS

With the discovery of penicillin in 1928 by Alexander Fleming, the human potential to tackle bacterial infections in both humans and animals grew immeasurably. Penicillin is made by the mold, *Penicillium notatum*, yet most antibiotics we know today are derived from Actinobacteria (also called actinomycetes), nature's most prolific antibiotic producers. They not only produce antibiotics in a huge variety but also produce chemicals that kill fungi, parasitic worms, and even insects.[4] Many other pharmaceuticals, such as antitumor agents and immunosuppressants, are also derived from actinomycetes. The antibiotics industry is valued at roughly $25 billion per year, underscoring the value of the knowledge of these bacteria.

Streptomyces, a genus of Actinobacteria, account for well over two-thirds of these commercially and therapeutically significant antibiotics, which are produced by means of complex secondary metabolic pathways. *Streptomyces* is, therefore, the most important source of antibiotics for medical, veterinary, and agricultural use. *Streptomyces* is a group of Gram-positive filamentous bacteria found in soils worldwide. The biomass per hectare of actinomycetes in 15 cm of topsoil comprises between 400 and 5000 kg.[5] They are among the most numerous and ubiquitous soil bacteria adapted to the utilization of plant remains and are extremely important in this environment because of their broad potential of metabolic processes and biotransformations. These include degradation of the insoluble remains of other organisms making *Streptomyces* imperative in carbon recycling.[6] *Streptomyces* is a member of the same taxonomic order as the causative agents of tuberculosis (TB) and leprosy (*Mycobacterium tuberculosis* and *Mycobacterium leprae*).

While secondary metabolites must in all likelihood confer an adaptive advantage, the roles of most are not fully understood. This is true even for antibiotics. Antibiotics offer a competitive advantage to antibiotic producers as compared to other ground-dwelling microorganisms that need to tap into the same biomass for nutrition. An ensnarement strategy was observed as well, where competing organisms are attracted and subsequently killed by excreted antibiotics and subsequently consumed supplying additional nutrition. A variety of antibiotics are produced by *Streptomyces* and their close relatives; some examples of medically important antibiotics are given in Table 1.[7]

Streptomycin was the first antibiotic to be discovered to be effective against TB. Chloramphenicol, now an outdated pharmacon in the Western world, was the first antibiotic to be synthetically produced and was shown to be effective against typhoid;[8] however, ophthalmic infections are still treated with chloramphenicol, and this antibiotic is still widely used in Asian countries, especially against typhoid fever.

Antibiotics, for the past six decades, have played a major role in human and animal health. Apart from curative use, antibiotics were extensively used as growth-promoting agents in animal husbandry. This practice is no longer allowed, at least within the EU, because of potential risks of transfer of bacterial resistance from animals to humans. Curative use of antibiotics in animal rearing is still a major issue. As an indication of the volume of veterinary antibiotics in use, total sales in the UK remained relatively steady between 1998 and 2003 at around 434 tons per year. Between 2004 and 2007 total sales dropped to 387 tons.[9] These figures suggest that antibiotics use in animal rearing is a daily reality that potentially impacts food quality. We will subsequently consider the regulatory and toxicological components of this use.

LEGISLATURE—THE EUROPEAN APPROACH

In Europe, the core regulatory framework in food law is Regulation 178/2002/EC.[10–12] According to this Regulation, "food" (or "foodstuff") denotes "any substance or product, whether processed, partially processed or unprocessed, intended to be, or reasonably expected to be ingested by humans." The scope of Regulation 178/2002/EC concerns "all stages of the production, processing and distribution of food..." and its general objective is to provide "a high level of protection of human life and health and the protection of consumers' interests, . . ." This Regulation thus sets general rules for all products that are brought to market. Importantly, the Regulation also constitutes the European Food Safety Authority (EFSA) and defines the Authority's task and fields of competence and authority.

Council Regulation EEC No. 2377/90 was implemented to establish maximum residue limits of veterinary medicinal products in foodstuffs of animal origin. This so-called "MRL Regulation" (maximum residue limit) introduced Community procedures to evaluate the safety of residues of pharmacologically active substances according to human food safety requirements. A pharmacologically active substance may be used in food-producing animals only if it receives a favorable evaluation. If it is considered necessary for the protection of human health, maximum residue limits ("MRLs") are established. They are the points of reference for setting withdrawal periods in marketing authorizations as well as for the control of residues in the member states and at border inspection posts.

EEC No. 2377/90 contains an Annex IV listing of pharmacologically active substances for which no maximum toxicological levels (tolerable daily intake—TDI) can be fixed, either from lack of toxicological or pharmacological data, e.g., the absence of a definable NOAEL (no observed adverse effect level) or LOAEL (lowest observed adverse effect level)—or because of genotoxic characteristics of the compound in question. These substances are consequently not allowed in the animal food-production chain. So-called zero-tolerance levels are in force for Annex IV for reasons that can be subsumed as follows:

- Lack of scientific data de facto makes the establishment of a TDI not feasible.
- The absence of a TDI and the subsequent impossibility to establish a maximum residue limit (MRL), in regulatory terms is understood as "dangerous at any dose" requiring zero-tolerance regulation.
- With the introduction of zero tolerance, a veterinary ban on Annex IV compounds (such as CAP) is in place, whereby the listed compounds, when producers compliance is achieved, would disappear from the food chain.

This approach has generated numerous problems. Technological advances in analytical equipement resulted in lower limits of detection whereby dwindling amounts of compounds can be detected. Toxicological relevance, and thereby food-safety, lost its significance in this development. One consequence thereof was that pharmaceutical companies lost interest in the European market; escalating regulatory stringency made it more difficult to obtain veterinary medicine authorisations.

As of the 6th of May 2009 Regulation EC No. 470/2009 has superseded EEC No. 2377/90.[13] This regulation aims among other issues to tackle the decreasing veterinary authorisations (preamble 14) and the issue of zero tolerance. The latter has been scrapped. Instead a risk analysis is required for illegal or non-authorised pharmacologically active substances (articles 15–19). As a result, the Annex IV list will be scrapped eventually. The reference point of action is related to Minimum Required Performance Limit ('MPRL'). MRPL is the concentration level that regulatory (reference) laboratories in the European Community should at least be able to detect and confirm. MRPLs since 2005 have been given some legal status in terms of *levels of concern*.[14] Levels of concern now increasingly seem a viable approach compared to the zero-tolerance as 'virtually safe' doses.[15] This articulates the toxicological issue of the dose-response curve with which we will conclude this discussion.

Table 1 Antibiotic-producing organisms, antibiotics, and targeted diseases.

Target disease	Antibiotic	Producing organism
Methicillin-resistant *Staphylococcus aureus* (MRSA)	Vancomycin	*Amycolatopsis* (*Streptomyces*) *orientalis*
Typhoid (*Salmonella enterica* serovar Typhi)	Chloramphenicol	*Streptomyces venezuelae*
Pathogens with transmissible penicillin resistance	Clavulanic acid	*Streptomyces clavuligerus*
Tuberculosis and leprosy	Rifampicin	*Amycolatopsis* (*Streptomyces*) *mediterranei*
Cancer	Daunomycin	*S. clavuligerus*

TOXICOLOGICAL MODELS OF ASSESSMENT—MATURING PERSPECTIVES

The question in what way high-dose and low-dose exposures correlate is a long-standing one, as one has to relate experimental data with real-life risks of exposure to antibiotics residues in food products. The age-old Paracelcus axiom, *Sola dosis facit venemum*, the dose makes the poison, does not address the shape of the curve linking both ends of the exposure scale (Fig. 1).

For the sake of simplicity two main toxicological linear models will be mentioned here. Model A depicts the "no-dose no-illness" approach when dealing with primarily genotoxic carcinogens. The fact that chemicals are capable to react with hereditary material makes the assumption that even one molecule might in theory generate cancer seemingly viable. Model A is usually referred to as the LNT model (linear non-threshold model). Model B assumes a threshold in the dose–response curve. So below the threshold the toxin is assumed not to generate any harmful effect in the exposed organism. Non-carcinogens are thought to usually exhibit such behavior. Model B is usually referred to as the LT model (linear threshold model).[16]

It is now argued, however, that the most fundamental shape of the dose–response is neither threshold nor linear, but U-shaped (C), and hence both current models A and B provide less reliable estimates of low-dose risks in antibiotic use. This U-shape is usually referred to as hormesis[16–20] and is best described as an adaptive response to low levels of stress or damage, resulting in enhanced robustness of some physiological systems for a finite period. More specifically, hormesis is defined as a moderate overcompensation to a perturbation in the homeostasis of an organism. The fundamental conceptual facets of hormesis are respectively: 1) the disruption of homeostasis; 2) the moderate overcompensation, 3) the re-establishment of homeostasis; and 4) the adaptive nature of the overall process. For example, heavy metals such as mercury prompt synthesis of enzymes called metallothioneins that remove toxic metals from circulation and probably also protect cells against potentially DNA-damaging free radicals produced through normal metabolism. Conversely, low doses of antitumor agents commonly enhance the proliferation of the human tumor cells, in a manner that is fully consistent with the hormetic dose–response relationship.[21]

High doses push the organism beyond the limits of kinetic (distribution, biotransformation, or excretion) or dynamic (adaptation, repair, or reversibility) recovery. This is the classical toxicological object of research usually required as a result of public and regulatory concerns whereby hormetic responses are by default regarded as irrelevant, or even contrary to policy interests, and therefore unlooked for. Acceptance of hormesis suggests that low doses of toxic/carcinogenic agents may reduce the incidence of adverse effects observed at higher dosages.

The future of toxicological risk assessments of antibiotics exposure in food possibly could best be described as follows: "With respect to hormesis it is ethically mandated that potential beneficial aspects of low exposure to potentially hazardous material are incorporated in the risk-benefit balancing procedure. The potential harm done by pollutants does not justify the invocation of a categorical principle. *Minimization of risk is not required if health benefits are also at stake....*"[22] Chemical substances, be they natural or synthetic, are neither bad nor good; they are both, depending on exposure levels and adaptive responses from the exposed organisms. This, as we know is especially true for antibiotics that have a wide range of activities, both for humans and animals. In terms of bacterial resistance, prudent and targeted use of antibiotics seems the best way forward. Here, "good" and "bad" go hand in hand as well.

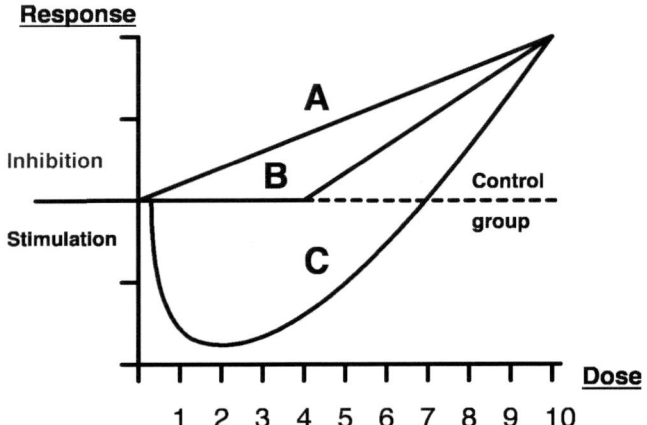

Fig. 1 Three toxicological dose–response models.

REFERENCES

1. Casewell, M.; Friis, C.; Marco, E.; McMullin, P.; Phillips, I. The European ban on growth-promoting antibiotics and emerging consequences for human and animal health. J. Antimicrob. Chemother. **2003**, *52*, 159–161.
2. Phillips, I.; Casewell, M.; Cox, T.; De Groot, B.; Friis, C.; Jones, R.; Nightingale, C.; Preston, R.; Waddell, J. Does the use of antibiotics in food animals pose a risk to human health? A critical review of published data. J. Antimicrob. Chemother. **2004**, *53*, 28–52.
3. Bezoen, A.; van Haren, W.; Hanekamp, J.C. *Emergence of a Debate: AGPs and Public Health*; Heidelberg Appeal Foundation: Amsterdam, The Netherlands, 1999.
4. Hopwood, D.A. *Streptomyces in Nature and Medicine. The Antibiotic Makers*; Oxford University Press: Oxford, 2007.
5. Brady, N.C.; Weil, R.R. *The Nature and Properties of Soils*, 13th Ed.; Prentice Hall: New Jersey, 2002.
6. Bentley, S.D.; Chater, K.F.; Cerdeño-Tárraga, A.-M.; Challis, G.L.; Thomson, N.R.; James, K.D.; Harris, D.E.; Quail, M.A.; Kieser, H.; Harper, D.; Bateman, A.; Brown, S.; Chandra, G.; Chen, C.W.; Collins, M.; Cronin, A.; Fraser, A.; Goble, A.; Hidalgo, J.; Hornsby, T.; Howarth, S.; Huang, C.H.; Kieser, T.; Larke, L.; Murphy, L.; Oliver, K.; O'Neil, S.; Rabbinowitsch, E.; Rajandream, M.A.; Rutherford, K.; Rutter, S.; Seeger, K.; Saunders, D.; Sharp, S.; Squares, R.; Squares, S.; Taylor, K.; Warren, T.; Wietzorrek, A.; Woodward, J.; Barrell, B.G.; Parkhill, J.; Hopwood, D.A. Complete genome sequence of the model actinomycete *Streptomyces coelicolor* A3(2). Nature **2002**, *417*, 141–147.
7. Chater, K.F. Streptomyces inside-out: a new perspective on the bacteria that provide us with antibiotics. Philos. Trans. R. Soc. B **2006**, *361*, 761–768.
8. Ehrlich, J.; Bartz, Q.R.; Smith, R.M.; Joslyn, D.A. Chloromycetin, a new antibiotic from a soil actinomycete. Science **1947**, *106*, 417.
9. Veterinary Medicines Directorate. *Sales of Antimicrobial Products Authorised for Use as Veterinary Medicines, Antiprotozoals, Antifungals, Growth Promoters and Coccidiostats, in the UK in 2007*; New Haw, Surrey, 2008.
10. Hanekamp, J.C.; Frapporti, G.; Olieman, K. Chloramphenicol, food safety and precautionary thinking in Europe. Environ. Liabil. **2003**, *6*, 209–219.
11. Hanekamp, J.C.; Kwakman, J. Beyond zero tolerance: a new approach to food safety and residues of pharmacological active substances in foodstuffs of animal origin. Environ. Liabil. **2004**, *1*, 3–39.
12. Hanekamp, J.C. Veterinary residues and new European legislation: a new hope? Environ. Liabil. **2005**, *1*, 52–55.
13. Regulation (EC) No 470/2009 of the European Parliament and of the Council of 6 May 2009 laying down Community procedures for the establishment of residue limits of pharmacologically active substances in foodstuffs of animal origin, repealing Council Regulation (EEC) No 2377/90 and amending Directive 2001/82/EC of the European Parliament and of the Council and Regulation (EC) No 726/2004 of the European Parliament and of the Council. *Official Journal of the European Communities* **L152**, 11–22. See for a discussion: Hanekamp, J.C. 2009. Neither Acceptable nor Certain Cold War Antics for 21st Century Precautionary Culture. *Erasmus Law Review 2* (2), 221–257.
14. Commission decision of 11 January 2005 laying down harmonised standards for the testing for certain residues in products of animal origin imported from third countries. Off. J. Eur. Commun. *L16*, 61–63.
15. *Comments on the Discussion Paper on Risk Analysis Principles and Methodologies in the CODEX Committee on Residues of Veterinary Drugs in Food*; Joint FAO/WHO Foods standards Programme CODEX committee on Residues of Veterinary drugs in Food, Thirteenth Session, Charleston, SC, December 4–7, 2001; CX/RVDF 01/9-Add.1.
16. Hanekamp, J.C.; Calabrese, E.J. Chloramphenicol, European legislation and hormesis. Dose Response **2007**, *5*, 91–93.
17. Calabrese, E.J.; Baldwin, L.A. Toxicology rethinks its central belief. Hormesis demands a reappraisal of the way risks are assessed. Nature **2003**, *421*, 691–692.
18. Moustacchi, E. DNA damage and repair: consequences on dose-responses. Mutat. Res. **2000**, *464*, 35–40.
19. Tubiana, M.; Aurengo, A. Dose–effect relationship and estimation of the carcinogenic effects of low doses of ionising radiation: the joint report of the Académie des Sciences (Paris) and of the Académie Nationale de Médecine. Int. J. Low Radiat. **2005**, *2* (3/4), 1–19.
20. Calabrese, E.J.; Staudenmayer, J.W.; Stanek, E.J.; Hoffmann, G.R. Hormesis outperforms threshold model in National Cancer Institute antitumor drug-screening database. Toxicol. Sci. **2006**, *94* (2), 368–378.
21. Calabrese, E.J. Hormesis and medicine. Br. J. Clin. Pharmacol. **2008**, *66* (5), 594–617.
22. Renn, O. An ethical appraisal of Hormesis: towards a rational discourse on the acceptability of risks and benefits. Am. J. Pharmacol. Toxicol. **2008**, *3* (1), 165–181.

Antimicrobial Packaging

Haiqiang Chen
Hudaa Neetoo
Department of Animal and Food Sciences, University of Delaware, Newark, Delaware, U.S.A.

Abstract

Surface growth of microorganisms is one of the leading causes of food spoilage and contamination. Food surfaces can be contaminated by spoilage and pathogenic microorganisms during handling, processing, and packaging. A variety of processing and preservation technologies have been used to control surface contamination. More recently, the concept of incorporating antimicrobials into packaging materials has been investigated to better control the growth of microorganisms by direct contact of the package with the surface of food. An overview of the biotechnological techniques used in the manufacturing of the various components of antimicrobial packaging materials is presented.

INTRODUCTION

Consumer preferences have forced many food companies to reduce levels of "chemical food additives" and consequently fueled much interest in "natural" microbial inhibitors. The application of "natural preservatives" on or into packaging materials represents one of the most promising avenues available to enhance food safety and extend the shelf life of foods. Antimicrobial surface treatments ensure that only low levels of the agent come into contact with the food as compared to the direct addition of preservatives.

Many natural antimicrobials, such as enzymes, organic acids, and bacteriocins, have been isolated from animal, plant, and microbial sources; however, their widespread use is often hampered by limited amounts that can be extracted or obtained in vivo, as well as non-optimal specific activities which may require large amounts for satisfactory results in foods. This is the point where biotechnology comes to play a pivotal role to optimize the production and specific activities of these antimicrobial compounds.

Although food packaging materials have traditionally been based on non-renewable materials, increasing attention is being given to sustainability and the replacement of non-renewable resources with renewable ones. Hence alternative packaging materials based on biological materials have been developed, and recently been made commercially available.[1]

This entry thus gives an exposure of the biotechnological techniques used in the production of natural antimicrobials as well as the packaging materials into which these agents are incorporated.

ANTIMICROBIAL COMPOUNDS

Bacteriocins

Much attention has been focused on the ability of bacteriocins to inhibit the growth of pathogens and their incorporation into packaging films has been widely examined.[2] Nisin is a polypeptide bacteriocin produced by some strains of *Lactococcus lactis* subsp. *Lactis*. Nisin exhibits antimicrobial activity against a variety of Gram-positive bacteria often associated with food. Toxicological studies have found nisin safe and it is currently allowed as a food preservative in over 50 countries worldwide.[3] As part of a hurdle system of preservation, nisin can also be used to inhibit the growth of pathogens such as *Listeria monocytogenes* and *Clostridium botulinum*. Production of nisin is usually carried out through batch (rather than continuous) fermentation processes.[3] Current biotechnological developments in nisin technology include the design of modified nisin molecules by genetic methods resulting in a wider microbial spectrum and activity as well as transfer of nisin production to bacteria that are capable of producing nisin in situ in food systems.[3]

Pediocin from several strains of *Pediococcus acidilactici* and *Pediococcus pentosaceus* have been explored for potential use as food preservatives. Yield of pediocin by *P. acidilactici* strains is relatively high and several methods have been tested for large-scale production.[4] Pediocin PA-1/AcH is currently being produced from *P. acidilactici* PAC1.0 by a commercial firm and the product is being sold as a flavor enhancer.[4] It is effective against *L. monocytogenes* and *C. botulinum*.

Organic Acids

Organic acids, such as benzoic acid, sorbic acid, propionic acid, acetic acid, and lactic acid and their salts, possess strong antimicrobial activity and have been incorporated into packaging materials.[5] These acids inhibit microbial growth by lowering the pH, affecting the proton gradient across cytoplasmic membranes, acidifying the cytoplasm, and interfering with chemical transport across membranes.

Lactic acid and its derivatives elicit a broad-spectrum of antimicrobial activity against Gram-positive bacteria including spore-forming bacilli and clostridia, as well as Gram-negative pathogens such as enterohemorrhagic *Escherichia coli* O157:H7 and *Salmonella*. Lactates also demonstrate antifungal activity against *Aspergillus* sp. Based on these attributes, lactic acid and lactates have been used for their ability to control the growth of pathogens and for shelf-life enhancement of fresh and semiprocessed foods. Fungi (e.g., *Rhyzopus oryzae*) and lactic acid bacteria are used worldwide to manufacture bulk quantities of lactic acid.[6] Its production is amenable to both batch and continuous fermentation systems although continuous fermentation techniques can shorten the non-productive operations associated with batch systems. Currently, the worldwide utility of lactic acid and its derivatives amounts to 100,000 metric tons per year.[6]

Citric acid is not used as a direct antimicrobial agent but it is recognized for enhancing the antimicrobial property of many other substances in foods, either by reducing pH or by chelating metal ions. Its use in controlling thermophilic bacteria in particular is well established. Currently, there are two organisms widely used in the production of citric acid, namely, yeast and *Aspergillus niger*.[7] Other organic acids that have also been successfully incorporated into packaging materials include benzoic acid, acetic acid, and propionic acid. These are all produced by traditional fermentation technologies, some of which have not changed for decades. Although it is conceivable that improvements in strain stability and product yields could be made using recombinant DNA technology, the relatively low cost of food acidulants makes it unlikely that biotechnological techniques will have a major impact on that business sector for the time being.

Lysozyme

Lysozyme is attractive as a natural food preservative because it is endogenous to many foods, specifically targets bacterial cell walls, is harmless to humans, and can be incorporated into or immobilized onto packaging materials. Singly or in combination with other natural antimicrobial compounds, it has desirable antimicrobial properties for use in minimally processed foods.[8] Lysozyme is more effective against Gram-positive than Gram-negative bacteria. It is commercially isolated almost exclusively from hen egg white. Systems for expression and secretion of lysozyme have been constructed using yeast and *E. coli*. Unfortunately until now, the amount of lysozyme secreted into the medium has been low. As more opportunities are found for lysozyme, there will be a need for genetic manipulation to enhance production yield.

EDIBLE AND BIO-BASED PACKAGING MATERIALS

Antimicrobials can be incorporated into a variety of plastic polymers (e.g., polyethylene [PE], polypropylene [PP], and Surlyn®) made from non-renewable resources or edible and bio-based polymers made from renewable resources. Edible biopolymers are typically used as films and coatings and can be consumed along with foods, while bio-based polymers are not edible. Edible biopolymers used in antimicrobial packaging are typically polysaccharide- or protein-based but they can also be lipid-based. The polysaccharide-based biopolymers include starch, cellulose derivatives, seaweed extracts (alginates and carrageenan), and chitosan. The protein-based biopolymers include whey protein isolate, collagen, gelatin, corn zein, wheat gluten, and soy protein isolate. Polylactic acid (PLA) and polyhydroxyalkanoates (PHAs) are two bio-based polymers that show great commercial potential. The following section focuses on edible and bio-based polymers produced from microorganisms (naturally occurring or genetically transformed) through fermentation processes.

Edible Biopolymers

Cellulose

Although the packaging field is still dominated by mineral oil-derived polymers, such as PE and PP, a notable exception is cellulose, which enjoys wide usage as an exterior packaging layer. Traditionally, cellulose has been harvested from plant resources; however, fungi and bacteria (*Acetobacter xylinum*) can also assemble cellulose by harnessing chemical energy from glucose or some organic substrate.[9] Direct synthesis of cellulose derivatives, such as carboxymethylcellulose and methylcellulose, by genetic modification techniques makes them attractive polymers for the food packaging industry. Recent successes with cloning and sequencing of genes for bacterial cellulose synthesis combined with new information on how these genes may function in vivo have given new insight into fermentation optimizations.[9]

Starch

Another biodegradable polysaccharide, which shows great promise for use by the food packaging industry is starch.

Although most researches aimed at enhancing the functional properties of starch have focused on incorporation of additives such as plasticizers, microorganisms present a unique and novel alternative because they produce a variety of enzymes that disintegrate or assimilate the polymers with unique properties. The molecular weight distribution of the final product can also be controlled by adjusting the duration of the fermentation process and other culture conditions. The purified thermoplastic starch developed by fungal treatment can be extracted, combined with glycerol and water, and cast to form packaging films. Fermentation of starch by fungal microorganisms to modify its structure and functionality presents initial steps towards a future where large-scale production of biopolymers for the food packaging industry would be possible.[10]

Chitosan

Chitosan is a natural polymer derived from chitin. It is the principal fiber component of the exoskeleton of shellfish and possesses inherent antimicrobial activity.[11] Chitosan is commercially produced from shrimp and crab shells through alkaline deacetylation of chitin;[11] however, supplies of raw materials are variable and seasonal and the process is laborious and costly. Moreover, the chitosan derived from such a process is heterogeneous with respect to their physicochemical properties[11] Recent advances in fermentation technology suggest that the cultivation of selected fungi can provide an alternative source of chitosan. The use of *Zygomycetes mycelia* has been proposed as a good alternative source of chitosan. Recently, the production of chitosan by culturing a *Rhizopus azygosporus* fungus or an *Actinomucor taiwanensis* fungus and its subsequent isolation from the culture has been patented.

Bio-Based Polymers

Polylactic acid (PLA)

Polylactic acid is a polyester polymerized from lactic acid monomers. The monomer itself is produced via fermentation of carbohydrate feedstock. Although not yet produced in bulk, large-scale production is being evaluated by at least two companies. Final polymer costs may well depend on the efficiency of the initial process to cheaply produce the lactic acid monomer by fermentation of biomass or waste products.[12] Recently, the use of PLA/pectin or PLA/chitosan composite films in antimicrobial packaging have been investigated and have shown great potential as novel antimicrobial food packaging materials.

Polyhydroxyalkanoates (PHA)

Polyhydroxyalkanoates are produced by fermentative microorganisms or genetically transformed bacteria. They function in microorganisms as energy substrates and for carbon storage. An advantage of these films is that their properties can be adjusted by manipulation of the growth media to obtain different polymers. Although there has been limited research on the incorporation of antimicrobials into PHA films, these show great promise as film substrates.[13]

ANTIMICROBIAL PACKAGING FILMS

The following methods are most commonly used to incorporate antimicrobials into packaging materials: 1) incorporation of antimicrobial agents directly into polymers; 2) coating or adsorbing antimicrobials onto polymer surfaces; 3) immobilization of antimicrobials to polymers via ionic or covalent linkages; and 4) use of polymers that are inherently antimicrobial. Antimicrobial packaging is an emerging and exciting area of food technology, which can confer many preservation benefits on a wide range of foods. Postpasteurization/prepackaging contamination of ready-to-eat (RTE) foods by foodborne pathogen *L. monocytogenes* is a significant problem currently faced by the food industry. Recent research has shown great promise for the use of antimicrobial packaging films to control *L. monocytogenes* in various RTE foods such as frankfurters and smoked salmon as well as to reduce the risk of surface contamination of spoilage microorganisms.

CONCLUSION

Antimicrobial packaging is an emerging technology that can contribute to the improvement of food safety and shelf-life extension. The burgeoning interest in the incorporation of "natural" antimicrobials into packaging materials for food preservation and protection coupled with advances in fermentation, enzymology, and recombinant DNA technologies have opened numerous avenues for novel production processes of antimicrobials. In addition, the new developments in polymer science and genetic engineering have also expanded the possibilities for improving the structure and function of existing polymers used by the food packaging industry. There is no doubt that future biotechnological research into a combination of microbially produced antimicrobial agents and biodegradable packaging materials will highlight a range of the merits of antimicrobial packaging in terms of food safety, shelf life, and environmental friendliness.

REFERENCES

1. Weber, C.J.; Haugaard, V.; Festersen, R.; Bertelsen, G. Production and applications of bio-based packaging materials for the food industry. Food Addit. Contam. **2002**, *19*, 172–177.

2. Cooksey, D.K.; Gremmer, A.; Grower, J. Characteristics of nisin-containing corn zein pouches for reduction of microbial growth in refrigerated pouches for reduction of microbial growth in refrigerated shredded cheddar cheese. In *Book of Abstracts IFT Annual Meeting, Dallas, TX*; Institute of Food Technologists: Chicago, 2000.
3. Thomas, L.V.; Clarkson, M.R.; Delves-Broughton, J. Nisin. In *Natural Food Antimicrobial Systems*; Naidu, A.S., Ed.; CRC Press: New York, 2000; 463–524.
4. Ray, B.; Miller, K.W. Pediocin. In *Natural Food Antimicrobial Systems*; Naidu, A.S., Ed.; CRC Press: New York, 2000; 525–566.
5. Han, J.H. Antimicrobial food packaging. In *Novel Food Packaging Techniques*; Ahvenainen, R., Ed.; CRC Press: Boca Raton, FL, 2000; 50–65.
6. Bogaert, J.C.; Naidu, A.S. Lactic acid. In *Natural Food Antimicrobial Systems*; Naidu, A.S., Ed.; CRC Press: New York, 2000; 613–636.
7. Sharma, R.K. Citric acid. In *Natural Food Antimicrobial Systems*; Naidu, A.S., Ed.; CRC Press: New York, 2000; 689–702.
8. Losso, J.N.; Nakai, S.; Charter, E.A. Lysozyme. In *Natural Food Antimicrobial Systems*; Naidu, A.S., Ed.; CRC Press: New York, 2000; 185–210.
9. Petersen, K.; Nielsen, P.V.; Bertelsen, G.; Lawther, M.; Olsen, M.B.; Nilsson, N.H.; Mortensen, G. Potential of bio-based materials for food packaging. Trends Food Sci. Tech. **1999**, *10*, 52–68.
10. Mohini Sain, M.; Jeng, R.; Saville, B.; Huang, C.B.; Hubbes, M.; Saville, B. Thermoplastic starch. Bioscienceworld, http://www.bioscienceworld.ca/ThermoplasticStarch (accessed August 27, 2007).
11. Knorr, D. Recovery and utilization of chitin and chitosan in food processing waste management. Food Technol. **1991**, *26*, 114–122.
12. Robertson, G.L. Edible and biobased food packaging materials. In *Food Packaging: Principles and Practice*, 2nd Ed.; CRC Press, Boca Raton, FL, 2006; 43–54.
13. Eldgenossiche materials Prufungs Anstalt (EMPA): Production of PHA from cheap waste resources, http://www.empa.ch/plugin/template/empa/*/87077 (accessed December 5, 2009).

Aroma Compounds

Adriane Medeiros
Debora Brand
Carlos R. Soccol
Bioprocess Engineering and Biotechnology Division, Federal University of Paraná, Curitiba, Brazil

Abstract

Aroma compounds produced biotechnologically are considered natural and have applications in the food, feed, cosmetic, chemical, and pharmaceutical industries. The biotechnological processes most frequently used for aroma compounds include microbial fermentation, enzymatic processes, and plant cell culture. Solid-state fermentation is a viable alternative to economically produce aroma compounds using advantageously agro-industrial residues as substrates for their production. Aroma compounds produced by plant cell cultures are characteristic to their origin and present very complex compositions; however, its production has several disadvantages when compared to fermentation technology.

INTRODUCTION

Biotechnological aroma compounds, also known as odorants, aromas, fragrances, or flavors, are considered as natural substances by European legislation on the sole condition that the precursors involved in their synthesis are natural.[1] In the United States, the Code of Federal Regulations (CFR) defines a natural aroma as: "... the essential oil, oleoresin, essence or extractive, protein hydrolysate, distillate, or any product of roasting, heating and enzymolysis, which contains the flavoring constituents derived from a spice, juice, vegetable or vegetable juice, edible yeast, herb, bud, bark, root, leaf or similar plant material, meat, seafood, poultry, eggs dairy products or fermentation products thereof, whose significant function is food flavoring rather than nutritional" (CFR 101.22.0.3).[2]

This definition of natural aroma comprises products that are converted by living cells or parts thereof, including enzymes. There is a supplementary regulation established by the Food and Drug Administration (FDA), in which compounds known originally as generally recognized as safe (GRAS) should still be classified as GRAS when produced by microbiological or enzymatic processes.

Fragrance and aroma compounds now have widespread applications in the food, feed, cosmetic, chemical, and pharmaceutical industries with a worldwide industrial size estimated at $16 billion in 2003.[3] According to The Freedonia Group, Inc.[4] the global claim for aroma compounds and fragrances increased 4.4% per year to $18.6 billion in 2008 reaching 7.3% in the Asia/Pacific region (excluding Japan). Food and beverage industries are among the major market segments for aroma compounds and fragrances accounting for 47% of total demand in 2003. A fast market growth is expected in soft drinks, snacks, convenience foods, health foods, and nutraceuticals; demand for these compounds will benefit from increased interest for natural and exotic compounds, which can be obtained biotechnologically.

Although a considerable amount of current research focuses on the production of food flavor and aroma compounds, only a few of them are obtained by biotechnological routes. The most relevant achievements are related to plant cell and microbial production, as well as enzyme-catalyzed reactions.[5]

Microbial Production

Many microorganisms are capable of synthesizing aroma compounds when grown on a culture medium. They have the ability to perform conversions, which would otherwise require multiple chemical steps. Microorganisms are used to catalyze specific steps. They are also an economical source of enzymes, which can be utilized to enhance or alter the flavors of many food products. In this way, biotechnological processes involved in the production of aromatic compounds can be divided in two groups, microbiological and enzymatic processes.[6] In fermentation, the production of aroma compounds starts from cheap and simple sources such as sugars and amino acids. The product is generated by the complex metabolism of the microorganism. When microorganisms are used to catalyze specific conversions of precursors and intermediates, the process is called biotransformation. While fermentation requires C- and N-sources, a specific substrate is necessary for microbial transformation. The enzymatic catalysis precedes a simple and specific transformation of the substrate molecule.

It is important to identify research whose objectives are to obtain complex products with natural characteristics. The first tendency imitates nature and develops a process with one or more microorganisms and enzymes. The second tendency tries to obtain the higher yield of characteristic components. The choice determines the methodology, which will be employed in vivo or in vitro and obtained by biosynthesis or bioconversions.[7]

Applications of Aroma Compounds

Most of the biotechnological processes that have been reported have not yet been applied in industrial production of aroma compounds. It is estimated that around 100 aroma compounds are produced industrially by microbial fermentation.[8] The major reason for the low number of products is the low yield. Microbial aroma compounds are often present in low concentrations in fermentation broths, resulting in high costs for downstream processing. Nevertheless, this fact is counterbalanced by the price of the naturally produced compounds, which is 10–100 times higher than that of synthetic compounds. For example, synthetic γ-decalactone, the impacting aroma compound of peach, costs \$150 for a kilogram (kg), while the same substance extracted from a natural source costs about \$6000/kg. The microbial production of γ-decalatone (peach aroma), involving the bioconversion by *Yarrowia lipolytica* of castor oil yields about 6 g/L.[8] Another example of fermentation process is the production of vanillin, which is widely used as a flavoring agent in a wide range of foods and fragrances. The use of biotechnological methods involving fungi to produce vanillin is being developed. Ferulic acid, used as a direct precursor in this bioconversion, is a product of the microbial oxidation of lignin, particularly by white-rot basidiomycetes.[9]

Attempts are being made by industries to produce natural aroma compounds by fermentation, which allows for the recovery of natural food additives preferred by the consumer. The aim of current biotechnological research is the development of low-cost processes with high yields. In order to achieve this, it is necessary to control the metabolic pathway and to develop alternative production techniques such as the use of solid-state fermentation (SSF), immobilized cells or genetically modified organisms.

Solid-State Fermentation

Agro-industrial residues such as sugarcane and cassava bagasse, apple pomace, and coffee husk and pulp have been used efficiently in several bioprocesses. The application of agro-industrial residues not only provides alternative substrates for SSF but also helps to solve pollution problems.[10]

Solid-state fermentation has received more and more interest from researchers. The development of novel and cheap production processes, such as SSF, may help overcome some of the current limitations of microbial aroma production, as well as widening the spectrum of biotechnologically accessible compounds.[11] Cassava and sugarcane bagasse, apple pomace, giant palm bran, and coffee husk have been used as substrates for aroma compound production in SSF.[12–14] Fruity flavors were detected in cultures of *Ceratocystis fimbriata* using coffee husk as substrate.[13] The authors found that the aroma detected in the headspace of the culture depended on the amount of glucose added to the medium. Increased levels of glucose decreased aroma intensity. Glucose concentration seemed to have a direct effect on metabolic pathways and on the nature of the volatile compounds produced. Among the compounds produced, ethanol and ethyl acetate were the most abundant, followed by isoamyl acetate, isopropyl acetate, and ethyl propionate.

Bramorski et al.[12] studied the production of volatile compounds by the edible fungus *Rhizopus oryzae* during solid-state cultivation on tropical agro-industrial substrates. When *R. oryzae* was grown on a medium containing cassava bagasse plus soybean meal (5:5 w/w), CO_2 production rate was at its highest (200 mL/L), whereas the highest volatile metabolite production was with amaranth grain as the sole substrate (282.8 mL/L). In the headspace, ethanol was the most abundant compound (more than 80%). Acetaldehyde, 1-propanol, ethyl propionate, and 3-methyl butanol were also present. CO_2 and volatile metabolite productions reached their maximum around 20 and 36 hours, respectively.

A strain of the yeast *Kluyveromyces marxianus* was used for the production of a fruity aroma in SSF using cassava bagasse as substrate.[14] Experiments were performed with a 2^5 statistical experimental design. The parameters studied were cultivation temperature, pH, initial water content, and carbon/nitrogen (C/N) ratio of substrate and inoculum size. The initial pH and the C/N ratio of the medium were statistically significant at 5% level for the production of volatile compounds. Aroma production increased in acidic pH (3.5) medium with a C/N ratio of 100. The results showed the feasibility of using cassava bagasse as substrate to produce a fruity aroma with *K. marxianus* in SSF. Bagasse and dessicated coconut have been used as solids substrates for the production of 6-pentyl-α-pyrone, an unsaturated δ-lactone with a strong aroma of coconut, by *Trichoderma harzanium*.[1]

Plant Cell Cultures

Plants produce a wide variety of volatiles, every cell of a plant contains the genetic information necessary to produce several components that constitute a natural aroma. Plant aromas can be considered as terminal metabolites containing abundant biochemical information, differing according to metabolism stages. Some reactions that naturally occur in plant cells are complex and cannot be easily achieved by synthetic pathways. The utilization of a plant cell or tissue culture process for the production of various aroma compounds will depend on its ability to compete with agricultural sources. To compete with the normal sources the plant culture process will, at least initially, involve low volume and high price compounds and high productivity, which requires a high yield of product.

Many differences occur when comparing plant and microbial cells for the production of aroma compounds. Plant cells are more fragile and susceptible to shear than microbial cells. Duplication time and product accumulation is much faster for microbial cells. Production costs are over 10-fold for cell cultures; however, efforts have been made to stimulate synthetic activities by selecting strains presenting high productivity, optimizing culture conditions, adding precursors, and using cell immobilization that may prolong culture viability.[15]

Several plants showing their characteristic aroma are good candidates for cell culture. For example, the culture of *Vanilla planifolia* produces vanilla as the major component. Vanilla is the most frequently used flavor ingredient in the food industry. Compounds such as p-hydroxibenzaldehyde, vanillic acid, and p-hydroxybenzyl methyl ether are also produced. Strawberry aroma is very complex; it is constituted by 278 volatile substances, comprising 33 acids, 39 alcohols, 17 aldehydes, 14 ketones, and 103 esters, among other compounds. Tissue culture for strawberry aroma was only performed for component groups such as esters. Other aroma compounds produced by plant cell cultures include apple, cinnamic acid, caryophyllen, cocoa, garlic, and onion.[5]

CONCLUSION

Aroma biotechnology is increasing rapidly and is becoming commercially competitive in the manufacturing of natural products. Microorganisms such as yeast and molds present several opportunities for use in biosynthetic and biotransformation processes. Perspectives in microbial aroma production can be found with the elucidation of the metabolic pathways and also with use of immobilized cultures. Additionally, as downstream processing is the major drawback in microbial production, future research in this area will open new avenues for technology transfer. Among fermentation technologies, SSF is a good choice for growth as it yields a more concentrated product.

REFERENCES

1. Sarhy-Bagnon, V.; Iozano, P.; Roussos, S. Coconut-like aroma production by Trichoderma harzanium in solid state fermentation. In : Roussos S., Lonsane B.K., Raimbault M., Viniegra Gonzales G. (editors), *Advances in Solid State Fermentation*. Kluwer Academic Publishers: Dordrecht, 1995.
2. Armstrong, D.W.; Gillies, B.; Yamazaki, H. In *Flavour Chemistry: Trends and Development*; Charalambous, G., ed.; Elsevier Science Publishers: New York, 1989.
3. Serra, S.; Fuganti, C.; Brenna, E. Biocatalytic preparation of natural flavours and fragrances. Trends Biotechnol. **2005**, *23*, 193–198.
4. http://www.freedoniagroup.com/pdf/1886smwe.pdf (accessed July 2007).
5. Longo, M.A.; Sanromán, M.A. Production of food aroma compounds: microbial and enzymatic methodologies. Food Technol. Biotechnol. **2006**, *44* (3), 335–353.
6. Welsh, F.W.; Murray, W.D.; Willians, R.E. Microbiological and enzymatic production of flavor and fragrance chemicals. Crit. Rev. Biotechnol. **1989**, *9* (2), 105–169.
7. Delest, P. Natural flavours: biotechnology limited... or unlimited? In *Bioflavour 95*, INRA, Ed.; Les colloques: Dijon, France, 1995; Vol. 75, 14–17.
8. Janssens, L.; De Poorter, H.L.; Vandamme, E.J.; Schamp, N.M. Production of flavours by microorganisms. Process Biochem. **1992**, *27*, 195–215.
9. Lomascolo, A.; Stentelaire, C.; Asther, M.; Lesage-Meessen, L. Basidiomycetes as new biotechnological tools to generate natural aromatic flavours for the food industry. Trends Biotechnol. **1999**, *17* (7), 282–289.
10. Soccol, C.R.; Vandenberghe, L.P.S. Overview of applied solid-state fermentation in Brazil. Biochem. Eng. J., **2003**, *13*, 205–218.
11. Feron, G.; Bonnarme, P.; Durand, A. Prospects for the microbial production of food flavours. Trends Food Sci. Technol. **1996**, *7*, 285–293.
12. Bramorski, A.; Christen, P.; Ramirez, M.; Soccol, C.R.; Revah, S. Production of volatile compounds by the fungus *Rhizopus oryzae* during solid state cultivation on tropical agro-industrial substrates. Biotechnol. Lett. **1998**, *20*, 359–362.
13. Soares, M.; Christen, P.; Pandey, A.; Soccol, C.R. Fruity flavour production by *Ceratocystis fimbriata* grown on coffee husk in solid-state fermentation. Process Biochem. **2000**, *35*, 857–861.
14. Medeiros, A.B.P.; Pandey, A.; Freitas, R.J.S.; Christen, P.; Soccol, C.R. Optimization of the production of aroma compounds by *Kluyveromyces marxianus* in solid-state fermentation using factorial design and response surface methodology. Biochem. Eng. J. **2000**, *6*, 33–39.
15. Hrazdina, G. Aroma production by tissue cultures. J. Agric. Food Chem. **2006**, *54*, 1116–1123.

Arthropod Host–Plant Resistant Crops

Gerald E. Wilde
Department of Entomology, Kansas State University, Manhattan, Kansas, U.S.A.

Abstract

Resistance of plants to pest attack is defined as the relative amount of heritable qualities possessed by the plant that influence the ultimate amount of damage caused by the pest. The use of plant resistance to manage arthropod pest populations provides an ideal approach to integrated pest management because it is biologically and economically sound, environmentally friendly, and generally compatible with other management tactics or strategies. The cultivar forms the foundation on which all pest management programs and tactics are applied, and its effects are specific, cumulative, and persistent.

INTRODUCTION

Because of the many advantages plant resistance offers, virtually every cultivated crop has been evaluated for this trait and one or more resistant sources have been identified. The challenge has been to incorporate these resistant sources into agronomically adapted and consumer acceptable, high-yielding cultivars. In addition to traditional breeding methods, the use of modern breeding techniques and genetic transformation of crops has opened the door to other ways of identifying, incorporating, and employing pest-resistance genes to effectively and economically manage arthropod pest populations. The use of resistant cultivars contributes significant economic and social benefits and sustainable agricultural systems to the world's farmers. The positive effects of resistant cultivars have been demonstrated repeatedly in crops as diverse as wheat, alfalfa, grape, sorghum, maize, rice, apple, and cotton.

PERCENTAGE OF CROPS THAT HAVE SOME DEGREE OF PEST RESISTANCE

Plant resistance has been employed to a greater or lesser degree in practically all of the major food, feed, and fiber crops. Table 1 lists a number of major crops grown in the world and the number of pests for which resistance has been employed to at least some extent in the field. Hectarage planted to resistant cultivars varies for each pest and crop and over time as new varieties and hybrids (both susceptible and resistant) are grown and, in some instances, as new pest biotypes (pest populations that are capable of damaging previously resistant sources) develop. For example, most of the modern rice varieties and hybrids grown in China, India, and other countries are resistant to one or more major pests.

Resistant American grape rootstocks have been used extensively over the world to control *Phylloxera vittifolae* (Fitch). A large percentage of the alfalfa planted in the United States is comprised of varieties resistant to aphid species. Sorghum hybrids with resistance to the greenbug have occupied up to 80% of the hectarage in the United States. Significant hectarages of wheat and barley in the United States, Canada, and North and South Africa have resistance to at least one pest. Most commercial soybean varieties are resistant to the potato leafhopper. Several cotton varieties carrying genes for resistance to jassids (*Empoasca* sp.) are grown widely in Africa, India, and the Philippines. In the United States, more than 65% of commercial maize hybrids have some resistance to corn leaf aphid, >90% have some resistance to first generation European corn borer, and >75% have some resistance to second generation corn borer.

However, many more resistance genes have been identified in all crops than have been used in modern commercial varieties and hybrids, because incorporating them into high yielding cultivars acceptable to growers has been difficult. Recently, transgenic crops have been utilized to combat major insect pests. Hybrids or varieties with insect-resistance genes have been developed in cotton, maize, and potato. An estimated 6.7 million hectares of transgenic corn resistant to the European corn borer, 2.5 million hectares of transgenic cotton resistant to several pests, and 20,000 hectares of transgenic potato resistant to Colorado potato beetle were grown in the world in 1998. The hectares planted to transgenic crops are likely to increase as additional countries register these products and this technology is used on additional crops. For example, specific biotechnology applications are being field tested for rice and wheat, which together occupy 400 million hectares globally.

Table 1 List of some major crops grown in the world and number of arthropod pests for which resistant cultivars have been used in the field by growers for pest management

Crop	No. of pests
Alfalfa	6
Apple	1
Asparagus	1
Barley	3
Bean	1
Cassava	2
Chickpea	0
Cotton	6
Grape	1
Lettuce	1
Maize	10
Millet	1
Oat	1
Pea	1
Peanut	4
Potato	1
Raspberry	1
Rice	14
Rye	1
Sorghum	6
Soybean	1
St. Augustine grass	1
Sugar beet	1
Sugarcane	3
Sunflower	1
Sweet clover	1
Sweet potato	1
Wheat	7

EFFECT OF PLANT RESISTANCE ON PEST POPULATIONS

The growing of pest-resistant cultivars can be used as a major control tactic or adjunct to other measures. Historically, the use of resistant cultivars combined with other tactics has resulted in a reduction of many pest species to subeconomic levels. Even small increases in resistance enhance the effectiveness of cultural, biological, and insecticidal controls. The extent to which growing resistant plants affects pest populations is dependent upon the level of resistance expressed, the mechanisms of resistance involved, and the number of hectares grown. The growing of resistant wheat on 50% of the hectarage in Kansas has been shown to reduce Hessian fly populations to extremely low levels. Resistance in wheat to wheat curl mite (ca. 25% of the hectarage) was effective in limiting the spread of wheat streak mosiac virus, which the mite transmits. The incorporation of leaf and stem pubescence into most commercial soybean varieties has resulted in population suppression of the potato leafhopper over the past 60 years. As the hectarage of sorghum resistant to the greenbug increased to >50%, the area of sorghum treated with insecticide was reduced by 50%. Tenfold reductions in pest populations have been observed where insect-resistant rice cultivars have been grown widely.

ECONOMIC AND SOCIAL BENEFITS

Assessing the economic benefits of plant resistance is difficult in the context of integrated pest management programs and is likely to be underestimated frequently and substantially. Even determining the obvious advantages (yield benefits and reduced production costs) may be difficult over a large area where pest populations vary from locality to locality and year to year. Other environmental benefits, such as cleaner water and food, reduced risks to farmers, more flexibility in planting and cropping systems, reduced disease transmission, and reduced secondary pest outbreaks, also are difficult to quantify. Nevertheless, some specific estimates are available. In the United States alone the estimated valued of using arthropod-resistant alfalfa, barley, corn, sorghum, and wheat cultivars is more than $1.4 billion each year. The net economic benefit of greenbug resistance in U.S. sorghum production is estimated at close to $400 million annually. The global economic value of arthropod-resistant wheat has been estimated at $250 million annually. The value of resistance to aphids in alfalfa in the major alfalfa-producing states of the United States is estimated at more than $100 million annually. Breeding for pest resistance in rice has been estimated to be responsible for one-third of recent yield increases and $1 billion of additional annual income to rice producers. The net return of insect-resistant Bt maize in the United States and Canada has been estimated in some studies at $42.00–$67.30 per hectare, but other studies have indicated less of an economic return. The average net economic return of insect-resistant Bt cotton in 1997 was $133 per hectare.

BIBLIOGRAPHY

Antle, J.M.; Pingali, P.L. Pesticides, productivity and farmer health: a Philippine case study. Am. J. Agric. **1994**, *76*, 418–430.

Global Plant Genetic Resources for Insect-Resistant Crops; Clement, S.L., Quisinberry, S.S.; Eds.; CRC Press: New York, 295.

Harvey, T.L.; Martin, T.J.; Seifers, D.L. Importance of plant resistance to insect and mite vectors in controlling virus diseases of plants: resistance to the wheat curl mite (Acari: Eriophyidae). J. Agric. Entomol. **1994**, *11*, 271–277.

Hyde, J.; Martin, M.A.; Preckel, P.V.; Edwards, L.R. The economics of Bt corn: valuing protection from the European corn borer. Rev. Agric. Econ. **1999**, *21*, 442–454.

James, C. *Global Review of Commercialized Transgenic Crops*; ISAAA Briefs No. 8, ISAA: Ithaca, NY, **1998**, 43. In *Insect Resistant Maize: Recent Advances and Utilization*, Proceedings of an International Symposium, International Maize and Wheat Improvement Center (CIMMYT), Nov 27–Dec 3, **1994**; Mihm, J.A., Ed.; CIMMYT: Mexico, D.F., **1997**; 302.

Painter, R.H. *Insect Resistance in Crop Plants*; University of Kansas Press: Lawrence, KS, **1968**, 520.

Smith, C.M. *Plant Resistance to Insects. A Fundamental Approach;* John Wiley & Sons: New York, **1989**, 286.

Smith, C.M.; Quisinberry, S.S. Value and use of plant resistance to insects in integrated pest management. J. Agric. Entomol. **1994**, *11*, 189–190.

van Emden, H.F. Host-Plant Resistance to Insect Pests. In *Techniques for Reducing Pesticides: Environmental and Economic Benefits;* Pimentel, D., Ed.; John Wiley: Chichester, England **1997**, 124–132.

In *Economic, Environmental, and Social Benefits of Resistance in Field Crops*, Proceedings, Thomas Say Publications in Entomology, Wiseman, B.R., Webster, J.A., Eds.; Entomological Society of America: Lanham, MD, **1999**, 189.

Contribution No. 00-252-B of the Kansas Agricultural Experiment Station.

Artificial Chromosomes in Plants

James A. Birchler
Division of Biological Sciences, University of Missouri, Columbia, Missouri, U.S.A.

Abstract
The addition of transgenes to crop plants has provided a means to introduce herbicide resistance for improved cultivation methods and insect resistance to increase yields. The amount of worldwide tillable land can no longer increase significantly, but world food production must increase; additional types of genetic modifications will be needed on many fronts. In order to facilitate additions of multiple genes that can modify crops for diverse properties, artificial chromosomes can act as a single entity to carry these genes. These chromosomes would optimize expression of transgenes, permit their accumulation on the same chromosome, and allow the transfer of a collection of genes from one variety to another with ease.

INTRODUCTION

For a chromosome to function, certain minimal parts are necessary. These include the centromere, telomeres and the origins of replication. The centromere is the site of attachment of spindle fibers for chromosomal movement. Specialized structures at the ends of chromosomes called telomeres, resolve replication and prevent fusion of the ends. In yeast, these elements were assembled several decades ago to produce so-called artificial chromosomes that have been used for a variety of practical applications as well as studies of chromosomal function.[1]

Artificial chromosomes have also been produced using human chromosome components by two different methods.[1] In one method, tissue culture cells are transformed with telomere sequences retaining just the centromeric region. Subsequently, additional genes and other engineered sequences are added by homologous recombination. In the second method, the centromere and other sequences are transformed into tissue culture cells. These sequences spontaneously form megabase-sized conglomerates that can function as a chromosome.

The greatest potential for artificial chromosome usage for practical applications may reside in the plant kingdom where this technology has the potential to provide a tool for agricultural and pharmaceutical biotechnology. Artificial chromosomes may provide a platform for reliable and predictable gene expression from a predetermined site into which many genes could be continually added to act as an independent chromosome. Under these conditions, multiple transgenes can be placed into a genome as a group that would not exhibit linkage to other genes, but would facilitate their transfer from one variety to another. In this entry, the initial production of plant artificial chromosome platforms will be described as well as the parameters that need to be developed to make them useful for crop improvement.

ARTIFICIAL CHROMOSOMES

The use of transformed centromere sequences as a means to produce artificial chromosomes in plants is complicated by the epigenetic component to centromere specification. In other words, the underlying DNA alone is insufficient to organize a centromere. Whereas plant centromeres typically have a very similar DNA repetitive array,[2-4] these sequences do not organize the kinetochore for chromosome movement as evidenced by the fact that centromeres can become inactive despite being intact.[5] (During division, the kinetochore is the chromosomal protein structure where the spindle fibers attach to pull the chromosomes apart.) Indeed, transformation of a centromere sequence-containing clone has been successful in rice and maize, but no kinetochore is organized.[6]

Engineered minichromosomes that can act as an artificial chromosome platform have been produced in maize.[7] The approach taken is to perform telomere truncation.[8] The telomeres of plants consist of many repeats of a simple sequence at the ends of each chromosome. When this sequence is transformed into maize via *Agrobacterium* transformation or via a biolistic approach, the presence of the telomere repeat at one end of the transformed sequence appears to convert the normal "non-homologous end-joining" mechanism that usually connects the introduced sequence into the genome into a de novo telomere-formation reaction. The creation of a telomere during transformation cleaves off the remainder of the chromosome arm.

There is currently no efficient means to perform homologous recombination to modify minichromosomes produced by telomere truncation. Therefore, it is necessary to place a site-specific recombination module on the transformed DNA in such a manner so as to keep it on the truncated chromosome.[7] Site-specific recombination involves the presence of target DNA sequences that are recognized by a recombinase enzyme for reciprocal exchange between a pair of target sites. The combination of a site-specific recombination cassette and the ability to perform telomere truncation confers the ability to add new sequences to the minichromosome in subsequent transformations.[1]

To minimize the chromosomal sequences present on an artificial chromosome platform, telomere truncation must be conducted for both chromosome arms, in some cases in consecutive operations. Alternatively, the starting material for truncation could be a supernumerary or B chromosome that is present in many species of plants including maize.[9] The maize B chromosome is basically inert and its centromere is at one end of the chromosome. Thus, truncation of the B chromosome with transformation cassettes that carry site-specific recombination modules form engineered minichromosomes in one step that can act as an artificial chromosome platform for further additions.[7] B chromosomes are available in maize and rye and have been studied extensively in these two species.[9] The maize B chromosome can also be transferred to oat, and the rye B chromosome has been added to several related species, in particular, wheat. They are maintained in populations despite the lack of vital genes by accumulation mechanisms that rely on non-disjunction of the centromere at specific stages of the life cycle.[9] For both the maize and rye B chromosomes, the very tip of the long arm is required to be in the same nucleus as the centromere for non-disjunction to occur. Thus, when telomere truncation occurs, the non-disjunction property is lost and the chromosome is stabilized. This observation shows that B chromosomes are likely to be excellent targets for the production of artificial chromosome platforms in those species in which they are present.

In the absence of B chromosomes, telomere truncation can be applied to lines that carry an extra copy of a particular chromosome. Following truncation the broken chromosome with a portion missing could be transmitted to the next generation by being covered by one of the remaining intact homologous chromosomes. Once the truncated chromosome is present as an extra chromosome, the second truncation to remove the other chromosome arm should be as easily accomplished as truncations of B chromosomes.

Artificial chromosome constructs in plants require some type of male gametophytic selection to have the transmission frequency at a high level. This is the case for two reasons. First, small chromosomes in plants tend to have reduced transmission through meiosis.[10] Second, all known cases of small chromosomes in maize show a failure of sister chromatid cohesion at meiosis I.[10] With normal-sized chromosomes, sister chromatids remain adhered to each other in meiosis I with the separation of the homologous chromosomes. The sisters then separate during meiosis II; however, this process is defective for small chromosomes such that sisters separate and proceed to opposite poles in meiosis I.

This issue with transmission should be overcome by including a gene on the artificial chromosome that will allow transmission under circumstances that would otherwise produce sterile pollen. One means of producing sterile pollen is to use lines with defective mitochrondria that produce cytoplasmic male sterility (CMS). This condition is transmitted genetically through the maternal parent because it is determined by a mitochondrial mutation but produces pollen abortion. For so-called gametophytic CMS, nuclear genes occur that will reverse sterility and restore viability in individual pollen grains. If such a restorer of CMS were introduced onto the artificial chromosome and placed in a background of CMS, then only the pollen grains carrying the minichromosome would be viable. With a single copy of such an engineered minichromosome being present, then near complete transmission should occur for the minichromosome from one generation to the next with continual crossing to a maternal line carrying the CMS characteristic. A second means to increase transmission would be to introduce a normal copy of a nuclear male sterility gene onto a minichromosome present in a line that is otherwise homozygous for the mutant allele; then, when this line is used as a male parent only the pollen grains carrying the minichromosome would function.

CONCLUSION

The use of telomere truncation to produce engineered minichromosomes sets the stage for future development of genetically modified plants. This technique bypasses the issue of the epigenetic component of centromere specification that complicates other approaches to the production of artificial chromosomes in plants. With the production of engineered minichromosomes that have the ability to add further additions to the artificial chromosome platform, it should be possible to add whole biochemical pathways to plants to confer new properties upon them or to use plants as factories for the large-scale production of useful metabolites. Combinations of transgenes can now be added to an independent chromosome that allows them to be inherited together. Useful properties include herbicide resistance, insect resistance, drought tolerance, efficient nitrogen utilization, nutritional improvements, and favorable quantitative traits.

ACKNOWLEDGMENT

Research work in the author's laboratory is sponsored by National Science Foundation grants DBI 041671, DBI 0423898, and DBI 071297.

REFERENCES

1. Yu, W.; Han, F.; Birchler, J.A. Engineered minichromosomes in plants. Curr. Opin. Biotechnol. **2007**, *18*, 425–431.
2. Ananiev, E.; Phillips, R.L.; Rines, H. Chromosome-specific molecular organization of maize (*Zea mays* L.) centromeric regions. Proc. Natl. Acad. Sci. U.S.A. **1998**, *95*, 13073–13078.
3. Zhong, C.X.; Marshall, J.B.; Topp, C.; Mroczek, R.; Kato, A.; Nagaki, K.; Birchler, J.A.; Jiang, J.; Dawe, R.K. Centromeric retroelements and satellites interact with maize kinetochore protein CENH3. Plant Cell **2002**, *14*, 2825–2836.
4. Ma, J.; Wing, R.A.; Bennetzen, J.L.; Jackson, S.A. Plant centromere organization: a dynamic structure with conserved functions. Trends Genet. **2007**, *23*, 134–139.
5. Han, F.; Lamb, J.C.; Birchler, J.A. High frequency of centromere inactivation resulting in stable dicentric chromosomes of maize. Proc. Natl. Acad. Sci. U.S.A. **2006**, *103*, 3238–3243.
6. Phan, B.H.; Jin, W.; Topp, C.N.; Zhong, C.X.; Jiang, J.; Dawe, R.K.; Parrott, W.A. Transformation of rice with long DNA-segments consisting of random genomic DNA or centromere-specific DNA. Transgenic Res. **2007**, *16*, 341–351.
7. Yu, W.; Han, F.; Gao, Z.; Vega, J.M.; Birchler, J.A. Construction and behavior of engineered minichromosomes in maize. Proc. Natl. Acad. Sci. U.S.A. **2007**, *104*, 8924–8929.
8. Yu, W.; Lamb, J.C.; Han, F.; Birchler, J.A. Telomere-mediated chromosomal truncation in maize. Proc. Natl. Acad. Sci. U.S.A. **2006**, *103*, 17331–17336.
9. Jones, N.; Houben, A. B chromosomes in plants: escapees from the A chromosome genome? Trends Plant Sci. **2003**, *8*, 417–423.
10. Han, F.; Gao, Z.; Yu, W.; Birchler, J.A. Minichromosome analysis of chromosome pairing, disjunction, and sister chromatid cohesion. Plant Cell **2007**, *19*, 3853–3863.

Artificial Insemination

William L. Flowers
Department of Animal Science, North Carolina State University, Raleigh, North Carolina, U.S.A.

Abstract

Artificial insemination (AI) is a biotechnology that allows for rapid genetic improvement, mating of animals housed at many different locations, improved biosecurity, and conservation of germ plasma from unique genetic lines. Semen collection, extension and storage of semen, and insemination are basic processes critical for the successful AI. Digital pressure is the preferred method of semen collection in swine and poultry, whereas an artificial vagina is used for horses and ruminants. Cryopreservation is the primary method for semen storage prior to insemination in cattle, while maintenance of fresh semen at temperatures between 5 and 20°C is common in swine, horses, sheep, goats, and poultry. Semen is deposited in the female reproductive tract with the use of specialized catheters and techniques in each of these species. If these aspects of AI are performed correctly, then fertility is equivalent, if not superior, to those achieved with natural service.

INTRODUCTION

Genetic improvement is the main reason for using artificial insemination (AI) in breeding programs. Ejaculates from most male livestock contain more spermatozoa than are needed to achieve adequate fertility. As a result, if semen is collected and diluted, then a greater number of females can be bred using AI than with natural service. In addition, because extended semen can be stored for varying lengths of time, its quality and health status can be evaluated prior to its use. Finally, conservation of germ plasma from unique genetic lines of animals for future propagation is also possible with AI. Successful AI in animals involves three basic processes: 1) semen collection; 2) semen extension and storage; and 3) insemination.

SEMEN COLLECTION

Digital pressure or massage and use of an artificial vagina are the two most common methods of semen collection in livestock species.[1] The decision as to which one is best for a given animal is based on several factors, including male reproductive physiology and behavior, quality of semen obtained, and the technical expertise required for a successful collection. The best collection procedure for a given animal is the one that most closely simulates the social and physiological stimuli that occur during natural mating (Table 1).

Digital Pressure

Pressure applied to the end of the penis is the most common collection technique for swine and poultry. Collection from boars is accomplished by sexually stimulating the boar to mount an object and achieve erection of the penis. The primary stimulus for mounting behavior in boars is an object that resembles an immobile sow and the primary stimulus for ejaculation is pressure on the end of the penis. Therefore, most boars can be trained to mount a collection dummy or dummy sow. After mounting, boars will begin to thrust forward until the penis protrudes from the sheath. Manual pressure is applied by grasping the tip of the penis with a gloved hand. The pressure on the tip of the penis mimics cervical contractions during natural mating and results in full extension of the penis and ejaculation. In male birds, the cloaca is the exit for the reproductive system and contains small nipple-like projections called papillae. Collection of semen is achieved by massaging the papillae. The area surrounding the vent is massaged while simultaneously stroking the back. These actions stimulate a spinal reflex arc that causes the male to become aroused sexually and causes the papillae to become erect. Once erect, the papillae are gently massaged until ejaculation occurs. Semen from both boars and male birds is collected directly into insulated containers to protect the sperm cells from temperature shock.

Table 1 Preferred collection and storage conditions for semen from cattle, chickens, horses, sheep and goats, swine, and turkeys.

Animal	Collection technique	Type of semen	Storage temperature	Maximum storage length
Cattle	Artificial vagina	Frozen	−196°C	>1 year
Chickens	Manual massage	Fresh	5°C	24–48 hr
Horses	Artificial vagina	Fresh	5–20°C	24–36 hr
Swine	Digital pressure	Fresh	15–18°C	2–5 days
Sheep/Goats	Artificial vagina	Fresh	5°C	4–6 days
Turkeys	Manual massage	Fresh	5°C with aeration	12–24 hr

Artificial Vagina

An artificial vagina is a device that imitates the female vagina and provides the appropriate thermal and mechanical stimulation, which is required for ejaculation in horses, sheep, goats, and cattle. Most artificial vaginas have the same basic design, which includes an outer casing, an inner lining, and an insulated pouch that is attached to one end of the outer casing. Warmed water is added to the space between the outer casing and inner lining prior to collection and is used to regulate the temperature and pressure to which the penis is exposed during collection. During collection of bulls, rams, or bucks, the male is allowed to mount a dummy, a female, or a neutered male. Stallions can be trained to mount a mare or a collection dummy called a phantom. When the male mounts, his sheath is gently deflected to the side and, as he thrusts forward to breed, the artificial vagina is placed over the end of his penis. The male ejaculates into the artificial vagina. Ejaculated semen flows through the artificial vagina and is collected in a small plastic tube connected to the end of the vagina.

SEMEN EXTENSION AND PRESERVATION

After collection, semen is diluted with semen extender that allows it to be stored prior to insemination (Table 1). For horses, swine, poultry, sheep, and goats, semen is stored at temperatures above 0°C and is referred to as fresh semen. For cattle, semen is stored at temperatures below 0°C and is referred to as frozen semen. The decision to use fresh or frozen semen is primarily based on fertility. In most species, fertility is considerably lower with frozen than fresh semen. Therefore, fresh semen is the preferred choice. In cattle, frozen semen is equivalent to that of fresh semen in terms of fertility. As a result, frozen semen is used because it allows for much longer periods of storage prior to insemination compared with fresh semen. However, if the primary goal of using AI is to produce a few live offspring from a genetically unique male, then fertility with frozen semen is sufficient to accomplish in all domesticated farm animal species.

Extenders used in the preservation of semen perform several basic functions, including providing nutrients for sperm metabolism; neutralizing metabolic wastes; stabilizing sperm membranes; preventing drastic changes in osmolarity of semen; and retarding bacterial growth. The chemical compositions of semen extenders are similar across species. Common ingredients in extenders include glucose, sodium bicarbonate, phosphate buffers, organic zwitterionic compounds, sodium citrate, potassium chloride, albumin and other proteins, and antibiotics.[2] Extenders used for preparing frozen semen include these ingredients but with cryoprotectants, which protect spermatozoa during the freezing and thawing process.

INSEMINATION

Decisions associated with the insemination process are considered to be the most important components of AI. These include the insemination dose, the insemination regimen (frequency and timing), and the insemination technique.

Insemination Dose and Insemination Regimen

The insemination dose consists of the number of spermatozoa and the total volume of extended semen. Semen-processing procedures, viability of spermatozoa, and the insemination technique are the main factors that influence the individual characteristics of insemination doses. Likewise, the insemination regimen is determined by the length of estrus, the timing of ovulation during estrus, and the viability of spermatozoa in the female reproductive tract. Characteristics of insemination doses and regimens most commonly used in cattle, sheep, goats, horses, chickens, turkeys, and swine are summarized in Table 2.

Insemination Technique

Female reproductive anatomy determines the type of insemination technique that is used.[2] From an anatomical perspective, the goal of insemination is to deposit semen near the posterior portion of the uterus.

Table 2 Characteristics of artificial insemination with semen for cattle, chickens, horses, sheep and goats, swine, and turkeys.

Animal	Insemination dose	Insemination location	Insemination strategies	Normal pregnancy rates (%)
Cattle	15–25 million sperm in 0.5–1.0 mL	Uterus	Once 12–18 hr after detected estrus	60–70
Chickens	150–200 million sperm in 0.2–0.5 mL	Caudal portion of oviduct	Once per week during breeding season	80–90
Horses	250–500 million sperm in 10–25 mL	Uterus or cervix	Every other day of estrus beginning on day 2 or 3	65–75
Swine	2–4 billion sperm in 60–80 mL	Cervix	Once each day of estrus	80–90
Sheep/Goats	100–300 million sperm in 0.1–0.5 mL	Cervix	Once each day of estrus	70–80
Turkeys	150–250 million sperm in 0.2–0.5 mL	Caudal portion of oviduct	Once per week during breeding season	80–90

Cattle

In cattle, a rectovaginal procedure is used. One hand is positioned so that it grasps the cervix through the ventral floor of the rectum. An insemination rod, which resembles a long, thin needle with a blunt end, is inserted into the vulva and guided through the vagina and cervix and into the uterus using the other hand. Gentle manipulation of the cervix through the rectum facilitates movement of the insemination rod through the cervical rings. Semen is deposited by pushing the plunger forward that physically pushed the semen out of the tip of the insemination rod. Gentle massaging of the clitoris for at least 10 seconds prior to insemination is a technique that has been shown to improve conception rates in beef cows and is a practice that is often used on repeat breeders.

Sheep and goats

In sheep and goats, a speculum and a long, narrow insemination rod are used to deposit semen in the cervix. The speculum is inserted into the vagina and positioned just caudal to the external opening of the cervix. Visualization of the cervical os is enhanced by the use of a standard medical head lamp. Once the cervix has been located, the insemination rod is inserted through the speculum and threaded into the cervix at least 3 cm. Once the insemination rod is correctly positioned a syringe or bottle containing semen is attached to the other end of the insemination rod and the semen is deposited.

Horses

The insemination catheter or pipette used in mares is a flexible plastic tube with a rounded tip. The insemination dose is loaded into a syringe with a plastic plunger and connected to the insemination rod with a short piece of pliable tubing. The tip of the catheter is placed in the palm of the hand with the index finger covering the tip. The gloved hand and insemination pipette are then passed into the cranial portion of the vagina, where the index finger is used to locate the cervix. The insemination catheter is advanced through the cervix into the uterine body and semen is slowly deposited.

Swine

In swine, the vagina becomes progressively smaller as it approaches the cervix and forms a smooth junction. The central canal of the cervix in the pig is a left-handed or counterclockwise spiral. Specialized catheters with left-handed spirals or compressible foam tips are used for insemination. Insemination doses are packaged in small bottles, tubes, or sealed plastic bags. With the tip pointed up, the catheter is passed through the vulva and into the vagina until resistance is encountered. The catheter is then rotated counterclockwise. When rotation becomes difficult, insertion is stopped and the dose of semen is attached to the rod and the semen is slowly deposited. Physical stimulation of the sow by massaging her back and rear enhances sperm deposition and transport of semen inside the sow.

Uterine or trans-cervical insemination catheters have been developed and have been used successfully in swine. The unique feature of these rods is that they contain a small tube that can be threaded through the large catheters that are commonly used for deposition of semen in the cervix. Once the conventional AI catheters have been inserted into the cervix, the smaller tube is threaded through the remainder of the cervix and into uterus.

Turkeys and chickens

For birds, the insemination rod resembles a medicine dropper or small plastic rod connected to a 1 mL syringe by flexible tubing. The female is held so that her vent is exposed and the area around the vent is massaged at the same time her back is stroked. As the hen become sexually aroused, the vent turns inside out and an opening resembling a rosette appears on the left side. The tip of the insemination rod is inserted about 1 cm into the rosette opening and semen is deposited.

CONCLUSION

Artificial insemination is used successfully in most livestock species for genetic improvement because semen from superior males can be used to breed increased numbers of females. For most species, fertility with artificial AI is equivalent or exceeds that normally achieved with natural service, if it is performed correctly.

REFERENCES

1. Hafez, E.S.E. Artificial insemination. In *Reproduction in Farm Animals*, 6th Ed.; Hafez, E.S.E., Ed.; Lea and Febiger: Philadelphia, PA, 1993; 424–439.
2. Flowers, W.L. Artificial insemination, in animals. In *Encyclopedia of Reproduction*; Knobil, E., Neill, J.D., Eds.; Academic Press: San Diego, CA, 1999; Vol. 1, 291–302.

Bacteria: Gram-Positive

Hua Wang
Department of Food Science and Technology, Ohio State University, Columbus, Ohio, U.S.A.

Abstract
The chapter introduced mechanisms for DNA transfer in Gram-positive bacteria. Common genetic tools, as well as traditional and advanced methods in genetic manipulation and analysis in representative Gram-positive bacteria were also illustrated.

INTRODUCTION

Gram-positive bacteria are of tremendous importance to biotechnology, food production and quality, and agriculture. They include food fermentation starter cultures and those involved in generation of energy and production of desirable chemical compounds through bioprocesses. Some members of the Gram-positive bacteria are dangerous pathogens such as *Bacillus anthracis*, *Clostridium botulinum*, *Listeria monocytogenes*, Group A streptococci ("flesh-eating bacteria"), and *Staphyloccocus aureus*. Genetic manipulation of Gram-positive organisms is essential for the generation of strains with improved properties and the development of effective treatment and disease prevention strategies. In general, Gram-positive bacteria are more difficult to manipulate genetically than their Gram-negative counterparts, but biological and physical techniques such as transformation, conjugation, transduction, and electroporation can be applied to both successfully, and virtually all of the newer techniques such as proteomics, PCR, and DNA hybridization are equally applicable to Gram-negative and Gram-positive bacteria. This entry will describe how these techniques are being utilized and how newer approaches such as genome sequencing and high-throughput technologies have the potential to exponentially increase our understanding of Gram-positive bacteria.

MECHANISMS FOR DNA TRANSFER INTO GRAM-POSITIVE BACTERIA

Conjugation

Conjugation involves the transfer of genetic elements such as plasmids and transposons from one cell to another through direct cell-to-cell contact. Some of these elements carry their own genetic determinants for DNA transmission (self-transmissible elements); others require assistance from proteins encoded elsewhere (mobilizable elements). The transfer of DNA via conjugation occurs frequently with a very low frequency, but some Gram-positive bacteria, such as certain strains of *Lactococcus lactis*, *Enterococcus faecalis*, *Bacillus* spp., and *Lactobacillus* spp., are capable of producing transconjugants with frequencies as high as 10^{-1} transconjugants/donor colony-forming units.[1] Traits such as bacteriocin production and immunity, bacteriophage resistance, lactose and citrate fermentation, and proteolytic activity are frequently associated with mobile elements in lactic acid bacteria. Recombinant genetic elements such as conjugative shuttle plasmids allow DNA transfer between different bacteria species such as *Escherichia coli* and *L. monocytogenes* for mutagensis and cloning purposes.

Transduction

Gene transfer mediated by bacteriophages is known as transduction. Transferable genetic material can be of recombinant origin and packaged into phage in vitro, or derived from host DNA accidentally packaged during phage infection. The resulting transducing phages can transmit their DNA to sensitive bacteria during the next round of the infection. In general, transduction has not played a large role in genetic research on Gram-positive bacteria, but it has been utilized for certain Gram-positive bacteria such as *Staphylococcus aureus*.

Natural Transformation

Some bacteria can take up "naked" DNA from the environment. This process is called transformation, and cells capable of DNA uptake are called "competent." Some Gram-positive bacteria, such as *Bacillus subtilis*, *Streptococcus pneumoniae*, and *Streptococcus mutans*, are naturally competent and transformation has been used extensively for genetic studies of these bacteria. Competency is frequently transient and dependent on the growth phase.[2]

Electroporation

The application of a short high-voltage electric pulse opens small pores in the membrane of bacteria allowing DNA to migrate into the cytoplasm. This technique is called electroporation and is now the preferred gene transfer method for many Gram-positive bacteria. Successful DNA transfer via electroporation requires cells suspended in a medium with very low conductivity. The number of cells that stably express the acquired DNA depends on the type of DNA (plasmid vs. linear DNA) and host factors such as restriction endonucleases.

Protoplast Transformation

Removal of cell wall components by enzymatic and other means results in the formation of so-called "protoplasts." Such cells are frequently able to take up DNA when maintained in an isotonic medium. Protoplast transformation is rarely done today, but generation of protoplasts for the purpose of fusing cells with different genetic properties is still being carried out to achieve "genome shuffling" through extensive recombination. For example, by fusing protoplasts from *Lactobacillus* sp. with different acid tolerance properties, it was possible to obtain more acid-tolerant variants.[3]

VEHICLES FOR THE TRANSFER OF DNA INTO GRAM-POSITIVE BACTERIA

Vectors are genetic vehicles, most often recombinant plasmids, which are able to introduce DNA into cells. A number of specialized vectors are available for genetic manipulation of Gram-positive bacteria.

Shuttle Vectors

The isolation of plasmids from Gram-positive bacteria is generally more complicated than the isolation of plasmids harbored by *E. coli*. Therefore, vectors have been constructed that can "shuttle" DNA between *E. coli* and Gram-positive bacteria. These shuttle vectors contain a plasmid origin of replication (ori) from *E. coli* and from the targeted Gram-positive bacterium. Often different selective markers are included in shuttle vectors since some markers expressed in *E. coli* are not expressed in Gram-positive bacteria and vice versa. Multiple cloning sites are usually included in the vector construction. The ori from the broad host range drug resistance plasmid pAMβ1 functions in many Gram-positive bacteria and has therefore been incorporated into many shuttle vectors. Shuttle vectors carrying DNA of interest are usually introduced into *E. coli* to obtain sufficient DNA for genetic manipulations such as site-directed mutagenesis and sequence determination. The vector carrying the modified DNA is then introduced into the Gram-positive bacterium for assessment of functionality.

Expression Vectors

Vectors that harbor an inducible promoter can be used to control the level and timing of transcription of DNA inserted downstream of the promoter. Such expression vectors facilitate the assessment of the functionality of cloned genes since it is possible to observe phenotype differences when transcription from the inducible promoter has been up- or down-regulated. An example is vector pMSP3535 harboring a promoter inducible by the bacteriocin, nisin, that functions in *L. lactis* and *E. faecalis*.[4]

Integration Vectors

When it is desirable that DNA transferred into a cell becomes part of the cell's chromosomal DNA, vectors are employed that cannot be maintained independently in the host cell. Plasmids without a Gram-positive ori or with an origin that becomes non-functional under certain conditions such as elevated temperatures are useful as integration vectors. The presence of homologous DNA in vector and chromosome can lead to the integration of the entire vector including the recombinant DNA or the exchange of the homologous DNAs. Transposons or insertion sequences located on the integration vector survive by inserting into the chromosome, usually at random sites. Genetic markers such as antibiotic resistance genes are employed to detect cells that have undergone an integration event. Genetic studies of *Listeria* spp. and other Gram-positive bacteria have greatly benefited from integration vector-mediated genetic exchange. Complementation of mutations, introduction of mutations within a specific gene, and replacement of mutated sequences with wild-type sequences are routinely carried out with the aid of integration vectors.

GENETIC MANIPULATION AND ANALYSIS OF GRAM-POSITIVE BACTERIA

The genomic content of Gram-positive bacteria can be analyzed and manipulated with the same methods used for Gram-negative bacteria. These methods include PCR techniques, mutagenesis procedures, genome, and microarray analyses.

Mutagenesis

Integration vectors and transoposons such as Tn*917* and Tn*916* are commonly used in Gram-positive bacteria to generate mutants.[5] "Artificial" transposon constructs based on the insertion sequences of transposon Tn*5*, and the transposition complex of phage Mu originating from Gram-negative bacteria are also successful in generating mutations in Gram-positive bacteria. Transposon derivatives such as Tn*phoZ* (containing a truncated alkaline phosphatase gene) are available for finding sites encoding proteins exported out of the cytoplasm of Gram-positive bacteria. Even general gene knockout systems such as those utilizing group II introns[6] have been applied to generate insertion mutations in *L. lactis, Clostridium perfringens,* and *S. aureus.*

Genome Sequencing

The first genome sequence of a Gram-positive bacterium, *B. subtilis*, was completed in 1997. Since then the complete genomes of many Gram-positive bacteria of biotechnological and medical interest have been published. These include genomes of pathogens and bacteria of biotechnological interest such as lactic acid bacteria and the solvent-producing *Clostridium acetobutylicum*. Open reading frames in the genome of one bacterium can frequently be assigned a putative function based on sequence similarity with genes of known function in other bacteria. The search for sequences encoding transcriptional regulators in different bacteria has been facilitated by genome comparisons. Genome data have also highlighted the ubiquitous nature of horizontal gene transfer between Gram-positive bacteria. It can be expected that genome sequence data will continue to increase our knowledge of the genetics of Gram-positive bacteria and will facilitate construction of useful strains of interest to biotechnology.

Microarray

On a DNA microarray, also called *gene chip, DNA chip,* or simply *chip,* a collection of as many as 30000 small single-stranded DNA fragments or oligonucleotides is immobilized on the surface of a glass, plastic, or silicon chip such that the location of specific sequences is known. These sequences serve as probes to which fluorescently labeled single-stranded nucleic acid sequences can bind via hybridization. The fluorescent signal from a particular spot on the chip is proportional to the amount of bound nucleic acid. Microarrays can thus provide information on the levels of specific RNA or DNA sequences in a sample.[7] Many bacterial genome microarrays, including those of Gram-positive bacteria such as *Listeria* spp., *Streptococcus* spp., *Staphylococcus* spp., *Bacillus* spp., *Lactococcus* spp., and *Lactobacillus* spp. have been manufactured. Applications of microarrays include comparison of gene expression levels under various environmental or physiological conditions or among various strains, and detection of target DNA or RNA sequences from specific bacteria. Examples of the application of microarrays for Gram-positive bacteria are the genetic analysis of the response of *S. mutans* to environmental stress, the detection of *B. anthracis*, and the identification of antibiotic resistance markers present in different bacteria.

Quantitative PCR and Real-Time PCR

In contrast to conventional PCR, quantitative and real-time PCR permit the determination of the initial copy number of amplifiable nucleic acid molecules in a sample. Technically these types of PCR are performed by measuring fluorescence originating from labeled probes included in the PCR or from dyes that reversibly bind to double-stranded DNA. Using PCR primers specific to a strain, species, or groups of bacteria, it is possible to calculate the number of these bacteria in a sample.[8] For example, the number of lactic acid bacteria or their bacteriophages present at various stages during fermentation can be monitored. Coupled with reverse-transcription, the numbers of specific mRNAs in a cell can also be estimated.

Fluorescent In Situ Hybridization (FISH)

Non-destructive techniques have been developed to study the microorganisms in a community in situ. Microscopic and digital image analysis techniques, particularly confocal laser-scanning coupled with fluorescent DNA probes, have been widely used in such studies.[9] The probes employed for studies of bacteria are primarily directed toward ribosomal RNAs because this type of RNA is relatively stabile and present in high copy numbers. For the hybridization event to occur inside the cell, the probes have to diffuse through membrane(s) and cell walls. For this reason, bacteria are fixed and treated with alcohol. For most experiments involving Gram-negative bacteria, probe penetration is rarely a problem. For some Gram-positives, it is necessary to treat fixed cells with lysozyme, protease, or detergents to ensure that probes reach their rRNA target. Detection of lactobacilli, for example, has been found to require such treatments to allow sufficient probe penetration. FISH has been applied to the detection and enumeration of Gram-positive members of bacterial communities in foods, intestinal, and environmental habitats.

Proteomics

Genome sequences and mRNA level data provide information on the genetic capacity of microorganisms, but this type of information can rarely be correlated with the content of functional proteins in cells. Qualitative and quantitative data on proteins are obtained primarily by 2-D protein gel electrophoresis. Coupled with mass spectrometry, proteins excised from 2-D gels can be identified by deduction of amino acid sequences of peptide fragments on the basis of their spectrometric patterns and comparison of the sequences with protein sequences in databases.[10] Proteomic approaches have been employed to compare protein expression patterns of several Gram-positive bacteria. An example is the comparison of proteins found in *L. monocytogenes* cells grown at room and refrigeration temperatures.

In summary, years of improvements in "conventional" genetic techniques for Gram-positive bacteria and the development of advanced molecular biology tools have removed many of the challenges experienced in the past genetic analyses of Gram-positive bacteria, and genetics studies and modification of the Gram-positive bacteria has become a fun and rewarding experience.

REFERENCES

1. Luo, H.; Wan, K.; Wang, H.H. A high frequency conjugation system facilitated biofilm formation and pAMβ1 transmission in *Lactococcus lactis*. Appl. Environ. Microbiol. **2005**, *71*, 2970–2978.
2. Solomon, J.M.; Grossman, A.D. Who's competent and when: regulation of natural genetic competence in bacteria. Trends Genet. **1996**, *12*, 150–155.
3. Patnaik, R.; Louie, S.; Gavrilovic, V.; Perry, K.; Stemmer, W.P.C.; Ryan, C.M.; del Cardayré, S. Genome shuffling of *Lactobacillus* for improved acid tolerance. Nat. Biotechnol. **2002**, *20*, 707–712.
4. Bryan, E.M.; Bae, T.; Kleerebezem, M.; Dunny, G.M. Improved vectors for nisin-controlled expression in Gram-positive bacteria. Plasmid **2000**, *44*, 183–190. Erratum in: Plasmid 2001 *45*:61.
5. Maguin, E.; Prevost, H.; Ehrlich, S.D; Gruss, A. Efficient insertional mutagenesis in lactococci and other Gram-positive bacteria. J. Bacteriol. **1996**, *178*, 93193–93195.
6. Karberg, M.; Guo, H.; Zhong, J.; Coon, R.; Perutka, J.; Lambowitz, A.M. Group II introns as controllable gene targeting vectors for genetic manipulation of bacteria. Nat. Biotechnol. **2001**, *19*, 1162–1167.
7. Lobenhofer, E.K.; Bushel, P.R.; Afshari, C.A.; Hamadeh, H.K. Progress in the application of DNA microarrays. Environ. Health Perspect. **2001**, *109* (9), 881–891.
8. Mackay, I.M. Real-time PCR in the microbiology laboratory. Clin. Microbiol. Infect. **2004**, *10*, 190–212.
9. Amann, R.; Fuchs, B.M.; Behrens, S. The identification of microorganisms by fluorescence *in situ* hybridization. Curr. Opin. Biotech. **2001**, *12*, 231–236.
10. Pandey, A.; Mann, M. Proteomics to study genes and genomes. Nature **2000**, *405* (6788), 837–846.

Bacteriophages: Pathogen Control

Clair Hicks
Department of Animal and Food Sciences, University of Kentucky, Lexington, Kentucky, U.S.A.

Abstract

Bacteriophage or phage (common term) are a type of bacterial virus that are small submicroscopic entities (around 0.1 micrometer in length) and can only be viewed using such procedures as scanning electron microscopy or transmission electron microscopy. Phage are approximately 100 times smaller than the bacteria they infect. Phage must use a bacteria host in their replication process. During the replication process, phage generates many copies of itself within the host cell and then lyses the cell to free the replicated copies into the environment where the replicated copies can then attach to new host cells. This scenario may be useful in reducing reservoirs of pathogens within an environment. The addition of pathogen-specific phage to foods has the potential to reduce the number of pathogens within a food or on the surface of a food.

INTRODUCTION

Many raw foods come into processing plants contaminated with low numbers of pathogenic organisms on their surfaces. Also, animals can harbor pathogenic organisms in their gut. When foods are processed, care must be taken to reduce the spread of pathogenic organisms to food or equipment surfaces. Processing procedures must include critical control points where these endemic pathogens can be destroyed. Many manufacturers use procedures that heat the product to temperatures that destroy pathogens in the food. Examples of this type of processing include pasteurization and thermal sterilization. Other methods that can destroy pathogens in foods include pickling, irradiation, and high-pressure processing; however, if the food is to be sold in its raw form, such as fresh fruits, vegetables, or meats, then only high-pressure processing and irradiation are effective as methods for destroying pathogens. Because of cost and image, food manufacturers are interested in developing alterative processing methods to reduce pathogenic loads of raw foods. A new processing method has been proposed for foods that are susceptible to contamination with a pathogen; this method relies on application of solutions containing suspensions of bacteriophage specific to known food pathogens. This bacteriophage is a type of virus that infects and kills the pathogen by a lytic process in which the cell of the pathogen is ruptured causing the cell to die. This entry will describe the specificity of phage, how they replicate and destroy their host, how the host can develop resistance by producing mutant offspring that are resistant to future phage attacks, and then discuss how this process can be optimized to reduce pathogens in raw food commodities within a processing plant.

BACTERIOPHAGE MECHANISMS

Phage are quite specific in that every type of phage has a specific host. Generally, all types of phage can only replicate within a fairly small group of bacterial host cell types (within a few strains) of a specific bacterium (single genera). Thus, a single type of phage may only be able to destroy one pathogenic strain of a single type of bacterium.[1] Although this sounds simple, the process is fairly complex and most host populations adapt rapidly to generate mutant cells that resist future phage infections. Phage attach to a host cell, first through a reversible attachment phase that allows the phage to locate a receptor-rich area on the host cell, and then an irreversible phase where the phage locks onto the membrane of the host cell.

Phage specificity is determined by: 1) the receptors on the host cell; 2) triggering mechanism of the phage; and 3) the compatibility of replication machinery between phage and host. The host cell membrane must have sites allowing the phage to first attach reversibly for a long enough period of time to locate receptors that support irreversible attachment. This process can best be visualized as both the host bacteria and phage being in a constant state of vibration (Brownian movement). Thus, the phage must have sufficient attraction (electrical potential) to the host cell membrane to keep it near the surface of the membrane (reversible attachment) as the phage moves (due to vibration) across the surface toward a receptor-rich area on the membrane. Once a receptor-rich area is located (usually a thin spot in the membrane) the phage can attach to the membrane irreversibly.[2] Other phage, such as T_1 coliphage, can attach anywhere on the membrane (Fig. 1). Since many bacteria have similar receptors

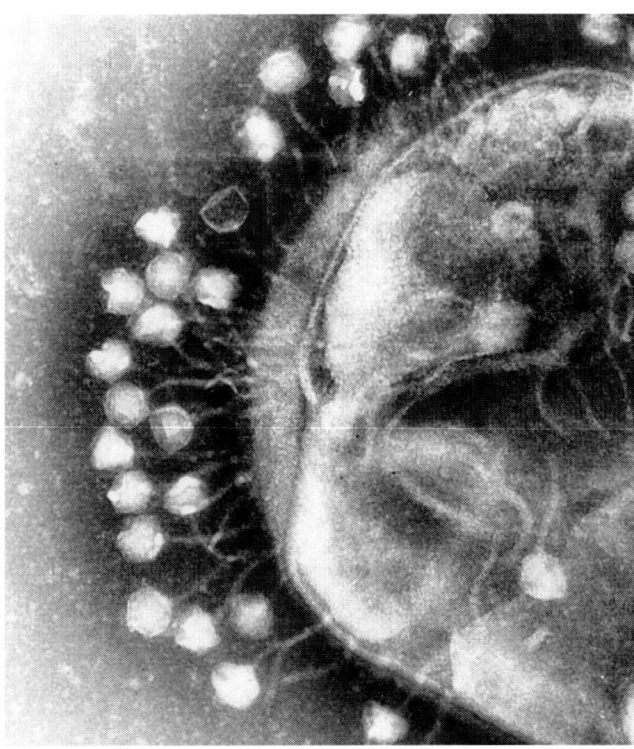

Fig. 1 Bacteriophage irreversibly attached to a bacterial host membrane. Scanning electron micrograph of a phage resembling coliphage T_1 attached to an *Escherichia coli* host. Note that many phages can attach to a single host and that a number of new phages have been replicated within the host cell. **Source:** From http://www.Wikipedia.org/wkki/bacteriophage.

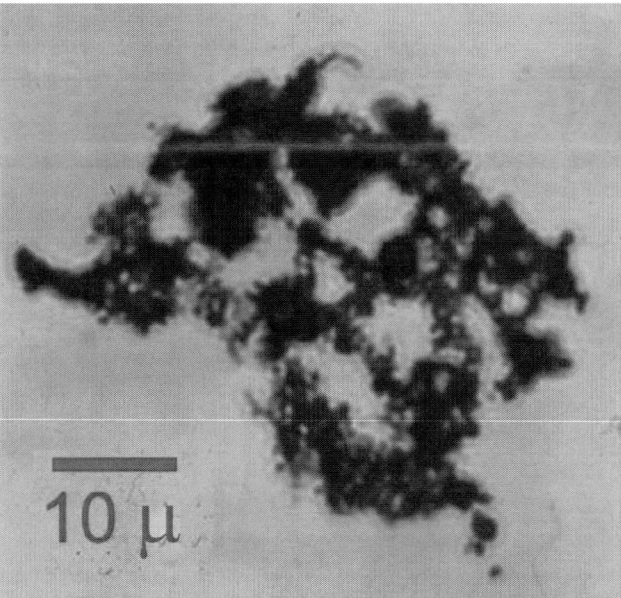

Fig. 2 Cellular debris left after lysis occurs. No viable cells remain in the debris field, although some cells retain their cellular shape. Debris fields aggregate because of the sticky nature of phage coat proteins. **Source:** Photograph by C.L. Hicks.

(e.g., various sugar moieties or functional surface groups), many types of phage can attach to common receptors.[3] Attachment can be thwarted if mutated daughter cells of the host alter their cell envelope. One common host mechanism is synthesis of a polysaccharide coat as part of its surrounding envelope.[1] For example, *Escherichia coli* is a bacterium that can rapidly mutate to inhibit coliphage infections.

The triggering mechanism is a defined chemical sequence usually involving a protein on the host cell's membrane that initiates a response in the phage indicating that it has found the correct host. The trigger mechanism causes the phage to release ATP (an energy source) that induces the phage to inject its DNA or RNA into the host cell.[4]

The DNA or RNA from the phage must be able to direct the replicating processes, which is first accomplished by allowing escape from the host's restriction/modification system.[1] If the DNA in the host cell already contains the genetic code of the virus (temperate or lysogenic state) then the host is resistant to any infections by that type of phage. Highly virulent phage (lytic phage) may produce 100–200 progeny where other phage types may produce fewer than 20 progeny before lysis occurs. Thus, specificity is only partially controlled by receptors, highly controlled by trigger mechanisms, and limited by replication processes.[1]

Once the components for a new phage are replicated, the parts are assembled in the host cell. After assembly, holin and endolysins (lytic enzymes) are produced, which degrade the cell membrane and cell wall, respectively, which allows the membrane to rupture (Fig. 2). When rupture occurs, the newly formed phage are liberated into the surrounding environment.[1] These newly liberated phage seek out a new host of the same cell type. In infections by lytic phage where high host cell concentrations exist, the phage can far outpace the growth of the host cells and can easily produce a quantity of phage that exceeds the number of host cells and destroys all host cells in the environment;[1] however, the host must be in a viable growth state for this to occur.

To employ this concept in a food material, the pathogens that most frequently contaminate the food must be identified and partially characterized. These pathogens are used to find or select types of phage that are highly successful in killing the selected pathogen. Phage must be collected (usually from places where the pathogen is found) from various sources and characterized to determine their suitability. A suitable phage has high specificity for the pathogen, produces high numbers of progeny (easy for manufacture), does not enter into the lysogenic state or select for mutants that are resistant to the phage, and is non-transducing.[1] The selected phage is then mass produced on an industrial

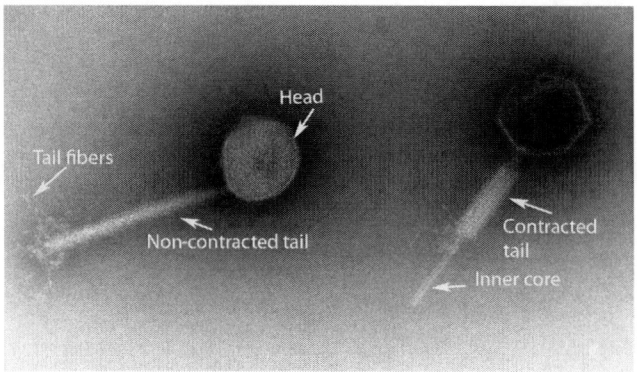

Fig. 3 Scanning electron micrograph of *Listeria* P100 bacteriophage. Left: P100 phage has a non-contracted tail with observable tail fibers. Tail fibers bind to host irreversibly. Right: P100 phage shows contracted tail with inner core extended. Inner core penetrates the host membrane similar to a hypodermic needle when DNA is injected. **Source:** Photograph by M.J. Loessner, Inst. Food Sci. and Nutr., Schmelzbergsstrasse 7, LFV B20, CH-8092 Zurich, Switzerland.

scale to prepare a phage/endolysin solution that can be used to spray on food and contact surfaces or be formulated with a food.

Pioneering research to use phage to eliminate or reduce a pathogen in foods, food contact surfaces or carcass surfaces was first patented for *Listeria monocytogenes* by Loessner in 2004 and 2007.[5,6] *L. monocytogenes* can produce a 25% or greater mortality rate when human infections occur. Mortality is particularly high in at-risk groups. Thus, the concept of using phage to reduce or limit the numbers of *Listeria* in some retail foods may become a new processing tool.[1] These patents propose that suspensions of *Listeria* phage and their endolysin could be applied to contact or food surfaces or added to formulated foods. When phage and endolysins are added at high numbers, lysis of contaminating *L. monocytogenes* cells would occur and enhance the safety of the product. Although these patents cover the use of many phage that can infect *L. monocytogenes*, it is the P100 phage that appears to most closely fit the selection criteria discussed previously (Fig. 3). The P100 phage is strictly lytic, produces high progeny numbers, effectively lyses all *Listeria* within its broad-host range, and does not select for mutant strains.[6]

Many foods are contaminated from food contact surfaces. *L. monocytogenes* is a pathogen that grows very well on some contact surfaces that are not properly washed and sanitized. Since some food types, such as cheese and meats products, are extremely good growth media for *Listeria*, Loessner[5,6] proposed that contact surfaces be sprayed with phage-containing solutions to reduce hot spots or reservoirs where *Listeria* contamination may exist. By spraying equipment with phage solutions, the phage can diffuse into surface cracks or crevices that harbor the pathogen, thus reducing the number of pathogens that may contaminate foods.

Phage could also be added to *Listeria*-prone foods, such as ready-to-eat meats and certain cheeses. Most *Listeria* outbreaks have been in ready-to-eat meats[7] and raw milk cheeses.[8,9] *L. monocytogenes* is a common bacterium, so its presence in raw milk and meats is expected. While most processing conditions (cooking, pasteurization, or sterilization) kill *L. monocytogenes*, this pathogen can also survive and grow in drains, on equipment and floor surfaces to provide a residual reservoir that can contaminate food surfaces or food-contact surfaces.[10] To incorporate phage into these types of foods could reduce or eliminate *Listeria* contaminations from direct and indirect reservoirs. The greatest problem of implementing this practice is that in the presence of its phage, *L. monocytogenes* may mutate to become resistant to the phage. Thus, a number of different phage types may require rotational use to reduce the emergence of host resistance. In some cases, cocktails of phage might be used to overwhelm the resistance of a pathogen or to enhance the number of pathogenic strains targeted, but rotation might be necessary to reduce selective pressure; however, some bacteria such as *E. coli* appear to produce resistance mutants so quickly that this procedure would be impractical.[1,10]

CONCLUSION

The use of phage to destroy a bacterial pathogen is quite possible; however, phage are quite host-specific, so phage selection becomes critical to the control of a pathogen in various food products. The use of several distinct phages may allow for the control of several select pathogens that commonly contaminate a particular food type. The use of phage to reduce or destroy all selected pathogens must coincide with a good sanitation program, because the addition of phage to a food will not inhibit spoilage organisms. Thus, the use of phage-processing protocols should be part of a total processing strategy that enhances the shelf life of a product and reduces the risk of pathogens being present in the finished food product. Phage that remain in the lytic stage such as P100 could become a successful processing tool to control *Listeria* in foods.

REFERENCES

1. Doyle, M.P.; Beuchat, L.R. *Food Microbiology*, 3rd Ed.; ASM Press: Washington, DC, 2007; 757–759, 775–778.
2. Hicks, C.L.; Clark-Safko, P.A.; Surjawan, I.; O'Leary, J. Use of bacteriophage-derived peptides to delay phage infections. Food Res. Int. **2004**, *37* (1), 115–122.
3. Klumpp, J.; Dorscht, J.; Lurz, R.; Bielmann, R.; Wieland, M.; Zimmer, M.; Calendar, R.; Loessner, M.J. The terminally redundant, nonpermuted Genome of *Listeria*

Bacteriophage A511: a model for the SPO1-like Myoviruses of Gram-positive bacteria. J. Bacteriol. **2008**, *190* (17), 5753–5765.
4. Kostyuchenko, V.A.; Navruzbekov, G.A.; Kurochkina, L.P.; Strelkov, S.V.; Mesyanzhinov, V.V.; Rossmann, M.G. The structure of bacteriophage T4 gene product 9: the trigger for tail contraction. Structure **1999**, *7* (10), 1213–1222.
5. Loessner, M.; Carlton, R.M. Virulent phages to control *Listeria monocytogenes* in foodstuff and in food processing plants. WO Patent: 004495, January 15, 2004.
6. Loessner, M.J. P100 Bacteriophage for control of *Listeria monocytogenes*. WO Patent: 093849, August 23, 2007.
7. http://www.fsis.usda.gov/FSIS_Recalls/Open_Federal_Cases/ index.asp (accessed December 3, 2009).
8. http://www.torontosun.com/news/canada/2008/09/07/6689011-sun.html (accessed December 3, 2009).
9. http://jds.fass.org/cgi/content-nw/full/87/13_suppl/E6/T3 (accessed December 3, 2009).
10. Hagens, S.; Offerhaus, M.L. Bacteriophage—new weapons for food safety. Food Tech. **2008**, *62* (4), 46–54.

Bananas: Somatic Cell Genetics

D.K. Becker
School of Life Sciences, Queensland University of Technology, Brisbane, Queensland, Australia

M.K. Smith
Queensland Department of Primary Industries, Brisbane, Queensland, Australia

Abstract

Bananas and plantains (*Musa* spp.) are grown throughout the humid tropics and subtropics, where they are of great importance both as subsistence crops and as sources of domestic and international trade. They are monocotyledonous perennial herbs consisting of sympodial rhizomes and large pseudostems composed of tightly clasping leaf sheaths, slightly swollen at the base. Suckers are freely produced. Bracts and flowers are inserted independently on a peduncle and are usually deciduous by abscission, except for functional female ovaries in the basal hands. Basal flowers are generally female with male flowers on distal hands.

INTRODUCTION

Edible, seedless banana, plantain cultivars are derived from intra- and interspecific hybridisation of the two seeded, wild diploid species, *Musa acuminata* (A genome) and *Musa balbisiana* (B genome). The haploid genome of both *M. acuminata* and *M. balbisiana* consist of 11 chromosomes, but the *acuminata* genome has been estimated as being slightly larger, 610 Mbp cf. 560 Mbp for *balbisiana*. Many edible hybrids are parthenocarpic, female sterile and triploid with the relative contribution of each species to the genome being annotated by either A or B. Dessert bananas are usually AAA, plantains AAB and cooking bananas ABB.

Banana breeding programs have largely focused on generating pest and disease resistant cultivars while at the same time retaining acceptable yield and fruit quality. The two major diseases of banana are Sigatoka leaf spots (*Mycosphaerella* spp.) and Fusarium wilt (*Fusarium oxysporum* f. sp. *cubense*). Due to infertility, triploidy and long generation time, very few conventionally bred cultivars have reached commercial release. As a result, a great amount of effort has been directed towards improving banana through the manipulation of somatic cells.

SOMACLONAL VARIATION AND MUTATION INDUCTION

Banana has benefited greatly from in vitro-induced mutation. This is largely due to the resources directed towards this approach in view of difficulties associated with breeding triploid cultivars. Genetic variation can arise as a result of the tissue culture process or be intentionally induced by the use of physical and chemical mutagens. All three approaches have been applied to banana.[1,2]

Banana plants regenerated from in vitro culture sometimes exhibit morphological and biochemical variation due to genetic changes, which is of serious concern for clonal micropropagation. These genetic changes, nonetheless, may be exploited as a source of genetic variation for banana improvement. Somaclonal variation in banana usually leads to undesirable characteristics. However, there are several examples of somaclonal variants with advantageous characteristics, some of which have been released commercially. Variants identified include Tai-Chiao No 1 (AAA), TC1-299 (AAA) and Mutiara (AAB) with *Fusarium* resistance, a variant of SH 3436 (AAAA) with greater resistance to Sigatoka diseases, JD Special (AAA) with larger fruit, and higher yielding variants of Agbagba (AAB).

Of the physical mutagens used in banana and plantain improvement, gamma rays have been most commonly used. Differences in radiosensitivity have been observed among different genotypes and ploidy levels. In one study,[3] the LD_{50} was 20–25 Gy for diploid (AA), 30–35 Gy for triploid (AAA), and 35–40 Gy for the tetraploid (AAAA). Plantain and cooking clones (AAB, ABB) had a radiosensitivity of 25–35 Gy. Among a population of plants that had regenerated from a 60 Gy irradiated explant of "Grande Naine" (AAA), an early-flowering plant was identified. The clone was extensively field tested, released as "Novaria," and entered into commercial production in Malaysia in 1993.

Ethyl methanesulphonate (EMS) has been widely used as a chemical mutagen in banana and has resulted, at least at an experimental level, in the generation of clones with increased *Fusarium* tolerance.[4] The concentration required is dependent on exposure time and whether a carrier is used. EMS has been shown to be effective at

8–16 mM with an exposure of 4–5 days and also at 200 mM with a much shorter exposure of 30 min. Dimethylsulphoxide (DMSO) has been used as a carrier agent as it facilitates greater uptake of EMS. Other mutagens such as sodium azide and diethylsulphate have also been shown to be effective in inducing variation from banana shoot apices.

Chemical methods have also been used to manipulate ploidy in *Musa*, with a view to incorporating the material into conventional breeding programs. By breeding and selecting elite diploid clones, and using them to generate autotetraploids and to cross with diploids, triploid bananas could be resynthesised. Tetraploidy has been readily induced with colchicine in *M. acuminata* and *M. balbisiana* seedlings and also in vitro shoot tips. The in vitro technique has the advantage of rapid multiplication and ease of distribution of new clones for evaluation and breeding.[5] Of importance is the number of in vitro multiplication cycles, usually more than three, following colchicine treatment in order to reduce chimerism. Treatment of shoot cultures with oryzalin has also been effective in producing high frequencies of tetraploids in banana.

SOMATIC EMBRYOGENESIS AND ORGANOGENESIS

Most reports of banana regeneration have been via embryogenic cell suspensions (ECSs). Potential uses of such cultures include micropropagation, as a source of regenerable protoplasts and for genetic transformation. There is also the potential for use as a source of cells for mutagenesis, although, to date, there have been no reports of this application. Banana somatic embryos are thought to be unicellular in origin and therefore offer the potential of generating nonchimeric plants from genetically altered cells.

ECSs have been generated using a range of explants including rhizome tissue, leaf bases, immature zygotic embryos, meristems, and immature male and female flowers. Currently, ECSs are most commonly derived from immature flowers or meristems.[6–7] Both these techniques appear to be applicable to a wide range of genotypes including diploids (AA, AB), triploids (AAA, AAB, ABB) and tetraploids (AAAB). ECSs contain a heterogeneous mix of cell types and their regenerative capacity depends on the proportion of cell types. Single isolated cells and large cell clumps (200 μm to 2 mm) in general do not give rise to somatic embryos, whereas small cell aggregates (50 to 100 μm), consisting of small cells with dense cytoplasms readily form somatic embryos.[8]

The ECSs from immature male flowers entail dissecting out clusters (hands) of immature flowers that are close to the apical meristem of the inflorescence and placing them on induction medium where they remain without subculture until embryogenic complexes appear. Embryogenic complexes are then transferred to liquid medium where a suspension of cells forms. Regeneration involves plating suspension cells on embryo development/maturation medium. Very large numbers of somatic embryos can be generated from these cells with as many as 3.7×10^5 embryos formed per 1 mL packed cell volume (PCV) being reported.[6] When embryos are transferred to germination medium, the germination rate varies depending on their maturity. Leaving embryos on embryo formation/maturation medium for long periods (up to four months) results in a germination rate of over 80%.

To generate ECSs from meristems, highly proliferating clumps of meristems are induced by placing in vitro shoot cultures on multiplication medium that, depending on the cultivar, contains varying concentrations of BAP. Over a series of subcultures on this medium, primordia gradually stop forming leaves and meristems enlarge, resulting in large white meristematic nodules. Following the formation of meristematic clumps, the upper 4–6 mm of these structures (termed "scalps") are used as explants for induction of embryogenesis followed by initiation of ECSs in a manner similar to that of flower-derived cultures.[7] Recent work suggests that "scalps" are not required, but rather ECSs can be generated directly from longitudinal sections of conventional shoot tip cultures.[9]

In the only reported field trial of plants derived from ECSs, measurements were taken for the growth characteristics of 500 "Grand Nain" (AAA) plants regenerated from flower-derived ECS.[10] Overall, there were no significant differences in growth characteristics. The frequency of off-types for ECS-derived plants was nil, while the frequency for those derived from micropropagation was between 1.8% and 3.3% depending on the accession. This study suggests that plants regenerated from ECSs may not be any more prone to somaclonal variation than plants derived from shoot-tip culture.

PROTOPLAST ISOLATION AND CULTURE

Somatic hybridization of bananas is possible and has potential to assist the breeding of edible bananas. In a similar manner to manipulation of ploidy with chemicals, tetraploids could be generated via protoplast fusion and triploids resynthesised by crossing with diploids. Allotetraploids created by the somatic hybridization of two elite diploid clones would be very different from autotetraploids produced by chromosome doubling. As there are now a number of examples of plant regeneration from protoplasts, fusion is now a technique that can potentially contribute to banana-breeding programs. Genotypes for which plants have been regenerated from protoplasts include breeding diploids (AA), dessert bananas (AAA), plantains (AAB), and cooking banana (ABB).[11] All reports of successful plant regeneration have two factors in common. Firstly, protoplasts were derived from ECSs and secondly, feeder cells and/or protoplast culture at high densities was required.

GENETIC TRANSFORMATION

As this entry is concerned with genetic manipulation of somatic cells, a brief mention should be made of genetic transformation. Both meristems and embryogenic cells have been targeted for transformation; however, due to greater efficiency and the greatly reduced risk of chimerism, embryogenic cells are the tissue of choice. Following the development of efficient regeneration systems for various cultivars, transformation systems using both microprojectile bombardment[12] and *Agrobacterium*-mediated transformation[9] of embryogenic cultures have been developed. Reports of transformation to date have focussed on the development of the technique itself and the examination of the strength and tissue specificity of various promoter elements. Work in this area has now moved on to conferring useful new traits including disease resistance and altered fruit characteristics. Indeed, potentially disease-resistant transgenic banana are currently undergoing glasshouse and field trials, and delayed fruit ripening using sense suppression of genes involved in ethylene biosynthesis has been reported.

CONCLUSION

The difficulties associated with conventional breeding of banana have been the driving force behind the genetic manipulation of somatic cells. In some cases, such as mutation breeding, new cultivars are already in commercial production. Other techniques, such as ploidy manipulation and protoplast fusion may provide assistance to conventional breeding or be utilised independently to generate new cultivars. Although at times some controversy is associated with genetic transformation, it does provide the potential to circumvent fertility problems and allow the transfer of genes within the banana gene pool or indeed introduce genes from completely unrelated organisms. The possibilities for banana improvement have entered a new and exciting phase based on our understanding of somatic cell genetics.

REFERENCES

1. Smith, M.K.; Hamill, S.D.; Becker, D.K.; Dale, J.L. Musacea. In *Biotechnology of Fruit and Nut Crops*; Litz, R.E., Ed.; CABI Publishing, New York: 2004, pp. 366–391.
2. Predieri, S. Mutation induction and tissue culture in improving fruits. Plant Cell, Tissue Organ Cult. **2001**, *64*, 185–210.
3. Novak, F.J.; Afza, R.; van Duran, M.; Omar, M.S. Mutation induction by gamma irradiation of in vitro cultured shoot tips of banana and plantain (*Musa* cvs.). Trop. Agric. (Trinidad) **1990**, *67*, 21–28.
4. Bhagwat, B.; Duncan, E.J. Mutation breeding of banana cv. highgate (*Musa* spp., AAA group) for tolerance to *Fusarium oxysporum* fsp. *cubense* using chemical mutagens. Sci. Hort. **1998**, *73* (1), 11–22.
5. Hamill, S.D.; Smith, M.K.; Dodd, W.A. In vitro induction of banana autotetraploids by cochicine treatment of micropropagated diploids. Aust. J. Bot. **1992**, *40* (6), 887–896.
6. Cote, F.X.; Domergue, R.; Monmarson, S.; Schwendiman, J.; Teisson, C.; Escalant, J.V. Embryogenic cell suspensions from the male flower of *Musa* AAA cv Grand Nain. Physiol. Plant. **1996**, *97* (2), 285–290.
7. Dhed'a, D.H.; Dumortier, F.; Panis, B.; Vuylsteke, D.; De Langhe, E. Plant regeneration in cell suspension cultures of the cooking banana cv. "Bluggoe" (*Musa* spp., ABB group). Fruits **1991**, *461* (2), 125–135.
8. Georget, F.; Domergue, R.; Ferriere, N.; Cote, F.X. Morphohistological study of the different constituents of a banana (*Musa* AAA, cv. Grande Naine) embryogenic cell suspension. Plant Cell Rep. **2002**, *19* (8), 748–754.
9. Ganapathi, T.R.; Higgs, N.S.; Balint-Kurti, P.J.; Arntzen, C.J.; May, G.D.; Van Eck, J.M. Agrobacterium-mediated transformation of embryogenic cell suspensions of the banana cultivar Rasthali (AAB). Plant Cell Rep. **2001**, *20* (2), 157–162.
10. Cote, F.X.; Folliot, M. Domergue, R.; Dubois, C. Field performance of embryogenic cell suspension-derived banana plants (*Musa* AAA, cv. Grande Naine). Euphytica **2000**, *112* (3), 245–251.
11. Assini, A.; Haicour, R.; Wenzel, G.; Cote, F.; Bakry, F.; Forought-Wehr, B.; Bakry, F.; Cote, F.X.; Ducreux, G.; Ambroise, A.; Grapin, A. Influence of donor material and genotype on protoplast regeneration in banana and plantain cultivars *Musa* spp. Plant Sci. **2002**, *162*, 355–362.
12. Becker, D.K.; Dugdale, B.; Smith, M.K.; Harding, R.M.; Dale, J.L. Genetic transformation of Cavendish banana (*Musa* spp. AAA group) cv. "Grand Nain" via microprojectile bombardment. Plant Cell Rep. **2000**, *19* (3), 229–234.

Beer Fermentation

Thomas H. Shellhammer
Department of Food Science and Technology, Oregon State University, Corvallis, Oregon, U.S.A.

Abstract

Beer production begins with malting, a process whereby grain (principally barley) is biochemically and physically converted to yield a starting material suitable for brewing. A fermentable substrate known as wort is produced within the brewhouse where starch is extracted from malted grain and enzymatically hydrolyzed to a range of saccharides, predominantly maltose. Hops are added during wort boiling for flavor and microbial preservation, and the sterile wort is fermented at conditions suitable for the yeast strain in question, *Saccharomyces cerevisiae* or *Saccharomyces uvarum*. Following fermentation, immature ("green") beer requires maturation, the objectives being vicinal diketone (diacetyl) reduction, clarification, carbonation, and in many instances, colloidal haze stabilization. At this point, the product is ready for packaging and release.

INTRODUCTION

Production of beer has existed in some form for over 8000 years. Its development and infiltration into various cultures was tied closely to the simultaneous development of agriculture. In it earliest forms, beer was a beverage of nourishment as well as pleasure, not to mention a source of potable hydration. Egyptians were sophisticated in their brewing practices and Osiris, the Egyptian god of agriculture, is credited by some as the "father" of beer.[1]

Beer most likely evolved from a series of events that currently would be considered accidents. Beginning with the production of coarse bread from pregerminated (sprouted) wheat, enzymes produced by the germinating wheat kernels resulted in partial saccharification of its starch thus producing the substrate necessary for an alcoholic fermentation. If this material was further hydrated and left to sit, it would have fermented as a result of the endogenous yeast present in the product. Despite the technological advances over the following 8000 years, the production of beer still resembles these "accidental" events. In fact, if one were to produce beer today without prior knowledge of or ties to its historical past, the production process would probably look quite different.

The key steps in manufacturing beer are malting, mashing, boiling, fermentation, finishing, and packaging. While malting does not happen in the brewery and is not technically considered brewing, it is so intrinsically linked to the entire process that it must be considered when discussing brewing.

MALTING

The main ingredient used to make beer, other than water, is malted barley. The malting process consists of three distinct events: 1) a steeping step where the moisture content of the kernel is raised from ~10 to ~44%; 2) a germination step where physical and biochemical changes occur due to the enzymatic breakdown of the starchy endosperm; and 3) a kilning stage where the kernel is dried to preserve the enzymes produced during germination and then heated to induce the Maillard reaction thereby producing color and characteristic biscuity, toasted, roasted flavors.

The biochemical sequence of events during germination of barley is initiated by water entering the kernel via the micropyle, the release of gibberellic acids via the migration of moisture across the scutellum, and activation and/or release of enzymes in the aleurone layer at the periphery of the endosperm (Fig. 1). Sequentially, β-glucanases, proteases, and amylases are produced within the aleurone and these slowly migrate through the endosperm toward the center of the kernel. In order to grow, the embryo must ultimately utilize amino acids produced via the action of exo- and endo-proteases on storage proteins plus sugars produced from the action of amylases on the starch granules. Enzymatic attack on these internal structures is only possible once the endosperm cell walls, which are composed of $(1\rightarrow3),(1\rightarrow4)$-$\beta$-D-glucan (~75%), arabinoxylan (~20%), and protein (~5%), have been sufficiently degraded to allow entry of this large molecular weight machinery. The maltster's objective is to achieve near complete destruction of

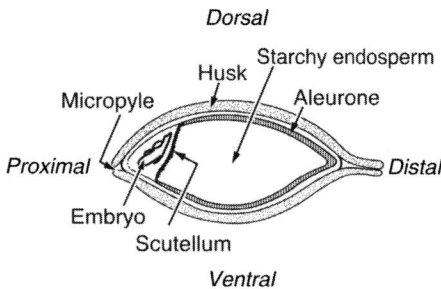

Fig. 1 Cross-section of a barley kernel. **Source:** From *Scientific Principles of Malting and Brewing*.[3]

the grain's endosperm cell walls, and hence minimize the amount of soluble β-glucans, without undue loss of starch via the amylases.

BREWHOUSE PROCESSES

While the malting process is generally carried out at a separate location, all subsequent steps happen within the confines of a brewery. First are those steps carried out in the brewhouse, namely milling, mashing, lautering, boiling, cooling, and aeration. Milling is a critical step toward effectively extracting the starch from the malted barley while at the same time allowing for efficient separation of the extracted and enzymatically converted material (wort) from the spent grains. Milling is often performed with a multi-roll (2, 4, 5, or 6 rolls) mill or in some instances, a hammer mill. The degree of particle reduction is dependent upon the technology selected for the wort separation process; for instance, a very coarse grind is used for a mash tun which is contrasted by a very fine grind for a mash filter. The milled grain is mixed with water and heated to achieve a desire time–temperature profile in a process referred to as mashing. The goal of this step is the enzymatic hydrolysis of the barley's starch, and in some cases with other starch substrates (referred to as adjuncts), via the endogenous α- and β-amylases that had been created or released during malting. The endo- and exo-amylases break down starch (a mixture of straight-chained amylase and branched-chained amylopectin) to yield a carbohydrate spectrum which is chiefly maltose but also includes oligosaccharides, including branched dextrins. Exogenous enzymes of fungal origin, such as debranching pullanses plus β-glucanses and arabinoxylanases, are sometimes used by brewers to increase yield, modify flavor, and improve downstream processing steps.

Once the hydrolysis of starch is complete, the wort must be separated from the spent grains in a process known as lautering or wort separation. Two general approaches may be taken to accomplish this task. A lauter tun is a vessel with a false bottom and rakes whereby the wort is strained through the spent grain bed (Fig. 2). Milling conditions earlier in the process are critical to the success of this step with small, uniform particles and an intact husk required. Alternatively, a mash filter can be used whereby wort is filtered through a modified plate and frame filter. In this case, milling conditions are set to completely pulverize the entire kernel, including the husk. The goal of the wort separation process is rapid and efficient separation of the wort from the spent grain without compromising wort quality. Hot (170°C) water is sparged through the grain bed to remove any remaining fermentable extract. This

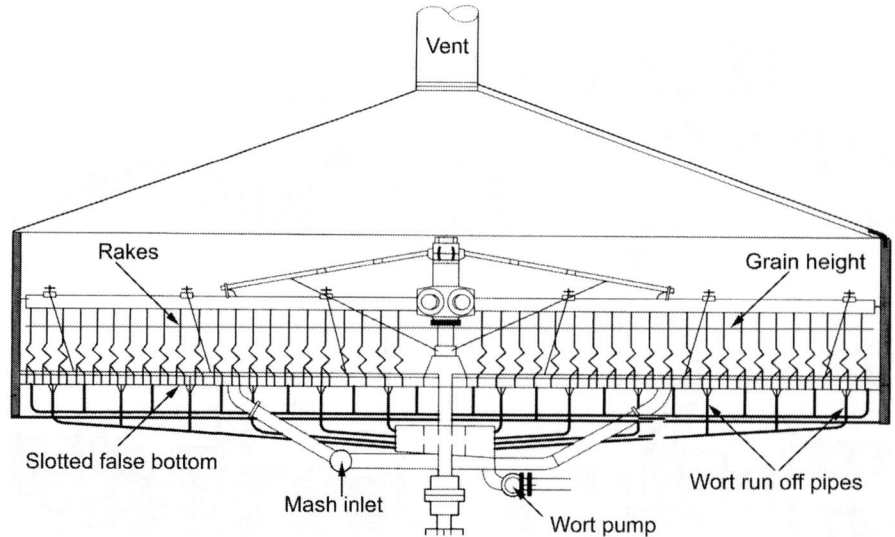

Fig. 2 Cross-section of a lauter tun. **Source:** Reproduced with permission of Huppmann AG.

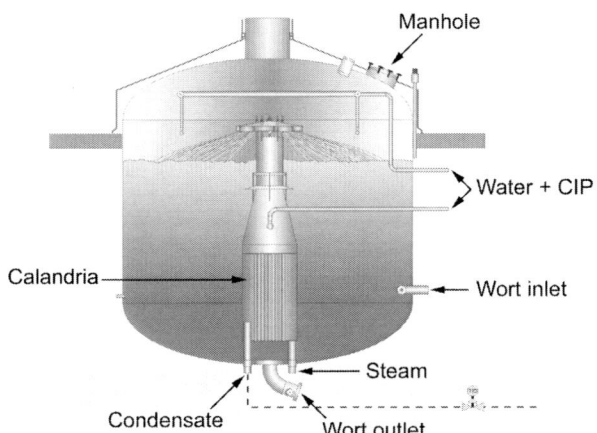

Fig. 3 Wort kettle with an internal calandria. **Source:** Reproduced with permission of Huppmann AG.

process is terminated with a small amount of residual sugars (<2% w/w) left in the spent grain bed since barley polyphenols, which can produce astringency in the final product, can be extracted from the husk as the pH of the liquid phase rises with the diluting sparge water.

Wort boiling takes place in a vessel that has ample heat transfer media (either in steam jackets on the sides and bottom of the vessel or with internal or external shell-and-tube heat exchangers [calandria]) to produce an evaporation rate of roughly 6% per hour (Fig. 3). The goal of the wort boiling phase is to sterilize the wort, isomerize hop acids, volatilize unwanted aromas (for instance, dimethyl sulfide, which is responsible for a cooked cabbage or creamed corn aroma), and coagulate protein–polyphenol–hop acid complexes (referred to as trub). Some of the consequences of this step are wort concentration from evaporation and color formation (most observable in lightly colored worts) from the Maillard reaction. It is in this stage of the process that hops are added.

Hops (*Humulus lupulus*) are climbing, perennial plants indigenous to Europe, Asia, and North America. It is the female plant of this diecious species that are cultivated for brewing needs. Resins, specifically the α-acids, and oils contained within the hop cones are of interest to the brewer. Hops are grown in northerly and southerly latitudes (35–55°C) because long days are required to delay flowering and produce acceptable yields.[2] Brewers may use dried, "whole" hops, pelletized hops (hammer-milled and extruded through a pelleting mill), or concentrated resin extracts. During boiling, the α-acids are extracted from the cone or pelleted material and isomerized because of the high temperature. The iso-α-acids are intensely bitter, substantially more soluble in wort and beer than α-acids, and have an antimicrobial (preservative) effect against lactic acid bacteria. The timing and amount of the hop addition will therefore determine how much bitterness is produced and how much volatile aroma is retained or lost. Most brewers use multiple hop additions at different times during wort boiling to achieve desired levels of bitterness and aroma from hops. Immediately following wort boiling, the hopped wort is separated from the trub (coagulated protein–polyphenol–hop acid complex) in a whirlpool separator or hop back. The sterile wort is cooled via a heat exchanger to fermentation temperature and oxygenated before pitching with yeast.

FERMENTATION

The dramatic transformation of sweet wort into alcoholic beer occurs as a result of yeast metabolizing simple sugars in the absence of oxygen. Strains of *Saccharomyces cerevisiae* and *Saccharomyces uvarum* differentiate ales and lager beer, respectively. The former species is well-suited for fermentation at 18–22°C, ferments in 3–4 days, and rises to the surface of the fermentor as it flocculates, while the latter ferments at cool temperatures (6–15°C), takes 7–10 days to attenuate and sediments when it flocculates. The two main byproducts of fermentation are ethanol and carbon dioxide, produced in nearly equal amounts by mass. Additionally, metabolic intermediates and byproducts such as organic acids, higher alcohols, esters, and vicinal diketones are produced in smaller amounts but contribute significantly to the flavor of the final product (Fig. 4).

Given that beer fermentations produce large quantities of viable yeast, each new fermentation is generally started with surplus yeast from previous fermentations, although pure cultures are propagated from single yeast cells every 8–15 generations (separate fermentations) to ensure yeast health and consistent fermentation performance. Yeast is added at a concentration of approximately 10^6 cells/mL/°P to sterile, aerated wort, where °P refers to degrees Plato (% wt./wt. soluble solids). Yeast requires some oxygen at the beginning of fermentation in order to produce sterols and unsaturated fatty acids; however, once attenuated, the immature ("green") beer is highly susceptible to oxidative damage and thus protected from further oxygen exposure.

MATURATION AND FINISHING

At the conclusion of fermentation, several physical and chemical changes take place that transform the product from being immature ("green") to a state that is ready for packaging, and these involve vicinal diketone (VDK) reduction, carbonation, (in many cases) colloidal haze stabilization, and clarification. This overall process is often referred to as lagering (i.e., cold storage for

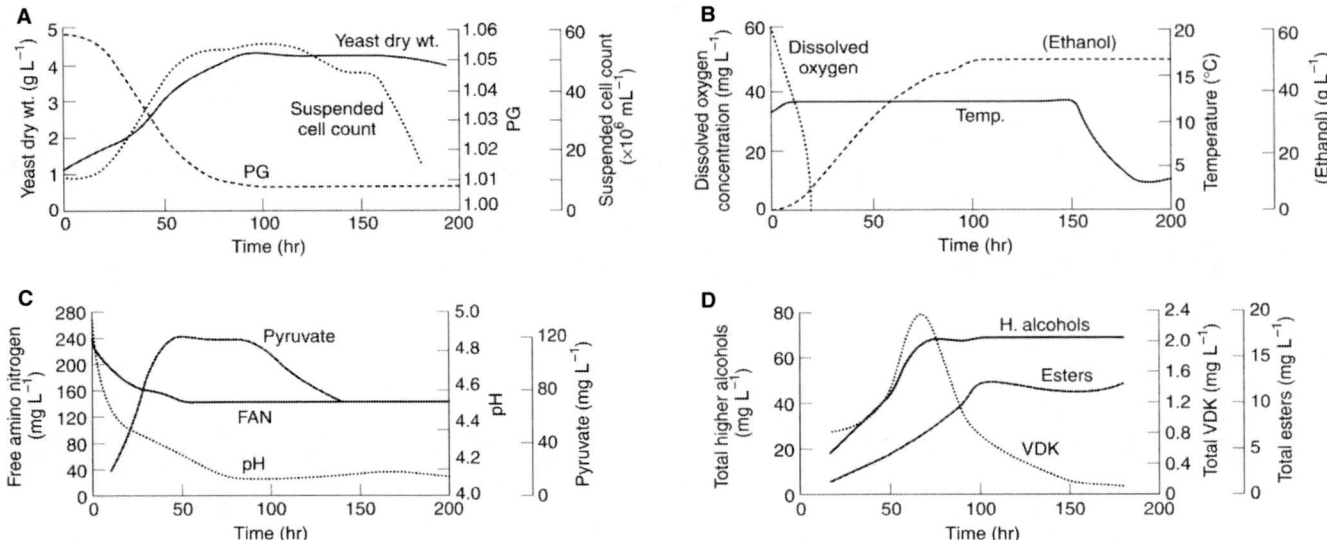

Fig. 4 Changes in present gravity (PG), yeast dry weight, suspended yeast cell count (A); ethanol concentration, temperature, dissolved oxygen concentration (B); free amino nitrogen concentration (FAN), pH, pyruvate concentration (C); total VDKs, total higher alcohols, total esters (D) during fermentation of a high-gravity (15° Plato) lager wort. The fermentation was maintained at a temperature of 12°C and terminated at 150 hours by cooling to 3°C. **Source:** From *Brewing Science and Practice*.[4]

the purpose of bringing about physical and biochemical changes).

Diacetyl (2,3-butanedione) and 2,3-pentanedione, the predominant VDKs in beer, are produced extracellularly from the oxidative decarboxylation of α-acetolactate and α-acetohydroxybutyrate, respectively, and have strong "buttery," "butterscotch," and "toffee" aromas. The acetohydroxy acids are excreted from the yeast cells as a byproduct of valine and isoleucine synthesis. Part of the maturation process involves waiting for live yeast to mop up the VDKs and convert them into more innocuous compounds, 2,3-butanediol and 2,3-pentanediol, which have much higher flavor thresholds.

The green beer in most cases does not have adequate carbonation since fermentation is carried out under atmospheric conditions. Carbonation can be adjusted by krausening the beer, which is adding an appropriate amount of fermenting wort to the lagering beer and then sealing the vessel thereby capturing the additional CO_2. Alternatively, CO_2 can be added directly to the beer to achieve the desired level.

Colloidal haze in beer results from the interaction between haze-active protein (proline-rich) and haze-active polyphenols. Haze is produced as flavonoids oxidize, polymerize, and bridge haze-active protein, via hydrogen bonding, to create a haze visible to the human eye. Preventing haze formation in the finished product is accomplished in a number of different manners. Adding tannic acid to force haze formation followed by filtration was a common method for many years. Today, most brewers use adsorptive treatments, which rely on silica hydrogel for haze active protein removal and polyvinylpolypyrrolidone (PVPP) for polyphenol removal.

Clarification was historically performed by allowing suspended material to settle over time. Brewers nowadays can speed this process up by using continuous centrifuges, adsorptive agents, or filtration. Once the beer has met specifications for ethanol level, bitterness, color, dissolved carbon dioxide, and oxygen, it is ready for release to packaging. In some instances, the beer is either sterile filtered, or thermally pasteurized, and then aseptically filled into kegs or small containers. Alternatively, it can be filled in small containers and then pasteurized via a tunnel pasteurizer. For those beers in which live yeast and fermentable sugars are added to the final beer before packaging, the finished packaged product is not pasteurized but is stored at room temperature to allow the residual sugars and oxygen to be consumed by the yeast in a process known as bottle conditioning.

CONCLUSION

Beer is produced via one of the oldest bioprocesses in existence. Its production method traces its roots many millennia back in time and the current production practices mimic many of the historically key steps albeit in a more technologically advanced and refined manner. The entire process is characterized by a sequence of enzymatically catalyzed biochemical transformations plus the combining of key ingredients and separation of their endproducts. Regardless of its style or historical age, it is a product that continues to capture people's attention and is enjoyed worldwide.

REFERENCES

1. Bamforth, C.W. From Babylon to Busch. In *Beer, Tap into the Art and Science of Brewing*, 2nd Ed.; Oxford University Press: New York, 2003; 7–48.
2. Atkinson, B. (chairman) for EBC Technology and Engineering Forum. Hop growing. In *European Brewery Convention: Hops and Hop Products, Manual of Good Practice*; Getranke-Fachverlag Hans Carl: Nurnberg, Germany, 1997; 7–24.
3. Bamforth, C.W. Barley and malting. *Scientific Principles of Malting and Brewing*; American Society of Brewing Chemists: St. Paul, MN, 2006; 21–44.
4. Briggs, D.E.; Boulton, C.A.; Brookes, P.A.; Stevens, R. Metabolism of wort by yeast. In *Brewing Science and Practice*; Woodhead Publishing Limited: Cambridge, UK, 2004; 401–468.

Bioactive Compounds in Mushrooms

Mark Shamtsyan
Department of Technology of Microbiological Synthesis, St. Petersburg State Institute of Technology, St. Petersburg, Russia

Abstract

For thousands of years mushrooms have been used not only for their nutritional value as food, but also for healing potential. At present there are at least 270 species of mushroom that are known to have therapeutic properties, which have the potential to be used as dietary supplements or in the fortification of foods as functional compounds. In addition to well-known immunomodulating and anticancer properties, mushrooms possess other important therapeutic properties including antioxidants, antihypertensive, cholesterol-lowering, liver protection, anti-inflammatory, antidiabetic, antiviral, antimicrobial, and some others. Mushrooms also can be a source of useful for enzymes for the food industry, such as milk-clotting enzymes.

INTRODUCTION

Mushrooms contain a large variety of bioactive compounds that are poorly understood and not well characterized. Their potential contributions to the food industry and agriculture are considerably underutilized. Bioactive compounds from mushrooms with immunomodulating, antibacterial, and antioxidant activities could be used as functional and antimicrobial food and animal feed supplements. As functional food supplements can be used mushroom derived substances with hypocholesterolic, hypolipidemic hepatoprotective, antidiabetic, and some other activities, while milk-clotting enzymes can find utilization in cheese making to substitute animal rennin.

Mushrooms are not a taxonomic group, but include about 14,000 species that have macroscopic fruiting bodies, seen by the naked eye. They are noted for their nutritional value and many are viewed as functional foods.

The protein content in mushrooms is higher than in most vegetables and somewhat less than in meat and milk. Mushrooms contain all the essential amino acids, but can be deficient in methionine and cysteine. They are rich in dietary fiber, and thus, calorific value of most of them is low. Mushrooms are a good source of vitamins, especially thiamine (B_1), riboflavin (B_2), niacin, biotin, and ascorbic acid. Undoubtedly, edible mushrooms represent a nutritious and tasteful source of food and can be an important dietary component for vegetarians.

Long ago people recognized that mushrooms could have valuable health benefits. Even shamans of Paleolithic peoples knew and used mushrooms to fight illnesses.

Pharaohs appreciated mushrooms as delicacies. According to hieroglyphics, 4600 years ago Egyptians believed that mushrooms were the "plant of immortality" and "a gift from the God Osiris." The delicious flavor of mushrooms intrigued the pharaohs so much that they declared mushrooms food for royalty and no commoner could ever touch them. This assured them the entire supply of mushrooms. Greeks turned to mushrooms as a source of strength for warriors before the battle. Ancient Romans considered them as the meal of gods and served them only for great feasts. Vikings ate mushrooms to place themselves in a state of "going berserk," while Native Americans used them for their magical hallucinogen effects to cross the barrier between the body and mind.[1]

Frozen in the Italian Alps 5000 years ago, "Ice-Man" Otzi, carried a medicine kit of dried mushrooms. Indeed, the oldest written record of mushrooms used as medicines is documented in the Indian medical treatise from 3000 B.C.[1]

The majority of traditional knowledge on the bioactive properties of mushrooms comes from East Asia.

Historically, mushrooms were gathered from the wild for consumption and for medicinal purposes. China has been the source of many mushrooms, since 600 (*Auricularia auricula*). *Agaricus bisporus* was first cultivated in France in the 17th century while *Pleurotus ostreatus* was first grown in the United States in the 20th century. While mushroom cultivation now spans many centuries, it is only over the last 3 to 4 decades that major expansions in basic research and practical knowledge has lead to mushroom production as a major global industry.

In the second half of the 20th century, technologies for mushroom cultivation were strongly developed, and at the beginning of 21st century the overall value of the world's mushroom production was estimated to be over $ 45 billion.[2]

Presently, there are at least mushroom 270 species of mushrooms that are known to have therapeutic properties.

Today, on the world pharmaceutical and food markets a large arsenal of mushroom compounds exist, which is representing a new class of nutriceuticals or "functional food additives." When used for therapeutic purposes, mushrooms are normally consumed as powdered concentrates or extracts in hot water. The liquid concentrates or dried powder can be placed in capsules drink, freeze-dried, or spray-dried to form granular powders, which allow for easier handling, packaging, transportation, and consumption.[3] These liquid concentrates or dried powdered mushrooms can be considered as dietary supplements or mushroom nutriceuticals with potential health benefits[4] (Table 1).

Naturally, during the passage of time human communities accumulated and assimilated the traditions of folk medicine with regard to mushrooms. While there is a great deal of attention focused on the various immunological and anticancer properties of mushrooms,[3,5] mushrooms can also offer other potentially important therapeutic properties, such as antioxidant activities, antihypertensive function, cholesterol-lowering ability, liver protection, anti-inflammatory effects, alleviation of diabetes, and antiviral and antimicrobial effects.[1,6–8]

IMMUNOMODULATING AND ANTITUMOR EFFECTS

Historically, mushrooms have played an important role in China treating many forms of immune disorders. Chinese pharmacopeias stores the information of more than 100 mushroom species used by practitioners of traditional Chinese medicine for a wide range of ailments. There is considerable folklore concerning many bioactive principles but it is important to distinguish between those with documented experimental clinical, mechanistic, or epidemiological data from those with unsubstantiated claims.

The main antitumor compounds presently isolated from mushroom fruit-bodies, submerged cultural mycelia, or liquid culture broth have been identified as either water-soluble β-D-glucans, β-D-glucans with heterosaccharide chains of xylose, mannose, galactose, or uronic acid or β-D-glucan-protein complexes—proteoglycans, which can induce immunomodulatory and therapeutic effect in animals and humans.[9]

The main chain of basic β-D-glucan is either β 1–3, β 1–4 or mixed β 1–3, β 1–4 with β 1–6 side chains of varying sizes occurring at different intervals (Fig. 1).[10] Levels of activity of these compounds can be related to their degree of branching, molecular weight, and solubility in water.

These substances are able to influence non-specific and specific immune responses of an organism and activate various kinds of immunocompetent cells or components, such as monocytes, neutrophils, cytotoxic macrophages, natural killer cells, dendritic cells, cytokines, interferons, and lymphocytes.

CARDIOVASCULAR AND CHOLESTEROL-LOWERING EFFECTS

Hypercholesterolemia increases the risk of cardiovascular diseases and is among the major health risk factors for human health in developed countries. Elevated levels of circulating cholesterol cause deposits to form inside blood vessels. These deposits can result in a disease process called arteriosclerosis.

Cholesterol has been divided into two major categories: low-density lipoprotein (LDL) and very low-density lipoprotein (VLDL), the so-called "bad" cholesterol, and high-density lipoprotein (HDL), the so-called "good" cholesterol.

The first step in the prevention and treatment of hypercholesterolemia and related cardiovascular diseases is the development of the nutritional regime with a diet low in fats and saturated fatty acids and rich in crude fibers. Mushrooms, due to their high fiber content, which can influence lipid digestion and inhibit intestinal absorption of cholesterol and other lipids, as well as low calorific value, are suitable for diets designed to prevent cardiovascular diseases.

Being rich in dietary fiber, some mushrooms also can produce substances inhibiting cholesterol syntheses. A major rate-limiting step in the biosynthetic pathway of cholesterol formation is the level of the microsomal enzyme, 3-hydroxy-3-methylglutaryl-coenzymeA reductase (HMG-CoA reductase) that catalyses the reductions of HMG-CoA into mevalonate.

Some species from the genus *Pleurotus* are capable to produce mevinolin (lovastatin), which was the first specific inhibitor of the HMG-CoA reductase to receive approval for the treatment of hypocholesteremia.

Eritadenine, a compound extracted from *Lentinus edodes*, is also able to lower blood serum cholesterol. Probably, eritadenine lowers cholesterol by decreasing the ratio of phosphatidylcholine to phosphatidylethanolamine in liver microsomes.

The addition of dried fruiting bodies or submerged mycelial growth to a high cholesterol diet effectively reduces cholesterol accumulation in the serum and liver of experimental rats redistributing cholesterol in favor of HLDL, reduced production of VLDL and LDL cholesterol, reduced cholesterol absorption, and reduced HMG-CoA reductase activity in the liver and triglyceride level in blood serum.[9,11–12]

HEPATOPROTECTIVE EFFECTS

Mushrooms are considered to be beneficial for a wide range of liver disorders, including hepatitis. Although, some mushrooms have been shown to be hepatotoxic, several compounds isolated from the fruiting body, mycelia, and spores of *Ganoderma lucidum*, such as, ganoderic acid, ganosporeric acid A were shown to have strong antihepatotoxic activity.

Table 1 Cross index of most studied higher basidiomycetes and their bioactive properties.

Taxa	BIOLOGICAL ACTIVITY														
	Immunomodulating	Antitumour	Antiviral	Antibacterial and Antiparasitic	Anti-inflammatory	Blood pressure regulation	Cardiovascular disorders, Hypercholesterolemia,	Antidiabetic	Kidney tonic	Hepatoprotective	Nerve tonic	Sexual potentiator	Chronic bronchitis	Antioxidant	Milk clotting
Auriculariales															
Auricularia auricula-judas (Bull.) Wettst	+	+				+	X						X		
Tremellales															
Tremella fuciformis Berk.	+	+		+			+	+		+			X		
Tremella mesenterica Rits.: Fr.	+	+			+								+		
Polyporales															
Schyzophyllum commune Fr.:Fr.	X	X		X	X			X	X					+	+
Dendropolyporus umbelatus (Pers. Fr.) Jul.	X	X							X			X			
Grifola frondosa (Dicks. Fr.) S.F. Gray	X	X	X	X		X	X	X		+	X		+		
Fomes fomentarius (L.) J.J. Kickx	+	+		+									+		
Fomitopsis pinicola (Schw.:Fr.) P. Karst.	+	+		+	+					+					
Trametes versicolor (L.:Fr.) Lloyd	X	X	X	X				X	X						
Trametes ochracea (Pers.) Gilb. & Ryvarden	+						+								
Piptopurus betulinus (Bull.:Fr.) P. Karst	+	+		+											
Hericium erinaceum (Bull.:Fr.) Pers.	X	+				X	X			X	X				
Cerrena unicolor (Bull.) Murrill.	+	+					+							+	
Irpex lacteus (Fr.) Fr.				+											X
Daedaleopsis confragosa (Bolton) J. Schröt				+	+									+	
Albatrellus ovinus (Schaeff.) Kotl. & Pouzar				+			+							+	
Trichapatum laricinum (P. Karst) Ryvarden														+	
Sparassis crispa (Wulfen) Fr.														+	
Laetiporus sulphureus (Bull.) Murrill		+								+				+	

(Continued)

Table 1 Cross index of most studied higher basidiomycetes and their bioactive properties. *(Continued)*

Species	1	2	3	4	5	6	7	8	9	10	11	12	13	14	15
Poria cocos (Schw.) Wolf	X	X	+						+						
Ganodermatales															
Ganoderma lucidum (Curt.:Fr.) P. Karst	X	X	X	X	X	X	X		X	X	X	X	X	+	
Ganoderma applanatum (Pers.) Pat.	+	+	+	+											
Hymenochaetales															
Inonotus obliquus (Ach. ex Pers.) Pilat		X		X									X		
Aphillophorales															
Phellinus linteus (L.ex Fr) Quel	X	X													
Phellinus robustus (P. Karst.) Bond. et Singer															+
Agaricales															
Pleurotacea															
Lentinus edodes (Berk.) Sing	X	X	X	X	X	X		X	X	X		X			
Pleurotus cornucopiae (Pau. Ex Pers.) Rollan	+	+											+		
Pleurotus ostreatus (Jack.:Fr.) Kumm	X	+	+	+						+			+		
Pleurotus pulmonarius (Fr.:Fr.) Quel		+													
Tricholomataceae															
Flamulina velutipes (Curt.:Fr.) P. Karst.	X	X	+	+	X								+		
Armillariella mellea (Vafl.:Fr.) P. Karst.						X	X			X					
Hypsizigus marmoreus (Peck) Bigel		X													
Marasmius androsaceus (L.:Fr.) Fr.						X				X					
Agaricaceae															
Agaricus blazei Murr.	X	X													
Agaricus bisporus (J. Lge) Imbach	X	+				+		X	X						
Coprinus comatus (O.F. Müll.) Persoon						+							+		
Coprinus domesticus (Bolton:Fr.) Grey													+	+	
Coprinellus micaceus (Bull.) Vilgalys		+		+									+		
Pluteaceae															
Volvariela volvacea (Bull.:Fr.) Sing.		+	+	+		+							+		
Bolbitaceae															
Agrocybe aegerita (Brit.) Sing.		+			+				+						
Strophariaceae															
Pholiota adiposa (Batsch) P. Kumm.			+	+	+								+		
Pholiota nameko (T. Ito) S. Ito & Imai	+	+													
Hypocreales															
Cordyceps sinensis (Berk) Succ.	X	X			X	X	X	X	X		X		X		

X - Commercially developed mushroom product
+ - Non-commercially developed mushroom product

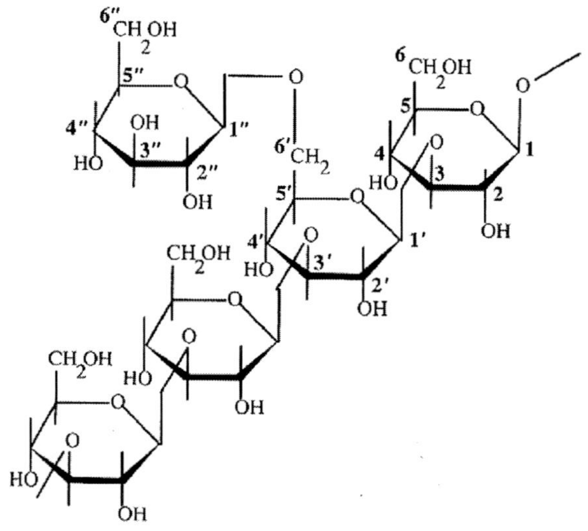

Fig. 1 Structure of β-1-3 glucan with 1–6 branching.

A polysaccharide fraction from *L. edodes* showed liver-protective action in animals together with improved liver function and an increased production of antibodies to hepatitis B.[3] There have been other interesting medical reports concerning distinct improvement with patients suffering from cirrhosis of the liver and chronic hepatitis B with extracts or polysaccharides from *Dendropolyprus umbellatus*, *Schizophyllum commune*, *Trametes versicolor*, *Poria cocos*, and *Tremella fuciformis*.

ANTIDIABETIC EFFECT

Extracts of several higher fungi, including *Tremella aurantia*, *Cordyceps sinensis*, *G. lucidum*, *A. auricula-judae*, *L. edodes*, *P. ostreatus*, and *Phellinus linteus* have been shown to lower blood glucose triglyceride levels. Such results strongly suggest that these mushrooms have potential preventive and therapeutic action for diabetes mellitus types I and II.

Antidiabetic activity of *Grifola frondosa* is related to the process of metabolism of adsorbed glucose. The blood glucose-lowering effect is thought to be a result of a high molecular weight glycoprotein.

ANTIMICROBIAL EFFECTS

Various antitumor polysaccharides from medicinal mushrooms would be expected to function by mobilizing the humoral immunity of the body to protect against viral, bacterial, fungal, and protozoal infections resistant to current antibiotics.

Several mushroom polysaccharides have shown antiviral activity against ectromelia virus and cytomegalovirus infections.[5] Lentinan, a commercial β-D-glucan preparation from *L. edodes*, has shown antiviral activity in mice against vesicular stomatis virus, encephalitis virus, Abelson virus, and adenovirus type 12; it has also stimulated non-specific resistance against respiratory viral infection in mice, and conferred complete protection against an LD75 challenge dose of virulent mouse influenza A/SW15. Against other microorganisms, lentinan has increased resistance to protozoal parasites, exhibited activity against *Mycobacterium tuberculosis* resistant to antituberculosis drugs, antagonism against *Bacillus subtilis*, *Staphylococcus aureus*, *Micrococcus lenuteus*, *Candida albicans*, and *Saccharomyces cerevisiae*, and increased host resistance to infections from potentially lethal *Listeria monocytogenes*.[9]

The sulfated polysaccharide schizophyllan, from *S. commune* has displayed strong anti-HIV activity, but antitumor effects have been inconsistent or absent. Schizophyllan has also been reported to enhance protection against *staphylococcal* infections.

Extensive examination of over 200 species of basidiomycetes in Spain demonstrated that almost 50% of the varieties examined had significant antibiotic activity against a range of test organisms. It is interesting to note that the bracket polypore *Piptoporus betulinus* carried by the historic Ice-man Otzi mentioned earlier displayed a strong broad-spectrum antibiotic activity.

L. edodes extracts can improve the beneficial intestinal biota of the gut. The active factor in the extract is considered to be trehalose. Trehalose can reduce the harmful effects of some bacterial enzymes such as α-glucosidase, α-glucuronidase, and tryptophanase, as well as function to reduce colon cancer formation.[9]

ANTIOXIDANT ACTIVITY

Antioxidants play a very important role in protecting the body from the harmful effects of free radicals.

Mushroom extracts and polysaccharides can decrease the production of oxygen-free radicals. Significant superoxide- and hydroxyl radical-scavenging activities have been demonstrated for several mushroom antitumor polysaccharides.[9]

NERVE TONIC ACTIVITY

Nerve growth factors (NGF) called erinacins (a series of diterpenoids) and hericenones (a class of benzyl alcohol) occur in the fruiting bodies, mycelia, and culture broth of *Hericium erinaceum*. These have been shown to stimulate NGF synthesis. The erinacines are the most powerful inducers of NGF synthesis among all currently identified natural compounds.

MILK-CLOTTING ACTIVITY

Higher basidial mushrooms are a source of biologically active supplements, but also are a source of useful enzymes.

Milk-clotting enzymes from higher basidiomycetes are promising substitutes for the proteolytic rennin used to form curds in cheese making.

The quality of cheese significantly depends upon the milk-clotting enzymes used. Usually rennet is extracted from calves' stomachs or produced by genetically engineering *Escherichia coli*. Requirements for rennet substitutes are strict and specific; enzymatic properties must optimally approach those of natural rennin. Together with strong curd-forming activity substitutes must possess insignificant general proteolytic activity, which can leads to proteolysis of casein.

Highly active proteases of rennet action have been isolated from the higher basidiomycetes and found practical use.

CONCLUSION

Mushrooms are a promising source of desirable compounds for the food industry, but their full potential is not yet fully developed. Their possibilities and utilization are tremendous, especially, for those products taken from submerged cultivation of higher mushrooms as bioactive food supplements and as a source of valuable enzymes.

REFERENCES

1. Chang, S.T. The world mushroom industry: trends and technological development. Int. J. Med. Mushr. **2006**, *8* (4), 297–314.
2. Mizuno, T.; Sakai, T.; Chihara, G. Health foods and medicinal usage of mushrooms. Food Rev. Int. **1995**, *11*, 69–81.
3. Chang, S.T.; Buswell, J.A. Mushroom nutriceuticals. World J. Microbial. Biotechnol. **1996**, *12*, 473–476.
4. Jong, S.C.; Donovick, R. Antitumour and antiviral substances from fungi. Adv. Appl. Microbiol. **1989**, *34*, 183–262.
5. Hobbs, C. *Medicinal Mushrooms: An Exploration of Tradition, Healing and Culture*; Botanica Press: Santa Cruz, CA, 1995.
6. Mizuno, T. The extraction and development of antitumor-active polysaccharides from medicinal mushrooms in Japan. Int. J. Med. Mushr. **1999**, *1* (1), 9–29.
7. Gunde-Cimerman, N. Medicinal value of the genus *Pleurotus* (Fr.) P. Karst. (Agaricales S.I., Basidiomycetes). Int. J. Med. Mushr. **1999**, *1* (1), 69–80.
8. Mau, J.L.; Lin, H.C.; Chen, C.C. Antioxidant properties of several medicinal mushrooms. J. Agric. Food Chem. **2002**, *50* (21), 6072–6077.
9. Bohn, J.A.; BeMiller, J.N. (1-3)-β-D-Glucans as biological response modifiers: a review of structure–functional activity relationships. Carbohydr. Polym. **1995**, *28*, 3–14.
10. Wasser, S.P.; Weis, A.L. Medicinal properties of substances occurring in higher Basidiomycete mushrooms: current perspective. Int. J. Med. Mushr. **1999**, *1* (1), 31–62.
11. Bobek, P.; Ginter, E.; Jurcovicova, M.; Ozdin, L.; Mekinova, D. Cholesterol lowering effect of the mushroom *Pleurotus ostreatus* in hereditary hypercholesterolemic rats. Ann. Nutr. Metab. **1991**, *35*, 191–195.
12. Popov, A.; Panchenko, A.; Denisova, N.; Petrischev, N.; Shamtsyan, M. Hypocholesterolic effect of some higher basidiomycetes. In *Research Progress in Biotechnology*; Zaikov, G.E., Ed.; Nova Science Publishers, Inc.: New York, 2008; 53–58.

Bio-Based Chemicals: Renewable Carbon for Biorefinery Production

Joseph J. Bozell
Biomass Chemistry Laboratories, Forest Products Center, University of Tennessee, Knoxville, Tennessee, U.S.A.

Abstract

Exploitation of domestic renewable carbon for the production of chemicals and fuels offers a means to reduce dependence on diminishing supplies of nonrenewable raw materials. The biorefinery, a renewables-based analog to the petrochemical refinery, has emerged as the organizing concept for the transition of renewable carbon to marketplace products. Increasing numbers of bio-based products and means for their cost effective manufacture within the biorefinery are appearing. This entry briefly overviews concepts associated with biorefinery development, describes technology for the conversion of renewable carbon, and lists several promising bio-based products that may be commercially useful.

INTRODUCTION

The interest in renewable feedstocks as raw materials for the production of strategic chemicals and fuels is at one of its highest levels in the last quarter century. Concern over the availability of supplies and wildly fluctuating price swings in crude oil, from a high of nearly $150/bbl in mid-2008 to less than $35/bbl in early 2009, have helped spur study of domestic feedstock sources demonstrating greater price stability and higher security.

The upsurge in R&D interest is also driven by the impact that carbon has on daily life in the industrialized world. The technological advances and standard of living enjoyed by modern society are the result of easy access to large, concentrated supplies of non-renewable carbon in the form of coal, crude oil, or natural gas. Two primary uses exist for non-renewable carbon. The majority (>90%) is used for energy extraction. By subjecting crude oil to a preliminary separation and subsequent combustion, energy is released and is used primarily for transportation, shelter, and power. Alternatively, a much smaller portion (<10%) of non-renewable carbon is subjected to cracking and conversion into the primary chemical building blocks of the petrochemical industry: aromatics (benzene, toluene, and xylene) and olefins (ethylene, propylene, and mixed butenes). Coal and natural gas provide additional building blocks in the form of syngas (a mixture of CO and H_2) and methane (CH_4). From this initial group of low molecular-weight organic compounds come the tens of thousands of products currently made by the chemical industry; however, the past 30 years have repeatedly experienced negative worldwide economic effects when non-renewable carbon supplies are threatened.

Accordingly, extensive effort is being directed at developing renewable carbon (biomass, such as agricultural crops or forest resources) as a source of strategic chemicals and fuels. The biorefinery is now widely recognized as the organizing concept to unify the transition of renewable carbon from its source in agricultural and silvicultural raw materials through intermediate monomeric and biopolymeric building blocks and ultimately to bio-based marketplace products. This brief entry will describe some of the major concepts regarding biorefinery development and renewable carbon for the production of new chemical intermediates and products.[1]

USE OF RENEWABLE CARBON IN THE BIOREFINERY

The biorefinery bears many similarities to the petrochemical refinery. For example, biorefinery operation faces three primary issues with renewable carbon that are also faced by the petrochemical industry when converting crude oil or other non-renewable carbon sources to fuels and chemicals.

Supply

Supply issues are primarily associated with the growing and collection of the biomass feedstock. Successful biorefinery operation will depend on the source of the starting feedstock, its availability, the best geographic location for supply production, and its sustainability, collection, transportation, densification, and processing.

Separation

As is the case with crude oil, biomass as collected from its source has limited utility and must be subjected to a preliminary upgrading before further conversion. Biorefinery separation technology is the rough equivalent of that used in the petrochemical industry to obtain the primary raw materials for conversion into other products. Thermochemical processes, analogous to crude oil distillation, apply heat to the biomass to effect an upgrading. In the absence of oxygen, heating of biomass induces pyrolysis, leading to the formation of the so-called bio-oils, a potential source of fuel or chemicals. Biomass can also be gasified to give syngas as the major product. These thermal conversions generally afford low-value intermediates that are complex mixtures, or the result of complete deconstruction of the biopolymers present in biomass. Alternatively, biomass can be subjected to more selective fractionation processes, analogous to petrochemical cracking. These processes separate the mixture of polymeric and monomeric structures that comprise biomass into individual process streams, each of which can serve as a starting point for the production of a wide range of chemical platforms, intermediates, or final products. Sources of renewable carbon within the biorefinery include: 1) carbohydrates, in the form of cellulose, hemicellulose, starch, or monomeric sugars; 2) aromatics, in the form of lignin; 3) hydrocarbons, in the form of plant triglycerides; and 4) smaller volume process streams such as protein or extractives.

Conversion

Once isolated, renewable building blocks must be converted into chemical intermediates and final products. It is often incorrectly assumed that biorefinery operation occurs within the single context of combining fermentable sugars with biological organisms. This is a common misconception, but one that must be dispelled as use of renewable carbon does not require biotechnology. Bioenergy and biorefinery development includes all aspects of the use of renewable carbon for the production of chemical products, fuels, and power, combined with conventional chemical, biochemical, or hybrid conversion technologies most appropriate for the process under consideration.

For the most part, issues of supply and separation are generally understood for both biochemical and petrochemical refining; however, conversion processes for renewable carbon suffer from a gap in technology development. The petrochemical industry has gained amazing transformational control over the behavior of their many crude oil-derived building blocks. By contrast, analogous use of renewable carbon suffers from a much narrower range of discrete building blocks, fewer methods to convert those building blocks to other materials, and a lack of information about the properties and performance available from those products. To address this gap, recent reports couple technology needs in the biorefinery with groups of product candidates potentially available from biorefinery carbohydrates and lignin (Table 1).[2,3]

THE CASE FOR RENEWABLE CARBON

Well into the 20th century, renewable carbon supplied a significant portion of the nation's chemical needs. The chemurgy movement of the 1930s, led by such notables as William Hale and Henry Ford, promoted the use of farm products as a source of chemicals, with the belief that *"anything that can be made from a hydrocarbon could be made from a carbohydrate."* It is only between 1920 and 1950 that we have witnessed a large-scale transition to a non-renewable-based economy.

Incorporation of domestic renewable carbon as part of the nation's raw material supply has been the topic of a large number of reviews suggesting advantages not available with petrochemical feedstocks. Renewable carbon sources afford industrial processes that are nearly CO_2 neutral. At the end of their life cycle, bio-based products release no more CO_2 than was originally metabolized in the biological production of the raw material. Biotechnology offers the ability to tailor plants for production of structurally defined intermediates or enhanced production of particularly useful biorefinery process streams. Biorefinery operation also offers benefits for the chemical industry by addressing several of the principles of green chemistry. For example, carbohydrate process streams within the biorefinery are well suited for transformations in aqueous media. Bio-based materials can be designed to give products that can break down in the environment at the end of their useful life, leading to environmentally beneficial processes when considered from the perspective of life cycle analyses, and heat and energy use in the chemical industry. New process technology based on renewable carbon offers a way to reduce industry's environmental footprint. In his review of progress on sustainable development, Metzger concluded, "renewables are the only workable solution."[4]

A vast amount of renewable carbon is produced in the biosphere; about 77×10^{10} tons are fixed annually. When measured in energy terms, the amount of carbon synthesized is equivalent to about 10 times the world consumption. Cellulose is the most abundant organic chemical on earth, with an annual production in the biosphere of about 1×10^{10} to 1×10^{11} tons. Lignocellulosic feedstocks provide a parallel lignin supply, which can comprise up to 25% by weight of biomass and is a promising source of aromatic chemicals. Lignin production by the pulp and paper industry is estimated to be 3.0 to 5.0×10^7 tons/yr. The U.S. agriculture excels at the production of carbohydrates through growing of annual domestic crops. In 2007, the corn industry produced 3.67×10^8 tons (1.31×10^{10} bu) of corn, containing 2.1×10^8 tons of carbohydrate as starch, and equivalent to over 5.0×10^8 bbl of crude oil. Soybean production reached

Table 1 The "Top 10" chemicals from carbohydrates and lignin.

Products from carbohydrates	Products from lignin
Succinic, fumaric, and malic acids	Thermochemical products (gasification, pyrolysis, and combustion)
2.5-Furandicarboxylic acid	Macromolecules (carbon fibers, polymer modifiers, resins, and adhesives)
3-Hydroxypropionic acid	Benzene, toluene, and xylene
Aspartic acid	Phenol
Glucaric acid	Substituted coniferols (propylphenol, eugenol, aryl ethers, etc.)
Glutamic acid	Oxidized lignin monomers (vanillin and syringaldehyde)
Itaconic acid	Diacids from chemical transformations
Levulinic acid	Diacids from biochemical transformations
3-Hydroxybutyrolactone	Aromatic polyols (cresols and catechols)
Glycerol	Cyclohexane and substituted cyclohexanes
Sorbitol	Quinones
Xylitol/arabinitol	

8.8×10^7 tons (2.97×10^9 bu), leading to 1.6×10^7 tons of plant oils in 2008. The pulp and paper industry converts over 2.4×10^8 tons/yr of wood, which is about 50% carbohydrate as cellulose, for the production of paper products.

Yet our chemical feedstock supply is overwhelmingly dominated by non-renewable carbon. Only about 2% comes from renewable sources, despite recent work that has identified a sustainable biomass supply of 1.3×10^9 tons/yr available in the United States without upsetting normal supplies of food, feed, and fiber and without requiring extensive changes in infrastructure or agricultural practices. Thus, except for the pulp and paper, corn-milling, and soybean-processing industries, relatively few examples of large-scale industrial bioprocessing exist. Further, the pulp and paper industry devotes only a small part of its production to chemicals, while the corn-milling industry is focused largely on starch and its commercial derivatives, ethanol and corn syrup. The current Renewable Fuels Standard has driven an increase in fermentation ethanol production in the United States by legislating the sustainable production of 16 billion gallons of fuel ethanol from cellulose by 2022. As a benchmark, 4.9 billion gallons of fuel ethanol was produced from cornstarch in 2006.[5–8]

ECONOMIC CONSIDERATIONS

Economic issues do not appear to present a major hurdle to the use of renewable carbon. Recent evaluations suggest that bio-based raw materials may be competitive with conventional sources of non-renewable carbon (Table 2). Lynd et al.[9] has compared the cost of non-renewable and renewable carbon sources on a cost/unit of energy basis. Cellulosic crops at $50/ton compare favorably with all non-renewable sources of carbon except coal; however, if coal-based processes are required to sequester generated CO_2, their relative cost/energy unit rises significantly. It is important to recognize that infrastructural costs (such as collection and transportation) associated with renewables are not included in these calculations and still not as well understood as those for non-renewables. Nonetheless, an energy basis calculation offers a promising initial comparison of renewables to non-renewables.

TECHNOLOGY FOR BIO-BASED PRODUCTS IN THE BIOREFINERY

Conversion of renewable carbon to chemical products falls into two general categories:

Conventional processing—renewables can be converted to marketplace chemicals using conventional chemical transformations. This approach was frequently used in the early part of the 20th century. It offers the advantage of using the vast knowledge possessed by today's chemical industry for the conversion of a new feedstock source to products; however, the technology development issues described earlier have hindered greater use of these kinds of approaches.

Biocatalysis—biocatalysis offers the benefits of proceeding under mild conditions, frequently in aqueous reaction media, and the use of readily available

Table 2 Projected raw material costs for the biorefinery.

Energy source	Price	$/GJ
Petroleum	$50/bbl	8.7
Gasoline	$1.67/gal	13.7
Natural gas	$7.50/ft^3	7.9
Coal	$20/ton	0.9
Coal/carbon capture	$106/ton	4.8
Electricity	$0.04/kWhr	11.1
Soy oil	$460/ton	13.8
Corn kernels	$82/ton ($2.30/bu)	6.6
Cellulosic crops	$50/ton	3.0

Source: From Nat. Biotechnol.[9]

renewable feedstocks as starting materials; however, issues of separation, expense, productivity, maintenance of organisms, and new capital investment have so far limited bioprocesses in the chemical industry, except where no other alternatives are available.

Significantly absent from the standard tools available for bioprocessing is the use of non-biological catalysis. Catalysis is a powerful technology for the petrochemical-based chemical industry with 80–90% of all chemical processes involving at least one catalytic step. It is likely that chemical catalysis tailored for renewables will see increasing use as a tool for within the biorefinery.

EXAMPLES OF CHEMICAL PRODUCTION IN THE BIOREFINERY

Conventional Processing

Conventional transformation of renewable feedstocks to products is not common in the chemical industry. In many cases, the products are those that can be isolated from existing renewable resources without further structural transformation. Examples include extractives from the pulp and paper industry used in the production of turpentine, tall oils, and rosins and the production of oils from corn or other oil crops or starch-based polymers. Other products are made by simple derivatization of the materials found naturally occurring in biomass. Industry produces a number of chemicals, including a wide range of cellulose derivatives such as cellulose esters and ethers, rayons, and cellophane. A few well-known routes exist for the conversion of renewables to low-molecular-weight monomeric products. Glucose is converted to sorbitol by catalytic hydrogenation, and to gluconic acid by oxidation. Furfural is manufactured by the acidic dehydration of corn by-products. Xylitol is also produced from xylose by hydrogenation. Vanillin and DMSO have been commercially produced from lignin.[10,11]

A number of promising laboratory-scale approaches to chemical products from renewable carbon have appeared. Chemical hydrolysis of oligo- and polysaccharides has been reported as an alternative to biochemical glucose formation from starch or cellulose. Catalytic dehydrations of carbohydrates to furans are known, such as the transformation of xylose to furfural and glucose or fructose to hydroxymethylfurfural (HMF). Further processing of HMF or furfural has been suggested as a route to bio-based diesel-grade fuels. A novel two-phase reaction system has been developed that dehydrates the starting carbohydrate feedstock, and can subsequently be used for hydrogenation of the intermediate furans. This system works effectively for fructose and has been adapted for use with several carbohydrate sources, including glucose, xylose, and polysaccharides. Direct conversion of cellulose to chlorinated furans as intermediates for biofuel production has also been reported.[12]

Certain transformations of carbohydrates afford products that show promise as platform chemicals. These products include: 1) levulinic acid and its derivatives, methyltetrahydrofuran, an automobile fuel extender; δ-aminolevulinic acid, a broad-spectrum biodegradable herbicide and insecticide used in cancer treatment; and diphenolic acid, a material for the production of polymers and other materials; 2) levoglucosan and levoglucosenone as products of sugar pyrolysis; 3) lignin as a chemical feedstock, e.g., in the production of quinones it has been widely suggested as a component in graft copolymers or polymer blends.

Biocatalysis

The most successful route so far for introducing renewables to the chemical industry has been through biotechnology, although its use for the production of high-volume chemicals is only starting to be realized. Most examples of the use of organisms or enzymatic steps in the production of chemicals have been limited to low-volume, high-value fine chemicals and pharmaceuticals such as oligosaccharides, amino acids, purines, vitamins, nicotine, and indigo. This is a sensible first application, given the strict structural requirements of many of these specialty materials. Biocatalysts are generally unchallenged in their ability to provide the stereo-, regio-, and enantioselectivity required by these specialty products. The development of more robust biological systems that can operate in extreme conditions (e.g., temperature, low-water levels, in organic solvents, and under high hydrostatic pressure) will also broaden their applicability. An important example of industrial biotechnology is the Mitsubishi Rayon process for acrylamide production, currently operating on a 3×10^4 metric tons/yr basis by the treatment of non-renewable acrylonitrile with nitrile hydratase.

Several examples of large-scale biotechnological processes are known (Table 3). Some operations have been used for many years, because there is no equivalent non-biological route. Ethanol and lactic acid are of particular

Table 3 Commercial uses of biotechnology.

Compound	Thousand tons/yr
EtOH	26,000
Fructose	1,000
Citric acid	1,000
Monosodium glutamate	1,000
Aspartame	600
L-lysine	350
L-lactic acid	250
L-ascorbic acid	80
Gluconic acid	50
Xanthan	30
Antibiotics	35

interest, because they represent chemicals whose original non-biological production has been almost totally replaced by biochemical manufacture. In addition, the pulp and paper industry has started to incorporate enzyme treatments into their pulping and bleaching sequences, and low-lactose milk (up to 250,000 L daily) can be produced by treating milk with β-galactosidase.[13]

A number of smaller scale processes have also been reported. Non-fermentative, high-yield production of higher alcohol biofuel candidates (such as propanol, butanol, and isobutanol) results from a genetically engineered strain of *Escherichia coli*. Biochemical transformations of cyclohexanol and cyclohexanone lead to the production of caprolactone as an intermediate in the production of adipic acid, a well-known component of nylon polymers offering a route to more selective conversions than the currently practiced nitric acid oxidation. Several organisms ferment glycerol to 3-hydroxypropionaldehyde (3-HPA). Although research on 3-HPA is still exploratory, it is an interesting chemical intermediate as the proposed central component of a network of several high-volume biorefinery products if the cost of glycerol continues to drop as a result of expanding biodiesel production. A potential large-scale use of 3-HPA is in the production of 1,3-PDO by catalytic hydrogenation. Alternatively, 3-HPA is a precursor to acrolein or 3-hydroxypropanoic acid, both precursors of acrylic acid.[14]

CONCLUSIONS

The United States possesses sufficient renewable resources to supply all domestic organic chemical needs without sacrificing traditional applications of renewables in the production of food, feed, and fiber. Further, the experience to date of those industries currently involved in biorefining shows that renewable carbon holds considerable promise as a feedstock complementary to those used by the chemical industry. Domestic sources of renewable carbon are numerous, and with careful attention paid to their sustainable use, improved methods of agriculture and biotechnology will ensure a continuing, environmentally friendly supply. Biorefinery development will play an increasingly important role in the future evolution of the chemical manufacturing. Progress in the use of conventional chemical processing and catalysis for the conversion of renewables to products are areas that will see significant growth as the world turns its attention to new carbon sources as nonrenewable crude oil feedstocks diminish.

REFERENCES

1. Bozell, J.J. Feedstocks for the future—biorefinery production of chemicals from renewable carbon. Clean **2008**, *36*, 641–647.
2. Werpy, T.; Petersen, G. *Top Value Added Chemicals from Biomass. Volume I—Results of Screening for Potential Candidates from Sugars and Synthesis Gas*; U.S. Department of Energy Report no. DOE/GO-102004–1992, U.S. Department of Energy, **2004**. http://www1.eere.energy.gov/biomass/pdfs/35523.pdf.
3. Bozell, J.J.; Holladay, J.E.; Johnson, D.; White, J.F. *Top Value Added Chemicals from Biomass. Volume II—Results of Screening for Potential Candidates from Biorefinery Lignin*; U.S. Department of Energy, Report PNNL-16983, U.S. Department of Energy, **2007**. http://www.pnl.gov/main/publications/external/technical_reports/PNNL-16983.pdf.
4. Eissen, M.; Metzger, J.O.; Schmidt, E.; Schneidewind, U. 10 years after Rio—concepts on the contribution of chemistry to a sustainable development. Angew. Chem. Int. Ed. **2002**, *41*, 414–436.
5. National Corn Growers Association. *The World of Corn*, **2007**. http://www.ncga.com.
6. American Soybean Association. *2008 Soystats*, **2008**. http://www.soystats.com.
7. *Pulp and Paper Factbook—North American Factbook 2001*; Paperloop Publications: San Francisco, 2002.
8. Perlack, R.D.; Wright, L.L.; Turhollow, A.F.; Graham, R.L.; Stokes, B.J.; Erbach, D.C. *Biomass as Feedstock for a Bioenergy and Bioproducts Industry. The Technical Feasibility of a Billion-Ton Annual Supply*; U.S. Department of Energy DOE/GO-102995–2135, ORNL/TM-2005/66; Renewable Fuels Association, U.S. Department of Energy, **2005**. http://www.ethanolrfa.org.
9. Lynd, L.R; Laser, M.S.; Bransby, D.; Dale, B.E.; Davison, B.; Hamilton, R.; Himmel, M.; Keller, M.; McMillan, J.D.; Sheehan, J.; Wyman, C.E. How biotech can transform biofuels. Nat. Biotechnol. **2008**, *26*, 169–172.
10. Fengel, D.; Wegener, G. *Wood: Chemistry, Ultrastructure and Reactions*; Walter DeGruyter: New York, 1984.
11. Murphy, D.J. *Designer Oil Crops. Breeding, Processing and Biotechnology*; VCH: Weinheim, 1994.
12. Chheda, J.N.; Dumesic, J.A. An overview of dehydration, aldol-condensation and hydrogenation processes for production of liquid alkanes from biomass-derived carbohydrates. Catal. Today **2007**, *123*, 59–70.
13. Wilke, D. Chemicals from biotechnology: molecular plant genetics will challenge the chemical and the fermentation industry. Appl. Microbiol. Biotechnol. **1999**, *52*, 135–145.
14. Thomas, S.M.; DiCosimo, R.; Nagarajan, A. Biocatalysis: applications and potentials for the chemical industry. Trends Biotechnol. **2002**, *20*, 238–242.

Biocontrol Agents: Genetic Improvement

Marjorie A. Hoy
Department of Entomology and Nematology, University of Florida, Gainesville, Florida, U.S.A.

Abstract

Biological control of pest arthropods or weeds involves the use of natural enemies to suppress pest populations. Many arthropod populations are effectively suppressed by a complex of naturally occurring parasitoids, predators, pathogens (viruses, bacteria, fungi, protozoa), or entomopathogenic nematodes. In other cases, natural enemies must be introduced (especially if an alien pest arthropod has invaded a new geographic area without its natural enemy complex) in a method called classical biological control.

INTRODUCTION

Natural enemies also can be mass reared and released to suppress pest arthropods, which is especially useful in ephemeral cropping systems in which the natural enemies are unable to suppress pests quickly enough to avoid economic injury; this tactic is called augmentation. It also is possible to modify cropping practices, such as crop rotation, maintenance of ground covers, or modification of pesticide application methods (concentrations applied or application sites), to preserve natural enemies so that they are more effective (conservation biological control).

Despite this diversity of tactics, the efficacy of natural enemies sometimes is limited by their intrinsic genetic characteristics.[1] Under these circumstances it may be appropriate to enhance their effectiveness by genetic manipulation. A primary example is the development of pesticide-resistant predators and parasitoids so that they can survive to suppress pests while other pests are controlled with a pesticide.[2] However, it will be ethically undesirable to develop resistant natural enemies if this increases the use of toxic pesticides with their undesirable environmental and human health effects. Thus, genetic manipulation requires that we understand why a natural enemy is ineffective and have a mechanism with which to alter it (Table 1).

HISTORY OF GENETIC IMPROVEMENT OF PREDATORS AND PARASITOIDS

Genetic manipulation of natural enemies (parasitoids or predators) was first discussed by Mally in 1918. He argued that, like the breeding of crops and domestic animals for artificial agricultural ecosystems, genetic selection could make natural enemies more effective. The first projects focused on determining whether it was, in fact, possible to select natural enemies for specific traits such as increased temperature tolerance, developmental rate, altered diapause and pesticide resistance, but the "improved" strains were not evaluated in the field or employed in practical pest management programs. Thus, the value of genetically modified natural enemies was doubted by most pest management specialists for many years. After 1973 several projects began to focus more on learning how to deploy the genetically modified natural enemies in pest management programs. Most involved selecting a predator or parasitoid of a secondary pest for resistance to pesticides and the resistant natural enemy strain was released so that it could establish and survive the pesticides that were applied to control a primary pest that was not controlled by any other method.[3]

TWO MAIN IMPLEMENTATION STRATEGIES

Implementation of genetically improved natural enemies typically has used one of two strategies: the improved natural enemy is deployed by inoculation, in which the new strain is released one or more times into the environment where it is expected to establish and persist. This strategy has the advantages of reducing rearing costs by requiring fewer individuals for release and the likelihood that fitness of the individuals reared in the field will be high. The inoculation strategy has at least three possible population genetic mechanisms by which it can be achieved: 1) the new strain is released into a new environment where no native populations occur (open niche), and it establishes and provides long-term control; 2) the new strain is released into the environment after the native population is greatly reduced (e.g., after pesticide applications or winter) and the new strain replaces the old, providing long-term control, especially if the wild population is unable to recolonize the release sites; or 3) the new strain is released into the environment and through interbreeding and selection in the field (perhaps with a pesticide), a new hybrid strain is produced that can persist.

The second strategy, augmentation, involves mass rearing the new strain and releasing it periodically. Augmentation is particularly appropriate for greenhouse systems or other crops of high value and short persistence so that the high costs of periodic releases can be justified and achieved effectively. However, it is difficult to imagine that natural enemy augmentation would be cost-effective or technically feasible over vast acreages of relatively low-value crops. Present-day rearing and release technologies would make such releases prohibitively expensive.

Several genetically modified parasitoid and predator species have been used in practical pest management programs, including the use of pesticide-resistant predatory mites to control spider mites in almonds, apples, citrus, and pears, and the use of pesticide-resistant parasitoids to control aphids in walnuts and California red scale insects in citrus.[1,2] The predatory mite (*Metaseiulus occidentalis*) implemented in a California almond IPM program resulted in a significant cost-benefit analysis because fewer pesticides were applied to control spider mites. Unfortunately, relatively few cost-benefit analyses have been conducted on genetically modified natural enemy programs. Genetic manipulation thus remains a minor component of biological control research tactics.

TRANSGENIC TECHNOLOGIES

A benefit of transgenic technology is the ability to insert cloned genes from any prokaryotic or eukaryotic species so that we are no longer limited by the intrinsic genetic variability within that species.[4,5] There also may be disadvantages to using transgenic methods because the steps involved in developing and implementing a pest management program must be modified to deal with the potential risks associated with the release of transgenic arthropods into the environment.[6–8]

Elegant molecular genetic methods have been and are being developed for inserting exogenous DNA into arthropods other than *Drosophila melanogaster*.[4] Most methods use viral or transposable element vectors to carry the foreign DNA into the chromosomes. Interesting genes and regulatory elements that have the potential for conferring useful traits upon specific transgenic arthropods in a tissue-specific manner have been developed for a few pest arthropods, but are lacking for arthropod natural enemies. Ideally, the completion of the *Drosophila* Genome Project will make identifying useful genes for natural enemy improvement much easier.

The breakthroughs in transforming arthropods by recombinant DNA technology have not been matched with significant breakthroughs in our ability to deploy them.[8] Significant efforts will have to be made to develop the data

Table 1 Important questions to answer when developing a genetic manipulation project if it is to be deployed successfully.

Phase I. Defining the problem and planning the project
  What genetic trait(s) limit effectiveness of beneficial species or might reduce damages caused by the pest?
  Can alternative control tactics be made to work effectively and inexpensively, and are they environmentally friendly?
  Can agencies be found to support the high costs and long duration of genetic manipulation projects?
  How will the genetically manipulated strain be deployed?
  What risk issues, especially of transgenic strains, should be considered in planning?
  What advice do the relevant regulatory authorities give regarding your plans to develop a transgenic strain?

Phase II. Developing the genetically manipulated strain and evaluating it in the laboratory
  Where will you get your gene(s)?
  Is it important to obtain a high level of expression in particular tissues or life stages?
  How can you maintain or restore genetic variability in your selection or transgenesis program after obtaining the pure lines?
  What methods can you use to evaluate "fitness" of the modified strains in artificial laboratory conditions that will best predict effectiveness in the field?
  Do you have adequate containment methods to prevent premature release of the transgenic strains into the environment?
  Do you have adequate rearing methods developed for carrying out field tests?
  What release rate will be required to obtain the goals you have set?
  Have you tested for mating biases, partial reproductive incompatibilities, or other population genetic problems?
  If the strain is transgenic, have you obtained approval from the appropriate regulatory authorities to release the strain in the environment or greenhouse?
  How will you measure effectiveness of the modified strain in the field trials?

Phase III. Field evaluation and eventual deployment in practical pest management project
  If the small-scale field trials were promising, what questions remain to be asked prior to deploying the manipulated strain?
  If permanent releases are planned, have all the risk issues been evaluated?
  How will the program be evaluated for effectiveness and potential environmental or other risks?
  Will the program be implemented by the public or private sector?
  What did the program cost and what are the benefits?
  What inputs will be required to maintain the effectiveness of the program over time?

and resources to deploy transgenic strains. Relevant issues include: What is the probability that the transgenic natural enemies (released permanently into the environment) will create future environmental problems? Will transgenes inserted into natural enemies somehow be transferred horizontally through currently unknown mechanisms to pest species? Can we develop mitigation methods or methods for retrieving transgenic natural enemies from the environment should they perform in unexpected ways? The concerns about potential risks will require both researchers and regulatory agencies to accept new responsibilities and conduct creative research.[8,9]

The first experimental field release of a transgenic natural enemy, a strain of the predatory mite *Metaseiulus occidentalis* carrying a molecular marker (lacZ construct with a *Drosophila* heatshock 70 promoter), was authorized in March 1996 by the U.S. Department of Agriculture Animal and Plant Health Inspection Service (USDA–APHIS). The temporary releases took place on the campus of the University of Florida after officials of the USDA–APHIS, Florida Department of Agriculture and Consumer Services, the University of Florida Biosafety Committee, and the U.S. Department of Fish and Wildlife also evaluated potential risks to threatened and endangered species and the environment. The releases took place without incident in March 1996 and no environmental harm was detected. Subsequent releases of transgenic arthropods may receive equivalent or greater scrutiny, especially if the goal is to establish the new strain permanently in the environment. Specific concerns include the potential for horizontal gene transfer (movement of the foreign gene from the transformed species to another species by a variety of little known mechanisms). Other potential risks include changes in host range or host specificity.[10,11]

If we assume that every field release of a transgenic arthropod should be conducted only after thorough peer review by scientists and regulatory agencies, appropriate efforts to contain transgenic insects and mites in the laboratory prior to their purposeful release into the environment should be made.[9] At present there are no guidelines regarding methods to contain transgenic arthropods in the laboratory, but we know that greenhouses or other general laboratory facilities usually are inadequate to prevent accidental releases. Suggestions have been made that the containment facilities for classical biological control projects, which are certified by the USDA–APHIS and state departments of agriculture, would be suitable for transgenic arthropods. Such facilities and procedures were developed to prevent the escape of undesired organisms, including pest arthropods, plant pathogens, and hyperparasitoids. Personnel working in these facilities adhere to specific handling and disposal procedures designed to prevent accidental releases.

CONCLUSIONS

The genetic manipulation of arthropod natural enemies remains a minor component of IPM tactics. Under certain circumstances, such programs have been effective and cost-effective. Recombinant DNA techniques offer exciting new opportunities for improving arthropod natural enemies, but also make genetic manipulation more complex and expensive, primarily because potential risks associated with releases of transgenic arthropods into the environment need to be considered.

The most readily implemented pest management projects employing genetically manipulated natural enemies are those where augmentative releases can be conducted and the organism used in relatively small areas such as temporary cropping systems, or where the natural enemy has a low dispersal rate and can be established in individual orchards, or where the natural enemy is released into a geographic region where the wild strain does not occur. The most difficult projects to implement are those in which the new biotype is expected to replace the endemic population. This is due to the difficulty in developing mass rearing technology, quality control methods, and lack of information on population structure and hidden partial reproductive isolation mechanisms.

One field release of a transgenic natural enemy has occurred but there are no guidelines yet as to how to evaluate the risks of permanent releases of transgenic arthropods and it could take five to ten years of evaluating short-term releases of transgenic arthropods before permanent releases are permitted if we follow the paths taken by transgenic plants and microorganisms.

It is difficult to anticipate the opportunities that might arise over the next few years as improved methods are developed for genetic manipulation of arthropods by recombinant DNA methods. However, deploying a transgenic arthropod natural enemy in a pest management program will remain a challenge, requiring risk assessments, detailed knowledge of the population genetics, biology, and behavior of the species in the field, as well as coordinated efforts between molecular and population geneticists, ecologists, regulatory agencies, and pest management specialists.

REFERENCES

1. Hoy, M.A. Pesticide Resistance in Arthropod Natural Enemies: Variability and Selection Responses. In *Pesticide Resistance in Arthropods*; Roush, R.T., Tabashnik, B.E., Eds.; Chapman and Hall: New York, **1990**; 203–236.
2. Hoy, M.A. Almonds: Integrated Mite Management for California Almond Orchards. In *Spider Mites, Their Biology, Natural Enemies, and Control*; Helle, W., Sabelis, M.W., Eds.; Elsevier: Amsterdam, **1985**; 1B, 299–310.
3. Hoy, M.A. In *Novel Arthropod Biological Control Agents, Biotechnology and Integrated Pest Management*, Proceedings of a Bellagio Conference on Biotechnology for

Integrated Pest Management, Lake Como, Italy, October **1993**; Persley, G.J., Ed.; CAB International: Wallingford, U.K., **1996**; 164–185.

4. Ashburner, M.; Hoy, M.A.; Peloquin, J. Transformation of arthropods—research needs and long term prospects. Insect Molec. Biol. **1998**, *7* (3), 201–213.

5. Hoy, M.A. Insect Molecular Genetics. In *An Introduction to Principles and Applications*; Academic Press: San Diego, **1994**; 1–540.

6. Hoy, M.A. Biological control of arthropods: genetic engineering and environmental risks. Biol. Control **1992**, *2*, 166–170.

7. Hoy, M.A. Criteria for release of genetically improved phytoseiids: an examination of the risks associated with release of biological control agents. Exp. Appl. Acarol. **1992**, *14*, 393–416.

8. Hoy, M.A. Impact of risk analyses on pest management programs employing transgenic arthropods. Parasitol. Today **1995**, *11* (6), 229–232.

9. Hoy, M.A.; Gaskalla, R.D.; Capinera, J.L.; Keierleber, C. Laboratory containment of transgenic arthropods. Amer. Entomol. **1997**, *43* (4), 206–209, 255–256.

10. Tiedje, J.M.; Colwell, R.K.; Grossman, Y.L.; Hodson, R.E.; Lenski, R.E.; Mack, R.N.; Regal, P.J. The planned introduction of genetically engineered organisms: ecological considerations and recommendations. Ecology **1989**, *70*, 298–315.

11. Purchase, H.G.; MacKenzie, D.R. *Agricultural Biotechnology Introduction to Field Testing*; Office of Agricultural Biotechnology, U.S. Department of Agriculture: Washington, DC, **1990**; 1–58.

Bioconversions: Fuel Ethanol

W.M. Ingledew
Lallemand Ethanol Technology, Parksville, British Columbia, Canada

Abstract
The fuel alcohol industry has become a significant energy-generating business based on production of liquid fuel for automobile use. There are a number of advantages that ethanol has over the use of petroleum, and these have led to the use of ethanol as both an oxygenate and a fuel extender to partially replace gasoline. The current processing of starch-containing grains through fermentation and ethanol purification is outlined and a glimpse of the future is provided.

ETHANOL: INTRODUCTION

The fermentative production of ethanol for fuel purposes in North America has accelerated in recent years such that since the early 1980s, production has soared from almost nothing to 12 billion U.S. gal/yr (45 billion L/yr). Over 2.6 billion U.S. gal of new capacity is under construction or idle. At this point in time (December 2009), 202 manufacturing plants exist in the United States, with 12 under construction or expansion.[1] Almost 25 billion liters are produced in Brazil from sugar cane juice, molasses, and sugar beets. In North America, the substrate is corn, with much smaller amounts made from sorghum, wheat, barley, hulless barley, rye, triticale, oats, hulless oats, sweet sorghum, distressed sugars and candy, whey, potatoes, and, in future, cellulose. Ethanol has established itself as the world's number one biotechnological commodity, and *Saccharomyces* yeasts are the most exploited microbes known to man. More than 185 billion L/yr (>13.2 billion U.S. gal/yr) of beverage, industrial, and fuel alcohol is now made around the world. Fuel alcohol makes up 38% of this alcohol.[2]

ETHANOL: THE FUEL

There are many advantages in using ethanol as a fuel. Each gallon of ethanol made displaces at least 2 gal of crude oil[3] and opens up markets for agricultural commodities. Ethanol from corn and all biomass is renewable energy. It improves the balance of payments and decreases the dependency upon foreign oil—leading to energy security. Moreover, ethanol is a liquid fuel that can be splash blended with gasoline and used in virtually any car at the 10% level. Many cars on the roads today are flex-fuel vehicles capable of burning any percentage of ethanol from 0% to 85%, and in some countries like Brazil, they can use up to 100%. Market prices then dictate the type of fuel used in many such cars. In Brazil, this means that many cars use 100% ethanol when petroleum prices are high.

The energy balance of ethanol manufacturing is now positive due to a number of technological advances including energy recycling, lowered costs of distillation, the use of the molecular sieve to remove the last 5% of water, fermentor and other equipment design, lowered costs of enzymes and yeasts, production of alcohol at higher concentrations using VHG technology,[4] the improvement and understanding of yeast nutrition, and plant diversification that leads to novel end products such as horticultural crops and fish, or livestock feedlots using distillers' wet grains and thin stillage as significant feed and water replacements. Novel innovative technologies such as grain fractionation, cold starch hydrolysis, water recovery methods, and the isolation of oils and valuable grain fractions also appear to be successful and lead to further economies.[5,6] As a result, the energy balance from the farm, crop production and transportation to the factory, processing, and purification of the alcohol and by-products is now considered to be positive.[7]

Moreover, ethanol has replaced MTBE (methyl tertiary butyl ether), the petroleum refiner's oxygenate for gasoline, and the second most commonly found chemical in groundwater now known to contaminate waters throughout North America. MTBE is thought to be carcinogenic and has now been banned in more than 30 U.S. states. Ethanol is the oxygenate replacement. Ethanol is also a fuel extender (reducing importation of oil) and has few obnoxious emissions or difficulties if spilled in the environment. Ethanol is environment friendly. The carbon dioxide liberated in fermentation of sugars to ethanol along with carbon dioxide liberated on burning ethanol for fuel closely balance the amount of carbon dioxide taken up by the corn plant in the previous growing season. Thus this fuel is more "Kyoto Accord friendly" than fossil fuels where use of every carbon molecule in oil, gasoline, or natural gas leads to carbon dioxide emissions that are implicated in global warming.

In addition to this, ethanol production is a rural activity. Most plants are located near the corn and employment is therefore created in these rural communities. More income is provided to farms that now claim to be producing energy not just food and feed. Rural revitalization then occurs, and millions of dollars are generated in local farm areas. As corn is the major substrate, it is not surprising to note that most fuel alcohol plants are located in the corn belt. Substrate is the major cost in making ethanol, and prices of corn and subsequent transportation are lowest near the point of harvest.

ETHANOL: THE NEED

To put the industry into perspective, it must be noted that the North American consumption of gasoline is now well over 567 billion L per year. To displace 10% of this, we will need 57 billion L of ethanol. At the current conversion of 2.7 U.S. gal ethanol per 56 lb (47 lb dry) bushel (401 L/tonne) of corn, we need ~141,000,000 metric tonnes of corn. Yearly production of corn is approximately 12.1 billion bushels or 298,000,000 tonnes, and therefore ~47% of the U.S. feed corn crop is needed to supply a 10% ethanol substitution in gasoline. Furthermore, the use of the entire feed corn crop would only provide about 20% of the liquid fuel needed for vehicular traffic. So, ethanol derived from grains will only be an octane enhancer and fuel extender—not a total replacement for gas. This, more than any other information, leads us to the inevitable conclusion that ethanol is only a partial answer for the dwindling stocks of petroleum and the political world's influences on gasoline supply. It is also clear, then, that further work is needed to provide the technologies necessary for the use of cellulosics as sources of glucose for additional ethanol. Cellulose is in excess and poorly used in today's world as an energy source. The job of converting cellulosics into mash suitable for ethanol production and free of inhibitors (acetic acid, furfural, hydroxymethyl furfural, etc.) and unusable substrates is formidable, and hundreds or millions of dollars have already been spent in research with only moderate progress. These are opportunities for biotechnologists—needed soon to alleviate the rising costs of the fuel of preference, gasoline.

Cellulose utilization will involve genetic engineering of plants and microorganisms as well as acceptance of these technologies in the world of tomorrow. More research will be needed on the isolation of valuable commodities from corn and other grains—prior to the use of the grains in the total "biorefining" process. Although there are patented microorganisms already, most grow poorly in industrial media, are influenced greatly by liberated chemicals from cellulosic residues, are unable to use effectively the pentose sugars found in wood, and do not make industrially significant levels of alcohol economically recoverable by distillation and molecular sieves. Noteworthy progress has been made in the development of cellulase enzymes at a price almost suitable for commercial use, but less understood is the total process, the economics, and the organism that will successfully work to produce enough ethanol at the right concentration.

ETHANOL: THE CURRENT PROCESS

The process of making ethanol is multidisciplinary. Knowledge of plant science, agricultural technologies, biochemistry, microbiology, engineering, food science, animal science, economics, accounting, and the environmental regulations of each state or province are all required. A distinct lack of people trained in the science of ethanol production and a lack of supporting resources during the current expansion of this industry are becoming a major impediment to new projects.

The process of making alcohol for fuel can be done in two major ways—dry grind and wet milling.[8,9] The dry grind process (Fig. 1) now makes up over 80% of production even though at one time wet millers made well over 50% of all alcohol destined for fuel. Most new plants are dry grind facilities. The dry grind process[8] begins with feed corn hopefully free from mold contamination and most of the dockage (weed seeds, chaff, and other materials). Corn is hammer-milled to a specific range of screen sizes, and then "mingled" with water and ammonia/ammonium hydroxide (pH adjustment and yeast nutrient). A high temperature alpha amylase enzyme is added and allowed to begin the process of starch conversion to dextrins. The mash is then heated—often in a jet cooker to over 100°C to solubilize (gelatinize) the starch, held for a defined period, and then cooled to allow a second alpha-amylase addition to convert all residual starch to dextrins. Mash is then cooled to about 30°C and passed to the fermentor (normally batch not continuous) where yeast and yeast nutrients are added. Glucoamylase enzyme is supplied to break down dextrins to glucose, which yeasts (simultaneous saccharification and fermentation) are able to take up and convert through a series of 12 internal enzymes that convert the sugar to ethanol and carbon dioxide at ~90% yield. Losses in yield result from the production of glycerol, new yeast cells, fusel (higher) alcohols, and other minor end products, and via losses in processing. The fermentation broth at end fermentation contains from 10% to 20% v/v ethanol (depending on the use of normal or very high gravity fermentation technology), and is passed to the beer still that leads to separation of whole stillage production from 100 proof alcohol (50% v/v). The whole stillage is centrifuged or screened to separate most of the solids from the thin stillage. The yeasts are found in both fractions and contribute almost a 5% increase in protein content to the resultant grain. The wet grain then goes for drying (or is fed wet to animals close to the factory). The thin stillage containing some yeast and small grain particles (proteins and cellulosic residues) can be recycled as backset replacing fresh water used in the next

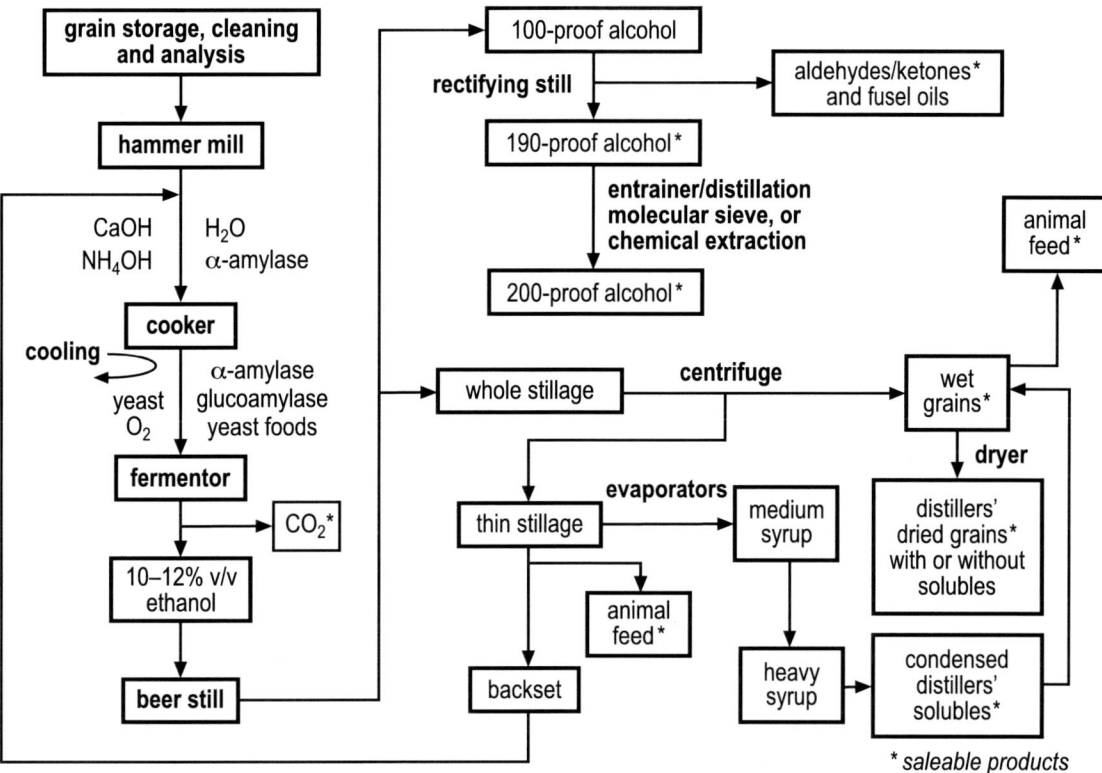

Fig. 1 Dry grind corn-to-ethanol production using continuous mash production upstream and continuous downstream processing but batch fermentation. Saleable products include ethanol (190 or 200 Proof), thin stillage, wet grain, distillers dried grain (with or without solubles), and perhaps carbon dioxide.[10]

slurry preparation. Some thin stillage is evaporated in the plant to form a syrup that is added to the wet grain and dried to form distiller's dried grain with solubles (DDGS) or the grain can be dried without thin stillage addition to make distiller's dried grain (DDG). DDG and DDGS are used as animal feed with their enriched protein contents but much lower starch than original grain. The 100 proof alcohol is passed into a second still where more water is removed forming 190 proof alcohol (95% v/v). The last 5% of water is then removed by passing distillate though a molecular sieve that, due to the pore size of the beads, is capable of separating water molecules from ethanol molecules. When one column is saturated with water, it is regenerated and a second (or third) column is then used to remove further water. It is important to note that most alcohol plants are continuous on the front end (mash production) and on the back end (alcohol purification and distiller's grain production) but almost 4/5 of these plants use a batch fermentation process where each fermentor is separately washed, sanitized, filled, fermented, emptied, and re-cleaned (Fig. 1). Enough fermentor volume must be supplied to take all the mash continuously produced over the time frame needed for fermentation, so multiple batch fermentors are employed. In other plants, particularly when wet milling is used, a train of fermentors is common such that the mash moves in sequence from fermentor 1 to 2 to 3 to 4—such that the medium residence time and fermentation rate match the flow rates of mash production and the flow rate of beer processing to collect the alcohol and the spent grains.

It should be noted that the carbon dioxide from most plants is not collected. In some cases where economics are best, carbon dioxide is collected as off gas, and processed to remove air and other gases such that the purified gas can be used in soft drink manufacture and for industrial uses that include oil recovery. This process is expensive and is probably only done when equipment costs and carbon dioxide supply and demand warrant the expense.

The wet milling process[9] varies significantly from dry grind ethanol technology. The process has been used for over 150 years to isolate corn starch for food purposes and for the production of syrups (corn syrup or glucose syrup, as well as high fructose corn syrup—a novel biotechnology of its own). Corn is milled to create process streams of steep water, germ (mostly oil), fiber, gluten, and starch. Oil and gluten are revenue-generating by-products to this industry. The process begins with a long steep of corn during which time soluble materials leach into the steep water. Fiber (bran), germ, and gluten (protein) are removed leaving a purity of starch of about 99%. The equipment needed in wet milling is far more sophisticated than in dry milling in order to process each of the five major components. Starch can then be used as food, as a substrate for producing high-fructose corn syrup, or as a substrate (with recycled corn steep liquor) for ethanol production. In today's world, the fate of starch is tied to economics of the possible end-use products. Wet milling companies almost always use continuous fermentors in trains of 4–5 that may or may not be operated in balanced growth and at optimal ethanol production rates.

The fuel alcohol industry has become an important part of energy security in the United States and has helped countries in their quest to reduce CO_2 emissions and to rejuvenate rural economies. The U.S. model is stimulating similar activities in other areas of the world. The industry will continue to grow over the next few years based on use of starch-containing grains, but future expansion will be limited due to the finite amounts of the feed-grade starchy substrates available. Fuel alcohol derived from grain will remain a significant bio-based part of the energy market, but in time will be augmented by novel cellulose technologies that will take this fuel to heights unimagined even today. The future of the industry and even greater reduction in petroleum usage will depend on cellulose-based biotechnologies.

REFERENCES

1. Renewable Fuels Association, http://www.ethanolrfa.org/industry/locations/ (accessed December 3, 2009).
2. Ingledew, W.M.; Lin, Y-H. Ethanol from starch-based fendstocks. Vol. 3. Industrial Biotechnology, Elsevier, Oxford, UK (in press).
3. The California Energy Commission, http://www.energy.ca.gov/gasoline/whats_in_barrel_oil.html (accessed November 17, 2006).
4. Ingledew, W.M. Alcohol production by *Saccharomyces cerevisiae*: a yeast primer. In *The Alcohol Textbook*, 3rd Ed.; Jacques, K.A., Lyons, T.P., Kelsall, D.R.; Eds.; Nottingham University Press: Nottingham, UK, 1999; 49–87.
5. Reidy, S. Beyond status quo. Broin introduces technology to improve corn to ethanol process. Biofuels J. **2005**, *4Q*, 98–100.
6. Singh, V.; Johnston, D.B.; Naidu, K.; Rausch, K.D.; Belyea, R.L.; Tumbleson, M.E. Comparison of modified dry-grind corn processes for fermentation characteristics and DDGS composition. Cereal Chem. **2005**, *82* (2), 187–190.
7. Farrell, A.E.; Plevin, R.J.; Turner, B.T.; Jones, A.D.; O'Hare, M.; Kammen, D.M. Ethanol can contribute to energy and environmental goals. Science **2006**, *311*, 506–508. Erratum published June 23, 2006.
8. Kelsall, D.R.; Lyons, T.P. Grain dry milling and cooking procedures: extracting sugars in preparation for fermentation. In *The Alcohol Textbook*, 4th Ed.; Jacques, K.A., Lyons, T.P., Kelsall, D.R., Eds.; Nottingham University Press: Nottingham, UK, 2003; 9–21.
9. May, J.B. Wet milling, process and products. In *Corn Chemistry and Technology*; Watson, S.A., Ramstad, P.E., Eds.; American Association of Cereal Chemists, Inc.: St. Paul, MN, 1999; 377–397.
10. Ingledew, W.H.; Austin, G.D.; Kelsall, D.R.; Kluhspies, C. The alcohol industry: How has it changed and matured? In *The Alcohol Textbook*, 5th Ed.; Ingledew, W.H.; Kelsall, D.R.; Austin, G.D.; Kluhspies, C. Eds.; Nottingham University Press: Nottingham, UK, 2009; 1–6.

Biodiesel: Enzymatic Production

John Birch
Martin Bell
Department of Food Science, University of Otago, Dunedin, New Zealand

Abstract

The alkali-catalyzed transesterification of lipids to produce biodiesel, thereby replacing fossil fuel usage with renewable sources, is a commercial reality; however, along with bioethanol production to replace petroleum fuel, large areas of arable land are needed, competing with food production resources and limiting biofuel production to approximately 10% of fuel demand. The biodiesel process also requires a relatively pure lipid source. Use of enzymes (lipases) to effect the conversion of triglycerides to acylesters (biodiesel) is an attractive alternative. Whole organisms along with immobilized extracellular and intracellular enzymes on a suitable support can utilize conventional and impure feedstocks, need not require expensive agricultural land, and potentially need less downstream operations to recover and separate the alcohol and glycerol present in the reaction mixture. Hurdles to industrial implementation of isolated enzymes and modified organisms are the cost of the enzyme and its reusability, and the yield and lipid composition of oleaginous organisms to suit biodiesel production.

INTRODUCTION

Production of biodiesel has gained importance as a renewable energy resource to replace conventional fossil fuels and provide environmental benefits. To date, commercialization of biodiesel production has largely been based on alkali transesterification processes that rely on feedstock purity, removal of glycerol by-product, and efficient recycling of the catalyst.[1,2] Enzymatic transformation of triglycerides to alcohol esters of fatty acids allows use of impure and waste feedstock and simplifies recovery of alcohol and glycerol;[2] however, commercialization requires solutions to issues such as cost-effective enzyme systems and maximization of yields. This entry considers feedstock sources, whole organisms, isolated lipase systems, and efficiencies of conversion with the potential to produce biodiesel from biomass that replaces the need for extensive oilseed production and potentially, lipid processing before alcoholysis.

BACKGROUND

In the early 1900s, Rudolf Diesel used peanut oil in his original motor;[3] however, this was not viable commercially as the oil did not perform well under cold conditions, clogging up the engine owing to its viscosity. Petroleum diesel fuel was used in the motor instead. During the Second World War, troops in South Africa used waste frying oil to power their trucks and tanks. The oil was not of high purity and clogged the engines; however, by converting the oil into methyl esters (alcoholysis), a suitable fuel could be made.[4] Thus, biodiesel was formed and the term has come to represent the alcohol esters derived from triglycerides (Fig. 1), rather than including the unprocessed fats. The pressure and heat that had to be used to make this early form of biodiesel was huge, being based on an acid-catalyst method. The process was performed batch-wise and was very dangerous.

The perception has existed linking diesel use with fossil fuel shortages and harm to the environment. Diesel fuel has high sulfur content and is sourced from organic stores deep underground, causing acid rain and adding to the Greenhouse Effect through the release of carbon dioxide. During the early 1980s, biodiesel and other alternative fuels had a resurgence of interest. The most successful preparations involved a base-catalyzed interesterification reaction using sodium hydroxide or sodium methoxide as catalysts (alkali catalysis, Fig. 1). There are now many types of processes that can be utilized to make biodiesel, which are safe and continuous.[1] This leads to more biofuel being made at reduced cost with the fuel available for retail sale typically as a blend with traditional diesel, notably in France, Germany, and the United States. Fuel standards have been established for biodiesel fuel in a number of countries. Many countries now subscribe to the Kyoto Protocol, set up in 1990, to help combat greenhouse gas emissions.

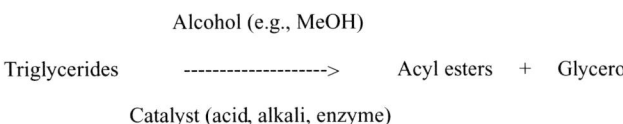

Fig. 1 Production of fatty acid esters from transesterification of lipids.

CONVENTIONAL FEEDSTOCKS AND ALKALI CATALYSIS

While process developments have seen the commercialization of biodiesel production, refined raw materials suitable for production can comprise up to 80% of the product cost.[2] The established method for conversion of plant and animal fats into biodiesel employing sodium hydroxide and methanol is severely compromised by the presence of free fatty acids and water, which destroys the catalyst. Only added-value rendered tallow is suitable as an animal-based feedstock. Commercially, rapeseed and soy oil crops are favored, requiring large tracts of agricultural land for production that can realistically only substitute for a fraction of existing transport fuels.[1]

ALTERNATIVE FEEDSTOCKS AND ENZYMATIC ALCOHOLYSIS

Many meat processing plants produce various grades of tallow (Fig. 2) and hard-to-treat lipid-rich wastes such as stick liquors, grease-trap wastes, and dissolved air flotation scums with no commercial value and requiring bioremediation treatment before disposal. If the process can handle impure feedstock, use of waste frying oils and microorganisms, such as microalgae, bacilli, molds, and yeasts, as lipid sources become feasible.[5]

Enzymatic catalysis for producing biodiesel is gaining popularity. As most enzymes work best at temperatures around 30–50°C, energy consumption costs are relatively low. Enzymes can work on degraded feedstock without significant loss of activity or the need for extra processing.

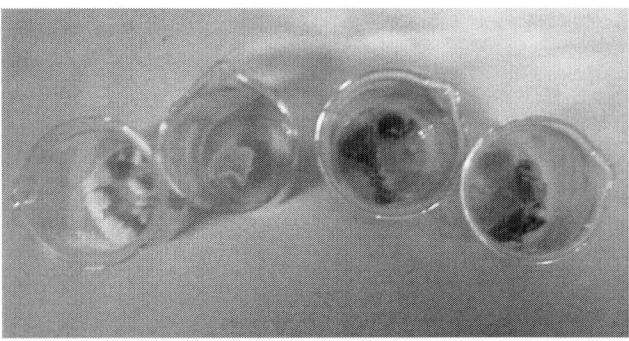

Fig. 2 Margarine fat and G1, G2, and G3 inedible tallow.

The catalyst used for the production of biodiesel is a triacylglycerol acylhydrolase (EC 3.1.1.3), a lipase. There are various biological sources of the lipase, though the enzyme is found in most organisms that hydrolyze fat.[6] These microorganisms include *Burkholderia cepacia* (formerly *Pseudomonas cepacia*), *Candida rugosa* (formerly *Candida cylindracea*), *Pseudomonas fluorescens*, *Mucor javanicus*, and *Rhizopus niveus*. Lipases from these microorganisms are the most studied of the biodiesel-producing lipases. All of these lipases have also been successfully cloned into *Escherichia coli*, enabling large amounts of enzyme to be produced. This enables the technology to become more viable, especially as some of the original organisms may be considered pathogenic.

Microorganisms for Biodiesel Production

The U.S. Department of Energy,[7] in the search for ameliorating CO_2 generation, investigated bubbling the gas produced from coal-burning power stations through algal ponds (Fig. 3). Under the correct conditions, algae will produce natural triglycerides, which then can be processed into biodiesel.

Trials performed in the middle of a desert showed that up to 50 g of fat can be produced per square meter per day. This equates to around 36,500 kg of fat for a 2000-m^2 pond per year; however, microalgal oils may need modification for biodiesel production owing to their high degree of unsaturation, which could potentially oxidize during storage. When oleaginous yeasts are grown with an excess of carbon and a limited quantity of nitrogen (high C:N), they may accumulate up to 60% of their cellular dry weight as intracellular lipid.[5] *E. coli* can also be genetically modified to produce ethyl esters from external glucose and fatty acid carbon sources.[1]

Lipases Used Directly with Lipid Feedstock

The main advantage of using a lipase system to produce biodiesel is that it does not discriminate among tri-, di-, and mono-acylglycerides and free fatty acids. The disadvantages of using the lipase system are the cost of producing the enzyme and the susceptibility of the enzyme to damage from the alcohol. Therefore, the enzyme must be immobilized to prevent its removal along with the biodiesel exiting the system; enzyme reuse is a necessity to make the process feasible.[8] The enzyme support must be porous to allow transfer of enough fat and alcohol to keep the reaction going, but not too much so as to deactivate the lipase from overexposure to methanol. Also, if there is a high water content, free fatty acids will form rather than methyl esters. The lipase works at the polar/non-polar interface.[8,9] A water activity between 0.5 and 0.6 is considered optimum for the reaction with a lipase from *Burkholderia cepacia* on

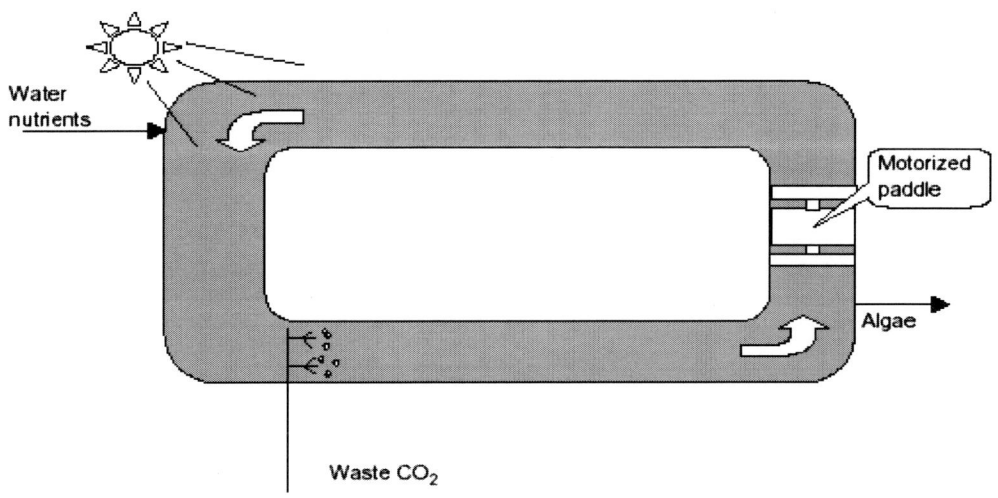

Fig. 3 Basic design of the algal pond. **Source:** From *A Look Back at the U.S. Department of Energy's Aquatic Species Program.*[7]

a sodium-rich phyllosilicate clay.[8] Addition of an alkylamine to occupy the charge sites of the clay and tetramethyl orthosilicate to aid the formation of a gel along with sodium fluoride improves enzyme activity in the matrix when conducted in a solvent-free environment at 40°C to transmethylate recycled restaurant grease in the presence of excess methanol.[9] Other lipase sources studied include *Thermomyces lanuginose* and *Candida antarctica*. Some of the immobilizing matrices that have been investigated include granulated silica, macroporous acrylic resin, Eupergit C, Celite, and phyllosilicate clay.

Changing the lipid source, enzyme source, and alcohol has shown that the lipases have higher activity with different alcohols and that solvent-free esterifications can be advantageous.[9] Novozyme 435 proved best for vegetable oils at 30–35°C with methanol when there was a five-fold excess of alcohol over lipid, while Lipozyme TL-IM had higher activity with ethanol and Lipozyme RM-IM had higher activity with butanol. When washing out the immobilized enzyme with hexane, the activity remained at 90% over seven cycles.[10] A study on transesterification of soybean oil using methanol and ethanol also found that the lipase from *Pseudomonas cepacia* was the best of nine lipases screened.[11] The lipase immobilized in a sol–gel support worked optimally under the following regimes: 35°C, 1:7.5 oil/methanol molar ratio, 0.5-g water, and 475-mg lipase for the reactions with methanol, and 35°C, 1:15.2 oil/ethanol molar ratio, 0.3-g water, and 475-mg lipase for the reactions with ethanol.

CONCLUSION

Industrial biodiesel production to replace fossil fuels is presently only possible using relatively pure lipid sources that compete with conventional food production. Oleaginous microorganisms, genetically modified microorganisms, and the use of purified, immobilized lipases have the potential to add to the commercial biodiesel-production armory in a more sustainable way. Costs of lipase production, the development of multiple reuse systems that generate high yields with extracellular lipases, and simpler downstream processing are continuing challenges; however, development of the technology to deliver lipid feedstocks employing whole cells for carbon dioxide capture is bringing cost-competitive alternatives to diesel fossil fuels ever closer. Integration of processes to utilize by-products, such as animal feed and bioethanol or methane production from residual biomass, along with continued development in glycerol and alcohol recovery and refinement will enhance competitiveness. Improvement of microalgal lipid composition, pre- or postprocessing, is advantageous for commercialization.

REFERENCES

1. Antoni, D.; Zverlov, V.; Schwarz, W. Biofuels from microbes. Appl. Microbiol. Biotechnol. **2007**, *77*, 23–35.
2. Vembanur, S.; Srinivasan, R.; Narasimhan, L.; Muthukumar, K. An overview of enzymatic production of biodiesel. Bioresour. Technol. **2008**, *99*, 3975–3981.
3. Ma, F.; Hanna, M. Biodiesel production: a review. Bioresour. Technol. **1999**, *70*, 1–15.
4. Wright, H; Segur, J.; Clark, H.; Coburn, S.; Langdon, E.; DuPuis, R. A report on ester interchange. Oil Soc. **1944**, *21*, 145–148.
5. Meng, X.; Yang, J.; Xu, X.; Zhang, L.; Nie, Q.; Xian, M. Biodiesel production from oleaginous microorganisms. Renewable Energy **2009**, *34*, 1–5.

6. Schmidt-Dannert, C. Recombinant microbial lipases for biotechnological applications. Bioorg. Med. Chem. **1999**, *7*, 2123–2130.
7. Sheehan, J.; Dunahay, T.; Benemann, J.; Roessler, P. *A Look Back at the U.S. Department of Energy's Aquatic Species Program—Biodiesel from Algae*; close-out Report. 325pp; NREL Report No. TP-580.24190
8. Hsu, A.; Foglia, T.; Shen, S. Immobilization of Pseudomonas cepacia lipase in a phyllosilicate sol–gel matrix: effectiveness as a biocatalyst. Biotechnol. Appl. Biochem. **2000**, *31*, 179–183.
9. Hsu, A.; Jones, K.; Foglia, T.; Marmer, W. Immobilized lipase-catalysed production of alkyl esters of restaurant grease as biodiesel. Biotechnol. Appl. Biochem. **2002**, *36*, 181–186.
10. Rodrigues, R.; Volpato, G.; Wada, K.; Ayub, M. Enzymatic synthesis of biodiesel from transesterification reactions of vegetable oils and short chain alcohols. J. Am. Oil. Chem. Soc. **2008**, *85*, 925–930.
11. Noureddini, H.; Gao, X.; Philkana, R. Immobilized *Pseudomonas cepacia* lipase for biodiesel fuel production from soybean oil. Bioresour. Technol. **2005**, *96*, 769–777.

Biological Control: Successes and Failures

Heikki Hokkanen
Department of Applied Biology, University of Helsinki, Helsinki, Finland

Abstract
Biological control of pests has been actively practiced for the control of pests, weeds, and plant diseases for more than 100 years. It has had some spectacular successes. However, the obtained successes are only the tip of the iceberg of all the work carried out in the field. A major ecological and economic challenge is to improve the ratio of successes in biological control, while retaining the excellent safety record of this approach to pest control.

INTRODUCTION

Biological control of pests has been actively practiced for the control of pests, weeds, and plant diseases for more than 100 years, and it has had some 150 spectacular successes,[1] which in economic terms have been just as impressive as in ecological terms: the calculated return for investment is 32:1, while for other control methods the ratio is around 2.5:1.[2,3] However, the obtained successes are only the tip of the iceberg of all the work carried out in the field. To date, more than 6000 introductions of alien natural enemies have been carried out, worldwide.[4] It is estimated that only about 35% of all introduced biocontrol agents have become ecologically established in the target ecosystem, and only 60% of these have provided any economic or biocontrol success.[3,5] Of all the individual biocontrol projects, only 16% have resulted in complete control of the target pest.[6] A major ecological and economic challenge is to improve the ratio of successes in biological control, while retaining the excellent safety record of this approach to pest control.

GENERAL PRINCIPLES

While it has been shown that biological control can be effective in any climate, ecosystem, and crop, the factors determining success or failure remain largely unknown, and often are economic rather than ecological in nature.[1] Very few general principles to improve the efficacy and predictability of biological control have emerged; these include better ecological background knowledge, genetic improvement (in particular, genetic engineering) of biocontrol agents, and the utilization of new ecological associations in selecting the biocontrol agents.[7] The genetic engineering of biocontrol agents, especially insects, is still in its infancy and cannot be expected to improve the success ratios in the foreseeable future. In contrast, the new association principle has—usually unknowingly—been used for a long time, and is increasingly employed to find more effective natural enemies for current biological control programs.

NEW ASSOCIATIONS

The standard biological control principle is to reestablish the ecological balance between an exotic pest and its natural enemies occurring in their country of origin (the "old association approach").[2,3] It has been argued, however, that this is an inefficient way of practicing biological control, because due to an evolved long-term equilibrium between the pest and the natural enemy, the control agent only seldom is very efficient.[6] To find more effective enemies one should search among agents that do not share an evolutionary history with the target pest (the "new association" approach). Such natural enemies can be found, for example, for the target pest in areas where the pest has been introduced only recently, or among enemies attacking related species in other geographical areas.[6]

EVIDENCE FOR IMPROVED EFFICACY

Analysis of past biocontrol successes and failures have indicated that when employing the new association principle it is possible to increase the success ratio by at least 75%.[6] More detailed studies showed that some natural enemy groups may be particularly attractive as new association agents (Table 1). Such analyses are, however, often confounded by the fact that new association agents seldom have been considered as the primary choice in biological control, and consequently, usually five to seven old association agents are introduced before a new association agent is tried. In addition, on average much greater numbers (two- to fourfold) of old association agents are normally introduced (Table 1), which further increases the probability

Table 1 Comparisons of biological control introductions with old and new association control agents utilizing Tachinidae, Braconidae, and Eulopidae.

	Proportion (%) of successes of all cases (introductions)		Total number of cases		Bias in the release numbers[a]
	Old	New	Old	New	
Tachinidae	10.9	17.1	92	41	3.8-fold
Braconidae	17.2	14.4	169	97	1.6-fold
Eulopidae	28.6	35.7	56	28	1.7-fold
Overall	17.4	18.7	317	166	

[a]Indicates how many more individuals on average of old association agents were released in the introduction projects, compared with new association agents. In the case of Tachinidae and Braconidae the mean number of released new association agents was below 5000 individuals, which is considered to be the necessary number to ensure a fair chance for the natural enemies to establish themselves. **Source:** From Hokkanen, H.M.T., unpublished data.

of biocontrol success, and biases the analyses against new association agents. Therefore, the estimate for improving the success ratio appears to be conservative.

Several spectacular, well-documented biocontrol successes that have employed new association control agents are known, and these include the complete control of serious pests such as the sugarcane borer *Diatraea saccharalis* in the Caribbean, coconut spike moth *Levuana iridescens* in Fiji, southern green stink bug *Nezara viridula* in Hawaii, the moth *Oxydia trychiata* in Colombia, and several scale insect species in California, Greece, and Australia.[7] Further, more detailed examples will be given below on new research with good prospects of success utilizing this approach.

RECENT CASES EMPLOYING NEW ASSOCIATIONS

Eurasian Watermilfoil

The Eurasian watermilfoil (*Myriophyllum spicatum*) was introduced into North America several decades, possibly 100 years, ago. It grows rapidly, forms a dense canopy on the water surface, and often interferes with recreation, inhibits water flow, and impedes navigation. Herbicides and mechanical harvesting have been used to control infestations, costing $150–$2000 per acre annually in Minnesota.[8]

Sometimes naturally occurring declines of the watermilfoil have been observed. The main causal agent proved to be a native beetle *Euhrychiopsis lecontei*, the milfoil weevil, which subsequently has shown control potential in controlled field experiments. The weevil is a specialist herbivore of watermilfoils, but prefers the Eurasian to its native host, the northern watermilfoil (*M. sibiricum*). Research is in progress to use the milfoil weevil effectively as a biocontrol agent against the Eurasian watermilfoil in North America.[8]

Lantana

Lantana camara is a serious weed of Mexican or Caribbean origin, affecting cropping lands and forest areas in 47 countries. Lantana was the focus of the first weed biocontrol effort in history (1902), and there is an enormous literature on Lantana biocontrol. Several complexes of herbivores have been credited for exerting some degree of biocontrol of the weed (e.g., in Hawaii), many employing new association agents jointly with old association agents. Latest research gives data on the good efficacy and release in Australia of the moth *Ectaga garcia* originating from South America, where it feeds on the related weed *Lantana montevidensis*.[9]

Triffid (Siam) Weed

The triffid weed (*Chromolaena odorata*) is a perennial shrub native of tropical America. In recent decades it has become a serious pest of humid tropics around the world.[10] It spreads rapidly in lands used for forestry, pasture, and plantation crops and can reach a height of three meters in open situations and up to eight meters in forests. For more than two decades the triffid weed has been the subject of intensive research as a target for biological control. However, so far all attempts at biocontrol of *C. odorata* have failed. Recently the new association biological control agent, arctiid moth *Pareuchaetes aurata aurata* collected from *C. jujuensis* in South America, was considered as more promising than the related moth *P. pseudoinsulata*, an old association control agent previously thought of as one of the best biocontrol candidates.[10]

Southern Green Stink Bug

The biological control of the southern green stink bug (the green vegetable bug) (*Nezara viridula*) in Australia,

New Zealand, and Hawaii has been heralded as a landmark example of classical biological control.[11] An egg parasitoid—old association agent—*Trissolcus basalis* and a tachinid fly—new association agent—*Trichopoda pennipes* have jointly provided these successes. Control by the fly has been considered as relatively more important, and indeed, in Australia where the fly has failed to establish, the control is poor and the bug remains a serious pest. Currently in Australia another new association tachinid fly, *Trichopoda giacomellii*, is being released after research showed it has excellent potential for control.[12]

Citrus Leafminer

Citrus agroecosystems have numerous potentially damaging pests often maintained under substantial to complete biological control by both old and new association agents. The citrus leafminer *Phyllocnistis citrella*, native to Asia, has spread rapidly throughout the citrus growing areas of the world in recent years.[13] It arrived in Florida in 1993 and in less than one year invaded and colonized the entire state. An old association parasitic wasp *Ageniaspis citricola* was introduced in 1994, and after establishment it has held the pest under control with significant help from native parasitoids such as *Pnigalio minio* (new association agent).[13] In some other areas native parasitoids similarly have shown significant control effect on the citrus leafminer (e.g., in Italy). This example illustrates well the fact that invading species often do not become pests, because effective local natural enemies keep them in check.

Tarnished Plant Bug

An ongoing study in the United States has identified as the most important parasitoid of the native pest *Lygus lineolaris*, the tarnished plant bug, the exotic species *Peristenus digoneutis*, originally introduced for the control of related introduced mirid plant bugs.[14] This example serves well to point out the importance of native pests, which in most if not all areas form the majority of all pest species. As old association biological control agents seldom can be utilized for the control of native pests, their biocontrol by introduced natural enemies has attracted relatively little attention and, indeed, only three decades ago was considered an impossible task. Several recent examples, usually utilizing new association control agents, show that biological control can work against native pests just as well as against exotic ones.

FUTURE PROSPECTS

Compared with chemical control, the success rates of biological control are outstanding. While only about one out of 15,000 tested chemicals ends up as a chemical pesticide meeting the requirements of efficacy and safety, approximately one out of seven introductions of natural enemies has been successful using old associations.[15] Using new association control agents this rate could still be increased to about one out of four, while the array of potential natural enemies is also substantially larger providing a wider choice. In addition, the potential uses for natural enemy introductions are broadened to include the control of native pests.

A major concern with respect to all biological control introductions is the question of nontarget safety. Biological control has an excellent record of safety[3,16] and it covers the new association agents as well: there have been some 1500–2000 introductions already (out of 6000) that have involved new association agents.[7] Those extremely few cases where a negative nontarget effect has been suspected as a result of biological control, all involve old association agents; therefore it is clear that new associations can safely be used to help obtain biological control successes at an increasing rate.

REFERENCES

1. Hokkanen, H.M.T. Success in classical biological control. CRC Crit. Rev. Plant Sci. **1985**, *3*, 35–72.
2. Cullen, J.M.; Whitten, M.J. Economics of Classical Biological Control: A Research Perspective. In *Biological Control: Benefits and Risks*; Hokkanen, H.M.T., Lynch, J.M., Eds.; Cambridge University Press: Cambridge, U.K., **1995**; 270–276.
3. *Biological Control: Benefits and Risks*; Lynch, J.M., Hokkanen, H.M.T., Eds.; Cambridge University Press: Cambridge, U.K., **1995**.
4. Waage, J. In *Agendas, Aliens and Agriculture*; Global Biocontrol in the Post UNCED Era, Cornell Community Conference on Biological Control, http://www.nysaes.cornell.edu/ent/bcconf/talks/waage.html (accessed January 5, 1999).
5. Hokkanen, H.M.T. Pest Management, Biological Control. In *Encyclopedia of Agricultural Science*; Academic Press, Inc.: San Diego, **1994**, *3*, 155–167.
6. Hokkanen, H.M.T.; Pimentel, D. New associations in biological control: theory and practice. Can. Entomol. **1989**, *121*, 829–840.
7. Hokkanen, H.M.T. New Approaches in Biological Control. In *CRC Handbook of Pest Management in Agriculture,* 2nd Ed., Pimentel, D., Ed.; CRC Press: Boca Raton, FL, **1991**; II, 185–198.
8. Newman, R.M. Biological control of Eurasian Watermilfoil. http://www.fw.umn.edu/research/milfoil/milfoilbc.html (accessed February 5, 1999).
9. Day, M.D.; Wilson, B.W.; Latimer, K.J. The life history and host range of *Ectaga garcia*, a biological control agent for

Lantana camara and *L. montevidensis* in Australia. BioControl **1998**, *43*, 325–338.
10. Kluge, R.L.; Caldwell, P.M. The biology and host specificity of *Pareuchaetes aurata aurata* (Lepidoptera: Arctiidae), a "New Association" biological control agent for *Chromolaena odorata* (Compositae). Bull. Entomol. Res. **1993**, *83*, 87–94.
11. Caltagirone, L.E. Landmark examples in classical biological control. Annu. Rev. Entomol. **1981**, *26*, 213–232.
12. Coombs, M. *Biological Control of Green Vegetable Bug in Australia and PNG*; Pest Management Current Programs and Projects. http://www.ento.csiro.au/research/pestmgmt/pmp16.htm (accessed October 5, 1999).
13. Timmer, L.W. *Citrus Leafminer Proves to be an IPM Success*; IPM Florida, **1996** Winter. http://www.ias.ufl.edu/~FAIRSWEB/IPM/IPMFL/v2n4/leafminer.htm (accessed February 5, 1999).
14. Day, W.H. Host preferences of introduced and native parasites (Hymenoptera: Braconidae) of phytophagous plant bugs (Hemiptera: Miridae) in alfalfa-grass fields in the northeastern USA. BioControl **1999**, *44*, 249–261.
15. Hokkanen, H.M.T. Role of Biological Control and Transgenic Crops in Reducing Use of Chemical Pesticides for Crop Protection. In *Techniques for Reducing Pesticide Use*; Pimentel, D., Ed.; John Wiley & Sons: New York, **1997**; 103–127.
16. *Evaluating Indirect Ecological Effects of Biological Control*; Scott, J.K., Quimby, P.C., Wajnberg, E., Eds.; CABI Publishing: Wallingford, U.K., **2001**; 261.

Biopesticides

Gavin Ash
School of Agriculture, Charles Sturt University, Wagga Wagga, New South Wales, Australia

Abstract

A biopesticide is a type of augmentative biological control agent in which an inundative application of a living organism is used to kill the target pest. Massive amounts of inoculum of the organisms are applied in an effort to manage the target. These chemicals, although derived from micro-organisms, are simply analogous to synthetic pesticides and do not contain a living organism as an active ingredient. Biopesticides have been used to control a range of pests including insects, weeds, and diseases. The success of this type of control revolves around the cost of production, the quality of the inoculum, and the field efficacy of the organism. Biopesticides are usually developed by commercial companies in an expectation that they will recoup their costs by sale of the product.

INTRODUCTION

A biopesticide is a type of augmentative biological control agent in which an inundative application of a living organism is used to kill the target pest. In this type of strategy, massive amounts of inoculum of the organisms (usually fungi, nematodes, or bacteria) are applied in an effort to manage the target. This is a general term applied to a range of pests and control organisms. Furthermore, this term does not include the use of toxins or secondary metabolites alone applied as pesticides.[1] These chemicals, although derived from micro-organisms, are simply analogous to synthetic pesticides and do not contain a living organism as an active ingredient. A parallel term to biopesticide, used to describe the suppression of the pest, is biopestistat (Fig. 1). These types of biological control agents, when applied to a pest, suppress the population to below an economic threshold or injury level. There is a growing interest in the use of these types of organisms in conjunction with competitive crops to control weeds.[2] However, they do not kill the target organism per se, and so are excluded from the remaining discussion of the term biopesticide.

Biopesticides have been used to control a range of pests including insects, weeds, and diseases. In all of these cases, the biopesticide is packaged, handled, stored, and applied in a fashion similar to that of traditional pesticides.[3] The success of this type of control revolves around the cost of production, the quality of the inoculum, and the field efficacy of the organism. In comparison to classical control, in which the cost of research and development is borne by the community, biopesticides are usually developed by commercial companies in an expectation that they will recoup their costs by sale of the product. This type of strategy can be used against both native and introduced pests.

Biopesticides may be further subdivided based on the type of target pest. For example, a bioherbicide is a biopesticide developed to kill weeds. Further subdivision based on the type of agent used is also common. The term mycoherbicide is used widely to describe the formulation of a fungal agent in a bioherbicide.

Bioinsecticides

More than 1500 species of pathogens have been shown to attack arthropods and include representatives from the bacteria, viruses, fungi, protozoa, and nematodes.[4] Diseases caused by insects have been known since the early 1800s with the first attempts at inundative applications of fungi to control insects being developed in 1884, when the Russian entomologist, Elie Metchnikoff, mass-produced the spores of the fungus *Metarhizium anisopliae*. The majority of entomopathogenic fungi belong to the Deuteromycotina, a group of fungi without known sexual stages. These fungi have been developed as biopesticides, primarily in tropical regions due to their requirements for high humidity for infection.

Bacteria that attack insects can be divided into nonspore-forming and spore-forming bacteria. The nonspore-forming bacteria include species in the Pseudomonaeae and the Enterobacteriaceae. The spore-forming bacteria belong to the Bacilliaceae and include species such as *Bacillus popilliae* and *Bacillus thuringiensis*. *B. thuringiensis* (Bt) has primarily been developed as a biopesticide to control Leptodopteran larva. However, other serotypes of Bt produce toxins that kill insects in the Coleoptera and Diptera as

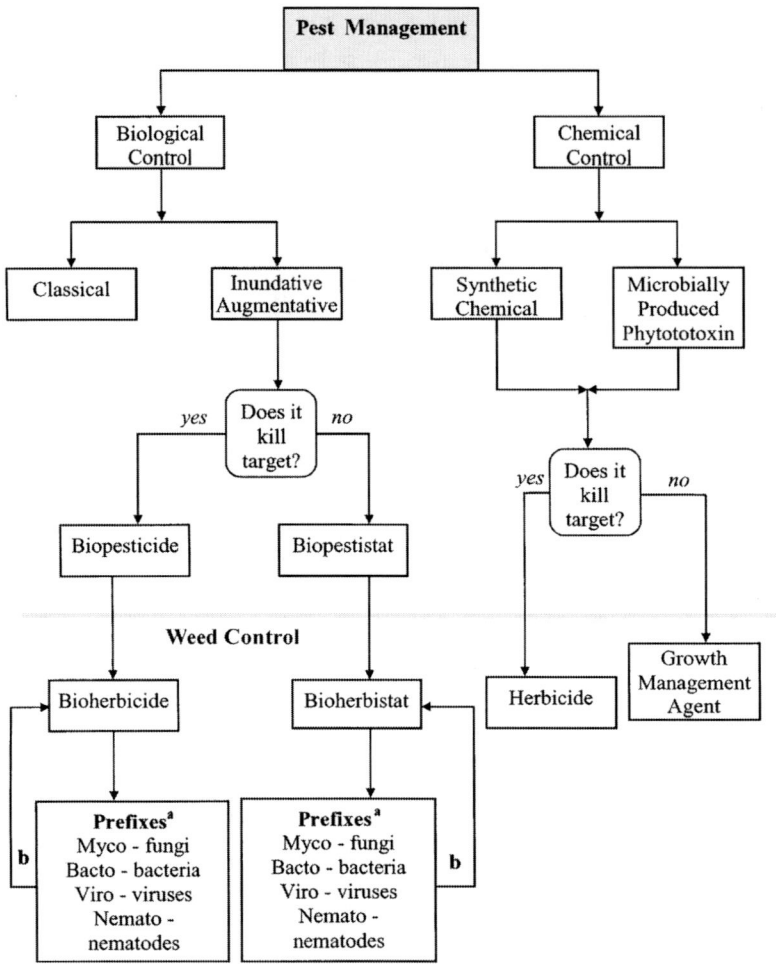

Fig. 1 A hierarachy of terminology in pest management with emphasis on weed control. **Source:** From Biocontrol Sci. Tech.[1]

well as nematodes. Commercial formulations of the bacteria contain living spores of the bacteria.

Entomopathogenic nematodes of the families Steinernematidae and Heterorhabditiae, in conjunction with bacteria of the genus *Xenorhabdus*, have been used successfully deployed as biopesticides, for example BioVECTOR.[4] They are usually applied to control insects in cryptic and soil environments. The nematodes harbor the bacteria in their intestines. The infective third-stage larvae enter the host through natural openings and penetrate into the haemocoel. The bacteria are voided in the insect and cause septicemia, killing the insect in approximately 48 h.

Biofungicides

Biological control of fungi that cause plant disease can be accomplished by a number of mechanisms including antibiosis, hyperparasitism or competition. Additionally, weak pathogens may induce systemic acquired resistance in the host, giving a form of cross-protection. Biofungicides have been used in both the phylloplane and rhizosphere to suppress disease. A biological control agent for the control of foliar pathogens in the phylloplane must have a high reproductive capacity, ability to survive unfavorable conditions, and the ability to be a strong antagonist or be very aggressive. A wide range of bacteria and fungi are known to produce antibiotics that affect other microorganisms in the infection court. Most often these organisms are sought from a soil environment, as this environment is seen as the richest source of antibiotic producing species. Species of *Bacillus* and *Pseudomonas* have been successfully used as seed dressings to control soil-borne plant diseases.[5] Fluorescent Pseudomonads are also often seen as a component of suppressive soils. These bacteria may prevent the germination of fungi by the induction of iron competition through the production of siderophores (iron chelating compounds). These are effective only in those soils where the availability of iron is low. Control of foliar and fruit pathogens such as *Botrytis cinerea*, a pathogen of strawberries, has been accomplished by the foliar application of the soil inhabiting fungus *Trichoderma viride*.[6] This fungus inhibits *Botrytis* using a combination of antibiosis and competition. On grapevines, *Trichoderma harzianum* competes with *B. cinerea* on senescent floral parts, thus preventing the infection of the ovary. It has also been shown to coil around the hyphae of the pathogen during hyperparasitism.[7]

Bioherbicides

Fungi are the most important group of pathogens causing plant disease. Therefore, fungi are most commonly used as the active ingredient in bioherbicides and as such the formulated organism is referred to as a mycoherbicide.[1] The aim of bioherbicide development is to overcome the natural constraints of a weed–pathogen interaction, thereby creating a disease epidemic on a target host(s).[8] For example, the application of fungal propagules to the entire weed population overcomes the constraint of poor dissemination. After removal of the host weed, the pathogen generally returns to background levels because of natural constraints on survival and spread.

The first commercially available biopesticide for the control of weeds was DeVine®, a mycoherbicide for the control of stranglervine in citrus groves in the United States. It was released in 1981.[8] In 1982, a formulation of *Colletotrichum gloeosporioides* f.sp. *aeschynomene*, was released to control northern jointvetch in soybean crops in the United States. Since these early examples there have been a range of organisms investigated with a view to the production of a commercial bioherbicide in Canada, United States, Europe, Japan, Australia, and South Africa. Necrotrophic or hemibiotrophic fungi are usually used as the basis of mycoherbicides, as they can be readily cultured on artificial media and so lend themselves to mass production. Other desirable characteristics of fungi under consideration as mycoherbicides include the ability to sporulate freely in artificial culture, limited ability to spread from the site of application, and genetic stability. In most cases these biopesticides are applied in a similar fashion to chemical herbicides using existing equipment, although the development of specialized application equipment and formulation may improve their efficacy and reliability.

The development of biopesticides relies on agent discovery and selection, development of methods to culture the pathogen (or nematode), creation of formulations that protect the organism in storage as well as aid in its delivery, studies of field efficacy, and methods of storage. Each biopesticide is unique, in that not only will the organism vary but so too will the host, the environment in which it is being applied, and economics of production and control.

There are a number of advantages of the use of biopesticides over the use of conventional pesticides, including the minimal residue levels, control of pests already showing resistance to conventional pesticides, host specificity, and the reduced chance of resistance to biopesticides. This indicates an emerging, strong role for biopesticides in any integrated pest management strategy and an important involvement in sustainable farming production systems in the future.

FUTURE DIRECTIONS

There have been some spectacular successes in the use of biopesticides, despite the perceived constraints to their deployment.[9] In the past, biopesticides have been expected to behave in the same way as synthetic pesticides. For the ultimate success of biopesticides, microorganisms developed for biological control must be viewed by researchers, manufacturers, and end-users in a biological paradigm rather than a chemical one.

The efficacy and reliability of biopesticides are largely determined by environmental parameters in the period between application and infection. In this period, the availability of moisture is of the utmost importance. Furthermore, the narrow host range of many pathogens may restrict their commercial attractiveness. Both of these issues can be addressed by research into the use of genetic engineering and formulation. As research into the molecular basis of host specificity and pathogenesis continues, it will become possible to produce more aggressive pathogens with the desired host range for biological control. The survival and efficacy of these pathogens will be enhanced through the use of novel formulations.

REFERENCES

1. Crump, N.S.; Cother, E.J.; Ash, G.J. Clarifying the nomenclature in microbial weed control. Biocontrol Sci. Tech. **1999**, *9*, 89–97.
2. Mortensen, K.; Makowski, R.M.D. Tolerance of strawberries to *Colletotrichum gloeosponriodes* F. sp. *malvae*, a mycoherbicide for control of round-leaved mallow (*Malva pusilla*). Weed Sci. **1995**, *43* (3), 429–433.
3. Van Drieshe R.G.; Bellows, T.S. *Biological Control*; Chapman & Hall: New York, NY, **1996**.
4. Kaya, H.K.; Gaugler, R. Entomopathogenic nematodes. Ann. Rev. of Entomol. **1993**, *38*, 181–206.

5. Johnsson, L.; Hokeberg, M.; Gerhardson, B. Performance of the *Pseudomonas chlororaphis* biocontrol agent ma 342 against cereal seed-borne diseases in field experiments. Eur. J. Plant Pathol. **1998**, *104* (7), 701–711.
6. Sutton, J.C.; Peng, G. Biocontrol of *Botrytis-cinerea* in strawberry leaves. Phytopathology **1993**, *83* (6), 615–621.
7. Oneill, T.M.; Elad, Y.; Shtienberg, D.; Cohen, A. Control of grapevine grey mould with *Trichoderma harzianum* t39. Biocontrol Sci. & Tech. **1996**, *6* (2), 139–146.
8. TeBeest, D.O.; Yang, X.B.; Cisar, C.R. The status of biological control of weeds with fungal pathogens. Ann. Rev. Phytopath. **1992**, *30*, 637–657.
9. Auld, B.A.; Morin, L. Constraints in the development of bioherbicides. Weed Tech. **1995**, *9* (3), 638–652.

FURTHER READING

Baker, K.F.; Cook, R.J. *Biological Control of Plant Pathogens*; W.H. Freeman and Company: San Francisco, CA, **1974**.

Chen, J.; Abawi, G.S.; Zuckerman, B.M. Efficacy of *Bacillus thuringiensis*, *Paecilomyces marquandii*, and *Streptomyces costaricanus* with and without organic amendments against *Meloidogyne hapla* infecting lettuce. J. Nematol. **2000**, *32* (1), 70–77.

Lacey, L.A.; Goettel, M.S. Current developments in microbial control of insect pests and prospects for the Early 21st Century. Entomophaga **1995**, *40* (1), 3–27.

Sticher, L.; Mauchmani, B.; Metraux, J.P. Systemic acquired resistance. Ann. Rev. Phytopathol. **1997**, *35*, 235–270.

Bioremediation

Ragini Gothalwal
P.S. Bisen
Institute of Microbiology and Biotechnology, Barkatullah University, Bhopal, India

Abstract
Rapid industrialization and urbanization during the past decades have revolutionized and affected society in many ways. However, ever-increasing industrial affluence has led to the generation of effluents that are posing a threat to the environment. Thus, it can be said that the boon of modernization has come along with the curse of pollution. The conventional methods employed for wastewater treatment include the use of chemicals and resins. However, these methods are not ecofriendly and in addition they are expensive and cumbersome. In such a scenario, the microbial biomass for bioremediation is an attractive option due to its easy operation and economic viability.

PESTICIDE POLLUTION

A pesticide is a double-edged sword. If used properly it is a boon to humanity but can involve hazards if carelessly handled. There has been an enormous increase in the use of various synthetic pesticides that contribute to the spectacular increase in the crop yield. The soil, groundwater, and sediments are the ultimate sink for the pollutants, where they either are broken down to simpler forms or remain persistent polychlorinated biphenyls (e.g., DDT). Some of the pesticides are susceptible to bioaccumulation/biomagnification and cause more danger to the environment. The safe and economical disposal of excess pesticide waste is a problem of considerable magnitude.

Some of the pesticides include nitroaromatic compounds, polycyclic aromatic hydrocarbons (PAH), derivatives of benzene, and phenolic compounds. Toxicity and bioaccumulation potential of chlorophenol increase with the degree of chlorination of lipophilicity. The constituents of many pesticides are generally derivatives of benzene. The aromatic ring has large negative resonance energy and for this benzene and its derivatives is a stable group of compounds.[1] Recent research has revealed a number of microbial systems capable of biodegradation of organic compounds.

MICROBIAL POTENTIAL OF PESTICIDE DEGRADATION

The natural capacity of microorganisms to degrade a huge variety of herbicides, pesticides, and some inorganic compounds[2] is the essence of the microbial method for degradation of soil contaminants (bioremediation)—the basis for "green" technologies.[3] Another very interesting feature of microorganisms is that they degrade organic substances that are produced only synthetically. Although degradation does not always lead to detoxification, in many cases the products are less hazardous and/or become susceptible to further degradation.

A relatively enormous catalytic power, large surface volume action, and rapid rate of reproduction of microorganisms contribute to the major role they play in the chemical transformation. Generally there are three possibilities by which the microbial community acquires the ability to degrade pesticide: 1) some organisms require adaptation before attacking a pollutant having degradative enzymes of a constitutive type; 2) some through random mutation; and 3) some have adaptive enzymes that are induced in the presence of a particular pollutant. The most important environmental factors affecting biodegradation are temperature, pH, water, and oxygen content. The biochemical processes induced by microorganisms under aerobic and anaerobic conditions are mineralization, detoxification, cometabolism, and activation. The microbes engaged in degradation of pesticides are bacteria, actinomycetes, fungi, algae, and cyanobacteria.

Bacteria and Actinomycetes

Twenty-eight genera of bacteria that utilize aliphatic hydrocarbon have been isolated (Table 1).

Fungi

There are two main ways by which fungi can be utilized to degrade environmental pollutants. The first approach is to modify sewage treatment systems, the second is

Table 1 Examples of bacteria, actinomycetes, and fungi engaged in pesticide degradation

Organism	Organic compounds or pesticides
Bacteria and Actinomycetes	
Alcaligens denitrificans	Fluoranthene (PAH)
Alcaligens faecalis	Arylacetonitriles
Archrombacter	Carbofuran
Arthrobacter	EPTC, glyphosate, pentachlorophenol
Bacillus sphaericue	Urea herbicides
Brevibacterium oxydans IH 35A	Cyclohexylamine
Burkholderia sp. P514	1,2,4,5-TeCB
Clostridium	Quinoline, glyphosate
Comomonas testosteroni	Arylacetonitriles
Corynebacterium nitrophilus	Acetonitrile, carboxylic acid, ketones
Dehalococcoides ethenogenes 195	Trichloroethylene (TCE)
Desulfitobacterium dehalogenes	Hydroxylated PCBs
Desulfovibrio sp.	Nitroaromatic compound
Flavobacterium	Pentachlorophenol
Geobacter sp.	Aromatic compound
Klebsiella pneumoniae	3&4 hydrobenzoate
Methylococcus capsulatus (Bath)	Trichloroethylene
Methylosinus trichosporium OB 3b	1,1,1-trichloroethane (TCA)
Moraxella	Quinoline, glyphosate
Nitrosomonas europaea	1,1,1-trichloroethane
Nocardia	Quinoline, glyphosate
Pseudomonas aeruginosa	Nitriles, biphenyl, parathion
Pseudomonas sp.	Quinoline, glyphosate
Pseudomonas stutzeri	Parathion
Pseudomonas cepacia	2,4,5-T
Pseudomonas paucimobilis	PCP
Pseudomonas putida 6786	Propane
Pseudomonas striata	Propham, chlorpham
Rhodococcus chlorophenolicus	Pentachlorophenol
Rhodococcus corallinus	S-triazines
Rhodococcus rhodochrous	Propane
Rhodococcus sp.	Propane, 1,1,1,-trichloroethane
Rhodococcus UM1	Pyrene
Fungi	
Aspergillus flavous	DDT
Aspergillus paraceticus	DDT
Aspergillus niger	2,4-D
Candida tropicalis	Phenol
Chrysosporium lignorum	3,4-dichloroaniline
Fusarium solani	Acylamilide
Fusarium oxysporum	DDT
Hendesonula toruleidea	2,4-D
Hydrogenomonas + *Fusarium* sp.	DDM, nitrile
Mucor alterans	DDT
Penicillium	Acylamilide
Penicillium megasporum	2,4-D
Phallinue weirii	DDT
Phanerochaete chrysosporium	PAH, 2,4,6-trinitrotoluene, pentachlorophenol, DDT, 2,4,5-T and lindane
Pleurotus ostreatus	DDT
Polyporue versicolor	DDT
Pullularia	Acylamilide
Rhodotorula	Benzaldehyde
Stereum hirsistum	Phenanthrene

(*Continued*)

Table 1 Examples of bacteria, actinomycetes, and fungi engaged in pesticide degradation. (*Continued*)

Organism	Organic compounds or pesticides
Trametes versicolor	Dieldrin
Trichoderma sp.	Nitrile
Trichoderma viride	DDT
Trichospron cutaneum	Phenol
Yeast	Paraquat
Ectomycorrhizal fungi	
Tylospora fibrillosa	Mefluidide
Thetophora terrestris	Mefluidide
Suillus variegatus	Mefluidide
Suillus granulatus	Mefluidide
Suillus luteus	Mefluidide
Hymenoscphur ericae	Mefluidide
Paxillus involutus	Mefluidide

"bioaugmentation." Fungi can utilize several cosubstrates for energy. Hence, fungi can effectively degrade several high-strength industrial wastes more efficiently than bacteria.[4]

The ecto and ericoid mycorrhizal fungi have been shown to degrade a wide range of persistent organic pollutants such as PCBs, atrazine, 2,4-D, chlorophenol, 2,4,4-trinitrotoluene, making the pollutants suitable target organisms for facilitating a bioremediation program. The sustainability of the organisms favor their use in bioremediation over white rot fungi that require oxidative enzymes[5] to facilitate remediation (Table 1).

Algae

Microalgae have received more attention in recent years as an alternative secondary biosystem for waste treatment. The success of an algal system relies on its ability to take up inorganic nutrients such as N and P from wastewater and assimilate them for growth, for example, *Chlorella vulgaris*, *Scendesmus* spp., *Chlamydomonas humicola*, and *Chlorella minesstissima* could rapidly utilize different organic compounds (acetate and glucose).

Cyanobacteria

Utilization of cyanobacteria in effluent treatment is a recent phenomenon. It has great potential to take up external nutrients, hence, it can be a good candidate for tertiary treatment of industrial effluents, in turn helping to solve the problem of eutrophication. The common cyanobacterial strain used in wastewater treatment includes *Anabaena doliolum*, *Anabaena* CH3, *Lyngbya gracilis*, *Phormidium faveolarum*, *Oscillatoria animalis*, *Oscillatoria pseudogeminata*, *Phormidium laminosum*, and *Spirulina maxima*. Cyanobacterial filters comprising of *Oscillatoria annae* and *Phormidium tenue* proved to be most suitable for sewage water treatment.

CONSTRUCTED STRAINS

In the natural environment a mixed culture plays an important role in the degradation of xenobiotic compounds. Some members of the culture might be able to provide important degradative enzymes whereas others supply surfactants or recombinant biocatalysts to ensure the bioavailability of pollutants.[6,7] A consortium of *Arthobacter ilicis* and *Agrobacterium radiobacter* mineralized ethylene glycol nitrate. The *Arthrobacter* strain was the actual degrading organism, although the second microbe facilitated the mineralization. The bacterial degradation of Benzene, Toluene, Ethylene, and Xylene (BTEX) has used microbial consortia since no pure culture is known to degrade all the components of BTEX efficiently.

The enzymes involved in the catalysis of many of these pesticides generally can be grouped into two classes, namely, hydrolases and oxidoreductases. Microorganisms have evolved the degradative capacity by altering the substrate specificity of an enzyme[6] already encoded in the genome, as seen in the evolution of the 3-chlorobenzoate pathway (Fig. 1). It is possible nonetheless, to use selective evolution (through the use of a chemostat) to develop a strain capable of degrading recalcitrant 2,4,5,-T.[6] This type of evolution is facilitated by the presence of a gene pool, so an organism can readily take up foreign DNA. However, recent advances in genetic techniques have opened up new avenues to move toward the goal of genetically engineered microorganisms to function as "designer catalysts," in which certain desirable biodegradation pathway or enzymes from different organisms are brought

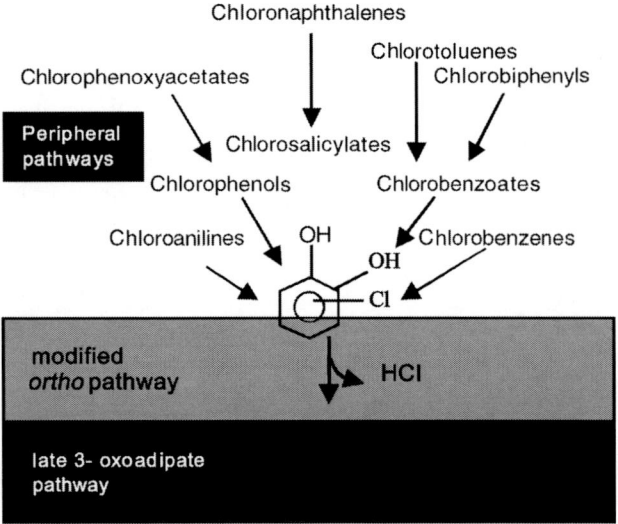

Fig. 1 Combination of pathway segments to develop a hybrid pathway for the mineralization of chloroaromatic compounds.

together in a single host with the aim of performing specific detoxification.[6,7] These applications will ultimately create "super biocatalysts" capable of degrading several pesticides rapidly and cost effectively.[8] The transfer of plasmid pJP4 from *Alcaligens eutrophus* JMP134 into *Pseudomonas cepacia* AC1100 was performed and constructed strain RHJ 1 was shown to efficiently degrade a mixture of 2,4-D and 2,4,5-T. Such strains in many instances may be better suited than microbial consortia for use in the degradation of certain toxic chemical mixtures (Table 2).

BIOREMEDIATION TECHNOLOGY

To ensure the success of any implemented bioremediation process, bench scale experiments are usually required to determine the natural and augmented microbial degrading activities in the contaminated media. Biotreatability tests determine the amount of specific microbial metabolic activity, stimulated by the addition of complementary

Table 2 Degradative plasmids with relevance to the construction of chloroaromatics-degrading hybrid strains.

Plasmid	Size (kb)	Substrate	Host
Peripheral pathways			
TOL	117	Xylenes, toluene, toluate	*Pseudomonas putida*
NAH7	83	Napthalene via salicylate	*Pseudomonas putida*
pWW60-1	87	Napthalene via salicylate	*Pseudomonas* sp.
pDTG1	83	Napthalene via salicylate	*Pseudomonas putida*
SAL1	85	Salicylate	*Pseudomonas putida*
pKF1	82	Biphenyl via benzoate	*Acinetobacter* sp. (reclassified as *Rhodococcus globerulus*)
pWW100	~200	Biphenyl via benzoate, methylbiphenyls via toluates	*Pseudomonas* sp.
pWW110	>200	Biphenyl via benzoate, methylbiphenyls via toluates	*Pseudomonas* sp.
pCITI	100	Aniline	*Pseudomonas* sp.
pEB	253	Ethylbenzene	*Pseudomonas fluorescens*
pRE4	105	Isopropylbenzene	*Pseudomonas putida*
pWW174	200	Benzene	*Acinetobacter calcoaceticus*
pHMT112	112	Benzene	*Pseudomonas putida*
pEST1005	44	Phenol	*Pseudomonas putida*
pVI150	Mega	Phenol, cresol, 3,4-dimethylphenol	*Pseudomonas* sp.
Central pathways			
pAC25	117	3-Chlorobenzoate	*Pseudomonas putida*
pJP4	77	3-Chlorobenzoate, 2,4-D	*Ralstonia eutropha* (formerly *Alcaligenes eutrophus*)
pBR60	85	3-Chlorobenzoate	*Alcaligens* sp.
pRC10	45	2,4-D	*Flavobacterium* sp.
pP51	100	1,2,4-Trichlorobenzene	*Pseudomonas* sp.
pMAB1s	90	2,4-D	*Burkholderia* (formerly *Pseudomonas*) *cepacia*

Source: From Annu. Rev. Microbiol.[10]

microorganisms and/or nutrients and the optimization of environmental factors.[9]

The immobilization of cells offers various advantages and the process is cost-effective. Though a majority of the work pertains to entrapment of cells in natural or synthetic polymers, a few are cross-linked with various compounds. The rotating biological contactor (RBC) is an effective reactor configuration and when operated sequentially, it can degrade even 500 mg/l phenol in 24 h.[5] Enhanced removal efficiency can be obtained by repeating anaerobic and aerobic treatment processes in combination.

Biocapsules have been tested for various applications and can be produced for site-specific application.[9] They are products that surround microorganisms and/or microbial components (protein or enzymes) with stabilizers, nutrients, and other materials to ensure their survival during storage and function on dispersal in various environments.

In situ groundwater biodegradation in the United States has been carried out through various processes by numerous companies, namely, Biogenesis Technologies, Biopim, ABB Environmental Service, Bioremediation System, Remediation Technologies Inc., IT Corporation, Geo-Microbial Technologies Inc., etc.[11] These companies use natural biological ingredients that are harmless. The contaminants that can be treated are BTEX, TPH, TCE, TCA, phenol, organic acid, monochlorobenzoate, and toxic radioactive materials (at contaminated military sites). The removal effectiveness can approach 100%.

FUTURE DEVELOPMENTS

The understanding of biotransformation pathways is necessary for designing bioremediation strategies and wastewater treatment processes. Enzymes with broad substrate specificity and microorganisms with degradative competence against a range of substrates offer the possibility that one or a few microbial cultures can degrade all the important wastes in a complex mixture. The simplest strategy is improving the biodegradation performance of a consortium through the addition of a "specialist" organism.

ACKNOWLEDGMENT

The authors gratefully acknowledge the financial assistance from University Grants Commission, New Delhi, under the SAP program.

REFERENCES

1. Paul, J.; Varma, A. Microbial Degradation of Xenobiotics—An Overview. In *Microbes: For Health, Wealth & Sustainable Environment*; Varma, A., Ed.; Malhotra Publishing House: New Delhi, **1998**; 413–432.
2. Eccles, H. Treatment of metal contaminated wastes: why select a biological process. Trends Biotechnol. **1999**, *17*, 462–465.
3. Karamarev, D.G. Biodegradion of soil contaminants. J. Sci. and Indust. Res. **1999**, *58*, 764–772.
4. Spain, J.C. Biodegradation of nitroaromatic compounds. Annu. Rev. Microbiol. **1995**, *49*, 523–555.
5. Manimekatai, R.; Swaminathan, T. Biodegradation of phenolic compounds using *Phaenrochate chrysosporium*. J. Sci. and Indust. Res. **1998**, *57*, 833–837.
6. Timmis, K.N.; Pieper, D.H. Bacteria designed for bioremediation. Trends Biotechnol. **1999**, *17*, 201–204.
7. Timmis, K.N.; Steffan, R.J.; Unterman, R. Designing microorganisms for treatment of toxic wastes. Annu. Rev. Microbiol. **1994**, *48*, 525–557.
8. Chen, W.; Mulchandani, A. The use of live biocatalysts for pesticide detoxification. Trends Biotechnol. **1998**, *16*, 71–75.
9. Bioremedation Applied Biosciences. http://www.bioprocess.com (accessed December 1999).
10. Reineke, W. Development of hybrid strain for the mineralization of chloroaromatics by patchwork assembly. Annu. Rev. Microbiol. **1998**, *52*, 287–331.
11. Bioremedation, **1999**. http://www.in weh.unu.edu/447/lectures/bioremedation.htm (accessed December 1999).

Biosensors

Leon Terry
Jordi Giné Bordonaba
Department of Food Security and Environmental Health, Cranfield University, Bedfordshire, U.K.

Abstract
Biosensors are analytical devices that can be operated on-site by unskilled personnel allowing rapid measurements and determinations to significantly enhance tasks such the monitoring of quality and safety aspects in foods. While most developments in the biosensor field have focused on the medical diagnostics industry, attention is now being trained on introducing the technology into other areas, such as the food and agricultural sector.

INTRODUCTION

The use of biosensors to monitor the condition of food and agricultural products, including both raw materials and processed items, should lead to a significant enhancement in both process and quality controls,[1,2] not only by enhancing the relevance of measurements but also by monitoring specific analytes that are key indicators of produce quality, safety, and consumer acceptability. Until recently these devices have been mainly targeted at the health-care industry, leading to a current market valuation for biosensors of approximately $5 billion/year;[3] however, based on the bedrock of technological advances that have been achieved to meet this current position, biosensor technology is now primed for expansion into other market sectors, particularly the agricultural and food industries. The design and application of biosensors can range from one-shot disposable devices to on-line systems designed to automatically monitor and measure components such as pesticide and herbicide residues, biochemical markers, and the presence of toxic compounds. For example, simple and robust biosensors that can be operated by non-skilled persons allowing "on-the-spot" determinations of fruit ripeness,[4] glucose content and perceived sweetness in strawberries,[5] or onion pungency[6,7] have been described. Using a multisensor array design simultaneously measuring a number of markers and coupled to a simple handheld meter, a biosensor system can be low cost and provide easy-to-understand results. Similarly, easy-to-use robust sensors could be designed for deployment and use in environments such as in-field operations and fresh produce distribution centers[2] and supermarkets.

Overall, the advantages for using biosensors in the food and agricultural industries, compared to conventional analytical tools, can be summarized as:

- Biosensors can be designed to be operated literally "in-the-field" by non-specialists. Hence, significantly reducing the cost and time delays incurred when using highly trained staff in centralized laboratories to carry out measurements on foods.
- Biosensors often have a high specificity for a wide range of target analytes.
- Many sensor systems can be manufactured using mass production techniques so that biosensors can be produced and sold at very low cost.

BIOSENSOR TECHNOLOGY

Practically all biosensors comprise several key components:

- A biological recognition element that acts to identify a specific target substrate. This identification is manifested by changes in one or more physiochemical properties, e.g., electron transfer or a change in pH. A number of such recognition elements have been used in various biosensor formats. The following list provides an overview of the most commonly used:
  - antibodies
  - enzymes
  - whole cells
  - subcellular organelles
  - tissue slices
  - DNA

  Conventionally, enzymes, antibodies and whole cells have tended to be the most frequently used biological recognition elements.
- A signal transduction method that acts to translate the biological recognition event (e.g., binding of an antibody to its corresponding antigen) into a signal output that can be relayed to the biosensor operator. As with the biological recognition elements, a number of transduction methods have been employed for the fabrication of

biosensors. The list shown below highlights the most common transduction methods used:
- electrochemical
- optical
- piezoelectric
- calorimetric
- acoustic

Electrochemical and optical transduction methods have proved to be the dominant approach used for biosensor design and fabrication. Broadly, electrochemical transduction methods are based on monitoring electroactive species that are either consumed or produced as a result of the interaction between the biological recognition element and its corresponding target analyte. In practice, the two main approaches to electrochemical transduction have tended to be either amperometric (i.e., monitoring a current output) or potentiometric (i.e., measuring a change in electric potential).

In contrast to the electrochemical transduction methods, optical biosensors operate by measuring responses to illumination or light emission. Typically, biosensors based on this approach tend to display higher levels of reproducibility. In addition, unlike most electrochemical biosensors, optical devices can be designed to operate without the use of reporter molecules (e.g., enzymes or other labels) as part of the recognition method (Fig. 1).

As described above, the overall aim of a biosensor is to convert a biologically induced recognition event into a usable signal. In order to achieve this function, biosensors incorporate a number of key supporting technologies, these include:

Sampling

An effective sampling system is an essential feature for any biosensor system, particularly in relation to carrying out measurements in agricultural and food items. Many of the components found in agricultural produce (e.g., fats, proteins, or starches) could have significant effect on the performance of the biosensor; either as a result of fouling the sensor surface or by producing erroneous signals (e.g., electrochemically active compounds such as ascorbic acid and phenylpropanoids).

Membranes

Almost all biosensor configurations include some membrane structures; they can be used for a number of applications such as:

- Retaining the biological recognition element, usually in close proximity to the transducer surface, helping to improve the efficiency of signal interpretation.
- Providing a protective barrier, preventing fouling of the sensor by components in the sample matrix.
- Helping to provide stability to the sensor, both for long-term storage and operation conditions.
- Providing, in many cases, a high degree of selectivity either through allowing only the target analyte through to the sensors surface or by eliminating interfering compounds that may affect the sensor signal.

Immobilization

Immobilizing the biological recognition element on or in proximity to the transducer is a major feature of many biosensor formats, allowing for the efficient transfer of signal from the biological recognition element to the transducer. A number of immobilization strategies have been adopted including physical adsorption, entrapment, and covalent binding. Each of these approaches has their own advantages and disadvantages; in practice, the biosensor

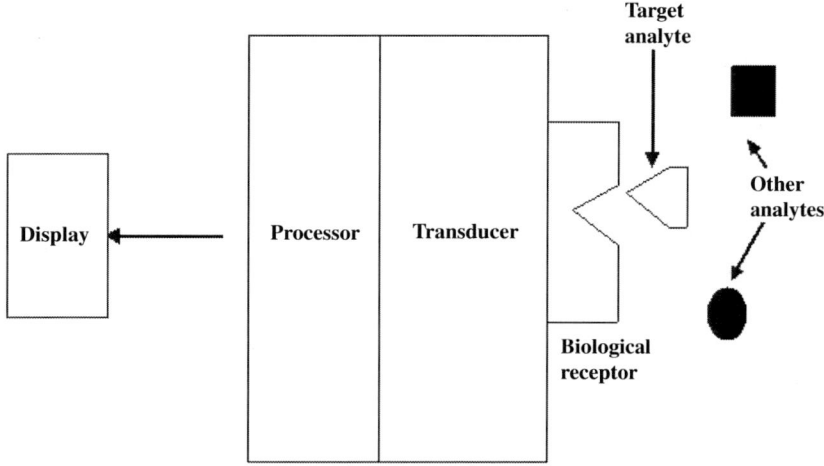

Fig. 1 A simple schematic showing the principal components of a biosensor.

technologist selects the immobilization process that is the most appropriate for a particular sensor format and application. For the interested reader there are a number of excellent publications that give an in-depth description of the basic technology that underpins biosensor research, development, and commercial exploitation.[2,8]

BIOSENSORS IN AGRICULTURE AND FOOD PRODUCTS: THE CURRENT PICTURE

Despite the obvious advantages that biosensors offer, their adaptation in the agricultural and food industries has been limited. There are several possible reasons for this. Most biosensor designers and manufactures have tended to concentrate on medical diagnostics, in particular the home-based blood glucose–monitoring market (self-testing carried out by diabetics). Secondly, the problems posed by sampling (as discussed earlier) have been overlooked in development of sensors aimed at the food and agricultural industries, thus preventing widespread introduction.

Nevertheless, in spite of this lack of widespread commercial adaptation, a significant amount of time and effort is being spent on developing biosensor systems for the food and agricultural industries. The following provides just a snapshot of recent research and development activity that has been carried out in various related aspects.

Pesticide Residues

The detection and quantification of pesticide residues on foodstuffs present a serious challenge to both food producers and retailers. Biosensor systems designed to meet this challenge have been described.[9,10]

Metabolites

The ability to easily measure a wide range of metabolites in food products on-site would provide a significant boost for applications in quality control. This is the ideal situation for using biosensors, given their reputation for being low-cost, rapid, and easy-to-use. Subsequently, there have been a significant number of reports describing the development of biosensors for such a task (Fig. 2).[4–7,11,12]

Pathogen Detection

The presence of foodborne pathogens (e.g., *Escherichia coli* O157:H7) can have a devastating effect, not only on the persons suffering the consequences of the contamination but also on the food industry itself, i.e., loss of confidence by the public leading to economic problems for the industry. Again, biosensor systems designed to rapidly and easily detect the presence of dangerous pathogens (without the

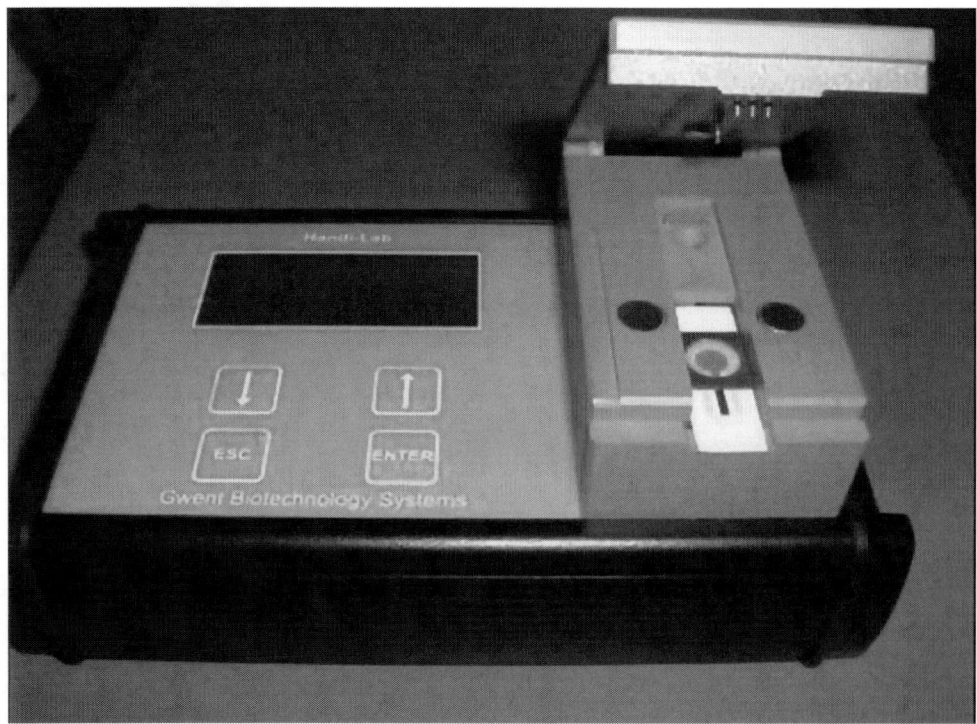

Fig. 2 A commercial biosensor device to measure pungency in onions that was fabricated by Gwent Electronic Materials in collaboration with Cranfield University.

recourse to using centralized testing at laboratories) have been developed.[13]

CONCLUSIONS

The picture is changing; there is increasing consumer demand for higher quality foods that incorporate traceability. Economically, this demand can only be met by using reliable and inexpensive methods of monitoring. Biosensors have the potential to significantly contribute to this area.

Looking to the future, the use of DNA probes as the bioreceptor component of biosensors will become more prominent. The high levels of sensitivity that can be achieved using DNA probes compared to the current (typically antibody-based) bio-receptors make it an attractive option particularly in the field of pathogen detection.

In contrast to the use of purely biological recognition elements, the use of synthetic receptors is currently being explored as a viable alternative for inclusion in biosensor systems.[14] Despite the high specificity inherent in many biological receptors, they can suffer from a number of drawbacks. Some of these shortfalls include poor stability and high costs for production. In contrast, synthetic receptors would have high stability and be very cheap to produce.

Other areas of future development will include miniaturization coupled with advanced fabrication techniques. This will lead to "lab-on-a-chip" type devices that will allow a multitude of analytes to be measured from one sample; all of which will be rapidly carried out on-site by non-specialist people using low-cost, robust biosensors. Linked to advances in sampling and extraction techniques the use of biosensors on farms, in warehouses, and food production plants is set to continue.

REFERENCES

1. Patel, P.D. (Bio)sensors for measurement of analytes implicated in food safety: a review. Trends Analyt. Chem. **2002**, *21* (2), 96–115.
2. Terry, L.A.; White, S.F.; Tigwell, L.J. The application of biosensors to fresh produce and the wider food industry. J. Agric. Food Chem. **2005**, *53*, 1309–1316.
3. Newman, J.D.; Tigwell, L.J.; Turner, A.P.F.; Warner, P.J. *Biosensors: A Clearer View*, Cranfield University, Cranfield Biotechnology Centre, 2004, 216.
4. Jawaheer, S.; White, S.F.; Rughooputh, S.D.D.V.; Cullen, D.C. Development of a common biosensor format for an enzyme based biosensor array to monitor fruit quality. Biosens. Bioelectron. **2003**, *18*, 1429–1437.
5. Giné Bordonaba, J.; Terry, L.A. Development of a glucose biosensor for rapid assessment of strawberry quality: relationship between biosensor response and fruit composition. J. Agric. Food Chem. **2009**, *57*, 8220–8226.
6. Abayomi, L.A.; Terry, L.A.; White, S.F.; Warner, P.J. Development of a disposable pyruvate biosensor to determine pungency in onions. Biosens. Bioelectron. **2006**, *21*, 2176–2179.
7. Abayomi, L.A.; Terry, L.A. A pyruvate dehydrogenase-based amperometric biosensor for assessing pungency in onions (*Allium cepa* L.). Sens. Instrumen. Food Qual. **2007**, *1*, 183–187.
8. Law, W.T.; Akmal, N.; Usmani, A.M., Eds.; *Biomedical Diagnostic Science and Technology*; Marcel Dekker: New York, 2002.
9. Andreescu, S.; Marty, J.-L. Twenty years research in cholinesterase biosensors: from basic research to practical applications. Biomol. Eng. **2006**, *23* (1), 1–15.
10. Pogacnik, L.; Franko, M. Detection of organophosphate and carbamate pesticides in vegetable samples by a photothermal biosensor. Biosens. Bioelectron. **2003**, *18* (1), 1–9.
11. Rotariu, L.; Bala, C.; Magearu, V. New potentiometric microbial biosensor for ethanol determination in alcoholic beverages. *Anal. Chim. Acta* **2004**, *513* (1), 119–123.
12. Adányi, N.; Tóth-Markus, M.; Szabó, E.E.; Váradi, M.; Sammartino, M.P.; Tomassetti, M.; Campanella, L. Investigation of organic phase biosensor for measuring glucose in flow injection analysis system. Anal. Chim. Acta **2004**, *501* (2), 219–225.
13. Ko, S.; Grant, S.A. A novel FRET-based optical fiber biosensor for rapid detection of *Salmonella typhimurium*. Biosens. Bioelectron. **2006**, *21* (7), 1283–1290.
14. Feng, L.; Liu, Y.; Tan, Y.; Hu, J. Biosensor for the determination of sorbitol based on molecularly imprinted electrosynthesized polymers. Biosens. Bioelectron. **2004**, *19* (11), 1513–1519.

Biotechnology

Manas R. Banerjee
Laila Yesmin
Research and Development Division, Brett-Young Seeds Limited, Winnipeg, Manitoba, Canada

Abstract
Biotechnology is an applied field that can be viewed from many perspectives. To some it is the age-old technique of selecting better cultivars and wine production, and to others it is the recombinant DNA technique that creates the basis of modern biotechnology. The term 'biotechnology' may suggest a single subject but in reality it is a multidisciplinary approach to utilize science for the benefit of our society. The word 'biotechnology' originates from Greek words 'bios' and 'technologos,' which means living and technical study, respectively. The literal meaning of biotechnology is, thus, the technical study of living things.

INTRODUCTION

The simplest way to look at biotechnology is as the use of living organisms to produce goods and services for industrial purposes. *Chambers Science and Technology Dictionary* defines biotechnology as "the use of organisms or their components in industrial or commercial process, which can be aided by the techniques of genetic manipulation in developing, e.g., novel plants for agriculture or industry."[1] Furthermore, the *Macmillan Dictionary of Biotechnology* defines the term as "the application of organisms, biological systems, or biological processes to manufacturing and service industries."[2] This definition actually includes any process in which organisms, tissues, cells, organelles, or isolated enzymes are used to convert biological or other raw materials to products of greater value, as well as the design and use of reactors, fermenters, downstream processing, analytical and control equipment associated with biological manufacturing processes.[2] Biotechnology has been used for centuries in fermentation, and cheese and bread making, but currently it involves innovative devices like molecular techniques, in vitro techniques, genetic engineering, cloning, protein manipulation, monoclonal antibodies, and stem cells. Biotechnology is no longer an applied branch of biological sciences rather it is a strategic technology having diverse application to the socioeconomic development and sustainability of the industrialized countries (Fig. 1).

APPLICATIONS AND PROSPECTS

Biotechnology has made remarkable progress over the last couple of decades and is well integrated with technical improvement of processing and instrumentations. For example, penicillin manufacturing has changed over the years with the development of efficient strain and improved operational techniques. The same is true for many pharmaceutical products, drugs, fine chemicals, biopolymers, energy, and processed food products. In developed world, major biotech industries are in pharmaceutical, food, agriculture, environment, and health sectors whereas in developing countries, food and agriculture biotechnology is progressing rapidly.

With the introduction of genetic engineering in the agricultural sector, biotechnology has made fabulous advances in crop yield and quality. Agronomic traits have been successfully developed or are being developed through genetic manipulation of crop plant. For instance, "roundup-ready" soybean and canola were developed to protect from herbicide and 'Bt' cotton and corn to provide protection against insect damage. Currently, genetically modified (GM) crops are commercially grown in more than 40 countries, with over 110 million hectares under cultivation.[3] This technology reached almost one billion acres of farmland in 2005 and millions of farmers are using it. Crops like soybean, oilseed rape, corn, and cotton have the largest acreages and are primarily grown in countries like the U.S., Canada, Argentina, and China. Biotechnological applications in agriculture have a potential market estimated at $67 billion per year.[4] Many developing countries are embracing agriculture biotechnology with the hope of increasing their food production to meet the demand of a growing population as food security is their main concern. The Food and Agriculture Organization of United Nations recognizes the potential of increasing production in agriculture, fisheries, and forestry through GM technology. This might not completely eradicate the world food shortage but it could definitely contribute to food security particularly in the developing world.

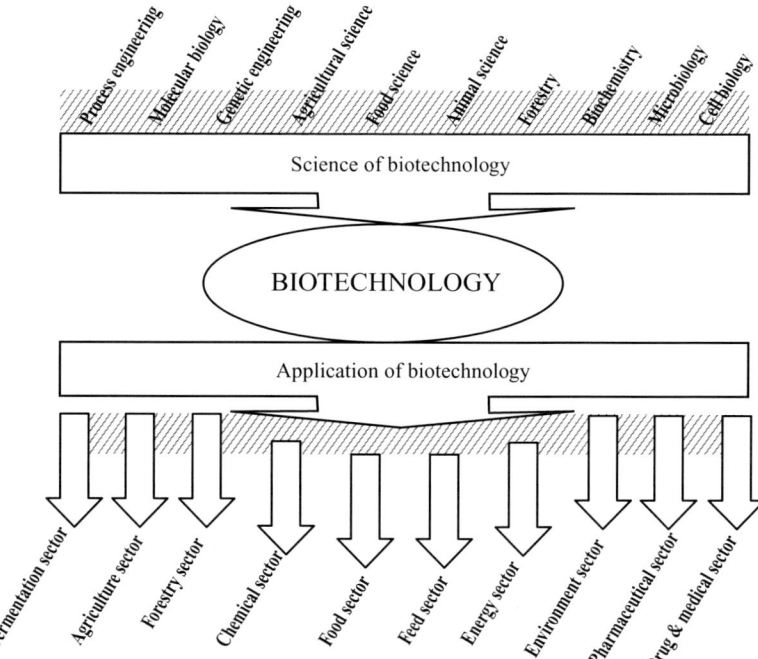

Fig. 1 Contribution of science towards the application of biotechnology.

With increasing yield, GM technologies have also been shown to reduce pesticide use and increase farm profitability. The technology now has been extended to local crops like banana, cassava, plantain, rice, and sorghum for abiotic stress tolerance and quality. "Golden rice" capable of producing Pro-vitamin A (beta-carotene) and iron-rich rice are also examples of this novel technology. As such, more challenging work in the areas of stress resistance, crop yield, and quality enhancement are on the horizon. Some microbiologically derived biopesticides and biofertilizers are already in the marketplace making strides in boosting agricultural production. Besides, biotechnology indirectly aids in carbon sequestration by reducing tillage that minimizes soil erosion, helps water retention, and reduces carbon escape from soil. Introducing myccorhizal fungi to increase plant growth that enhances moisture and nutrient uptake resulting in plant capacity to sequester more carbon is another example. Biotechnology is also contributing to the improvement of biodiversity and restoring ecological health via the reduction of carcinogenic chemical use. Recent advances on effective identification of genes, and genomic and proteomic research play a tremendous role in improving biodiversity as well. Preservation of the terrestrial ecosystem is an important step to maintain biodiversity as ecological damage by deforestation has already wiped out and/or threatened the existence of many valuable species. If GM technology could produce more food per unit of existing farmland, then more land will be protected for wildlife and forest species. Biotech research based on a gene called the "cellulose binding domain" that accelerates tree growth (superfast GM tree) would give hope to reforestation and damage control. GM technology also opens the door for plant-based edible vaccines for preventable human diseases like tetanus, diphtheria, measles, and cholera. This cost-effective vaccine would have tremendous impact on global population health especially in developing and under-developed countries. For example, potato-derived hepatitis B vaccine would likely be the first edible vaccine that could save millions of lives.[5]

As the public's awareness of "green technology" is increasing so is its reluctance to use harmful products. With biotech innovation, nondegradable toxic products are being replaced with biodegradable and less toxic products, and the applications are extended into plastics, food processing, textiles, polymers, and energy and mining industries. At present, the environmental sector uses biotechnology in pollution control, waste management, and renewable resource areas. Although the present use of transgenics in bioremediation is limited, the potential could be substantial in the foreseeable future. Biotechnology can also be applied in the conventional fossil fuel and renewable energy sector. For instance, a nontoxic biopolymer of xanthan gum and cellulose avoids the problems of conventional drilling mud via better viscosity and binding in the drilling process. Production of renewable fuel-like ethanol can be achieved more efficiently by using a modified cellulase enzyme that maximizes the conversion of cellulose into fermentable sugar, and in turn helps in reducing greenhouse gas emissions.

Biotechnology is extensively used in the food and food processing industry. A range of essential food products and food additives require microbes or microbial source.

Lactobacillus, *Steptococcus*, and *Leuconostoc* are commonly used in dairy industry for flavor, consistency, and reducing spoilage time. Various techniques have been used to modify the functional properties of foods such as spreadability of butter and enhancement of fruit and vegetable qualities. Fermentation technology is the other key area of this applied science. Although pharmaceutical and food industries rely heavily on the advancement of this technology, almost every sector has its use. Many important drugs, antibiotics, steroids, vaccines, hormones, enzymes, diagnostic products, solvents, food and food processing products, animal feed, animal vaccines, biopesticides, and inoculants are the results of this technology. Improvements on bioreactors, computer-controlled fermentation processes, and downstream processing have added a new dimension to the biotech industry. With the evidence of new diseases and many recurring diseases, the drug and pharmaceutical sector is rapidly expanding in areas of antibiotics, vaccines, hormones, enzymes, and diagnostic tests using monoclonal antibodies. Biotechnology also makes it possible to mass-produce these substances that might otherwise be cumbersome and expensive to produce. New methods allow producing purer drugs like human insulin from GM bacteria that does not cause any allergic reactions in patients. Evidently, biotechnology is becoming more diverse in application and beginning to make substantial contributions to the health and well being of our society.

DILEMMAS AND CONCERNS

Like any other science, biotechnology is also not free from some obvious concerns. Environmental release of transgenes can impact food safety, human health, and the environment because of its vast applications in the food and agricultural industry. There is also public perception that the use of GM products or continued biotech research is ethically or morally wrong. Whether they are scientifically based or not, it is important that all these issues be carefully considered.

Herbicide and Insecticide Resistance

The possibility of transferring transgenes from GM crops to wild relatives to create "superweeds" and insects that have developed tolerance to insecticides to create "superbugs" is not an unreal threat. Herbicide-resistant crops are voluntarily coming up in different fields creating huge problems. However, agronomic practices like crop rotation, hybrid rotation, and integrated pest management could reduce these risks.

Transfer of Allergens

To develop soybeans with higher methionine content, a Brazil nut gene was inserted, which is responsible for methionine-rich protein, into soybean. The transgenic soybean was supposed to be utilized as animal feed. But the product was found to be allergenic and the soybean was never put on the market.[6] StarLink corn was released for animal feed only. The corn contains Cry9C protein, a potential allergen. The genes were found in human food and seed corn causing many corn-based foods to be recalled, although no health problems associated with the corn were reported.[4] However, there is a real concern that an accidental mix-up with the human foods could exist.

Antibiotic-Resistant Genes

Antibiotic-resistant genes originally isolated from bacteria are used as selectable markers for transforming plants. By the use of such genes, cells that have been modified can grow in the presence of a specific antibiotic, for instance, kenamycin-resistant corn. If these genes are transferred into human or other bacteria, they may increase antibiotic resistance in the human or bacterial population. As such worldwide, an estimated nine billion dollars are wasted each year on ineffective use of antibiotics.[7]

Gene Escape and Genetic Pollution

Preventing the movement of transgenes to the other plants is a concern for biotechnologists to avert genetic pollution. There are some gene containment techniques like male sterility, terminator technology, apomixis, cleistogamy, chloroplast transformation, and transgenic mitigation that can reduce gene transfer,[8] but presently these are used for theoretical or laboratory purposes only.[9]

Social, Moral, and Ethical Issues

Multinationals have heavily invested in biotechnology and they would naturally want a return on their investment. It is a legitimate concern to think that these companies will either enforce technology fees or monopolize the marketplace. However, studies have revealed that farmers also enjoy a financial benefit from GM crops.[6] This is mainly because of less pesticide use and enhanced production. In Canada, GM canola farmers experienced a 40% reduction in herbicide costs and a 10% increase in yield.[10] Besides, other issues like increased reliance on developed countries by developing countries, control of food production by a special few, foreign exploitation of natural resources, and infringement on the natural organism's inherent traits are of great concern.[3] Is it morally right to do gene transfer from bacteria to plant, animal to another animal, human to animal? Are these ethically correct? These dilemmas warrant a rational answer.

CONCLUSIONS

Although there are several GM products commercially available, biotechnology has no magic solution to all our problems. It has tremendous promise but to attain its full potential it must be environmentally compatible, economically viable, and socially responsible. This is an ever-emerging technology that needs proficient management of its development. However, biotechnology may be the single most sought-after tool at our disposal capable of making a significant contribution towards food security and sustainable development.

REFERENCES

1. *Chambers Science and Technology Dictionary*, New Ed.; Walker, M.B., Ed.; W&R Chambers Ltd: Edinburgh, **1991**; 1–1008.
2. Coombs, J. *Macmillan Dictionary of Biotechnology*; Macmillan Press: London, **1986**; 1–330.
3. Venter, B. Modified crops: the pros and cons. In *Pretoria News; Pretoria, 2004*; http://www.pretorianews.co.za/index.php?SectionID=665&ArticleID=2150709 (accessed September 2004).
4. Borem, A.; Santos, F.R.; Bowen, D.E. *Understanding Biotechnology*; Prentice Hall, PTR: New Jersey, **2003**; 1–216.
5. Marsa, L. Garden variety vaccines may be edible alternative. In *Los Angeles Times*; Los Angeles, 2005. Today in AgBioView from; http://www.agbioworld.org (accessed April 2005).
6. *Biotechnology Myths & Facts*. Global Knowledge Center on Crop Biotechnology: Manila, 2002; 1–37. CropBiotechnet; http://www.isaaa.org/kc (accessed August 2004).
7. Grace, E.S. Biotechnology of the body. In *Biotechnology Unzipped Promises & Realities*; Trifolium Books Inc.: Toronto, **1997**; 55–96.
8. Daniell, H. Molecular strategies for gene containment in GM crops. Nat. Biotechnol. **2002**, *20*, 581–586.
9. Slatter, A.; Scott, N.; Fowler, M. Future prospects of GM crops. In *Plant Biotechnology, The Genetic Manipulation of Plants*; Oxford University Press: Oxford, New York, **2003**; 305–333.
10. *An Agronomic and Economic Assessment of Transgenic Canola*. Canola Council of Canada, 2001; 1–95. Canola Council of Canada; http://www.canola-council.org/production/gmo_main.htm (accessed September 2004).

Biotechnology: Ethical Aspects

Kathrine Hauge Madsen
Centre for Bioethics and Risk Assessment, Royal Veterinary and Agricultural University, Frederiksberg, Denmark

Peter Sandoe
Department of Agricultural Sciences (Weed Science), Royal Veterinary and Agricultural University, Frederiksberg, Denmark

Abstract

Genetic engineering of crops and other food products has resulted in widespread opposition, particularly in Europe. Partly in response to the concerns of the general public, a system has been set up to make scientifically based risk assessments prior to the release of genetically modified organisms (GMOs) and the marketing of GM products. However, these risk assessments do not seem fully to address worries commonly felt by members of the general public. The aim of this entry is to define the issues that are at stake in a broad ethical discussion concerning GM food.

PUBLIC CONCERNS

The public's concerns about the genetic modification of crops and other applications of gene technology to food production have been voiced in some parts of the world. In Europe, a large-scale survey, the so-called "Eurobarometer," conducted in 1999 showed that on average only 31% of those asked "mostly or totally agreed" that using modern biotechnology in the production of foods should be encouraged. Compared to the previous survey, made in 1996, this represents a fall of 13%.[1] Public concern in Europe has led to a temporary ban on new releases of transgenic crops until new and stricter legislation is in place. In the United States, by contrast, the technology has been positively received and widely adopted with less of public debate.

The extensive Eurobarometer surveys strongly suggest that lack of knowledge about gene technology does not explain this opposition to GM food. Indeed increased knowledge serves to merely polarize those who have not yet made up their minds into positive or, more frequently, negative attitudes. Furthermore, the surveys show that resistance is not directed towards the technology as such. Most of those interviewed welcomed the medical progress brought about by developments in genetic engineering. Thus, it is not the process of genetic engineering per se that engenders opposition to it, but rather specific applications of it.

The authors of the Eurobarometer surveys give the following interpretation of what will be required if the people of Europe are to accept applications of gene technology outside the medical sphere:

> First, usefulness is a precondition of support; second, people seem prepared to accept some risk as long as there is a perception of usefulness and no moral concern; but third, and crucially, moral doubts act as a veto irrespective of people's views on use and risk.

Consumer choice is a further requirement that would seem to play a role in encouraging acceptance of the technology.

The remainder of this entry looks in more detail at these four requirements for acceptability: usefulness, low risk, consumer choice, and no "moral" objections. However, a few words should first be said about the involvement of ethics in this area.

THE ROLE OF ETHICS

To understand ethical concerns is to open an agenda and encourage discussion. To set out the ethical issues concerning the development and use of gene technology, a distinction needs to be made between two types of ethical question. The first and most fundamental sort of question raises the issue: *What sorts of concern are morally relevant when deciding whether or not it is morally acceptable for us to use gene technology in agriculture*? Suppose, for instance, that we are trying to decide whether it is ethically acceptable to create herbicide-resistant crops. Some people think that as long as these crops do not give rise to environmental hazards and are safe to eat there is no ethical problem. Others, however, think that, besides environmental and food safety issues, other concerns are morally relevant. For example, some people object to the creation of genetically modified crop plants because they think that modifying living organisms in this way violates the order of nature or involves an attempt to modify the work of the Creator, thus amounting to an objectionable exercise in "playing God."

When discussing ethical issues in biotechnology, it is crucial that an effort should be made to distinguish these sorts of concerns. In too many discussions, those involved simply fail to see that their disagreement arises from the fact that they take different concerns to be morally relevant. Where this happens, the moral debate about biotechnology is apt to make less progress than it otherwise might.

Of course, recognition that the biotechnology debate involves a range of distinct concerns is no guarantee that further discussion will result in consensus. But parties to the argument will at least know what it is that they disagree about at the fundamental level, and thus gain some sort of understanding of why other people reject views that they take to be settled truths. It may, furthermore, be possible to argue rationally that some concerns genuinely are more important than others, or that some alleged moral concerns are not morally relevant at all.

The second sort of ethical question addresses the problem of *how to weigh the different ethical concerns against each other*. This presupposes that the first question has already been answered and it is known what the relevant concerns are. Because these issues are separable, two people may agree entirely on what the relevant concerns are and nevertheless continue to disagree over the use of gene technology in food production, because they disagree over what weight to assign to their agreed concerns.

USEFULNESS

The usefulness of gene technology is obvious to the agrochemical industry and plant breeding companies, as a way of extending the market of existing products by developing herbicide-resistant crops rather than searching for new products. For crop seed companies, the technology has reduced the time frame for development and enabled the production of varieties with an entirely new set of traits, something, which could not as easily have been achieved using traditional breeding methods.

As far as farmers are concerned, the genetically engineered crops currently in production have already displayed a number of advantages, such as increased yield levels, reduced pesticide use, reduced input costs, and more flexibility. However, many farmers are concerned that it may become increasingly difficult to handle and sell genetically engineered products as a result of increasing opposition towards these crops.

To date, the usefulness of gene technology in food production has not been obvious to the general public in Europe. Cheaper food is not really an issue in this part of the world. Many people think that food prices are already low: for example, only 8% of Danes are willing to accept the case for genetically modified food products on the basis of cost savings alone.[2] Rather, people tend to be concerned about the undesirable effects of modern intensive farming. However, some adherents of gene technology argue that herbicide-resistant crops may be useful from the point of view of environmental protection because herbicides may be applied at a later stage of growth leaving food for insects and small mammals in the field, thus enhancing wildlife.

One reason why the environmental argument has not gained public support in Europe and among influential nongovernment organizations is that in Europe there is widespread scepticism about the usefulness of pesticides in general because of possible unwanted side effects on human health and the environment. Many see the development towards nonchemical farm management practices as the very same thing as sustainable farming. The introduction of herbicide-resistant crops is based on the use of chemicals, and it is therefore automatically viewed as a nonsustainable path for the future.

Scepticism may change if benefits of applying gene technology to food production become more obvious. If so, the next issue to be discussed is whether there are significant risks involved in the application of gene technology to food production, and whether these risks are worth taking.

RISK

Many biotechnologists see gene technology as a precise method by which one or more identifiable genes are inserted in plants or animals. This contrasts with conventional breeding methods, where thousands of random genes are transferred. It is therefore easier to predict the toxicological and ecological effects of inserting the genes. Opponents of the technology, however, argue, that inserting genes from distant organisms, which would never interbreed under natural circumstances, might have completely unknown effects that may only be identified through long-term monitoring. Furthermore, ecological risk assessment has not yet been able to quantify the risk of growing genetically engineered crops because biological systems are complex.

In order to address public concerns about environmental risks, the so-called "Precautionary Principle" has gradually been adopted into international environmental legislation. Within the European Union, the principle creates actionable rights in connection with products or processes carrying unacceptable risks, provided that potential hazards have been identified, and that it is not possible to make a scientific risk assessment and thereby define risk with adequate precision.

Despite the general doubts already mentioned, it is fair to say that a number of GM crops appear to be safe in terms of ecological risk assessment, judged by ordinary scientific standards. Furthermore, from the point of view of health, no serious hazards have been identified as regards the use of GM crops for food production. However, it is important to be aware that the general population and its spokespeople may have a broader notion of what counts as a hazard than the one taken for granted in ordinary risk assessment.

There may also be hazards of a socio-economic nature. One argument often heard in the debate draws attention to the monopoly enjoyed by the multinational agrochemical industry. These companies have bought several plant-breeding companies to link seed production to agrochemicals. Many people resent this development towards monopolization. Instead they wish to secure influence on development at the community level—a development that is based on "need, not greed," and that allows the consumer free choice.

CONSUMER CHOICE

Consumers object to food produced by genetic engineering if they have no freedom of choice not to buy the product. Such an opportunity can be achieved only by labeling genetically modified food products—something already regulated by the novel food EU-directive. From the consumer's point of view, it may be desirable to extend the labeling requirements to include food, which, while it no longer contains genetically engineered compounds, was produced using this technology. Stringent labeling requirements could, however, discourage retailers from stocking genetically engineered food products. This is because, following the imposition of such labeling requirements, there would probably be increased costs, because now two selections of each product may have to be available: one produced with, and one produced without, the use of genetic engineering. At this point it is difficult to predict which strategy will be more profitable in the long run to the producer.

MORAL OBJECTIONS

Many people see genetically modified food as a "threat the to the natural order of things," because the processes, or results of gene technology are somehow "unnatural," and for that reason objectionable. Some people have also felt that "scientists should not play God." According to this allegation, scientists involved in biotechnology are playing God in the sense that they are trying to modify the work of the Creator. In so doing, they are, perhaps, assuming the role of divine creators themselves. If this allegation were taken quite literally, it would force mankind to stop doing many things that have been essential to human survival. For several thousand years agriculture has been based on the principle of selection, and thus on deliberately changing species. Few think this is a bad thing, even if it involves modifying the work of the creator, so it is not really clear why more advanced techniques that aim at similar ends should be considered objectionable exercises in "playing God."

The first allegation, that the processes or results of gene technology are "unnatural" faces problems as well. First of all, it will have to be explained what is meant by "unnatural." And this explanation needs to rule out the inappropriate kinds of "advanced" biotechnology but not the more traditional farming methods—methods which almost everyone accepts and which cannot plausibly be claimed to be morally dubious, such as the use of ordinary selective breeding.

Secondly, even if such an explanation could be provided, it would still have to be shown that unnatural in the specified sense is also immoral. This point is particularly important since, as is well known, throughout history the characterization of certain acts or practices as unnatural has often revealed no more than stubborn prejudice. Discrimination against homosexuals is a case in point.

Even though these allegations are difficult to sustain, the fact that they are widely treated with sympathy is significant. It shows, among other things, that there is a need for continuing public discussion of nature and humankind's place in it.

REFERENCES

1. Anonymous. Eurobarometer 52.1, *The Europeans and Biotechnology,* Report by INRA (Europe)—ECOSA; Directorate-General for Research, Directorate B—Quality of Life and Management of Living Resources Programme and Directorate-General for Education and Culture "Citizens Centre" (Public Opinion Analysis Unit); Brussels, Belgium, **2000**; 85.
2. Anonymous. *Negative Holdning (Negative Attitude),* A Survey by the Danish Technology Council; AC Nielsen and AIM A/S; Politiken 15 November **1998**; 2 (in Danish).

Bovine Embryos: In Vitro Culture

Karen Moore
Department of Animal Sciences, University of Florida, Gainesville, Florida, U.S.A.

Abstract

Culture of bovine embryos has transitioned over the last two decades from requiring in vivo culture to the successful production of viable embryos cultured completely in vitro. This has been achieved by gaining knowledge into the changing needs of the early preimplantation embryo. Attempts to produce a culture system that more closely mimics the maternal environment and meets these changing needs are essential for embryonic survival in vitro. The aim of this short entry is to provide a summary of key requirements for successful in vitro culture of the bovine embryo and methods utilized for assessing their efficacy for maintaining embryo viability.

INTRODUCTION

The quality and survival of in vitro–produced (IVP) bovine embryos have historically lagged behind that of in vivo–derived embryos. Although 70–80% of bovine oocytes are successfully fertilized in vitro, only 20–40% survive the first week of life. While oocyte competence, maturation, and fertilization can impact survival, this entry will focus only on in vitro culture conditions that are crucial to the survival of the early bovine embryo. Culture conditions must be improved so that embryo survival is not compromised. This will help make the artificial reproductive technologies, such as IVP, ovum pick up, somatic cell nuclear transfer, cryopreservation, and others more feasible for use by the cattle industry.

In vitro culture of bovine embryos has improved dramatically over the last two decades. As the embryo develops from the zygote to the blastocyst, it changes in morphology, physiology, biosynthetic activity, and metabolism. During this first week of life, the embryo must go through cleavage, activate its embryonic genome, become polarized and go through compaction, form a blastocoele, and begin the process of differentiation. Attempts to understand the embryo's needs and the maternal environment more fully have led to the development of more successful culture systems. These systems more closely mimic the reproductive environment of the cow, which in turn allows these embryonic processes to occur more naturally. The challenge is to develop a system or systems that meet the changing needs of the developing embryo, minimize cellular stress, and prevent loss of viability. This entry will address key ingredients for successful in vitro culture of bovine embryos and potential methods for assessing their efficacy for maintaining viability.

CULTURE SYSTEMS

Culture systems for bovine embryos have evolved over time, initially requiring short-term in vivo culture to obtain viability, which was time-consuming and expensive. Next, attempts to mimic this environment led to the utilization of embryo coculture with tissue explants or epithelial monolayers from the oviduct, uterus, or other somatic cell lines. Although these improved survival over culture medium alone, the results were variable. In an attempt to provide a more defined, reliable environment and remove variability, in vitro embryo culture has now progressed to either a semi-defined system or to sequential culture, which will be addressed below. Each system has attempted to refine the culture medium, supplements, and the environment to achieve development similar to what is achieved in vivo.

MEDIUM

Culture medium has changed from complex tissue culture media to those media more specific for culture of bovine embryos. Earlier research utilized complex media designed for somatic cells when coculture was more prominent. More recently though, a strategy has been used to develop media that more closely resemble the maternal environment and meet the changing needs of the embryo. Table 1 gives a comparison of the ingredients found in six commonly used bovine embryo culture media: SOFaa, CR1aa, KSOMaa, MBMOC, G1.2, and G2.2.[1–4] They all contain a basal salt solution, a buffer, an energy source, and possibly supplements of amino acids, vitamins, minerals, heavy metal cation chelators, macromolecules, proteins, and nucleotides. Although it is clear that these media vary substantially, several things are common to all: use

Table 1 Composition of commonly used media for bovine embryo culture.

Ingredient (mM)	Culture medium					
	SOFaa	CR1aa	KSOM	MBMOC	G1.2	G2.2
NaCl	107.7	114.7	95.0	81.62	90.08	90.08
$NaH_2PO_4 \cdot 2H_2O$	—	—	—	—	0.25	0.25
KCl	7.16	3.1	2.5	4.83	5.5	5.5
KH_2PO_4	1.19	—	0.35	1.19	—	—
$MgCl_2$	0.49	—	—	—	—	—
$MgSO_4$	—	—	0.2	1.19	1.0	1.0
$CaCl_2 \cdot 2H_2O$	1.71	—	1.7	1.7	1.8	1.8
Lactate	3.30	5.0	10.0	31.30	10.5	5.87
Pyruvate	0.33	0.40	0.20	0.27	0.32	0.10
Glucose	1.50	—	0.20	—	0.50	3.15
BSA	8 mg/mL	3 mg/mL	8 mg/mL	6 mg/mL	2 mg/mL	2 mg/mL
$NaHCO_3$	25.07	26.2	25.0	25.12	25.0	25.0
L-Glutamine	—	1.0	1.0	1.0	0.5	1.0
Taurine	—	—	—	—	0.1	—
Non-essential amino acids[a]	All	All	All	All	All	All
Essential amino acids[a]	—	—	—	—	—	All
EDTA	—	—	0.01	0.11	0.01	—
Na acetate	—	—	—	0.61	—	—
Ca pantothenate	—	—	—	—	—	0.0042
Choline chloride	—	—	—	—	—	0.0072
Folic acid	—	—	—	—	—	0.0023
Inositol	—	—	—	—	—	0.01
Niacinamide	—	—	—	—	—	0.0082
Pyridoxal	—	—	—	—	—	0.0049
Riboflavin	—	—	—	—	—	0.0003
Thiamine	—	—	—	—	—	0.003

[a]See Table 2.

of ultrapure water (18 mOhm resistivity) free of pyrogens, an osmolarity of 260–295 mOsm/kg, and a pH of 7.2–7.4. These points are critical for the basal medium, but in and of themselves do not produce optimal results. It is the addition of supplements to these basal media that help to optimize embryo development in culture and these supplements will be addressed briefly below.

SUPPLEMENTS

Serum was widely used in the earlier works for culturing bovine embryos, as it was known to bind growth factors, chelate heavy metal cations, and serve as an osmolyte. However, serum gives variable results due to its source and the undefined components it contains. Use of serum in culture can also alter the metabolism of the early embryo, can lead to increased fatty acid content in blastocysts, can increase the frequency of mixoploid embryos, and may result in alterations in gene expression profiles over in vivo–derived embryos. Moreover, serum can lead to problems later in development, such as poor cryosurvival and reduced survival following embryo transfer, including large offspring syndrome, making its use less than optimal for today's culture systems.

To circumvent the problems related to serum, work has been done to develop semi-defined culture systems. Bovine serum albumin (BSA) is the most abundant protein in the female reproductive tract and has been the protein of choice for achieving development to blastocyst, increased cell numbers, improved embryo quality, and increased pregnancy rates. However, BSA is still an undefined component in culture leading to issues with batch variability due to its undefined levels of fatty acids, growth factors, and other small molecules. Commercially available serum replacers have also been tested with some benefit over serum in embryo culture but are still not completely defined. Hyaluronan, one of the most abundant glucosaminoglycans found in follicular fluid and bovine reproductive tract fluids, can also be used in embryo culture to increase development to blastocyst, increase cell number, and increase survival post thaw. More recently though, the use of recombinant human albumin has shown promise for effective culture of bovine embryos.

Current efforts to develop completely defined culture systems without the loss of viability are underway. Synthetic macromolecules, such as polyvinyl alcohol and polyvinylpyrrolidone, are increasingly used in these defined systems. These synthetic polymers have surfactant qualities similar to BSA, but their use in culture is still

Table 2 Composition of non-essential and essential amino acid cocktails (Invitrogen).[a]

Non-essential amino acids	Concentration (mM)	Essential amino acids	Concentration (mM)
Alanine	0.1	Arginine	0.6
Asparagine	0.1	Cystine	0.1
Aspartate	0.1	Glutamine	2.0
Glutamate	0.1	Histidine	0.2
Glycine	0.1	Isoleucine	0.4
Proline	0.1	Leucine	0.4
Serine	0.1	Lysine	0.4
		Methionine	0.1
		Phenylalanine	0.2
		Threonine	0.4
		Tryptophan	0.05
		Tyrosine	0.2
		Valine	0.4

[a]Concentrations specified by Eagle.[8]

suboptimal, often yielding lower development to blastocyst, decreased cell numbers, and altered amino acid profiles. These problems suggest that beneficial components normally present in BSA-containing systems are missing and need to be added to make these defined environments more conducive to embryo development and survival.

Energy sources for bovine embryo culture are another area of concern for in vitro culture and depend upon the stage of the embryo.[5] Embryos rely on oxidative phosphorylation for the generation of ATP during early preimplantation development. The early embryo utilizes pyruvate and lactate through the tricarboxylic acid cycle for ATP production. Glucose is known to be inhibitory to early cleavage stage embryos; however, in small amounts glucose is still beneficial for nucleic acid and lipid biosynthesis. High glucose can also skew the sex ratio in favor of male embryos.[6] At compaction and blastocyst formation the embryo's metabolism changes, so that glucose becomes the preferred energy source. These are important points to consider when culturing embryos and have led to the development of sequential culture systems[3] that can accommodate these changing metabolic needs, making them more similar to what occurs in vivo. By moving embryos to a new medium (either at compaction or every 48 hours), embryo viability is enhanced.

Amino acid supplementation also appears to be the key to improving development in vitro. Amino acids can serve as osmolytes, act as buffers of intracellular pH, can chelate heavy metal cations, serve as energy sources, reduce apoptosis, and act as anabolic precursors. Both utilization and secretion of amino acids by embryos change during development and attempts to provide concentrations similar to those found in the oviduct or uterine environment have proven beneficial.[7] Early cleavage stages benefit most from non-essential amino acids and L-glutamine, whereas essential amino acids can be detrimental at this stage (Table 2). However, following compaction both essential and non-essential amino acids help in development. A word of caution is warranted here though, as amino acids can be deaminated at high temperatures (37°C) and through embryo metabolism, resulting in high ammonium production. Embryos are highly sensitive to ammonium, which can result in a developmental block at the 8–16 cell stage if severe or produce later developmental defects and may be one of the culprits in the production of large offspring syndrome. These problems can be partially circumvented by moving embryos to fresh culture-drops every 48 hours.[9]

Antioxidants are very useful in the culture of bovine embryos since reactive oxygen species (ROS) production increases over time in culture. Scavengers of oxygen-free radicals are present in the more complex culture systems, but as culture conditions are refined, ROS can prove problematic. Addition of cysteine, glutathione, EDTA, or β-mercaptoethanol has been shown to reduce free radicals in these defined systems and improve blastocyst formation.

Finally, no discussion of embryo culture medium would be complete without the mention of growth factors.[5] Embryos derived in vivo are exposed to several growth factors, including insulin, insulin-like growth factor-1, leukemia inhibitory factor, epidermal growth factor, platelet activating factor, and transforming growth factor α, to name a few. Several of these have been shown to be beneficial for development in culture and for increasing pregnancy rates. However, defining the appropriate conditions for their use in vitro is often difficult, and further research is warranted.

ENVIRONMENT

Sensitivity of embryos to their culture environment can lead to reduced development in culture, as well as long-term alterations in fetal and postnatal development. Culture of bovine embryos has evolved from strictly in vivo culture

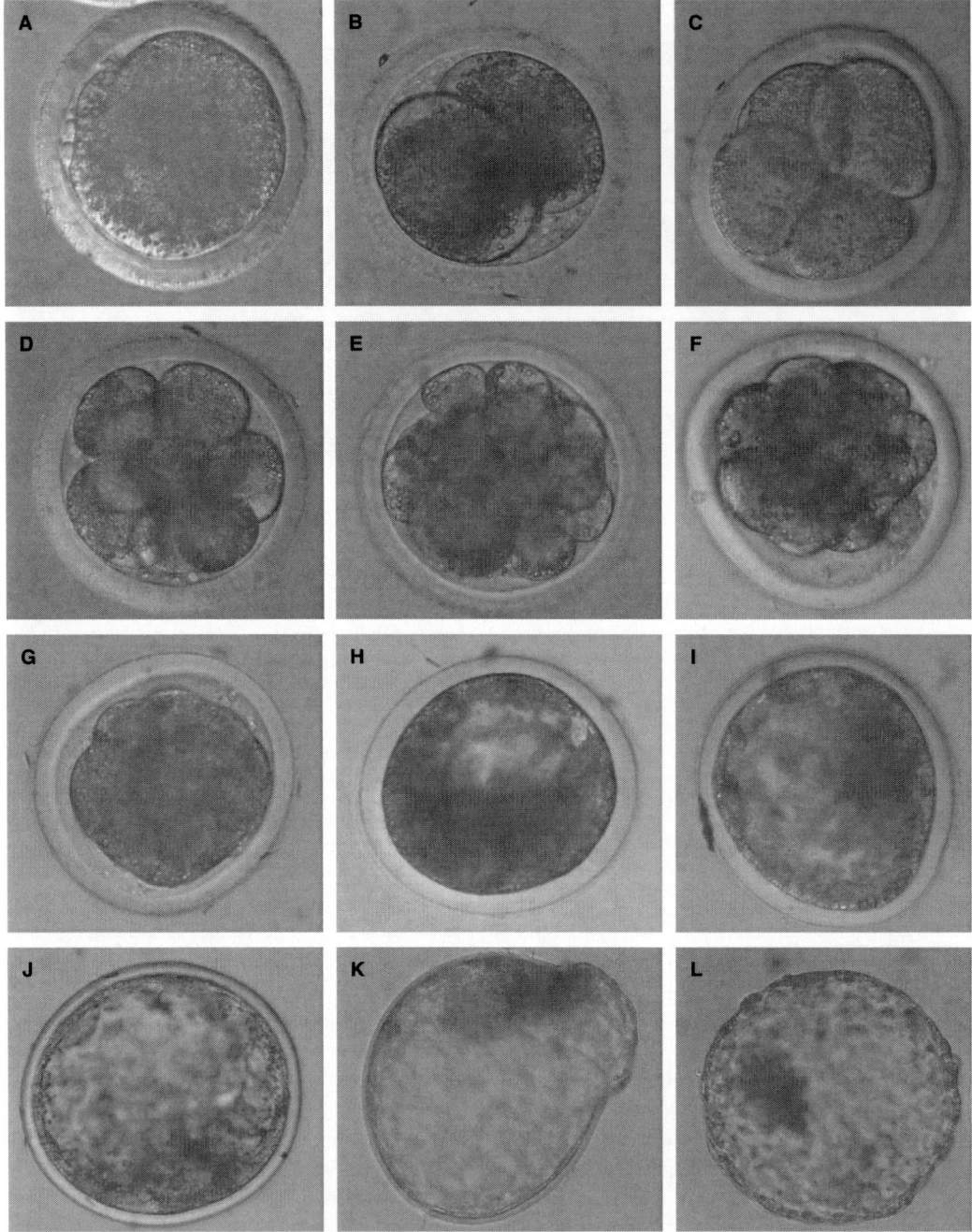

Fig. 1 Bovine embryo developmental stages and appropriate ages of attainment postfertilization (pf). (A) 1-cell zygote, day 1 pf; (B) 2-cell, day 1–2 pf; (C) 4-cell, day 2–3 pf; (D) 8-cell, day 3–4 pf; (E) 16-cell, day 4–5 pf; (F) morula, day 4–5 pf; (G) compact morula, day 5–7 pf; (H) early blastocyst, day 6–8 pf; (I) blastocyst, day 7–9 pf; (J) expanded blastocyst, day 7–9 pf; (K) hatching blastocyst, day 7–9 pf; (L) hatched blastocyst, day 7–9 pf.

to more refined in vitro systems used today. Embryos have been cultured in a variety of vessels including gassed tubes, organ culture dishes, 4-well plates (Nunc), terasaki plates (Greiner Bio-One), microdrops, hanging drops, and Well of the well (WoW; Minitube). Culturing in microdrops, however, is still the most common method in use at this time. Different culture volumes have been investigated (embryo: medium ratio of 1:1 to 1:30), with reduced volumes being more beneficial, most likely due to paracrine and autocrine stimulation. Reduced volumes result in increased development to blastocyst, increased cell numbers, improved cryosurvival and survival post transfer. Microdrops must be covered with oil (mineral, paraffin, or silicone) in order to maintain both osmolarity and pH of the medium and should be prepared several hours prior to use to ensure equilibration of pH and temperature.

A lot of emphasis has been placed on culture medium to this point, but there are additional components that are critical to the embryo culture environment. Humidity is crucial for maintaining the correct osmolarity of media by avoiding dehydration. Temperature in culture incubators is traditionally set at 38.5–39°C, similar to that of the maternal environment. When embryos are brought out of the incubator, attempts to maintain this temperature through the use of heated stages can be beneficial.

The gas atmosphere in room air is not what an embryo would normally experience. The in vivo-derived embryo normally encounters an O_2 range of 2–8% in the oviduct, which is reduced further in the uterus to 1.5–5%.[3] Most culture is performed either in 5% CO_2 in air or in a reduced oxygen environment of 5% CO_2, 5% O_2, and 90% N_2. The latter more closely mimics that found in the reproductive tract, and in most cases is more beneficial, resulting in increased cell numbers, cryosurvival, higher pregnancy rates, and lower birth weights for IVP embryos. A low oxygen environment can be achieved using specially equipped incubators, through the use of modular chambers or even appropriately gassed desiccators. High oxygen environments, however, increase ROS that can result in lipid peroxidation and cause DNA damage, lowering embryo development and viability. Most culture systems today take advantage of the benefits of growing embryos in low oxygen.

pH also has a critical impact on embryo development in culture. This includes pH of the medium, as well as the intracellular pH (pHi). Prior to compaction, the early embryo is impaired in its ability to regulate pHi. pH can be altered by components in the medium, such as lactic acid, amino acids, buffers (bicarbonate, HEPES, MOPS), as well as the gaseous environment. Acidic or basic conditions can lead to alterations in mitochondrial function and distribution, as well as microfilament organization.[9] Therefore, great care must be exercised when working with embryos in bicarbonate-based media under atmospheric conditions to avoid rapid increases in pH.

Finally, technical skill should not be discounted when working with embryos. Embryos should be handled with care and speed, while moving them in small volumes. This can be done using either a fine drawn Pasteur pipet attached to a syringe or a mouth pipet, or more commonly by using a positive displacement pipet. All equipment and media must be sterile and handled with good sterile technique. Attention to these details will result in shorter periods of suboptimal temperature, pH, and lighting, and will improve development in culture substantially.

ASSESSING CULTURE EFFECTIVENESS

Pregnancy and calving are the ultimate measures of successful culture of IVP embryos. However, this can be cost-prohibitive and time-consuming, so several assays have been utilized for evaluating viability at the preimplantation stage. Most common is the use of embryo staging (Fig. 1) and quality grading. Embryos should be within one stage of the appropriate developmental stage for their age, since embryos further delayed are usually not viable. Quality grades are very useful for determining which embryos can be frozen (grade 1 and 2), but otherwise have limited value for predicting longevity. Other more recent assays include viability staining, counting total cells, immunocytochemistry for various protein markers, electron microscopy for ultrastructural changes, PCR for appropriate gene expression,[10] evaluation of cleavage rates, respiration and/or metabolite screening.[3] The problem with several of these methods is that they are either technically demanding and/or lethal to the embryo. Future efforts to define viability markers that are not harmful to the developing embryo, are easily measured, and closely resemble what is seen for in vivo-derived embryos as a great asset for further improving bovine embryo culture.

CONCLUSIONS

The days of static culture conditions are becoming obsolete, as sequential culture systems continue to improve and more closely mimic the metabolic requirements normally found in vivo. Efforts to minimize stressors and regulate homeostasis in culture are crucial for maintaining survival both in vitro, as well as later in development post transfer. This entry has summarized the key areas for successful bovine in vitro embryo culture today. However, further efforts for improving culture conditions are warranted in order to improve development, ultimately reducing embryonic losses in utero, and allowing for greater use of embryos produced by the artificial reproductive technologies.

ACKNOWLEDGMENT

This entry was supported by the Florida Agricultural Experiment Station.

REFERENCES

1. Tervit, H.R.; Whittingham, D.G.; Rowson, L.E. Successful culture in vitro of sheep and cattle ova. J. Reprod. Fertil. **1972**, *30* (3), 493–497.
2. Rosenkrans, C.F., Jr.; First, N.L. Effect of free amino acids and vitamins on cleavage and developmental rate of bovine zygotes in vitro. J. Anim. Sci. **1994**, *72* (2), 434–437.

3. Gardner, D.K.; Lane, M. Culture of mammalian preimplantation embryos. In *A Laboratory Guide to the Mammalian Embryo*; Gardner, D.K., Lane, M., Watson, A.J., Eds.; Oxford University Press: New York, 2004; 41–61.
4. Moore, K.; Bondioli, K.R. Glycine and alanine supplementation of culture medium enhances development of in vitro matured and fertilized cattle embryos. Biol. Reprod. **1993**, *48* (4), 833–840.
5. Thompson, J.G. In vitro culture and embryo metabolism of cattle and sheep embryos — a decade of achievement. Anim. Reprod. Sci. **2000**, *60–61*, 263–275.
6. Kimura, K.; Spate, L.D.; Green, M.P.; Roberts, R.M. Effects of d-glucose concentration, d-fructose, and inhibitors of enzymes of the pentose phosphate pathway on the development and sex ratio of bovine blastocysts. Mol. Reprod. Dev. **2005**, *72* (2), 201–207.
7. Li, R.; Wen, L.; Wang, S.; Bou, S. Development, freezability and amino acid consumption of bovine embryos cultured in synthetic oviductal fluid (SOF) medium containing amino acids at oviductal or uterine–fluid concentrations. Theriogenology **2006**, *66* (2), 404–414.
8. Eagle, H. Amino acid metabolism in mammalian cell cultures. Science **1959**, *130* (3373), 432–437.
9. Lane, M.; Gardner, D.K. Understanding cellular disruptions during early embryo development that perturb viability and fetal development. Reprod. Fertil. Dev. **2005**, *17*, 371–378.
10. Wrenzycki, C.; Herrmann, D.; Lucas-Hahn, A.; Korsawe, K.; Lemme, E.; Niemann, H. Messenger RNA expression patterns in bovine embryos derived from in vitro procedures and their implications for development. Reprod. Fertil. Dev. **2005**, *17*, 23–35.

Cattle Embryo Transfer

Curtis R. Youngs
Department of Animal Science, Iowa State University, Ames, Iowa, U.S.A.

Robert A. Godke
Department of Animal Sciences, Louisiana State University Agricultural Center, Baton Rouge, Louisiana, U.S.A.

Abstract

Embryo transfer is a reproductive biotechnology that was successfully performed for the first time in cattle more than 50 years ago. However, it was not until the mid-1970s that a commercial embryo transfer industry emerged following the development of non-surgical embryo collection and transfer methods. More than half a million embryos are transferred each year throughout the world as a means of improving the overall genetic quality of cattle that produce meat, milk, and other products for human use. Embryo cryopreservation (freezing) is a routine part of the commercial embryo transfer industry, and recent advances in embryo sexing, in vitro fertilization, and nuclear transfer ("cloning") have created opportunities for novel application of embryo transfer technologies in cattle not only in production agriculture but also in biomedical research.

INTRODUCTION

The first successful mammalian embryo transfer was conducted using rabbits in 1890. This was followed several years later by successful embryo transfer in sheep and goats (1930s) and in pigs and cattle (1950s). Commercial use of embryo transfer in cattle did not occur in North America until the early 1970s, and initial methods involved time-consuming and labor-intensive surgical collection and transfer techniques. The popularity of embryo transfer increased markedly when non-surgical embryo collection and transfer procedures became commercially available in 1976. Currently, more than half a million cattle embryos are transferred each year on a global basis. The goal of this entry is to review the current use of embryo transfer as a biotechnological tool for genetic enhancement of cattle.

REASONS TO PERFORM EMBRYO TRANSFER IN CATTLE

Embryo transfer capitalizes on genetically superior females in much the same way that artificial insemination takes advantage of genetically elite males. A mature bull can sire ≥50 calves per year through natural mating but can sire as many as 65,000 calves per year through artificial insemination. It has been estimated that ≥200,000 ova exist in the ovaries of a female calf at the time of her birth, yet a typical cow produces fewer than 10 calves in her lifetime. Thus, tremendous opportunity exists to increase the number of offspring from any given cow through the use of embryo transfer technology. In many instances, embryo transfer enables a genetically valuable cow to produce more calves in one year than she would in her normal reproductive lifetime, facilitating more rapid genetic improvement. The more rapid genetic improvement is achieved by producing more calves from cows with the best genetics in the herd and fewer calves from cows with the poorest genetics in the herd (because the latter group of cows serves as recipient females for embryos obtained from the genetically superior cows). Embryo transfer may also be used as a means to transport cattle (in the form of embryos) from one location to another while minimizing the risk of disease transmission. Embryos can be washed to eliminate pathogens, whereas live animals pose greater risk of carrying disease.

EMBRYO TRANSFER PROCEDURES

The process of embryo transfer involves the selection of genetically superior cows that then undergo a procedure known as superovulation. Superovulation is induced by treatment with follicle-stimulating hormone (a hormone which the cow naturally produces each day) to cause follicles in the ovaries that would normally undergo atresia to grow, enabling the ovary to release multiple ova at the time of ovulation instead of one ovum. The superovulated donor female is artificially inseminated with semen from a genetically elite bull and approximately 7 days later the embryos are recovered from the donor cow's uterus by a non-surgical, transcervical flushing technique. Harvested embryos are located, evaluated microscopically, and then either transferred to recipient females whose estrous cycles were synchronized with that of the donor female or cryopreserved (frozen) for subsequent transfer.

The commercial embryo transfer industry has evolved from a few large centralized embryo transfer businesses to many smaller companies offering on-farm embryo collection and transfer services. Although the basic methodology has changed little since the 1970s,[1] an enhanced understanding of the pattern of ovarian follicular growth (obtained via ultrasonography) has facilitated the development of improved superovulatory treatment protocols. Experienced embryo transfer technicians often attain embryo recovery rates >75% with an average of six transferable quality embryos obtained per donor collection. Pregnancy rates following embryo transfer are often ≥70% under excellent conditions.

EMBRYO CRYOPRESERVATION

In the early 1970s researchers devised methodology to cryopreserve (freeze) embryos. Initial success with mouse embryos was followed shortly thereafter by success with cattle embryos. Currently, more than half of all cattle embryos transferred in the commercial embryo transfer industry are embryos that have been cryopreserved. The methodologies have changed over the years, leading not only to less time-consuming procedures but also to improved pregnancy rates following embryo transfer.

Embryos must be placed into a cryoprotective solution before they are slowly cooled to a temperature of $-196°C$ (the temperature of liquid nitrogen). The cryoprotective solution facilitates dehydration of the embryo so that ice crystal formation is reduced inside the embryonic cells during freezing (which would destroy them). Although the conventional procedure requires that the cryoprotective compound be removed from frozen-thawed embryos prior to transfer, an increasingly popular method involves the direct transfer of embryos into recipient cows without cryoprotectant removal. This direct transfer method involves the use of highly permeating cryoprotectant compounds that cause little osmotic stress on embryonic cells during cryoprotectant addition and removal. More recently, a novel method known as vitrification has enabled the ultra-rapid cryopreservation of embryos without the use of a specialized embryo freezing machine. By utilizing high concentrations of cryoprotectant compounds and ultra-rapid cooling rates, the solution containing the embryo changes from a liquid to a solid (referred to as "glass") without forming ice crystals. Pregnancy rates following transfer of excellent-quality frozen-thawed embryos are often similar to those obtained after transfer of fresh embryos, except for those cryopreserved via vitrification which typically yield slightly lower pregnancy rates.[2]

EMBRYO SEXING

Many of the larger commercial embryo transfer companies offer a DNA (deoxyribonucleic acid) test to determine the biological sex of an embryo prior to its transfer. A small number of cells are mechanically removed from the embryo, and the DNA from those cells is amplified using the polymerase chain reaction (PCR). Thousands of copies of DNA may be made via PCR from a single embryonic cell. Amplified DNA from both the Y chromosome (which indicates a male embryo) and either an autosome (a non-sex chromosome) or the X chromosome is analyzed. Results, which can be obtained in as few as 2 hours, are obtained from more than 95% of embryos, and sexing accuracy typically exceeds 98%. The capability of sexing embryos gives cattle producers the option of selectively producing bull calves (ideal for meat production) or heifer calves (ideal for milk production). However, sexed embryos typically do not survive cryopreservation as well as intact embryos, so most sexed embryos are transferred fresh with little reduction in pregnancy rate compared with transfer of unsexed embryos.

IN VITRO FERTILIZATION

One of the most rapidly expanding areas of embryo transfer technologies is that of in vitro fertilization, the so-called "test tube baby" technology. The first calf produced from in vitro fertilization (IVF) was born in the early 1980s, yet widespread adoption of this technology did not occur until the mid-1990s because IVF is a complex multistep process that requires a well-equipped laboratory and a skilled technician. Substantial research efforts have led to procedures that are commercially viable, and more than 300,000 IVF bovine embryos are being produced annually throughout the world. However, IVF is labor-intensive and likely to be too expensive to use in cows that respond well to standard "in vivo" production protocols.

The IVF procedure offers an alternative to cattle producers who have genetically valuable cows that for some reason are unable to produce viable embryos through standard embryo collection and transfer procedures. This technology can be used with ova harvested from older non-ovulatory cows, females with physical injuries (e.g., lameness), and problem-breeding cows (e.g., scarred cervix, cystic ovarian disease). Ova are collected from donor females via transvaginal ultrasound-guided aspiration, a procedure in which a stainless steel needle is inserted through the wall of the vagina and directly into the ovarian follicles to retrieve the ova.

With IVF, the potential exists for more embryos to be produced in a shorter period of time because the collection procedure can be repeated on the same cow 6 or more times per month. Ova can be harvested from an early postpartum (<25 days) cow before she resumes normal reproductive cyclicity, enabling her to produce one or more extra calves before she is mated to establish a pregnancy. Ova also can be collected from the cow's ovaries during the first 3 mos of pregnancy. In addition, transvaginal ultrasound-guided aspiration can be used to harvest ova from prepubertal heifers.

From high quality ova harvested from cattle ovaries, one would expect a 90% in vitro maturation rate and 80–90% fertilization and cleavage rates. Development of IVF zygotes to the morula and blastocyst stage typically is 35–50%. The resultant IVF embryos are then non-surgically transferred into recipient females at the appropriate stage of their estrous cycle, with pregnancy rates from excellent quality embryos ranging from 50 to 65%. Although viable healthy calves have been produced from frozen-thawed IVF embryos, the success rate is generally lower than that with unfrozen IVF embryos. Vitrification appears promising to help overcome this reduced posttransfer pregnancy rate. The problem of "large offspring syndrome" (IVF calves born with extremely high birth weights) has been mostly overcome due to refinements in embryo culture methodology.

EMBRYO "CLONING"

In the early 1980s, procedures were developed to produce genetically identical twin offspring ("clones") by bisecting (splitting) embryos. Pregnancy rates obtained after the transfer of a half-embryo (or demi-embryo) were nearly as good as those obtained following transfer of intact embryos. A variety of methods are available to produce demi-embryos, and some are relatively inexpensive and practical for use within small embryo transfer companies.

Embryo bisection offers the potential of doubling the number of viable embryo transfer offspring produced from valuable donor females. For example, 100 high quality, intact embryos may result in the birth of 65 embryo transfer calves. In contrast, 100 similar quality embryos divided into halves would yield 200 demi-embryos which may result in 120 embryo transfer calves (120% pregnancy rate from 100 embryos), assuming pregnancy rate of demi-embryos is reduced by 5%. Of course, twice as many recipient females are needed when transferring demi-embryos unless both demi-embryos from a single embryo are transferred into the same recipient to induce twinning.

In the mid-1980s, a different method of "cloning" called embryonic cell nuclear transfer emerged.[3] Nuclear transfer involves the transfer of individual undifferentiated embryonic cells to enucleated oocytes (ova whose chromosomes have been removed). Unlike embryo bisection where usually two (or a maximum of four) genetically identical offspring can be produced, nuclear transfer gives the theoretical opportunity to produce numerous genetically identical offspring from each multi-cell embryo. Embryos produced during the "cloning" process can themselves be used as donor cells for repeated nuclear transfer in a process known as serial nuclear transfer. Nuclear transfer-derived offspring have been produced over four generations of "cloning" from a single cattle embryo. Unexpectedly, some nuclear transfer offspring are afflicted with abnormally large birth weights and an increased incidence of developmental defects. The specific reason(s) for these problems is not clear, although laboratory embryo culture conditions have been implicated.

In the mid-1990s yet another form of "cloning" was reported when the world was introduced to "Dolly."[4] Dolly was created via somatic cell nuclear transfer, a procedure in which an ordinary body cell (a differentiated cell) is used as the donor cell for nuclear transfer instead of a cell from an early embryo (an undifferentiated cell). Cells used to produce Dolly were harvested from the mammary gland of a mature sheep. The mammary cells were cultivated in the laboratory to produce a larger population of mammary gland cells, thus enabling researchers to have an essentially unlimited supply of donor cells. The production of Dolly was an important scientific breakthrough because it was the first mammal produced in the world from a differentiated body cell (somatic cell).

To construct "cloned" embryos with this new approach, a somatic cell obtained from a developing fetus or an adult animal (male or female) is transferred to an enucleated oocyte. The enucleated oocyte with the newly introduced foreign somatic cell is activated via electrical pulses, and the reprogrammed cell nucleus directs development into a

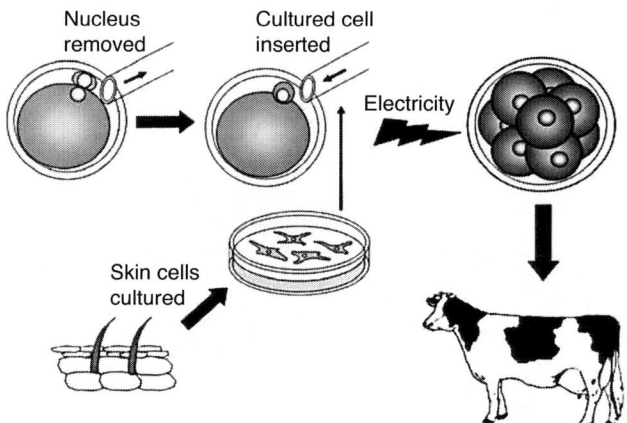

Fig. 1 The basic scheme for somatic cell nuclear transfer (SCNT). Skin cells are removed from the donor animal and are propagated in culture. Chromosomes are removed from an ovum to create an enucleated oocyte. An individual somatic cell is injected into an enucleated cow oocyte, followed by pulses of electricity (electrofusion) to activate the reconstructed embryo. The DNA inside the nucleus of the skin cell directs the development of the fetus that is produced following the transfer of the SCNT embryo into a recipient female. **Source:** Drawing courtesy of A.M. Landry.

"cloned" embryo which may then be transferred to a recipient female (see Fig. 1). Once the donor somatic cell population has been multiplied, hundreds of "cloned" embryos can be produced in the laboratory on a weekly basis, using ova extracted from abattoir ovaries or ovaries of living cows utilizing transvaginal ultrasound-guided oocyte aspiration.

"Cloning" provides cattle producers with an opportunity to reproduce valuable seed stock animals, animals that have suffered an injury and can no longer reproduce, or males that have been castrated. Biomedical researchers are also using "cloning" to create additional copies of animals genetically enhanced through the gene transfer process known as transgenesis. However, one must recognize that "cloning" procedures reproduce that which already exists but does not directly result in genetic improvement.

CONCLUSION

Advances in development and application of embryo transfer technologies have accelerated in the past decade. Increased knowledge of ovarian follicular growth patterns has led to improved methods of superovulation and reproductive management. Enhanced procedures for cryopreservation of embryos have led to greater adoption of this technology, including for embryos produced via in vitro fertilization technology. Refined approaches for embryo sexing have led to pretransfer embryonic sex determination in the commercial embryo transfer industry. Somatic cell nuclear transfer has opened new avenues to propagate genetically valuable animals. Although the availability and cost effectiveness of some of these new technologies remain in question, there is little doubt about their potential impact on future livestock production. These new technologies, if economically practical, will provide producers with the opportunity to change the genetic merit of their cattle at a faster rate than is now possible by conventional breeding methods.

REFERENCES

1. Hasler, J.F. Current status and future of commercial embryo transfer. Anim. Reprod. Sci. **2003**, *79* (3/4), 245–264.
2. Seidel, G.E., Jr.; Walker, D.J. Pregnancy rates with embryos vitrified in 0.25-ml straws. J. Reprod. Dev. **2006**, *52*, S71–S76.
3. Willadsen, S.M. Nuclear transplantation in sheep embryos. Nature **1986**, *320* (6057), 63–65.
4. Wilmut, I.; Schnieke, A.E.; McWhir, J.; Kind, A.J.; Campbell, K.H.S. Viable offspring derived from fetal and adult mammalian cells. Nature **1997**, *385* (6619), 810–813.

Cereal Foods: Starter Cultures

Claude P. Champagne
Pierre Gélinas
Edward R. Farnworth
Food Research and Development Centre, Agriculture and Agri-Food Canada, Saint-Hyacinthe, Quebec, Canada

Abstract
Starter cultures for cereals are those which are the most extensively used for fermented foods. Bread and beer are the major fermented foods prepared with starter cultures but can include other traditional foods such as soy sauce or miso as well. Cereal starters are primarily yeasts, but lactic acid bacteria and molds also serve as starters for the production of specialty cereal-based foods.

INTRODUCTION

Some of the most popular fermented foods, such as bread and beer, are produced by microbial starters. Wheat, barley, rice and, to a lesser extent, corn (maize), and rye are major raw materials for such basic foods. Historically, crushed grain was simply mixed with small or large amounts of water to produce bread or beer. Dough inflation or bubbling of the cereal extract was the first indication of fermentation. Over time, humans have domesticated microorganisms. At the turn of the 20th century, industries had developed to produce highly specialized fermentation starter cultures. Today, microbial strains are tailored for specific food applications. This entry briefly presents major microorganisms involved in the fermentation of cereal-based foods, as well as cultures for lesser-known fermented cereal-based foods.

ENDOGENOUS MICROFLORA

Originally, a mixture of crushed grain and water was incubated, allowed to bubble, and later turn sour, owing to the action of endogenous microorganisms present on the surface of grains. By trial and error, our ancestors developed the art of preparing fermented foods with improved sensory appeal and enhanced shelf life. During 1860–1870, Louis Pasteur confirmed the importance of isolating and using pure microorganisms to prepare fermented foods, particularly beer, to improve the chances of producing foods with desirable qualities. Even today, cereal-based foods are produced with endogenous microbiota in small-scale production, such as for some artisan sourdough-type breads or in the production of regional specialties, mainly in less industrialized countries. However, commercial yeast and, to a lesser extent, lactic acid bacteria (LAB) and mold concentrates are now widely available for preparing standardized foods such as bread or beer.

BAKER'S YEAST

Basic Characteristics and Production

Baker's yeast is the most renowned microorganism for cereal fermentation. *Saccharomyces cerevisiae* is still the most popular yeast species for bread applications with several strains being used for the same purpose. In essence, yeast strains for baking applications are selected according to: 1) their potential for high gas production in dough; 2) fermentation rate at various temperatures; 3) ability to assimilate sugars other than sucrose (maltose, e.g., to enable its use in non-sugared dough); 4) constant gas production during long fermentation periods; and most importantly, 5) tolerance to stresses such as drying, freezing, high sugar concentrations, and preservatives (acids, propionate).

A strong gas producer from carbohydrates, baker's yeast becomes very active when enough water is mixed with flour at optimal temperatures (20–35°C). Basically, in the bread and brewery fermentations, carbohydrates are converted into ethanol and CO_2. In the traditional bread-making process, sugars such as glucose or maltose, liberated by the hydrolysis of flour starch serve as substrates for fermentation. Alternatively, sugar concentrates (mainly sucrose) may be added to activate yeast fermentations, which is commonly done in bread dough manufactured in the United States and England. However, if the sugar concentration is too high, yeast gas production will diminish; this can occur in specialty pastries.

In commercial baker's yeast production, a fed-batch fermentation process under high aeration is used to optimize biomass production. In such a process, the sugar

Table 1 Some flavor compounds formed by yeasts in bakery products, alcoholic beverages, and some spirits, soy sauces, and dairy products.

Fatty acids	Esters	Aldehydes	Alcohols
Acetic acid	Ethyl acetate	Acetaldehyde	Ethanol
Butyric acid	Isoamyl acetate	2-Nonenal	2-Methyl butanol
3-Methyl butyric acid	Ethyl caproate	3-Methyl butanal	3-Methyl butanol
	Ethyl caprylate		2-Phenyl butanol
	Ethyl laurate		2-Phenyl ethanol
	Ethyl hexanoate		Isoamyl alcohol
			Octanol

Source: From *Microbiology of Fermented Foods*.[2]

concentration in the growth medium (cane and beet molasses) is kept at a constant low level to prevent a Crabtree effect, which is evidenced by alcohol production even in the presence of oxygen. The fermentation process also has the particularity of generating intracellular accumulation of trehalose. This is obtained by growing cells under conditions of limited nitrogen availability to enhance survival of yeast during refrigerated storage or the drying process. Baker's yeast technology has been reviewed by Gélinas.[1]

Commercial Formats

Yeast concentrates are available in three forms: 1) the liquid fresh state; 2) the compressed fresh form; and 3) the dry form. Large bakeries mostly use the fresh liquid or compressed forms, while small operations and household users prefer the dry format. The liquid fresh product is obtained after concentrating yeast solids from 5% to 18–20% by centrifugation. This cream yeast is easy to pump and blend with other major baking ingredients. Compressed yeast is the most widely used yeast product. It is commonly made from taking yeast paste from a vacuum drum filter, forming it into cakes or bricks, and then packaging. Shelf life of fresh yeast is about 2–3 weeks under refrigerated conditions; its gassing power slowly diminishes during that period. Glutathione leaking from dead cells is notoriously deleterious to gluten (dough weakens).

Between 1940 and 1970, major efforts were made to screen drying-tolerant strains and optimize drying conditions so that baker's yeast would remain active for several months at room temperature. If the packaging for dry yeast is not opened, these commercial yeast products may keep for more than a year. Around 1970, a new dehydrated yeast process was developed to avoid the necessity of rehydration before incorporation into dough. These yeast products are unique in having a very fine granulation, which facilitates water uptake in dough. If the packages are not opened, vacuum- or nitrogen-packaging also offers some protection against the oxidation of yeast lipids. Also, survival following rehydration is improved through the use of specific surfactants.

Flavors

Besides CO_2 production, baker's yeast forms substantial amounts of ethanol which enhance bread aroma that is often associated with Maillard-type reactions formed during bread baking. Baker's yeast also forms small amounts of aroma compounds, especially during long dough fermentation periods. Numerous flavor compounds can be produced by yeast (Table 1) and are critical components of the sensory properties of fermented cereals. However, compared to brewer's yeast, the bread-making process is normally too short to allow the production of large amounts of aroma compounds from yeast cells. This has now become an exciting area of research. For example, aroma-forming yeasts have been screened for possible use in breads where shortening of the length of time for dough fermentations can result in a lack of desirable bread aromas and flavors.

BREWER'S YEAST

Beer manufacturing from barley and rice lasts for several days under specific conditions depending on the type of beer. Hence, the selection of brewer's yeast strains is commonly based on: 1) their optimal fermentation temperature; 2) sedimentation properties; 3) capacity to form specific aroma components (Table 1); and 4) ethanol tolerance. Unlike baker's yeast which is widely available in the marketplace, most of the brewer's yeast strains are kept at the brewing plant. While baker's yeast is intended to be embedded once in the dough and killed by baking, brewer's yeast is recovered in the brewing process and reused several times for manufacturing of subsequent batches of beer.

There are two types of brewer's yeast: 1) ale (also called top-fermenting); and 2) lager (bottom-fermenting). Ale yeasts are typically classified as *S. cerevisiae*, while lager strains are often termed as *Saccharomyces carlsbergensis* (now classified as *Saccharomyces pastorianus*).[3] In the past, surplus brewer's yeast was very popular for bakery applications. Today, brewer's yeast is considered a poor baker's yeast because of its weak gassing power in dough and production of off-flavors in bread. Conversely, baker's yeast may also be used to prepare beer, but the taste and foaming properties of the final product are inferior to a brewer's yeast–fermented beer.

LACTIC ACID BACTERIA

Lister isolated the first lactic acid bacterium in 1873, which is about the same time that Pasteur made his discoveries on the fermentation of beer with pure yeast cultures.[4] Lactic acid bacteria have a very different fermentation behavior compared to yeasts and form organic acids instead of alcohol from sugars as their major metabolic products. Historically, the use of sourdough was necessary for rye bread production. The inhibition of endogenous amylases by acidification prevented excessive starch degradation during baking and this was a prerequisite for an acceptable bread volume.[5] Most species used in the fermentation of cereal foods are members of the genus *Lactobacillus* (Table 2). There are two types of LAB: 1) homofermentative, which mainly produce lactic acid during fermentation; and 2) heterofermentative that produce gas, acetic acid, and a variety of other products in addition to lactic acid.[8] Although less popular than yeast for the preparation of fermented cereal foods, pure LAB starters are now available for sourdough-type bread.[6] A microbial starter culture that is a mixture of LAB and yeasts, such as kefir grains, has been suggested as an alternative starter to produce bakery products with a distinctive aroma.[9] Like yeasts, LAB (especially heterofermentative types) produce numerous aroma compounds which add to the complexity and richness of sourdough products.[10] The majority of sourdough LAB do not produce cell wall-associated proteinases[8] which can lead to bitterness in other fermented foods. The lactobacilli starters used in cereal fermentations are mainly marketed in a freeze-dried form, but as with yeast, the liquid form may be useful for some large-scale applications.[11]

In addition to *Lactobacillus*, the most commonly found LAB used to produce fermented grain foods are members of the *Leuconostoc* and *Pediococcus* genera (Table 2). Some *Leuconostoc* strains produce exopolysaccharides (EPS), particularly dextran. The sourdoughs obtained with EPS-producing strains demonstrate improved freshness, crumb structure, mouthfeel, and softness in baked goods (wheat-rich dough products to rye sourdough breads).[12] EPS production by LAB also has the potential to replace more expensive hydrocolloids used as bread improvers.[13]

Although the technology of producing LAB starters is well developed and commercial products are available, in many cases (Table 2), fermentations are spontaneous, resulting from the contaminating microbiota and particular fermentation conditions (e.g., temperature, salt, and water activity). One can expect that as production volumes increase, inoculation of cereal-based ingredients with selected LAB starters will occur. It is noteworthy that at least two specific oatmeal-based foods contain probiotic cultures (Table 2). Most fermented products containing lactobacilli are potential carriers for these health-beneficial cultures.

MOLDS

Foods produced with molds are less important in America and Europe than foods produced with yeasts and bacteria starters. Some of the most popular cereal foods developed in Asia are prepared by mold fermentation;[14] saké and miso are probably the most famous of these foods. Generally prepared from a paste made from rice, barley, or soybeans, miso is consumed as a condiment. Miso is the result of a fermentation process using *Aspergillus* spp. which may last from a few days to 1–12 months. Made from rice, saké is somewhat similar to beer production except that koji molds (*Aspergillus oryzae*) break down rice starch into

Table 2 Microbial cultures involved in lesser known fermented cereal products.

Food	Country	Ingredients	Microorganisms[a]
Adai	India	Cereal, legume	*Pediococcus* sp., *Streptococcus* sp., *Leuconostoc* sp.
Anarshe	India	Rice	Lactic acid bacteria
Arroz fermentado	Ecuador	Rice	Lactic acid bacteria, yeasts
Aya-bisbaya	Mexico	Rice	Lactic acid bacteria
Balao balao	Philippines	Rice, shrimp	*Lb. brevis*, *Ln. mesenteroides*, *P. cerevisiae*
Banku	Ghana	Maize, cassava	Lactic acid bacteria, molds
Bhatura	India	Wheat	Lactic acid bacteria, yeasts
Bogobe	Botswana	Sorghum	*Lactobacillus* sp.
Burukutu	Nigeria	Sorghum, cassava	Lactic acid bacteria, *Candida* sp., *S. cerevisiae*
Chicha	Peru	Maize	*Aspergillus* sp., *Penicillium* sp., yeasts
Dokla	India, Sri Lanka	Rice, chick pea, Bengal gram, fenugreek	Lactic acid bacteria, yeasts
Dosa, dosai	India, Sri Lanka	Rice, black gram	*Leuconostoc* sp., *Lb. fermentum*, *Saccharomyces* sp.

(*Continued*)

Table 2 Microbial cultures involved in lesser known fermented cereal products. (*Continued*)

Food	Country	Ingredients	Microorganisms[a]
Enjera	Ethiopia	Tef	*Candida guilliermondii*
Fermented oatmeal[b] (ProViva)	Sweden	Oatmeal	*Lb. plantarum*
Hopper	Sri Lanka	Rice or wheat	Lactic acid bacteria, *S. cerevisiae*
Idli	India	Rice, black gram	*Ln. mesenteroides, Sc. faecalis, Lb. delbrueckii, Lb. fermenti, Lc. lactis, P. cerevisiae, Geotrichum candidum, Torulopsis* sp.
Llambazi, lakubilisa	Zimbabwe	Maize	Lactic acid bacteria, yeasts, molds
Injera	Ethiopia	Sorghum, tef, corn, millet, barley, wheat	*Lb. plantarum, Aspergillus* sp., *Penicillium* sp., *Rhodotorula* sp., *Candida* sp.
Jalebi	India	Wheat, dahi	*Lb. fermentum, Lc. lactis, Lb. buchneri, Sc. faecalis*
Kanga kopiro	New Zealand	Maize	*Leuconostoc* sp., *Clostridium* sp.
Kanji	India	Rice, carrot	*Hansenula anomala*
Kenkey	Ghana	Maize	*Lb. fermentum, Lb. reuteri, Lb. plantarum, P. pentosaceus, Lb. brevis, Candida* sp., *Saccharomyces* sp., *Penicillium* sp., *Aspergillus* sp., *Fusarium* sp.
Khanomjeen	Thailand	Rice	*Lactobacillus* sp., *Streptococcus* sp.
Kichudok, takju	Korea	Rice	*Saccharomyces* sp.
Kishk, kushuk, trahanas	Egypt, Syria, Lebanon	Milk (yoghurt), wheat	*Lb. casei, Lb. plantarum, Lb. brevis, B. subtilis, B. licheniformis, B. megaterium*, yeasts
Kisra	Sudan, Iraq, Arabian Gulf	Sorghum, millet	*Lactobacillus* sp., *Lb. brevis, Lb. fermentum, E. faecium, Acetobacter* sp., *S. cerevisiae*
Kulcha	India, Pakistan	Wheat	Lactic acid bacteria, yeasts
Kwuna-zaki	Nigeria	Millet	Lactic acid bacteria, yeasts
Lao-chao	China, Indonesia	Rice	*R. oryzae, R. chinensis, A. oryzae, Saccharomycopsis* sp.
Mantou	China	Wheat	*Saccharomyces* sp.
Mahewu, mogou	South Africa	Maize	*Lc. lactis*
Mawe	South Africa	Maize	*Lb. brevis, Lb. fermentum, Ln. mesenteroides, Lc. lactis, P. pentosaceus, W. confusa*, yeasts
Me	Vietnam	Rice	Lactic acid bacteria
Minchin	China	Wheat gluten	*Paecilomyces* sp., *Aspergillus*, sp., *Cladosporium* sp., *Fusarium* sp.
Mutwiza	Zimbabwe	Maize	*P. pentosaceus*
Nan	India, Pakistan, Afghanistan, Iran	Wheat	Lactic acid bacteria
Nasha	Sudan	Sorghum, pearl millet	*Streptococcus* sp., *Lactobacillus* sp., *Candida* sp., *S. cerevisiae*
Ogi	Nigeria	Maize, sorghum, millet	*Lb. plantarum, Lb. brevis, Lb. fermentum, Ln. mesenteroides, W. confusa, Saccharomyces* sp., *Candida* sp.
Pozol	Mexico	Maize	*Lc. lactis, Lb. plantarum, Lb. casei, Lb. delbrueckii, Lb. fermentum, Clostridium* sp.
Puda, pudla	India	Bengal gram, mung bean, wheat	Lactic acid bacteria, yeasts
Puta	Philippines	Rice	*Ln. mesenteroides, Sc. faecalis, S. cerevisiae*
Rabdi	India	Maize, buttermilk	*P. acidilactici, Bacillus* sp., *Micrococcus* sp.
Togwa	Tanzania	Maize, sorghum	*Lb. plantarum, Lb. brevis, Lb. fermentum, Lb. cellobiosus, P. pentosaceus, W. confusa, S. cerevisiae, C. tropicalis*
Trahanas, tarhanas, kishk	Greece, Turkey	Wheat, sheep milk (yoghurt)	Lactic acid bacteria
Tsukemono	Japan	Vegetables, rice	Lactic acid bacteria
Uji	Kenya, Uganda, Tanzania	Maize, sorghum, cassava	*Ln. mesenteroides, Lb. plantarum*
Yosa®[b]	Finland	Oat bran	Lactic acid bacteria, *Bifidobacterium* sp.

[a]A., *Aspergillus*; B., *Bacillus*; C., *Candida*; Lb., *Lactobacillus*; Lc., *Lactococcus*; Ln., *Leuconostoc*; P., *Pediococcus*; R., *Rhizopus*; S., *Saccharomyces*; Sc., *Streptococcus*; W., *Weissella*.
[b]Contains documented probiotic strains.
Source: Adapted from Decock and Cappelle[6] and Farnworth.[7]

fermentable sugars which are transformed into alcohol by yeasts able to tolerate high levels of alcohol. Typical saké contains about 12–16% alcohol and is closer to a rice wine than a beer. Other examples of mold-containing fermented foods are presented in Table 2.

CONCLUSION

Originally, endogenous microbiota in ground cereals (flour) initiated cereal fermentations. Particularly with baker's and brewer's yeasts, starter cultures currently play a key role in the preparation of fermented cereal foods which are among the most popular worldwide. Microbial starter cultures have a major influence on the production of safe and uniform fermented bakery foods, and beer. As for the wine industry, new advances in brewery may pass through the limited use of genetically-modified strains.[15] However, these traditional and highly conservative food sectors are not likely to be among the first to adopt these technologies.

REFERENCES

1. Gélinas, P. Yeast. In *Bakery Products: Science and Technology*; Hui, Y.H., Corke, H., De Leyn, I., Nip, W.-K., Cross, N., Eds.; Blackwell Publishing: Ames, IA, 2006; 173–192.
2. Stam, H.; Hoogland, M.; Laane, C. Food flavours from yeast. In *Microbiology of Fermented Foods*; Wood, B.J.B., Ed.; Blackie Academic & Professional: London, 1998; 506–542.
3. Josephsen, J.; Jespersen, L. Starter cultures and fermented products. In *Handbook of Food and Beverage Fermentation Technology*; Hui, Y.H., Goddik, L.M., Hansen, A.S., Josephsen, J., Nip, W.-K., Stanfield, P., Toldra, F., Eds.; Marcel Dekker: New York, 2004; 23–49.
4. Bamford, C.W. *Food, Fermentation and Micro-organisms*; Blackwell: Oxford, 2005; 28–34.
5. Brand, M.J. Sourdough products for convenient use in baking. Food Microbiol. **2007**, *24*, 161–164.
6. Decock, P.; Cappelle, S. Bread technology and sourdough technology. Trends Food Sci. Technol. **2005**, *16*, 113–120.
7. Farnworth, E. The beneficial health effects of fermented foods—potential probiotics around the world. J. Nutraceuticals, Funct. Med. Foods **2004**, *4* (3/4), 93–117.
8. Gänzlea, M.G.; Vermeulen, N.; Vogel, R.F. Carbohydrate, peptide and lipid metabolism of lactic acid bacteria in sourdough. Food Microbiol. **2007**, *24*, 128–138.
9. Plessas, S.; Pherson, L.; Bekatorou, A.; Nigam, P.; Koutinas, A.A. Bread making using kefir grains as baker's yeast. Food Chem. **2005**, *93*, 585–589.
10. Rehman, S.; Paterson, A.; Piggott, J.R. Flavour in sourdough breads: a review. Trends Food Sci. Technol. **2006**, *17*, 557–566.
11. Carnevali, P.; Ciati, R.; Leporati, A.; Paese, M. Liquid sourdough fermentation: Industrial application perspectives. Food Microbiol. **2007**, *24*, 150–154.
12. Lacaze, G.; Wick, M.; Cappelle, S. Emerging fermentation technologies: development of novel sourdoughs. Food Microbiol. **2007**, *24*, 155–160.
13. Arendt, E.K.; Ryan, L.A.M.; Dal Bello, F. Impact of sourdough on the texture of bread. Food Microbiol. **2007**, *24*, 165–174.
14. Gélinas, P.; McKinnon, C. Fermentation and microbiological processes in cereal foods. In *Handbook of Cereal Science and Technology*, 2nd Ed.; Kulp, K., Ponte, J.G. Jr., Eds.; Marcel Dekker: New York, 2000; 741–754.
15. Schuller, D.; Casal, M. The use of genetically modified *Saccharomyces cerevisiae* strains in the wine industry. Appl. Microbiol. Biotechnol. **2005**, *68*, 292–304.

Cereal-Based Grain Products: Fermented Indigenous Grains

Neela Badrie
Department of Food Production, University of the West Indies, St. Augustine, Trinidad and Tobago

Abstract

By utilizing microorganisms and their enzymes, cereal grains can be transformed into value-added products. The applications, processing, microbial types, and characteristics of some indigenous fermented foods and beverages are presented in this entry. Fermented foods include kenkey, kishk, tarhana, ogi, medida, and togwa. The alcoholic beverages described are boza, pito, takju, kaffir, sake, and chicha.

INTRODUCTION

Fermented foods and beverages contribute diverse flavors, aromas, and textures to the human diet. In these products, preservation is normally enhanced through the syntheses of lactic acid, alcohol, acetic acid, and carbon dioxide. The applications, processing, fermentation, microbial types, and characteristics of fermented foods are presented in the section "Fermented Foods" and those of alcoholic beverages are presented in the section "Fermented Alcoholic Beverages."

FERMENTED FOODS

Kenkey (Kenky)

Kenkey is a cooked, sour-fermented white maize corn dumpling, wrapped in leaves and boiled; it is a popular dish in Ghana. It is used as a weaning food for babies.

The traditional kenkey-making process takes 4–6 days. Maize grains are soaked in water at 4–25°C for 1–2 days to allow for endogenase activity of proteases and carbohydrases. The resulting meal is fermented as stiff dough. An accelerated option (24 hours) involves manufacture of kenkey in sausage casings. By precracking the kernels and hydrating the maize to 40% moisture (w/w), time is shortened from 48 to 10 hours. Also, fermentation is shortened from 3–4 days to 12 hours by incorporating aflata (gelatinized maize paste) into the dumpling and using starter dough.[1]

Penicillium, *Aspergillus*, *Fusarium*, *Candida*, *Saccharomyces*, *Trichosporon*, *Kluveromyces*, and *Debaryomyces* sp. are found in raw maize during steeping and early fermentation. After 24–48 hours of fermentation, *Candida krusei* and *Saccharomyces cerevisiae* dominate. At the advanced stage of fermentation, more than 96% of the bacteria present are obligate heterofermentative lactobacilli, such as *Leuconostoc mesenteroides* and *Lactobacillus fermentum*. Acid-forming streptococci multiply between 24 and 36 hours of the fermentation.

The pH of the dough falls from 6.4–6.8 to 3.5–4.1 owing to fermentation. The flavor and aroma comprise a mixture of diacetyl, acetic acid, and butyric acid.

Kishk

Kishk (Fugush) is a dried product from yoghurt and parboiled "cracked" wheat, which is consumed by the Bedouins of North Africa, the Middle East, and the Indian subcontinent. It is made from a dough containing salt, milk (unfermented, acidified with glucono-δ-lactone, or fermented with yoghurt starter), rolled oats, oat flour, parboiled "cracked" wheat (Burghol), and/or wheat flour.[2] The mixture is dried (10–13% moisture), ground into a powder, and stored in the form of round balls.[3]

The principal fermenting microorganisms include *Lactobacillus casei*, *Lactobacillus plantarum*, *Lactobacillus brevis*, and yeasts. Aspergilli, penicillia, and *Rhizopus* have also been isolated. Three thermophilic species of fungi isolated from kishk are *Malbranchea sulfurea*, *Rhizomucor pusillus*, and *Thermomyces lanuginosus*. Lactic acid and acetic acid are the major organic acids found in the food.

Tarhana

Tarhana (trahana) is a Turkish dry-fermented form of a yoghurt–cereal mixture. Tarhana soup is used as a part of any meal. Tarhana is prepared by mixing wheat flour (two parts) and yoghurt (one part) from various milk types with yeast and other ingredients followed by fermentation for 1–7 days. The yoghurt contains *Streptococcus thermophilus* and *Lactobacillus bulgaricus*. The ingredients are mixed, kneaded, fermented, dried, and sieved. Vacuum-drying retains more active aroma components compared to sun-drying.

The low moisture (10%) and low pH (3.4–4.2) have a bacteriostatic effect on the resident microbiota. The amounts of lactic, acetic, and propionic acids increase while citric acid decreases during 3-day tarhana fermentations. Tarhana is used as a high-protein dietary supplement (~15% protein). Aldehydes are the predominant class of aroma compounds in sun-dried and vacuum-dried tarhanas.

Ogi

Ogi is a generic name. In most states of Nigeria, it is called maize ogi. Cooked ogi porridge is called pap to which sugar and different forms of milk are added to give about 8% solids. It is used to thicken soups and stews and is easily digestible.

In the traditional process of ogi-making, whole grains are water-steeped for 24–72 hours, wet-milled, sieved, and fermented (no addition of inoculum) in slurry troughs for 24–72 hours. In a shorter modified process (1–5 days), the two-stage fermentation (i.e., soaking and souring stages) is reduced to a single-stage process with better total protein recovery. A mixed culture of *L. plantarum*, *Lactococcus lactis*, and *Zygosaccharomyces rouxii* is used for the fermentation. Use of a high-lysine producing mutant of *L. plantarum* increases available lysine content from 228.5 ± 12.0 mg/100 g to 525.1 ± 25.8 mg/100 g.[4] The relevant molds are *Rhizopus*, *Oospora*, *Fusarium*, *Aspergillus*, and *Penicillium* sp. Isolatable yeasts are *S. cerevisiae*, *Rhodotorula* sp., and *Candida mycoderma*. Members of the genus *Corynebacterium* hydrolyze the starch of maize to form organic acids, while *S. cerevisiae* and *C. mycoderma* contribute to flavor. Ogi has a pH of 3.6–3.7 with relatively large amounts of lactic and acetic acids and trace amounts of formic and butyric acids.

Medida

Medida is a thin, fermented porridge from the Sudan. It can be prepared from malted flour using bifidobacteria. The product has high total solids content (21%) with a flowing consistency.

Brown rice (after removal of husk) and skim milk are used in the formulation. The initial pH of 6.7 is lowered to a final pH of 4.4 by addition of *Bifidobacterium longum*. Under refrigerated storage, counts of *B. longum* remain stable during the first week of fermentation (9.7 ± 0.10 log cfu mL^{-1}), then subsequently decrease to 0.9 log cfu mL^{-1} the following week. The final levels of lactic and acetic acids are 56.8 ± 0.80 and 56.3 ± 2.00 μmol mL^{-1}, respectively.[5]

Togwa

Togwa is a cereal-based, starch-saccharified, non-alcoholic, lactic acid-containing gruel and weaning food consumed in sub-Saharan Africa. It is made mostly from maize, sorghum, or finger millet flour. The flour is mixed with water (one part of flour and nine parts of water) and cooked for 10–20 min. The gruel is cooled to 35°C. Ten percent (v/v) of a previous togwa culture is mixed into the gruel with 5% (w/v) malt flour for the 9- to 24-hours lactic acid fermentation. The gruel is pasteurized at 97°C for 10 min.

The cultures isolated from native togwa are *L. brevis*, *L. fermentum*, *L. plantarum*, *Pediococcus pentosaceus*, *Weissella confusa*, *S. cerevisiae*, *Hansenula anomala*, *Candida tropicalis*, and *Issatchenkia orientalis* in coculture with either *L. brevis* or *L. plantarum*. Proteinase and aminopeptidase activities are higher in naturally fermented togwa than those made with starter cultures (14–30% and 12–70%, respectively).[6] Togwa has a pH of 3.2–4.0, which is mainly due to lactic acid fermentation. Ethanol is the predominant volatile organic compound.

FERMENTED ALCOHOLIC BEVERAGES

Boza

Boza is a very dense, indigenous, opaque, and creamy beer of Egypt, Bulgaria, and Turkey. It can be made by yeast and lactic acid bacteria fermentation of millet, cooked maize, wheat, sorghum, or rice semolina/flour.

Grains are cleaned and conditioned in water to soften the endosperm. The grits are solubilized by cooking, diluted with water, and strained to produce "sugarless crude boza." In fermentation, 20–25% (w/v) sucrose together with 2–3% (w/v) yeast (back slopping) are added to crude boza and incubated at 18–20°C for 12 hours.

The lactic acid bacteria isolated during the fermentation included *Leuconostoc paramesenteroides*, *Leuconostoc mesenteroides* subsp. *mesenteroides*, *Leuconostoc mesenteroides* subsp. *dextranicum*, *Lactobacillus sanfranciscensis*, *L. coryniformis*, and *L. fermentum*. The yeasts are *Saccharomyces uvarum* and *S. cerevisiae*. Other microorganisms include *Torulopsis candida*, *Streptococcus* sp., *Kocuria varians*, and *Bacillus cereus*.

Volatile esters and higher alcohols such as ethyl acetate and isoamyl acetate are found in boza. The final pH is 3.7–4.0 and the alcohol concentration is 3.8–4.2%.

Pito

Pito is a creamy, yellow-to-brown alcoholic beer with a sweet to bitter taste. It is produced and consumed in Nigeria and Ghana.

Fig. 1 Production of pito.

Cereal grains (maize, sorghum, or a combination) are soaked in water for 2 days and malted for 5 days in baskets. The contents are ground, mixed with water, and concentrated by boiling, filtration, and fermentation. In the traditional methods of fermentation, a woven belt serves as entrapment of the microbial starter from a previous fermentation, or cooled concentrate or dried scum (foam) is added to the mixture to start the fermentation (Fig. 1).

Pito is produced using a combination of *L. plantarum* and *S. cerevisiae*.[7] Important yeasts are *Hanseniaspora uvarum* and *Kluyveromyces africanus*. The molds are *Rhizopus oryzae*, *Aspergillus flavus*, *Penicillium culosum*, and *Penicillium citrinum*. Pito typically has a pH of 3.5, 1.5–3.5% v/v ethanol, and 0.7–1.0% w/w lactic acid.

Takju

Takju is a Korean beer produced from rice or wheat. It is turbid with suspended insoluble solids and active yeasts. The starter for takju is called nuruk. The nuruk contains fungal amylolytic enzymes to saccharify starch and yeasts for the alcoholic fermentation. It is made from solid-state fermentation of moistened wheat flour using *Aspergillus usamii* and is ready for application in 2–3 months. *Rhizopus*, *Aspergillus niger*, *Debaryomyces hansenii*, *Debaryomyces occidentalis*, *Pichia anomala*, *Pichia fabianii*, and *Saccharomyces fibuligera* have been isolated in nuruk. In the traditional takju process, steamed and cooled rice is inoculated with nuruk of 12% moisture for fermentation. Fermentation is 90% completed within 3–4 days. After 3 days of fermentation, takju has 1% titratable acidity, a pH of 4.0, and 7–10% ethanol.

Kaffir

Kaffir beer (Bantu beer) is an opaque alcoholic beer with a yeasty odor and fruity flavor. It is pink because of the presence of sorghum tannins. The beer is made from malted sorghum, unmalted maize, and millet. The main steps in kaffir beer-brewing are shown in Fig. 2. The brewing

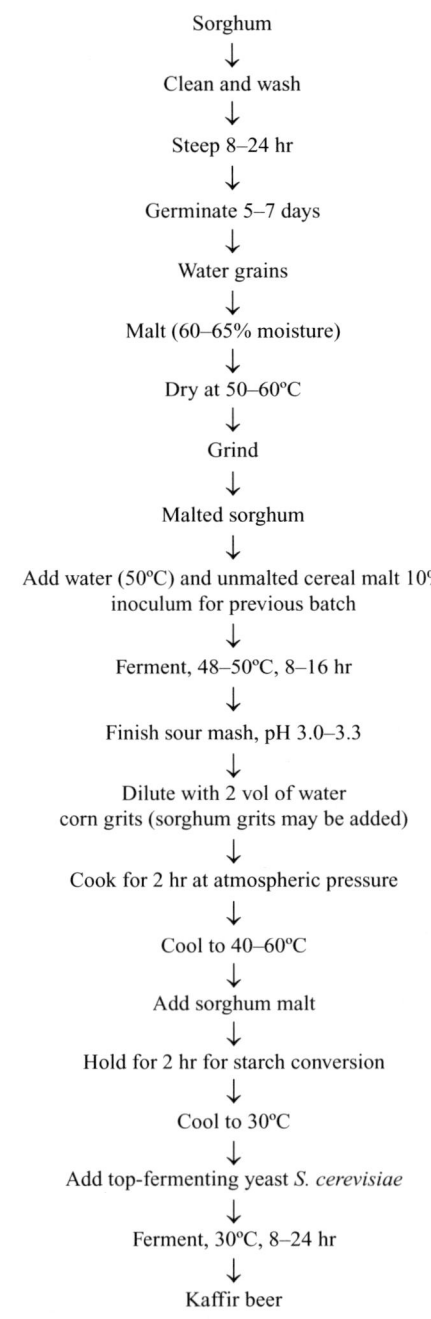

Fig. 2 Steps in the production of kaffir beer.

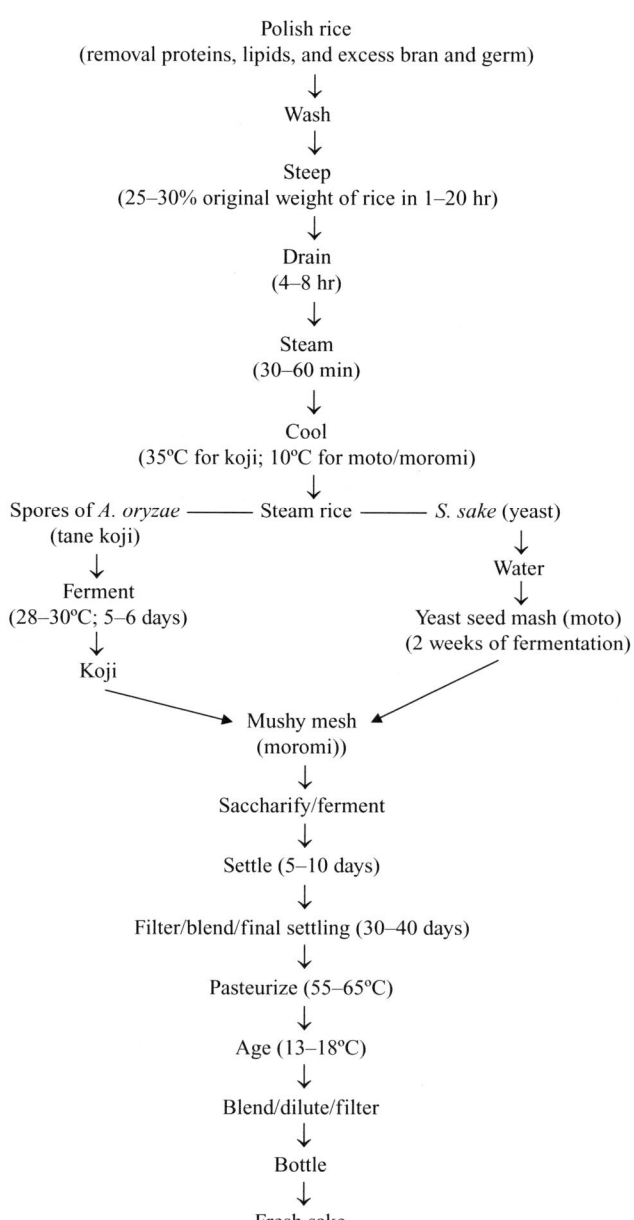

Fig. 3 Steps in sake brewing.

process involves both lactic acid and alcohol fermentations. In commercial breweries, the top-fermenting yeast (*S. cerevisiae*) is added. In the traditional process, the yeasts are introduced with the malt. The beer has a pH of 3.3–3.5, 0.3–0.6% lactic acid, and 2–4% alcohol content with an approximate shelf life of 40 hours. *Acetobacter* has the potential to spoil kaffir beer owing to the production of acetic acid.

Sake

Sake (clarified rice wine) is a traditional clear, pale yellow rice wine, consumed in Japan and China. Sake is prepared by digesting a mixture of cooked rice and koji with concurrent fermentation with lactobacilli and yeasts. Koji is a concentrate of fungal amylases, proteases, and other enzymes obtained by overgrowing steamed rice or barley with selected strains of *A. oryzae* (Fig. 3). Saccharification and fermentation by *Saccharomyces sake* proceed for over 30–40 days in the mushy mesh (moromi). Some defects in sake can be improved by accelerated maturation through suitable dosage with gamma radiation.[8] Sake has approximately 0.3% lactic acid and 12–15% alcohol.

Chicha

Chicha is a clear, yellowish South American alcoholic maize beverage which resembles cider. It has been consumed by Andes Indians for centuries. In traditional chicha fermentation, saliva serves as source of amylase to convert maize starch into fermentable sugars. Amylases are also available in the malting (germination) of the maize kernels or as with the addition of commercial enzymes. The germinated dried maize (jora) is ground to form pachucho, which is mixed with water and boiled for 3.5 hours. Sugar or molasses may be added in the second boiling to increase alcoholic content in the finished product. Following cooling, the hulls and starch are separated by rubbing and the pachucho is further boiled for 4 hours. The cool filtrate is inoculated with *S. cerevisiae* and lactobacilli and fermented for 1 day. Other active microorganisms include *Leuconostoc*, *Acetobacter*, *Aspergillus*, *S. uvarum*, *Mycoderma vini*, *Endomyces lactis*, and *Monilia candida*. The chicha has a semisharp flavor with an alcohol content of 2–12% v/v.

CONCLUSION

As discussed in this entry, most of the examples of indigenous fermented cereals are fermented by natural fermentation with native microbiota or by the practice commonly known as back-slopping. Usually lactic acid bacteria initiate the fermentation followed by the activities of yeasts and molds. Lactic acid and alcohol are the two major components contributing to the sensory qualities of these products.

REFERENCES

1. Nout, M.J.R.; Kok, B.; Vela, E.; Nche, P.F.; Rombouts, F.M. Acceleration of the fermentation of kenkey, an indigenous fermented maize food of Ghana. Food Res. Int. **1995**, *28* (6), 599–604.

2. Tamime, A.Y.; Muir, D.D.; Khaskheli, M.; Barclay, M.N.I. Effect of processing conditions and raw materials on the properties of kishk. Compositional and microbiological qualities. Lebensm. Wiss. Technol. **2000**, *33* (6), 444–451.

3. Blandino, A.; Al-Asseri, M.E.; Pandiella, S.S.; Cantero, D.; Webb, C. Cereal-based fermented foods and beverages. Food Res. Int. **2003**, *36* (6), 527–543.

4. Adebawo, O.O.; Akingbala, J.O.; Ruiz-Barba, J.L.; Osilesi, O. Utilization of high lysine-producing strains of *Lactobacillus plantarum* as starter culture for nutritional improvement of ogi (Nigerian fermented gruel). World J. Microbiol. Biotechnol. **2000**, *16* (5), 451–455.

5. Kabeir, B.M.; Abd-Aziz, S.; Muhammad, K.; Shuhaimi, M.; Yazid, A.M. Growth of *Bifidobacterium longum* BB536 in medida (fermented cereal porridge) and their survival during refrigerated storage. Lett. Appl. Microbiol. **2005**, *41* (2), 125–131.

6. Mugula, J.K.; Nnko, S.A.M.; Narvhus, J.A.; Sorhaug, T. Microbiological and fermentation characteristics of togwa, a Tanzanian fermented food. Int. J. Food Microbiol. **2002**, *80* (3), 187–199.

7. Orji, M.U.; Mbata, T.I.; Anicne, G.N.; Anonkhai, I. The use of starter cultures to produce "pito," a Nigerian fermented alcoholic beverage. World J. Microbiol. Biotechnol. **2003**, *19* (7), 733–736.

8. Chang, A.C. The effects of gamma radiation on rice wine maturation. Food Chem. **2003**, *83* (3), 323–327.

Cheese Production: Nonstarter Culture Bacteria

Giorgio Giraffa
Agricultural Research Council, Research Center for Forage and Dairy Productions (CRA-FLC), Lodi, Italy

Abstract

Cheese production is based on the use of added starter cultures of lactic acid bacteria that initiate rapid milk acidification; however, many cheeses, especially traditional cheeses, obtain their flavor intensity from the presence and activity of non-starter, secondary biota. Recently, new bacterial cultures with industrially important functionalities are under development. Such cultures, defined as "starter adjuncts," can also contribute to microbial safety or offer organoleptic, nutritional, or health benefits. This entry provides an overview of the ecology, claimed role, and application in cheese of non-starter bacteria.

INTRODUCTION

Microorganisms are essential during both cheese production and ripening. They can be grouped into starter and secondary cultures. Lactic fermentation begins with the addition of starter cultures of lactic acid bacteria (LAB) to milk. The starter LAB (SLAB) are responsible for acid development during cheese manufacture. Later, the secondary biota, along with the starter bacteria, promote a series of biochemical events which are extremely important for proper development of both cheese flavor as well as texture during ripening.

The secondary biota is composed of complex mixtures of bacteria, yeasts, and molds. The bacterial community is composed of non-starter LAB (NSLAB) such as lactobacilli, pediococci, leuconostocs, and enterococci; propionic acid bacteria; and "smear bacteria" such as coryneforms, micrococci, and staphylococci.[1] Given the acknowledged importance of non-starter bacteria in the dairy industry, the objective of this entry is to give an overview of the taxonomy, ecology, technological role, and application to cheese production of this heterogeneous group of bacteria.

TAXONOMY, PHYSIOLOGY, AND HABITAT

Non-Starter LAB

Non-starter LAB are widespread and can be isolated from many vegetal and animal sources. The non-starter *Lactobacillus* species, most often encountered in cheese are the facultatively heterofermentative lactobacilli (FHL). They ferment hexoses to lactic acid and may produce CO_2 from gluconate. They are aerotolerant and mesophilic (i.e., with optimal growth temperature of 25–30°C).

The genus *Pediococcus* is composed of homofermentative bacteria which produce lactic acid from glucose. Some species exhibit extreme tolerance to temperature, pH, and NaCl concentration.

The genus *Leuconostoc* contains obligately heterofermentative bacteria which produce CO_2 from glucose as well as lactic acid and ethanol or acetic acid. They are facultatively anaerobic and mesophilic.

The genus *Enterococcus* consists of homofermentative and facultatively anaerobic bacteria with an optimum growth temperature of 35°C. They can grow in the presence of 6.5% NaCl and survive heating at 60°C for 30 min.

Propionic Acid Bacteria

Propionic acid bacteria (PAB) are Gram-positive, non-motile, non-sporeforming, anaerobic to aerotolerant, pleomorphic, and mesophilic. Propionic acid bacteria ferment sugars and organic acids, such as lactate, to a mixture of propionic, acetic and succinic acids, and CO_2 as major metabolic end products. Two major groups within the genus *Propionibacterium* are recognized: 1) the "cutaneous" (clinical); and 2) the "classical or dairy" PAB. Clinical PAB are isolated from human or animal skin. Dairy PAB are commonly found in milk and dairy products, but have also been isolated from soil, silage, brines for olive fermentation, and rum distilleries.

Smear Bacteria

Coryneforms are a heterogeneous group of Gram-positive bacteria from the genera *Arthrobacter*, *Brevibacterium*, *Corynebacterium*, *Microbacterium*, and *Rhodococcus*. Strictly speaking, the term "coryneform" has no taxonomic meaning except that these bacteria are irregularly shaped rods and produce carotenoid pigments when grown on solid surfaces. Their preferred habitat is mammalian skin and soil. *Brevibacterium linens* is the major coryneform growing on the surface of smear-ripened cheeses. *B. linens* is rod-shaped, obligately aerobic, mesophilic, and usually halotolerant (growth in the presence of 5% NaCl).

Staphylococci and micrococci are Gram-positive cocci that are generally halotolerant. They also occur primarily on mammalian skin and in soil. Staphylococci are facultatively anaerobes with optimum growth temperature of 30–37°C. Enterotoxin production and coagulase activity are the two generally accepted characteristics used to separate pathogenic from non-pathogenic staphylococci. Micrococci are strictly aerobic with respiratory metabolism and optimum growth temperature of 25–37°C. Colonies are usually pigmented in shades of yellow or red.

ECOLOGY IN CHEESES

Facultatively heterofermentative lactobacilli, pediococci, and enterococci form a significant portion of the NSLAB of most cheese varieties during ripening and consequently, their detection has been recommended as a microbial fingerprint for traditional cheeses. Leuconostocs are also included within the NSLAB, although they are often present in dairy starter cultures.

Facultatively heterofermentative lactobacilli, pediococci, and enterococci are typical adventitious NSLAB that are able to grow in cheese after manufacture. They originate from milk and although pasteurization usually inactivates them, recontamination from the manufacturing environment may easily occur. Therefore, it is normal for these bacteria to be present at high levels in cheeses derived from both raw and pasteurized milk. A relatively limited number of species are able to survive adverse cheese process conditions, such as extreme pH, temperature, and salinity. The most frequently found species are *Lactobacillus casei* (or *L. paracasei*), *Lactobacillus plantarum*, *Lactobacillus rhamnosus*, *Lactobacillus curvatus*, *Pediococcus acidilactici*, *Pediococcus pentosaceous*, *Enterococcus faecium*, and *Enterococcus faecalis*.[1–4]

Although little is known about the effect of stress on *Leuconostoc* cells, the presence of this bacterium in numerous cheese varieties made without addition of *Leuconostoc* starter is frequent, particularly in raw milk cheeses.[5] *Leuconostoc mesenteroides* and *Leuconostoc lactis* are the most frequently encountered species in cheeses.

Propionic acid bacteria constitute the essential secondary microbiota in Swiss-type cheeses such as Emmental, Gruyère, and Comté. The raw milk environment represents the main source of PAB; however, with the introduction of pasteurization, PAB are now added to the milk to ensure their presence in cheese at the beginning of ripening. The dairy PAB most widely encountered in cheeses are *Propionibacterium freudenreichii*, *Propionibacterium thoeni*, *Propionibacterium jensenii*, and *Propionibacterium acidipropionici*.[1,3]

Coryneforms and staphylococci/micrococci are present on the surface of many cheeses. Generally, these bacteria are isolated from soft, smear-ripened cheeses such as Camembert or Taleggio, but also from blue and hard cheeses. Their alkalophilic and salt-tolerant nature explain their selection on the surface of smear-ripened cheeses, where a washing of the rind with a brine solution is usually applied during ripening.[1,3]

BIOCHEMICAL PROPERTIES OF TECHNOLOGICAL INTEREST

Facultatively heterofermentative lactobacilli show a large diversity of properties. The peptidase activities of *L. casei* contribute to the hydrolysis of bitter peptides to non-bitter peptides with the release of free amino acids.[3] FHL are also involved in amino acid catabolism, which is a major process for flavor formation in cheese. In particular, glutamate dehydrogenase (GDH) activity is one of the key properties for selecting FHL as starter adjuncts. Natural GDH activity has been reported in *L. plantarum* and *L. casei/L. paracasei*.[6] Strains of *L. curvatus*, *L. plantarum*, and *L. rhamnosus* produce bacteriocins active against clostridia, *Staphylococcus aureus*, and *Listeria* spp.[3] Many strains of *L. casei/L. paracasei*, *L. plantarum*, and *L. rhamnosus* isolated from dairy products possess the ability to survive at low pH and in the presence of bile salts, which are important prerequisites of probiotic cultures.[3] Less is known about useful dairy characteristics of pediococci and no studies have identified their importance in cheese ripening.

The characteristics of technological interests in leuconostocs rely essentially on their sugar and organic acid metabolism. In addition to the heterofermentative utilization of glucose, leuconostocs metabolize citrate with the formation of diacetyl, acetate, acetoin, and 2,3-butanediol.[5]

The positive influence that enterococci may have on cheese is owing to their acidifying, proteolytic and lipolytic activities, citrate utilization, and production of aromatic volatile compounds. Moreover, enterococci are capable of producing bacteriocins with activity against pathogenic bacteria such as *Listeria monocytogenes* and *S. aureus*.[4]

For selection of PAB as adjuncts, gas production, peptidase, and lipase activities, and the ability to catabolize amino acids to different flavor compounds are desirable properties.[1,3] In addition to producing distinctive red-orange pigments, smear bacteria demonstrate peptidolytic and lipolytic activities which are important for the production of aroma compounds in cheese, such as methane thiol from *B. linens*.

Table 1 Non-starter bacteria used or potentially applicable in cheese production.

Bacterial group/genus/species	Biochemical properties of technological interest	Application as starter adjuncts
FHL	Peptidase activity	Ripening cultures in Cheddar cheese
	Amino acid catabolism (GDH activity)	Ripening cultures in Cheddar cheese
	Antimicrobial activity	Late blowing prevention in Emmental cheese
	Probiotic properties (ability to survive to low pH; bile salt resistance)	Probiotic cultures in Cheddar and Argentinean Fresco cheeses
Leuconostoc spp.	Citrate utilization	Aromatic cultures in raw milk cheeses
	Gas production	Cultures involved in cheese curd openness
Enterococcus faecium/E. faecalis	Proteolytic and lipolytic activities	Ripening cultures in soft cheeses
	Citrate utilization	Aromatic cultures in cheeses
	Antimicrobial activity	Anti-*Listeria* cultures in soft cheeses
Dairy PAB	Proteolytic and lipolytic activities	Ripening cultures in Swiss-type cheeses
	Amino acid catabolism	Ripening cultures in Swiss-type cheeses
	Gas production	Cultures involved in cheese curd openness in Swiss-type cheeses
Smear bacteria/*Brevibacterium linens*	Production of carotenoid pigments	Ripening cultures in soft- and semihard, surface-ripened cheeses
	Peptidase and lipolytic activities	Ripening cultures in soft- and semihard, surface-ripened cheeses
	Antimicrobial activity	Anti-*Listeria* cultures in soft, surface-ripened cheeses

Acronyms: FHL, facultatively heterofermentative lactobacilli; PAB, propionic acid bacteria; GDH, glutamate dehydrogenase.

FUNCTIONALITY AND ROLE IN CHEESE PRODUCTION

In cheese technology, starter adjuncts are selected cultures added for purposes other than acid formation. Acidification is exclusively performed by the primary starter culture. Starter adjuncts can be used as ripening cultures (e.g., as cultures added to accelerate ripening or produce desirable flavors), or to act as probiotic or protective cultures. Protective cultures are starter adjuncts showing specific antimicrobial activity against pathogenic or spoilage organisms. A recent trend is the use of probiotic adjuncts with the capability to produce biologically active peptides (such as the angiotensin-converting enzyme-inhibitory peptides) as a result of their proteolytic activity during cheese ripening. Principal applications of non-starter bacteria as culture adjuncts in cheese technology are summarized in Table 1.

Several trials have been done with *L. casei/paracasei* and/or *L. rhamnosus* as ripening cultures in the manufacture of many cheeses, especially Cheddar cheese, but the indications appear rather controversial.[1,2] The main reason for this statement is the difficulty to select the correct strain combination which links the required properties to concomitant lack of imperfections. The ability of many strains of *L. casei/paracasei* and *L. plantarum* to survive in the gastrointestinal tract and colonize the intestine is initiating the development of probiotic cheeses containing FHL. Strains of *L. casei/paracasei* have been used as probiotic cultures in Cheddar cheese.[3,7]

Leuconostocs have been tested as flavor or gas-producing ("opening") cultures in different raw milk cheeses or to inhibit off-flavor producing bacteria in Cheddar.[5]

Cheese trials carried out to evaluate the feasibility of enterococci as ripening cultures indicated that they positively influence taste, aroma, color, and structure, as well as the overall sensory profile, of the ripened cheeses. Bacteriocin-producing enterococci have also been used as anti-*Listeria* cultures in soft cheeses.[4]

Propionic acid bacteria are used as ripening cultures in Swiss-type cheeses. The CO_2 produced during the propionic fermentation is responsible for the typical holes in these cheeses. Propionic acid bacteria also are involved in flavor formation.

Smear bacteria are used as ripening cultures in surface-ripened cheeses. The addition of *B. linens* either to the milk or the cheese after brining has been suggested and several strains are presently marketed as selected starter adjuncts.[1–3] The wide phenotypic diversity of smear bacteria might offer further opportunities for the selection of appropriate starter formulations for specific cheeses. Future efforts in defining starter adjuncts for surface-ripened cheeses should include the other coryneforms as well as micrococci and coagulase-negative staphylococci.

CONCLUSION

To respond to the increasing demand for product diversification, the use of bacterial strains with expanded characteristics (such as flavor production, probiotics traits, or antimicrobial activity) is promising. Several pools of nonstarter bacteria are now available for the cheesemaker to select for cheese innovation. More specifically, the interest in the microbiota of traditional cheeses, which often contains heterogeneous pools of bacterial strains, should be encouraged for potential future applications. The increase in our understanding of complex microbial consortia and the interactions, which take place in cheese, will ultimately result in a better control of both cheese production and ripening.

REFERENCES

1. Beresford, T.P.; Fitzsimons, N.A.; Brennan, N.L.; Cogan, T.M. Recent advances in cheese microbiology. Int. Dairy J. **2001**, *11* (4–7), 259–274.
2. Wouters, J.T.M.; Ayad, E.H.E.; Hugenholtz, J.; Smit, J. Microbes from raw milk for fermented dairy products. Int. Dairy J. **2002**, *12* (2–3), 91–109.
3. Chamba, J.F.; Irlinger, F. Secondary and adjunct cultures. In *Cheese. Chemistry, Physics and Microbiology*, 3rd Ed.; Elsevier Academic Press: London, UK, 2004; Vol. 1, 191–206.
4. Giraffa, G. Enterococci and dairy products. *Handbook of Food Production Manufacturing*, 1st Ed.; John Wiley & Sons Inc.: Hoboken, New Jersey, USA, 2007; Vol. II, 85–97.
5. Hemme, D.; Foucaud-Scheunemann, C. *Leuconostoc*, characteristics, use in dairy technology and prospects in functional foods. Int. Dairy J. **2004**, *14* (6), 467–494.
6. Tanous, C.; Kieronczyk, A.; Helinck, S.; Chambellon, E.; Yvon, M. Glutamate dehydrogenase activity: a major criterion for the selection of flavour-producing lactic acid bacteria strains. Antonie van Leeuwenhoek **2002**, *82* (1–4), 271–278.
7. Ong, L.; Henrikkson, A.; Shah, N.P. Development of probiotic Cheddar cheese containing *Lb. acidophilus*, *Lb. paracasei*, *Lb. casei* and *Bifidobacterium* spp. and the influence of these bacteria on proteolytic patterns and production of organic acids. Int. Dairy J. **2006**, *16* (5), 446–456.

Cheese: Yeasts and Molds

Steve Labrie
Department of Food Science and Nutrition, University of Laval, Quebec City, Quebec, Canada

Abstract
This entry provides biotechnological approaches to understanding and controlling yeasts and molds during cheese ripening. The main focus is on molecular tools used for strain genotyping and community analyses in cheeses.

INTRODUCTION

Cheeses are living foods composed of a protein matrix containing fat, residual sugars, and minerals that is colonized by microorganisms that metabolize available substrates in order to survive as individuals and as a community. During the ripening process, coordinated microbial activities influence cheese texture and flavor as well as the composition and dynamics of the microbial community. The surface microbiota of mold-ripened and smear-ripened cheeses add another level of complexity. In addition to their contribution to the general appearance of the cheese, yeasts and molds produce desirable compounds that contribute to the final flavor and quality of the cheese.

This entry describes the advantages of using molecular biology techniques to understand the contribution of fungi in cheese ripening. First, the predominant yeasts and molds in cheese will be briefly reviewed, with a focus on their genetics. The main molecular techniques used for strain selection and typing will then be discussed. Lastly, we will review research trends and the new molecular tools being used to investigate cheese ripening, flavor formation, and the dynamics of the microbial community.

PREDOMINANT YEASTS AND MOLDS IN CHEESES AND THEIR GENETICS

Surface-ripened and veined cheeses are inoculated with carefully selected strains of yeasts and molds. At last count, at least 18 different genera encompassing more than 60 species have been isolated from cheeses.[1–4] The main mold species are *Penicillium camemberti* (syn. *Penicillium caseicolum* and *Penicillium candidum*) for Camembert-type cheeses and *Penicillium roqueforti* for blue-veined cheeses. They work in synergy with yeasts, of which the most common species are *Geotrichum candidum*, *Debaryomyces hansenii*, *Kluyveromyces lactis*, *Kluyveromyces marxianus*, and *Yarrowia lypolitica*.[1,2] A number of strains have been developed for various cheese applications and are commercially available worldwide.

Given their economical value, it is surprising that so little is known about their genetics.

Compared to lactic acid bacteria, little effort has been devoted to investigating the genomics of dairy fungi and to characterizing their genes. The entire genomes of three dairy yeasts have been sequenced to date (*D. hansenii*, *K. lactis*, and *Y. lypolitica*), but no complete sequences are available for *Penicillium* spp.[5] While 21 complete genome sequences of lactic acid bacteria have been published, only 10 *P. camemberti* genes, 9 *P. roqueforti* genes, and 11 *G. candidum* genes are listed in public databases. Assuming that the genomes of these fungi code for 5000 to 10,000 genes, we thus have a severely limited picture of their genetic organization and potential.[5,6] The limited use of molecular biology techniques in studying cheese fungi explains, in part, why so little information on their genetics is available and why there has been so little innovation in this field compared, for example to lactic acid bacteria.

MOLECULAR TOOLS FOR TYPING YEASTS AND MOLDS

Strain typing and tracking are of the utmost importance for the cheese industry since starter strains coexist with native microorganisms that also grow during cheese ripening. These strains must be accurately identified to be able to determine their roles in the ripening process and monitor their biological diversity. The selection of a typing method depends on the information needed. Different tools are used to type newly isolated strains, identify commercial strains protected under license, track individual strains in a complex ecosystem, and profile the microbial community as a whole (Table 1).

GENERAL METHODS FOR IDENTIFYING GENERA AND SPECIES

Despite the lack of information on the genetics of these species, a number of molecular techniques have been

Table 1 Molecular techniques for typing dairy fungi and studying cheese ecology.

Method	Principles	Advantages	Disadvantages	Level of typing	Reproducibility	Reference
Ribosomal DNA sequencing	Partial sequencing of the 28 S, 18 S, ITS, and IGS regions of rDNA	Makes rapid and accurate identification of fungal species possible	Rarely discriminates between strains	Genus and species	High	[7,8]
Karyotyping	Electrophoretic analysis of intact chromosomes	Provides information on cell organization	Requires prior knowledge of genus/species	Strain	Fair/good	[7]
RAPD-PCR	Random PCR amplification of genomic DNA	Makes high-throughput screening possible without knowing the DNA sequence; makes SCAR selection possible	Inter-laboratory reproducibility is often low	Genus, species, or strain; depends on the primers used	Fair/good	[3,8,9]
Locus-RFLP	Locus-specific PCR amplification	Is highly reproducible	Needs photodocumentation tools to interpret and analyze the profiles	Subspecies and strain	High	[9,10]
SSCP	Amplification of 18 S rDNA and electrophoretic resolution based on size and conformation	Useful for analyzing microbial communities and determining the relative proportions of strains	Needs specialized equipment and only provides relative quantification	Genera and species in whole samples	High	[11]
Real-time PCR	Quantification of target genes using fluorescent molecules detected during PCR amplification	Allows quantitative analysis of fungi in a given sample	Needs expensive equipment and reagents, and is difficult to apply when diversity is high	Quantifications of genera, species, and strains; depends on the primers used	High	[12]

successfully employed to type and characterize them. Ribotyping is the most common molecular method used to identify the genus/species of unknown isolates because of its moderate cost and high precision. By comparing the sequences of portions of the ribosomal RNA-coding operon (18 S, 5.8 S, and 28 S subunits interspersed with the ITS1 and ITS2 internal spacers) to published sequences, it is possible to rapidly assign a genus and a species to a particular strain.[7,8] However, strains cannot be discriminated using the rDNA sequencing approach. A PCR-based approach can also be used. Seven primer sets are currently available to precisely identify predominant yeast species involved in smear-cheese ripening.[2] This method is effective for high-throughput screening of unknown isolates and can be applied directly to whole DNA extracted from smear-ripened cheese.

STRAIN DISCRIMINATION AND TRACKING

A number of molecular methods are available to help discriminate between isolates when precise strain typing or tracking is important. Since dairy yeasts exhibit inter-strain chromosome length polymorphisms, pulsed-field gel electrophoresis (PFGE) of intact chromosomes can be used to determine their karyotypes.[7] However, PFGE only provides limited resolution and also requires good technical skills to ensure pattern reproducibility. As such, it is often difficult to interpret differences between strains. Restriction fragment length polymorphism (RFLP) analyses of loci containing enough variations can also be used to differentiate between fungal strains. Mitochondrial DNA (mt-DNA) and ribosomal intergenic spacers (ITS and IGS) regions have been proposed as suitable loci for differentiating strains and species, respectively, of dairy fungal isolates and for conducting ecological surveys of cheeses.[9,10]

The random amplified polymorphic DNA—polymerase chain reaction (RAPD-PCR) method has been adapted to identify the genera of dairy yeasts and to differentiate them to the subspecies and strain level.[3,4,8] RAPD-PCR uses 10-bp oligonucleotides to randomly amplify DNA fragments from genomic DNA at low annealing temperatures (36–40°C). Band patterns are then used to discriminate between genera, species, and strains. Since knowing the DNA sequence of the strain of interest is not required, this approach is very useful for studying dairy fungi. The main drawback is the low level of interlaboratory reproducibility due to the varying skill levels of researchers and the different equipment they use, especially the thermocycler. Other PCR-profiling techniques use either universal primers that hybridize with conserved regions (primer N21) or microsatellite primers (RAM-PCR) that can be employed at higher annealing temperatures (50°C) and thus provide improved reproducibility.[11,12] RAPD and RAM-PCR generate band profiles that can be digitalized and analyzed by computer. Sequence-characterized amplified region (SCAR) markers can be identified by cloning and sequencing conserved bands common to all samples. SCAR markers are sequences present in all strains that can be used as templates to design DNA probes to identify specific yeast strains or to develop specific PCR primer sets to detect yeast strains with the common marker.[13]

MOLECULAR TOOLS FOR CULTURE-INDEPENDENT FUNGAL COMMUNITY ANALYSES

Cheeses have diverse ecosystems that evolve throughout the ripening period. Studying the various steps of this evolutionary process (colonization, development, cooperation, competition, metabolism, etc.) will lead to a better understanding of the roles of each species. Various methods have been proposed to study the microbial communities of mold-ripened and smear-ripened cheeses. Single-strand conformation polymorphism (SSCP) analysis was originally developed to screen mutations and is based on the theory that the electrophoretic mobility of a single-stranded DNA (ssDNA) molecule is dependent on its size and conformation under non-denaturing conditions. The 18 S rDNA of fungi can be used as template to generate ssDNA molecules with a unique conformation due to interspecies heterogeneity within the sequences. When the total DNA of a given ecosystem is extracted (the metagenome), a band profile is generated by SSCP where each band represents a different 18 S rDNA that can be linked to a given species. SSCP has recently been optimized and has been used to generate a detailed fingerprint of the complex yeast community of Salers cheese.[14] Based on this information, it is possible to determine the composition of the community and to follow population shifts during cheese ripening.

Enumerating fungi on cheese surfaces is often difficult because of the mycelium of filamentous fungi. Real-time quantitative PCR (qPCR) has been proposed as a way to detect and quantify major yeasts species involved in smear cheese ripening.[15] This method can rapidly identify and quantify four yeast species from complex microbial ecosystems without prior isolation. qPCR has been used to show that *G. candidum* is the dominant yeast species in Livarot cheese.[15] It can also be used to follow yeast community dynamics at different stages of the ripening process.

LIMITATIONS OF CULTURE-INDEPENDENT METHODS

Molecular tools can be useful for typing strains and analyzing microbial communities. However, as pointed out by many authors, isolating DNA from complex matrices (like food) is a limiting step in all culture-independent microbiota analyses because it is more difficult to extract DNA from some microorganisms than others.[15] Many validation phases are needed to ensure that community profiles closely resemble reality.

FUNCTIONAL GENOMICS OF DAIRY FUNGI

The identification of biochemical pathways and the attribution of functions to dairy fungi genes should lead to a better understanding of flavor formation and control of the ripening process. However, very few studies have investigated gene function and expression in dairy fungi. Most research on flavor formation has focused on enzyme activities, substrate consumption, and catabolite production.[16] However, the availability of the genome sequences of *D. hansenii*, *K. marxianus*, and *Y. lipolytica* has opened the way to new approaches for studying the genetics of dairy fungi. For example, microarrays have been used to identify genes expressed during multistrain growth on -methionine, lactate, and lactose in a synthetic cheese medium.[16] An analysis of the transcriptional profiles of co-cultured yeasts has also shown that microarrays can also be used to study metabolic interactions between species in complex microbial communities.

CONCLUSION

The availability of molecular methods for detecting, identifying, and quantifying microorganisms has made it possible to track fungal species and strains throughout the ripening process. An integrated molecular analytic approach for studying the genomes and transcriptomes of cheese fungi will provide further information on the evolution of these dynamic communities during ripening and help researchers to identify the crucial steps involved in flavor development. In addition, it will be possible to determine the roles of individual cheese fungi as more complete sequences and the results of the functional analyses of unknown genes become available. Based on a good understanding of the genetic background of each strain, it will then be possible to develop new starter culture combinations that will optimize the activities of each strain. In addition, technologically relevant genes could be used as "biomarkers" to selectively quantify cheese microorganisms by qPCR or to conduct transcriptomic analyses using microarrays. These tools, combined with conventional biochemical analyses, will improve our understanding of cheese ripening and open the way to predictive analysis schemes to help manage dairy fungi and flavor development.

REFERENCES

1. Spinnler, H.E.; Gripon, J.-C. Surface mould-ripened cheese. In *Cheese: Chemistry, Physics and Microbiology: Major Cheese Groups*, 3rd Ed.; Fox, P.F., McSweeney, P.L.H., Cogan, T.M., Guinee, T., Eds.; Elsevier Academic Press: Amsterdam, 2004; Vol. 2, 157–174.
2. Gente, S.; Larpin, S.; Cholet, O.; Guéguen, M.; Vernoux, J.P.; Desmasures, N. Development of primers for detecting dominant yeasts in smear-ripened cheeses. J. Dairy Res. **2007**, *74* (2), 137–145.
3. Prillinger, H.; Molnár, O.; Eliskases-Lechner, F.; Lopandic, K. Phenotypic and genotypic identification of yeasts from cheese. Antonie Van Leeuwenhoek **1999**, *75* (4), 267–283.
4. Vasdinyei, R.; Deák, T. Characterization of yeast isolates originating from Hungarian dairy products using traditional and molecular identification techniques. Int. J. Food Microbiol. **2003**, *86* (1–2), 123–130.
5. Dujon, B.; Sherman, D.; Fischer, G.; Durrens, P.; Casaregola, S.; Lafontaine, I.; De Montigny, J.; Marck, C.; Neuvéglise, C.; Talla, E.; Goffard, N.; Frangeul, L.; Aigle, M.; Anthouard, V.; Babour, A.; Barbe, V.; Barnay, S.; Blanchin, S.; Beckerich, J.M.; Beyne, E.; Bleykasten, C.; Boisramé, A.; Boyer, J.; Cattolico, L.; Confanioleri, F.; De Daruvar, A.; Despons, L.; Fabre, E.; Fairhead, C.; Ferry-Dumazet, H.; Groppi, A.; Hantraye, F.; Hennequin, C.; Jauniaux, N.; Joyet, P.; Kachouri, R.; Kerrest, A.; Koszul, R.; Lemaire, M.; Lesur, I.; Ma, L.; Muller, H.; Nicaud, J.M.; Nikolski, M.; Oztas, S.; Ozier-Kalogeropoulos, O.; Pellenz, S.; Potier, S.; Richard, G.F.; Straub, M.L.; Suleau, A.; Swennen, D.; Tekaia, F.; Wésolowski-Louvel, M.; Westhof, E.; Wirth, B.; Zeniou-Meyer, M.; Zivanovic, I.; Bolotin-Fukuhara, M.; Thierry, A.; Bouchier, C.; Caudron, B.; Scarpelli, C.; Gaillardin, C.; Weissenbach, J.; Wincker, P.; Souciet, J.L. Genome evolution in yeasts. Nature **2004**, *430* (6995), 35–44.
6. Fungal Genomes Central, http://www.ncbi.nlm.nih.gov/projects/genome/guide/fungi/ (accessed June 2008).
7. Corredor, M.; Davila, A.M.; Casaregola, S.; Gaillardin, C. Chromosomal polymorphism in the yeast species *Debaryomyces hansenii*. Antonie Van Leeuwenhoek **2003**, *84* (2), 81–88.
8. Lopandic, K.; Zelger, S.; Bánszky, L.K.; Eliskases-Lechner, F.; Prillinger, H. Identification of yeasts associated with milk products using traditional and molecular techniques. Food Microbiol. **2006**, *23* (4), 341–350.
9. Belén Flórez, A.; Álvarez-Martín, P.; López-Díaz, T.M.; Mayo, B. Morphotypic and molecular identification of filamentous fungi from Spanish blue-veined Cabrales cheese, and typing of *Penicillium roqueforti* and *Geotrichum candidum* isolates. Int. Dairy J. **2007**, *17* (4), 350–357.
10. Petersen, K.M.; Møller, P.L.; Jespersen, L. DNA typing methods for differentiation of *Debaryomyces hansenii* strains and other yeasts related to surface ripened cheeses. Int. J. Food Microbiol. **2001**, *69* (1–2), 11–24.
11. Gente, S.; Desmasures, N.; Panoff, J.M.; Guéguen, M. Genetic diversity among *Geotrichum candidum* strains from various substrates studied using RAM and RAPD-PCR. J. Appl. Microbiol. **2002**, *92* (3), 491–501.
12. Naumova, E.S.; Smith, M.T.; Boekhout, T.; de Hoog, G.S.; Naumov, G.I. Molecular differentiation of sibling species in the *Galactomyces geotrichum* complex. Antonie van Leeuwenhoek **2001**, *80* (3), 263–273.
13. Corredor, M.; Davila, A.M.; Gaillardin, C.; Casaregola, S. DNA probes specific for the yeast species *Debaryomyces hansenii*: useful tools for rapid identification. FEMS Microbiol. Lett. **2000**, *193* (1), 171–177.
14. Callon, C.; Delbès, C.; Duthoit, F.; Montel, M.C. Application of SSCP-PCR fingerprinting to profile the yeast community in raw milk Salers cheeses. Syst. Appl. Microbiol. **2006**, *29* (2), 172–180.
15. Larpin, S.; Mondoloni, C.; Goerges, S.; Vernoux, J.P.; Guéguen, M.; Desmasures, N. *Geotrichum candidum* dominates in yeast population dynamics in Livarot, a French red-smear cheese. FEMS Yeast Res. **2006**, *6* (8), 1243–1253.
16. Cholet, O.; Hénaut, A.; Casaregola, S.; Bonnarme, P. Gene expression and biochemical analysis of cheese-ripening yeasts: focus on catabolism of L-methionine, lactate, and lactose. Appl. Environ. Microbiol. **2007**, *73* (8), 2561–2570.

Chemoprevention with Dietary Phytopharmaceuticals

Young-Joon Surh
College of Pharmacy, Seoul National University, Seoul, South Korea

Chang Yong Lee
Department of Food Science and Technology, Cornell University, Geneva, New York, U.S.A.

Abstract

Numerous epidemiological, clinical, and experimental research conducted over the past few decades has provided convincing data suggesting that dietary constituents may prevent or even treat several forms of human diseases, including cancer. The rapid progress in our understanding of the cellular signal transduction pathways that mediate unique metabolic and other physiologic processes has paved the way to unveiling the molecular milieu of cellular homeostasis. A wide variety of nutraceuticals derived from plant-based diet, termed "*phytopharmaceuticals*," can alter or correct cellular malfunctions caused by aberrant signal transmission, which often occur during malignant transformation. Alternatively, some phytopharmaceuticals, by inducing transcription of antioxidant genes, can potentiate cellular capacity against oxidative injury implicated in multistage carcinogenesis. Modulation of cellular signaling pathways represents an essential component of molecular target-based cancer prevention with dietary phytopharmaceuticals.

THE PROMISE OF FRUITS AND VEGETABLES—RESERVOIR OF PHYTOPHARMACEUTICALS

There is ample epidemiologic evidence, together with data from animal and cell culture studies, supporting an inverse association between the frequency and/or amount of fruits and vegetables consumed and the risk of several forms of chronic ailments, including cancer, cardiovascular disease, stroke, neurodegenerative disorders, hypertension, diabetes, and cataracts. These observations have led to the establishment of public health promotion programs, such as "Five-a-Day for Better Health" (now expanded to eating five to nine a day for better health) campaign in the United States. Increasing fruit and vegetable intake has become a global priority in the prevention of many chronic diseases.[1]

Diets rich in plant foods, including fruits, vegetables, and whole grains, contain substantial amounts of vitamins, minerals, and fiber that provide desirable health benefits beyond basic nutrition; however, plant-based foods are also full of other biologically active substances, collectively termed phytochemicals ("*phyto*" in Greek means plant). Although not considered essential nutrients, some phytochemicals have the potential to prevent diseases or to mitigate their symptoms, and are hence called "phytopharmaceuticals." The majority of phytopharmaceuticals are synthesized by plants as secondary metabolites. Phytopharmaceuticals represent a chemical arsenal that protects plants from ultraviolet light, insects, bacteria, fungi, viruses, pollutants, and drought. Such cytoprotective activities of phytopharmaceuticals are likely to contribute to chemopreventive effects and other health benefits of fruits and vegetables. As illustrated in Fig. 1, some bioactive phytopharmaceuticals or their metabolites directly modulate the expression of target genes or the activity/stability of their protein products. Alternatively, phytopharmaceuticals can indirectly influence gene expression through regulation of intracellular signaling cascades.

PHYTOPHARMACEUTICALS AS GUARDIANS OF OUR HEALTH

Numerous laboratories around the world have been mining botanicals for biologically active phytochemicals that may be used as medicines. The terms, phytopharmaceuticals and nutraceuticals, have been coined to name identify bioactive phytochemicals with therapeutic value. While the former term is preferred for use by health professionals, the latter is more commonly used by food scientists and dieticians. Phytopharmaceuticals can be classified as carotenoids, polyphenols, terpenoids, isothiocyanates, indoles, isoflavones, and organosulfur compounds (Table 1). We can benefit from the majority of these bioactive phytochemicals as well as other nutrients by eating five to nine servings of colorful fruits and vegetables per day in combination with sufficient amount of whole grains, legumes, and nuts (http://www.5aday.gov/index.html). It is

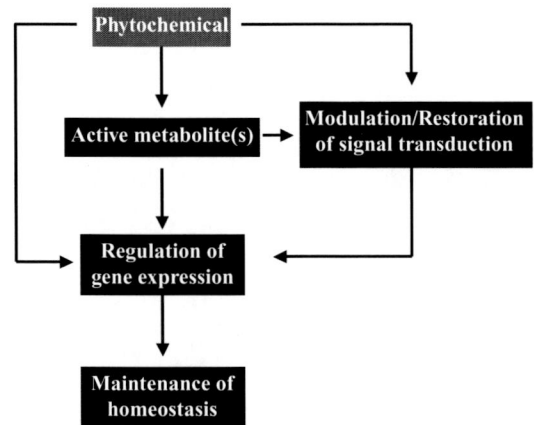

Fig. 1 Phytopharmaceutical gene interactions. Phytopharmaceuticals or their active metabolites can directly regulate target gene expression. Alternatively, some phytochemicals may modulate or restore the particular cellular signal transduction events, thereby indirectly influence gene expression.

estimated that there may be more than 100 different phytopharmaceuticals in just a single serving of vegetable or fruit. The identification of phytopharmaceuticals and elucidation of mechanisms of their action are exciting and emerging areas of recent research.

HOW CAN PHYTOPHARMACEUTICALS FIGHT DISEASES?

Phytopharmaceuticals have multifaceted actions. Among the diverse beneficial effects exerted by phytopharmaceuticals is their potential to inhibit, reverse, or retard multistage carcinogenesis, which constitutes an important part of current cancer chemoprevention strategies.[1–4] Chemopreventive effects that certain dietary phytopharmaceuticals exert are likely to be the sum of several distinct mechanisms that include blockage of metabolic activation, DNA binding of carcinogens, stimulation of detoxification, repair of DNA damage, suppression of cell proliferation and metastasis or angiogenesis, and induction of differentiation or apoptosis of precancerous or malignant cells (Fig. 2).

Examples of chemopreventive dietary phytopharmaceuticals are epigallocatechin gallate from green tea, curcumin from turmeric, genistein from soybeans, brassinin from cabbage, resveratrol from grapes, lycopene from tomatoes, diallyl sulfide from garlic, gingerol from ginger, and caffeic acid phenethyl ester from honey bee propolis.[1]

INTRACELLULAR SIGNAL NETWORK AS A NOVEL TARGET OF CHEMOPREVENTIVE PHYTOPHARMACEUTICALS

Despite enormous progress in clarifying the biochemical events that are associated with the pathophysiology of many human malignancies, identification of molecular and cellular targets of chemopreventive phytopharmaceuticals is still incomplete. Research directed toward elucidating underlying molecular mechanisms of chemoprevention with dietary phytopharmaceuticals has recognized key components of signal transduction pathways as potential targets.[1,5–10] A vast variety of molecules and events are involved in relaying intracellular signals. Both external

Table 1 Representative phytopharmaceuticals derived from vegetables and fruits.

Phytochemicals	Sources
Carotenoids	
β-Carotene	Carrots, pumpkins, cantaloupe, sweet potatoes, spinach, kale
Lycopene	Tomatoes, red peppers, watermelon
Lutein	Kale, spinach, kiwifruit, broccoli, Brussels sprouts
Polyphenols	
Quercetin	Apples, pears, cherries, grapes, onions
Resveratrol	Grapes, red wine, red grape juice
Epigallocatechin gallate	Green tea
Hesperidin	Citrus fruits (oranges, lemon, lime, tangerine, grapefruit)
Ellagic acid	Berries, grapes, kiwifruits
Terpenoids	
D-limonene	Citrus fruit peel
Iosthiocyanates	
Sulforaphane	Broccoli, kale, Brussels sprouts
Indoles	
Indol-3-carbinol	Cruciferous vegetables (cabbage, watercress, broccoli, turnips)
Isoflavones	
Genisteine and daidzein	Soy products
Organosulfur compounds	
Allyl sulfides and allicin	Garlic, onion, chives, scallions

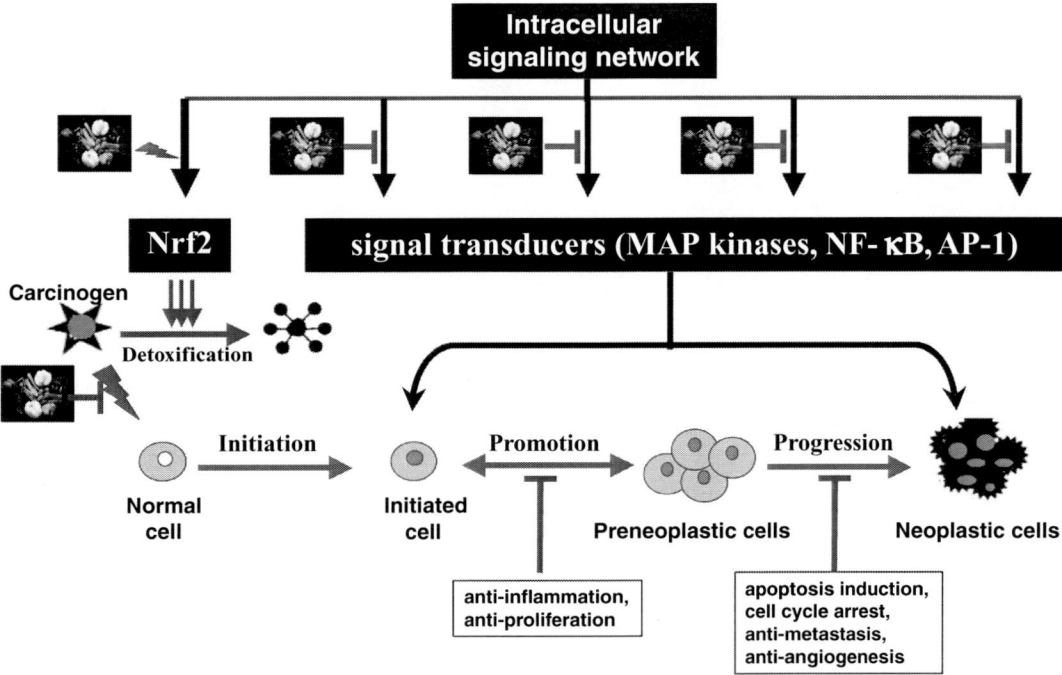

Fig. 2 Intracellular signaling network involved in multistage carcinogenesis as a molecular target for chemopreventive phytopharmaceuticals. Some phytopharmaceuticals block the interaction of a carcinogen with the target cell DNA or stimulate the detoxification of a carcinogen by activating the transcription factor Nrf2 (see details of Nrf2 activation described in the legend to Fig. 3). Others can also exert chemopreventive effects by suppressing the signal transduction mediated by other transcription factors, such as NF-κB and AP-1 or their upstream protein kinases including MAP kinases that are frequently overactivated during the promotion and progression stages of carcinogenesis.

and endogenous stimuli turn on or switch off critical events of this relay, thereby transmitting proper signaling to diverse downstream target molecules in a highly sophisticated fashion for fine-tuning of cellular homeostasis. Components of upstream or cytoplasmic signaling networks include protein kinases, such as the family of proline-directed serine/threonine kinases specifically named mitogen-activated protein (MAP) kinases, protein kinase C, phosphatidylinositol-3-kinase, protein kinase B/Akt, and glycogen synthase kinase. Inappropriate or aberrant activation or silencing of any of the aforementioned kinases or downstream transcription factors can result in disruption of cellular homeostasis. Therefore, targeted modulation or restoration of the intracellular signaling network by phytopharmaceuticals offers a unique strategy for preventing abnormal cell proliferation and other malfunctions linked to cancer formation; however, in contrast to molecular-target-based anticancer drugs (e.g., Gleevec and Iressa) developed after extensive activity-guided or mechanism-based screening for target specificity as well as for efficacy, the majority of chemopreventive phytopharmaceuticals so far investigated have pleiotropic effects on intracellular signaling, and perhaps with few exceptions, no single pathway can account for the tumor-preventive activities of a specific phytopharmaceutical. Such complexity hampers the identification of critical target molecules of chemopreventive phytopharmaceuticals.

CELLULAR SIGNALING FOR ANTIOXIDANT ENZYME INDUCTION BY CHEMOPREVENTIVE PHYTOPHARMACEUTICALS

Oxidative stress induced by reactive oxygen species (ROS) has been implicated in multistage carcinogenesis. Thus, oxidative DNA damage can initiate malignant transformation, whereas ROS-mediated oxidative or inflammatory tissue injury may stimulate the proliferation of initiated cells during the promotion stage and even facilitate the progression of carcinogenesis. Many phytopharmaceuticals have capabilities to rescue cells from oxidative assault, by scavenging excess ROS, and/or more actively potentiating cellular built-in defenses through induction of de novo synthesis of antioxidant enzymes, such as glutathione S-transferase, NAD(P)H:quinone oxidoreductase 1, UDP-glucuronyltransferase, glutamate cysteine ligase, glutathione synthetase, glutathione peroxidase, heme oxygenase-1, and thioredoxin reductase. While some of these antioxidant enzymes catalyze the reactions responsible for producing endogenous antioxidants, such as reduced glutathione, bilirubin, and thioredoxin, others can indirectly maintain cellular redox-status. The 5'-flanking region of genes encoding most antioxidant enzymes harbors the common cis-acting sequences, termed "antioxidant response elements (ARE)" or "electrophile response elements (EpRE)." The ARE- or EpRE-driven transcription of

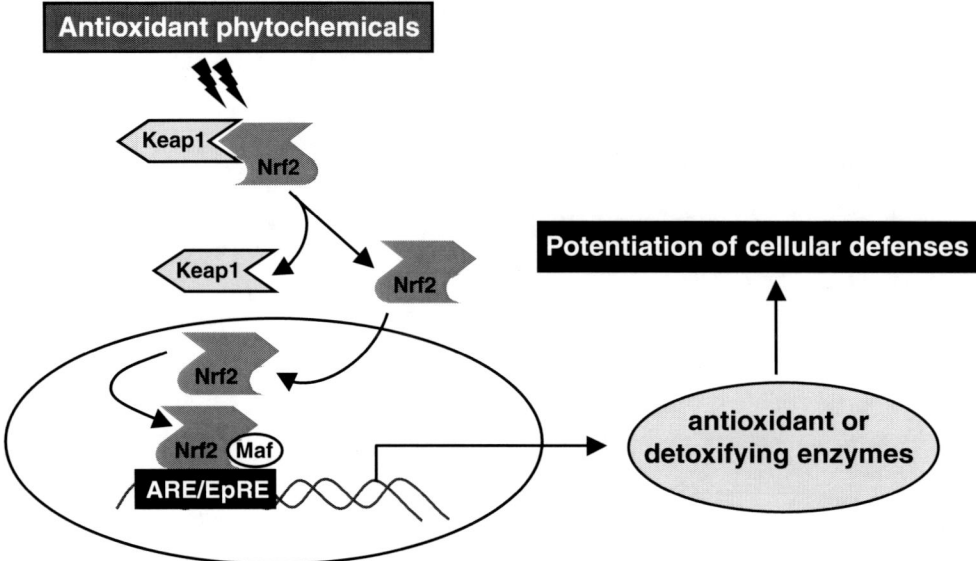

Fig. 3 Upregulation of antioxidant/detoxifying enzyme gene expression via Nrf2-ARE signaling. Nrf2 is present as an inactive complex with the cytosolic protein Keap1. Some phytopharmaceuticals stimulate dissociation of Nrf2 from Keap1 and facilitate the translocation of Nrf2 to nucleus where it binds together with small Maf protein to the antioxidant response elements (ARE) or electrophile response elements (EpRE). This leads to transcriptional induction of an array of antioxidant and phase-2 detoxifying enzymes, such as glutathione S-transferase, NAD(P)H:quinone oxidoreductase 1, glutamate cysteine ligase, glutathione synthetase, heme oxygenase-1, and thioredoxin reductase, thereby fortifying cellular antioxidant defenses.

antioxidant genes is mainly mediated by a redox-sensitive transcription factor, Nrf2, that is sequestered in the cytoplasm as an inactive complex with an actin-binding protein, "Kelch-like ECH-associated protein 1 (Keap1)." Some antioxidant phytochemicals stimulate dissociation of Nrf2 from Keap1 and facilitate subsequent nuclear translocation and ARE-binding of Nrf2, thereby inducing antioxidant/carcinogen detoxifying gene expression[1,6,8,11,12] as illustrated in Fig. 3. The regulation of signal transduction pathways mediating antioxidant or phase-2 detoxification enzyme induction may comprise mechanistic basis for the chemopreventive effects of phytopharmaceuticals.

DEVELOPMENT OF FUNCTIONAL FOODS CONTAINING CHEMOPREVENTIVE PHYTOPHARMACEUTICALS: NUTRIGENOMIC PERSPECTIVES

While chemoprevention with phytopharmaceuticals might be best acquired by daily intake of whole foods containing phytopharmaceuticals, dietary supplements of active ingredients are recommended for those who have limited access to fruits, vegetables, and fibers, or for high-risk populations who may need extra amounts of phytonutrients. The global functional food market is expanding enormously, as consumers are turning to dietary products that promise health benefits. The term "nutrigenomics" or "nutragenetics" has been coined, and much attention is being focused on this exciting new wave of nutrition research. Nutrigenomics or nutragenetics can also help us understand how diseases such as cancer can be affected or alleviated with dietary components. With advances in molecular biological techniques to assess single nucleotide polymorphisms (SNPs), we are now more aware of the specific genes that can directly or indirectly contribute to individual differences in susceptibility to carcinogenesis and pathogenesis of other ailments. When high-risk individuals are identified, health practitioners or dieticians may recommend specific dietary supplements that may modulate, correct, or restore cellular signaling events that are likely to be disrupted in these people. So, it would be possible to formulate a cocktail containing chemopreventive ingredients in consideration of an individual's family history to a specific cancer. While such tailored supplementation with designer foods to match a person's genome appears fascinating and theoretically plausible, it is not that simple, and there is a long way to go before we can be certain of the relative benefits versus the risks of increased consumption of individual food ingredients or their mixtures. The fact that a food with a certain ingredient is beneficial to health does not necessarily mean the ingredient taken in a pill form will be. Moreover, undesired side effects or even deleterious effects may be caused as a consequence of long-term intake of formulated supplements. In this context, we should be reminded by a Finnish ATBC study and a U.S. CARET study: Although β-carotene had been listed as one of the elite class cancer-fighting agents since a U.S. National

Academy of Science report in 1982, high-dose β-carotene supplements were linked to an extra-high rate of lung cancer and death, according to results of intervention trials.

CONCLUDING REMARKS

The cellular signaling network, which is regulated by a wide array of kinases, transcription factors, and other proteins, often goes awry in various pathophysiologic processes. The resultant effects appear as a disturbance in homeostasis and consequently an abnormal cellular function that can lead to cancer and other disorders. As disruption of cellular signaling pathways is considered a common denominator in the pathogenesis of various ailments, it is therefore worthwhile to identify and better understand the mechanism of signal-transducing molecules that can be affected by individual phytopharmaceuticals. Such elucidation will allow for better assessment of the underlying molecular and cellular mechanisms; however, extreme complexity of the cellular signaling network, especially various unidentified cross-talk between signaling molecules, appears to be a major hurdle that hampers characterization of the underlying mechanisms. In the era of "-omics," such as nutritional genomics, proteomics, and metabolomics, a state-of-the-art approach utilizing high-throughput technologies will enable researchers to simultaneously analyze clusters of individual genes and their complex expression profiles that can be altered by specific phytopharmaceuticals.[10] Such an effective multidisciplinary approach will facilitate molecular analysis of phytopharmaceuticals and their effects on global gene expression in a more integrated way.

Although it seems reasonable to suggest that proper dietary manipulation might substantially affect rates of cancer and other diseases, the general veracity of this hypothesis remains unproven. Furthermore, unraveling the intricacies of the genetic variability in the context to the multitude of phytochemicals and other influences to which humans are exposed requires a concerted and long-term research effort.

ACKNOWLEDGMENTS

This work was supported by the Functional Food Research grant from the Ministry of Education, Science and Technology, Republic of Korea (awarded to YJS).

REFERENCES

1. Surh, Y.-J. Cancer chemoprevention with dietary phytochemicals. Nat. Rev. Cancer **2003**, *3* (10), 768–780.
2. Lee, K.W.; Lee, H.J.; Lee, C.Y. Vitamins, phytochemicals, diets, and their implementation in cancer chemoprevention. Crit. Rev. Food Sci. Nutr. **2004**, *44* (6), 437–452.
3. Bagchi, D.; Preuss, H.G. *Phytopharmaceuticals in Cancer Chemoprevention*; CRC Press: Boca Raton, FL, 2005.
4. Birt, D. Phytochemicals and cancer prevention: from epidemiology to mechanism of action. J. Am. Diet. Assoc. **2006**, *106* (1) 20–21.
5. Kundu, J.K.; Surh, Y.-J. Breaking the relay in deregulated cellular signal transduction as a rationale for chemoprevention with anti-inflammatory phytochemicals. Mutat. Res. **2005**, *591* (1–2), 123–146.
6. Surh, Y.-J.; Kundu, J.K.; Na, H.-K.; Lee, J.-S. Redox-sensitive transcription factors as prime targets for chemoprevention with anti-inflammatory and antioxidative phytochemicals. J. Nutr. **2005**, *135* (12), 2993S–3001S.
7. Bode, A.M.; Dong, Z. Targeting signal transduction pathways by chemopreventive agents. Mutat. Res. **2004**, *555* (1–2), 33–51.
8. Chen, C.; Kong, A.N. Dietary cancer-chemopreventive compounds: from signaling and gene expression to pharmacological effects. Trends Pharmacol. Sci. **2005**, *26* (6) 318–326.
9. Dorai, T.; Aggarwal, B.B. Role of chemopreventive agents in cancer therapy. Cancer Lett. **2004**, *215* (2) 129–140.
10. Go, V.L.W.; Butrum, R.R.; Wong, D.A. Diet, nutrition, and cancer prevention: the postgenomic era. J. Nutr. **2003**, *133* (11), 3830S–3836S.
11. Lee, J.-S.; Surh, Y.-J. Nrf2 as a novel target for chemoprevention. Cancer Lett. **2005**, *224* (2) 171–184.
12. Yu, X.; Kensler, T. Nrf2 as a target for cancer chemoprevention. Mutat. Res. **2005**, *591* (1–2) 93–102.

Chitosan: Produced by Microorganisms

Renuka Karuppuswamy
Department of Food Science and Technology, University of New South Wales, Sydney, New South Wales, Australia

Abstract

In our daily life, biodegradable polymers have attained much commercial importance because of their pollution-free environmental biocompatibility. Chitin, the second most abundant naturally occurring biopolymer after cellulose, is found in the exoskeleton of crustacea, fungi, and some insects. Chitosan, the deacetylated chitin, is used in various industrial sectors because of its unique properties compared to chitin. However, chemical conversion of chitin to chitosan is mostly achieved from crustacea waste, which reduces the quality of chitosan and, at the same time, generates chemical waste. Enzymatic deacetylation of chitin to chitosan using chitin deacetylase is found to be an effective alternative to chemical methods. However, this way of producing chitosan depends on the seasonal availability of crustacea waste. Production of chitosan from various microorganisms not only improves the quality of chitosan but also produces higher yields. Therefore, high-quality chitosan can be produced all year round using microorganisms without relying on the seasonal shellfish industry.

INTRODUCTION

Polymers have become part of our everyday activities. Recently, natural biodegradable polymers have gained great attention due to their safe environmental biocompatibility. Chitin, the shell material of crustacea, fungi, and some insects, is one among them. Chitosan, a chitin-derivative, has been evaluated for numerous applications in agricultural, food, cosmetics, pharmaceutical, biomedical, papermaking, and wastewater treatment because of its better properties suited to various applications. The production of chitin and chitosan is currently based on crustacea waste discarded by seafood-processing industries and fresh food markets. Recovery of these valuable compounds from crustacea waste not only solves the waste disposal problem, but also adds additional income. Chemical deacetylation of chitin to chitosan is highly effective. However, it produces an enormous amount of chemical waste that causes major threat to the environment. This method also affects the product quality by side reaction and polymer chain breakdown. Moreover, the use of strong alkali during deacetylation process corrodes the installation. Therefore, an alternative approach is necessary to overcome these problems. This entry discusses the use of microorganisms in obtaining chitosan.

CHITIN DEACETYLASES

Chitin consisting of poly-*N*-acetyl-D-glucosamine units is the second most abundant naturally occurring polymer after cellulose. Chitosan is derived from chitin by deacetylation using hot sodium or potassium hydroxide solution.[1] The biomedical market for chitin, chitosan, and their derivatives is estimated at $1.25 billion per year because of their biocompatibility, biodegradability, and low toxicity. Pure chitin and chitosan are worth $10 per gram in wholesale.[2] Therefore, there is a strong demand for chitosan in commercial industrial sectors.

Chitosan exists as a major component of the cell wall in zygomycetes. It is synthesized through the tandem action of chitin synthetase and chitin deacetylase (CDA). In this pathway, chitin synthetase polymerizes *N*-acetyl glucosamine precursors to chitin and CDA catalyzes the deacetylation of the nascent chitin into chitosan. Therefore, CDA plays an important role in the fungal growth of zygomycetes.[3] In order to overcome the drawbacks associated with chemical deacetylation of chitin to chitosan, studies have been focusing to find an alternative approach.

As a result, enzymatic deacetylation/conversion of crustacea chitin to chitosan using CDA has been extensively studied, and several organisms producing CDA have been screened out. CDA has been first identified and partially purified from *Mucor rouxii* fungal extract.[4] Subsequently, numerous CDAs have been extracted, purified, and characterized from various fungal strains, including *Absidia*, *Rhizopus*, and *Aspergillus* species. These studies mainly concentrated on the increased production of chitosan by applying some suitable modifications both in the chitin structure and in the CDA production conditions.

Later in a study, several fungal strains were reviewed for chitosan production using fungal fermentation.[5] This study also screened 33 fungal strains as chitosan-producing microorganisms. Of the 33 strains, there are 9 zygomycetes, 1 basidiomycetes, 5 ascomycetes, and 18 deuteromycetes. They are: *A. coerulea, Gongronella butleri, Agaricus bisporus, Aspergillus oryzae, Aspergillus terreus, Aspergillus terricola, Aspergillus usamii, Blakeslea trispora, Aspergillus flavus, Cladosporium ladosporioides, Gibberella fujikuroi* var. intermedia (*ATCC 42052*), *M. rouxii, R. oryzae, Trichoderma viride, Trichothecium roseum, Gliocladium catenulatum, Aspergillus niger, Ashbya gossypii, Penicillin chrysogenumare, Absidia glauca, Aspergillus clavatus, A. nidulans, Botrytis cinerea, Ceratocystis ips, Cladosporium cucumerinum, Epicoccum nigrum, Humicola grisea, Mucor hiemalis, Myrothecium verrucaria, Penicillium digitatum, Phycomyces blakesleeanus, Rhizopus stolonifer,* and *Sclerotinia sclerotiorum*. This study concludes that chitosan can be extracted from some common industrial waste fungal mycelia, such as *P. chrysogenum* (used in penicillin production), *Aspergillus species* (used in lipase, pentosanase, riboflavin, citric, gluconic acids, b-glucanase, cellulose, and proteases production), and *Mucor* species (used in rennin production). This will also eliminate the incineration of the industrial waste fungal mycelia. Fungi of no or little industrial importance, for example, *A. glauca, A. nidulans,* and *M. rouxii* can also be the chitosan producers because of their high chitosan content in cell walls.[5]

Nitrogen is one of the important factors for the production of fungal chitosan because it acts as a nutrient source to synthesize chitin/chitosan for their cell wall.[6] The increase in molecular weight of fungal chitosan can be obtained by using more nitrogen source in the submerged fermentation medium.[7] The type of fungal strain and fermentation conditions also affect the quantity and the quality of chitosan extracted from the fungal media.[7] Fungal chitosan from *G. butleri* USDB 0201 grown on sweet potato pieces supplied with different amount of urea has been investigated to find out the effect of nitrogen in chitosan production. The maximum yield of chitosan (11.4 g/100 g mycelia) has been obtained from the fungal mycelia grown on the solid state fermentation medium supplied with 14.3 g urea/kg of solid substrate. The optimum pH for fungal chitosan production is 5.5–6.5.[7]

Few studies have also been focused to reduce the high cost associated with the culturing medium. Therefore, various fungal strains, such as *Absidia atrospora IF09471, A. coerulea IF04011, A. coerulea IF04012, A. glauca IF04002, A. glauca IF04003, A. glauca IF04004, A. glauca* var. *pardoxa IF04007, A. glauca* var. *paradoxa IF04431, G. butleri IF08080,* and *G. butleri IF08081* have been grown on the wastewater containing a mixture of barley-*shochu* (Japanese distilled spirit) distillery wastewater, buckwheat-*shochu* distillery wastewater, and sweet potato-*shochu* wastewater. *G. butleri IF08081* grown in the sweet potato-*shochu* wastewater is most suitable for chitosan production. This method is cost effective and requires no addition of any nutrients to the medium.[8] It will be economical if chitosan can be produced from microorganisms using industrial organic waste. Similarly, several studies have been investigated to produce fungal chitosan using the industrial waste such as *A. niger* mycelia obtained as a waste from a citric acid production plant[9] and *R. oryzae* grown on beet molasses,[10] and on the industrial waste whey medium added with growth hormones.[11] These chemical alternative methods produce high-purity chitosan with higher recovery rate. Therefore, extraction of chitosan from microorganisms not only offers an attractive way to waste management but also leads to commercial feasibility of producing high-quality chitosan in larger scale.

CONCLUSION

Generally production of chitosan using microorganisms gives low yields but highly deacetylated forms of chitosan than obtained by deacetylation of crustacean chitin. However, low yields may be improved by modifying the processing conditions. There are several advantages using microorganisms, especially fungi, to produce chitosan. Chitosan is extractable from some common industrial waste fungal mycelia, which is economical and aids for effective waste management. The physiochemical properties of the fungal chitosan can be controlled and standardized by changing the processing parameters. Chitosan can be produced under controlled conditions all year round and is independent of seasonal shellfish industry. Chitosan obtained from microorganisms can be used in medical and pharmaceutical industries where there are concerns associated with seafood residues that can cause allergic reaction in sensitive individuals.

ACKNOWLEDGMENT

I would like to thank Professor Willem F. Stevens, Mahidol University, Thailand, for his continuous support in giving valuable ideas and suggestions.

REFERENCES

1. Andrady, A.L.; Xu, P. Elastic behaviour of chitosan films. J. Polym. Sci. Part B: Polym. Phys. **1997**, *35*, 517–521.
2. Steward, P. New uses for cull spuds may be on horizon. Capital Press Agriculture Weekly, Washington, 2004; *67* (792), Article No. 13517.
3. Shimahara, K.; Takiguchi, Y.; Kobayashi, T.; Uda, K.; Sannan, T. Screening of mucroaceae strains suitable for chitosan products. In *Chitin and Chitosan*; Skjak-Braek, G., Anthonsen, T., Sandford, P., Eds.; Elsevier, Essex, UK, 1989; 171–178.

4. Araki, Y.; Ito, E. A pathway of chitosan formation in *Mucor rouxii*. Enzymatic deacetylation of chitin. Eur. J. Biochem. **1975**, *55* (1), 71–78.
5. Hu, K.J.; Hua, J.L.; Hob, K.P.; Yeunga, K.W. Screening of fungi for chitosan producers, and copper adsorption capacity of fungal chitosan and chitosanaceous materials. Carbohydr. Polym. **2004**, *58*, 45–52.
6. Moore-Landecker, E. *Fundamentals of the Fungi*, 4th Ed.; Prentice-Hall: Upper Saddle River, NJ, 1996; 251–278.
7. Nwe, N.; Stevens, W.F. Effect of urea on fungal chitosan production in solid substrate fermentation. Process Biochem. **2004**, *39*, 1639–1642.
8. Yokoi, H.; Aratake, T.; Nishio, S.; Hirose, J.; Hayashi, S.; Takasaki, Y. Chitosan production from shochu distillery wastewater by funguses. J. Ferment. Bioeng. **1998**, *85* (2), 246–249.
9. Cai, J.; Yang, J.; Du, Y.; Fan, L.; Qiu, Y.; Li, J.; Kennedy, J.F. Enzymatic preparation of chitosan from the waste *Aspergillus niger* mycelium of citric acid production plant. Carbohydr. Polym. **2006**, *64*, 151–157.
10. Goksungur, Y. Optimization of the production of chitosan from beet molasses by response surface methodology. J. Chem. Technol. Biotechnol. **2004**, *79* (9), 974–981.
11. Chatterjee, S.; Chatterjee, S.; Chatterjee, B.P.; Guha, A.K. Enhancement of growth and chitosan production by *Rhizopus oryzae* in whey medium by plant growth hormones. Int. J. Biol. Macromol. **2008**, *42* (2), 120–126. Online accepted on October 4, 2007.

Cloning: Breeding

Rodomiro Ortiz
Intensive Agroecosystems Program, International Center for Maize and Wheat Improvement (CIMMYT), Juarez, Mexico

Abstract
Mutation and genetic recombination allow breeding asexual crops such as potato, cassava, sweet potato, yam, plantain/banana, sugar cane, and fruit trees. Analytical breeding (through ploidy manipulations) allows the broadening of the genetic base of asexual crops. This entry indicates that biotechnology, through tissue culture, transgenics, and genomics, offers new tools for completing other methods used for breeding clones.

Clones are propagules arising from asexual (or vegetative) propagation, which ensures that plant propagules are genetically alike. The most important vegetatively propagated food crops are potato, cassava, sweet potato, yam, plantain/banana, sugar cane, and fruit trees. Other crops with asexual propagations are some ornamentals, grasses, and forages. Among the most common planting materials are tubers (e.g., potatoes and yams), vines (e.g., sweet potatoes), stem cuttings (e.g., cassava), and suckers (e.g., plantains and bananas). In vegetatively propagated crops, the common origin (or the same source) of planting materials is crucial to have uniform and meaningful trials. Tissue culture–derived plantlets are also promising planting materials to achieve propagule uniformity in some of these food crops.

CROSSBREEDING

Breeding clones result from selecting among mutant plants or hybrids derived from the union of male and female gametes, which can be either natural or through hand crossing. If none of this occurs, the genotypes of the clones will remain constant. Such distinct breeding system offers the advantage of fixing immediately any desirable genetic combination and propagating it indefinitely. For example, "Russet Burbank," which was selected as a clone in the nineteenth century, still remains as one of the most popular potato cultivars grown by North American farmers. However, the potential lack of genetic recombination may be a major challenge in some asexual crops because of this breeding system. When recombination occurs the same genetic principles apply as for sexual (or seed) crops.

Crossbreeding methods for vegetatively propagated crops rely on sexual hybridization (i.e., seeds produce new genotypes after crossing selected parents). However, low fertility as a result of natural and human selection may impede crossbreeding of asexual crops, which may be further exacerbated by polyploidy. Both environment and genotype appear to control fertility in asexual crops. Special protocols are used to maximize flowering in some vegetatively propagated crops. Time (i.e., photoperiod), temperature, and intensity of light are the most important factors affecting flowering in these crops.

The main goal of breeding clones is to obtain genotypes that are phenotypically uniform (homogeneous) but often highly heterozygous, particularly if non-additive gene action controls the commercial trait(s) of interest. The non-additive gene action may arise from intra- or interallelic (epistasis) interactions. The conventional plan for breeding clones consists of 1) selecting appropriate parents for crossing schemes; 2) early or late selection in clonal generations, which is determined by the heritability of the targeted trait(s); and 3) adequate environmental sampling (i.e., evaluation across a number of locations and years) for testing advanced breeding materials, leading to cultivar development. The steps followed in the most common breeding scheme are shown in Fig. 1.

ANALYTICAL BREEDING

Genetic manipulations of complete chromosome sets are called ploidy manipulations, i.e., scaling up and down chromosome numbers of a species within a polyploid series. The most important vegetatively propagated food crops (e.g., potato, sweet potato, yam, plantain/banana, and some fruit trees) possess well-endowed genetic resources from their wild relatives, which are often of lower ploidy.[1] Chromosome sets are manipulated with haploids, 2n gametes, and through interspecific–interploidy crosses. Analytical breeding schemes mainly rely on ploidy manipulations to "capture" diversity from exotic (wild or non-adapted)

Source population (5,000–100,000 seedlings) after crossing selected parents
↓ Defect elimination or mild-selection for specific attributes

Single plots (100–3,000 selected clones) for clonal evaluation
↓ Screening for specific attributes as per breeding plan

Preliminary yield trial (25–100 clones) with 2 replications
↓ Screening to confirm attributes and early yield assessment

Advanced yield trial (10–25 clones) with 3–4 replications in at least 3 locations
↓ Further yield assessment

Uniform yield trial (5–15 best clones) with 4 replications in many locations
↓ Yield assessment and testing stability across location range

On-farm participatory testing of elite materials (2–5 clones)
↓ Farmer (and sometimes end-users') testing

Multiplication of selected clone(s) and cultivar release
↓ Through appropriate national committee

Fig. 1 Common breeding scheme for vegetatively propagated tropical crops at the International Institute of Tropical Agriculture. Propagule numbers are crop-dependent.

germplasm and use $2n$ gametes to incorporate this genetic diversity through unilateral ($n \times 2n$ or $2n \times n$) or bilateral ($2n \times 2n$) polyploidization.[2] Haploids are propagules with the gametophytic chromosome number (n), and $2n$ gametes possess the sporophytic chromosome number of the parental source. The most interesting examples of analytical breeding are in vegetatively propagated species such as potato,[3] sweet potato,[4] and cassava[5] among roots and tubers, and plantain/banana[6] among fruit crops. This breeding approach appears promising in sugar cane,[7] blackberry,[8] blueberry,[9] strawberry,[10] and other fruit crops.[11]

Potato may be regarded as the model crop either for breeding clones by conventional methods[12] or for broadening the genetic base of crop production, particularly through analytical breeding.[13] In potato ploidy manipulations, chromosome sets are easily managed with wild species, maternal haploids obtained through parthenogenesis, $2n$ gametes arising from meiotic mutants, and the endosperm balance number (EBN). This endosperm dosage system, also common to other angiosperm genera, requires a 2:1 maternal to paternal genetic contributions for normal seed development after hybridization.[14] The wild species (mostly diploids) bring new genetic variation to the breeding pool, whereas haploids "capture" this genetic diversity when they are crossed with diploid wild species. The resulting haploid-species hybrids producing $2n$ gametes and the EBN needed for the successful gene transfer (Fig. 2) are the means for broadening the genetic base of the cultivated potato through unilateral or bilateral polyploidization, which is confirmed by a recent analysis with genetic markers.[15] Furthermore, such analysis suggests that the need for broadening the genetic base of potato may be for specific chromosomes or regions within chromosomes.

SOMACLONAL VARIATION

Irrespective of the advantages of tissue culture for vegetatively propagated crops, such as high-throughput production of healthy propagules, somaclonal variation may affect

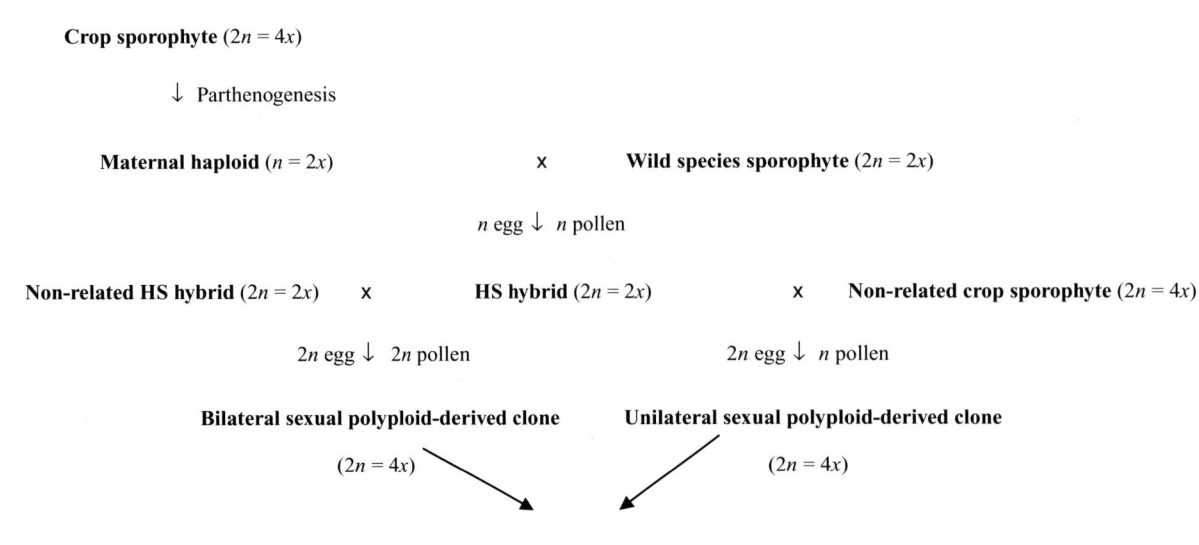

Fig. 2 Ploidy manipulations (or scaling up and down chromosome numbers) for breeding clones with haploids and $2n$ gametes in a crop within a polyploid series. The example refers to a tetraploid ($2n = 4x$) crop with diploid wild species. HS, haploid species.

the true to type during micropropagation.[16] Somaclonal variation refers to genetic variation among plants from the same original genotype arising from tissue culture regeneration. Careful plant breeders minimize somaclonal variation by 1) deliberately selecting stable sources of materials for primary explants; 2) limited subculturing and multiplication (e.g., less cycles, short time for subculturing, and a few hundred plants per primary explant); and 3) nursery screening to detect and rogue off-types. Somaclonal variation may provide a potential source for genetic improvement of some vegetatively propagated crops. However, in most crops, the range of somaclonal variants recovered through shoot-tip culture seems to be narrow, mimics naturally occurring variation, or produces defective genotypes.

FUTURE PROSPECTS OF BIOTECHNOLOGY IN IMPROVING CLONES

Plant breeders determine their genetic response to selection (R) using the following equation:

$$R = i h^2 \sigma_P$$

where i is the selection intensity, expressed in standardized units; h^2 is the narrow-sense heritability in the reference population; and σ_P is the phenotypic standard deviation of the selected characteristic. For example, in vitro selection techniques may provide a means for achieving a higher intensity of selection, whereas molecular markers are used to tag specific chromosome segments bearing the desired gene(s) to be transferred (or incorporated) into bred clones.[17] In this way, indirect selection with codominant molecular markers tightly linked to the gene(s) controlling the characteristic(s) of interest improves R, because codominant markers have an h^2 equal to 1. Transgenic plants offer new genes to the breeding pool, thus enhancing the phenotypic standard deviation of the population. Thus, biotechnology offers new means for incorporating wild and other exotic alleles in the genetic betterment of vegetatively propagated crops by tagging chromosomes with molecular markers as tools of marker-aided breeding or by broadening the genetic endowment through transgenics. For example, gene order conservation between genomes (or synteny) allows the use of the tomato (a sexual crop) genetic map to locate potato genes.

Biotechnology applications currently in use for improving vegetatively propagated food crops include tissue culture, genetic transformation, DNA fingerprinting, marker-aided analysis of genetic diversity, and DNA markers for new diagnostic tools.[18] For example, transgenic potatoes with resistance to Colorado potato beetle or viruses such as potato virus Y (PVY) and potato leaf roll virus (PLRV) as well as other transgenic clones of banana, cassava, and sweet potato are currently undergoing field testing to verify their transgenic expression for desired traits.[19] One of the most successful engineering of host-plant resistance in clonal crops species includes the use of a gene from the virus that encodes the virus coat protein into the plant genome. The transgenic plant made makes this protein and disrupts virus development. Other approach consists of gene silencing; i.e., a transgenic plant with a virus' genes produces mRNA or other RNA intermediates that form double-stranded RNAs, which leads to shutting down the expression of both host and virus copies of that gene.

Likewise, advances in potato genomics, which ensued from the extensive genetic mapping using diploid stocks in the 1990s,[3] may assist in the marker-aided incorporation or introgression of *Solanum* genetic resources into the tetraploid breeding populations. So far, molecular marker-aided genetic analysis allowed the segregation of complex multigenic characteristics into their discrete genetic factors and confirmed early hypotheses about the transmission of heterozygosity through $2n$ gametes. More research undertakings are needed to assess what biotechnology tools should be included in breeding clones; e.g., genomics appears to be a powerful tool for such an endeavor aiming to better use the sources of variation available in gene banks, particularly in genetically ill-endowed clonal crops, and to enhance the efficiency of gene transfer throughout the breeding process.

Similarly, such advances are also modifying the breeding of other crops such as banana/plantain, cassava, sweet potato, and yam.[19] The use of DNA markers as a selection aid was advocated in the 1990s for facilitating and speeding asexual crop breeding. For example, plantain and banana researchers elsewhere aimed to identify genetic markers for traits they could select at the seedling stage in hybrid populations of a giant perennial plant where the first bunch emerged between 12 and 18 months after planting. In the process, they identified rapid amplified polymorphic DNA markers for A and B putative donor genomes in *Musa* species and adapted a fluorescent in situ hybridization technique to determine *Musa* distinct genomes (Fig. 3). They also assessed germplasm variation in *Musa* germplasm with many DNA marker systems or used microsatellites for genetic-aided analysis and cultivar registration. However, *Musa* breeders have tried, without great success, to predict heterosis with microsatellites, but their research indicated that pedigree-based analysis might still prove useful for selecting parents of prospective *Musa* hybrid populations. More of their work led to the finding of an amplified fragment length polymorphism band likely to be associated with fruit parthenocarpy (or production of fruit without fertilization), but they still rely on field testing and selection for getting new, elite plantain, and banana hybrids. Perhaps, on balance so far, the main public good from this investment in plantain and banana genomics will be the abundant knowledge gathered. However, as discussed above, the promise of genomics to assist the genetic enhancement of plantain, banana, and other asexual crops still remains, and therefore "the jury at present is out."[19]

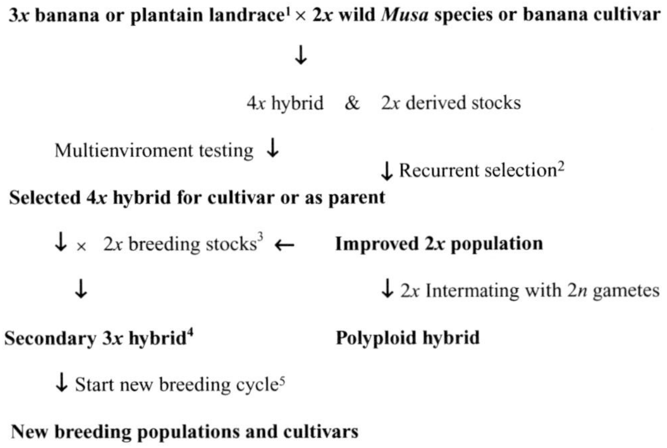

3x banana or plantain landrace[1] × 2x wild *Musa* species or banana cultivar
↓
4x hybrid & 2x derived stocks

Multienviroment testing ↓
↓ Recurrent selection[2]

Selected 4x hybrid for cultivar or as parent

↓ × 2x breeding stocks[3] ← **Improved 2x population**

↓   ↓ 2x Intermating with 2n gametes

Secondary 3x hybrid[4]   **Polyploid hybrid**

↓ Start new breeding cycle[5]

New breeding populations and cultivars

Fig. 3 Ploidy manipulation for the genetic enhancement of the *Musa* genome.
[1] Transgenic clones can also be parents.
[2] Genetic research and DNA marker-aided selection may be undertaken.
[3] After progeny testing to select for combining ability.
[4] Arising from hand pollination or through open pollination from polycrosses. A secondary 3x hybrid may be a potential new cultivar after selecting for yield, quality, and other desired attributes, e.g., pest resistance.
[5] Following above schemes from the top of this diagram.
Source: From Adv. Banana Plantain R&D Asia Pacific.[20]

ACKNOWLEDGMENT

The author thanks the late Dirk R. Vuylsteke, a meticulous *Musa* scientist and humanitarian, who dedicated his short but outstanding professional life to improve African agriculture and with whom he had the privilege to work on the genetic enhancement of banana and plantain clones using plant breeding and biotechnology.

REFERENCES

1. Ortiz, R. Germplasm enhancement to sustain genetic gains in crop improvement. In *Managing Plant Genetic Diversity*; Engels, J.M.M., Brown, A.H.D., Jackson, M., Eds.; IPGRI: Rome, Italy, **2002**; 275–290.
2. Peloquin, S.J.; Ortiz, R. Techniques for introgressing unadapted germplasm to breeding populations. In *Plant Breeding in the 1990s*; Stalker, H.T., Murphy, J.P., Eds.; CAB International: Wallingford, **1992**; 485–507.
3. Ortiz, R. Potato breeding via ploidy manipulations. Plant Breed. Rev. **1998**, *16*, 15–86.
4. Iwanaga, M.; Freyre, R.; Orjeda, G. Use of *Ipomoea trifida* (HBK) G. Don germplasm for sweet potato improvement. 1. Development of synthetic hexaploids of *I. trifida* by ploidy manipulations. Genome. **1991**, *24*, 201–208.
5. Hahn, S.K.; Bai, K.V.; Asiedu, R. Tetraploids, triploids, and 2n pollen from diploid interspecific crosses with cassava. Theor. Appl. Genet. **1990**, *79*, 433–439.
6. Vuylsteke, D.; Ortiz, R.; Ferris, R.S.B.; Crouch, J.H. Plantain improvement. Plant Breed. Rev. **1998**, *16*, 267–320.
7. Bremer, G. Problems in breeding and cytology of sugar cane. Euphytica **1961**, *10*, 59–78.
8. Hall, H.K. Blackberry breeding. Plant Breed. Rev. **1990**, *8*, 244–312.
9. Ortiz, R.; Bruederle, L.P.; Vorsa, N.; Laverty, T. The origin of polysomic polyploids via 2n pollen in *Vaccinium* section *Cyanococcus*. Euphytica **1992**, *61*, 241–246.
10. Bringhurst, R.S.; Voth, V. Breeding octoploid strawberries. Iowa State J. Res. **1984**, *58*, 371–382.
11. Sanford, J.C. Ploidy manipulations. In *Methods in Fruit Breeding*; Moore, J.N., Janick, J., Eds.; Purdue University Press: West Lafayette, IN, **1983**; 100–123.
12. Tarn, T.R.; Tai, G.C.C.; de Jong, H. Breeding potatoes for long-day, temperate climates. Plant Breed. Rev. **1992**, *9*, 217–332.
13. Ortiz, R. The state of use of potato genetic diversity. In *Broadening the Genetic Bases of Crop Production*; Cooper, H.D., Spillane, C., Hodgkin, T., Eds.; Food and Agriculture Organization of the United Nations (FAO): Rome, Italy, **2001**; 181–200.
14. Ehlenfeldt, M.K.; Ortiz, R. On the origins of endosperm dosage requirements in *Solanum* and other angiosperma genera. Sex. Plant Reprod. **1995**, *8*, 189–196.
15. Ortiz, R.; Huamán, Z. Allozyme polymorphism in tetraploid potato gene pools and the effect of human selection. Theor. Appl. Genet. **2001**, *103*, 792–796.
16. Jain, S.M.; Brar, D.S.; Ahloowalia, B.S. *Somaclonal Variation and Induced Mutations in Crop Improvement*; Springer-Verlag: New York, **1998**; 615 pp.
17. Ortiz, R.; Watanabe, K. Genetics contributions to breeding polyploid crops. Rec. Res. Dev. Genet. Breed. **2004**, *1*, 269–286.
18. Ortiz, R. Biotechnology with horticultural and agronomic crops in Africa. Acta Hort. **2005**, *642*, 43–56.
19. Ortiz, R.; Dochez, C.; Moonan, F.; Asiedu, R. Breeding vegetatively propagated crops. In *Plant Breeding*; Lamkey, K., Lee, M., Eds.; Blackwell Publishing: Ames, IA, **2006**; 251–268.
20. Tenkouano, A. Current issues and future directions for *Musa* genetic improvement research at the International Institute of Tropical Agriculture. Adv. Banana Plantain R&D Asia Pacific. **2001**, *10*, 11–23.

Cloning: Nuclear Transfer

X. Yang
M.G. Carter
S.R. Gao
F. Du
J. Xu
L.Y. Sung
Center for Regenerative Biology, University of Connecticut, Storrs, Connecticut, U.S.A.

Abstract

Nuclear transfer allows the cloning of animals by transferring a nucleus from an early embryo, embryonic stem cell, or somatic cell into an oocyte from which the nuclear DNA has been removed. This technique has now been used to clone 13 animal species, including farm animals, laboratory research animals, pets, and endangered species. Here, we follow the history of nuclear transfer cloning technology and the species in which it has been carried out to date.

INTRODUCTION

The concept of nuclear transfer (NT), as a method of demonstrating that cellular differentiation factors present in the cytoplasm can "reprogram" a differentiated nucleus, dates back to the writings of Yves Delage in 1895,[1] but its technical feasibility was unproven until 1952, when Briggs and King produced normal tadpoles via transfer of embryonic nuclei into frog eggs.[2] Nuclear transfer techniques were useful solely as research tools until the mid- to late 1980s, when they were used in sheep and later in cattle to produce high-quality, identical calves, marking the beginning of their commercial use. In the latter case, nuclei from early embryos were inserted into enucleated oocytes collected from the bovine oviduct or from cow ovaries obtained from an abattoir. After reaching the blastocyst stage, these reconstructed eggs were implanted into the uterus of a surrogate mother.

A breakthrough which changed the scope of potential applications for NT occurred when Ian Wilmut and his colleagues at the Roslin Institute in Glasgow, Scotland, produced the first animal cloned from the nucleus of a cell from adult tissue: Dolly the sheep.[3] This method, called somatic cell nuclear transfer (SCNT), inserts the nucleus from a somatic cell into an enucleated oocyte. This is commonly done through membrane fusion between a nuclear donor cell and an enucleated oocyte by electrical shock or by injection of a donor nucleus directly into an enucleated oocyte.

Since Dolly was born, scientists have used SCNT to produce clones of 13 animal species, including cats, a dog, gaur, and livestock. Although there are and will continue to be stringent legal controls or prohibitions on using SCNT for human reproductive cloning, the method is potentially an excellent tool for therapeutic cloning, or the generation of human embryonic stem (hES) cells for the production of non-immunogenic cells and tissues for patient transplantation therapy.[4]

The remainder of this entry will discuss the use of SCNT for reproductive cloning in animals and therapeutic cloning for humans.

NT IN FARM ANIMALS

Successful SCNT has resulted in live farm animal clones, including sheep, cattle, goats, pigs, mules, and horses. For effective reprogramming of the genome of a differentiated somatic cell nucleus, the donor nucleus is introduced into the cytoplasm of an oocyte that has had its own DNA removed. Oocytes are capable of nuclear remodeling/reprogramming, which can reestablish the totipotency of an introduced nucleus. Functionally, this process is similar to what occurs during normal embryonic development. Although the mechanisms responsible for this type of nuclear reprogramming are not known, they are hypothesized to be part of the system that normally reprograms the nascent zygotic genome during fertilization and early development. Hence, SCNT holds great promise as a basic research tool for studying the normal reprogramming of embryonic DNA.

Cloning, however, is not yet a standard practice in animal agriculture. Overall, cloning efficiency remains low, ranging from 1 to 10%. "Dolly," the first farm animal clone, was produced by Ian Wilmut's Roslin group after

Fig. 1 Famous Japanese black beef breeding bull, Kamitakafuku, and his clones. The first two clones were born in December 1998. Ear skin cells of one of the clones were used as nucleus donors for nuclear transfer cloning, thus producing a second-generation clone. **Source:** Adapted from Cozzarelli et al.[6] and Kubota et al.[7]

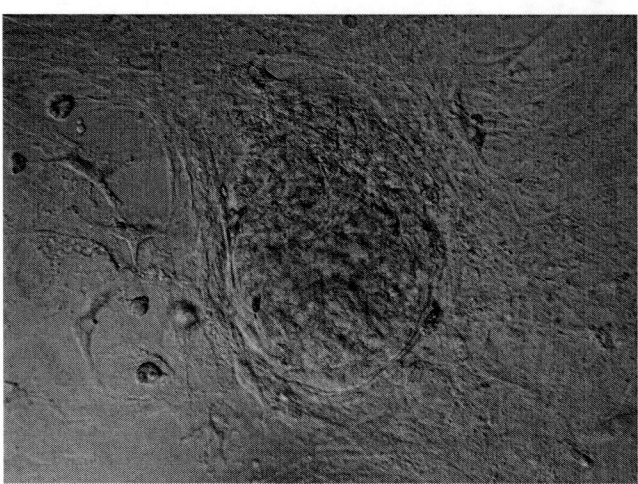

Fig. 2 Bovine nuclear transfer embryonic stem cell colonies. **Source:** Photo courtesy of X. Cindy Tian, Center for Regenerative Biology, University of Connecticut.

277 attempts.[3] Moreover, in most species studied so far, many clones die shortly after birth, neonates may suffer physiological deformities, and fetuses and young may suffer from large offspring syndrome. These all have a negative impact on the economic applications for cloning farm animals.

SCNT allows opportunities for genetic engineering, genome preservation, and genetic improvement in farm animals, including the use of these animals as biomedical reactors for producing pharmaceuticals, such as vaccines or human polyclonal antibodies, in transgenic animal clones. The method may also be used to generate animal models for numerous human diseases and for clinical drug and toxicology testing.

The purpose of cloning in the animal industry is to produce "perfect" animals with high genetic merit and economic value. For example, a cloning group led by one of the authors of this entry, Xiangzhong Jerry Yang, generated first- and second-generation clones in 1998 and 2000 from a 17-year-old Japanese black beef champion stud bull by using bull skin cells as nuclear donors[5] (Fig. 1). To enhance milk composition and milk processing efficiency, New Zealand scientists introduced additional copies of bovine casein genes into female bovine fibroblasts cells, cloning transgenic calves that produced milk with additional protein, improving its functional properties. More recently, James Robl and colleagues produced cloned cattle via SCNT technology with the donor cell's bovine spongiform encephalopathy (BSE) gene (the gene responsible for mad cow disease) knocked out completely.[8] Environmental changes may have a significant impact on the productivity and longevity of resultant clones. Furthermore, technical improvements in cloning processes are necessary to provide a toolbox for agricultural and biomedical applications (Fig. 2).

The use of genetically modified cloned pigs to produce organs and tissues for xenotransplantation is one such application in the medical toolbox of the future. Pig organs are commonly held as the most appropriate match in terms of function and size to replace diseased human organs, but the major hurdle to using porcine organs and tissues clinically has been immunologic rejection, largely because of antiporcine Galα1,3 antibodies in primates. These antibodies will trigger complement-mediated destruction of porcine vascular endothelium, where Galα1,3-Gal epitopes are expressed, leading to hyperacute rejection (HAR) of the transplant.[9] Investigators have used SCNT to create genetically modified swine which lack functional copies of the α1,3-Galactosyltransferase gene, and cannot express the epitopes which trigger HAR.[10] Because porcine embryonic stem cells have yet to be derived, SCNT currently offers the only way to modify the pig genome, and more recent studies have used it to express transgenic human glycotransferases in porcine tissues, in an effort to improve xenograft-host compatibility further.[11]

RABBIT CLONING

Rabbits are a valuable scientific research resource, so cloned rabbits would be valuable research organisms. The first successful production of a rabbit clone by using adult somatic cells was reported by Chesne et al. in 2002.[12] Rabbit cloning employs similar procedures to those used to clone other species; however, there are major unique differences: 1) reconstructed embryos are transferred to a surrogate's oviducts at the 2- to 4-cell stages instead of in the blastocyst stages, as is done in many other species; and 2) there is a narrow temporal window to transfer the embryos into recipient surrogate animals.

The rabbit is considered to be a valuable animal model for many human diseases, including diabetes, cardiovascular disease, and cystic fibrosis. Nuclear transfer will potentially provide a means to produce gene-targeted rabbits for the study of these human diseases. Previously, this was not possible, because no ES cell lines (essential for gene targeting before SCNT technology became available) have yet been reported.

NT IN PETS

Cloning pets has had rare success. Thus, pet cloning remains very expensive and will be limited to a few people who can afford to pay to have their beloved pets cloned. The first cat clone, CC, was produced by a research team led by Mark Westhusin at Texas A&M University in 2002.[13] In 2005, the first cloned dog, Snuppy, produced by Woo Suk Hwang's group in South Korea,[14] attracted significant media coverage and created quite a sensation with the general public. Both CC and Snuppy were produced by SCNT cloning.

Currently, cat cloning is commercially available at approximately $50,000 per clone and the cost is likely to be even higher for commercial dog cloning. The cloning technology, however, is at an early stage of development and is inefficient. Some cloned pets died during embryonic or fetal development or shortly after birth, although many clones apparently are normal and healthy a few days after birth. Although the common belief is that the clones are genetically identical to the original donors, they also may have different coat color patterns, different behaviors, and different physiological conditions than the nucleus donor animals, perhaps as a result of epigenetic effects.

ENDANGERED SPECIES CLONING

Nuclear transfer is a potential means of conserving endangered species. In general, the procedure is similar to the regular cloning process, including the sampling and storage of somatic cells from individual animals, preparation of somatic cells for NT into enucleated oocytes of the same species, or a closely related species, embryo culture, and embryo transfer. One notable difference is that the oocyte cytoplasm used to create embryos is often derived from common domesticated species. For example, oocytes from *Bos taurus* (cow), *Capra hircus* (goat), *Oryctolagus cuniculus* (rabbit), or *Felis cattus* (domestic cat) may be used because of the insufficient number of available oocyte donor animals from the endangered species and the lack of understanding of the reproductive physiology of those species. A Japanese team of researchers has even proposed that this approach can be used to resurrect extinct species like the Wooly Mammoth. Concerns exist that the use of eggs from another species eliminates the genetic information contained in the oocyte mitochondrial DNA. Cloning of several endangered species, including the gaur, the banteng, and the African wild cat have been reported.[15] There are reported to be ongoing efforts to clone other species, such as the panda[16] and the Tasmanian tiger, but they are not yet successful. Two problems that need solving before NT can be used as an effective method for the conservation of endangered species are the poor cloning efficiency and the lack of availability of the oocytes from the same or related species.

THERAPEUTIC SCNT FOR HUMAN THERAPY

SCNT can produce blastocysts that will yield cells that can be differentiated into specific cell lines. Unlike reproductive cloning, as is carried out in animals, the intent of human therapeutic SCNT is not to implant the SCNT-generated preembryo into a human uterus. Instead, SCNT preembryos would be harvested at the blastocyst stage and used to derive individual, specific ES cell lines. Successfully derived ES cells are then propagated in vitro. Genetic defects may be corrected by homologous recombination during in vitro culture before SCNT.

Nuclear transfer ES (ntES) cell lines were first established from cloned mouse blastocysts.[17] Despite the potential technical and scientific success of using SCNT to establish human cell lines, ethical and religious debates on the concept of using SCNT-generated preembryos for therapeutic purposes continue in many countries.[18] Furthermore, there are other ethical and legal issues, such as ownership of the eggs and any subsequent commercial products derived from the eggs. National and local regulation of this research is likely to become the norm.

CONCLUSION

NT technology has allowed the cloning of animals and generation of ES cells from animals and humans in the absence of fertilization. As we write this, attempts are being made to clone animals other than the 13 species we discussed above. NT provides a means of dispersing high-quality livestock worldwide without the complexities of shipping large animals, producing genetically identical research animals that will all respond similarly to the same protocol, propagating endangered species even if the appropriate same-species surrogate mother for the species is unavailable, and recreating a facsimile of someone's beloved pet. In the near future, it may help to provide therapeutic tissues and organs, for both human and veterinary medicine, to cure or prevent many physically, emotionally, and financially devastating diseases.

Although researchers and society are faced with ethical and moral dilemmas because of the rapidity at which the technology is moving and the potential consequences of SCNT, the positive value of the technology is evident. Legal constraints must regulate how this technology is used and the public must be educated to understand the potential value of this work.

REFERENCES

1. Beetschen, J.C.; Fischer, J.L. Yves Delage (1854–1920) as a forerunner of modern nuclear transfer experiments. Int. J. Dev. Biol. **2004**, *48* (7), 607–612.
2. Briggs, R.; King, T.J. Transplantation of living nuclei from blastula cells into enucleated frogs' eggs. Proc. Natl. Acad. Sci. U. S. A. **1952**, *38* (5), 455–463.
3. Wilmut, I.; Schnieke, A.E.; McWhir, J.; Kind, A.J.; Campbell, K.H. Viable offspring derived from fetal and adult mammalian cells. Nature **1997**, *385* (6619), 810–813.
4. Wilmut, I. Human cells from cloned embryos in research and therapy. Br. Med. J. **2004**, *328* (7437), 415–416.
5. Tian, X.C.; Kubota, C.; Yang, X. Cloning of aged animals: a medical model for tissue and organ regeneration. Trends Cardiovasc. Med. **2001**, *11* (8), 313–317.
6. Cozzarelli, N.R.; Fulton, K.R.; Sullenberger, D.M. Liberalization of PNAS copyright policy: noncommercial use freely allowed. Proc. Natl. Acad. Sci. U. S. A. **2004**, *101* (34), 12399.
7. Kubota, C.; Yamakuchi, H.; Todoroki, J.; Mizoshita, K.; Tabara, N.; Barber, M.; Yang, X. Six cloned calves produced from adult fibroblast cells after long-term culture. Proc. Natl. Acad. Sci. U. S. A. **2000**, *97* (3), 990–995.
8. Richt, J.A.; Kasinathan, P.; Hamir, A.N.; Castilla, J.; Sathiyaseelan, T.; Vargas, F.; Sathiyaseelan, J.; Wu, H.; Matsushita, H.; Koster, J.; Kato, S.; Ishida, I.; Soto, C.; Robl, J.M.; Kuroiwa, Y. Production of cattle lacking prion protein. Nat. Biotechnol. **2007**, *25* (1), 132–138.
9. Yang, X.; Tian, X.C.; Dai, Y.; Wang, B. Transgenic farm animals: applications in agriculture and biomedicine. Biotechnol. Annu. Rev. **2000**, *5*, 269–292.
10. Phelps, C.J.; Koike, C.; Vaught, T.D.; Boone, J.; Wells, K.D.; Chen, S.H.; Ball, S.; Specht, S.M.; Polejaeva, I.A.; Monahan, J.A.; Jobst, P.M.; Sharma, S.B.; Lamborn, A.E.; Garst, A.S.; Moore, M.; Demetris, A.J.; Rudert, W.A.; Bottino, R.; Bertera, S.; Trucco, M.; Starzl, T.E.; Dai, Y.; Ayares, D.L. Production of alpha 1,3-galactosyltransferase-deficient pigs. Science **2003**, *299* (5605), 411–414.
11. Ramsoondar, J.J.; Machaty, Z.; Costa, C.; Williams, B.L.; Fodor, W.L.; Bondioli, K.R. Production of alpha 1,3-galactosyltransferase-knockout cloned pigs expressing human alpha 1,2-fucosyltransferase. Biol. Reprod. **2003**, *69* (2), 437–445.
12. Chesne, P.; Adenot, P.G.; Viglietta, C.; Baratte, M.; Boulanger, L.; Renard, J.P. Cloned rabbits produced by nuclear transfer from adult somatic cells. Nat. Biotechnol. **2002**, *20* (4), 366–369.
13. Shin, T.; Kraemer, D.; Pryor, J.; Liu, L.; Rugila, J.; Howe, L.; Buck, S.; Murphy, K.; Lyons, L.; Westhusin, M. A cat cloned by nuclear transplantation. Nature **2002**, *415* (6874), 859.
14. Lee, B.C.; Kim, M.K.; Jang, G.; Oh, H.J.; Yuda, F.; Kim, H.J.; Shamim, M.H.; Kim, J.J.; Kang, S.K.; Schatten, G.; Hwang, W.S. Dogs cloned from adult somatic cells. Nature **2005**, *436* (7051), 641.
15. Gomez, M.C.; Pope, C.E.; Giraldo, A.; Lyons, L.A.; Harris, R.F.; King, A.L.; Cole, A.; Godke, R.A.; Dresser, B.L. Birth of African Wildcat cloned kittens born from domestic cats. Cloning Stem Cells. **2004**, *6* (3), 247–258.
16. Chen, D.Y.; Wen, D.C.; Zhang, Y.P.; Sun, Q.Y.; Han, Z.M.; Liu, Z.H.; Shi, P.; Li, J.S.; Xiangyu, J.G.; Lian, L.; Kou, Z.H.; Wu, Y.Q.; Chen, Y.C.; Wang, P.Y.; Zhang, H.M. Interspecies implantation and mitochondria fate of panda-rabbit cloned embryos. Biol. Reprod. **2002**, *67* (2), 637–642.
17. Munsie, M.J.; Michalska, A.E.; O'Brien, C.M.; Trounson, A.O.; Pera, M.F.; Mountford, P.S. Isolation of pluripotent embryonic stem cells from reprogrammed adult mouse somatic cell nuclei. Curr. Biol. **2000**, *10* (16), 989–992.
18. Anonymous. A world of approaches to stem cells. Sci. Am. **2005**, *293*, A20–A21.

Cloning: Stem Cells of Different Developmental Potency

Björn Oback
Ruakura Research Centre, AgResearch, Hamilton, New Zealand

Abstract

Following nuclear transfer (NT), the most stringent measure of donor cell reprogramming is development into fertile adults. This is referred to as cloning efficiency. Cloning efficiency depends on the ability of the nuclear donor cell to be fully reprogrammed. Donor reprogrammability decreases with developmental potency of the donor cell, i.e., from totipotent blastomeres via pluripotent stem cells to multipotent somatic stem cells. Within different somatic lineages, however, no conclusive correlation between differentiation status and cloning efficiency was found. This may reflect technical limitations of the NT-induced reprogramming assay. Alternatively, differentiation status and reprogrammability might be unrelated within each category of cell potency.

INTRODUCTION

Nuclear transfer (NT) cloning has demonstrated that it is possible to completely reverse cell differentiation. The efficiency of this process depends 1) on uncharacterized factors in the recipient oocyte or zygote carrying out the reprogramming reactions; and 2) on the ability of the nuclear donor cell to be fully reprogrammed ("reprogrammability"). Reprogrammability can be measured at different levels (Fig. 1). At the molecular level, it manifests itself as changes in DNA methylation, histone modifications, and chromatin proteins. At the RNA level, it influences termination of somatic and initiation of embryonic gene expression. At the cellular and organismal level, molecular changes affect functionality and phenotype during development. Early events, such as blastocyst formation rate and frequency of deriving embryonic stem (ES) cells, are not strongly correlated with progression to later developmental milestones, such as implantation, placentation, organogenesis, and birth. The most definitive measure of complete reprogramming is the development into fertile adults. This is referred to as cloning efficiency and quantified as the proportion of embryos transferred into surrogate mothers that develop into viable offspring. It has been postulated that mammalian cloning efficiency is inversely correlated with donor cell differentiation status. If there was a hierarchical relationship between differentiation and reprogrammability, undifferentiated stem cells should have increased cloning efficiency. This hypothesis was systematically addressed in mammalian NT experiments where donor cell type was the only parameter that varied.

TOTIPOTENT STEM CELLS

The zygote and its early cleavage products, called blastomeres, are not conventionally regarded as stem cells, even though all other cells stem from them. Following blastomere NT in mouse, there is a gradual restriction in cloning efficiency from the 1- to the 4-cell stage, followed by a steep decrease from the 4- to the 8-cell stage[1,2] (Fig. 2). This is consistent with the notion that individual 4- but not 8-cell mouse blastomeres are still totipotent (able to give rise to all cell types on their own). ES cell derivability also decreases at the 4-cell stage,[3] coinciding with emerging epigenetic differences between blastomeres.[4] In cattle, a similar restriction point in cloning efficiency occurs one cell division later,[5] consistent with loss of totipotency after the 8-cell stage in this species. Overall, cloning efficiencies with blastomeres are one order of magnitude higher than with somatic cells.[1]

PLURIPOTENT STEM CELLS

Pluripotent stem cells are immortal and capable of giving rise to all cells of an animal, including the germ line. They are only available in rodents and originate from early embryos (ES cells) or from delivering pluripotency-inducing genes into somatic cells (induced pluripotent stem or iPS cells). They can also be derived from the germ line, either from primordial germ cells (embryonic germ or EG cells) or testis (multipotent germ stem or mGS cells).

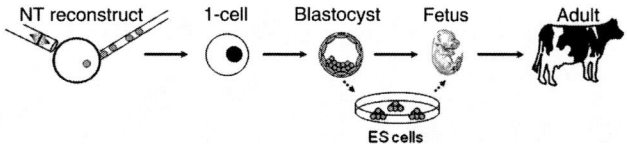

Donor reprogrammability assays at different developmental stages:
1. Molecular (stage-specific DNA, RNA and protein profile)
2. Functional (stage-specific physiology and developmental potency)
3. Phenotypic (stage-specific histology and morphology)

Fig. 1 Donor cell reprogrammability assays. In order of increasing stringency, these assays measure reacquisition of pluripotency at different developmental stages, using molecular, functional, and phenotypic evaluation.

In direct comparisons, ES cells did not perform significantly better than cumulus or immature Sertoli cells.[6] Considering that many targeted ES cell lines and subclones have non-reprogrammable karyotypic abnormalities, however, true ES cell reprogrammability might surpass somatic cells. In addition, ES cells are transcriptionally and epigenetically heterogeneous. DNA-methylation patterns of imprinted genes as well as gene expression vary widely among subclones of a given ES cell line and even among individual cells of an ES subclone after in vitro culture. This makes it difficult to distinguish between reprogramming errors and pre-existing epigenomic errors in the ES donors.

iPS cells are a complementary experimental system to NT-clones that allows studying the reacquisition of pluripotency through epigenetic reprogramming. Some somatic cell types may show higher reprogrammability into iPS cells, however, a conclusive comparison of different genotype-, sex-, and cell cycle-matched iPS donors has not yet been undertaken. Clearly their applicability for NT cloning will be very important to evaluate, emphasizing the importance of deriving these cells in livestock.

Germ-line derived pluripotent stem cells from fetal or postnatal sources have not been used as NT donors either, mainly because they are likely to be androgenetically imprinted and thus incompatible with full-term development.

MULTI- AND UNIPOTENT STEM CELLS

Tissue-specific somatic stem cells are capable, at the single cell level, of proliferation, self-renewal, and the production of one or multiple types of daughter cells within a lineage. Their reprogrammability has been systematically compared to that of their isogenic differentiated progeny in several different lineages (Fig. 3).

The first lineage to be studied in detail was the neuronal. Neural stem cells (NSC) have similar cloning efficiency to ES cells. Provided that NSCs are genetically and epigenetically more stable and homogenous, their true reprogrammability should thus be lower than ES cells. Compared to neurons, NSCs show higher cloning efficiency, however, these results are not quite significant and mature neurons can harbor non-reprogrammable karyotypic alterations. Hematopoietic stem cells (HSCs) are among the longest-studied and best-defined stem cells. For HSCs of the same genetic background and sex as NSCs, cloning efficiency is 0.7% per transferred 4-cell,[7] compared to

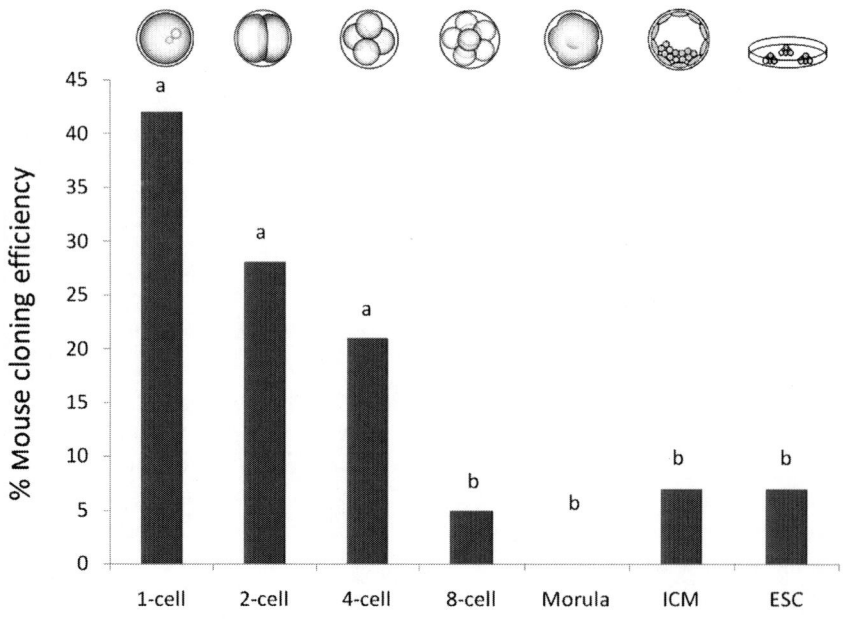

Fig. 2 Mouse embryonic cell cloning efficiency. All donor cells were from F1 crosses of C57BL/6 with either CBA or C3H. Cloning efficiency was calculated as surviving offspring per number of morulae/blastocysts transferred; a, b: different letters denote significant differences ($P < 0.05$) using the Fisher 2×2 exact test. ICM, inner cell mass of the blastocyst; ESC, embryonic stem cell.

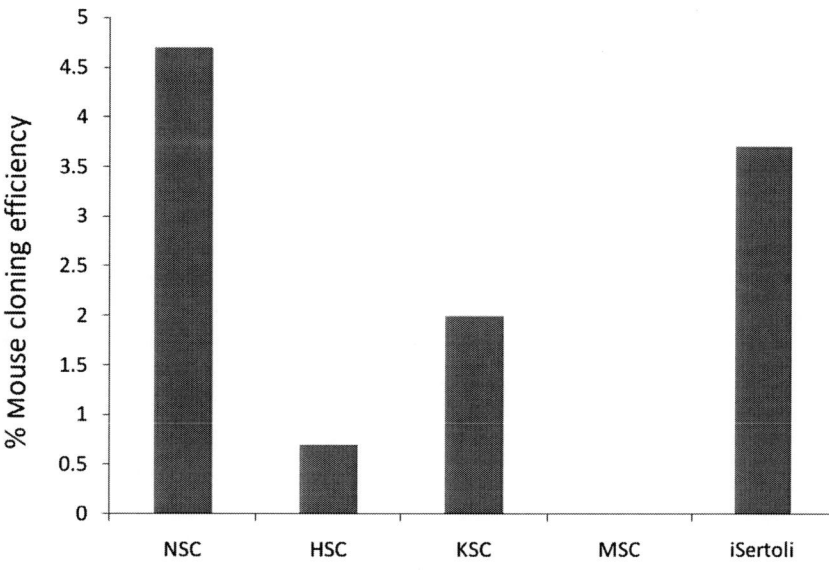

Fig. 3 Mouse somatic stem cell cloning efficiency. All donor cells were from male F1 crosses of 129/Sv with either C57BL/6 (B6), B6 hybrids, or CD1. Cloning efficiency was calculated as surviving offspring per number of morulae/blastocysts transferred. HSC, hematopoietic stem cell; iSertoli, immature Sertoli cell; KSC, keratinocyte stem cell; MSC, mesenchymal stem cell; NSC, neural stem cell.

0.5% per transferred 2- to 4-cell using NSC.[6] When HSCs are compared to their differentiated progeny, granulocytes, there are no differences in cloning efficiency. The findings in HSCs are supported by data from using bone marrow-derived mesenchymal stem cells (MSCs). Following NT with FACS-sorted clonally-derived MSCs of proven multilineage potential, only 34% of embryos cleave into 4-cell and in vivo development arrests before implantation.[8] Adult keratinocyte stem cells (KSCs) and more differentiated transiently amplifying keratinocyte progenitors (KPCs), isolated from the murine skin epithelium by FACS-sorting, show a similar trend. For both male and female KSCs and KPCs, there are no significant differences in blastocyst development or cloning efficiency (2% vs. 0.4% for male and 1% vs. 0% for females), although KSC-derived embryos tend to survive better.[9]

In cattle, increasingly differentiated cells from the skeletal muscle lineage were used as donors.[10] Myogenic precursors and in vitro-differentiated mononucleated myotubes were individually analyzed for cell-type-specific antigens. Isogenic muscle fibroblasts lacking muscle-specific antigens served as a control group. After manual size selection, molecularly characterized cells were used for NT. Despite significant differences in blastocyst development, cloning efficiency did not depend on donor cell type.

In deer, cells from the antlerogenic periosteum (AP) represent a unique population of functionally defined putative stem cells. These cells are anatomically and histologically well-described and give rise to all different antler lineages (e.g., skin, blood, nerve, cartilage, bone, and connective tissue), even after ectopic transplantation into deer or the skull bone of nude mice. Their removal from the presumptive antler growth region abolishes future antler formation. In vitro, AP cells are highly proliferative and can be differentiated into several mesodermal cell lineages, such as bone, cartilage, and adipocytes. Quiescent AP cells were compared to their in vitro-differentiated progeny, adipocytes. Implantation rates and development into adulthood were not significantly different between AP- vs. adipocyte-derived blastocysts.[11]

Whilst not as well-defined on the molecular level as the previous stem cells, immature Sertoli progenitors are to-date the only convincing example of a relatively undifferentiated cell resulting in high reprogrammability. Sertoli cells are a major somatic cell type in the testis and central to male gonad formation and spermatogenesis. In mice, immature Sertoli cells are the most efficient somatic donors by far, achieving up to 15% cloning efficiency with some genotypes[5] and matching the cloning efficiency of isogenic ES cells.

CONCLUSION

Cloning efficiency decreases with the developmental stage of the donor cell. This decrease is not gradual but appears to occur at two discrete restriction points, first during early-cleavage stages and second during the establishment of somatic lineages in the mammalian embryo (Fig. 4). It is hypothesized that these events correlate with the loss of totipotency and pluripotency respectively, during development. Within the somatic differentiation continuum, over 20 well-defined cell types, representing 10 different lineages, have been tested for their reprogrammability. No lineage of consistently high cloning efficiency has emerged from

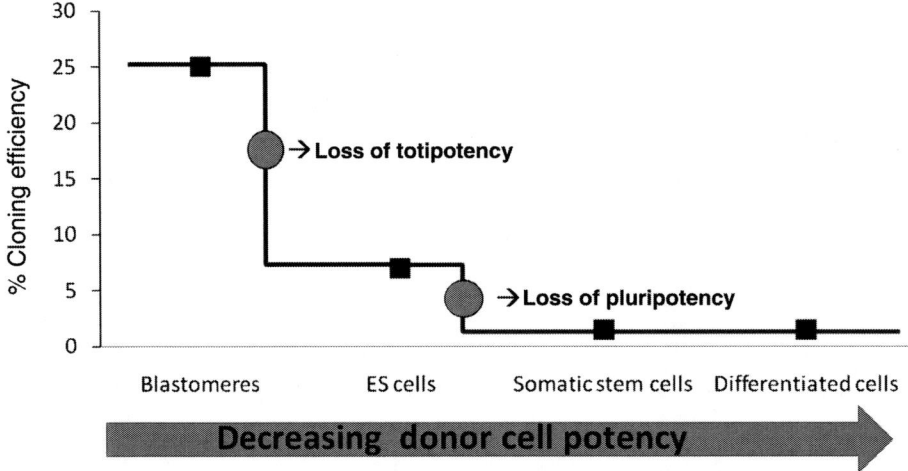

Fig. 4 Hypothetical restriction points in donor cell reprogrammability as a function of decreasing donor cell potency. Within each category of donor potency, cloning efficiency remains relatively constant.

these comparisons. When reprogrammability was compared using undifferentiated somatic stem cells and their differentiated isogenic progeny, no conclusive correlation was found. It would be informative to repeat such experiments with pluripotent stem cells and their differentiated progeny. Given the small number of viable offspring in a typical NT experiment, this assay may simply not be sensitive enough to detect subtle hierarchical relations between differentiation and reprogrammability. Alternatively, differentiation status and reprogrammability might be unrelated within each category of cell potency. In order to elucidate the mechanism underlying differences in embryonic and somatic reprogrammability, we need a clearly defined structural correlate for cellular potency. This correlate may be found in the dynamics of chromatin structure, which forms the molecular basis for epigenetic variation between donor cell genomes.

REFERENCES

1. Hiiragi, T.; Solter, D. Reprogramming is essential in nuclear transfer. Mol. Reprod. Dev. **2005**, *70* (4), 417–421.
2. Ono, Y.; Kono, T. Irreversible barrier to the reprogramming of donor cells in cloning with mouse embryos and embryonic stem cells. Biol. Reprod. **2006**, *75* (2), 210–216.
3. Wakayama, S.; Hikichi, T.; Suetsugu, R.; Sakaide, Y.; Bui, H.T.; Mizutani, E.; Wakayama, T. Efficient establishment of mouse embryonic stem cell lines from single blastomeres and polar bodies. Stem Cells **2007**, *25* (4), 986–993.
4. Torres-Padilla, M.E.; Parfitt, D.E.; Kouzarides, T.; Zernicka-Goetz, M. Histone arginine methylation regulates pluripotency in the early mouse embryo. Nature **2007**, *445* (7124), 214–218.
5. Oback, B.; Wells, D.N. Donor cell differentiation, reprogramming, and cloning efficiency: elusive or illusive correlation? Mol. Reprod. Dev. **2007**, *74* (5), 646–654.
6. Mizutani, E.; Ohta, H.; Kishigami, S.; Van Thuan, N.; Hikichi, T.; Wakayama, S.; Kosaka, M.; Sato, E.; Wakayama, T. Developmental ability of cloned embryos from neural stem cells. Reproduction **2006**, *132* (6), 849–857.
7. Inoue, K.; Ogonuki, N.; Miki, H.; Hirose, M.; Noda, S.; Kim, J.M.; Aoki, F.; Miyoshi, H.; Ogura, A. Inefficient reprogramming of the hematopoietic stem cell genome following nuclear transfer. J. Cell Sci. **2006**, *119* (Pt 10), 1985–1991.
8. Inoue, K.; Noda, S.; Ogonuki, N.; Miki, H.; Inoue, S.; Katayama, K.; Mekada, K.; Miyoshi, H. Ogura, A. Differential developmental ability of embryos cloned from tissue-specific stem cells. Stem Cells **2007**, *25* (5), 1279–1285.
9. J.; Greco, V.; Guasch, G.; Fuchs, E.; Mombaerts, P. Mice cloned from skin cells. Proc. Natl. Acad. Sci. U.S.A. **2007**, *104* (8), 2738–2743.
10. Green, A.L.; Wells, D.N.; Oback, B. Cattle cloned from increasingly differentiated muscle cells. Biol. Reprod. **2007**, *77* (3), 395–406.
11. Berg, D.K.; Li, C.; Asher, G.; Wells, D.N.; Oback, B. Red deer cloned from antler stem cells and their differentiated progeny. Biol. Reprod. **2007**, *77* (3), 384–394.

Cocoa Fermentation: Chocolate Flavor Quality

Emmanuel Ohene Afoakwa
Nutrition and Food Science, University of Ghana, Legon-Accra, Ghana

Alistair Paterson
Centre for Food Quality (SIBPS), University of Strathclyde, Glasgow, U.K.

Abstract

Chocolate aroma and flavor attributes not only originate in precursor compounds present in cocoa beans, but are generated during fermentation and drying and transformed into desirable odor notes in the manufacturing process. Complex biochemical modifications of bean constituents are further altered by thermal reactions in the roasting and conching steps and during alkalization leading to the development of the finished flavor character; however, the extent to which fermentation component influences chocolate flavor formation and the relationships contributing to final flavor quality are not clear. With increasing specialty niche products in chocolate confectionery area, greater understanding of factors contributing to variations in flavor character during cocoa fermentation has significant commercial implications.

INTRODUCTION

The main varieties of the cocoa tree *Theobroma cacao* (family Sterculiacae) are: 1) *Criollo*, rarely grown because of disease susceptibility; 2) *Nacional* with fine flavor grown in Ecuador; 3) *Forastero* from the Amazon region; and 4) *Trinitario*, a hybrid of *Forastero* and *Criollo*. *Forastero* varieties form most of the bulk or basic cocoa market. World annual cocoa bean production is approximately 3.7 million metric tons and major producers are the Ivory Coast, Ghana, Indonesia, Brazil, Nigeria, Cameroon, and Ecuador. There are also a number of smaller producers, particularly of fine cocoa, which forms less than 5% world trade.[1]

Chocolate has a distinctive flavor character with specific notes related to bean genotype, growing conditions, and processing factors.[2] Fermentation is a key processing stage that causes the death of the bean and facilitates removal of the pulp and subsequent drying. During this stage, there is formation of flavor precursors and color development, and a significant reduction in bitterness. The biochemistry of cocoa beans in fermentation is still under study, as are contributions from roasting, alkalization, and conching.[3] The biochemical and chemical processes leading to chocolate flavor formation and development, and their relationships contributing to the final character and perceptions of quality are not fully understood. This entry discusses the general fermentation practices, the resultant chemical and biochemical changes, and their effects on chocolate flavor.

GENERAL PRACTICES AND EFFECTS OF BIOCHEMICAL CHANGES DURING COCOA FERMENTATION

Fermentation is essential for development of appropriate flavors from precursor compounds. After pod harvest, beans and adhering pulp are transferred to heaps, boxes, or baskets for fermentation lasting from five to six days for *Forastero* beans or one to three days for *Criollo* (Table 1). In the first day, the adhering pulp liquefies and drains off with steady rises in temperature. Native microorganisms produce acetic acid and ethanol under anaerobic conditions that inhibit seed germination and contribute to structural changes, such as removal of compartmentalized enzymes and substrates with migration of cytoplasmic components through the cocoa cotyledon generally between 24 and 48 hours of bean fermentation. By the third day, the mass of the beans will reach approximately 45°C, remaining at 45–50°C until fermentation is complete.[3] Mucilaginous pulp of the beans undergoes ethanolic, acetic, and lactic fermentations with the consequently greater heat and acid levels terminating seed germination with notable swelling and key changes in cell membranes that facilitate enzyme and substrate migration.

During fermentation, the rate of diffusion of organic acids into the cotyledons, the timing of initial entry, and the duration at optimal and final pHs are crucial for optimal flavor development. Beans of higher pH (5.5–5.8) are considered unfermented with a low fermentation index and cut test score, and those of lower pH (4.75–5.19)

Table 1 Origin, cocoa variety, and fermentation duration effects on flavor character.

Origin	Cocoa type	Duration (days)	Special flavor character
Ecuador	Nacional (Arriba)	2 Short	*Aromatic*, *floral*, *spicy*, and *green*
Ecuador	Criollo (CCN51)	2	*Acidic*, *harsh*, and *low cocoa*
Ceylon	Trinitario	1.5	*Floral*, *fruity*, and *acidic*
Venezuela	Trinitario	2	*Low cocoa* and *acidic*
Venezuela	Criollo	2	*Fruity* and *nutty*
Zanzibar	Criollo	6 Medium	*Floral* and *fruity*
Venezuela	Forastero	5	*Fruity*, *raisin*, and *caramel*
Ghana	Forastero	5	*Strong basic cocoa* and *fruity*
Malaysia	Forastero/Trinitario	6	*Acidic* and *phenolic*
Trinidad	Trinitario	7–8 Long	*Winey*, *raisin*, and *molasses*
Grenada	Trinitario	8–10	*Acidic*, *fruity*, and *molasses*
Congo	Criollo/Forastero	7–10	*Acidic* and *strong cocoa*
Papua New Guinea	Trinitario	7–8	*Fruity* and *acidic*

are considered well fermented. Fermentation techniques can reduce acid notes and maximize chocolate flavors.[4] Ziegleder[5] compared natural acid (pH 5.5–6.5) and alkaline (pH 8.0) cocoa extracts obtained by direct extraction; the former possessed a more intense chocolate aroma than the latter, which was attributed to high contents of aromatic acids and sugar degradation products with persistent sweet aromatic and caramel notes. Cocoa beans of lower pH (4.75–5.19) and higher pH (5.50–5.80) were scored lower for chocolate flavor and higher for off-flavor notes, respectively. Chocolate from intermediate pH (5.20–5.49) beans was scored more highly for desired chocolate flavor.[6]

During fermentation, sucrose and proteins are partially hydrolyzed, phenolic compounds are oxidized, and glucose is converted into alcohols as well as oxidized to acetic and lactic acids. Cocoa beans subsequently undergo an anaerobic hydrolytic phase, followed by aerobic condensation. Timing, sequence of events, and degree of hydrolysis and oxidation varies among fermentations. The concentration of flavor precursors is dependent of enzymatic mechanisms. Color changes also occur with hydrolysis of phenolic components by glycosidases accompanied by bleaching that influences the final flavor character.[7]

The nitrogenous flavor precursors formed during the anaerobic phases are predominately amino acids and peptides made available for non-oxidative carbonyl–amino condensation reactions that are promoted during the elevated temperatures that occur during fermentation, drying, roasting, and grinding. Although degraded to flavor precursors, residual protein is also diminished by phenol–protein interactions. During the aerobic phases, oxygen-mediated reactions occur, such as oxidation of protein–polyphenol complexes that were formed anaerobically. Such processes reduce astringency and bitterness; oxidized polyphenols influence subsequent degradation reactions.[8]

The fermentation method determines the final quality of the products produced, especially flavor. Previous studies on postharvest pod storage and bean spreading have shown marked improvement in chocolate flavor and reductions in sourness, bitterness, and astringency.[9] In commercial production, similar effects have been obtained through combinations of pod storage, pressing, and air-blasting.[10] Variations in such factors as pod storage and duration affect the pH, titratable acidity, and temperature achieved during fermentation, influencing enzyme activities and flavor development.[9]

Important flavor-active components produced during fermentation include: ethyl-2-methylbutanoate, tetramethylpyrazine, and certain pyrazines. Bitter notes are evoked by theobromine and caffeine, together with diketopiperizines formed from roasting through thermal decompositions of proteins. Other flavor precursor compounds derived from amino acids released during fermentations include 3-methylbutanol, phenylactaldehyde, 2-methyl-3-(methyldithio)furan, 2-ethyl-3,5-dimethyl- and 2,3-diethyl-5-methylpyrazine.[11] Immature and unfermented beans develop little chocolate flavor when roasted and excessive fermentation yields unwanted hammy and putrid flavors.[12]

EFFECTS OF ENZYMATIC CHANGES

During fermentation, microbial activity on the cocoa pulp generates heat and produces ethanol, acetic, and lactic acids that kills the bean. Pulp fermentation products penetrate slowly into beans causing swelling and stimulating enzymatic reactions that yield desirable flavor precursors, and upon roasting, characteristic flavor and aroma notes. Fresh beans with low contents of flavor precursors have limited commercial usage and any activities in the fermentation are unable to rectify this shortcoming.[3]

Subcellular changes in the cotyledons release key enzymes that affect reactions among substrates preexisting in unfermented beans.[8] Enzymes exhibit different stabilities during fermentation and may be inactivated by heat, acids, polyphenols, and proteases. Aminopeptidase, cotyledon invertase, pulp invertase, and polyphenol oxidase are significantly inactivated, carboxypeptidase partly inactivated, whereas endoprotease and glycosidases remain active during fermentation.[11] Hansen et al.[12] noted differences in enzyme activities that can be partly explained by pod variation and genotype, but in general, activities present in unfermented beans do not appear to be a limiting factor for optimal flavor precursor formation in fermentation. Significant fermentation effects may relate to factors such as storage protein sequence and accessibility, destruction of cell compartmentalization, enzyme mobilization, and pulp and testa changes.

Proteases affect multiple cellular processes in plants, such as protein maturation and degradations associated with tissue restructuring and cell maintenance. Key aspartic proteinases (EC 3.4.23) have been characterized in a number of *Theobroma cacao* gymnosperms, and activity in seeds of *Theobroma cacao* has been extensively studied.[11] Partially purified aspartic proteinase had optimal activity at 55°C and pH 3.5. A processing sequence is required to produce cocoa beans with good flavor. Pulp sugar fermentation should yield high levels of acids, particularly acetic acid. As seed pH decreases, cell structure is disrupted, which triggers mobilization and activation of primary aspartic proteinase activity with massive degradation of cellular protein.[11] Fermentation proteinase and peptidase activities seem critical for good flavor quality.[12]

EFFECTS OF DRYING

Flavor development from cocoa bean precursors continues during drying with development of the characteristic brown color. Major polyphenol oxidizing reactions are catalyzed by polyphenol oxidases, giving rise to new flavor components, and loss of membrane integrity, inducing brown color formation. Use of artificial drying can increase cotyledon temperatures and cause case hardening that restricts loss of volatile acids that leads to detrimental effects to the final chocolate flavor.[7]

After fermentation and drying, the moisture-content target for cocoa beans is ca. 6–8%. For storage and transport, moisture content should be <8% or mold growth is possible. Indicators of well-dried, quality beans are good brown color, low astringency and bitterness, and an absence of off-flavors, such as smoky notes and excessive acidity. Sensory assessment of cocoa beans dried using different strategies, i.e., sun-drying, air-blowing, shade-drying, and oven-drying suggested sun-dried beans are rated higher in chocolate development with fewer off-notes.[10] Off-notes from incomplete drying or rain-soaking may result in high levels of water activity and mold contamination, producing high concentrations of strongly flavored carbonyls, leading to alterations in bean flavor, producing hammy off-flavors, which is also correlated with over-fermentation.[7]

CONCLUSION

Cocoa bean fermentation is crucial to the formation of key volatile fractions (alcohols, esters, and fatty acids) and the provision of flavor precursors (amino acids and reducing sugars) (Table 1), which are important notes contributing to chocolate character. Drying reduces levels of acidity and astringency in cocoa nibs by decreasing the volatile acids and total polyphenols enhancing the color and taste of finished chocolate. Direct relationships are thus observed between the initial composition and postharvest treatments (fermentation and drying) of cocoa beans and technological effects on flavor development and character in chocolate.

REFERENCES

1. Afoakwa, E.O.; Paterson, A.; Fowler, M. Factors influencing rheological and textural qualities in chocolate—a review. Trends Food Sci. Tech. **2007**, *18* (4), 290–298.
2. Clapperton, J.F. A review of research to identify the origins of cocoa flavour characteristics. Cocoa Growers' Bull. **1994**, *48*, 7–16.
3. Beckett, S.T. *Industrial Chocolate Manufacture and Use*, 5th Ed.; Blackwell Science: Oxford, UK, 2009; 405–428, 460–465.
4. Holm, C.S.; Aston, J.W.; Douglas, K. The effects of the organic acids in cocoa on flavour of chocolate. J. Sci. Food Agric. **1993**, *61*, 65–71.
5. Ziegleder, G. Composition of flavour extracts of raw and roasted cocoas. Z. Lebensm. Unters. Forsch. **1991**, *192*, 512–525.
6. Jinap, S.; Dimick, P.S.; Hollender, R. Flavour evaluation of chocolate formulated from cocoa beans from different countries. Food Control **1995**, *6* (2), 105–110.
7. Lopez, A.S.; Dimick, P.S. Cocoa fermentation. In *Biotechnology: A Comprehensive Treatise*. Vol. 9: *Enzymes, Food and Feed*, 2nd Ed.; Reed, G., Nagodawithana, T.W., Eds.; VCH: Weinheim, 1995; 563–577.
8. Afoakwa, E.O.; Paterson, A.; Fowler, M.; Ryan, A. Flavor formation and character in cocoa and chocolate: a critical review. Crit. Rev. Food Sci. Nutr. **2008**, *48* (9), 840–857.
9. Biehl, B.; Meyer, B.; Said, M.B.; Samarakoddy, R.J. Bean spreading: a method of pulp preconditioning to impair strong nib acidification during cocoa fermentation in Malaysian. J. Food Agric. **1990**, *51*, 35–45.
10. Said, M.B.; Jayawardena, M.P.G.S.; Samarakoddy, R.J.; Perera, W.T. Preconditioning of fresh cocoa beans prior to fermentation to improve quality: a commercial approach. The Planter **1990**, *66*, 332–345.
11. Voigt, J.; Biehl, B.; Heinrich, H.; Kamaruddin, S.; Marsoner, G.; Hugi, A. *In vitro* formation of cocoa-specific aroma precursors: aroma-related peptides generated from cocoa seed protein by co-operation of an aspartic endoprotease and a carboxypeptidase. Food Chem. **1994**, *49*, 173–180.
12. Hansen, C.E.; Manez, A.; Burri, C.; Bousbaine, A. Comparison of enzyme activities involved in flavour precursor formation in unfermented beans of different cocoa genotypes. J. Sci. Food Agric. **2000**, *80*, 1193–1198.

Cold Plasmas Used for Food Processing

H.C. Mastwijk
M.N. Nierop Groot
Wageningen UR Food and Biobased Research, Wageningen, The Netherlands

Abstract
Application of cold plasma has been reported in agriculture, food, and bioscience literature as an effective, non-chemical, gas-phase disinfection agent that can be applied at moderate temperatures. The unusual thermodynamic properties of these gases are discussed with focus on nitrogen-based atmospheric plasma. Typical concepts, such as electron temperature and charge-exchange processes with surfaces are explained in more detail. Finally, some general phenomena related to microbial inactivation are presented, including a survey on outstanding issues concerning research and development efforts aimed at utilization of cold plasma disinfection.

INTRODUCTION

Plasma refers to a gaseous-like state where a portion of the molecules present is excited or ionized.[1] A cold plasma is subject to the unusual condition that the fraction of excited-state molecules (Fig. 1) exceeds the predicted fraction by the Boltzmann factor[2] when evaluated at the observed gas temperature. Under these conditions the valence electrons in, e.g., a cold nitrogen plasma may have a temperature of 54,000 K (Fig. 2), whereas the gas temperature is maintained at 40°C. These remarkable properties of a cold plasma gas allow for charge-mediated modifications at moderate temperatures that cannot be realized by intense heating, as observed in Fig. 3.

The situation where a small fraction of excited-state molecules can coexist in a mixture with a relatively large fraction of gas in the ground state, suggests that thermodynamic laws are violated;[2] however, the interaction between molecules and the exchange of energy in the process of binary molecular collisions are subject to selection rules.[1] Spin selection rules prevent the loss of energy of excited-state molecules by natural decay or exchange with ground-state molecules. Both these effects are known to occur in molecular nitrogen[3] owing to the specific electronic structure of N_2 that allows a triplet state (where electron spins are aligned) as the first excited state,[4] whereas the ground state is a singlet state (electron spins in opposition). Transitions between a triplet and a singlet state are unlikely to occur as a result of spin selection rules.[1,4] Therefore, nitrogen gas that contains a small fraction of triplet N_2 cannot lose energy that is stored in this state, neither by decay to the singlet ground state[1,4] nor by triplet–singlet interconversion collisions.[3,4] This effect, where a molecular gas seemingly disobeys the laws of thermodynamics, is well known (the Overhauser effect) for nuclear ensembles that are spin-polarized.[5] A plasma that contains triplet nitrogen is stabilized by this phenomenon for more than a second, resulting in a formidable afterglow that may be several meters in length.[3,4]

The only effective way for triplet nitrogen to lose its energy is by interconversion reactions.[1,4] In the case of nitrogen, interconversion reactions may occur when triplet nitrogen reacts with any molecule other than nitrogen. This may be arranged by exposure of the plasma gas to a surface.

PRODUCTION OF COLD PLASMA

Plasma that can be sustained at standard normal conditions is usually referred to as atmospheric plasma. Atmospheric plasma generated in an environment of inert gas at standard normal conditions using electric discharges were first reported by Lord Rayleigh in 1916.[4] The three basic requirements for production of a cold plasma are an electrode system, a feed of carrier gas, and an auxiliary high voltage power supply.[6,7] These relatively basic requirements for the production of cold plasma, the low thermal load of the gas to the product, and its typical non-thermal signature (Fig. 3–Fig. 5) are the main reasons for a common belief that cold plasma technology can be utilized as a large-scale disinfection method.[8]

NITROGEN-BASED ATMOSPHERIC PLASMA

The particular importance of molecular nitrogen as a carrier gas for plasma that operates near standard normal conditions is nitrogen's great abundance (~78%) in air.

Secondly, it is fairly easy to derive pure nitrogen gas from air using modern membrane technology, and thirdly,

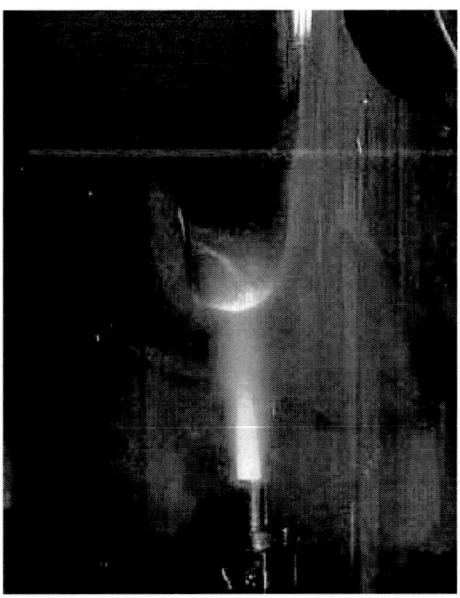

Fig. 1 Characteristic afterglow of a cold plasma produced by an electric discharge in a jet of molecular nitrogen. The gas temperature of the flame was 40°C.

molecular nitrogen is a safe and proven food-grade gas that has been used in large scale for a long time within many areas of food processing and biosciences, e.g., in modified atmospheric packaging.

INTERACTION OF COLD PLASMA WITH SURFACES AND USE IN DIAGNOSTICS

Cold plasma disposes its energy by interconversion reactions when brought into contact with a surface. For ions

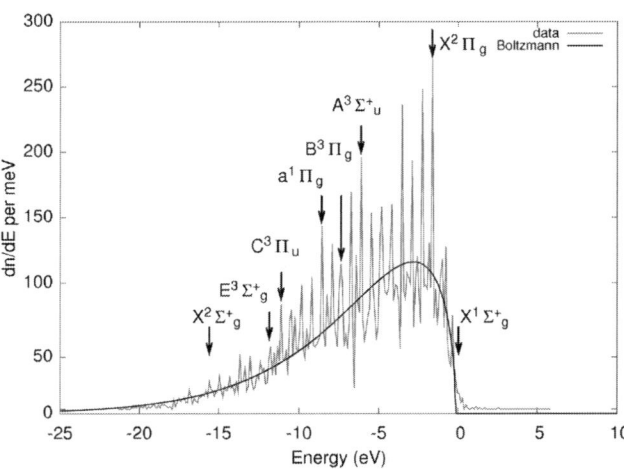

Fig. 2 The energy distribution of electrons in a jet of nitrogen plasma at a temperature of 40°C. The coarse shape is a Boltzmann distribution of a free electron gas at a temperature of 54,000 ± 2000 K. The peaked structure is owing to vibrational excitations of N_2. The lowest vibrational levels for the specific electronic states in molecular nitrogen are shown.

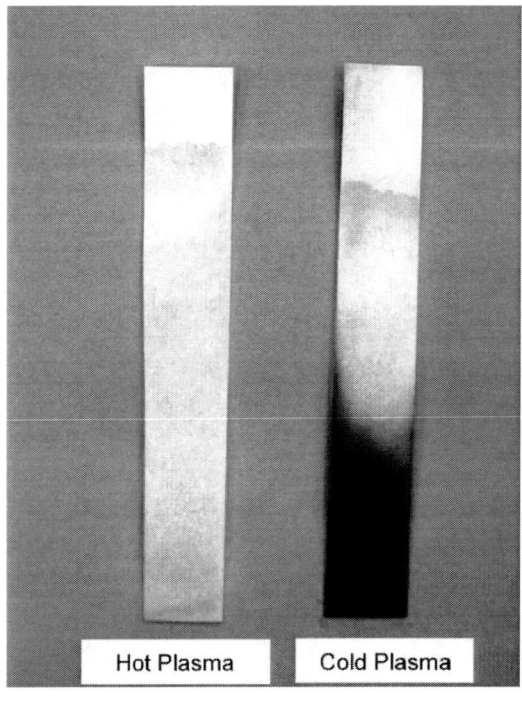

Fig. 3 Electron transfer reactions invoked by plasma treatment on starch–iodine paper. Redox reactions result in a purple decolouration of the indicator paper. Redox reactions are absent in a hot plasma that was obtained by combustion of butane at a temperature of 1200°C but are observed in a cold atmospheric nitrogen plasma at a gas temperature of 30°C for electron temperatures of approximately 10,000 K.

and excited-state molecules this process is known as Auger de-excitation.[9] It governs a binary charge exchange process between one molecule in the plasma and one molecule at the target surface. These charge-exchange processes have been known and studied for many decades, but exclusively under high-vacuum conditions.[9] Auger de-excitation allows the detailed recording of electron transfer reactions (electron spectroscopy) used for the characterization of chemical bonds at the surface under consideration.[9] Recently, advancements have been made to establish a

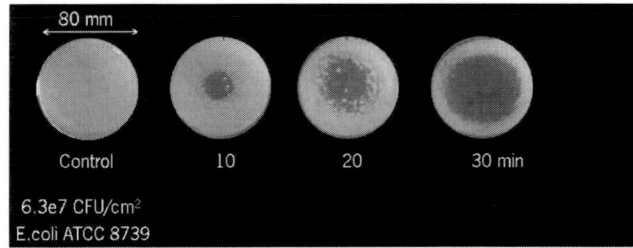

Fig. 4 Loss in viability of *Escherichia coli* after cold plasma treatment. Cells were initially prepared on tryptic soy agar dishes for confluent growth (control). The sterile zones that appear on the treated plates after exposure for 10, 20, and 30 min indicate the loss of ability to produce colony-forming units (CFU).

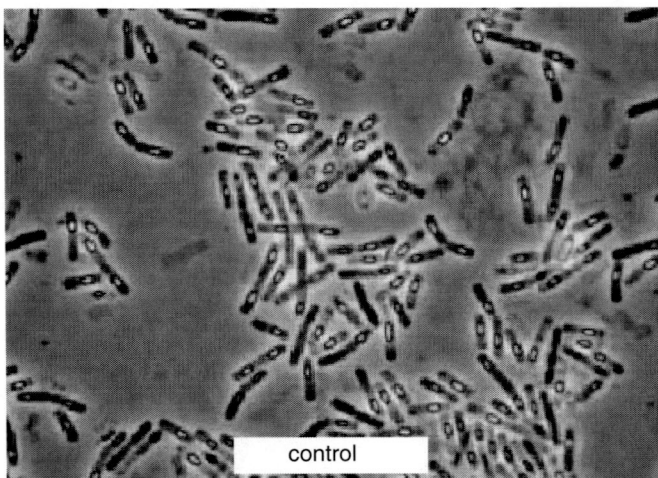 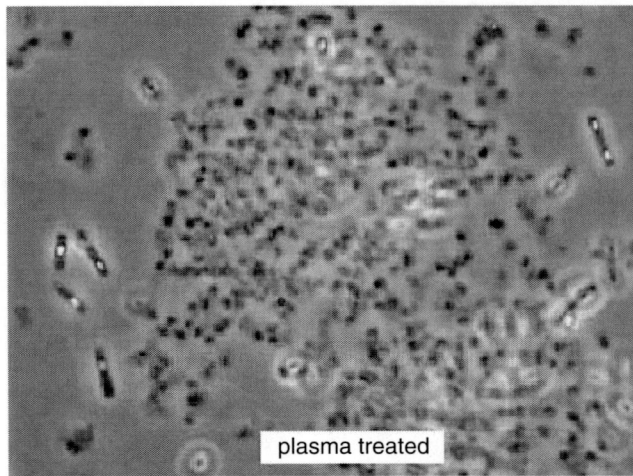

Fig. 5 Light microscope images of cells of *Bacillus cereus* treated by cold plasma. The loss of viability is probably related to the structural damage that is usually observed.

method for applying Auger electron spectroscopy at ambient conditions.[3] This technique resembles in some aspects the technique of Langmuir probes,[6,10] which is a standard tool for plasma diagnostics to derive electron- and ion-energy spectra; however, Auger electron spectroscopy is known to deliver much more detailed information; charge exchange between different vibrational states can be observed.[3,9,10] In addition, it allows for in situ analysis of ongoing charge transfer reactions that occur at the surface of the target under investigation. Other diagnostic tools that are usually applied to detect and analyze the active constituents in a plasma are photon spectroscopy,[3,4,6] titrations, and current–voltage recordings using Langmuir probes.[6,10]

KNOWN APPLICATIONS OF COLD PLASMA

Electrical discharges that proceed in air are widely used in agricultural, food, and bioscience applications. The production of ozone and singlet oxygen can be considered as an example of oxygen-based cold plasma. Ozone is an oxidizing agent[11] that is used for disinfection of water and removal of volatile organic compounds in air. Corona discharges in air[6] are used for the treatment of polymer materials to increase surface energy. By breaking the cross-linking of aligned polymers, the surface characteristics of the materials under investigation can be efficiently altered at moderate temperatures without the risk of deformation.

SURFACE DECONTAMINATION

Cold plasma technology is currently considered a realistic option to inactivate pathogens that can be found on the surface of perishable foods.[12–15] The main motivations for considering this technology is that decontamination is effective at moderate temperatures in the range of 20–50°C.[12–15] Secondly, in principle, few chemical residues are to be expected using cold plasma.[8,16]

Many reports have been made demonstrating the effectiveness of particular plasma treatments against various target organisms, including bacteria, fungi, yeasts, and moderate inactivation of spores.[8,16] In typical decontamination experiments, as depicted in Fig. 4 and Fig. 5, inert carrier gases, such as nitrogen, helium, argon, air, and mixtures are used. Such plasma applications that have been reported usually rely on electric discharges in an efflux of these carrier gases close to ambient conditions.[6,7,10]

Most often, the large number of different plasma generators used, the variety in operating conditions adapted, and the different types of microorganisms used, have restricted comparison of reported results.[8,16] No systematic studies have been devoted to retrieve critical cold plasma conditions for microbial inactivation. Hence, the interaction of cold plasma gas with microorganisms and the possible mechanisms of inactivation remain largely unspecified.[8,12–16] It has been suggested that UV light, ions, and the occurrence of free radicals are three key elements that are responsible for microbial inactivation;[8,16] however, dehydration and other stresses to microorganisms caused by exposure to cold plasma treatment, such as the temperature shock invoked by evaporative cooling of wet samples in a jet of gas, are important factors that cannot be excluded.[17] Observation of complex inactivation kinetics[16] is a strong indication that treatment conditions are constantly changing and probably induce a multifactorial stress response in microorganisms. These factors make it difficult to draw firm conclusions on general mechanisms of microbial inactivation invoked by cold plasma from the scientific literature.

Major causes of difficulties in interpretation of published data are found in the inherently large degrees of freedom in the design of equipment, treatment conditions,[8] and choice in available diagnostics.[6]

Ideally, the inactivation kinetics of microorganisms by cold plasma are studied by varying a single, isolated parameter without the interference of secondary stresses to the target microorganism caused by the choice of a particular experimental set-up. Secondly, critical parameters of a given plasma treatment should be quantified, recorded, and understood to a level that reproducible decimal reduction rates (reaction constants) and z values (activation energy) can be assigned to the critical parameters of a treatment analogous to heat pasteurization and sterilization processes.[18] Validation of processing conditions can only be achieved when the critical parameters in the process that is utilized are known to assure that critical levels, required for inactivation, are actually met during processing. Thirdly, methodologies to identify relevant molecular processes involved in microbial inactivation using standardized conditions are desirable.[8,16] Recently, commercially available laboratory-scale equipments have been developed to maintain and secure a traceable standard on the production and molecular diagnostics of cold plasma technology.[19] It has become evident that future research and development efforts in the field of cold plasma decontamination should not only focus on development of the technology but should be combined with basic sciences, including biotechnology.

CONCLUSION

Cold plasma technology is an emerging disinfection method that relies on the exposure of electronic activated gases to surfaces. The action of cold plasma is non-thermal and non-chemical based if inert carrier gases are used. Microbial inactivation has been demonstrated, but critical parameters in treatment have not been identified. Scientific evidence for the proposed mechanisms of interaction with microorganisms and substrates is limited. There is an urgent need for the standardization of the production and diagnostics of cold plasma and the collection of kinetic data on microbial inactivation.

ACKNOWLEDGMENTS

This work was financially supported by the Dutch Ministry of Economic Affairs: EOS (Energie Onderzoek Subsidie) through contract LT-01044 and by the Commission of the European Communities, Framework 6, Priority 5 "Food Quality and Safety" through contract FP6-CT-2006–015710.

REFERENCES

1. Joachain, C.J. *Introduction to Quantum Mechanics*; Longman: Essex, 1989.
2. Reif, F. *Fundamentals of Statistical and Thermal Physics*; McGraw-Hill: Singapore, 1965.
3. Mastwijk, H.C.; van Dijk, C.; Wichers, H.J. Observation of free hole gases, http://arxiv.org/abs/0711.4737 (accessed February 2009).
4. Lofthus, A.; Krupenie, P.H. The molecular spectrum of nitrogen. J. Phys. Chem. Ref. Data **1977**, *6*, 113.
5. Ebert, M.; Grossmann, T.; Heil, W. Otten, W.E.; Surkau, R.; Leduc, M.; Bachert, P.; Knopp, M.V.; Schad, L.R.; Thelen, M. Nuclear magnetic resonance imaging with hyperpolarised helium-3. Lancet **1996**, *347*, 1297–1299.
6. Iza, F.; Kim, G.J.; Lee, S.M.; Lee, J.K.; Walsh, J.L.; Zhang, Y.T.; Kong, M.G. Microplasmas: sources, particle kinetics, and biomedical applications. Plasma Processes Polym. **2008**, *5* (4), 322–344.
7. Okazaki, S.; Kogoma, M.; Uehara, M.; Kimura, Y. Appearance of stable glow discharge in air, argon, oxygen and nitrogen at atmospheric pressure using a 50 Hz source. J. Phys. D Appl. Phys. **1993**, *26*, 889–892.
8. Moreau, M.; Orange, N.; Feuilloley, M.G.J. Non-thermal plasma technologies: new tools for bio-decontamination. Biotechnol. Adv. **2008**, *26*, 610–617.
9. Hotop, H. *Experimental Methods in the Physical Sciences*; Dunning, F.D., Ed.; Academic Press: San Diego, 1996; Vol. 29B, 191–215.
10. Anders, A.; Kuhn M. Characterization of a low-energy constricted-plasma source. Rev. Sci. Instrum. **1998**, *3* (69), 1340–1344.
11. Kim, J.-G.; Yousef, A.E.; Khadre, M.H. Ozone and its current and future application in the food industry. In *Advances in Food Science and Nutrition*; Taylor, S., Ed.; Elsevier Sci. Ltd.: London, UK, 2003; Vol. 45, 167–218.
12. Critzer, F.J.; Kelly-Wintenberg, K.; South, S.L.; Golden, D.A. Atmospheric plasma inactivation of foodborne pathogens. J. Food Prot. **2007**, *70* (10), 2290–2296.
13. Deng, S.; Ruan, R.; Kyoonmok, C.; Huang, G.; Lin, X.; Chen, P. Inactivation of *Escherichia coli* on almonds using nonthermal plasma. J. Food Microbiol. Saf. **2007**, *72* (2), M62–M66
14. Niemira, B.A.; Sites, J. Cold plasma inactivates *Salmonella stanley* and *Escherichia coli* O157:H7 inoculated on golden delicious apples. J. Food Prot. **2008**, *71* (7), 1357–1365.
15. Perni, S.; Liu D.W.; Shama G.; Kong M.G. Cold atmospheric plasma decontamination of pericaps of fruit. J. Food Prot. **2008**, 71 (2), 302–308.
16. Moison, M.; Barbeau, J.; Moreau, S.; Pelletier, J.; Tabrizian, M.; Yahia, L.H. Low-temperature sterilization using gas plasmas: a review of the experiments and an analysis of the inactivation mechanisms. Int. J. Pharm. **2001**, 226.
17. Niemira, B.A.; Lelieveld, H.L.M. In *Cold Plasma: An Emerging Technology for Food Processing*, IFT symposium, New Orleans, June 30, 2008.
18. Schlegel, H.G. *General Microbiology*; 7th edition; Press Syndicate of the University of Cambridge: Melbourne, 1993.
19. http://www.omve.com/img//files/Cold-Plasma Demonstrator CP121.pdf (accessed May 10, 2009).

Corn Sweeteners: Enzyme Use

Christine Scaman
Food, Nutrition, and Health, University of British Columbia, Vancouver, British Columbia, Canada

Abstract

The use of enzymes for production of sweeteners from corn starch is the largest sector of the food enzyme industry. Major steps in sweetener production include starch liquefaction by endo-amylases, followed by saccharification by exo-amylases and debranching enzymes, and glucose conversion to fructose by isomerase. Each step yields industrially important sweeteners, with high fructose corn syrup as the major product in terms of volume and economic impact. In addition to these well-established sweeteners, a variety of products are derived using enzymes with unique hydrolytic activity, alone or in combination with transferases. Examples of these include trehalose, gentio-oligosaccharides, and branched-chain oligosaccharides. The focus of research on enzymes with desirable stability and catalytic activities from natural sources or through genetic modification will continue to drive advances in sweetener production from corn starch.

INTRODUCTION

Starch is composed of two polymers of glucose: amylose, a predominantly linear molecule with α-1,4 bonds and 0.2–0.7% α-1,6 bonds, and amylopectin, a branched molecule with approximately 5% α-1,6 bonds in addition to α-1,4 bonds. Starch is the primary energy storage compound in photosynthetic plants, and many organisms have evolved enzymes to take advantage of this energy store. These sources provide food technologists with an array of starch-acting enzymes that can be used to produce sweeteners to meet the needs of the food industry.

The major enzymatic steps in corn starch hydrolysis to produce glucose and maltose syrups and high fructose corn syrup are summarized. A brief overview of low-intensity and alternative sweeteners produced by enzymatic modification of starch hydrolyzates is also presented. The physical, chemical, and health properties of these products are highlighted.

ENZYMATIC STARCH HYDROLYSIS

The production of the most commonly used sweeteners from corn starch arises from liquefaction, saccharification, and isomerization reactions (Fig. 1).[1] Details of these reactions are summarized below. Glucosidases commonly used in these steps are given in Table 1.

Liquefaction

α-Amylases reduce the viscosity of a starch solution by producing dextrins that are 2–10 glucose units in size through a random hydrolysis of amylose and amylopectin, hence the term liquefaction. In a typical operation, the pH of a starch slurry is adjusted to pH 6 with NaOH, and enzyme and calcium (added to enhance enzyme thermostability) are added. The temperature is increased to 105°C for 5 min, allowing starch gelatinization and disassociation of lipid–amylose complexes. After dropping the temperature to 95–100°C, the solution is held for 1–2 hours until the desired dextrose equivalent (DE), or reducing power, is obtained. Pure dextrose (glucose) has a DE of 100 as every glucose molecule has one reducing group while starch has a DE of effectively 0. A solution with a DE of 8–12 is commonly produced, although a DE of up to 40 may be required for malto-oligosaccharides production. Sources of industrially important thermophillic α-amylases are *Bacillus licheniformis* and *Bacillus stearothermophilus*. The fungal amylase from *Aspergillus oryzae*, used at 55–70°C, results in the accumulation of the disaccharide maltose (4-*O*-α-D-glucopyranosyl-D-glucose) and dextrins with 1,6 branches for the production of maltose syrups. The fungal amylase produces a syrup with a higher amount of larger dextrins than β-amylase, which is also used in maltose syrup production (see below).

Saccharification

Amyloglucosidase removes glucose residues from the non-reducing end of starch polymers and dextrins. Given an extended reaction time, a solution of approximately 95% glucose can be obtained in this process, known as saccharification. The enzyme can quickly hydrolyze α-1,4 linkages, although di- or trisaccharides, and 1,6 linkages in branched starch are hydrolyzed more slowly. A commercial source of

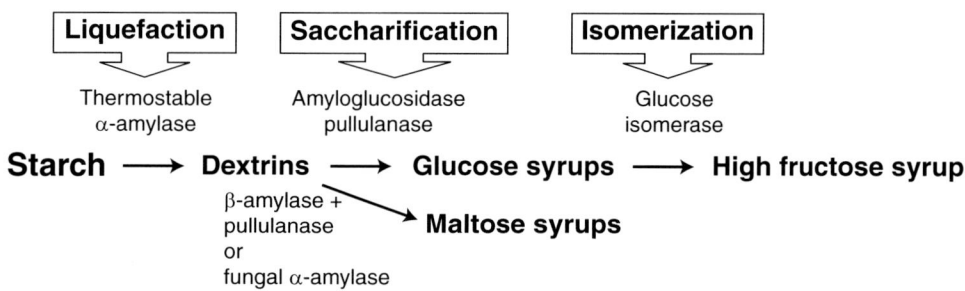

Fig. 1 Overview of enzymes used in production of common sweeteners from starch.

Table 1 Glucosyl hydrolases acting on starch.

Common name(s)	EC#	Activity	Sources
α-Amylase	3.2.1.1	Endo-hydrolase acting on (1,4)-α-D-glucosidic linkages; acts on starch, and related polysaccharides and oligosaccharides in a random manner; reducing groups are liberated in the α-configuration	*Bacillus stearothermophilus* *Bacillus licheniformis* *Bacillus subtilis* *Aspergillus oryzae*
β-Amylase	3.2.1.2	Exo-hydrolase acting on (1,4)-α-D-glucosidic linkages; removes successive maltose units from the non-reducing end of starch, glycogen, and related poly- and oligosaccharides producing β-maltose	Malted barley *Bacillus polymyxa*
Amyloglucosidase or glucoamylase	3.2.1.3	Exo-hydrolase cleaving terminal (1,4)-α-D-glucose residues successively from non-reducing ends of the chains with release of β-D-glucose; most forms of the enzyme can rapidly hydrolyze 1,6-α-D-glucosidic bonds when the next bond in the sequence is α-1,4	*Aspergillus niger* *Aspergillus awamori* *Rhizopus oryzae*
Pullulanase	3.2.1.41 Formerly 3.2.1.69	Endo-hydrolase acting on (1,6)-α-D-glucosidic linkages in pullulan, amylopectin and glycogen; maltose is the smallest sugar that it can release	*Bacillus acidopullulyticus* *Bacillus cereus* *Klebsiella planticola*
Group II Pullulanase	3.2.1.135	Endo-hydrolase acting on both α-1 to α-4 and α-1 to α-6 glycosidic bonds; main degradation products are maltose and maltotriose	*Bacillus* species
Isoamylase	3.2.1.68	Endo-hydrolase acting on 1,6-α-D-glucosidic branch linkages in glycogen, amylopectin and their β-limit dextrins; differs from EC 3.2.1.41 by its inability to hydrolyze pullulan	*Pseudomonas amyloderamosa*
α-Glucosidase	3.2.1.20	Exo-hydrolase acting on terminal, non-reducing 1,4-linked α-D-glucose residues with release of α-D-glucose; acts most efficiently on short malto-oligosaccharides	*Aspergillus niger* *Pyrococcus furiosus* *Sulfolobus solfataricus*
Maltogenic α-amylase	3.2.1.133	Exo-hydrolase acting on 1,4-α-D-glucosidic linkages in polysaccharides so as to remove successive α-maltose residues from the non-reducing ends of the chains	*Bacillus stearothermophilus* *Bacillus subtilus*
Maltotriose-forming amylase	3.2.1.116	Exo-hydrolase acting on 1,4-α-D-glucosidic linkages in amylase to remove successive maltotriose residues from the non-reducing chain ends	*Microbacterium imperiale*
Maltotetraose-forming amylase	3.2.1.60	Exo-hydrolase acting on 1,4-α-D-glucosidic linkages in amylase to remove successive maltotetraose residues from the non-reducing chain ends	*Pseudomonas stutzeri*
Maltopentaose-forming amylase	EC # not assigned	Exo-hydrolase acting on 1,4-α-D-glucosidic linkages in amylase to remove successive maltopentaose residues from the non-reducing chain ends	*Bacillus licheniformis*
Maltohexaose-forming amylase	3.2.1.98	Exo-hydrolase acting on 1,4-α-D-glucosidic linkages in amylase to remove successive maltohexaose residues from the non-reducing chain ends	*Bacillus subtilus* *Klebsiella pneumoniae*

Sources: van der Maarel et al.;[1] http://ca.expasy.org/enzyme/; http://www.brenda-enzymes.org/index.php4; http://www.cazy.org/index.html.

the enzyme is *Aspergillus niger*, which has a pH optimum near 4.5 and is stable at 60°C. Therefore, after liquefaction, the substrate solution must be acidified. The amount of enzyme and reaction time must be carefully controlled as the enzyme can resynthesize glycosidic linkages in a process known as reversion which occurs near the end of the reaction resulting in decreased yield.

Saccharification is much more efficient if a debranching enzyme, such as pullulanase or isoamylase, is used with amyloglucosidase. Both of these enzymes hydrolyze α-1,6 bonds, preventing the accumulation of α-1,6 branched oligomers near the end of the reaction.

The products of extensive saccharification are glucose and maltose syrups. Glucose syrups are distinguished from maltose syrups as mixtures that contain more glucose than maltose. Glucose syrups can be classified as low conversion (20–35 DE), intermediate conversion (35–55 DE), high conversion (55–70 DE), and very high conversion (70–98 DE). Low conversion glucose syrup, with a relative sweetness of 30–35% sucrose (Table 2), provides viscosity or body in products and acts as a humectant. Very high conversion syrups are used in the bakery and beverage industries, and provide the starting material for the manufacture of high fructose syrups.

β-Amylases are used to produce maltose syrups. The enzyme, obtained from barley or soybeans, is expensive, so use of β-amylases is limited to specialty products. This enzyme hydrolyzes maltose units from the non-reducing ends of the starch polymer. Hydrolysis is inhibited several glucose units away from α-1,6 branch points, so when the enzyme is used alone a maximum of 50–60% maltose (dry basis) can be obtained; although, up to 85% maltose can be reached when pullulanase or isoamylase are also used. Maltose syrups (40–45 DE), high maltose syrups (48–52 DE), and extra high maltose syrups (50–60 DE) show little tendency to crystallize due to the low glucose content (typically <5%), and are relatively non-hygroscopic. Extra high conversion maltose syrup typically contains 30–35% glucose, 30–45% maltose, and 8–13% maltotriose. Syrups composed of specific oligomers of maltose (i.e., maltotriose, maltotetraose, maltopentaose, and maltohexaose) can be obtained using microbial amylases with these specificities instead of β-amylase (Table 1).

Glucose–Fructose Isomerization

Xylose or glucose isomerase (EC 5.3.1.5) catalyzes the isomerization of monomeric keto sugars such as glucose to their enol isomers.[2] Magnesium, manganese, and cobalt are enzyme activators and stabilizers; calcium is an inhibitor. Therefore, sufficient metal must be added to displace calcium remaining from the α-amylase liquefaction step. The enzyme requires pH 7.5 or higher for good stability and activity, and the reaction is usually carried out at 60°C. At higher temperatures, enzyme stability suffers and decomposition by-products are produced, while at lower temperatures microbial growth may be a problem and a lower fructose:glucose equilibrium is obtained. The reaction is carried out using enzyme immobilized in a fixed-bed reactor with glucose syrup flowing continuously through it.

Enzymatic isomerization of glucose to fructose represents one of the largest commercial applications of immobilized enzyme. Immobilization techniques developed to offset the expense of the isomerase use cell-free enzyme or whole microbial cells. An example of whole cell immobilization consists of disrupting bacterial cells, cross-linking them with glutaraldehyde, followed by concentration, extrusion, drying, and sieving. A commercial enzyme preparation, Sweetzyme IT (Novozyme AS, Franklinton, NC, USA), can convert 18,000 kg of syrup dry matter for every 1 kg of isomerase.[3] The operating lifetime of a batch of enzyme can be 200–360 days. Cell-free enzyme systems, such as enzyme immobilized on an anion exchange resin, have improved flow characteristics compared to immobilized cells.

High fructose corn syrup (HFCS) is a popular sweetener, particularly in the United States, where it has a cost advantage over other sweeteners due in part to the artificially low market price for corn. High fructose syrups containing either 42 or 55% fructose (dry basis) are commonly used. These are standardized using chromatographically purified 90%-fructose syrup. The most significant application of HFCS is soft drinks as 55% HFCS is equal in sweetness to sucrose syrups (Table 2). The 42% syrup, which is less sweet than sucrose, finds applications in the baking and other industries.

There is controversy over the relationship of fructose consumption and adverse health effects, especially obesity. It has been argued that HFCS with its similar composition to other natural sweeteners, such as honey and fruit concentrates, is not related to obesity development except when consumed in excess.[4] Adverse metabolic effects can be noted when pure fructose is consumed, rather than the more even ratio of glucose and fructose present in HFCS.

Table 2 Relative sweetness of carbohydrates.

Compound	Relative sweetness
Sucrose	1.0
Glucose	0.6–0.8
Fructose	1.1–1.7
Maltose	0.3–0.6
Trehalose	0.4
Glucose and maltose syrups	0.3–0.5
High fructose syrup (42–55%)	0.9–1.0
Sucromalt	0.6–0.7
Isomalto-oligosaccharides	0.3–0.6
Gentio-oligosaccharides	0.3–0.6

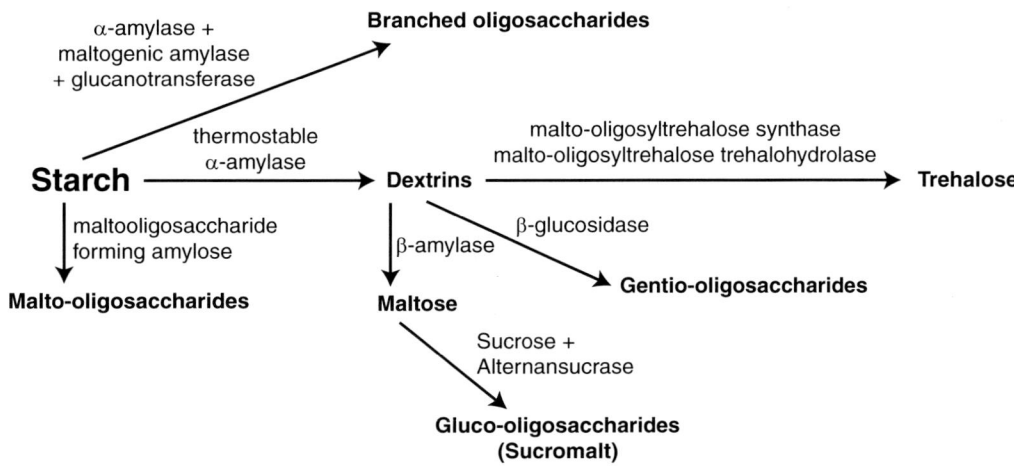

Fig. 2 Overview of enzymes used in production of low intensity and alternative sweeteners from starch.

LOW-INTENSITY AND ALTERNATIVE SWEETENERS

The enzymatic hydrolysates of corn starch provide the basis for production of products with mild sweetness and health benefits, derived from further chemical or enzymatic modifications (Fig. 2). Trehalose (α-D-glucopyranosyl α-D-glucopyranoside) is a non-reducing, non-cariogenic sugar with wide pH stability and cryoprotectant activity. Three enzymatic approaches to trehalose production from starch or hydrolysis products have been reported, including the use of a novel transferase (malto-oligosyl trehalose synthase, EC 5.4.99.15) and α-amylase (malto-oligosyl trehalose trehalohydrolase, EC 3.2.1.141) from an *Arthrobacter* species,[5] and utilization of an α-glucosidase from *Chaetomium thermophilum* var. *coprophilum*.[6]

Sucromalt (Cargill, Inc., Minneapolis, MN, United States) is a nutritive sweetener produced by treating sucrose and corn syrup with *Leuconostoc mesenteroides* alternansucrase (EC 2.1.4.140).[7] The enzyme hydrolyses sucrose, and transfers a glucose onto the non-reducing end of maltose to form dextrans with α-1,6 bonds and α-1,3 side chains. Sucromalt contains 41% fructose, 52% gluco-oligosaccharides, and minor amounts of a reversion by-product, leucrose.

Branched oligosaccharides (isomalto-oligosaccharides), containing α-1,6 as well as α-1,4 linkages (isomaltose and panose), can be used as low-intensity/low-calorie sweeteners and humectants. Traditionally, these have been produced from starch using α-amylase, followed by β-amylase, and then α-glucosidase. A more efficient method uses α-amylase, followed by maltogenic amylase and α-glucanotransferase.[8] Gentio-oligosaccharides, composed of up to five glucose molecules linked by β-1,6, are produced using β-glucosidase. Although they have a bitter taste, gentio-oligosaccharides will become sweet or bland depending on the degree of enzymatic glucosylation using alternansucrase.[9]

There is strong evidence that some oligosaccharides have health benefits.[10] These include a favorable change in intestinal microbiota (prebiotic effect), prevention of constipation, inhibition of pathogenic bacterial adhesion to the intestinal surface, and beneficial effects on blood lipids profiles.

FUTURE PROSPECTS AND CONCLUSION

Corn starch is one of the most important sources of material for production of sweeteners and will remain so in the future. Although glucose and high fructose corn syrup compose the bulk of the corn starch-derived sweeteners, increased production of other products with specific sensory, chemical, physical, and health benefits is likely.

Research efforts will continue to identify enzymes with novel activities or engineer improvements in those that are well known in order to optimize sweetener production.[11] Improvements include greater thermal and pH tolerances, lack of metal dependence, and better overall stability and catalytic efficiency.[12]

REFERENCES

1. van der Maarel, M.J.E.C.; van der Veen, B.; Uitdehaag, J.C.M.; Leemhuis, H.; Dijkhuizen, L. Properties and applications of starch-converting enzymes of the alpha-amylase family. J. Biotechnol. **2002**, *94* (2), 137–155.

2. Bhosale, S.H.; Rao, M.B.; Deshpande, V.V. Molecular and industrial aspects of glucose isomerase. Microbiol. Rev. **1996**, *60* (2), 280–300.

3. Olsen, H.S. Enzymes in starch modification. In *Enzymes in Food Technology*; Whitehurst, R.J., Law, B.A., Eds.; CRC Press: Boca Raton, FL, 2002; 200–228.

4. White, J.S. Straight talk about high-fructose corn syrup: what it is and what it isn't. Am. J. Clin. Nutr. **2008**, *88* (suppl), 1716S–1721S.

5. Schiraldi, C.; Di Lernia, I.; De Rosa, M. Trehalose production: exploiting novel approaches. Trends Biotechnol. **2002**, *20* (10), 420–425.
6. Giannesi, G.C.; de Lourdes Teixeira de Moraes Polizeli, M.; Terenzi, H.F.; Jorge, J.A. A novel alpha-glucosidase from *Chaetomium thermophilum* var. *coprophilum* that converts maltose into trehalose: Purification and partial characterisation of the enzyme. Process Biochem. **2006**, *41* (8), 1729–1735.
7. Eapen, A.K.; Chengelis, C.P.; Jordan, N.P.; Baumgartner, R.E.; Zheng, G.H.; Carlson, T. A 28-day oral (dietary) toxicity study of sucromalt in Sprague–Dawley rats. Food Chem. Toxicol. **2007**, *45* (11), 2304–2311.
8. Lee, H.S.; Auh, J.H.; Yoon, H.G.; Kim, M.J.; Park, J.H.; Hong, S.S.; Kang, M.H.; Kim, T.J.; Moon, T.W.; Kim, J.W.; Park, K.H. Cooperative action of alpha-glucanotransferase and maltogenic amylase for an improved process of isomaltooligosaccharide (MO) production. J. Agric. Food Chem. **2002**, *50* (10), 2812–2817.
9. Cote, G.L. Acceptor products of alternansucrase with gentiobiose. Production of novel oligosaccharides for food and feed and elimination of bitterness. Carbohydr. Res. **2009**, *344* (2), 187–190.
10. Mussatto, S.I.; Mancilha, I.M. Non-digestible oligosaccharides: a review. Carbohydr. Polym. **2007**, *68* (3), 587–597.
11. Kelly, R.M.; Dijkhuizen, L.; Leemhuis, H. Starch and alpha-glucan acting enzymes, modulating their properties by directed evolution. J. Biotechnol. **2009**, *140*, 184–193.
12. Olempska-Beer, Z.S.; Merker, R.I.; Ditto, M.D.; DiNovi, M.J. Food-processing enzymes from recombinant microorganisms—a review. Regul. Toxicol. Pharm. **2006**, *45* (2), 144–158.

Crops: Feral De-Domestication

Jonathan Gressel
Department of Plant Sciences, Weizmann Institute of Science, Rehovot, Israel

Abstract

Although domestication of crops was a slow process governed by selection of recessive traits, de-domestication to weedy, feral forms by selecting for abundant dominant back mutations (endo-ferality) can be rapid. Feral forms can also appear quickly owing to hybridization with wild or weedy relatives (exo-ferality). Examples of such evolution are described for 10 crops. If biotechnology leads to more monoculture, the possibilities of evolution of feral forms increase, requiring diligence and transgenic failsafe mechanisms to delay the phenomenon.

INTRODUCTION: DOMESTICATION AND FERALITY

Domestication of crops has been a long, unending process requiring centuries or millennia (Fig. 1) of selecting mainly rare recessive traits that define the "domestication syndrome," Domestication is being accelerated by marker-assisted breeding and by transgenic introduction of novel traits not found in the crop genomes or in those of interbreeding relatives.[1] The back mutation reversion of a recessively domesticated species to the dominant feral form (de-domestication) can be rapid (evolution of endo-ferality), e.g., it took a few generations to select and breed foxes to be docile pets, but in fewer generations without continued selection, they become feral.[2] The same probably occurs in plants, where there are less data.[3] The process of becoming feral can be hastened when and if the domesticated species can hybridize with its wild progenitor or a wild relative (exo-ferality). The feral form is a partially de-domesticated form (not fully de-domesticated to the wild form). Feral forms, unlike crops, can exist independently without being dependent on managed cultivation. The feral forms are usually weedy versions of the crop. Ferality is a graded continuum process, and thus the term may be disliked by those who prefer more definitive, "stop the evolutionary clock," definitions. As most crops are highly domesticated and can exist only under cultivation, feral populations will be expected to evolve first in human-disturbed situations; e.g., cultivated fields and roadsides, and not in natural (wild) habitats. Ferality may occur first as "volunteer" weeds (emanating from crop seeds not harvested in the previous season) bearing mutations to feral forms that later continue accumulating feral genes.

Many crops are ultimately domesticated to polyploidy (or amphiploidy from genomic combination of related species). The selection for the recessive traits likely occured before polyploidization or interspecific genome combinations, as it would take much longer to select for recessive domestication traits in redundant genomes. The reversion to dominant feral forms is under no such constraints in polyploids.

DNA evidence has countered the presumption of speciation during domestication by "splitter" taxonomists. Many cases where wild and cultivated forms had been given different Latin binomials are unspeciated (i.e., can cross breed), justifying "lumper" taxonomists.

WHAT IS KNOWN ABOUT FERAL PLANTS?

More than 95% of >2100 documents in a database search for "feral" referred to feral animals. Of the 115 documents in a search for "feral and plants," 92 referred to feral animals that devour or pollinate plants. The 23 abstracts on feral plants appeared about one per year. Five more citations came forth when specific crop names were used instead of "plant." Still, some authors use "weedy" or other adjectives for feral forms of crops, e.g., weedy wheat, red rice, wild rice, etc., for what is both weedy and feral.

The possibility that transgenic monoculture might lead to more gene flow and ferality led is author to convene a symposium to ascertain common as well as species-specific evolutionary processes that occur when a crop becomes feral. The analyses were mainly limited to implications to agriculture from the evolution of ferality within agroecosystems. The published book[3] is the source of much of the information herein.

The evolution of feral forms was thought to be predominantly due to gene flow from adjacent ruderal (human disturbed) or wild ecosystems (exo-ferality). Many had studied gene flow to the wild; the F_1 of crop X wild backcrossing to the wild species. To achieve exo-feral forms, the F_1 backcrosses to the crop, re-imbuing it with feral traits. Endo-ferality, where the crop back mutates to a feral form, may be more prevalent than had been thought.

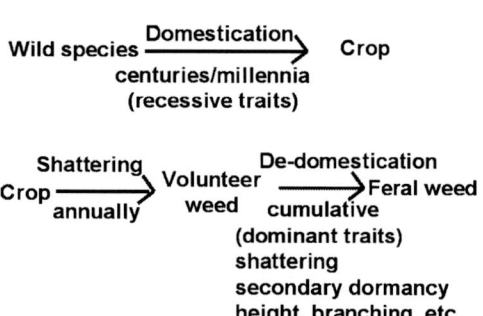

Fig. 1 The processes of domestication of crops (mainly selection for recessive traits from wild progenitors), and de-domestication to feral forms.

VOLUNTEERS: THE FIRST STEP TO FERALITY

The first step to ferality may be via "volunteer weeds" — offspring of crop seed that "shattered" (prematurely dropped seeds) before harvest in previous seasons (Fig. 1). If these volunteers can remain continuously in the same fields for extended periods, there are many chances for selection of back mutations of various feral traits, and then for them to recombine.

Not many genes need to mutate for a "volunteer" weed to become feral weed. Some of the initial volunteer weeds may have a higher frequency of the ability to shatter, filling the soil seed bank for the next growing season. There will then be selection for those individual mutants that have enhanced secondary dormancy, the ability to germinate non-uniformly during the following season and over a few years, filling the soil seed bank for many seasons to come. This allows incipiently feral forms to "bypass" early herbicide treatments and cultivations, and for some, to remain dormant under rotational crops, only to come up and interbreed with the crop itself.

Many crops have been "dwarfed" (a recessive trait) to enhance the ratio of seed to straw (harvest index). De-domesticating volunteer weeds that are taller and branch more will provide a competitive advantage on the volunteers. Other adaptive minor traits will soon appear as the feralized volunteers continue to evolve. Thus, it is important to assure that "volunteer" weeds be controlled each season to prevent such an evolutionary scenario. It is hard to control volunteer weeds in monoculture crops, forcing farmers to use rational rotations. Volunteer individuals mutating to feral weedy traits having a competitive selective advantage can cross with others bearing other feral traits, becoming more intractable than the volunteer weed parents.

There is less chance of ferality evolving when certified seed is cultivated. The provider of certified seed culls off-types in production fields — the back mutations are quickly removed from the population. Farmer-saved seed will propagate feral forms at the expense of the crop.

DOCUMENTED FERALITY

Many crops have feral weedy forms (Table 1), some clearly endo-feral, some exo-feral, some sequential, some either, in different places. Much is surmised, as it is rare that one can be fully certain of evolutionary pasts. Some documented cases are described below to demonstrate the variety of factors involved.

Sugar and table beets were domesticated from annual wild beets *Beta vulgaris* ssp. *maritima*, by selection for types with roots that thickened throughout a season, and did not move resources into a flowering (bolting) stem until vernalized (cold temperature treated). Non-bolting is recessive. The exo-feral evolution of weedy feral beets that bolt without vernalization is the recurring motif in an excellent

Table 1 Examples of crops that have de-domesticated to feral forms by back mutation (endo-ferality) or by crossing with progenitors or relatives (exo-ferality).

Crop	Latin binomial	Endo/exo-feral	Feral traits in weedy types	Reference[a]
Sugar beet	*Beta vulgaris*	Endo and exo[b]	Annual bolting	[4]
Oilseed rape	*Brassica napus*	Exo	Antifeedants, dormancy	[5]
Foxtail millet	*Setaria italica*	Exo	Shattering, tillering	[6]
Sorghum	*Sorghum bicolor*	Endo	Shattering	[8]
		Exo	Shattering/rhizomes	[8]
Wheat	*Triticum aestivum*	Endo	Shattering, colored grain, non-threshing	[11]
Rye	*Secale cereale*	?	Shattering, at times perennial, small seeds, late flowering	[12]
Radish	*Raphanus sativus*	Endo and exo[c]	Bolting, shattering, thick and segmented siliques, dormancy	[13]
Sunflower	*Helianthus annuus*	Mainly exo[d]	Multi-stemmed, small flower, shattering, tiny seeds	[14]
Oats	*Avena sativa*	Endo	Shattering	[15]
Rice	*Oryza sativa*	Endo and exo	Shattering, red grain	[16–21]

[a]References in [ ] refer to chapter numbers in Gressel.[3]
[b]In seed production areas exo- with progenitor, and in some fields, endo; a question mark denotes that there is no straightforward guess.
[c]In different locations.
[d]Endo-ferality cannot be excluded in Europe.

book on gene flow.[4] Sugar beets are cultivated for seed in southern France near sea beets, and genes flow. Endo-feral weedy forms also occur in Italy, where it is clearly a function of back mutation to the dominant bolting form (Chapter 4).[3]

Oilseed rape (*Brassica napus*) can cross with weedy *B. rapa* = *B. campestris* yielding exo-feral forms that contain glucosinolate and erucic acid, antifeedants bred out of oilseed rape. Secondary dormancy can set in owing to gaining a light requirement for germination (Chapter 5).[3] Most exo-feral hybrids and backcrosses are unfit, but rare backcrosses bearing a transgene encoding herbicide resistance can establish even in the absence of the herbicide selector.[5] Volunteer oilseed rape had not been allowed to remain in the field owing to crop rotation. This could change if there is a shift to monoculture after the disease, insect, and weed problems are solved transgenically. Then one might expect endo-feral, highly shattering forms to appear.

Wild *Setaria viridis* evolved in two directions, to weedy *S. viridis* (green foxtail) and to cropped *S. italica* (foxtail millet). At times the weed and crop have crossed, yielding a worse weed, giant green foxtail. It has also exo-ferally crossed with the related *S. adhaerens* to produce an allotetraploid weed, *S. faberii* (Chapter 6).[3]

Ornamental tree species often escape into the wild displacing native species (Chapter 2)[3] but one cannot determine what/if genetic feral changes have taken place. Feral olives, spread by birds eating cultivated olives, have displaced wild species in Australia and have a form that appears like wild oleaster. There are no basic genetic differences between cultivated olive and wild oleaster other than pruning. Olive groves left unpruned revert to oleaster, pruning transforms them back to olive (Chapter 15).[3]

Sorghum in the field can endo-ferally back mutate to a very weedy shattering "shattercane." At some point the crop or its progenitor crossed with *Sorghum propinquum* yielding *S. halepense* (Johnsongrass), one of the world's worst weeds because it propagates both by seed and by underground rhizomes forming perennial clumps (Chapter 8).[3]

Wheat (*Triticum aestivum*) is an allohexaploid derived from three *Aegilops* progenitor species. Hybrids between wheat and weedy and ruderal *Aegilops* species have been reported through the ages, including in herbaria (Chapter 3).[3] Most of the hybrids are not fertile, and even when fertile, there is no evidence of real feral forms persisting. A feral weedy form of hexaploid wheat was found in Tibet, with colored, shattering seeds that could not be threshed. No wild *Aegilops* species are known in Tibet, and it is surmised that three dominant back mutations occurred in wheat cultivated there for 2–4 millennia ago and abandoned (Chapter 11).[3]

The crop rye (*Secale cereale*) was introduced to North America, where it is now hardly grown. A feral form is a pernicious weed. In most places weedy rye has many crop traits and the feral forms are not fully wild. This could be due to endo-feral back mutations after abandonment or to exo-feral crosses, with its cointroduced perennial relative *S. strictum*, so which type of evolution occurred is not clear (Chapter 12).[3] The feral rye types are rapidly diverging. It is proposed that the lack of continual crossing with the crop owing to lack of cultivation is enhancing divergence.[6]

Raphanus raphanistrum, the progenitor of radish *R. sativus* is mainly a wild species, but weedy radish (*R. sativus*) can be quite a problem. Whether the weed evolved endo-ferally or with the help of its progenitor is an open question, with both answers feasible. It is possible but hard to experimentally achieve weedy feral populations by abandoning radishes, and it is easier to get them from crop x wild hybrids, but that does not answer the question (Chapter 13).[3]

Cultivated, feral weedy, and wild sunflowers are distinct types of *Helianthus annuus*, originating in North America, yet feral weedy forms occur throughout the world where sunflowers are cultivated. The endo- or exo-feral origin of the feral weedy form is unclear, as feral forms often contaminate crop seeds (Chapter 14).[3] Between 1% and 6% of genes in weedy sunflowers had significantly less variation suggesting involvement in weediness.[7]

Oats (*Avena sativa*) and weedy oats (*A. sterilis* and *A. fatua*) are interbreeding yet very different morphotypes that can and do transfer genes. "Fatuoids," a common feral type of oats that shatters with same mechanism as *A. fatua* are very common in oat fields and are considered to be endo-feral back mutations (Chapter 15).[3]

Feral, red, or weedy rice has become the major weed problem in rice cultivation wherever farmers have switched to direct seeding of rice from transplanting month-old seedlings in freshly cultivated paddies. Direct-seeded rice does not have a "head start" allowing it to compete out the feral form. Feral rice is exceedingly variable, even in the same field, suggesting multiple back-mutational evolutionary events occurring in parallel. The only common feature is shattering, while seed color, awn type, and length, and other features vary. This suggests an endo-feral origin, especially where the progenitors do not exist. Still, in Asia, where the perennial progenitor *O. rufipogon* and the annual progenitor *O. nivara* exist in ruderal areas near paddies, hybrid swarms with cultivated rice are often visible at the border. In feral weedy rice there may have exo-ferally introgressed genes from the progenitors (Chapters 16–21).[3]

WILL TRANSGENICS HASTEN THE EVOLUTION OF FERAL FORMS?

Many claim that transgenics are no different from traditionally bred crops, which is not be the case for two interrelated reasons:

1. Many traits being transformed do not exist in the crop genome. Such traits might provide a selective

advantage over previous volunteers of the crop or against competing weeds;
2. The use of transgenics may change agronomic practices in ways that will enhance the rate of de-domestication.

Consider the scenario for oilseed rape described in the Documented Ferality section. Many growers may prefer monoculture of the valuable crop, and the volunteers could remain. The implications of whether less rotation will hasten ferality must be assessed. This can be only done after we have more basic information on the evolution of feral forms from crops.

TRANSGENICALLY PREVENTING/MITIGATING FERALITY

It is clear that ferality is a possibility and there are cases where it may be exacerbated by introduced transgenic traits. This is especially the case in rice, where transgenic herbicide resistance would allow the control of feral rice, until the herbicide-resistant rice endo-ferally becomes a weed, or exo-ferally introgresses the gene. Most methodologies proposed for gene containment are leaky, and once a gene introgresses, spread can occur. Gene containment cannot prevent endo-feral evolution of weediness. Transgenic mitigation, where mitigating transgenes that render weedy forms unfit to compete are tandemly linked with the gene of choice, can prevent the establishment of endo- and exo-feral forms. This was demonstrated in oilseed rape where dwarfing was done using a mitigator gene linked to herbicide resistance. This form could neither compete with the wild type, and its hybrid with *B. rapa* could not compete with the weed (Chapter 22).[3] A dominant transgene preventing bolting (e.g., an antisensed kaurene oxidase) would be an appropriate mitigator to be tandemly linked with genes of choice in sugar beets and other root crops, as both endo- and exo-feral bolting would be suppressed. Antishattering mitigator genes would have the same effect in rice and other grains.[1]

CONCLUSIONS

The case studies demonstrate the nuances and differences in the domestication of crops and their de-domestication to feral forms. The possibility of evolution of ferality in maize is extremely remote because of the complexity of the domestication process (Chapter 10),[3] yet feral/weedy wheat evolved in a crop with an even more complex evolutionary background. One cultivated species needed no mutations to become feral; other species require mutations, and maize is so highly domesticated that it may never become feral as it has yet to survive more than a generation or two. Other crops can hybridize with their weedy progenitors giving rise to more pernicious weedy forms, especially on ruderal roadsides, where seeds from fields and seeds from transported crops can mix, and are vulnerable to evolution of ferality.[8] One cannot delineate a "unified theory of ferality," just unified suggestions to appreciate nature, understand its intricacies, use history to predict the future, and use biotechnology to prevent/mitigate the evolution of ferality.

ACKNOWLEDGMENTS

The author learned much from *Crop Ferality and Volunteerism*[3] and apologizes to them because space limitations precluded referencing each entry. The OECD and the Rockefeller Foundation supported the workshop culminating in *Crop Ferality and Volunteerism*.[3]

REFERENCES

1. Gressel, J. *Genetic Glass Ceilings—Transgenics for Crop Biodiversity*; Johns Hopkins University Press: Baltimore, 2008.
2. Trut, L.N. Early canid domestication; the fox farm experiment. Am. Sci. **1999**, *87*, 160–169.
3. Gressel, J., Ed. *Crop Ferality and Volunteerism*; CRC Press: Boca Raton, FL, 2005.
4. Ellstrand, N.C. *Dangerous Liaisons—When Cultivated Plants Mate with Their Wild Relatives*; Johns Hopkins University Press: Baltimore, 2003.
5. Warwick, S.I.; Legere, A.; Simard, M.J.; James, T. Do escaped transgenes persist in nature? The case of an herbicide resistance transgene in a weedy *Brassica rapa* population. Mol. Ecol. **2008**, *17*, 1387–1395.
6. Burger, J.C.; Holt, J.M.; Ellstrand, N.C. Rapid phenotypic divergence of feral rye from domesticated cereal rye. Weed Sci. **2007**, *55*, 204–211.
7. Kane, N.C.; Rieseberg, L.H. Genetics and evolution of weedy *Helianthus annuus* populations: adaptation of an agricultural weed. Mol. Ecol. **2008**, *17*, 384–394.
8. Garnier, A.; Deville, A.; Lecomte, J. Stochastic modelling of feral plant populations with seed immigration and road verge management. Ecol. Model. **2006**, *197*, 373–382.

Cyclodextrins

Estrella Nuñez-Delicado
Department of Science and Food Technology, San Antonio Catholic University of Murcia, Guadalupe, Spain

José Antonio Gabaldón-Hernández
National Technology Center for Preservatives and Food, Molina de Segura, Spain

Abstract

Cyclodextrins are natural cyclic oligosaccharides that consist of (α-1,4)-linked α-D-glucopyranose units, shaped like truncated cones with a lipophilic central cavity and a hydrophilic outer surface, which can dissolve in water. The most important property of cyclodextrins is their ability to include into their hydrophobic cavity a wide range of solid, liquid, and gaseous compounds, such as aldehydes, ketones, alcohols, organic acids, fatty acids, aromatics, gases, halogens, oxyacids, and amines, by a molecular complexation. This inclusion process exerts a profound effect on the physicochemical properties of guest molecules. Because of the molecular complexation and their negligible cytotoxic effects, cyclodextrins are widely used in the biotechnology field in many industrial processes for the development of new food products, technologies, and analytical methods. Cyclodextrins may be used to enhance a substrate's aqueous solubility and reduce toxicity, thus increasing the conversion of lipophilic substrates in industrial processes and in diagnostic reagents. Therefore the actual and potential uses of cyclodextrins in biotechnological applications for food and agriculture are extensive.

INTRODUCTION

Cyclodextrins are natural macrocyclic non-reducing oligosaccharides, discovered about 100 years ago, produced as a result of intramolecular transglycosylation reaction from degradation of starch by the cyclodextrin glucanotransferase (CGTase) enzyme. Cyclodextrins consist of (α-1,4)-linked α-D-glucopyranose units shaped like truncated cones with a lipophilic central cavity and a hydrophilic outer surface that is water-soluble. The apolar internal cavity provides a hydrophobic matrix that enables cyclodextrins to form inclusion complexes with a wide variety of hydrophobic guest molecules. The most common natural cyclodextrins consist of six, seven, or eight glucopyranose units, identified as α-, β-, and γ-cyclodextrins, respectively.

All toxicity studies have demonstrated that orally administered cyclodextrins are practically non-toxic owing to a lack of absorption in the gastrointestinal tract. Moreover, inclusion in cyclodextrins exerts a profound effect on the physicochemical properties of guest molecules as they are temporarily locked within the host cavity, giving rise to beneficial modifications of guest molecules. These chemical interactions make cyclodextrins suitable for applications in the fields of analytical chemistry, pharmaceutical/industrial catalysis, environment protection, fermentations, agriculture, and food production.

USE OF CYCLODEXTRINS

As a consequence of the conformation of the glucopyranose units forming α-, β- and γ-cyclodextrins (Fig. 1), all secondary hydroxyl groups (C_2 and C_3) are situated on one of the two edges of the ring, whereas the primary ones (C_6) are placed on the other edge. In reality, the ring is a conical cylinder, often characterized as a doughnut or wreath-shaped truncated cone (Fig. 1). The internal cavity is lined by hydrogen atoms and glycosidic oxygen bridges. The non-bonding electron pairs of the glycosidic–oxygen bridges are directed toward the inside of the cavity that produces a high electron density with Lewis base characteristics.

The C_2-OH group of one glucopyranose unit can form a hydrogen bond with a C_3-OH group of an adjacent glucopyranose unit. In the case of the β-cyclodextrin molecule, a complete secondary belt is formed by these H bonds, therefore, the β-cyclodextrin is a rather rigid structure with the lowest water solubility of all cyclodextrins (18 mg/cm^3 in water at 25°C). In the case of α-cyclodextrin, the H-bond belt is incomplete because one glucopyranose unit is in a distorted position. So instead of six possible H-bonds, only four can be established simultaneously making α-cyclodextrin more soluble than β-cyclodextrin (140 mg/cm^3 in water, at 25°C). The molecule of γ-cyclodextrin is non-coplanar and has a

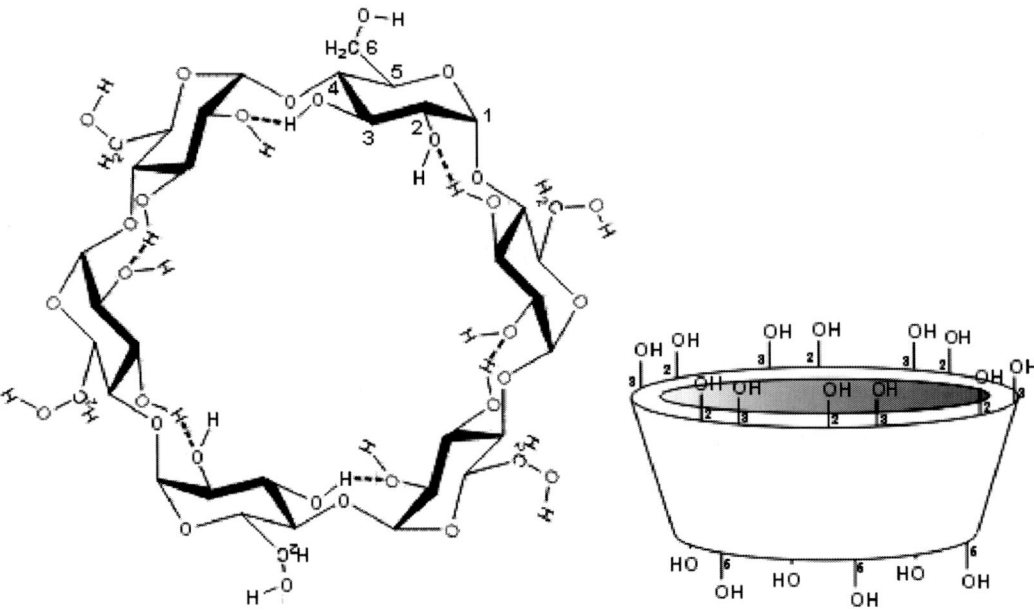

Fig. 1 Chemical structure and shape of α-cyclodextrin.

more flexible structure; therefore, it is the most water-soluble of the three cyclodextrins (220 mg/cm^3 in water at 25°C).[1]

The reactivities of C_2-OH, C_3-OH, and C_6-OH depends on the reaction conditions (pH, temperature, reagents), and can be modified by substituting a hydrogen atom or hydroxyl group by aminations, esterifications, or etherifications; all of them prepared by chemical or enzymatic reactions. The aim of such derivatizations may be: 1) to improve the solubility of cyclodextrins and its complexes; 2) to improve the fitting and association between cyclodextrins and its guest; 3) to attach specific catalytic groups to the binding site; or 4) to form insoluble, immobilized cyclodextrin-containing structures. Only a few of the thousands of cyclodextrins derivatives described can be taken into account for industrial-scale synthesis and utilization.[2]

Cyclodextrins can specifically link covalently or noncovalently to other cyclodextrins. Because of this capability, cyclodextrins can be used as building blocks for construction of supramolecular complexes that can be employed for the separation of complex mixtures of molecules and enantiomers.[3]

Cyclodextrin molecules are relatively large with a substantial number of hydrogen donors and acceptors, and consequently are poorly absorbed through biological membranes. Toxicity studies have demonstrated that orally administered cyclodextrins are non-toxic owing to the lack of absorption from the gastrointestinal tract.[4] The natural α- and β-cyclodextrins, unlike γ-cyclodexrin, cannot be hydrolyzed by human salivary and pancreatic amylases, but all the three are subjected to fermentation by intestinal microbiota.[5]

The most notable feature of cyclodextrins is their ability to form inclusion complexes with a very wide range of solid, liquid, and gaseous compounds, including aldehydes, ketones, alcohols, organic acids, fatty acids, aromatics, gases, halogens, oxyacids, and amines by molecular complexation. The lipophilic cavity of cyclodextrin molecules provides a microenvironment into which appropriately sized non-polar moieties can enter to form inclusion complexes. No covalent bonds are broken or formed during formation of the inclusion complexes.[6] The binding of guest molecules within the host cyclodextrin is not permanent but rather is in a dynamic equilibrium governed by a complex stability or equilibrium constant, K_c. The simplest and most frequent case is the host:guest ratio 1:1; however, 2:1, 1:2, 2:2, or even more complicated associations and higher-order equilibria can simultaneously exist.

Complexes can be formed by a variety of techniques that depend on the properties of the active material, the equilibrium kinetics, the formulation ingredients and processes, and the final dosage desired. Among the methods used are simple dry mixing, mixing in solutions and suspensions followed by a suitable separation, the preparation of pastes, and several thermo-mechanical techniques.

Inclusion in cyclodextrins exerts a profound effect on the physicochemical properties of guest molecules as they are temporarily locked within the host cavity, giving rise to beneficial modification of guest molecules. These properties are:

- Solubility enhancement of highly insoluble guest.
- Modified spectral properties of the guest.
- Modified reactivity of included molecules. In most cases, the guest is stabilized against degradative effects,

but in many cases cyclodextrin behaves as an artificial enzyme that accelerates reactions.
- Strong decrease of diffusion and volatility of the included guest.
- Modified chromatographic mobility of the guest.
- Taste modification by masking off flavors and unpleasant odors, and controlled release of drugs and flavors.

Measurements of equilibrium constants, K_c, of the complexes are important since these constants represent an index of changes in physicochemical properties of a compound upon inclusion. In theory, any methodology that can be used to observe these changes in guest physicochemical properties (solubility, chemical reactivity, UV/Vis absorbance, fluorescence, drug retention time in liquid chromatography, pKa values, potentiometric measurements, chemical stability, nuclear magnetic resonance, and effects on drug permeability through artificial membranes) may be used to determine the stoichiometry of the complexes formed and the numerical value of their stability constants.[7]

As a result of molecular complexation phenomenon, cyclodextrins are widely used in many industrial products, technologies, and analytical methods. The negligible cytotoxic effects of cyclodextrins and their beneficial properties are important attributes that make them suitable for applications in analytical chemistry, the pharmaceutical and catalysis fields, environmental protection, fermentations, agriculture, and foods.[8]

One of the classical applications of cyclodextrins is in the field of pharmaceuticals. The pharmaceutical industry is always in need of new formulating aids to enhance the solubility and bioavailability of active ingredients while diminishing any adverse effects. Moreover, cyclodextrin acts as a controlled delivery system, increasing the bioavailability of many drugs; however, less than 10% of all produced cyclodextrins are consumed by the pharmaceutical industry. The largest cyclodextrin consumers are the food and cosmetic industries; however, in this entry, we centered on the biotechnological applications of cyclodextrin in agriculture and food.

Cyclodextrins form complexes with a wide variety of agricultural chemicals including herbicides, insecticides, fungicides, repellents, pheromones, and growth regulators with similar results as in the case of drugs. The dose of environmentally polluting synthetic agrochemicals could probably be significantly reduced when complexed by cyclodextrins, and/or their effectiveness could be enhanced.[1] Cyclodextrin complexation also results in the increase of water solubility of benzimidazole-type fungicides making them more available in soil. In addition to its ability to increase the solubility of the hydrocarbon for biodegradation and bioremediation, cyclodextrins also decrease the toxicity resulting in an increase in microbial and plant growth.[2] Cyclodextrins can also be applied to delay germination of seeds because some of the amylases that degrade the starch of the seeds are inhibited. Initially the plant grows more slowly, but later on this is largely compensated by improved plant growth yielding a 20–45% larger harvest.[3]

Cyclodextrins can also play an important role in environmental science in terms of solubilization of organic contaminants, and the enrichment and removal of organic pollutants and heavy metals from soil, water, and the atmosphere.[9] Cyclodextrins are also applied in water treatment to increase the stabilizing action, encapsulation, and absorption of contaminants. Using cyclodextrins, highly toxic substances can be removed from industrial effluent by the formation of inclusion complexes. Environmentally unacceptable aromatic compounds, such as phenol, p-chlorophenol, and benzene, are considerably reduced in wastewaters after treatment with β-cyclodextrins. Cyclodextrins are used to scrub gaseous effluent from organic chemical industries. Low cost, biocompatibility, and effective degradation make β-cyclodextrins a useful tool for bioremediation processes.[10]

Cyclodextrins have also found numerous applications in the food industry. β-cyclodextrin has been on the GRAS list from U.S. FDA since 1998, as a flavor carrier and protectant at a level of 2% in numerous food products.[11] Cyclodextrins can be used in foods mainly as carriers for molecular encapsulation of flavors and other sensitive ingredients, such us fats, colors, and bioactive compounds. Its application offers the following advantages: 1) protection of active ingredients against oxidation; 2) elimination or reduction of undesired tastes, odors, and microbial contaminations; 3) technological advantages that include stability, standardizable compositions, simple dosing, and handling of dry powders; 4) and improvement of the shelf life of food products.

Cyclodextrins act as complexating agents to protect flavor compounds in foods exposed to processing methods such us freezing, thawing, and microwaving. Complexation with sweetening agents such as aspartame stabilizes and improves taste. Moreover, cyclodextrin itself is a promising new sweetener. The bitter taste of grape and citrus juices caused by the presence of limonoids (limonin) and flavonoids (naringin) is a major problem in the food industry. Cross-linked cyclodextrin polymers are useful to remove these bitter components by formation of stable inclusion complexes. Cyclodextrins are also used to control the bitterness of tannins, plant and fungal extracts, milk casein hydrolysates, and overcooked teas and coffees.[11]

Flavonoids and terpenoids are beneficial to human health because of their antioxidative and antimicrobial properties, but these compounds cannot be used in foodstuffs owing to their very low solubility in water and bitter taste. Combination of flavonoids and terpenoids with cyclodextrins has been recently studied by many research groups[12] in order to solve the solubility and flavor problems. The most prevalent use of cyclodextrins in process aids is the removal of cholesterol from animal products such as eggs and dairy products. Cyclodextrin-treated materials show 80% removal of cholesterol. There are many low-cholesterol milk products produced by this technology, including butter, cheese, and cream.[11] Free fatty acids can also be removed from fats using cyclodextrins, thus improving the frying properties of the fat.[8] Fruits and vegetable juices have also treated with cyclodextrins to remove phenolic compounds in order to lessen or control enzymatic browning.[13,14]

An interesting application of cyclodextrins in foods is production of cyclodextrin-containing food packaging materials. Cyclodextrin or antimicrobial agents complexed with cyclodextrin have been incorporated into food packaging films to reduce the loss of the aroma compounds and improve antimicrobial effectiveness during storage.[11]

In the biotechnology field, cyclodextrins may be used to avoid difficulties that arise from enzyme-catalyzed transformations of substrates in aqueous media. Cyclodextrins and their derivatives enhance the solubility of substrates in water and reduce their toxicity; cyclodextrins increase the conversion of lipophilic substrates in industrial processes and in diagnostic reagents. The yield of product-inhibited fermentations can be improved, organic toxic compounds can be better tolerated by microbial cells at higher concentrations, and compounds in small amounts can be isolated simply and economically from complicated mixtures.[1]

CONCLUSION

In conclusion, the actual and potential uses of cyclodextrins in biotechnological applications for food and agriculture are extensive. While a series of cyclodextrin-containing products or cyclodextrin-using technologies is widely known in the food industry, in future, significant new applications are expected from the use of cyclodextrins in agriculture, biotechnology, and environmental protection.

ACKNOWLEDGMENTS

This work was partially supported by the Fundación Séneca (03025/PPC/05) and Ministerio de Educación y Ciencia (AGL2006-08702/ALI).

REFERENCES

1. Szejtli, J. Past, present and future of cyclodextrin research. Pure Appl. Chem. **2004**, *76*, 1825–1845.
2. Martin del Valle, E.M. Cyclodextrins and their uses: a review. Process Biochem. **2004**, *39*, 1033–1046.
3. Szejtli, J. Introduction and general overview of cyclodextrin chemistry. Chem. Rev. **1998**, *98*, 1743–1753.
4. Irie, T.; Uekama, K. Pharmaceutical applications of cyclodextrins. III. Toxicological issues and safety evaluation. J. Pharm. Sci. **1997**, *86*, 147–162.
5. Loftsson, T.; Brewster, E.; Másson, M. Role of cyclodextrins in improving oral drug delivery. Am. J. Drug Deliv. **2004**, *2*, 1–15.
6. Loftsson, T.; Brewster, E. Pharmaceutical applications of cyclodextrins: 1. Drug solubilisation and stabilization. J. Pharm. Sci. **1996**, *85*, 1017–1025.
7. Hirose, K. A practical guide for the determination of binding constants. J. Incl. Phenom. Macroc. Chem. **2001**, *39*, 193–209.
8. Singh, M.; Sharma, R.; Banerjee, U.C. Biotechnological applications of cyclodextrins. Biotechnol. Adv. **2002**, *20*, 341–359.
9. Hedges, R.A. Industrial applications of cyclodextrins. Chem. Rev. **1998**, *98*, 2035–2044.
10. Bardi, L.; Mattei, A.; Steffan, S.; Marzona, M. Hydrocarbon degradation by a soil microbial population with beta-cyclodextrin as surfactant to enhance bioavailability. Enzyme Microbiol. Technol. **2000**, *27*, 709–713.
11. Szente, L.; Szejtli, J. Cyclodextrins as food ingredients. Trends Food Sci. Technol. **2004**, *15*, 137–142.
12. Lucas-Abellán, C.; Fortea, M.I.; López-Nicolás, J.M.; Núñez-Delicado, E. Cyclodextrins as resveratrol-carriers system. Food Chem. **2007**, *104*, 39–44.
13. Hicks, K.B.; Haines, R.M.; Tong, C.B.S.; Sapers, G.M.; El-Atawy, Y.; Irwin, P.L.; Seib, P.A. Inhibition of enzymatic browning in fresh fruit and vegetable juices by soluble and insoluble forms of β-Cyclodextrin alone or in combination with phosphates. J. Agric. Food Chem., **1996**, *44*, 2591–2594.
14. López-Nicolás, J.M.; Núñez-Delicado, E.; Sánchez-Ferrer, A.; García-Carmona, F. Kinetic model of apple juice enzymatic browning in the presence of cyclodextrins: the use of maltosil-β-cyclodextrin as secondary antioxidant. Food Chem. **2007**, *101*, 1164–1171.

Dairy Lactococci

Baltasar Mayo
Department of Microbiology and Biochemistry, Dairy Institute of Asturias (CSIC), Villaviciosa, Spain

Abstract
Lactococci are commonly dominant in natural dairy fermentations and in majority components of starter cultures used in large-scale cheese making. Strains of *Lactococcus lactis*, of both *lactis* and *cremoris* subspecies, and the diacetyl-forming biovariety diacetylactis are of industrial significance. The growth of lactococci in milk induces a rapid fall in the pH, prevents the growth of undesirable microorganisms, and provides optimal conditions for ripening. Lactococci are also responsible for the formation of cheese texture and flavor through their proteolytic and amino acid conversion pathways. Physiological and genetic data accumulated over the past 25 years allowed lactococci to be used for the expression of heterologous proteins, the synthesis of food-grade additives and nutraceuticals, and the delivery of vaccines.

INTRODUCTION

Lactococci are adenine- and thiamine-rich, Gram-positive, lactic acid bacteria cocci (Fig. 1) commonly dominant in natural habitats such as spontaneous milk fermentations, plant materials, cattle mucosa, and the intestinal tract of fish. First included within *Streptococcus* as members of Lancefield serological group N, the lactococci were later assigned to an independent genus on the basis of their nucleic acid hybridization characteristics, enzyme immunorelationships, and lipid and cell wall composition. Now, the genus *Lactococcus* encompasses five species: 1) *Lactococcus garvieae*; 2) *L. lactis*; 3) *L. piscium*; 4) *L. plantarum*; and 5) *L. raffinolactis*. *L. lactis* includes three subspecies (*cremoris*, *lactis*, and *hordniae*) and a diacetyl-forming biovariety (*L. lactis* subsp. *lactis* biovar diacetylactis).[1] All are non-pathogenic, except for *L. garvieae*, which causes lactococcosis in fish and subclinical mastitis in cattle.

In the modern industry, the growth of adventitious lactococci in milk has been replaced by the deliberate addition of carefully selected strains (i.e., starter cultures). Fermented products manufactured with *L. lactis* starters include many varieties of soft and hard cheeses, fermented butter, buttermilk, and Scandinavian ropy milks (e.g., viili and langfil)—products that require an annual supply of some 100 million tons of milk.[2] The growth of *L. lactis* in milk is associated with the rapid production of lactic acid, which provides flavor, assists in curd formation, prevents the growth of pathogenic and spoilage bacteria, and creates the optimal biochemical conditions for ripening. Lactococci also contribute to the final texture (moisture, softness) and further flavor of dairy products through their proteolytic and amino acid conversion pathways.[3] All these functions determine quality, safety, and shelf life of dairy-fermented products.

L. lactis is the only technologically important species, and both the *lactis* and the *cremoris* subspecies are widely used as starter cultures. *L. lactis* subsp. *lactis* is distinguished from subsp. *cremoris* according to five phenotypic criteria: 1) the ability to grow at 40°C; 2) the ability to grow in 4% NaCl; 3) the ability to grow at pH 9.2; 4) the ability to ferment maltose; and 5) the ability to deaminate arginine. *L. lactis* subsp. *lactis* biovar diacetylactis is distinguished by its capability to assimilate citrate, from which it produces diacetyl and acetoine. Starter cultures are probably derived from a small number of cheese-making strains isolated from successful fermentations and maintained by back-sloping techniques over the years. Currently, there is growing interest in dairy and non-dairy wild isolates to complement industrial strains accompanied with a continued search for strains harboring unique flavor-forming capabilities or production of novel broad-range antimicrobial compounds.

This entry provides a general overview of recently reported physiological and genetic data of relevance in the industrial exploitation of lactococci and discusses the use of *L. lactis* in new biotechnological settings made possible by the development of new genetic tools and the amenability of these bacteria to genetic engineering.

FUNCTIONALITY OF DAIRY LACTOCOCCI

The rapid production of lactic acid demands a rapid growth of lactococci in milk up to high cell densities. To reach high numbers, lactococci use lactose and milk proteins as

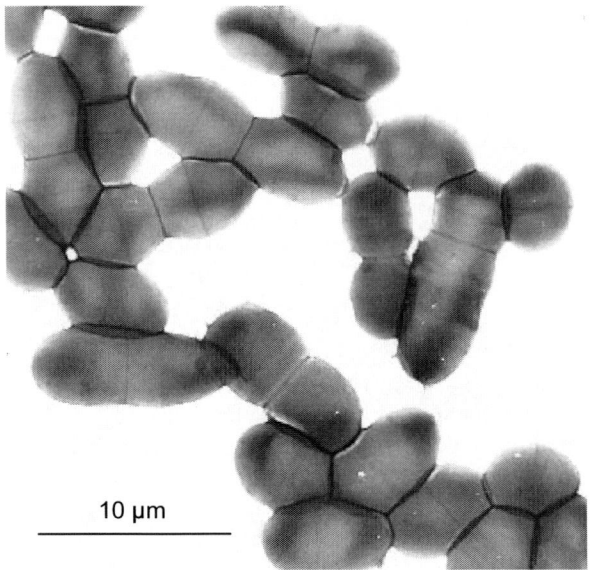

Fig. 1 *Lactococcus lactis* cells grown in M17 medium under the scanning electron microscope.

a source of energy and nutrients. These abilities were soon associated with plasmids, whose loss correlated with poor growth in milk. Most strains of *L. lactis* harbor between 4 and 7 plasmids ranging in size from 2 to 70–80 kbp (see Fig. 2). Though some plasmids are cryptic, phenotypes of paramount industrial importance other than lactose assimilation and proteinase activity have frequently been reported associated with plasmids, for example, phage resistance, bacteriocin production, and resistance to these molecules; exopolysaccharide production; and citrate metabolism in the biovar diacetylactis. The location of key functions

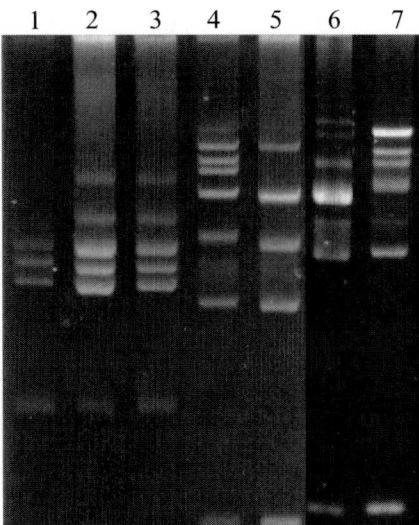

Fig. 2 Plasmid profiles of wild *Lactococcus lactis* strains isolated from artisan starter-free cheeses. Related and unrelated profiles can be observed. Up to seven different plasmid bands are present in sample 4. **Source:** From FEMS Microbiol. Rev.[4]

in plasmids is detrimental since these elements are nonessential and can be lost; however, most plasmids can be mobilized, providing possibilities for a controlled dissemination of desirable properties.

Lactose Metabolism

Starters assimilate lactose efficiently by an ATP-driven translocase system (PTS), in which lactose resulted phosphorylated during transport and then broken down by a β-phospho-galactosidase (β-PGal) enzyme.[5] Both the lactose-specific PTS components and β-PGal genes are located in plasmids.[4] The lactose operon of *L. lactis* has been found homologous to that of *Staphylococcus aureus*, suggesting the recent spread of lactose metabolism among Gram-positives. By contrast, *L. lactis* strains of plant origin do not normally have β-PGal activity; instead, they make use of a specific permease, and the unmodified intracellular lactose is hydrolyzed by a β-galactosidase (β-Gal).

Proteolytic System

Amino acids are either essential or stimulatory in *L. lactis* growth; however, free amino acids and short assimilable peptides are not found in sufficient quantity in milk for *L. lactis* to attain the high cell densities. The configuration and abundance of the caseins make them the best choice for supplying the required amino acids. For the degradation of caseins, lactococci possess a complete proteolytic system consisting of proteinases, peptidases, and amino acid and peptide uptake systems.[3] An extracellular serine-proteinase, PrtP, encoded by a plasmid-located gene (*prtP*) is essential in this process. PrtP is exported through the cell membrane via an N-terminus signal peptide, bound to the cell wall by its C-terminal region, and activated by autolytic cleavage of a proregion. Its activation also requires the action of a lipoprotein chaperone (PrtM) encoded next to PrtP and transcribed in the opposite orientation. PrtP produces more than 100 peptides from caseins, some of which are taken up by the oligopeptide uptake system (OppA). Inside the cells, the peptides are further degraded by a pool of around 20 peptidases.

The proteolytic system is also involved in the production of amino acid-derived aroma compounds.[3] The process starts by the transaminases, which transfer of the amino groups to α-ketoglutarate converting amino acids into their corresponding α-keto acids. α-ketoglutarate is a rate-limiting molecule, therefore, glutamate dehydrogenase is a key enzyme in this process.

Citrate Metabolism

Diacetyl and acetoine (2,3-butanediol) are important flavor compounds in the bouquet of fermented butter, buttermilk, and soft cheeses.[3] The production of these compounds is significant only in the presence of citrate and excess of pyruvate. In co-metabolism with lactose, citrate is converted by citrate lyase to oxaloacetate, which is further transformed to pyruvate and CO_2 by oxaloacetate decarboxylase. Citrate in milk (~1.5 mg/mL) is transported by a citrate permease encoded on a 7- to 9-kbp plasmid present in only *L. lactis* subsp. *lactis* biovar diacetylactis strains.

Phage and Phage Resistance

The susceptibility of starter cultures to bacteriophages remains the biggest problem in the dairy industry; the lysis of starters retards or halts dairy fermentations, occasionally causing severe economic losses.[6] In addition, lactococcal strains harbor temperate phages that, under certain circumstances, can undergo a lytic cycle. Four categories of plasmid-encoded natural resistance mechanisms have been identified in *L. lactis*: 1) interference with phage adsorption; 2) interference with phage DNA injection; 3) DNA restriction/modification (R/M); and 4) abortive infection. Strain rotation and the spread of resistance by natural transfer (conjugation) have worked well in minimizing phage attack.

Nisin and Other Bacteriocins

Bacteriocins are a heterogeneous group of antibacterial proteins that vary in their mode of action, molecular weight, genetic organization, and biochemical properties.[7] Strains of *L. lactis* are known to produce different bacteriocins; the most well known is nisin (an approved food preservative [E-234]). Nisin is a 34 amino acid-long hydrophobic peptide containing five characteristic β-methyl-lanthionine rings (lantibiotic) formed by posttranslational modifications. A cluster of 11 genes arranged in three operons and encoded in a 70-kbp conjugative transposon is involved in nisin biosynthesis, resistance, and regulation. Other bacteriocins (lactococcin A, B, M, 972 [Fig. 3], lacticin 3147) with antilisterial and/or anticlostridial properties are being investigated as bioprotective agents.

Extracellular Polysaccharides

Strains of *L. lactis* strains that produce extracellular polysaccharides (EPS) are naturally present in Scandinavian ropy milks. EPS are of technological interest for the in situ improvement of textural characteristics of dairy products, especially those of low-fat yogurts and cheeses. Potentially they could replace EPSs produced by non-food-grade bacteria. Plasmid and chromosomal EPS gene clusters have been characterized in different strains.

GENE CLONING AND EXPRESSION

Cryptic lactococcal plasmids have provided the platform for the development of sophisticated genetic tools (e.g., vectors

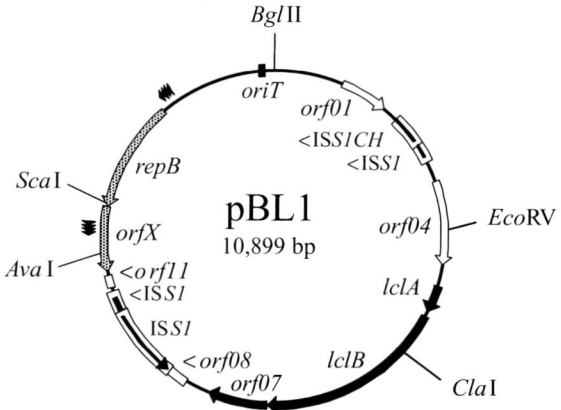

Fig. 3 Physical and genetic map of the bacteriocinogenic plasmid pBL1 from *Lactococcus lactis* IPLA972 encoding lactococcin 972. Position of unique restriction enzyme sites are indicated. Arrows and boxes show length and direction of open reading frames (ORFs); near by inside the plasmid are their names or numbers. Genes implicated on bacteriocin production and immunity are indicated by black arrows. Genes involved in replication are marked by doted arrows. Other complete and incomplete ORFs are denoted by empty arrows or boxes, respectively. Intact and remnant copies of IS elements are indicated by boxes with a mark inside. Direct repeats in front of *repB*, repeats in the coding region of *orfX*, and the position of the functional origin of transference (*oriT*) are also depicted.

for cloning [including food-grade vectors], expression, and integration).[4] Of note is the NICE (NIsin-Controlled gene Expression) system, which has been in use for more than 10 years. Genetic tools have allowed the engineering and rerouting of *L. lactis* metabolism to obtain the efficient conversion of sugars into diacetyl.[8] The formation of alanine as the only fermentation end product has also been achieved, and *L. lactis* has even been converted into a mannitol producer. More recently, plant-derived odorous compounds such as octyl acetate and linalool have been produced by recombinant *L. lactis*. Owing to its "Generally Regarded As Safe" (GRAS) status, *L. lactis* is receiving increased attention for the expression of therapeutic proteins, such as human interleukin-10 and interferon-gamma,[9] and for the presentation of antigens for mucosal immunization.

GENOMICS AND PROTEOMICS

Genome sequencing, genomics, and proteomics opened up a new era of research in the physiology and genetics of microorganisms. All the components required for aerobic respiration were observed in the genome of the first *L. lactis* strain to be fully sequenced (*L. lactis* subsp. *lactis* bivoar diacetylactis IL1403). Thus, *L. lactis* can undertake oxidative phosphorylation if an exogenous protoporphyrinogen is provided, which has been exploited in a new industrial starter culture-producing process.[10] The adaptation of *L. lactis* to the nutrient-rich environment of milk occurred by gene decay and the acquisition of key functions by horizontal transfer.[10] Decay is notable in metabolic genes of carbohydrates and amino acids. Conversely, acquisition of a copy of *metC*, endowing lactococci with the ability to synthesize methionine (which is rare in milk), was crucial.

CONCLUSION

The adventitious lactococci present in milk have been used unknowingly through the ages for manufacturing a series of fermented dairy commodities still much appreciated for their desirable sensory and nutritive properties. Nowadays, large-scale industrial facilities make use of carefully selected *L. lactis* strains to control the fermentations, thus ensuring the quality and safety of the fermented products. Robust, engineered strains appear to have a future in the manufacture of these traditional foods but can also serve as cellular factories for the production of flavors and ingredients, and as probiotics for the presentation of antigens and the delivery of therapeutic agents to the human gut mucosa.

REFERENCES

1. http://www.bacterio.cict.fr/l/lactococcus.html (accessed June 2008).
2. Cogan, T.M.; Accolas, J.-P. *Dairy Starter Cultures*; John Wiley and Son: Chichester, UK, 1995.
3. Smit, G.; Smit, B.A.; Engels, W.J.M. Flavour formation by lactic acid bacteria and biochemical flavours profiling of cheese products. FEMS Microbiol. Rev. **2005**, *29*, 591–610.
4. Mills, S.; McAuliffe, O.E.; Coffey, A.; Fitzgerald, G.F.; Ross, R.P. Plasmids of lactococci. Genetic accessories or genetic necessities? FEMS Microbiol. Rev. **2006**, *30*, 243–273.
5. Kowalczyk, M.; Bardowski, J. Regulation of sugar catabolism in *Lactococcus lactis*. Crit. Rev. Microbiol. **2007**, *33*, 1–13.
6. Coffey, A.; Ross, R.P. Bacteriophage-resistance systems in dairy starter trains: molecular analysis to application. Antonie van Leeuwenhoek **2002**, *82*, 303–321.
7. de Vuyst, L.; Leroy, F. Bacteriocins from lactic acid bacteria: production, purification, and food applications. J. Mol. Microbiol. Biotechnol. **2007**, *13*, 194–199.
8. de Vos, W.M.; Hugenholtz, J. Engineering metabolic highways in lactococci and other lactic acid bacteria. Trends Biotechnol. **2004**, *22*, 72–79.
9. Steidler, L.; Rottiers, P. Therapeutic drug delivery by genetically modified *Lactococcus lactis*. Ann. N. Y. Acad. Sci. **2007**, *1072*, 176–186.
10. Kok, J.; Buist, G.; Zomer, A.l.; van Hijum, S.A.; Kuipers, O.P. Comparative and functional genomics of lactococci. FEMS Microbiol. Rev. **2005**, *29*, 411–433.

Dairy: Fermented Products

Elaine P. Black
Department of Food and Nutritional Sciences, University College Cork, Cork, Ireland

Abstract

The preservation of milk by fermentation has been practiced for thousands of years, creating a myriad of products such as cheeses, cultured butter, yoghurts, and fermented drinks. The elevation of traditional "farmhouse" manufacture of such foods to large-scale industrial production has made use of many tools of modern biotechnology. Defined starter cultures create consistency and convenience in modern dairy factories previously unknown with the use of traditional techniques. Recombinant chymosin is used widely in the production of cheese; this allows cheese-makers to keep up with the demand for their product without having to rely on a limited supply of calf rennet. Along with the benefits of the nutritional aspects of fermented dairy foods, it is now well documented that the by-products of fermentation of lactic acid bacteria (LAB) or starter cultures can lead to increased microbial safety by the production of bacteriocins or other antagonistic metabolites. More recently, fermented dairy products have become an important delivery vehicle for probiotic LAB or bioactive peptides that have beneficial effects on human health and well-being.

INTRODUCTION

The fermentation of milk was once, and remains in many parts of the world, a purely artisanal way of preserving food. Today the manufacture of cheese, yoghurt, and other fermented dairy products is a multimillion dollar business. Fermented dairy products are microbiologically safer, have an extended shelf life, and can have enhanced flavor and texture characteristics compared to the raw material, milk. This entry will review some of the developments in biotechnology that have had an impact on fermented dairy foods.

Modern times see the use of recombinant DNA techniques in the production of chymosin, the enzyme responsible for clotting of cheese milk, the development of starter and adjunct cultures, and the enhancement of probiotic bacteria. The successful sequencing of the genomes of bacteria involved in dairy fermentations open up new opportunities for the understanding and improvement of the survival and activities of these bacteria. The importance of food safety and the health-promoting benefits of fermented dairy foods are also addressed with a brief look at bacteriocins, probiotics, and bioactive peptides.

STARTER CULTURES

Starter cultures are essential in the manufacture of fermented dairy products. Traditionally, a method known as "back-slopping" was used to begin the fermentation. This involved the transfer of the starter or inoculum, usually a portion of the previous batch of fermented product, to a fresh batch.[1] While such methods are still used for many traditional fermented foods, large-scale commercial manufacture of products such as cheese and yoghurt relies almost exclusively on the many scientific developments that have been made over the past few decades to produce consistent, safe, and economically viable products.

Starter cultures, with few exceptions, are members of the group of bacteria known as lactic acid bacteria (LAB). The chief role of the primary starter culture is the production of lactic acid from lactose. Starter cultures can also produce metabolites such as diacetyl, acetaldehyde, and CO_2, which contribute to the flavor of fermented dairy products. Their proteolytic systems can also be involved in developing flavor and aroma during cheese ripening.[1] The reduction in pH, competition with spoilage and pathogenic bacteria, and the production of antimicrobials by LAB contribute greatly to the safety of fermented dairy products. Secondary and adjunct cultures are often used in cheese-making for the development of specific organoleptic properties such as propionibacteria for the production of holes in Emmental cheese or *Penicillium roquefortii* for the development of mold in blue-mold varieties of cheese.[1]

The past three decades have seen extensive research into the physiology and genetics of LAB, especially those that are of industrial relevance, e.g., starter cultures and probiotics. The plasmids of starter strains encode most of the traits essential for fermentation and contribute to the individuality of a starter culture.[2] These traits include hydrolysis of casein, transport and metabolism of lactose, citrate metabolism, production of bacteriocins, and phage resistance. It is possible to transfer plasmids between strains via the natural mechanism of conjugation. For example, conjugation has been used successfully for the development of bacteriophage-resistant strains with desirable fermentative traits that are used routinely in the cheese industry.[3] Another food-grade method of starter culture improvement involves screening and natural selection. By screening for spontaneous mutants of *Lactobacillus bulgaricus* with little or no β-galactosidase activity, i.e., strains that produce flavor compounds but have a reduced ability to ferment lactose, yoghurt with desirable flavor attributes without postacidification problems can be produced.[4] Advances in molecular biology, such as genome sequencing, microarray technology, and proteomics, have enabled a more thorough investigation and understanding of the role of LAB in fermentation.

One of the most important advances in the area of starter cultures was the determination of the complete genome sequence of the chromosome of *Lactococcus lactis* IL1403.[2] At least 19 other industrially important starter strains or probiotic bacteria have been fully sequenced or are in the process of being sequenced worldwide.[1] The driving forces of such research are not just the problems encountered by manufacturers, e.g., bacteriophage infection, but also the opportunities such as the technological advantages and health benefits offered by many of these bacteria. A range of genetic tools has been developed over the past 30 years to allow the introduction of phage-resistance mechanisms[3] and the manipulation the metabolic pathways of LAB.[2,4] A number of innovative studies have demonstrated the ability to engineer starter strains to overproduce desirable metabolites, such as vitamins and compounds of importance in cheese and yoghurt flavor. *L. lactis* MG1363 can be manipulated to overproduce vitamin B_2 (riboflavin).[5] Efforts have been made to overproduce diacetyl and mannitol (a possible low-calorie sweetener in yoghurt) and exocellular polysaccharides (EPS) in yoghurt starter cultures for the enhancement of flavor and texture.[4]

The future of starter strain development must marry old and new technologies. Traditional fermented dairy products remain an important source of new bacteria to be exploited for possible beneficial effects or metabolites. The bacteriocin, lacticin 3147, e.g., was discovered in a strain of *L. lactis* from Irish kefir grain.[6] A multifaceted approach, involving enhanced screening and selection techniques, conjugation, genome sequencing, comparative genetics, proteomics, genetic engineering, and metabolic engineering, is currently employed to further starter strain development and enhance their role in dairy fermentations.

RECOMBINANT CHYMOSIN

Rennet or calf chymosin is the milk-clotting enzyme responsible for the coagulation of milk during cheese-making. The enzyme acts by the cleavage of κ-casein resulting in destabilization of the casein micelle and subsequent clotting in the presence of calcium. Alternatives to rennet include bovine pepsin, fungal proteinases, and other proteolytic enzymes. These enzymes generally have higher levels of non-specific proteolytic activity and higher heat stabilities resulting in poor flavor development and lower yields in some cheese.[7] The limited supply of calf rennet and inadequate rennet substitutes are two reasons why calf chymosin was one of the first mammalian enzymes to be cloned and expressed in a microorganism.[8]

The production of recombinant chymosin is a perfect example of the exploitation of modern biotechnological tools for developing fermented foods. The gene for calf prochymosin has been cloned in various microorganisms, e.g., *Escherichia coli*, *Saccharomyces cerevisiae*, *L. lactis*, *Bacillus subtilis*, *Kluyveromyces lactis*, and *Aspergillus niger*.[7] It has even been expressed in human cell culture (HeLa cells), secreted and successfully activated to chymosin. In *E. coli*, the proenzyme is produced as inclusion bodies that must be extracted from the host cells, after which the enzyme can be solubilized and activated. The proenzyme produced in fungi such as *K. lactis* and *A. niger* can be secreted into the culture medium and subsequently activated. Recombinant calf chymosin is more pure than traditional calf rennet and is identical to its natural counterpart in its ability to coagulate milk for cheese-making. It is produced commercially in large quantities in *E. coli* (California Biotechnology and DSM Food Specialities, the Netherlands) *K. lactis* (DSM Food Specialities, the Netherlands) and mammalian cells (Upjohn, United States, MI, USA) and is reportedly have up to 80% of the global market share for rennet.[7,8]

PROBIOTICS IN FERMENTED DAIRY PRODUCTS

Fermented dairy products have been long associated with good health, one reason for which is the presence of health-promoting microorganisms within these foods. The concept of probiotics has been attributed to Metchnikoff,[9] who proposed in 1908 that the longevity of the Caucasian people was associated with the frequent consumption of fermented milks. A growing body of scientific evidence demonstrates the benefits of the ingestion of probiotic bacteria. Probiotic consumption can benefit such conditions as irritable bowel syndrome, gastroenteritis, lactose intolerance, diarrhea, depressed immune function, cancer of the intestine, and genitourinary tract infections.[5]

Fermented probiotic products include fermented milks such as acidophilus milk, yoghurts, and recently, probiotic cheeses have been developed.[10] Common probiotic starters are *Lactobacillus acidophilus* and species of *Bifidobacterium* supplemented with traditional inocula such as *Streptococcus thermophilus* and *L. bulgaricus*, which provide acceptable flavors and textures. The slow-growing

Table 1 Genome-sequencing projects of starter and probiotic bacteria.

Microorganism	Fermented dairy product	Genome size	Institute
Bifidobacterium breve	Probiotic products	In progress	University College Cork, Ireland
Bifidobacterium longum bv. *Infantis* str. ATCC 15697	Probiotic products	In progress	DOE Joint Genome Institute
Bifidobacterium longum DJO10A	Probiotic products	2 Mb (draft assembly)	DOE Joint Genome Institute
Bifidobacterium longum NCC2705	Probiotic products	2 Mb	Nestlé Research Centre, Switzerland
Brevibacterium linens BL2	Mold- and smear-ripened cheese	4 Mb (draft assembly)	DOE Joint Genome Institute
Lactobacillus acidophilus NCFM	Yoghurt	1 Mb	California Polytechnic State University/North Carolina State University
Lactobacillus casei ATCC 334	Cheese	2 Mb	DOE Joint Genome Institute
Lactobacillus delbrueckii subsp. *bulgaricus* ATCC BAA-365	Yoghurt and Swiss and Italian type cheeses	1 Mb	Genoscope
Lactobacillus helveticus	Cheese	In progress	University of Wisconsin/Utah State University
Lactobacillus helveticus DPC4571	Cheese	In progress	Teagasc, Moorepark Food Research Centre, Ireland
Lactobacillus johnsonii NCC 533	Probiotic	1 Mb	Nestlé Research Centre, Switzerland
Lactobacillus rhamnosus	Cheese	In progress	New Zealand Dairy Board
Lactobacillus salivarius subsp. *salivarius* UCC118	Probiotic	2 Mb	University College Cork, Ireland
Lactococcus lactis subsp. *cremoris* SK11	Hard cheese	2 Mb	DOE Joint Genome Institute
Lactococcus lactis subsp. *lactis* Il1403	Soft cheese	2 Mb	INRA, France
Lactococcus lactis subsp. *lactis* MG1363	Cheese	2.5 Mb	Institute of Food Research, Norwich, UK/Alimentary Pharmabiotic Centre, Cork, Ireland/University of Groningen, The Netherlands/University of Bielefeld, Germany
Propionibacterium freudenreichii subsp. *shermanii* CIP 103027	Swiss cheese	In progress	Genoscope
Streptococcus thermophilus CNRZ1066	Yoghurt	1 Mb	INRA, France
Streptococcus thermophilus LMD-9	Yoghurt mozzarella	1 Mb	DOE Joint Genome Institute
Streptococcus thermophilus LMG 18311	Yoghurt	1 Mb	Catholic University of Leuven, Belgium

Abbreviations: DOE, Department of Energy; INRA, Institut National de la Recherche Agronomique.

nature of some probiotic bacteria has lead to addition of preprepared cultures in the form of Direct Vat Set (DVS) to yoghurt and milk. Probiotic microorganisms are live cultures and must survive numerous hurdles in the journey through the human body with colonization of the intestine; however, they must first survive long enough in the delivery food to reach their host. Bacteria are susceptible to many stresses such as temperature, oxygen, and acidity.[5] Fermented milks and yoghurts are good delivery vehicles as they have a short shelf life. Current and future knowledge of the genetic makeup of probiotic cultures allow scientists to identify new ways of exploiting these cultures to survive the stresses encountered in the delivery vehicle and in the human host. Consumer awareness of functional foods that contain probiotics has increased a great deal giving a significant boost to the popularity of associated fermented dairy products.

BIOACTIVE PEPTIDES

A further advantage of the fermentation of milk by LAB is the production of bioactive peptides. These peptides are physiologically active peptides derived from milk proteins such as α-, β-, κ-caseins and whey proteins.[11]

While bioactive peptides are inactive within the parent protein molecule, they can become active once liberated by milk fermentation: 1) by proteolytic starter cultures; 2) by digestion with gastrointestinal enzymes; or 3) artificially by hydrolysis using proteolytic enzymes. Peptides derived from milk proteins have been shown to display various health-promoting activities, affecting the digestive, immune, cardiovascular, and nervous systems.[5,11]

The highly proteolytic nature of many dairy starter cultures allows the release of a range of bioactive peptides into common fermented dairy products. A number of industrially important starter cultures such as *Lactococcus lactis*, *Lactobacillus helveticus*, and *Lactobacillus delbreuckii* have been shown to produce bioactive peptides from β- and κ-casein that have antihypertensive, antioxidative, and immunostimulatory properties.[11] Peptides that exhibit antimicrobial, antioxidative, and antihypertensive activities have been found in traditional fermented dairy products, e.g., cheese, yoghurt, and fermented milk.[11] Furthermore, these peptides or their precursors are susceptible to further degradation and hence activation by gastrointestinal enzymes such as pepsin and trypsin. The health benefits of bioactive peptides found in traditional fermented dairy products have yet to be established; however, following the current trend of functional food production, bioactive peptides evaluated in human clinical trials have been incorporated into some dairy products.[5,11] Fermented milk drinks, e.g., supplemented with the peptides, Val-Pro-Pro and Ile-Pro-Pro, derived from β-casein and κ-casein and claiming to reduce blood pressure are already commercially available.

BACTERIOCINS IN FERMENTED DAIRY PRODUCTS

Bacteriocins are small peptides produced by various bacteria, which inhibit the growth of other bacteria. The active site of antagonism by bacteriocins is the cell membrane. Bacteriocins produced by Gram-positive bacteria are generally active only against other Gram-positive bacteria.[6] The production of bacteriocins by LAB in dairy products results in safer and more stable foods. Nisin, a bacteriocin produced by *L. lactis*, is used commercially in more than 50 countries as a preservative to control spoilage and pathogenic bacteria such as *Listeria monocytogenes*, *Clostridium* spp., *Bacillus* spp., and *Staphylococcus* spp. Bacteriocins can be added directly as a food additive or as a bacteriocin-producing adjunct culture.[6] The production of Gouda cheese with a nisin-producing strain has been found to be effective in controlling the outgrowth of *Clostridium tyrobutyricum* as well as any subsequent spoilage of the cheese, and also inhibits growth of *L. monocytogenes* in cottage and Camembert cheeses.[12] Furthermore, bacteriocins have shown promise as technological aids in cheesemaking. Nisin added at a precise level can increase the lysis of starter cultures to accelerate cheese ripening and improve flavor. The genetic determinants of the bacteriocin, lacticin 3147, are located on a conjugative plasmid and have been transferred to Cheddar cheese-making strains with the resultant cheese being free of non-starter bacteria (NSLAB).[6]

CONCLUSIONS

While many exciting and innovative technological and organoleptic advances are possible by genetic manipulation, the future focus of research and development in fermented dairy products is more likely to be on the health-promoting benefits that can be achieved. As research continues, we can gain a better understanding of the microorganisms concerned and how they, or their metabolites, can have an impact on human health. Future prospects will include further exploitation of bioactive peptides and probiotics but may also see the arrival of custom-made foods for the delivery of drugs and vaccines. There are also opportunities to harness the information garnered from complete genome sequences of starter bacteria to improve culture performance and activity during fermentation.

REFERENCES

1. Parente, E.; Cogan, T. Starter cultures: general aspects. In *Cheese: Chemistry, Physics and Microbiology*, 3rd Ed.; Fox, P.F., McSweeney, P.L.H., Cogan, T.M., Guinee, T.P., Eds.; Chapman and Hall: London, 2004; 123–148.
2. Callanan, M.J.; Ross, R.P. Starter cultures: genetics. In *Cheese: Chemistry, Physics and Microbiology*, 3rd Ed.; Fox, P.F., McSweeney, P.L.H., Cogan, T.M., Guinee, T.P., Eds.; Chapman and Hall: London, 2004; 149–162.
3. McGrath, S.; Fitzgerald, G.F.; van Sinderen, D. Starter cultures: bacteriophage. In *Cheese: Chemistry, Physics and Microbiology*, 3rd Ed.; Fox, P.F., McSweeney, P.L.H., Cogan, T.M., Guinee, T.P., Eds.; Chapman and Hall: London, 2004; 163–190.
4. Mollet, B. Genetically improved starter strains: opportunities for the dairy industry. Int. Dairy J. **1999**, *9*, 11–15.
5. Stanton, C.; Ross, R.P.; Fitzgerald, G.F.; Van Sinderen, D. Fermented functional foods based on probiotics and their biogenic metabolites. Curr. Opin. Biotechnol. **2005**, *16*, 198–203.
6. Ross, R.P.; Morgan, S.; Hill, C. Preservation and fermentation: past, present and future. Int. J. Food Microbiol. **2002**, *79*, 3–16.

7. Crabbe, M.J.C. Rennets: general and molecular aspects. In *Cheese: Chemistry, Physics and Microbiology*, 3rd Ed.; Fox, P.F., McSweeney, P.L.H., Cogan, T.M., Guinee, T.P., Eds.; Chapman and Hall: London, 2004; 19–46.
8. Johnson, M.E.; Lucey, J.A. Major technological advances and trends in cheese. J. Dairy Sci. **2006**, *89*, 1174–1178.
9. Metchnikoff, E. The prolongation of life. In *Optimistic Studies*; William Heinemann: London, 1908; 1–343.
10. Oberman, H.; Libudzisz, Z. Fermented milks. In *Microbiology of Fermented Foods*, 2nd Ed.; Wood, B.J.B., Ed.; Blackie: London, 1998; 209–247.
11. Korhonen, H.; Pihlanto, A. Bioactive peptides: production and functionality. Int. Dairy J. **2006**, *16*, 945–960.
12. Caplice, E.; Fitzgerald, G.F. Food fermentations: role of microorganisms in food production and preservation. Int. J. Food Microbiol. **1999**, *50*, 131–149.

Dairy: Starter Cultures

Robert W. Hutkins
Department of Food Science and Technology, University of Nebraska, Lincoln, Nebraska, U.S.A.

Abstract

Modern manufacture of cheese and cultured dairy products relies on the consistency provided by dairy starter cultures. These cultures contain species of *Lactococcus*, *Lactobacillus*, *Leuconostoc*, and *Streptococcus* that are selected on the basis of their proven and consistent performance under actual manufacturing conditions. Dairy cultures are grown in large fermentors, usually under optimized conditions. Depending on the particular product, and the temperatures to which these products are exposed, dairy cultures are classified as mesophilic or thermophilic. They are also distinguished on the basis of their intended use—frozen bulk cultures are added to an intermediate vessel and propagated such that the total amount of available culture is increased several fold. Manufacturers also produce highly concentrated cultures (frozen or lyophilized) that can be added directly to the production tanks or vats. Dairy starter bacteria are vulnerable to bacteriophage, and several molecular strategies have been devised to reduce or prevent infections. The genomes of several strains of dairy lactic acid bacteria have been sequenced and used to modify and improve their metabolic properties; however, their application awaits regulatory and consumer acceptance.

INTRODUCTION

For the first 5000 or so years when fermented foods were manufactured, the fermentation, without exception, was initiated and performed by wild microorganisms. These organisms were introduced into the starting material in one of the two ways. They were simply part of the indigenous microbiota of the raw material, such that under the correct set of environmental circumstances, these organisms were able to grow and outcompete the other members of the local microbiota to produce appropriate metabolic end products. Alternatively, a portion of successfully fermented product from a previous batch, that presumably contained the relevant organisms, was added back (thus, this process is referred to as back-slopping) to fresh substrate to start a new fermentation. Although these approaches worked for thousands of years (and still do), they suffer from several disadvantages, ranging from the relatively minor occurrence of a slow fermentation to very serious situations as would occur if pathogens grow or produce toxins directly in the food. Moreover, in the 21st century, food industry has other requirements that make these older systems untenable. For most fermented foods, especially cheese and other fermented dairy products, modern production methods rely on high-throughput processing, consistent production schedules and product quality, and food safety standards that cannot be achieved using natural or back-slopping approaches. Therefore, the industry has become dependent on starter cultures to perform most food fermentations.[1] In the dairy industry, in particular, the use of starter cultures is now almost universally accepted. Given the huge volume of culture that is necessary for modern cheese manufacturing, concentrated starter cultures have become essential today.[2] Moreover, the ability to modify and improve the organisms that comprise dairy starter cultures is now possible, thanks to significant advances in molecular genetics and microbial genomics.

MICROORGANISMS AND THEIR PROPERTIES

Except for some specialized cheese cultures, most dairy starter cultures contain selected strains of lactic acid bacteria (LAB). The LAB form a cluster of low G + C, non-sporeforming, catalase-negative, anaerobic or facultative Gram-positive rod-, or coccus-shaped bacteria. Although the cluster consists of 12 genera, the dairy culture LAB belong to one of only four phylogenetically-related genera, *Lactococcus*, *Streptococcus*, *Leuconostoc*, and *Lactobacillus*; however, only a few species within these genera are suitable as starter cultures (Table 1). In general, the dairy LAB are most often described according to their metabolism and temperature optimum. For example, the optimum growth temperature for *Lactococcus lactis* subsp. *lactis* and *Lactococcus lactis* subsp. *cremoris* is around 28–32°C, thus starter cultures containing these organisms are referred to as mesophilic cultures. These bacteria are also homofermentative, meaning that the main end product (more than 90%) from glucose metabolism is lactic acid. Therefore, in some applications, these bacteria may also be referred to as

Table 1 Lactic acid bacteria used as dairy starter cultures.

Organism	General properties	Application
Lactococcus lactis subsp. *lactis*	Homofermentative and mesophilic	Cheese, cultured dairy products
Lactococcus lactis subsp. *cremoris*	Homofermentative and mesophilic	Cheese, cultured dairy products
Lactococcus lactis subsp. *lactis* biovar. *diacetylous*	Homofermentative and mesophilic	Cheese, cultured dairy products
Lactobacillus helveticus	Homofermentative and thermophilic	Cheese
Lactobacillus delbrueckii subsp. *bulgaricus*	Homofermentative and thermophilic	Cheese, yogurt
Streptococcus thermophilus	Homofermentative and thermophilic	Cheese, yogurt
Leuconosotoc lactis	Heterofermentative and mesophilic	Cheese, cultured dairy products
Leuconosotoc mesenteroides subsp. *cremoris*	Heterofermentative and mesophilic	Cheese, cultured dairy products

acid-producers. Similarly, *Streptococcus thermophilus* is another obligate homofermenter, but its optimum growth temperature is somewhat higher (37–45°C). The genus, *Lactobacillus*, is metabolically and physiologically much more diverse compared to other LAB, but only two species, the homofermentative *Lactobacillus delbrueckii* subsp. *bulgaricus* and *Lactobacillus helveticus*, are generally used as starter cultures. Usually, these organisms are paired with *S. thermophilus*, and although neither the lactobacilli nor *S. thermophilus* are true thermophiles, dairy cultures containing these bacteria are often referred to as thermophilic cultures. Finally, species of *Leuconostoc* are mesophilic and heterofermentative, such that lactic acid, as well as acetic acid, ethanol, and carbon dioxide can be formed from glucose. Because the leuconostocs also produce other volatile end products, they are often referred to as flavor producers.

Although they serve no technological function, another group of bacteria, called probiotics, are often included in yogurt and other cultured dairy products. Probiotic bacteria are defined as "live microorganisms which when administered in adequate amount confer a health benefit on the host." In other words, they are added for their nutritional function. The main probiotic organisms used in dairy applications include *Lactobacillus acidophilus* and other species of *Lactobacillus* and *Bifidobacterium*. In general, these organisms are produced industrially in the same manner as starter cultures (i.e., under conditions that maximize cell density), and are added as direct-to-vat type cultures (see below). In contrast to starter cultures and probiotics, other LAB are used to improve or accelerate cheese ripening. The adjunct cultures usually contain species of *Lactobacillus*. Finally, there are several other non-LAB and fungi that comprise specialized cultures (Table 2). These include *Propionibacterium freudenreichii* subsp. *shermani* and *Brevibacterium linens* (used in the manufacture of Swiss and Limburger cheeses, respectively) and *Penicillium roqueforti* and *Penicillium camemberti* (for blue-veined and cheese with surface mold, respectively).

FUNCTIONS AND APPLICATIONS

Dairy cultures perform three main functions in fermented milk products. First, they ferment lactose to lactic acid. The ensuing decrease in pH is then responsible for coagulation, in the case of yogurt and sour cream, and for enhancing syneresis and calcium solubilization in cheese. Lactic acid and low pH also serve a major preservation function. Second, starter cultures generate a myriad of flavor or flavor precursors. Lactic acid itself provides a pleasant tart flavor, but other culture-produced compounds, such as diacetyl and acetaldehyde, contribute important flavor and aroma characteristics in sour cream and yogurt, respectively. Finally, dairy cultures can modify the texture of cheese, mainly by producing proteolytic enzymes that degrade casein, and also by producing polysaccharides that add viscosity and water-binding capacity to various culture dairy products.

Starter culture suppliers offer a variety of products in several different forms. As noted above, cultures are usually classified as either mesophilic or thermophilic. The mesophilic dairy cultures are used for most hard cheeses, including Cheddar and Cheddar-type cheeses. These cultures are also used for cottage cheese, sour cream, cultured buttermilk, and other fermented milk products. By contrast, thermophilic cultures are used for many Swiss and Italian cheese varieties, as well as for yogurt; however, it is not unusual for a strain of *L. lactis* to be included in thermophilic culture or for *S. thermophilus* to be added to mesophilic cultures.

Regardless of the type of culture (i.e., mesophilic vs. thermophilic), most modern starter cultures contain specific strains having defined metabolic, physiological, and genetic properties. Strain selection is based on several cell functions relevant for the particular product, such as those

Table 2 Other organisms used as starter cultures.

Organism	Application
Bacteria	
Brevibacterium linens	Cheese: pigment, surface
Propionibacterium freudenreichii subsp. *shermanii*	Cheese: eyes in swiss
Mold	
Penicillium camemberti	Cheese: surface ripens white
Penicillium roqueforti	Cheese: blue veins, protease, lipase

related to fermentation rate, flavor formation, and texture development. The ability to resist infection by bacteriophage (viruses that attack bacteria) is particularly important and, in many cases, dictates the choice of strain in a given culture. For example, it is now common for a culture to contain two or more defined strains, each having a different sensitivity or resistance to the most common bacteriophage. Thus, if a bacteriophage capable of infecting a single starter culture strain was to emerge within a given cheese plant, the companion strain or strains, that had a different phage sensitivity, would still be able to perform the fermentation. Likewise, the culture itself could be "rotated" such that a specific culture would be used for a limited period of time, then, it would be removed from the rotation and replaced by a different culture containing strains unrelated (on a phage sensitivity basis) to the original culture.

INDUSTRIAL PRODUCTION

For producing highly concentrated starter cultures, controlled fermentation conditions are used (see Fig. 1). The medium, which must be food grade, typically contains a source of hydrolyzed protein, readily fermented carbohydrate, and growth factors. Importantly, the medium pH is maintained during growth, within a prescribed range (depending on the organism), by addition of ammonia or other alkaline material to prevent overacidification and loss of cell viability. Catalase may also be added to inactivate hydrogen peroxide. The cells are eventually collected and either packaged directly into cans and then frozen or concentrated first and then packaged and frozen. Rapid freezing is performed in liquid nitrogen ($-196°C$) to promote high cell viability. Alternatively, cells can be harvested from fermentors and lyophilized; the resulting freeze-dried and highly concentrated cells can then be maintained at normal freezing or refrigeration temperatures. Another means by which cultures can be frozen is to add the cells dropwise into liquid nitrogen, forming small pellets that can easily be dispensed. Viability of frozen or lyophilized cells can generally be enhanced by addition of cryoprotective agents.

There are two types or ways in which starter cultures are used by the dairy industry. The culture can be added directly to the vat (containing the milk), provided the concentration of cells is high enough to initiate the fermentation within the expected time. These "direct-to-vat" cultures are convenient, require no special handling or hardware costs, and avoid many of the problems associated with phage proliferation (see below). By contrast, for large operations, these cultures can be expensive, and thus, bulk cultures are usually used instead. Bulk cultures are inoculated into tanks (containing culture media) and are then incubated to produce a large volume of viable cells. The bulk culture can then be added to multiple cheese vats during the course of one or more days. It is critical that bulk cultures be grown under pH control to prevent acid damage, and both

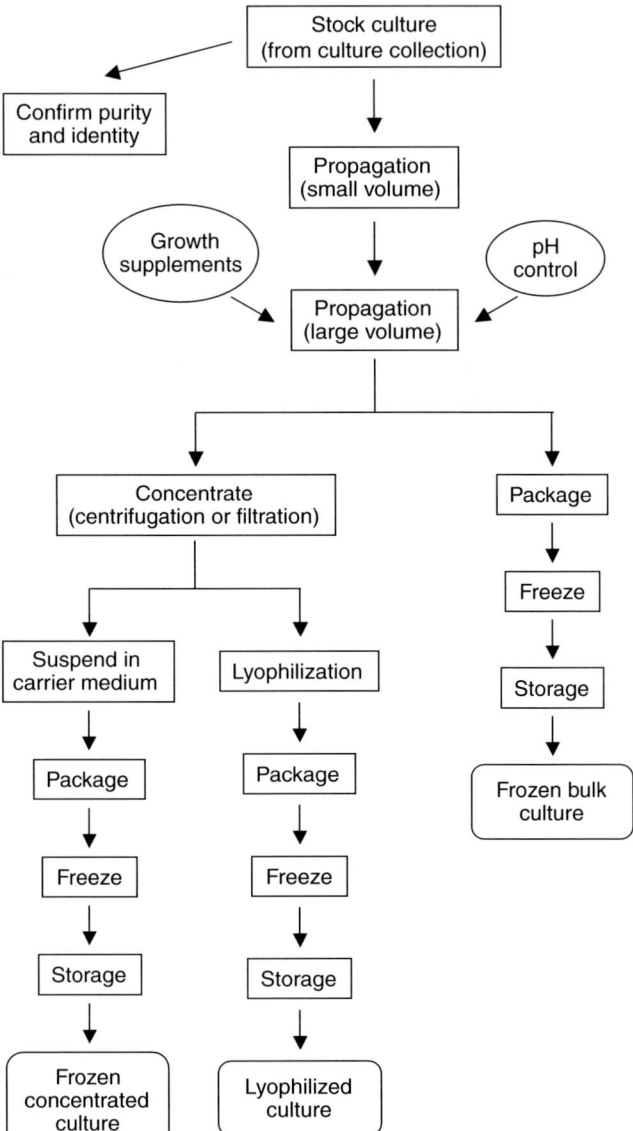

Fig. 1 Major production steps for dairy starter cultures. **Source:** From *Dairy Starter Cultures*.[3]

external pH control (via addition of a neutralizing agent) and internal pH control (via incorporation of buffer salts in the culture medium) are available. Preclusion of bacteriophage by sanitation, air handling, and other means, is also essential during bulk culture preparation.

BACTERIOPHAGE AND THEIR CONTROL

In cheese manufacture or any "open" environment where lactic fermentations occur, bacteriophage capable of infecting and killing starter culture bacteria will invariably be found. Infection of a single cell by a single bacteriophage may ultimately lead to more than a billion new bacteriophage, within just a few hours. This is more than enough bacteriophage to decimate the starter culture. Less severe infections may still cause slow fermentations, resulting in production delays and inferior product quality. Bacteriophage have long been a problem for manufacturers of cheeses made using mesophilic cultures (containing *L. lactis* susbsp. *lactis*); however, the incidence of phage infections against thermophilic cultures (mainly *S. thermophilus*) has increased significantly, because of the increase in the production of Mozzarella cheese, yogurt, and other products that rely on thermophilic starter cultures. Thus, the bacteriophage problem is one that affects the entire fermented dairy foods industry. It is worth noting that bacteriophage problems have been a major driving force for studying the molecular biology of dairy LAB (see below).

As noted above, sanitation, air handling, plant design, and phage exclusion are the first lines of defense against bacteriophage. Phage infections can also be minimized by using phage inhibitory media to propagate bulk cultures. These media contain chelating agents that bind calcium and thereby interfere with phage attachment to cells. A third and more direct approach for reducing infection by bacteriophage is to isolate starter culture strains that harbor genes encoding for one or more bacteriophage-resistance systems. Several such systems exist and have been characterized. Examples include cell-mediated mechanisms in which: 1) initial bacteriophage adsorption is inhibited or blocked; 2) bacteriophage DNA is hydrolyzed within the cytoplasm; and 3) the bacteriophage infection and replication process is aborted (abortive infection). Because there is a genetic basis for these systems, it is possible to transfer, using naturally occurring gene transfer techniques, the relevant genes from one organism to another.[4] Also, a single strain can accommodate more than one bacteriophage resistance system; thus, strains can be constructed that are resistant to several types of bacteriophage.

ENGINEERED PHAGE RESISTANCE

In addition to exploiting the innate phage resistance properties of LAB, it is also possible to engineer phage resistance by molecular manipulation. In general, these strategies are similar to the approaches described above, in which they rely on blocking adsorption and inhibiting replication. For example, it is possible to prevent phage adsorption by genetically inactivating the gene that encodes for a protein located on the host cell surface that is recognized by the phage as a receptor-binding site. In the absence of this protein, phage adsorption is inhibited.[5] Another strategy involves introducing excess copies of a specific DNA sequence that are used by the phage to initiate the phage genome replication process.[6] The more "decoy" copies of these origin of replication or *ori* sites there are in the cells, the more the DNA replication machinery is diluted and the fewer phage will be made. A so-called suicide method has also been devised in which a DNA-degrading enzyme (i.e., a restriction enzyme) is turned on upon exposure of the cell to bacteriophage.[7] Although this results in the "programmed" death of the infected cell, cell death occurs before the phage had replicated. Thus, phage proliferation does not occur, and the remaining cell population is not affected. A final example involves engineering starter culture bacteria such that the cells transcribe the non-coding or antisense region of mRNA whose sense strand encodes for an essential phage gene.[8] Although the exact mechanism by which this antisense approach inhibits phage multiplication has not yet been established, it has been suggested that the antisense and complementary sense mRNA hybridize, preventing translation and synthesis of the essential protein and thereby reducing phage replication.

DAIRY STARTER CULTURES IN THE 21ST CENTURY

The phenotypic properties of starter culture LAB are well known and have long formed the basis of strain selection, classification, and identification; however, many of their relevant biochemical, as well as physiological traits, are now defined at the gene level. Among the first genes characterized in LAB were plasmid-borne genes, encoding for proteins and enzymes involved in lactose metabolism, casein utilization, bacteriocin production, and bacteriophage resistance. More recently, several of the genomes of dairy starter LAB, including *L. lactis* subsp. *lactis*, *L. delbrueckii* subsp. *bulgaricus*, and *S. thermophilus* have been sequenced, revealing detailed information on central metabolism, stress response, biosynthesis of polysaccharides, and flavor generation.[9] Importantly, gene transfer technologies now exist that make it possible to construct novel strains having unique functional properties. Metabolic engineering strategies could be used in numerous ways to improve dairy fermentations. For example, diverting pyruvate away from lactate formation and toward synthesis of the flavor compound, diacetyl, has been accomplished by genetically disrupting genes in lactococci coding for lactate dehydrogenase or α-acetolactate decarboxylase (and such strains are now in commercial use). Cheese ripening can be accelerated by increasing expression of genes involved in proteolysis and peptide degradation. Increased synthesis of texture-forming exopolysaccharides can be achieved by overexpressing enzymes responsible for formation of exopolysaccharide precursors. Ultimately, regulatory issues regarding the use of such strains must, however, be resolved before the widespread application of these strains is realized.[10]

REFERENCES

1. Caplice, E.; Fitzgerald, F. Food fermentations: role of microorganisms in food production and preservation. Int. J. Food Microbiol. **1999**, *50* (1–2), 131–149.

2. Cogan, T.M. History and taxonomy of starter cultures. In *Dairy Starter Cultures*; Cogan, T.M., Accolas, J.-P., Eds.; VCH Publishers, Inc.: New York, 1996; 1–23.
3. Sandine, W.E. Commercial production of dairy starter cultures. In *Dairy Starter Cultures*; Cogan, T.M., Accolas, J.-P., Eds.; VCH Publishers, Inc.: New York, 1996; 191–206.
4. Sanders, M.E.; Leonhard, P.J.; Sing, W.D.; Klaenhammer, T.R. Conjugal strategy for construction of fast acid-producing, bacteriophage-resistant lactic streptococci for use in dairy fermentations. Appl. Environ. Microbiol. **1986**, *52* (5), 1001–1007.
5. Garbutt, K.C.; Kraus, J.; Geller, B.L. Bacteriophage resistance in *Lactococcus lactis* engineered by replacement of a gene for a bacteriophage receptor. J. Dairy Sci. **1997**, *80* (8), 1512–1519.
6. Hill, C.; Miller, L.A.; Klaenhammer, T.R. Cloning, expression, and sequence determination of abacteriophage fragment encoding bacteriophage resistance in *Lactococcus lactis*. J. Bacteriol. **1990**, *172* (11), 6419–6426.
7. Djordjevic, G.M.; O'Sullivan, D.J.; Walker, S.A.; Conkling, M.A.; Klaenhammer, T.R. A triggered-suicide system designed as a defense against bacteriophage. J. Bacteriol. **1997**, *179* (21), 6741–6748.
8. Kim, S.G.; Batt, C.A. Antisense mRNA-mediated bacteriophage resistance in *Lactococcus lactis*. Appl. Environ. Microbiol. **1991**, *57* (4), 1109–1113.
9. Klaenhammer, T.; Barrangou, R.; Buck, B.L. Genomic features of lactic acid bacteria effecting bioprocessing and health. Antonie van Leeuwenhoek **2005**, *29*, 393–409.
10. Pedersen, M.B.; Iversen, S.L.; Srrensen, K.I.; Johansen, E. The long and winding road from the research laboratory to industrial applications of lactic acid bacteria. FEMS Microbiol. Rev. **2005**, *29*, 611–624.

Drought and Drought Resistance

Salvatore Ceccarelli
Biodiversity and Integrated Gene Management, International Center for Agricultural Research in the Dry Areas (ICARDA), Aleppo, Syria

Abstract
Drought has affected humankind since the beginning of agriculture causing the collapse of several civilizations. The entry starts with describing how plant breeding and biotechnology have traditionally attempted to improve drought resistance of crops and then discusses innovative avenues to cope with drought in the near future when the frequency of drought is likely to increase due to climate changes.

INTRODUCTION

Drought is the most serious constraint to agricultural production, and is still one of the most difficult challenges to agricultural scientists. After defining drought and drought resistance, the entry describes what has been done in the past, explains the reasons for the lack of success, and discusses how a new paradigm may be able to increase the level of drought resistance in view of climate changes. The objective is to describe drought and drought resistance from an historical perspective, current research, and strategies to cope with the consequences of climate changes.

BREEDING FOR DROUGHT RESISTANCE

Drought, or more generally, limited water availability has historically been the main factor limiting crop production. Water availability has been associated with the rise of multiple civilizations, while drought has caused the collapse of empires and societies, such as the Akkadian Empire (Mesopotamia, ca. 4200 calendar years B.P.), the Classic Maya (Yucatan Peninsula, ca. 1200 calendar years B.P.), the Moche IV–V Transformation (coastal Peru, ca. 1500 calendar years B.P.)[1] and the early bronze society in the southern part of the Fertile Crescent.[2]

Although in the past drought reached the front pages of the media only when it caused famine and death, in the last few years drought warnings have become more and more frequent. More people are now aware that drought is a permanent constraint to agricultural production in many developing countries, and is increasingly causing crop losses in developed countries. The development of drought-resistant cultivars would be a major breakthrough.[3] However, from a genetic point of view drought resistance is a very elusive trait because its occurrence, severity, timing, and duration vary from location to location and in the same location from year to year. Cultivars successful in one dry year may fail in another, or cultivars resistant to terminal drought may not be resistant to intermittent drought, or to drought occurring early in the season.[4] To make matters worse, drought seldom occurs in isolation; it often interacts with other abiotic (particularly temperature extremes) and biotic stresses (for example, root diseases and nematodes).[5] Moreover, areas with high risk of drought generally have low-input agriculture.[6] Thus, breeding for drought tolerance is made more difficult by its interactions with other stresses. Because of its complexity, drought means different things to different people in different areas and for different crops.[7]

The complexity of drought resistance also emerges from molecular studies in which quantitative trait loci (QTL) for characters associated with drought resistance in a number of crops are located on many chromosomal regions. One exception is represented by regulatory genes such as DREB, which, when introduced into wheat, has been shown in *Arabidopsis thaliana* to play a crucial role in promoting the expression of drought tolerant genes. Transformed wheat plants, expressing DREB, demonstrated substantial resistance to water stress manifested by a 10-day delay in wilting in plastic house conditions.[8]

Two contrasting philosophies have been used to breed drought-tolerant varieties. In the first, selection is done under optimum growing conditions assuming that an increased yield potential will have a carryover effect in less favorable conditions. In the second, selection is done in the presence of drought, and can take two forms: 1) selection for physiological or developmental traits (analytical breeding); and 2) direct selection for grain yield (empirical or pragmatic breeding).

The first philosophy has been one of the main concepts upon which the Green Revolution was based,[9] which, in high-input environments, not only achieved large production increases but also decreased agricultural biodiversity and increased environmental degradation because the varieties it produced were able to express their superiority only if supplied with chemical inputs (fertilizers, pesticides, and herbicides). On the other hand, in the regions where farmers were too poor to afford chemical inputs and in the regions chronically affected by drought, it failed to improve drought resistance. Therefore, two billion people still lack reliable access to safe, nutritious food, and 800 million of them are chronically malnourished.[10] One corollary of this philosophy is that, because favorable environments tend to be similar in time and space, one or few varieties *widely adapted* can perform well in all of them, hence the reduction in agricultural biodiversity.

The second philosophy recognizes the complexity of drought and is based on breeding for *specific adaptation*. In the analytical approach physiologists classify mechanisms for drought resistance into drought escape, drought avoidance, and drought tolerance. Several traits have been associated with those mechanisms: the physiological or biochemical traits include osmotic adjustment, proline content, stomatal conductance, canopy temperature, relative water content, leaf turgor, abscisic acid content, transpiration efficiency, water use efficiency, carbon isotope discrimination, and retranslocation; the developmental and morphological traits include leaf emergence, early growth vigor, leaf area index, leaf waxiness, stomatal density, tiller development, flowering and maturity time, cell membrane stability, cell wall rheology, and root characteristics.

Many of these traits have been analyzed with molecular techniques in a number of crops and mapped on several chromosomal regions. However, although biotechnology has revolutionized the development of new varieties with genes being identified for engineering or marker-assisted selection for a range of "simple" traits, and despite a substantial investment in the study of more complex traits, there are not yet success stories based on the identification of specific genes and their utilization for drought resistance.[11]

The analytical approach has been very successful in understanding which traits are associated with drought resistance but much less successful in developing new cultivars with improved drought resistance under field conditions where drought varies in occurrence, severity, timing, and duration. Therefore, it is the interaction among traits that determines the overall crop response to drought, rather than the expression of any specific trait.[12]

Although breeding for drought resistance based on direct selection for grain yield in the target environment (empirical or pragmatic breeding) appears intuitively as the most obvious solution, it has faced criticism because the chances of progresses appear slow and remote. Two issues with selection in stress environments are precision of selection and existence of several target environments.

On the first issue, a literature review[13] has shown the absence of a consistent relationship between grain yield and the magnitude of heritability and the possibility of combining precision and relevance by conducting trials in the target environment even when it is a stress environment.

The second issue, how to deal with the multitude of target environments, is intimately associated with broad and specific adaptation, two concepts that have been debated in plant breeding since the early 1920s and are still highly controversial.[14] Because of the heterogeneity of the environments affected by drought, breeding for specific adaptation is essential to increase the yield of crops grown in these environments. Breeding for specific adaptation to drought conditions is often considered an undesirable breeding strategy, because it is usually associated with lower yields under favorable conditions and produces too many varieties for the seed companies to handle.

The issue of the heterogeneity and the multitude of the environments affected by drought has been addressed by developing an innovative type of breeding program for drought-prone areas based on the collaboration between farmers and scientists known as participatory plant breeding.[15] Participatory plant breeding is based on breeding for specific adaptation to both the climatic and social environment and for wide adaptation over time (also called stability or dependability) and is conducted independently in areas that are consistently different. An additional strategy is to replace breeding for uniform varieties (pure lines or hybrids) with population (evolutionary) breeding, which produces heterogeneous varieties better able to cope with the year-to-year variability expected to increase with climate changes. Evolutionary breeding[16] is based on populations with large genetic variability, such as mixtures of F_2, which can be handled directly by farmers in a multitude of environments while they slowly evolve and adapt.

Both analytical and empirical breeding for drought resistance can benefit from the increasing knowledge on the genetic control of the several traits known to be associated with drought resistance. However, molecular biologists should not forget that ultimately a new variety has to prove its superior drought tolerance in the field and in the hands of farmers.

CONCLUSION

Drought is a complex problem that affects the livelihoods of millions of poor every year and most likely will become more frequent with the climate changes. Current knowledge on the molecular basis of drought resistance is not yet sufficient to develop more drought resistant crops or varieties. Therefore, for the time being, conventional breeding based on specific adaptation and evolutionary strategies, and with the collaboration of farmers, appears as the only realistic avenue to develop drought-resistant varieties in the short term. As our knowledge of the genetic basis of drought resistance improves, biotechnological techniques such as transformation might offer the possibility of introducing new traits and regulatory mechanisms that would otherwise be inaccessible.

ACKNOWLEDGMENTS

The collaboration of Dr. Stefania Grando who for more than 20 years has conducted breeding for drought resistance at the International Center for Agricultural Research in Dry Areas (ICARDA), in Aleppo (Syria) is gratefully acknowledged.

REFERENCES

1. deMenocal, P.B. Cultural responses to climate change during the late holocene. Science **2001**, *292*, 667–673.
2. Rosen, A.M. Environmental change at the end of early Bronze Age Palestine. In *L'urbanisation de la Palestine à l'âge du Bronze ancient*; De Miroschedji, P., Ed.; BAR International: Oxford, UK, **1990**; 247–255.
3. Baum, M.; van Korff, M.; Guo, P.; Lakew, B.; Hamwieh, A.; Lababidi, S.; Udupa, S.M.; Sayed, H.; Choumane, W.; Grando, S.; Ceccarelli, S. Molecular approaches and breeding strategies for drought tolerance in Barley. In *Genomic Assisted Crop Improvement. Vol. 2: Genomics Applications in Crops*; Varshney, R., Tuberosa, R., Ed.; Springer, Netherlands, **2007**; 51–79.
4. Turner, N.C. Optimizing water use. In *Crop Science: Progress and Prospects*; Nösberger, L., Geiger, H.H., Struick, P.C., Eds.; CAB International Publishing: Wallingford, UK, **2002**; 119–135.
5. Ceccarelli, S.; Grando, S.; Baum, M.; Udupa, S.M. Breeding for drought resistance in a changing climate. In *Challenges and Strategies for Dryland Agriculture*; Rao, S.C., Ryan, J., Eds.; CSSA Spec. Publ. 32. ASA and CSSA: Madison, WI, **2004**; 167–190.
6. Cooper, P.J.M.; Gregory, P.J.; Tully, D.; Harris, H.C. Improving water use efficiency of annual crops in the rainfed farming systems of West Asia and North Africa. Exp. Agric. **1987**, *23*(2), 113–158.
7. Passioura, J. Increasing crop productivity when water is scarce—from breeding to field management. Agric. Water Manage. **2006**, *80*, 176–196.
8. Pellegrineschi, A.; Reynolds, M.; Pacheco, M.; Brito, R.M.; Almeraya, R.; Yamaguchi-Shinozaki, K.; Hoisington, D. Stress-induced expression in wheat of the *Arabidopsis thaliana* DREB1 A gene delays water stress symptoms under greenhouse conditions. Genome **2004**, *47*, 493–500.
9. Hazell, P.B.R. The green revolution. In *Oxford Encyclopedia of Economic History*; Mokyr, J., Ed.; Oxford University Press: Oxford, UK, **2003**; 478–480.
10. Fresco, L.O.; Baudoin, W.O. Food and nutrition security towards human security. In *ICV Souvenir Paper*, International Conference on Vegetables: World Food Summit Five Years Later, Rome, June 11–13, 2002; Nath, P., Gaddagimath, P. B., Dutta, O. P. Dr. Prem Nath Agricultural Science Foundation, Bangalore, India.
11. Chapman, S.C. Use of crop models to understand genotype by environment interactions for drought in real-world and simulated plant breeding trials. Euphytica **2008**, *161*, 195–208.
12. Ceccarelli, S.; Acevedo, E.; Grando, S. Breeding for yield stability in unpredictable environments: single traits, interaction between traits, and architecture of genotypes. Euphytica **1991**, *56*, 169–185.
13. Ceccarelli, S. Adaptation to low/high input cultivation. Euphytica **1996**, *92*, 203–214.
14. Ceccarelli, S.; Grando, S. Plant breeding with farmers requires testing the assumptions of conventional plant breeding: lessons from the ICARDA barley program. In *Farmers, Scientists and Plant Breeding: Integrating Knowledge and Practice*; Cleveland, D.A., Soleri, D., Eds.; CAB International Publishing: Wallingford, UK, 2002; 297–332.
15. Ceccarelli, S.; Grando, S.; Baum, M. Participatory plant breeding in water-limited environments. Exp. Agric. **2007**, *43*, 411–435.
16. Suneson, C.A. An evolutionary plant breeding method. Agron. J., **1956**, *48*, 188–191.

Electric Field Stress on Plant Systems

Ana Balasa
Anna Janositz
Dietrich Knorr
Department of Food Biotechnology and Food Process Engineering, Berlin University of Technology, Berlin, Germany

Abstract

Plant foods have beneficial effects on the prevention of civilization-related diseases. Numerous epidemiological studies documented an inverse association between fruit and vegetable consumption and different types of cancer and cardiovascular ailments. The components responsible for this protective impact are phytochemicals, plant secondary metabolites, which play a major role in plant adaptation to strained environmental conditions and contribute to color, flavor and taste of the foods. Secondary metabolites can be found in a vast variety and are widely distributed in plants, whereas their dynamical increase can occur after external stress stimulation by biotic and abiotic factors. The application of mild to low energy pulsed electric field treatment (PEF) on plant systems comprises these stress factors. PEF processing includes the exposure of plant material to short repeated pulses of a high voltage leading to membrane permeabilization. Accompanied events of reversible and/or irreversible cell poration can be used as a stimulus to target plants for the production of nutritionally valuable phytochemicals. Such a process allows the development of new innovative foods that might play an important role in the prevention of cancer and cardiovascular diseases.

SECONDARY METABOLITES

Interactions between plants and their environment have a very elaborate character due to complexity of plant biochemistry. The term "secondary" was introduced for the first time in the late 19th century by biochemist Albrecht Kossel, when secondary metabolites were regarded as nonessential plant substances. Currently, they are considered to play a vital role in plants' defense mechanisms with various functions necessary for plant survival. In contrast to primary metabolites, they do not have an apparent function in plant metabolism on growth and development, so their absence will not cause immediate death. However, as metabolic intermediates they are needed for plant existence and survival. They may be present as inactive precursors of the functional enzymes that are converted to biological active form after induced stress or may be located in their biological active form within the cell. Secondary metabolism and its regulation are not completely elucidated on both biochemical and genetic levels; however, it is known that numerous enzymes are involved in the production of secondary compounds.

STRESS RESPONSES IN PLANT SYSTEMS IN NATURE

Biological plant stress is induced by changes in external environmental conditions, which can either inhibit or promote plant growth and development. Environmental stress involves different biotic (result of interaction with other living organisms) and abiotic factors (water deficiency or excessiveness, temperature alterations, light intensity, and radiation). Their impact depends on the quantity of aforementioned elements.[1] External conditions (stress) may trigger a wide range of possible plant responses, which induce changes on different levels: macroscopic, cellular, and genetic level. Changes in growth rates, plant size, and crop yield can be observed.[2] Altering cellular metabolism and re-routing of metabolic pathways can cause accumulation or additional production of secondary metabolites within the cell. Plants are using a large diversity of mechanisms to produce or release secondary compounds into the surroundings or to start synthesis of proteins and phytochemicals that remain in the cell. This behavior can be seen as a defense mechanism that plants evolved as a response to extreme circumstances.[3,4] Each response is unique and cannot be directly extrapolated for different stress factors, and it even varies with diversity of plant growth phase and its development. Phytochemical defenses found in plants can be described through production or accumulation of different bioactive compounds, pigments, flavors, and other low molecular substances. These compounds are also recognized as effective antioxidant and anti-inflammatory substances, which are at certain levels contributing to the nutritional value of food and, therefore, are recognized as dietary constituents with anticarcinogenic effects. Plant adaptation to many different stress conditions in the nature could be substituted with a stimulated defense response triggering a response to a particular stress situation.

PULSED ELECTRIC FIELD PROCESSING

During the past decade, industrial interest in developing gentle food technologies to replace the currently common thermal processing has increased substantially. The non-thermal application of pulsed electric fields (PEF) is included among these emerging processes and has received considerable relevance in bio- and food technology. It involves the exposure of biological cell material to short repeated pulses (μs–ms) of a high voltage with the result of either temporary or permanent pore formation in cell membrane leading to membrane permeabilization. Several models that describe mechanism of permeabilization have been suggested. The most accepted one "the dielectric breakdown model" was proposed by Zimmerman in 1986.[5] It considers the cell membrane as a dielectric barrier that separates ionic species and free charges on both sides of the membrane (Fig. 1). The application of an external electric field induces polarization of the different charged ions and therefore the charging of the membrane. Pores are formed above a critical membrane potential which can be followed by pore expansion. When cells are exposed to a high-intensity or a continuous stream of electrical field pulses, irreversible pore formation occurs. Reversible pore formation takes place when the electric field is applied while the transmembrane potential is reaching the critical level of 1 V and the pores formed are small in relation to the whole membrane surface. After removal of the electric field, pores reseal and cells restore their vitality. The benefit of electroporation is an important aspect of process and product development because it is aimed at maintaining fresh-like physical and chemical characteristics of food products, as well as controlling the microbial safety with minimal or no change during processing. Besides the main research field, the pasteurization of electrical conductive food products due to bacterial cell destruction, PEF technology can be also implemented in several areas concerning the permeabilization of plant cell material. Mass transfer processes such as extraction, expression, drying, infusion, as well as operations to gently transfer desirable compounds are facilitated by PEF treatment.[6–8] In addition to permanent pore formation after electric field processing at high intensities, the application of mild, sublethal PEF treatment causes reversible membrane disintegration with the maintenance of cell viability.[9] This effect provides a potential for targeted influenced plant cells toward the generation of health related compounds as a stress response to the electric field.[10] Thus, mild PEF applications can stimulate cells to act as "bioreactors" by producing desirable secondary metabolites.

PULSED ELECTRIC FIELDS INDUCED STRESS

Application of PEF treatment with low to mild treatment intensities (when energy inputs between 0.01 and 5 kJ/kg are used) induces temporary pore development in the cell membrane, which allows disturbance of the phospholipid bilayer. This method is routinely used in biotechnology to facilitate the delivery of foreign material such as DNA, drugs, medicines, etc. into the cell. Conductive channels across the cell membrane occur due to quick electrical shock, whereby electrically insulating properties of the membrane are quickly restored, and the cell is reestablishing its vital functions.[9] Since metabolic activity is recovered, low-intensity treatments offer a potential to induce stress reactions in plant systems that have been observed in nature. Biosynthesis of plant metabolites after PEF treatment may be stimulated through a broad range of factors. Changes in pH during the treatment, altered membrane transport and osmotic imbalance are among those factors. These events can undergo a change in accordance with induced electric potential, which cells recognize as a stress and respond to with accumulation and/or production of secondary metabolites. The initial applications of PEF treatment, on the borderline between reversible and irreversible membrane permeabilization, in an attempt to induce artificial stress response within plant cell cultures were realized by Dörnenburg and Knorr.[11] Electric field pulses in a range of 0 up to 1.6 kV/cm were applied under sterile conditions to suspension cultures placed in the electroporation chamber. The effect of different treatment intensities was observed through sustainable release of pigments (anthraquinones and amaranthin) from cultured plant tissue (*Morinda citrifolia* and *Chenopodium rubrum*, respectively) while maintaining cell viability. Reversible permeabilization and extent of the pulse-induced changes in the structural properties of the cell system have been studied by Angersbach, Heinz, and Knorr.[9] Similar breakdown phenomena in different cell systems (potato, apple, and fish tissue and plant cell suspension) were observed after a single pulse with critical field amplitude was applied. Further investigation on PEF processing to induce secondary metabolite production in plant systems was introduced by Guderjan et al.[10] An increase in isoflavonoids diadzain (up to 20%) and genistein (up to 21%) from soybeans was observed after electric field treatments of different intensities. The highest amount of diadzain was achieved after 50 pulses

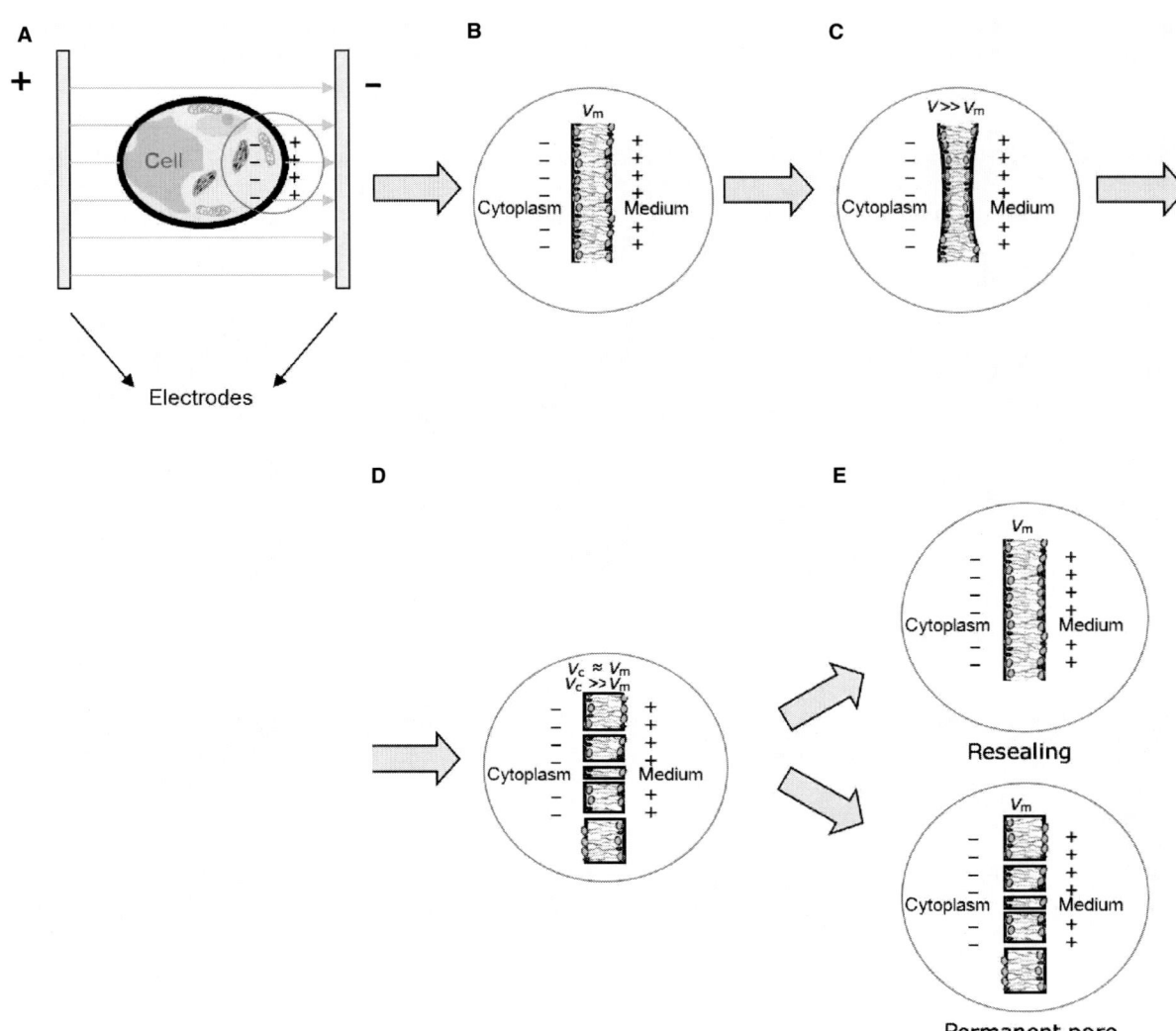

Fig. 1 Schematic diagram of PEF-induced membrane permeabilization (A) cell in an electric field; (B) enlarged cell membrane with potential V_m; (C) polarization of intracellular and extracellular charges induces increased potential difference V across the membrane; attraction of charges with different polarities on both sides of membrane lead to membrane compression; (D) critical membrane potential V_c reaches and overcomes membrane potential which leads to a permeabilization of cell membrane with (E) reversible and irreversible pore formation.

of 1.3 kV/cm, which resulted in total energy input of 1.857 kJ/kg, whereas only 20 pulses at 1.3 kV/cm (energy input of 0.743 kJ/kg) were necessary for the maximal amount of genistein. Balasa and Knorr[12] have investigated the impact of PEF as a pretreatment method in traditional wine-making process and its influence on total polyphenolic content in grape juice and grape pomace. They reported an increase of 13% of total polyphenolic content in grape juice and 24% in grape pomace after 50 pulses of electric field treatment at 0.5 kV/cm (energy input of 0.1 kJ/kg). Current research activities reveal that biological systems can be affected by PEF-induced stress reactions (Table 1) and can offer new possibilities of targeted modifications of functional food properties.

CONCLUSION

Pulsed electric field treatment was found to be an effective pretreatment method for enhancement of secondary metabolites production in plant systems and cultured plant cells. Increased phenolic content was observed in plant material after low-intensity PEF application, when cell membranes were not completely disintegrated.

Ongoing research activities are concentrating on the stress application of pulsed electric fields to investigate biochemical and molecular mechanisms by which plants tolerate external impact and therefore are able to withstand the stress with increased biosynthesis of secondary metabolites.

Table 1 Plant secondary metabolites stimulated with PEF treatment.

Secondary metabolite	Plant material	PEF treatment	Reference
Amaranthin	Plant cell suspension *Chenopodium rubrum*	Electric field strength 0–1.6 kV/cm, 0–30 pulses; Treatment chamber: electrodes 300 × 70 mm, distance 77 mm, parallel plate configuration, volume 1.6 L	[11]
Diadzain	Soybeans	Electric field strength 1.3 kV/cm, 50 pulses, pulse duration of 280 μsec, total energy input 1.857 kJ/kg; Treatment chamber: parallel plate configuration	[10]
Genistein	Soybeans	Electric field strength 1.3 kV/cm, 20 pulses, pulse duration of 280 μsec, total energy input 0.743 kJ/kg; Treatment chamber: parallel plate configuration	[10]
Total polyphenolics	Grapes *Vitis vinifera*	Electric field strength 0.5–2.4 kV/cm, 50 pulses, pulse duration of 280 μsec, total energy input 0.1–2.3 kJ/kg; Treatment chamber: parallel plate configuration	[12]

ACKNOWLEDGMENTS

Parts of this work have been supported by an EU-funded Integrated Project NovelQ "Novel Processing Methods for the Production and Distribution of High-Quality and Safe Food," FP6-CT-2006–015710, Priority 5 "Food Quality and Safety."

REFERENCES

1. Beck, E.; Fettig, S.; Knake, C.; Hartig, K.; Bhattarai, T. Specific and unspecific responses of plants to cold and drought stress. J. Biosci. **2007**, *32*, 501–510.
2. Bohnert, H.J.; Nelson, D.E.; Jensen, R.G. Adaptations to environmental stresses. Plant Cell **1995**, *7*, 1099–1111.
3. Dixon, R.A.; Paiva, N.L. Stress-induced phenylpropanoid metabolism. Plant Cell **1995**, *7* (7), 1085–1097.
4. Namdeo, A.G. Plant cell elicitation for production of secondary metabolites: a review. Pharmacognosy Rev. **2007**, *1* (1), 69–79.
5. Zimmermann, U. Electrical breakdown, electropermeabilization and electrofusion. Rev. Physiol. Biochem. Pharmacol. **1986**, *105*, 176–257.
6. Jaeger, H.; Balasa, A.; Knorr, D. Food industry applications for pulsed electric fields. In *Electrotechnologies for Extraction from Food Plants and Biomaterials*; Food Engineering Series, Lebovka, N., Vorobiev, E., Eds.; Springer: New York, **2008**; 181–216.
7. Lebovka, N.I.; Praporscic, I.; Ghnimi, S.; Vorobiev, E. Temperature enhanced electroporation under the pulsed electric treatment of food tissue. J. Food Eng. **2005**, *69*, 177–184.
8. Ade-Omowaye, B.I.O.; Angersbach, A.; Taiwo, K.A.; Knorr, D. Use of pulsed electric field pre-treatment to improve dehydration characteristics of plant based foods. Trends Food Sci. Technol. **2001**, *12*, 285–295.
9. Angersbach, A.; Heinz, V.; Knorr, D. Effects of pulsed electric fields on cell membranes in real food systems. Innov. Food Sci. Emerg. Technol. **2000**, *1*, (2), 135–149.
10. Guderjan, M.; Toepfl, S.; Angersbach, A.; Knorr, D. Impact of pulsed electric field treatment on the recovery and quality of plant oils. J. Food Eng. **2005**, *67* (3), 281–287.
11. Dörnenburg, H.; Knorr, D. Cellular permeabilization of cultured plant cell tissues by high electric field pulses or ultra high pressure for recovery of secondary metabolites. Food Biotechnol. **1993**, *7*, 35–38.
12. Balasa, A.; Knorr, D. Extraction of total polyphenolics from grapes in correlation with the degree of membrane poration. In *COST Meeting 928–300606*; Reykjavik, Island, June 30–July 1, 2006.

ELISA Assays: Microorganisms and Toxins in Foods

Robert Levin
Department of Food Science, Massachusetts Agricultural Experiment Station, University of Massachusetts, Amherst, Amherst, Massachusetts, U.S.A.

Abstract

The mechanism of operation of the major ELISA protocols for detection of toxins and foodborne infectious bacteria are presented in this entry. Included are details involving conjugation of haptens to carrier proteins, the use of enzymes in ELISA assays, advantages of microtiter plate formats, and detection of specific toxins and infectious bacteria in foods via ELISA assays.

INTRODUCTION

ELISA assays by definition are enzyme-linked immunosorbent assays, and are presently among the most widely used serological assays to detect and quantitate a wide variety of toxins and infectious bacteria in foods. Advantages of ELISA assays encompass high specificity due to the immunological nature of the assays, extraordinary sensitivity, frequently in the picogram (pg) range of detection, and small sample size. All ELISA assays require at least one antibody (Ab). Most such antibodies involve the immunoglobulin G (IgG) fraction of antisera but are not necessarily restricted to the G fraction. Antisera are produced by the injection of suitable animals with the immunogenic molecule or microorganism of interest. The rabbit is the conventional animal of choice for production of polyclonal antibodies (pAbs); mice are used for production of monoclonal Abs (mAbs) in conjunction with hybridoma methodology.[1] Antisera to specific antigens are often available commercially from additional animals such as the goat, cow, horse, sheep, cat, dog, rabbit, poultry, rat, and mouse. In general, the larger the animal, the less expensive is the antiserum. The industry also produces a wide variety of anti-immune sera, e.g., rabbit antisera that will react with all IgGs from the goat and vice versa. The availability of such a wide variety of anti-immune sera greatly facilitates the development of ELISA assays. This entry deals specifically with the mechanism of operation of ELISA assays, the choice of enzymes and antibodies, conjugation mechanisms, and how one goes about establishing an ELISA assay for detecting toxins and infectious bacteria associated with foods.

HAPTENS AND THEIR CONJUGATION TO PROTEINS

Many organic molecules such as toxins in foods having a molecular weight of less than 5000 (haptens) are usually not immunogenic. This problem is overcome by covalently conjugating the hapten to a large protein molecule or carrier. Bovine serum albumin (BSA, mol.wt. 67,000) is most frequently used as a protein carrier because of its high degree of purity, availability, and low cost. Keyhole limpet hemocyanin (KLH, mol.wt. ~7 million) and polylysine (mol.wt. 70,000–150,000) are alternate protein carriers. Because of its large size, KLH may precipitate during cross-linking, which can create handling problems. BSA is readily soluble but is itself highly immunogenic. Ovalbumin (OVA, mol.wt. ~45,000) is another suitable carrier and is ideal for a second carrier (as in antiglobulin ELISA sandwich assays) where BSA is used for primary antibody production.

If a free amino group is available on the hapten it can be readily conjugated to BSA using formaldehyde or gluteraldehyde as a covalent bridge. Reaction mechanisms are also available for conjugating through available –OH, –CHO, –COOH, and –SH groups.[2,3]

Injection of a mammal with such a purified conjugate will result in a family of Ab molecules, each directed against a specific domain of the conjugate. Some Ab molecules will be directed against overlapping domains of the BSA while others will be capable of recognizing the hapten. Such a family of Abs is referred to as polyclonal in that each Ab molecular species is produced by an individual B lymphocyte. Higher levels of immunological specificity can often be obtained with the use of mAbs that are homogeneous, in that all the Ab molecules present are identical and will recognize the same domain of an antigen. Methodology involving the production of monoclonal antibodies has been described in detail.[1]

THE USE OF ENZYMES IN ELISA ASSAYS

The second crucial component of ELISA assays involves the use of an enzyme reaction to produce a colorimetrically detectable indication of the presence or absence of the target antigen. Enzymes most frequently used with ELISA assays are alkaline phosphatase, β-galactosidase, and horseradish peroxidase (HRP). The substrate for alkaline phosphatase is usually p-nitrophenylphosphate (PNPP). The substrate for β-galactosidase is usually o-nitrophenyl-β-D-galactopyranoside) ONPG. HRP is cheaper than the other two enzymes but requires two substrates, H_2O_2 plus a proton donor such as 2,2'azino-di-[3-ethylbenzthiazoline sulphonate] (ABTS), orthophenylene diamine (OPD), or tetramethylbenzidine (TMB). Among these three proton donors, TMB is preferred because it is non-mutagenic. For quantitative ELISA assays, the product of the enzyme reaction must be soluble.

The enzymes are usually conjugated to specific or anti-immune antibodies. Conjugation reactions retaining enzyme activity usually involve the use of gluteraldehyde in a one-step reaction[4] or a two-step reaction.[5] The two-step gluteraldehyde method results in enzyme and antibody being present in equal amounts in the conjugate resulting in enhanced sensitivity.[6] Periodate conjugation[7] is also widely used.

THE USE OF MICROTITER PLATES IN ELISA ASSAYS

Most ELISA assays are performed using 72- or 96-well microtiter plates with wells having a capacity of 300 μL, facilitating assay of many samples and considerable saving of antibody preparations and other costly reagents. The availability of microplate readers with variable wavelengths and internal incubation at a variety of temperatures greatly enhances the efficiency of processing multiple samples.

Polystyrene microtiter plates harbor electrostatic binding sites on their surface that will bind proteins in a non-specific manner. This is utilized by the addition of diluted antiserum (or the hapten conjugated to a carrier protein) to wells (usually 100 μL) to achieve non-specific binding of Abs or the hapten to the bottom of the well over several hours. The unbound Ab (or hapten–protein conjugate) molecules are then rinsed out and blocking solution (300 μL) is then added. Blocking solutions consist of any suitably inexpensive protein that will bind non-specifically to the surface of the wells so as to prevent further non-specific binding of ELISA reagents thereafter. After 15 min., the blocking agent is rinsed out and thereafter only immunospecific binding will occur on the surface of the wells. Blocking agents consist of 0.1% BSA, 0.1% fat-free skimmed milk (FFSM), or 0.1% gelatin. If FFSM is used it is first prepared as a 1.0% (w/v) solution in 0.1 M phosphate buffer, pH 7.5 containing 8.5% NaCl. Boiling in the presence of phosphate solubilizes the casein. It is then diluted 10-fold with d.H_2O for use. BSA should not be used as a blocking agent if the original hapten conjugate used for injection to produce Abs involved BSA as the protein carrier.

The sensitivity of ELISA assays can be significantly increased if the Ab portion of antiserum is first purified. This is readily achieved with the use of affinity chromatography involving protein A bound to agarose beads in miniature columns and frequently results in $\sim$100-fold purification of the IgG fraction of the antiserum.

It is important to keep in mind that all ELISA assays consist of a series of additions of reagents, and that after the addition of each reagent and following the required incubation time, the wells are thoroughly washed with a washing solution consisting usually of 0.02% Tween 20 in 0.05 M phosphate buffer (pH 7.5) containing 0.85% NaCl. In addition to Tween 20, Triton X-100 and Nonidet-40 are also employed. These are non-ionic detergents that are used to prevent the hydrophobic adsorption of proteins to the surface of the wells and to decrease the hydrophobic aggregation of proteins.[8] There are five fundamental ELISA protocols: direct, indirect, direct competitive, indirect competitive, and double antibody sandwich. Three commonly used ELISA formats using microtiter plates are presented in Fig. 1. In the direct ELISA (Fig. 1A), antigen or a hapten–protein conjugate is bound to the surface of the well followed by the addition of Ab–enzyme conjugate, and then substrate for the enzyme is added. The intensity of the color developed after a predetermined incubation time is then directly related to the amount of antigen bound to the well. In the indirect ELISA (Fig. 1B) antigen or a hapten–protein conjugate is first bound to the well followed by the addition of primary antibody, then secondary antibody, then enzyme, and finally substrate. The advantage here involves the use of commercially available secondary Ab already conjugated to the enzyme and does not require the primary antibody to be conjugated to the enzyme. The indirect sandwich ELISA (Fig. 1C) is particularly useful if a hapten by itself is not capable of binding to the surface of the well Here, primary Abs are bound initially to the well, followed by the addition of antigen (sample), then secondary Abs conjugated to the enzyme, and then the substrate. The competitive ELISA assays are the most sensitive. In the direct competitive ELISA (Fig. 2A), Abs are first bound to the surface of the well followed by the simultaneous addition and incubation of the sample (antigen) and antigen–enzyme conjugate. These two components then compete for binding to the Abs bound to the well. The intensity of the final color developed is inversely related to the amount of antigen in the sample. In the indirect competitive ELISA (Fig. 2B), antigen or hapten–protein conjugate is first bound to the surface of the well. This is then followed by the simultaneous addition and incubation of the sample (antigen) and primary Abs against the antigen. The primary Abs then compete for the soluble antigen in the sample and for the antigen bound to the well. After rinsing, secondary Ab–enzyme conjugate is then added followed by rinsing and the addition of substrate. The intensity of the final color developed is inversely related to the amount of antigen in the sample.

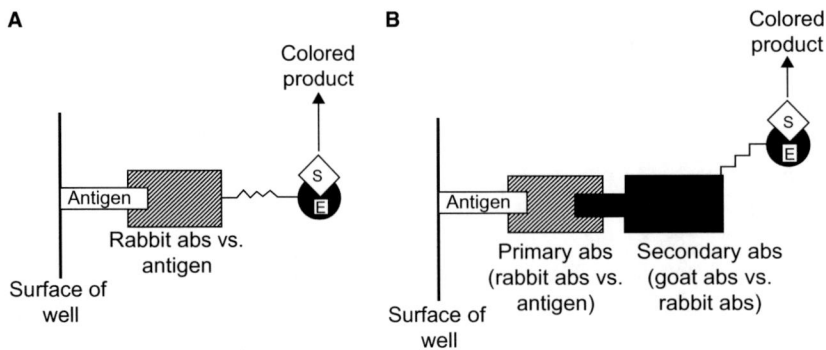
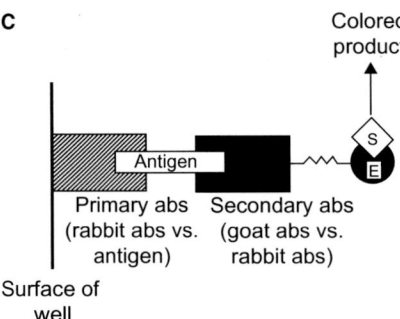

Fig. 1 Fundamental ELISA techniques.

Application of ELISA Assays for Detection of Toxins in Foods

ELISA assays have been developed for the detection and quantitation of a wide variety of toxins in foods. These toxins include paralytic shellfish toxins, ciguatera toxins, brevetoxin, domoic acid, okadaic acid, aflatoxins, ochratoxin, trichothecenes, and ricin, in addition to allergens in certain foods.[9]

Application of ELISA Assays for Detection and Identification of Foodborne Bacterial Pathogens

The application of ELISA assays to foodborne pathogenic bacteria is particularly useful in quantitating such organisms from selective enrichment cultures involving most probable number (MPN) assays. In addition, ELISA assays are frequently used to detect toxins produced by pure cultures of isolates of infectious organisms from foods to

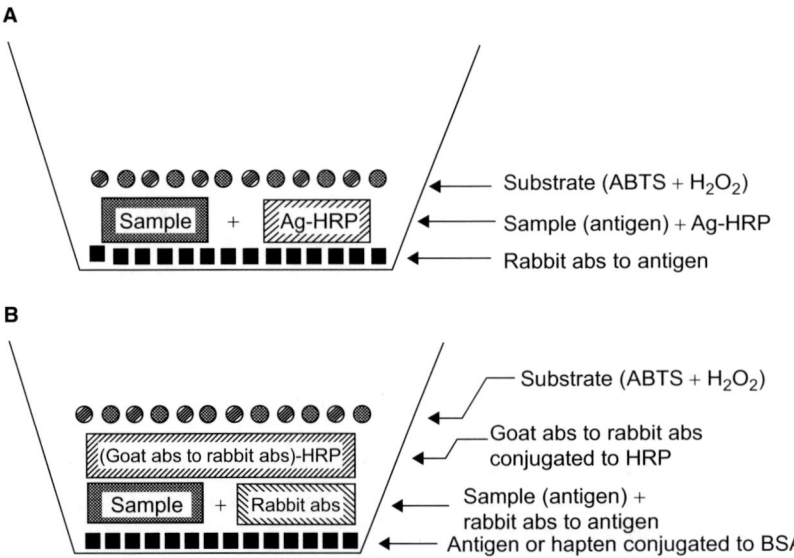

Fig. 2 Direct and indirect competitive ELISA techniques.

assess their pathogenic potential.[10] ELISA assays have been developed for foodborne pathogenic bacteria such as *Salmonella*, *Escherichia coli* O157:H7, *Vibrios*, *Listeria monocytogenes*, and *Campylobacter jejuni*.[11]

CONCLUSIONS

ELISA assays have the advantages of extreme sensitivity and high specificity. In addition, the small assay volumes involved reduce the consumption of antibodies and conjugates so as to reduce the cost per assay and also greatly facilitates the simultaneous processing of large numbers of samples. Robotic systems (although quite costly) are now available for complete automation of ELISA assays. ELISA assays are being increasingly used to greatly enhance the sensitivity of PCR assays for detection of low numbers of infectious microorganisms in foods.[12]

REFERENCES

1. Zolla, H. *Monoclonal Antibodies: A Manual of Techniques*; CRC Press: Boca Raton, FL, 1987; 214.
2. Harlow, E.; Lane, D. *Antibodies—A Laboratory Manual*; Cold Spring Harbor Laboratory: Cold Spring Harbor, NY, 1988; 726 pp.
3. Ishikawa, E. Labeling of antibodies and antigens. In *Immunoassay*; Diamandis, E.P., Christopoulos, T.K., Eds.; Academic Press: New York, 1996; 191–204.
4. Avrameas, S. Coupling of enzymes to proteins with gluteraldehyde. Use of conjugate for detection of antigens and antibodies. Immunochemistry **1969**, *6*, 43–52.
5. Avrameas, S.; Ternyck, T. Peroxidase labeled antibody and Fab conjugates with enhanced intracellular penetration. Immunochemistry **1971**, *8*, 1175–1179.
6. Van Weemen, B.K.; Schuurs, A. Immunoassay using antibody–enzyme conjugates. FEBS Lett. **1974**, *43*, 215–218.
7. Nakane, P.K.; Kawaoi, A. Peroxidase labeled antibody. A new method of conjugation. J. Histochem. Cytochem. **1974**, *22*, 1084–1091.
8. Butler, J.E. Solid phases in immunoassay. In *Immunoassay*; Diamandis, E.P., Christopoulos, T.K., Eds.; Academic Press: New York, 1996; 191–204.
9. Levin, R.E. Application of ELISA assays for detection and quantitation of toxins in foods. In *Food Biotechnology*, 2nd Ed.; Shetty, K., Paliyath, G., Pometto, A., Levin, R. Eds.; Marcel Dekker: New York, 2005; 1543–1565.
10. Erdenlig, S.; Ainsworth, A.; Austin, F. Production of monoclonal antibodies to *Listeria monocytogenes* and their application to determine the virulence of isolates of channel catfish. Appl. Environ. Microbiol. **1999**, *65* (7), 2827–2832.
11. van Knapen, F.; Elgersma, A.; Notermans, S.H.W.; Ruitenberg, E.J. *Applications of ELISA in Food Hgiene*, Rapport van hete Institute voor veeteelkundig Onderzoek "Schoonoord" (B-237), 1983; 27–54.
12. Hong, Y.; Berrang, M.; Tongrui, L.; Liu, T.; Hofacre, C.L.; Sanchez, S.; Wang, L.; Maurer, J.J. Rapid detection of *Campylobacter coli*, *C. jejuni*, and *Salmonella enterica* on poultry carcasses by using PCR-enzyme-linked immunosorbent assay. Appl. Environ. Microbiol. **2003**, *69* (6), 3492–3499.

Embryo Transfer Traits: Genetic Parameter Estimation

F.A. di Croce
A.M. Saxton
J.L. Edwards
Department of Animal Science, Institute of Agriculture, University of Tennessee, Knoxville, Tennessee, U.S.A.

D.E. Casanova
Faculty of Veterinary Science, National University of Buenos Aires, Buenos Aires, Argentina

F.N. Schrick
Department of Animal Science, University of Tennessee, Knoxville, Tennessee, U.S.A.

Abstract

The need to improve multiple ovulation/embryo transfer (MOET) programs through approaches such as genetic selection motivated this arena of new research. These methods for estimating genetic parameters (and the results shown) indicate the feasibility of including MOET traits in future breeding strategies. Understanding of procedures for estimating quantitative genetic parameters, along with the use of molecular genetic techniques, should accelerate improvements in the efficiency of MOET technology.

INTRODUCTION

The use of statistical parameters for genetic improvement of multiple ovulation/embryo transfer (MOET) technology in cattle, primarily through quantitative genetics and selective breeding, is an area of rapidly developing research. The MOET technology rapidly increases genetic progress, reduces risk of disease transmission, eliminates cost/difficulty of animal transport, and expands the number of progeny from genetically superior parents. However, critical limitations of MOET include variability in superovulatory response of donor animals and pregnancy success from transferred embryos. These unpredictable responses create logistical problems that reduce availability of embryos and profitability.

Although significant genetic variation for fertility is generally accepted, development of breeding values for fertility traits has mostly been ignored. Strategies that elevate fertility in cattle by means of genetic (genomic) selection will become increasingly important.

LIMITATIONS OF MOET

Primarily due to the reduction of generation intervals, producers can rapidly acquire superior genetics utilizing MOET. Several studies have revealed a potential of 10–20% in additional genetic gain compared with traditional breeding programs.[1]

Utilization of MOET in the cattle industry has grown over the last 30 years. Annually, more than 500,000 embryos are produced around the world.[2] Despite large growth during the past decade, this technology has not increased in efficiency as related to number of embryos produced per superovulated donor.[2] Mean embryo production is <6 embryos per superovulated donor (range of 0 to >60) with 20% of donors failing to produce any embryos.[2] Thus, genetic improvement through selection becomes a possible approach to solve these limitations.

METHODOLOGY FOR ESTIMATION OF GENETIC PARAMETERS

Multiple equations have been developed over time to aid in estimating genetic parameters and must be introduced to assist in understanding the methodology. In simplistic terms, the observed or phenotypic trait (P) is described by additive genetics (A; effects transmitted to progeny), all other genetics (G), and environment (E) resulting in the equation $P = A + G + E$. Corresponding variances components can be defined as $Vp = Va + Vg + Ve$. How one trait affects another is important, represented by a model of covariance components ($C_{1,2} = A_{1,2} + G_{1,2} + E_{1,2}$).

From these, one can calculate heritability = Va/Vp, repeatability = $(Va + Vg)/Vp$, and genetic = $A_{1,2}/\sqrt{(Va_1 \times Va_2)}$ and phenotypic = $C_{1,2}/\sqrt{(Vp_1 \times Vp_2)}$ correlations between traits 1 and 2.

Variance and covariance "components" are estimated using statistical mixed models, which include fixed and random effects. Maximum likelihood is the current estimation method of choice; in particular, restricted maximum likelihood (REML). Packages used for MOET genetic parameter estimation include ASReml[1] and DFReML (now Wombat).[3–6] These programs must be able to model single traits (univariate analysis) to estimate variance components, and model two or more traits (multivariate analyses) for the estimation of covariance components.

The most common mixed model used is the "animal" model. In this model, all animals are included as a random effect, while other explanatory terms are fixed effects (such as year and herd). Animal models use pedigrees to track relationships among animals, and related individuals will share additive genetic (co)variance. Animals within the same herd will share environmental variance. In this manner, variances are separated into components of interest.

Phenotypic trait and pedigree data are required to estimate genetic parameters using the animal model. Several phenotypic traits are being studied to evaluate efficiency and to estimate genetic parameters of MOET schemes. These traits include number of palpable corpora lutea (CLN), recovered ova (RO), transferable embryos (TE), degenerated embryos (DE), unfertilized oocytes (UFO), percentage of transferable embryos (PTE = TE × 100/RO), and pregnancy rates following embryo transfer (PR).[1,5]

Fixed effect data of importance to the MOET process include embryo transfer technician and/or company, origin (location) and breed of the donor, month and/or season of superovulation, dose of FSH (follicle-stimulating hormone), number of previous collections, status of transferred embryos (frozen and fresh), embryo quality, and development.[1,3] Pedigree data, another required input, are a list of animals under analysis with their corresponding sires and dams.

From a genetic standpoint, special care is needed to correctly attribute a trait to the responsible genotype (individual, maternal, or paternal). For example, number of RO is a result of the donor's genetics.[1] In traits describing embryonic survival (TE, DE, and UFO), the paternal effect must also be taken into account.[1] Furthermore, when evaluating PR of transfer, one must consider the genotype of the recipient, donor, and sire of embryos. A representation of genetic contribution (synergistic model) for MOET traits is depicted in Figure 1.

Provided estimates of the variance components, predictions of individual additive genetic effects, or breeding values can be estimated from Henderson's mixed model equations. Best linear unbiased predictions (BLUP) are

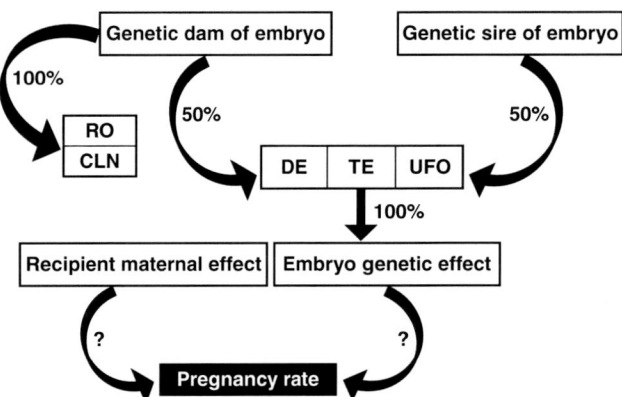

Fig. 1 Synergistic model for MOET technology illustrating genetic attribution of MOET traits to genotypes (dam, sire, and recipient). MOET traits in the model are CLN, number of palpable corpora lutea; RO, number of recovered ova; TE, number of transferable embryos; DE, number of degenerated embryos; UFO, number of unfertilized oocytes; and PR, pregnancy rate following embryo transfer. Percentages and arrows refer to the degree of attribution between genotypes and MOET traits.

produced by these equations. BLUPs are breeding values, predictions of the value of that individual as a parent, since additive genetic effects are transmitted to progeny.

Given the predicted breeding values determined by BLUP, genetic improvement is accomplished by selecting superior animals as parents. Estimating the rate of genetic gain depends upon three factors: heritability (h^2), selection differential, and generation interval. Selection differential is determined by selection intensity (i = standardized mean deviation of selected parents) and phenotypic variation (SD) present in the population. Utilizing this methodology, Di Croce et al.[5] estimated a genetic gain of 1.16% per generation in PR following embryo transfer assuming 50% selection for PR, estimated by the formula ($i = 0.8$) × (SD = 49.962) × ($h^2 = 0.029$).

RELEVANT LITERATURE REGARDING GENETIC ESTIMATES FOR MOET TRAITS

Heritability, the fraction of variation available for genetic improvement, is relatively low in MOET traits. This is a consequence of large environmental variation, and does not necessarily indicate a lack of genetic variation.[7] Generally, studies have reported heritabilities ranging from 0.23 to 0.34 for number of RO, and from 0.00 to 0.59 for number of TE.[1,3–5] Similar to non-MOET PR results, heritability for recipient PR is below 0.06.[1,5] Repeatability measures predictability of future performance of an individual. Estimates have ranged from 0.11 to 0.57 for TE and 0.11 and 0.51 for RO.[3,4,8] These values are high enough to indicate that culling of low performing cows should result in higher performance in the remaining herd.

König et al.[1] reported the importance of direct selection of TE or indirect selection of RO as a breeding strategy to increase TE. The authors developed a complete description of the genetic and phenotypic correlation of MOET traits where RO demonstrated a positive genetic correlation with TE (0.74). However, RO also demonstrated a positive genetic correlation with the number of DE (0.89) and the number of UFO (0.76) indicating that despite attempts to improve the number of TE, increases in undesirable traits may also occur.[1] An interesting negative genetic correlation was found between RO and PTE (−0.17) indicating a possible increase in reproductive losses associated with increased RO.[1]

König et al.[1] also illustrated the negative genetic correlation between milk yield and TE (−0.27). This may explain why high producing cows fail to give an average number of TE.[1] Furthermore, an increase of 1000 kg in milk production decreased the numbers of RO and TE by 0.94 and 0.46, respectively.[1] Additionally, a negative association between somatic cell count (SCS) and RO or TE was observed.[1]

GENOMIC SELECTION: A NEW APPROACH FOR GENETIC IMPROVEMENT

Traditionally, selection for economically important traits has been based on phenotypic records of the individual and its relatives (as described above). However, research suggests that information at the DNA level will lead to faster genetic gain than achieved, based on phenotypic data only.[9] Single nucleotide polymorphisms (SNPs), single DNA "letter" changes, have been utilized as markers to identify genetic differences in animals. SNPs are densely located throughout the genome, and effects of these locations (not necessarily the SNP itself) on a trait can be estimated in a model that considers them as random factors.[10]

Informative markers can be found by combining SNP and phenotypic information for any trait. Meuwissen et al.[9] developed methods to predict genetic parameters based on a genome-wide dense marker map with a limited number of phenotypic records using marker haplotypes. Authors observed that selection based on breeding values predicted from markers could substantially increase the rate of genetic gain in animals, especially in those traits where accuracy of selection is low. Low accuracy values are mostly due to traits with a limited number of phenotypic records or with low heritability.[9]

Genomic estimated breeding value (GEBV) can be obtained by combining the total estimated effects for that genotype.[10] The high accuracy of GEBV (80% estimated by the correlation between GEBV and true breeding value) and the possibility that GEBV can be calculated at birth are major advantages of the genomic approach compared to traditional breeding values.[10] Genome-wide selection could produce twice the genetic change compared to current progeny testing schemes. Therefore, complex genetic traits with low heritability, such as fertility and MOET, are potential applications for genomic improvement.

CONCLUSION

Multiple ovulation embryo transfer schemes have the potential for enabling rapid genetic change by increasing reproductive capacity of animals. However, growth of MOET has become limited by high variability in donor response and the high percentage of donors which fail to produce any progeny. Although few studies have been performed, genetic selection of donors appears to be a potential approach to improve efficiency of MOET traits. The promising results of moderate heritability and repeatability shown in this entry indicate the feasibility of including MOET traits in future breeding goals. Furthermore, the use of molecular genetics in parameter estimation may aid in identifying individuals who perform well in both arenas. Utilizing genome-wide estimated breeding value could produce a genetic change greater than current progeny testing schemes. To apply this strategy, it will be necessary to identify informative markers (SNPs), especially in the case of MOET and fertility, which may create an even more effective approach for improving the efficiency of MOET schemes and the overall fertility of the livestock industry.

REFERENCES

1. König, S.; Bosselmann, F.; von Borstel, U.U.; Simianer, H. Genetic analysis of traits affecting the success of embryo transfer in dairy cattle. J. Dairy Sci. **2007**, *90*, 3945–3954.
2. Hasler, J.F. The current status and future of commercial embryo transfer in cattle. Anim. Reprod. Sci. **2003**, *79*, 245–264.
3. Peixoto, M.G.C.D.; Pereira, C.S.; Bergmann, J.A.G.; Penna, V.M.; Fonseca, C.G. Genetic parameters of multiple ovulation traits in Nellore females. Theriogenology **2004**, *62*, 1459–1464.
4. Tonhati, H.; Lôbo, R.B.; Oliveira, H.N. Repeatability and heritability of response to superovulation in holstein cows. Theriogenology **1999**, *51*, 1151–1156.
5. Di Croce, F.A.; Saxton, A.M.; Rohrbach, N.R.; Schrick, F.N. Genetic parameter estimation for embryo transfer traits. Reprod. Fertil. Dev. **2008**, *21*, 169.

6. Meyer, K. WOMBAT – a tool for mixed model analyses in quantitative genetics by restricted maximum likelihood (REML). J. Zhejiang Univ. Sci. **2007**, *B8*, 815–821.
7. Veerkamp, R.F.; Beerda, B. Genetics and genomics to improve fertility in high producing dairy cows. Theriogenology **2007**, *68*, 266–273.
8. Bastidas, P.; Randel, R.D. Effects of repeated superovulation and flushing on reproductive performance of *Bos indicus* cows. Theriogenology **1987**, *28*, 827–835.
9. Meuwissen, T.H.E.; Hayes, B.J.; Goddard, M.E. Prediction of total genetic value using genome-wide dense marker maps. Genetics **2001**, *157*, 1819–1829.
10. Schaeffer, L.R. Strategy for applying genome-wide selection in dairy cattle. J. Anim. Breed. Genet. **2006**, *123*, 218–223.

Enzymes

Geoffrey Smithers
Geoffrey Smithers Consulting Group, Highett, Victoria, Australia

Peter Roupas
Brad Woonton
Chakra Wijesundera
Food and Nutritional Sciences, Commonwealth Scientific and Industrial Research Organisation (CSIRO), Werribee, Victoria, Australia

Abstract

Enzymes are proteins that catalyze the biochemical reactions necessary for life, and can be found in all animal, plant, and microbial food raw materials. For millennia, man has harnessed the naturally occurring and ubiquitous enzymes found in microorganisms—bacteria, yeasts, and molds—to prepare foods such as bread, cheese, beer, and wine. For example, in cheese making, the enzyme chymosin is used to modify the casein protein in milk allowing it to coagulate and ultimately form the curd that, with maturity, becomes cheese.

BACKGROUND

In modern food processing, enzymes are used for a wide range of applications, including cheese making, dairy processing, baking, brewing, starch processing, oil processing, and in the production of beverages such as fruit juices (Table 1). In these applications, the enzyme is often a critical processing aid, but can also influence the texture, appearance, and nutritional value of the final product, and may generate desirable flavors and aromas. Food enzymes are occasionally derived from animal or plant sources, but the vast majority originate from microorganisms, primarily species of *Bacillus*, *Aspergillus*, *Streptomyces*, and *Kluyveromyces*.

Modern biochemistry and biotechnology techniques are being used to improve the productivity, variety, efficiency, cost-effectiveness, and specificity of enzymes available to the food processor. Some examples are shown in Table 2. However, the global debate on genetic modification is slowing progress.

In food processing, enzymes are often welcomed as alternatives to traditional physical and/or chemical processes and their associated undesirable side reactions, facilitating enhanced environmental friendliness (e.g., energy conservation and reduced pollution), fewer waste byproducts, and greater yield of a higher quality product. Enzymes also facilitate processing that otherwise would not be possible by other synthetic means (e.g., use of pectinase to produce clear apple juice concentrate).

The value of the global food and beverage enzyme market is expected to reach $1.2 billion by 2011 with growth of 8% annually.[1]

BRIEF HISTORY OF ENZYMES AND THEIR USE IN FOOD PROCESSING AND PRODUCTION

- 2000 B.C.—Egyptians and Sumerians develop fermentation for use in brewing, bread baking, and cheese making.
- 1000–800 B.C.—Calves' stomachs used to transport milk. Through the action of the naturally occurring enzyme chymosin, the milk was coagulated spawning the start of the cheese industry.
- 1878—Components of yeast cells that cause fermentation are identified and the term "enzyme" is first used, derived from the Greek term meaning "in yeast."
- 1926—Enzymes are first demonstrated to be proteins.
- 1982—First food application of an enzyme (α-amylase) produced using gene technology.
- 1988—Recombinant chymosin approved for use in Switzerland, although other countries slow to adopt.
- 1990—Enzyme food processing aids produced using gene technology, and approved for limited use in cheese making in the United States, and in baking in the United Kingdom.

Table 1 Applications of enzymes in modern food processing and production.

Food industry sector	Enzyme	Function
Dairy	Chymosin (rennin)	Coagulant in cheese making
	β-Galactosidase (lactase)	Hydrolysis of lactose to give lactose-free/reduced milk and whey products
	Proteases (various)	Hydrolysis of whey proteins in manufacture of hydrolysates and enzyme-modified cheese
	Catalase	Removal of hydrogen peroxide
	Lipase	Enzyme-modified cheese, infant formula
Baking	α-Amylases	Breakdown of starch, maltose production
	Amyloglycosidases	Saccharification
	Maltogen amylase (Novamyl)	Delays process by which bread becomes stale
	Proteases	Breakdown of proteins
	Pentosanase	Breakdown of pentosan, leading to reduced gluten production and also improved baking performance
	Glucose oxidase	Stability of dough
Wine and fruit juice	Pectinase	Increase of yield and juice clarification
	Glucose oxidase	Oxygen removal
Meat	Proteases (e.g., papain)	Meat tenderizing. Protein solubilization for sauces (fish)
Brewing	Cellulases, β-glucanases, α-amylases, proteases, maltogenic amylases	Liquefaction, clarification and supplement to malt enzymes
Alcohol production	Amyloglucosidase	Conversion of starch to sugar
Starch	α-Amylase, glucoamylases, hemicellulases, maltogenic amylases, glucose isomerases, dextranases, β-glucanases, cyclodextrin glycotransferase	Modification and conversion of carbohydrates (e.g., to dextrose or high fructose syrups)
Inulin	Inulinases	Production of high fructose syrups
Protein	Proteases, peptidases	Breakdown of proteins for feed

Table 2 Industrially-relevant enzymes produced using gene technology and their primary areas of application in the food industry.

Enzyme	Food application
α-Acetolactate decarboxylase	Brewing
α-Amylase	Baking, brewing, distilling, starch
Catalase	Mayonnaise
Chymosin (rennin)	Cheese
β-Glucanase	Brewing
α-Glucanotransferase	Starch
Glucose isomerase	Starch
Glucose oxidase	Baking, egg mayonnaise
Hemicellulase	Baking
Lipases	Fats, oils
Maltogenic amylase	Baking, starch
Phytase	Starch
Proteases	Baking, brewing, dairy, alcohol production, fish, meat, starch, vegetable
Pullulanase	Brewing, starch
Xylanase	Baking, starch

The following entries will provide an overview of the use of amylases, lipases, and proteases in food processing and transformation, including the occurrence and commercial production of relevant enzymes in these classes, their biological significance, applications in the food industry, and future trends.

REFERENCE

1. The Freedonia Group (2009) World Enzymes – Industry study with forecasts for 2013 and 2018, *Study 2506*. http://www.freedoniagroup.com (accessed January 4, 2010).

Enzymes: Amylases

Geoffrey Smithers
Geoffrey Smithers Consulting Group, Highett, Victoria, Australia

Peter Roupas
Brad Woonton
Chakra Wijesundera
Food and Nutritional Sciences, Commonwealth Scientific and Industrial Research Organisation (CSIRO), Werribee, Victoria, Australia

Abstract

Enzymes that convert starch belong to the amylase family. Because of their importance in food processing, amylases hold the maximum market share of food related enzyme sales world wide. The major amylases employed in the food industry are α-amylase, β-amylase, glucoamylase, limit dextrinase, and cyclodextrin glycosyltranferase. This entry summarizes the occurrence, biological significance, commercial production and application of amylases in the food industry.

DEFINITION

Based on amino acid sequence homology, the majority of the enzymes that convert starch belong to one family, the α-amylase family or "family 13" glycosyl hydrolases. This group of enzymes: 1) hydrolyzes α-glucoside bonds to produce α-anomeric mono- or oligosaccharides from α-1,4 or α-1,6 glycosidic linkages; 2) possesses a $(\beta/\alpha)_8$ or TIM barrel structure at the catalytic site; and 3) have four highly conserved regions in their primary sequence. By this system, there are at least 21 enzymes that can be classified as either starch-hydrolyzing or starch-modifying (transglycosylating) amylases.[1]

The most commonly used amylases in the food industry are those that degrade starch (amylose and amylopectin) by catalyzing the cleavage of the α-1,4- and α-1,6-glycosidic linkages to produce dextrins, oligosaccharides, maltose, and D-glucose. The main enzymes that break down starch are α-amylase, β-amylase, glucoamylase, and limit dextrinase. α-Amylase (1,4-α-D-glucan glucanohydrolase) is a metalloenzyme containing at least one calcium ion for stability. It has been assigned the number 3.2.1.1 by the Enzyme Commission of the International Union of Biochemistry and Molecular Biology (IUBMB) and is an endohydrolase that randomly attacks the α-1,4-glycosidic linkages of amylose to produce dextrins and oligosaccharides. β-Amylase (1,4-α-D-glucan maltohydrolase) has been assigned the number 3.2.1.2 by IUBMB and is an exohydrolase that cleaves the α-1,4-glycosidic linkages from the non-reducing end of amylose to remove maltose units successively. Glucoamylase (glucan 1,4-α-glucosidase) has been assigned the number 3.1.1.3 by IUBMB and is an exohydrolase that cleaves the α-1,4-glycosidic linkages from the non-reducing end of amylose to remove glucose units successively. Although the natural substrate of glucoamylase is the α-1,4-glycosidic linkage, it can catalyze the hydrolysis of α,β-1,1-, α-1,6-, α-1,3-, and α-1,2-glycosidic linkages. At high glucose concentrations, glucoamylases can catalyze the reverse reaction, reforming glycosidic bonds to produce isomaltose, isomaltotriose, kojibiose, nigerose, maltose, α,β-trehalose, panose, and isomaltotetraose in decreasing equilibrium concentrations. Pullulanase (pullulan α-1,6-glucanohydrolase) and limit dextrinase (dextrin α-1,6-glucanohydrolase) have been assigned the numbers 3.2.1.41 and 3.2.1.142 by IUBMB, respectively. These enzymes catalyze the hydrolysis of α-1,6-D-glucosidic linkages in pullulan, amylopectin, and the α- and β-amylase limit dextrins of amylopectin and glycogen when there is a maltose residue on either side of the α-1,6-linked glucose linkage. Cyclodextrin glycosyltransferase (CGTase; 1,4-α-D-glucan 4-α-D-(1,4-α-D-glucano)-transferase) has been assigned the number 2.4.1.19 by IUBMB. It is a unique member of the amylase family as its main products when hydrolyzing starch are cyclodextrins and high molecular weight dextrins.

OCCURRENCE, BIOLOGICAL SIGNIFICANCE, AND COMMERCIAL PRODUCTION

α-Amylase is widely distributed in nature, playing a key role in the metabolism of carbohydrates by microorganisms, plants, and animals. α-Amylase can easily be isolated from microorganisms, plants, and animals; however, the most economical means of production is by the fermentation of microorganisms such as *Bacillus stearothermophilus* and subsequent recovery of the enzyme using standard precipitation, membrane concentration, and chromatographic procedures. β-Amylase is not as widely distributed as α-amylase and can be isolated from higher plants and bacteria. It can be found in high concentrations in germinated barley, soybean, and sweet potato. Like α-amylase, β-amylases are commercially produced by fermentation of microorganisms from the genus *Bacillus*. Glucoamylase is produced by a number of bacteria, yeasts, and filamentous fungi, and is often produced commercially by the growth of *Aspergillus* species on suitable substrates. Bacterial pullulanases are commonly found in and purified from microorganisms of the genera *Klebsiella* and *Bacillus*; however, new sources of these enzymes are being found in hyperthermophilic bacteria such as *Thermococcus*.[2] Limit dextrinases are known to occur in several plants, but only the enzyme found in barley (*Hordium* spp.) is of commercial importance as it is found within the malted barley kernel and plays a significant role during brewing. CGTase has been identified in bacteria belonging to the genera *Bacillus*, *Clostridium*, *Klebsiella*, *Micrococcus*, *Thermoanaerobacter*, *Thermococcus*, and *Anaerobranca*. CGTases are usually produced commercially from *Bacillus macerans*;[3] however, heat- and acid-stable CGTases can be commercially produced from *Thermoanaerobacter*.

APPLICATIONS IN THE FOOD INDUSTRY

Amylases are employed extensively in the manufacture of food products, in the chemical analysis of food, and for thermal process control. Due to their wide usage within the food industry, amylases hold the maximum market share of enzyme sales worldwide.

Syrups

Apart from using starch-containing plant parts directly as a food source, starch is often harvested and chemically or enzymatically processed into starch hydrolysates such as glucose syrups, fructose syrups, starch or maltodextrin derivatives, and cyclodextrins.[1] The amount of starch hydrolysis and glucose (dextrose) in the syrup is given by the dextrose equivalents (DE), a measure of the amount of reducing equivalents expressed as glucose per unit dry weight. Glucose has a DE value of 100, maltose of 53, maltotriose of 36, and starch of almost 0.

Glucose containing syrups were traditionally produced by high temperature acid hydrolysis of starch. With the commercial production of thermophilic amylases, glucose, and maltose syrups are now produced by the addition of a heat stable α-amylase to a starch slurry, followed by heating to 95–100°C for approximately 90 min. During this process, the α-amylase hydrolyzes the starch to produce a solution containing linear and branched dextrins with a DE between 8 and 12. An exo-acting enzyme such as β-amylase or glucoamylase is then added that hydrolyzes the α-1,4-glycosidic bonds from the non-reducing end to produce glucose or maltose units, respectively. The high maltose syrups are sometimes preferred, as they have a higher viscosity and lower hygroscopicity, fewer tendencies to crystallize, and more resistance to browning than glucose syrups. To increase the yield of glucose and/or maltose, bacterial pullulanases can be added to supplement glucoamylase in hydrolyzing the α-1,6-glycosidic linkages of the branched dextrins. For each of the enzymatic catalyzed steps, the pH, calcium ion concentration, and the temperature of the starch solution is optimized for the enzyme employed. Novel applications of screw extruders or membranes[4] for the continuous hydrolysis and production of starch with enzymes have been developed and are currently employed by some manufactures. There is also a move towards entrapment of amylases in alginate beads for recovery and repeated use.[5]

Cyclodextrins

Cyclodextrins are cyclic α-1,4-linked oligosaccharides of 6, 7, or 8 glucose units (α-, β-, or γ-cyclodextrins, respectively) that are arranged so that the inside is hydrophobic and the outside hydrophilic.[3] Cyclodextrins have a number of potential applications in the food industry, including the masking of undesirable flavors,[6] reducing cholesterol adsorption, fat adsorption, and weight gain,[7] to deliver bioactive food components to various sites within the intestinal tract and to protect sensitive hydrophobic compounds from oxidation. In the production of cyclodextrins, starch is first liquefied by a heat-stable α-amylase followed by cyclization with a CGTase.[1] CGTase can also be used to produce novel glycosylated compounds, e.g., the intense sweetener, stevioside, has been glycosylated to improve solubility and decrease bitterness.[8]

Bread and Bakery Products

The baking industry is a large consumer of starch and starch-modifying enzymes from microbial and cereal sources. During bread manufacture, amylases are added to achieve two main outcomes. Firstly, during bread manufacture they are added to the dough to degrade the damaged starch into smaller dextrins, which are subsequently fermented by the yeast to produce carbon dioxide and ethanol. This increased fermentation activity results in increased loaf volume and improved crumb texture.[1] Directly after the manufacture of bread and bakery products, a series of changes start to occur that eventually result in the deterioration of quality, a term called staling. Staling causes an increase in the crumb firmness, a loss of crust crispness, a decrease in crumb moisture content, and a loss of flavor. The retrogradation of the amylopectin in the bread is considered to be one of the major factors contributing to the staling effect. Addition of α-amylase during the production process is thought to alter starch structure, particularly amylopectin and reduce retrogradation of bakery products. However, over-addition can lead to excess production of low molecular weight dextrins, and bread with a sticky and gummy texture. The use of de-branching enzymes such as pullulanase in combination with α-amylases are thought to overcome the problems associated with using α-amylase alone by rapidly hydrolyzing the low molecular weight dextrins. The addition of exoamylases such as β-amylase, amyloglucosidase, and other maltogenic amylases during bread making results in the cleavage of maltose, maltotriose, maltotetraose, or glucose molecules from amylopectin side chains, which is suggested to result in reduced amylopectin retrogradation.[9]

Beverages and Beer

Amylases find use in the beverage industry to remove starch, and to decrease beverage viscosity that leads to improved transmembrane flux rates during fruit juice clarification processes.[10] Amylases also find use in the production of malt-based beverages such as beer and whisky. During the brewing process, endogenous α-amylase, β-amylase, and limit dextrinase from malt break down the malt starch into fermentable sugars, which are then metabolized by yeast into carbon dioxide and ethanol. In the absence of malted grain in the grist, exogenous enzymes from microbial sources such as α-amylase are often added to improve the rate and quantity of fermentable sugar production. Although not widely employed, α-amylase, β-amylase, and limit dextrinase can be successfully utilized to produce alcoholic beverages from a range of unmalted cereals, including sorghum and barley.[11]

Infant Formula

α-Amylases from germinated cereal grains have been employed to reduce the viscosity and enhance the digestibility of gruels made from cowpea, sorghum, and other plant species. These partly digested meals significantly increase the energy intake of malnourished children following shigellosis. Although germinated cereals are the preferred and most economical source of α-amylase, the use of exogenous enzymes from microbial sources has also been reported as a successful means to enhance the viscosity and digestibility of similar gruels.[12]

Food Chemical Analysis

Amylases, in particular α-amylase, have found extensive use in the extraction of components and chemical analysis of food. For example, amylases are routinely employed to release fiber from food before quantification.[13] They are also used to determine the content of vitamin K1 (phylloquinone), B1 (thiamine), B2 (riboflavin), B3 (niacin), C (ascorbic acid), and B9 (folic acid) in a range of food matrices.[14] Amylases have found wide use in the analysis of starch, fructose, inulin, and elements in food. Amylases have also found use in the determination of added annatto coloring to food and in the extraction and analysis of lipids from extruded products.

Other Uses

A novel application of heat-stable α-amylases is in the development of time temperature indicators (TTI). Once the deactivation kinetics of α-amylase attached to glass beads or entrapped in gel beads have been studied, the TTIs can be subsequently employed to assess the extent of thermal processing of foods.[15]

Amylases have also been employed to significantly enhance the yield and quality of oil extracted from coconut and rice bran. They have also been reported to be useful in the manufacture of corn-based biodegradable desiccants, the production of food acids such as citric and lactic acid, and in the extraction of fiber and protein components on a large scale from corn and rice matrices.[16]

FUTURE APPLICATIONS AND INNOVATION

Since their discovery and commercial production, amylases have proven to be very valuable in the manufacture and chemical analysis of food products. Future trends may include the use of specific amylases to produce α-1,6-glucosidic linked oligosaccharides that reduce dental caries and improve intestinal microbiota. There may also be an increase in the use of biotechnology to not only improve enzyme production yields from microorganisms, but also to produce enzymes with enhanced catalytic properties. Such enzymes may have greater temperature, pH, and salt stability.

REFERENCES

1. Van Der Maarel, M.J.E.C.; Van der Veen, B.; Uitdehaag, J.C.M.; Leemhuis, H.; Dijkhuizen, L. Properties and applications of starch-converting enzymes of the alpha-amylase family. J. Biotechnol. **2002**, *94*, 137–155.
2. Duchiron, F.; Legin, E.; Gantelet, H.; Ladrat, C.; Barbier, G. New thermostable enzymes for crop fractionation. Ind. Crops Prod. **1997**, *6*, 265–270.
3. Biwer, A.; Antranikian, G.; Heinzle, E. Enzymatic production of cyclodextrins. Appl. Microbiol. Biotechnol. **2002**, *59*, 609–617.
4. Sarbatly, R.; England, R. Critical review of membrane bioreactor system used for continuous production of hydrolyzed starch. Chem. Biochem. Eng. Q. **2004**, *18*, 155–165.
5. Konsoula, Z.; Liakopoulou-Kyriakides, M. Starch hydrolysis by the action of an entrapped in alginate capsules alpha-amylase from *Bacillus subtilis*. Process Biochem. **2006**, *41*, 343–349.

6. Szejtli, J.; Szente, L. Elimination of bitter, disgusting tastes of drugs and foods by cyclodextrins. Eur. J. Pharm. Biopharm. **2005**, *61*, 115–125.
7. Somogyi, G.; Posta, J.; Buris, L.; Varga, M. Cyclodextrin (CD) complexes of cholesterol—their potential use in reducing dietary cholesterol intake. Pharmazie **2006**, *61*, 154–156.
8. Pedersen, S.; Dijkhuizen, L.; Dijkstra, B.W.; Jensen, B.F.; Jørgensen, S.T. A better enzyme for cyclodextrins. Chemtech **1995**, *25*, 19–25.
9. Wursch, P.; Gumy, D. Inhibition of amylopectin retrogradation by partial beta-amylolysis. Carbohydr. Res. **1994**, *256*, 129–137.
10. Ceci, L.N.; Lozano, J.E. Amylase for apple juice processing: Effects of pH, heat, and Ca^{2+} ions. Food Technol. Biotechnol. **2002**, *40*, 33–38.
11. Goode, D.L.; Halbert, C.; Arendt, E.K. Optimization of mashing conditions when mashing with unmalted sorghum and commercial enzymes. J. Am. Soc. Brew. Chem. **2003**, *61*, 69–78.
12. Chakravarthi, S.; Kapoor, R. Development of a nutritious low viscosity weaning mix using natural ingredients and microbial amylases. Int. J. Food Sci. Nutr. **2003**, *54*, 341–347.
13. Champ, M.; Langkilde, A.M.; Brouns, F.; Kettlitz, B.; Le Bail-Collet, Y. Advances in dietary fibre characterisation. 2. Consumption, chemistry, physiology and measurement of resistant starch; implications for health and food labelling. Nutr. Res. Rev. **2003**, *16*, 143–161.
14. Konings, E.J.M. A validated liquid chromatographic method for determining folates in vegetables, milk powder, liver, and flour. J. AOAC Int. **1999**, *82*, 119–127.
15. Tucker, G.S.; Lambourne, T.; Adams, J.B.; Lach, A. Application of a biochemical time-temperature integrator to estimate pasteurisation values in continuous food processes. Innov. Food. Sci. Emerg. Technol. **2002**, *3*, 165–174.
16. Tang, S.; Hettiarachchy, N.S.; Eswaranandam, S.; Crandall, P. Protein extraction from heat-stabilized defatted rice bran: II. The role of amylase, celluclast, and viscozyme. J. Food Sci. **2003**, *68*, 471–475.

Enzymes: Chymosin and Other Milk Coagulants

P.L.H. McSweeney
N. Bansal
Department of Food and Nutritional Sciences, University College Cork, Cork, Ireland

Abstract
The enzymatic coagulation of milk involves limited proteolysis of the caseins by enzymes in preparations called rennets, followed by aggregation of the rennet-altered casein micelles in the presence of calcium ions and at temperatures greater than ~18°C. In this entry, the milk protein system is reviewed briefly and the mechanism of rennet coagulation of milk is discussed. The different types of rennets and rennet substitutes are outlined and the role of residual rennet in cheese during ripening is also considered.

THE MILK PROTEIN SYSTEM

Various aspects of the milk protein system were comprehensively described by Swaisgood;[1] a brief summary is given here. Bovine milk, and probably the milk of all other species, contains two major and distinctly different categories of protein: 1) the caseins; and 2) whey proteins. The principal whey proteins in milk (~20% of total N, TN) are β-lactoglobulin and α-lactalbumin, with lesser amounts of blood serum albumin and immunoglobulins, and trace amounts of numerous other proteins, including at least 60 enzymes. The caseins, which constitute ~80% TN in bovine milk, are phosphoproteins, insoluble at pH 4.6 and 20°C and comprise four principal proteins, α_{s1}-, α_{s2}-, β-, and κ-caseins, in the approximate ratio of 40:10:35:12. Minor casein fractions are also present in milk, including the γ-caseins and most proteose peptones, derived from β-casein, and the so-called "λ-caseins" derived from α_{s1}-casein, both formed by post-translational proteolysis by plasmin, the principal indigenous proteinase in milk. More than 95% of the caseins in milk exist in the form of casein micelles. Casein micelles are spherical, 50–300 nm in diameter, with a M_r of ca. 10^8 kDa (i.e., on average, micelles contain ca. 5000 casein monomers, each of mass 20–24 kDa). On dry matter basis, micelles contain ~94% protein and 6% low molecular mass species, mainly Ca and PO_4 with some Mg and citrate, collectively known as colloidal calcium phosphate (CCP).

Although the precise structure of the casein micelle is a subject of debate, it is widely believed that α_{s1}-, α_{s2}- and β-caseins, which are sensitive to calcium, are protected from precipitation by κ-casein, which is located predominantly at the micelle surface. κ-Casein stabilizes the micelle by contributing to its zeta (surface) potential (~-20 mV) and by steric hindrance caused by its protruding C-terminal segments, which give the micelles a "hairy" appearance.

κ-Casein contains 169 amino acid residues and the N-terminal approximately two-thirds of the molecule is quite hydrophobic while the C-terminal region is hydrophilic and is usually gylcosylated; the Phe_{105}-Met_{106} bond is conventionally thought to be the division between these parts of the molecule. Recent studies of casein structure and function were reviewed by De Kruif and Holt.[2]

RENNETS AND RENNET SUBSTITUTES

"Rennet" is the generic term for an animal enzyme preparation used to coagulate milk. Originally, rennet was extracted from the abomasum of the young of the dairy animal (most commonly, the calf) which contains aspartyl proteinases; the principal enzyme in calf rennet is chymosin but smaller amounts of pepsin are also present (typically ~88–94% chymosin and 6–12% pepsin), depending on the age of the animal at slaughter; as the age at slaughter increases, the proportion of chymosin decreases and that of pepsin increases. Rennets extracted from the abomasa of calves, kids, or lambs were the principal commercial coagulants until recently. Other than animal rennets, natural coagulants from vegetable sources such as *Cynara cardunculus*, *Ficus carica*, *Arctium minus*, and *Solanum dobium* are also used traditionally in some parts of the world.

Calf chymosin exists as three different genetic variants: A, B, and C. Chymosin A and B are allelic variants, differing in only one amino acid (at position 243). Chymosin C is three amino acids shorter than the A and B variants. The A variant has slightly higher specific activity for cleaving κ-casein but is less stable than the B form; the C form has the lowest specific activity against κ-casein. The molecular weight of chymosin is in the range of 32–39 kDa. The pH optimum for chymosin varies with the substrate. For example, the pH optimum for chymosin on Na caseinate is ~3.5 and that for first stage of rennet action in milk is ~6.0. The optimum temperature for first phase of rennet action in milk is ~45°C at pH 6.6.[3]

With an increase in cheese production, it became necessary to find sources of rennets other than the abomasa of young dairy mammals. Although many, or perhaps most proteinases coagulate milk, only a few can be used successfully as rennet substitutes. For production of good quality cheese, it is necessary that the rennet substitute should have weak general proteolytic activity, narrow specificity on κ-casein, and is easily denatured in whey; otherwise, excessive proteolysis by the residual enzyme in the cheese curd during ripening may cause yield losses during manufacture and a bitter flavor in the final product. Aspartyl proteinases have a narrow specificity with a preference for peptide bonds to which a bulky hydrophobic residue supplies the carboxyl group. Among the available and potentially useful milk coagulants, pepsins from the stomach of adult ruminants or other animals and aspartyl proteinases of fungal origin were the first to be used as rennet substitutes; the most acceptable of these are bovine pepsin and the aspartyl proteinases from *Rhizomucor miehei*, *R. pusillus*, and *Cryphonectria parasitica*. Chicken and porcine pepsins have also been assessed for their suitability for cheese making but these enzymes are not used widely. The pH and temperature optima of the substitutes are not identical to that of calf chymosin. The pepsins and fungal proteinases are most stable at pH ~ 2–3 and 3.5–7.0, respectively. The temperature optimum for proteinases from *R. pusillus* and *R. miehei* is higher than that of chymosin. More recently, the gene for calf chymosin has been cloned into microorganisms permitting the production of chymosin B by fermentation. Such "fermentation-produced chymosins" are now widely used for cheese manufacture in many countries and they give excellent results.

MECHANISM OF RENNET COAGULATION OF MILK

Rennet is used to coagulate milk for the production of rennet-coagulated cheeses (including most important varieties and about 75% of total production) and rennet casein. The rennet coagulation of milk can be divided into two distinct but overlapping stages: 1) enzymatic proteolysis; and 2) aggregation and gelation of the renneted micelles in the presence of Ca^{2+} and at temperatures above ~18°C.

$$\text{Casein} \xrightarrow[\text{Rennet}]{\text{1st Stage}} \textit{para}\text{-Casein} + \text{(glyco)macropeptides} \xrightarrow[\text{Ca}^{2+}, >18°C]{\text{2nd Stage}} \text{Gel}$$

During primary stage of rennet action, κ-casein is hydrolyzed at (usually) or near its Phe_{105}-Met_{106} bond, thus removing a polypeptide (usually residues 106–169). Chymosin, pepsins, and the acid proteinases of *R. meihei* and *R. pusillus* hydrolyze the Phe–Met bond of κ-casein during coagulation, whereas the acid proteinase of *C. parasitica* cleaves the Leu_{104}–Phe_{105} bond. The hydrolysis of the Phe_{105}–Met_{106} bond in κ-casein by chymosin is several orders of magnitude faster than any other bond in the casein system; it appears that the conformation of the protein in the region of this bond (98–111) renders it highly susceptible to hydrolysis by aspartyl proteinases.[4] The hydrolysis of κ-casein releases a hydrophilic peptide, the (glyco)macropeptide, which is lost in whey; the remainder of the molecule (residues 1–105) remains attached to the casein micelle and is known as *para-κ*-casein. Cleavage of κ-casein leads to destabilization of the colloidal casein system in milk and aggregation of casein micelles at $\geq 18°C$. This aggregation then leads to the formation of a gel-like matrix which is processed further to yield cheese or rennet casein. The physiological role of chymosin is to coagulate milk in the stomach of the neonate thus delaying discharge into the small intestine thereby increasing the efficiency of digestion.

ROLE OF RENNET IN CHEESE RIPENING

While the major role of chymosin or other enzymes in rennet is to destabilize the casein micelle and thus bring about the gelation of milk, a step essential in the production of rennet-coagulated cheese, the small amount of coagulant trapped in the curds contributes to the ripening of many cheese varieties. Proteolysis in cheese during ripening is very complex and has been reviewed extensively.[5–10] Proteolysis involves enzymes from the milk (particularly plasmin), the coagulant, and the cheese microflora although only the contribution of the coagulant will be discussed here.

About ~0–15% of the rennet activity added to milk is retained in the cheese curd, depending on factors related to the milk or associated with the cheese manufacture and the coagulant itself. The factors related to milk and cheese manufacturing which influence rennet retention include the concentration of caseins in milk, casein micelle size, ratio of different caseins in micelles, ionic strength of milk, heat treatment of milk, extent of acidification of milk prior to addition of rennet, pH at whey drainage, cooking temperature, and method and the level of moisture in the final cheese.[11,12] Factors related to coagulant itself are the type and ratio of enzymes in the rennet blend, their stability to pH and temperature during cheese making, and the influence of pH on their ability to bind to the caseins.

Studies on cheese with a controlled microflora have shown that the residual coagulant is responsible for the level of proteolysis during ripening detectable by gel electrophoresis and for most of the nitrogen soluble in water or at pH 4.6; however little trichloroacetic acid (TCA)- or phosphotungstic acid (PTA)-soluble nitrogen (containing mainly short peptides and amino acids, respectively) is produced by residual coagulant.

Chymosin generally prefers to hydrolyze peptide bonds containing hydrophobic and aromatic amino acid residues. Relative to many other proteinases, chymosin is weakly proteolytic; indeed, limited proteolysis is one of the characteristics to be considered when selecting proteinases for use as rennet substitutes. As mentioned above, the primary chymosin cleavage site in the bovine milk protein system is the Phe_{105}-Met_{106} bond in κ-casein; α_{s1}-, α_{s2}- and β-caseins are not hydrolyzed during milk coagulation but may be hydrolyzed in cheese during ripening. In buffer at pH 5.4, chymosin cleaves β-casein at seven sites: Leu_{192}-Tyr_{193} > Ala_{189}-Phe_{190} > Leu_{165}-Ser_{166} = Gln_{167}-Ser_{168} = Leu_{163}-Ser_{164} > Leu_{139}-Leu_{150} = Leu_{127}-Thr_{128}. NaCl inhibits the hydrolysis of β-casein by chymosin to an extent dependent on pH; hydrolysis is strongly inhibited by salt which promotes hydrophobic interaction of the susceptible C-terminal regions of the protein. Extensive hydrolysis of β-casein by chymosin during cheese ripening (as might occur at low salt-in-moisture values) is highly undesirable as the peptides produced are very hydrophobic and bitter.[10]

The primary site of chymosin action on α_{s1}-casein is Phe_{23}-Phe_{24}; in 0.1 M phosphate buffer, pH 6.5, chymosin also cleaves α_{s1}-casein at Phe_{28}-Pro_{29}, Leu_{40}-Ser_{41}, Leu_{149}-Phe_{150}, Phe_{153}-Tyr_{154}, Leu_{156}-Asp_{157}, Tyr_{159}-Pro_{160} and Trp_{164}-Tyr_{165} and at pH 5.2 in the presence of 5% NaCl (i.e., similar to the conditions in many cheese varieties), in addition at Leu_{11}-Pro_{12}, Phe_{32}-Gly_{35}, Leu_{101}-Lys_{102}, Leu_{142}-Ala_{144} and Phe_{179}-Ser_{180}. The hydrolysis of α_{s1}-casein by chymosin is influenced by pH and ionic strength. α_{s2}-Casein appears to be relatively resistant to proteolysis by chymosin; cleavage sites are restricted to the hydrophobic regions of the molecule (residues 90–120 and 160–207).[10]

In cheeses with a high cooking temperature, e.g., Emmental, and in *pasta filata* varieties (e.g., Mozzarella), the curds for which are stretched at high temperatures, chymosin is extensively denatured and makes relatively little contribution to ripening. Similarly, if the pH of cheese increases during ripening (e.g., Camembert), chymosin activity is less important although substantial activity remains.

Proteolysis by residual rennet also influences cheese flavor. Some peptides produced by rennet action are small enough to influence flavor directly. These peptides are further hydrolyzed by microbial proteinases and peptidases to small peptides and amino acids. Peptides produced by the coagulant during ripening contribute at least to background flavor, and perhaps, to bitterness if the activity of such enzymes is excessive. Residual rennet may also be associated with softening of cheese texture through hydrolysis of its casein matrix; subsequent proteolysis by plasmin or bacterial proteinases further modifies texture. It was thought that hydrolysis of the Phe_{23}-Phe_{24} bond of α_{s1}-casein caused the marked softening characteristic of the early stages of the ripening of Cheddar cheese. However, recent work[13] suggests that the solubilization of calcium phosphate during the early stages of ripening is of more significance.

REFERENCES

1. Swaisgood, H.E. Chemistry of the caseins. In *Advanced Dairy Chemistry: Proteins*, 3rd Ed., Part A; Fox, P.F., McSweeney, P.L.H., Eds.; Kluwer Academic/Plenum Publishers: New York, 2003; Vol. 1, 139–201.
2. De Kruif, C.G.; Holt, C. Casein micelle structure, functions and interactions. In *Advanced Dairy Chemistry: Proteins*, 3rd Ed., Part A; Fox, P.F., McSweeney, P.L.H., Eds.; Kluwer Academic/Plenum Publishers: New York, 2003; Vol. 1, 233–276.
3. Crabbe, M.J.C. Rennets: general and molecular aspects. In *Cheese: Chemistry, Physics and Microbiology*, 3rd Ed.; Fox, P.F., McSweeney, P.L.H., Cogan, T.M., Guinee, T.P., Eds.; Elsevier Academic Press: Amsterdam, 2004; Vol. 1, 19–45.
4. Visser, S.; van Rooijen, P.J.; Slangen, C.J. Peptide substrates for chymosin (rennin). Isolation and substrate behaviour of two tryptic fragments of bovine κ-casein. Eur. J. Biochem. 1980, 108 (2), 415–421.
5. O'Keeffe, A.M.; Fox, P.F.; Daly, C. Proteolysis in Cheddar cheese: role of coagulant and starter bacteria. J. Dairy Res. 1978, 45 (3), 465–477.
6. Grappin, R.; Rank, T.C.; Olson, N.F. Primary proteolysis of cheese proteins during ripening. J. Dairy Sci. 1985, 68 (3), 531–540.
7. Sousa, M.J.; Ardo, Y.; McSweeney, P.L.H. Advances in the study of proteolysis in cheese during ripening. Int. Dairy J. 2001, 11 (4–7), 327–345.
8. Fox, P.F.; Law, J.; McSweeney, P.L.H.; Wallace, J. Biochemistry of cheese ripening. In *Cheese: Chemistry, Physics and Microbiology*, 2nd Ed.; Fox, P.F., Ed.; Chapman & Hall: London, 1993; Vol. 1, 389–438.
9. Fox, P.F.; Singh, T.K.; McSweeney, P.L.H. Proteolysis in cheese during ripening. In *Biochemistry of Milk Products*; Andrews, A.T., Varley, J., Eds.; Royal Society of Chemistry: Cambridge, 1994; 1–31.
10. Upadhyay, V.K.; McSweeney, P.L.H.; Magboul, A.A.A.; Fox, P.F. Proteolysis in cheese during ripening. In *Cheese: Chemistry, Physics and Microbiology*, 3rd Ed.; Fox, P.F., McSweeney, P.L.H., Cogan, T.M., Guinee, T.P., Eds.; Elsevier Academic Press: Amsterdam, 2004; Vol. 1, 392–433.
11. Holmes, D.G.; Duersch, J.W.; Ernstrom, C.A. Distribution of milk clotting enzymes between curd and whey and their

survival during cheddar cheese making. J. Dairy Sci. 1977, 60 (6), 862–869.
12. Garnot, P.; Molle, D.; Piot, M. Influence of pH, type of enzyme and ultrafiltration on the retention of milk clotting enzymes in Camembert cheese. J. Dairy Res. 1987, 54 (2), 315–320.
13. O'Mahony, J.A.; Lucey, J.A.; McSweeney, P.L.H. Chymosin-mediated proteolysis, calcium solubilization, and texture development during the ripening of Cheddar cheese. J. Dairy Sci. 2005, 88 (9), 3101–3114.

BIBLIOGRAPHY

1. Fox, P.F.; McSweeney, P.L.H. *Dairy Chemistry and Biochemistry*; Blackie Academic and Professional Publishers: London, 1998.
2. Horne, D.S.; Banks, J.M. Rennet-induced coagulation of milk. In *Cheese: Chemistry, Physics and Microbiology*, 3rd Ed.; Fox, P.F., McSweeney, P.L.H., Cogan, T.M., Guinee, T.P., Eds.; Elsevier Academic Press: Amsterdam, 2004; Vol. 1, 47–70.

Enzymes: Lipases

Geoffrey Smithers
Geoffrey Smithers Consulting Group, Highett, Victoria, Australia

Peter Roupas
Brad Woonton
Chakra Wijesundera
Food and Nutritional Sciences, Commonwealth Scientific and Industrial Research Organisation (CSIRO), Werribee, Victoria, Australia

Abstract
Lipases are a versatile class of enzymes that catalyze the hydrolysis of lipids. Lipases are responsible for a range of bioconversion reactions that are important in food processing and that influence food quality. This entry summarizes the occurrence, biological significance, commercial production and application of lipases in the food industry, and their influence on food quality.

DEFINITION

Lipases are enzymes that catalyze the hydrolysis of lipids. The reaction occurs at the lipid/water interface and is referred to as lipolysis. Since triacylglycerols (TAGs) are the most abundant class of lipids in nature, free fatty acids (FFA) and partial acylglycerols are the common products of lipolysis. However, under appropriate conditions, lipases also catalyze synthetic reactions, such as esterification, transesterification, alcoholysis, and acidolysis. The amount of water in the reaction mixture determines the direction of lipase-catalyzed reactions.[1] Excess water favors hydrolysis while the presence of little or no water favors esterification and transesterification. Many lipases function in organic solvents where they catalyze a number of useful reactions including esterification.[2] Lipases have been assigned the official name triacylglycerol acylhydrolases and the number 3.1.1.3 by the Enzyme Commission of the International Union of Biochemistry and Molecular Biology (IUBMB). Enzymes that hydrolyze esters of short-chain acids and water-soluble esters are called esterases.

OCCURRENCE AND COMMERCIAL PRODUCTION

Lipases occur in plants, animals, and microorganisms. They are prepared either by extraction from animal or plant tissues or by cultivation of microorganisms. Microorganisms are preferred to plant and animal as sources of lipases because they can be obtained in bulk, are generally cheaper to grow, and their lipase contents are more predictable and controllable.[3] The process involves fermentation followed by purification. The microbial lipases can be of fungal or bacterial origin. Pancreatic lipase of porcine origin is one of the earliest recognized lipases. Other important sources of animal lipases are the pancreas of cattle, sheep, hogs, and pigs. The current global enzyme market is estimated to be approximately $ 3.7 billion and lipases account for around 5% of this market.[1]

BIOLOGICAL SIGNIFICANCE

Lipases are indispensable for the bioconversion of lipids in nature such as in the biosynthesis of milk fat in the mammary gland. Several different lipases occur in the human body. They are primarily produced in the pancreas but also in the mouth and stomach. Some lipases are located inside living cells and degrade lipids, while others such as pancreatic lipases are located outside of cells and are involved in metabolism, adsorption, and transport of lipids throughout the body. Low levels of pancreatic lipase have been implicated in diseases such as pancreatitis, cystic fibrosis, and celiac disease. As a consequence, lipase supplements have been used to treat such conditions. Lipase inhibitors have been offered for the treatment of obesity as non-hydrolyzed fat cannot be absorbed by the gastrointestinal tract and, therefore, does not contribute to fat deposition.[2]

EFFECTS ON FOOD QUALITY

Lipolysis can have detrimental effects on the quality of food, and in particular, is a constant concern in the dairy industry. The significance of lipolysis in milk is two-fold: flavor production and altered functionality. Short- and medium-chain length FFA resulting from lipolysis have strong and rancid flavors that, in most cases, are considered undesirable. FFA are more susceptible to oxidation than intact TAGs, thus lipolysis exacerbates rancidity problems. Lipolysis causes depression of the foaming ability of milk when injected with steam, thus the difficulty in producing acceptable foam when making cappuccino coffee. The effect is due to the production of partial acylglycerols, which are surface active and displace the foam-stabilizing proteins at the air–water interface of the foam bubbles. Some other functionality defects, such as impaired creaming ability during separation and increased churning time in the manufacture of butter have also been reported.[4]

The lipolytic activity in bovine milk is due mostly, if not entirely, to lipoprotein lipase (LPL). It is involved in the synthesis of milk fat TAGs in the mammary gland and its presence in milk is due to a "spill over" from this gland. Although human milk lipase appears to have a major role in digestion of milk fat by the newborn, bovine LPL has no known biological role in milk.[4] Milk fat in freshly secreted milk is in the form of globules, which are enveloped in the protective milk fat globule membrane (MFGM). Although there is enough lipase activity in milk to cause rapid hydrolysis of a large portion of the milk fat, in reality this does not happen unless lipolysis is induced by physical damage to the MFGM during milk handling or processing. LPL is relatively unstable, and high-temperature, short-time pasteurization (72°C for 15 seconds) inactivates most, if not all, of the enzymes present in milk. LPL is also unstable to acid and hence would normally be inactivated in the stomach.

While lipolysis is undesirable in fresh milk, the same can be desirable, and in some cases necessary, for flavor development in cheese. In certain cheese varieties such as Blue and hard Italian cheeses (e.g., Romano, Parmasan, and Provolone) lipolysis reaches high levels and is a major pathway for flavor generation; however, the role of lipolysis in the flavor of Cheddar cheese and similar varieties is less apparent. Even though FFA may not directly contribute to the flavor of these cheese varieties, they can be converted to highly flavored compounds such as esters, methyl ketones, and lactones. Lipases catalyze not only the generation of FFA in cheese by lipolysis but also the conversion of the FFA to esters. For example, a lipase from *Candida cylindracea* can first hydrolyze milk fat and then esterify the released FFA with ethanol to produce ethyl esters in both aqueous and non-aqueous media. It has also been suggested that lipases can catalyze the biosynthesis of thioester flavor compounds in dairy systems provided that the thiols and FFA are available in sufficient amounts and that the water activity is favorable.[5]

APPLICATIONS IN THE FOOD INDUSTRY

Although the cost of production and scale up technologies continues to hinder wider commercial exploitation, lipases have tremendous potential in food processing and manipulation because of their versatility and ability to catalyze industrial processes under mild conditions, without the use of undesirable chemicals such as strong acids and bases and without generating high amounts of waste.

Enhancement of Cheese Flavor and Production of Cheese Flavorings

Lipase-mediated flavor generation in cheese occurs via lipases derived from starter, non-starter, or adjunct bacteria in cheese. Such reactions may be accelerated by addition of exogenous lipases.[5] Encapsulation technology has been used to regulate the reaction to prevent texture and flavor defects that can result from excessive lipolysis.[6] Another important application of lipases is in the manufacture of enzyme modified cheeses (EMCs). Compared to natural cheeses, EMCs have very high flavor intensities and are used as cheese-like flavorings in various food products such as soups, snacks, salad dressings, dips, and frozen foods. EMCs are produced by lipolysis of fresh cheese curd with selected lipases.

Infant Formulae

Most lipases are regiospecific and exclusively release the fatty acids attached to the *sn*-1 and *sn*-3 positions (the first and third hydroxyl group of glycerol) of the TAG. This regiospecificity of lipases has been utilized to produce structured lipids with particular biological or processing properties. In infant formulae, it is important to match human milk not only for the fatty acid composition but also for the positional distribution of fatty acids on the milk TAG molecules. Palmitic acid, which is the main saturated fatty acid in human milk, is located predominantly at the *sn*-2 position (the central hydroxyl group of glycerol) of the milk TAGs. During digestion, pancreatic lipases specifically hydrolyze fatty acids from the *sn*-1 and *sn*-3 positions, producing a monoacylglycerol with palmitic acid in the *sn*-2 position, which is more readily absorbed than free palmitic acid. Free palmitic acid binds to calcium ions and forms poorly absorbed insoluble soaps that cause constipation in infants.[6] The required structural configurations have been achieved in commercial infant formulae by acidolysis of tripalmitin in the presence of *sn*-1/*sn*-3 specific lipases.

Medium-Chain Triglycerides (MCTs) and Omega-3 Concentrates

TAGs with medium-chain fatty acids (MCFAs) at the *sn*-1 and *sn*-3 positions and a longer chain, polyunsaturated fatty acid (PUFA) at the *sn*-2 position have applications as highly digestible energy supplements for patients requiring artificial nutrition. In these applications, MCTs serve as a ready energy source and the PUFA provide the essential FA. Such specialty fats are also prepared by lipase catalyzed acidolysis of common fats and oils such as fish oils with MCFAs. Since the bioactive long-chain omega-3 acids such as eicosapentaenoic acid (EPA) and docosahexaenoic acid (DHA) are generally concentrated at the sn-2 position of fish oil TAGs, the product from this reaction is a mixture of TAGs in which omega-3 acids occupy the sn-2 position and medium-chain acids occupy the sn-1 and sn-3 positions.

The sn-1/sn-3 specificity as well as the relative lack of reactivity of lipase toward longer chain highly unsaturated fatty acids have been utilized in the preparation of long-chain omega-3 concentrates. Omega-3 acids are an important class of bioactive compounds that provide protection against a range of degenerative diseases including cardiovascular disease and are also essential for brain and vision development in infants. Fish oils and other oils of marine origin are the main source of long-chain omega-3 acids and in these oils the omega-3 acids are concentrated at the sn-2 position of the TAG. Hydrolysis of such oils in the presence of lipase removes acids other than omega-3 acids from the sn-1/sn-3 positions leaving behind a mixture of acylglycerols enriched in omega-3 acids, which can be further processed.

Diglyceride Oils

A vegetable oil–derived product predominantly composed of diacylglycerols (DAGs) has recently been introduced into the market.[7] DAGs have reduced absorption and consequently not transformed as much as TAG oils into body neutral fat. DAG oils are produced by lipase catalyzed partial hydrolysis of vegetable oils.

Cocoa Butter Substitutes

The sn-1/sn-3 regiospecificity of lipases is also utilized in the production of cocoa butter substitutes. Cocoa butter is a valued ingredient in the manufacture of chocolate primarily because of its unique melting characteristics resulting from special TAG molecules in which the sn-2 is predominantly occupied by oleic acid and the sn-1 and sn-3 positions by palmitic or stearic acids. Cocoa butter is expensive because of low availability. Cheaper alternatives have been produced from more abundant oils by introduction of palmitic or stearic acid to the sn-1 and sn-3 positions via lipase catalyzed reactions.

Trans Fat Alternatives

The solid fat content and melting characteristics required of margarine fats and bakery fats have traditionally been obtained by partial hydrogenation of liquid oils. This process also produces *trans* fats, and such industrially produced *trans* fats have been shown to exert the same detrimental effects on health as do saturated fats. Recent food labeling regulations in several countries, including Denmark, Canada, and the United States, aimed at discouraging the consumption of *trans* fat have hastened the search for alternatives to partially hydrogenated fats. Interesterification of liquid oils with more saturated or fully hydrogenated oils using immobilized lipase is being developed for this purpose.

FUTURE INNOVATIONS AND APPLICATIONS

The growth in the use of lipases in the food industry is likely to continue with the development of better and more diverse applications. This may be achieved by discovery/development of lipases with improved specificity toward TAG positions and/or fatty acids with known health implications. Better understanding of mechanisms of lipase action in food emulsion systems will assist in the development of food products with improved flavor, shelf life, and nutrition. Also, knowledge of the role of lipases in intestinal lipid absorption and lipoprotein metabolism will assist in the development of strategies for reducing the risk of obesity and atherogenesis.

REFERENCES

1. Vakhlu, J.; Kour, A. Yeast lipases: enzyme purification, biochemical properties and gene cloning. Electron. J. Biotechnol. **2006**, *9*, 69–85.
2. Schmid, R.D.; Verger, R. Lipases: interfacial enzymes with attractive applications. Angew. Chem. Int. Ed. **1998**, *37*, 1608–1633.
3. Sharma, R.; Chisti, Y.; Banerjee, U.C. Production, purification, characterization, and application of lipases. Biotechnol. Adv. **2001**, *19*, 627–662.
4. Deeth, H.C. Lipoprotein lipase and lipolysis in milk. Int. Dairy J. **2006**, *16*, 555–562.
5. Collins, Y.F.; McSweeney, P.L.H.; Wilkinson, M.G. Lipolysis and free fatty acid catabolism in cheese: a review of current knowledge. Int. Dairy J. **2003**, *13*, 841–866.
6. Houde, A.; Kademi, A.; Leblanc, D. Lipases and their industrial applications. Appl. Biochem. Biotechnol. **2004**, *118*, 155–170.
7. Bradford, L.L.; Egbert, W.R.; Gottemoller, T.V.; Lane, D.B.; Mount, M. (2006) Compositions containing protein and DAG oil and methods for making them. US Patent No. US 2006/0286280.

Enzymes: Molecular Aspects of Chymosin

Rodney J. Brown
College of Life Sciences, Brigham Young University, Provo, Utah, U.S.A.

Abstract

Chymosin (EC 3.4.23.4) is an aspartyl protease found in the stomachs of bovine calves. It is the enzyme of choice for milk coagulation in cheese making. Chymosin was one of the first enzymes purified and cheese making was one of the earliest applications of biotechnology or applied biochemistry. Because of the scarcity of chymosin, substitute enzymes from many sources have been found, none of which is as acceptable as chymosin. Production of chymosin by recombinant techniques has made pure chymosin available in commercial quantities.

INTRODUCTION

Chymosin (EC 3.4.23.4), formerly rennin, is an aspartyl protease found in the mucosa of the fourth stomach (abomasum) of young milk-fed bovine calves. Although chymosin can be found in the stomachs of many young ruminants, domestic bovine calves are the main source. The natural function of chymosin is coagulation of milk proteins during digestion; however, its use in the coagulation of milk during cheese making is of more interest here.

PREPARATION

Chymosin is prepared from calf stomachs by salt extraction followed by clarification and filtration.[1] Such preparations contain substances other than chymosin, including other proteolytic enzymes. As calves mature, enzymes extracted from their stomachs change from almost exclusively chymosin in young calves to predominantly pepsin (EC 3.4.23.1), another aspartic protease, in adult animals. Those preparations from young ruminants, containing mostly chymosin, are called rennet or calf rennet. Preparations from older animals, containing mostly pepsin, are called bovine rennet. Further complicating this issue, pepsin is sometimes added to calf rennet to make a less-expensive product. Chymosin can be further purified or crystallized for laboratory use.

The gene product for chymosin is preprochymosin, an inactive precursor form of the enzyme. As with many proteolytic enzymes, initial inactive forms exist to prevent the enzyme from being active too soon and functioning detrimentally. Proteolysis by enzyme or acid removes a 16-amino acid signal peptide to convert preprochymosin to prochymosin that later is cleaved again to remove a 42-amino acid propeptide to form active chymosin.[2] The active site is in a large groove on the side of the bean-shaped molecule. The two catalytic aspartic acid residues are buried at the bottom of the active site near the center of the molecule.

CHEESE MAKING

Cheese making is one of the earliest examples of biotechnology or applied biochemistry. Living organisms have long been used in fermentation processes, but the clotting of milk in cheese making requires soluble enzymes. Chymosin was one of the first enzymes purified.[3] Standardization of rennet in Denmark in the 1870s started the conversion of cheese making from an art to a science.[4]

Milk is a concentrated source of nutrients for young mammals. The challenge of accommodating a large amount of protein (ca. 3% of cow's milk or ca. 25% of milk solids) in a liquid medium is partially answered by the arrangement of ca. 80% of milk protein (α_{s1}, α_{s2}, β, and κ-caseins) in large suspended particles called casein micelles. Though they are more like multisubunit proteins than micelles in the strict chemical sense, casein micelles continue to be called micelles as they were originally named. The micelles vary in size but can occur as large complexes of 150,000 protein molecules each. In addition to casein, the micelles contain a high level of calcium phosphate. The κ-casein tends to be located on the surfaces of casein micelles and serves an important role in stabilizing them.

Milk coagulation is a complex process, involving a series of overlapping but distinct steps.[5] Chymosin initiates milk coagulation by specifically cleaving the Phe^{105}–Met^{106} bond in κ-casein, yielding *para*-κ-casein and a large peptide called glycomacropeptide.[6] As the name implies, the glycomacropeptide is the most hydrophilic portion of κ-casein. When it is released into the medium, *para*-κ-casein stays with the other caseins, which aggregate to form a network of protein strands that trap fat and other milk components to make cheese.[7]

Enzymatic activity in cheese making does not end with coagulation.[8] Residual enzymes in the cheese curd continue to hydrolyze proteins to produce cheese flavor compounds and influence curd texture during the long, slow process of cheese ripening. Since the cheese-making process follows a course of changes in temperature, acidity, and moisture, the stability and activity of enzymes over the course of these changes is important. Chymosin appears to be made for this process.[9]

CHYMOSIN SUBSTITUTES

Most proteolytic enzymes are able to clot milk. By most criteria, chymosin is the best enzyme for coagulation of milk in cheese making; however, a steady increase in cheese consumption and accompanying decreases in veal calf consumption over the past 50 years caused a shortage of calf stomachs. This, plus other factors such as religious dietary restrictions, motivated an intense search for enzymes to mimic and therefore replace or supplement chymosin in cheese making. Enzymes from animal, microbial (usually fungal), and plant sources have been used, but chymosin has never lost its place as the reference against which all others are compared.

Pepsin (EC 3.4.23.1) is the predominant enzyme in adult bovine stomachs and can be found in the stomachs of other mammals as well. When the pepsin source is swine stomachs, the term porcine pepsin is used. Chicken pepsin has been prepared for kosher use.[10]

Microbial milk-clotting enzymes have been extracted from many fungal sources, the most successful for cheese making being *Cryphonectria parasitica* (EC 3.4.23.22), *Mucor pusillus* (EC 3.4.23.23), and *Mucor miehei* (EC 3.4.23.23). These have been produced by several companies and sold under a number of different trade names. They are all proteases and they all coagulate milk, but they vary in general proteolysis and stability during the cheese-making process.

Plants have been used as sources of milk-clotting enzymes too. This has been limited to areas where consumption of animal products is prohibited for religious or other reasons.

Finally, genetic engineering has produced pure recombinant chymosin that is chemically and functionally identical to chymosin from bovine calves. Many organisms, including *Escherichia coli*, *Saccharomyces cerevisiae*, *Aspergillus niger*, *Aspergillus oryzae*, *Kluyveromyces marxianus*, and *Kluyveromyces lactis* have served as hosts for the chymosin gene. The choice of which organism provides the most commercially available chymosin is still being made in the market place; however, the shortage of chymosin is over and pure chymosin products are competing with the earlier chymosin substitutes with the former predominating over the latter. Recombinant chymosin has been accepted for use in cheese making even in areas where genetically modified organisms are generally not used. Genetically engineered chymosin was first used in commercial cheeses in the United States in 1991 and has remained popular ever since.

All milk-clotting enzymes have general proteolytic capability in addition to their ability to cleave κ-casein to initiate coagulation. The degree of activity varies, but most of the chymosin substitutes exhibit more general proteolysis than does chymosin. These enzymes also vary in their ability to survive the high temperature and acidity of cheese making, and their pH and temperature optima for activity also vary.[11] For example, porcine pepsin is destroyed during cheese making. These factors affect the degree to which the milk-clotting enzymes affect development of flavor and texture in cheese during aging. For this reason, the presence of some non-chymosin enzymes can be beneficial. This is one criticism aimed at use of pure, recombinant chymosin in cheese making.

CONCLUSIONS

The discovery that the stomachs of calves are able to coagulate milk led to the early isolation of chymosin and to the early application of applied biochemistry to the cheese industry. The cheese industry became a forerunner of modern biotechnology. Not too long ago, the desirable cheese-making characteristics of chymosin and the increasing popularity of cheese caused a shortage of chymosin. This increased attention paid to chymosin and chymosin substitutes eventually pushed chymosin to become one of the first commercial products of recombinant technology. Aside from its own importance as an enzyme, chymosin serves as a model for the development of other enzymes for industrially important applications of biotechnology.

REFERENCES

1. Berridge, N.J. Rennin and the clotting of milk. In *Advances in Enzymology and Related Subjects of Biochemistry*; Nord, F.F., Ed.; Interscience: New York, 1954; Vol. 15, 423–448.
2. Harris, T.J.R.; Lowe, P.A.; Lyons, A.; Thomas, P.G.; Eaton, M.A.; Millican, T.A.; Patel, T.P; Bose, C.C.; Carey, N.H.; Doel, M.T. Molecular cloning and nucleotide sequence of cDNA coding for calf preprochymosin. Nucleic Acids Res. **1982**, *10*, 2177–2187.
3. Foltmann, B. General and molecular aspects of rennets. In *Cheese: Chemistry, Physics and Microbiology*, 2nd Ed.; Fox, P.F., Ed.; Elsevier Applied Science: Barking, Essex, England, 1999; Vol. 1, 37–68.

4. Scott, R. Introduction to cheesemaking. In *Cheesemaking Practice*, 2nd Ed.; Elsevier Applied Science: London, 1986, 37–43.
5. Dalgleish, D.G. The enzymatic coagulation of milk. In *Advanced Dairy Chemistry-1; Proteins*; Fox, P.F. Ed.; Elsevier Applied Science: London, 1992; 579–619.
6. Foltmann, B. A review of prorennin and rennin. C. R. Trav. Lab. Carlsberg **1966**, *35*, 143–231.
7. Brown, R.J.; Ernstrom, C.A. Milk-clotting enzymes and cheese chemistry. Part I—Milk-clotting enzymes. In *Fundamentals of Dairy Chemistry*, 3rd Ed.; Wong, N.P.; Jenness, R.; Keeney, M.; Marth, E.H., Eds.; Van Nostrand Reinhold: New York, 1988; 609–633.
8. McMahon, D.J.; Brown, R. Effects of enzyme type on milk coagulation. J. Dairy Sci. **1985**, *68*, 628–632.
9. Brown, R.J. Dairy products. In *Enzymes in Food Processing*, 3rd Ed.; Nagodawithana, T.; Reed, G., Eds.; Academic Press: New York, 1993; 347–361.
10. Cogan, U.; Kopelman, I.J.; Schab, R. Combined temperature-concentration effects on the clotting rate of chicken pepsin. J. Dairy Sci. **1981**, *65*, 1130–1134.
11. Walsh, M.K.; Xiaosham, L. Thermal stability of acid proteases. J. Dairy Sci. **2000**, *67*, 637–640.

Enzymes: Proteases

Geoffrey Smithers
Geoffrey Smithers Consulting Group, Highett, Victoria, Australia

Chakra Wijesundera
Brad Woonton
Peter Roupas
Food and Nutritional Sciences, Commonwealth Scientific and Industrial Research Organisation (CSIRO), Werribee, Victoria, Australia

Abstract

Protease enzymes catalyze the hydrolysis of peptides bonds. Proteases are usually classified by the nature of the hydrolysis, their mechanism of catalytic action and their source. These enzymes are quite ubiquitous and have been used in food processing transformations for thousands of years. The major proteases employed in the food industry include chymosin, trypsin, and papain. This entry summarizes the occurrence, biological significance, commercial production and applications of proteases in the food industry.

DEFINITION

Proteases are enzymes that catalyze the hydrolysis of proteins and peptides by the addition of a water molecule across the peptide bond resulting in breakage of the amino acid chain at this point. They are usually classified according to the nature of their catalytic action on proteins and peptides (exo/endo proteases/peptidases), their source (microbial, plant, or animal), and the mechanism by which they catalyze the proteolytic reaction (i.e., molecular detail of the enzyme active site).

Proteases have been assigned the official name peptide hydrolases by the Enzyme Commission (EC) of the International Union of Biochemistry and Molecular Biology (IUBMB). They fall into subclass 3.4, and have been further defined according to their catalytic action as exopeptidases (EC 3.4.11–3.4.19) or endopeptidases (EC 3.4.21–3.4.24). Exopeptidases sequentially cleave one amino acid (or occasionally a dipeptide) from the N- (aminopeptidases) or C- (carboxypeptidases) terminus of the susceptible protein. By contrast, endopeptidases cleave the amino acid chain at susceptible points throughout the primary sequence of the protein substrate, such points being dependent upon the specificity of the particular endopeptidase. Endopeptidases are the most widely used proteases in food processing.[1,2]

Endopeptidases are usually classified on the basis of the functional amino acid or coenzyme at the active site of the enzyme, and include serine (EC 3.4.21) (e.g., trypsin, chymotrypsin), cysteine (EC 3.4.22), aspartic (EC 3.4.23) (e.g., pepsin), and metallo (EC 3.4.24) proteases. Blocking of the functional amino acid side-chain or removal of the coenzyme (e.g., Zn^{2+}) in the case of the metalloproteases leads to total inactivation. Activity of the endopeptidases is highly dependent upon pH. In general, the serine proteases show maximum activity at alkaline pH, the cysteine proteases at neutral pH, and the aspartic proteases at acid pH.

Aminopeptidases (EC 3.4.11) (e.g., pronase) are found widely throughout nature as intracellular or membrane-bound enzymes. They are used widely in biochemistry and biotechnology for extensive hydrolysis of proteins at bench-scale, but not in commercial food processing. Carboxypeptidases have been classified according to the functionality at the active site, namely, as serine (EC 3.4.16), metallo (EC 3.4.17), and cysteine (EC 3.4.18) carboxypeptidases. Dipeptide hydrolases (EC 3.4.13) have also been included in the exopeptidase subgroup and, as the name suggests, are specific for dipeptide substrates.

OCCURRENCE AND COMMERCIAL PRODUCTION

Proteases are present in all life forms, and animal and plant tissues and microorganisms have been used as raw materials for the manufacture of protease preparations for use in food processing and food component transformation. Examples include trypsin from ox or pig pancreas, papain from papaya, bromelain from pineapple, alcalase from *Bacillus licheniformis*, general protease from *Aspergillus oryzae*, chymosin from calf stomach and bacterial/fungal sources, and thermolysin from *Bacillus stearothermophilus*. In recent times, commercial production has focused more on the use of microbes because quality of the raw material is easier to control, cost of production is minimized, enhanced quantities of the target enzyme can be most easily engineered, and elimination of "carryover" of any troublesome disease agents can be assured. Fungal (e.g., from *Aspergillus* species) and bacterial (e.g., from *Bacillus* species) protease preparations are used widely in the food industry. New food-approved enzymes, many with unique characteristics (e.g., thermal/pH stability and unusual/targeted specificity), are being developed via natural selection or genetic engineering and are produced via genetically modified organisms (GMO).

A number of membrane-bound or intracellular exopeptidases are used extensively for complete hydrolysis of proteins on a small-scale,[3] but are too expensive for industrial-scale use such as in food processing, because they are difficult to isolate cost effectively.

Commercial proteases are supplied as liquid "ready-to-use" preparations or as stabilized solids containing excipients as carriers and/or stabilizers. Crystalline purified protease preparations are rarely used in food processing for handling and safety reasons, the latter associated with possible inhalation of "protease dust" and the resultant undesirable physiological effects and possible allergic reactions. Proteases account for ~40% of the total commercial enzymes in the United States[4] and approximately 24% (as a percentage of revenue) in food applications in Europe.[5]

BIOLOGICAL SIGNIFICANCE

Proteases are indispensable in the bioconversion of dietary proteins into their constituent peptides and amino acids. Examples include pepsin in the stomach, and trypsin and chymotrypsin in the intestine. These proteases act on their protein substrates releasing peptides and constituent amino acids that are absorbed by the body primarily for tissue structural growth and repair and other physiological functions (e.g., certain amino acids act as neurotransmitters).

APPLICATIONS IN THE FOOD INDUSTRY

The manufacture of many food products requires the judicious use of proteases. Examples include cheese, bread, and beer, since proteolysis in the raw materials of these products is a critical reaction during their production. In general, proteases exploited in food processing originate from i) the food material itself (e.g., autolysis of fish protein in the manufacture of Asian fish sauce); ii) microbes growing in the food (e.g., manufacture of some alcoholic beverages); iii) extraneous enzyme preparations added to the food (e.g., manufacture of hydrolysates for clinical nutrition products); and iv) a combination of one or more of the above [e.g., cheese making (ii) and maturation (iii)].[2] An undesirable consequence of the use of proteases in food processing has often been the generation of peptide-based bitterness originating from short-medium chain length peptides containing hydrophobic amino acids. Flavourzyme®, an exopeptidase preparation from *Aspergillus oryzae*, has shown promise in reducing the bitterness of a range of hydrolyzed protein preparations.[3]

Proteases Used as Processing Aids

Proteases are used to increase productivity for specific transformations leading to traditional food products. Perhaps the best-known and most longstanding example is the application of chymosin (rennet) from animal sources in the treatment of milk as an early step in the cheese-making process. Chymosin (an endopeptidase) cleaves κ-casein leading to destabilization of the casein micelles and formation of the curd. The origins of this process can be traced back to at least 800 B.C., and it is thought to have originated from the observation that milk transported in calves' stomachs coagulated through the action of chymosin in the stomachs. In the mid-1800s, calf rennet became a standardized product for use in commercial cheese-making. Nowadays, microbial and fungal chymosin preparations, many from engineered organisms, are widely used in the dairy industry for cheese-making.[6]

Proteases (e.g., mixed protease preparations of pepsin, trypsin, and chymotrypsin) are used in the baking industry to modify wheat gluten and wheat flour in order to enhance their functionality in various baked goods applications.[7]

In the brewing industry, a number of proteases ensure processing efficiency and quality of the final product. Examples include treatment of beer with proline-specific proteases to eliminate haze formation during cold storage,[8] and the use of neutral proteases from bacterial sources during mashing of malt and grain to enhance the nitrogen content of the wort.[9]

Papain has been used for hundreds of years to tenderize meat and is still the most widely used protease for this purpose. It can be applied as a powdered enzyme or an enzyme solution that is injected into the carcass. The proteolytic reaction has to be controlled as excessive protein degradation can lead to severe loss of meat quality.[3] By contrast, total degradation of fish protein into soluble amino acids and short peptides is the aim during manufacture of Asian fish sauces. This process employs a combination of fish autolysis and the action of introduced proteases/exopeptidases.[3]

Proteases Used in the Manufacture of New Functional Protein and Peptide Ingredients/Products

Applications of the vast and growing array of proteases can provide a means to prepare new and novel protein and peptide ingredients and foods for everyday and specialized purposes. For example, carefully controlled and limited proteolysis of food protein sources can enhance existing or reveal new functional properties, including solubility, aeration (foam formation), emulsification, and bioactivity.[10]

An enzymatic hydrolyzate of soybean proteins was first used some 50 years ago, most notably as an egg white substitute, and more recent improvements to the hydrolysis process have provided soluble soybean hydrolyzate preparations for fortification of beverages, in nutritional applications, and as a food functional (emulsification, foaming) ingredient.[11]

Dairy proteins have been used as substrates for the manufacture of a range of casein and whey protein hydrolysates. The latter, for example, are preferred by elite athletes and body builders over egg protein, because of their amino acid compositional balance and the high content of branched chain amino acids that are thought to be important in muscle recovery after exercise. Dairy protein hydrolysates are available on the market as stand-alone ingredients and as "ready-to-make" powdered beverage products. In situ hydrolysis of dairy proteins through the action of proteases from fermentative lactobacilli forms the basis for the manufacture of "bioactive" fermented dairy drinks available in Europe and in Japan. These products contain bioactive tripeptides derived from casein, and have clinically proven effects in reducing hypertension in the consumer.[12]

FUTURE INNOVATIONS AND APPLICATIONS

Of the myriad proteases that have been identified by scientists, only a few are widely used in food transformation mainly due to non-ideal reaction conditions, instability, or prohibitive cost. These restrictions are set to be addressed through future innovations in biotechnology and biochemistry, notably genetic and protein engineering. Application of these disciplines will allow the "tailoring" of proteases for specific purposes and the cost-effective production of stable enzymes with optimum activity under the desired food processing conditions. Furthermore these techniques can aid in acquiring a detailed understanding of the structure–function relationship of proteases that will permit enhancement of specificity, stability, and activity of these enzymes. They can also facilitate production of proteases from unusual plant, animal, and microbial sources with desirable characteristics (e.g., stability and specificity) for a nominated purpose in food processing and transformation.

REFERENCES

1. Perlmann, G.E.; Lorand, L. *Methods in Enzymology*; Vol 19: *Proteolytic Enzymes*; Academic Press: New York, 1970.
2. Adler-Nissen, J. *Proteases in Enzymes in Food Processing*, 3rd Ed.; Academic Press: New York, 1993; 159–203.
3. Raksakulthai, R.; Haard, N.F. Exopeptidases and their application to reduce bitterness in food: a review. Crit. Rev. Food Sci. Nutr. **2003**, *43* (4), 401–445.
4. Frost and Sullivan. US Commercial Enzyme Markets, Report 5113-39, 1998, http://www.frost.com (accessed January 6, 2010).
5. Frost and Sullivan. Advances in Industrial Enzymes—A Global Technology Assessment, Report D324, 2004, http://www.frost.com (accessed January 6, 2010).
6. Neelakantan, S.; Mohanty, A.K.; Kaushik, J.K. Production and use of microbial enzymes for dairy processing. Curr. Sci. **1999**, *77* (1), 143–148.
7. Song, K.A.; Koh, B.K. Effect of enzymatically hydrolyzed vital wheat gluten on dough mixing and the baking properties of wheat flour frozen dough. Food Sci. Biotechnol. **2006**, *15* (2), 173–176.
8. Lopez, M.; Edens, L. Effective prevention of chill-haze in beer using an acid proline-specific endoprotease from *Aspergillus niger*. J. Agric. Food Chem. **2005**, *53* (20), 7944–7949.
9. Goode, D.L.; Halbert, C.; Arendt, E.K. Mashing studies with unmalted sorghum and malted barley. J. Inst. Brew. **2002**, *108* (4), 465–473.
10. Haard, N.F. Enzymatic modification of food proteins. In *The Chemical and Functional Properties of Food Proteins*; Sikorski, Z., Ed.; Technomic Publishing: Lancaster, 2001, 155–190.
11. Moure, A.; Sineiro, J.; Dominguez, H.; Parajo, J.C. Functionality of oilseed protein products: a review. Food Res. Int. **2006**, *39* (9), 945–963.
12. Takano, T. Milk derived peptides and hypertension reduction. Int. Dairy J. **1998**, *8* (5–6), 375–381.

Estrus Synchronization

Pietro S. Baruselli
Department of Animal Reproduction, University of Sao Paolo, Sao Paolo, Brazil

Gabriel A. Bó
Institute of Animal Reproduction, Cordoba (IRAC), Córdoba, Argentina

Abstract

Our expanding knowledge of the control of follicular wave dynamics during bovine estrous cycle has resulted in new prospects of precisely controlling the occurrence of estrus and ovulation. Follicular wave development and ovulation can be synchronized by treatments with prostaglandin and GnRH or a combination of estradiol and progestogen/progesterone. Synchronization protocols dramatically improve reproductive success by offering the possibility of planning the application of assisted reproductive technologies and allowing producers to breed more cattle in less time.

INTRODUCTION

Reproductive biotechnologies such as artificial insemination (AI), embryo transfer, and in vitro embryo production have been widely studied and globally utilized in a successful manner. These technologies enabled a faster and a more efficient genetic improvement of beef and dairy herds. However, estrous detection failure is one of the major problems that affects breeding efficiency in cows and impairs the widespread use of reproductive biotechnologies. Therefore, an alternative to increase the number of animals submitted to these biotechnologies is the use of treatments that synchronize the time of estrus and ovulation.

The goal of the synchronization program is to control the luteal phase and the follicular growth and the time of ovulation, thus allowing the application of biotechnologies to a large number of animals. Hormones used for estrus synchronization programs mimic those that occur naturally during estrous cycles.

OVARIAN FOLLICULAR DYNAMICS DURING THE ESTROUS CYCLE

In order to understand the benefits of the use of estrus synchronization protocols and how they work, it is necessary to first comprehend the concept of ovarian follicular waves. The existence of waves of follicle development during estrous cycles has been documented in cattle.[1] By definition, a wave of follicle development consists of the simultaneous growth of a large number of small antral follicles (follicular recruitment) followed by the selection of a dominant follicle and regression of subordinate follicles. The dominant follicle is defined as the largest follicle of the follicular wave and it has the ability to suppress the growth of smaller follicles. This phenomenon is regulated by a follicular selection mechanism which causes atresia of all the other follicles in the cohort.[2]

Several studies showed that the recruitment of follicular waves and the selection of the dominant follicle are based on differential responsiveness of follicles to FSH and LH. As reviewed by Mapleptoft et al.,[3] surges in plasma FSH concentrations are followed 1–2 days later by the emergence of a new follicular wave, while FSH is subsequently suppressed by hormones produced by the growing follicles (e.g., estradiol and inhibin). In each wave, the dominant follicle acquires LH receptors and continues to grow while subordinate follicles (that continue to depend on FSH) undergo atresia. Suppression of LH, as a consequence of progesterone secretion by the corpus luteum (CL), causes the dominant follicle to cease its metabolic functions and regress, leading to a new FSH surge and emergence of a new follicular wave.[3] Luteal regression allows LH pulse frequency to increase, and the largest follicle present at that time increases its growth and secretes relatively high amounts of estradiol that results in positive feedback on the hypothalamus, LH surge, and ovulation.

SYNCHRONIZATION OF ESTRUS AND OVULATION USING PROSTAGLANDIN (PGF$_{2\alpha}$)

Prostaglandin (PGF$_{2\alpha}$) is the most commonly used treatment for synchronizing estrus in cattle.[4] Both PGF$_{2\alpha}$ and its analogues induce regression of the CL in cycling animals with a functional CL at the time of treatment, and this can be followed by estrous detection and breeding. The major limitation of PGF$_{2\alpha}$ is the lack of efficiency in females without a CL, e.g., females within 5–6 days of a previous estrus, prepubertal heifers, and postpartum anestrous cows. In cows with a CL, luteolysis of the CL and estrus can be distributed over a 6-day period after $PGF_{2\alpha}$ treatment.[5] The interval from $PGF_{2\alpha}$ treatment to ovulation is determined by the stage of development of the dominant follicle at the time of treatment.[5] If $PGF_{2\alpha}$ is given when the dominant follicle is in the late growing or early static phase, ovulation will occur within 3–4 days. On the other hand, $PGF_{2\alpha}$ treatment given when the dominant follicle is in the mid- to late-static phase (i.e., when it is no longer viable) will result in ovulation of the dominant follicle from the next follicular wave 5–7 days later.[5] The interval is a reflection of the time required for the dominant follicle of the new wave to grow and develop to a preovulatory state. This emphasizes that both luteal and follicular control are required to obtain satisfactory synchronization of estrus.

SYNCHRONIZATION OF ESTRUS AND OVULATION USING PROSTAGLANDIN ($PGF_{2\alpha}$) AND GnRH

Currently there are many estrus synchronization protocols that use gonadotropin-releasing hormone (GnRH) in conjunction with $PGF_{2\alpha}$. An injection of GnRH in any phase of the estrous cycle results in a peak of LH that promotes the ovulation of the dominant follicle and the emergence of a new follicular wave 2–3 days later.[6] With the ovulation of the dominant follicle, concentrations of progesterone remain high; therefore, when $PGF_{2\alpha}$ is given 7 days later, it induces luteolysis providing ovulatory conditions to the dominant follicle of this follicular wave. Another GnRH injection should be administered 48–56 hours after the $PGF_{2\alpha}$ injection for better synchronization of ovulation and fixed timed AI (FTAI) 16 hours later. This protocol is known as the Ovsynch protocol.[6] Use of the Ovsynch protocol has resulted in acceptable pregnancy rates after FTAI in cycling cows, but pregnancies are typically lower in heifers and also cows in postpartum anestrus.[6,7]

SYNCHRONIZATION OF ESTRUS AND OVULATION USING PROGESTERONE

For more than 30 years, studies have been performed to evaluate different progesterone/progestogen treatments in various estrus synchronization schemes.[4] Results have shown that progestogen treatments, if given long enough to allow normal regression of the CL (≥ 14 days), induce synchronous estrus. However, these treatments are associated with reduced fertility.[8]

Currently, recommended doses of progestogen/progesterone used to control the estrous cycle in cattle (MGA feeding, norgestomet ear implant, and progesterone-releasing intravaginal devices) are associated with greater secretion of LH secretion compared with the normal luteal phase, and increased-LH supports the development of persistent dominant follicles.[8] Therefore, the lowered fertility associated with progestogen treatment for the synchronization of estrus has been attributed to prolonged persistence of the dominant follicle and consequently the ovulation of a subfertile oocyte.[8] Treatments that induce regression of the persistent follicle have resulted in emergence of a new follicular wave and improved pregnancy rates. This has led to the development of estrus synchronization regimens that involve a short-term progestogen treatment combined with an induced regression of the persistent follicle at the beginning and a prostaglandin injection shortly before or at the termination of the progestogen treatment.[8]

Various progesterone-releasing intravaginal devices impregnated with different amounts of progesterone (0.5–1.9 g) are commercially available. These devices contain natural progesterone, and blood concentrations around 4–5 ng/mL of progesterone are reached shortly after insertion. Ear implants contain norgestomet, which has a greater biological potency than natural progesterone. Thus, lower doses of norgestomet compared with progestogen are required (3–6 mg). Melengestrol acetate (MGA) is another synthetic form of progesterone. MGA is mixed in feed and the consumption amount per day is around 0.5 mg. Most estrus synchronization programs recommend 14 days of MGA feeding. During MGA treatment, estrus and ovulation are suppressed and after removal of MGA from feed, females show estrus within 2–6 days. It is not recommended to inseminate at the estrus immediately following MGA feeding because of poor fertility (persistent dominant follicle). Acceptable fertility has been reported when $PGF_{2\alpha}$ is administered 17 days after MGA feeding and heifers are bred 12 hours after estrus.[9]

SYNCHRONIZATION OF ESTRUS AND OVULATION USING PROGESTERONE AND ESTRADIOL

The use of progesterone-releasing intravaginal devices and estradiol benzoate together is currently one of the most popular treatments for estrus and ovulation synchronization in cattle in South America.[7,8,10] The use of estradiol in combination with progestogen/progesterone treatments was introduced in the 1960s when it was found that estradiol induced luteal regression,[11] and therefore, this hormone was incorporated into treatments for estrus synchronization in cattle. However, the effects of estradiol on ovarian follicular growth were elucidated more than 25 years later. During the 1990s, it was observed that estradiol suppressed antral follicle growth.[8] The administration of 5 mg estradiol-17β in progestogen-implanted cattle synchronized the emergence of a new follicular wave, on average, 4.3 ± 0.2 days later.[8] The mechanism of estrogen-induced suppression of follicular growth appears to be systemic, and involves suppression of FSH. Once the estradiol was metabolized, there was an FSH surge, and a new follicular wave emerged. A lower dose of estradiol is normally given at the time of or 24 hours after progestogen removal to induce a synchronous LH surge and ovulation.[10] This has permitted fixed-time AI with relatively high pregnancy rates. Unfortunately, estradiol cannot be used in some countries.

Some estrus synchronization protocols have the ability to shorten the anestrous postpartum interval. The use of equine chorionic gonadotropin(eCG) at the time of progesterone-releasing device removal resulted in increased pregnancy rates in suckled cows treated during postpartum anestrus.[7] The beneficial effect of eCG treatment seems to be through stimulation of dominant follicle growth and maturation, resulting in increased ovulation rate in anestrous animals and greater progesterone production by the subsequent CL.[7]

More recent studies have revealed that it is not only possible to synchronize the timing of ovulation for fixed-time AI in single ovulating animals, but also in superstimulated donors and embryo recipients.[3]

CONCLUSION

The widespread use of emerging reproductive technologies requires precise methods of estrous cycle control. Effective control of the estrous cycle requires the synchronization of both luteal and follicular functions. The estrus synchronization treatments may facilitate the application of assisted reproductive biotechnologies, thus allowing a more efficient genetic improvement in herds all over the world.

REFERENCES

1. Ginther, O.J.; Kastelic, J.P.; Knopf, L. Composition and characteristics of follicular waves during the bovine estrus cycle. Anim. Reprod. Sci. **1989**, *20* (3), 187–200.
2. Ginther, O.J.; Wiltbank, M.C.; Fricke, P.M.; Gibbons, J.R.; Kot, K. Selection of the dominant follicle in cattle. Biol. Reprod. **1996**, *55* (6), 1187–1194.
3. Mapletoft, R.J.; Bó, G.A.; Baruselli, P.S. Control of ovarian function for assisted reproductive technologies in cattle. Anim. Reprod. **2009**, *6* (1), 114–124.
4. Odde, K.G. A review of synchronization of estrus in postpartum cattle. J. Anim. Sci. **1990**, *68* (3), 817–830.
5. Kastelic, J.P.; Ginther, O.J. Factors affecting the origin of the ovulatory follicle in heifers with induced luteolysis. Anim. Reprod. Sci. **1991**, *26* (1–2), 13–24.
6. Pursley, J.R.; Mee, M.O.; Wiltbank, M.C. Synchronization of ovulation in dairy cows using $PGF_{2\alpha}$ and GnRH. Theriogenology **1995**, *44* (7), 915–923.
7. Baruselli, P.S.; Reis, E.L.; Marques, M.O.; Nasser, L.F.; Bó, G.A. The use of hormonal treatments to improve reproductive performance of anestrous beef cattle in tropical climates. Anim. Reprod. Sci. **2004**, *82/83*, 479–486.
8. Bó, G.A.; Baruselli, P.S.; Moreno, D.; Cutaia, L.; Caccia, M.; Tríbulo, R.; Tríbulo, H.; Mapletoft, R.J. The control of follicular wave development for self-appointed embryo transfer programs in cattle. Theriogenology **2002**, *57* (1), 53–72.
9. Patterson, D. J.; Kojima, F. N.; Smith, M. F. A review of methods to synchronize estrus in replacement beef heifers and postpartum cows. J. Anim. Sci. 2003 81 (E. Suppl. 2) : E166–177.
10. Ayres, H.; Martins, C.M.; Ferreira, R.M.; Mello, J.E.; Dominguez, J.H.; Souza, A.H.; Valentin, R.; Santos, I.C.; Baruselli, P.S. Effect of timing of estradiol benzoate administration upon synchronization of ovulation in suckling Nelore cows (*Bos indicus*) treated with a progesterone-releasing intravaginal device. Anim. Reprod. Sci. **2008**, *109* (1–4), 77–87.
11. Wiltbank, J.N.; Ingalls, J.E.; Rowden, W.W. Effects of various forms and levels of estrogens alone or in combination with gonadotrophins on the estrus cycle of beef heifers. J. Anim. Sci. **1961**, *20* (1), 341–346.

Farm Animals: Embryo Transfer

John F. Hasler
Bonner Peak Ranch, Laporte, Colorado, U.S.A.

Abstract

Embryo transfer is one step in the process of removing one or more embryos from the reproductive tract of a donor female and transferring them to one or more recipient females. Embryos can also be produced in the laboratory via techniques such as in vitro fertilization (IVF) or more sophisticated technologies such as cloning or production of transgenic embryos. These technologies are of little value without adding the embryo transfer step. Embryo transfer has been used extensively in farm animals including cattle, horses, sheep, goats, and pigs and to a lesser extent in buffaloes and most species of camelids. Embryo transfer also has been successful in domestic dogs and cats and in virtually all species of laboratory rodents and in several species of primates. Lastly, embryo transfer has been successful on a limited basis in numerous wild mammals. However, availability of suitable recipients for rare or endangered species has limited the application of the technology. This entry focuses on embryo transfer in farm animals.

BACKGROUND

The first successful embryo transfers were conducted in 1890 in rabbits at Cambridge, England (for review see Hasler).[1] Subsequently, there were only a few reports of embryo transfers in any species until the first calf was born in 1951. Commercial embryo transfer in farm animals started with cattle in North America in the early 1970s. The initial primary economic driving factor was the high prices being paid for various breeds of so called "exotic" beef cattle imported into North America in small numbers from Europe. Over the past 35 years, a large international bovine embryo transfer industry has developed. It was recently estimated that in North America, approximately 1 out of every 300 dairy calves is born from embryo transfer.[2]

Embryo transfer is usually used to increase the number of genetic offspring that a donor female can produce in a given unit of time because recipients gestate the pregnancies to term. On the average, dairy cattle in North America produce only about three calves in their lifetime. Under some circumstances, with repeated superovulation, individual donor cows have produced more than 100 offspring in 2–3 years.

EMBRYO TRANSFER INDUSTRY

The exact size of the embryo transfer industry is difficult to determine, but it is truly international in scope. The International Embryo Transfer Industry conducts an annual assessment of the number of embryos recovered and transferred worldwide. A summary of the results for in vivo-derived embryos in cattle and horses for 2005 is shown in Table 1 and for in vitro-derived embryos in cattle in Table 2.[3] Clearly, cattle represent the greatest amount of commercial activity and the largest portion of the industry is centered in North America. A growing percentage of cattle embryos in North America and Europe are frozen instead of being transferred fresh after collection, whereas in South America, a very high percentage is transferred fresh. This probably represents a combination of differences in culture, export markets, and confidence in the technology of embryo freezing.

SUPEROVULATION

Embryos can be recovered from donor females following a normal estrous cycle that does not involve any hormonal stimulation. However, this limits the potential number of embryos available, especially in cattle. Consequently, in most cases donors are superovulated. Superovulation is defined as the treatment of a female with gonadotropins so that more ova than normal are ovulated. The most common gonadotropin employed is follicle-stimulating hormone, which is derived from either porcine or ovine pituitary glands. Regulation of the estrous cycle of the donor with prostaglandin F2α and progesterone usually is done concurrently with superovulation. Superovulated cattle are usually inseminated artificially, and produce approximately 10 ova of which 6 are viable embryos. The range of responses, however, is very large in all species; cattle on rare

Table 1 The number of in vivo-derived cattle and horse embryos collected and transferred in different regions of the world in 2005.

Region	Bovine				Equine
	No. recoveries	No. embryos	No. fresh transfers	No. frozen transfers	No. recoveries[a]
Africa	1,893	12,612	3,453	3,223	77
Asia	19,811	135,633	49,814	65,745	NA[b]
North America	65,520	392,232	130,523	146,223	12,000[a]
South America	26,052	150,434	110,817	14,433	12,100[c]
Europe	16,995	96,581	36,500	48,787	NA
Oceania	590[d]	2,480	1,300	1,360	NA
Total	130,861	789,972	332,407	279,771	24,177

[a]Partially estimated.
[b]Not available.
[c]Includes only Brazil and Columbia.
[d]Highly underreported.
Source: From Embryo Transf. Newsl.[3]

occasions produce as many as 100 ova yielding 70 or more viable embryos. Superovulation has proven to be reliable and effective in sheep, goats, and pigs. In contrast, the mare has proven to be quite difficult to superovulate. Recently, however, a commercially available equine pituitary extract (eFSH) has proven moderately effective in increasing ovulation rates.[4]

EMBRYO RECOVERY

Embryos were recovered from cattle by mid-ventral surgery in the early days of the commercial industry. Non-surgical recovery utilizing flexible Foley catheters replaced surgery in cattle in the late 1970s. With this technique, the uterus is repeatedly irrigated about a week after estrus with media composed primarily of NaCl and physiological salts under gravity flow. In most cases the outflow is directed through a filter, trapping the embryos. Equine embryos are recovered similarly. Laparoscopic surgical recovery of embryos, with the donor under intravenous anesthesia, is the primary method in swine, sheep, and goats.

EVALUATION OF EMBRYOS

Preimplantation embryos from all domestic farm animals are microscopic in size, ranging from 150 to 300 microns in diameter; they, due to their density, sink rapidly to the bottom of their container, usually a petri dish, and are identified and evaluated with a stereomicroscope at 10× to 50× magnifications. Pig embryos do not survive chilling (below 15°C), but the embryos from most farm animals can be maintained in a variety of holding/culture media for periods of up to 24 hours at room temperature.

EMBRYO TRANSFER

Embryos are transferred into surrogate cattle and horses by non-surgical methods similar to artificial insemination. Most transfers into sheep, goats, and pigs are via a surgical approach similar to that used for embryo recoveries. It is important that estrous cycles of surrogates be closely synchronized within ±24 hours of the age of the embryo.

FREEZING

Cryopreservation of embryos plus storage in liquid nitrogen (temperature, −196°C) has several important advantages. First, it eliminates the immediate need for recipients that are closely synchronized in estrus with the donor at the time of embryo recovery. Frozen embryos can be thawed when recipient(s) of suitable breed, size, and estrous synchrony are available. In addition, freezing

Table 2 The number of in vitro-produced cattle embryos transferred in different regions of the world in 2005.

Region	No. embryos transferred	
	Fresh	Frozen
Africa	NA[a]	8
Asia	49,099	78,396
North America	1,451	18
South America	129,340	68
Europe	2,689	3,127
Oceana	898	897
Total	183,477	82,514

[a] Not available.
Source: From Embryo Transf. Newsl.[3]

allows convenient movement of embryos over long distances, including internationally and, in contrast to the transport of live animals, virtually eliminates the chances of transporting infectious microbes.[5] Also, their surrogate mothers endow calves resulting from embryo transfer with significant immunity to local infectious microbes via colostrum.

Currently, ethylene glycol is used as a cryoprotectant, and embryos are transferred immediately following thawing. Embryos of most farm animals can be frozen with high survival rates, but pig embryos have proven very difficult to freeze successfully, and there is currently no commercially practical protocol. Horse embryos are frequently chilled to 4°C and transported in periods of 12–24 hours to suitable surrogates. Equine embryos more than 300 microns in diameter do not survive cryopreservation, but increasing numbers less than 300 microns are being successfully vitrified and transferred directly after thawing.[6]

MICROMANIPULATION

One application of micromanipulation is the splitting of embryos into two demi-embryos. This technique has limited commercial value in pigs, sheep, and goats and has been successful in horses in only a few cases. Splitting has been used commercially on a small scale in cattle since the mid 1980s because more calves are produced due to doubling the number of embryos. Of course, identical twins result if both halves go to term. It has probably not been used widely because the technique requires a high degree of skill. The equipment used for splitting embryos is also used by some commercial cattle practitioners to remove embryo biopsies consisting of a few cells for determination of sex using the polymerase chain reaction.

IN VITRO FERTILIZATION

Most donors of embryos are artificially inseminated, so fertilization occurs in vivo. With IVF, sperm are added to the oocytes in a test tube or petri dish. Usually the processes of in vitro maturation, fertilization, and culture go together. IVF has been successful in all species of farm animals, but it has proven especially difficult in horses, and has been commercially applied most frequently to cattle, especially to females with infertility problems.[7] To obtain oocytes from cows or mares, a long needle is inserted into the ovary through the vaginal wall in the same way as oocytes are recovered from infertile women for IVF. Ultrasonography is used to guide the needle so oocytes can be aspirated from ovarian follicles. An often used, alternative approach is to obtain oocytes from slaughterhouse ovaries to produce relatively inexpensive IVF-derived embryos for research, specific export markets, and for getting dairy cattle pregnant in heat-stressed environments. The increasing availability of sex-selected semen has created a potential commercial market for embryos produced from slaughterhouse-derived oocytes fertilized with enriched for X-bearing sperm by a flow cytometric cell sorter.[8]

CLONING AND TRANSGENICS

Domestic sheep, represented by the now famous "Dolly," were the first mammalian species to be cloned by the nuclear transfer of somatic cells of an adult donor animal. All the other farm animal species covered by this entry have subsequently been cloned, with the most recent success being the cloning of one horse and three mules in 2003. Several companies in North America operate commercial cattle cloning programs and hundreds of clones have been born. However, as of early 2006, the United States Food and Drug Administration had not yet made a ruling regarding the safety of cloned animals for human consumption. As a consequence, commercial cloning of cattle has slowed considerably.

Animals carrying foreign genes, or transgenics, have been produced from cattle, sheep, goats, and pigs. Animals carrying valuable transgenes are then cloned. A number of commercial enterprises have produced hundreds to thousands of transgenic goats, sheep, and pigs. Goals for these transgenics include increased disease resistance, improved feed efficiency and growth characteristics, and modifications in milk composition. There are also large commercial programs to make genetically engineered farm animals that produce pharmaceutical products and even industrial products from the milk of these animals.[9]

RESEARCH

Embryo transfer has been used for studying reproductive mechanisms such as the intrauterine spacing and migration of pig embryos, embryo-uterine interactions in several species, and aging of the reproductive tract. A significant amount of research is being conducted in the areas of superovulation, embryo freezing, IVF, interspecific embryo transfer, and cloning.

CONCLUSION

Embryo transfer, whether thought of as a single step, or a process including superovulation and embryo recovery and storage, is an important research tool. In addition, there are numerous commercial applications of embryo transfer. The process is more complicated and expensive than artificial insemination, so it is usually used to amplify the reproductive rates of the top few percent of genetically valuable females.

REFERENCES

1. Hasler, J.F. The current status and future of commercial embryo transfer in cattle. Anim. Reprod. Sci. **2003**, *79* (3–4), 245–264.
2. Seidel, G.E., Jr.; Elsden, R.P.; Hasler, J.F. *Embryo Transfer in Dairy Cattle*; W.D. Hoard & Sons Company: Fort Atkinson, 2003.
3. Thibier, M. Transfers of both in vivo derived and in vitro produced embryos in cattle still on the rise and contrasted trends in other species in 2005. Embryo Transf. Newsl. **2006**, *24* (4), 12–18.
4. Niswender, K.D.; Alvarenga, M.A.; McCue, P.M.; Hardy, Q.P.; Squires, E.L. Superovulation in cycling mares using equine follicle stimulating hormone (eFSH). J. Equine Vet. Sci. **2003**, *23*, 497–500.
5. Wrathall, A.E.; Sutmöller, P. Potential of embryo transfer to control transmission of disease. In *Manual of the International Embryo Transfer Society*, 3rd Ed.; Stringfellow, D.A., Seidel, S.M., Eds.; IETS: Savoy, IL, 1998; 17–44.
6. Eldridge-Panuska, W.D.; Caracciolo de Brienza, V.; Seidel, G.E. Jr.; Squires, E.L.; Carnevale, E.M. Establishment of pregnancies after serial dilution or direct transfer by vitrified equine embryos. Theriogenology **2005**, *63*, 1308–1319.
7. Hasler, J.F. The current status of oocyte recovery, in vitro embryo production, and embryo transfer in domestic animals, with an emphasis on the bovine. J. Anim. Sci. **1998**, *76* (Suppl. 3), 52–74.
8. Wheeler, M.B.; Rutledge, J.J.; Fischer-Brown, A.; VanEtten, T.; Malusky, S.; Beebe, D.J. Application of sexed semen technology to in vitro embryo production in cattle. Theriogenology **2006**, *65*, 219–227.
9. Forsberg, E.J. Commercial applications of nuclear transfer cloning: three examples. Reprod. Fertil. Dev. **2005**, *17*, 59–68.

Feeds: Genetically Modified

Fred Owens
Matthias Liebergesell
Pioneer Hi-Bred International, Inc., Johnston, Iowa, U.S.A.

Abstract

Currently marketed transgenic crops have traits to protect those crops from insect damage and to simplify weed management. The livestock industry has benefited indirectly from such traits through stabilization of the feed supply and more consistent feed quality. The next generation of transgenic crops that improves nutrient or energy content and availability are discussed here. Extensive collection of safety data and review by regulatory agencies will assure that the commercial transgenic crops and feedstuff derived from such crops will produce animal products and human foods that are as safe or safer than from crops developed by other selection procedures.

INTRODUCTION

Genetic modification (GM) is defined as the transfer of genes from one organism to another through modern biotechnology. Transfer of genes within or across species, genera, and kingdoms results in transgenic organisms, a type of "genetically modified" organisms (GMO). By changing the agronomic characteristics or nutritive value of a crop, transgenic crops can improve both the supply and quality of food and feed and thereby have potential to enhance animal performance.

BACKGROUND

Transgenic crops were cultivated on 309 million acres globally in 2008[1] with expansion continuing (Fig. 1). Although the majority (62%) of acres planted to transgenic crops were in industrial countries, the growth rate for biotech crops is greater for the developing than the developed countries and for the Southern than the Northern hemisphere. In 2006, of the total worldwide acreage with transgenic crops, 57% was planted to soybeans (herbicide tolerant), 25% to maize (insect resistant and/or herbicide tolerant), 13% to cotton (insect resistant and/or herbicide tolerant), and 5% to canola (herbicide tolerant). Similar to growth worldwide, the land area devoted to transgenic crops in the world (1) and the United States[2] has increased steadily since 1995 (Fig. 2[3,4]).

IMPACTS OF TRANSGENIC CROPS ON LIVESTOCK FEED QUALITY AND COST

Production (Input) Traits

Most transgenic plants currently marketed were developed to improve crop production (often called "input" traits). Specific traits have increased the plant's tolerance to specific herbicides (e.g., glufosinate, glyphosate, or sulfonyl urea) and provided resistance to insects that attack the herbage or roots of plants (e.g., European corn borer and western corn rootworm for maize).[5] Such traits enable changes in crop production procedures that can reduce feed contaminants typically involved with production agriculture (e.g., pesticide and herbicide residues). By reducing the risk of crop damage from insect pests, viruses, and fungi, transgenic crops have helped stabilize the feed supply and modulate volatility in feed prices. By reducing the need for soil tillage, transgenic crops also help to conserve water and soil and expand crop production into new regions.[4] Transgenic crops used as feed for livestock include two cereal grains (maize, wheat), five oilseeds (canola, soybean, flax, sunflower, cotton), two forage crops (alfalfa, bentgrass), and two crops whose by-products are fed to livestock (potato, sugar beet). More details about specific traits that have been approved for food and feed are available elsewhere.[5,6]

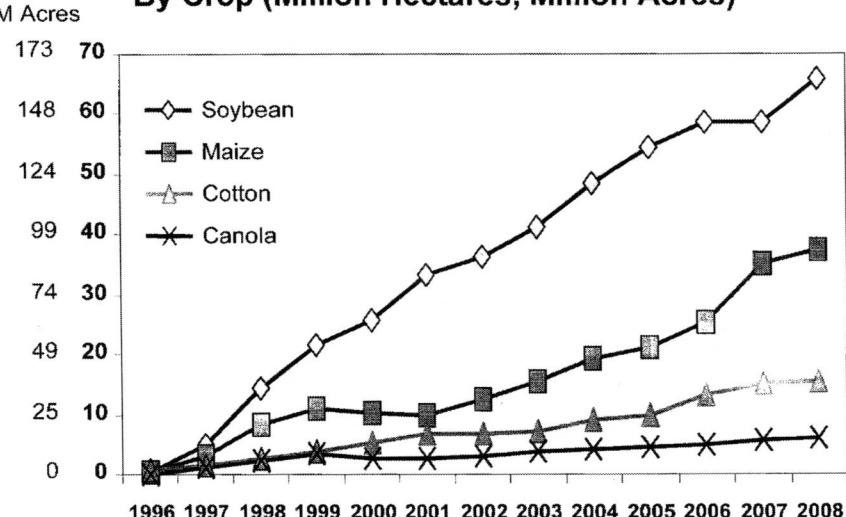

Fig. 1 Worldwide area planted to biotech crops from 1996 to 2008. **Source:** From *Global Status of Commercialized Biotech/GM Crops*.[1]

Product (Output) Traits

Transgenic crops developed with improved product (also called "output") traits have targeted the crop's value for food, feed, or industrial use. Only a limited number of nutritionally enhanced crops, improved either through biotechnology or conventional breeding, have been commercialized (Table 1). Most products have been developed with improved nutritive or health value for humans (e.g., oil fatty acid composition). However, specific modifications have included alterations in the concentrations or availability of limiting amino acids (e.g., lysine), energy (e.g., oil), nutrients (e.g., phosphorus), or of other compounds that can improve the quality of animal products (improved meat flavor, tenderness, or extended shelf life). For example, maize and soybean products with increased phosphorus availability and improved amino acid balance will help both to reduce the dietary need for protein and phosphorus

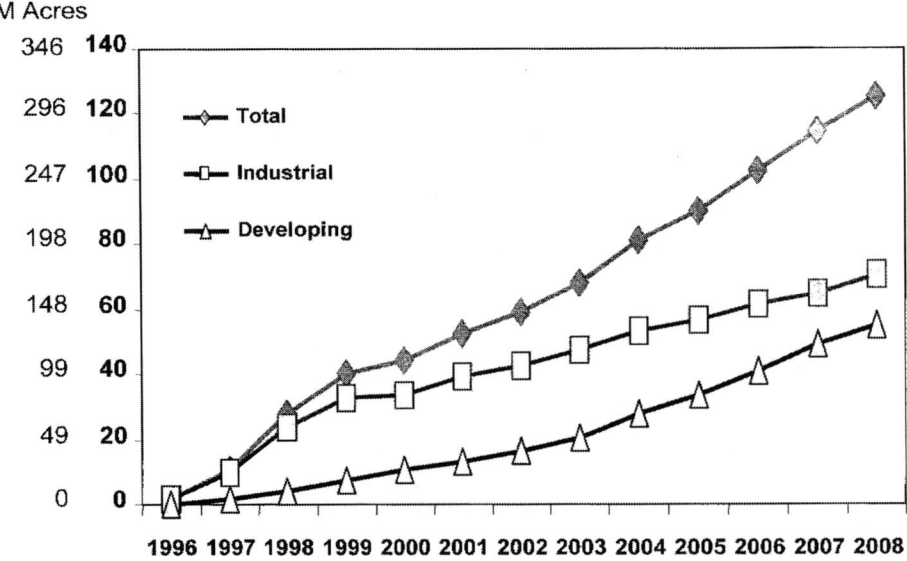

Fig. 2 Adoption of genetically engineered crops in the U.S. **Source:** From *Global Status of Commercialized Biotech/GM Crops*.[1]

Table 1 Currently marketed crops with modified nutrient composition.

Nutrient	Crop	Mechanism
Oil: enriched in lauric and myristic acids	Canola	Transgenic
Oil: enriched in oleic acid, low linoleic acid	Canola	Chemical mutagenesis
Oil: enriched in oleic acid	Soybean	Transgenic
Oil: low linoleic acid	Soybean	Traditional plant breeding
Protein: enriched in lysine	Maize	Transgenic
Phosphorus: phytate for increased P digestibility	Canola	Transgenic

supplements and the excretion of nutrients by non-ruminants. Although selected primarily through screening rather than genetic modification, maize hybrids with higher available energy have been identified for selected use as non-ruminant feed. When the genes responsible for such differences are identified, transgenic maize with even greater energy availability, developed specifically for these markets, can be expected.

In order to preserve the full value of crops with improved output traits, transgenic crops must be kept separated (identity preserved) from commodity grain during production, handling, and marketing. Such separation presents both physical and economic challenges to traditional grain handlers who often blend "commercial" grains. When the economic advantage of grain with a specific trait is recognized by industrial or agricultural end users, greater linkage within the marketing channel for grains will occur. Finally, appropriate assays need to be available throughout the grain marketing channel to verify the presence of the desired trait.

SAFETY ASPECTS OF TRANSGENIC CROPS FOR FOOD AND FEED

Public Safeguards

In the United States, the safety of transgenic crops is reviewed by three governmental regulatory agencies: 1) the Food and Drug Administration (FDA); 2) the Environmental Protection Agency (EPA); and 3) the Department of Agriculture (USDA). The degree of involvement of each agency depends on the trait modified. The U.S. regulatory agencies have specific responsibilities,[7] and a list of completed regulatory reviews is available elsewhere.[6]

Transgenic Crop Safety Assessment

Before a transgenic crop can reach the market, it undergoes extensive testing for food safety, feed safety, and environmental safety. Regulatory agencies request, receive, and review data concerning the genes and proteins used to develop the trait, crop characteristics, food and feed composition, and potential environmental impacts that have been compiled over several years and from multiple locations. A molecular analysis determines the functional integrity of the introduced genetic material. Stability of the trait, its efficacy, and agronomic performance of the crop are monitored over multiple generations and environments. Novel proteins are tested for potential adverse effects on humans including toxicity and allergenicity. Nutrient analysis of forage and grain (proximate analysis, protein composition, oil composition, major minerals, major vitamins, secondary metabolites, and antinutrients), as well as animal nutritional performance studies (small animals), provide data to assure that the transgenic crop is "substantially equivalent" to the conventional counterpart that has not been modified genetically. The environmental risk assessment includes a diverse study of numerous factors such as the impact on non-target organisms (for all insect tolerant traits), weediness potential, gene flow via pollination, and horizontal gene transfer (e.g., gene transfer from crop to soil microorganisms).

LIVESTOCK FEEDING STUDIES

Crop developers and academic groups conduct livestock feeding trials to document that the feed derived from transgenic crops is nutritionally adequate and will support normal growth and well-being. In numerous trials and field tests with poultry, swine, dairy, and beef cattle, no unexpected effects of transgenic plants or plant products on livestock productivity have been detected.[8–11] A comprehensive list of livestock feeding studies with traits currently on the market is available.[12] Future transgenic crops are widely expected to enhance both the quantity and quality of feed for animals and food for mankind.[13,14]

REFERENCES

1. International Service for the Acquisition of Agri-biotech Applications. Global Status of Commercialized Biotech/GM Crops: ISAAA Brief 39-2008: Executive Summary, 2009. http://www.isaaa.org/resources/publications/briefs/39/executivesummary/default.html (accessed January 2010).
2. Fernandez-Cornejo, J. Adoption of genetically engineered crops in the U.S. Economic Research Service, USDA, 2009. http://www.ers.usda.gov/data/biotechcrops/ (accessed January 2010).
3. USDA. Biotechnology FAQ, 2009. http://www.usda.gov/wps/portal/!ut/p/_s.7_0_A/7_0_1OB?contentidonly=true&

navid=AGRICULTURE&contentid=BiotechnologyFAQs.xml (accessed January 2010).
4. AgBios. GM Crops used in Animal Feeds. http://www.agbios.com/cstudies.php?book=COMM&ev=GMFEEDS&chapter=GMCrops&lang= (accessed January 2010).
5. AgBios. Search the GM crop data base. http://www.agbios.com/dbase.php?action=ShowForm (accessed January 2010). 2009b.
6. USDA. Search the U.S. Database of Completed Regulatory Agency Reviews, 2009 http://usbiotechreg.nbii.gov/database_pub.asp (accessed January 2010).
7. USDA. United States Regulatory Agencies Unified Biotechnology Website. http://usbiotechreg.nbii.gov/ (accessed January 2010).
8. Flachowsky, G., A. Chesson, and K. Aulrich. Animal nutrition with feeds from genetically modified plants. Archives of Animal Nutrition **2005**, *59* (1), 1–40.
9. Owens, F.N. Corn genetics and animal feeding value. Minnesota Nutrition Conference Proceedings, pp. 2–25. University of Minnesota, St. Paul. http://www.ddgs.umn.edu/articles-proc-storage-quality/2005-Owens%20(MNC)%20Corn%20genetics.pdf (accessed January 2010).
10. Federation of Animal Science Societies. FASS Facts: Are the milk, meat, and eggs from livestock fed biotech feeds safe to eat? Savoy, IL. http://www.fass.org/geneticcrops.pdf (accessed January 2010).
11. Council for Agricultural Science and Technology. Safety of meat, milk, and eggs from animals fed crops derived from modern biotechnology. Animal Agriculture's Future through Biotechnology, Part 5. http://www.cast-science.org/websiteUploads/publicationPDFs/feedsafety_ip.pdf (accessed January 2010).
12. Federation of Animal Science Societies. References – Feeding Transgenic Crops to Livestock http://www.fass.org/references/feeding_transgenic_crops_to_livestock.htm (accessed January 2010).
13. Hartnell, G.F., Hatfield, R.D., Mertens, D.R., and Martin, N.P. Potential benefits of plant modification of alfalfa and corn silage to dairy diets. In: Proceedings of the 20th Southwest Nutrition and Management Conference, University of Arizona, Tucson. pp. 156–172. http://cals-cf.calsnet.arizona.edu/animsci/ansci/swnmc/papers/2005/Hartnell_SWNMC%20Proceedings%202005.pdf (accessed January 2010).
14. PEW Initiative on Food and Biotechnology. Emerging challenges for biotech specialty crops. http://www.pewtrusts.org/uploadedFiles/wwwpewtrustsorg/Summaries_-_reports_and_pubs/WorkshopReport.pdf (accessed January 2010).

Fish Roe: Fermentation

Alaa El-Din A. Bekhit
Department of Food Science, University of Otago, Dunedin, New Zealand

Abstract

Fermented fish roe products are available in different forms and under different names around the world. Salted dried (e.g., karasumi), salted fermented (e.g., karashi mentaiko and jeotgal), or salted roes (e.g., suijiko and tarako) are some of the most popular fermented roe products that contain significant amounts of n-3 fatty acids. The development of desired texture, flavor, and color of these products is regulated by factors such as the fish species, maturity of the roe, age of the fish, season, the environmental and processing conditions, and the extent of protein degradation and lipid oxidation. Proteases, from endogenous and exogenous sources, and lipoxygenases from the roe contribute to the generation of different taste and flavor compounds depending on processing and storage conditions. Fermentation of fish roe may also have a beneficial impact on the nutrition by degrading cholesterol and decreasing the level of certain trace elements without comprising the n-3 fatty acids in the roe.

INTRODUCTION

Fermented roe products are widely available in different forms as traditional delicacies, but our knowledge of the biochemical changes that take place during processing and storage, the biological systems behind these changes, and the nutrition of these products remain very limited as most of the research published on these products are in languages other than English. In the present entry, a brief discussion on the utilization of fish roe and the major fermented roe products is presented. The two main biochemical events (protein degradation and lipid hydrolysis) and the source of enzymes involved are discussed to highlight their roles in the development of the characteristic attributes of fermented roe. A summary of new information of the beneficial role of fermentation to reduce cholesterol and trace element content as well as potential health benefits from fermented roe products is provided.

FISH ROE UTILIZATION

Fish roe products that are processed by means of salting, fermenting, drying, or a combination of these steps have been traditionally consumed as delicacies in many countries and are important in international trade. These products were originally developed to fully utilize and preserve the catch, but now they are considered to be prime products of certain species. Well-known examples include: salted dried mullet roe, known as *karasumi* in Japan; *avgotaraho* in Greece; *bottarga* in Italy; *batarekh* in Egypt; *yu chian-cha* in China, and *wuoze* in Taiwan; and salted fermented fish roe (such as *karashi mentaiko* in Japan and *jeotgal* in Korea).

Caviar (salted and cured individual fish eggs) from sturgeon and several other fish species is the most desired form of fish roe due to its sensory properties and high value. The natural biological variation in the egg characteristics (e.g., size, color, and flavor) that contribute to the hedonic attributes required in caviar products limits the fish species that can be used in the production of these products. Several processing methods can modify the color and flavor through the use of food-grade dyes and flavoring compounds. The grain size may be of particular importance for the production of caviar and caviar-like products due to desired mouth-feel characteristics. Most the fish species that have been used for caviar-like products have egg diameters of ≥ 2 mm (Table 1). Thus, egg size seems to be the limiting factor for the generation of caviar-like products from the roe of many fish species (Table 1). Egg size is not important in fermented roe products; therefore, fermentation as a process represents a better way to enhance the utilization of the roe from many fish species.

FERMENTED FISH ROE PRODUCTS

A widely accepted definition of fermentation is "the transformation of organic compounds into simpler entities by either the actions of microorganisms or the actions of enzymes which may be intrinsically available in the raw materials or added exogenically." With regard to fish roe products this includes: salted dried (e.g., karasumi and

Table 1 Egg size (mm diameter) for selected fish used and not used in caviar and caviar-like products.

Fish	Species	Diameter (mm)
Used for caviar and caviar-like products		
Salmon	Chum (*Oncorhynchus keta*)	3.5–4
	Pink (*O. gorbuscha*)	3.5–4
	Coho (*O. kisutch*)	4–4.5
	Sockeye (*O. nerka*)	4–4.5
	Chinook (*O. tshawytscha*)	6–7
Sturgeon	Beluga (*Huso huso*)	2.83–4.51
	Sevruga (*Acipenser stellatus*)	2.1–2.85
	Osetra (*A. gueldenstaedti*)	2.36–3.94
	A. baerii	3.06–4.03
	A. persicus	2.98–4.46
	A. nudiventris	2.98–3.46
Paddle	*Polyodon spathula*	2.21–3.22
Pike	*Esox lucius*	2.5–2.8
Lumpfish	*Cyclopterus lumpus*	≈ 2.0
Orange roughy	*Hoplostethus atlanticus*	2.1–2.3
Not used for caviar and caviar-like products		
Herring	*Clupea harengus*	0.9–1.2
Mackerel	*Scomber scombrus*	0.8–1.0
Carp	*Cyprinus carpio*	0.8–1.6
Whitefish	*Coregonus huntsmani*	0.9–1.4
Flounder	*Pseudopleuronectes americanus*	0.8–1.2
Mullet	*Mugil cephalus*	0.46–0.68
Atlantic cod	*Gadus morhua*	0.53–0.57
Rohu	*Labeo rohita*	1.30–1.48

kazunoko), salted fermented (e.g., jeotgal), salted marinated (e.g., karashi mentaiko) as well as salted (e.g., tarako and suijiko) fish roe products since several biochemical changes occur in the organic components during these processes leading to desired and characteristic products.

Salted Dried Fish Roe

Mullet roe has traditionally been the main roe used in salted dried fish roe; however, roe from several other fish species such as tuna and ling have also been used. Research on the use of roe from species such as hoki, hake, salmon, and southern blue whiting to produce salted dried roe products have been reported, but no information regarding the acceptability of the products to consumers is available; however, it is apparent that there are vast differences in the organoleptic and sensorial characteristics of fermented products from these fish species (Fig. 1). The major biochemical changes in these products during processing/maturation are mainly due to the action of endogenous enzymes in the roe, although the native microbiota of the raw materials and equipment can contribute to these changes. The final product, which has a firm waxy texture desired by consumers, is achieved at a moisture content of approximately 30%. The product is normally coated with beeswax to maintain the moisture level and reduce oxidative processes during storage.

Salted Fermented Fish Roe

Two main products are well known under this category, karashi mentaiko in Japan and jeotgal in Korea. These are regarded as similar products, varying in the degree of metabolic changes (texture, flavor development, and appearance characteristics) and the process responsible for these changes during fermentation (endogenous enzymes or microorganisms). Karashi mentaiko is the specific name for the salted spiced Alaska pollack or cod roe that is marinated in a mixture of saké, konbu (kelp), uzu (citrus fruit), salt, and red chili. The marination lasts for a week and starts with an initial soaking step in saké to firm and reduce the microbial load of the roes. During karashi mentaiko processing, migration of moisture and salt takes place and changes in the proximate composition occur until equilibrium is reached. The membrane of the eggs changes the orientation of its molecular structure irreversibly and it becomes semipermeable. Other ingredients such as katsuo (fish flakes), sugar, and soy sauce are used for flavoring.

Jeotgal is a general term used to describe a fermented product generated using a variety of seafood sources including roe. The fermentation process relies on the actions of

Fish Roe: Fermentation

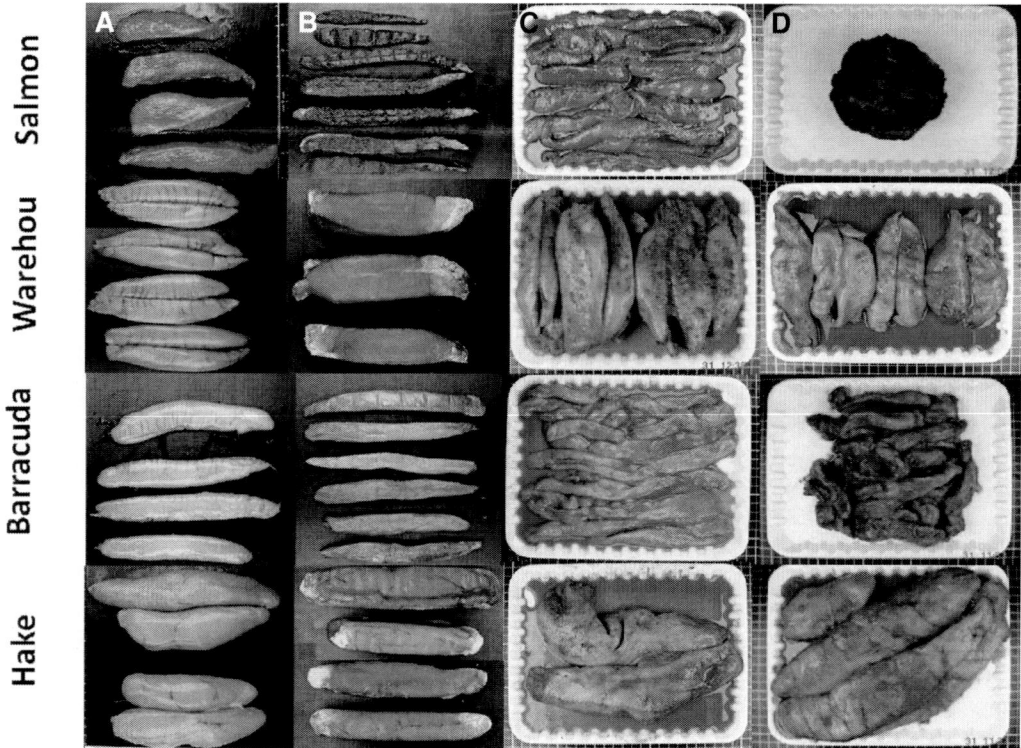

Fig. 1 Impact of processing (karasumi-like and jeotgal at 4 and 25°C) on the sensory properties of roe products. A = Fresh roe; B = Karasumi-like product, C = Jeotgal (4°C); D = Jeotgal (25°C).

microorganisms as well as the endogenous enzymes in the roe. The processing time for jeotgal can vary from several weeks to about a year.

Salted Fish Roe

This class of products is included because several changes in the organic compounds (e.g., generation of free amino and organic acids) and organoleptic changes have been reported,[1,2] which are attributed to the actions of endogenous enzymes. This category includes: sujiko (salmon roe), tobiko (flying fish roe), kazunoko (herring roe), and tarako (Alaska pollack roe), which are very popular products in Japan.

BIOCHEMICAL CHANGES

The transformation of organic compounds during fermentation, which lead to the development of the desired texture and flavor, is caused by the actions of endogenous enzymes in the roe and the microorganisms in the materials used during processing. Protein degradation and lipid oxidation are of great importance in developing the required characteristics of the products, with the activity of the proteases and lipoxygenases being dictated by the processing and storage conditions. The generation of free amino and fatty acids, volatile compounds, and the utilization of sugars available

in roe cause the formation of the characteristic flavor notes in fermented roe products and the reactions among these compounds will influence the color of the product through the Maillard reaction.

Protein Degradation

The extent of protein hydrolysis in the roe during processing depends on the source of the proteases (endogenous versus microbial), the processing conditions (salt content, additives, and the processing time), the fish species, and its maturity. Proteolysis is particularly important for the color of karasumi since inhibiting the proteolytic activity through protease inhibitors and heating at high temperature reduces the browning of karasumi.[3]

Endogenous proteolytic enzymes

Several proteases found in fish roe are important in the formation and degradation of the yolk during the development of the oocyte. The dominant protease and its activity level in the ovary vary with the developmental stage. Several natural protease inhibitors, which are expressed at different development stages, are involved in the regulation of these proteases.[4] Trypsin-like, chymotrypsin-like, and aminopeptidases are found in the eggs of mullet and Alaska pollack.[3,5] Cathepsins B, L, and D are found in the eggs of scad, salmon, perch, and killifish. The optimal activities of these enzymes collectively cover a wide range of pH values (2.6–8.2) and temperatures (35–60°C). Protease activity in mullet and Alaska pollack roes is either unaffected or slightly activated by sodium chloride up to 150 mM and by processing into karasumi or tarako.[6] These proteases seems to play a greater role in salted products (e.g., sujiko and tarako) where the generation of free amino acids is not related to microbial activity,[1] and similar actions may be important in marinated products such as karashi mentaiko. The exposure of the egg wall to the action of these proteases during harvesting can have detrimental effects on the quality of the final product. For example, these enzymes can affect the wall strength (altering texture) and wall permeability that consequently affects salt uptake during processing. For these reasons, fresh and intact roes are required for the production of consistent premium quality fish roe products.

Microbial proteolytic enzymes

Several halophilic bacteria and archaea are able to produce proteases that are active under moderate and high-salt concentrations. Extracellular proteases have been purified from *Archaebacterium* 172P1, *Halobacterium mediterrane*, *Natrialba magadii*, *Pseudoalteromonas* sp. CP76, and *Salinivibrio* sp. AF-2004. Several *Bacillus* species (*B. subtilis*, *B. pumilus*, *B. horti*, and others) of marine origin were found to grow over a wide range of pH values (5.7–11.5) and salt concentrations from 7% to 15% NaCl.[7] A halotolerant intracellular protease from *B. subtilis* FP-133 isolated from fish paste was active over a salt range from 0% to 12.5% with a maximum activity at 5%. Other protease-producing bacteria such as *Halobacillus* sp. SR5-3 and *Halobacterium* sp. PB 407 have been isolated from fish sauce. The microbiota of jeotgal is dominated by Gram-positive, sporeforming bacilli, and the contribution of these microorganisms to protein hydrolysis during ripening seems to be very similar to that found during fish sauce production. A similar protease was reported earlier from *Bacillus* isolated from salt,[8] thus the raw material used in processing (e.g., salt, chili, and other ingredients) may be very important in supplying essential ripening microorganisms.

Aerobic bacterial and fungal counts increase during karasumi processing and subsequent storage. The growth kinetics of the microorganisms is dependent on the time and the temperature of the storage period. Thus, a wide variation in the aerobic microbial load in karasumi (<1 to 7.1 logs CFU/g) can be expected depending on production and storage conditions. The microbiota present in karasumi and the contribution of the microorganisms to the biochemical changes during processing and storage remain to be elucidated.

Lipid Oxidation

Both lipase and lipoxygenases are present in fish eggs. Lipoxygenases in particular have been the focus of much research due to their physiological roles in the development and maturation of the ovary, and its involvement in the generation of active flavor compounds.[9] Lipid hydrolysis occurs in salted roe products and roe treated with antibiotics,[1] and is reduced in frozen and pasteurized roe, indicating the significant role that endogenous enzymes play in lipid hydrolysis during processing and storage. The addition of NaCl seems to activate lipoxygenase in herring roe.

Wax esters are dominant in mullet roe and are quite stable during the processing of karasumi. The decomposition of triglycerides is the main contributor to free fatty acid formation in karasumi. The major n-3 fatty acids do not change significantly over 15 days of drying and the oxidation of lipids is not significant during processing but increases during subsequent storage dependent on storage conditions.

IMPACT OF FERMENTATION ON TRACE ELEMENT AND CHOLESTEROL CONTENTS

It is well known that fermentation delivers preservative and nutritional benefits (e.g., by breaking down proteins to more easily digested amino acids and peptides) as well as enhancement of the organoleptic characteristics of the food. In the case of fish roe, several additional nutritional changes take place depending on the final product and the processing conditions. For example, salted dried fish roe products per unit weight will have higher levels of trace elements as compared to raw roes, whereas salted fermented fish roe will tend to have lower concentrations of certain trace elements. Another important biochemical change is the reduction in cholesterol concentration during fermentation. Fish eggs have a high cholesterol concentration (200–560 mg/100 g) that is roughly ten times the concentration of cholesterol in the corresponding muscle. This has been a concern to consumers and a limiting factor in the consumption of roe products.[10] Cholesterol-degrading bacteria have been found in fermented fish and one isolate was identified as *B. subtilis* SFF34. Two extracellular cholesterol oxidases are present in *B. subtilis* SFF34 that convert cholesterol into 4-cholesten-3-one. Metabolism by *B. subtilis* SFF34 leads to the generation of campesterol without increasing levels of cholesterol-oxidized derivatives. The changes in cholesterol concentration and in two of its derivatives (cholesta-3,5-diene (α) and cholesta-3,5-diene (β)) in hoki roe during fermentation at 25°C over 24 days is shown in Fig. 2. Gradual decreases in cholesterol, cholesta-3,5-diene (α), and cholesta-3,5-diene (β) concentrations occur over a 24-day fermentation with 33%, 100%, and 75.53% reductions in cholesterol, cholesta-3,5-diene (α), and cholesta-3,5-diene (β), respectively. Thus, fermentation of fish roe can decrease cholesterol concentration. If *B. subtilis* SFF34 is added, the potential to generate

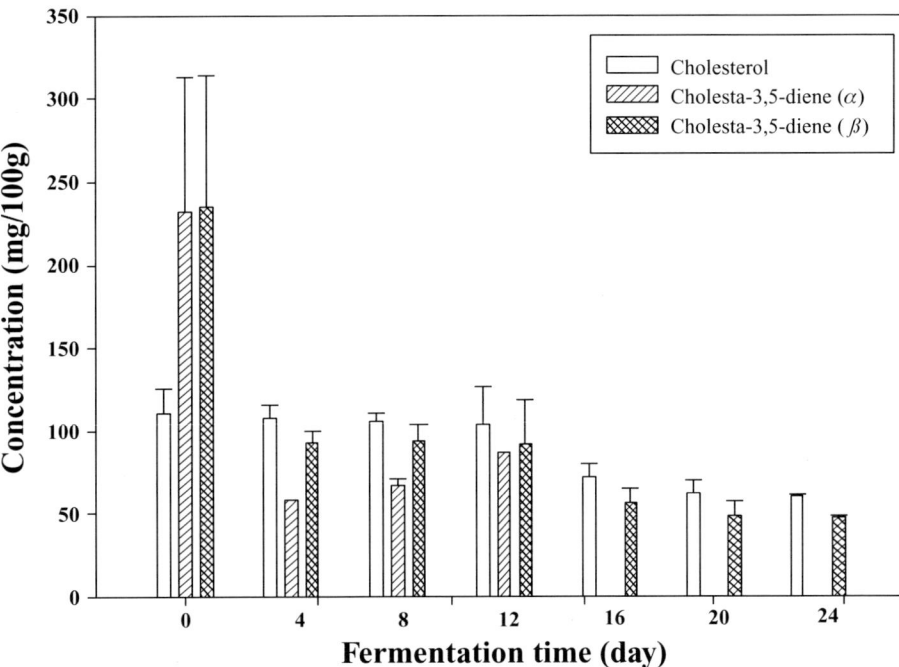

Fig. 2 Impact of fermentation on cholesterol, cholesta-3,5-diene (α) and cholesta-3,5-diene (β) concentrations in hoki roe during 24 days of fermentation at 25°C.

compounds that have hypocholesterolemic activity, such as campesterol, is feasible.

HEALTH BENEFITS OF FERMENTED FISH ROE PRODUCTS

Fish roe contains a high n-3 fatty acid concentration. These valuable compounds are not affected significantly during processing with a final concentration range of 23–38.8% of the total fatty acids in karasumi and kazunoko. The consumption of fermented fish roe products increases n-3 fatty acids (especially docosahexaenoic and eicosapentaenoic fatty acids) in plasma and hepatic lipids in mice and humans. Despite high cholesterol levels, consumption of fish roes and their products decreases total plasma cholesterol, phospholipids, and glucose concentrations, and increases plasma adiponectin in mice.[10,11] This paradox may be related to the high wax ester concentrations in fish roe products (e.g., karasumi, kazunoko, and sujiko) that have been reported to be in the range of 63.7–84.3% of total lipids. Very long-chain fatty acids alcohols, which constitute about 50% of the wax esters of fish roe products, have been reported to lower plasma cholesterol in humans through regulation of cholesterol metabolism in the peroxisome and by altering low-density lipoprotein uptake and metabolism.[12] Also, the high concentrations of docosahexaenoic and eicosapentaenoic fatty acids (32–38% of the total fatty acids) in fish roe used in fermented products can have plasma cholesterol-lowering effects.[10]

Several other potential health benefits for fermented roe products have recently been reported. Lipid extracted from Greek avgotaracho strongly inhibits platelet activating-factors and thrombin in vitro with the polar fraction of the oil being the most active. Kazunoko lipids are more effective than fish oil in lowering the level of arachidonic acid in the brain lipids of mice and improve their learning capacity.

CONCLUSION

Development of the characteristic attributes of fermented fish roe is governed by endogenous and exogenous proteolytic and lipolytic enzymes. Fermentation of roe can have beneficial impacts by reducing concentrations of cholesterol and certain trace elements. There is growing evidence for health-promoting benefits from consuming certain fermented roe products. With the rapid increase in fish farming and the potential to harvest fish year-round, the amounts of immature fish roe that is not suitable for caviar-like products is increasing, leading to by-products that are poorly utilized. Fermentation can be a useful and cheap technology to utilize these by-products.

ACKNOWLEDGMENT

The author would like to thank Dr. Phil Bremer for his helpful comments on the final drafts.

REFERENCES

1. Miyaji, T.; Nakagawa, T.; Tomizuka, N. Assessment of the microbial contribution to the processing of salted salmon roe (Sujiko). Biocontrol Sci. **2007**, *12*, 111–113.
2. Chiou, T.K.; Matsui, T.; Konosu, S. Comparison of extractive components between raw and salted Alaska Pollack roe (Tarako). Nippon Suisan Gakkaishi **1989**, *55*, 515–519.
3. Pan, B.S.; Huang, W.L.; Tsai, C.H.; Chang, K.L.B. Effects of protease inhibitors from rice bran and sweet potato on browning of dried mullet roe. Food Sci. **1997**, *24* (1), 75–85.
4. Yamashita, M.; Konagaya, S. Cysteine protease inhibitor in egg of chum salmon. J. Biochem. **1991**, *110*, 762–766.
5. Chiou, T.K.; Matsui, T.; Konosu, S. Purification and properties of an aminopeptidase from Alaska Pollack *Theragra chalcogramma*, roe. J. Biochem. **1989**, *105*, 505–509.
6. Chiou, T.K.; Matsui, T.; Konosu, S. Proteolytic activities of mullet and Alaska Pollack roes and their changes during processing. Nippon Suisan Gakkaishi **1989**, *55*, 805–809.
7. Ivanova, E.P.; Vysotskii, M.V.; Svetashev, V.I.; Nedashkovskaya, O.I.; Gorshkova, N.M.; Mikhailov, V.V.; Yumoto, N.; Shigeri, Y.; Toguchi, T.; Yoshikawa, S. Characterization of *Bacillus* strains of marine origin. Int. Microbiol. **1999**, *2*, 267–271.
8. Kamekura, M.; Onishi, H. Protease formation by a moderately halophilic *Bacillus* strain. Appl. Microbiol. **1974**, *27*, 809–810.
9. Pan, B.S.; Tsai, J.R.; Chen, L.M.; Wu, C.M. Lipoxygenase and sulfur-containing amino acid in seafood flavor formation. In *Flavor and Lipid Chemistry of Seafoods*; Shahidi, F., Cadwallader, K.R., Eds.; ACS Symposium Series No. 674; American Chemical Society: Washington, DC, 1997, 64–75.
10. Moriya, H.; Kuniminato, T.; Hosokawa, M.; Fukunaga, K.; Nishiyama, T.; Miyashita, K. Oxidative stability of salmon and herring roe lipids and their dietary effect on plasma cholesterol levels of rats. Fish. Sci. **2007**, *73* (3), 668–674.
11. Higuchi, T.; Shirai, N.; Suzuki, H. Effects of herring roe on plasma lipid, glucose, insulin and adiponectin levels and hepatic lipid contents in mice. J. Nutr. Sci. Vitaminol. **2008**, *54*, 230–236.
12. Hargrove, J.L.; Greenspan, P.; Hartle, D.K. Nutritional significance and metabolism of very long chain fatty alcohols and acids from dietary waxes: review. Exp. Biol. Med. **2004**, *229*, 215–226.

Food Labeling

Thomas P. Wilson
Department of Food and Animal Sciences, Alabama A&M University, Normal, Alabama, U.S.A.

Barbara Rasco
Department of Food Science and Human Nutrition, Washington State University, Pullman, Washington, U.S.A.

Abstract

The labeling of food involves a wide range of issues driven by product safety considerations, along with consumer demand for more information about the composition, healthful properties, type of food production technologies, and sources of food products and dietary supplements. To date, the United States has requirements for labeling of the following: listing of ingredients; weight or fill; manufacturer location and contact information; storage instructions; nutrient labeling, including of specific macronutrient components such as carbohydrate, salt, fat (most recently trans fat) and specific food ingredients such as whole grains, milk and milk products; content of various micronutrient components (most recently omega-3 fatty acids); provisions for nutrient content and health claims for foods and dietary supplements; specifications for organic foods; requirements for allergen labeling and related claims, such as "gluten free," and country of origin labeling. Under current labeling regulations, genetically modified foods require labeling only if the food has a significantly different nutritional property than its conventional counterpart; if a new food includes an allergen that consumers would not expect to be present or if a food contains a toxicant beyond acceptable limits. This entry provides an overview of food labeling regulation in the United States only. Most developed countries have labeling requirements covering these same content areas in addition to provisions for multilingual labeling, common provisions showing compliance with government technical specifications, and mandatory storage or handling information or expiration date labeling.

INTRODUCTION

The history of food labeling is a reflection of the current political climate and has evolved to address a variety of issues and challenges since the passage of the Pure Food and Drugs Act of 1906. This act established basic legal requirements for the labeling of food products in the United States, partly to address concerns with foods being misrepresented to the consumer as to their contents and to address other issues of economic fraud such as container fill. Initially, food labeling was administered by the U.S. Department of Agriculture (USDA). In the successor legislation, the 1938 Federal Food, Drug and Cosmetic Act (FFDCA), food labeling provisions were amended, expanded, and refined. The 1938 act also contained the enabling legislation creating the Food and Drug Administration (FDA), which gave this new agency jurisdiction and responsibility for food products labeling for all foods with the exception of meat and poultry which remained the responsibility of the USDA. There has been a plethora of amendments to this act since its adoption, a few of the relevant ones presented below.

Subsequently, the Fair Packaging and Labeling Act of 1966 amended the FFDCA and required FDA to promulgate regulations requiring that all "consumer commodities" be labeled to disclose net quantity of contents, common and usual name, and name and place of business of the product's manufacturer, packer, or distributor, or whomever is considered to be the most responsible person for that particular food. In 1990, the Nutrition Labeling and Education Act (NLEA) again amended the FFDCA and mandated nutritional content information labeling for most food products. In 1994, the Dietary Supplement Health and Education Act (DSHEA) was enacted to regulate the labeling of dietary supplements, and in 2003, the FDA amended its regulations to require that *trans* fatty acids content also be declared on the nutrition facts panel of conventional foods and dietary supplements. In 2004, the Food Allergen Labeling and Consumer Protection Act amended the FFDCA again to bring food allergens within the scope of FDA regulatory authority requiring that food labels declare the presence of specific food allergens. Finally, in response to concerns about the safety of imported food products, the 2002 and 2008 Farm

Bill included requirements that certain foods are required to have country of origin labeling.

Generally, food products' labels must contain, at a minimum, the name and address of the manufacturer, net content, fill or weight, the common name of the food product, and a list of its ingredients. Failure to include all required information or for false or misleading statements on a food product label is referred to as "misbranding" and constitutes a violation of the labeling laws and regulations punishable by administrative sanctions and in some cases, criminal penalties. In addition to these general provisions, the FDA has promulgated regulations that define exactly what constitutes a food label, where it should be located on a food package and precisely what information should be on particular labeling and how it is to be disclosed. Additionally, specific regulations have been promulgated for the labeling of specific food product categories such as genetically modified foods, organic foods, medical foods, foods for special dietary purposes such as weight loss, bottled water, infant formula, hypoallergenic foods. In the case of dietary supplements, specific regulations also govern the use of health claims that promote beneficial food/health relationships that have been approved by the FDA.

FOOD LABELING ISSUES: THE HISTORY OF FOOD PRODUCT LABELING IN THE UNITED STATES

The stated purpose of food label legislation in the United States is to prevent fraud, deception, misleading statements (misbranding), and to provide information about important product properties so that the consumer is able to make an informed judgment about the product that would drive purchasing decisions. The label of a food product has been defined by regulation to include any display of written and graphic matter on the immediate container of the food product and any information that appears on the outside container or wrapping along with any other printed information that accompanies the food product. The 1906 Pure Food and Drug Act represented the first national attempt to regulate food products including food labels. It was enacted in part to prohibited false or misleading statements on food product labels and to proscribe the sale of food products sold with no indication of contents or ingredients or which made outrageous claims about their dubious heath benefits. It also prohibited false and misleading claims, fraudulent marketing of imitation products, and labeling that misrepresented the content of important constituents such as fat in butter, milk, oils and fat, eggs, or meat. The FFDCA expanded the scope of the labeling requirement of the 1906 Act by promulgating definitions and standards of identity or technical product specifications for certain foods that greatly facilitated FDA enforcement authority over misbranded food products. The FFDCA also increased the penalties available to FDA for misbranding violations.

In 1966, Congress enacted the Fair Packaging and Labeling Act (FPLA) the stated purpose of which is to facilitate value comparisons and to prevent unfair or deceptive packaging and labeling of many household consumer commodities. The FPLA expanded the scope of labeling authority beyond food products to include "consumer commodities," covering the majority of products packaged for sale in interstate commerce and consumed or "used up" by the consumer such items as aluminum foil, candles, household cleaning fluids, and light bulbs. The FPLA required household consumer commodities to bear a label statement identifying the product, e.g., detergent, sponges, etc.; the name and place of business of the manufacturer, packer, or distributor; and the net quantity of contents in terms of weight, measure, or numerical count. The act also authorizes additional regulations with respect to descriptions of ingredients, slack fill of packages, use of "cents-off" or lower price labeling, and characterization of package sizes. A metric labeling requirement was added in 1992 which took effect in 1994.

In 1990, in response to increased consumer demand for more accurate and relevant information about the nutritional content of food products, the Nutrition Labeling and Education Act (NLEA) was passed. It amended the FFDCA and mandated nutritional content information labeling for all foods products except those made by the smallest manufacturers based upon set serving sizes and provided that a food would be misbranded, and therefore subject of FDA enforcement authority unless it stated the following mandatory and voluntary components in the correct order and format, under the label's "Nutrition Facts" panel:

- *total calories*
- *calories from fat*
- calories from saturated fat
- *total fat*
- *saturated fat*
- *trans fat*
- polyunsaturated fat
- monounsaturated fat
- *cholesterol*
- *sodium*
- potassium
- *total carbohydrate*
- *dietary fiber*
- soluble fiber
- insoluble fiber
- *sugars*
- *protein*
- *vitamin A*
- percent of vitamin A present as beta-carotene
- *vitamin C*
- *calcium*
- *iron*
- other essential vitamins and minerals

(The italicized items are not required unless added or if a product claim is made.)

The NLEA-exempted food products served for immediate consumption, such as in hospital cafeterias and airplanes, and food sold in food-service operations, such as at mall cookie counters, or by sidewalk vendors, ready-to-eat food prepared primarily on site, deli, and candy store items, food products shipped in bulk, as long as it is not for sale in that form to consumers. Other exempt products included medical foods used to address the nutritional needs of patients with certain diseases that must comply with specific nutrient requirements under other regulations. Also foods that have no significant nutrient content such as plain coffee and tea, some spices were exempt. Recently, local governments have required nutrition labeling of foods sold in food service establishments within their jurisdiction.

The NLEA also addresses the concerns of Congress that health claims were becoming increasingly common in the marketplace in the 1980s but that these were not regulated. A health claim is a claim made on the label or in labeling that expressly or by implication characterizes the relationship of any substance to a disease or health-related condition. In contrast, structure function claims describe the role of a nutrient or dietary ingredient intended to affect normal structure or function in humans, for example, "calcium builds strong bones." A nutrient content claim is one that characterizes the level of a nutrient in a food, such as "free," "low," "high," or "reduced" for fat, cholesterol, sodium, and calories.

Until the NLEA, the FDA had not issued clear, enforceable rules to regulate such health-related claims. Until this law was passed, FDA may have lacked the statutory authority to permit health claims on foods without also requiring that the claim meets the premarket approval requirements applicable to drugs. The NLEA treats health claims for foods and dietary supplements separately. Health claims on foods may be made without FDA approval as a new drug, as long as there is "significant scientific agreement" regarding the validity of the claim. No premarket approval is required for structure/function or nutrient content claims.

The Labeling of Dietary Supplements

In 1994, the Dietary Supplement Health and Education Act (DSHEA) defined dietary supplements as products that contain a "dietary ingredient" intended to supplement the diet. The term "dietary ingredients" refers to vitamins, minerals, herbs or other botanicals, amino acids, and substances such as enzymes, organ tissues, glandular materials, and metabolites. The label of a dietary supplement must include: a descriptive name of the product stating that it is a "supplement"; the net quantity of contents statement; the ingredient list; the name and place of business of the manufacturer, packer, or distributor, expiration date, and the nutrition labeling information in the form of a "Supplement Facts" panel that must also identify each dietary ingredient contained in the product. Ingredients not listed on the "Supplement Facts" panel must be listed in an "other ingredient" statement beneath the panel. Additionally, DSHEA authorizes the use of FDA approved health claims on the labels or labeling or dietary supplements that advertise a beneficial relationship to a disease or health-related condition in manner typically reserved for "drugs" without the product being classified as a drug. In 1999, the issue of allowable health claims received judicial attention in the case of *Pearson v. Shalala*, which permitted supplement manufacturers to include a disclaimer on a supplement product label stating that the FDA had not approve health claims made about the product. As for foods, structure/function and nutrient content claims do not require premarket authorization by the FDA.

The Labeling of Food Allergens

In 2004, the Food Allergen Labeling and Consumer Protection Act (FALPA) was passed to help consumers with food allergies and intolerances to identify and avoid foods that contain major food allergens. FALPA requires that the label of a food that contains an ingredient or a protein from a major food allergen declare the presence of the allergen in the manner described by the law and applies to food products that are labeled on or after January 1, 2006. FALCPA identifies eight foods or food groups as the major food allergens. Similar legislation in Europe covers 14 potentially allergenic food components. In the United States, labeling is required for milk, eggs, fish, shellfish, tree nuts, peanuts, wheat, and soybeans. FALCPA provides manufacturers with two options regarding allergen labeling. The first option is to include the name of the food source in parenthesis following the common or usual name of the major food allergen in the list of ingredients in instances when the name of the food source of the major allergen does not appear elsewhere in the ingredient statement. The second option is to place the word "contains" followed by the name of the food source from which the major food allergen is derived, immediately after or adjacent to the list of ingredients, in type size that is no smaller than the type size used for the list of ingredients. FALCPA also provides mechanism to petition the Secretary of Health and Human Services for an exemption from this labeling requirement either through a petition process or a notification process that demonstrates that a food ingredient does not cause an allergic response that poses a risk to human health.

The Labeling of Genetically Modified Food Products

Both the FDA and USDA require labeling of a food produced using modern methods of biotechnology food if the food's composition differs significantly from that of its conventional counterpart. This is known as the doctrine of "substantial equivalence." However, most biotech foods on the market have been found to be the substantial equivalent of their conventional counterparts, hence, most biotech foods including genetically modified food products are not required to be labeled as such unless: the food has a significantly different nutritional property, if a new food product includes an allergen that consumers would not expect to be present, or if a food contains a toxicant beyond acceptable limits.

The Labeling of Alcohol Beverages

The 1998 Alcoholic Beverage Labeling Act requires that the labels of alcoholic beverages to carry a "government warning," which must state that according to the Surgeon General, women should not drink alcoholic beverages during pregnancy because of the risk of birth defects, that consumption of alcoholic beverages impairs your ability to drive a car or operate machinery, and may cause health problems. However, information about the composition and nutritional value of alcoholic beverages, although proposed, has not been mandated.

The Labeling of Meat Products

The labeling of meat products is the responsibility of the Food Safety and Inspection Service (FSIS) of the USDA. USDA's regulatory authority over meat product labels is derived from several statutes including the Federal Meat Inspection Act (FMIA), Wholesome Meat Act, the Poultry Products Inspection Act (PPIA), the Egg Products Inspection Act (EPIA), the Agricultural Marketing Act (AMA), the FFDCA, and the FPLA. Broadly defined, the USDA through FSIS is authorized under these acts to regulate marking, labeling, or packaging of meat, poultry, or processed parts to prevent the use of any false or misleading mark, label, or container. Both the USDA and the FDA have similar labeling requirements; however, USDA requires that meat product labels be approved by the agency prior to the food being introduced into interstate commerce and USDA may withhold this approval at its discretion. Additionally, USDA requires that the official inspection legend (Fig. 1) and establishment number appear on product labels as well as the basic information regarding the identity, ingredients, quantity and location.

In 1993, following the outbreak of *E. coli* O157:H7, a consumer group sued the USDA to prevent the use of the official inspection seal on meat and poultry products unless it was accompanied by a label warning (Fig. 2) to the consumer that the product may contain bacteria capable of causing infection, disease, or death and giving safe handling instructions. In the court-approved settlement, the USDA agreed to publish a regulation requiring safe handling labels for raw meat and poultry products by August 15, 1993.

Fig. 1 USDA inspection label.

The Labeling of Transesterified Fats

Trans fats are defined by the FDA as all unsaturated fatty acids containing one or more isolated double bonds in a trans configuration. In 1994, the Center for Science in the Public Interest, a consumer advocacy group, filed a petition with FDA requesting that the agency take steps to require inclusion of trans fat content on nutrition labels. In response, the FDA issued a proposal in 1999, implemented in 2006, requiring manufacturers of conventional foods and some dietary supplements to list trans fats as a separate item on the nutrition facts panel. Dietary supplement manufacturers must also list trans fat information on the Supplement Facts panel when their products contain reportable amounts (0.5 g or more) of trans fat.

Country of Origin Labeling

Following a string of food-borne illness outbreaks, new concerns over the safety of food imports and some of the largest meat recalls in history, the 2002 Farm Bill required country of origin labeling (COOL) for beef, lamb, pork, fish and other aquatic foods, perishable agricultural commodities, peanuts, ginseng, pecans, and macadamia nuts be provided

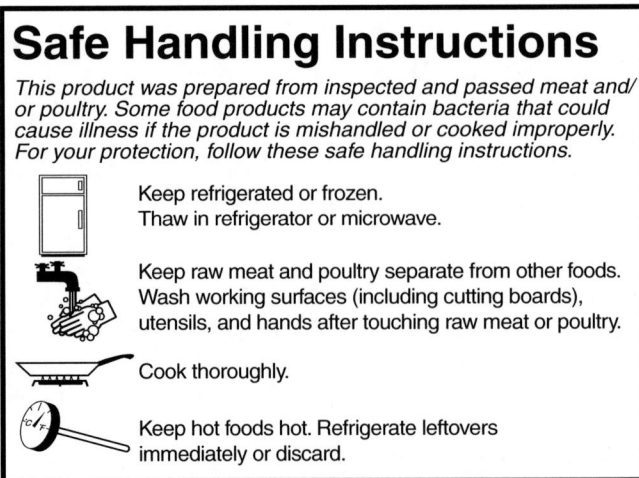

Fig. 2 USDA safe handling instruction label.

on retail packages by March 2009. This information is to provide shoppers with information to select products from countries perceived to have higher safety standards or who wish to purchase local products.

Organic, Eco-labeling, Fair Trade, and Sustainability

Although not under FDA jurisdiction, food producers can participate in fee for service and licensing programs to label products for specific market niches. Foods meeting production standards for USDA organic certification can label their foods or ingredients as organic. Similarly, licenses are issued by various groups for agriculture and fisheries that meet their criteria for ecologically sustainable or fair trade practices.

CONCLUSION

The regulatory scheme for the labeling of food and meat products in the United States has evolved from the requirement to provide only basic information about the food products we consume, to specific labeling requirements for specific food categories, nutrient content information, health claims, country of origin, and information regarding specific classes of ingredients such as trans fats and food allergens. Future labeling challenges include the labeling addressing the following issues: food consumed away from home, labeling of nutritive components such as omega-3 fatty acids and antioxidants, and more stringent labeling of carbohydrate components. Carbohydrate labeling, eco-labeling, and sustainable practices, "gluten free," whole grain statements, mail order products, Internet sales of food products, food service products, club stores, specialty foods, and Voluntary Labeling of Milk and Milk Products. Eco-labeling and sustainable agricultural practices, and labeling of products for mail order and internet sale and in food service.

BIBLIOGRAPHY

1. "*The Food Label*" – FDA Backgrounder, U.S. Food and Drug Administration, May 1999.
2. Hutt, P.; Merrill, R.; Grossman, L. *Food and Drug Law*; NY University Casebook Series, 3rd Ed.; Foundation Press: New York, 2007.
3. 21 CFR §101 (2001).
4. Pure Food and Drug Act of 1906 United States Statutes at Large (59th Cong., Sess. I, Chp. 3915, pp. 768–772).
5. Federal Food Drug and Cosmetics Act United States Code (U.S.C.) Title 21, Chapter 9 §341.
6. Fair Packaging and Labeling Act, 15 U.S.C. §§1451–1461 (1967).
7. Nutrition Labeling and Education, Public Law §§101–535 (1990).
8. 21 CFR §101.13q (2001).
9. Food Allergen Labeling and Consumer Protection Act, Public Law §§108–282 (2004).
10. "A Food Labeling Guide" – Guidance for the Industry, U.S. Food and Drug Administration, 2008.

Food Regulations: Global Harmonization

Yasmine Motarjemi
Quality Management, Nestlé, Vevey, Switzerland

Abstract
In the globalized world of the 21st century, the harmonization of food safety legislation is of highest importance. The primary purpose of laws and regulations in the area of food safety, be it at a national or international level, is the protection of consumers' health and of their right to access accurate information; however, the harmonization of food legislation also plays a crucial role in the promotion and facilitation of trade and economic development. Although rules and codes for food production date back to ancient times, global harmonization of food legislation has received heightened attention during the past few decades; a major milestone was the establishment of the Codex Alimentarius Commission from 1961 to 1963 and the coming into force of the Agreement of the World Trade Organization on Sanitary and Phytosanitary Measures in 1995. Despite all the efforts made by the international community and the achievements so far, global harmonization of food legislation remains a huge challenge, particularly considering the gap between the industrialized and the developing countries.

INTRODUCTION

Food legislation covers a broad range of subjects, including safety, nutrition, characteristics of the food, labeling, claims, marketing of novel foods or products derived from novel technologies, requirements for operations of food establishments, and procedures for the administration of food control. In the context of the harmonization of legislation at global or regional level, it is often matters related to safety aspects that receive the most attention. Therefore, this text focuses more on the harmonization of food safety legislation.

At the outset, it should also be pointed out that harmonization of food legislation does not necessarily mean of developing one law for all nations. Rather, it is a process by which appropriate national laws are established and administered in view of ensuring food quality and safety standards that protect the health of consumers, and at the same time by which trade is facilitated at the domestic and international level, without applying arbitrary or unjustifiably discriminating requirements.[1]

PURPOSE OF FOOD LEGISLATION AND REGULATIONS

Food legislation is the core function of governments. The term encompasses basic food laws and regulations. In an ideal system, food laws should set the basis for compliance and administrative policy and the operational programs to be implemented by the competent national food control authority. On the contrary, food regulations provide specific and detailed information needed to comply with the provision of the law.[1] They also provide the basis for specific administrative, procedural and technical requirements to ensure food quality and safety.

The main purpose of food legislation is to

- Protect public health;
- Protect consumers from products that are spoiled and fraudulent, or otherwise unfit for consumption; and
- Provide consumers with relevant and accurate information so that they can make an informed choice.

Although not often duly acknowledged, food regulations are also of great importance to the food and associated industries,[2] as they:

- Promote fair trade by ensuring a consistent standard among competing businesses;
- Increase the confidence of consumers in the food supply;
- Provide guidance on matters related to food safety; and
- Ensure that all stakeholders of the food chain, i.e., both suppliers and customers, fulfill their role.

Additionally, food regulations provide food industry norms that are used in designing and validating food safety assurance systems and in reassuring consumers that their products are safe and meet the agreed safety and quality standards.

IMPORTANCE OF GLOBAL HARMONIZATION

Since the latter part of the 20th century, the world has increasingly become globalized, as demonstrated by the financial crisis that swept the world in 2008. Globalization has also had consequences on food safety, and many food safety incidents such as the melamine crisis in 2008 demonstrated how malpractice in one country can have global implications.[3]

With the increase in international trade in food and feed, as well as international travel and migration, consumers are increasingly exposed to foods that are produced and processed in distant countries, hence a greater potential risk. Rapid circulation of information through modern means of communication is also influencing consumer expectations in terms of food safety and quality and raising questions on their national food safety and nutritional standards.

For many countries, food export is a major source of foreign exchange earnings and an important factor of economic development. Underdevelopment is a major cause of poverty, which itself is a source for society's ills, including foodborne diseases and malpractices affecting food safety. The latter exacerbate the food safety situation and jeopardize food export. Therefore, while food standards should be health-based and adequately protect the health of consumers, food standards that are too high and cannot be justified on health reasons will be unfavorable for society as they can create barriers to trade, damage the economy of countries, and sustain underdevelopment. In a study where the regulatory requirements of the European Union (EU) were compared with those set by the Codex Alimentarius Commission (CAC), it was found that EU standards resulted in a considerable loss of revenue from exports of cereal, edible nuts and preserved fruit for African countries and decreased the African export revenue by $670 million.[4] Thus, too stringent a food safety standard by importing countries can create unjustified trade restrictions, as well as underdevelopment and poverty in some other countries. Nevertheless, food standards have to be health-based and should ensure adequate protection of consumers. Consequently, global harmonization of food safety regulations has become of paramount importance in the 21st century.

For food businesses, the harmonization of food legislation is also of primary importance as producing food according to the legislation of different countries further complicates operations that are already inherently very complex. It will also decrease efficiency in production and increase operational risks, as management of change often creates difficulties.

There is also a fundamental reason for which global harmonization of food regulations is important.[5] This relates to the moral and ethical obligation that human beings have towards each other and towards the observation of the *Universal Declaration of Human Rights* (1948). In Article 1, it proclaims that:

> All human beings are born free and equal in dignity. They are endowed with reason and conscience and should act towards one another in a spirit of brotherhood.

Article 25 stipulates the following:

> Everyone has the right to a standard of living adequate for the health and well-being of himself and of his family, including food, clothing, housing, medical care and necessary social services.

Although the term *food safety* was not explicitly mentioned at the time, it was implicitly understood that the term food means "safe food." In its world Declaration on Nutrition, The Food and Agriculture Organization/World Health Organization (FAO/WHO) International Conference also recognized that access to nutritionally adequate safe food is the right of each individual. By this token, and considering that all human beings, regardless of race color, sex, language, religion, and social origin have equal rights, the above, globally accepted principles imply that all individuals have a right to the same standard of food safety and the same degree of health protection. Global harmonization of regulations, particularly in matters related to food safety, thus becomes a major step in recognizing this right. In other words, it contributes to ensuring that the population around the world benefits from the same degree of health protection from foodborne hazards, and therefore, the same standard of food safety. If we fail to do this, we will be creating opportunities for dumping food, i.e., a food that is considered as unfit for human consumption in one country is exported to another country with a lower standard.

For international food companies, it is also an ethical concern to market foods with the same standard of food safety all over the world. In the absence of a harmonized legislation, it will be difficult for these companies to operate and to locally source raw material and meet the same standard. This principle is of particular importance in contemporary issues such as products derived from biotechnology. An unjustified rejection of products derived from biotechnology in one country or region can make its export to the other countries unethical and unacceptable, and may deprive these countries of technological solutions that can perhaps address their nutrition, food safety, or food security issues.

TABLE 1. ACHIEVEMENTS IN THE GLOBAL HARMONIZATION OF FOOD SAFETY REGULATIONS.[7]

The Codex Alimentarius Commission is an intergovernmental body operating under the auspices of FAO and WHO. Its work includes a collection of standards, codes of practices, guidelines, and other recommendations, among which:

186	Commodity standards
46	Commodity-related texts
9	Food labeling provisions
5	Food hygiene provisions
3	Food safety risk assessment
15	Methods of sampling and analysis
8	Inspection and certification procedures
6	Animal food production provisions
12	Maximum level for contaminants including detection and prevention methods
1112	Food additives provisions, covering 292 food additives
7	Food additives-related texts
2930	Maximum limits for pesticide residues, covering 218 pesticides
441	Maximum limits for veterinary drugs, covering 49 veterinary drugs
3	Regional guidelines

The work of Codex is based on and driven by science. Three main FAO/WHO expert bodies provide scientific advice to Codex Alimentarius:

(a) The Joint FAO/WHO Expert Committee on Food Additives (JECFA), created in 1955, to examine chemical and toxicological aspects of food additives, contaminants, and veterinary drugs in foods for human consumption;

(b) The Joint FAO/WHO Meetings on Pesticide Residues (JMPR), established in 1963, to recommend maximum residue limits (MRLs) for pesticides and environmental contaminants as well as methods of analysis in specific food products; and

(c) The Joint FAO/WHO Expert Meetings on Microbiological Risk Assessment (JEMRA) set up in the year 2000 to provide advice on the microbiological aspect of food safety.

PROGRESS IN THE HARMONIZATION OF FOOD LEGISLATION

Historical Background

The history of the harmonization of food regulations followed a path parallel to that of food regulation at community and national levels. The development of food regulations started in the early period of human civilization, as many cultures and religions made provisions with regard to foods that were considered as hygienic or fit for human consumption. These were often based on empirical knowledge. As civilization advanced and trade flourished, the concern for food adulteration and fraud grew. For instance, in Mesopotamia, the Sumer civilization (5300 B.C.) made specific provisions for the preparation of cheese. One of the earliest codes, *The Code of Hammurabi*, was developed in 1760 B.C. in Babylon to protect the population from adulteration and fraud. For many centuries, the concerns for fraud and adulteration remained the focus of legislation. Luxury products such as tea, coffee, sugar, oil, and wine were easy targets. Efforts to prevent deceptive practices continue to date. Addition of textile dyes to foods and spices, mixing mineral and vegetable oils, and adulteration of milk with melamine are a few examples. While in the industrialized countries minute amounts of adulterated ingredients are frequently detected thanks to sophisticated analytical methods and instruments, it is to be presumed that a large number of cases of adulterated ingredients go unnoticed in the developing countries.

Some countries began reviewing their food legislation only since the industrialization era. The period from the 1800s to the present time is referred to as the legislative period.[6] Different factors were behind this development. Advances in food science and technology led to new, sophisticated, or complex food products, such as canned foods that were invented in the early 1800s. Urbanization increased the number of steps in the food chain, and with this change, new opportunities for fraud or contamination were created. International trade, facilitated by transport and also new food processing technologies, started to grow. Changes in lifestyle boosted the demand for industrialized food products and commercial food establishments. These trends all underpinned the need for strengthening food legislation; the first general food laws were adopted and enforcement agencies were established in this period.[7]

As for food legislation, the history of the harmonization of food regulations goes back to ancient times. As religions conquered new populations, they introduced their food code into the communities converted to the new religion. In this way, they represent the first attempts in harmonizing food laws. The need for developing international standards and for harmonizing food legislation at a global level received recognition in the 20th century, and it gained momentum toward the end of the century with the increase of international food trade and travel.

Some of the earliest international standards were those of milk and dairy products elaborated by the International Dairy Federation in 1903. The establishment of the FAO by the United Nations and of the WHO in 1945 and 1948, respectively, were also important milestones. The former received the mandate to set up international standards for quality and composition of products while the WHO set up health standards.

In the late 1940s, there were several attempts to set up regional food codes. In 1949, Argentina proposed a regional Latin American Food Code, the "Codigo Latino-Americano de Alimentos." Between 1954 and 1958, Austria actively pursued the creation of a regional food code, the Codex Alimentarius Europeaus or the European Codex Alimentarius.

The Codex Alimentarius Commission

The CAC, as we know it today, was created during the period from 1961 to 1963 (Table 1). Until the mid-1990s, the work of Codex was mainly recommendations to its member states that were free to accept or reject these recommendations (of Codex). Nevertheless, the work of Codex was often used as a reference for national legislation, particularly in countries with less developed food legislation.

Codex and WTO Agreements

In 1994, the Uruguay Round of Multilateral Negotiations was concluded in Marrakesh and the World Trade Organization (WTO) was established in 1995. As a consequence, two agreements that had major implications for the harmonization of food legislation came into force. These were the *Agreement on Sanitary and Phytosanitary Measures* (SPS) and the *Agreement on Technical Barrier to Trade* (TBT).

The SPS recognizes the governments' right to take sanitary measures, but it specifies that the measures should be based on science and applied to the extent necessary to protect human health, and that it should not discriminate unjustifiably between members where identical or similar conditions prevail. Both agreements encourage WTO member states to base their measures on international standards, guidelines, and recommendations of the CAC. It is also expected that WTO member states accept the sanitary and phytosanitary measures of others as being equivalent if the exporting country demonstrates to the importing country that its measures meet the importing countries' appropriate level of health protection. To determine what appropriate level of health protection corresponds to food safety and facilitates the principle of the "equivalence" of the SPS Agreement, new terms such as *food safety objectives* or *performance objectives* have been coined. Although these concepts are still being debated internationally, the SPS Agreement has, nevertheless, played a major role in encouraging countries to bring their legislation in line with international food standards, guidelines, and recommendations.

The TBT Agreement is also important to food trade. It provides member states with the right to consider other legitimate factors in their decision-making process. Examples of such factors are considerations regarding environment, animal welfare, and consumer preferences.

In the years 1995, 1997, and 1998, during three consecutive consultations of FAO and WHO, the concept of risk analysis and risk management was developed and introduced in the work of the CAC. The concept stipulates a functional separation between risk assessment and risk management, yet with adequate interaction between the two. It also defines the process of food safety management, from scientific evaluation to the risk management decision, taking into consideration other legitimate factors as well as communication to stakeholders (Fig. 1). This development was also reflected at national and regional levels and is presently the backbone of decision-making at governmental level in most countries. On the basis of this model, the *European Food Safety Authority* was created in Europe and entrusted with the responsibility of conducting risk assessment, while the risk management decision remained with the European Commission. While these principles have been adopted nationally in many countries, notably in North America and Europe, there are differences in their implementation, particularly with regard to risk management decisions and how uncertainties in risk assessment and other legitimate factors should weigh in the decision-making process.

In Europe, past experience with bovine spongiform encephalopathy (BSE) and consumer reaction to the BSE crisis have influenced the process of risk analysis; therefore, in the decision-making process:

1. There is a conservative approach toward the application of the "precautionary principle" in times of uncertainties in the risk assessment; and
2. Greater attention is paid to consumer perception and preference regarding food safety risk, ethical, and/or environmental issues.

In North America, the general public is less averse to modern methods of food production and processing; however, large-scale foodborne disease outbreaks, particularly outbreaks of *Escherichia coli* O157:H7 affecting children, have led to the fact that the public is more sensitive to the microbiological risk.

The WTO and Dispute Settlement

In addition to the international agreements on trade, the WTO also provides judgment in disputes related to the SPS and requires members who violate the SPS Agreement to modify or withdraw their non-compliant sanitary measures. The WTO can authorize countries affected by the violation of the SPS agreement to take retaliatory measures. The intervention of the WTO in the US-EU hormone-treated beef dispute is a historical example that showed the difficulties in harmonizing food regulations.[8,9]

Other Related Instruments

In the context of harmonizing food regulations, two other developments merit mentioning. The first relates to the development of International Organization for Standardization (ISO) standards and the second is International Health Regulations (IHR).

International Organization for Standardization

ISO has been operating since 1947. Its role is to facilitate the international coordination and harmonization of industrial standards to promote international exchange of goods and services.[10] ISO is a non-governmental organization, and as such, the standards are voluntary and not enforceable as regulations. Nevertheless, some of its standards, particularly those related to health, safety, and environment, have been adopted in some countries as part of the regulatory framework or are referred to in legislation. ISO Standards are developed by experts in the industry or business sector. They are, therefore, referred to as private standards. Standards with regard to food are developed under the Technical Committee (TEC) on Agriculture (ISO/TEC 34). Within the scope of this committee, over 500 international standards exist and over 100 more are under development. With regard to food safety, the ISO 22000 standard entitled *Food Safety Management Systems—Requirements for Any Organization in the Food Chain*[11] has received attention in regulating the customer–supplier relationship.

International Health Regulations

Concomitant with the industrialization of the world, international travel and migration also expanded along with a greater risk of spreading infectious diseases. Between 1830 and 1847, two major cholera epidemics struck Europe. These were catalysts for international cooperation in public health and led to the first International Sanitary Conference in Paris in 1851. In 1951, 3 years after the establishment of WHO, its member states adopted the International Sanitary Regulations, which were replaced by the International Health Regulations in 1969. IHR is an international legal instrument, binding 194 countries. The aim of IHR is to prevent and respond to public health risks that have the potential to cross borders and threaten people worldwide. The latest revision of IHR dates back to 2005 and came into force on June 15, 2007. The IHR requires countries to report certain disease outbreaks and public health events to the WHO. Under this requirement, WHO member states are requested to report a food product that may be or is suspected to be contaminated and has been imported from or exported from other member states.[12] To support the exchange of information on contaminated food products entering international trade, in 2004, in collaboration with the FAO, the WHO developed the International Food Safety Authorities Network (INFOSAN), including a food safety emergency

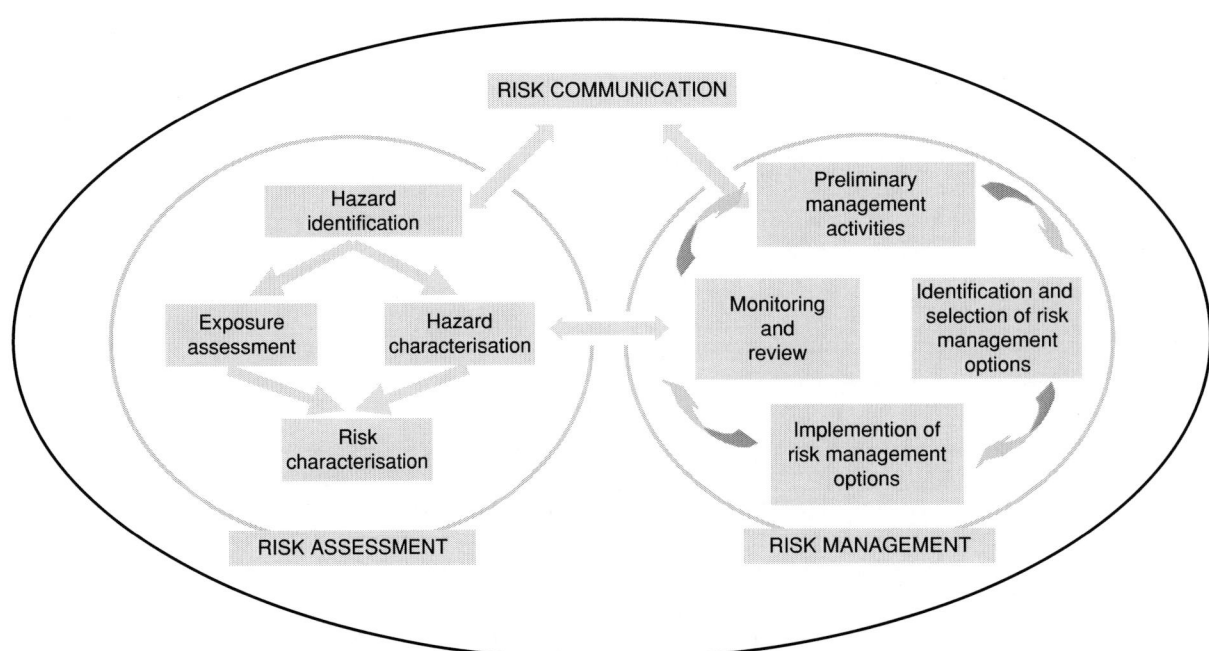

Fig. 1 Generic framework for risk analysis.

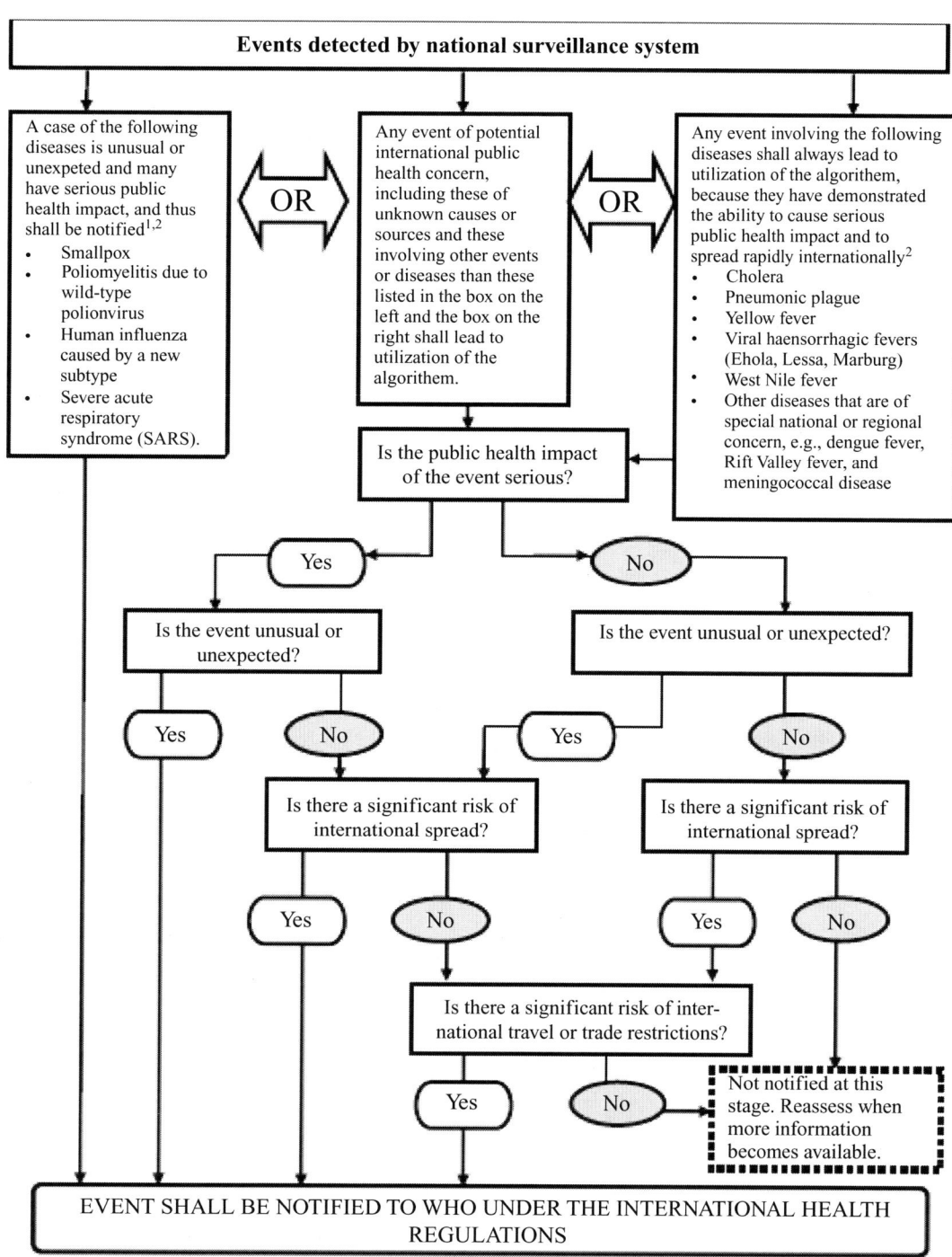

Fig. 2 Decision instrument for the assessment and notification of events that may constitute a public health emergency or international concern. **Source:** From *International Health Regulations*.[12]

component that is associated with processes under IHR (see Fig. 2).

FUTURE CHALLENGES

Despite its importance, the global harmonization of food legislation and regulations will remain a major challenge for the 21st century. The international community will have to reach agreements that provide adequate health protection and receive consumer acceptance, but at the same time allow: 1) economic growth; 2) progress in science and technology; and 3) fair competition between large and small industries as well as industrialized and developing countries.[5] An essential factor for the harmonization of food regulations is assistance to developing countries and small businesses in improving their standard of food safety as well as their infrastructure and capabilities, together with the education of consumers and of the general public.[14] In relation to the global harmonization of food regulations, the recent "Global Harmonization Initiative" merits mentioning. In the context of this initiative, since 2004 food scientists from all over the world have been working to obtain a global consensus among scientists on food safety issues and provide tools to facilitate the global harmonization of food regulations.[15]

REFERENCES

1. Whitehead, A.J. International food trade, harmonisation and mutual acceptance. In *International Food Safety Handbook*; van der Heijden, K., Younes, M., Fishbeen, L., Miller, S., Eds.; Marcel Dekker: New York, 1999; 747–766.
2. Motarjemi, Y.; Mortimore, S. Industry's need and expectations to meet food safety. Food Control **2005**, *16*, 523–529.
3. WHO/FAO. *Toxicological and Health Aspects of Melamine and Cyanuric Acid*, Report of a WHO Expert Meeting in collaboration with FAO, December 1–4, 2008; World Health Organization: Geneva, Switzerland, 2009; http://www.who.int/foodsafety/publications/chem/Melamine_report09.pdf (accessed April 21, 2009).
4. Wilson, J.; Otsuki, T. Balancing risk reduction and benefits from trade in setting standards. In *Food Safety in Food Security and Food Trade*; Unnevehr, L.J., Ed.; International Food Policy Research Institute: Washington, DC, 2003, 11–12.
5. Motarjemi, Y.; van Schothorst, M.; Käferstein, F. Future challenges in global harmonisation of food safety legislation. Food Control **2001**, *12*, 339–346.
6. Scheuplein, R.J. History of food regulations. In *International Food Safety Handbook*; van der Heijden, K., Younes, M., Fishbeen, L., Miller, S., Eds.; Marcel Dekker: New York, 1999; 647–659.
7. Codex Alimentarius. In *Understanding the Codex Alimentarius*; Food and Agriculture Organization of the United Nations and the World Health Organization: Rome, Italy, 2008; ftp://ftp.fao.org/codex/Publications/understanding/Understanding_EN.pdf (accessed April 21, 2009).
8. Kastern, J.J.; Pawsey, R.K. Harmonising sanitary measures and resolving trade disputes through the WTO-SPS framework. Part 1: case study of the US-EU hormone-treated beef dispute. Food Control **2002**, *13*, 49–55.
9. Lugard, M.; Smart, M. The role of science in international trade law. Regul. Toxicol. Pharmacol. **2006**, *44*, 69–74.
10. Petró-Turza, M.; Guissé, C. The International Organization for Standardization. In *International Handbook of Foodborne Pathogens*; Milliotis, M.D., Bier, J.W., Eds.; Marcel Deckker: New York, 2003; 753–769.
11. ISO. In *ISO 22000: Food Safety Management Systems—Requirements For Any Organization in the Food Chain*; International Organization for Standardization: Geneva, Switzerland, 2005.
12. WHO. In *International Health Regulations (2005)*; World Health Organization: Geneva, Switzerland, 2008.
13. Kirk, M.; Musto, J.; Gregory, J.; Fullerton, K. Obligations to report outbreaks of foodborne diseases under the International Health Regulations (2005). Emerg. Infect. Dis. **2008**, *14* (9), 1440–1442.
14. Motarjemi, Y.; Gorris, L. Safety foremost. Int. Food Ingredients **2009**, (2), 32–33.
15. Lelieveld, H. Progress in the global harmonization initiative. Trends Food Sci. Technol. **2009**, *20* (1), S82–S84.

Functional Foods

Michael Schweizer
School of Life Sciences, Heriot-Watt University, Edinburgh, U.K.

Abstract

Functional foods may play an important role in managing risk of chronic diseases if consumed on a regular basis. Consumers are likely to purchase health-promoting foods and therefore, it is essential that health claims are well substantiated. There is a growing body of evidence that the consumption of specific food components is associated with a positive effect on diagnostic biomarkers for conditions such as cardiovascular disease, diabetes, and obesity.

INTRODUCTION

Food is regarded as functional if, in addition to fulfilling nutritional requirements, it proves to be beneficial in some ways, e.g., decreases the risk of diseases, optimizes health, or has a positive impact on the ageing process. The evidence for benefits can be provided by either of the two means: 1) epidemiological studies, which measure a specific endpoint, e.g., level of serum cholesterol or a reduced disease risk; or 2) by the results of molecular biological studies demonstrating the influence of a food component on the expression of a specific gene or metabolic event. Ideally, the health benefit(s) of a functional food would be confirmed by both of these criteria thus providing the consumer with information which has a scientifically sound basis.

In this entry, some of the evidence, peer-reviewed and published within the past 5 years relating to functional foods and their efficacy will be discussed. The selected functional food targets are: coronary heart disease (CHD), diabetes, cancer, and obesity, all known to be diet-related diseases which are recognized as enormous social and financial burdens in developed countries. Therefore, any positive impact of diet on these diseases would be of benefit to both the well-being and the economy of the countries involved. The concept of functional foods could also be a means of dealing with the problems of malnutrition globally by breeding mineral- and vitamin-dense varieties of rice, wheat, maize, beans, and cassava for agricultural use.[1]

Functional foods are regarded differently from country to country. The Japanese government defines FOSHU ("foods for specified health issues") as "foods that are expected to have certain health benefits, and have been licensed to bear a label claiming that a person using them for a specified health use may expect to obtain the health use through the consumption thereof"; however, this classification has no status outside Japan. The acceptance of functional foods worldwide is based on psychological rather than physiological or biochemical criteria since the vast majority of consumers lack the knowledge for critical assessment of the last two types of data. It is, therefore, essential that research data supporting health claims are rigorous and presented in such a way that it appeals to, but does not overwhelm, the discerning consumer.

The connection between CHD and the consumption of fish oil high in ω-3 unsaturated fatty acids has been well studied. It is now emerging that the ratio of ω-3/ω-6 fatty acids in the human diet may be increased by a diet enriched in fish oil, which in addition to being cardioprotective,[2] may also be regarded as anti-inflammatory and anticarcinogenic.

A novel approach to providing a functional food is the development of a biscuit recipe reduced in salt and sugar contents, containing additional vitamin B_{12} and vitamin C as well as folic acid and prebiotic fiber. Although the biomedical assessment was carried out on only 12 subjects, it was found that two biomarkers: 1) serum homocysteine, elevated levels of which are a strong risk factor for cardiovascular disease; and 2) blood glucose levels decreased significantly following the consumption of four biscuits per day for 24 days.[3]

Obesity is a worldwide problem which is a difficult target for functional foods. There are several compounds, including conjugated linoleic acid (CLA), green tea, and calcium, which have the potential to act as functional food ingredients in the fight against obesity. These compounds have been shown to influence substrate utilization or thermogenesis.[4,5] The evidence to date for CLA is conflicting, but as CLA-supplemented food products become increasingly available, further research and monitoring is essential. Green tea has been shown to have a short-term effect on energy expenditure and fat oxidation; however, it is unclear whether this is because of caffeine or the catechins contained in green tea. More information on long-term weight control is required. It has been suggested that dietary calcium may suppress 1,2,5-dihydroxyvitamin D production, thus inhibiting lipogenesis and stimulating lipolysis. Another explanation for the role of calcium in weight management is the formation of non-absorbed complexes of calcium and fat which are excreted in the feces. Again additional information is required, but the feasibility of green tea and calcium as functional food ingredients is acceptable given their long history of safe consumption.

Another functional food approach to combat obesity is based on epidemiological evidence that by increasing the amount of soluble fiber in the diet, it is possible to influence weight gain negatively over an extended period of time. Data from intervention studies suggest that increasing dietary fiber leads to a reduction in food intake and subsequent weight loss. Potential mechanisms underlying these observations are increased satiety, reduced absorption of macronutrients, and alterations in the hormonal status of the gut. There are a number of natural foods, e.g., fruit, vegetables, whole grains, and legumes, which can be regarded as functional in the fight against obesity.[6]

Epidemiological studies have highlighted the chemoprotective properties of broccoli.[7] Broccoli and other cruciferous vegetables are excellent sources of antioxidants and essential minerals as well as glucosinolates, which, following hydrolysis to isothiocyanate, are responsible for induction of detoxification enzymes; however, the content of these components varies considerably from variety to variety making it difficult to quantify the amount of broccoli which should be consumed. Bioavailability, defined as the fraction of ingested nutrients used to meet functional demands in target tissues, differs from source to source. In fact, polyphenols, the most abundant in the diet, do not always lead to the highest concentration of active metabolites in the target tissue.[8] Therefore, although vegetables may be regarded as the ideal functional food, it would only be by means of extensive, specific plant breeding programs that it would be possible to create varieties with precise ratios of phytochemicals.

An interesting development in reprogramming plant metabolism is the overexpression of anthocyanidin synthase of rice, a multifunctional dioxygenase which causes the increase of flavonols and anthocyanins at the expense of procyanidins. This strategy of pathway engineering has been exploited by overexpressing rice anthocyanidin synthase in a specific mutant strain of rice totally lacking anthocyanins. This approach illustrates the possibility of creating a transgenic plant enriched for antioxidants without the introduction of any "foreign" DNA.[9] This approach could be extended to different food crops to increase their content of natural products of nutritional value.

The science of "nutrigenomics"—the study of the response of humans to food and food components—is capable of revealing the molecular targets of dietary compounds including prebiotics and probiotics and thus, identifies potential biomarkers.[10] Even though the long-term goal of nutrigenomics is to provide information which can be used to create individual tailor-made diets, the identification of defined metabolic responses to the application of a dietary compound could be useful in the manufacture and marketing of functional foods.

Prebiotics can represent a substantial portion of functional foods. In addition to improving the composition of intestinal microbiota, prebiotics are capable of acting as immunomodulators and providing protection against infections. In contrast to probiotics which are live organisms whose consumption at a sufficient level can confer a health benefit on the consumer, prebiotics are substances which act by stimulating growth and/or activity of specific beneficial bacteria found in the human colon (e.g., bifidobacteria) to improve the health of the host.[11] Probiotics can be delivered in functional foods as live bacterial cultures in yogurt and other types of foods, or in the form of dried preparations prepared by freeze- or spray-drying. An important consideration is the survival of the probiotic cultures following passage through the gastrointestinal tract; here the food matrix may play a role in overcoming this hurdle. For example, it has been shown that cheddar cheese has a more protective effect against gastric juice on probiotic bacteria than yogurt. Furthermore, fermented dairy products, in particular yogurt, contain the nutrients, calcium, potassium, and protein, and the non-nutrients, sphingolipids, butyric acid, and conjugated linoleic acid; therefore, yogurt may be regarded as a prime example of a functional food.

Peptides released in vivo or in vitro from either plant or animal proteins may be bioactive, i.e., have an effect on living tissue. Bioactive peptides have a wide range of effects. These effects can be antihypertensive, antimicrobial, antioxidant, or hypocholesterolemic. Occurring naturally in traditional foods, these peptides are obviously ideal as ingredients in functional foods. The nature of these bioactive peptides means that their influence on cellular metabolism can be tested readily in cell culture, thus providing scientific evidence for functional foods; however, health claims must be substantiated by human studies carried out over extended time periods since the effect of diet on health is chronic rather than acute. For screening purposes, the identification of biomarkers predicting beneficial effects which can be measured in vivo is of vital importance.[12]

The evidence for the efficacy of a functional food is generally obtained in a randomized clinical trial (RCT). If multiple RCTs have been conducted for a specific food or substance, there is the potential for meta-analysis (systematic review) which treats the trials as if they were a single factor of greater statistical power, although some bias may occur, e.g., publication bias, since studies with positive outcomes are more likely to be published. One such meta-analysis based on articles published between January 1999 and April 2003 showed that phytosterols and soluble fiber had a cholesterol-lowering effect whereas ω-3 fatty acids lowered tri-acylglycerols but increased cholesterol; both results were dependent on the length of treatment and were independent of dose, number of patients per study, age, and body mass index.[13]

CONCLUSION

The examples given, although by no means complete, represent a cross-section of functional foods and their targets. It is vitally important that health claims made for manufactured functional foods adhere to standards set by national and international governing bodies, e.g., the FDA. As stated in the text, it is more difficult to set standards for free-growing or farmed crops, such as broccoli, where the content of the bioactive substance fluctuates from variety to variety; however, this situation could be addressed by targeted plant breeding. Until such time that individually tailored diets are available, consumers should aim for a balanced diet including functional foods for the prevention and management of any specific disease.

REFERENCES

1. Bouis, H.E. Micronutrient fortification of plants through plant breeding: can it improve nutrition in man at low cost? Proc. Nutr. Soc. **2003**, *62*, 403–411.
2. Murphy, K.J.; Meyer, B.J.; Mori, T.A.; Burke, V.; Mansour, J.; Patch, C.S.; Tapsell, L.C.; Noakes, M.; Clifton, P.A.; Barden, A.; Puddey, I.B.; Beilin, L.J.; Howe, P.R. Impact of foods enriched with n-3 long-chain polyunsaturated fatty acids on erythrocyte n-3 levels and cardiovascular risk factors. Br. J. Nutr. **2007**, *97*, 749–757.
3. Boobier, W.J.; Baker, J.S.; Davies, B. Development of a healthy biscuit: an alternative approach to biscuit manufacture. Nutr. J. **2006**, *5*, 7.
4. Kovacs, E.M.; Mela, D.J. Metabolically active functional food ingredients for weight control. Obes. Rev. **2006**, *7*, 59–78.
5. Diepvens, K.; Kovacs, E.M.; Vogels, N.; Westerterp-Plantenga, M.S. Metabolic effects of green tea and of phases of weight loss. Physiol. Behav. **2006**, *87*, 185–191.
6. Slavin, J.L. Dietary fiber and body weight. Nutrition **2005**, *21*, 411–418.
7. Moreno, D.A.; Carvajal, M.; Lopez-Berenguer, C.; Garcia-Viguera, C. Chemical and biological characterisation of nutraceutical compounds of broccoli. J. Pharm. Biomed. Anal. **2006**, *41*, 1508–1522.
8. Wittemer, S.M.; Ploch, M.; Windeck, T.; Müller, S.C.; Drewelow, B.; Derendorf, H.; Veit, M. Bioavailability and pharmacokinetics of caffeoylquinic acids and flavonoids after oral administration of Artichoke leaf extracts in humans. Phytomedicine **2005**, *12*, 28–38.
9. Reddy, A.M.; Reddy, V.S.; Scheffler, B.E.; Wienand, U.; Reddy, A.R. Novel transgenic rice overexpressing anthocyanidin synthase accumulates a mixture of flavonoids leading to an increased antioxidant potential. Metab. Eng. **2007**, *9*, 95–111.
10. Sutton, K.H. Considerations for the successful development and launch of personalised nutrigenomic foods. Mutat. Res. Fund. Mol. Mech. Mutagen. **2007**, *622*, 117–121.
11. Pineiro, M.; Stanton, C. Probiotic bacteria: legislative framework—requirements to evidence basis. J. Nutr. **2007**, *137*, 850S–853S.
12. Hartmann, R.; Meisel, H. Food-derived peptides with biological activity: from research to food applications. Curr. Opin. Biotechnol. **2007**, *18*, 163–169.
13. Castro, I.A.; Barroso, L.P.; Sinnecker, P. Functional foods for coronary heart disease risk reduction: a meta-analysis using a multivariate approach. Am. J. Clin. Nutr. **2005**, *82*, 32–40.

Gametes and Embryos: Sublethal Hydrostatic Pressure Treatment

Csaba Pribenszky
Department of Animal Breeding and Genetics, St. Istvan University, Budapest, Hungary

Abstract

Assisted reproductive techniques require carefully established in vitro environment and finely adjusted procedures. Gametes and embryos are sensitive to environmental stresses, so in vitro procedures aim to minimize the inevitable harmful conditions. Applying stress to precondition cells has only been investigated recently. As for the initial stressor high hydrostatic pressure (HHP) was selected because of its unique physical features like: 1) it acts immediately and equally at every point of the sample without gradient effect; and 2) it can be applied with the highest precision and safety. Studies clearly demonstrated that by utilizing a well-defined and properly applied HHP stress treatment to reproductive cells before insemination, cryopreservation, enucleation, or in vitro maturation/culture, post-thawing survival, continued in vitro development, blastocyst ratio, and cell number, pregnancy, and/or farrowing rates improved considerably compared to untreated controls. The optimal stress treatment may be different according to species, cell types, and developmental stages. Although the biological mechanism is still unclear, several processes including alteration of the protein profile and biochemical balance of cells, production of shock proteins, and temporary arrest of the cell cycle might explain the observations. The birth of healthy piglets from treated fresh or frozen-thawed semen or from the transfer of embryos reconstructed from treated recipient eggs for SCNT demonstrates the in vivo safety of the procedure. In mice, two-generation studies have demonstrated that offspring from HHP treated embryos have normal reproductive function. Improving cells' stress tolerance by a defined sublethal stress treatment may confirm a completely new strategy in assisted reproductive technologies.

INTRODUCTION

Industrial meat or milk production incorporates assisted reproductive techniques, such as semen cryopreservation and artificial insemination. The quantity of the production is highly dependent on the efficacy of the utilized technologies.

Introducing sublethal stress treatment of semen before the process of insemination or cryopreservation has been published to enhance cell survival and increase the lifespan of spermatozoa after cryopreservation, thawing, or cooled in vitro storage. This results in the production of a higher number of insemination doses (number of straws) per bull, or higher litter size in swine. Sublethal high hydrostatic pressure (HHP) treatment has also been shown to substantially increase the post-thawing survival rate of vitrified oocytes in pig, and vitrified blastocysts in cattle, sheep and mice. The treatment of porcine immature (GV) or matured (MII) oocytes has also increased blastocyst yield and quality after in vitro embryo production or somatic cell nuclear transfer (Table 1).

The purpose of this entry is to show the establishment of an in vitro procedure that increases the general resistance, fertilizing ability, and/or developmental competence of gametes and embryos, to present the use, efficacy, safety, possible biological background, and perspectives of sublethal HHP stress treatment of cells in the process of routine assisted reproductive techniques.

SUBLETHAL STRESS

The way how organisms adopt to stress conditions is not yet understood well. Studies describe that sublethal stresses (e.g., heat shock, cold shock, oxidative stress, hydrostatic pressure) induce cellular stress reaction that includes the alteration of the protein profile of the cells, depending on the intensity and type of the sublethal stress itself, as well as on the type of the stressed cell. Other cellular stress reactions include cell cycle control, redox regulation, DNA damage sensing and repair, lipid and energy metabolism, as well as protein degradation, and apoptosis if stress impact goes over a specific limit.[11]

Depending on the cell type and the nature or magnitude of the environmental stress, cells react with stress reaction that actually increases their stress resistance, or at higher

Table 1 Success rates achieved in different ART procedures by the sublethal hydrostatic pressure treatment of gametes and embryos.

Biological material treated with HHP (Reference)	System, reason	Outcome (treated vs. control)	Reference institute
Spermatozoa (bull, boar)[1–5]	To facilitate cryopreservation:	Ratio of motile cells post-thaw: Bull: 54% vs. 45% Boar: 59.7 ± 2.6% vs. 37.4 ± 2.2% Number of live piglets born/sow (pregnancy rate ~70% both groups): 9.4 ± 1.0 vs. 4.4 ± 1.1	St. István University, Faculty of Veterinary Science, Budapest, Hungary; A.I. Station Klessheim, Austria; A.I. Station Vieselburg, Austria; Bos Genetic, Hungary; Animal Technology Institute, Taiwan
	To improve insemination:	Number of live piglets born/sow (pregnancy rate ~ 80% both groups): 11.9 ± 3 vs. 11.0 ± 2.9	St. István University, Faculty of Veterinary Science, Budapest, Hungary
	Basic research: proteomics (2-D, mass-spec), protein profile:	Ubiquinol-cytochrome C reductase complex core protein 1, perilipin, and carbohydrate-binding protein AWN precursor levels elevated significantly after treatment, and remained elevated after thawing	Animal Technology Institute, Taiwan
MII oocyte (porcine)[6–8]	To facilitate vitrification:	Blastocyst rate from vitrified eggs: 14.1 ± 1.4% vs. 1.3 ± 1.3%	Institute of Genetics and Biotechnology, University of Aarhus, Denmark
	To facilitate somatic cell nuclear transfer:	Blastocyst rate from reconstructed embryos: 57.0 ± 2.4% vs. 28.9 ± 3.3% (Yucatan donor cell line) (68.2 ± 4.1% vs. 46.4 ± 4.2% (LW1–2 donor cell line)	Institute of Genetics and Biotechnology, University of Aarhus, Denmark
	To facilitate in vitro system and vitrification:	Survival rate of cryopreserved reconstructed embryos: 61.6 ± 4.0% vs. 30.2 ± 3.9%	Institute of Genetics and Biotechnology, University of Aarhus, Denmark
GV oocyte (porcine)	To facilitate oocyte maturation and in vitro system:	Blastocyst formation rate: 54% vs. 39% Mean cell number of blastocysts: 59.9 vs. 47.3	Institute of Genetics and Biotechnology, University of Aarhus, Denmark
In vitro produced embryo (bovine)[9]	To facilitate vitrification:	Survival rate of cryopreserved embryos: 77% vs. 61% embryos hatched	EMBRAPA, Brazil; UFMG, Brazil
In vivo produced embryo (mouse)[10]	To facilitate cryopreservation: Basic research: gene expression:	Survival rate of cryopreserved embryos: 94% vs. 46% embryos Upregulation of selected genes after treatment or even after 120 min	St. István University, Faculty of Veterinary Science, Budapest, Hungary; Animal Biotechnology Centre, Gödöllö, Hungary

Results are presented as mean ± SE; or effect size, as they were published in the referred publications.

impacts, loss of cell viability or cell death (apoptosis or necrosis) occurs. For this reason the choice of the sublethal stressor to precondition cells is vital, as well as the fine-tuning of the stress treatment according to the species and the type of cell.

Applying sublethal hydrostatic pressure stress treatment to gametes and embryos to improve their resistance interestingly comes from the teachings of former food microbiology studies.

To reduce the microbial contamination but preserve food quality, sequential combination of relative mild treatments including hydrostatic pressure has been investigated.

However, in contrast with the supposed effect, significantly increased proliferation of *Listeria monocytogenes* as the consequence of sequential treatment with cold shock and HHP was reported. Biological effects of the first sublethal treatment have protected the bacteria from the detrimental effects of the second sublethal treatment.[12]

Based on former studies and on its unique physical features, HHP was chosen as a sublethal stressor for the preconditioning of reproductive cells and tissues.

HHP as a Sublethal Stressor

Hydrostatic pressure, unlike all other environmental stresses, acts immediately and uniformly at each point of the sample. It is not related to penetration problems or gradient effects and can be applied with the highest precision, consistency, and reliability.

Hydrostatic pressure is transferred instantly to the cells. The change of the pressure parameters during the treatment also takes effect immediately, such as decompression to atmospheric pressure (0.1 MPa) ceases the direct HHP effect in the cells right at the moment, when the set value reaches 0.1 MPa.

Experiments revealed that unlike with temperature, where even a 2–3°C increase in the physiological temperature might reduce viability, hydrostatic pressure acts with a wide safety margin, e.g., treatment of bovine blastocysts was effective at 40 MPa, but even an 80 MPa/30 min, treatment caused no loss in the continued in vitro development and hatching rate.

Treatment, that is executed by a purpose-made, programmable hydrostatic pressure device, is inserted into the routine procedures before the cryopreservation, insemination, enucleation, or in vitro maturation step.

HHP Treatment in ART

For the treatment, spermatozoa, eggs, embryos, or other cells or tissues are loaded in their optimal media into straws (0.25 mL or 0.5 mL ministraws, 5 mL maxistraws) or insemination bags, then sealed (heat, plastic/iron ball, plug, whatever is applicable), and placed into the pressure chamber of a custom made, computer controlled HHP machine (Cryo-Innovation Ltd., Budapest, Hungary). The pressure chamber is filled with distilled water. The cell specific program is run, executing hydrostatic pressure treatment in the range of 20–60 MPa (surprisingly 200–600 times the physiological value), for the duration of 30–90 min. Treatment temperature may influence optimal pressure parameters and the efficacy of the treatment as well. Treatment of porcine eggs or semen at body temperature yielded better post-thawing survival rates than treatment at room temperature,[1,7] though room temperature treatments provided significant improvement compared to controls as well.[1–9] The temperature of the HHP treatment has to be adjusted to the state-of-the-art temperature management of the cryopreservation processes (e.g., if semen sample was cooled down to room temperature, then HHP treatment shall not be done at 37°C).

At the first phase of the experiments, cells were exposed to different levels (5–80 MPa) of hydrostatic pressures applied for various durations (30–120 min) to determine the sublethal zone with no irreversible damage. At the second phase, samples were first exposed to this sublethal dose, and then incubated for 10–120 min under normal culture conditions for recovery. Later, the required intervention (cryopreservation, insemination, parthenogenetic activation, in vitro maturation or culture, enucleation followed by somatic cell nuclear transfer) was performed. Eventually, effects were assessed in vitro by investigating morphology, functional parameters including motility, membrane integrity, and developmental competence, respectively, or in vivo by pregnancy rate, litter size, and study of the offspring.

In each experiment, treatment groups were compared with a control group where HHP treatment was omitted.

Surprisingly, the pressure tolerance limit of mammalian gametes and embryos was found in the 20–80 MPa zone, where cells survived 30–120 min treatments without any loss of their viability, although the highest hydrostatic pressure that these cells normally encounter is less than 0.2 MPa.

The following paragraphs summarize HHP effects according to cell types in different ART procedures.

Spermatozoa

HHP induced stress tolerance was shown in bovine and porcine spermatozoa. Porcine ejaculated semen was extended and treated with 30 MPa pressure for 90 min, then cryopreserved. Significantly improved cell survival was achieved after thawing.[1,2] The treatment did not alter pregnancy or farrowing rates, but a significant increase of litter size was achieved following the insemination of frozen-thawed semen.[3] Other studies have demonstrated that HHP treatment increased litter size when treated semen was used for fresh semen artificial insemination.[4] As a result of the insemination of HHP treated fresh or cryopreserved spermatozoa, 491 and 155 healthy piglets were born, respectively, not being different from the control in sex ratio, weight, stillbirth, and malformations.[3,4]

Increased post-thawing motility and membrane integrity was also achieved by the HHP treatment of bull spermatozoa.[5]

HHP treatment has induced a significant increase in the production of several proteins that are supposed to play a crucial role in the process of fertilization. Ubiquinol-cytochrome C reductase complex core protein 1, perilipin, and carbohydrate-binding protein AWN precursor were identified as HHP response proteins being significantly higher in HHP treated samples measured after treatment, following 5 hours of equilibration, and also post-thawing.[2] As spermatozoa are regarded as transcriptionally inactive cells, the above proteins may be the result of posttranscriptional stabilization of the mRNA by the sublethal stress, as it was described earlier in chondrocytes.

Oocytes

Porcine oocytes were found relatively sensitive to HHP, accordingly a 20 MPa pressure for 60 min proved to be the optimal treatment to increase stress tolerance.[6] Subsequent vitrification and parthenogenetic activation resulted in a tenfold increase in blastocyst developmental rates in vitro.[7] In another experiment pressure-treated oocytes were enucleated and used as recipients for somatic cell nuclear transfer. Both blastocyst rates and the survival of these blastocysts after vitrification have increased significantly. The strongest effect was observed when 1–2 hours recovery time was applied between the end of the HHP treatment and the initiation of vitrification or enucleation. Transfer of cloned embryos derived from HHP treated oocytes has resulted in two healthy piglets.[8] Similarly, sublethal HHP stress treatment of immature, GV stage oocytes resulted in increased blastocyst rate and higher blastocyst cell number following in vitro maturation, parthenogenetic activation, and in vitro culture. On the basis of these achievements, experiments involving mouse and human oocytes have also been started recently. According to the preliminary results, the HHP stress tolerance of murine and human oocytes is comparable to that observed in pigs.

Preimplantation stage embryos

Treatment of in vivo-derived mouse blastocysts with 60 MPa for 30 min has significantly improved their survival rates after traditional freezing.[10] Bovine embryos produced in vitro could be treated with even higher pressures (80 MPa, 30 min). The treatment of 60 MPa for 1 hour before vitrification resulted in higher survival, re-expansion, and hatching rates.[9]

HHP induced transcriptional changes of mouse blastocysts included the upregulation of Azin1, Gas5, Gadd45g, and Sod2 immediately after HHP, and Gadd45g after 120 min culture. The study demonstrated that HHP activates short- and long-term growth arrest and oxidative stress related genes.

HHP stress treated blastocysts, after transfer, has resulted in 240 fit mouse pups, 12 of which were further bread to test lifetime and reproductive parameters, not found to be different from the results of untreated controls.

CONCLUSION

Experiments have proved that a well-defined sublethal HHP stress treatment offers a solution to improve the overall quality of gametes and embryos (i.e., stress tolerance (cryotolerance), fertilizing ability, and developmental competence). Although the understanding of the exact molecular mechanism requires further research, accumulating data demonstrate the beneficial effect and safety of the "high pressure technology" in ART.

ACKNOWLEDGMENTS

To Gabor Vajta, to all the participating institutes (see Table 1), and to the several Hungarian research grants from NKTH(EGG_CARE), ITDH, and GKM.

REFERENCES

1. Pribenszky, C.; Molnar, M.; Horvath, A.; Harnos, A.; Szenci, O. Hydrostatic pressure induced increase in post-thaw motility of frozen boar spermatozoa. Reprod. Fertil. Dev. **2005**, *18* (2), 162–163.
2. Huang, S.H.; Pribenszky, C; Kuo, Y.H.; Teng, S.H.; Chen, Y.H.; Chung, M.T.; Chiu, Y.F. Hydrostatic pressure pretreatment affects the protein profile of boar sperm before and after freezing-thawing. Anim. Reprod. Sci. **2009**, *112*, 136–149.
3. Kuo, Y.H.; Pribenszky, C.; Huang, S.Y. Higher litter size is achieved by the insemination of high hydrostatic pressure-treated frozen-thawed boar semen. Theriogenology **2008**, *70*, 1395.
4. Pribenszky, C; Molnár, M.; Kútvölgyi, G.; Harnos, A.; Horváth, A.; Héjja, I. Sublethal hydrostatic pressure treatment improves fresh and chilled boar semen quality in vitro and in vivo. Reprod. Fertil. Dev. **2009**, *21*, 107.
5. Pribenszky, C.; Molnár, M.; Horváth, A.; Kútvölgyi, G.; Harnos, A.; Szenci, O.; Dengg, J.; Lederer, J. Improved post-thaw motility, viability and fertility are achieved by hydrostatic pressure treated bull semen. Reprod. Fertil. Dev. **2007**, *19* (1), 181–182.
6. Pribenszky, C.; Du, Y.; Molnár, M.; Harnos, A.; Vajta, G. Increased stress tolerance of matured pig oocyte by high hydrostatic pressure impulse. Anim. Reprod. Sci. **2008**, *106* (1–2), 200–207.
7. Du, Y.; Pribenszky, C.; Molnár, M.; Zhang, X.; Yang, H.; Kuwayama, M.; Pedersen, A.M.; Villemoes, K.; Bolund, L.; Vajta, G. High hydrostatic pressure (HHP): a new way to improve in vitro developmental competence of porcine matured oocytes after vitrification. Reproduction **2008**, *135* (1), 13–17.
8. Du, Y.; Lin, L.; Schmidt, M.; Bøgh, I.B.; Kragh, P.M.; Sørensen, C.B.; Li, J.; Purup, S.; Pribenszky, C.; Molnár, M.; Kuwayama, M.; Zhang, X.; Yang, H.; Bolund, L.; Vajta, G. High hydrostatic pressure treatment of porcine oocytes before handmade cloning improves developmental competence and cryosurvival. Cloning Stem Cells. **2008**, *10* (3), 325–330.
9. Pribenszky, C.; Siquiera, E.; Molnár, M.; Rumpf, R. Improved post-warming developmental competence of open pulled straw-vitrified in vitro-produced bovine blastocysts by sublethal hydrostatic pressure pre-treatment. Reprod. Fertil. Dev. **2008**, *20* (1), 125.
10. Pribenszky, C.; Molnar, M.; Cseh, S.; Solti, L. Improving post-thaw survival of cryopreserved mouse blastocysts by hydrostatic pressure challenge. Anim. Reprod. Sci. **2005**, *87*, 143–150.
11. Kültz, D. Molecular and evolutionary basis of the cellular stress response. Annu. Rev. Physiol. **2005**, *67*, 225–257.
12. Wemekamp-Kamphuis, H.H.; Karatzas, A.K.; Wouters, J.A.; Abee, T. Enhanced levels of cold shock proteins in *Listeria monocytogenes* LO28 upon exposure to low temperature and high hydrostatic pressure. Appl. Environ. Microbiol. **2002**, *68*, 456–463.

Gene Expression Patterns: In Vitro Techniques in Oocytes and Preimplantation Embryos

Christine Wrenzycki
Clinic for Cattle, Reproductive Medicine Unit, University of Veterinary Medicine, Hannover, Germany

Abstract
Numerous factors such as environmental changes have been shown to have a profound effect on oocyte and embryo quality and subsequently preimplantation development. Messenger RNA (mRNA) analysis has become a powerful tool to identify key molecules associated with developmental competence. Before the major activation of the embryonic genome, the bovine preimplantation embryo is controlled by maternal genomic information that is accumulated during oogenesis. The early steps of development, including the timing of the first cleavage, activation of the embryonic genome, compaction, and blastocyst formation, can be affected by the culture media and conditions as well as the production procedure itself. As the in vitro culture conditions do not fully mimic the in vivo situation and as it has been shown that morphological and physiological differences exist between in vivo and in vitro cultured embryos, the gene expression levels of in vitro produced embryos should be compared with those of in vivo embryos, the golden standard embryos. Changes in gene expression patterns between in vitro and in vivo embryos may help to assess the normality of in vitro produced embryos and to optimize the in vitro culture conditions for all mammalian species, including humans.

INTRODUCTION

Current in vitro production (IVP) systems and related biotechnologies, i.e., somatic cell nuclear transfer (SCNT), have been significantly improved in cattle. But the in vivo situation still cannot be mimicked sufficiently well. Morulae and blastocysts can be produced with reasonable efficiency. However, generated embryos display a number of marked differences compared to their in vivo counterparts including different gene expression patterns. It has been suggested that persistent alterations from the normal expression patterns may have detrimental effects on embryonic development and contribute to the incidence of the "large offspring syndrome (LOS)," of which a high birth weight and an extended gestation length are the predominant features.[1–3] The underlying mechanisms are largely unknown today, but alterations of epigenetic modifications of embryonic and fetal gene expression patterns, primarily caused by alterations in DNA methylation, are thought to be involved.[4]

The preimplantation bovine embryo is initially under the control of maternal genomic information that is accumulated during oogenesis. Soon, the genetic program of development becomes dependent on new transcripts derived from activation of the embryonic genome. The early steps in development including timing of first cleavage, activation of the embryonic genome, compaction, and blastocyst formation can be affected by the culture media and conditions as well as the production procedure itself. These perturbations can possibly result in a dramatic decrease in the quality of the resulting blastocysts, and may even affect the viability of offspring born after transfer.

Analysis of mRNA expression patterns of developmentally important genes essential in early development provides a useful tool to assess the normality of the embryos produced and to optimize assisted reproductive technologies. Messenger RNA studies in bovine oocytes and embryos have emerged as a rapidly moving field and a large body of literature exists.[1–7] Numerous aberrations have been found including (complete) suppression of gene expression, (incorrect) de novo gene expression, or more frequently to a significant up- or downregulation of an already expressed gene.[4] Defining the expression patterns of specific genes critically involved in preimplantation development will aid in selecting markers for determining embryo quality. The identification and characterization of the short-term effects raise the question about long-term consequences and safety of assisted reproductive technologies. Recent reports show that in vitro culture of murine embryos can have irreversibly long-term consequences of postnatal development, growth, physiology, and behavior in resulting offspring.[8,9]

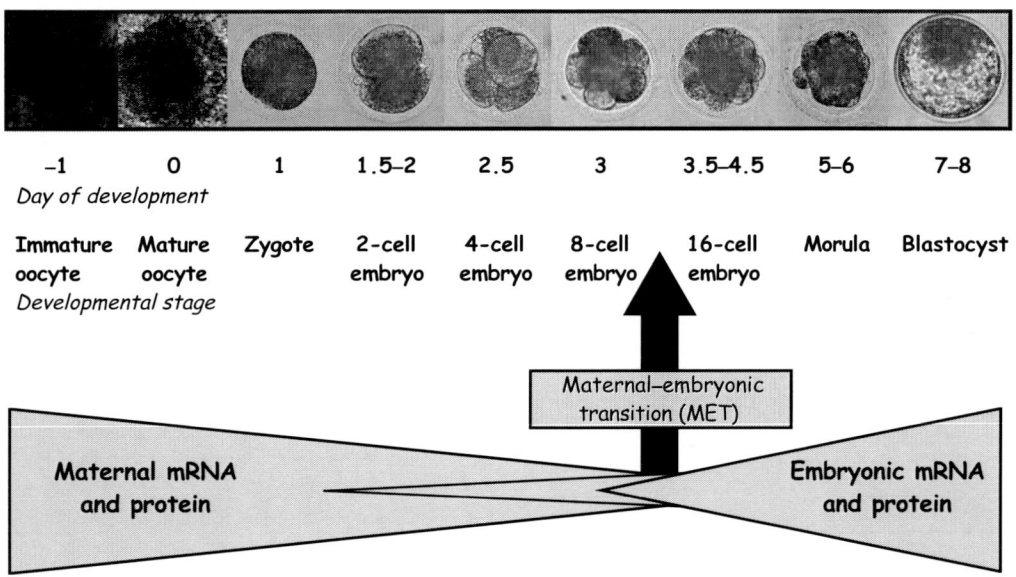

Fig. 1 Messenger RNA and protein expression in preimplantation bovine embryos.

The molecular deviations observed in studies with farm and laboratory animals emphasize the need for further studies to gain insight into the expression patterns correlated with an undisturbed embryonic and fetal development. Understanding the molecular mechanisms will aid to improve biotechnologies applied to early embryos in all species, including humans. The bovine embryo could serve as a useful model as there is growing evidence that the bovine is a good model for human preimplantation development.

BODY OF TEXT

The first major developmental transition that occurs following fertilization is the maternal to embryonic transition (MET) in which the developmental program that is initially directed by maternally inherited transcripts and proteins is replaced by a new program as the consequence of the expression of new genes. This occurs in bovine embryos at the 8- to 16-cell stage and in humans at the 4- to 8-cell stage.[10] The MET has at least three functions that are required for normal development.[11] The first function is to destroy oocyte-specific transcripts that are not subsequently expressed. The second function is to replace maternal transcripts that are common to the oocyte and early embryo with embryonic transcripts and the third is to promote the dramatic reprogramming in the pattern of gene expression that is coupled with the generation of novel transcripts that are not expressed in the oocyte.

Currently, relatively little is known regarding the protein production of mammalian embryos. The lack of sensitivity has been the major obstacle for the application of proteomics into the field of mammalian embryology (at a minimum, hundreds of oocytes or embryos are needed).

However, new developments in technology have increased the sensitivity. Now, it is feasible to profile the proteome of individual mammalian oocytes and embryos.[12] This will allow studies to compare mRNA and protein profiles in single early bovine embryos. There is already some evidence that mRNA and protein levels are not always related.

The developmental program is controlled by genetic and epigenetic mechanisms. Epigenetic marks are heritable through mitosis and meiosis and are not dependent on changes in DNA sequence. The precise timing of gene activation/inactivation is crucial for normal preimplantation development and any delay or lack of expression leads to abnormalities. Thus, the reprogramming needs to attain the correct timing of activation and the well-orchestrated pattern of gene expression to initiate and continue embryonic development. Fig. 1 shows the gene expression patterns in the preimplantation bovine embryo. In general, in vitro-produced embryos exhibit many different alterations and nearly all of them show some level of abnormality, even those of high morphological quality. The exact mechanisms by which in vitro procedures induce abnormal epigenetic modifications responsible for the deregulation of gene expression are not known.

In human ART, there have been reports of an increased incidence of IVF or ICSI conceptions amongst children with Beckwith–Wiedemann syndrome (BWS) and Angelman syndrome (AS). BWS and AS are model imprinting disorders which result from altered expression or mutations in imprinted genes that are critical for normal growth and development. The main characteristics of BWS are pre- and/or postnatal overgrowth, anterior abdominal wall defects, macroglossia, and neonatal hypoglycaemia. AS is characterized by severe mental retardation, delayed motor development, poor balance, jerky movements, absence of speech, and a happy disposition. Importantly, more than

90% of children with BWS that were born after ART had imprinting defects, compared with 40–50% of children with BWS and conceived without ART. Moreover, there have been other reports suggesting an association between AS and ART. The majority of children with AS born after ART had an imprinting defect as the underlying etiology, specifically loss of methylation of the maternal allele.[13,14]

CONCLUSION

Knowledge of the mechanisms regulating the preimplantation development of mammalian embryos generated by different biotechnologies has produced an increasing number of approaches to evaluate the normality of these embryos. Comparison with their in vivo counterparts is important in this respect. There is a need for further research with animal models to understand the developmental plasticity of early embryos and to reduce the potential of adverse consequences. Improvements of methodology during all the steps of in vitro manipulation in such a way as the produced embryos will have a similar well-orchestrated gene expression pattern as their in vivo counterparts will make it possible to produce more transferable embryos of higher quality.

ACKNOWLEDGMENTS

The financial support of the Deutsche Forschungsgemeinschaft (DFG), the European Commission, and the H.W. Schaumann Foundation is gratefully acknowledged.

REFERENCES

1. Farin, P.W.; Piedrahita, J.A.; Farin, C.E. Errors in development of fetuses and placentas from in vitro-produced bovine embryos. Theriogenology **2006**, *65* (1), 178–191.
2. Lonergan, P.; Fair, T. In vitro-produced bovine embryos: dealing with the warts. Theriogenology **2008**, *69* (1), 17–22.
3. Wrenzycki, C.; Herrmann, D.; Niemann, H. Messenger RNA in oocytes and embryos in relation to embryo viability. Theriogenology **2007**, *68* (Suppl 1), 77–83.
4. Wrenzycki, C.; Herrmann, D.; Lucas-Hahn, A.; Korsawe, K.; Lemme, E.; Niemann, H. Messenger RNA expression patterns in bovine embryos derived from in vitro procedures and their implications for development. Reprod. Fertil. Dev. **2005**, *17* (1–2), 23–35.
5. Kues, W.A.; Sudheer, S.; Herrmann, D.; Carnwath, J.W.; Havlicek, V.; Besenfelder, U.; Lehrach, H.; Adjaye, J.; Niemann, H. Genome-wide expression profiling reveals distinct clusters of transcriptional regulation during bovine preimplantation development in vivo. Proc. Natl. Acad. Sci. U.S.A. **2008**, *105* (50), 19768–19773.
6. Goossens, K.; Van Soom, A.; Van Poucke, M.; Vandaele, L.; Vandesompele, J.; Van Zeveren, A.; Peelman, L.J. Identification and expression analysis of genes associated with bovine blastocyst formation. BMC Dev. Biol. **2007**, *7*, 64.
7. Fair, T.; Carter, F.; Park, S.; Evans, A.C.; Lonergan, P. Global gene expression analysis during bovine oocyte in vitro maturation. Theriogenology **2007**, *68* (Suppl 1), 91–97.
8. Mahsoudi, B.; Li, A.; O'Neill, C. Assessment of the long-term and transgenerational consequences of perturbing preimplantation embryo development in mice. Biol. Reprod. **2007**, *77* (5), 889–896.
9. Thompson, J.G.; Mitchell, M.; Kind, K.L. Embryo culture and long-term consequences. Reprod. Fertil. Dev. **2007**, *19* (1), 43–52.
10. Telford, N.A.; Watson, A.J.; Schultz, G.A. Transition from maternal to embryonic control in early mammalian development: a comparison of several species. Mol. Reprod. Dev. **1990**, *26* (1), 90–100.
11. Schultz, R.M. The molecular foundations of the maternal to zygotic transition in the preimplantation embryo. Hum. Reprod. Update **2002**, *8* (4), 323–331.
12. Katz-Jaffe, M.G.; Gardner, D.K. Embryology in the era of proteomics. Theriogenology **2007**, *68* (Suppl 1), 125–130.
13. Manipalviratn, S.; DeCherney, A.; Segars, J. Imprinting disorders and assisted reproductive technology. Fertil. Steril. **2009**, *91* (2), 305–315.
14. Amor DJ, Halliday J. A review of known imprinting syndromes and their association with assisted reproduction technologies. Hum. Reprod. **2008**, *23* (12), 2826–2834.

Genetic Engineering and Biotechnology: Biosafety and Environmental Impact

Martina Newell-McGloughlin
UC Systemwide Biotechnology Research and Education Program (UCBREP), University of California–Davis, Davis, California, U.S.A.

Abstract

Regulatory oversight of the products of modern biotechnology is undertaken within a framework that ensures adequate protection of the consumer and the environment while not stymieing innovation. The United States uses health and safety laws written prior to the advent of modern biotechnology to review genetically engineered products. To date, the United States has not issued any new legislation for these products. The current system was delineated under the 1986 Coordinated Framework for Regulation of Biotechnology. Over 150 products have been reviewed with just 120 attaining unregulated status under the relevant regulatory agencies. The first generation of biotechnology products commercialized were crops focusing largely on input agronomic traits. The coming generations, although appearing more complex, represent many of the same issues of risk. Animals will also enter the regulatory net and while they may parallel plants at some level of application, namely disease resistance, yield, food quality and as "pharming" factories, they have clear departures on many significant levels. The issues of ensuring biosafety for the consumer and the environment are the prime concern whatever the organism, microbe, plant, arthropod, or animal. For all organisms continuing improvements in molecular and genomic technologies are contributing to the acceleration of product development. These new products and new approaches on the horizon are the subject of continuous reassessment by the relevant agencies to determine appropriate criteria to manage risk while insuring that the development of innovative technologies and processes is encouraged to provide value-added commodities for the consumer.

INTRODUCTION

Agriculturalists have been genetically modifying animals and crop plants through cross breeding, mutation selection, and culling those with undesirable characteristics for hundreds of years. Thus, from a scientific perspective the term "genetically modified/manipulated organism" is not an accurate descriptor of the products solely of modern biotechnology, as virtually all domesticated crops and animals have been subjected to varying degrees of genetic modification. Risk assessment should be undertaken within a regulatory framework that ensures adequate protection of the consumer and the environment while not stymieing innovation. The first generation of biotechnology products commercialized were crops focusing largely on input agronomic traits. Twenty-three countries, including twelve less developed countries (LDCs), now have biosafety protocols in place and grew 114.3 million ha (282.4 million acres) of biotech crops in 2007. The coming generations of crop plants can be grouped into four broad areas each presenting what, on the surface, may appear as unique challenges to regulatory oversight. The present and future focus is on continuing improvement of agronomic traits such as yield and abiotic stress in addition to the biotic stress tolerance of the present generation; crop plants as biomass feedstocks for biofuels and "bio-synthetics"; value-added output traits such as improved nutrition and food functionality; and plants as production factories. Animals will also enter the regulatory net and although they may parallel plants at some level of application, namely disease resistance, yield, food quality, and as pharming factories, they have clear departures on many significant levels. The issues of ensuring biosafety for the consumer and the environment are the prime concern whatever the organism, microbe, plant, arthropod, or animal. For all organisms continuing improvements in molecular and genomic technologies are contributing to the acceleration of product development. These new products and new approaches on the horizon require a reassessment of appropriate criteria to manage risk while insuring that the development of innovative technologies and processes is encouraged to provide value-added commodities for the consumer (Fig. 1).

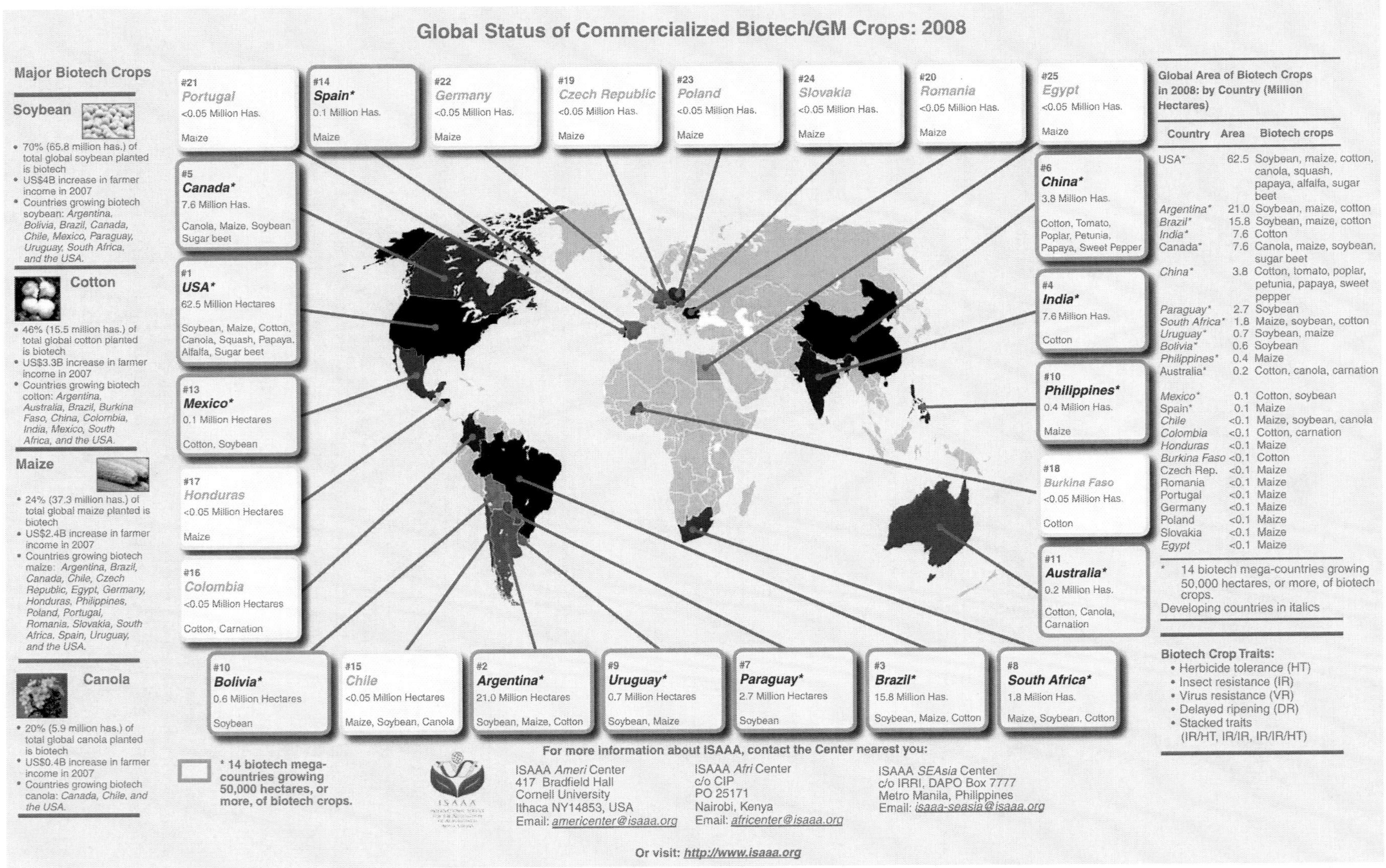

Fig. 1 Biotech crop countries and mega-countries*, 2008. Biotech crops 2008: 125 million ha (310 million acres); 25 countries (15 LDC) 12% increase over 2007, 13.3 million farmers (12 million, 2007); 90% resource-poor LDC farmers (12.3–11 million, 2006) most Bt cotton; Spain lead country in Europe planting 100,000 ha. **Source:** From *Global Status of Commercialized Biotech/GM Crops.*[1]

Table 1 Oversight authority for agricultural biotechnology products.

Agency	Products regulated
U.S. Department of Agriculture	Plant pests, plants, veterinary biologics, animals, fish, and arthropods
Environmental Protection Agency	Microbial/plant pesticides, new uses of existing pesticides, and novel microorganisms
Food and Drug Administration	Food, feed, food additives, veterinary and human drugs, and medical devices

CURRENT U.S. REGULATIONS

The United States has been grappling with these regulatory issues considerably longer than most other countries. It also has the largest stake in the technology. Since regulations in other jurisdictions, including the European Union (EU), Australia/New Zealand Authority (ANZA), India, Japan, and China are still evolving and are being influenced by the U.S. experience.

At the federal level the United States has attempted, and largely succeeded, in developing a coordinated, risk-based system to ensure that new biotechnology products are safe for the environment and human and animal health. The United States uses health and safety laws written prior to the advent of modern biotechnology to review genetically engineered products. To date, the United States has not issued any new legislation for these products. Established as a formal policy in 1986, the Coordinated Framework for Regulation of Biotechnology[2] describes the federal system for evaluating products developed using modern biotechnology. The coordinated framework is based upon health and safety laws developed to address specific product classes. The move from the laboratory to the field proved a challenge for the relevant agencies. In the short time that elapsed after the initial development of the capabilities of recombinant DNA technology, the controversy over this research shifted focus from the presumed risks associated with the possible escape of GMOs from research laboratories to the nature of the long-term environmental impact of GMOs that are intentionally released.

Under the coordinated framework, agencies that were responsible for regulatory oversight of certain product categories or for certain product uses are also responsible for evaluating those same kinds of products developed using genetic engineering. Three agencies share primary responsibility for regulating the organisms, products, and processes of recombinant DNA technology, whether they be designed for closed systems or for environmental release: the Food and Drug Administration (FDA), the United States Department of Agriculture (USDA), and the Environmental Protection Agency (EPA) (Table 1). Regulations under the Occupational Safety and Health Administration (OSHA) cover those working with rDNA, while the Department of Health and Human Services (DHHS) oversees the health of the general public. Their responsibilities are complementary and, in some cases, overlapping. USDA-APHIS has jurisdiction over the planting of genetically engineered plants. EPA has jurisdiction over planting and food and feed uses of pesticides engineered into plants. These are referred to as plant-incorporated protectants or PIPs. FDA has jurisdiction over food and feed uses of all foods from plants. The agencies have developed a coordinated Web site to keep the public informed.[3]

This framework has allowed the United States to build upon agency experience with organisms and products developed using conventional techniques. Accordingly, the agencies base their analysis of oversight not principally on any presumed theoretical "exotic" risks but rather exploit the vast cache of accumulated knowledge that has been amassed over years of research using highly developed scientific procedures for assessing field tests and planned introductions. Using such a scientific approach deduced from earlier principles is probably the most effective mechanism regulatory bodies can take in crafting effective regulations. The laws currently used to regulate the products of modern biotechnology are the Plant Protection Act (PPA), the Federal Food, Drug, and Cosmetic Act (FFDCA), the Federal Insecticide, Fungicide, and Rodenticide Act (FIFRA), and the Toxic Substances Control Act (TSCA). New regulations have been developed under these statutes as needed to address genetically engineered products developed. New regulations, policy statements, and guidelines will continue to be developed as needed.

The details of what items are regulated, such as organisms and processes, and how both may be regulated (time frames, permitting processes, and penalties) are written by each agency that has the appropriate authority. All formal federal regulations are published in the Federal Register and also in the Code of Federal Regulations, a large multivolume series. For example those regulations for agriculture and the USDA comprise fifteen volumes and those governing biotechnology as overseen by APHIS-BRS are found in Volume 7, Section 340.

Many states and local authorities were not satisfied with existing oversight, and some 14 states and several municipalities enacted their own biotechnology legislation many of which have since been modified or revoked. Agencies must also adhere to the umbrella National Environmental Policy Act (NEPA), which is binding on all federal agencies. Depending on its intended use, a product may or may not be reviewed by all three regulatory agencies. A food crop plant developed using genetic engineering to produce a pesticide in its own tissue provides an example that is reviewed by all three regulatory agencies.

THE UNITED STATES DEPARTMENT OF AGRICULTURE (USDA)

Scope for coverage is food and fiber products. While its authority is as equally applicable to genetically modified animals, plants, and microorganisms, it is in one sense outside the circle of ecological safety in that its primary concerns are the safety of crop plant and food animals and the safety and wholesomeness of food products.

The USDA has nine divisions that deal with biotechnology:

- The Agricultural Research Service (ARS)
- The Food Safety and Inspection Service (FSIS)
- The Animal and Plant Health Inspection Service (APHIS)
- The Agricultural Marketing Service (AMS)
- The Cooperative State Research Service (CSRS)
- Extension Service
- The National Agricultural Library (NAL)
- The Forest Service (FS)
- The Economic Research Service (ERS)

USDA policy on the regulation of biotechnology, consistent with the overall federal policy, does not view GMOs as fundamentally different from those produced using traditional methods. The USDA considered that the products of the new techniques of biotechnology were in principle covered by regulations that had been implemented for existing technologies. They did, however, consider that the assessment of the products of the new technologies in some instances required specific information that necessitated the introduction of some new regulations and the updating of some existing ones.

Within USDA, the Animal and Plant Health Inspection Service (APHIS) is responsible for protecting agriculture from pests and diseases. The Federal Plant Protection Act (7 USC. §§7701 et seq.) is the primary statute under which APHIS regulates agricultural biotechnology. Enacted in 2000, this statute replaced the former Federal Plant Pest Act. Originally intended to prevent the introduction and interstate movement of plant pests, the Plant Pest Act had been adapted by APHIS to regulate genetically engineered plants so that they do not become "plant pests." The section within APHIS with prime responsibility is APHIS Biotechnology Regulatory Services (APHIS-BRS). Under the Plant Protection Act, USDA-APHIS has regulatory oversight over products of modern biotechnology that could pose such a risk. Accordingly, USDA-APHIS regulates organisms and products that are known or suspected to be plant pests or to pose a plant pest risk, including those that have been altered or produced through genetic engineering. These are called "regulated articles." A regulated article is defined in APHIS regulations as "any organism which has been altered or produced through genetic engineering" if the donor organism, recipient organism, vector, or vector agent is a "plant pest" (7 C.F.R. §340.1). The APHIS defines a plant pest broadly to include "any living stage" of insects, bacteria, fungi, viruses, or various other organisms that can damage or cause injury to plants or plant parts (7 C.F.R. §340.1). Many plant pathogens commonly used as vectors or promoters in agricultural biotechnology, such as *Agrobacterium* spp. and cauliflower mosaic viruses, are considered plant pests under APHIS regulations [7 C.F.R. §340.2(a)]. Use of any of these plant pests to make a transgenic plant makes that plant a regulated article. The agency may also designate as a regulated article any product of genetic engineering that the agency determines or has reason to believe is a plant pest (7 C.F.R. §340.1). USDA-APHIS regulates the import, handling, interstate movement, and release into the environment of regulated organisms that are products of biotechnology, including organisms undergoing confined experimental use or field trials. Regulated articles are reviewed to ensure that, under the proposed conditions of use, they do not present a plant pest risk through ensuring appropriate handling, confinement, and disposal.

In 1993, with modifications in 1995 and 1997, USDA introduced the final rule notification in lieu of permit process for GMOs that are field tested in accordance with specific safety criteria. The amendment simplifies procedures for the introduction of certain genetically engineered organisms, expedites review for certain determinations of non-regulated status, and adjusts procedures for the reporting of field tests conducted under notification to the biology of the test organisms. This enables APHIS, when appropriate, to extend the existing determination of non-regulated status for new products that do not raise new risk issues. USDA-APHIS regulations provide a petition process for the determination of non-regulated status. If a petition is granted, that organism will no longer be considered a regulated article and will no longer be subject to oversight by USDA-APHIS. The petitioner must supply information such as the biology of the recipient plant, experimental data and publications, genotypic and phenotypic descriptions of the genetically engineered organism, and field test reports. The agency evaluates a variety of issues including the potential for plant pest risk; disease and pest susceptibilities; the expression of gene products, new enzymes or changes to plant metabolism; weediness and impact on sexually compatible plants; agricultural or cultivation practices; effects on non-target organisms; and the potential for gene transfer to other types of organisms. A notice is filed in the Federal Register and public comments are considered on the environmental assessment and determination written for the decision on granting the petition. Copies of the USDA-APHIS documents are available to the public.

One exception to the notification process described above is for plants that have been genetically modified to produce pharmaceutical or industrial products [40 C.F.R. §340.3(b)(4)(iii)]. These plants require a permit before field testing or interstate shipment. The APHIS policy is to inspect all field trials involving such organisms at least annually, and the agency has stated that such organisms will never be eligible for "deregulation."[5] Currently, test plots for crops engineered to produce are regulated by the APHIS permit system. USDA regulations require developers to have clearly written procedures, to handle wastes properly, and to maintain production and control records. Plant-made pharmaceuticals also come under the U.S. Food and Drug Administration's (FDA) purview. The FDA will monitor the manufacturing process as well as the purity and consistency of the products under its "good manufacturing practices" guidelines. Nevertheless, both USDA and FDA recognize that this is a new application of the technology and are in the process of coming up with new guidance specifically for pharmaceutical plants.

The growing and harvesting of the crop itself must comply with principles of confinement that essentially means keeping the crop and its products on the land where it was grown until removed for processing with no inadvertent exposure to the public and minimal exposure of products to workers and the environment. Adequate analytical methods for detection of expression (i.e., protein) products must be demonstrated and all confinement systems and procedures must be based on sound scientific principles. Identity preservation within a closed loop system to prevent co-mingling of pharmaceutical crops with food crops will be a prime directive for industry as well as regulatory agencies.

In addition to the this, under NEPA, the USDA has a responsibility for ensuring ecological safe utilization of crops, livestock, and veterinary products produced both from traditional and recombinant DNA methods. FSIS assures the safety and wholesomeness of food products.

Over 9000 field tests have been analyzed by USDA involving more than 180 organisms at 40,000 sites throughout the United States. The agency has assessed the biotech plants for their efficacy, performance, and suitability for release in the environment. Globally, approximately 30,000 field trials have been conducted on 100 organisms in 45 countries.[6] There has not been a single report of any unexpected or unusual outcome. The APHIS also issues licenses and permits for production, importation, sale, and experimental use of various types of biological products. Specific information that must be submitted for GMOs and products includes detailed information on stability, genetic constructs and vectors, and the effects of any insertions and deletions on the organism.

In 2008, the USDA stated that new rules will be phased into use. Products currently under review for non-regulated status will not be affected. They propose a multitiered permit system for field testing bioengineered plants wherein higher risk products are subject to tighter controls; they also considered conditional decisions and post-marketing monitoring requirements for commercialization of certain bioengineered plants; they recommended more coordination with the FDA and the EPA to include early food safety assessments to address low level intermittent presence of unapproved traits in the food supply; and proposed new policies for field testing and commercializing plant-made pharmaceuticals (PMPs) and plant-made industrial products (PMiPs). However, although there was a comprehensive and thorough prologue that quite clearly delineated the history, issues, and process involved in promulgating the proposed regulations the suggested amendments, which appeared in the Federal Register[7] in October 2008, have little basis in science or proportionality. The proposed regulations place little value on precedent, experience, or history basically reducing the regulatory process to assessment on a case-by-case basis. This could greatly increase the burden of proof and place such an onus on meeting regulatory requirements that it may very well have a detrimental effect on all agricultural biotech research.

THE ENVIRONMENTAL PROTECTION AGENCY (EPA)

The EPA claims that its biotechnology regulatory program is based on five important principles that guide their decision-making policy. They are using sound science, ensuring transparency of the decision-making process, maintaining consistency and fairness, collaborating with their regulatory partners, and building public trust.

Under NEPA the EPA has broad jurisdiction over environmental impact. It mostly considers GMOs under the aegis of the Toxic Substances Control Act (TSCA) and the Federal Insecticide, Fungicide, and Rodenticide Act (FIFRA). In general, the TSCA will not be triggered if the microbes are used to produce foods, additives, drugs, vaccines, cosmetics, or medical devices if they are regulated by the FDA or the USDA. Other than those listed exemptions all other new or new-use chemical end products from microorganisms, which themselves may be considered new, are subject to the TSCA. Consent orders and negotiated agreements are used as the mechanisms to cover environmental introductions.

The FIFRA mandates the registration of pest-control products and "economic poisons" prior to production and sale. In order to test a product, an applicant must submit complete data. Product evaluation includes the equivalent to an environmental assessment or environmental impact statements. Since 1984, notification of intent to field test GMOs and non-engineered non-indigenous pathogenic and non-pathogenic microorganisms was required, regardless of testing site size.

The EPA currently regulates chemical and bio pesticides that are externally applied to plants. In the 1990s, the agency expanded its federal regulatory powers over the characteristics of plants that help plants resist diseases and pests. The agency coined a new term for these characteristics calling them "plant-pesticides." All plants are able to prevent, destroy, repel, or mitigate pests or diseases. That ability occurs naturally, and some crops have been bred for resistance to specific pests. The EPA proposed to single out for regulation those pest-resistant qualities that were transferred to the plant through recombinant DNA technology. Eleven professional scientific societies took grave objection to the suggested regulations and they responded by developing a document entitled "Appropriate Oversight for Plants with Inherited Traits for Resistance to Pests: A Report." They determined that evaluation of the safety of substances in plants should be based on the toxicological and exposure characteristics of the substance and not on whether the substance confers protection against a plant pest. A compromise was reached, and the EPA now regulates PIPs.

The PIPs rules were issued January 17, 2001. The rules consider the impact on beneficial insects and non-target organisms, toxicity of pesticidal compound, safety for human consumption, ecological hazards, and insect resistance. The specific tests for FIFRA PIPs clearance are wildlife exposure studies for PIPs on non-target organisms; feeding studies on beneficial insects, including honeybees, ladybird beetles, and lacewings, earthworms and birds, fish and mammals at ten to one hundred times the concentration; environmental fate—how long it takes for the PIP protein to degrade in plant material left on or in the soil. Reviews are conducted case-by-case based on the product and the risk and not the means by which the organism was created. However, interestingly enough, despite the assertion that no experimental use permits (EUPs) are required for undirected mutagenesis, most transconjugants and plasmid-cured strains, EUPs are required for all live rDNA GMOs irrespective of product or risk.

Under the TSCA, the EPA acquires information in order to identify and regulate potential hazards and exposures. The TSCA applies to the manufacturing, processing, importation, distribution, use, and disposal of all chemicals in commerce, or intended for entry into commerce, that are not specifically covered by other regulatory authorities (e.g., substances other than food, drugs, cosmetics, and pesticides). The TSCA's applicability to the regulation of products of biotechnology is based on the interpretation that organisms are chemical substances under the TSCA. The EPA's TSCA Biotechnology Program of the Office of Prevention and Toxic Substances currently regulates microorganisms intended for general industrial uses. The program conducts a pre-market review of "new" microorganisms, i.e., those microorganisms formed by deliberate combinations of genetic material from organisms classified in different taxonomic genera. Developers must notify the EPA 90 days prior to manufacture or 60 days prior to field testing of a product regulate by the TSCA.

The EPA also sets tolerance limits for residues of pesticides on and in food and animal feed, or establishes an exemption from the requirement for a tolerance, under the Federal Food, Drug and Cosmetic Act.

An economic analysis[8] shows that in the first 11 years of GM crop cultivation, global net farm income increased by $33.8 billion since 1996; the environmental footprint associated with pesticide use was reduced by 15.4%; there was a reduction in carbon dioxide emissions in 2006 equivalent to taking nearly 6.6 million cars off the road for a year. Reduced-till agriculture means healthier soil, with reduced erosion and far less carbon dioxide release. In general, cultivation is not a sustainable practice. It is energy intensive, exposes soil to wind and water erosion. It allows rain to compact the soil, increases the oxygen content of the soil, allowing organic matter to oxidize away. In turn, lower organic matter in the soil allows more compaction and more nutrient loss. Pesticide use fell by over 286 million kg (−7.8%: equivalent to about 40% of the annual volume of pesticide active ingredient applied to arable crops in the European Union). Less spraying means fewer tractor passes, contributing to lower CO_2 emissions. Insect-resistant maize also has a collateral effect—less insect damage results in much less infection by fungal moulds that reduces mycotoxins that are known health risks, causing such problems as liver cancer to humans and animals. The only "natural" way to control those fungi is the use of copper sulfate that has one of the highest toxic hazard ratings of acceptable pesticides and selects for antibiotic resistant bacteria in the soil.

FOOD AND DRUG ADMINISTRATION (FDA)

The FDA regulates biotechnology under the authority of the Food, Drug, and Cosmetic Act (FDCA) and the Public Health Services Act (PHSA). The agency has a mandate to ensure efficacy and safety of food and pharmaceutical products. The agency has a major responsibility in biotechnology in that the majority of the current market share of biotechnology products passes through the agency for review, and it has already reviewed thousands of biotechnology products.

The FDA is responsible for ensuring the safety and proper labeling of all plant-derived foods and feeds, including those developed through bioengineering. All foods and feeds, whether imported or domestic and whether derived from crops modified by conventional breeding techniques or by genetic engineering techniques, must meet the same rigorous safety standards. Under the FDCA, it is the responsibility of food and feed manufacturers to ensure that the products they market are safe and properly labeled. Generally, whole foods, such as fruits, vegetables, and grains, are not subject to pre-market approval. The primary legal tool that the FDA has successfully used to ensure the safety of foods is the adulteration provisions of section 402(a)(1).

The act places a legal duty on developers to ensure that the foods they present to consumers are safe and comply with all legal requirements. The FDA has authority to remove a food from the market if it poses a risk to public health. Foods derived from new plant varieties developed through genetic engineering are regulated under this authority. A second section of the act that the FDA relies on is the food additive provision (section 409). Under this section, substances that are intentionally added to food are food additives, unless the substance is generally recognized as safe (GRAS). Food additives are subject to review and approval by the FDA before they may be used in food. When requested to do so, the FDA also reviews and affirms the GRAS status of food ingredients.

In May 1992, the FDA issued a policy statement on regulating biotechnology food products. The FDA requires pre-market review only for foods into which substances are intentionally introduced, significantly changing the structure, function, or amount currently found in the food. If a new food product developed through biotechnology does not contain substances that are significantly different from those already in the diet, it does not require pre-market approval. To help sponsors of foods and feeds derived from genetically engineered crops comply with their obligations, the FDA encourages them to participate in its voluntary consultation process. All foods and feeds from genetically engineered crops currently on the market in the United States have gone through this consultation process. With one exception, none of these foods and feeds were considered to contain a food additive, and so did not require approval prior to marketing. On a number of occasions, the FDA stated they would publish a proposed rule mandating that developers of bioengineered foods and animal feeds notify the agency when they intend to market such products and that they would require that specific information be submitted to help determine whether the foods or animal feeds pose any potential safety, labeling, or adulteration issues. However consultation still remains voluntary.[9]

On the "labeling" issue, FDA policy guidelines state that foods produced through biotechnology will be subject to the same labeling laws as all other foods and food ingredients. Labeling would be required for biotech products in some instances, but not because the products were made using biotech. The FDA requires that labeling of a food or food ingredient or additive be truthful and not misleading and that the product be declared by its common or usual name. In general, the information on the label pertains to the composition and attributes of the food or food ingredients or additive but not to the details of agricultural practices or the manufacturing process.

In 2006, to address the possibility that material from a new plant variety intended for food use might inadvertently enter the food supply before its sponsor has fully consulted with the FDA, the FDA announced the availability of a draft guidance document entitled "Guidance for Industry: Recommendations for the Early Food Safety Evaluation of New Non-Pesticidal Proteins Produced by New Plant Varieties Intended for Food Use." It discussed the early food safety evaluation of new proteins in new plant varieties, particularly in new bioengineered varieties that are under development for possible use as food for humans or animals. The draft guidance also described procedures for communicating with the FDA about this evaluation. The issuance of draft guidance was proposed in August 2002 in a Federal Register Notice (67 FR 50578) published by the Office of Science and Technology Policy (OSTP) as part of proposed federal actions to update field test requirements and to establish early voluntary food safety evaluations for new proteins produced by bioengineered plants.

They opined that rapid developments in genomics are resulting in dramatic changes in the way new plant varieties are developed and commercialized. Scientific advances are expected to accelerate leading to the development and commercialization of a greater number and diversity of bioengineered crops. As the number and diversity of field tests for bioengineered plants increase, the likelihood that cross-pollination due to pollen drift from field tests to commercial fields and commingling of seeds produced during field tests with commercial seeds or grain may also increase. This could result in low-level presence in the food supply of material from new plant varieties that have not been evaluated through FDA's voluntary consultation process for foods derived from new plant varieties (referred to as a "biotechnology consultation" in the case of bioengineered plants). The FDA believes that any potential risk from the low level presence of such material in the food supply would be limited to the possibility that it would contain or consist of a new protein that might be an allergen or toxin.

While the FDA has not found and does not believe that new plant varieties under development for food and feed use generally pose any safety or regulatory concerns, this guidance is consistent with the FDA's policy of encouraging communication early in the development process for a new plant variety. Such communication helps to ensure that any potential food safety issues regarding a new protein in such a new plant variety are resolved prior to any possible inadvertent introduction into the food supply of material from that plant variety.

Depending on its intended use, a product may or may not be reviewed by all three regulatory agencies. A food crop plant developed using genetic engineering to produce a pesticide in its own tissue provides an example that is reviewed by all three regulatory agencies. A common example of this type of product is Bt corn that has incorporated a gene isolated from the soil bacterium, *Bacillus thuringiensis* (Bt). The Bt gene encodes a pesticidal protein (Cry). The USDA-APHIS under the Plant Protection Act (PPA) regulates the corn as a regulated article based on the procedures used to introduce or express the Bt gene. USDA-APHIS oversight begins early in the development cycle and continues until the developer applies for and is granted non-regulated status for the plant because it can be shown that the plant is not a plant pest. Until such time as the developer is granted non-regulated status for the plant, the USDA-APHIS oversees the transportation (including importation), field-testing, and disposal of the plant. Developers of Bt crops also consult with the FDA about possible other, unintended changes to the food or feed, for example, possible changes in nutritional composition or levels of native toxicants. Although this consultation is voluntary, all of the food/feed products commercialized to date have gone through the consultation process. The consultation with the FDA serves to ensure that safety or other regulatory issues that fall within the agency's jurisdiction, including appropriate labeling of the food, are resolved prior to commercial distribution.

When an unregulated event falling under all three jurisdictions is detected in commercial distribution, all three coordinate to mitigate the misadventure. For example, in February 2008, the three agencies coordinated efforts following notification by Dow AgroSciences' subsidiary Mycogen that the company detected extremely low levels of an unregistered genetically engineered (GE) PIP in three of its commercial GE hybrid corn seed lines. They quickly determined that the unregistered product produces proteins that are identical to a registered product and issued a statement that the USDA, the EPA, and the FDA have concluded that there are no public health, food, or feed safety concerns. Additionally, the USDA and the EPA have determined that the unregistered GE corn PIP poses no plant pest or environmental concerns.[10]

Coexistence

The issue is what, if any, are the economic consequences of adventitious presence of material from one crop system within another based on the notion that farmers should be able to cultivate freely the crops of their choice using whichever production system works best in any given context (GM, conventional, or organic). It is never a food or environmental safety issue but rather a production and marketing matter. The heart of the issue is assessing the likelihood of adventitious presence of material from one production system affecting another and the potential impacts. This requires consistency when dealing with adventitious presence of any unwanted material including, but most definitely not limited to, biotech-derived material. It is unrealistic to expect 100% purity for any crops, or products derived therefrom, so thresholds that are consistent across all materials should be set and should not discriminate (e.g., thresholds for adventitious presence of biotech material should be the same as applied to thresholds for other unwanted material and vice versa).[11]

Historically, the worldwide market has adequately addressed economic liability issues relating to the adventitious presence of unwanted material in any agricultural crop. For example, for certified seed the onus is on the producers, who require isolation from undesired pollination for the purity of their product, to insure such purity. By extension, the onus is on growers of any specialty crops to take action to protect the purity of their crops since these are self-imposed standards for and by that market. They cannot hold accountable their non-certified neighbors for contamination in their seed fields since they chose that course—it was not imposed on them. Such growers usually are rewarded by higher prices for taking such actions. Existing legislation in North America and the EU is more than adequate to protect all grower and consumer interests but if new regulations were considered to address economic liability provisions for any negative economic consequences of adventitious presence of unwanted material, the same principle should apply to all farmers regardless of their chosen production methods. All measures should be proportionate, non-discriminatory, and science-based. Within the EU, provision has been made for a *de minimis* threshold for unavoidable presence of GMOs but no actual threshold has been set. Therefore, the default state of the 0.9% on labeling and traceability is the one enforced. In the United States, organic products cannot be (legally) downgraded or the producer decertified by unintentional presence and none have to date.

ANIMAL BIOTECHNOLOGY

Biotechnology as it relates to animals largely comprises three main thrust areas, namely biologics (including vaccines and "therapeutics"), agricultural, and biomedical applications.

As with plants the two principal agencies that oversee animal biotech applications are the USDA and the FDA. Within the USDA primary jurisdiction for biologics lies under the Virus, Serum, Toxin Act. USDA-APHIS Veterinary Services inspects biologics production establishments and licenses veterinary biological substances, including animal vaccines that are products of biotechnology. Within FDA the Center for Veterinary Medicine (CVM) is the primary center responsible for approving recombinant products. For example recombinant bovine somatotropin (rBST), a growth hormone injected into dairy cows, fell under CVM jurisdiction. The CVM considered the additions of rBST (or the genetic modifications involved in producing rBST) to constitute the addition of an animal drug, making the product subject to its pre-market approval authority.

While neither transgenic fish nor animals have yet been approved for human consumption, regulations are in place to cover their use. As with plants the principal federal regulation of transgenic animals and fish is governed by the Food, Drug and Cosmetics Act.

As far back as 1986, the FDA asserted jurisdiction over transgenic animals and fish on the grounds that the transgene and any expressed proteins affect the "structure and function" of the receiving animal analogous to the modalities of alternative veterinary drug formulations. This gene-based modifications of animals for production or therapeutic claims fall under CVM regulation as new animal drugs. They work in consultation with the other FDA centers for food and feed safety evaluation. The FDA jurisdiction has been upheld by the federal courts. Perhaps more so than for plants, the principal focus for transgenic animal production is their use as bioreactors for the production of biotherapeutics in milk, serum, and other bodily fluids. So it makes logical sense that they would be regulated as Institutional New Drugs under CVM. However, faced with regulatory decision-making regarding the ever-evolving GM and cloning pipeline, the FDA asked the National Research Council (NRC) to identify and prioritize science-based concerns posed by animal biotechnology. The NRC released its report, Animal Biotechnology: Science-Based Concerns to the public on August 21, 2002, drawing the attention of scientists, the biotech industry, non-governmental organizations, and the national media.

Their concerns fell into three areas: medical, food, and environmental. On the medical side the principal concerns were modification of animals for biomedical purposes especially xenotransplantation mobilization of new infectious agents. On food issues they focused on new proteins and food safety concerns posed by biological activity, allergenicity, or toxicity, which they determined should be evaluated on a case-by-case basis. The key issue regarding cloned animals is whether and to what degree the genomic reprogramming results in altered gene expression that raises food safety concerns. They concluded that although it is difficult to quantify concerns without data comparing the composition of food products from cloned and non-cloned animals, there is no current evidence that food products derived from adult somatic cell clones or their progeny present a safety concern. Their greatest concern was surprisingly environmental, specifically the potential for GM organisms to escape and become established in the natural environment. They specifically considered that the existing regulatory framework might not prove adequate, particularly with regard to transgenic arthropods. Understandably there were concerns regarding the potential for this technology to cause pain, physical and physiological distress, behavioral abnormality, and health problems but noted that also the potential existed to alleviate or reduce those problems. Broadly, they expressed concern about the technical capacity of the agencies to address potential hazards

As noted under CVM, investigational applications are filed for genetic modifications under the authority of an investigational new animal drug exemption (INAD) or a similar provision. The INAD regulations are published in the Code of Federal Regulations, Title 21, Part 511.1(b). As part of the INAD submission, researchers must document their plans regarding the disposition of all investigational animals after their participation is completed. This is important in the case of food animal species where, with a showing of adequate safety data, the sponsor may request disposition of animals by slaughter for food or for processing into animal feed components. Although no transgenic animals have been approved for use as human food, a very limited number have been approved for rendering into animal feed components. In September 2008, FDA published for public comment proposed regulations that crystallize this notion of that every transgenic animal expressing a novel trait will be subject to the procedures and regulations for drug approval. In January, 2009, FDA issued a final guidance for industry on the regulation of genetically engineered (GE) animals. The FDCA defines "articles (other than food) intended to affect the structure or any function of the body of man or other animals" as drugs. An rDNA construct that is in a GE animal and is intended to affect the animal's structure or function meets the definition of an animal drug, whether the animal is intended for food, or used to produce another substance. Under this guidance developers of these animals must demonstrate that the construct and any new products expressed from the inserted construct are safe for the health of the GE animal and, if they are food animals, for consumption. Many consider that this "new drug" paradigm does not fit transgenic animals well. Miller[12] proposes that a better model is the approach taken by the Center for Food Safety and Nutrition that places the burden of ensuring the safety of foods and food ingredients on those who produce them.

In December 2003, California found itself in the national spotlight when it became the only state to ban the sale of a genetically engineered pet fish called the GloFish based on the potential for negative ecological impact.

Clones

On January 15, 2008,[13] after years of detailed study and analysis, the FDA issued an advisory that concluded that meat and milk from clones of cattle, swine, and goats, and the offspring of clones from any species traditionally consumed as food, are as safe to eat as food from conventionally bred animals. Interestingly, they considered that there was insufficient information for the agency to reach a conclusion on the safety of food from clones of other animal species such as sheep. The release was advisory rather than regulatory as withholding clones from the food and feed supply was voluntary since no regulations exist banning this practice.

The advisory consisted of three documents on animal cloning outlining the agency's regulatory approach: a risk assessment; a risk management plan; and guidance for industry. The documents were originally released in draft form in December 2006. Since that time, the risk assessment has been updated to include new scientific information. That new information reinforces the food safety conclusions of the drafts, namely that milk and meat products from cloned cattle, pigs, and goats are safe for consumers to eat since cloned cattle between 6 and 18 months of age are "virtually indistinguishable" from their conventional parents and can give birth to healthy offspring. The agency said that any risk was small given that the group that could pose problems, live neonatal clones, is unlikely to enter the food supply; it poses an extremely limited risk for consumption. The proposed plan outlines measures that FDA might take to address the risks that cloning poses to animals involved in the cloning process. These risks all have been observed in other assisted reproductive technologies currently in use in common agricultural practices.

The draft guidance for industry addresses the use of food and feed products derived from clones and their offspring. The guidance is directed at clone producers, livestock breeders, and farmers and ranchers purchasing clones. It provides the agency's current thinking on the use of clones and their offspring in human food or animal feed.

In the draft guidance, the FDA does not recommend any special measures relating to human food use of offspring of clones of any species. Because of their cost and rarity, clones will be used as are any other elite breeding stock—to pass on naturally occurring, desirable traits such as disease resistance and higher quality meat to production herds. Because clones will be used primarily for breeding, almost all of the food that comes from the cloning process is expected to be from sexually reproduced offspring and descendents of clones, and not the clones themselves.

INTERNATIONAL PERSPECTIVE

Biotechnology is a worldwide industry. In the case of genetically modified crops alone, some 35,000 field trials have been done on more than 100 crops in 45 countries, including many of the 27 countries of the EU. Despite the fact that almost 80% of these trials take place in the United States, the regulatory issues are of concern to all, and biotechnology issues are front and center in many international fora. For instance, several nations are working on the United Nations' Biosafety Protocol. The purpose of the protocol is to protect biological diversity from potential adverse effects resulting from the transboundary movement of living modified organisms, including those made through biotechnology. It will also address the needs of developing countries to develop the capacity to assess and manage potential risks. The Codex Alimentarius Commission has established the Codex Committee on Food Labeling to arrive at a common international position on labeling.

The Organization for Economic Cooperation and Development and the World Health Organization have embraced the concept of substantial equivalence as the cornerstone of safety assessment for genetically modified foods and crops. The most complex regulatory and political situation is in the EU, especially with regard to labeling for consumers.

The OECD's Working Group on Harmonization of Regulatory Oversight in Biotechnology decided at its first session, in June 1995, to focus its work on the development of consensus documents that are mutually recognized as among member countries. These consensus documents contain information for use during the regulatory assessment of a particular product. The consensus documents comprise technical information for use during the regulatory assessment of products of biotechnology and are intended to be mutually recognized among OECD member countries. These documents focus on the biology of organisms (such as plants, trees, or microorganisms) or introduced novel traits.

Variations due to breeding and the application of modern biotechnology have been studied frequently by scientific experts sponsored by organizations such as the United Nations (UN) Food and Agriculture Organization (FAO), the European Commission, the Royal Society, and the U.S. National Academy of Sciences. In each case, the conclusions were that modern biotechnology is no more likely than conventional breeding to produce unintended effects. Indeed, many expert reviews have concluded that the greater precision and more defined nature of the changes introduced into crops via modern biotechnology may actually be safer than changes produced by conventional plant breeding.

Recent reports have demonstrated that GM crops are often more closely related to the isogenic parental strain used in their development than to other members of the same genus and species. For example, metabolomic studies in *Solanum tuberosum* have shown that conventional plant breeding produces both intended and unintended effects and that insertion of transgenes can occur with little apparent effect on composition, even when the GM variety produces significant quantities of a new metabolite (e.g., inulin). Indeed, when the introduced gene product (DP2-3 fructans) was removed from the analysis parameters, multivariate statistical analysis showed no significant variation in the metabolic phenotype, including harmful glycoalkaloids, between the GM crop and the progenitor lines, whereas other conventionally bred cultivars showed clearly separated metabolic phenotypes.[14] Similar results have been observed at the proteome level for other plant species.

The consensus of scientific opinion and evidence is that biotechnology-derived foods and feeds present no new or unusual dangers to the environment or human health (Food and Agriculture Organization FAO/WHO, OECD, Seven Academies Report, French Medical Association, Royal Society of London, National Research Council, Society of Toxicology). The U.S. National Research Council (NRC) in "Genetically Modified Pest-Protected Plants: Science and Regulation" determined that no difference exists between crops modified through modern molecular techniques and those modified by conventional breeding practices. The NRC emphasized that it was not aware of any evidence suggesting foods on the market today are unsafe to eat as a result of genetic modification. In fact, the scientific panel concluded that growing such crops could have environmental advantages over other crops. An EU Commission Report[15] that summarized biosafety research of 400 scientific teams from all the 15 original EU countries conducted over 15 years stated that research on biotechnology-derived plants and derived products so far developed and marketed, following usual risk assessment procedures, has not shown any new risks to human health or the environment beyond the usual uncertainties of conventional plant breeding. Indeed, they concluded, the use of more precise technology and the greater regulatory scrutiny probably make them even safer than conventional plants and foods. A declaration signed by over 3,500 scientists including 25 Nobel Laureates reiterates this position.

The medical community has also supported the introduction of biotechnology-derived plants and foods. The American Medical Association states that "it is the policy of the AMA to endorse or implement programs that will convince the public and government officials that genetic manipulation is not inherently hazardous and that the health and economic benefits of recombinant DNA technology greatly exceed any risk posed to society." Although France has been among the most skeptical countries about this technology, the French Academy of Medicine report, Les plantes génétiquement modifiées "Genetically Modified Plants" called for an end to the European moratorium on genetically modified crops as there is no demonstrated concern from a human health or environmental impact perspective. These crops have been approved for import and sale in the dozens of countries and for growth as crops in more than 30. For example, Bt maize, herbicide-tolerant soybeans, and herbicide-tolerant canola among others have been approved for import into the EU and Japan.

The main principles of the international consensus approach are listed below. They serve to illustrate the variety of principles that have been at the center of the discussions and that are continuously being updated:

- Substantial equivalence: This is the guiding principle for the safety assessment. In short, substantial equivalence involves the process of comparing of the GM product to a conventional counterpart with a history of safe use. Such a comparison commonly includes agronomic performance, phenotype, expression of transgenes and composition (macro- and micronutrients) and identifies the similarities and differences between the GM product and the conventional counterpart. Based on the differences identified, further investigations may be carried out to assess the safety of these differences. These assessments include any protein(s) that are produced from the inserted DNA.
- Potential gene transfer: Where there is a possibility that selective advantage may be given to an undesirable trait from a food safety or environmental impact perspective, this should be assessed, for example in the highly unlikely event of a gene coding for a plant made pharmaceutical is transferred to commodity to corn. Where there is a possibility that the introduced gene(s) may be transferred to other crops, the potential environmental impact of the introduced gene and any conferred trait must be assessed.
- Potential allergenicity: Since most food allergens are proteins, the potential allergenicity of newly expressed proteins in food must be considered. A decision-tree approach introduced by ILSI/IFBC in 1996 has become internationally acknowledged and recently updated by Codex FAO/WHO, 2003. The starting point for this approach is the known allergenic properties of the source organism for the genes. Other recurrent items in this approach are structural similarities between the introduced protein and allergenic proteins, digestibility of the newly introduced protein(s), and eventually if needed, sera-binding tests with either the introduced protein or the biotechnology-derived product.
- Potential toxicity: Some proteins are known to be toxic, such as enterotoxins from pathogenic bacteria and lectins from plants. Commonly employed tests for toxicity include bioinformatic comparisons of amino acid sequences of any newly expressed protein(s) with the amino acid sequences of known toxins with those of introduced proteins, as well as rodent toxicity tests with acute administration of the proteins. In addition to purified proteins, whole grain from GM crops has been tested in animals, commonly in subchronic (90-day) rodent studies.
- Unintended effects: Besides the intended effects of the modification, interactions of the inserted DNA sequence with the plant genome are possible sources of unintended effects. Another source might be the introduced trait unexpectedly altering plant metabolism. Unintended effects can be both predicted and unpredicted. For example, variations in intermediates and endpoints in metabolic pathways that are the subject of modification while undesirable are predictable, while switch on of unknown endogenous genes through random insertion in control regions is both unintended and unpredictable, The process of product development that selects a single commercial product from hundreds to thousands of initial transformation events eliminates the vast majority of situations that might have resulted in unintended changes. The selected commercial product candidate event undergoes additional detailed phenotypic, agronomic, morphological and compositional analyses to further screen for such effects.
- Long-term effects: It is acknowledged that the pre-market safety assessment should be rigorous to exclude potentially adverse effects of consumption of foods or feeds derived from GM crops. Nevertheless, some have insisted that such foods should also be monitored for long-term effects by post-market surveillance. No international consensus exists as to whether such surveillance studies are technically possible without a testable hypothesis in order to provide meaningful information

regarding safety, and a GM crop with a testable safety concern would most likely not pass regulatory review. The notion of using measurable biomarkers has been suggested but these need to be determined for all foods and feeds whatever the source and the question of reasonable economic burden arises.

Besides the international organizations such as FAO/WHO, OECD, ILSI, and IFBC, other organizations have also formulated their views and recommendations on oversight of GM crops.

The data that are currently used for the safety assessment of GM crops has focused on the potential perceived risks associated with modern biotechnology. There are now worldwide data from more than 10 years of commercial use of GM crops and over two decades of research experience and no verified adverse consequences have been reported. These positive results have allayed the fears of many and have resulted in the greater acceptance of GM foods. Indeed, some scientists have begun to question the painstaking pre-market safety assessment of GM crops as practiced in some countries, and recommend that the extent and type of data that is part of a current safety assessment be updated to reflect this long-term safe experience with GM crops coupled with new information about plant genome plasticity.[16-20] They suggest that refinements to the process could include incorporation of factors such as "familiarity" (e.g., for commonly used proteins such as CP4 EPSPS, Cry1Ab, and PAT) and the source of the gene (e.g., when the gene is from the same crop species or is one with a history of safe use) into the overall safety assessment, influencing the extent to which event-specific data are needed.

REFERENCES

1. James, C. *Global Status of Commercialized Biotech/GM Crops: 2008*, ISAAA Briefs No. 39; ISAAA: Ithaca, NY, 2009.
2. Federal Register. *Coordinated Framework for Regulation of Biotechnology*; Office of Science and Technology Policy, June 26, 1986, 51 FR 23302, http://usbiotechreg.nbii.gov/CoordinatedFrameworkForRegulationOfBiotechnology1986.pdf (accessed December 5, 2008).
3. Unified Biotechnology Website. U.S. Regulatory Agencies Unified Biotechnology Website, http://usbiotechreg.nbii.gov/ (accessed December 5, 2008).
4. USDA Petition, http://www.aphis.usda.gov/brs/ (accessed December 5, 2008).
5. White, J. *Regulatory Considerations Overview*, Presentation at the USDA/CBER Plant-Derived Biologics Meeting (Transcript), Ames, IA, 2000, http://www.fda.gov/cber/minutes/plnt2040600.pdf (accessed December 5, 2008).
6. International Field Test Sources. *Information Systems for Biotechnology*, 2007, http://www.isb.vt.edu/cfdocs/globalfieldtests.cfm (accessed December 5, 2008).
7. Federal Register. *Importation, Interstate Movement, and Release into the Environment of Certain Genetically Engineered Organisms*, Oct 9, 2008, 7 CFR Part 340 [Docket No. APHIS-2008-0023], http://www.regulations.gov/fdmspublic/component/main?main=DocketDetail&d=APHIS-2008-0023 (accessed December 5, 2008).
8. Brooks, G.; Barfoot, P. *Co-Existence of GM and Non-GM Arable Crops: The Non-GM and Organic Context in the EU*; PG Economics Ltd.: Dorchester, UK, 2008.
9. The Food and Drug Administration, Center for Food Safety and Applied Nutrition (FDA/CFSAN). FDA to strengthen pre-market review of bioengineered foods. HHS News. (May 3, 2000). [pamphlet].
10. Event "32". *USDA, EPA and FDA Statement on Genetically Engineered Corn "Event 32,"* 2008, http://www.aphis.usda.gov/newsroom/content/2008/02/ge_corn_e32.shtml.
11. Brooks, G.; Barfoot, P. *Co-Existence of GM and Non-GM Arable Crops: The Non-GM and Organic Context in the EU*; PG Economics Ltd.: Dorchester, UK, 2004.
12. Miller, H.I. Will animal biotech bring home the bacon? In *Foundation for Biotechnology Awareness and Education*, Oct 2008, http://fbae.org/2009/FBAE/website/news_08_10_will-animal-biotech.html.
13. CVM and Animal Cloning. A risk based approach to animal cloning, 2008, http://www.fda.gov/cvm/cloning.htm.
14. Chassy, B.; Egnin, M.; Gao, Y.; Glenn, K.; Kleter, G.A.; Nestel, P.; Newell-McGloughlin, M.; Phipps, R.H.; Shillito, R. Nutritional and safety assessments of foods and feeds nutritionally improved through biotechnology: case studies. Compre. Rev. Food Sci. Food Saf. **2008**, *7* (1), 50–99.
15. EU Commission Report. *EC-sponsored Research into the Safety of Genetically Modified Organisms*. Fifth Framework Programme—External Advisory Groups "GMO research in perspective," Report of a workshop held by External Advisory Groups of the "Quality of Life and Management of Living Resources" Programme, 2001, http://europa.eu.int/comm/research/quality-of-life/gmo/index.html; http://europa.eu.int/comm/research/fp5/eag-gmo.html.
16. Bradford, K.J.; Gutterson, N.; Parrott, W.; Van Deynze, A.; Strauss, S.H. Reply to "Regulatory regimes for transgenic crops." Nat. Biotechnol. **2005**, *23*, 787–789.
17. Bradford, K.J.; Van Deynze, A.; Gutterson, N.; Parrott, W.; Strauss, S.H. Regulating transgenic crops sensibly: lessons from plant breeding, biotechnology and genomics. Nat. Biotechnol. **2005**, *23*, 439–444.
18. Kalaitzandonakes, N.J.; Alston, J.M.; Bradford, K. Compliance costs for regulatory approval of new biotech crops. Nat. Biotechnol. **2007**, *25* (5), 509–510.
19. McHughen, A. Fatal flaws in agbiotech regulatory policies. Nat. Biotechnol. **2007**, *25*, 725–727.
20. McHughen, A.; Smyth, S. US regulatory system for genetically modified [genetically modified organism (GMO), rDNA or transgenic] crop cultivars. Plant Biotechnol. J. **2008**, *6* (1), 2–12.

BIBLIOGRAPHY

1. Brent, P.; Bittisnich, D.; Brooke-Taylor, S.; Galway, N.; Graf, L.; Healy, M.; Kelly, L. Regulation of genetically modified foods in Australia and New Zealand. Food Control **2003**, *14*, 409–416.

2. Federal Register. *Guidance for Industry; Recommendations for the Early Food Safety Evaluation of New Non-Pesticidal Proteins Produced by New Plant Varieties Intended for Food Use*, 2006; Vol. 71, No. 119, 35688–35689, http://www.cfsan.fda.gov/~dms/bioprgu2.html.
3. Food and Drug Administration, Center for Food Safety and Applied Nutrition (FDA/CFSAN). *Foods Derived from New Plant Varieties Derived through Recombinant DNA Technology: Final Consultations under FDA's 1992 Policy*; Office of Premarket Approval: Washington, DC, 1999, http://vm.cfsan.fda.gov/~lrd/biocon.html.
4. Food and Drug Administration, Center for Food Safety and Applied Nutrition (FDA/CFSAN). *Guidance for Industry; Recommendations for the Early Food Safety Evaluation of New Non-Pesticidal Proteins Produced by New Plant Varieties Intended for Food Use*; Office of Food Additive Safety: College Park, MD, 2006, http://www.cfsan.fda.gov/~dms/bioprgu2.html.
5. Committee on Genetically Modified Pest-Protected Plants, National Research Council, *Genetically Modified Pest-Protected Plants: Science and Regulation* (Washington, D.C.: National Academy Press, 2000). http://www.nap.edu/openbook.php?isbn=0309069300.
6. ADSF. Les plantes génétiquement modifiées. Rapport sur la science et la technologie n°13. Académie Des Sciences Française, Paris, France, 2003. http://www.academie-sciences.fr/publications/rapports/rapports_html/RST13.htm.
7. Food and Drug Administration, Center for Veterinary Medicine (CVM), *Regulation of Genetically Engineered Animals Containing Heritable Recombinant DNA Constructs Guidance for Industry Final Guidance*, Center for Veterinary Medicine (HFV-100), Food and Drug Administration, 7500 Standish Place, Rockville, MD January 15, 2009 http://www.fda.gov/AnimalVeterinary/DevelopmentApprovalProcess/GeneticEngineering/GeneticallyEngineeredAnimals/default.htm (accessed January 2009).

Genetic Engineering: Evolution

John Davison
Yves Bertheau
National Institute for Agronomic Research (INRA), Versailles, France

Abstract
This entry traces the early evolution of genetic engineering from its beginnings, in unrelated fields of basic science, to the present day. The results of this fundamental scientific research have led to the development of a multibillion-dollar biotechnology industry, producing innovative human and animal health care products and improved varieties of cultivated plants. The public debates, controversies, and ethical issues related to genetic engineering are also briefly discussed.

INTRODUCTION AND BACKGROUND

By the 1970s, impressive progress had been made in the then new fields of molecular genetics and molecular biology. The bacterial and bacterial virus groups at Cold Spring Harbor Laboratory (New York, U.S.A.), the Pasteur Institute (France), and Caltech (U.S.A.) had recruited many new members across the United States and Europe. Thus, the genetic structures and functions of bacterial viruses were already well understood. Similarly, the genetic map of the model bacterium *Escherichia coli* had been extensively refined; there was a good understanding of the gene structure and regulation; and there was considerable knowledge of bacterial plasmids (small, circular, and self-replicative DNA molecules that are important in bacterial antibiotic resistance and horizontal gene transfer). The mechanisms of DNA replication, RNA synthesis, and protein synthesis as well as the related coding problem, of how the four-nucleotide-based DNA molecule uses a triplet code to specify the 20 amino acids present in proteins, had been elucidated. The 3-D structure of DNA had been determined and attempts at nucleotide sequencing had been initiated.

In contrast to these impressive advances in bacterial molecular genetics, the study of plants and animals was mostly limited to genetic approaches using a variety of model systems (e.g., yeast, fruit fly, and mouse). With the exception of the study of smaller plant and animal viruses, none of these researches was at a molecular level since the genomes were complex and, at that moment in time, intractable to molecular analysis. To progress further in the molecular biology of plants and animals, a major breakthrough was needed. This breakthrough was achieved in 1973 by S.N. Cohen and H.W. Boyer,[1] and became known as genetic engineering, or genetic manipulation.

THE BEGINNINGS OF GENETIC ENGINEERING IN MICROORGANISMS

In 1973, a number of independent, different lines of scientific research, necessary as the basis of genetic engineering, had been successful. Restriction enzymes were independently discovered by W. Arber, H.O. Smith, and D. Nathans, and became so important that in the following years, new biotechnology companies were founded simply to purify and supply them. Restriction enzymes (endonucleases) are a kind of molecular scissors, and are important since they enable genetic engineers to cleave DNA molecules at specific and predictable sites (defined by the nucleotide sequence at that site). Specific DNA cleavage is essential for genetic engineering, but having cleaved the DNA, it is necessary to be able to reseal (ligate) it in new combinations in vitro. This was achieved using newly characterized enzymes called DNA ligases, which were discovered independently in the laboratories of Lehrman, Gellert, Hurwitz, and Richardson. Finally, another step was necessary before genetic engineering could be achieved: it was necessary to be able to reintroduce the DNA molecules into *E. coli*, which is normally refractory to DNA uptake. Mandel and Higa found that *E. coli* cells treated with high concentrations of calcium chloride became permeable to large DNA molecules. Using these three new discoveries, Cohen and Boyer were able to perform a (nowadays) simple experiment. They cleaved a small plasmid (pSC101) carrying a gene for resistance to the antibiotic tetracycline, using the restriction enzyme *EcoRI*. They found that pSC101 was cleaved at a single site in a region essential neither for plasmid duplication (replication) in *E. coli* nor for the expression of the tetracycline resistance gene. These two facts were serendipitous, since they could not be known in advance, but were nonetheless essential for the success of the experiment. The cleavage of pSC101 thus produced a linear DNA molecule that did not lack essential genetic information. Cohen and Boyer[1] then mixed this DNA molecule with another DNA molecule derived from a plasmid carrying a gene for resistance to the antibiotic kanamycin and added the enzyme DNA ligase, which joins DNA fragments together at random. The ligase reaction produced a mixture of joined DNA molecules, which were then introduced into calcium chloride-treated *E. coli*. The bacteria were then spread on bacteriological petri dishes in the presence of both tetracycline and kanamycin. Most of the bacteria were killed by the antibiotics. However, rare bacteria were able to grow, being resistant to both antibiotics. These bacteria were shown to contain a single plasmid that contained DNA segments containing the two antibiotic resistance genes from the two parental plasmids. In this way, the first man-made DNA molecule had been created, and genetic engineering technology was born. The early discoveries leading to genetic engineering have been documented in detail.[2]

On the surface, the experiment of Cohen and Boyer[1] represents nothing more than the transfer of an antibiotic resistance gene from one plasmid to another—something that had been observed in bacterial genetics and clinical epidemiology many years ago. However, the transfer had taken place, for the first time, in vitro and the observations had very profound and long-reaching implications. It immediately became obvious that, if a bacterial gene could be inserted into a plasmid, then, in principle, any DNA molecule (bacterial, viral, animal, plant, or human) could also be inserted into the plasmid. Soon afterward, the ribosomal gene from a toad was inserted into a plasmid of *E. coli*. Then, a human sequence coding for the hormone insulin (used for the treatment of type I diabetes) was expressed in *E. coli* to produce the first human therapeutic protein made in bacteria. Today, more than 40% of human therapeutic proteins (antigens for vaccines and industrial enzymes) are produced in microorganisms, such as *E. coli* or the baker's yeast *Saccharomyces cerevisiae*.

The beginning of genetic engineering was not without difficulties. This was an unknown territory and it was necessary to proceed with caution, with due care for potential risks. P. Berg and M. Singer,[3] eminent scientists of the time, quickly called for a meeting of interested scientists, which took place at Asilomar, California, U.S.A., in 1975. After much discussion, it was decided to hold a moratorium on the construction of genetically engineered organisms, pending a period of reflection as to the possible risks and appropriate guidelines for the physical and biological containment conditions for these experiments. The Asilomar guidelines eventually evolved into the National Institutes of Health guidelines. Berg later admitted that "we overestimated the risks, but had no basis for deciding, and it was sensible to choose the prudent approach."[2]

THE BEGINNINGS OF GENETIC ENGINEERING IN PLANTS

The possibility of performing genetic engineering in plants came to light during a completely unrelated study by M.D. Chilton, in 1977, who was studying the plant pathogenicity of the bacterium *Agrobacterium tumefaciens*. She came to the remarkable conclusion, accepted only with difficulty at that time, that during infection of the plant, a small specific part (the T-DNA) of the very large Ti plasmid was transferred to the plant, which became incorporated into the plant genome. We now know that the T-DNA contains two kinds of genes: those concerned with tumorigenicity and those concerned with the synthesis of peculiar amino acids called opines. During bacterial infection of a plant wound, the T-DNA is transferred to the plant cell where it induces plant tumors, which in turn act as factories for the synthesis of opines that are exported from the plant cells to the extracellular bacteria. The bacterium thus creates an ecological niche for its own use by turning the plant into a factory creating a source of carbon and nitrogen. It was later shown that the tumorigenic genes and the opine synthesis genes are not necessary for the DNA transfer process, and several groups realized that it might be possible to replace these genes by foreign genes and thus perform genetic engineering in plants. This idea took several years to evolve, largely because of the time needed to define the basic scientific parameters of the system and to construct vector plasmids able to transport the foreign gene. In 1983, three groups, directed by J. Schell and M. van Montagu (University of Gent), M.D. Chilton (University of Washington), and the Monsanto Company (St. Louis, Missouri, U.S.A.), showed that the Ti plasmid could be used to transfer foreign DNA, carried between the T-DNA borders, into plants (Fig. 1).[1–3] It was rapidly shown that plants carrying bacterial genes encoding enzymes for the degradation of herbicides, such as glufosinate (Basta®, Aventis Crop Science Pty. Ltd., Glen Iris, Victoria, Australia) or glyphosate (Roundup®, Monsanto Company, St. Louis, Missouri, U.S.A.), became tolerant to the herbicides. The use of such transgenic plants impeded weeding and low-till agriculture, and thus soil erosion. Similarly, plants expressing an insect toxin (from the bacterium *Bacillus thuringiensis*) became insect resistant by killing the insect pests that eat them. The use of herbicide-tolerant plants and insect-resistant plants reduces herbicide and pesticide use, and thereby environmental pollution and farm-worker exposure. Although these examples still represent the most commercialized transgenic crops (such as maize, rapeseed, soy, and cotton), new generations with other advantages should soon be available. These include, e.g., rice containing enhanced levels of vitamin A, designed to combat the vitamin A deficiency that causes blindness in about 500,000 undernourished children in developing countries. Other examples include iron-fortified rice and potatoes with improved protein content. Future genetically modified (GM) crops will also contain genes

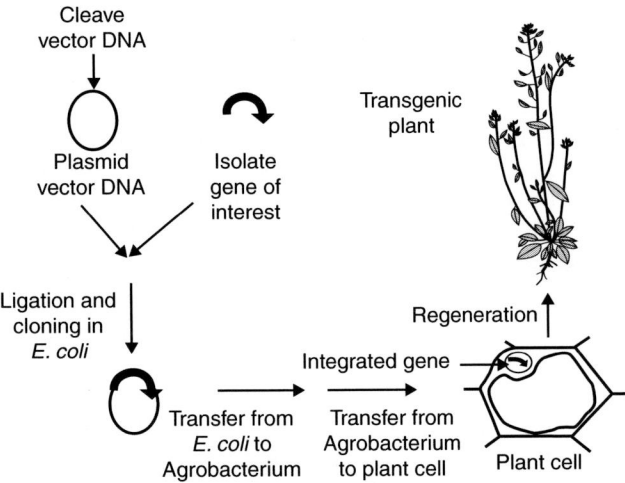

Fig. 1 Example of construction of a transgenic plant. Foreign DNA carrying the gene of interest (bold arrow) is inserted between the T-DNA borders (not shown) of the circular plasmid vector DNA molecule that has been cleaved with an appropriate restriction enzyme. The recombinant construction is transferred into *E. coli*, and then further transferred to *Agrobacterium tumefaciens* containing the Ti plasmid that is capable of transferring it to plant cells. The T-DNA, with its accompanying transgene, integrates into the plant DNA and is stably inherited. Cells containing the recombinant DNA may be selected according to the genetic marker (often antibiotic resistance) it carries. The entire plant is then regenerated by the use of medium containing suitable plant growth hormones. Expression of the gene of interest in the plant is achieved using plant regulatory signals.

for improved tolerance to drought, salinity, heat, and cold. In the future, other non-food crops may be used to produce pharmaceuticals, may be used as industrial raw materials, or may have applications in phytoremediation of polluted soils.

THE BEGINNINGS OF GENETIC ENGINEERING IN ANIMALS

Immediately following the discovery of genetic engineering in bacteria, attempts were made to introduce DNA into model animals, which was a novel way to understand the molecular basis of animal development, differentiation, and disease. Rapid success was obtained with nematode worms, fruit flies, toads, and fish, and later with mice. The initial technique consisted of injecting the recombinant DNA directly into the fertilized oocyte at an early stage of development.[7] The recombinant DNA integrates into the host DNA at random, often with multiple copies at the same site. If the integration takes place into the DNA of the fertilized oocyte, then the resulting animal will be transgenic. If it takes place in a postzygotic stage, then the animal will be a mosaic possessing two different cell types: one transgenic and the other normal.

The mouse as an animal model is of particular interest, because of the availability of a well-developed database resulting from many years of sophisticated genetic and physiological studies. The mouse is also phylogenetically closer to a human being, and is thus a suitable model for studying several human diseases. In order to obtain suitable mouse mutants for the study of genetic diseases, it is necessary to be able to specifically inactivate, or modify, known mouse genes. This is impossible using injected oocytes, where the transgenic DNA inserts at random into the mouse genomic DNA. It was necessary that the transgenic DNA be specifically directed to the host site with the same (homologous) nucleotide sequence (as can be easily done in yeast and bacteria, but not in plants). The breakthrough came from a different line of research: the study of stem cells that can be grown in vitro and that have the capacity to differentiate into any cell type, including the germ line. It was shown that the transgenic DNA can be introduced into embryonic stem cells, cultured in vitro, and that the resulting cells (some of which are transgenic) can then be screened for the presence of the transgene inserted into the homologous integration site. Once the correct transgenic cell line has been found, this can be grown and its characteristics verified. This cell line can then be injected into a fertilized oocyte.[8] In this situation, the transgenic embryonic stem cells will grow and differentiate normally and the resulting animal will be a chimera, in which some cells are transgenic and some are normal. Since the transgenic stem cells enter into the germ cell line, mating this chimeric animal with normal mice will produce some animals (called heterozygotes) carrying a single copy of the transgene (Fig. 2). In these animals, one copy of the gene will be knocked out, but the animals will normally survive because of the presence of the other normal gene. Mating two heterozygotic mice together will give some normal mice, some heterozygotic mice, and some homozygotic mice (containing two copies of the transgene). If the knocked-out gene is not essential for survival, then the homozygotic mouse will be viable and capable of reproduction and may prove to be a useful model to study the role of that particular gene in human genetic diseases. A large number of knock-out mice have been generated, providing models for studying a variety of human diseases, including cancer, cystic fibrosis, β-thalassemia, hypercholesterolemia, atherosclerosis, and Creutzfeldt–Jakob disease.

Genetically modified animals (e.g., cows, goats, and rabbits) are used for the production of human pharmaceutical proteins (antigens for vaccines, immunomodulators, and hormones), which are present in their milk. Genetically modified pigs that have been "humanized" (i.e., modified to prevent rejection by the human immune system) have also been used as potential organ donors. The possibility of the transfer of pig viruses to humans raises ethical and health concerns.

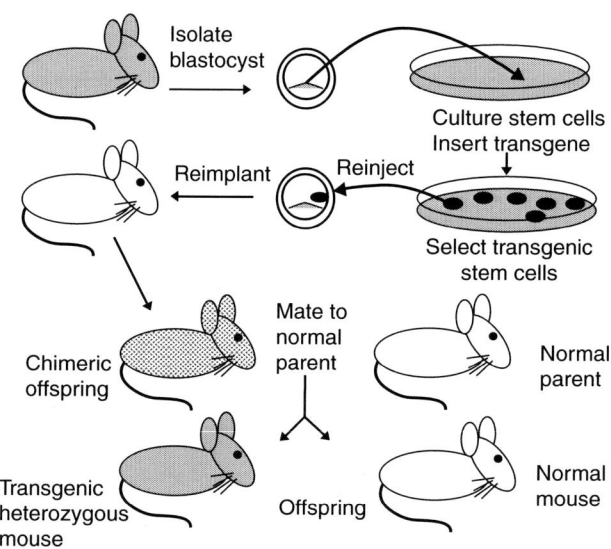

Fig. 2 Example of construction of a transgenic mouse. Stem cells are removed from a mouse blastocyst (developing embryo) and are cultured in vitro. The transgenic DNA is introduced into the stem cells and those containing the recombinant DNA may be selected according to the genetic marker it carries. The insertion of the transgene into the correct (homologous) site may also be verified. The appropriate stem cell line is then reinjected into a new blastocyst and develops as if it is an integral part of the new embryo, participating in all cell types, including germ cells. The initial animal is a chimera (shown stippled), containing normal as well as transgenic cells. However, pure heterozygotic transgenic (containing one transgene per diploid genome) animals may be obtained by mating with a normal mouse. If viable, homozygotic (containing two transgenes per diploid genome) animals may be obtained by further crosses (not shown).

HUMAN GENE THERAPY

Another important, but still highly experimental, line of research studies the possibility that some human genetic diseases can be cured by the insertion of a functional transgene corresponding to the mutated gene in the patient. Naturally, because of ethical concerns, such experiments do not affect the germ line, and thus the transgene cannot be passed on to new generations. The first such trial took place in 1990, which attempted to alleviate the symptoms of a human genetic disease caused by the deficiency of the enzyme adenosine deaminase. Such patients suffer from acute immunodeficiency and can normally be cured only by bone marrow transplantation from a compatible donor. The experiment attempted to infect the patient's own T lymphocytes, grown in vitro, with a recombinant retrovirus carrying a correct copy of the adenosine deaminase gene. This virus should integrate into the host genome of the T cells and express the enzyme, thus alleviating the symptoms. Similar attempts have been made with cystic fibrosis and severe combined immunodeficiency (SCID)-X1 disease.[9] Some clinical success in the alleviation of the symptoms was observed in SCID-X1, but health problems (such as leukemia) arose because of the viral vectors that carry the transgene. For the moment, these techniques are experimental and improved protocols are needed.

CONCLUSIONS

The beginnings of genetic engineering in bacteria, plants, and animals provide a fascinating story and it is interesting to reflect on what general conclusion may be drawn. One point that clearly arises from this short history of genetic engineering is the way in which scientific discoveries are made. In almost all cases, the discoveries on which genetic engineering is based were made by scientists working in unrelated fields of basic science, and these discoveries had no prospect of immediate application. None of the initial scientists had the objective of creating a transgenic bacterium, plant, or animal. However, the sum of these basic discoveries has created a worldwide industry with enormous human health benefits and commercial impacts.

Ethics, public opinion, and the possible need for regulation must always be taken into account for any new invention. Public opinion had been adverse to the introduction of many previous new technologies (e.g., printing press, electricity, trains, automobiles, chlorination of water, and vaccination) and genetic engineering was no exception. This human characteristic that resists change has been referred to as "Luddite," named after the leader of protestors who destroyed the new textile looms in Great Britain in the early nineteenth century. The Asilomar conference was effective in providing guidelines for researchers, and public opinion was calmed. This was aided by the obvious medical benefits of microbial genetic engineering.

Genetic engineering in animals also caused initial public outcry, particularly surrounding the patented Harvard mouse or Oncomouse® (a mouse prone to develop cancer and designed to study the mechanisms of cancer and to test the effect of environmental carcinogens; E. I. du Pont de Nemours and Company, Wilmington, Delaware, U.S.A.). However, given the obvious medical benefits and stringent regulations governed by such agencies as the U.S. Food and Drug Agency (FDA), the Environmental Protection Agency (EPA), and the U.S. Department of Agriculture (USDA), public outcry subsided. [A curious exception is a fluorescent transgenic fish, GloFish® (Yorktown Technologies, Austin, Texas, U.S.A.), which escapes regulation by the FDA, USDA, and EPA, since it does not fall under the category of food, agriculture, or environment.]

In contrast to transgenic animals, the main purpose of GM crops was to grow them on a large scale in the agricultural environment and to use them to feed animals and humans. This caused a huge public outcry in Europe, but to a much smaller extent in the United States, and was augmented by unscientific scare reports from nongovernmental organizations. Genetically modified plants arrived at a moment when European consumer confidence in governments, regulators, and even scientists was at a low point because of a series of food health scares. Many European consumers found GM plants to be unnatural and dangerous and objected to being forced to consume them. In addition, poor public relations policies of the major producers of GM plants tended to dissimulate information from the public; however, an active campaign to educate public might have produced better results. Avoidable accidents involving mixing of GM and non-GM plants made the situation worse. The arguments for and against GM crops have been extensively analyzed.[10] Today, Europe produces almost no GM food and feed, and has instituted complex regulations on the labeling, detection, and coexistence of GM and non-GM crops. European supermarkets presently refuse to sell GM products. In contrast, the anti-GM crop movements have held less sway in the United States, Canada, Argentina, Brazil, and China, which have become the world's major producers and consumers of GM crops without demonstrable ill effects on the consumers or on the environment.

It is difficult and hazardous to try to predict the future. Eventually, the unfavorable public opinion of GM crops is likely to subside (as was the case with genetic engineering in bacteria) when the benefits are realized and the potential risks are better understood and appropriately regulated. Many scientists, including N. Borlaugh and M.S. Swaminathan (the fathers of the "green revolution" that has been feeding developing countries since the 1960s), feel that genetically engineered crops will be the only way to sustainably feed the ever-increasing world population in the face of dwindling agricultural land surfaces and climate change.

REFERENCES

1. Cohen, S.N.; Chang, A.C.; Boyer, H.W.; Helling, R.B. Construction of biologically functional bacterial plasmids in vitro. Proc. Natl. Acad. Sci. U.S.A. **1973**, *70* (11), 3240–3244.
2. Echols, H. *Operators and Promoters: The Story of Molecular Biology and Its Creators*; University of California Press: Berkeley, CA, **2001**, 318–351.
3. Berg, P.; Baltimore, D.; Brenner, S.; Roblin, R.O.; Singer, M.F. Asilomar conference on recombinant DNA molecules. Science. **1975**, *188* (4192), 991–994.
4. Shaw, C.H.; Leemans, J.; van Montagu, M.; Schell, J. A general method for the transfer of cloned genes to plant cells. Gene **1983**, *23* (3), 315–330.
5. Barton, K.A.; Chilton, M.D. Agrobacterium Ti plasmids as vectors for plant genetic engineering. Methods Enzymol. **1983**, *101*, 527–539.
6. Fraley, R.T.; Rogers, S.G.; Horsch, R.B.; Sanders, P.R.; Flick, J.S.; Adams, S.P.; Bittner, M.L.; Brand, L.A.; Fink, C.L.; Fry, J.S.; Galluppi, G.R.; Goldberg, S.B.; Hoffmann, N.L.; Woo, S.C. Expression of bacterial genes in plant cells. Proc. Natl. Acad. Sci. U.S.A. **1983**, *80* (15), 4803–4807.
7. Palmiter, R.D.; Brinston, R.L. Transgenic mice. Cell. **1985**, *41* (2), 343–345.
8. Bradley, A.; Zheng, B.; Liu, P. Thirteen years of manipulating the mouse genome: a personal history. Int. J. Dev. Biol. **1998**, *42* (7), 943–950.
9. Buckley, R.H. Molecular defects in severe combined immunodeficiency and approaches to immune reconstitution. Annu. Rev. Immunol. **2004**, *22*, 625–655.
10. Miller, H.I.; Conko, G. *The Frankenfood Myth: How Protest and Politics Threaten the Biotech Revolution*; Praeger Publishers: New York, **2004**.

Genetically Modified Foods: Consumer Attitudes

Christine M. Bruhn
*Center for Consumer Research, Department of Food Science and Technology,
University of California–Davis, Davis, California, U.S.A.*

Abstract
Despite food-related applications of biotechnology being in the marketplace for over a decade, consumers in the United States and Europe remain relatively uninformed about this technology. United States consumers do not realize the type or number of applications in the supermarket. Concern about biotechnology is volunteered by few consumers and the majority indicates that they will purchase biotechnology-modified products with appealing benefits. European consumers express concern and are less optimistic about the likelihood of experiencing benefits. Both U.S. and European consumers are more comfortable with applications to plants as compared to animals; however, animal applications are viewed positively if the reason for the application is considered beneficial. Beneficial applications should be well-received with the appropriate communications program.

INTRODUCTION

Genetic modification can be applied to a broad range of areas, offering advantages in food production and processing, environment stewardship, and human health. While the range of promising applications is broad, regulatory differences in the United States and Europe have given provisions to unlabeled use in one location and restricted use in another. This entry will summarize consumer attitudes based upon surveys, not marketplace behavior.

The use of words influences consumer response. Studies repeatedly demonstrate a change in response depending on the word used to describe the technology with consumers more accepting of "biotechnology" over "genetic engineering." Consumer response is also influenced by the implied benefit or threat conveyed through the wording of a question and by responses that are a forced choice among specific options or open-ended.

UNITED STATES CONSUMERS AWARENESS AND ATTITUDES

Despite media coverage for over a decade, in 2008, 35% of the consumers said that they had heard little and 30% said that they had heard nothing about biotechnology or genetic engineering.[1] Only 9% of the consumers in 2008 stated that they had heard much about this area. This low level of awareness has been consistent across the last decade.[2] Similarly, 66% of the consumers do not know if foods produced using biotechnology are available in the marketplace.[1] Of the 23% who recognized that there are genetically modified foods in the supermarket, 40% volunteered that these modified foods are vegetables, 23% mentioned fruits, and 22% specified corn products.[1] Only 3% mentioned soy as a modified product, yet corn and soy are the primary modified crops grown in the United States.

Consumer surveys provide general information on consumer attitudes toward product innovations. In 2008, 30% of the U.S. consumers believed that biotechnology would benefit their family within next 5 years with 53% of consumers uncertain.[1] In an open-ended response, consumers specified that 34% of them expected nutritional or health benefits, 27% expected improved quality or taste, and 22% expected economic benefits. In past years, over 60% of the consumers believed that they would receive benefits within 5 years. The decrease in percentage of those expecting benefits may reflect lack of consumer awareness of the economic impact of current applications or the fact that most applications focus on agronomic traits.

Consumers are more interested in change when it is tied to a specific application, as compared to use of technology in general. When applications are identified, United States consumers have a positive attitude toward many applications. The majority of the consumers said that they would purchase genetically modified products to improve taste or freshness with 27% being very likely and 51% somewhat likely to buy products with this benefit.[1] Consumers also supported agronomic applications with a total of 78% of consumers suggesting that they would purchase products modified to reduce pesticide use. Health-related applications were also endorsed with 75% reporting that they were very or somewhat likely to buy foods modified to reduce saturated fat, 76% would buy to avoid trans fat, and 78%

would buy if the fat were more healthful with omega-3 fatty acids. Although few have heard about using biotechnology to produce medicines from food crops, 41% favored this use, 25% were neither favorable nor unfavorable, and 25% stated that they did not know enough about the application to express an opinion.

Few U.S. consumers perceive that modifications by biotechnology as risky. When asked in an open-ended question about food safety concerns, only 1% volunteered concerns about genetically modified food.[1] By contrast, concerns about foodborne disease and safe handling were mentioned by 50% and 29%, respectively.

Labeling of genetically modified foods is not a high priority for most of the consumers. When asked if there was any information currently not on a food label that they would like to see added, only 1% asked for information on genetic modification.[1] When consumers were asked to select one item from a list of potential label additions, 17% chose labeling if the product was genetically altered, 33% selected if pesticides were used in production, 8% favored if the product was imported, 16% responded that they needed no additional information and 15% said that they did not know.[3] By contrast, when the consumers were asked to agree or disagree with the statement that all foods modified by biotechnology should be labeled, 80% agreed.[4] It is unlikely that consumers are aware of the U.S. Food and Drug (FDA) policy on labeling genetically modified foods. When told that FDA requires labeling if modification changes nutritional value or introduces an allergen, 60% supported this policy with an additional 27% neither supporting nor opposing. Only 8% were somewhat opposed and 5% were strongly opposed to the FDA labeling policy.[1]

In addition to the impact on food safety, consumers have also been asked to rate the importance of potential environmental risks. A majority of consumers, 64%, considered the potential risk of contaminating a plant species by genetic transfer to be very important.[5] Other potential risks and the percentage of consumers considering the risk to be very important included the potential to create super weeds, 57%; to develop pesticide resistant insect, 57%; to reduce genetic diversity, 49%; and the potential that modified plants could harm others, 48%. Consumers also responded to the importance of potential environmental benefits. Almost three-quarters of the consumers, 74%, rated the cleaning-up of toxic pollutants as very important.[5] Other potential benefits and the percentage of consumers considering the benefit very important included reducing soil erosion, 73%; using less fertilizer, 72%; developing drought-resistant plants, 68%; developing disease-resistant trees, 67%; and using less pesticides, 61%.

Few consumer studies have focused on animal applications of biotechnology. In a survey in 2008, only 22% of the consumers were favorable toward using biotechnology with animals to produce food crops; however, when phrased another way, that animal biotechnology allows movement of beneficial traits from one animal to another in a precise way, 34% viewed the application as favorable. Hearing why an application was being considered also had a positive effect on consumer attitudes.[1] The fact that an application could reduce the environmental impact of animal waste had a positive effect on 52% of the consumers and a negative effect on 16%. An application that increased farm efficiency and decreased the amount of feed needed by animals had a positive effect on 55% and a negative effect on 19%. An application that improved quality and safety of food through improved animal health or improved nutritional quality had a positive effect on 62% of consumers and a negative effect on 15%. These results suggested that animal applications are viewed negatively by 15% of consumers with the remaining either neutral or positive.

The term cloning appears to be viewed more negatively than animal biotechnology or genetic engineering with only 23% indicating that they have a very favorable or somewhat favorable impression of cloning.[1] Even though they were told that the FDA considers products from cloned animals to be safe, willingness to purchase meat, milk, or eggs from cloned animals was expressed by only 48% of consumers. Similarly, only 48% of consumers would buy products from the offspring of cloned animals, compared to 65% who would purchase products from animals enhanced by genetic engineering.

European Consumers

A survey of European consumer attitudes toward diverse social issues was published in the *Eurobarometer*. Findings in the 2002 study on biotechnology showed that few (11%) Europeans consider themselves informed about biotechnology.[6] Questions related to the facts of biology suggested that basic knowledge was lacking among the general population. In 1999, only 35% of the consumers correctly responded that the following statement is false, "Ordinary tomatoes do not contain genes while genetically modified tomatoes do." Furthermore, only 42% recognized that eating genetically modified fruits does not change your personal genes, where 24% believed human genes would be changed and the remaining were uncertain. Responses varied by country with consumers from the Netherlands having the largest percentage with correct responses. A comparison of responses in 1996 and 1999 showed the little increase in public knowledge. In fact, fewer consumers responded correctly in 1999 to the statement that eating genetically modified fruits changes human genes when compared to 1996.

Only about half of European consumers were aware of many of the applications of biotechnology.[6] Slightly over half, 56%, were aware that genetic modification could be used to make plants resistant to insect attack. About half were aware that these tools could be used to detect inherited diseases, prepare human medicine or vaccine from animals, or clone human cells to replace those that are not functioning. Slightly fewer (44%) were aware that tools of biotechnology could be used to introduce human genes into bacteria to make medicine such as insulin for diabetics. Only 28% knew that biotechnology could be used to clean toxic spills.

When asked to rate if a biotechnology application is useful, risky, or should be encouraged, medical applications were considered most valuable, including using tools of biotechnology to detect hereditary disease, using human genes in bacteria to produce medicine, and replacing non-functional human tissue.[6] Environmental remediation, described as developing genetically modified bacteria to clean toxic spills, was also highly regarded. Usefulness rating for modifying food to obtain high protein content, longer shelf life, or changed taste was lower than other applications with a usefulness rating at the midpoint.

Europeans considered all applications of genetically engineering somewhat risky.[6] The applications considered most risky were food production and the cloning of animals. Using tools of biotechnology to detect hereditary disease received the lowest risky rating, but was still at the midpoint of risk.

When asked to rate the moral acceptability of various biotechnology applications, developing genetically modified bacteria for environmental remediation, using biotechnology to detect hereditary disease, and cloning animals to produce medicine or vaccine were rated highest in acceptability.[6] Making plants resistant to insects and modifying food to increase protein, shelf life, or taste, were below the midpoint of moral acceptability.

Those applications European consumers felt should be encouraged, corresponded to perceptions of usefulness and moral acceptability.[6] Gaskell[7] observed that as the perceived usefulness of an application declines, perceived risk increases and moral acceptability and support declines. Perception of the value or worth of potential benefits is key to acceptance of the application of biotechnology and the products produced.

Attitude Change

A review of attitudes toward genetic modification over the last 10 years suggests that U.S. consumers have become less optimistic about the value of potential applications while Europeans are more optimistic.[8] Attitudes in Europe have not been linear. Between 1991 and 1999 consumers were quite negative; however, willingness to buy modified products under certain conditions increased between 2002 and 2005.

The less optimistic perspective in the United States could have been influenced by negative media coverage and the perception that potential risks were not under control. Media coverage of biotechnology increased from less than 1% of articles in 1997 to 6% in 1999 and 12% in 2001.[9] A content analysis of media coverage of biotechnology articles in 1999 found claims of harm were expressed in 70% of the articles, while discussions of benefits were only in 30% of the stories.[10]

People can change their attitudes in response to information from a credible source. A university-sponsored program delivered in California and Indiana brought information about biotechnology to community groups.[11] Participants who believed that biotechnology offers society benefits increased from 68% before the program to 96% afterwards. Similarly those believing biotechnology presents society with risks increased from 46 to 65%. In an assessment of the overall impact of biotechnology on human health, those believing the impact would be positive increased from 71% initially to 90% after the program.

CONCLUSION

The key to consumer acceptance of genetic modification is perceived benefit. Consumers expect human, animal, and environmental safety to be protected. Consumer attitudes are more likely to be positive when people understand why animals or plants are modified, why the potential benefit is important, and that neither the animal nor the environment is harmed.

Since consumer awareness of potential applications and factors related to safety assessment is low, a comprehensive communication program is appropriate. Messages about potential benefits should be presented multiple times using a variety of media. Messages about biotechnology which highlight potential benefits, address consumer concern, and are delivered by trusted, knowledge sources are critical to long-term acceptance and realization of the benefits this technology can facilitate.

REFERENCES

1. Cogent Research. Food biotechnology: a study of U.S. consumer attitudinal trends, 2008 report. 2008. http://www.ific.org/research/biotechres.cfm (accessed January 10, 2001).
2. Cogent Research. Support for food biotechnology holds with increased recognition of benefits. 2002. http://ific.org/proactive/newsroom/release.vtml/id=19981.
3. Bruskin. National opinion polls on labeling of genetically modified foods. 2001. http://www.cspinet.org/new/poll_gefoods.html (accessed January 6, 2001).
4. Pew Initiative on Biotechnology. Overview of findings: 2004 Focus groups and polls. 2004. http://pewagbiotech.org/research/2004update/overview.pdf (accessed January 5, 2009).
5. Pew Initiative. Environmental savior or saboteur? Debating the impacts of genetic engineering. 2002. http://www.pewtrusts.com (accessed January 9, 2009).

6. INRA. *Eurobarometer 52.1—The Europeans and Biotechnology*. 2002. http://europa.eu.int/comm/public_opinion/archives/eb/ebs_134_en.pdf.
7. Gaskell, G. *Agricultural Biotechnology and Public Attitudes in the European Union*. 2000. http://www.agbioforum.org/ (accessed May 2002).
8. Bonny, S. How have opinions about GMOs changed over time? The situation in the European Union and the USA. CAB Reviews: Persp. Agric. Vet. Sci. Nutr. Nat. Res. **2008**, *3* (093), 1–17.
9. Center for Media and Public Affairs. Food for thought IV. Reporting on diet nutrition and food safety news. 2002. http://www.ific.org/.
10. International Food Information Council. Food for thought III. Reporting of diet, nutrition, and food safety. In *Executive Summary 1999 vs 1997 vs 1995*. 2000; p. 7.
11. Bruhn, C.M.; Mason, A. Community leader response to educational information about biotechnology. J. Food Sci. **2002**, *67*, 399–403.

Genetically Modified Organisms (GMOs): Authorized, Unauthorized and Unknown

John Davison
Yves Bertheau
National Institute for Agronomic Research (INRA), Versailles, France

Abstract
Although the introduction of GMO plant and plant material to the U.S. market has been met with little opposition; in Europe, it has led to considerable polemic, destruction of field trials, boycott by supermarkets, and even violence. In response to this, the European community has developed the most severe GMO regulations in the world. The need for traceability and detection of GMO has led to some technical difficulties that have been discussed by the European Network of GMO Laboratories, which gives advice to the European Commission. In particular, unauthorized and unknown GMOs, which may contaminate shipments destined for Europe, pose considerable difficulties for current detection methods. The increasing proportion of GMO being produced, with the consequent increasing admixing, may pose problems of animal feed supply, upon which Europe is dependent. This may, in turn, lead to food price increases in Europe.

INTRODUCTION

In 1983, three reports from the University of Ghent, the University of Washington, and the Monsanto Company showed that the Ti plasmid of *Agrobacterium tumefaciens* could be used to transfer foreign DNA into plant genome, thus producing the first genetically modified (GM) plants. This discovery had enormous implications for plant genetics and agriculture. In the last 20 years, plant biotechnology has grown into a multibillion-dollar international industry.[1] In the process, the Monsanto Company transformed itself from a chemical company into the world's leading biotechnology company. The earliest and still most important commercialized transgenic plants were corn (maize), cotton, soybean, and canola, and contained transgenes conferring tolerance to herbicides, or resistance to insects and pests. These result in enhanced benefits for the farmers and have a positive environmental effect, due to a decrease in the application of herbicides and pesticides, a reduction in exposure of farm workers, and permitting low-tillage agriculture that reduces soil erosion. The next generation of GM crops will offer enhanced food quality to the consumer, such as increased content of vitamin and trace elements and reduced content of allergens.[2] Other plant traits in the pipeline include virus resistance, drought tolerance, and salt tolerance. Still other GM crops may be better designed for biofuel production, for less-polluting paper manufacture, or for the production of pharmaceutical products (such as human proteins, antigens, and antibodies).[3] It is evident that some of these products must not mix with the food and feed chains, and that strict rules for coexistence, identity preservation, traceability, and detection are necessary.

The cultivation of GM crops is presently limited to a few countries. The United States grows 55% of GM crops, followed by Argentina (19%), Brazil (10%), Canada (7%), and China (4%). Europe cultivates almost no GM crops, except for a relatively small amount of maize in Spain. In the next few years, this situation is likely to change dramatically. China is expected to dramatically increase its transgenic crop cultivation. Similar increases in GMO cultivation are foreseen in India, South Africa, Australia, and even Europe.[4]

This entry will present European GMO regulations, with special focus on GMO detection and traceability and the difficulties of their application to unauthorized and unknown GMOs.

GMO REGULATIONS IN THE EUROPEAN COMMUNITY

Attitudes to GM food and feed vary in different societies and cultures. Relatively little attention has been paid to GM food by the American public, and unlabeled GM foods are available in the local supermarkets. In contrast, the European attitude has been hostile, perhaps due to a number of unrelated food scandals that have left the general population mistrustful of the governments and food safety authorities. The often-violent attitude of the anti-GMO movements, toward what they call "Frankenfoods," has increased public distrust in plant biotechnology. In consequence, Europe has adopted the most severe GMO legislation in the world, with specific directives and regulations governing the cultivation, import, and marketing of GM products. A major difference between the European regulatory attitude and that of the United States lies in the concepts of "substantial equivalence" and of "methods of production." The U.S. regulations do not require that an approved GM product is labeled for marketing, provided that it is "substantially equivalent" to the non-GM product. This is true for the vast majority of GM food and feed, but, for example, golden rice, which contains elevated levels of vitamin A, would need to be labeled. However, European regulations (e.g., Directive 2001/18 EEC and Regulations EC 1829/2003 and EC 1830/2003[5,7]) are concerned with the "method of production" and reflect the belief that the consumers must be informed so that they can express their freedom of choice as to whether, or not, they wish to consume GM food. Thus under European regulations, highly processed foods, in which the DNA can no longer be detected, still need to be labeled if they are derived from GM material. These regulations also deal with the possibility that non-GM food or feed may become accidentally admixed with GM food and feed and define the labeling threshold for such contamination (discussed later in the entry). Regulation 1829/2003 also obliges the plant biotechnology companies to provide detection methods, and pay for the validation thereof, for products that they wish to commercialize in Europe. The detection methods are then validated by the EC Joint Research Centre, Community Reference Laboratory (JRC/CRL), with the support of European Network of GMO Laboratories (ENGL), and subsequently published on its Web site. Food and feed containing authorized GMOs, or having an accidental (or adventitious) presence of authorized GMOs, of greater that 0.9% must be labeled. The burden of showing that the GMO presence is unavoidable, and that all suitable precautions were taken to avoid admixing, remains with the companies. Only authorized GMOs are concerned by the 0.9% labeling threshold. Unauthorized GMOs are not permitted, at any level, in the European food and feed chain. Finally, there is currently no legally defined threshold for seeds.

The meaning of % GM is not defined in Regulation 1829/2003 and has been a subject of great discussion. The experimental measurements lead to DNA ratios; the farmers, shippers, and regulatory authorities work in mass/mass ratios in contrast, and the conversion between the two systems cannot easily be made.[7,11]

FOOD SAFETY OF GMO FOOD AND FEED

It should be stressed, since it is a frequent source of misunderstanding, that the EU GM traceability detection and labeling regulations are not concerned with food safety, but simply with consumer choice for the food they wish to eat. Under Regulation 178/2002 ("General Food Law"), the safety of food and feed, including GMOs and environmental impact, is determined by an independent scientific body; the European Food Safety Authority (EFSA). A favorable EFSA recommendation is necessary before the European Commission and member states can take a decision on the applicant's request. GMO detection, labeling, and traceability are not part of EFSA's remit, although traceability may serve to clarify unforeseen incidents (such as accidental or deliberate food contamination) in the food chain.

AUTHORIZED GMOS

In the EU, authorized GMOs are those that have received a positive EFSA assessment and a positive decision from the Commission. Authorized GMOs necessarily have a complete EFSA dossier, including a description of the transgenes incorporated. In addition they require a JRC/CRL validated method of detection, which usually includes DNA sequence information on the site of integration of the transgene in the genome. Reference material (later in the entry) must also be provided by the applicant. Such GMOs can then be detected by the national enforcement laboratories using such validated protocols. These invariably use the quantitative real-time PCR (Q-PCR) reaction. In Q-PCR, specific regions of DNA are amplified using short DNA primer oligonucleotides to initiate the DNA polymerase reaction. PCR amplification thus specifically amplifies the region between these two primer binding sites. In GMO identification methods, the primers are generally designed to bind, on the one hand, to part of the transgene cassette, and on the other, to the plant genomic region close to the site of insertion. Since the site of transgene insertion within the genome is unique to that transgenic event, the amplification is specific and can differentiate different transgenic plants even though these may carry the same transgene cassette. However for EC regulations, specificity is obligatory but insufficient, since quantification of the copy number is also necessary. To obtain quantification of GMO content, the GMO target region is always amplified simultaneously with a standard reference gene (e.g., for maize, the alcohol dehydrogenase 1, *adh1*, gene is often used) able to identify and quantify the species DNA being tested. The GMO percentage is obtained from the ratio of the number of target GMO DNA sequences to the target taxon specific sequences calculated in terms of haploid genomes.[11] These results may be converted to mass/mass units by comparison with the simultaneous amplification of "Certified Reference Material (CRM)," which is series of Institute for Reference Materials and Measurements (IRMM) standardized samples, each containing a certified percentage (currently mass/mass) of GMO and non-GMO material. A decision tree for GMO detection and labeling is given in Fig. 1.

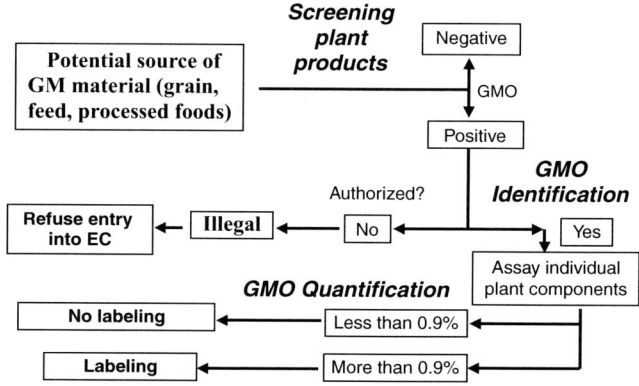

Fig. 1 A decision tree for detection, acceptance, labeling, or rejection of GMO food and feed in the European Community.

Q-PCR is, for the moment, the only method that combines a sufficient high accuracy and low limit of detection. The disadvantage is that such methods are slow, costly, require trained personnel, and can only be performed in specialized laboratories using expensive computerized machines. Their application for a continuously increasing number of GMOs in budget-restricted enforcement laboratories may thus be difficult.

Sampling (e.g., how to obtain representative samples from a large cargo that may contain one or several GMOs in heterogeneous distribution) is a major, and still unresolved, problem that may have costly implications in legal disputes.[7,12]

MIXTURES OF GM PLANT MATERIAL

GMO quantification becomes difficult when the sample lot contains mixtures of GMOs, and particularly so when the mixtures are not of the same species; for example, when a cargo of non-GM maize is contaminated by GM-soybean. In this case, the EU regulations specify that threshold for labeling of 0.9% is measured per ingredient (analytically translated into taxon), independent of its proportion in the total sample. In extreme situations, the presence of trace botanical impurities of, for example, traces of 100% GM soybean in a cargo of entirely non-GM maize would result in it being labeled as containing 100% GMO.

Another complication arises in the case of stacked genes, which refers to the presence of two or more different transgene cassettes in the same cultivar (e.g., herbicide tolerance and insect resistance). Such stacked GMOs plants are obtained by standard genetic crosses, as routinely performed in plant breeding. Stacked GM crops are currently cultivated in several countries, and thus may become accidentally mixed with cargoes destined for the EU. However, using current PCR detection methods it is very difficult to differentiate between samples containing stacked genes and those that are a simple mixture non-stacked transgenes.

UNAUTHORIZED GMOS

GMOs that are unauthorized in Europe may nonetheless be approved in other countries (asynchronous approval). Alternatively, they may be completely unknown (next paragraph). Regulation 1929/2003 obliges the notifying company to provide the method of detection at the time that it makes an application for the GMO commercialization in EU. However, there is a major weakness in the European regulations: in contrast to authorized GMOs, unauthorized GMOs do not usually have a CRL-validated detection method. Indeed, many unauthorized GMOs are simply experimental varieties that may never be commercialized. Furthermore, the method applied in the GMO construction belongs to the intellectual property portfolio of the company. Thus, while European regulations[5,7] specify that there is zero tolerance for non-authorized GMOs, the authorities have no means to detect the unauthorized GMOs, other than by presupposing that these may contain certain common elements (Pnos, Tnos, herbicide-tolerance genes, etc.). In contrast to the EU, the U.S. Department of Agriculture's Animal and Plant Health Inspection Service (USDA/APHIS), in close collaboration with the Food and Drug Administration (FDA) and the U.S. Environmental Protection Agency (EPA), shows more tolerance for trace contaminants, provided that they pose no significant risk to health or the environment, contain well-known transgenes, and are similar to other GM plants already deregulated.

It is a fact that non-authorized GMOs do accidentally enter the food and feed chains. Thus, in 2000, Aventis "Starlink" corn, authorized in the United States as animal feed but not for human consumption, was found to have contaminated food chains. In 2002, Prodigene corn, genetically modified to produce a pig vaccine, contaminated a field of soybean, which was then harvested and sent to the grain elevator. The entire harvest of 500,000 bushels of soybean was destroyed at a total cost, including the fine, exceeding 1 million dollars. In 2005, Syngenta produced and distributed a limited quantity of Bt10 corn (instead of the approved variant Bt11). All Bt10 plants were destroyed by APHIS and the seed stock quarantined. Nonetheless, some Bt10 corn found its way to Europe and Japan. In 2006, imported noodles bought in Asian stores, were found to test positively for Chinese insect-resistant transgenic rice; Xianyou Bt63. In 2006, Bayer Crop Science found that trace amounts of an experimental variety of GM rice (LLrice601) had contaminated non-GM long-grain rice. After a long study, the USDA was unable to determine the source of this contamination. LLrice601 contains the glufosinate herbicide-tolerance gene already present in other varieties of GM rice, and was quickly deregulated in the United States by USDA/APHIS. A preliminary EFSA report, supported by similar reports from Canada, UK, Germany, and Holland, suggested the probable food safety of LLrice601. These accidents impacted the revenues of the companies, since entire shipments from the United States contaminated by

Starlink, Bt10, or LLrice601 were prevented from entering Europe. Detection methods had to be rapidly developed, for both Bt10 and LLrice601, by the JRC/CRL, in collaboration with the responsible firms. Similar actions were taken in other parts of the world including Japan and Korea.

The EC DG Agri has recently issued a report on the economic effects of trace amounts of unapproved GMOs on European feed and livestock production. Under the worst-case scenario, the European refusal of shipments carrying low levels of GMOs could result in feed shortages in Europe, with severe consequent price rises for meat and poultry.[13]

UNKNOWN GMOS

Another class of GMOs is referred to as unknown GMOs since, in principle, nothing is publicly known about them, including their molecular characteristics and safety aspects. Such GMOs are not tolerated at any level in most countries, including the United States. In addition, unknown GMOs cannot be routinely detected using present technologies, due to the absence of molecular information concerning their construction. Some new methods are being developed under the EC-funded Co-Extra program,[14] which attempt to make an educated guess as to which transgenes are likely to be present, to detect combinations of these transgenes in the sample, and to compare these combinations to databanks of known GMOs. Most of these methods assume that unknown GMOs would contain at least some already-known transgenes and regulatory elements, since it would be commercially too costly to develop GMO inserts with new genes and controlling elements. Nonetheless, GMOs containing only new transgenic elements would remain undetectable by any presently available detection procedure in routine analyses.

CONCLUSION

Several factors are likely to enhance the problem of GMO detection, and thus complicate the traceability and detection of GM plants in the coming years. First, worldwide planting of transgenic crops has continually increased over the last 10 years, for example, GM soybeans now represent more that 80% of world soybean production, and further increases are likely. The preponderance of GM crops increases the probability that admixing of GM and non-GM crops will occur, even in the presence of identity preservation systems. Second, although the number of genetic traits in use is presently limited, the next few years will bring a considerable increase in new transgenes, corresponding to new genetic traits, requiring new detection methods. Third, the number of kinds of crops will also increase.

For example, GM variants of a number of crops (including rice, wheat, mangos, poplar, and potatoes) are nearing commercialization, the major obstacles often being political and market acceptance (see for instance the Canadian refusal of Monsanto's herbicide-tolerant wheat), rather than food safety. In addition, it is clear that there is urgent need for international agreement on asynchronous approval of GMOs, whereby a GMO, authorized as safe in one country, is unauthorized and thus not tolerated in another. Such situations inevitably lead to incidents of admixing. Some countries, such as Argentina, which exports extensively to Europe, appear to base their GMO approval procedures not only on biosafety, but also on the possibility of approval in the European marketplace. Nonetheless, the illegal use of non-approved GMOs by individual farmers could undermine these policies. All of these factors will pose additional strains on the European regulatory system and on the regulatory enforcement laboratories and possibly produce food and feed shortages and price increases.

ACKNOWLEDGMENTS

Part of this work was funded by the EC FP6 research project Co-Extra (contract 007158) and by the EC FP6 research project PETER (contract number 031717).

REFERENCES

1. USDA Report. *Economic Information Bulletin*, The first decade of genetically engineered crops in the United States; U.S. Department of Agriculture, 2006, http://www.ers.usda.gov/publications/eib11/eib11.pdf (accessed August 2008).
2. Unnevehr, L.; Pray, C.; Paarlberg, R. Addressing micronutrient deficiencies: alternative interventions and technologies. AgBioForum **2007**, *10* (3), 124–134, http://www.agbioforum.org (accessed August 2008).
3. Fischer, R.; Stoger, E.; Schillberg, S.; Christou, P.; Twyman, R.M. Plant-based production of biopharmaceuticals. Curr. Opin. Plant Biol. **2004**, *7*, 152–158.
4. Runge, C.F.; Ryan, B. *Global Diffusion of Biotechnology: Diffusion and Research in 2004*, A report prepared for the Council on Biotechnology Information; Washington, DC, 2004, http://www.apec.umn.edu/faculty/frunge/globalbiotech04.pdf (accessed August 2008).
5. Regulation (EC) No 1829/2003 of the European Parliament and of the Council of 22 September 2003 on genetically modified food and feed, 2003, http://europa.eu/scadplus/leg/en/lvb/l21154.htm (accessed August 2008).
6. Regulation (EC) No 1830/2003 of the European Parliament and of the Council of 22 September 2003 concerning the traceability and labelling of genetically modified organisms and the traceability of food and feed products produced from genetically modified organisms and amending Directive 2001/18/EC, 2003, http://europa.eu/scadplus/leg/en/lvb/l21170.htm (accessed August 2008).

7. Davison, J.; Bertheau, Y. European regulations on genetically modified organisms: their interpretation, implementation and difficulties in compliance. Abbreviation unknown. CAB reviews: perspectives in agriculture, veterinary science. Nutrition and Natural Resources **2007**, *2* (7), 1–12.
8. Holst-Jensen, A. Sampling, detection, identification and quantification of genetically modified organisms (GMOs). In *Food Toxicants Analysis. Techniques, Strategies and Developments*; Pico, Y., Ed.; Elsevier: Amsterdam, The Netherlands, 2007; 231–268.
9. Holst-Jensen, A.; De Loose, M.; Van Den Eede, G. Coherence between legal requirements and approaches for detection of genetically modified organisms (GMOs) and their derived products. J. Agric. Food Chem. **2006**, *54* (8), 2799–2809.
10. Weighardt, F. European GMO labeling thresholds impractical and unscientific. Nat. Biotechnol. **2006**, *24* (Pt 1), 23–25.
11. ENGL Explanatory document on the use of "Percentage of GM-DNA copy numbers in relation to taxon specific DNA copy numbers calculated in terms of haploid genomes" as a general unit to express the percentage of GMOs, http://engl.jrc.it/docs/HGE%20release%20version%201.pdf (accessed August 2008).
12. Paoletti, C.; Donatelli, M.; Kay, S.; Van Den Eede, G. Simulating kernel lot sampling: the effect of heterogeneity on the detection of GMO contaminations. Seed Sci. Technol. **2003**, *31* (Pt 3), 629–638.
13. EC-DGAgri Report. Economic impact of unapproved GMOs on EU feed imports and livestock production, 2004, http://ec.europa.eu/agriculture/envir/gmo/economic_impactGMOs_en.pdf (accessed August 2008).
14. EC Co-Extra integrated project Web site. GM and non-GM supply chains: their CO-EXistence and TRAceability, 2004–2009, http://www.coextra.eu/ (accessed August 2008).

Genomic Resources: Genetic Conservation

David A. Kudrna
Rod A. Wing
Department of Plant Sciences, University of Arizona, Tucson, Arizona, U.S.A.

Abstract
Conservation of plant genetic resources for food and agriculture is an important undertaking for future generations. The advent of biotechnology in crop science has produced an explosion of genetic data and molecular clone resources from all major crop and model plants as well as many minor and orphan crop species. Molecular resources are produced from genomic DNA and maintained as large insert clones [bacterial artificial chromosomes (BACs)] or from transcribed mRNA in the form of complementary DNA (cDNA) clones, representing expressed genes. The millions of clones available as resources are used as substrates to sequence entire genomes, to clone agriculturally important genes, and to investigate global gene expression patterns with the goals of preventing crop failures, pathogenic outbreaks, and famine. Importantly, these genomic gene-bank resources also provide powerful tools to work in concert with established germ plasm conservation methods for the necessary evaluation and preservation of biological diversity.

CONSERVATION AND BIOTECHNOLOGY

In 1996, the Food and Agriculture Organization (FAO) of the United Nations established a global plan of action for the conservation and sustainable utilization of plant genetic resources for food and agriculture. Beginning in 1998, and at frequent intervals thereafter, FAO publishes updates[1] on various aspects of genetic conservation of foods and related agricultural materials. This surge of awareness for genetic conservation of foods has moved to aquaculture, to domesticated animals, and to all organisms used for fiber, shelter, medicine, and raw materials. Multinational centers involved with research and in situ and ex situ gene banking of germ plasm were established for the purpose of preserving biological diversity for future generations.[2]

With the development of molecular biological tools,[3] the usefulness of biotechnology in agriculture appears a perfect match.[4–6] As scientific teams unravel the genetic basis of many physiological systems, likewise, other teams discover the genetic reasons for specific crop failures, poor yields, and pathogenic outbreaks. Biotechnology is the mix of genetics from multiple systems applied to the utility of a living organism. What began as small gene cloning and sequencing experiments in the 1980s has now exploded with thousands of molecular projects aimed at discovering cures for disease, famine, scientific understanding, and product development. Such projects involve living organisms at all phylogenetic branches of life. These far-reaching efforts, to be feasible, have created millions of clones that, when put together, represent the genomes of those organisms from which they came. These genome clone libraries are an ex situ genetic conservation of the organism at the molecular level.

The framework of genetics rests on the four nucleotide bases of DNA and how they are ordered, expressed, regulated, and coordinated. These DNA sequences, either in the form of cloned genes or as whole organisms, are the sources of genes needed for both basic research and plant breeding. For many plant species, it is common to introduce (transform) genes, add regulatory sequences, and measure protein or gene products. Transformation offers a very exact use of genetic conservation of genomic resources and may avoid genetic disruption of a host genome and allow the use of genes from foreign organisms.[7]

GENOMIC RESOURCES

Genomic resource clones produced in laboratories can be classified into two types: genomic (whole genome) and expressed. Genomic, whole-genome, clones are currently the backbone of genome studies because they contain an organism's complete genetic makeup, every chromosome from telomere to telomere. These clones are produced from a particular organism's whole-genome DNA that is extracted from tissue or purified nuclei, precisely cut and selected for uniform size, and ligated into a predetermined plasmid vector capable of successful replication of the cloned DNA followed by introduction into the bacterial host, *Escherichia coli*, by transformation. These cloned DNA fragments are quite large (100–300 kb), exist as single copy clones in bacterial host cells, and are referred to as bacterial artificial chromosomes or BACs. Each BAC clone contains a single fragment of the source DNA such that a specified number of these clones represent the source organism's genome in what is called a BAC library. The statistical representation of the source organism's genome may be determined by analyzing the size of the inserted DNA fragments, the number of clones in the library, and the size of the source organism's genome.

Expressed clones use a source organism's mRNA as the starting point for the purpose of obtaining complementary DNA (cDNA) only from expressed genes. The transcribed DNA is cloned into a high-copy replicating vector and transformed into *E. coli*. DNA pieces of expressed genes are called expressed sequence tags (ESTs), while full-length complementary DNA (FLcDNA) clones have the complete nucleotide sequence of a gene. Tremendous efforts have been undertaken to identify hundreds of thousands of ESTs from major plant species from mRNA extracted from several tissues of plants grown under constricting growth conditions. Complete genome representation FLcDNA libraries for most major crop species are under development. The DNA sequences of BAC and EST clones are deposited into public databases. Three principal databases that contain all publicly available sequence information from around the world are National Center for Biotechnology Information (NCBI, http://www.ncbi.nlm.nih.gov), European Bioinformatics Institute (EMBL-EBI, http://www.ebi.ac.uk/embl), and DNA Database of Japan (DDBJ, http://www.ddbj.nig.ac.jp). As huge amounts of additional data are generated (genetic and physical maps, DNA and protein sequences, discoveries, tools and methods, collaborators, etc.), centers for compiling, and more importantly displaying in an informative manner, data for major plant projects have been developed. A concise list of major plant databases for the most popular species is listed in Table 1, while a more comprehensive list identifying germ plasm, clone resources, sequence information, collaborators, bioinformatics, genetic conservation, crop improvement, links, etc. is available.[8]

GENE BANKING OF GENOMIC RESOURCES

Bacterial artificial chromosome and expressed sequence tag clone resource centers (Table 1) have been created to handle the vast amounts of molecular resources available today. For example, the Arizona Genomics Institute (AGI, http://www.genome.arizona.edu) and Clemson University Genomics Institute (CUGI, http://www.genome.clemson.edu) collaborate to maintain and construct the largest collection of plant agriculture BAC and EST libraries in the world. Researchers worldwide have also donated large numbers of libraries created by their labs for the purpose of archiving and distributing these resources to the scientific community. Libraries are named based on genus, species, specific plant variety or accession, and library type, and individual clone addresses are based upon a standard 384-well microtiter dish. Once these libraries are produced or obtained, robotic manipulations and computerized bar code databases are used to assemble, archive, handle, and distribute the large numbers of clones. Efficient handling of libraries for fast, accurate, and collaborative resource distribution is facilitated with laser-reading-equipped robots (Fig. 1). Long-term storage of the libraries in storage freezing media (buffered growth broth and glycerol) is used for banking clones in bar-coded, databased ultracold freezers at $-80°C$. A standard 20 ft^3 ultracold freezer houses more than 1 million clones. Although this storage technology has been known for decades, the exact biological life of stored *E. coli* cultures is unknown; thus bacterial cultures are replicated and refreshed every 5 years. Library copies are maintained in triplicate at either AGI or CUGI with backup copies at the opposite location. Resources are distributed via ordering from Web sites allowing for quick and accurate distribution of the clones. Because *E. coli* is the organism containing the source DNA, worldwide movement of resources presents no concerns for the transfer of agricultural pathogens or deleterious germ plasm and is easily performed by overnight shipping with dry ice. In the past year, AGI and CUGI distributed over 175,000 individual clones, 80 whole libraries, and 1750 screening filters; produced 46 BAC libraries containing over 3 million clones, 10 cDNA libraries (over 250,000 clones), and 20 subclone libraries; accepted donated libraries for archiving and distribution (15 BAC and 49 EST) comprising nearly 3 million clones. Distribution for all resources is on a cost-recovery basis.

Additional valuable resources and tools are accessible with utilization of molecular genomic resources. Macroarray, high-density, hybridization screening filters can be produced from all BAC and cDNA libraries to identify clones containing desirable nucleotide sequences. Each filter contains up to 23,000 duplicate spotted clones and is reusable multiple times. Subclone libraries can be produced by a variety of methods and are used for sorting specific sequences within larger clones. High-throughput sequencing can be performed on BACs and cDNAs for in silico (via computer) screening of deposited orthologs or chromosome assembly. Deposited sequence data may be used for cross-species comparisons and a multitude of organism research projects. DNA fingerprinting and physical map assembly by Finger Print Contig (FPC)[9] software can be used to develop physical maps of a genome. Molecular genetic (linkage) maps may be assembled using a variety of methods (such as restriction fragment length polymorphism, RFLP; amplified fragment length, AFLP; simple sequence repeat, SSR) involving both genomic or cDNA clones. Microarray screening of cDNAs can be used to identify a specific gene(s) (clone DNA) expressed under

Table 1 Major plant databases

Name	Web site	Source species	Data description
Arabidopsis Information Resource, The (TAIR)	http://www.arabidopsis.org/	*Arabidopsis*	Comprehensive resource for *Arabidopsis*, germ plasm, clones, databases, maps, collaborators, public resource
Arizona Genomics Institute (AGI) Computational Laboratory (AGCoL)	http://www.genome.arizona.edu	Agriculture and world plant species	Clones, genomic tools, biocomputing and bioinformatics, sequencing, physical mapping, links, collaborators
Consultative Group on International Agricultural Research (CGIAR)	http://www.cgiar.org/	World agriculture	Portal for world agriculture centers, leadership for world foods, policies, collaborators, links
Laboratory for plant genomics and genefinder genomic resources	http://hbz7.tamu.edu/index.htm	Agriculture and world plant species	Clones, genomic tools, bioinformatics, maps, links, collaborators
Munich Information Center for Protein Sequences (MIPS)	http://www.mips.biochem.mpg.de/	Proteins of plants	Protein information, bioinformatics, databases, genomes, proteomes, collaborators
Plant genome database	http://www.plantgdb.org/	Many plant species	Databases, EST annotation, bioinformatics, links, collaborators
The Institute for Genomic Research (TIGR)	http://www.tigr.org	Many plant species	Genomics (functional, applied, informatics), sequencing, collaborators
CAMBIA intellectual property resource	http://www.cambiaip.org/Home/welcome.htm	World agriculture	Intellectual property issues relevant to biotechnology in international agriculture
Graingenes	http://wheat.pw.usda.gov/index.shtml	Wheat, barley, rye, triticale, oat	Molecular maps, phenotypic info, collaborators, germ plasm information, links
Gramene	http://www.gramene.org	Grass, grains	Comparative genome analysis of grasses, bioinformatics, databases, maps, rice resources
Plant array database	http://www.univ-montp2.fr/%7Eplant.arrays/index.html	Many plant species	Microarray (expressed genes) data monitoring of plants, collaborators
Arabidopsis information on the world wide web	http://weeds.mgh.harvard.edu/atlinks.html	*Arabidopsis*	Maps, bioinformatics, stocks, clones, links, collaborators
UK cropnet	http://www.ukcrop.net	World crops	Bioinformatics, databases, collaborators
Maize genetics and genomics database	http://http://www.maizegdb.org	Maize	Maize genome gateway, molecular maps, clones, collaborators

Fig. 1 Laser-equipped robots, such as the "Q-bot" (Genetix LTD, Hampshire, UK) shown here, provide fast and accurate handling of the huge numbers of biological samples.

strict experimental conditions such as pathogen infection, salt stress, temperature or light alteration, growth development, etc.

Publicly generated genomic resources must be maintained and made available to the scientific community to avoid unnecessary re-creation of resources. As example, to reap the enormous U.S. federal investment in the generation of these (and future) genomic resources, the National Science Foundation (NSF) is requiring investigators to include a 5-year plan (including funds) for maintenance and distribution of resources. The majority of investigators deposit their resources in facilities designed to handle such large projects. Caution should be noted that several "for profit" companies have attempted to provide services for clone archiving and distribution; however, the majority have stopped providing such services because of profitability or business plan issues.

CONCLUSION

Conservation of genomic resources offers considerable value to the overall goal of preserving biological diversity. However, efforts of this magnitude do require careful consideration because costs for operation and maintenance of a resource facility are not trivial. Expenses for freezers, robots, incubators, buildings, personnel, and electricity are expensive, but when strategically incorporated into a functioning molecular research institution, the utility of the resources for the various applications quickly becomes necessary to the operation. Other considerations are clone storage life, biological consequences as a result of human errors, mechanical dependence, and intellectual property issues. Future developments in ambient temperature storage and full robotic handling (see http://www.genvault.com) are promising for decreasing expenses, improving efficiency, and extending culture life.

These current and expanding genomic gene-bank resources and sequencing databases represent a significant economic and scientific investment. With the usefulness of these resources impacting today's research, agriculture, and ecology communities and advancing knowledge and practical agriculture more dramatically in the future, long-term availability issues must be considered. Sufficient attention by the global scientific community to management, preservation, and availability must occur for these technology tools to be beneficial for our future.

REFERENCES

1. http://www.fao.org (accessed April 2003).
2. Koo, B. The price of conserving agricultural biodiversity. Nat. Biotechnol. **2003**, *21*, 126–128.
3. Sambrook, J.; Fritsch, E.F.; Maniatis, T. *Molecular Cloning: A Laboratory Manual*, 3rd Ed.; Cold Spring Harbor Laboratory Press: Cold Spring Harbor, NY, **2001**.
4. Phillips, R.L. Biotechnology and agriculture in today's world. Food Nutr. Bull. **2000**, *21* (4), 457–459.
5. Briggs, S.P. Plant genomics: More than food for thought. Proc. Natl. Acad. Sci. U.S.A. **1998**, *95*, 1986–1988.
6. Phillips, R.L.; Freeling, M. Plant genomics and our food supply. Proc. Natl. Acad. Sci. U.S.A. **1998**, *95*, 1969–1970.
7. Committee on Managing Global Genetic Resources: Agricultural Imperatives, Board on Agriculture, National Research Council *Agricultural Crop Issues and Policies—Managing Global Genetic Resources*; National Academy Press: Washington, DC, **1993**; 189–204.
8. http://www.genome.arizona.edu/agi/berc/pub/major_plant_databases.html (accessed April 2003).
9. Soderlund, C.; Longden, I.; Mott, R. FPC: A system for building contigs from restriction fingerprinted clones. CABIOS **1997**, *13*, 523–535.

Genomics Research: Livestock Production

Monika Sodhi
Lawrence B. Schook
Department of Animal Sciences, University of Illinois, Urbana, Illinois, U.S.A.

Abstract

The major emphasis of animal genomic research has been to increase the productivity and disease resistance ability of economically important farm animal breeds to meet the ever increasing demands of animal products. Traditionally, the approach of livestock improvement was based on the classical breeding where in the parents of the next generation were selected based on their phenotype without knowing the genetic variation underlying the traits of interest. To this end, genomic tools have helped to move away from this "Black box" approach and understand the complex genetic components of phenotypic variation. Through structural and comparative genomics it had become possible to identify the genomic regions/quantitative trait loci (QTL) harboring the performance altering genes. However, the QTL search followed by fine mapping could prove useful to a limited extent as detection of casual mutations underlying QTL's was time consuming and cost extensive. These limitations restricted the extensive implementation of marker assisted selection technology in the commercial breeds. Recent advances in genomic research including whole genome sequencing, expression array, and highly multiplexed and high throughput single nucleotide polymorphism (SNP) genotyping platforms have a large impact on the ability for genome mining and pinpoint genes that influence biological and economically important traits. In future, selection decision based on the genomic breeding values complemented with other quantitative genetic evaluation approaches can very well lead to improve genetic gain in livestock species. In this chapter, an effort has been made to describe the issues, scope and needs those confront the animal genomics community in today's industry driven era.

INTRODUCTION

Livestock science has an increased challenge to meet the food demand for an ever-expanding world population. Livestock production started ~8000–10,000 years ago with the domestication of various livestock species. Along with domestication, livestock improvement was attempted following a classical breeding approach where parents of the next generation were selected on the basis of their phenotypes. This was somewhat a black-box approach as nothing in specific (DNA sequence, gene, or underlying traits?) was known and thus, it had limitations to improve the livestock production. Simultaneously, with the rise in income and urbanization, the demand/consumption of milk and meat has been increasing tremendously (Table 1, Fig. 1). During the 1996/1998–2020 period, the projected growth rates in developing countries for milk and meat are projected to be 3.0 and 2.9% annually, respectively. In addition, by 2020–2030, world demand for meat and dairy products is expected to increase by 40–50%.[2] To increase the production of meat and milk, it is not recommended to add on the number of livestock because of the associated pollution risks. Hence, introduction of new technologies to meet the ever-increasing demand for the livestock products is of great urgency.

The biological mechanisms linking genetic variation with phenotype have been the major challenge for researchers. Understanding and capturing the genetic component of phenotypic variation could play an important role in productivity enhancement in the livestock species. The molecular genetics revolution in the 1980s and 1990s led to the emergence of a new scientific discipline, genomics, resulting from the convergence of genetics, molecular biology, and bioinformatics. The objective of genomics is systematic structural and functional analysis of complex genomes to understand what individual genes do and how they interact to control biological processes. The genomic tools have helped to know how to predict "sequence to consequence" and harness this genetic variation to facilitate stable changes in production, fertility, and health through selection. Advances in genomics have contributed significantly toward the quality and efficiency of livestock production such as commercial strains of chickens growing at implausible rates or pigs growing considerably faster, leaner, and utilizing much less feed and with increased litter size.

Table 1 Projected food consumption trends of various livestock products to the year 2020 in developing world.

	Projected growth of consumption (%/yr)	Total consumption (million metric tons)		Percent of total world consumption	Per capita consumption (kg)	
	1997–2020	1997	2020	2020	1997	2020
Beef	2.9	27	47	61	6	7
Pork	2.4	47	81	67	10	13
Poultry	3.9	29	49	64	7	8
Meat	3.0	111	188	65	25	30
Milk	2.9	194	391	57	43	62

Sources: From Delgado;[1] FAOSTAT statistical database. FAO,[2] http://faostat.fao.org/ default.aspx.; ftp://ftp.fao.org/docrep/nonfao/lead/x6155e/x6155e00.pdf.
Note: Consumption indicates direct use as food, measured as uncooked weight, bone in. Meat includes beef, pork, mutton, goat, and poultry. Milk represents cattle and buffalo milk and milk products in liquid milk equivalents. Metric tons and kilograms are three-year moving averages centered on the years shown.

Genomics can be grossly divided into two basic areas: 1) structural genomics; and 2) functional genomics. Structural genomics corresponds to characterization of the physical nature of whole genome and functional genomics is the understanding of the genes transcribed and translated into protein products, their interactions, and also the regulation of these processes.

STRUCTURAL GENOMICS

Structural genomics encompasses construction of genetic maps, physical maps, comparative maps, and ultimately, the determination of the complete DNA sequence of the genome of interest with a goal to understand the genome organization and locate DNA sequence variations, i.e.,

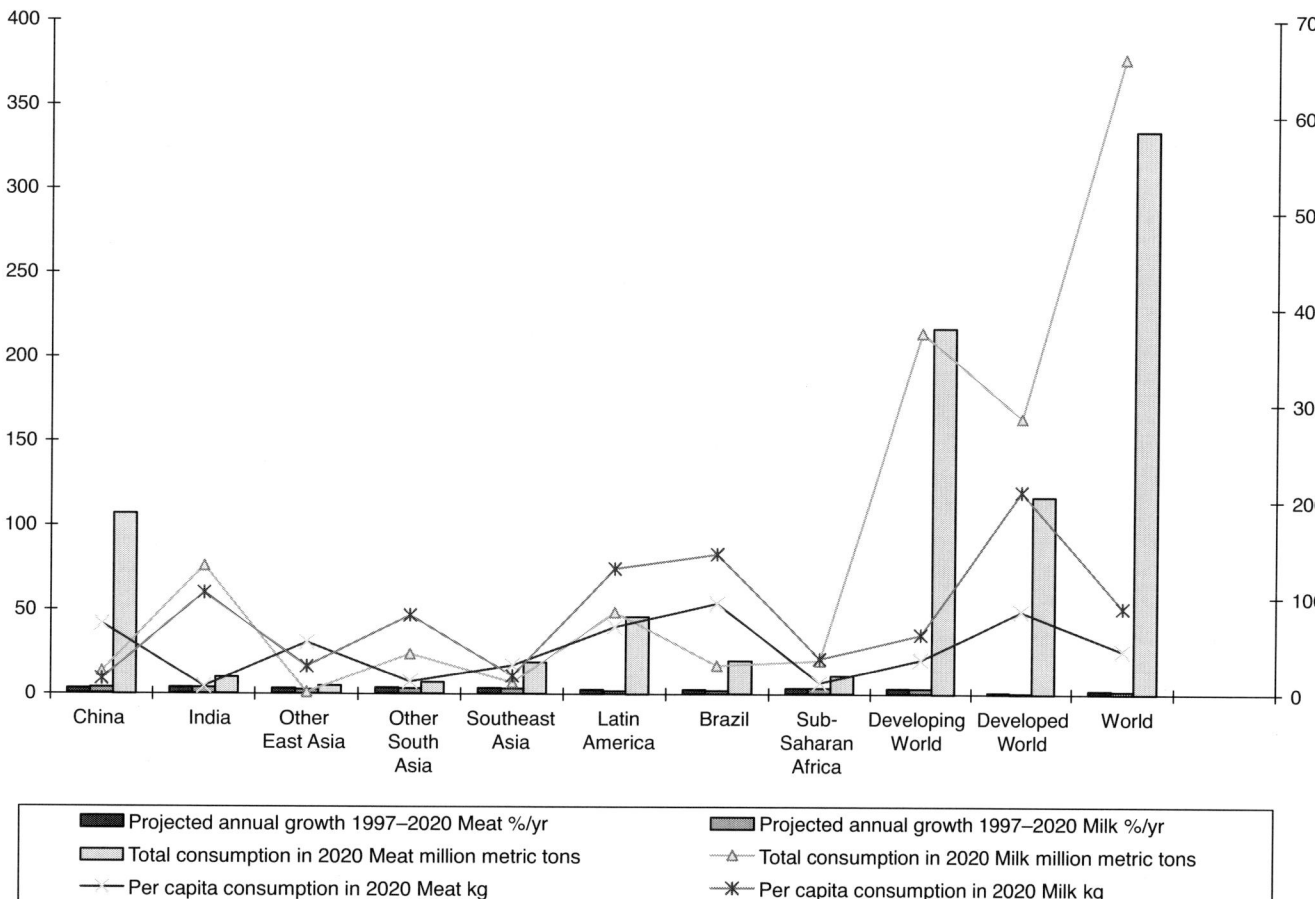

Fig. 1 Trend of demand vs. projected production of meat and milk. **Sources:** From Delgado.[1] FAO annual data 2002, total and per capita meat consumption for 1997 are annual average of 1996–1998 values. Projections are from the July 2002 version of IMPACT.

Table 2 Status of quantitative trait loci identified in different livestock species.

Species	QTLs	Traits
Pig	1831	316
Cattle	1123	101
Chicken	657	112
Sheep	53	28

Source: From http://www.animalgenome.org/QTLdb.

polymorphisms. Structural genomics started in 1980s with the genetic mapping to identify the locations on the chromosome that might contain the performance altering genes, which however, was not a very successful approach as the sequence of the DNA was not known. The development of new techniques enabled identification of genes that underlie the genetic variation of production traits in livestock species.

Although scientists have been searching for quantitative trait locus (QTL, a region that contains a gene [or locus] affecting a quantitative trait) in livestock for over 40 years, genome-wide QTL studies could not be conducted until genetic markers that spanned the entire genome were developed. For construction of high-density genetic maps, microsatellite markers (repetitive sequences) were the markers of choice because of their high polymorphism and abundant distribution throughout the genome. Microsatellite-based high-density genetic maps facilitated the fine mapping of hereditary traits of agricultural importance, characterization of meiosis, and provided a foundation for physical map construction. Resource populations required for construction of genetic maps were developed by crossing widely divergent breeds representing different alleles having large effects on QTLs. Worldwide resource populations are available for most of the livestock species and has been successfully used to identify numerous economically important trait loci (ETL) in a variety of livestock species (Table 2). The information related to livestock QTL is available in the animal quantitative trait locus database (AnimalQTLdb, http://www.animalgenome.org/QTLdb).

Over the years, QTL studies have contributed considerably to the development of more efficient selection procedures employing genetic markers, a strategy called marker-assisted selection (MAS). Presently several gene tests are being used in MAS programs to improve milk and meat quality, growth and reproduction in various livestock species (Table 3). QTL studies have improved the swine industry the most with several gene tests being used under MAS program for various production traits in over 60% of the pigs produced in United States.[4] However, mapping of QTL is not a very straightforward approach. It has been very difficult to search for a gene or mutation causing a QTL effect in an area surrounded by millions of unknown DNA sequences in the region. To locate these genes, the concept of comparative genomics was applied.

Comparative genomics utilizes the DNA sequence similarities between species. Although the arrangement of DNA sequences is not identical across species, large regions of the genome/functional elements have been conserved throughout evolution. Hence, it is possible to harness the information available in the gene-rich mammalian species and extrapolate the same in livestock.

Comparative mapping has been very effectively used to identify a number of genes in livestock.[5] The best example of the use of comparative genomics in livestock is the localization of chromosomal region containing the double muscling gene. Two research groups were able to localize the QTL effect to bovine chromosome 2 in the mid-1990s based on its location in the mouse genome. However, the specific gene was annotated in cattle by comparative mapping after the identification of "myostatin" gene in mice having a large impact on muscle development. The myostatin or GDF8 gene was mapped to a region of the human genome syntenic to that of bovine chromosome 2 and a single nucleotide switch from guanine to adenine at codon 313 in the gene that caused the double muscling effect was identified. Grobet et al.[6] were able to inactivate the GDF8 gene postnatally and demonstrated that these animals achieved the same level of muscle hypertrophy as animals with inactive myostatin genes. Such information is beneficial to the cattle industry as the percentage of lean in the carcass could be increased without the associated problems of dystocia because of the heavy muscled fetus. A few other genes that have been mapped in cattle through the "QTL search followed by comparative fine mapping" approaches include the thyroglobulin and calpastatin genes affecting the degree of marbling and meat tenderness; diacylglycerol acetyltransferase (DGAT) having an affect on fat deposition in milk and leptin, a protein important in energy metabolism.

Linkage maps, QTL searches, comparative mapping, and fine mapping were useful, but time-consuming and cost extensive. Most of the economically important traits in livestock are under the control of numerous genes simultaneously, and interact both with one another as well as the production environment. Searching these genes one by one and establishing the genotype–phenotype association was a gigantic task. Later, in the last half of 1990s, the efforts started for the whole genome sequencing of human and mouse using high-throughput sequencing (National Human Genome Research Institute, NHGRI) and whole genome shotgun (Celera genomics) approach. The animal scientists took advantage of infrastructure built by the NHGRI and efforts were initiated for sequencing the major livestock species.

The first drafts of the sequence of chicken and cattle genomes were completed in 2005 (http://www.genome.gov/12512874). In 2006, the sequencing of the horse and the pig genomes was also begun, with the first draft of the horse sequence completing in 2007, and the

Table 3 Commercial DNA tests available for different livestock species.

	Species			
Trait category	Cattle—dairy	Cattle—Beef	Sheep	Pig
Congenital defects	BLAD	CMS		RYR
	Citrulinaemia	Citrulinaemia		
	DUMPS	Congenital myasthenic syndrome		
	Freemartinism	Freemartinism		
	MSUD	MSUD		
	Mannosidosis	Mannosidosis		
	CVM			
Genetic disorder	α-Mannosidosis	α-Mannosidosis		
	β-Mannosidosis	β-Mannosidosis		
	Bovine hereditary zinc deficiency	Bovine hereditary zinc deficiency		
	Dwarfism	Dwarfism		
	Factor XI deficiency	Pompes disease (E7)		
	Factor IX deficiency	Pompes disease (E13)		
	α-Mannosidosis	Myophosphorylase deficiency		
	Hypotrichosis and oligodontia	Inherited congenital myoclonus		
	Deficiency of uridine Monophosphate synthetase			
	Weaver syndrome	Protoporphyria		
	Platelet bleeding	Platelet bleeding		
	Albinism			
Disease	Muscular hypertrophy	Muscular hypertrophy	Prp	F18
Appearance	MC1R/MSHR	MC1R/MSHR		MC1R/MSHR
	Red coat color	Red coat color		CKIT
	Black coat color	Black coat color		
	TYR or Albinism	MGF or Roan		
Milk quality	κ-Casein			
	β-Lactoglobulin			
	FMO3			
Milk yield and composition	κ-Casein			
	GRH			
	DGAT			
Meat quality	TG			RYR
	CAST			RN/PRKAG3
	CAPN1			>15 PICmarq
	CAPN3			
	RN/PRKAG3			
	RYR			
	GH1			
Growth and composition		Myostatin	Callipyge	MC4R
				IGF-2
Reproduction			Booroola	
			Inverdale	
			Hanna	
Feed intake				MC4R

Source: From J. Anim. Sci.[3]

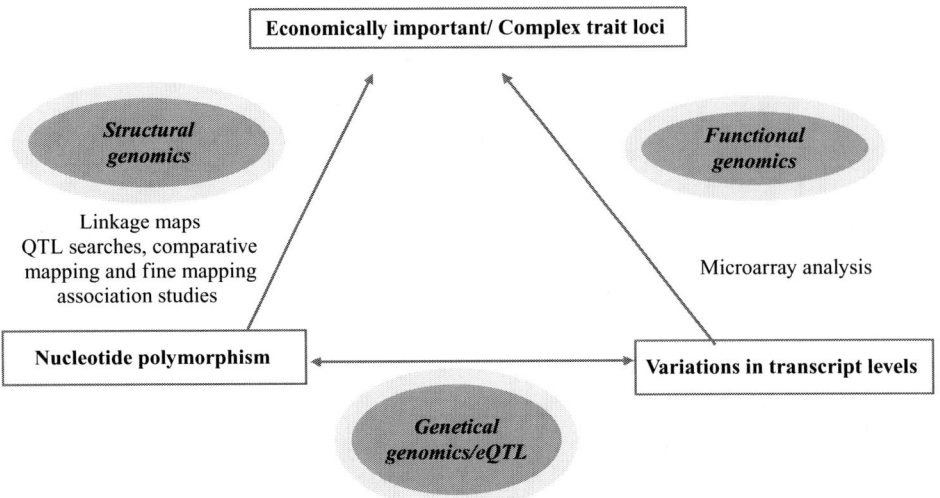

Fig. 2 Integrative approaches applied to decipher economically important traits.

pig sequence draft has also become available recently (http://www.ensembl.org/Sus_scrofa). Taking advantage of the availability of sequence draft, single nucleotide polymorphism (SNP) arrays have already been developed for chicken (Illumina™ 18K iSelect chip), cattle (Bovine 50K Illumina™ iSelect chip), pig (Illumina™ iSelect pig DNA chip, 60K), and sheep (Illumina™ OvineSNP50) and are under development for other livestock species. SNP chip provides a platform where hundreds or thousands of markers can be genotyped in a cost effective manner. SNP array-based association studies are being extended to traits like behavior, disease resistance, and structural and environmental soundness.[7]

With the availability of species-specific SNP chips, it is now possible to undertake whole genome scans[8] and genomic selection approaches can be accomplished to correlate phenotype and genotype across the whole genome simultaneously, rather than by one locus at a time. Thus, SNP-based platforms are available to predict the molecular genetic value of individual animals for specific traits, viz., marbling and tenderness in beef cattle. SNP chips also provide a higher resolution map of the genome that can be used to detect region of chromosomal abnormalities, e.g., aneuploidies, microdeletions, microduplications, and loss of heterozygosity (LOH). As a variety of diseases are linked to such chromosomal abnormalities, SNP chips have the power to identify such regions and provide insight about the diseases and further suggest targets for intervention.[9]

FUNCTIONAL GENOMICS

QTL mapping is a powerful method and a number of QTLs have been reported in various livestock species, but only a few of these have been localized at the gene level.[10] A gap still exists in the knowledge to localize QTL and then, to identify the underlying genes and their function which reduces the ability to harness the potentials of genomics. With the recent completion of the genome sequence of the major livestock species and availability of high-throughout technology, a functional genomics approach can now be used for the assignment of function to the identified genes and also for the organization and control of genetic pathways combining to determine the complex phenotype. The use of microarray technology for global profiling of gene expression is the primary method of functional genomics and has made it possible to compare expression levels for active genes under a variety of genetic and environmental conditions. Examples showing the power of this technology include identification of set of genes conferring resistance to Marek's disease in chicken and *Trypanosoma congolense* infection in cattle.

Another powerful application of microarray technology is to synergistically harness the powers of recombination and functional analyses in a combined approach known as "Genetical Genomics," or expression QTL (eQTL) mapping (Fig. 2). Such an approach has the potential to identify *cis*-acting and *trans*-acting genes in the region of the functional QTL and master regulators of quantitative trait variation as well, wherein polymorphisms at a single locus may regulate/control variation in gene expression of many other genes. Genetical genomics approach allows one to understand the groups of genes and governing variation in complex traits. Integrated information on phenotypic traits, pedigree structure, molecular markers, and gene expression can be used for estimating heritability of mRNA transcript abundances so as to infer the regulatory gene network and map eQTL. Such studies will help in answering the questions concerning genes involved in meat and milk production, meat and milk quality and composition, and response to production systems/husbandry stressors. Eventually, this knowledge will further aid in selection and management of livestock.

IMPLICATIONS OF GENOMICS

The most important utilization of the genomic data toward livestock production would be the use in selective breeding. Traditionally, the selection methods are based on the estimated breeding value (EBV) for traits of economic importance that are estimated from phenotypic records, genetic covariances among traits, and pedigree relationships in populations. In most cases, however, the actual location, identity, and functional role of the QTL remain unknown. In addition, efficiency of traditional selection practices is low when the traits are difficult to measure and have low heritability.

Solution to these problems is gene-assisted selection, which is easily applied to traits that are difficult to measure, and have low heritability and slow genetic change. With the availability of the high-throughput platforms (SNP array), it is possible to incorporate DNA information from markers into EBVs. Selection decisions based on genomic breeding values (GEBV) are calculated as the sum of effects of dense genetic markers or haplotypes across the entire genome and thus, all the QTLs contributing to variation in a trait are captured. The QTL effects are first measured on a large reference populations with phenotypic records and then, only marker information is required to calculate GEBV in subsequent generations. GEBV holds special importance in developing countries where structured resource populations are not available for most of the livestock species. GEBV approach has already been evaluated in United States, New Zealand, Australia, and the Netherlands and is expected to double the rate of genetic gain in the dairy industry.[11]

SUMMARY

The demand for livestock products is projected to grow significantly in the developing countries. Traditional breeding methods are not well equipped to meet this increasing demand of meat and dairy products. The integrated use of structural and functional genomics has the potential to speed up the livestock production progress through understanding the genotype–phenotype relationship and genotype × environment interaction of livestock genomes as a whole. Selection decisions based on GEBV are made possible with the advances in the whole genome sequencing and high-throughput SNP panels. This approach of GEBV to select animals with desired allelic combinations is cost effective and can even target non-measurable traits.

REFERENCES

1. Delgado, C.L. Rising consumption of meat and milk in developing countries has created a new food revolution. J. Nutr. **2003**, *133*, 3907S–3910S.
2. Food and Agriculture Organization of the United Nations. *FAOSTAT Statistical Database*. FAO, 2009, http://faostat.fao.org/default.aspx; ftp://ftp.fao.org/docrep/nonfao/lead/x6155e/x6155e00.pdf.
3. Dekkers, J.C.M. Commercial application of marker- and gene-assisted selection in livestock: Strategies and lessons. J. Anim. Sci. **2004**, *82*, E313–E328.
4. Rothschild, M.F. Porcine genomics delivers new tools and results: this little piggy did more than just go to market. Genet. Res. **2004**, *83*, 1–6.
5. O'Brien, S.J.; Menotti-Raymond, M.; Murphy, W.J.; Nash, W.G.; Wienberg, J.; Stanyon, R.; Copeland, N.G.; Jenkins, N.A.; Womack, J.E.; Marshall Graves, J.A. The promise of comparative genomics in mammals. Science **1999**, *286*, 458–462, 479–481.
6. Grobet, L.; Pirottin, D.; Farnir, F.; Poncelet, D.; Royo, L.J.; Brouwers, B.; Christians, E.; Desmecht, D.; Coignoul, F.; Kahn, R.; Georges, M. Modulating skeletal muscle mass by postnatal, muscle-specific inactivation of the myostatin gene. Genesis **2003**, *35*, 227–238.
7. Rothschild, M.F.; Plastow, G.S. Impact of genomics on animal agriculture and opportunities for animal health. Trends Biotechnol. **2007**, *26*, 21–25.
8. Meuwissen, T.H.E.; Hayes, B.J.; Goddard, M.E. Prediction of total genetic value using genome-wide dense marker maps. Genetics **2001**, *157*, 1819–1829.
9. Scharpf, R.B.; Ting, J.C.; Pevsner, J. BIOINFORMATICS SNP chip: R classes and methods for SNP array data. Bioinformatics **2008**, *23*, 627–628.
10. Andersson, L.; Georges, M. Domestic animal genomics: deciphering the genetics of complex traits. Nat. Rev. Genet. **2004**, *5*, 202–212.
11. Hayes, B.J.; Bowman, P.J.; Chamberlain, A.J.; Goddard, M.E. Invited review: genomic selection in dairy cattle: progress and challenges. J. Dairy Sci. **2009**, *92*, 433–443.

Genomics: Animal Agriculture

Juan Loor
Department of Animal Sciences, University of Illinois, Urbana, Illinois, U.S.A.

Abstract

One of the greatest challenges in animal agriculture is to understand the genetic basis underlying muscle, adipose, and bone growth, lactation, and disease susceptibility. High-throughput sequencing and transcriptomics technologies have dramatically accelerated the rate at which biological and genetic information can be collected from agricultural animals. The classical definition of genomics has been expanded in recent years to encompass the study of regulation of all the genes of a cell or tissue at the level of DNA (genotype), mRNA (transcriptome), protein (proteome), and metabolite (metabolome). Among genome-enabled technologies, microarrays have been most-widely used in animal agriculture, thus allowing the simultaneous analysis of the expression of thousands of genes in tissues or cells. Together with whole-animal level information, large-scale DNA and mRNA data are poised to accelerate accumulation of knowledge of the genetic regulation in agricultural species. Through the use of bioinformatics, clustering, and gene network analysis, transcriptomics is beginning to allow the identification of regulatory mechanisms that are associated with functional development of tissues of agricultural animals. Continued application of genome-enabled technologies (genotyping, microarrays, proteomics, gene silencing) will contribute to our understanding of regulatory points at different stages of muscle growth and lactation as well as during disease incidence. This information will provide new insights into opportunities for enhancing the efficiency of animal production and well-being.

INTRODUCTION

Availability of DNA sequence information for model organisms (e.g., mouse, rat) has already facilitated the characterization of the behavior of molecular networks at multiple points of growth, development, and disease. Although the scientific method in animal agriculture research has allowed us to amass substantial amounts of information at the tissue and animal level, major gaps in knowledge of molecular adaptations in domestic animals remain. This manuscript is focused on the use of transcriptomics to study the respective functions of skeletal muscle, fat, and liver during postnatal growth and lactation in cattle. The goal is to provide specific examples from published studies where transcriptomics has advanced our understanding of animal function and how information gathered could impact livestock production in the next 10–20 years. Perspectives for a systems biology approach in animal agriculture through the use of transcriptomics, proteomics, and metabolomics are presented.

FUNCTIONAL GENOMICS THROUGH DNA MICROARRAYS

Insights into Cattle Muscle and Adipose Tissue Development

Functional genomics is generally defined as the study of the transcriptome.[1] The transcriptome (i.e., mRNA) encodes genetic information about a protein and changes in its expression exert a major influence on physiological conditions. DNA microarray technology allows the semiquantitative and simultaneous monitoring of the expression of thousands of genes in tissues or cells. A cattle muscle/fat cDNA microarray was used recently to evaluate transcript profiles of skeletal muscle tissue of 11-month-old Japanese Black and Holstein steers.[2] The former have a greater inherent capacity for intramuscular fat deposition (i.e., marbling), driven partly by a greater rate of preadipocyte differentiation relative to other breeds such as Angus. Genes expressed preferentially in Japanese Black muscle tissue (total of 17) included lipogenic transcription factors (sterol regulatory element binding factor 1, *SREBF1*; thyroid hormone responsive SPOT14, *THRSP*) as well as lipogenic enzymes (stearoyl-CoA desaturase, *SCD*; fatty acid binding protein 4, *FABP4*). Holstein skeletal muscle was characterized by genes associated with energy metabolism (myosin genes; pyruvate dehydrogenase kinase 4, *PDK4*) and muscle contraction (ATPase, *ATP2A1*). Despite the well-defined effects of age and growth rate on cattle lipid deposition, which differ across breeds, differences in gene expression profiles in this study provided molecular evidence for the greater propensity of Japanese Black to produce beef with greater marbling.

Another study[3] explored temporal (3, 7, 12, 20, 25, and 30 months of age) gene expression profiles of skeletal muscle from Wagyu × Hereford (WH) and Piedmontese × Hereford cattle (PH), which exhibit contrasting amounts of intramuscular fat content (ca. 10.7% vs. 4.3% for WH and PH). Two animals from each breed with the most extreme intramuscular fat content (16.5% vs. 3.4% for WH and PH) were selected for microarray analysis using the same platform as Wang et al.[2] Microarrays were coupled with quantitative RT-PCR of 17 genes on the entire group of animals ($n = 6$–7). Results revealed contrasting mRNA expression patterns for several genes associated with adipogenesis and lipogenesis between breeds but also across stages of growth. For example, the transcription factor peroxisome proliferator-activated receptor γ (*PPARG*) had greater expression as early as seven months of age, i.e., close to weaning, in WH vs. PH steers. A battery of genes associated with lipogenesis (e.g., *FABP4*, *SCD*, fatty acid synthase [*FASN*], adiponectin [*ADIPOQ*]) and energy metabolism (e.g., *PDK4*) also had greater expression in WH steers at seven months of age. Unlike previously thought, gene expression patterns from microarrays and RT-PCR suggested[3] that the predisposition of WH cattle to accumulate fat is well-developed close to the time of weaning. Data from this study point to a time-frame during which early nutritional intervention could be used as a practical means to initiate precocious intramuscular fat development and, potentially, increase the likelihood of achieving greater marbling at slaughter age. Several of these putative regulators of adipogenesis could potentially be used as markers of marbling in breeding programs.

Insights into Metabolic Disease from Liver Transcriptomics in Dairy Cattle

Prepartum energy nutrition

After parturition, dairy cows enter a period of severe negative energy balance that places them at higher risk of developing metabolic disorders such as fatty liver and ketosis. In the first study of its kind, a cDNA microarray was used to characterize hepatic gene expression patterns in dairy cows fed with different levels of energy prepartum.[4] Compared with cows restricted to 80% of energy requirements during the dry period, cows that were overfed throughout the dry period (intake of >160% of energy requirements) had sharply decreasing energy balance between the last week prepartum and the first week postpartum. After parturition, overfed cows also had higher triacylglycerol in liver, higher blood non-esterified fatty acids and β-hydroxybutyrate (a product of liver fatty acid oxidation) around and after parturition, and higher insulin prepartum.[4]

From the set of liver expressed genes, a total of 111, 2270, and 200 genes were differentially expressed according to day × diet, day, and diet, respectively. Using k-means clustering, a number of distinct expression patterns for genes affected due to day × diet interaction were discerned. Among those, cows fed with restricted energy prepartum clearly had a greater level of upregulation of 54 genes during the periparturient period most of which were associated with important biological functions such as inflammatory responses, fatty acid oxidation, signal transduction, and gluconeogenesis.[4] These genes represent putative metabolic markers that could be targeted via nutrition to minimize the risk of cows developing liver-related metabolic disorders.

Ketosis

The biochemistry of metabolic adaptations in liver due to ketosis has been known for several decades.[5] Work with rodents, however, indicates that metabolic regulation relies partly on transcriptional control as a long-term mechanism affecting the level of expression of key enzymes.[6] Large-scale adaptations in metabolic and cell signaling gene networks in liver tissue from cows induced to develop ketosis via feed-restriction early postpartum were recently evaluated via microarrays.[7] Several target genes of the nuclear receptor PPARα, which in non-ruminants controls liver fatty acid oxidation, were differentially expressed by ketosis including carnitine palmitoyltransferase 1A (liver) (*CPT1A*), acyl-Coenzyme A oxidase 1, palmitoyl (*ACOX1*), acetyl-Coenzyme A acyltransferase 1 (*ACAA1*), and apolipoprotein A-I (*APOA1*). These were upregulated with ketosis as a necessary response of the liver to cope with the increased influx of non-esterified fatty acids from blood. Furthermore, it was suggested that upregulation of the PPARα co-activator *PPARGC1A* might be required for activation of the transcription machinery controlled by PPARα. From additional gene network analysis, it appeared that liver possesses additional nuclear receptors (e.g., *PPARD*, hepatocyte nuclear factor 4, alpha [*HNF4A*]) that respond to stimuli from the outside (e.g., non-esterified fatty acids) and allow the tissue to adapt to a new physiological state. Those nuclear receptors have received less attention than PPARα in terms of liver metabolism but represent novel markers for future studies searching for

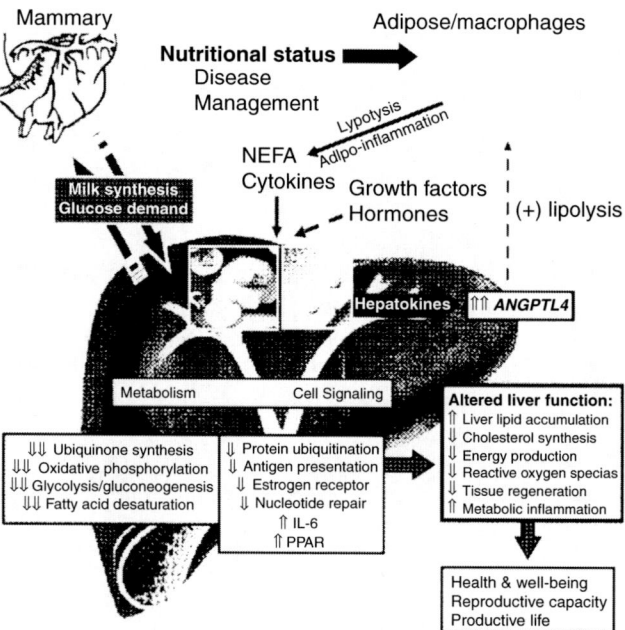

Fig. 1 Interrelationships between nutritional status and mammary demands for milk synthesis on physiological adaptations in adipose tissue and liver early postpartum. Copious milk production by the postpartum cow takes precedence in the utilization of the animal's resources, i.e., demands for milk synthesis place a large toll on carbohydrate resources. Metabolic changes underlying onset of ketosis are associated primarily with a shortage of carbohydrate precursors and, as a consequence, rates of gluconeogenesis. Clearly, management and/or disease incidence are likely to drastically influence the nutritional status of an early postpartum cow. Thus, if either of these factors prevents the animal from achieving the expected rates of feed intake, the likelihood of developing ketosis increases. Upon withdrawal of feed, metabolic signals (e.g., glucocorticoids, catecholamines) enhance adipose tissue lipolysis and non-esterified fatty acid delivery to liver. Resident macrophages in adipose could contribute to influx of cytokines and other proinflammatory mediators into liver. It is also likely that both hepatocytes and Kupffer cells can directly respond to influx of non-esterified fatty acids (e.g., palmitic acid, oleic acid) and/or adipokines, which might lead to inflammation (e.g., *IL6* upregulation). The ketotic liver reduces energy production and metabolism, perhaps due to decreased milk production, but maintains fatty acid oxidation through upregulation of peroxisome proliferator-activated receptor (PPAR) signaling pathways. Reduced energy production would also ensure that generation of reactive oxygen species is lower, thus lessening tissue damage during ketosis. Nutrition-induced ketosis in an early postpartum cow could affect the animal's health and well-being and in the long term have negative effects on reproduction and productive life.

preventatives of liver-metabolic diseases. Authors used transcriptomics data as well as available measurements of metabolic state (e.g., blood non-esterified fatty acids, β-hydroxybutyrate, insulin) to develop an integrative model of adaptations at several levels in cows undergoing ketosis (Fig. 1).

SYSTEMS BIOLOGY APPROACH IN ANIMAL AGRICULTURE

General Considerations

The biological complexity of agricultural animals unavoidably requires a systems biology approach, i.e., a way to systematically study the complex interactions in biological systems using a method of integration instead of reduction.[8] One of the goals of systems biology is to discover new emergent properties that may arise from examining the interactions between all components of a system to arrive at an integrated view of how the organism functions.[9] Work in model organisms during the past 15 years has demonstrated the applicability of high-throughput methods to discern regulatory and metabolic networks.[10] Fig. 2 outlines a systems biology scheme that would be amenable to high-throughput studies of animal tissue at the level of transcriptomics, proteomics, and metabolomics. The systems approach might lead to the discovery of regulatory targets that could be tested further (i.e., model-directed discovery) or help address a broader spectrum of basic and practical applications including interpretation of phenotypic data, metabolic engineering, or interpretation of lactation phenotypes.

Bioinformatics

Bioinformatics involves the use of mathematics and biochemistry to solve biological problems at the molecular level.[9] The core principle of bioinformatics is utilizing computer resources to solve problems on scales of magnitude far too great for human discernment. A bioinformatics approach, e.g., through the use of gene ontology (GO) analysis, will allow discerning the biological functions in tissues at specific points of development or under a particular nutritional management. One of the aims of the GO Consortium is to provide a controlled vocabulary that can be used to describe *any* organism. However, it is intuitive that many functions, processes, and components are not common to all life forms. Annotation with respect to the biological context of livestock would be an important undertaking. A recent entry provides an in-depth overview of the development of an animal trait ontology, which is essential for annotating genes/proteins to biological functions.[11]

CONCLUSIONS

The use of high-throughput technologies, such as microarrays, has allowed a biological holistic view of the complex system represented by the major tissues in agricultural animals. Genome-enabled technologies have already uncovered molecular functions and pathways that are key factors during growth and lactation. Particularly important is the discovery of interacting networks of genes, which is suggestive of functional interactions as well as common-regulated

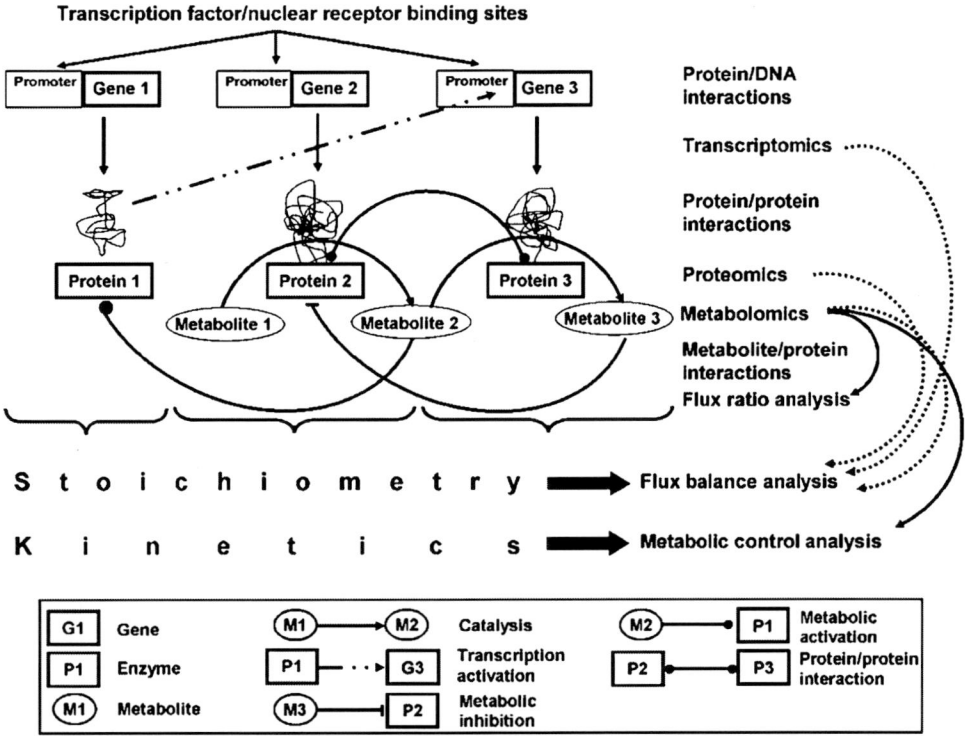

Fig. 2 Systems biology scheme to study livestock tissue function in the context of growth and lactation. Transcriptomics, proteomics, and metabolomics will result in an initial characterization of tissues (e.g., skeletal muscle, adipose, liver, mammary gland) across developmental and mature physiological stages. Ultimately, there will be a need to integrate data obtained at each omics level with currently-available enzyme kinetic and stoichiometry data so that more updated models of livestock tissue function can be developed.

functions. Greater understanding of the significance and regulation of those interactions is the next challenge for livestock biologists. Besides microarray technology, other techniques will need to be implemented in the near future to more-fully understand these complex interactions. The discovery and functional characterization of transcription factors involved in tissue adaptations to a new physiological state could result in long-term practical applications. These molecules can be controlled by effectors such as nutrients, hormones, and/or growth factors. Therefore, controlling (increasing, decreasing) the availability of these effectors in tissues of economic importance might allow for fine regulation of the system.

REFERENCES

1. Schoolnik, G.K. Functional and comparative genomics of pathogenic bacteria. Curr. Opin. Microbiol. **2002**, *5* (1), 20–26.
2. Wang, Y.H.; Byrne, K.A.; Reverter, A.; Harper, G.S.; Taniguchi, M.; McWilliam, S.M.; Mannen, H.; Oyama, K.; Lehnert, S.A. Transcriptional profiling of skeletal muscle tissue from two breeds of cattle. Mamm. Genome **2005**, *16* (3), 201–210.
3. Wang, Y.H.; Bower, N.I.; Reverter, A.; Tan, S.H.; De Jager, N.; Wang, R. McWilliam, S.M.; Cafe, L.M; Greenwood, P.L.; Lehnert, S.A. Gene expression patterns during intramuscular fat development in cattle. J. Anim. Sci. **2009**, *87* (1), 119–130.
4. Loor, J.J.; Dann, H.M.; Guretzky, N.A.; Everts, R.E.; Oliveira, R.; Green, C.A.; Litherland, N.B.; Rodriguez-Zas, S.L.; Lewin, H.A.; Drackley, J.K. Plane of nutrition prepartum alters hepatic gene expression and function in dairy cows as assessed by longitudinal transcript and metabolic profiling. Physiol Genomics **2006**, *27* (1), 29–41.
5. Baird, G.D.; Heitzman, R.J.; Hibbitt, K.G. Effects of starvation on intermediary metabolism in the lactating cow. A comparison with metabolic changes occurring during bovine ketosis. Biochem. J. **1972**, *128* (5), 1311–1318.
6. Desvergne, B.; Michalik, L.; Wahli, W. Transcriptional regulation of metabolism. Physiol. Rev. **2006**, *86* (2), 465–514.
7. Loor, J.J.; Everts, R.E.; Bionaz, M.; Dann, H.M.; Morin, D.E.; Oliveira, R.; Rodriguez-Zas, S.L.; Drackley, J.K.; Lewin, H.A. Nutrition-induced ketosis alters metabolic and signaling gene networks in liver of periparturient dairy cows. Physiol. Genomics **2007**, *32* (1), 105–116.
8. Loor, J.J.; Cohick, W.S. ASAS centennial paper: lactation biology for the twenty-first century. J. Anim. Sci. **2009**, *87* (2), 813–824.
9. Bruggeman, F.J.; Westerhoff, H.V. The nature of systems biology. Trends Microbiol. **2007**, *15* (1), 45–50.
10. Feist, A.M.; Palsson, B.O. The growing scope of applications of genome-scale metabolic reconstructions using *Escherichia coli*. Nat. Biotechnol. **2008**, *26* (6), 659–667.
11. Hughes, L.M.; Bao, J.; Hu, Z.L.; Honavar, V.; Reecy, J.M. Animal trait ontology: the importance and usefulness of a unified trait vocabulary for animal species. J. Anim. Sci. **2008**, *86* (6), 1485–1491.

Genomics: Captive Breeding and Wildlife Conservation

Alfred L. Roca
Lawrence B. Schook
Department of Animal Sciences, University of Illinois, Urbana, Illinois, U.S.A.

Abstract

The genomes of many domestic and wild mammals are being sequenced in order to enhance agricultural productivity, provide insights into biomedical model organisms, and identify conserved functional regions of the human genome. Sequencing of these species "genome-enables" other species within the same genus, family, or order by facilitating conservation genetic studies on wild, and often endangered, relatives of the sequenced species. Sequenced genomes have enabled researchers to examine whether interbreeding with domestic species has affected the germline of related wild species. Genomic sequences improve the ability of researchers to make inferences about neutral genetic variation within species; to find deleterious alleles, which can be purged through selective breeding programs; and to identify adaptive alleles. The sharply dropping cost of sequencing technology may allow the genomes of thousands of species to be sequenced, assisting conservation and captive breeding efforts on their behalf.

INTRODUCTION

Sequencing the human genome has greatly assisted researchers seeking to comprehend the genetic basis of health and disease. In order to properly understand, annotate, and interpret human genomic information, a comparative approach using the genomic sequences of other mammals is being used to elucidate the structure, function, and regulation of genes and other genomic elements.[1] For example, genomic or phylogenetic "shadowing" has been proposed as a means of identifying functional elements within the human genome.[2] In phylogenetic shadowing, DNA sequences from multiple species are aligned and regions that are conserved as well as non-conserved regions are identified (Fig. 1).[2] Regions conserved across species tend to have a functional role, such as the exons that code for the amino acids of proteins, or other elements that play an important role in regulating the transcription or expression of genes. Thus, genomic or phylogenetic shadowing has been used to discover and characterize functional components of the human genome, such as novel gene regulatory elements.[2]

The initial selection of mammalian species recommended for sequencing (Fig. 2) largely relied on the role of the species as important agricultural animals (livestock); as model organisms for biomedical studies (rodents, primates, livestock); as veterinary models for human diseases (cat, dog, horse); or as species important for comparative studies of recent human evolution (primates and their relatives) or of evolutionary events affecting all placental mammals (afrotherians such as the savanna elephant or xenarthrans such as the sloth).[1] In addition, some species were recommended for genomic sequencing in order to further studies on basic evolutionary questions, such as bats (reduced genome size) or insectivores (retention of primitive morphological traits[1]). Fig. 2 lists the species of mammals identified by the National Center for Biotechnology Information as having genome sequencing projects that are complete, under assembly, or in progress. Although many species were selected based on agricultural, biomedical, or evolutionary criteria, the genome sequences of each species will have a profound impact on conservation and captive breeding research, by allowing the development of markers for use in conservation genetics, and by permitting immediate access to sequences of candidate genes for hereditary traits or diseases that may affect the fitness of free-ranging or captive species.

GENOME-ENABLED SPECIES

Genomic sequencing of a species does not only facilitate genetic research within the sequenced species. It also facilitates genetic studies of closely related species within the same genus, family, or even order of mammals.[3] Many species of mammals being sequenced are either themselves endangered or part of the same genus or family as other species that are endangered.[3] Even mammals chosen because they are common species of livestock, companion

Genomics: Captive Breeding and Wildlife Conservation

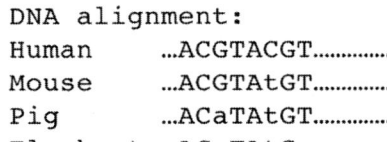

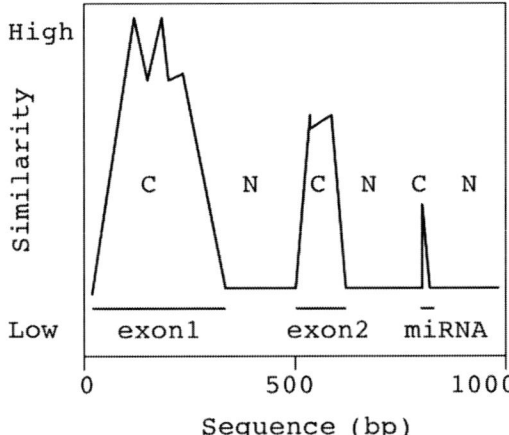

Fig. 1 Phylogenetic or genomic shadowing provided a major justification for sequencing non-human mammalian genomes.[2] Alignment of genomic sequences from a diverse set of organisms (top) will reveal regions (bottom) where the DNA sequence is conserved (C) or non-conserved (N). Functional regions of the genome tend to show greater sequence conservation across species. Thus non-human genomes can be used to annotate the human genome by identifying conserved regions of known or unknown functional importance for human health and disease studies. In the illustrative example depicted, conserved regions identify functional exons (coding regions of proteins) and a microRNA (which regulates gene expression).

animals, or model organisms in biomedical research (Fig. 2) can "genome-enable" a broad range of related wild or endangered species.[3] For example, the genome of the domestic dog was recently sequenced[4] since the species is a very common companion animal, and there are an estimated four hundred million dogs worldwide. Yet in addition to its impact on dog veterinary research, sequencing of the domestic dog genome will also enable the development of genetic markers and facilitate genetic studies among other members of the dog family Canidae, which includes some 35 species, several of which are endangered or threatened (Fig. 3). Thus, genetic studies on endangered African wild dogs or on the threatened maned wolf of South America (Fig. 3) will be greatly facilitated by the domestic dog genome project.[4]

Like the canids, other domesticated species will "genome-enable" their wild relatives. At the family level of classification, the genome sequence of the pig will enable studies on wild suids such as pygmy hogs and babirusas;[5] the domestic cat genome will enable studies on cheetahs, lions, tigers, and other wild felids.[6] Sequencing of the horse genome will enable genetic studies of wild horses, wild asses, and zebras; in one study of 20 short tandem repeat (STR or microsatellite) markers developed in the domestic horse, 15 proved useful in two species of zebras.[7] In all, the current mammalian genome sequencing projects will "genome-enable" almost a thousand endangered species of mammals.[3]

INTROGRESSIVE HYBRIDIZATION

Introgressive hybridization, in which the germ line of a wild species is threatened due to interbreeding with a domestic species, is a threat to the conservation of many wild species, notably those that are closely related to common domestic animals. For example, introgressive hybridization by domestic cats is a threat to African and European wild cats; hybridization with domestic dogs is a major threat to the highly endangered Simian wolf of Ethiopia; while hybridization with domestic horses threatens the Przewalski's or Mongolian wild horse. Genetic methods can assess the extent of hybridization present among wild populations, and genomic sequencing of domestic animal genomes has greatly facilitated such studies. For example, the genome sequence of domestic cattle was used to generate closely linked short tandem repeat (STR) markers for 14 chromosomal segments, examined in 14 populations of American bison (Fig. 4).[8] Although cattle and bison will preferentially mate with members of their own species, male bison were deliberately crossed with female domestic cattle in many of the private ranches that saved the bison from extinction in the late 1800s.[8] These small herds were the sources later used to stock protected populations across North America, and recent genetic tests have detected the introgression of cattle alleles in many bison populations. However, the plains bison population in Yellowstone National Park in the United States and the wood bison population in Wood Buffalo National Park in Canada are known from historical records to have been continuously free-ranging.[8] Thus bison populations from these two parks were used to identify STR loci with species-specific alleles that distinguish bison from cattle.[8] Using this information, along with previous studies of mtDNA and nuclear markers, six additional bison populations were identified for which no evidence was detected of introgression of either mitochondrial or nuclear markers, although several other bison populations carried high frequencies of domestic cattle alleles.[8] This effective use of genomic markers to detect interspecies hybridization has important consequences for conservation. For example, translocations of bison to establish new populations can rely on source herds that are free of introgression by domestic cattle, while hybrid source populations can be avoided in captive breeding of plains or wood bison.

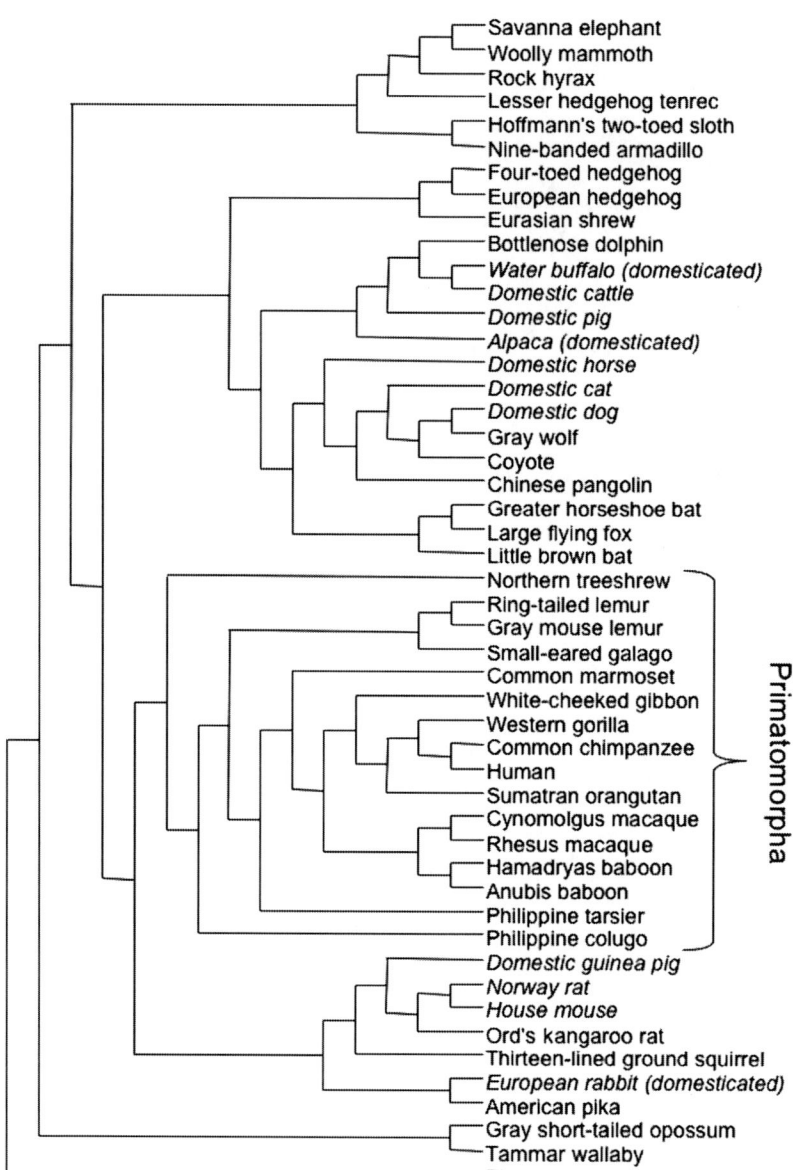

Fig. 2 Mammalian species for which genome sequencing projects are complete, under assembly, or in progress. Many primates and their close relatives (Primatomorpha) were chosen for sequencing due to their close evolutionary relationship to humans, while many other species are domestic or laboratory animals (italicized). Other criteria for selecting species included primitive or derived morphology, unique genomic attributes, or great evolutionarily distance from human.[1] The phylogenetic relationships among the species are shown. Branch lengths are not proportional to genetic distances or chronological age.

ADAPTIVE AND DETRIMENTAL VARIATION

Many endangered species have gone through population bottlenecks that led to reduced genetic variation within the species, while zoo populations often relied on a small number of founder individuals, with subsequent inbreeding of individuals in their collections. Zoos have instituted captive breeding programs that attempt to adjust the contributions of individual founder animals to the captive gene pool in order to maximize genetic diversity. Most conservation genetic studies of endangered species have largely relied on genetic markers presumed to be neutral, i.e., that do not affect the fitness of the species. For example, both mitochondrial DNA and short tandem repeats (STRs) are commonly used in analyses of population structure because these markers are assumed to be free of selective pressure. While genomic data enables more comprehensive surveys of neutral variation, genomic methods should also facilitate identification of alleles that contribute either to positive adaptive variation within a species or to negative deleterious effects on organismal health,[3] and may reveal how different taxa have adapted to different climatic or other conditions. Adaptive alleles could be targeted by selective breeding efforts, to increase the fitness of zoo or of managed

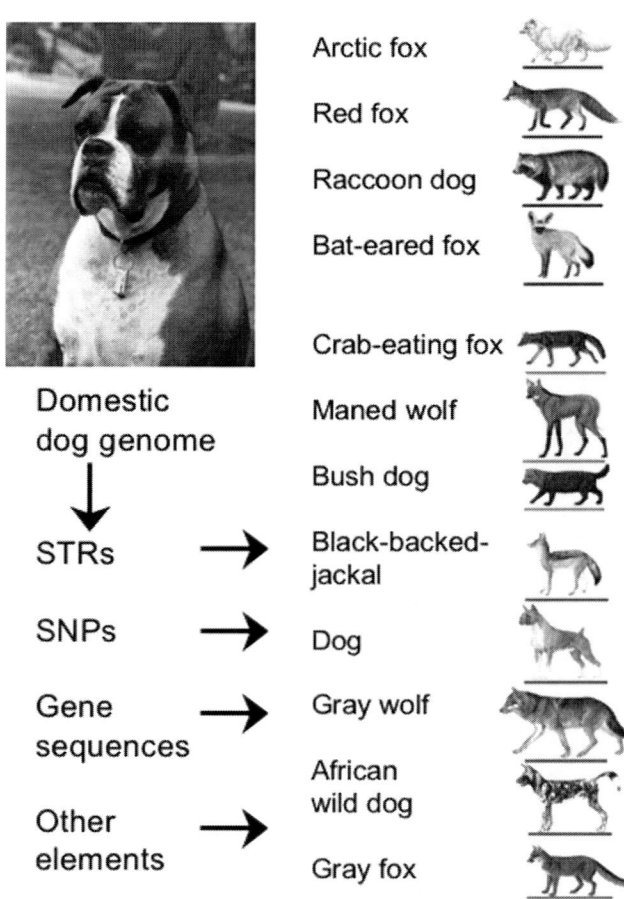

Fig. 3 Genomic sequences from one species will allow many related species to become "genome-enabled." For example, molecular markers derived from genomic sequencing of the domestic dog (left)[4] can be tested and used for research studies on wild species of the dog family Canidae (right). These include endangered or threatened species such as the African wild dog and the maned wolf. Thus although the domestic dog was selected for genomic sequencing based on the utility of its DNA sequences for veterinary and biomedical research, the sequence of the dog genome will also facilitate studies on related wild species, including genetic studies involving short tandem repeats (STRs), single nucleotide polymorphism (SNPs), gene sequences or other genomic elements. **Source:** Images courtesy of the NHGRI and The Broad Institute of MIT and Harvard; and of Elsevier.

wild populations. A number of diseases in zoo animals have been proposed for investigation using genomic approaches, including iron overload disorders in lemurs, cardiovascular diseases in bonobos, and susceptibility to pathogens in a broad range of species.[9] Susceptibility to many diseases will involve the major histocompatibility complex (MHC), a gene-dense region of the genome that plays a major role in immune function. The generation of genomic sequences has already enabled the comparative study of MHC structure across mammals,[6] and genomic data may reveal immunogenetic networks involved in host-pathogen interactions in this era of zoonotic diseases.

In wild populations, genomics may enable more extensive studies of the effects of fishing or hunting on heritable traits,[10] including quantitative traits. For example, the sequence of the elephant genome[11] could be used to examine how hunting for the ivory trade has impacted loci involved in determining tusk growth; or used to develop better genetic markers for determining, using DNA from confiscated tusks, the source population of illegally poached ivory. A wild population that may be especially amenable to genomic identification of loci that determine fitness is the Florida panther (Fig. 4), a subspecies of the puma that suffered the deleterious effects of substantial inbreeding following its isolation and decline in South Florida.[12] An increasing number of wild panthers were found to suffer from heart defects and cryptorchidism that resulted from substantial inbreeding, as the population numbered fewer than 50 individuals for many generations. In an attempt to reverse the increase in hereditary diseases due to inbreeding, several Texas pumas were introduced into Florida to increase the genetic diversity of the panthers. With the recent sequencing of a domestic cat genome (Fig. 4),[6] the Florida panther has become a "genome-enabled" taxon.[3] DNA has been sampled from generations of panthers, and pedigrees are well established, so the Florida panther could become one of the first wild species in which alleles affecting fitness can be identified. Likewise, since newly developed sequencing platforms permit genomic sequencing of ancient DNA from museum samples, current allele frequencies could be compared to historical patterns within the subspecies.[12] Thus, new genomic technologies will permit fine-tuning of the genetic restoration efforts for subspecies, guided by genetic patterns found among outbred specimens collected by museums before any recent genetic bottlenecks.

CONCLUSION

One of the most striking technological advances of recent years is the decline in the cost of DNA sequencing (Fig. 5), analogous in speed to the decline in the price of computing power or the rise of the internet.[10] This trend has been accelerated by the development of next generation sequencing platforms (Fig. 5).[10] The decline in sequencing costs will make genomic studies more feasible in wildlife conservation and captive breeding research. The limiting factor will no longer be sequencing costs but the cost of personnel and ability to collect samples.[10] The price decline has led to discussions on possibly sequencing the entire genomes of 10,000 different vertebrate species. Declining sequencing costs may also expand the scope of zoo research into new areas, such as measuring gene expression in individual animals to determine the effects of stress or other factors. It will also be possible to compare large numbers of individuals in captive collections to those in natural or historic populations, to determine whether, to what degree, and at which

Fig. 4 Genome sequencing projects enable the study of introgressive hybridization and genetic fitness in wild species. An intensive study[8] of the degree to which surviving populations of the American bison (A) have experienced introgression of cattle alleles due to hybridization was enabled after the genome of a domestic cow (B) was sequenced. Bison populations that have remained genetically intact will be especially important for conserving the species. (C) The Florida panther is a wild puma subspecies that has suffered the deleterious hereditary effects of substantial inbreeding.[12] It has been intensively studied with well-characterized pedigrees. Sequencing of the domestic cat (D) genome[6] may permit identification of deleterious alleles in the genome of the panther. **Source:** Photos courtesy of Melissa Dowland; Michael MacNeil, USDA; Robert C. Garrison; and Dr. Kristina Narfstrom, University of Missouri–Columbia.

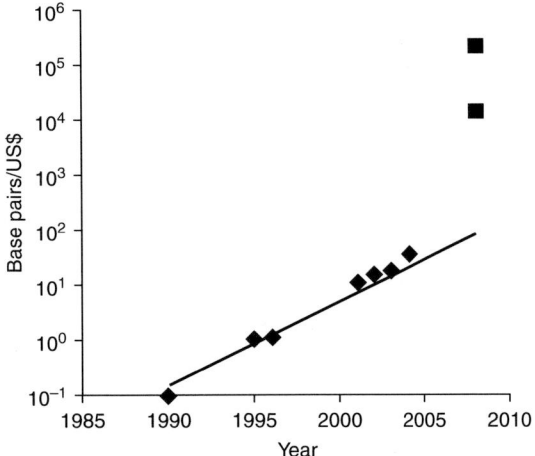

Fig. 5 The number of base pairs that can be sequenced for $1 has consistently doubled every 2 years (diamonds). Moreover, the recent development of next generation sequencing platforms has enabled an even faster increase in efficiency (squares). The plummeting cost suggests that genomic sequencing may soon become possible for thousands of wild and endangered species. **Source:** From Fish Fish.[10]

loci captive breeding may have inadvertently changed allele frequencies. Thus genomic technologies may revolutionize the conservation and captive breeding management of species.

REFERENCES

1. O'Brien, S.J.; Eizirik, E.; Murphy, W.J. Genomics. On choosing mammalian genomes for sequencing. Science **2001**, *292* (5525), 2264–2266.
2. Boffelli, D.; McAuliffe, J.; Ovcharenko, D.; Lewis, K.D.; Ovcharenko, I.; Pachter, L.; Rubin, E.M. Phylogenetic shadowing of primate sequences to find functional regions of the human genome. Science **2003**, *299* (5611), 1391–1394.
3. Kohn, M.H.; Murphy, W.J.; Ostrander, E.A.; Wayne, R.K. Genomics and conservation genetics. Trends Ecol. Evol. **2006**, *21* (11), 629–637.
4. Wayne, R.K.; Ostrander, E.A. Lessons learned from the dog genome. Trends Genet. **2007**, *23* (11), 557–567.
5. Schook, L.B.; Beever, J.E.; Rogers, J.; Humphray, S.; Archibald, A.; Chardon, P.; Milan, D.; Rohrer, G.;

Eversole, K. Swine genome sequencing consortium (SGSC): a strategic roadmap for sequencing the pig genome. Comp. Funct. Genomics **2005**, *6* (4), 251–255.

6. Pontius, J.U.; Mullikin, J.C.; Smith, D.R.; Agencourt Sequencing Team; Lindblad-Toh, K.; Gnerre, S.; Clamp, M.; Chang, J.; Stephens, R.; Neelam, B.; Volfovsky, N.; Schäffer, A.A.; Agarwala, R.; Narfström, K.; Murphy, W.J.; Giger, U.; Roca, A.R.; Antunes, A.; Menotti-Raymond, M.; Yuhki, N.; Pecon-Slattery, J.; Johnson, W.E.; Bourque, G.; Tesler, G.; NISC Comparative Sequencing Program, O'Brien, S.J. Initial sequence and comparative analysis of the cat genome. Genome Res. **2007**, *17* (11), 1675–1689.

7. Moodley, Y.; Baumgarten, I.; Harley, E.H. Horse microsatellites and their amenability to comparative equid genetics. Anim. Genet. **2006**, *37* (3), 258–261.

8. Halbert, N.D.; Ward, T.J.; Schnabel, R.D.; Taylor, J.F.; Derr, J.N. Conservation genomics: disequilibrium mapping of domestic cattle chromosomal segments in North American bison populations. Mol. Ecol. **2005**, *14* (8), 2343–2362.

9. Ryder, O.A. Conservation genomics: applying whole genome studies to species conservation efforts. Cytogenet. Genome Res. **2005**, *108* (1–3), 6–15.

10. Hauser, L.; Seeb, J.E. Advances in molecular technology and their impact on fisheries genetics. Fish Fish. **2008**, *9* (4), 473–486.

11. Roca, A.L.; O'Brien, S.J. Genomic inferences from Afrotheria and the evolution of elephants. Curr. Opin. Genet. Dev. **2005**, *15* (6), 652–659.

12. Culver, M.; Johnson, W.E.; Pecon-Slattery, J.; O'Brien, S.J. Genomic ancestry of the American puma (*Puma concolor*). J. Hered. **2000**, *91* (3), 186–197.

Herbicide-Resistant Crops

Stephen O. Duke
Natural Products Utilization Research Unit (NPURU), U.S. Department of Agriculture/Agricultural Research Service (USDA/ARS), University, Mississippi, U.S.A.

Abstract

The term herbicide-resistant crops (HRC) (sometimes termed herbicide-tolerant crops) has come to mean crops that have been genetically altered by biotechnology to be resistant to herbicides to which they are normally susceptible. Until the advent of plant biotechnology, selective herbicides were designed to kill important weed species while causing limited injury to major crops. Nonselective herbicides that kill almost all plant species were used only at times and places where crop injury was not a concern or with complicated application methods that avoided contact with the crop. There was little success in breeding crops for herbicide resistance especially to nonselective herbicides. Crops with genetics altered by biotechnology to impart herbicide resistance now offer the farmer valuable new tools for weed management. During the past few years, herbicide-resistant canola, cotton, maize, and soybeans have been widely adopted in North America and a few countries outside of North America. This technology has had many critics who have pointed out an array of environmental, toxicological, and societal risks.

CURRENT IMPACT ON WEED MANAGEMENT

The largest segment of the transgenic crop market has been HRCs. Several HRCs are currently available in North America (Table 1). At this time, the most widely utilized HRCs are those that are resistant to two nonselective herbicides, glyphosate (e.g., Roundup®) or glufosinate (e.g., Basta®). Glyphosate-resistant crops in particular have been widely adopted in cotton, soybean, maize, and canola in North America. In 1999, 55 and 37% of the soybean and cotton acreage, respectively, in the United State was planted with glyphosate-resistant varieties. An even larger proportion of the soybean crop in Argentina was glyphosate resistant. The use of glyphosate-resistant maize grew from 950,000 acres in 1998 when it was introduced to 2.3 million acres in 1999. The rapid adoption of glyphosate-resistant crops in the United State (Fig. 1) indicates that farmers find this trait to be very valuable. Other HRCs, such as bromoxynil-resistant cotton (Fig. 1), have been useful in situations with special weed problems. At this time, HRCs are not available to European farmers because of public resistance to their use.

RISKS AND BENEFITS

Generalities regarding risks and benefits of HRCs are difficult to make, as what is true for one HRC can be quite different for another, and even different for the same crop at another place or time. Furthermore, risks and benefits must be considered within the context of current and predicted future farming practices. These products are relatively new, and there are relatively few data to support predicted risks and benefits. Nevertheless, an attempt will be made to point out likely potential benefits and risks of particular HRCs. Many of these risks are being considered by regulatory agencies in their regulation of HRCs.

Benefits and Risks for the Farmer

A major benefit of the HRCs that are resistant to nonselective herbicides (glyphosate and glufosinate) is that the herbicide kills all or almost all weeds. Thus, in these crops, one herbicide can substitute for several selective herbicides that were needed to manage an array of weed species. Furthermore, glufosinate and glyphosate are used as foliar sprays after the weeds have appeared. Theoretically, the farmer can avoid prophylactic herbicide treatments and only rely on the nonselective herbicide after the weed problem appears. Some weed species, however, require relatively high rates of glyphosate for adequate control. In these cases, farmers are finding that the most efficacious weed management with glyphosate-resistant crops sometimes requires use of a selective herbicide with glyphosate.

Perhaps one of the most attractive features of being able to apply nonselective herbicides directly on the crop is that it greatly simplifies weed management, eliminating or reducing the need for tilling, for applying preemergence herbicides, and for decisions as to which selective postemergence herbicides should be used. Management simplicity favors

Table 1 Herbicide-resistant crops available in North America.

Herbicides	Crop	Year	Resistance mechanism
Bromoxynil	Cotton	1995	Enhanced degradation
Sethoxydim[a]	Maize	1996	Altered target site
Glufosinate	Maize	1997	Altered target site
	Canola	1997	Altered target site
Glyphosate	Soybean	1996	Altered target site
	Canola	1997	Altered target site and enhanced degradation
	Cotton	1997	Altered target site
	Maize	1998	Altered target site
Imidazolinones[a]	Maize	1993	Altered target site
	Canola	1997	Altered target site
Sulfonylureas	Soybean	1994	Altered target site
Triazines[a]	Canola	1984	Altered target site

[a]Not transgenic.

the small farmer who cannot afford crop protection consultants.

Despite the fact that U.S. farmers have had to pay for both the herbicide and a technology premium for the HRC seeds, glyphosate-resistant crops have significantly lowered the cost of weed management. In soybeans, these costs have been lower than conventional weed management costs, resulting in the prices of herbicides for use in non-HRC soybeans being substantially lowered. Thus, HRCs have lowered the cost of weed management for all soybean farmers, whether or not they plant a glyphosate-resistant crop.

Many selective herbicides are not entirely selective, causing some phytotoxicity to the crop at certain doses under some conditions. Farmers have learned to accept this because the crop usually outgrows the effect, and there is rarely any significant crop loss. Nevertheless, farmers prefer to have no crop injury from herbicides. HRCs eliminate or greatly reduce crop injury by herbicides at early stages of development. Whether occasional developmental abnormalities in later stages of glyphosate-resistant cotton are due to glyphosate or not has been a contentious issue.

Both glyphosate and glufosinate have activity against some fungi and microbes. There have been reports that, in addition to killing weeds, glufosinate can reduce certain plant pathogen damage to some HRCs. This type of unpredicted benefit has been understudied. There are some potential problems for farmers with HRCs. If a farmer rotates HRC crops (e.g., maize after soybeans) that are resistant to the same herbicide, the unharvested seed of the previous crop can result in a serious weed problem. Evolution of resistance to herbicides is a growing problem, although not to the extent of insecticide or fungicide resistance. Evolution of resistance to glyphosate has not been a significant problem, despite the heavy use of this herbicide over a long period. A bigger problem has been weed species shifts in glyphosate-resistant crops to those species that require higher doses for adequate management (e.g., *Amaranthus rudis* in soybeans). In some crops, the transgene may be introduced into a sexually compatible weedy relative (introgression), creating the need to use additional herbicides. This has not been reported yet, but it will occur eventually if reproductive barriers are not incorporated into certain HRCs.

Glyphosate and glufosinate are commonly sprayed over the tops of the HRCs as a foliar spray. Spray drift to nontarget plants, including other crops, has been a problem since selective herbicides such as 2,4-D were introduced. The potential adverse impact of spray drift is increased when nonselective herbicides are used, in that only the transgenic cultivars of the crop are resistant. The potential of a severe herbicide application error is compounded when the HRC and non-HRC varieties are grown in close proximity.

Adoption of HRCs largely has been driven by short-term economic advantage for the farmer. As mentioned above, the replacement of other herbicides by glyphosate has reduced the value and price of competing herbicides. Furthermore, the price of glyphosate has steadily declined due to the expiration of its patent. Herbicides are the largest segment of the pesticide market. Thus, a major portion of the pesticide market has been significantly devalued, resulting in an escalation of the horizontal integration of the pesticide industry. Fewer companies and the devalued herbicide market will ultimately result in fewer herbicides from which to choose. The impact of this situation on farmers' abilities to cope with new weed problems and on the development of nonchemical weed management alternatives is difficult to predict.

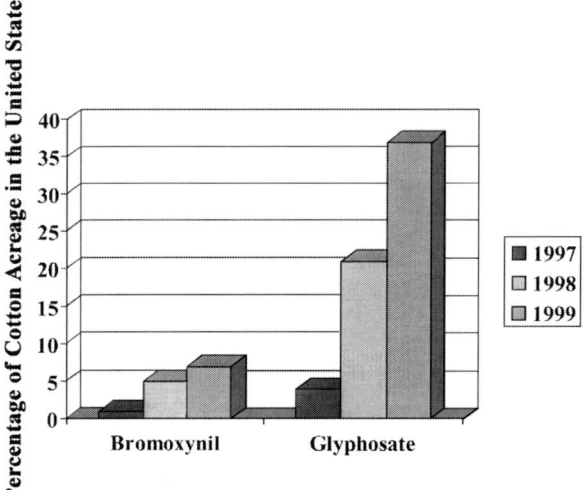

Fig. 1 Adoption of bromoxynil-resistant and glyphosate-resistant cotton in the United States during the last three years of the twentieth century.

Environmental Benefits and Risks

Other than removing land from its natural state, the primary long-term environmental damage of most agriculture has been soil erosion due to tillage. The soil that moves from plowed fields during rainfall events is often contaminated by pesticides, contributing to surface water contamination. The biggest hindrance to adoption of reduced and no-tillage agriculture has been inadequate weed management. The postemergence herbicides to which HRCs have been engineered allow farmers to reduce or, in some cases, eliminate tillage, thereby reducing soil erosion.

The leading HRCs are those resistant to glyphosate and glufosinate. These herbicides are among the most environmentally benign herbicides available. Both are amino acid analogues that degrade rapidly in the environment. Glyphosate is virtually inactivated upon contact with soil, due to its ability to bind soil components strongly. Both are toxicologically safer than most of the products that they replace, despite being relatively high dose rate herbicides.

A transgene that confers herbicide resistance represents a new potential threat to the environment. In some crops such as canola (*Brassica napus* L.), the transgene can introgress into weedy relatives. The herbicide resistance transgene confers no advantage to the weedy relative in a natural ecosystem. It can, however, favor the introgression of other transgenes of clear value in the wild (e.g., Bt toxin) that are packaged with the herbicide resistance gene. In the agricultural setting, all offspring of the weedy relative that are not a result of a cross are killed by the herbicide. Such a process could eventually lead to movement of trangenes of great survival value into natural populations, leading to significant natural ecosystem disruption. Reproductive barriers can be engineered into crops with introgression potential.

The vast majority of cropland in North America is devoted to agronomic crops and weed management in these crops is almost completely dependent on herbicides. Herbicides are expensive, but there are no economical alternatives to herbicides for weed management in these crops on the horizon. Thus, HRCs have simply substituted one herbicide for another. Studies done so far show little or no overall reduction in herbicide use rate (mass per unit area) with HRCs. However, the herbicides used with the most accepted HRCs are generally less environmentally suspect than the herbicides that they replace. For example, in maize and soybeans, glyphosate and glufosinate replace herbicides such as triazines and cloroacetamides that have generated environmental and toxicological concern.

THE FUTURE

Without public opposition, availability of currently available and future HRCs would eventually result in almost universal use of these products in all major crops. Growth of the adoption of HRCs will, however, depend on more than their value to the farmer. In a world economy, the rejection of transgenic crops by the European public could have a profound influence on their utilization in exporting countries such as the United States, even if there is relatively little opposition to their use where they are grown. Public opinion where they are now accepted could change. Whether HRC use increases or decreases is unlikely to significantly influence reliance on herbicides for weed management in major crops. New technologies such as precision agriculture and decision aid programs for weed management will, however, reduce both the volume of herbicides used and their environmental impact.

BIBLIOGRAPHY

Dekker, J.; Duke, S.O. Herbicide-resistant field crops. Adv. Agron. **1995**, *54*, 69–116.

Herbicide-Resistant Crops—Agricultural, Environmental, Economic, Regulatory, and Technical Aspects; Duke, S.O., Ed.; CRC Press: Boca Raton, FL, **1996**.

Dyer, W.E.; Hess, F.D.; Holt, J.S.; Duke, S.O. Potential benefits and risks of herbicide-resistant crops produced by biotechnology. Hort. Rev. **1993**, *15*, 367–408.

Hess, F.D.; Duke, S.O. Genetic Engineering in IPM: A Case Study. Herbicide Tolerance. In *Emerging Technologies for Integrated Pest Management. Concept, Research, and Implementation;* Kennedy, G.G., Sutton, T.B., Eds.; American Phytopathology Society Press: St. Paul, **2000**; 126–140.

High Pressure Processing and Enzymatic Reactions in Food

Alan Kelly
Department of Food and Nutritional Sciences, University College Cork, Cork, Ireland

Thom Huppertz
NIZO Food Research, Ede, The Netherlands

Abstract
The quality of many food products, and changes therein during storage, depends on the activity of indigenous enzymes; in addition, food allergenicity may be correlated with the resistance of certain ingested proteins to proteolytic digestion in the human body. High-pressure processing, a relatively new processing strategy that is gaining acceptance in the food industry, has interesting applications in controlling enzymatic reactions related to food quality. This entry considers applications of this technology for mediating enzyme action in dairy products, as well as fruits, vegetables and seafood.

INTRODUCTION

Although first described as a method of food preservation over a century ago by the remarkably pioneering Bert Hite, the use of high-pressure (HP) processing for food treatment has only become a commercial reality in the last 20 years. Since 2000, the number of food companies using the technology has risen dramatically, and the spread of HP applications has become global.

The basic principle of HP treatment is that a food product, in a flexible package, is immersed in a suitable pressure-transmitting medium within a strong metal vessel, usually cylindrical in shape; a pump, piston, or other mechanical force is then used to subject the medium to intense pressures of the orders of 1000–10,000 times atmospheric pressure (i.e., 100–1000 MPa), which is transmitted uniformly and instantaneously to the food product. The process may be applied at any desired temperature provided the vessel has a suitable temperature-control system, but is most commonly performed at ambient temperatures; the increase in pressure does result in a modest heating effect due to adiabatic heating, and the reverse happens on decompression.

Under pressure, many desirable effects associated with heat treatment of food are also found; microorganisms are inactivated (yeasts and molds being generally sensitive, spores being very resistant), proteins are denatured, and some enzymes become inactivated;[1] however, and perhaps most significantly, many undesirable effects of heat treatment, such as the destruction of nutrients and loss of sensory quality, are avoided, as small molecules with little higher-order structure are unaffected by HP. This ability to induce a set of effects that share many advantages with heat treatment, without some of the disadvantages, has been the primary driver for its industrial adoption. However, its high cost, combined with relatively small vessel scale, means that it has been adopted for the most part only in niche markets where its benefits outweigh and thereby negate these not inconsiderable drawbacks. Products for which HP treatment has been successfully applied include whole-shell oysters, fruits and vegetable products and juices, processed meat products, and guacamole, sauces, and salsas.

HP treatment can inactivate enzymes, and, in fact, this inactivation of enzymes is a key factor in its actual or potential application for certain food products; this will be the subject of this entry. Indeed, as will be seen, the effects of pressure on food enzymes and the reactions they control are complex and, depending on the enzyme, conditions applied, and the substrate (and changes therein under pressure), acceleration, deceleration, or inactivation may occur, with accordingly variable consequences for the food product concerned.

EFFECTS OF PRESSURE ON ENZYMES IN DAIRY PRODUCTS

One of the earliest reports of a possible application for HP treatment for dairy products was a Japanese patent that claimed that holding fresh Cheddar cheese curd at a pressure of 50 MPa for three days resulted in the development of a free amino acid content and flavor equivalent to that of a 6-month-old ripened cheese. As the development of cheese flavor is the result of a very complex set of biochemical pathways catalyzed by a heterogeneous pool of enzymes from the milk, the starter (and non-starter) bacteria, and the coagulant (e.g., rennet), this suggested that enzymatic processes in food could potentially be affected, and perhaps even accelerated, under pressure. Subsequent scientific studies did find increased proteolysis in cheese treated under conditions described by the Japanese researchers, and the effect could be magnified by increasing pressure while significantly reducing treatment time; however, the maximum extent of acceleration achieved remained substantially less than that suggested in the original patent, and of a magnitude unlikely to be of commercial interest.[2,3] A possible mechanism for the acceleration involved changes in the casein matrix induced by pressure rendering the protein substrate more accessible to enzyme action.

Later studies showed that proteolysis in milk due to the indigenous enzyme, plasmin, could be increased relative to that found in raw milk following certain treatments that resulted in disruption of the casein micelle (the complex aggregated structures within which the molecules of casein are found suspended in milk) while not being severe enough to induce significant inactivation of the enzyme; treatment at more severe conditions resulted in net decreases in proteolysis due to increasing enzyme inactivation.[4] Similar studies of enzyme inactivation in cheese by direct measurement of residual enzyme activity characterized the pressure sensitivities of plasmin and chymosin,[5] and raised the potential application of HP treatment for the arresting of cheese ripening by inactivation of ripening enzymes, thereby offering a potential novel means of fixing the quality of cheese at its optimum, which may be of particular interest for fast-ripening (e.g., high moisture) cheese types. This opportunity has indeed been the subject of industrial patents. The future perspective for utilization of HP technology in the dairy industry has been recently reviewed.[6]

EFFECTS OF HIGH PRESSURE ON ENZYMES IN FRUIT AND VEGETABLES

Indigenous enzymes can cause undesirable changes in fruit and vegetable products. Prime examples of such enzymes are polyphenol oxidase (PPO) and pectin methylesterase (PME). PPO causes browning of fruits, particularly during postharvest handling and storage, whereas PME causes reductions in the viscosity and cloud formation in fruit and vegetable juices. As a result, PPO and PME must be inactivated for preservation of premium quality fruit and vegetable products. Thermal treatment is often unsuitable for this purpose as the heat-load required to inactivate PPO or PME causes non-enzymatic browning of the product and a loss of nutrients and sensory properties; however, HP processing has potentially great opportunities for controlling enzymatic deterioration of various fruit and vegetables byproducts. Even though quite high pressures, e.g., 400–600 MPa, may be required to cause significant inactivation of PPO and, particularly, PME, such treatment does not cause any undesirable side effects on color, sensory and nutrients properties of the product. HP-induced inactivation of PPO and PME increases with increasing pressure, holding time, and temperature at which the HP treatment is carried out.[7]

Several applications based on the HP-induced inactivation of enzymes in fruit and vegetables have been commercialized. These include the use of HP processing for guacamole and avocado products to give a shelf life greater than 1 month, as well as premium orange and other fruit juice products, yielding excellent physical stability without compromising on sensory and nutritional properties.

EFFECTS OF HIGH PRESSURE ON ENZYMES IN SEAFOOD

High pressure has been suggested to improve the quality of some seafood products, e.g., cold-smoked salmon. This occurs through the inactivation of some of its enzymes, e.g., cathepsin B, cathepsin L, and calpain. These proteolytic enzymes can cause extensive softening, particularly in cold-smoked salmon and related products. HP processing of cold-smoked salmon, even at moderate pressures, e.g., 300 MPa, results in considerable inactivation of the aforementioned proteases and can thus prevent, or at least reduce, deterioration of texture of the product during storage.[8]

ENHANCING PROTEOLYSIS BY HIGH-PRESSURE PROCESSING

The hydrolysis of proteins is a process that occurs in the digestive tract in the body, but is also widely used industrially to prepare protein products with improved functional properties, improved absorption in the body, and reduced allergenicity. Some proteins show significant resistance to proteolysis, and when these proteins are also major components of food classes and display considerable allergenicity, problems can arise. A prime example is the whey protein, β-lactoglobulin, which is stable to hydrolysis in the stomach. High pressure has been investigated as a means of improving the digestibility of β-lactoglobulin by means of a range of enzymes. Both pretreatment of the protein with high pressure prior to enzymatic digestion and the digestion of β-lactoglobulin by enzymes under pressure have been studied. In general, the digestibility of HP-treated β-lactoglobulin increases with increasing degree of denaturation of the protein, which increases with pressure, treatment time, and temperature.[9–11] This suggests that the altered structure of the HP-denatured protein facilitates access of the enzyme to cleavage sites.

In terms of proteolysis under pressure, the effect of pressure is strongly dependent on the type of enzyme used, as this approach is influenced by the effect of pressure on both enzyme activity and protein confirmation.[9,10] As such, a maximum in the digestion of whey proteins by alcalase occurs at ~200 MPa, whereas digestion by neutrase, corolase, and pepsin increases with pressure up to 300 MPa.[10] In the case of whey proteins, proteolysis under pressure not only leads to improved digestibility, but can also result in reduced allergenicity.[11] Such aspects may be of interest for hypoallergenic infant formula, although the economics of such processes may provide considerable challenges. Moreover, digestion of the milk protein β-casein by plasmin is not affected under pressure at pressures up to 300 MPa, but reduced at higher pressures, probably due to enzyme inactivation.[4,12] Better digestion of soy proteins by pepsin at relatively low pressure, e.g., 100 MPa, has also been observed.

CONCLUSIONS

In conclusion, HP processing can, as briefly described in the above sections, enable the production of premium food products that cannot be achieved by traditional processing technologies. The benefits of HP processing for several commercialized applications are based on the fact that deteriorative enzymes can be inactivated without undesirable side effects on the organoleptic and nutritional properties of the product. Particularly for the area of fruit and vegetable products, several more applications based on such principles are likely to be developed in the near future. Other opportunities may arise in the controlled hydrolysis of potentially allergenic protein sources; however, as with all potential applications, the cost and limited scale associated with the process may prove to be limitations. The economical factors associated herewith will probably result in HP processing remaining a viable opportunity in niche markets but limit its "mainstream" applications.

REFERENCES

1. Hogan, E.; Kelly, A.L.; Sun, D.-W. High pressure processing of foods: an overview. In *Emerging Technologies for Food Processing*; Sun, D.-W., Ed.; Elsevier: London, 2005; 3–32.
2. O'Reilly, C.E.; Kelly, A.L.; Murphy, P.M.; Beresford, T.P. Effect of high pressure of 50 MPa on Cheddar cheese ripening. Innov. Food Sci. Emerg. Technol. **2000**, *1*, 109–117.
3. O'Reilly, C.E.; Kelly, A.L.; Murphy, P.M.; Beresford, T.P. High pressure treatment: applications in cheese manufacture and ripening. Trends Food Sci. Technol. **2001**, *12*, 51–59.
4. Huppertz, T.; Fox, P.F.; Kelly, A.L. Plasmin activity and proteolysis in high pressure-treated bovine milk. Lait **2004**, *84*, 297–304.
5. Huppertz, T.; Fox, P.F.; Kelly, A.L. Susceptibility of plasmin and chymosin in Cheddar cheese to inactivation by high pressure. J. Dairy Res. **2004**, *71*, 496–499.
6. Patel, H.A.; Carroll, T.; Kelly, A.L. Nonthermal preservation technologies for dairy applications. In *Dairy Processing and Quality Assurance*; Chandan, R.C., Kilara, A., Shah, N.P., Eds.; Wiley-Blackwell: Ames, IA, 2008; 465–482.
7. Ahmed, J.; Ramaswamy, H.S. High pressure processing of fruits and vegetables. Stewart Postharvest Review **2006**, *2*, 1–8.
8. Laksmhanan, R.; Pigott, J.R.; Patterson, A. Potential applications of high pressure for improvement in salmon quality. Trends Food Sci. Technol. **2003**, *14*, 354–363.
9. Chobert, J.M.; Briand, L.; Dufour, E.; Dib, R.; Dalgalarrondo, M.; Haertle, T. How to increase β-lactoglobulin susceptibility to peptic hydrolysis. J. Food Biochem. **1997**, *20*, 439–462.
10. Penas, E.; Snel, H.; Floris, R.; Prestamoa, G.; Gomez, R. High pressure can reduce the antigenicity of bovine whey protein hydrolysates. Int. Dairy J. **2006**, *16*, 969–975.
11. Zeece, M.; Huppertz, T.; Kelly, A.L.; Effect of high-pressure treatment on in-vitro digestibility of beta-lactoglobulin. Innov. Food Sci. Emerg. Technol. **2008**, *9*, 62–69.
12. Scollard, P.J.; Beresford, T.P.; Murphy, P.M.; Kelly, A.L. Barostability of milk plasmin activity. Lait **2000**, *80*, 609–619.

Horses: Commercial Oocyte Technologies

Elaine Carnevale
College of Veterinary Medicine and Biomedical Sciences, Colorado State University, Fort Collins, Colorado, U.S.A.

Abstract

Commercial uses of oocyte technologies are rapidly advancing for the horse, primarily driven by the demand to obtain offspring from mares and stallions with poor fertility. Oocytes for commercial use are primarily collected from the preovulatory follicles of estrous donors. For the subfertile mare, oocyte transfer has proven to be a successful method to obtain offspring. Donors' oocytes are transferred into the oviduct of an inseminated recipient; and sperm transport, fertilization, and embryo and fetal development occur in the healthy reproductive tract of the recipient. If a valuable mare dies, her ovaries can be shipped to a laboratory for the collection, maturation, and transfer of oocytes. For subfertile stallions with low sperm numbers or poor sperm quality, a single sperm can be injected into an oocyte in a procedure called intracytoplasmic sperm injection. While in vitro fertilization has not been successful in the horse, the development of other oocyte technologies has provided the equine industry with methods to obtain offspring from mares and stallions that would otherwise be considered infertile.

INTRODUCTION

The refinement of methods to collect and manipulate the equine oocyte has resulted in assisted reproductive technologies for commercial use in the horse. Although in vitro fertilization has been successful for many domestic animals, this procedure has not been successful for the horse.[1] Therefore, other oocyte technologies have been developed and modified for commercial purposes, with the primary use of these procedures being to obtain pregnancies from mares and stallions with poor fertility.

The collection and transfer of a donor's oocyte into an inseminated recipient's oviduct have resulted in clinical procedures to obtain foals from mares with oviductal, uterine, or cervical problems.[2] After the deaths of valuable mares, ovaries have been transported to laboratories for the collection, maturation, and transfer of oocytes, with the resulting births of healthy foals.[3] Development of oocyte technologies has also provided new methods for obtaining offspring from stallions with poor quality sperm or limited sperm numbers. Low numbers of sperm can be transferred into the oviduct with an oocyte in a procedure called gamete intrafallopian transfer; and even fewer sperm are used for intracytoplasmic sperm injection, with a single sperm injected into the oocyte.[1,4] In this entry, oocyte technologies, currently in commercial use for the horse, will be reviewed.

OOCYTE COLLECTIONS

Commercial oocyte technologies are dependant on the ability to successfully collect and handle oocytes. Oocytes are easier to remove from preovulatory follicles rather than immature follicles in the live mare. Therefore, most oocytes are collected from follicles that have been hormonally induced to mature. Oocyte and follicular maturation can be initiated in the estrous mare by administration of human chorionic gonadotropin (hCG) or a GnRH analog (desloreslin acetate). Two primary methods of oocyte collection have been used. Oocytes can be collected through the mare's flank after a trocar is placed through the abdominal muscles. The ovary is positioned per rectum, and a needle is inserted through the trocar and into the follicular antrum. For transvaginal, ultrasound-guided follicular aspirations,[5] an ultrasound transducer housed in a plastic casing is placed in the anterior vagina. The mare's ovary is manipulated per rectum to position the preovulatory follicle over the face of the transducer. A needle can be advanced, through a needle guide in the casing, to puncture the vaginal and follicular walls. The oocyte is then collected using alternate suction and lavage of follicular contents.

OOCYTE TRANSFER

A highly successful method for oocyte transfer was reported in 1995,[6] and similar techniques are still used for commercial purposes.[2,5] For commercial oocyte transfer, a donor's oocyte is transferred into the oviduct of an inseminated recipient mare. The donor's oocyte can be collected just before ovulation of the dominant follicle, at approximately 36 hours after administration of hCG, and rapidly transferred into the recipient's oviduct. However, to help in timing of transfers and to avoid the possibility of ovulation before oocyte collection, the donor's oocyte is often collected approximately 12 hours before the expected time of ovulation and cultured for completion of maturation. Oocytes have been cultured in different media, but one standard medium is Tissue Culture Medium 199 with additions of 10% fetal calf serum, 0.2 mM pyruvate, and 50 μg/mL gentamicin.[6] Oocytes are cultured at a temperature between 38.5°C and in an atmosphere of 6% CO_2 and air.

Cyclic and non-cyclic recipients have been used for oocyte transfer.[5] If a cyclic recipient is used, her oocyte must be removed prior to the transfer of a donor's oocyte. An artificial estrus is induced in non-cyclic recipients by the administration of estrogen. Recipients are inseminated within the uterus with semen from the desired stallion. Recipients are often inseminated before or before and after the transfers. Exposure of the oviduct is obtained during a standing surgery. A small incision is made through the skin in the flank area; muscle layers are bluntly dissected; the peritoneum is punctured; and the ovary is exposed through the incision. In the author's laboratory, oocytes are transferred approximately 3 cm past the infundibular os using a fire-polished glass pipette.

Oocyte transfers were first used for commercial purposes during the late 1990s. The primary use for oocyte transfer was to obtain offspring from mares that were not successful embryo donors because of uterine, oviductal, or cervical problems. Results of the oocyte transfer program at the Colorado State University were recently reviewed.[2] Over a period of 4 years, 86 donors were included in the program with a mean age of 19 years and histories of infertility using standard breeding methods or embryo transfer. Oocytes were collected from 77% (548/710) of aspirated follicles. On average, oocytes were collected 21 hours after hCG administration to donors and cultured for 16 hours before transfer into recipients' oviducts. Pregnancy rates were 40 and 32% at 16 and 50 days after transfer, respectively. One or more recipients were pregnant at 50 days for 71% of donors. Results of the program demonstrate the potential for oocyte transfer as a commercial procedure to obtain offspring from mares that would otherwise be considered infertile.

COLLECTION AND TRANSFER OF OOCYTES FROM DEAD MARES

When a valuable mare dies, her ovaries contain potentially viable oocytes that can be collected, matured, and transferred to produce offspring. Currently, limited facilities are available to collect and transfer the oocytes, requiring ovaries to be transported to a suitable facility for processing. Prolonged transportation intervals require that oocytes be shipped at a reduced temperature to minimize tissue necrosis. Although reduced temperatures will affect oocyte viability, research results suggest that ovaries can be shipped at 12 or 22°C with similar success.[7]

For the harvesting of oocytes from commercial cases, we try to minimize the transportation interval. Upon arrival at our laboratory, ovaries are rinsed in a physiological saline solution at approximately the same temperature during transportation. Follicles on the ovary are identified, sliced, and scraped using a bone curette. Oocytes are identified under a dissecting microscope and placed into a culture medium. In our laboratory, oocytes are cultured in Tissue Culture Medium 199 or a SOF-based medium; media are supplemented with 10% fetal calf serum, 0.2 mM pyruvate, 25 μg/mL gentamicin sulfate, 1 μg/mL LH, 15 ng/mL FSH, 1 μg/mL E_2, 500 ng/mL P_4, 10 ng/mL IGF, and 100 ng/mL EGF. Oocytes are cultured at 38.5°C in 6% CO_2 and air for 24–30 hours. One or multiple oocytes are transferred into a recipient's oviduct as described above.

In 2001, the first commercial offspring was produced after the collection and transfer of oocytes from ovaries that had been transported a long distance. Between 2001 and 2004, oocytes were transferred from the ovaries of 25 mares of various light-horse breeds. These ovaries were transported to our laboratory after the mares had died ($n = 4$) or were euthanized ($n = 21$) for medical reasons.[3] Mares were euthanized by attending veterinarians using their preferred procedures, and the ovaries were collected after euthanasia. The ovaries from six mares were transported in less than 1 hour to our laboratory at approximately 37°C. Nineteen mares died in distant locations, and their ovaries were transported in 8–26 hours, with ovarian temperatures ranging from 10 to 23°C upon arrival at our laboratory.

An average of 11 oocytes were collected per mare, with more oocytes collected from the ovaries of younger mares (4 to 19 years, mean of 12 oocytes) than from older mares ($\geq$20 years, mean of 5 oocytes). Approximately 5 oocytes were transferred into each of the 46 recipients. Embryonic vesicles resulted from 15% of the transferred oocytes as detected using ultrasound at 16 days after transfer. More embryos developed from the oocytes that were transported in less than 1 hour versus 8–26 hours to the laboratory (36% and 10% rates of embryo development, respectively). Although 30% of recipients were diagnosed as pregnant by Day 16, six of the 14 pregnancies were lost by Day 60. One or multiple foals were produced for 24% (6/25) of the donors.

GAMETE INTRAFALLOPIAN TRANSFER

Gamete intrafallopian transfer (GIFT) involves the transfer of oocytes and sperm into the oviduct. The advantage of GIFT is that low numbers of sperm (200,000–500,000) can be used; therefore, it has the potential to be used when numbers of viable sperm are limited. Although experimental use of GIFT was successful when fresh semen was used, transfer of cooled or frozen semen into the oviduct resulted in significantly lower pregnancy rates (83%, 25%, and 8%, respectively).[8] Although limited attempts at GIFT have been made for commercial purposes, the procedure is restricted by the need for fresh semen.

INTRACYTOPLASMIC SPERM INJECTION

Intracytoplasmic sperm injection (ICSI) has provided a method to achieve fertilization in vitro of equine oocytes.[1,4] For ICSI, a single sperm is selected and injected into the ooplasm of a mature (metaphase II) oocyte. Ideally, embryos resulting from ICSI could be allowed to develop in vitro until the late morula or blastocyst stages, when they would be transferred into recipients' uteri. However, development of equine embryos in vitro is not optimal. Therefore, different tactics have been used for the transfer of sperm-injected oocytes. In our laboratory, we transfer injected-oocytes and early-cleavage embryos into the oviducts of recipients to minimize any detrimental effect of embryo culture in vitro. However, in other laboratories, injected oocytes have been transferred into sheep oviducts for early embryo development or successfully cultured in vitro prior to uterine transfers.[4]

Although the first foal was reported from intracytoplasmic sperm injection in 1996,[1] the procedure has only recently been used for commercial purposes. This procedure provides a method to achieve pregnancies when the number of sperms or their viability is limited. Because a limited quantity of frozen semen is available for some stallions, methods to conserve sperm are being developed; these methods include cutting a small section of the straw for thawing or thawing and refreezing sperm at lower concentrations. At this time, ICSI provides the best option for obtaining foals from stallions with poor sperm quality or limited sperm numbers.

CONCLUSIONS

Because in vitro fertilization is not successful in the horse and because of the demands for offspring from valuable mares and stallions with poor fertility, the horse provides a unique animal model for assisted reproductive technologies. During the last 10 years, rapid advances have occurred in our ability to collect and manipulate equine oocytes. Commercial applications of oocyte technologies in the horse have resulted in clinical procedures to obtain offspring from mares and stallions with poor fertility; however, further research is needed to optimize the success and efficiency of oocyte technologies.

REFERENCES

1. Squires, E.L.; Carenvale, E.M.; McCue, P.M.; Bruemmer, J.E. Embryo technologies in the horse. Theriogenology **2003**, *59*, 151–170.
2. Carnevale, E.M.; Coutinho da Silva, M.A.; Panzani, D.; Stokes, J.E.; Squires, E.L. Factors affecting the success of oocyte transfer in a clinical program for subfertile mares. Theriogenology **2005**, *64*, 519–527.
3. Carnevale, E.M.; Coutinho da Silva, M.A.; Preis, K.A.; Stokes, J.E.; Squires, E.L. Establishment of pregnancies from oocytes collected from the ovaries of euthanatized mares. *Proceedings of 50th Annual Convention of the American Association of Equine Practitioners*, Denver, CO, December 8, 2004; 531–533.
4. Hinrichs, K. Update on equine ICSI and cloning. Theriogenology **2005**, *64*, 535–541.
5. Carnevale, E.M. Oocyte transfer. In *Current Therapy in Equine Medicine*, 5th Ed.; Saunders: Philadelphia, 2003; 285–287.
6. Carnevale, E.M.; Ginther, O.J. Defective oocytes as a cause of subfertility in old mares. Biology of Reproduction **1995**, *1*, 209–214.
7. Preis, K.A.; Carnevale, E.M.; Coutinho da Silva, M.A.; Caracciolo di Brienza, V.; Gomes, G.M.; Maclellan, L.J.; Squires, E.L. In vitro maturation and transfer of equine oocytes after transport of ovaries at 12 or 22°C. Theriogenology 2004, *61*, 1215–1223.
8. Coutinho da Silva, M.A.; Carnevale, E.M.; Maclellan, L.J.; Preis, K.A.; Seidel, G.E., Jr.; Squires, E.L. Oocyte ransfer in mares with intrauterine or intraoviductal insemination using fresh, cooled, and frozen stallion semen. Theriogenology **2004**, *61*, 705–713.

Insects and Mites: Biological Control

Ann E. Hajek
Department of Entomology, Cornell University, Ithaca, New York, U.S.A.

Abstract

Biological control is defined as the use of natural enemies to suppress a pest population, making the pests and their associated damage less abundant. Natural enemies were first used to control insect pests when farmers in ancient China and Yemen moved colonies of predaceous ants to control pests of tree crops. Today, the natural enemies used to control insect and mite pests include a diversity of predators, parasitoids, and pathogens. Specific strategies have been developed for release of natural enemies or enhancement of their persistence and activity. Biological control has been used very successfully for permanent suppression of introduced pests. Among natural enemies applied for shorter-term control, in 1990 even the most widely used biological control agent, *Bacillus thuringiensis*, accounted for < 1% of the insecticide market. However, biological control agents are widely used for control in environmentally sensitive areas or controlled environments and constitute important components of integrated pest management programs.

STRATEGIES FOR USING BIOLOGICAL CONTROL

Natural enemies can be used in a variety of very different ways. The first major uses of natural enemies for pest control were directed at control of introduced insect pests. Natural enemies from the land of origin of introduced pests were released in areas of pest introduction. This strategy, called classical biological control, now also includes introduction of exotic natural enemies to control native pests. In all cases, a high degree of host specificity is required in the natural enemies to be introduced. After the exotic natural enemy is established in the new location, its effectiveness is based on population increases in response to increasing densities of pest populations. Classical biological control can be dramatically effective, with 34% of insect natural enemies that are released becoming established and 17% completely controlling devastating pests. Classical biological control is known to be extremely cost effective with cost benefit estimates of up to 200:1, if a program is successful at establishing an effective natural enemy.

A second strategy, augmentation, involves releasing natural enemies for pest control, usually in instances where natural enemies can be effective but are not sustained in the environment at high enough densities to provide control. Inundative augmentation is used when only the natural enemies that are released in high numbers are expected to exert control. Under inoculative augmentation, control effects are more delayed and are predominantly exerted by the progeny of the released organisms. Natural enemies used for augmentation are often mass reared, so understanding requirements for mass production of high quality natural enemies that are healthy and active after shipping and release is critical for the use of this strategy.

The third major strategy, conservation, involves manipulations to enhance the persistence and activity of natural enemies already occurring in the environment. This strategy takes on a diversity of forms based on requirements of the individual natural enemies. To cause less mortality of natural enemies, use of synthetic chemical pesticides that kill natural enemies can be altered in different ways ranging from eliminating their use to selecting pesticides with less impact on natural enemies to timing pesticide applications to minimize the effect on natural enemies. Alternatively, natural enemy populations can be increased by maintaining or improving the environment to provide ideal conditions. For example, irrigating, strip-harvesting, intercropping, retaining vegetation adjacent to crops, and planting cover crops all have been shown to provide favorable habitats and food to maintain or increase populations of natural enemies. In a program to control the brown planthopper on rice in south and southeast Asia, the activity of a suite of native natural enemies, aided by host plant resistance and application of insecticides only when absolutely necessary, provided better control than pesticides alone.

TYPES OF NATURAL ENEMIES

Predators

Predators are generally larger than their prey and each usually consumes several prey individuals either for growth of immatures or for subsistence and reproduction of adults.

Fig. 1 An adult of the multicolored Asian lady beetle (*Harmonia axyridis*), which was introduced from Asia for control of aphids and scales (length ca. 1 cm). Both larvae and adults are predatory. (Photo by J. Ogrodnick.)

Fig. 2 An adult of the parasitoid *Muscidifurax raptor* (length ca. 2 mm) parasitizing house flies. Females lay eggs in fly puparia, larvae grow while consuming the fly pupae and winged adults then emerge to mate and find more hosts. These flies can be purchased and released to augment naturally occurring populations. (Photo by S. Long.)

The predatory life style is very common among insects and mites but predators with the most importance for biological control belong to four insect orders (Hemiptera, Coleoptera, Diptera, and Hymenoptera) and eight mite families (Fig. 1). Predators feed on a diversity of prey life stages, from eggs to adults, and display a range of host specificity, from feeding only on one prey species to generalized feeding on many prey species. One of the most famous examples of classical biological control is the introduction of the highly host specific Vedalia beetle that was imported from Australia and released against outbreak populations of the introduced cottony cushion scale threatening the southern California citrus industry. After the 1888–1889 releases of this predator, cottony cushion scale populations decreased dramatically and, by 1890, scale populations had been decimated. The immense success of this early program was instrumental in building interest in use of natural enemies for biological control. As a second example, in more recent years, phytoseiid mites attacking tetranychid spider mites have been developed for mass release in greenhouses or on some outdoor crops. Pesticide-resistant mite strains have also been developed for use against spider mites attacking tree crops.

Parasitoids

Parasitoids develop at the expense of a single host and usually kill their hosts. Parasitoids have been used extensively for biological control because, in contrast to predators, the impressive degree of host specificity often characteristic of parasitoids leads to sensitive responses to changes in host density. Parasitoids used for biological control are predominantly in the Order Hymenoptera (Fig. 2) with less common use of Diptera. The immature parasitoid is usually a featureless larva associated with the host while winged adults disperse to mate and find new hosts. To enable their close association with hosts, parasitoids have adopted amazing and diverse life cycles. Different species of parasitoids attack different life stages of hosts (egg through adult) and can develop either externally on hosts or internally within hosts. One to many parasitoid individuals of one or more species can develop within a host. Parasitoids have been more widely used for classical biological control than either predators or pathogens. In recent years, the tiny wasp *Epidinicarsis lopezi* was released by land and air in 34 countries in Africa to control the introduced cassava mealybug. Due to the activity of this wasp, cassava mealybug is no longer considered a problem, saving African farmers hundreds of millions of dollars in reduced crop losses. Some parasitoids widely used for augmentative biological control are tiny species of *Trichogramma* attacking eggs of Lepidoptera and *Encarsia formosa*, a member of the Aphelinidae that attacks whiteflies. Both of these tiny wasps are mass produced in insectaries and shipped to users for release against pest populations threatening crops.

Pathogens

Microorganisms that are parasitic, referred to as pathogens, are masters at exploiting insect and mite hosts (Fig. 3). Pathogens important for biological control include a diversity of viruses, bacteria, fungi, protozoa, and nematodes. This range of types of pathogens exhibits a comparable medley of diverse interactions with their hosts. Of primary importance, viruses, bacteria, and most protozoa must be ingested by hosts in order to infect, while fungi and some

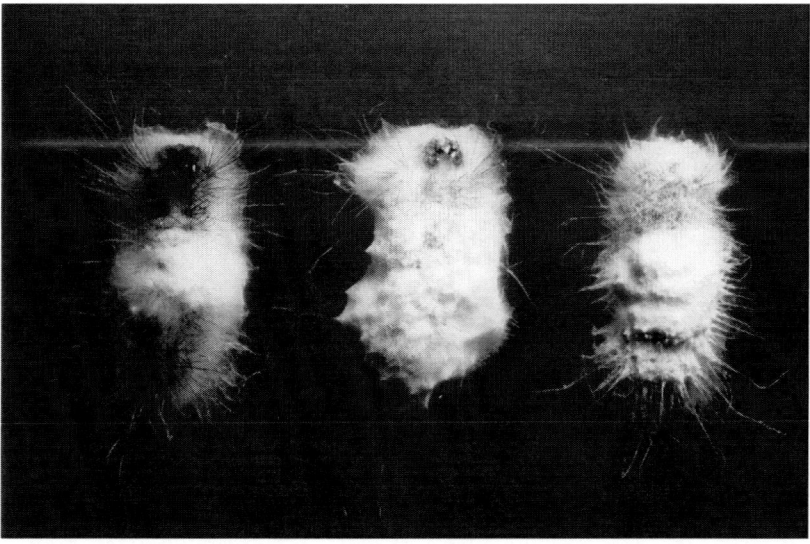

Fig. 3 Gypsy moth (*Lymantria dispar*) larvae killed by the entomopathogenic fungus *Paecilomyces farinosus* (each larva ca. 3 cm in length). When infecting, this fungus penetrates externally through the larval cuticle, then increases within the host and, after host death, grows out through the integument to produce spores that will infect healthy hosts. (Photo by T. Ebaugh.)

protozoa can penetrate directly through the host cuticle. The nematodes of greatest importance to biological control, *Steinernema* and *Heterorhabditis*, can enter hosts through body openings although some possess the ability to penetrate directly through the cuticle. While some pathogens have mechanisms for active dispersal to find new hosts, host finding is generally not directed and these pathogens rely principally on their production of huge numbers of progeny in order to be assured of locating healthy hosts. Associations between pathogens and hosts range from facultative to obligate but pathogens important for biological control are all specialized for infecting only insects and mites. Pathogens have been used for classical biological control relatively infrequently although some programs have provided complete control. Much of the development of pathogens has been directed toward inundative augmentation of mass-produced microbes. The bacterium *Bacillus thuringiensis* is applied more than any other biological control agent. Strains of this bacterium predominantly kill Lepidoptera, Diptera, and Coleoptera through the activity of a toxin destroying the integrity of the gut. For many years, this bacterium was applied principally as a spray but recently several crop plants have been engineered to express genes encoding the toxin.

BIOLOGICAL CONTROL IN PRACTICE

There is great demand for use of biological control programs to eliminate insect and mite pests, especially in environmentally sensitive areas and areas where humans live. Use of natural enemies to control pests can be highly effective due to the diversity of types of natural enemies and approaches. However, because biological control involves management of living organisms, it can be somewhat unpredictable. Therefore, biological control programs generally are tailored to specific pest systems and to optimize control, often require knowledge of the biology and ecology of the insect or mite host. Biological control has proven to be most effective under certain conditions (Table 1), although these generalities should not prevent investigations of use of biological control for alternative situations.

Introduced pests are not insignificant, comprising 39% of the 600 major arthropods pests in the United States. By 1990, more than 4300 introductions of exotic parasitoids and predators had been made to control insect pests, many of which were introduced. Due to its low cost and permanent effectiveness when successful, classical biological control continues as one of the first control strategies to be investigated after a new pest has been introduced.

Augmentation and conservation are now often employed as important parts of integrated pest management programs. Although since the late 1940s (the start of the DDT era), for most pests synthetic chemical pesticides have been the first control strategy considered, use of natural enemies for

Table 1 Characteristics of systems and conditions more commonly associated with successful biological control.

Highly efficient natural enemy
Less mobile pest living in an exposed location
Perennial crop, natural habitat, or controlled environment, for example, a greenhouse
Crop or environment where some pest damage is tolerated
Controls for other pests do not interfere with the activity of natural enemies

control is increasing, especially for specific applications and systems.

BIBLIOGRAPHY

Barbosa, P., Ed.; *Conservation Biological Control*; Academic Press: San Diego, **1998**; 396.

Bellows, T.S., Fisher, T.W., Caltagirone, L.E., Dahlsten, D.L., Gordh, G., Huffaker, C.B., Eds.; *Handbook of Biological Control: Principles and Applications of Biological Control*; Academic Press: San Diego, **1999**; 1046.

Evans, H.F., Ed.; *Microbial Insecticides: Novelty or Necessity?*; British Crop Protection Council: Surrey, UK, **1997**; 301.

Flint, M.L.; Dreistadt, S.H. *Natural Enemies Handbook: An Illustrated Guide to Biological Pest Control*; University of California Press: Berkeley, CA, **1998**; 154.

Greathead, D., Waage, J., Eds.; *Insect Parasitoids*; Academic Press: London, **1986**; 389.

Lacey, L.A.; Goettel, M.S. Current developments in microbial control of insect pests and prospects for the early 21st century. Entomophaga **1995**, *40* (1), 3–27.

Lynch, J.M., Hokkanen, H.M.T., Eds.; *Biological Control: Benefits and Risks*, Cambridge University Press: Cambridge, UK, **1995**; 304.

New, T.R. *Insects As Predators*; New South Wales University Press: Kensington, Australia, **1991**; 178.

Van Driesche, R.G.; Bellows, T.S., Jr. *Biological Control*; Chapman & Hall: New York, **1996**; 539.

Intellectual Property and Plant Science

Susanne Somersalo
John Dodds
Dodds & Associates, Washington, District of Columbia, U.S.A.

Abstract
In this entry, we will briefly go through various means for protection relevant to plant sciences. Thereafter we will deal with the current issues of intellectual property and modern plant sciences and provide an overview on development of intellectual property legislation and international treaties affecting it.

INTRODUCTION

Intellectual property refers to creations of the human mind and it may be inventions, artistic works, designs, names, images, and so on. Generally speaking the prime form of intellectual property to protect technical innovations is a patent, and the most well-known means to protect plant varieties is plant breeders' rights. Recently, other forms of intellectual property such as, copyrights, trademarks, and trade secrets have become important not only in other fields of life sciences but also in plant breeding.

Even if the intellectual property rights are by large national rights, there are several international treaties, the most important being the Trade Related Aspects of Intellectual Property Rights (TRIPS) agreement, regulating the contents of the national intellectual property laws. In addition to international treaties, there are intergovernmental organizations such as World Intellectual Property Organization (WIPO) and International Union for the Protection of New Varieties of Plants (UPOV) administering certain intellectual property practices.

A more detailed overview of the current issues of intellectual property rights in agricultural biotechnology is given in a recent review.[1] An excellent overview of issues related to plant breeders' rights is available in a recent case study.[2]

MEANS TO PROTECT INTELLECTUAL PROPERTY

The basic means to protect intellectual property are patents, copyrights, trademarks, trade secrets, and plant breeder's rights. There are variations in the national laws, and some countries provide broader protection than others. For example, the concept of plant patent is specific for the protection system of the United States. A basic understanding of these mechanisms is essential for anyone whose research may lead to an invention.

Patents

Historically a patent was a grant made by a sovereign that would allow for the monopoly of a particular industry, service, or goods. Over time the concept has been refined from a public policy perspective, and it has evolved to an agreement between the government and the inventor/creator.

In return for the right to exclude others from the practice of the invention the government requests the inventor to fully disclose the enablement of the invention. Furthermore, the monopoly is now limited by time and clearly is only applicable in the territory under the jurisdiction of the government.

In exchange for a limited-term right (usually 20 years) to exclude others from making, using, or selling the invention, the inventor must provide a complete and accurate public description of the invention and the best mode of "practicing" it. This provides others with the ability to use that information to invent further, thus promoting technology development for the benefit of society.

Markedly, one can have an issued patent and still have no right to practice the invention due to lack of approval of some government instance. An example related to plant breeding is an inventor having an issued patent for use of transgenic plant in a country where the government has not approved cultivation of transgenic plants.

Copyrights

A copyright is a type of intellectual property protection for "authors" of original works. Basically, a copyright protects an original work and allows the author an exclusive right to reproduce the work, prepare derivatives, distribute copies of the work, and perform the copyrighted work publicly.

Historically copyrights have been important in protecting the rights for artists and authors. Today, copyrights are becoming more and more important in protecting the rights of database developers. In relation to plant science, copyrights may be relevant means of protecting databases, such as GIS databases or databases containing gene sequences.

Trademarks

Trademark is a word, phrase, symbol, design, or a combination of those that distinguishes the source of one's goods or services from those of others. A trademark can be valid only when it is used in or on connection with the goods or services in commerce. Trademarks are important means for distinguishing a product or a technology. As in other industries, trademarks are becoming increasingly important for seed industry to brand its products. A remarkable advantage of a trademark is that it is valid as long as it is used in commerce, i.e., as long as the product is on market.

Geographical Indications

Geographical indications are defined in the TRIPS agreement as a type of intellectual property. A geographical indication is a sign used on goods that have a specific geographical origin and possess qualities or a reputation that are due to that place of origin.

Most commonly, a geographical indication consists of the name of the place of origin of the goods. Agricultural products typically have qualities that derive from their place of production and are influenced by specific local factors, such as climate and soil. An example is IDAHO for potatoes grown in the state of Idaho.

Markedly, geographical indications—just as trademarks—protect neither the information embodied in the goods, nor any method of producing the goods. Geographical indications serve as an assurance of source or quality, and they are important in similar sense as trademarks.

Trade Secrets

Trade secret may be a cheap way to protect one's intellectual property: having a trade secret simply requires that the intellectual property is kept secret. An example of a trade secret could be composition of a culture medium or a method to transform a plant species.

The positive aspect in trade secrets in addition to its potential for a lower-cost approach is that there is no expiration date. However, the negative side is that once the secret is disclosed, the protection is gone if the legal process cannot contain it through an effective litigation strategy.

Plant Breeders' Rights

Trade Related Aspects of Intellectual Property Rights agreement obliges the member countries to provide protection for plant varieties either through patents or by an effective *sui generis* system or a combination of them. The most common *sui generis* system is Plant Breeders' Rights (PBR) under UPOV convention, called plant variety protection (PVP). International Union for the Protection of New Varieties of Plants is a separate intergovernmental organization established in 1961. Presently, UPOV has 68 member countries including all the important agricultural producer countries.

International Union for the Protection of New Varieties of Plants provisions allow the developers of new plant varieties to control multiplication and sale of the reproductive material of a new variety but leave certain exemptions for further breeding and use in non-commercial purposes.

Plant Patent

Plant patent is a specific type of a patent granted in the United States to an inventor who has invented or discovered and asexually reproduced a distinct and new variety of plant species. The provision excludes tuber-propagated plants or plants found in an uncultivated state.

CURRENT ISSUES

Patenting Life

Intellectual property system is a dynamic system that needs to be reevaluated in the course of developing technologies and trade practices. Originally patents were typically granted for various kinds of mechanical and chemical inventions. The emergence of modern biotechnology more than 20 years ago prompted the U.S. Supreme Court to reevaluate the concept of patentability. In 1980, the Court issued its famous decision[3] stating that a genetically modified bacteria qualifies as patentable subject matter and "everything under the sun made by man is patentable." Five years later the Board of Appeals and Interferences of the U.S. Patent and Trademark Office held that genetically modified plants are patentable.[4]

However, as recently as 2001 the U.S. Supreme Court still discussed whether utility patents may be issued for plants.[5] The ruling held that newly developed plant breeds fall within utility patent protection. A sexually reproduced plant variety may receive a double protection by PVP and by utility patent.

Today the international treaties set forth the frames for minimum protection of intellectual property, but no intellectual property treaty regulates how far a member country may extend the protection. Accordingly, there are variations among the countries as to what extend living organisms can be protected. The rulings of the U.S. courts, even if having effect only in the jurisdiction of the Untied States, have been important because they have set a new tone for discussion of patentability of life forms everywhere in the world.

Access to Germplasm

Plant and animal breeding is different from any other field of technology in the sense that it has generally been impossible to make progress in terms of inventions without having access to "prior art," that is, access to germplasm.

Despite of this essential difference between art of breeding and other technology fields, inventions related to plant breeding may still be protected by various means of intellectual property rights in similar way as inventions in mechanics, where one can make an important invention without having any prior art available. This is an issue that is raised time after time, because of concern that intellectual property rights might prevent access to germplasm and thereby affect the capacity to breed, research, and provide better varieties for food and feed.

Today various international treaties have provisions that are aimed to ease access to germplasm. Most importantly, Food and Agricultural Organization (FAO) has adopted the International Treaty of Plant Genetic Resources (ITPGR) in 2001. Among other provisions, ITPGR provides ex situ collections of most of the important food and feed plants. Based on this treaty Consultative Group on International Agricultural Research (CGIAR) holds over 600,000 accessions of crop, forage, and agro forestry genetic recourses. International Treaty of Plant Genetic Resource requires a standardized material transfer agreement (MTA) to guarantee that no intellectual property right shall be claimed for material received from the system.

Wild germplasm, on the other hand, is an important part of the art of plant breeding, and wild germplasm might not be as well represented in gene banks as cultivation based germplasm.[6] This argument would inevitably lead to a very broad issue of compatibility of existing plant intellectual property system with the rights of indigenous people's traditional knowledge.

Traditional Knowledge

Traditional knowledge refers to the knowledge, innovations, and practices of indigenous and local communities around the world. Traditional knowledge has been transmitted orally from generations to generations, and it tends to be collectively owned. Among many other things, agricultural practices, including the development of plant species and animal breeds, may be traditional knowledge. This kind of traditional knowledge is based on genetic recourses. Today the relationship between genetic resources, traditional knowledge, and intellectual property rights is one of the most important issues on the agenda of several international treaties.

Convention of biological diversity constitutes the central instrument concerning biodiversity at the international level. Convention of biological diversity affirms the sovereign rights of states to exploit their own resources pursuant to their own environmental policies.

The convention also provides a broad framework for member states' policies concerning access, development, and transfer of technologies and acknowledges the necessity for all parties to recognize and protect intellectual property rights in this field. It points to the need for equitable sharing of benefits arising from the use of traditional knowledge, innovations, and practices relevant to the conservation of biodiversity and the sustainable use of its components.

The issues related to traditional knowledge are also closely related to IP protection in developing world. People in many developing countries are asset poor but knowledge rich, and poor communities seldom gain from that knowledge. As it might be difficult for indigenous people to invest in conventional IP protection, let alone litigations to protect their IP, new IP protection and contract models may be needed to protect such knowledge and to bring the benefit to the poorest. Fingers and Schuler[7] discuss the issues related to intellectual protection and indigenous knowledge.

CONCLUSIONS

Even if the concept and means for intellectual property are old, the issues emerging with the rapid development of life sciences together with ethical concerns have made it necessary to reevaluate the concept. Presently, there are several intergovernmental organizations trying to create harmonized guidelines for protection of intellectual property in industrialized as well as in developing countries. Especially interesting issues are those related to the protection of traditional knowledge.

REFERENCES

1. Erbish, F.H.; Maredia, K.M., Eds. *Intellectual Property Rights in Agricultural Biotechnology,* 2nd Ed.; Michigan State University: East Lansing, MI, **2003**.
2. Louwaars, N.P.; Tripp, R.; Eaton, D.; Eaton, D.J.F.; Hu, R.; Pal, K.; Tripp, R.; Henson-Apollonio, V.; Mendoza, M.; Muhhuku, F.; Wekundah, J. *Impacts of Strengthened Intellectual Property Rights Regimes on the Plant Breeding Industry in Developing Countries: A Synthesis of Five Case Studies*, Report Commissioned by the World Bank; Wageningen University and Research Center: The Netherlands, **Feb 2005**; 176.
3. *Diamond, v. Chakrabarty*, 447 U.S. 303, 65 (**1980**).
4. *In Re Hibberd,* 227 U.S.P.Q 443, 444 (BNA, **1985**).
5. *J.E.M. Ag. Supply, Inc. v. Pioneer Hi-Bred Int'l, Inc.,* 534 U.S. 124 (**2001**)
6. Gepts, P. **2004**. Who owns biodiversity, and how should the owners be compensated? Plant Phys. *134*, 1295–1307.
7. Fingers, J.M.; Schuler, P. *Poor People's Knowledge: Protecting Intellectual Property in Developing Countries*; Oxford University Press, New York, **2004**; 249.

Interspecies Embryo Transfer

Duane Kraemer
Department of Veterinary Physiology and Pharmacology, Texas A&M University, College Station, Texas, U.S.A.

Abstract

Interspecies embryo transfer in animals has been used for agricultural, conservation, and basic scientific purposes. At least 24 different combinations of taxa have been involved in intertaxon embryo transfer. Although relatively little is known about the mechanisms that control compatibility between species, the ability to produce fertile hybrids is probably the most useful predictor of successful intertaxon embryo transfer. Renewed interest in the use of interspecies embryo transfer has been prompted by the advent of nucleus transfer cloning, in which interspecies recipient ova and embryo recipients are often desired. Chimerism and inner-cell-mass transfer are being explored as ways to broaden the range of compatibility between various taxa.

INTRODUCTION

The movement of preimplantation embryos between animals of different taxonomic classes is usually termed interspecies embryo transfer. The objectives of this entry are to define the process, state how it may be useful, and list the taxa in which the process has been successful. Criteria are presented that may be helpful in predicting donor and recipient combinations in which interspecies embryo transfers may be possible, as well as approaches, which may be useful for expanding the range of compatibility between different taxa.

TEXT

Embryo transfer is the movement of preimplantation embryos from the reproductive tract of the genetic mother (donor) to the reproductive tract of the host (recipient) mother. If donor and recipient are of the same species, it would be referred to as intraspecies (intraspecific) embryo transfer. In special situations, such as in most human embryo transfers, the donor and recipient may be the same individual. If the donor and recipient are each from a different genus, the process would be designated as intergeneric embryo transfer. A general term that is sometimes used for this topic is intertaxon embryo transfer. The concept of intertaxon embryo transfer is not new. It is interesting that the first successful intraspecies embryo transfers in livestock were controls for an experiment on intergeneric transfer of embryos between domestic fine wool sheep and Angora goats.[1]

The most common intertaxon embryo transfer in agricultural food production is between *Bos taurus taurus* and *Bos taurus indicus* cattle. This type of transfer is so widely accepted that data are not available on the relative efficiency of transfers in either direction between these two subspecies. The motivations for such transfers are usually to introduce the reciprocal subspecies into an area where the numbers of one or the other of these subspecies is in short supply and is needed for meat or milk production. Often the offspring of these transfers are used for the production of hybrids. In these situations the hybrids are often sufficiently valuable to warrant the use of embryo transfer for their propagation, and either subspecies or even the hybrids can serve as the embryo recipients. Availability and economics are usually more important than biology in determining which of these subspecies to use as recipients, although adaptability to the prevailing climate should also be considered.

Other bovine interspecies embryo transfers have been performed (see Table 1) but the main motivation is conservation rather than meat and milk production. However, there has been some interest in using the Banteng in crossbreeding experiments for meat production, and Gaur[2] hybridization and embryo transfer have been used in gene mapping research. There is interest in embryo transfer of bison, water buffalo, and Eland embryos to domestic cattle recipients although to date no live offspring have been produced in these experiments. The Eland has served as recipient for Bongo embryos demonstrating the usefulness of interspecies embryo transfer for preservation of the endangered Bongo. Numerous different equine[3] and feline[4] intertaxon embryo transfers have been performed (see Table 1). Although the motivation for this work has been for basic research and conservation, it should be recognized that they are food-producing animals in some cultures, so there might be some agricultural use for such transfers in these species.

Table 1 Interspecific and intergeneric embryo transfers that have resulted in live offspring.

Donor		Recipient	
Genus, species, and subspecies	Common name	Genus and species	Common name
Ovis orientalis musimon	European mouflon	*Ovis aries*	Domestic sheep
Ovis orientalis gmelini	Armenian red sheep	*O. aries*	Domestic sheep
Ovis musimon	Domestic mouflon	*O. aries*	Domestic sheep
O. aries	Domestic sheep	*Capra hircus*	Domestic goat
C. hircus	Domestic goat	*O. aries*	Domestic sheep
Bos taurus taurus	Domestic cattle of European origin	*Bos taurus indicus*	Domestic cattle of Indian origin
Bos taurus indicus	Domestic cattle of Indian origin	*Bos taurus taurus*	Domestic cattle of European origin
Bos gaurus gaurus	Gaur	*Bos taurus taurus*	Domestic cattle of European origin
Bos javanicus	Banteng	*Bos taurus taurus*	Domestic cattle of European origin
Mus caroli/Mus musculus	Wild mouse	*M. musculus*	Laboratory mouse
Tragelaphus euryceros	Bongo	*Tragelaphus oryx*	Eland
Capra pyrenaica	Spanish ibex	*C. hircus*	Domestic goat
C. pyrenaica	Spanish ibex	*C. pyrenaica* × *C. hircus*	Ibex-goat hybrid
Mustela lutreola	European mink	*Mustela putorius hybrid*	European polecat
Felis sylvestris lybica	African wild cats	*Felis catis*	Domestic cats
Felis silvestrus ornate	Indian desert cat	*F. catis*	Domestic cat
Equus zebra	Grants zebra	*Equus caballus*	Domestic horse
Equus przewalski	Przewalski's horse	*E. caballus*	Domestic horse
Equus assinus	Donkey	*E. caballus*	Domestic horse
E. caballus	Domestic horse	*E. assinus*	Donkey
E. caballus	Domestic horse	*E. calabus* × *E. assinus*	Mule
E. assinus	Donkey	*E. calabus* × *E. assinus*	Mule
Lama pacos	Alpaca	*Lama glama*	Llama
Macaca mulatta	Rhesus monkey	*Macaca nemestrina*	Pigtail macque

Numerous other intertaxon embryo transfers have been performed for conservation or basic research, and they are presented here to illustrate the variety of relationships that can support intertaxon embryo transfer.

Unfortunately, there are still insufficient data to permit the development of a set of criteria that would permit prediction of which taxon relationship would support successful embryo transfer. Clearly, identical chromosome number is not required. The ability to hybridize is a helpful criterion, especially if the hybrids are fertile, but does not assure success in unassisted intertaxon embryo transfer. Although more data are needed to be certain, it appears that even though American bison and domestic cattle, and Desert Bighorn and Armenian Red sheep[5] can hybridize, they do not support each other's embryo transfer pregnancies to term. It appears that placental structure and physiology are the limiting factors, but the specifics of these criteria are not defined as suggested by sheep to goat and goat to sheep embryo transfers.[3] For some of the combinations in which no pregnancies are diagnosed, the signal between the conceptus and the ovary may be the limiting factor (Fig. 1).

Several approaches have been developed to assist intertaxon embryo transfers. Several of these approaches are explained by Anderson.[3] They include either active or passive immunization of recipient mares with donkey lymphocytes, which increased the normal development of donkey to horse pregnancies. Chimera formation has resulted in successful goat to sheep[6] pregnancies and in mouse interspecies embryo transfers.[7] Although most of the offspring are chimeras, a few of the offspring appeared to originate from one of the embryos. A specialized type of chimeric embryo in which the trophectoderm is that of the recipient species has been shown to support gestation to term in mouse interspecies transfers[6] and in goat to sheep

Fig. 1 An Armenian Red sheep (hair sheep) offspring produced by in vitro fertilization and interspecies transfer to the fine wool domestic sheep recipient.

transfers, although the efficiency is low, and this approach was not effective for supporting gerbil to mouse intergeneric embryo transfers. Much more work is needed to determine the biological limits of this procedure, which is usually referred to as "inner-cell-mass" transfer. Rorie et al. have published a relatively simple protocol for this procedure in sheep and goats.[8] The fact that the usually infertile mare mule is capable of carrying both horse and donkey pregnancies to term has prompted the testing of other hybrids for use as intertaxon embryo transfer recipients. Clearly, Bos taurus taurus and Bos taurus indicus cattle produce fertile hybrids, which can support each subspecies' pregnancies to term and can also support twin pregnancies, one of each species. Hybrids between Spanish ibex (Capra pyrenaica hispanica) and domestic goats (Capra hircus) proved to be much more effective recipients of Spanish ibex embryos than the pure domestic goat. Previously, the only term pregnancies from this interspecific transfer were when the goat recipient carried her own fetus, plus the transferred Spanish ibex fetus.[9]

Embryos for intertaxon embryo transfer can be obtained using embryos produced by a variety of procedures. The highest conception rates are usually obtained from in vivo-produced embryos, by either natural service or artificial insemination. They may also be produced by in vitro oocyte maturation and in vitro fertilization[10] or by intracytoplasmic sperm injection. Recently, intertaxon embryo transfer has been used in conjunction with somatic cell nucleus transfer for the production of offspring from endangered species. This might also prove to be useful in food and fiber producing animals when the desired species is in short supply, as is sometimes the case in international shipment of germplasm. Short-term (1–5 days) embryo transfers to host species recipients have been performed for culture of micromanipulated embryos before they are collected and retransferred to the definitive host. Many of these experiments have been described previously.[11]

CONCLUSION

Interspecies (intertaxon) embryo transfer is a useful practice in distribution and conservation of germplasm. However, there are relatively narrow limits of compatibility between various taxa. Much remains to be learned about the biology of the process, and how best to predict and improve the results. The International Embryo Transfer Society and its members are excellent sources of information on this fascinating topic.

ACKNOWLEDGMENT

The author wishes to acknowledge the assistance of Ms. Mary Garza for valuable assistance in the preparation of this manuscript and the many authors whose research is included but could not be referenced.

REFERENCES

1. Warwick, B.L.; Berry, R.O. Intergeneric and intraspecific embryo transfers. J. Hered. **1949**, *40*, 297–303.
2. Hammer, C.J.; Tyler, H.D.; Loskutoff, N.M.; Armstrong, D.L.; Funk, D.J.; Lindsey, B.R.; Simmons, L.G. Compromised development of calves (*Bos gaurus*) derived from in-vitro generated embryos and transferred interspecifically into domestic cattle (*Bos taurus*). Theriogenology **2001**, *55*, 1447–1455.
3. Anderson, G.B. Interspecific pregnancy: barriers and prospects. Biol. Reprod. **1988**, *38*, 1–15.
4. Gomez, M.C.; Pope, C.E. Current concepts in cat cloning. In *Epigenetic Risks of Cloning*, 1st Ed.; Inui, A., Ed.; CRC Press, Taylor & Francis Group: Boca Raton, FL, 2006; 111–151.
5. Flores-Foxworth, G.; Coonrod, S.A.; Moreno, J.F.; Byrd, S.R.; Kraemer, D.C.; Westhusin, M. Interspecific transfer of IVM IVF-derived red sheep (*Ovis orientalis gmelini*) embryos to domestic sheep (*Ovis aries*). Theriogenology **1995**, *44*, 681–690.
6. Fehilly, C.B.; Willadsen, S.M.; Tucker, E.M. Interspecific chimerism between sheep and goat. Nature (London) **1984**, *307*, 634–638.
7. Croy, B.A.; Rossant, J.; Clark, D.A. Effects of alteration in the immunocompetent status of *Mus musculus* females on the survival of transferred *Mus caroli* embryos. J. Reprod. Fertil. **1985**, *74*, 479–489.
8. Rorie, R.W.; Pool, S.H.; Prichard, J.F.; Betteridge, K.J.; Godke, R.A. A simplified procedure for making reconstituted blastocysts for interspecific and intergeneric transfers. Vet. Rec. **1994**, *135*, 186–187.
9. Fernandez-Arias, A.; Alabart, J.L.; Folch, J.; Beckers, J.F. Interspecies pregnancy of Spanish ibex (*Capra pyrenaica*) fetus in domestic goat (*Capra hircus*) recipients induces abnormally high plasmatic levels of pregnancy-associated glycoprotein. Theriogenology **1999**, *51*, 1419–1430.
10. Pope, C.E.; Gelwicks, E.J.; Wachs, K.B.; Keller, G.; Maruska, E.J.; Dresser B.L. Successful interspecies transfer of embryos from the Indian desert cat (*Felis silvestrus ornata*) to the domestic cat (*Felis catus*) following in vitro fertilization. Biol. Reprod. Suppl. **1989**, *40*, 61.
11. Adams, C.E. Egg transfer in carnivores and rodents, between species, and to ectopic sites. In *Mammalian Egg Transfer*, 1st Ed.; Adams, C.E., Ed.; CRC Press: Boca Raton, FL, 2006; 49–61.

Intracytoplasmic Sperm Injection (ICSI)

Albert L. Smith
Fertility Lab Consulting, Deming, New Mexico, U.S.A.

Abstract
Intracytoplasmic sperm injection, or ICSI, has been developed not only as a research tool but has had commercial application both in the livestock industry and in the human infertility field. ICSI allows fertilization of an egg without sperm capacitation in the female tract, and bypasses the physical barriers of the zona pellucida and the vitelline membrane. ICSI allows fertilization of eggs by sperm, which might not normally penetrate and fertilize an egg.

INTRODUCTION

ICSI is a procedure attempted by many scientists over the past half century. Many scientists destroyed many embryos of different species in a vain effort to inject sperm directly into an oocyte. Technology was the factor that prevented most scientists from being successful in this endeavor. In the early days of ICSI, no cheap, reliable, and commercially available instruments were available for making pipettes for injection and for holding the oocyte during the procedure. Much time was lost making tools, which were of incorrect size or broke while mounting to the injection tool. Improvements in technology have made injection and holding pipettes commercially available, along with media for injection, and special injection instruments to provide a steady pressure during the procedure.

BASICS OF ICSI

Although the technology for performing ICSI has changed over the years, the basic concept of ICSI has remained the same. An oocyte is held by a slight vacuum pressure holding pipette to keep the egg stationary. The holding pipette must hold the oocyte in such a way that the egg does not move and with a vacuum that is neither too high nor too low: if the vacuum is too low, the egg can drop off the holding pipette and if the vacuum is too high, the egg itself can be sucked into the vacuum line. Before injection, the egg is denuded of cumulus and corona cells, which surround it. The cytoplasm of the oocyte is visible and the breakdown of germinal vesicle and the first polar body extrusion can be determined. The quality of the egg itself can be determined from the appearance of the cytoplasm. The injection pipette is used for injecting the sperm into an egg. This consists of a glass or plastic pipette, which has been pulled down to a small diameter. The injection pipette is ground to a fine bevel to facilitate the penetration of the tip of the pipette into the egg itself. A slight vacuum pressure is used to aspirate the sperm into the pipette, and slight positive pressure is used to inject the sperm into the egg. Pressure can be critical during the injection procedure, since a pressure too high can rupture the cytoplasm of the egg, while a pressure too low will not expel the sperm.

Both the holding pipette and the injection pipette tips are held in a drop of buffered medium. The medium is necessary to maintain the osmotic pressure of the egg and sperm during the procedure. The drop may be in a petri dish overlaid with oil, or may be a hanging drop, which together with the egg is placed on a microscope slide and the slide is inverted over the light opening of the stage. Both holding and injection pipette holders are attached to separate vacuum sources via small diameter plastic tubing. The tubing (e.g., Tygon) is flexible enough to bend around the path between the pipette and the vacuum source, but stiff enough not to collapse when vacuum is placed on the tubing. The vacuum source may either be a disposable syringe, a specially designed syringe, an adjustable electric air pump of some kind, or a piezo injector. The procedure itself may be done at room temperature, about 18°C, or on a heated stage or even in a temperature controlled room to keep the sperm and oocytes near 37°C.

The medium used for the micromanipulation process can vary from lab to lab. The medium used may be a modified Dulbecco's PBS, a HEPES buffered medium, or a sodium bicarbonate buffered medium used in a 5% CO_2 in air chamber, which surrounds the microscope stage. The procedure itself sounds simple enough: 1) aspirate a sperm into the injection pipette; 2) hold the egg secure with the other pipette; 3) move the injection pipette to the egg; 4) penetrate the zona pellucida and vitelline membrane of the oocyte with the injection pipette; 5) deposit the sperm in the cytoplasm of the egg; 6) remove the injection pipette from the egg; and 7) retrieve the egg and place it in culture medium for observation and development.

EARLY ATTEMPTS OF ICSI

Early attempts of ICSI include the work of Uehara and Yanagamachi[1] who injected human sperm into hamster oocytes to determine the ability of the sperm to undergo decondensation after injection. The sperm used were lyophylized, freeze-dried sperm.[1] Early work by Markert[2] involved injection of sperm heads only, after separation of sperm head and tails with sonication. Injection of motile sperm into the cytoplasm was difficult because the sperm were motile and could swim out the injection pipette, or further into the injection pipette. Motile sperm injected into the cytoplasm would also swim around the cytoplasm and never begin to decondense.

The use of polyvinyl pyrrolidine (PVP), at concentrations of 3–5%, was instrumental in advancing the development of the ICSI procedure.[3] PVP, because of its high viscosity could effectively slow sperm down to the point that the sperm could be aspirated into an injection pipette. The increased viscosity of PVP more closely resembled the viscosity of the egg itself. Injection of a conventional medium into an egg was usually disastrous because the cytoplasm became diluted and began to seep out of the egg or a bubble of medium might become trapped within the egg. Sperm, once injected, could begin swimming normally again once they were exposed to normal viscosity medium, or start swimming around in the egg cytoplasm once it had been injected. The technique for immobilizing the sperm by touching or crushing the sperm tail prevented this situation.

THE ULTIMATE SUCCESS OF ICSI

ICSI has been successfully applied to a number of species, including cattle, horses, swine, humans, etc. ICSI has revolutionized the animal IVF (in vitro fertilization) industry by ensuring fertilization with the direct injection of a sperm into an egg. Since timing of fertilization is controlled, the potential for gamete aging is reduced and this ultimately increases the fertilization rate and survivability of embryos.[4] Because of the use of ICSI in human IVF, micropipettes, injectors, and other pieces of equipment necessary for the procedure became commercially available. Animal scientists were thus able to buy their instruments instead of spending hours making pipettes.

Although sperm need to capacitate in vitro or in vivo before fertilization, sperm, which are injected directly into the cytoplasm, do not need to go through these processes. Tailless sperm seem to be problematic with human ICSI;[5] however, results of ICSI in other species, e.g., mouse, do not seem affected by the presence of a tail.[6] The orientation of the oocyte during the injection procedure has been discussed as a factor in the ultimate success of the procedure. A meiotically mature oocyte will have extruded the first polar body creating a perivitelline space. Since the oocyte DNA is closely aligned to the position of the first polar body, injections are made to maximize the distance between the first polar body and the injection site. By orienting the polar body at a position of 12:00 or 6:00 (on an imaginary clock dial with the holding pipette at 9:00), the sperm can be injected into the oocyte at a position a safe distance from the area of the oocyte DNA. The human oocyte has an extremely tough vitelline membrane so that the injection tip must be pushed deeply into the cytoplasm before actual penetration occurs, while mice oocytes are extremely susceptible to rupture.[7] The tough membrane of the human oocyte is probably the reason that ICSI has been adopted so readily into the array of assisted reproduction techniques even though about 10% of human eggs do not survive the sperm injection procedure.[8]

One widely used technique for the human oocyte is to push the injection tip into the cytoplasm at least one half the diameter of the oocyte, with the immobilized sperm at the very tip of the pipette. The membrane is aspirated until the membrane ruptures at the point of aspiration. The cytoplasmic contents can be seen being pulled into the pipette after the membrane ruptures. The membrane can be seen returning to its normal position along the zona pellucida. The sperm is then injected into the oocyte along with as little medium as possible, and the pipette is quickly removed. This ensures that the sperm is placed well into the cytoplasm, and at a distance from any microtubules involved in the second meiotic division of the oocyte. If the pipette is pushed too far into the cytoplasm and touches the membrane on the opposite side, rupture of the cytoplasm is almost immediate.

By picking up several immobilized sperm simultaneously in the injection pipette, it is possible to inject several oocytes in rapid succession, although care must be used to ensure that only one sperm is injected per oocyte. Newer technologies allow oocytes to be stripped of their cumulus and corona cells so mature oocytes can be injected immediately. This actually decreases the total "fertilization time" so that oocytes may undergo pronuclear formation earlier than anticipated. In this case, second polar body extrusion or cell division may be used as a positive indicator that fertilization has occurred when ICSI is performed.

ICSI IN HUMAN INFERTILITY

ICSI was originally used in the human infertility field as a solution to the problem of male factor infertility.[8] The sperm of patients who had low sperm numbers, low motility, or a high number of abnormal sperm types could be used to fertilize oocytes. Unlike animal species where sperm concentration and motility are not a problem, human beings are different. Human sperm is considered "normal" with up to 86% abnormal sperm types. Human sperm concentrations, motility, and volumes are much lower compared to livestock species. Consequently, there have been situations where more oocytes are collected than visible motile sperm.

In situations where no sperm are available in the ejaculate, epididymal or testicular sperm are aspirated and used for the injection procedure. Motile sperm from the epididymis, testes, and cryopreserved sperm have been utilized for the ICSI procedure in humans.[9]

The focus of the ICSI procedure in human infertility treatment has changed from male factor infertility to include a number of conditions including unexplained infertility and combined male and female factors. Of 122,872 reported Assisted Reproductive Technology procedures performed in the United States in 2003,[10] 56% involved ICSI, even though only 13% of cases were male factor alone. Unexplained infertility, combinations of infertility problems, the presence of antisperm antibodies, all have been treated with ICSI. The downside to the extensive use of ICSI in the human infertility field may be the possible consequences on the children. Genetic defects that affect sperm production may be transmitted to children born from ICSI, thus creating another generation with reproductive problems to face.

ICSI has been an invaluable tool in the animal sciences industry. The use of ICSI has permitted scientists to study fertilization in a more controlled manner. ICSI has been applied to cattle,[11] lab animals,[12] and has been suggested as a method for increasing the numbers in endangered species.[13] ICSI, combined with flow cytometry, may allow sex selection of embryos.[11] What was once considered an almost impossible procedure holds promise for both animal and human reproduction.

REFERENCES

1. Uehara, T.; Yanagamachi, R. Microsurgical injection of spermatozoa into hamster eggs with subsequent transformation of sperm nuclei into male pronuclei. Biol. Reprod. **1976**, *15*, 467–470.
2. Markert, C.L. Fertilization of mammalian eggs by sperm injection. J. Exp. Zool. **1983**, *228*, 195–201.
3. Heuwieser, W.; Yang, X.; Jiang, S.; Foote, R.H. Activation of in vitro matured bovine oocytes after microinjection of immobilized sperm. Proceedings of the Annual Meeting of the International Embryo Transfer Society, Denver, Colorado, January 12–13, 1992; Theriogenology, *37* (1), 221.
4. Smith, A.L.; Lodge, J.R. Interactions of aged gametes: in vitro fertilization using in vitro-aged sperm and in vivo-aged oocytes. Gamete Res. **1987**, *16*, 47–56.
5. Sauias-Magnan, J.; Metzler-Guillmaria, C.; Mercier, G.; Carles-Marcorelles, F.; Grillo, J.M.; Guichaoua, M.R. Failure of pregnancy after intracytoplasmic sperm injection with decapitated spermatozoa: a case report. Hum. Reprod. **1999**. *14* (8), 1989–1992.
6. Kurokawa, M.; Fissore, R.A. ICSI generated mouse zygotes exhibit altered calcium oscillations, inositol 1,4,5 triphosphate receptor-1-down regulation and embryo development. Hum. Reprod. **2003**, *19*, 523–533.
7. Yoshida, N.; Perry, A.C.F. Piezo actuated mouse intracytoplasmic sperm injection (ICSI). Nat. Protoc. **2007**, *2*, 296–304.
8. Van Steirteghem, A.; De Vos, A.; Staessen, C.; Verheyen, G.; Aytoz, A.; Bonduelle, M.; Tournaye, H.; Devroey, P. Is ICSI the ultimate ART procedure? In *Fertility and Reproductive Medicine*, Proceedings of the XVI World Congress on Fertility and Sterility, San Francisco, California, October 4–9, 1998; Kemper, R.D.; Cohen, J.; Haney, A.F.; Younger, B.J., Eds.; Elsevier: Amsterdam, 1998; 27–38.
9. Witt, M.A. ICSI and the biological sperm reservoir. In *Fertility and Reproductive Medicine*, Proceedings of the XVI World Congress on Fertility and Sterility, San Francisco, California, October 4–9, 1998; Kemper, R.D.; Cohen, J.; Haney, A.F.; Younger, B.J., Eds.; Elsevier: Amsterdam, 1998; 429–434.
10. 2003 Assisted Reproductive Technology Success Rates. National Summary and Fertility Clinic Report. http://www.apps.nccd.gov.ART2003/nation03.asp pp 75 (accessed January 23, 2006).
11. Hamano, K.; Li, X.; Qian, X. Gender preselection in cattle with intracytoplasmically injected, flow cytometrically sorted sperm heads. Biol. Reprod. **1999**, *60*, 1194–1197.
12. Horiuchi, T. Application study of intracytoplasmic sperm injection for golden hamster and cattle production. J. Reprod. Dev. **2006**, *52* (1), 13–21.
13. Wirtu, G.W.; Pope, C.E.; Gomez, M.C.; Cole, A.; Paccamonti, D.L.; Dresser, B.L. 249 intracytoplasmic sperm injections of Eland (*Taurtotragus oryx*) and Bongo (*Tragilaphus eurycerus*) antelope oocytes. Reprod. Fertil. Dev. **2006**, *19* (1), 241.

Kimchi Fermentation

Jae-Kun Chun
*Department of Food and Animal Biotechnology, College of Agriculture and Life Sciences,
Seoul National University, Seoul, Korea*

Abstract

Kimchi is a Korean fermented food made of brined vegetables and seasonings having an acidic and spicy taste. The earliest form of kimchi consisted of only salted vegetables, but kimchi has evolved to incorporate seasonings such as red pepper, leek, garlic, and fish to create versatile tastes, textures, colors, and shapes. Kimchis are categorized into about 64 varieties, and *Baechu* kimchi is probably the most popular. Kimchi can be considered a health food containing a high level of dietary fiber, vitamins, and other nutrients synthesized during fermentation. Kimchi fermentation is a natural lactic acid fermentation developed for the long-term storage of vegetables. The method to prepare kimchi as well as the necessary raw materials, fermentative microorganisms, and storage conditions for kimchi are described.

INTRODUCTION

Kimchi is a traditional fermented Korean food made of vegetables such as Chinese cabbage and Korean radish. Principally, kimchi can be made from almost all vegetables and fruits found in Korea. Kimchi has an acidic taste with spicy and strong flavors, and also has a cool mouth feeling due to volatile organic acids and dissolved CO_2 in the liquid. Kimchi has been consumed since the 12th century in the Korean peninsula.[1] The earliest form of kimchi consisted of only salted vegetables, but has evolved to incorporate seasonings such as red pepper, leek, garlic, and fish to create versatile tastes, colors, and shapes.[1–3]

Kimchi is produced from natural fermentation associated primarily with lactic acid bacteria. The lactic acid bacteria come from natural sources, that being the vegetables, seasoning ingredients, and the surrounding environment.[4] Technologically, kimchi is one of the oldest foods derived from biotechnology with long-term storage of vegetables in mind. Kimchi enhances the nutrient value of vegetables by fermentation with seafood products such as oyster, shrimp, and fish. Increased production of commercial kimchi and advances in storage technology of kimchi have affected various business and industrial fields such as vegetable cultivation and supply, the food industry, distribution systems, and the home appliance market.

VARIETIES OF KIMCHI

Kimchi is categorized by various criteria: the ratio of primary and secondary vegetables used in formulation, the growing season, the methodology used, and the region where kimchi is prepared.[2] Primary vegetables include Chinese cabbage, Korean radish, pony radish, cucumber, Chinese leek, green pepper, cabbage, and eggplant. Minor vegetables are Korean radish (shredded), Indian mustard greens, carrot, and dropwort. The seasoning ingredients include salted fermented fish sauces (shrimp, oyster, Alaskan pollack, squid, flounder, and yellow corvina), red pepper powder, green onion, garlic, ginger, Chinese leek, onion, pine nut, gingko nut, and sugar.[2–3]

Generally, kimchis are named after the primary vegetables used, for example, *Baechu* kimchi prepared with Chinese cabbage (*Baechu* in Korean), *Moo-kimchi* made of Korean radish (*Moo* in Korean), and *Oyi*-kimchi prepared with cucumber (*Oyi* in Korean). Some types of kimchi are named after the seasons of preparation as such *Bom-kimchi* prepared in the spring, *Kimjang-kimchi* made at a kimchi-making-event; *Bom* and *Kimjang* are Korean words meaning spring and kimchi-making-event in autumn, respectively. Among 64 varieties of kimchi, *Baechu* kimchi is the most popular variety representing about 70% of all kimchi consumed.[2–3]

Kimchi from the cooler climates of northern Korea has less salt as well as lower contents of red pepper and brined seafood. In contrast, kimchi from the southern districts with warmer temperatures is salty along with increased levels of spices. Kimchi from the coastal areas contains increased quantities of fermented fish. Generally the content of red pepper, fermented fish, and liquid greatly affects the sensorial acceptance of kimchi.

DIETARY IMPORTANCE OF KIMCHI

Kimchi is a low-calorie food (18 kcal/100 g) containing sources of probiotic lactic acid bacteria (10^8–10^9

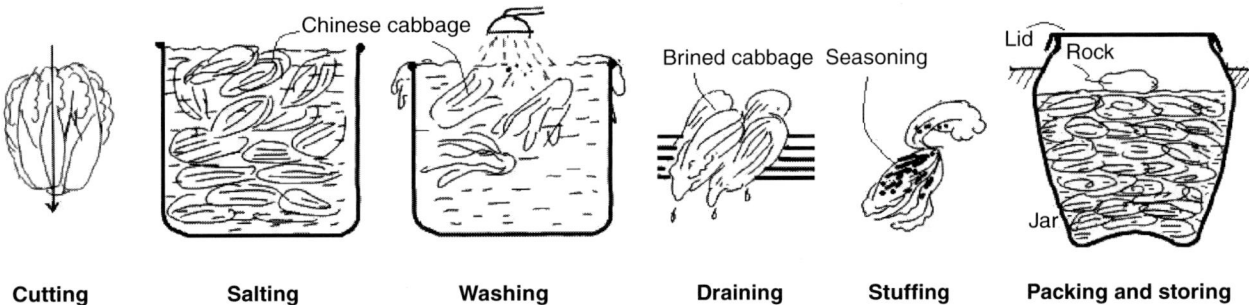

Fig. 1 The kimchi-making process.

colony-forming units/g), vitamins, organic acids, and minerals originating from the raw vegetables and biosynthesized during the fermentation process.[3] Born from the need to preserve vegetables throughout the country's long and bitter winter months, kimchi has been served as an essential side dish at any Korean meal. Kimchi is a unique food technology to fortify the nutrient value of vegetables with animal protein from oyster, shrimp, and fish. Solids and juices of kimchi are consumed, and widely used as cooking materials for various Korean derivative dishes such as kimchi-stew, kimchi-rice, and kimchi-noodles.

KIMCHI FERMENTATION

Kimchi Preparation Process

Unless otherwise specified, kimchi means *Baechu*-kimchi made of Chinese cabbage as the primary vegetable with secondary vegetables including Korean radish, leek, garlic, red pepper, salt, fruit, and fermented fish. The preparation process of *Baechu*-kimchi consists of four major processes: the salting process of Chinese cabbage, preparation of seasoning with secondary vegetable, stuffing seasoning, and packing and storing as shown in Fig. 1.

The salting process is carried out by first cutting whole Chinese cabbage into two or four parts length-wise. The cabbage is salted using a 10–15% salt solution for several hours until the cabbage leaves soften. The salted cabbage is then washed in running water and thoroughly drained. The degree of salting determines the salinity of the kimchi that is also affected by the temperature of the salt solution.[5,6] The preparation of seasoning is done by mixing in the secondary vegetables: slices of Korean radish, leek and fruit, mashed roots of garlic and ginger, chopped green onions, and fish such as fermented anchovies and pickled baby shrimp. This process determines the fundamental taste of the kimchi.

Stuffing is an operation in which the prepared seasoning is inserted between leaves of brined Chinese cabbage. The amount and uniformity of the seasonings in the stuffing influence color, taste, and the spiciness of the kimchi.

Finally, the stuffed Chinese cabbage is packed into an earthenware jar, and the jar is stored in a cool place. Traditionally, the packed kimchi jar is buried (at least 80–90% of the container) underground to keep it from freezing during the winter season.[3,7] Kimchi is cured at 0–20°C, for days or months, depending on the climate. At the early stage of fermentation, kimchi juice is produced from the stuffed cabbages and covers most of the packed kimchi, thus providing favorable conditions for the anaerobic kimchi fermentation. A flat rock placed above the packed kimchi is necessary to maintain a tightly packed state in the jar during the increasing levels of juice (Fig. 1).

Preparation methods vary depending on the types of kimchi. Modern kimchi fermentation is different from the classical type due to use of mechanical refrigeration that automatically controls fermenting and storage environments.[6]

Microorganisms Associated with the Kimchi Fermentation

The kimchi fermentation is a natural lactic acid fermentation associated with many microorganisms. Concentrations of aerobic and anaerobic microorganisms reach 10^3–10^7 and 10^4–10^6 cells/mL, respectively (Fig. 2). Important bacteria isolated from kimchi include *Lactobacillus plantarum*, *L. brevis*, *Enterococcus faecalis*, *Leuconostoc mesenteroides*, *Pediococcus pentosaceous*, and *Weissella koreanis*.[4,8,9]

L. mesentroides helps suppress the growth of anaerobic microbes, and thus provides a favorable anaerobic environment for lactic acid bacteria, primarily growth of *Lactobacillus* spp. At a later stage, yeasts such as *Saccharomyces* and *Torulopsis* grow.

Physicochemical Change during Kimchi Fermentation

Various changes occur both in chemical composition and physical properties during fermentation. Concentrations of acids, sugars, organic acids, vitamins, amino acids, and CO_2 changes as shown in Fig. 3.[6,8]

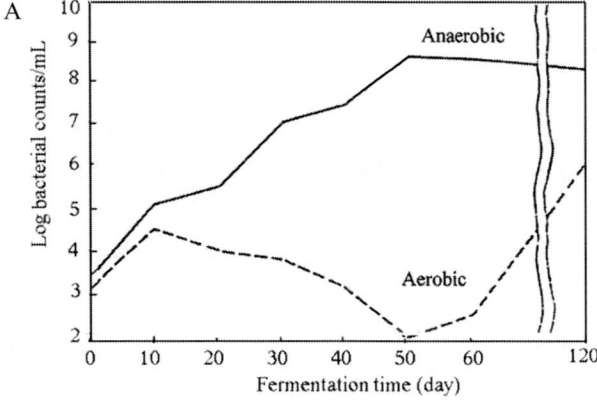

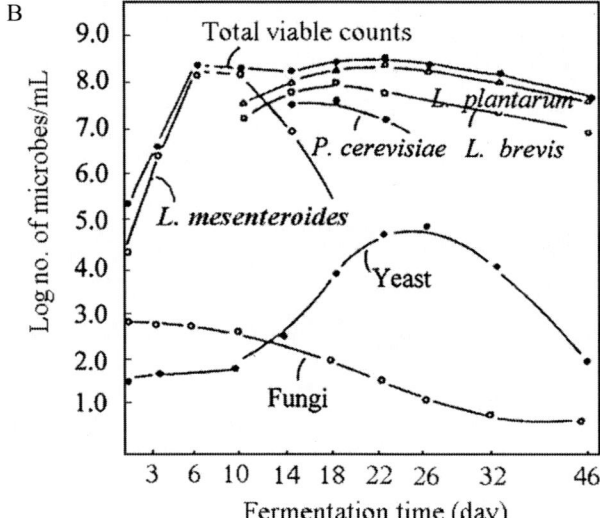

Fig. 2 Change in microorganisms during the kimchi fermentation. (A) Changes in aerobes and anaerobes. **Source:** *Studies on the dynamic changes of bacteria during the Kimchi fermentation*.[4] (B) Overall microbial changes. **Source:** From *Determination of microbial community as an indicator of Kimchi fermentation*.[8]

Factors Affecting Kimchi Fermentation

Raw materials

Since the identity of kimchi comes from the type and quality of vegetables used, vegetable cultivars and the degree of vegetable ripeness are important factors. For example, various cultivars of Chinese cabbage (*Brassica campestris (syn. rapa)* L. spp. *pekinensis (Lour.) Rupr.*) have been developed with optimization of seasonal supply, disease tolerance, and adaptation to low and high-altitude cultivations as breeding targets. The most popular cultivars for kimchi processing have common desirable characteristics such as tight heading, late bolting, and yellow inner leaves that result in soft texture with high sugar content in the kimchi.[7,10]

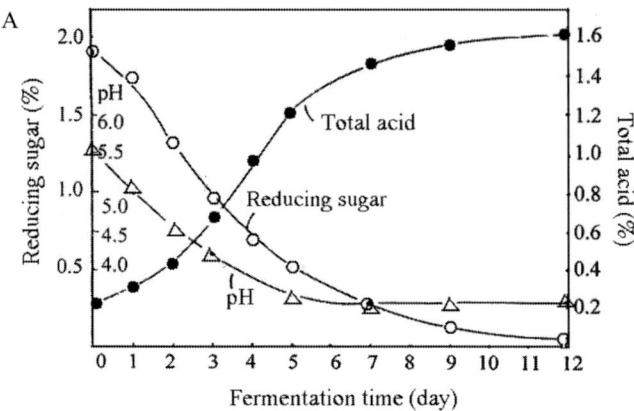

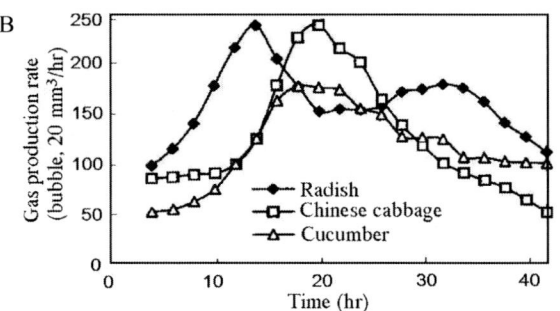

Fig. 3 Physicochemical changes during the kimchi fermentation. (A) Changes of acidity and sugar. **Source:** From *Traditional fermented food products in Korea*.[9] (B) Gas production rate. **Source:** From *Automation of Kimchi fermentation based on pattern analysis*.[6]

Salt concentration and temperature

The salt content of kimchi falls in narrow range from 2% to 6%.[5,6,11] The salt content provides favorable conditions for halophytic bacteria, primarily lactic acid bacteria, and inhibits undesirable microorganisms. Another prime factor is the temperature of the fermentation over time. The optimum temperature range for the kimchi fermentation is 4–10°C.[6,11]

Acidity and storage of kimchi

The best degree of acidity is obtained when the pH reaches 4.3. The monitoring methods to determine the completion of the kimchi fermentation are the measurement of end products of the fermentation, such as lactic acid content, the pH of kimchi juice and the extent of gas bubble of CO_2.[6,8,9]

Since kimchi retains its fermentative microbial populations, it is an ongoing fermenting food; the acceptability of kimchi changes with time.[8,11] Acidic deterioration due to over-ripening is one of the least desirable situations. This condition is best prevented using cold storage with possible incorporation of thermal pasteurization and preservatives. Many Koreans store kimchi in a kimchi-refrigerator that automatically controls the fermenting and storage temperatures of kimchi based on a microprocessor program.[6] Unlike the classical storage method using a kimchi jar, these days raw kimchi is commonly packed in a rectangular plastic box of 12-L capacity and stacked in a kimchi refrigerator with a capacity of 120–500 L to ferment in a fermenting mode, and following adequate ripeness, a switch to storage mode until consumption. In this fashion, kimchi can be consumed after storage for up to 1 year.

Distribution of kimchi

With increased demand for kimchi, the kimchi industry has expanded in market volume and the diversity of products. *Baechu*-kimchi is easily available in most of cities for domestic and world markets. Commercially, kimchi is produced in manufacturing lines facilitated with unit operations, including cutting, brining, washing, mixing, stuffing, filling, fermenting, and packaging. Kimchi is filled in a pre-cured state in various packages, such as cans, glass jars, flexible plastic pouches. Kimchi that is not commercially sterile is maintained under constant refrigeration.

CONCLUSION

Since the kimchi fermentation is a natural one, not employing starter cultures, the taste of kimchi is not always consistent. Differences come from the abovementioned factors. The quality criteria of optimum ripeness of kimchi are difficult to define with scientific data such as chemical analysis and physical measurement, because individual acceptance and preference are also deeply associated with the culture and lifestyle of the consumer. This means that there is no criterion on ripeness because the preference of taste is different among individuals. With the increasing demand for commercial kimchi, quality consistency, standard recipes, use of starter cultures, and automation of labor-intensive stuffing operations become important issues for future incorporation into the production of kimchi. Detailed descriptions of kimchi can be found in several review articles.[1–3,11,12]

REFERENCES

1. Chang, J.H. Studies on the origin of Korean vegetable pickles. Thesis Collection of Sung-Sim Womens College, **1975**, *6*, 149–174.
2. Cho, J.S. *Research on Korean Fermented Foods*; Kichonyonkusa: Seoul, 1981; 91.
3. Cheigh, H.S.; Park, K.Y. Biochemical, microbiological, and nutritional aspect of Kimchi. Crit. Rev. Food Sci. Nutr. **1994**, *34* (2), 175–203.
4. Kim, H.S.; Chun, J.K. Studies on the dynamic changes of bacteria during the Kimchi fermentation. J. Korean Nucl. Sci. **1966**, *6*, 112–118.
5. Park, W.S.; Lee, I.S.; Han, Y.S.; Koo, Y.J. Kimchi preparation with brined Chinese cabbage and seasoning mixture stored separately, Korean J. Food Sci. Technol. **1994**, *26*, 231.
6. Chun, J.K.; Kim, K.M.; Woo, D.H. Automation of Kimchi fermentation based on pattern analysis. Food Eng. Prog. **1999**, *3* (3), 181–185.
7. Chun, J.K. Chinese cabbage utilization in Korea; Kimchi processing technology. In *Proceeding of the 1st International Chinese Cabbage*; Talekar, N.S., Griggis, T.D., Eds.; AVRDC: Tainan, Taiwan, 1981; 49–54.
8. Han, H.U.; Lim, C.R.; Park, H.K. Determination of microbial community as an indicator of Kimchi fermentation. Korean J. Food Sci. Technol. **1990**, *22* (1), 26–32.
9. Mheen, T.I.; Kwon, T.W.; Lee, C.H. Traditional fermented food products in Korea. Korean J. Appl. Microbiol. Biotechnol. **1981**, *9* (4), 253–261.
10. Chun, C.H. Chinese cabbage. In *Horticulture in Korea*; Lee, J.M., Choi, G.-W., Janick, J., Eds.; Korean Society for Horticultural Science: Suwon, Korea, 2007; 117–121.
11. Hui, Y.H.; Goddick, L.M.; Hansen, A.S., Eds. *Handbook of Food and Beverage Fermentation Technology*; Marcel Dekker: New York, 2004; 621–855.
12. Rehm, H.J.; Reed, G. Indigenous fermented food. Biotechnology **1988**, *9*, 645–661.

Laboratory Animals: Cryopreservation of Oocytes and Sperm

Eric Walters
National Swine Research and Resource Center, University of Missouri, Columbia, Missouri, U.S.A.

Abstract

Recently there has been a huge demand for improved cryopreservation protocols in laboratory animals. Part of this demand for improved protocols is a result of a demand for genetically modified (GM) animals. With this demand for GM animals there is an additional increase in the exchange of animals/animal tissue between institutions domestically as well as internationally. As the number of models increase either by the addition of new models or the modification of existing models, the need to preserve the gametes and embryos of these models increases. There are many reasons why cryopreservation is important, and they include: 1) maintaining genetic diversity; 2) genetic banking of endangered species; 3) enhancing the distribution of genetically superior lines; 4) treating illness-induced infertility such as cancer; and 5) genetic banking of genetically modified animals. With the development of cryopreservation techniques, the application to preserve the germplasm of the animals initially began with embryos from several species. However, embryo banks proved to be a costly venture when compared to the cost associated with banking spermatozoa. Despite this decreased cost, one of the drawbacks to the preservation of sperm was banking only haploid gametes. Recently, the development of more sophisticated assisted reproductive technologies, such as intracytoplasmic sperm injection, has alleviated some of the difficulties with utilizing cryopreserved sperm for the rederivation of genetically modified organisms. Despite the advancement in assisted reproductive technologies, the efficiencies of gamete cryopreservation in many species remain relatively low.

CRYOBIOLOGY

Cryobiology refers to the study of low temperature on biological systems such as their cellular components. It is believed that as early as the 17th century, researchers have been studying cryobiology. For a more historical perspective of cryobiology refer to the history of sperm cryopreservation in the *Sperm Banking: Theory and Practice* book[1] as well as many other good reviews in recent years.

In cryobiology, it is known that species and cell types have different "death points" or as Luyet and Gehenio stated, "the definite temperature at which an organism passes from the living to the dead state."[2] Much of the first cryobiology work was done in the 17th and 18th century with spermatozoa from amphibians, rabbits, and dogs. Prior to the late 1940s, most of the cryopreservation research was empirical, which also encompassed competing intracellular ice avoidance hypotheses. However, in 1949, a serendipitous discovery by Polge and co-workers showed that glycerol had a protective affect on fowl spermatozoa during the cryopreservation procedure.[3] Afterwards, other chemicals such as ethylene glycol and propylene glycol were found to have protective properties similar to glycerol and were defined as permeating chemical protecting agents (CPAs). With these discoveries, a new empirical era of cryopreservation was on the horizon and much of the work was investigating the interactions between CPAs, osmotic tolerance limits or ability to shrink and swell, and cooling and warming rates for specific species. For example, there are significant differences in the osmotic tolerance limits between cell types and species (Fig. 1).

RODENTS

Spermatozoa

Gamete cryopreservation in the rat is still inferior in terms of post-thaw survivability when compared to the mouse; however, it is comparable to the pig (Table 1). Nearly 20 years ago mouse sperm were successfully cryopreserved.[4,5] However, the post-thaw survival of mouse sperm is highly variable among strains with inbred strains being more difficult to successfully cryopreserve. Similarly, rat sperm cryopreservation has been reported but the efficiency of establishing pregnancies and maintaining litter size after intrauterine insemination with cryopreserved sperm is extremely low. Compared to mouse sperm, rat sperm from different strains remain difficult to cryopreserve with sperm from inbred strains being the most difficult ones to cryopreserve. In addition to the extreme difficulty of cryopreserving rodent sperm, there is an added sensitivity

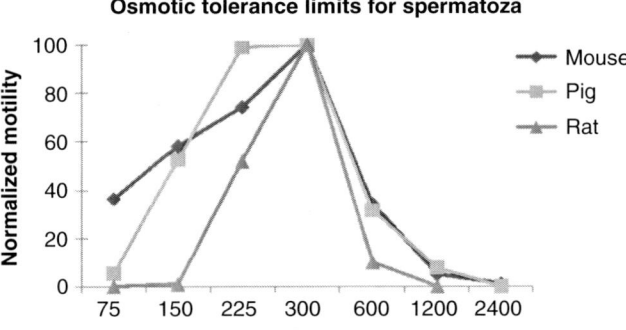

Fig. 1 Osmotic tolerance limits for the mouse, rat, and pig. Spermatozoa were exposed to various osmolalities for 5 min and returned to isosmotic (300 mOsm) then motility was analyzed. Differences can be seen in the osmotic tolerance limits which correlates with the sperm's ability to shrink and swell in response to the addition and removal of CPAs.

of mechanical stress. Mechanical stress can result from centrifugation, shear force, and extracellular ice formation. More recently, the movement in rodent sperm cryopreservation has been toward utilizing freeze-drying technology; i.e., the samples are stored at ambient temperatures. This technology facilitates shipping of rodent spermatozoa as the use of liquid nitrogen is eliminated. Despite the advantages to freeze-dried spermatozoa, the need for more sophisticated training and equipment is required. With the freeze-drying technology, the spermatozoa are unable to recover motility when rehydrated. However, DNA integrity is maintained, thus allowing freeze-dried sperm to be used in combination with assisted reproductive technologies such as intracytoplasmic sperm injection.

Embryos and Oocytes

Embryo cryopreservation for the mouse and the rat is very successful and a well-established technique. Typically rodent embryos are frozen at two different developmental stages, zygote and morula. Cryopreservation of both developmental stages is successful in terms of recovery of the genetic line. Despite the success with both developmental stages, most investigators cryopreserve zygotes instead of morula. The use of zygotes increases the amount of starting material, improves post-thaw survival and pregnancy rates. However, a disadvantage of cryopreserving rat zygotes is the inconsistency with the rat embryo culture system. This can be overcome by transferring the embryo directly to the recipient after thawing. Similar to the rat, the mouse is routinely frozen at the zygote stage as well, but the mouse culture system is more robust than the rat.

Human infertility affects approximately 5–15% of the worldwide population. Even though embryo cryopreservation is routine in many species, it introduces several practical and ethical issues when discussing human infertility. Due to the practical and ethical problems associated with embryo cryopreservation, there has been a demand for cryopreservation of immature or mature oocytes; thus alleviating some of the practical and ethical problems. Oocyte cryopreservation, to date has had limited success in the mouse and the rat; including both immature and mature oocytes. Both methods of slow- and ultrafast cooling have been utilized for successful oocyte cryopreservation; however, both have led to decreased viability resulting in reduced post-thaw maturation, fertilization, and culture compared to fresh oocytes.[7] There are several reasons for this decreased viability post-thaw; 1) irreversible changes in the meiotic spindle; 2) slow permeability of the cryoprotectant into the oocyte; and 3) alternations of the oocyte's organelles.

SWINE

In this entry the pig was included as a lab animal since it is often used as a biomedical model. The pig makes an excellent model for the biomedical community with its anatomical and physiological similarities to the human. Movement of swine biomedical models, domestically and internationally, will rely heavily on the ability to successfully cryopreserve gametes and embryos. Despite the high demand for the cryopreservation of pig gametes, the efficiency remains low. In addition to the low efficiency, the pig is in the infancy stage of the development of assisted reproductive technologies when compared to the mouse or even the rat.

Spermatozoa

Boar sperm cryopreservation has limited success depending on whether you are addressing the agricultural or research community. For the agricultural community, sperm cryopreservation has limited use due to the fact that there

Table 1 Efficiency of germplasm cryopreservation by species and germplasm.[a]

Species	Spermatozoa	Embryos	Oocytes	Ovarian tissue	Testicular tissue	Embryonic stem cells
Mouse	Fair	Good	Fair	Fair to Poor	Poor	Good
Rat	Fair to Poor	Good	Poor	Fair to Poor	Fair to Poor	n/a
Swine	Fair to Poor	Poor	Poor	Poor	Poor	Poor

[a]**Source:** From Biol. Reprod.[6]

is reduced economics when compared to fresh or liquid-cooled semen. Frozen-thawed semen results in a 1.5–3 pig/litter reduction and a 50% decrease in the pregnancy rate. However, the biomedical community is accepting of these results as they can recover the genetic line with one or two animals. Boar sperm cryopreservation still needs improved cryopreservation protocol to address this limited success. As the swine industry becomes a global industry, the demand for movement of genetics will increase. In combination with the high regulations set forth by USDA for importation and exportation of live animals, the use of cryopreservation will dramatically increase.

Embryos and Oocytes

Porcine embryos and oocytes are similar to sperm as they are sensitive to cryoinjuries at temperatures below 15°C. Despite this sensitivity to reduced temperatures, embryo cryopreservation has been successful. There are a couple of differences between pig and rodent embryo cryopreservation: 1) the pig embryos have a higher lipid content; and 2) cooling rates for cryopreservation are much different. Pig embryos have a much higher lipid content when compared to rodent embryos, and this lipid content is believed to be one of the stumbling blocks for embryo cryopreservation. There has been a significant amount of research focused on removing the intracellular lipid content at various timepoints in embryonic development with successful live offspring production. Another difference between porcine and rodent embryo cryopreservation is the cooling rate at which the embryos are frozen. In the rodent, cooling rates would be considered slow cooling due to the fact that cooling rate ranges from 0.5–2°C/min. However, in the pig the most successful method is vitrification. Vitrification is ultrafast cooling, the embryos goes from room temperature to liquid nitrogen ($-196°C$) instantly. The true definition of vitrification is the formation of a glass-like phase with no formation of crystalline structures.

Regardless of the accomplishments of embryo cryopreservation in the pig, oocyte cryopreservation has yet to produce live offspring. Similar to the embryo, the oocyte's lipid content as well as many other organelles such as the mitochondria, meiotic spindle, and cortical granules are altered as a result of the preservation of oocytes. Research in oocyte cryopreservation has begun to investigate some of these obstacles with items such as delipidation of the oocyte, and stabilization/relaxation of the meiotic spindle. As stated before, one of the disadvantages of oocyte cryopreservation is the limited success in the development of assisted reproductive technologies.

CONCLUSION

In conclusion, cryopreservation of lab animal gametes and embryos ranges from very successful to very limited depending on cell type and species of interest. There is still a lot of work that needs to be done to improve the success rate of cryopreservation of lab-animal sperm, oocytes, and embryos. We believe it will take the combination of empirical approaches as well as fundamental cryobiology to design cell- and species-specific cryopreservation protocols.

REFERENCES

1. Walters, E.M.; Woods, E.; Benson, J.; Critser, J.K. The history of sperm cryopreservation. In *Sperm Banking: Theory and Practice*; Pacey, A., Tomlinson, M., Eds.; Cambridge University Press: Cambridge, NY, 2009; 1–17.
2. Luyet, B.; Gehenio, P. *Life and Death at Low Temperatures*. Biodynamica: Normandy, MO, 1940.
3. Polge, C.; Smith, A.U.; Parkes, A.S. Revival of spermatozoa after vitrification and dehydration at low temperatures. Nature **1949**, *164*, 666.
4. Okuyama, M.; Isogai, S.; Hamada, H.; Ogawa, S. In vitro fertilization (IVF) and artificial insemination (AI) by cryopreserved spermatozoa in mouse. J. Fertil. Implant (Tokyo) **1990**, *7*, 116–119.
5. Tada, N.; Sato, M.; Yamonoi, J.; Mizorgi, T.; Kasai, K.; Ogawa, S. Cryopreservation of mouse spermatozoa in the presence of raffinose and glycerol. J. Reprod. Fertil. **1990**, *89*, 511–516.
6. Mazur, P.; Leibo, S.; Seidel, G. Cryopreservation of the germplasm of animals used in biological and medical research: importance, impact, status, and future directions. Biol. Reprod. **2008**, *78*, 2–12.
7. Hosu, B.G.; Mullen, S.; Critser, J.K.; Forgacs, G. Reversible disassembly of the actin cytoskeleton improves the survival rate and developmental competence of cryopreserved mouse oocytes. Plosone **2008**, *3* (7), 1–9.

Lactic Acid Fermentation: Direct

Rojan P. John
National Institute for the Scientific Research of the Water Ground Environment, Quebec City, Quebec, Canada

G.S. Anisha
Department of Zoology, Government College, Kerala, India

K. Madhavan Nampoothiri
Ashok Pandey
Biotechnology Division, National Institute for Interdisciplinary Science and Technology (CSIR), Kerala, India

Abstract

The idea of cost-effective production of lactic acid leads to the process of direct lactic acid fermentation from agro-industrial residues instead of the utilization of refined sugars or double- or triple-step fermentation. Agricultural wastes and other cost-effective industrial wastes can be utilized for single-step conversion to lactic acid by several methods, such as: 1) co-culture fermentation using starch-degrading organisms and lactic acid-producing organisms; 2) enzyme-mediated saccharification of starch and its fermentation to lactic acid by lactic acid-producing fungi such as *Rhizopus*; 3) lactic acid fermentation by amylolytic lactic acid bacteria; and 4) simultaneous conversion of starch to glucose and glucose to lactic acid by addition of α-amylase and glucoamylase along with lactic acid bacteria. These direct fermentations can reduce the cost of production by utilizing complex industrial residues without power-consuming liquefaction and saccharification processes. Utilization of amylase-producing strains can avoid the use of hydrolyzing enzymes and can increase productivity as there is no inhibition of enzymes by higher substrate concentrations.

INTRODUCTION

Lactic acid fermentation is one of the oldest fermentations. In practice since ancient times, lactic acid fermentation is still gaining increased attention owing to its versatile applications in the food, textile, cosmetic, pharmaceutical, and chemical industries. Recently there has been a considerable increase in research interest for lactic acid fermentation owing to its use as monomer for the synthesis of polylactic acid, which has received much attention as a versatile biodegradable plastic. Lactic acid synthesis can be done by chemical processes from petrochemical by-products or by biological fermentation of simple and complex sugars. The use of fermentation to produce lactic acid is now more than 90% of total lactic acid production because of the availability of raw materials, need for economic processes, ease in handling, eco-friendliness, and desirability in selecting an isomeric form.

The conventional method of lactic acid production is by fermentation from refined sugars; however, refined substrates increase the cost of production. Refined sugars are still in use because of: 1) the non-availability of potential amylolytic strains for lactic acid fermentation; 2) the need to develop a commercial strain for high yield efficiency of lactic acid production; 3) the inability of microorganisms to utilize alternate substrates with high yields; and 4) the inability of microorganisms to use easily available, inexpensive, and renewable agricultural materials. As there is a need for an alternative substrate for lactic acid fermentation, the use of hydrolyzates from raw agricultural biomass came into practice. It solved many environmental issues related to disposal of many agro-food residues, but the need for high temperatures in hydrolysis created additional costs of production, and so alternative or modified processes for the production of lactic acid were investigated. The direct conversion of agro-residues came as an economic alternative from the development of new processes and microorganisms.

ADVANTAGES OF DIRECT FERMENTATION OF LACTIC ACID

Fermentation of lactic acid from agro-industrial residues is widely appreciated owing to low-cost, high production rates, high yield, little or no by-product formation, the ability to be fermented with little or no pretreatment, and year-round availability.[1] The production of lactic acid from starchy biomasses commonly involves a two- or three-step

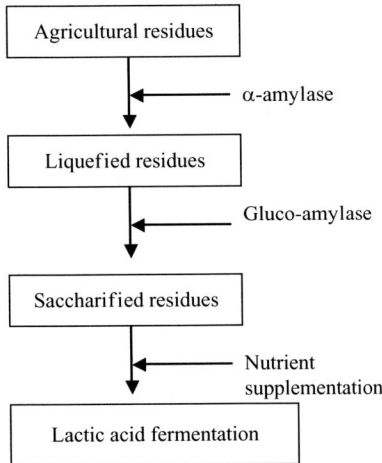

Fig. 1 Schematic diagram of multistepped conventional fermentation.

process: 1) liquefaction of starch by α-amylase; 2) enzymatic saccharification of the liquefaction products to produce glucose (both steps can be replaced with acid hydrolysis); and 3) fermentation of glucose to lactic acid (Fig. 1). The pretreatment of substrates makes the whole process cost-inefficient, whereas a reduction in cost of lactic acid production can be achieved if the starchy materials can be fermented directly.

Woiciechowski et al.[2] studied the hydrolysis of cassava bagasse starch by acid and enzymatic hydrolysis. Both methods were found to be quite efficient when considering parameters such as the percentage of hydrolysis, cost of the chemicals and energy consumption, and the time required for the process. Although acid hydrolysis is time-saving and cost-effective, an increase of salt levels occurs during neutralization that affects bacterial growth, product yield, and lactic acid productivity. Enzymatic hydrolysis of starch is better as it yields a high percentage of reducing sugars. For example, enzymatic hydrolysis of 150 kg of cassava bagasse costs approximately $2470 mostly for the enzymes and power in the lengthy saccharification process.[2] Direct fermentation reduces the cost of energy consumption for liquefaction and saccharification and thus is very cost-effective and time-saving.

The enzymatic liquefaction and saccharification steps can be eliminated by using symbiotic co-cultures of amylolytic and lactic acid-producing microorganisms. There have been many attempts to produce lactic acid directly from starch using wild amylolytic lactic acid bacteria.[3,4] The utilization of lactic acid-producing fungi, such as *Rhizopus* sp., is a cost-effective single-step conversion of agro-residues to lactic acid.[5] Simultaneous conversion of starch to glucose and glucose to lactic acid from the addition of α-amylase and gluco-amylase along with lactic acid bacteria is a recently tested technique for lactic acid production.[3,6] The different processes of lactic acid fermentation in direct conversion are shown in Fig. 2.

DIFFERENT TYPES OF DIRECT LACTIC ACID FERMENTATION

Co-Cultures of Amylase and Lactic Acid-Producing Organisms

Co-immobilization of starch-degrading organisms, such as *Aspergillus awamori* and the lactic acid-producing

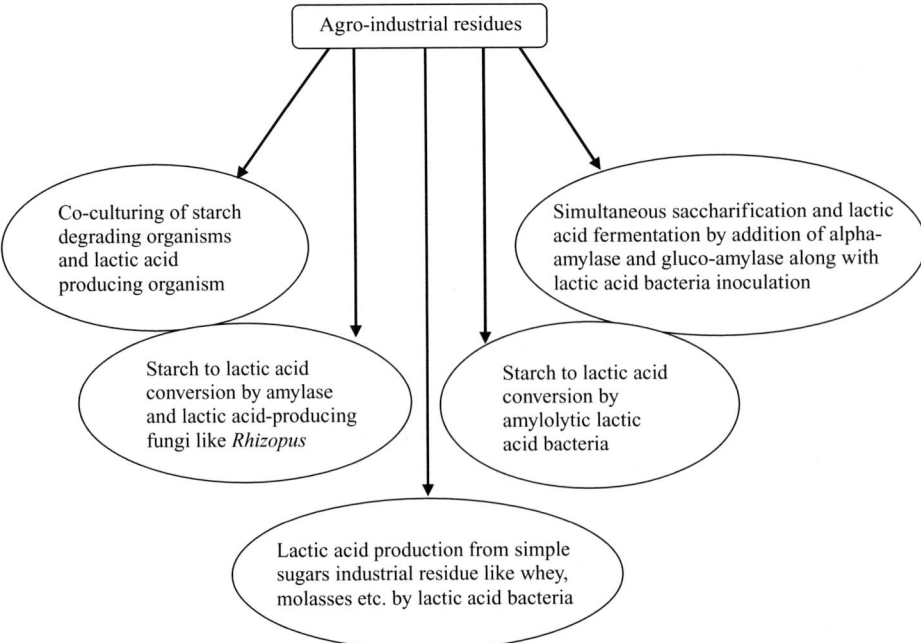

Fig. 2 The different processes of direct lactic acid fermentation.

bacterium, *Lactococcus lactis*, is done for simultaneous saccharification and lactic acid production.[7] *A. awamori* grows within the aerobic outer layer of the starch gel and *L. lactis* within the anaerobic inner core of the gel. The drawback of this system of mixed cultures is that the strains do not always have similar optimal process conditions. Here, one organism is aerobic and the other (*L. lactis*) is preferably anaerobic or microaerophilic. Simultaneous enzyme production, starch saccharification, and lactic acid fermentation from raw cassava starch in a circulating-loop bioreactor with *A. awamori* and *L. lactis* spp. *lactis* immobilized in loofa sponge have been reported by Roble et al.[8] Two columns with different conditions were used such that one aerated raiser column contained the fungal culture and starch, and another column was unaerated column with the bacterial culture in which the flow of substrate was downward. The column with the fungal culture was connected through an upper pipe and a lower pipe to the second column containing the bacterial culture. The substrate was saccharified by the fungal culture in the first column, which released sugars for use by the bacterial culture in the second column for production of lactic acid. The unused substrate was again circulated back to the first column for further saccharification. By controlling the valve opening and closing frequency, the circulation of liquid medium and its components, including oxygen concentration, was controlled (Fig. 3). Repeated fed-batch lactic acid production was performed for more than 400 hours and the average lactic acid yield and productivity from raw cassava starch were 0.76 g lactic acid per gram of starch and 1.6 g lactic acid per liter per hour, respectively. Since separate systems were used for starch saccharification by immobilized fungal culture and lactic acid fermentation by immobilized lactic acid bacteria, the effect of oxygen on lactic acid production could be significantly reduced.

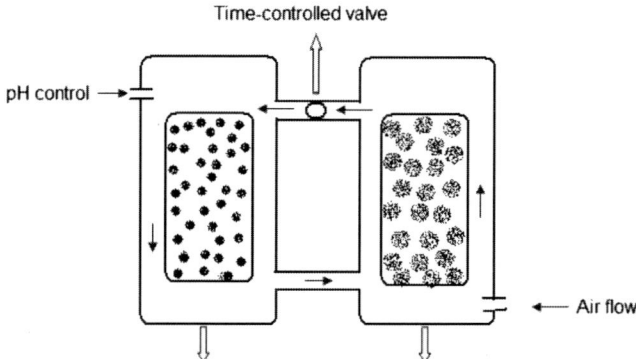

Fig. 3 The diagram shows a circulating-loop bioreactor with an amylase-producing fungus and lactic acid bacteria immobilized in loofa sponge. **Source:** From Biotechnol. Lett.[8]

Direct Lactic Acid Fermentation by Lactic Acid-Producing Fungi

Certain fungi, such as *Rhizopus* sp., can produce lactic acid. The best-known fungal sources of lactic acid are *Rhizopus oryzae* and *Rhizopus arrhizus*. Single-step simultaneous saccharification and fermentation of lactic acid by *R. arrhizus* 36017 and *R. oryzae* 2062 using potato, corn, wheat, and pineapple waste streams as production media have been reported.[9] *R. arrhizus* gave a high lactic acid yield up to 0.94–0.97 g per gram of starch or sugars, while *R. oryzae* had a lactic acid yield of 0.65–0.76 g per gram in 36–48 hours of fermentation. Supplementation with $2\,g\,L^{-1}$ of ammonium sulfate, yeast extract, and peptone, stimulated an increased yield of 8–15% lactic acid.[9] Even though homofermentative hexose-utilizing lactic acid bacteria have much greater efficiencies than fungi to convert sugars to lactic acid, research on lactic acid production by *Rhizopus* sp. has continued because of the ease of product separation and purification, and the ability of *Rhizopus* to utilize both complex carbohydrates and pentose sugars.[9] Research continues on enhancing lactic acid yield by strain improvement, substrate selection, and the addition of extra enzymes, such as pectinase, for increased accessibility of substrate by microorganisms.

Direct Lactic Acid Fermentation by Amylolytic Lactic Acid Bacteria

Isolation of amylolytic lactic acid bacteria have been reported from different tropical starchy fermented foods prepared from tubers and cereals. Important strains of *Lactobacillus* are *L. plantarum*, *L. manihotivorans*, *L. fermentum*, *L. amylovorus*, and *L. amylophilus*. Reddy et al.[4] has reviewed different types of lactic acid fermentation by amylolytic lactic acid bacteria and different substrates for direct lactic acid production. Lactic acid production using amylolytic lactic acid bacteria is still an interesting area of research because of the need for efficient methods and strains for direct utilization of starchy waste.

Direct Lactic Acid Fermentation by Simultaneous Saccharification and Fermentation

Most lactic acid bacteria are mesophiles, whereas conventional starch hydrolysis is done at elevated temperatures. To circumvent this problem much research has been carried out on simultaneous enzymatic saccharification and fermentation. John et al.[3] described reports on hydrolysis at low (37°C) and high (60°C) temperatures. The percentage of saccharification was almost the same at low and high temperatures, but the former takes longer than the latter. In simultaneous enzymatic saccharification and fermentation there is no inhibition of substrate on lactic acid production as it is in multistepped hydrolysate fermentation. As soon as the fermentable sugar is released,

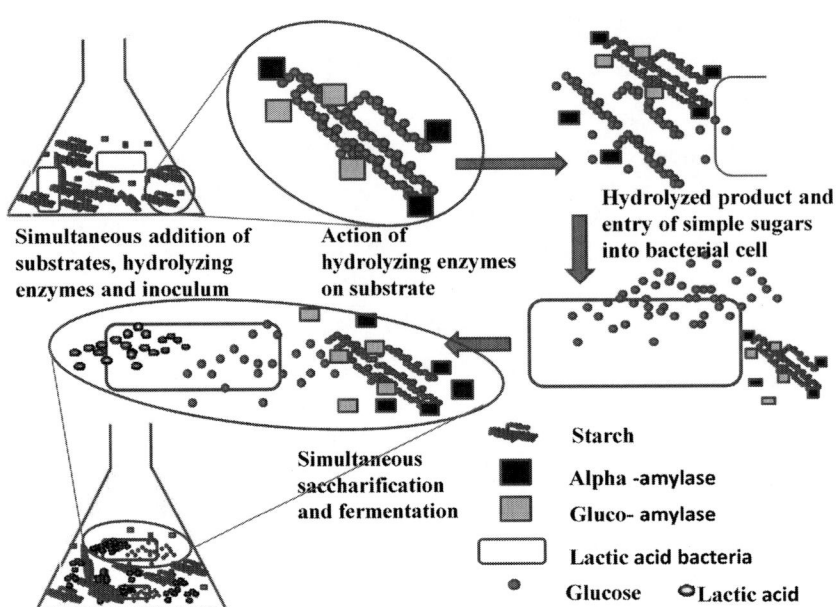

Fig. 4 Schematic diagram of simultaneous saccharification and fermentation processes.

the microorganism can metabolize the sugar for energy production and formation of lactic acid as by-product. The process of simultaneous saccharification and fermentation from starchy residues is represented in Fig. 4.

Usually starchy substrates are used as raw materials for lactic acid production in simultaneous saccharification and fermentation, and other direct lactic acid fermentations. There are some reports on the utilization of cellulosic industrial residues, such as Kraft pulp sludge for simultaneous saccharification and fermentation, but use of cellulose in simultaneous saccharification and fermentation is not economic because the cost of cellulase is very high and requires higher temperature for optimal action. Patel et al.[10] isolated a strain of *Bacillus* for the conversion of cellulosic and hemicellulosic hydrolysate at 50°C, which possibly could be an optimum temperature for cellulase activity. Although lactic acid bacteria, such as *L. delbrueckii*, *L. acidophilus*, and *L. casei* can produce lactic acid from glucose at higher concentrations, these bacteria lack the ability to ferment pentoses. So newly isolated strains, such as *Bacillus* sp. 36D1[9] can consume xylose and arabinose along with glucose and a complete bioremediation of the cellulosic waste can be achieved. Besides these complex residues, simple sugar-containing by-products, such as whey and molasses, can be utilized for the production of lactic acid using lactic acid bacteria.

CONCLUSION

Lactic acid bacteria have received extensive attention because of their easy use, high rate of growth, lactic acid productivity, safety, and conversion efficiency in terms of lactic acid yield. Lactic acid bacteria require many nutrients and lack the ability to utilize many complex nutrients. In this regard, fungal species have gained great interest, and have been recognized as suitable organisms for lactic acid production from complex substrates; however, most strains of these fungi produce other organic acids and by-products, which reduce the purity of the lactic acid and have lower conversion efficiency. Even though productivity is much reduced in the direct conversion to lactic acid by amylolytic strain, the costs of enzymes for saccharification can be reduced. These lactic acid bacteria can be used for the direct fermentation of glycogen waste from shellfish-processing industries. Cellulosic and starchy residues can be converted to lactic acid by simultaneous acid or enzyme saccharification and fermentation and accomplish the conversion of natural biopolymer waste to a multipurpose biodegradable polymer component, lactic acid. The cost for nitrogen and carbon supplementation in the production of lactic acid can be reduced by the utilization of industrial residues, which can also be an eco-friendly way for the disposal of industrial wastes. Strain improvements can be achieved by classical and modern methods for better yield and productivity.

REFERENCES

1. Wee, Y.-J.; Kim, J-N.; Ryu, H.W. Biotechnological production of lactic acid and its recent applications. Food Technol. Biotechnol. **2006**, *44* (2), 163–172.
2. Woiciechowski, A.L.; Nitsche, S.; Pandey, A.; Socool, C.R. Acid and enzymatic hydrolysis to recover reducing sugars from cassava bagasse: an economic study. Braz. Arch. Biol. Technol. **2002**, *45* (3), 393–400.
3. John, R.P.; Anisha, G.S.; Nampoothiri, K.M.; Pandey, A. Direct lactic acid fermentation: focus on simultaneous

saccharification and lactic acid production. Biotechnol. Adv. **2009**, *27* (2),145–152.
4. Reddy, G.; Altaf, M.D.; Naveena, B.J.; Venkateshwar, M.; VijayKumar, E. Amylolytic bacterial lactic acid fermentation—a review. Biotechnol. Adv. **2008**, *26* (1), 22–34.
5. Zhang, Z.Y.; Jin, B.; Kelly, J.M. Production of lactic acid from renewable materials by *Rhizopus* fungi. Biochem. Eng. J. **2007**, *35* (3), 251–263.
6. John, R.P.; Nampoothiri, K.M.; Pandey, A. Fermentative production of lactic acid from biomass: an overview on process developments and future perspectives. Appl. Microbiol. Biotechnol. **2007**, *74* (3), 524–534.
7. Kurusava, H.; Ishikawa, H.; Tanaka, H. L-lactic acid production from starch by coimmobilized mixed culture system of *Aspergilus awamori* and *Streptococcus lactis*. Biotechnol. Bioeng. **1988**, *31* (2), 183–187.
8. Roble, N.D.; Ogbonna, J.C.; Tanaka, H. L-lactic acid production from raw cassava starch in a circulating loop bioreactor with cell immobilized in loofa (*Luffa cylindrica*). Biotechnol. Lett. **2003**, *25* (13),1093–1098.
9. Jin, B.; Yin, P.; Ma, Y.; Zhao, L. Production of lactic acid and fungal biomass by *Rhizopus* fungi from food processing waste streams. J. Ind. Microbiol. Biotechnol. **2005**, *32* (11–12), 678–686.
10. Patel, M.A.; Ou, M.S.; Ingram, L.O.; Shanmugam, K.T. Simultaneous Saccharification and co-fermentation of crystalline cellulose and sugar cane bagasse hemicellulose hydrolysate to lactate by a thermotolerant acidophilic *Bacillus* sp. Biotechnol. Prog. **2005**, *21* (5), 1453–1460.

Lactobacillus plantarum in Foods

Maria Marco
Food Science and Technology, University of California–Davis, Davis, California, U.S.A.

Abstract

Lactic acid bacteria (LAB) have been integral to the production, preservation, and protection of food supplies throughout human history. The lactic acid bacterium, *Lactobacillus plantarum*, is of particular importance owing to its involvement in fruit, vegetable, milk, and meat fermentations and potential roles in improving human health. *L. plantarum* is a rapid producer of lactic acid; this capacity is necessary and beneficial or results in spoilage and off-flavors depending on the food product. Recent advancements in molecular and genetic analyses of this species have provided new information on its adaptations for growth and survival in food matrices and mammalian digestive tracts and its distinguishing biochemical and health-modulatory characteristics. This information opens opportunities to optimize and control *L. plantarum* for improvements in food quality and performance in the gut.

INTRODUCTION

Lactobacillus plantarum is a member of the lactic acid bacteria (LAB), a group of microorganisms essential for the production of numerous fermented food products. *L. plantarum* is distinct from other LAB as it is a highly versatile and flexible species encountered in a variety of niches including plant surfaces and mammalian digestive tracts. From these environments, *L. plantarum* enters into food supply chains resulting in either improvements or impairments in food quality. The capacity of *L. plantarum* to convert sugars to significant amounts of lactic acid by fermentation in low-pH environments is a distinguishing property of this species and is important for the development of food flavors, texture, shelf life and preservative properties. This species also produces a variety of other extracellular proteins and metabolites, which affects the safety and quality of food products and potentially human health. This entry will examine the contributions of *L. plantarum* to plant, dairy, and meat products and application of strains of this species as probiotics.

L. PLANTARUM DIVERSITY AND METABOLISM

Strains of *L. Plantarum* are low G + C content Gram-positive bacteria known for production of organic acids from carbohydrates in the absence of oxygen. This species is most closely related to *L. brevis* and *Pediococcus pentosaceus*, forming a distinct lineage among the order, *Lactobacillales*.[1] The genome sequence is known for the human isolate, *L. plantarum* WCFS1, and its coding capacity illustrates the functional significance of this species in food fermentations.[2] Specifically, *L. plantarum* WCFS1 possesses a large repertoire of genes involved in responding to environmental stresses including high acidity (pH $\leq$ 4.5) and salt concentrations (2–8%) and temperature extremes (2–40°C). Other food-relevant properties of *L. plantarum* WCFS1 include the production of small peptides, which inhibit the growth of other Gram-positive bacteria, including human pathogens, and a large number of surface-anchored proteins suggesting the ability to bind to numerous surfaces and substrates for growth.[2]

L. Plantarum WCFS1 is also illustrative of this species as it possesses the capacity to transport and consume diverse carbohydrates, including mono- and di-saccharides and other oligosaccharides with unknown composition.[2] Variation in carbohydrate metabolism genes among *L. plantarum* strains isolated from different plant and animal environments suggests substantial niche-specific adaptations.[3] *L. plantarum* WCFS1 is also typical of this species regarding its capacity to perform heterofermentation and homofermentation. In the presence of glucose, *L. plantarum* undergoes homofermentation producing D- and L-lactate as the only end products. This organism produces high amounts of lactate (over 1% weight/volume), in a manner that has the potential to be uncoupled from culture growth rates. While lactate is generally regarded as the dominant end product of *L. plantarum* during food fermentations, this organism also possesses the phosphoketolase pathway enabling heterolactic fermentation and production of ethanol, acetate, lactate, and CO_2 from other carbon sources including ribose, a pentose sugar found in meats and plants. These compounds impart flavor characteristics and inhibit the growth of spoilage and pathogenic bacteria.

L. PLANTARUM IN PLANT-BASED FOODS AND BEVERAGES

Before harvest, L. plantarum and several other LAB species are typically found in low but persistent amounts on plant surfaces including fruits and vegetables. After harvest and in conditions containing low amounts of oxygen, L. plantarum grows to high population sizes (10^9 cells/g) on plant materials. This activity is desirable and required for the production of food products such as sauerkraut and olives, economically relevant fermented foods consumed in the West, as well as other indigenous fermented foods produced globally. In sauerkraut fermentations, a strict heterofermentative LAB, Leuconostoc mesenteroides, initiates the fermentation and is followed by L. plantarum, which becomes the dominant fermenter during the latter half of the 2-week production period.[4] Although the mechanisms behind the appearance of L. plantarum during the latter stages of fermentation are not fully understood, this organism is particularly adept at growth and lactic acid production under anaerobic and low-pH conditions (e.g., sauerkraut final pH is 3.7) as compared to most other LAB. This species is also important in Spanish- and Greek-style olive fermentations. L. plantarum is able to grow despite the presence of high salt concentrations and plant phenolic compounds in the olive brines. Members of this species are able degrade some of olive phenols including oleuropein, a bitter-tasting compound, as well as perform other flavor- and aroma-forming reactions, which influence olive organoleptic properties.[5]

In olives and other food fermentations, LAB is often present with yeasts and other microorganisms. Depending on the desired product, fermentation processes have been designed either by trial-and-error or by directed efforts to favor the growth of one or more of these groups of microorganisms, sometimes in succession to each other. For example, L. plantarum is one of the main LAB that influences sourdough fermentations. Although yeasts are essential in bread-making, L. plantarum provides benefits including functional attributes such as the ability to hydrolyze phytic acid, an antinutritional factor present in flours.[6] Similarly, in cocoa fermentations, L. plantarum is an important member of the LAB community, which develops in response to the growth of yeasts and contributes to the properties of chocolate by the production of lactic acid and consumption of citric acid.[7] In yeast-fermented beverages such as wine and beer, attempts are made at limiting the growth of L. plantarum as it is associated with reductions in shelf life and production of off-flavors. While this species has the potential to convert malate to lactate (malolactate fermentation), a desirable reaction in some wines, it is a relatively minor contributor to this process and because some strains produce undesirable products including biogenic amines and a precursor, ethyl carbamate, a putative carcinogen, this species is generally regarded to have limited value in wine production.[8]

L. Plantarum in Dairy and Meat Products

L. plantarum is found in the intestinal tracts of animals and is associated with farms and livestock environments where it comes into contact with animal products including milk and meat. In fermented dairy products, L. plantarum has relatively minor roles compared to some other LAB. An exception is the involvement of L. plantarum as a nonstarter LAB (NSLAB) in cheese production. In comparison to starter LAB (e.g., Lactococcus lactis), which divides rapidly in milk, L. plantarum generally does not reach high population sizes in fresh milk owing to its inability to hydrolyze milk proteins (e.g., casein) and obtain sufficient amounts of essential amino acids for growth. Instead adventitious NSLAB like L. plantarum grow during cheese-ripening and become dominant members of the mature cheese microbiota, probably in response to proteolytic capacities of starter cultures and potentially other modifications to the cheese matrix including reductions in pH. Although NSLAB can be associated with cheese defects, L. plantarum, along with other LAB, are being added to some cheeses as adjunct cultures to better control the release of free amino acids and fatty acids affecting sensory characteristics. For example, glutamate dehydrogenase-producing strains of L. plantarum were shown to accelerate and intensify aroma and flavors during maturation of cheddar cheese.[9]

Some fermented meats are also produced through the actions of L. plantarum. This species along with L. sakei and L. curvatus are the main LAB isolated from European-style sausages and salami.[10] Between these species, L. sakei typically dominates meat fermentations, although strains of L. plantarum are used as starter cultures in commercial sausage production. The function of L. plantarum starter cultures is to obtain rapid declines in pH assuring the safety and quality characteristics of final product.[10]

L. PLANTARUM AND HEALTH BENEFITS

L. plantarum is a normal inhabitant of the human gut and strains of this species are commercially marketed as probiotics. Probiotics are defined as living microorganisms, which upon consumption in sufficient amounts confer a health benefit on the host.[11] L. plantarum 299v, a commercially used strain, has shown potential for alleviation of irritable bowel syndrome and protection against enteropathogenic bacterial infection.[12] This strain and L. plantarum NCIMB8826 were shown to survive passage through the human gut.[12] Cell products of the highly related strains L. plantarum WCFS1 and L. plantarum NCIMB8826 have been identified, which confer gut persistence or modulate human immune and epithelial cell responses.[13] These genes encode for the transport and degradation of certain carbohydrates as well as surface-localized compounds, which might mediate direct effects on the intestinal epithelium.[13]

L. plantarum 299v and other probiotic strains are increasingly incorporated into foods for human consumption. Although dairy is the most common food matrix for probiotic bacteria, plant- and meat-based matrices are attractive options for delivery of viable probiotics to the human gut; however, contributions of the food matrix on the health benefits conferred by L. plantarum and other probiotic strains remains to be determined. Considering that foods are composed of micromolecular and macromolecular compounds, vitamin, and minerals, each with known roles in human health, it is very likely that the choice of diet will affect probiotic function and overall efficacy in the gut.

CONCLUSIONS

L. plantarum is an important species in the production and preservation of healthy foods. Its application in plant, meat, and milk fermentations is intended to provide improvements in the sensory and safety attributes of products made from these matrices. Biochemical properties of this species also provide opportunities for designing foods with health benefits. These health benefits can be obtained through transformations of the chemical constituents of the foods and by the inclusion of probiotic strains. The application of molecular genetics to study the diversity and functional capacities of this species is rapidly accelerating the development of L. plantarum starter cultures, probiotics, and the control of indigenous L. plantarum populations in food matrices for optimal food quality and health benefits.

REFERENCES

1. Makarova, K.; Slesarev, A.; Wolf, Y.; Sorokin, A.; Mirkin, B.; Koonin, E.; Pavlov, A.; Pavlova, N.; Karamychev, V.; Polouchine, N.; Shakhova, V.; Grigoriev, I.; Lou, Y.; Rohksar, D.; Lucas, S.; Huang, K.; Goodstein, D.M.; Hawkins, T.; Plengvidhya, V.; Welker, D.; Hughes, J.; Goh, Y.; Benson, A.; Baldwin, K.; Lee, J.H.; Díaz-Muñiz, I.; Dosti, B.; Smeianov, V.; Wechter, W.; Barabote, R.; Lorca, G.; Altermann, E.; Barrangou, R.; Ganesan, B.; Xie, Y.; Rawsthorne, H.; Tamir, D.; Parker, C.; Breidt, F.; Broadbent, J.; Hutkins, R.; O'Sullivan, D.; Steele, J.; Unlu, G.; Saler, M.; Klaenhammer, T.; Richardson, P.; Kozyavkin, S.; Weimer, B.; Mills, D. Comparative genomics of the lactic acid bacteria. Proc. Natl. Acad. Sci. U.S.A. **2006**, *103* (42), 15611–15616.

2. Kleerebezem, M.; Boekhorst, J.; van Kranenburg, R.; Molenaar, D.; Kuipers, O.P.; Leer, R.; Tarchini, R.; Peters, S.A.; Sandbrink, H.M.; Fiers, M.W.; Stiekema, W.; Lankhorst, R.M.; Bron, P.A.; Hoffer, S.M.; Groot, M.N.; Kerkhoven, R.; de Vries, M.; Ursing, B.; de Vos, W.M.; Siezen, R.J. Complete genome sequence of *Lactobacillus plantarum* WCFS1. Proc. Natl. Acad Sci. U.S.A. **2003**, *100* (4), 1990–1995.

3. Molenaar, D.; Bringel, F.; Schuren, F.H.; de Vos, W.M.; Siezen, R.J.; Kleerebezem, M. Exploring *Lactobacillus plantarum* genome diversity by using microarrays. J. Bacteriol. **2005**, *187* (17), 6119–6127.

4. Plengvidhya, V.; Breidt, F., Jr.; Lu, Z.; Fleming, H.P. DNA fingerprinting of lactic acid bacteria in sauerkraut fermentations. Appl. Environ. Microbiol. **2007**, *73* (23), 7697–7702.

5. Maria, J.; Jose, L.; Curiel, A.; Rodriguez, H.; Rivas, B.D.L.; Munoz, R. Study of the inhibitory activity of phenolic compounds found in olive products and their degradation by *Lactobacillus plantarum* strains. Food Chem. **2008**, *107* (1), 320–326.

6. Pepe, O.; Blaiotta, G.; Anastasio, M.; Moschetti, G.; Ercolini, D.; Villani, F. Technological and molecular diversity of *Lactobacillus plantarum* strains isolated from naturally fermented sourdoughs. Syst. Appl. Microbiol. **2004**, *27* (4), 443–453.

7. Schwan, R.F.; Wheals, A.E. The microbiology of cocoa fermentation and its role in chocolate quality. Crit. Rev. Food Sci. Nut. **2004**, *44* (4), 205–221.

8. Spano, G.; Massa, S. Environmental stress response in wine lactic acid bacteria: beyond *Bacillus subtilis*. Crit. Rev. Microbiol. **2006**, *32* (2), 77–86.

9. Williams, A.G.; Withers, S.E.; Brechany, E.Y.; Banks, J.M. Glutamate dehydrogenase activity in lactobacilli and the use of glutamate dehydrogenase-producing adjunct *Lactobacillus* spp. cultures in the manufacture of cheddar cheese. J. Appl. Microbiol. **2006**, *101* (5), 1062–1075.

10. Ammor, M.S.; Mayo, B. Selection criteria for lactic acid bacteria to be used as functional starter cultures in dry sausage production: an update. Meat Sci. **2007**, *76* (1), 138–146.

11. Sanders, M.E. Probiotics: considerations for human health. Nut. Rev. **2003**, *61* (3), 91–99.

12. de Vries, M.C.; Vaughan, E.E.; Kleerebezem, M.; de Vos, W.M. *Lactobacillus plantarum*—survival, functional and potential probiotic properties in the human intestinal tract. Int. Dairy J. **2006**, *16* (9), 1018–1028.

13. Marco, M.L.; Pavan, S.; Kleerebezem, M. Towards understanding molecular modes of probiotic action. Curr. Opin. Biotechnol. **2006**, *17* (2), 204–210.

Leavened Breads

Neela Badrie
Department of Food Production, University of the West Indies, St. Augustine, Trinidad and Tobago

Abstract
Fermented doughs may be leavened and in most cases are baked or steamed prior to consumption as bread. The entry describes wheat gluten and the physical and biochemical changes associated with leavened dough (yeast, acid, and frozen) development. The applications, processing technologies, fermentation, and major microorganisms are presented for various leavened breads.

INTRODUCTION

Dough fermentation is based on the metabolic activities of yeast and lactic acid bacteria (LAB) in which carbohydrates are broken down primarily to lactic acid, acetic acid, ethanol, and CO_2. Two main factors associated with leavened breadmaking are CO_2 gas and the gluten network. These factors along with wheat gluten, the physical and biochemical changes in dough development, and characteristics and fermentation technologies of various leavened breads are discussed in this entry.

WHEAT GLUTEN

Wheat flour is the preferred raw material for production of yeast-leavened bread because of its gas retention capacity. Hard wheat, which is high in protein (12–14%), forms more elastic dough for high-leavened breads.[1] Doughs from other cereals lack gluten proteins. Gluten refers to a set of proteins that are grouped as glutenins and gliadins.[2] The ability to make leavened breads depends on the viscoelasticity of the dough, which allows for the entrapment of CO_2 released during yeast fermentation. The glutenin subunits (HMW-GS) having a high molecular weight are largely responsible for the elasticity of gluten.

LEAVENING

The yeast, *Saccharomyces cerevisiae*, used in leavening the dough for baking, exists in its four main commercial forms, namely, liquid yeast, compressed yeast, active dry yeast, and instant active dry yeast.[3] The production of a defined cellular structure in leavened-baked breads depends on the number, size, and retention of gas bubbles in the dough as the protein sets during baking. This process is influenced by the dough formation and the mixing conditions. As the yeast produces CO_2 during anaerobic fermentation, the gas goes into the solution in aqueous phase within the dough. Entrapped nitrogen gas plays a major role in dough conditioning by providing bubble nuclei into which CO_2 can diffuse.

DOUGH DEVELOPMENT

Dough development is an undefined term in breadmaking which covers a number of complex changes in bread ingredients (Fig. 1). In the process of developing bread dough, changes in the physical properties of the dough are crucial in improving its ability to retain CO_2. These changes are brought about by both the hydration of the gluten proteins in the flour and the application of energy through the process of kneading.[4] The viscoelastic properties of the dough depend on the HMW-GS fraction, whose proteins are stretched into linear chains forming elastic sheets under the gas bubbles in a 3-D network. The higher proportions of glutenins to gliadins result in stronger doughs which require more mixing to produce loaves of greater volume resulting in better dough-handling characteristics.[3] An underdeveloped gluten network can lead to reduced volume and increased bread firmness.

YEAST-LEAVENED DOUGHS

Straight Dough Bulk Fermentation

The mechanism for dough development in bulk fermentation depends on the metabolic activity of the yeast. The basic steps are:[4]

1. Mixing of the ingredients to form homogeneous dough.
2. Resting of the dough for a prescribed time period (minimum 1 hour) depending on flour quality,

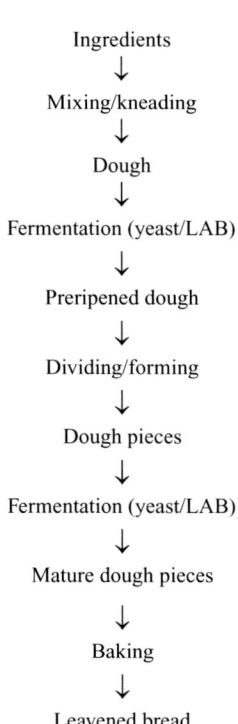

Fig. 1 Major steps in the production of bread from fermented doughs.

yeast level, dough temperature, and type of bread produced.
3. Remixing ("knocking-back") of the dough part way through the fermentation.

Sponge and Dough

In the sponge-and-dough process of leavened breadmaking, a part of the ingredients is thoroughly fermented before being added back to the remainder of the ingredients for further mixing to form the final dough.[4] The key features are:

- A two-stage process in which the total quantity of flour, water, and other ingredients are mixed to form a homogenous soft dough-sponge.
- Resting of the sponge for a prescribed time period (4–16 hours) depending on flavor requirements.
- Mixing of the sponge with the remainder of the ingredients to form a homogeneous dough.
- Immediate processing of the final dough, although a short period of bulk fermentation may be given.

Chorleywood Bread Process (CBP)

Traditional baking processes require prolonged periods (up to 3 hours) of fermentation before the dough is ready for the oven; however, the CBP relies on an increase in yeast level (50–100%), use of ascorbic acid as an improver/oxidant, and high-speed continuous mixing of the dough to accelerate dough development. The main difference between the CBP and the bulk fermentation lies in the rapid development of the dough in the mixer rather than through a prolonged resting period. The modification of the protein structure in dough to retain gas, due to yeast fermentation, occurs within 5 min of starting the mixing stage.

FROZEN-LEAVENED DOUGH

Dough Development and Stability

Yeast-leavened frozen doughs require flours milled from hard wheat varieties (~12–14% protein). In frozen dough development, the mixing stage, along with subsequent mechanical make-up operations (dividing, rounding, sheeting, and forming) are more important than in conventional bread processes. The dough must possess optimal rheological properties to avoid excessive proofing times, decreased bake volumes, and poor textural properties. Gas retention problems can result from ice-crystallization damage of the 3-D gluten protein network which is responsible for gas retention. It is recommended to thaw frozen dough at 5°C for 16 hours. Storage should be limited to 10–12 weeks.

Frozen-Dough Yeast

Higher levels of yeast (5–6% flour basis) are normally used in frozen dough formulations to compensate for losses in yeast-gassing power during extended storage. Unlike the typical dry yeasts which have a moisture level of 5–6%, frozen-dough dry yeasts have about 25% moisture. Compressed yeast is the most preferred form of yeast. Ethanol-tolerant strains of *S. cerevisiae* and *Saccharomyces chevalieri* have been used in frozen doughs owing to their osmotic and freeze tolerances.

ACID-FERMENTED LEAVENED DOUGH

Sourdough

Sourdough is defined as dough in which the microorganisms, mainly LAB and yeasts, originate from sourdough or a sourdough starter; these microorganisms are metabolically active or need to be reactivated. These bacteria and yeasts are responsible for leavening capacity, primarily through CO_2 production.[5] The sourdough microbiota perform various functions, such as: 1) increased acidification for technological quality of doughs from rye flour; 2) generation of flavor and aroma compounds such as acetic acid, lactic acid, and alcohols; 3) provision of synergistic effects on yeast; and 4) antagonistic action toward spoilage fungi.[6]

Traditionally, the starter for sourdough is a mixture of cereal flour (wheat of rye) and water for dough consistency

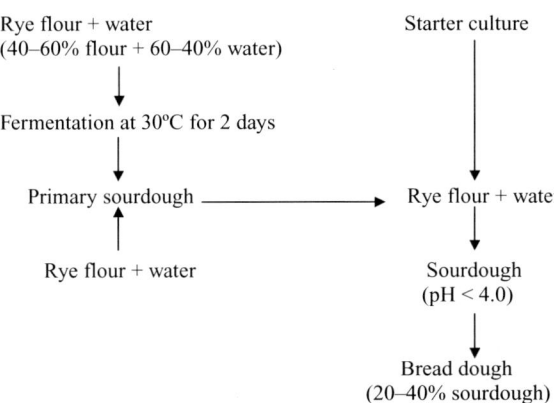

Fig. 2 Steps in the preparation of rye sourdough and rye bread dough. **Source:** From *Handbook of Fermented Functional Foods*.[7]

Table 1 Microorganisms isolated from sourdoughs.

Lactic acid bacteria (LAB)	*Lactobacillus sanfranciscensis* (sanfrancisco)
	L. plantarum
	L. acidophilus
	L. alimentarius
	L. amylovorus
	L. buchneri
	L. casei
	L. crispatus
	L. delbrüeckii
	L. farciminis
	L. fermentum
	L. fructivorans
	L. johnsonii
	L. kimchii
	L. leichmannii
	L. panis
	L. mindensis
	L. paralimentarius
	L. pontis
	L. reuteri
	Pediococcus acidilactici
	P. pentosaceus
	Weisella confusus
Yeasts	*Candida krusei*
	C. guillermondii
	C. milleri
	C. norvegensis
	Endomycopsis fibuliger
	Hansenula anomala
	H. subpelliculosa
	Pichia polymorpha
	P. saitoi
	Saccharomyces cerevisiae
	S. chevalieri
	S. curvatus
	S. exiguous
	S. fructum
	S. inusitatus
	S. panis fermentati
	Torulopsis delbrüeckii
	T. stellata
	T. unisporus

with incorporation of the natural biota of saved starter. Fig. 2 shows the steps in production of rye sourdough.[7] The process in San Francisco-style (French bread) sourdough is similar to other sourdough breads, as the starter is built about every 8 hours, 7 days a week. The predominant LAB in sourdough belongs to the genus *Lactobacillus* (Table 1). The most common yeast is *Candida milleri*. The dough is usually fermented at 27°C for 48 hours. An association between LAB and *S. cerevisiae* results in higher relative percentage of yeast fermentation products, such as 1-propanol, 2-methyl-1-propanol, 3-methyl-1-butanol, and ethanol) in sourdough. Using fungi-specific PCR detects *Candida humilis, Debaryomyces hansenii, S. cerevisiae,* and *Saccharomyces uvarum* in sourdoughs.[8]

In rye breads, the acidification of the sourdough is essential for the requirements of the bread (containing >20% rye flour) because water-soluble proteins of rye flour do not form gluten.[9] The pH of these doughs is kept within 4.2–4.7; a higher pH is advisable for doughs with lower rye content and lower enzyme activity. The pH values of German commercial rye breads range from 4.2 to 5.4.

ACID-LEAVENED BREADS

Enjera (Injera)

Enjera is a pancake-like acid-fermented leavened bread which is consumed as staple in Ethiopia and Sudan. It is made from either white or red tef (*Eragrostis tef*) or sorghum or wheat.[10]

Tef or other cereal flour or combinations of flours are mixed with water and "irsho" (inoculum). The thin watery batter ferments for 17–72 hours and is then incorporated in the main part of the batter and fermented for another $1/2$–2 hours. The acidic leavened batter is steam-baked.

Pullaria, Aspergillus, Penicillium, Rhodotorula, Hormodendrum, and *Candida* have been isolated from enjera samples. *Candida guilliermondii* is commonly found; this yeast produces acid and gas; however, the initial fermentation is carried out by Gram-negative anaerobic rods such as *Enterobacter, Hafnia, Citrobacter, Klebsiella, Escherichia,* and *Proteus* in which the pH falls to 5.0–5.5. Later, *Streptococcus, Leuconostoc,* and *Lactobacillus* develop and lower the pH to below 4.1.

Idli and Dosa

Idli is a naturally fermented white acid-leavened steamed pancake-like bread of Southern India and Sri Lanka.[10] Its

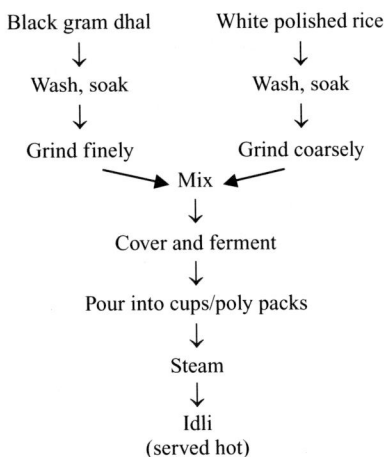

Fig. 3 Steps in the processing of idli.

appeal as a breakfast food is due to its moist, soft, and spongy texture and sour flavor. Idli is prepared from a fermented batter of white polished rice (*Oryza sativa* L.) and black gram dhal (*Phaseolus mungo*) of varying proportions. Dosa batter is very similar to idli batter except that it is finely ground and thinner.[10] It is quickly fried as a thin, fairly crisp pancake.

Fig. 3 outlines the major processing steps for idli. The batter, which is fermented overnight or longer at 30°C, is poured into small cups or a special idli pan having cups to facilitate in situ fermentation; poly packs have also been used. Product shelf life is 3–5 days when held at $30 \pm 2°C$ and 10 days under refrigerated conditions.

Back-slopping is the traditional technique in idli fermentation. *Leuconostoc mesenteroides* is primarily responsible for production of CO_2 and acid as well as strengthening the bread dough by production of dextrans. Other active LAB are *L. fermentum* and *Lactobacillus delbrueckii*. Yeasts isolated from idli include *S. cerevisiae*, *Torulopsis glabrata*, *Debaryomyces hansenii* var. *hansenii*, *S. exiguus*, and *Guehomyces pullulans*. The initial pH of around 6.0 falls to 4.3–5.3.

Kisra

The traditional preparation of the Sudanese staple bread, kisra is from sorghum, which is nutritionally inferior to other blends of flours. Protein and lysine contents of kisra increase by 73% as a result of supplementation of sorghum with 30% groundnut flour.

L. fermentum, *Lactobacillus reuteri*, and *Lactobacillus amylovorus* are the dominant bacteria in doughs inoculated with wet or dry sorghum dough preparations. After inoculation with the dry dough preparation, *Candida krusei* is commonly detected. During fermentation, the pH declines from 5.5 to 3.4, as lactic acid predominates.

CONCLUSION

Yeast and acid-leavened breads are produced mainly by *Saccharomyces* spp. and *Lactobacillus* spp. respectively. The trends in leavened bread technology are to add starter cultures, shorten the period for bulk fermentation and processing, apply more mechanical mixing to dough, and incorporate more functional ingredients.

REFERENCES

1. Coultate, T.P. *Food: The Chemistry of Its Components*, 2nd Ed.; The Royal Society of Chemistry: Cambridge, UK, 1995; 85–125.
2. Johnson-Green, P. *Introduction to Food Biotechnology*; CRC Press: Boca Raton, FL, 2002; 81–135.
3. Bonjean, B.; Guillaume, L.-D. Yeasts in bread and baking products. In *Yeasts in Food*; Boekhout, T., Robert, V., Eds.; Woodhead Publishing Limited: Abington, Cambridge, England, 2003; 289–307.
4. Cauvain, S.P. Breadmaking processes. In *Technology of Breadmaking*; Cauvain, S.P., Young, L.S., Eds.; Blackie Academic and Professional: London, 1998; 18–44.
5. Hansen, A.-S. Sourdough bread. In *Handbook and Beverage Fermentation Technology*; Meunier-Goddik, L., Hansen, A.-S., Josephsen, J., Nip, W.-K., Stanfield, P.S., Toldrá, F., Eds.; Marcel Dekker: Basel, Switzerland, 2004; 729–755.
6. Spicher, G. Preparation of stable sourdoughs and sourdough starters by drying and freeze-drying. In *Frozen and Refrigerated Doughs and Batters*, 2nd Ed.; Kulp, K., Lorenz, K., Brümmer, J., Eds.; American Association of Cereal Chemists, Inc.: St. Paul, MN, 1998; 53–61.
7. Molin, G. The role of *Lactobacillus plantarum* in foods and in human health. In *Handbook of Fermented Functional Foods*; Farnworth, E.R., Ed.; CRC Press: Boca Raton, FL, 2003; 305–342.
8. Meroth, C.B.; Hammes, W.P.; Hertel, C. Identification and population dynamics of yeasts in sourdough fermentation processes by PCR-denaturing gradient gel electrophoresis. Appl. Environ. Microbiol. **2003**, *69* (12), 7453–7461.
9. Lönner, C.; Ahrné, S. *Lactobacillus*: baking. In *Food Biotechnology Microorganisms*; Hui, Y.H., Khachatourians, G.G., Eds.; VCH Publishers Inc.: Eureka, CA, 1995; 797–844.
10. Steinkraus, K.H. Indigenous fermented foods involving an acid fermentation- preserving and enhancing organoleptic and nutritional qualities of fresh foods. In *Handbook of Indigenous Fermented Foods*, 2nd Ed.; Marcel Dekker: New York, 1996; 111–347.

Lemon Essential Oils: Bioactive Properties in Fermented Foods and Beverages

M.A. Del Nobile
Department of Food Science, University of Foggia, Foggia, Italy

B. Speranza
M. Sinigaglia
A. Conte
Institute for Biotechnology Research and Applications for Security and Promotion of Local Products and Quality (BIOAGROMED), University of Foggia, Foggia, Italy

Abstract

This entry focuses on the content, composition, and antimicrobial and antioxidant properties of lemon essential oils as natural food preservatives. The most recognized in vitro assays and model food systems used to assess their effectiveness are discussed. A few applications to fermented foods and beverages are also reported.

INTRODUCTION

Modern consumer trends and food legislation require high-quality foods that are preservative-free, safe, and mildly processed with extended shelf life. Society appears to be experiencing a trend of "green" consumerism, favoring fewer synthetic food additives and products with smaller impact on the environment. Consequently in the last decade, research on natural compounds from plants has notably increased. Antiseptic qualities of plants and their extracts have been recognized since the early 1900s. Antimicrobial and antioxidant properties of derived essential oils have been extensively reviewed recently due to their potential role in medical procedures, cosmetics, and pharmaceutical and food applications.[1] The principal reason essential oils are used by the food industry is to enhance palatability of food products; however, another reason is that spices help control undesirable microorganisms in foods and thereby contribute to health. Essential oils from plants such as basil, cumin, caraway, thyme, rosemary, and coriander have inhibitory effects on microorganisms.[2,3] Citrus fruits and their essential oils represent another source of natural compounds to prolong the shelf life and improve the safety of food.[4,5] Commercial lemon essential oils are generally obtained from different parts of the lemon plant by cold pressing (peels) or distillation (leaves) and can be used as ingredients in the pharmaceutical industry, as well as in perfumes. In the food industry, lemon essential oils and their essences are used as flavor enhancers in soft drinks, alcoholic beverages, and fruit-based products. In this entry, the antimicrobial and antioxidant properties of lemon essential oils are overviewed and attention is focused on the most recognized mechanism of action and on their effectiveness as demonstrated in model food systems and food products.

LEMON ESSENTIAL OILS

Isolation of functional compounds from citrus fruits can be of interest to the food industry for use in the reduction of oxidative changes in food to improve quality and nutritional value.[6] Fernandez-Lopez et al.[7] reported that the presence of functional dietary fiber and antioxidants added in citrus by-products improves the nutritional quality of cooked and dry-cured sausage. Lemon oils can also have a very pronounced antimicrobial activity, even if their complexity and variability make it difficult to correlate their action to a specific component. Lemon oils are characterized by a wide range of volatile compounds, some of which are important flavor quality factors. In addition to such compounds, a large number of flavonoids, responsible for the characteristic fruit color, possess not only antimicrobial potential but also anti-inflammatory and antitumor activities. In general, the antimicrobial effects of lemon oils have been explained through the presence of C10 and C15 terpenes with aromatic rings, and phenolic hydroxylic groups capable of forming hydrogen bonds with active sites of target enzymes. Moreover, other active terpenes, as well as alcohols, aldehydes, and esters, can contribute to the overall antimicrobial effect.

In fact, such molecules can interact with bacterial cell structures causing the inhibition of cell growth or cell death. Belletti et al.[8] evaluated the headspace composition of lemon essence by using the SPME-GC technique (solid phase microextraction—GC mass spectrometry). Limonene was the main constituent (42.30%); high quantities of β-pinene (21.28%), γ-terpinene (14%), α-pinene (7.79%), β-myrcene (4.04%), and p-cymene (3.4%) were also found. Other components were present in negligible amounts: 4-carene, β-thujene, cis-β-ocimene, and terpinolene. Oxygenated monoterpenes were measured as citronellal, neral, geranial, and linalool (1.73% in total). Traces of sesquiterpenes (farnesene and caryophyllene), ethanol, aliphatic aldehydes (octanal and non-anal), esters (ethyl butanoate, neryl acetate, and geranyl acetate), and ketones were also reported. Considering the large number of chemical groups present in lemon essential oils, each of them with a potential mode of action, it is most likely that their antibacterial activities are not attributable to one specific mechanism. An important characteristic of these components is their hydrophobicity, which enables them to partition in the lipids of the bacterial cell membrane, thus disrupting membrane structure and rendering them more permeable. Leakage of ions and other cell contents can then occur and as a consequence, extensive loss of critical molecules leads to cell death. It is also generally known that these components can interfere with the proton motive force, electron flow, and active transport of the membrane.[1]

Screening experiments have reported lemon oils to be one of the best broad-spectrum candidates for inhibition of foodborne pathogens and spoilage pathogens organisms. Most research work was carried out by in vitro assay. Conte et al.[9] investigated the efficacy of purified lemon extract against foodborne spore-forming bacteria, lactic acid bacteria, and yeasts. Lemon oil was a very effective antagonist. The minimum inhibitory concentration (MIC) to prevent microbial growth was very low, ranging around 200 mg/L. Very few applications of lemon oils to model food systems or actual food products have been reported in the literature. Conte et al.[10] assessed the potential use of lemon extract to inhibit the growth of *Oenococcus oeni* and *Lactobacillus plantarum*, bacteria involved in the malotactic fermentation of wine. Evolution of carbon dioxide was used to measure metabolic activity in a model food system and the MIC and the non-inhibiting concentration (NIC) were recorded. Results suggested lemon extract exhibited a non-linear dose-related inhibitory effect on bacterial growth. Concentrations slightly higher than NIC levels appreciably reduced the growth rate as well as MIC level. In a follow-up study in wine,[11] depletion of L-malic acid, production of L-lactic acid, and gas evolution in the headspace of sealed vials indicated that lemon oil prevented the malolactic fermentation in wine, suggesting that lemon oil could be used to control the malolactic fermentation in wine.

Suhr and Nielsen[12] tested the activity of several essential oils on the growth of rye bread spoilage fungi and found that the antifungal effects of plant extracts depended on the method of application. In fact, they used two methods: Rye bread-based agar medium supplemented with each active agent and actual rye bread exposed to volatile oil in air. Larger phenolic compounds, typical active agents of thymol and eugenol, have the best effect applied directly to a medium, whereas smaller compounds, such as citral and allyl isothiocyanate contained in lemongrass oil, are more effective when added as volatiles.

CONCLUSIONS

The development of more specific assays to provide information related to the deterioration of foods and biological systems should be a goal of future research. Modern consumers ask for natural products, free of synthetic additives. Therefore, the application of natural compounds should continue in the future in greater detail. Scientists will explore for new sources of natural components with potent antioxidant and antimicrobial properties. Of course, these extracts, their mixtures, isolates, and concentrates have to meet all requirements for human health and safety.

REFERENCES

1. Burt, S. Essential oils: their antibacterial properties and potential applications in foods—a review. Int. J. Food Microbiol. **2004**, *94*, 223–253.
2. Dorman, H.J.D.; Deans, S.G. Antimicrobial agents from plants: antibacterial activity of plant volatile oils. J. Appl. Microbiol. **2000**, *88*, 308–316.
3. Lambert, R.J.W.; Skandamis, P.N.; Coote, P.; Nychas, G.J.E. A study of the minimum inhibitory concentration and mode of action of oregano essential oil, thymol and carvacrol. J. Appl. Microbiol. **2001**, *91*, 453–462.
4. Abdalla, A.E.; Roozen, J.P. Effect of plant extracts on the oxidative stability of sunflower oil and emulsion. Food Chem. **1999**, *64*, 323–329.
5. Conte, A.; Scrocco, C.; Sinigaglia, M.; Del Nobile, M.A. Innovative active packaging systems to prolong the shelf life of mozzarella cheese. J. Dairy Sci. **2007**, *90*, 2126–2131.
6. Ponce, A.G.; Del Valle, C.E.; Roura, S.I. Natural essential oils as reducing agents of peroxidase activity in leafy vegetables. LWT **2004**, *37*, 199–204.
7. Fernandez-Lopez, J.; Fernandez-Gines, J.M.; Aleson-Carbonell, L.; Sendra, E.; Sayas-Barbera, E.; Perez-Alvarez, J.A. Application of functional citrus by-products to meat products. Trends Food Sci. Technol. **2004**, *15*, 176–185.

8. Belletti, N; Ndagijimana, M.; Sisto, C.; Guerzoni, M.E.; Lanciotti, R.; Gardini, F. Evaluation of the antimicrobial activity of citrus essences on *Saccharomyces cerevisiae*. J. Agric. Food Chem. **2004**, *52*, 6932–6938.
9. Conte, A.; Speranza, B.; Sinigaglia, M.; Del Nobile, M.A. Effect of lemon extract on foodborne microorganisms. J. Food Prot. **2007**, *70*, 114–118.
10. Conte, A.; Sinigaglia, M.; Del Nobile, M.A. Use of lemon extract as natural preservative in wine applications. J. Food Prot. **2007**, *70*, 1896–1900.
11. Conte, A.; la Gatta, B.; Brescia, I.; La Notte, E.; Del Nobile, M.A. Control of a malolactic bacterium by lemon extract. Adv. Food Sci. **2008**, *3*, 145–149.
12. Suhr, K.I.; Nielsen, P.V. Antifungal activity of essential oils evaluated by two different applications techniques against rye bread spoilage fungi. J. Appl. Microbiol. **2003**, *94*, 665–674.

Lipases in Foods

Hugo S. Garcia
Food Investigation Unit (UNIDA), Technological Institute of Veracruz, Veracruz, Mexico

Charles G. Hill, Jr.
Department of Chemical Engineering, University of Wisconsin-Madison, Madison, Wisconsin, U.S.A.

Abstract

Lipases are of interest to researchers because of their unique biocatalytic properties. This entry treats the nature of these enzymes, the reactions that they catalyze in either aqueous or organic media, and important kinetic aspects of their behavior. Applications of lipases involving modification of food systems, especially fats and oils, are also discussed.

INTRODUCTION

Lipases (EC 3.1.1.3, triacylglycerol hydrolases) are enzymes that catalyze the hydrolysis of glycerides as well as other esters. The products of the hydrolysis of glycerides are fatty acids and lower glycerides (or glycerol). These enzymes may also be referred to as acylglycerolases, acyl hydrolases, or triacylglycerol hydrolases.

At the interface of lipid micelles with water, lipases are converted from an inactive molecular configuration to a configuration with biocatalytic activity for cleavage or synthesis of ester bonds. Until recently, two criteria were used to define a "true" lipase: 1) activation of these enzymes requires the presence of an interface between different phases; and 2) the tertiary structure of the lipase should contain a "lid" covering the active site of the enzyme. On contact with the interface, the lid is displaced to provide substrates with access to the active site. Because exceptions (enzymes that have a lid but exhibit no interfacial activation) have been identified, lipases are currently defined as carboxylesterases that catalyze either the hydrolysis or synthesis of acylglycerols containing long-chain (≥ 10 carbon atoms) fatty acid residues. Hydrolysis of acylglycerols with constituent fatty acid residue chain lengths below 10 carbon atoms usually indicates the presence of an esterase. However, most lipases are also capable of hydrolyzing these esterase substrates.

Lipases may also be viewed as serine hydrolases containing a Ser-His-Asp catalytic triad. When bound to an interface, these enzymes exhibit reorientation of a short α-helix to expose the active site. Most lipases are acidic glycoproteins with molecular weights from 20 to 60 kDa. The structures, mechanisms, and genetic modifications of lipases from mammalian and microbial sources have been successfully compiled.[1,2]

OVER-EXPRESSION AND MODIFICATION OF LIPASES

Molecular biology has developed tools to facilitate improvements in biocatalysts in terms of stability, activity, and specificity. Usually these improvements involve over-expression of the lipase gene through cloning. Many lipases have been successfully cloned and over-expressed in *E. coli* using different vectors. The amount of enzyme produced via this route is usually greater and subsequent purification is easier than when the original source is employed. Molecular modifications of enzymes usually involve either a rational design approach based on molecular models to effect changes in individual amino acids by site-directed mutagenesis to modify specific residues that could lead to profound changes in activity and/or stability or a directed evolution approach involving random mutations on the lipase gene.

Detailed knowledge of the three-dimensional structures of lipases is important for achieving successful changes based on these strategies. Most lipases whose structures are known contain a substrate-binding site near the edge of a central β sheet. The size, shape, and hydrophobicity of the site are correlated with differences in selectivity for binding fatty acid chains. Thus, modification of specific amino acid residues may increase or decrease the hydrophobicity of the site with concomitant effects on substrate specificity, enantioselectivity, and regioselectivity. These properties depend on the nature and structure of the substrate, as well as its interactions with the active site. The thermostabilities of several lipases (e.g., *Candida antarctica* fraction B) have also been modified by protein engineering and structural changes involving directed evolution.

REACTIONS CATALYZED BY LIPASES

The classic reaction that provided the stimulus for studies of lipases is the hydrolysis of ester bonds in insoluble triacylglycerols. During hydrolysis one water molecule is consumed for each ester bond that is broken. However, lipases can also catalyze the reverse reaction that forms ester bonds from a hydroxyl moiety and a carboxylic acid with concomitant release of a water molecule. Lipases can also catalyze the cleavage and formation of ester bonds in a sequential fashion (interesterification) with no net consumption or formation of water. The generic group of reactions known as interesterification reactions consists of acidolysis in which the acyl group of a fatty acid is exchanged for a fatty acid residue in an ester; transesterification in which two acyl groups are interchanged between two ester molecules; and alcoholysis in which an acyl group is exchanged between an ester and an alcohol. Lipases also catalyze aminolysis reactions in organic solvents. The overall stoichiometries of several generic reactions mediated by lipases are depicted in Fig. 1.

Hydrolysis
$R_1COOR^* + H_2O \longrightarrow R_1COOH + R^*OH$

Esterification
$R_1COOH + R^*OH \longrightarrow R_1COOR^* + H_2O$

Acidolysis
$R_1COOR^* + R_3COOH \longrightarrow R_1COOH + R_3COOR^*$

Alcoholysis
$R_1COOR^* + R_3OH \longrightarrow R_1COOR_3 + R^*OH$

Transesterification
$R_1COOR^* + R_3COOR_4 \longrightarrow R_1COOR_4 + R_3COOR^*$

Aminolysis
$R_1COOR^* + R_3NH_2 \longrightarrow R_1CONHR_3 + R^*OH$

For an sn – 1,3 specific lipase:
$$R^* \equiv \begin{array}{c} CH_2- \\ | \\ CHR_x \\ | \\ CH_2R_x \end{array}$$

For an sn – 2 specific lipase:
$$R^* \equiv \begin{array}{c} CH_2R_x \\ | \\ CH- \\ | \\ CH_2R_x \end{array}$$

For a non-specific lipase R^* encompasses both of the above definitions. R_1, R_3, R_4 and R^* are alkyl groups while the R_x are either acyl or hydroxyl groups.

Fig. 1 Reactions catalyzed by lipases.

KINETICS OF REACTIONS CATALYZED BY LIPASES

The activity of lipases increases when the substrate concentration reaches its solubility limit so that a second phase is formed. This trait identifies the primary fundamental difference between lipases and esterases. Esterases have no catalytic activity for hydrolysis of insoluble esters. For soluble substrates (esters) the kinetics of the esterase-mediated reactions obey conventional Michaelis-Menten rate expressions. By contrast, lipases have no activity on soluble substrates.

Mathematical models of the kinetics of lipase-mediated reactions are generally based on manipulation of generalized Michaelis-Menten rate laws to arrive at mathematical expressions containing only the concentrations of species that can be experimentally determined. These models are then fit to experimental data. Irrespective of the type of reaction catalyzed by the lipase, the best description of the reaction kinetics is generally provided by a rate expression based on a Ping Pong Bi Bi mechanism. This mechanism involves two key molecular processes: nucleophilic attack on the substrate ester bond by the hydroxyl group of the serine moiety of the lipase, followed by subsequent hydrolysis of the acylated enzyme complex. Paiva et al.[3] have reviewed the literature concerning this mechanism and the associated rate expressions. Quantitative studies of the reaction kinetics have involved the use of single and multiple response models to account for the release of either total fatty acids or of each individual fatty acid during lipase-catalyzed hydrolysis of butteroil to produce lipolyzed butteroils.

Relevant efforts have also been made to model the rates of a variety of lipase-mediated synthesis reactions involving triacylglycerols including acidolysis, glycerolysis, alcoholysis, and transesterification reactions, as well as the synthesis of sugar esters. Control of the water activity (a_w) in the reaction medium for these synthesis reactions is critical because the dependence of enzyme activity on a_w is an intrinsic property of the enzyme-catalyzed reaction and not a function of the solvent or other solutes.[4]

LIPASES IN NON-AQUEOUS MEDIA

Lipases are effective biocatalysts in organic as well as in aqueous media. Some water must always be present if the enzyme is to retain the molecular configuration responsible for its biocatalytic activity. However, the high activity and the degree of specificity characteristic of the behavior of lipases in nearly anhydrous media significantly enhance prospects for commercial biotechnological applications of lipases. Associated advantages include increased solubility of non-polar substrates and products, shift of the thermodynamic equilibrium position to favor synthesis over hydrolysis, better yields, ease of product and enzyme recovery, suppression of water-dependent side reactions such as hydrolysis, elimination of microbial contamination, reduced inhibition by lipophylic substrates/products, ease of manipulation of substrate- and enantio-selectivities, enhanced thermostability, and elimination of the need for immobilization because of insolubility of the enzyme.

FOOD APPLICATIONS OF LIPASES

Baking

In recent years, reports have suggested that phospho(lipases) can be used as substitutes or supplements for traditional emulsifiers, as these enzymes degrade polar wheat lipids to produce emulsifying lipids "in situ." Addition of lipases decreases the surface tension of the batter and increases bulk viscosity to reduce batter aeration time. Commercial lipases are currently available for this particular application (see Table 1).

Fats for Infant Formulas

Infants rely on breast milk fat as a primary source of nutrients and dietary energy (50–60%). This fat is rich in palmitic acid located predominantly (>60%) in position sn-2, and may account for 20–25% of total fatty acid intake. Positions sn-1 and sn-3 of breast milk fat are occupied by unsaturated fatty acid residues. *Betapol* is a commercially available product of Unilever (Loders Croklaan) prepared using an sn-1,3 specific lipase. This product matches both the fatty acid composition and the positional distribution of fatty acid residues on the glycerides present in breast milk. Sahin and coworkers[5] have designed a human milk fat substitute containing 5% n-3 fatty acids and 40% oleic acid while retaining ca. 77% palmitic acid residues at the sn-2 position.

Dairy Products

The dairy industry employs lipases and pregastric esterases to impart desired flavor characteristics in the manufacture of several cheese varieties, to prepare concentrated dairy flavors that are used as natural food ingredients and to produce enzyme-modified cheeses. Several microbial lipases are available and these enzymes have also been used by the dairy flavors industry (see Table 1). Traditional pregastric (calf, lamb, and kid goat) esterases have been immobilized and used for the production of lipolyzed butter oils.[6]

Designer Fats and Oils

The vast majority of lipase applications in food products involve modifications of fats and oils. Total or partial transesterification of the triacylglycerides of a fat or oil with those from a different oil can be employed to modify the physico-chemical, sensory, and nutritional properties of the starting material. Major developments in the fats and oil industries have been driven by searches for equivalents for cocoa butter and breast milk fat, as well as for margarines or fats with improved spreadability (or that contain no *trans* fatty acid residues) and structured lipids with beneficial nutritional properties.

Cocoa butter is a major constituent of chocolate products and melts at roughly human body temperature. This substance is composed primarily (>70%) of quasi-symmetrical triacylglycerols (SOS, POP, and POS) that contain unsaturated oleic acid residues at the sn-2 position and saturated fatty acid residues at the sn-1,3 positions. Inconsistencies in market demand and price volatility have led to searches for routes for production of cocoa butter equivalents (CBE) from cheaper vegetable oils. Several companies in Europe and Japan now produce CBE using lipase-catalyzed interesterification reactions.[7]

Industrial processes for production of margarines and shortenings originally involved partial hydrogenation of oils over inorganic catalysts. However, these processes invariably lead to formation of fats containing *trans* isomers of fatty acids. Recent documentation of the link between dietary ingestion of *trans* fatty acids and coronary heart disease has generated interest in the production of zero-*trans* versions of margarines and shortenings, in particular the use of lipases to interesterify natural hard fats (e.g., palm stearin) or fully hydrogenated oils, with naturally occurring oils.[8] Although the enzymatic process poses no major technical difficulties, matching the physical properties of the resulting semisolid fats to those of traditional products has been a real challenge. In order to comply with regulations by different governments concerning labeling and the maximum levels of *trans* fatty acids allowed, research and practical efforts have focused on four technological options: modification of the hydrogenation process, use of interesterification, use of fractions high in solids from natural oils, and use of trait-enhanced oils. The goal is to develop options that provide equivalent functionality, are economically feasible, and do not greatly increase saturated fat content.[9]

Flavors and Fragrances

Lipases can be employed by the food industry for the organic synthesis of low-molecular-weight esters that are responsible for fruity flavors and odors. Lipase-catalyzed processes include synthesis of aliphatic, terpene, and aromatic esters in solvent-free media, preparation of (z)-3-hexen-1-yl acetate (a model compound for green flavor notes) using the same lipases, and synthesis of butyl acetate and isoamyl propionate. Enzymatic processes for the synthesis of capsaicin analogs (olvanil) involve amidation of vanillyl amine with fatty acid derivatives; this approach allows one to control the pungency and irritating effects of capsaicinoids and thus better utilize their beneficial

Table 1 Commercially available food-grade lipases, phospholipases, and esterases.

Manufacturer	Source	Proposed use
Amano Enzyme Inc. (http://www.amano-enzyme.co.jp)	*A12*, from *A. niger* *AY30*, from *C. rugosa* *G50*, from *P. camembertii* *R*, from *P. roqueforti* *F*, from *R. niveus*	All listed lipases can be used for processing fats and oils
Biocatalysts Ltd. (http://www.biocatalysts.com)	*Penicillium roqueforti* *Rhizopus* sp. Porcine pancreas *Mucor javanicus* *Candida rugosa* *Rhizopus niveus*	All listed lipases can be used to produce enzyme-modified cheese *Candida rugosa* can be employed for the general hydrolysis of fats *Rhizopus niveus* can be used to hydrolyze chicken fat
Cargill Bioactives US, LLC (http://www.cargilltexturizing.com)	Kid-goat PGE (pregastric esterase) Lamb PGE Calf PGE	All listed lipases can be used to produce enzyme-modified cheese
DANISCO (http://www.danisco.com) Genencor Division	Kid-goat PGE Lamb PGE Calf PGE *Aspergillus oryzae* Glycolipase from *Fusarium heterosporum* Phospholipase from *Streptomyces violaceoruber*	All listed PGE can be used to produce enzyme-modified cheese Enzyme-modified cheese Lysogalactolipids for the baking industry Lysophosoholipids from eggs
DSM (http://www.dsm.com)	Piccantase K, Kid-goat PGE Piccantase L, Lamb PGE Piccantase C, Calf PGE Piccantase A, Microbial Piccantase R, Microbial	All listed enzymes can be used to produce enzyme-modified cheese
NOVOZYMES (http://www.novozymes.com)	Lecitase Lecitase Ultra Lipopan Lipozyme TL IM Lipozyme TL 100 L Noopazyme Novozym 435 Palatase	Oil degumming Oil degumming Baking industry Oils and fats industry Prehydolysis of fats Pasta/Noodles Oil based specialties Dairy industry
RENCO (http://www.renconz.com)	Kid-goat PGE Lamb PGE Calf PGE	All listed PGE can be used to produce enzyme-modified cheese
Specialty Enzymes and Biochemicals, Inc. (http://www.specialtyenzymes.com)	*Aspegillus niger* *Candida cylindracea*	Both enzymes can be used to produce dairy flavors
Verenium (http://www.verenium.com)	Phospholipase from *Pichia pastoris*	Degumming of vegetable oil

physiological activities as analgesics and stimulants of the cardiovascular and respiratory systems.

Surfactants

Mono- and diacylglycerols (DAG) are important food-grade emulsifiers. Although these compounds have traditionally been prepared by chemical glycerolysis of fats and oils, the corresponding enzymatic processes have been thoroughly investigated. A wide variety of oils, fatty acids, reaction schemes, and lipases have been studied in both solvent-based and solvent-free media.[10] Approaches include partial hydrolysis, glycerolysis, and esterification. Lipase-catalyzed processes are used by Archer Daniels

Midland and the KAO Corporation (Japan) to manufacture oils rich in DAG to be consumed by individuals interested in the associated health claims. Phospholipids and their partial hydrolysis products (lysophospholipids) are categorized as surfactants because of their amphiphylic nature. By varying the molecular structures of these emulsifiers, one can manipulate their physico-chemical properties and emulsifying capabilities. The necessary structural changes are catalyzed by different phospholipases (A_1, A_2, C or D).[11] Olestra® is a sugar (poly)ester that is the best-known representative of a new generation of non-ionic surfactants with interesting nutritional properties because these esters are poorly absorbed in the gastrointestinal tract. These fat substitutes can be utilized as frying oils and as replacements for fats in food products such as ice cream, cheese, margarine, baked goods, and fried foods. Sugar esters can be synthesized using either biocatalysts (lipases) or traditional inorganic catalysts.

Refining (Degumming) of Edible Oils

Novozymes has developed a process for degumming edible oils based on a phospholipase A_1 from *Fusarium oxysporium* (Lecitase®). This phospholipase hydrolyzes phospholipids to the corresponding lysophospholipids that are easy to separate because they migrate to the aqueous phase. Phospholipases have also been employed to effect transformations of natural phospholipids.[11]

Nutraceuticals

Dietary ingestion of those fatty acids categorized as nutraceuticals produces not only nutritional benefits, but also physiological benefits in the form of either therapeutic or preventive medicinal effects. These fatty acids include α-linolenic, γ-linolenic, arachidonic, conjugated linoleic acid, pinolenic acid, and polyunsaturated omega-3 fatty acids such as docosahexaenoic and eicosapentaenoic acids. Lipases can be employed to incorporate residues of these fatty acids in naturally occurring fats and oils to produce structured or designer lipids for nutraceutical applications. In some cases residues of more than one physiologically beneficial fatty acid can be incorporated in the designer triglyceride.[12] Lipase-mediated acidolysis, glycerolysis, and transesterification reactions can be employed for this purpose.

CONCLUSIONS

Lipases constitute a group of highly versatile biocatalysts. They are characterized by unique abilities to effect both hydrolytic and synthesis reactions involving a broad range of substrates and products. They function as biocatalysts in both aqueous and organic solvents, and possess high regio- and substrate specificities. (The regioselectivity of a lipase refers to its ability to catalyze reactions at various positions on the glycerol backbone of substrates, e.g., sn-1,3; sn-2; or non-selective.) Hence lipases are an extremely attractive enzyme for use in selective modification of fats and oils.

REFERENCES

1. Alberghina, L.; Schmid, R.D.; Verger, R. *Lipases: Structure, Mechanism and Genetic Engineering*, GBF Monographs; VCH: Weinheim, Germany, 1990; Vol. 16.
2. Wooley, P.; Petersen, S.B. *Lipases, Their Structure, Biochemistry and Application*; Cambridge University Press: Cambridge, England, 1994.
3. Paiva, A.L.; Balcao, V.M.; Malcata, F.X. Kinetics and mechanisms of reactions catalyzed by immobilized lipases. Enzyme Microb. Technol. **2000**, *27*, 187–204.
4. Lee, C.H.; Parkin, K.L. Effect of water activity and immobilization on fatty acid selectivity for esterification reactions mediated by lipases. Biotechnol. Bioeng. **2001**, *75*, 219–227.
5. Sahin, N.; Akoh, C.C.; Karaal, A. Human milk fat substitutes containing omega-3 fatty acids. J. Agric. Food Chem. **2006**, *54*, 3717–3722.
6. Garcia, H.S.; Qureshi, A.; Lessard, L.; Ghannouchi, S.; Hill, C.G., Jr. Immobilization of pregastric esterases in a hollow fiber reactor for continuous production of lipolyzed butteroil. Lebensm. Wiss. U. Technol. **1995**, *28*, 253–258.
7. Xu, X. Production of specific-structured triacylglycerols by lipase-catalyzed reactions: a review. Eur. J. Lipid Sci. Technol. **2000**, *102*, 287–303.
8. Otero, C.; Lopez-Hernandez, A.; Garcia, H.S.; Hernandez-Marin, E.; Hill, C.G., Jr. Continuous enzymatic transesterification of sesame oil and a fully hydrogenated fat: effects of reaction conditions on product characteristics. Biotechnol. Bioeng. **2006**, *94*, 877–887.
9. Hunter, J.E. Dietary trans fatty acids: review of recent human studies and food industry responses. Lipids **2006**, *41*, 967–992.
10. Gunstone, F.D. Review. Enzymes as biocatalysts in the modification of natural lipids. J. Sci. Food Agric. **1999**, *79*, 1535–1549.
11. Ulbrich-Hofmann, R. Phospholipases used in lipid transformation. In *Enzymes in Lipid Modification*; Uwe, T.B., Ed.; Wiley-VCH Verlag GmBH: Weinheim, Germany, 2000, 219–262.
12. Garcia, H.S.; Arcos, J.A.; Ward, D.J.; Hill, C.G., Jr. Synthesis of glycerides containing n-3 fatty acids and conjugated linoleic acid by solvent-free acidolysis of fish oil. Biotechnol. Bioeng. **2000**, *70*, 587–591.

Maize: Durable Resistance Breeding

Randall Wisser
Department of Plant and Soil Science, University of Delaware, Newark, Delaware, U.S.A.

Abstract

Ephemeral versus durable plant disease resistance is classified retrospectively. The isolation of genes controlling ephemeral resistance has revealed a common mechanism of action: recognition-mediated resistance. The few genes underlying durable resistance that have been cloned appear to represent different mechanisms of action. Thus, knowledge of a genes mechanism of resistance may provide a characterizable link to its longevity. Breeding for non-recognition-mediated resistance would provide the path to *most* durable resistance. Identifying the genes underlying diverse resistance mechanisms in maize may be achieved by working with its tremendous natural variation, which in combination with new technologies is now being exploited for more efficient analysis of natural variation. As these genes are identified and characterized, other technologies may be used to facilitate breeding. It is expected that new perspectives on maize plant–pathogen interactions will emerge from the detailed characterization of natural resistance variation, and it is hoped that this will help shift breeding for durable resistance from a retrospective to predictive science.

INTRODUCTION

Breeding for host resistance is challenged by evolving pathogens against which we seek stable and long-term resistance. Largely due to their research tractability, emphasis has been placed on understanding the function of resistance genes that have large effects, which happen to have ephemeral utility in agricultural production systems. The classification of specific genes as durable is difficult since most durable resistance is characterized as a phenomena pertaining to an individual or a population—not an individual gene. Advances in plant research are now allowing for the detailed characterization of durable resistance. If specific genes can be classified as durable they may be emphasized in breeding programs. This entry presents a perspective on ephemeral versus durable resistance in the context of currently available evidence, some approaches being used to characterize durably associated resistance, and some techniques available for the development of resistant cultivars.

THE BILATERAL VIEW OF PLANT DISEASE RESISTANCE

Gene-for-gene (GFG) resistance is manifested as a qualitative response (i.e., resistant versus susceptible) arising from the interaction of specific plant and pathogen genotypes. The qualitative nature of GFG resistance has allowed for its detailed characterization across many plant–pathogen systems. GFG resistance is typically found for biotrophic pathogens that attempt to evade detection while deriving nutrition from the living host. If the pathogen's presence is detected by a resistance gene (R-gene) product in the host, a localized cell death response ensues (referred to as a hypersensitive response), which limits disease development. GFG resistance is less frequently effective against necrotrophic pathogens that reproduce on dead plant tissue. Substantial empirical evidence has shown that GFG resistance is ephemeral in agricultural production systems.

Horizontal and vertical resistances are defined in statistical terms, as a two-way system with host and pathogen sources of variation[1] (Fig. 1). Significant host main effects indicate horizontal resistance and significant interaction effects between the host and pathogen indicate vertical resistance. Likewise, significant pathogen main effects indicate aggressiveness and significant pathogen-by-host interaction effects indicate virulence. There are several interesting properties of this definition: 1) variation in horizontal and vertical resistance, and similarly, variation in aggressiveness and virulence may coexist (i.e., the host and pathogen may have significant main and interaction effects; Fig. 1); 2) vertical resistance in the host and specific virulence in the pathogen are inseparable; and 3) a basis (e.g., genetic, physiological, etc.) for horizontal and vertical resistance is not implicit. Horizontal resistance is pathogen-independent and thus considered durable. It has been generally associated with quantitative mulitgenic resistance (i.e., resistance varying on a quantitative scale controlled by multiple genes). Vertical resistance is pathogen-dependent and thus considered ephemeral. It has been generally associated with GFG resistance and qualitative monogenic resistance.

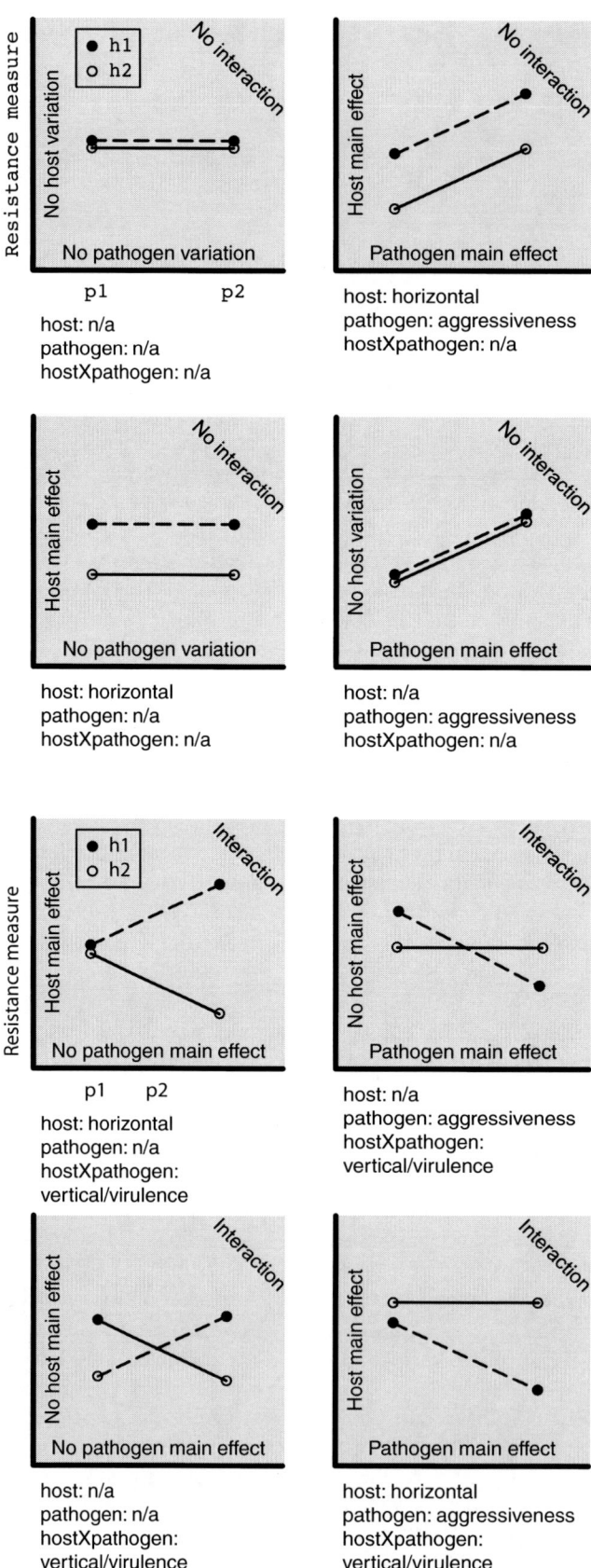

Fig. 1 Vertical and horizontal resistance defined in statistical terms, as a two-way system with host and pathogen sources of variation.

Since the concepts of GFG resistance[2] and vertical and horizontal resistance[1] were introduced, a bilateral view of plant resistance has developed whereby aspects of resistance have been classified as either vertical or horizontal, narrow- or broad-spectrum, qualitative or quantitative, and monogenic or multigenic (Fig. 2a). In the same bilateral way, each term pair has been associated with ephemeral versus durable host resistance; however, current evidence reveals crossover between the divide and breakdown of a bilateral view[3] (Fig. 2b). There exists the notion that the spectrum of resistance (defined in terms of whether resistance is vertical/narrow-spectrum versus horizontal/broad-spectrum) is the major determinant host factor to explain and predict the durability of resistance—this may be justifiable, but the spectrum of resistance can never be fully ascertained since not all plant and pathogen genotypes can be tested. However, as the mechanisms of action have been revealed for the genes underlying resistance loci classified as ephemeral or durable, a more useful relationship is thus far observed between the mechanism of action underlying resistance and durability. This relationship has been suggested by others[4,5] and is an important distinction with implications for breeding for durable resistance for any pathosystem. One must keep in mind, however, that host variables alone do not fully explain resistance durability: host, pathogen, and environment variables and interactions thereof represent a more complex and realistic explanation for the longevity of a host's resistance. Here, only host aspects are discussed.

A BREAKDOWN OF THE DIVIDE

There is a good amount of evidence to indicate that the path to *least* durable resistance in breeding is to emphasize the incorporation of GFG resistance conditioned by single genes with qualitative effects that confer vertical/narrow-spectrum resistance involved in pathogen recognition. The ability for the pathogen to quickly overcome GFG resistance has been attributed to the large selection pressure imposed on its population by this type of resistance. In some cases, individual aspects of GFG resistance (e.g., qualitative resistance and monogenic resistance) have separately been associated with ephemerality, which does not always hold true. For instance, durable qualitative monogenic resistance (e.g., the *mlo* gene in barley) and durable partial (i.e., resistance that is not complete) monogenic resistance (e.g., the *Lr34* gene in wheat) are known for some pathosystems, albeit these are rare cases. Quantitative multigenic resistance has been associated with durability, and there is no field evidence to dispute this. A major explanation for its durability is based on the observation that quantitative multigenic resistance is predominantly horizontal. Interestingly, some of the individual genomic loci containing the genes underlying quantitative multigenic resistance (quantitative resistance loci or QRLs) have been found to have differential effects against different pathogen isolates.[6] Thus, if the spectrum of resistance is predictive of durability, each gene underlying multigenic resistance may not be equally durable.

THE GENETIC BASIS OF DURABLE RESISTANCE

A few durable resistance genes have been cloned and their mechanism of action has been characterized. These include the barley *mlo* gene that confers resistance to *Erysiphe graminis* f. sp. *hordei*[7] and the wheat *Lr34* gene that confers resistance to *Puccinia triticina*.[8] The sequences of these genes are distinct from each other and those commonly found to underlie GFG resistance. More importantly, the mechanisms of action for these genes are distinct from those known for GFG R-genes.

It is the expectation that quantitative multigenic resistance is durable that makes its characterization important. Numerous studies on maize diseases have reported the map locations for QRL (see a synthesis of 437 maize QRL by Wisser et al.[9]). The use of these mapped QRL in resistance breeding, e.g., by marker-assisted introgression, is less apparent. This is at least in part due to the moderate and variable effects of QRL leading to imprecision in QRL mapping. Advances in genetic mapping designs, statistics, and sequencing and genotyping are leading to remarkable increases in QRL mapping resolution in maize, which should improve the use of marker-assisted introgression. The eventual cloning and thorough characterization of many QRL will add value to our understanding of the associative network of plant disease resistance-related phenomena as they pertain to durability. Awaiting direct evidence of the genes underlying QRL, a review of the literature led[3] to conclude that QRL likely draw from diverse mechanisms of resistance, which could explain the apparent durability of quantitative multigenic resistance.

At least four approaches are being applied in attempts to identify the genes underlying naturally occurring quantitative trait variation. A positional cloning approach was used to dissect a large effect QRL, *Rcg1*, conditioning resistance to *Colletotrichum graminicola* that causes anthracnose stalk rot.[10] While this represents the first QRL cloned in maize, it is an extreme case of a QRL in terms of its unusually large effect size. The resistance mechanism for *Rcg1* is unknown; however, two genes underlie the functionality of this QRL, and it is noteworthy that these genes have sequence motifs similar to GFG R-genes. The *Rcg1* locus is at low frequency in temperate U.S. maize and its durability is not known. The three other approaches that are allowing for more efficient fine mapping of QRL and the identification of genes controlling QRL are: 1) advanced

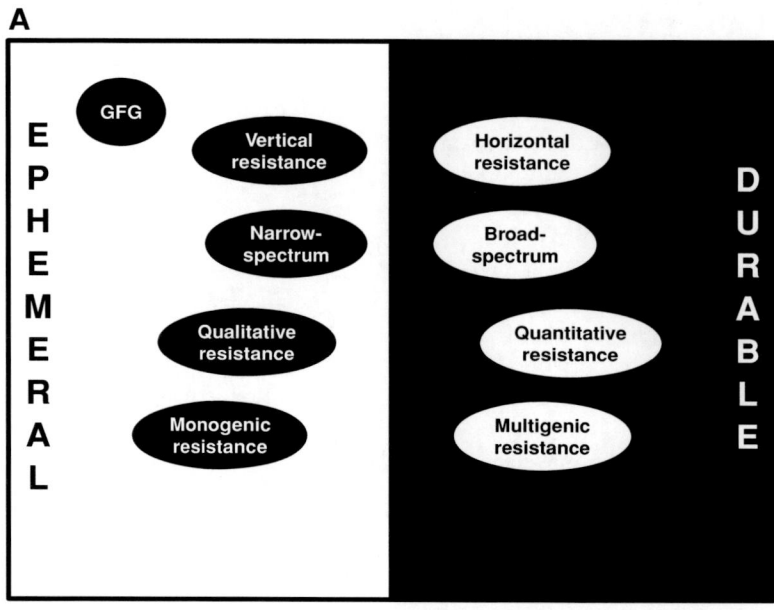

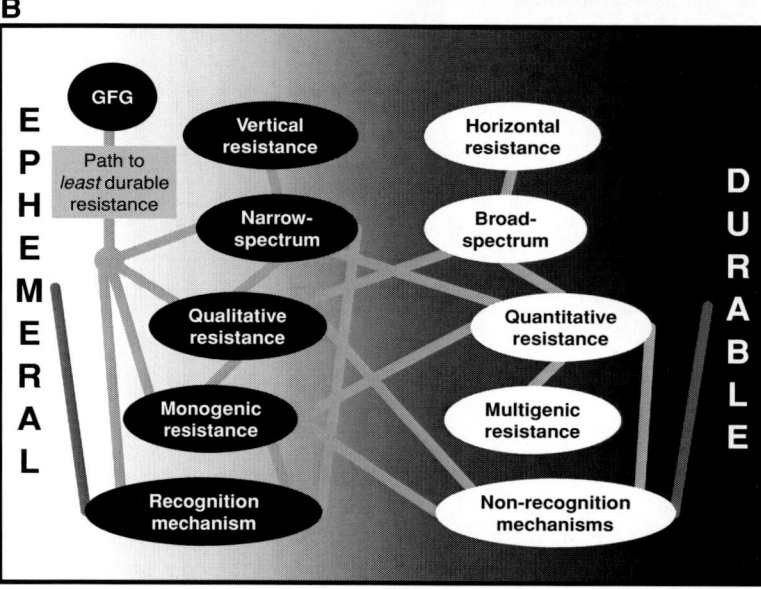

Fig. 2 (A) Classical versus (B) contemporary perspectives on aspects of plant disease resistance and their relationship to ephemerality and durability in agricultural production systems. The generalized associations of resistance-related phenomena with ephemeral versus horizontal resistance are indicated by background shading (e.g., horizontal resistance, broad-spectrum, quantitative resistance, multigenic resistance, and non-recognition based resistance have all been associated with durable resistance). In the classical view relationships are perceived to be sharply divided. In the contemporary view the divide is blurred. Resistance-related phenomena are indicated in black or white filled ovals. Lighter lines indicate evidence based relationships and darker lines indicate hypothesized relationships.

intercross recombinant inbred line mapping;[11] 2) high-resolution association mapping;[12] and 3) nested association mapping.[13] These approaches are now being applied to study maize disease resistance including resistance to southern leaf blight, northern leaf blight, and gray leaf spot (R.J. Nelson and P. Balint-Kurti, personal communication). A technological explosion of genotyping and genome sequencing capabilities is facilitating these QRL mapping endeavors, and certainly new approaches and facilitating technologies will follow. Ultimately, many genes will be identified that explain natural variation in quantitative resistance. Their use in breeding is a separate issue.

DEVELOPING DURABLY RESISTANT PLANTS

Strategies to engineer resistance through genetic transformation have been considered or applied. The host defense response may be envisaged as a series of cascading molecular and biochemical processes triggered and regulated, in some cases, by a few key genes. Activating these cascades through the controlled expression of key genes has been considered for engineering resistance. For instance, Stuiver and Custers[14] describe the constitutive expression of genes that induce the hypersensitive response independent of the pathogen's presence. This essentially amounts to the conversion of vertical resistance to horizontal resistance. At least one major issue with this approach is the consequent constitutive reduction of photosynthetic tissue and thus yield. A possible remedy could be to bring gene induction under the control of a spatially and/or temporally dependent promoter or a non-race-dependent pathogen inducible promoter, which will take time to develop. The application of this approach may be limited to certain pathogens (e.g., obligate biotrophs), but could be a durable approach for them.

The incorporation of specific resistance alleles is central to breeding and at least three techniques are possible: 1) introgression-by-hybridization with phenotypic or marker-assisted selection; 2) genetic transformation (discussed above); and 3) gene replacement via homologous recombination. The latter two approaches require knowledge of the gene(s) of interest; whereas introgression-by-hybridization benefits from this knowledge but it is not required (linked molecular markers can be use). Unlike genetic transformation that randomly inserts a sequence of interest into the genome, homologous recombination allows specific sequences in the genome to be replaced. The development of routine homologous recombination in plants would be a significant technological advance that could leverage the findings of genetically dissected QRL to facilitate resistance breeding through more precise genome manipulation. As the genes underlying resistance variation are discovered, specific superior alleles that confer high levels of resistance may be subsequently identified and incorporated into breeding germplasm. This targeted allele swapping technique could be applied to either ephemeral or durable pathosystems and may draw on either natural or synthetic alleles. Synthetic alleles can be generated by in vitro techniques, including error-prone PCR and DNA shuffling. These developments will take time and offer no guarantees for success in increasing the longevity of resistance. Access and use of the rich diversity of natural maize alleles are still currently the most apparent path to ensure long-term resistance and requires the curation and preservation of natural and irreplaceable maize diversity. Following an emerging theme for natural plant durable resistance gene mechanisms, it seems logical that breeding to incorporate diverse mechanisms of resistance into the same culitvar would provide the breeding path to *most* durable resistance.

CONCLUSION

It should go without saying that the primary interest in breeding for plant resistance is to attain sufficient (i.e., does not limit production) and durable performance under a variety of production conditions. Our ability to deliberately breed for durable resistance is limited since we do not have the ability to predict the longevity of a host's resistance. Current and building evidence suggests that resistance durability is most directly tied to the mechanism of action of the underlying resistance gene(s). Thus, GFG or recognition-based resistance mechanisms should be avoided and diverse non-recognition-based mechanisms should be exploited. The exploitation of natural genetic diversity is paramount to understanding resistance mechanisms and as a reservoir of superior alleles for breeding. Fortunately, new and improving technologies are expediting analyses of the vast wealth of natural maize genetic diversity, which will further facilitate resistance breeding.

REFERENCES

1. Vanderplank, J.E. *Disease Resistance in Plants*; Academic Press: New York, 1968.
2. Flor, H.H. Host-parasite interaction in flax rust—its genetics and other implications. Phytopathology **1955**, *45*, 680–685.
3. Poland, J.A.; Balint-Kurti, P.J.; Wisser, R.J.; Pratt, R.C.; Nelson, R.J. Shades of gray: the world of quantitative disease resistance. Trends Plant Sci. **2009**, *14* (1), 21–29.
4. Lindhout, P. The perspectives of polygenic resistance in breeding for durable resistance. Euphytica **2002**, *124*, 147–156.
5. Parlevliet, J.E. Durability of resistance against fungal, bacterial, and viral pathogens; present situation. Euphytica **2002**, *124*, 217–226.
6. Marcel, T.C.; Gorguet, B.; Ta, M.T.; Kohutova, Z.; Vels, A.; Niks, R.E. Isolate specificity of quantitative trait loci for partial resistance of barley to *Puccinia hordei* confirmed in mapping populations and near-isogenic lines. New Phytol. **2008**, *177* (3), 743–755.
7. Büschges, R.; Hollricher, K.; Panstruga, R.; Simons, G.; Wolter, M.; Frijters, A.; van Daelen, R.; van der Lee, T.; Diergaarde, P.; Groenendijk, J.; Töpsch, S.; Vos, P.; Salamini, F.; Schulze-Lefert, P. The barley Mlo gene: a novel control element of plant pathogen resistance. Cell **1997**, *88* (5), 695–705.
8. Krattinger, S.G.; Lagudah, E.S.; Spielmeyer, W.; Singh, R.P.; Huerta-Espino, J.; McFadden, H.; Bossolini, E.; Selter, L.L.; Keller, B. A putative ABC transporter confers durable resistance to multiple fungal pathogens in wheat. Science **2009**, *323* (5919), 1301–1302.
9. Wisser, R.J.; Sun, Q.; Hulbert, S.H.; Kresovich, S.; Nelson, R.J. Identification and characterization of regions of the rice genome associated with broad-spectrum, quantitative disease resistance. Genetics **2006**, *169*, 2277–2293.
10. Borglie, K.; Butler, K.H. *Polynucleotides and Methods for Making Plants Resistant to Fungal Pathogens*. E.I. Du Pont de Nemours & Company, Pioneer Hi-Bred International, Inc., University of Delaware: Wilmington, DE. World

Intellectual Property Organization Patent. WO 2008/157432 A1. See: http://www.wipo-int/pctdb/en/wo.jsp?wo=2008157432&IA=US2008067021&DISPLAY=STATUS. US: 15-Jun-07, Number: 60/944, 209, International: 13-June-0, Number: PCT/US2008/06702.

11. Balint-Kurti, P.J.; Zwonitzer, J.C.; Wisser, R.J.; Carson, M.L.; Oropeza-Rosas, M.A.; Holland, J.B.; Szalma, S.J. Precise mapping of quantitative trait loci for resistance to southern leaf blight, caused by *Cochliobolus heterostrophus* race O, and flowering time using advanced intercross maize lines. Genetics **2007**, *176* (1), 645–657.

12. Harjes, C.E.; Rocheford, T.R.; Bai, L.; Brutnell, T.P.; Kandianis, C.B.; Sowinski, S.G.; Stapleton, A.E.; Vallabhaneni, R.; Williams, M.; Wurtzel, E.T.; Yan, J.; Buckler, E.S. Natural genetic variation in lycopene epsilon cyclase tapped for maize biofortification. Science **2008**, *319* (5861), 330–333.

13. Yu, J.; Holland, J.B.; McMullen, M.M.; Buckler, E.S. Genetic design and statistical power of nested association mapping in maize. Genetics **2008**, *178*, 539–551.

14. Stuiver, M.H.; Custers, J.H.H.V. Engineering disease resistance in plants. Nature **2001**, *411*, 865–868.

Male Gametogenesis

David Honys
Institute of Experimental Botany, Academy of Sciences of the Czech Republic, Prague, Czech Republic

David Twell
Department of Biology, University of Leicester, Leicester, U.K.

Abstract

This entry describes the sequential phases of angiosperm pollen development—microsporogenesis and microgametogenesis—emphasizing the vital role of the cell wall interface and sporophytic-gametophytic interactions. This entry further describes recent progress in genomewide studies of haploid gene expression, genetic approaches that are being used to identify genes required for key cellular processes, and aspects of pollen biotechnology in crop improvement.

INTRODUCTION

The haploid male gametophytes of higher plants play a vital role in plant fertility and crop production through the generation and transport of the male gametes to ensure fertilization and seed set. There has been an evolutionary tendency toward reduction of the male gametophyte and its increasing functional dependence on the sporophyte. This trend is most acute within flowering plants, such that the male gametophyte consists of just two or three cells when shed as pollen grains. Despite its diminutive form, the functional specialization of the male gametophyte is thought to be a key factor in the evolutionary success of flowering plants through mechanisms that promote rigorous selection of superior haploid genotypes and outbreeding.

POLLEN DEVELOPMENT: FROM MICROSPOROCYTE TO MATURE POLLEN

Microsporogenesis and microgametogenesis take place inside the anther loculi that are lined by the tapetal cell layer (tapetum). Microsporogenesis is initiated upon meiotic division of the diploid pollen mother cell (microsporocyte) that produces four haploid microspores composing a tetrad[1,2] (Fig. 1). During microgametogenesis, microspores released from the tetrads undergo cell expansion, cell wall synthesis, asymmetric division, and differentiation of the vegetative and generative cells before partial dessication and release from the anther. The tapetal cells play a major role in pollen development through their contribution to microspore release, nutrition, pollen wall synthesis, and pollen coat deposition. Disturbance of tapetal cell functions usually results in reduced pollen fertility or male sterility through a variety of mechanisms, including arrest of microgametogenesis at the microspore stage or altered pollen hydration through modified pollen coat composition.

Microsporogenesis

A unique feature of the walls surrounding the microsporocytes and newly formed microspores within the tetrad is that they consist largely of callose, a β-1-3-glucan. The callose wall is secreted by microsporocytes before meiosis I and separates the microspores within the tetrad following meiosis II (Fig. 1). Microspores begin to synthesize the first elements of the sculptured outer pollen wall layer (exine), starting with primexine that functions as a template for subsequent exine elaboration. When young microspores are still developing the exine within the tetrad, an enzyme complex (callase) is secreted by the tapetal cells, allowing individual microspores to be released from the tetrads. Correct timing of callase secretion is critical because premature or delayed dissolution of the callose wall results in male sterility.[3]

Microgametogenesis

Once released, free microspores increase in size and their multiple small vacuoles enlarge and fuse into a single large vacuole, occupying most of the volume of the cell. In concert, the microspore nucleus migrates to a peripheral position that is required for the subsequent asymmetric division at pollen mitosis I (PMI)[4,5] (Fig. 1). PMI results in two morphologically and functionally distinct cells—a large vegetative cell and a small generative cell. The generative

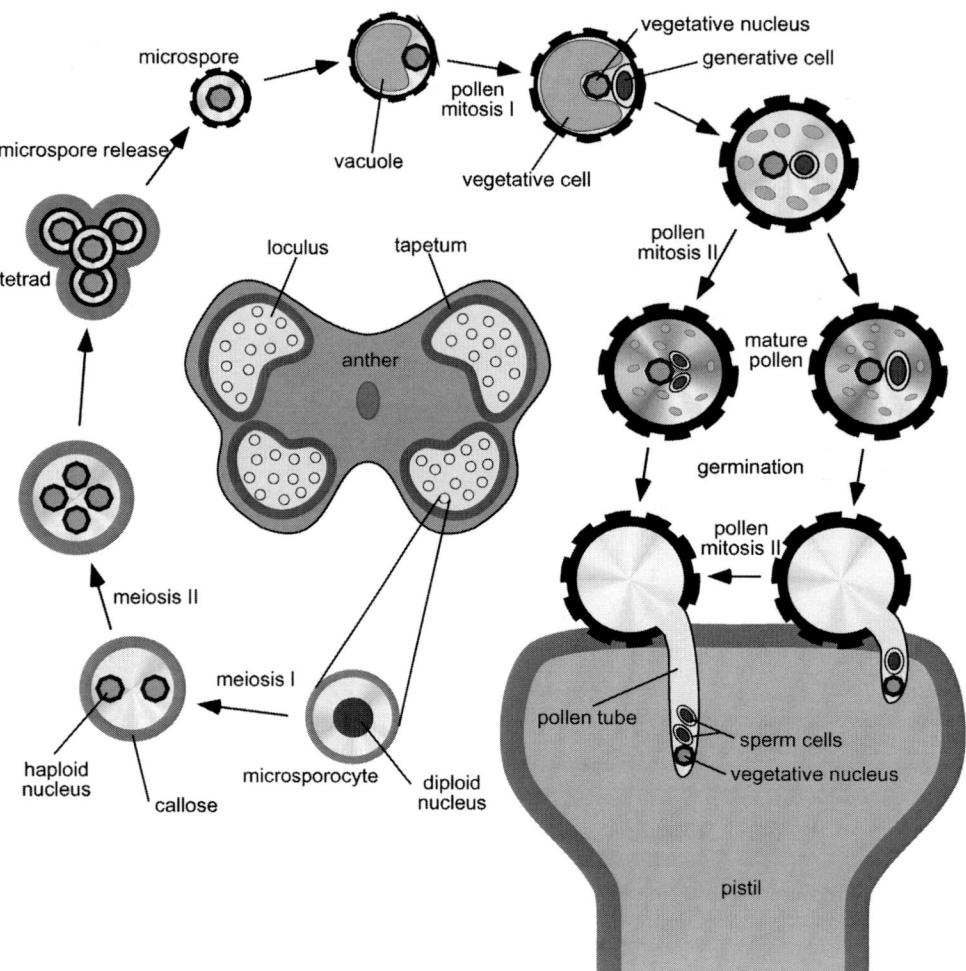

Fig. 1 Schematic diagram illustrating pollen development.

cell subsequently becomes engulfed within a membrane-bound compartment in the cytoplasm of its vegetative sister. This involves dissolution of the hemispherical callose wall separating the vegetative and generative cells, inward migration, and membrane fusion events. The asymmetric division at PMI is a key determinative event in generative cell fate.[6]

In microspores and immature pollen cultivated in vitro, gametophytic development can be switched to a sporophytic pathway by heat stress and/or starvation treatment, leading to microspore embryogenesis and haploid plant formation.[2] Such techniques are routinely applied to accelerate breeding programs through the rapid generation of double haploid plants and selection among large numbers of homozygous lines.

The generative cell undergoes further mitotic division at pollen mitosis II (PMII) to produce the two sperm cells. In tricellular pollen this division occurs within the anther, whereas in bicellular pollen it occurs within the growing pollen tube. Although the majority of flowering plants produce bicellular pollen, many important food crop plants such as rice, wheat, and maize produce advanced but often short-lived tricellular pollen grains (Fig. 2).[2]

During pollen maturation the vegetative cell accumulates considerable carbohydrate and/or lipid reserves that are transient or are stored in the mature pollen grain. Transient reserves are thought to provide metabolites for energy-demanding developmental events such as asymmetric division and pollen cell wall (intine) synthesis. Osmoprotectants including proline also accumulate in mature pollen grains to protect vital membrane and proteins from damage. In mature pollen grains the extensive stores of lipids and polysaccharides are required to supply the extensive demands for plasma membrane and pollen tube wall synthesis.

During dehydration, the final phase of pollen maturation, pollen grains are finally prepared for release from the anthers. This represents an adaptation to survive exposure to the hostile terrestrial environment. The extent of dehydration varies widely in different species; for example, in poplar the water content is reduced to only 6%, maize loses 50%, whereas cucumber pollen remains fully

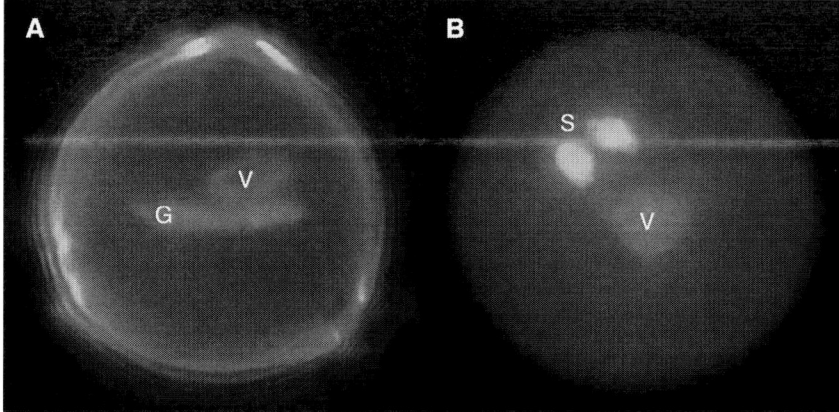

Fig. 2 Bicellular and tricellular pollen. (A) Bicellular tomato and (B) tricellular oilseed rape pollen-stained with the DNA stain DAPI. Nuclear DNA within the vegetative (V) and generative (G) or vegetative (V) and sperm cells (S) are highlighted.

hydrated. The degree of dehydration and the levels of cytoplasmic reserves positively correlate with pollen fitness and viability. Hydrated pollen is very susceptible to dehydration stress and generally survives only a few hours, whereas fully dehydrated pollen may survive for months or even years under certain conditions.

THE POLLEN WALL—A VITAL INTERFACE

The unique activities and biological role of the pollen grain are reflected in the unique composition of the pollen wall. The pollen wall and its coatings isolate and protect the male gametophyte and its associated gametes and mediate the complex communication with the stigma surface. The pollen wall consists of an inner intine and outer exine layer. Its synthesis begins at the microspore stage, when the pectocellulosic intine and the primexine are formed. The primexine serves as a matrix for subsequent deposition of sporopollenin. Sporopollenin is a highly resistant biopolymer containing fatty acids and phenylpropanoids; its synthesis involves tight cooperation between microspore cytoplasm and tapetal cells. The exine is not evenly distributed over the pollen grain surface, and regions lacking sporopollenin form apertures that are used as sites for pollen tube emergence. The number and size of apertures and exine patterning are under strict sporophytic control.

The formation of pollen coatings is completed at later stages of microgametogenesis. Remnants of degenerating tapetal cells are deposited onto the pollen grain surface creating the pollen coat. The pollen coat is involved in pollen–pistil signalling, self-incompatibility, pollen hydration, adhesivity, color, and odor. The yellow or purple colors of mature pollen grain results from the presence of both carotenoid and phenylpropanoid compounds. These features, as well as the elaborate patterning of sporopollenin, are highly variable among different plant species. In animal-pollinated species, pollen is often decorated with elaborate structures that facilitate vector adhesion, whereas in wind-pollinated species, pollen lacks such sculpturing or may be decorated with air sacs to increase buoyancy (Fig. 3).

HAPLOID GENE EXPRESSION

Development of the male gametophyte is associated with an extensive haploid gene expression program. To date, approximately 150 pollen-specific genes falling into 50 functional classes have been cloned from various species.[4][7] Recent genomewide studies using microarray hybridization technology have comprehensively demonstrated the scale and diversity of haploid gene expression in *Arabidopsis thaliana*.[8] Mature *Arabidopsis* pollen grains express approximately 5000 different mRNA species out of more than 27,000 predicted from the *Arabidopsis* genome. Approximately 40% of these transcripts are predicted to be preferentially or specifically expressed in pollen. Moreover, there is significant overlap between sporophytic and male gametophytic gene expression that reflects the large proportion of genes that are required for basic cellular functions. The most abundant classes of pollen-specific genes are predicted to have functions associated with transcriptional regulation, signal transduction, cytoskeleton organization, and cell wall synthesis. This highlights the importance of these functions for the unique cellular specialization required for pollen differentiation and function.

Interestingly, some of these pollen-specific genes encode proteins representing major allergens, the cause of hayfever and allergic asthma. Recent work has shown that the expression of one class of allergens from ryegrass pollen can be reduced without significant impact on plant fertility,[9] indicating that the genetic engineering of hypoallergenic cultivars is possible.

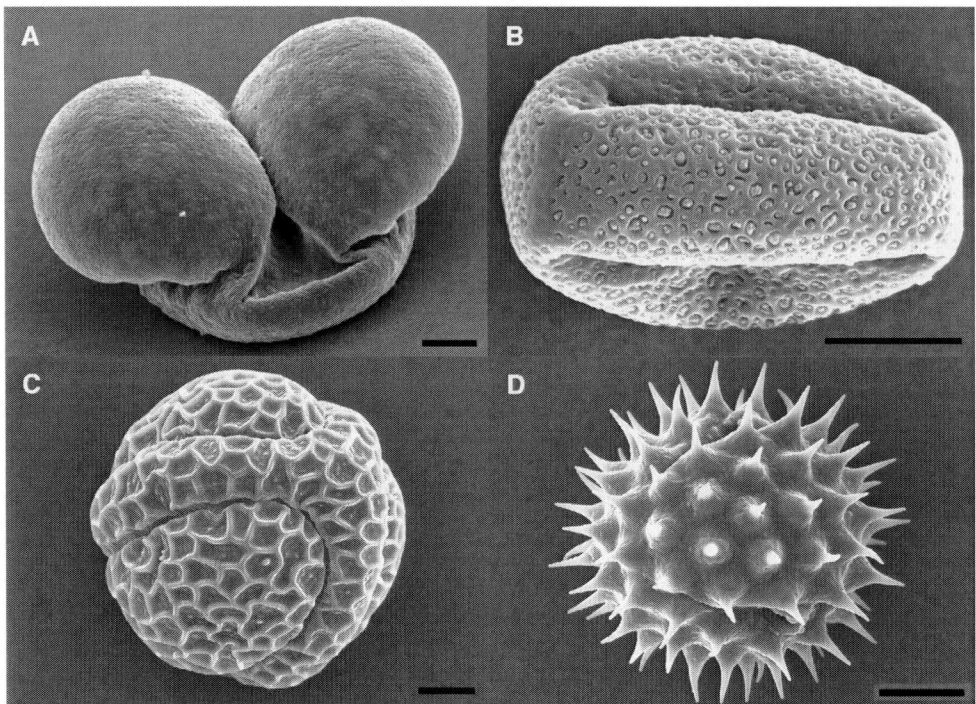

Fig. 3 Pollen morphology and wall patterning. Scanning electron micrographs of (A) pine, (B) papaya, (C) passion flower, and (D) sunflower pollen. Bar length = 10 μm.

Pollen-expressed genes have also been classified according to their temporal regulation. In addition to constitutively expressed genes, groups of early genes active in microspores and late genes acting after PMI have been identified. The late-pollen genes and their regulation have been most intensively studied, resulting in the functional dissection of several pollen-specific transcriptional and translational regulatory elements.[7]

GENETIC APPROACHES: MUTANTS AND DEVELOPMENT

Renewed interest in genetic analysis and the use of the model species A. thaliana have had a significant impact on the identification of gametophytic genes controlling pollen development. Successful mutant selection approaches have been devised that involve screening for aberrant pollen morphogenesis or marker-segregation ratio distortion in populations of plants treated with chemical or physical mutagens or by T-DNA or transposon-mediated mutagenesis.[4]

Morphogenesis screens have led to the identification of a number of mutants with novel phenotypes that affect processes throughout microgametogenesis (Table 1). These include mutants that disturb asymmetric cell division at pollen mitosis I,[6,10,11] division of the generative cell division,[4,6] the positioning and structure of the male germ unit,[12] and the repression of pollen germination within the anther.[13] Among these, the *gemini pollen1* (*gem1*) mutation has been tracked to a microtubule-associated protein with strong similarity to the human chTOG and *Xenopus* XMAP215 family of microtubule-associated proteins that stimulate plus-end microtubule growth.[11] The GEM1 protein plays a vital role in microspore polarity and cytokinesis through its involvement in microtubule assembly, and is associated with interphase, spindle, and phragmoplast microtubule arrays.[11] Further exciting discoveries are expected to arise from the identification of the mutant genes responsible for other gametophytic phenotypes described in Table 1.

Screening for gametophytic mutants using marker-segregation ratio distortion has also been used to identify genes involved in microgametogenesis and/or postpollination events including pollen germination, tube growth, and guidance to the ovule. These screens have mostly employed T-DNA or transposon insertion populations that harbor dominant antibiotic or herbicide resistance markers. For example, if a DNA insertion inactivates an essential male gametophytic gene, then the ratio of resistant to sensitive progeny will deviate significantly below the expected 3:1 ratio toward 1:1. A number of tagged mutants have now been identified that affect microgametogenesis, pollen germination or pollen tube growth.[14–16] Continued effort in screening and the isolation of tagged genes is expected to lead to a sophisticated genetic map of gametophytic proteins that function during microgametogenesis and postpollination events.

Table 1 Gametophytic mutations and genes that affect pollen development in *Arabidopsis thaliana*.

Gene	Mutant	Mutant phenotype	Function	Protein identity	Ref.
BOD	both male and female gametophytic defective (1,2,3)	Pollen arrested during bicellular development with pleiotropic effects	Pollen maturation after pollen mitosis I	ND	[17]
DUO	duo pollen (1,2,3,4,5,6)	Pollen fails to enter or complete generative cell division	Generative cell morphogenesis and cell cycle progression	ND	[4]
EFP	emotionally fragile pollen	Pollen shows diffuse callosic staining	Repression of callose synthesis during pollen maturation	ND	[13]
GEM	gemini pollen (1,2)	Twin-celled and binucleate pollen formed as a result of abnormal division at pollen mitosis I	Regulation of microspore polarity and cytokinesis through microtubule organization	GEM1: Homologous to chTOGp/XMAP215 family of microtubule-associated proteins	[4,11,12]
GUM	male germ unit malformed	Sperm cells separated from the vegetative nucleus	Association of vegetative nucleus and sperm cells	ND	[12]
GWP	gift wrapped pollen	Pollen contains internal callosic tube-like structures	Regulation of callose synthesis	ND	[13]
HAM	halfman	Pollen degenerates during bicelluar stage	Pollen maturation after pollen mitosis I	~150 kb deletion including 38 predicted genes	[18]
LIP	limpet pollen	Generative cell remains attached to pollen wall	Required for GC internalization	ND	[15]
MAD	male gametophytic defective (1,2,3)	Pollen arrested during bicellular development	Pollen maturation after pollen mitosis I	ND	[17]
MUD	male germ unit displaced	Male germ unit displaced to the cortical cytoplasm in mature pollen	Regulation of nuclear-cytoplasmic organization	ND	[12]
PDP	polka dot pollen	Pollen shows localized callose staining	Regulation of callose synthesis	ND	[13]
RTG	raring-to-go	Precocious germination of pollen within the anther locule	Regulation of pollen hydration status	ND	[13]
SCP	sidecar pollen	Pollen with an extra vegetative cell resulting from early symmetric microspore division	Control of microspore division timing or polarity establishment	ND	[10]
TDT	T-DNA transmission defective (6,17,38,40)	Aborted mature pollen	Pollen maturation	ND	[14,16]
TIO	two-in-one	Microspores fail to initiate or complete cytokinesis following pollen mitosis I	Regulation of phragmoplast structure and/or function	ND	[4]

Only gametophytic mutants that show phenotypes that are detectable before pollen germination are included. Mutations affecting male meiosis and postpollination events are excluded.
In the column labeled 'mutant,' the numbers in parenthesis refer to individual mutants with the same mutant symbol prefix.
ND = not determined.

CONCLUSION

The quest to achieve a comprehensive description of the cellular, molecular, and genetic events that control pollen development is not only of fundamental interest, but may find application in the production and genetic improvement of crops. Current knowledge is applied in several areas of crop production, including crop yield optimization and hybrid seed production, and in the production of economic products such as honey and pharmaceuticals. The recent progress in mutational and genome-wide analysis now provides enormous potential to develop new strategies for the molecular dissection and modification of pollen development and functions.

ACKNOWLEDGMENTS

We thank Stefan Hyman at the University of Leicester Electron Microscopy Facility for scanning electron microscopy and the University of Leicester Harold Martin Botanical Gardens for pollen samples. We gratefully acknowledge financial support for our research from the NATO/Royal Society Fellowship Programme and the Grant Agency of the Czech Republic (DH) and the UK Biological and Biotechnology Research Council (DT).

REFERENCES

1. Shivanna, K.R.; Cresti, M.; Ciampolini, F. In Pollen Development and Pollen–Pistill Interaction. In *Pollen Biotechnology for Crop Production and Improvement*; Shivanna, K.R., Sawhney, V.K., Eds.; Cambridge University Press: Cambridge, **1996**; 15–39.
2. Touraev, A.; Vicente, O.; Heberle-Bors, E. Initiation of microspore embryogenesis. Trends Plant Sci. **1997**, *2*, 297–302.
3. Goldberg, R.B.; Beals, T.P.; Sanders, P.M. Anther development: Basic principles and practical applications. Plant Cell **1993**, *5*, 1217–1229.
4. Twell, D. *Pollen Developmental Biology. Plant Reproduction*; O'Neill, S.D., Roberts, J.A., Ed.; Annual Plant Reviews, Sheffield Academic Press: Sheffield, **2002**; Vol. 6, 86–153.
5. Bedinger, P. The remarkable biology of pollen. Plant Cell **1992**, *4*, 879–887.
6. Twell, D.; Park, S.K.; Lalanne, E. Asymmetric division and cell-fate determination in developing pollen. Trends Plant Sci. **1998**, *3* (8), 305–310.
7. Hamilton, D.A.; Mascarenhas, J.P. In Gene Expression During Pollen Development. In *Pollen Biotechnology for Crop Production and Improvement*; Shivanna, K.R., Sawhney, V.K.; Ed.; Cambridge University Press: Cambridge, **1996**; 40–58.
8. Honys, D.; Twell, D. Comparative analysis of the *Arabidopsis* pollen transcriptome. Plant Physiol. **2003**, *203*, 640–652.
9. Bhalla, P.; Swoboda, I.; Singh, M.B. Antisense-mediated silencing of a gene encoding a major ryegrass pollen allergen. Proc. Natl. Acad. Sci. U. S. A. **1999**, *96*, 11676–11680.
10. Chen, Y.-C.; McCormick, S. *Sidecar pollen*, an *Arabidopsis thaliana* male gametophytic mutant with aberrant cell divisions during pollen development. Development **1996**, *122*, 3243–3253.
11. Twell, D.; Park, S.K.; Hawkins, T.J.; Schubert, D.; Schmidt, R.; Smertenko, A.; Hussey, P.J. MOR1/GEM1 plays an essential role in the plant-specific cytokinetic phragmoplast. Nat. Cell Biol. **2002**, *4*, 711–714.
12. Lalanne, E.; Twell, D. Genetic control of male germ unit organization in *Arabidopsis*. Plant Physiol. **2002**, *129*, 865–875.
13. Johnson, S.A.; McCormick, S. Pollen germinates precociously in the anthers of *raring-to-go*, an *Arabidopsis* gametophytic mutant. Plant Physiol. **2001**, *126*, 685–695.
14. Bonhomme, S.; Horlow, C.; Vezon, D.; de Laissardiere, S.; Guyon, A.; Ferault, M.; Marchand, M.; Bechtold, N.; Pelletier, G. T-DNA mediated disruption of essential gametophytic genes in *Arabidopsis* is unexpectedly rare and cannot be inferred from segregation distortion alone. Mol. Genet. **1998**, *260*, 444–452.
15. Howden, R.; Park, S.K.; Moore, J.M.; Orme, J.; Grosslinklaus, U.; Twell, D. Selection of T-DNA—Tagged male and female gametophytic mutants by segregation distortion in *Arabidopsis*. Genetics **1998**, *149*, 621–631.
16. Procissi, A.; de Laissardiere, S.; Ferault, M.; Vezon, D.; Pelletier, G.; Bonhomme, S. Five gametophytic mutations affecting pollen development and pollen tube growth in *Arabidopsis thaliana*. Genetics **2001**, *158*, 1773–1783.
17. Grini, P.E.; Schnittger, A.; Schwarz, H.; Zimmermann, I.; Schwab, B.; Jurgens, G.; Hulskamp, M. Isolation of ethyl methanesulfonate-induced gametophytic mutants in *Arabidopsis thaliana* by a segregation distortion assay using the multimarker chromosome 1. Genetics **1999**, *151*, 849–863.
18. Oh, S.-A.; Park, S.-K.; Jang, I.; Howden, R.; Moore, J.M.; Grossniklaus, U.; Twell, D. Halfman, an Arabidopsis male gametophytic mutant associated with a 150 kb chromosomal deletion at the site of transposon insertion. Sex Plant Reprod. **2003**, *16*, 99–102.

Malolactic Fermentation in Wine

Maret du Toit
Department of Viticulture and Oenology, Institute for Wine Biotechnology, Stellenbosch University, Matieland, South Africa

Abstract
Malolactic fermentation is discussed with regard to the biochemical reaction, starter cultures used, factors that will influence the process, the time of inoculation, and the impact it has on wine aroma.

INTRODUCTION

Winemaking is a complex process that involves microorganisms such as yeasts, lactic acid bacteria (LAB), acetic acid bacteria, and fungi. These microbes originate from the grapes, and the cellar environment or certain species can be introduced in the form of commercial starter cultures. The production of wine comprises two main fermentation processes, namely alcoholic fermentation that is conducted mainly by *Saccharomyces cerevisiae* and the secondary fermentation, called malolactic fermentation (MLF), performed by LAB, especially *Oenococcus oeni*.[1,2] The genera *Lactobacillus*, *Leuconostoc*, *Pediococcus*, and *Oenococcus* are the LAB associated with winemaking. The concentration of viable LAB populations is approximately 10^2–10^4 cells/mL in must from healthy grapes, and can increase to 10^6 cells/mL depending on the degree of rot and maturity of the grapes. During fermentation, the LAB microflora evolves not only in numbers but also in the variety of species.[2,3] LAB modifies the sensory characteristics of the wine through the production of certain flavor compounds (both desirable and undesirable) released during the fermentations. Some LAB can also produce secondary metabolic products that can affect the wholesomeness of the wine such as biogenic amines.[3–6] This entry will focus on the process of MLF, the impact of MLF starter cultures, and the influence of this process on wine quality.

MALOLACTIC FERMENTATION

Malolactic fermentation is the decarboxylation of the dicarboxylic acid, L-malic acid, one of the main organic acids in wine, to the monocarboxylic acid, L-lactic acid, and CO_2. The reaction is catalyzed by malate decarboxylase (*mleA*) or the so-called malolactic enzyme, and requires the coenzyme NAD^+ and manganese (Mn^{2+}). This reaction results in a pH increase of 0.1–0.3 units and a decrease in titratable acidity from 1 to 3 g/L. MLF is primarily performed to reduce wine acidity, especially in cooler climate regions where the malic acid can be as high as 8 g/L.[4,6] Secondly, owing to their diverse enzyme profiles, LAB can alter the flavor and aroma profile of wine and ultimately wine quality.[7,8] MLF also provides microbial stability, since malic acid, which can serve as a carbon source to support the growth of potential spoilage LAB, is degraded. There have been several studies in the past two decades that investigated the benefits of MLF in wine. MLF is mostly conducted in red wines and occasionally in white wines (most common is Chardonnay). MLF consists of three steps: 1) the uptake of malic acid from the external environment by a specific transporter, malate permease (*mleP*); 2) the decarboxylation of the malic acid to lactic acid with an increase in internal pH; and 3) the transport of lactic acid from the cell with one H^+ and an increase in the proton motive force (Δp) over the cell membrane. This increase in Δp will activate the ATPase in the cell membrane to produce ATP that can be used for transport of certain processes in the cell.[1–3] Using genetic tools, it has been shown that many wine-associated LAB species have the malolactic enzyme and they are very homologous.

STARTER CULTURES

Today *O. oeni* is still the preferred species for commercial starter cultures used in the wine industry. This is due to its tolerance to alcohol concentrations of as high as 15%, low pH (<3.5), low temperature, and sulphur dioxide (SO_2) levels used in winemaking. Spontaneous MLF is unreliable and can cause problems since the winemaker cannot predict which LAB species will be performing MLF. This can lead to a loss in wine quality, especially at pH >3.6 where *Lactobacillus* and *Pediococcus* can be the dominating microbes in the wine. Risks associated with spontaneous MLF are 1) increased volatile acidity, 2) degradation of glycerol to increase bitterness in wine, 3) increase in viscosity or ropiness, 4) production of volatile phenols, 5) mousiness,

6) production of biogenic amines, 7) degradation of arginine to produce ethyl carbamate, and 8) masking of varietal aromas. The advantages of using commercial MLF starter cultures are: 1) inoculation at cell concentration of $>10^6$ cells/mL is achieved, 2) a short lag phase is insured and this will increase the rate of MLF, 3) the impact of spoilage microbes is reduced, 4) better control over the fermentation is ensured, 5) the attributes of the inoculated strain is known, and 6) that strains have been selected that do not contain negative characteristics.[2,3,6] MLF can be induced in wine by different techniques. The lees of wine that have undergone MLF can be used to inoculate wine that still has to undergo MLF, but the biggest risk is that other spoilage microbes could also be inoculated into the wine. The same disadvantage is experienced with the use of mother tanks, where one tank is inoculated and after two-thirds of the malic acid has been degraded in the mother tank this is then transferred to another tank. MLF with commercial starter cultures is the best way to ensure a good and clean MLF and are normally inoculated after alcoholic fermentation, and if correctly implanted it will dominate and outnumber the undesirable indigenous flora. MLF starter cultures are mainly selected for their viability after being freeze-dried, and for performing malic acid degradation and survival under harsh winemaking conditions. Warmer climate regions provide harsher conditions for MLF compared to cooler climate areas owing to higher sugar concentrations that will subsequently lead to high alcohol levels ($>14\%$). Also the nitrogen status of the grapes is lower, and therefore nutrient management is important. The pH is mostly above 3.6, which leads to higher indigenous LAB cell numbers that will compete with the starter culture for survival.[3,6] Today strains are also selected for their potential impact on the sensorial quality of the wine. The past decade has seen great improvements in the efficiency and diversity of MLF starter cultures available in the market.

TIME OF INOCULATION

There are three phases in the winemaking process where inoculation of MLF can occur. The advantage and disadvantage of each phase will be discussed below.

Simultaneous Inoculation of Yeast and Bacteria

This practice is not commonly performed to avoid the risk of acetic acid and D-lactic acid being produced from glucose by heterofermentative LAB, such as *O. oeni* en *Lactobacillus* spp. High levels of acetic acid are toxic to wine yeasts and can cause stuck or sluggish alcoholic fermentations. Research conducted during the past 4 years in different wine-producing countries (including Canada, Germany, Italy, Australia, and South Africa) have shown that there is no significant increase in volatile acidity of wines that have been coinoculated when compared to wines that were inoculated after alcoholic fermentation.[6,9,10] One advantage of coinoculation is that there is no ethanol produced that will have an impact on the growth of the LAB. Also there are ample nutrients available for the LAB to use and therefore the addition of complex bacterial nutrients is unnecessary. Both alcoholic fermentation and MLF will be completed sooner than consecutive processes, which will allow the winemaker to stabilize the wine sooner and proceed with the ageing process. In general, this process is three weeks faster than inoculating after alcoholic fermentation.[3,6]

Inoculation during Alcoholic Fermentation

This practice has the advantage that the alcohol levels are not so high yet and that the LAB can adapt during the completion of alcoholic fermentation. During this phase of winemaking most of the free SO_2 is bound to carbonyl compounds and therefore the antimicrobial activity of the SO_2 is very low. The MLF bacteria can also still use the heat of the fermentation to increase their growth. The biggest risk for inoculation at this stage is the intense antagonistic activity of yeast-producing compounds such as decanoic acid.[6,11]

Inoculation after Alcoholic Fermentation

Inoculation after alcoholic fermentation does not pose the risk of acetic acid production as all sugars have been utilized. The two major problems experienced are stuck or sluggish MLF owing to high alcohol levels and a depletion of nutrients. Conducting MLF on the yeast lees can provide essential nutrients for the MLF bacteria as yeast autolysis progresses. All commercial MLF starter culture producers also have complex bacterial nutrients that can be added to ensure that the bacteria have optimal conditions in which to grow and perform MLF.[3,6]

FACTORS INFLUENCING MLF

The four most important factors that will influence the growth and MLF are pH, alcohol concentration, SO_2, and temperature. These factors have been shown to act synergistically and might have a greater effect together than alone. Generally, wines with a pH below 3.5 do not support the growth of *Pediococcus* and *Lactobacillus* spp. and, invariably, *O. oeni* dominates in these wines. The sensitivity of LAB to ethanol and free molecular SO_2 increases at a low pH. A total SO_2 concentration of >100 mg/L is sufficient to inhibit the growth of LAB. These values vary depending on the species, wine pH and the amount of insoluble solids present in wine. In general, strains of *O. oeni* appear to be the most sensitive. In wine, a temperature of 18–20°C is optimal for the growth of LAB and the induction of MLF. The optimal decarboxylation of malic acid occurs between 20 and 25°C. Growth and MLF are strongly inhibited by a low temperature and only a few strains of *O. oeni* can conduct MLF below 15°C. In general, *Lactobacillus* species are more alcohol tolerant than either *Pediococcus* or *Oenococcus* species, since strains have been isolated from dessert and fortified wines. Alcohol tolerance decreases with an increase in temperature and at low pH values.[1,3,6]

Other factors that can influence MLF are the yeast strain used for alcoholic fermentation; phenols and tannins; nutritional status of the wine; fungicide residues; oxygen; medium chain fatty acids; and lysozyme levels.[3,6]

IMPACT ON WINE AROMA

It has been proved in the past decade that MLF is more than just deacidifying the wine and that it does affect wine quality positively or negatively. The era of genomics and metabolomics has shown that the metabolic profile of LAB is complex, diverse, and strain dependent.[7,8] Attributes imparted by MLF are described as nutty, buttery, lactic, nutty, oaky, and sweaty. The degradation of malic acid will impact on the flavor and aroma of wine since malic acid has a strong green, metallic taste, whereas lactic acid is much softer on the palate. In general, MLF will increase the buttery character, reduce the vegetative character, improve the mouth feel, and improve the overall balance of the wine. The perception of MLF flavors is dependent on the cultivar. Neutral cultivars, such as Chardonnay, benefit from MLF more than aromatic cultivars like Chenin blanc. MLF in red wines significantly decreases free anthocyanins and astringency. The increase in pH can also reduce the color of red wine.[3,5,6]

The most important aroma compound produced by LAB from citric acid metabolism is diacetyl. Diacetyl imparts the buttery/nutty characteristic in wine. This metabolic pathway is closely linked to MLF. Diacetyl contributes positively to wine complexity when present at 1–4 mg/L, and is regarded as a spoilage characteristic when present at >5 mg/L. Bartowsky and Henschke has published a comprehensive review on diacetyl and wine.[12]

MLF can affect the glycosidically bound grape-derived flavor compounds through hydrolysis by the β-glycosidase enzyme. This will lead to an increase in norisoprenoids and monoterpenes. MLF also interacts with oak compounds and it has been shown that oak lactones and vanillin increases, whereas eugenol and furfural decreases. Esters are important aroma compounds responsible for fruity aromas. The most important esters associated with MLF are ethyl lactate, ethyl acetate, ethyl hexanoate, and ethyl octanoate and their synthesis is due to the esterase activity of LAB. Apart from the above-mentioned compounds, LAB can also metabolize phenolic acids, polyols, amino acids, lipids, peptides/proteins, aldehydes, and polysaccharides that will lead to compounds that can positively or negatively affect the sensorial properties of the wine.[7,8]

CONCLUSION

LAB and MLF are important for improving the quality of wine. With fundamental knowledge gained through omics technologies in the future, we should understand more about the contribution of MLF to wine aroma, flavor, and quality. Technological advances and further research will also assist us in understanding and therefore control MLF better in practice.

REFERENCES

1. Ribéreau-Gayon, P.; Dubourdieu, D.; Donèche, B.; Lonvaud, A. *Handbook of Enology, Volume 1: The Microbiology of Wine and Vinifications*; Wiley: Chichester, England, **2006**.
2. Fugelsang, K.C. *Wine Microbiology*; Chapman & Hall: New York, **1997**.
3. Henick-Kling, T. Malolactic fermentation. In *Wine Microbiology and Biotechnology*; Fleet, G.H., Ed.; Taylor & Francis Inc.: New York, **2003**; 289–326.
4. Lonvaud-Funel, A. Lactic acid bacteria in the quality improvement and depreciation of wine. Antonie Van Leeuwenhoek **1999**, *76*, 317–331.
5. Swiegers, J.H.; Bartowsky, E.J.; Henschke, P.A.; Pretorius, I.S. Yeast and bacterial modulation of wine aroma and flavour. Aust. J. Grape Wine Res. **2005**, *11*, 139–173.
6. Morenzoni, R. *Malolactic Fermentation in Wine*; Lallemand Inc.: Montréal, Canada, **2006**.
7. Matthews, A.; Grimaldi, A.; Walker, M.; Bartowsky, E.; Grbin, P.; Jiranek, V. Lactic acid bacteria as a potential source of enzymes for use in vinification. Appl. Environ. Microbiol. **2004**, *70*, 5715–5731.
8. Malolactic fermentation in wine—beyond deacidification. A review. J. Appl. Microbiol. **2002**, *92*, 589–601.
9. Jussier, D.; Morneau, A.D.; De Orduña, R.M. Effect of simultaneous inoculation with yeast and bacteria on fermentation kinetics and key wine parameters of cool-climate Chardonnay. Appl. Environ. Microbiol. **2006**, *72*, 221–227.
10. Rosi, I.; Fia, G.; Canuti, V. Influence of different pH values and inoculation time on the growth and malolactic activity of a strain of *Oenococcus oeni*. Aust. J. Grape Wine Res. **2003**, *9*, 194–199.
11. Alexandre, H.; Costello, P.J.; Remize, F.; Guzzo, J.; Guilloux-Benatier, M. *Saccharomyces cerevisiae–Oenococcus oeni* interactions in wine: current knowledge and perspectives. Int. J. Food Microbiol. **2004**, *93*, 141–154.
12. Bartowsky, E.J.; Henschke, P.A. The "buttery" attribute of wine—diacetyl—desirability, spoilage and beyond. Int. J. Food Microbiol. **2004**, *96*, 235–252.

Mammalian Sperm Sexing

Duane L. Garner
GametoBiology Consulting, Graeagle, California, U.S.A.

Abstract
Sex can be predetermined for mammals by separating the sex-determining sperm prior to insemination. A flow cytometer/cell sorter is used to separate the sex-determining gametes, X- and Y-chromosome-bearing sperm, using their DNA content differences. This sperm sexing technology has been commercialized in cattle with impending applications to other species. Successful flow cytometric sexing of mammalian sperm will likely stimulate the development of more efficient approaches to predetermining sex in mammalian offspring.

INTRODUCTION

Sex can be predetermined for mammalian offspring by separating the sex-determining sperm prior to insemination. The separation procedure, which is based on sperm DNA content, uses a flow cytometer/cell sorter to separate the sex-determining gametes with an accuracy of 85–95%.[1–3] Precise measurement of mammalian sperm DNA content was first established at Lawrence Livermore National Laboratory.[4,5] This sperm measurement system was enhanced at the USDA Beltsville Agricultural Research Center, thereby enabling sorting of live mammalian sperm according to their DNA content.[6] This sexing technology has advanced to commercialization in cattle[1,2,7–9] with impending marketable applications in other domestic and wildlife species. Flow cytometric sorting has demonstrated that sexing sperm can effectively alter the sex ratio of mammalian offspring.

BIOLOGICAL BASIS OF SPERM SEXING

Mammalian semen can be sexed because the X-chromosome-bearing sperm that produce females contain about 4% more DNA than do the Y-chromosome-bearing sperm that produce males.[1–6] Freshly collected sperm are stained with a DNA-specific bisbenzimidazole dye, Hoechst 33,342, for approximately 1 hour prior to sorting.[3–6] Hoechst 33,342-stained sperm fluoresce bright blue when exposed to a laser beam of short wavelength light and the X-bearing sperm are differentiated from the Y-bearing sperm because the former fluoresce brighter than the latter due to their greater DNA content. The fluorescence of each stained sperm is measured in a stream of fluid as it passes in front of a photomultiplier tube (PMT).[1–11] The resultant data are integrated using a powerful computer. Only the DNA content of properly oriented sperm can be measured accurately because the flat surface of each sperm head must be properly oriented relative to the PMT.[1,2,4,10]

FLOW SORTING SYSTEM

As the liquid stream containing sperm exits the sorter nozzle, it is vibrated at about 80,000 oscillations/sec to break the stream into individual droplets.[1] Although not all of these formed droplets contain sperm, those that do are electrically charged, either positive or negative, according to the DNA content information previously provided by the PMT detectors. Droplets containing improperly oriented sperm, more than one sperm, or dead sperm, as determined by the uptake of a dead cell stain, are disposed of as waste because no charge is applied to these droplets. As the droplets pass by an oppositely charged plate they are deflected into a collection vessel according to the DNA content of the sperm. Those droplets containing Y-sperm are simultaneously directed to a separate collection vessel by applying an opposite charge to those droplets to deflect them toward the oppositely-charged plate. Three streams of droplets containing X-sperm, Y-sperm, or no sperm as well as too many sperm are collected into separate vessels. Approximately 40% of the living sperm going through the sorter at a speed of approximately 120 km/hr can be accurately sexed and collected. Thus, at an event rate of 28,000 total sperm/sec, nearly 5000 live sperm/sec of each sex can be sorted simultaneously with 85–95% accuracy.[9–13] The current system can produce approximately 10 to 15 × 10^6 live bovine sperm/hr of each sex-determining gamete. Considerable numbers of sperm are, however, lost in the centrifugation and other postsorting steps making actual yields somewhat lower.[1,2,7] Sperm of other species tend to sort at somewhat slower rates due to differences in sperm head shape, size, and DNA content differences.[9,10]

UTILIZATION OF SEX-SORTED SPERM

The first sex-selected mammalian offspring were rabbits born following surgical insemination of sexed sperm.[6] Later, in a milestone achievement, a similar system was used to produce sex-selected piglets, demonstrating the utility of the system for preselecting the sex of offspring in domestic livestock.[11] The first calves born were from embryos derived from in vitro fertilization (IVF) with sex-sorted sperm, but most of the recent work in cattle has been done using artificial insemination (AI).[1,2,8,9]

Cattle

Sex-sorted bovine sperm must be reconcentrated by centrifugation[3,8–10] before they can be properly packaged into 0.25 mL French straws at doses of 1 to 6×10^6 sperm/straw. This is contrasted to conventional AI procedures that use at least 20×10^6 sperm/straw, making the insemination dose for sex-sorted bovine sperm about 1/20 to 1/3 that of a normal AI dose. This is necessary because it takes some time to sort each dose, even at the high sort rates now being achieved. Fortunately, sexed bovine sperm can be cryopreserved, thereby allowing efficient use in AI.[14] Current sexing technology has been applied mainly in heifers due to their inherently higher fertility. Several regimens have been used successfully to synchronize estrus to optimize insemination with low doses of sexed sperm.[3] These include: 1) a single injection of 25 mg of prostaglandin $F_{2\alpha}$ ($PGF_{2\alpha}$); 2) feeding 500 mg of melengestrol acetate (MGA) daily for 14 days followed by injection of 25 mg $PGF_{2\alpha}$ 17–19 days after last feeding of MGA; 3) 20–25 mg $PGF_{2\alpha}$ injected at 12 days intervals; and 4) injecting 50–100 μg of GnRH followed by 25 mg of $PGF_{2\alpha}$ 7 days later.[1,3,8,9]

Pregnancy rates from the use of sexed, cryopreserved sperm have been 60–80% of control inseminations with about 10 times greater sperm doses.[3,8] The lower pregnancy rates with sex-sorted sperm are seen even when numbers of sex-sorted sperm inseminated are equal to those of controlled insemination.[15] Inseminations with sex-sorted sperm revealed differences in pregnancy rates among bulls,[2] suggesting that only sperm from highly fertile bulls should be sorted for predetermining sex of their offspring. Differences among bulls in pregnancy rates were found 30 days after transfer of embryos fertilized by sex-sorted sperm.[16] Consistent with this, blastocyst development rates tend to be slower with sex-sorted sperm.[16,17] Field studies showed no incidence of abnormalities or increased abortion rates when calves from sexed sperm were compared to calves from control semen from the same bulls.[18]

Until recently, field trials with sexed sperm were conducted by sorting the gametes in fluid pressurized to 50 psi. This pressure level, however, was found to be detrimental to sperm membrane integrity.[10,19] With the latest procedures, where the pressure has been lowered to 40 psi, pregnancy rates in heifers inseminated with sexed sperm have been nearly 80% of controls, depending on the particular bull,[2] inseminator skills,[9] and management level of the herd.[9,12]

Swine

The first sexed piglets were from sows surgically inseminated with sex-sorted sperm.[11] Even with recent technological improvements in sorting efficiency, production of adequate numbers of boar sperm for uterine insemination of swine is limiting because normally 3 to 5×10^9 sperm/dose are used with the insemination being repeated two to three times during each estrus.[20–22] Delivery of fewer sperm deeper into the porcine reproductive tract is necessary before sex-sorted sperm can be used effectively with swine AI. Such an insemination approach has been developed using a flexible catheter for deep non-surgical insemination of a limited number of sperm.[21,22] A 1.3-m deep uterine insemination catheter was used to deposit 7.5 mL containing 150×10^6 unsexed sperm in the upper end of the uterine horn of sows in natural postweaning estrus. Sows inseminated one time every 32 hours postonset of estrus achieved pregnancy rates of 83.3% (50/60 sows) for sexed sperm and 87.3% (48/55) unsexed sperm.[21,22] Acceptable fertility rates were achieved with only 2–5% of the sperm used conventionally for AI in pigs.[21,22] These preliminary data suggest that such an approach could efficiently produce single sex litters. Although use of Hoechst 33,342 staining and ultraviolet lasers is a concern, exposure to these entities did not induce genotoxic effects in flow-sorted boar sperm.[23]

Sheep

Early efforts to predetermine the sex of lambs used low doses of freshly sorted ram sperm that were surgically deposited directly into the uteri of estrous-synchronized ewes.[24] Pregnancy rates were low, but demonstrated that live lambs could be produced with sex-sorted ram sperm. Recent trials in Australia with sex-sorted ram sperm resulted in pregnancy rates of 25% for ewes inseminated with X-sorted sperm and 15% for Y-sorted sperm at 4×10^6 sperm/dose. Control inseminations with 140×10^6 cryopreserved, thawed sperm yielded 54% pregnant ewes.[24] In this trial, thawed sperm were deposited either by standard laproscopic methods or placed into the oviduct with a catheter after minilaparotomy. It is likely that pregnancy rates with sexed sperm could be improved by increasing sperm/dose and precisely controlling the time of insemination relative to ovulation.

Horses

The first sex-selected filly was produced by surgical insemination with a limited number of flow sorted X-sperm.

This effort was followed by non-surgical artificial insemination of mares with flow-sorted sperm. Reasonable conception rates with sorted stallion sperm have been achieved only if the timing of insemination was optimized relative to ovulation induced with hCG or GnRH. Although stallion sperm do not sort as efficiently as those from bulls, semen from some stallions can be sorted at rates greater than 2000 sperm/sec producing nearly 5×10^6 live sperm/hr of each sex. Thus, with low-dose insemination, several doses of sexed sperm can be produced with a sorter each day.[9]

A hysteroscopic insemination technique, which arose from clinical examination of the mare's reproductive tract using the video endoscope, has been used to inseminate mares with fewer sperm than used with conventional equine artificial insemination.[25] Very low numbers of sperm are placed directly onto the oviductal papillus at the uterotubal junction with the video endoscope. Application of this technique to deposit small numbers of sex-sorted sperm deep into the reproductive tract appears to make predetermination of sex a practical possibility in horses. When mares were hysteroscopically inseminated with 5×10^6 fresh, sex-sorted motile sperm, 38% became pregnant compared to 40% pregnancy rate in mares who were inseminated with 5 million non-sorted motile sperm.[26]

Other Mammals

Living sperm from several other mammalian species have been successfully sorted according to their DNA content including sperm from humans, rabbits, bison, buffalo, elk, cats, and even sperm from white-sided dolphins.[2,7,10]

CONCLUSION

Additional refinements in the system are needed, but sexing sperm using flow sorting has been commercialized for cattle in Great Britain, the United States and some other countries. It is being developed for commercial application in horses, swine, sheep, and some exotic mammalian species. Successful flow cytometric sexing of mammalian sperm will likely stimulate development of more efficient approaches to sperm sexing.

REFERENCES

1. Seidel, G.E., Jr.; Garner, D.L. Current status of sexing mammalian spermatozoa. Reproduction **2002**, *124*, 733–743.
2. Garner, D.L.; Seidel, G.E., Jr. History of commercializing sexed semen for cattle. Theriogenology **2008**, *69*, 886–895.
3. Seidel, G.E., Jr.; Schenk, J.L.; Herickhoff, L.A.; Doyle, S.P.; Brink, Z.; Green, R.D.; Cran, D.G. Insemination of heifers with sexed sperm. Theriogenology **1999**, *52*, 1407–1420.
4. Pinkel, D.; Lake, S.; Gledhill, B.L.; Van Dilla, M.A.; Stephenson, D.; Watchmaker, G. High resolution DNA content measurements of mammalian sperm. Cytometry **1982**, *3*, 1–9.
5. Garner, D.L.; Gledhill, B.L.; Pinkel, D.; Lake, S.; Stephenson, D.; Van Dilla, M.A.; Johnson, L.A. Quantification of the X- and Y-chromosome-bearing sperm of domestic animals by flow cytometry. Biol. Reprod. **1983**, *28*, 312–321.
6. Johnson, L.A.; Flook, J.P.; Hawk, H.W. Sex preselection in rabbits: live births from X and Y sperm separated by DNA and cell sorting. Biol. Reprod. **1989**, *41*, 199–203.
7. Garner, D.L. Sex-sorting mammalian sperm: concept to application in animals. J. Androl. **2001**, *22*, 519–526.
8. Seidel, G.E., Jr.; Brink, Z.; Schenk, J.L. Use of heterospermic insemination with fetal sex as a genetic marker to study fertility of sexed sperm. Theriogenology **2003**, *59*, 515.
9. Seidel, G.E., Jr. Sexing mammalian sperm-intertwining of commerce, technology, and biology. Anim. Reprod. Sci. **2003**, *79*, 145–156.
10. Garner, D.L. Flow cytometric sexing of mammalian sperm. Theriogenology **2006**, *65*, 943–957.
11. Johnson, L.A. Gender preselection in domestic animals using flow cytometrically sorted sperm. J. Anim. Sci. **1992**, *70* (Suppl. 2), 8–18.
12. Amann, R.P. Issues affecting commercialization of sexed sperm. Theriogenology **1999**, *52*, 1441–1457.
13. Johnson, L.A.; Welch, G.R. Sex preselection: high-speed flow cytometric sorting of X and Y sperm for maximum efficiency. Theriogenology **1999**, *52*, 1323–1341.
14. Schenk, J.L.; Suh, T.K.; Cran, D.G.; Seidel, G.E., Jr. Cryopreservation of flow-sorted bovine sperm. Theriogenology **1999**, *52*, 1375–1391.
15. Bodmer, M.; Janett, F.; den Daas, N.; Reichert, P.; Hässig, M.; Thun, R. Fertility in heifers and cows after low dose insemination with sex-sorted and non-sex sorted sperm under field conditions. Theriogenology **2005**, *64*, 1647–1655.
16. Zhang, M.; Lu, K.H.; Seidel, G.E., Jr. Development of bovine embryos after in vitro fertilization of oocytes with flow cytometrically sorted, stained and unsorted sperm from different bulls. Theriogenology **2003**, *60*, 1657–1663.
17. Lu, K.H.; Seidel, G.E., Jr. Effects of heparin and sperm concentration on cleavage and blastocyst development rates of bovine oocytes inseminated with flow cytometrically sorted sperm. Theriogenology **2004**, *62*, 819–830.
18. Tubman, L.M.; Brink, Z.; Suh, T.K.; Seidel, G.E., Jr. Characteristics of calves produced with sperm sexed by flow cytometry/cell sorting. J. Anim. Sci. **2004**, *52*, 1029–1036.
19. Suh, T.E.; Schenk, J.L.; Seidel, G.E., Jr. High pressure flow cytometric sorting damages sperm. Theriogenology **2005**, *64*, 1035–1048.
20. Rath, D.; Long, C.R.; Dobrinsky, J.R.; Welch, G.R.; Schreier, L.L.; Johnson, L.A. In vitro production of sexed embryos for gender preselection: high speed sorting of X-chromosome-bearing sperm to produce piglets after embryo transfer. J. Anim. Sci. **1999**, *77*, 3346–3352.
21. Vazquez, J.M.; Martinez, E.A.; Parrilla, I.; Roca, J.; Gil, M.A.; Vasquez, J.L. Deep intrauterine insemination in

natural post-weaning estrus sows. In *Proceedings of the Sixth International Conference on Pig Reproduction*; University of Missouri: Columbia, MO, 2001; 134.

22. Martinez, E.A.; Vazquez, J.M.; Roca, J.; Lucas, X.; Gil, M.A.; Vazquez, J.L. Deep intrauterine insemination and embryo transfer. In *Proceedings of the Sixth International Conference on Pig Reproduction*; University of Missouri: Columbia, MO, 2001; 129.

23. Parrilla, I.; Vazquez, J.M.; Cuello, C.; Gil, M.A.; Roca. J.; Di Berardino, D.; Martinez, E.A. Hoechst 33342 stain and u.v. laser exposure do not induce genotoxic effect in flow –sorted boar spermatozoa. Reproduction **2004**, *128*, 615–621.

24. Hollinshead, F.K.; O'Brien, J.K.; He, L.; Maxwell, W.M.C.; Evans, G. Pregnancies after insemination of ewes with sorted, cryopreserved ram spermatozoa. Proc. Soc. Reprod. Biol. **2001**, *32*, 20.

25. Morris, L.H.A.; Hunter, R.H.F.; Allen, W.R. Hysteroscopic insemination of small numbers of spermatozoa at the uterotubal junction of preovulatory mares. J. Reprod. Fertil. **2000**, *118*, 95–100.

26. Lindsey, A.C.; Morris, L.H.A.; Allen, W.R.; Schenk, J.L.; Squires, E.L.; Bruemmer, J.E. Hysteroscopic insemination of mares with low numbers of nonsorted or flow sorted spermatozoa. Equine Vet. J. **2002**, *34*, 128–132.

Manipulated Embryos: Cryopreservation

Andras Dinnyes
BioTalentum Ltd., Godollo, Hungary

T.L. Nedambale
Germplasm Conservation and Reproductive Biotechnologies, Agricultural Reserve Council, Animal Production Institute, Irene, South Africa

Jun Liu
Molecular Animal Biotechnology Laboratory, Szent Istvan University, Godollo, Hungary

Abstract

Cryopreservation of manipulated embryos is complicated because such embryos often have a reduced viability. Cryopreservation methods, including the current main strategies like interrupted slow freezing and vitrification (which includes traditional and ultrarapid vitrification methods) are explained together with the associated cryoinjuries. Some cryopreservation strategies are recommended to overcome the "double jeopardy" of the combined adverse effects of manipulation and cryopreservation.

INTRODUCTION

Cryopreservation of manipulated embryos is becoming more frequent with the advances in manipulations directly on mammalian embryos for specific purposes or indirectly on animals or gametes to produce embryos. Such embryos often have an altered viability and their cryopreservation might have medium- or long-term consequences. In this entry, three typical cryopreservation methods: 1) interrupted slow freezing; 2) vitrification; 3) and ultrarapid vitrification; and physical events occurring during cryopreservation are explained shortly together with the associated cryoinjuries such as osmotic damage, chilling injury, intracellular ice formation, and recrystallization. Some cryopreservation strategies are recommended for the given manipulations based on recent research progress and reports.

EMBRYO MANIPULATION

Embryo manipulation is a common practice in modern agricultural and laboratory animal research and breeding. Assisted reproductive technologies (ART), in particular artificial insemination (AI), in vitro fertilization (IVF), and intracytoplasmic sperm injection (ICSI), supported by gamete and embryo cryopreservation have accelerated the progress of genetic improvement and enabled the distribution of germplasm worldwide for domestic species and supporting conservation efforts for endangered species and captive breeding programs.

Embryo manipulations are modifying the "natural" production of embryos; they range from simple hormonal treatments that change the timing and extent of ovulation to sophisticated procedures of ART, including the in vitro production (IVP) of embryos or micromanipulation for nuclear transfer (NT) combined with transgenic (TG) modifications.[1]

In vitro systems to produce viable mammalian embryos have become increasingly defined over the past 15 years in several species. The process can be broken down into three main steps: 1) IVM (in vitro maturation), during which the oocyte is matured under the influence of hormones until it is capable of being fertilized; 2) IVF, when a spermatozoon penetrates the oocyte and the second polar body is extruded; and 3) IVC (in vitro culture), the progression from the one-cell zygote to the desired developmental stage suitable for embryo transfer. Recent reports indicate that genes that are differentially expressed in IVP embryos, compared to embryos resulting from artificial insemination or produced in vivo, may be affected by in vitro oocyte maturation, fertilization, and culture conditions compared to embryos resulting from artificial insemination or produced in vivo. The IVP embryos differ from in vivo-produced embryos in many ways, including the timing of development, morphology, membrane composition, intercellular communication, metabolism, cell number, the incidence of mixoploidy, and gene expression. These differences might contribute significantly to embryo survival following cryopreservation.

Some embryo manipulations are mechanically and/or biologically invasive. Embryo biopsy has been widely utilized in preimplantation genetic diagnosis (PGD) and embryo-sexing. The procedure involves opening of the zona pellucida (ZP) and removal of one or more blastomeres, thus, usually results in reduction in the capacity of the biopsied embryo to survive subsequent cryopreservation. Embryo splitting or bisectioning remains an efficient technique for increasing the number of offspring from a single embryo; however, split embryos do not survive cryopreservation well. With the progresses in modern molecular embryology, more embryo manipulation technologies have been developed to create TG embryos by chimera or somatic cell NT approaches applying genetically modified cells.

The physical effects of these manipulations or the altered in vitro environment can change gene function patterns, cell physiology, and, subsequently, the long-term health and viability of the embryos and the developing progeny. The embryos will try to adapt to the stress to which they are subjected, and will be able to compensate to a certain extent for the suboptimal conditions. However, there are some long-term changes that still may occur, including epigenetic modifications, altered intracellular signaling, metabolic stresses, gene expression changes, apoptotic changes, and disturbed cell proliferation. This multifactorial process, whereby short-term epigenetic, metabolic, and proliferative conditions (possibly coupled with an altered maternal physiology) impose homeostatic changes in gene expression and settings of the neuro-endocrine axis during later gestation, may result in long-term effects such as abnormal placentation, altered maternal nutrient provisions, abnormal fetal growth rate, abnormal birth weight and postnatal growth, and, ultimately, adulthood diseases, including cardiovascular and metabolic syndromes.[1]

EMBRYO CRYOPRESERVATION

Embryo cryopreservation is a well-established procedure for banking of genetically valuable strains of livestock; it allows easy international trade of female genetics and long-term gene banking. In endangered breeds and species, gamete and embryo banks can ensure the preservation of genetic diversity. In laboratory animals, often with TG modifications, the large number of strains can be economically and safely managed with the help of cryostorage for banking embryos from lines that are crucially important for biomedical research.[2,3] Technically, there are two current main strategies: 1) interrupted slow freezing (ISF); and 2) vitrification, which includes traditional vitrification (mostly in straws) and ultrarapid vitrification methods (Table 1).

Interrupted Slow Freezing

The most common embryo cryopreservation method is the ISF procedure, consisting of an initial slow, controlled-rate cooling to subzero temperatures followed by rapid cooling as the sample is plunged into liquid nitrogen (LN_2) for storage. During the controlled slow cooling, extracellular ice formation is induced (seeding) at a temperature just below the solution's freezing point, and then the cooling continues at a given rate in the presence of a growing ice phase, which

Table 1 Comparison of the three approaches for embryo cryopreservation.

Parameters	Cryopreservation techniques		
	Interrupted slow freezing	In-straw vitrification	Ultrarapid vitrification
Embryo storage container	Straw, cryovial	Straw	Special devices (such as open-pulled straw, cryoloops, cryotop) or no container (such as solid surface)
CPA concentration (additional sucrose concentration)	1.3–1.5 M (0–0.3 M)	5.5–7.5 M (0–0.3 M)	3.5–5.5 M (0–0.3 M)
Time in the equilibration CPA solution	N/A	2–5 min	2–5 min
Time in final CPA concentration for cooling	15–20 min	>1 min	>10 sec
Time required for cooling to final storage temperature	90–120 min	2–3 min	<0.1 sec
Ice crystal formation in embryo suspension	Yes	No	No
Osmotic injury	Low risk	High risk	High risk
Toxic injury	Low risk	High risk	High risk
Chilling injury	High risk	Low risk	Low risk
Warming rate	Low to moderate	Moderate to high	High
Cost of cooling apparatus	High	Low	Low
Commercial applications	Extensive	Limited	Limited

Source: From Reprod. Fertil. Dev.[1]

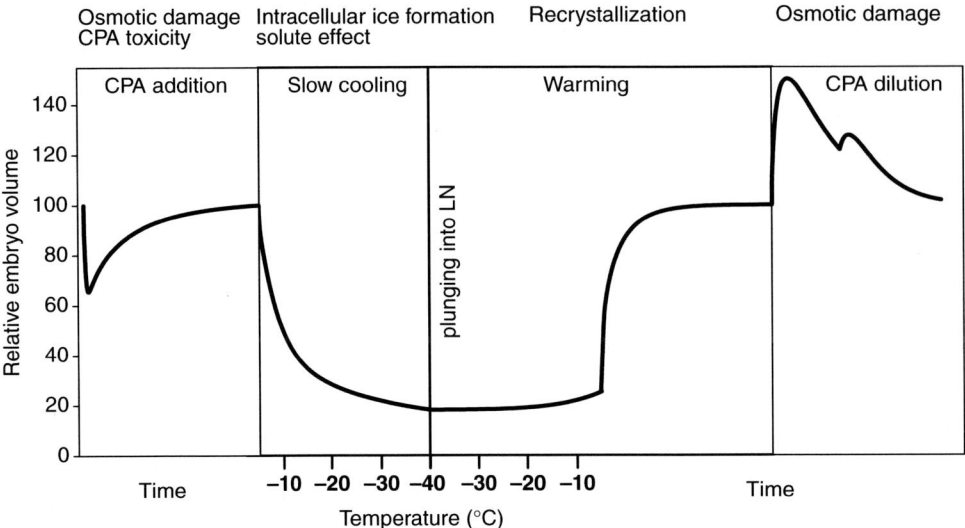

Fig. 1 Embryo volume changes during a typical ISF cryopreservation procedure; in this case, ISF terminated at −40°C and a two-step dilution without non-permeating CPA (such as sucrose) was used. Time and temperature scales were used for the x-axis.

raises the extracellular solute concentration in the unfrozen fraction and results in water being removed from the cell via exosmosmosis. Permeating cryoprotective agents (CPAs), such as glycerol, dimethyl sulfoxide (DMSO), ethylene glycol (EG), or propylene glycol (PG) are typically included in the freezing medium, to protect the cells against injury from the high concentrations of electrolytes that develop as water is removed from the solution as ice. These CPAs become increasingly concentrated intracellularly as the cell dehydrates. The slow cooling step is terminated at an intermediate temperature (in the range of −30 to −80°C) then plunging, a rapid cooling step, is initiated in which the remaining intracellular water either vitrifies or form small non-damaging ice crystals.

Control of the cooling and warming rates is crucial. If cells are cooled too rapidly during the controlled slow cooling process, water does not exit the cells fast enough to maintain equilibrium and, therefore, the embryos freeze intracellularly, resulting in death in most cases. If cooling is too slow, the long duration can cause "solution effects" injury resulting from the high concentration of extra-and intracellular solutes, probably because of the effects of the solutes on the cellular membrane or through osmotic dehydration.[4] Embryo cryopreservation protocols commonly use controlled cooling rates of 0.3–0.6°C/min.

During thawing, the small intracellular ice crystals might subsequently undergo recrystallization, forming bigger ice crystals that rupture the cell membrane, thus leading to fatal damage. Rapid thawing can prevent recrystallization.

In general, we can expect coupled flows of water and CPAs when CPAs are added, during freezing, thawing, and when CPAs are removed from the embryos,[5,6] resulting in a series of anisosmotic conditions. During freezing the embryo's cells dehydrate and shrink and remain shrunken during storage, but return to their isosmotic volume on thawing. Finally, the cells are subjected to potentially lethal swelling upon CPA dilution and removal. The addition/dilution of permeating CPAs and non-permeating CPAs (e.g., sucrose, glucose, or trehalose) to/from the embryos are usually conducted in a multiple-step manner to minimize the magnitude of volume excursion (Fig. 1).

Vitrification

Vitrification, cryopreservation without ice formation, is a process of converting a solution into a glass-like amorphous solid state that is free of any crystalline structures. This can be achieved by utilizing high cooling rates upon embryos loaded with high concentrations of CPAs.[7] Vitrification eliminates ice crystal formation-related embryo damages and can reduce chilling injuries (damages owing to low-temperature exposure of the cells, without ice formation) by reducing the time the embryos spend in suboptimal temperature zones compared to that of freezing procedures. Vitrification has the advantage, compared to ISF, that the need of time and expensive equipment is reduced.

A typical vitrification protocol requires a high concentration of CPAs in the medium (30–50% compared with 5–10% for ISF) at a cooling rate easily achievable for conventional laboratory conditions and setups. Such high CPA concentrations can be detrimental to embryos, causing both biochemical alterations and lethal osmotic injury. Various strategies have been described to counter the potential toxicity of solutions including: 1) the use of a combination of CPA solutes, each of which is below a toxic concentration, yet in combination would allow vitrification; 2) addition of non-permeating CPAs such as disaccharides (e.g., sucrose or trehalose) or high molecular weight molecules (e.g., Ficoll, polyvinylalcohol, or polyvinylpyrrolidone) can significantly reduce the amount of permeable CPA required;

3) the use of compounds which counteract the toxicity of other agents (e.g., acetamide with DMSO); and 4) reducing the time for which and/or the temperature at which the embryos are exposed to CPAs.

Reducing the solute concentrations in combination with higher cooling rates is an efficient strategy employed to achieve vitrification of embryos. In the ultrarapid vitrification methods, very high cooling rates are achieved while the volume of the vitrification solution is minimized by using specially designed containers, such as open-pulled straws, electron microscope copper grids, cryoloops, cryotops, gel-loading tips, and other similar devices, or without any containers in a few microliter-size drops into liquid nitrogen or on cooled surfaces (solid surface vitrification, SSV).[8] In the processes of warming of vitrified solutions, high warming rates are essential to prevent devitrification, a detrimental event that converts vitrified state into a crystallized structure resulting in intracellular ice formation and cellular damages.

Many of these novel ultrarapid technologies use direct exposure of samples to LN_2 that often precludes effective biocontainments.

Optimal cryopreservation methodology varies by species and breeds, mostly because of the variations in membrane permeability and embryo sensitivity to CPAs. Cryosensitivity is altered by the developmental stage of the embryo, reflecting its changing membrane permeability toward water and CPAs, and effective cell repair and protective mechanisms.[1]

CRYOPRESERVATION OF MANIPULATED EMBRYOS

Manipulation steps can result in aberrant physical (membrane composition, cytoskeletal, and chromosomal changes) and physiological (gene expression, protein production) patterns in embryos. The cryopreservation process exposes embryos to significant changes in their thermal, chemical, and physical environment. These changes would evidently introduce additional stress and epigenetic modifications to manipulated embryos, further reducing their viability.

Tolerance toward cryopreservation is affected by inherited vigor (e.g., expressed in the kinetics of early cleavages and further development, cell numbers in a given developmental stage) and numerous external factors such as IVP conditions, serum and lipid content of the media, postthaw recovery steps, and even the oil used to cover the microdrops in culture. Manipulated mammalian embryos may be more sensitive to the environment in which they develop, either in vitro or in vivo. Differences of IVP embryos compared to in vivo-derived ones include darker cytoplasm, lower density, more lipids (specifically, more triglycerides and fewer lipids from other classes), swollen blastomeres, more fragile ZP, differences in intercellular communication, and a higher incidence of chromosomal abnormalities.[9] These factors contribute to the higher sensitivity to cryoinjury in IVP embryos. Some species are particularly problematic, including pig embryos, reflecting their high sensitivity to chilling.

Remedy for these problems include improved IVC conditions and improved cryopreservation methods. The serum and lipid composition of IVC media can play a major role in the cryotolerance of embryos. The IVP medium supplemented with proteins or sera from various sources could alter the chemical composition of the embryos and their susceptibility to cryopreservation. In bovine IVP embryos improving culture conditions and replacing serum with albumin or synthetic molecules improved cryosurvival.

It has been also shown that IVP embryos incorporate more fatty acids into triglycerides than what they do in vivo-derived embryos, with possible effects on membranes. These intracellular lipids may be responsible for the low tolerance of IVP embryos to cryopreservation and for their lower buoyant density.

In several reports, the survival of IVP blastocysts was significantly higher after vitrification than after controlled freezing indicating that the chilling sensitivity of IVP embryos can be circumvented by vitrification. The ultrarapid embryo vitrification has been successful for species that are particularly susceptible to cryodamage, including the pig.

There are special cases of embryo manipulations aimed to improve cryotolerance. When the lipid content of porcine or bovine zygotes have been removed by centrifugation and then micromanipulation following IVC, the resulting blastocysts were much more tolerant toward vitrification.[10] The addition of hyaluronan or linoleic acid to the IVC media or reduction of the cytoplasmic lipid content with phenazine ethosulfate (PES), a compound that oxidizes NADPH, was beneficial for IVP bovine embryo cryotolerance.

The oil overlay on IVC medium microdrops can be an important factor in cryosurvival, as it will take part in the exchanges of components between the oil layer and the medium and ultimately, its components would influence the composition of the embryos.

Invasive micromanipulation methods, including embryo biopsy or splitting, usually damage the integrity of the ZP. It is postulated that the ZP plays an important role in preserving the embryo integrity through the steps of dehydration, shrinkage, and rehydration. However, hatched bovine and pig blastocysts have been successfully vitrified, demonstrating that the intact ZP is not necessary in advanced preimplantation stages.

CONCLUSION

In conclusion, further breakthroughs are needed to overcome the existing limitations for the cryopreservation of manipulated embryos. The key to the success would be the reduction of adverse epigenetic effects owing to the manipulations. Better controlled IVP environments, close to physiological handling conditions, improved ET and pregnancy maintenance methods together could contribute to better overall results. The right choice of the method and specific steps of the cryopreservation can overcome many manipulation related "double jeopardy" problems. The advances in TG technology will provide new tools to manipulate the physiology of mammalian embryos to enhance cryosurvival of TG embryos.

Selection of the "best" embryos for cryopreservation based not only on their morphological characteristics but also on their developmental dynamics, metabolic and molecular[11] profiles would increase survival rates. Increased stress tolerance by exposure of embryos to a limited amount of stress (i.e., high pressure) can improve cryotolerance.[12]

Most importantly, further improvements beyond the empirical approaches are needed. Fundamentally new avenues will open when accumulated molecular biological data coupled with membrane characteristics and CPA toxicities would allow to model the embryo mathematically as a "system" and to make reliable predictions for cryobehavior, design optimal cryoprocedures, and take into account long-term biological consequences as well.

ACKNOWLEDGMENTS

Short extracts of the text and Table 1 were reproduced with permission from A. Dinnyes and T.L. Nedambale (2009). Cryopreservation of manipulated embryos: tackling the double jeopardy. Reprod. Fertil. Dev. 21 (1) 45–59. Copyright IETS 2009. Published by CSIRO PUBLISHING, Melbourne, Australia. http://www.publish.csiro.au/nid/45/issue/4842.htm.

The authors' work reported herein was supported by the European Union FP6 (MEDRAT-LSHG-CT-2005–518240 and CLONET MRTN-CT-2006–035468), COST Action FA0702; by the Hungarian Government (NKFP_07_1-ES2HEARTHU (OM-00202-2007) project) and a Hungarian–South African Bilateral Scientific and Technological (TET No. OMFB-00302/2008 & NRF-RT24000) collaborative project.

REFERENCES

1. Dinnyes, A.; Nedambale, T.L. Cryopreservation of manipulated embryos: tackling the double jeopardy. Reprod. Fertil. Dev. **2009**, *21* (1), 45–59.
2. Leibo, S.P. Cryopreservation of oocytes and embryos: Optimization by theoretical versus empirical analysis. Theriogenology **2008**, *69* (1), 37–47
3. Mazur, P.; Leibo, S.P.; Seidel, G.E., Jr. Cryopreservation of the germplasm of animals used in biological and medical research: importance, impact, status, and future directions. Biol. Reprod. **2008**, *78* (1), 2–12.
4. Mazur, P.; Leibo, S.P.; Chu, E.H. A two-factor hypothesis of freezing injury. Evidence from Chinese hamster tissue-culture cells. Exp. Cell. Res. **1972**, *71* (2), 345–355.
5. Kedem, O.; Katchalsky, A. Thermodynamic analysis of the permeability of biological membranes to non-electrolytes. Biochim. Biophys. Acta **1958**, *27* (2), 229–246.
6. Mazur, P. Kinetics of water loss from cells at subzero temperatures and the likelihood of intracellular freezing. J. Gen. Physiol. **1963**, *47*, 47–69.
7. Rall, W.F.; Fahy, G.M. Ice-free cryopreservation of mouse embryos at -196 degrees C by vitrification. Nature **1985**, *313* (6003), 573–575.
8. Dinnyes, A.; Dai, Y.; Jiang, S.; Yang, X. High developmental rates of vitrified bovine oocytes following parthenogenetic activation, in vitro fertilization, and somatic cell nuclear transfer. Biol. Reprod. **2000**, *63* (2), 513–518.
9. Dinnyes, A.; Liu, J.; Nedambale,T.L. Novel gamete storage. Reprod. Fertil. Dev. **2007**, *19* (6), 719–731.
10. Nagashima, H.; Kashiwazaki, N.; Ashman, R.J.; Grupen, C.G.; Seamark, R.F.; Nottle, M.B. Removal of cytoplasmic lipid enhances the tolerance of porcine embryos to chilling. Biol. Reprod. **1994**, *51* (4), 618–622.
11. Boonkusol, D.; Gal, A.B.; Bodo, S.; Gorhony, B.; Kitiyanant, Y.; Dinnyes, A. Gene expression profiles and in vitro development following vitrification of pronuclear and 8-cell stage mouse embryos. Mol. Reprod. Dev. **2006**, *73* (6), 700–708.
12. Bock, I.; Mamo, S.; Polgar, Zs; Pribenszky, Cs. Changes in gene expression of mouse blastocysts treated with high hydrostatic pressure pulse. Reprod. Domest. Anim. **2008**, *43* (Suppl. 3), 145–146.

Meat Fermentation

Immaculada Franco
Sonia Fonseca
Maria C. Garcia Fontán
Javier Carballo
Food Technology Area, University of Vigo, Ourense, Spain

Abstract

The manufacturing of fermented meats involves the action of naturally present or added lactic acid bacteria, *Micrococcaceae*, and *Staphylococcaceae* (and, in some cases, external molds). In order to drive the manufacture process, the knowledge and control of these microorganisms is essential. In this entry, the applications of molecular biology to this field are reviewed.

INTRODUCTION

Meat fermentation/ripening is mainly due to endogenous enzymes (mainly proteases and lipases) and enzymatic activity exerted by the autochthonous microbiota or added starter cultures. Classical methods for the identification and control of these microorganisms are time-consuming and usually lead to ambiguous results. Molecular biology offers new alternatives in the identification and characterization of microbial strains and in the investigation of microbial groups or specific microorganisms in fermented meats.

IMPLICATIONS OF THE BIOTECHNOLOGY IN THE MANUFACTURE AND QUALITY CONTROL OF THE FERMENTED MEAT PRODUCTS

Since raw-cured meat products made from entire pieces (e.g., hams, dry-cured pork foreleg, and jerky) do not experience a true lactic fermentation during their manufacture processes, we will focus on the genuine fermented meat products, the dry fermented sausages. It is clear that the sensory and compositional changes that occur in fermented sausages during their manufacture are mainly due to endogenous catalytic enzymes (mainly proteases and lipases) and the autochthonous microbiota or added starter cultures, which develop in the product during its fermentation and ripening. Modern industry uses starter cultures with specific functions to assure the quality and safety of fermented sausages (see Table 1). In the fermented meats industry, biotechnology can be applied for the identification and genotypic characterization of the microorganisms used in the manufacturing processes.

IDENTIFICATION AND CHARACTERIZATION OF MICROORGANISMS

The study of the microbiota of dry fermented sausages and the search for new starter cultures require identification of the strains involved with particular attention to lactic acid bacteria group and the *Micrococcaceae* and *Staphylococcaceae*. Classical identification methods can provide valuable information on a variety of technologically relevant properties of specific strains (e.g., growth at different temperatures and NaCl concentrations, acetoin production, and nitrate reduction); however, the classical culture methods are time-consuming and can lead to bacterial misidentification. Culture methods also present limitations in differentiating species that are similar taxonomically, e.g., strains of *Lactobacillus curvatus* and *Lactobacillus sakei* isolated from meat products can be difficult to differentiate. Comparable issues arise within the genus *Lactobacillus* for the species *plantarum*, *paraplantarum*, and *pentosus*.

Therefore, several rapid methods for bacterial identification have been developed based on molecular techniques in recent years. They include: SDS-PAGE of whole cell proteins,[1] species-specific PCR or hybridization,[2] restriction fragment length polymorphism (RFLP) analysis of the 16S rRNA gene,[3] pulsed-field gel electrophoresis (PFGE;[4]), random amplification of polymorphic DNA (RAPD)-PCR analysis,[5] and restriction analysis of amplified 16S rDNA gene (ARDRA;[6]). These techniques allow identification at intraspecies level which may help to select, within the same species, strains with particular and desirable technological properties.

Because of the enterotoxigenic character of some staphylococcal strains, the investigation of the ability for enterotoxin production is essential in the selection of starter cultures. PCR methods of detection of enterotoxin genes

Table 1 Primary microbial species used as starter cultures in the manufacture of the raw-fermented sausages.

Species	Performance
Bacteria	
Lactobacillus curvatus	Formation of lactic acid
	Production of aroma
	Lipolytic activity
	Proteolytic activity
	Production of bacteriocins
Lactobacillus sakei	Formation of lactic acid
	Production of aroma
	Lipolytic activity
	Production of bacteriocins
Lactobacillus plantarum	Formation of lactic acid
	Production of bacteriocins
Pediococcus acidilactici	Formation of lactic acid
	Production of bacteriocins
Pediococcus pentosaceus	Formation of lactic acid
	Production of bacteriocins
Staphylococcus xylosus	Reduction of nitrate and nitrite
	Color fixation
	Production of catalase (antioxidant effect)
	Proteolytic activity
	Lipolytic activity
	Formation of flavor
Staphylococcus carnosus	Reduction of nitrate and nitrite
	Color fixation
	Production of catalase (antioxidant effect)
	Proteolytic activity
	Lipolytic activity
	Formation of flavor
Kocuria varians	Reduction of nitrate
Yeasts	
Debaryomyces hansenii	Lipolytic activity
	Oxygen consumption
Molds	
Penicillium camemberti	Lipolytic activity
	Proteolytic activity
	Formation of flavor
	Oxygen consumption
Penicillium nalgiovense	Lipolytic activity
	Proteolytic activity
	Formation of flavor
	Oxygen consumption

have been ascertained;[7] this is a valuable tool in the strain selection of staphylococci.

INVESTIGATION OF MICROBIAL GROUPS OR SPECIFIC MICROORGANISMS

The investigation of microbial groups or specific microorganisms in the manufacturing process or in the final products has been traditionally made by plate-counting on a selective culture medium. This method requires the use of high quantities of media, which usually are not entirely selective for the enumeration of a specific microbial group. Moreover, it has become evident that classical culture techniques are not valid in assessing the microbial diversity present in traditionally fermented foods, since these procedures tend to underestimate bacterial biodiversity and can lead to misidentification.

Using both classical and molecular methods, Fontana et al.[8] studied the dynamics of the microbial groups and species responsible for the fermentation of an artisanal dry-sausage produced in Argentina. In general, information obtained by denaturing gradient gel electrophoresis (DGGE) analysis agreed reasonably with results from classical, RAPD, and 16S rDNA analyses. DGGE analysis revealed the presence of additional species that were not detected while using culture-dependent microbiological procedures. The authors concluded that molecular characterization techniques can be complementary to traditional microbiological methods in the analysis of microbial systems such as a fermented sausage.

Molecular methods have been successfully applied to the study of the specific composition of commercial starter cultures and performance of the different species during sausage fermentation.[9] In fact, they have been used to determine starter species not declared on the product label.

Recently, Luxananil et al.[10] monitored the starter culture *Lactobacillus plantarum* BCC 9546 during the fermentation of a traditional Thai pork sausage. *L. plantarum* BCC 9546 was transformed with a recombinant plasmid pRV85 to produce the recombinant strain, *L. plantarum* G11, which is resistant to erythromycin and emits green fluorescence. *L. plantarum* G11 was used as a starter culture for sausage fermentation and its growth was easily monitored by plating on a selective medium and assaying for fluorescent activity. The growth of the recombinant strain during fermentation was very similar to that of the parental strain, and the acidity, texture, and color of the fermented sausages inoculated either with *L. plantarum* G11 or *L. plantarum* BCC 9546 were similar.

GENETIC KNOWLEDGE OF THE MICROBIAL STARTERS

Most of the genetic knowledge of *L. sakei* derives from the complete genomic sequence of *L. sakei* 23 K,[11] a strain originally isolated from a dry fermented sausage. The genome was 1884661 base pairs, encoding 1883 predicted genes. Genomic sequencing revealed a specialized metabolic repertoire, including purine nucleoside scavenging that may contribute to the ability of this species to successfully compete on raw meat products. Many genes appear to be responsible for robustness during the stresses encountered with meat processing (e.g., NaCl, spices, smoke, low temperatures, changes in redox, and oxygen levels). Genes potentially responsible for biofilm formation and cellular aggregation that may help the organism to colonize meat surfaces have been identified. Regarding the role of *L. sakei* as starter culture in dry-fermented sausage manufacturing, genomic analysis confirmed that the main function of *L. sakei* is to ferment carbohydrates into lactic acid.

L. sakei contains a variable number of plasmids, which may encode either beneficial (e.g., production of bacteriocins) or non-desirable (e.g., antibiotic resistance) traits. Therefore, they should be investigated during the selection of potential starter or adjunct cultures.

Pfeiler and Klaenhammer[12] reviewed the genomes of 20 lactic acid bacteria species, including *Pediococcus pentosaceus* ATCC 25745 which has been widely usually used as a starter in fermented sausages. The genome of this strain has a size of 1.8 Mbp; it does not contain plasmids but harbors two prophages integrated into its chromosome.

Regarding *Staphylococcus* species, most of the genetic knowledge of *S. carnosus* was obtained from strain TM300. The size of the circular chromosome was estimated to be 2590 kpb.[13] Wagner et al.[13] constructed a physical and genetic map of *S. carnosus* TM300. The analysis of single, multiple, and partial digests in combination with hybridization experiments with genetic markers and cross-Southern hybridization allowed the alignment of 31 fragments. Eighteen genetic loci were placed on the genome map. From the technological point of view, the most important genes were the *nir/nar* genes involved in nitrite and nitrate reduction.

Genetic knowledge of *S. xylosus* was obtained from strain C2a. A physical and genetic map of *S. xylosus* C2a was established by locating 47 restriction fragments and 33 genetic markers. The identified loci mainly concerned carbohydrate utilization and antioxidant capacities. The sequencing of the complete genome of *S. xylosus* is actually in progress (http://www.genoscope.cns.fr).

FUTURE CHALLENGES AND CONCLUSION

The main future challenge of meat starter culture technology is the genetic engineering of commercial starter cultures with genes conferring new metabolic capabilities. This would allow creation of the "ideal starter culture" that in one single species would gather all the metabolic traits needed to carry out the desired functions; however, most of the mechanisms of DNA transference are not yet described or explored in the species used as starter cultures in the fermented meat products. Additionally, in the global marketplace, there still exist many regulatory issues restricting the use of genetically modified starter cultures in foods; these concerns will have to be dealt with for strain improvements to continue for industrial applications.

Molecular techniques provide useful assistance to the fermented meat industry in the identification and characterization of microbial strains, in the investigation of microbial groups or individual microbial types, and in the genetic knowledge of the microbial starters. One can expect potential applications of metagenomics, transcriptomics, proteomics, and metabolomics to the field of starter cultures as well.

REFERENCES

1. Samelis, J.; Tsakalidou, E.; Metaxapoulos, J.; Kalantzopoulos, G. Differentiation of *Lactobacillus sake* and *Lactobacillus curvatus* isolated from naturally fermented Greek dry salami by SDS-PAGE of whole-cell proteins. J. Appl. Bacteriol. **1995**, *78*, 157–163.
2. Yost, C.K.; Nattress, F.M. The use of multiplex PCR reactions to characterize populations of lactic acid bacteria associated with meat spoilage. Lett. Appl. Microbiol. **2000**, *31*, 129–133.
3. Sanz, Y.; Hernández, M.; Ferrús, M.A.; Hernández, J. Characterization of *Lactobacillus sake* isolates from dry-cured sausages by restriction fragment length polymorphism analysis of the 16S rRNA gene. J. Appl. Microbiol. **1998**, *84*, 600–606.
4. Singh, S.; Goswami, P.; Singh, R.; Heller, K.J. Application of molecular identification tools for *Lactobacillus*, with a focus on discrimination between closely related species: a review. LWT—Food Sci. Technol. **2009**, *42*, 448–457
5. Berthier, F.; Ehrlich, S.D. Genetic diversity within *Lactobacillus sakei* and *Lactobacillus curvatus* and design of PCR primers for its detection using randomly amplified polymorphic DNA. Int. J. Syst. Bacteriol. **1999**, *49*, 997–1007.
6. Bonomo, M.G.; Ricciardi, A.; Zotta, T.; Parente, E.; Salzano, G. Molecular and technological characterization of lactic acid bacteria from traditional fermented sausages of Basilicata region (Southern Italy). Meat Sci. **2008**, *80*, 1238–1248.
7. Blaiotta, G.; Ercolini, D.; Pennacchia, C.; Fusco, V.; Casaburi, A.; Pepe, O.; Villani, F. PCR detection of staphylococcal enterotoxin genes in *Staphylococcus* spp. strains isolated from meat and dairy products. Evidence for new variants of *se*G and *se*I in *S. aureus* AB-8802. J. Appl. Microbiol. **2004**, *97*, 719–730.
8. Fontana, C.; Cocconcelli, P.S.; Vignolo, G. Monitoring the bacterial dynamics during fermentation of astisanal Argentinean sausages. Int. J. Food Microbiol. **2005**, *103*, 131–142.
9. Cocolin, L.; Urso, R.; Rantsiou, K.; Cantoni, C.; Comi, G. Multiphasic approach to study the bacterial ecology of fermented sausages inoculated with a commercial starter culture. Appl. Environ. Microbiol. **2006**, *72* (1), 942–945.
10. Luxananil, P.; Promchai, R.; Wanasen, S.; Kamdee, S.; Thepkasikul, P.; Plengvidhya, V.; Visessanguan, W.; Valyasevi, R. Monitoring *Lactobacillus plantarum* BCC 9546 starter culture during fermentation of Nham, a traditional Thai pork sausage. Int. J. Food Microbiol. **2009**, *129*, 312–315.
11. Chaillou, S; Champomier-Vergès, M.C.; Cornet, M.; Crutz-Le Coq, A.M.; Dudez, A.M.; Martin, V.; Beaufils, S.; Darbon-Rongére, E.; Bossy, R.; Loux, V.; Zagorec, M. The complete genome sequence of the meat-borne lactic acid bacterium *Lactobacillus sakei* 23 K. Nat. Biotechnol. **2005**, *23*, 1527–1533.
12. Pfeiler, E.A.; Klaenhammer, T.R. The genomics of lactic acid bacteria. Trends Microbiol. **2007**, *15* (12), 546–553.
13. Wagner, E.; Doskar, J.; Götz, F. Physical and genetic map of the genome of *Staphylococcus carnosus* TM300. Microbiology **1998**, *144*, 509–517.

Meats: Proteomics

Macdonald Wick
Department of Animal Sciences, Ohio State University, Columbus, Ohio, U.S.A.

Abstract
The use of biotechnology as applied to meat science strives to increase the quantity and quality of meat produced for human consumption. The recent advances in the fields of animal genomics and functional genomics, also known as animal proteomics, are beginning to allow meat animal producers to identify those genes and gene products affecting muscle growth and meat quality. This entry will briefly describe current methodologies associated with the study of the animal genome and proteome as they relate to their use for meat animals.

INTRODUCTION

The term "biotechnology" refers to the use of molecular techniques to improve the quality of biological systems. Techniques currently fundamental to the biotechnology of meat are derived from and often unite the disciplines of genetics, physiology, biochemistry, molecular biology, chemistry, immunology, statistics, bioinformatics, and image analysis. This entry will briefly discuss the most common current technologies being used to advance the quality of meat.

Meat is usually defined as the edible constituents of postmortem skeletal muscle tissue of animals. Although skeletal muscle is designed by nature to be the primary locomotor organ of land and sea animals, it is what occurs to skeletal muscle after the animal dies that generates meat. The quality of that meat is often defined in terms of consumer satisfaction, that quality is the result of fundamental biological mechanisms attendant with muscle growth, development, and repair that continue postmortem to produce the product we recognize as meat. Those mechanisms are based on the fundamental biological paradigm that states that all information for animal structure from cell to complete organism is contained in the genome (DNA). That DNA acts as a template for the transcription of messenger RNA (mRNA) that acts as the blueprint for the translation or manufacture of proteins, the functional molecules of the cell, tissue, organ and ultimately, the animal. Transcription and translation are affected by the environment, both internally (intrinsic) and externally (extrinsic). Muscle is, therefore, the result of the interaction between the genome and the environment.

Meat biotechnology focuses on elucidating those mechanisms that are under genetic and environmental control. It therefore seems reasonable that by identifying the gene(s) and, more importantly, the product(s) of those genes controlling the phenotype of muscle or meat and standardizing the environment, we could produce predictably high-quality muscle and meat. This is the goal of the meat industry and meat biotechnology.

One of the complicating factors in meat biotechnology is that muscle and meat phenotypes are not the result of a single gene product but the cooperative interaction of multiple temporal and spatially specific gene expression events. Meat biotechnology seeks to identify those events by quantifying and or controlling the gene expression products. The identification of the genes underlying meat quality is the result of classical genetics and molecular genetics, genomics. The study of the combined and often simultaneous gene expression events underlying meat quality is functional genomics or proteomics. This entry focuses on the more widely used genetic and functional genomic technologies used for investigating the genome and the proteome of agricultural animals. The information gained by these technologies may be used in the future to control those genes and gene products through selective breeding.

EMERGENT TECHNOLOGIES

Gene manipulation to increase muscle growth and quality through transgenic methodologies is possible. Although the U.S. Food and Drug Administration has concluded that "meat and milk from clones of cattle, swine and goats and the offspring of clones from any species traditionally consumed as food are as safe to eat as food from conventionally bred animals,"[1] to date, no transgenic animals have been approved for use as human food. A very limited number of transgenic organisms have been approved for rendering into animal feed components.[2,3]

Meat scientists as well as the meat industry have spent considerable resources to reduce the ante- and postmortem environmental challenges that negatively affect the quality characteristics of tenderness, juiciness, and flavor. The effect of the environment on gene expression is a topic on to itself, and no more space will be devoted to its discussion. The remainder of this entry will present a brief overview of relatively few but predominant "gene-based and protein-based biotechnologies" of meat production and quality.

DNA-BASED TECHNOLOGIES

Although usually associated with the technologies of the later 20th and early 21st century biotechnology, genomics are rooted in the domestication of livestock in the southwest Fertile Crescent around Mesopotamia near current Iraq around 9000 years ago.[5] People soon realized that animals could be grown larger by selecting parental lines or breeding stock that tended to grow more rapidly. Unwittingly, these early biotechnologists were, in fact, selecting for the passage of genes that interacted with the environment in specific ways to increase muscle mass growth rates. The progeny that did not put on more muscle or meat were eliminated from the gene pool. This type of genetic selection saw its greatest use in the last 50 years when the greatest advances in the enhancements of quantitative trait selection resulted in pigs and cattle with reduced intramuscular fat (marbling) and turkeys with reduced juiciness as a result of lowered water-holding capacity.

SEXING SPERM

From the beginning of meat animal agriculture, the sex of the animal played an important economic role. An excellent in-depth review of the technologies and economics of sexing sperm in cattle has recently been published.[6] In the cattle industry, heifers are valued based on their milk production and their ability to give birth. Steers are valued based on their increased muscle mass accumulation. As a result, steers are worth around $60 more than heifers.[7] Therefore, in the United States it would be desirable to control the sex of the offspring of meat-producing animals if it were economically practical. The most efficacious methodology is based on the fact that bovine X-sperm contains 3.8% more DNA than Y-sperm. Sperm is incubated in a solution containing the DNA-binding fluorescent dye, Hoechst 33342. The fluorescently decorated sperm is then processed through a cell sorter called a flow cytometer that counts the sperm. Once the sperm count has been determined, it is introduced by artificial insemination (AI) procedures. Accuracies of 90% have been reported.[7] Sperm-sexing is predominantly being studied and employed in cattle and horses, but other species could potentially benefit from sexing sperm as well. Currently, sperm-sexing in pigs have been shown to be efficacious, although not widely employed.[8]

DNA ARRAYS

Advances in animal biotechnology have been facilitated by recent progress in sequencing and analyzing animal genomes, identification of molecular markers (e.g., microsatellites, expressed sequence tags [ESTs] derived from sequence analysis of cDNA, quantitative trait loci [QTLs]), and a better understanding of the mechanisms that regulate gene expression.

The use of DNA array technology allows for the quantification of RNA transcripts present in muscle tissue at a selected time point. Although the concentration of a given mRNA transcript is not necessarily directly correlated with the concentration of its translated gene product, the relative concentration of mRNA can give great insight into the effect of a given genome by environment interaction.[9] A generalized flow diagram of a typical DNA array experiment is presented in Fig. 1. Patrick Brown's group at Stanford was the first to print arrays of polymerase chain reaction (PCR) fragments amplified from DNA libraries on a glass surface the size of a standard microscope slide using a robotic printing device.[9,10] The technology remains essentially the same today and is referred to as cDNA microarrays. To conduct a microarray analysis of gene expression, RNA is purified from tissues or cells of interest and labeled with fluorescent dyes. After hybridization of the labeled RNA to the array, the slides are scanned and the fluorescent signal in each cDNA element on the slide provides a measure of the expression of the corresponding gene. A more comprehensive review of DNA

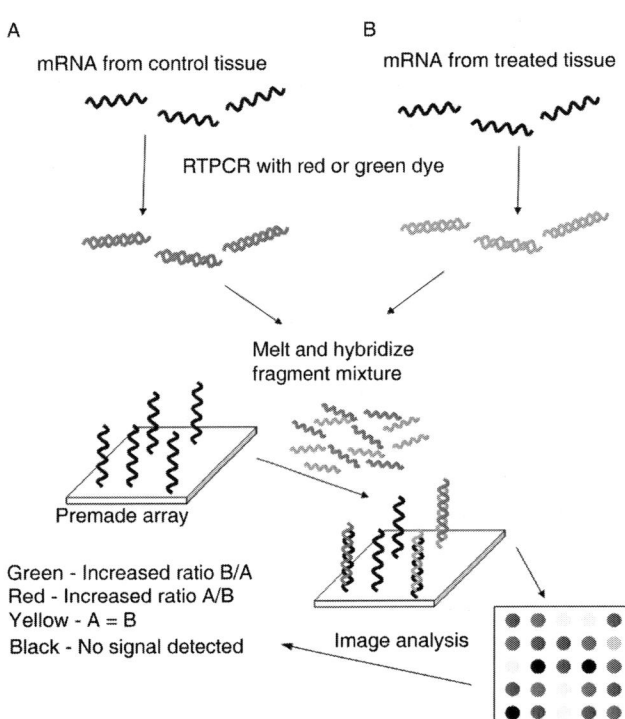

Fig. 1 Flow diagram for a DNA-based microarray experiment.

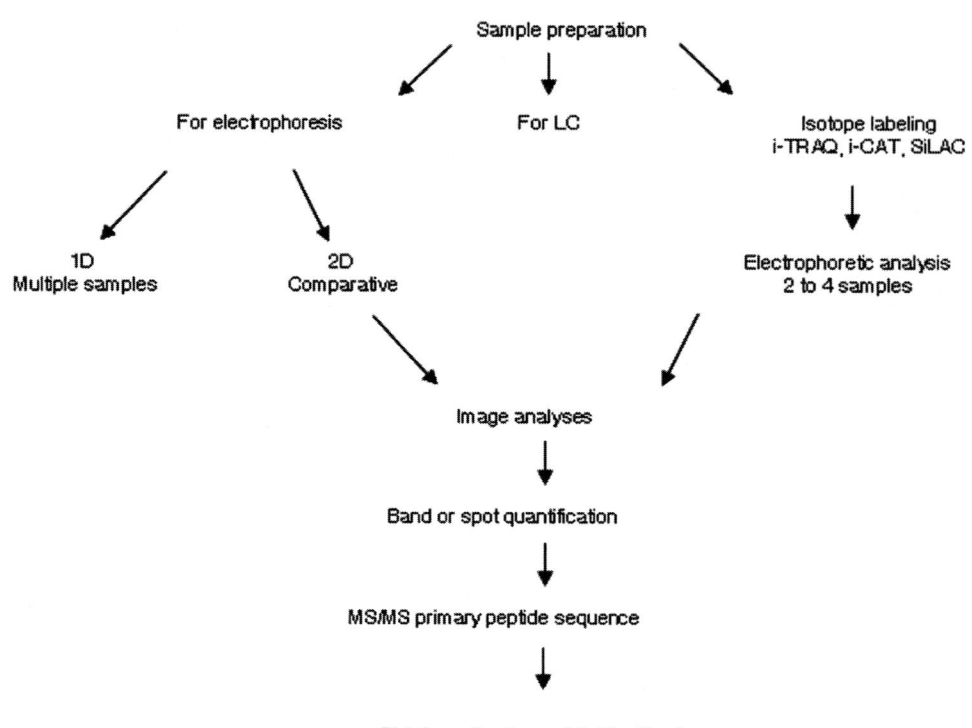

Fig. 2 Flow diagram for a typical proteomic experiment. Decision tree exists at the 1D vs. 2D electrophoresis.

microarray technology is given by Bendixen et al.[11] and at the website, http://kbrin.a-bldg.louisville.edu/Workshop/NCBIWorkshop_files/frame.htm.

PROTEIN ARRAYS

Currently protein arrays are more commonly being applied to human disease and have not gained wide use in meat biotechnology. Protein arrays are comprised of a library of proteins or antibodies immobilized on a grid in a similar fashion to the DNA array shown in Fig. 1. Because the amount of RNA transcript may or may not be directly related to the amount of the protein that the transcript codes for, the use of protein chip or protein arrays is beginning to gain favor. The principles for the protein array are similar to those of the DNA array. That is, proteins from a control sample and treatment sample are differentially labeled with dye. The difference in protein array technology is that the capture agent is generally an antibody specific for the given protein of interest. Therefore, the researcher must have reason to believe that that protein or set of proteins is unique to the sample. A visual review of protein microarray technology is presented on the Purdue University website.[12] A more comprehensive review of the subject is reviewed elsewhere.[13,14]

PROTEOMICS

Knowledge of the temporal and spatial translation of the genome into mRNA has the limitation that the presence or absence of mRNA is not necessarily directly related to the quantity of its translation product.[15] The translation products of the genome in a cell or tissue comprise the proteome. Proteins, the major component of the proteome, are isolated and characterized employing a series of methodologies generally based on the separation of the proteins by comparative one-dimensional (1D) or two-dimensional (2D) electrophoresis and the use of downstream image, sequence, and bioinformatics technologies. The complex interrelationship of the multiple methods used in proteomics is diagramed in Fig. 2. The decision to employ 1D or 2D electrophoresis for subsequent protein identification is based predominantly on whether comparative experiments (2D) or multiple samples or being used in the design. This will be discussed in the following section.

1D VS. 2D ELECTROPHORESIS IN PROTEOMIC ANALYSIS

Two-dimensional electrophoresis resolves proteins by their isoelectric point in the first dimension and by their

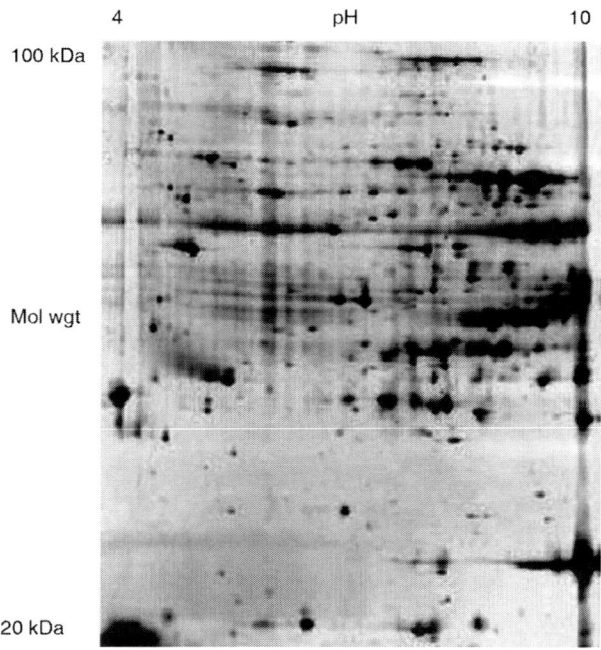

Fig. 3 Typical 2D electrophoretic separation of the extract from a single muscle sample. Broad isoelectric separation in a pH gradient from 4 to 10 and molecular weight range from ~100 to ~20 kDa.

molecular weight (kDa) in the second. A typical 2D electrophoretic resolution of muscle tissue is presented in Fig. 3. Each of the electrophoretic techniques has its own advantages and disadvantages. For instance, a common proteomic analysis based on 2D electrophoresis can resolve up to 1000 protein/peptide spots in a single run. The most common way to determine the changes in a control vs. a "test" sample is to perform DIGE. This technique is designed to determine which of the 1000 protein/peptide spots is statistically different between a control and test sample. Commonly, the experiment must be performed three times per sample. The disadvantage of 2D is its relative low differential molecular weight range and the high cost per sample.

On the other hand, 1D electrophoretic separation (Fig. 4) can resolve up to 100 protein/peptide bands. The advantage of this technique is its ability to be high-throughput. That is, 10–15 samples can be run on the same gel. One of the major advantages of this strategy is that it can be used for linear regression analysis to identify which bands are associated with changing characteristics of a given tissue or organ.[16,17] In addition, this strategy reduces the gel-to-gel variation that is inherent in 2D electrophoresis. The major disadvantage of 1D is the high potential of not resolving comigrating proteins.

CONCLUSION

Biotechnology is a complement—not a substitute—for many areas of conventional agricultural research. It offers

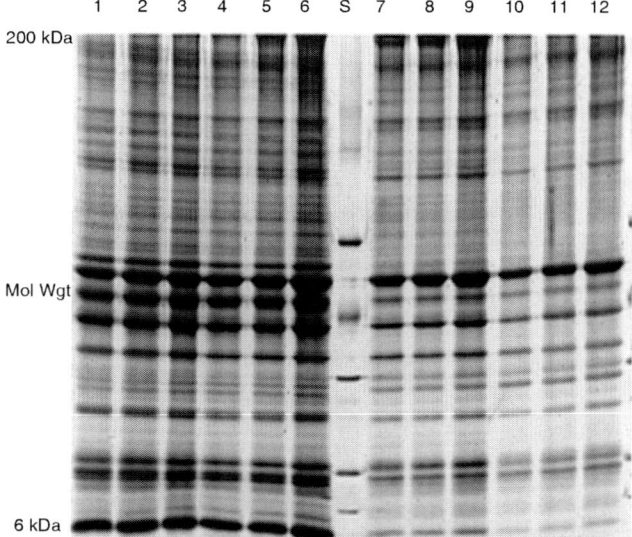

Fig. 4 Typical 1D electrophoretic separation of the extract from 12 different muscle samples each with different muscle growth characteristics. Molecular weight range from ~200 to ~6 kDa.

a range of tools to improve our understanding and management of genetic environmental impacts on meat animals. These tools are already making a contribution to breeding and conservation programs and to facilitating the diagnosis, treatment, and prevention of plant, human, and animal diseases. The application of biotechnology will provide meat scientists the tools that enable meat animal producers to make informed breeding decisions to increase the production level and quality of muscle and meat. It should be noted that, currently, biotechnology-based approaches are an extension and not a replacement of conventional approaches to food animal production and fabrication.

REFERENCES

1. Transcript of FDA Press Conference on Cloning Risk Assessment – Afternoon Media Telecon, January 16, 2008.
2. U. S. Food and Drug Administration, Center for Veterinary Medicine, http://www.fda.gov/cvm/transgen.htm (accessed May 17, 2008).
3. U. S. Food and Drug Administration, Center for Veterinary Medicine. *Appendix D: Transgenic Clones*, http://www.fda.gov/cvm/CloningRA_AppendixD.htm (accessed May 2008).
4. Brown, P.; Meyer, R.; Cardone, F.; Pocchiari, M. Ultra-high-pressure inactivation of prion infectivity in processed meat: a practical method to prevent human infection. Proc. Natl. Acad. Sci. **2003**, *100* (10), 6093–6097.
5. Diamond, J. To farm or not to farm. In *Guns, Germs and Steel: The Fates of Human Societies*, pp 86–89, 1st Ed.; W. W. Norton: New York.
6. Seidel, G.E., Jr. Overview of sexing sperm. Theriogenology **2007**, *6*, 443–446

7. Seidel, G.E., Jr. Economics of selecting for sex: the most important genetic trait. Theriogenology **2003**, *59*, 585–598.
8. Grossfeld, R.; Klinc, P.; Seig, B.; Rath, D. Production of piglets with sexed semen employing a non-surgical insemination technique. Theriogenology, **2005**, *63*, 2269–2277.
9. Southern, E.M. DNA microarrays. History and overview. Method Mol Biol. **2001**, *170*, 1–15.
10. DeRisi, J.; Penland, L.; Brown, P.O.; Bittner, M.L.; Meltzer, P.S.; Ray, M.; Chen, Y.; Su, Y.A.; Trent, J.M. Use of a cDNA microarray to analyse gene expression patterns in human cancer [see comments]. Nat. Genet. **1996**, *14* (4), 457–460.
11. Bendixen, C.; Hedegaard, J.; Horn, P. Functional genomics in farm animals – microarray analysis. Meat Sci. **2005**, *71*, 128–137.
12. http://www.stat.purdue.edu/sbc/protein_array_analysis.html (accessed May 2008).
13. Zhu, H.; Snyder, M. Protein chip technology. Curr. Opin. Chem. Biol. **2003**, *7* (1), 55–63.
14. Mitchell, P. A perspective on protein microarrays. Nat. Biotechnol. **2002**, *20* (3), 225–229.
15. Gygi, S.P; Rochon, Y.; Franza, B.; Abersold, R. Correlation between protein and mRNA abundance in yeast. Mol. Cell Biol. **1999**, *19*, 1720–1730.
16. Sawdy, J.C.; Kaiser, S.A.; St-Pierre, N.R.; Wick, M.P. Myofibrillar 1-D fingerprints and myosin heavy chain MS analyses of beef loin at 36 h postmortem correlate with tenderness at 7 days. Meat Sci. **2003**, *67* (3), 421–426.
17. Updike, M.S.; Zerby, H.N.; Sawdy J.C.; Lilburn M.S.; Kaletunc G.; Wick, M.P. Proteins associated with thermally-induced gelation of Turkey breast meat. J. Food Sci. **2006**, *71*, E398–E402.

Metabolite Extraction from Plant Tissues

Iryna Smetanska
Department of Methods in Food Technology, Berlin University of Technology, Berlin, Germany

Abstract
Extraction of one or more components from a complex mixture is a requirement for many operations in the food and biotechnology industries. This technology is used for recovery of active constituents from raw materials and for removal of undesirable constituents. Extraction methods rely on exploiting differences in physical or chemical properties of the mixture of components (e.g., molecular size and shape, density, solubility, and electrostatic charge).

SOLVENT EXTRACTION

Solvent extraction is a separation process that involves the removal of individual constituents from a mixture of solids or liquids upon addition of a solvent in which the original constituents have different solubilities. The solubilizing ability and selectivity of liquid solvents (mainly based on water or hydrocarbons like hexane, or alcohol) are used to leach or extract the components from the applied source material.

In liquid extraction, a liquid solvent is used to separate liquids by preferentially dissolving one of them. In leaching or solid extraction, a liquid solvent is used to dissolve soluble material from its mixture with an insoluble solid.

Liquid–liquid extraction is used in food processing to either selectively recover valuable or to remove undesirable components. The selection of solvent is based on solvent capacity, selectivity, chemical inertness, thermophysical properties (e.g., density, viscosity, and boiling point), flammability, toxicity, costs, and availability.

Extraction using water has obvious advantages of low cost and safety and is used to extract coffee and tea. Oils and fats require an organic solvent and as these are highly flammable, great care is needed in operating procedures: the equipment should be gas-tight and spark-proof.

The equipment used for liquid–liquid extraction can be classified based on the manner in which the phases are contacted: by gravity and by centrifugal force. In the class of equipments where the countercurrent flow is produced by gravity, they are further classified by the type of agitation as mostly spread static and mechanically agitated, but also rotary-agitated and reciprocating columns.

Static extraction columns are the simplest and cheapest, but not very efficient because of the large axial mixing or dispersion that occurs in the continuous phase as a direct result of the droplet rising through the continuous phase causing a large-scale circulation or axial mixing.

Mechanically agitated columns have the ability to control drop size independently of flow rates, so that losses of solvent can be minimized while maintaining efficiency. The agitation increases the interfacial area and column mass transfer efficiency. The mechanically agitated columns may be rotary devices or reciprocating devices.

Solid–liquid extraction or leaching is a separation process affected by a fluid involving the transfer of solutes from a solid matrix to a solvent. It is an extensively used unit operation, e.g., to recover proteins in oilseed meals. Solid–liquid extraction is easily automated, faster, and in general more efficient than liquid–liquid extraction.

The two main operations include the contact of liquid solvent with the solid for transfer of solute from the solid to the solvent and separation of the extract from the residual solid.

The preparation of the material for extraction is important to make the solute more accessible to the solvent by size reduction of solid. This increases surface area per unit volume of solids. The preparation of the material for extraction involves crushing, grinding, flaking, or cutting.

The effect of rupture of cell membranes offers various applications to support extraction process. In order to release the products from cells stored in vacuoles of plant cells, two membrane barriers (plasma membrane and tonoplast) have to be penetrated. Cell permeabilization depends on the formation of pores in one or more of the membrane systems of the plant cell, enabling the passage of various molecules into and out of the cell. To the methods causing membrane permeabilization belong high electric field pulses, high hydrostatic pressure, ultrasound, etc. Application of the high electric field pulses is based on the principle of development of membrane pores under external electric fields. Extraction of caffeine from coffee using water could be increased by the application of high pressure as well as increase in temperature. Also, treatment with high hydrostatic pressure (250 MPa) causes permeabilization of tonoplast. Both procedures could become effective tools for product recovery from tissues with minimum effects on product composition.

Short chain alcohols (methanol and ethanol), n-hexane, ketone (acetone), esters (ethyl acetate, n-butyl acetate), chlorinated hydrocarbons (methylene dichloride and ethylene dichloride), and liquid CO_2 can be used as solvents for leaching.

Temperature plays an important role in solid extraction. Higher temperatures give higher solubility of solute in solvents, permitting higher rates of extraction. However, higher temperatures may also mean high solvent losses, extraction of some undesirable constituents, and damage to some sensitive components.

Solid–liquid extractors may be either single- or multistage. *Single-stage extractors* are closed tanks, fitted with a mesh base to support the solid particles of food. Heated solvent percolates down through the particles and is collected below the mesh base, with or without recirculation. They are used to extract oils or to produce coffee or tea extracts. *Multi-stage extractors* comprise a series of up to 15 tanks, linked together so that solvent emerging from the base of one extractor is pumped to the next. These are used to extract sugar from beet.

Absorption may be incorporated into the extraction process. By extraction the substances pass through a solid material that selectively adsorbs some components. These compounds are then stripped from the solid phase by a suitable solvent and concentrated by removal of the solvent. Adsorption collects and concentrates selected compounds according to their affinities to a solid (e.g., silica gel, charcoal, bentonite, or aluminum oxide). For elution, water, water–alcohol, organic bases, or hydrochloric acid are used.

SUPERCRITICAL FLUID EXTRACTION

Supercritical fluid extraction (SFE) is the process of separating a mixture in solid as well as in liquid state by contacting it with a fluid maintained under conditions of pressure above its critical point.

Supercritical fluids have relatively high diffusivity; they can penetrate into porous solid materials more effectively than liquid solvents, and, consequently, it may render much faster mass transfer resulting in faster extractions. The extraction time could be reduced from hours or days in a liquid–solid extraction to a few minutes in SFE.

The extraction is usually performed at low temperatures, so it may be an ideal technique for thermally labile compounds. It can be applied to systems of different scales: from analytical (less than a gram to a few grams of samples) to large industrial scale (tons of raw materials). SFE uses no or significantly less quantities of organic solvents. By use of CO_2 as the solvent, there are no environmental risks and no fire hazards. It is a powerful solvent, relatively inert, inexpensive, non-toxic, non-flammable, recyclable, readily available in high purity, and leaves no residues. With a critical point at $31.1°C$ and 7.38 MPa, supercritical CO_2 can be used at temperatures and pressures that are relatively safe, convenient, and particularly appropriate for the extraction of volatile and heat-labile compounds.

Extraction with supercritical fluids is also a unit operation that could be employed for a variety of applications including the extraction and fractionation of edible fats and oils, separation of tocopherols, clean-up of herb medicines and food products from pesticides. However, nowadays the application of SFE still remains restricted due to high investment costs.

AQUEOUS TWO-PHASE EXTRACTION

Aqueous two-phase extraction (ATPE) generally involves use of two incompatible water-miscible polymers (normally polyethylene glycol and dextran), or a water-miscible polymer and a salt (Na_2SO_4), to form two immiscible aqueous phases each containing 75% water.

This technology provides mild conditions for recovery of proteins and other biomolecules with minimal loss of activity. The advantages of the ATPE are biocompatibility, easy processing, high capacity, easy and precise scale-up, high product yields, high potential for continuous processing, and low investment cost.

LIQUID MEMBRANE EXTRACTION

This separation technique includes three-phase system: two phases of a similar nature but different composition (aqueous–aqueous, organic–organic, and gas–gas) separated by a third phase of a different nature, insoluble into the other two. This phase can be the emulsion liquid membrane, and the supported liquid membrane.

In the **emulsion liquid membrane** (also called the surfactant liquid membrane) configuration, the liquid membrane is formed by dispersing into the feed (phase 1) an emulsion of the stripping phase (phase 3) in an organic phase (phase 2) containing an emulsifying agent. Here the liquid membrane is the continuous phase of the emulsion and the viability of the process depends primarily on the stability of the emulsion.

In the **supported liquid membrane** process, the liquid membrane phase impregnates a microporous solid support placed between the two bulk phases. The membrane is stabilized by capillary forces making unnecessary the addition of stabilizers to the membrane phase.

As the distribution ratio between phases 1 and 3 is the product of those in the two pairs of fluids, the potential effectiveness of the liquid membrane process is considerably greater than that of conventional solvent extraction.

Thus, the liquid membrane process is particularly suitable for the treatment of dilute feeds.

The liquid membrane method allows higher partitioning of the substances as by the conventional solvent extraction process, thus allowing a high separation percentage. However, the process has its disadvantages; one is that the need to produce a stable emulsion requires the use of additives that slow down the rate of extraction and, even if their solubility is negligible, they may contaminate the raffinate. Another disadvantage is the emulsion rupture due to its swelling caused by the transport of the external phase into the emulsion.

REVERSE MICELLE EXTRACTION

This method involves use of microscopic water-in-oil micelles formed by surfactants and suspended within a hydrophobic organic solvent to isolate proteins from an aqueous feed. The micelles are microdroplets of water and provide a compatible environment for the protein, allowing its recovery from a crude aqueous feed without significant loss of protein activity. The presence of the aqueous microphase in the extracting phase may enhance the extraction of hydrophilic solutes by solubilizing them in the reverse micellar cores. However, this seems to vary with the characteristics of the system and the type of solute. Furthermore, in many instances the mechanism of extraction enhancement is not simply solubilization into the reverse micellar cores.

CONCLUSIONS AND PERSPECTIVES

Solvent extraction is less energy intensive compared to other separation processes. By choosing a suitable solvent having a high selectivity and distribution coefficient for the solute, the energy requirement in the separation step of the extraction process may be significantly reduced.

Supercritical fluids provide good reaction media due to their capacity to homogenize a reaction mixture, high diffusivity, and controlled phase separations and distribution of products. The application of this technology remains restricted to few products due to high investment costs. Future technology should permit reduced cost of equipment and lower pressure operations.

Aqueous two-phase extraction is a versatile technique for the downstream processing of biomolecules, such as flavoring substances, dipeptides, and nucleotides from acid hydrolysis. This technology has potential to achieve the desired purification and concentration of the product in a single step.

Liquid membrane extractions as novel separation techniques offer some advantages over conventional solvent extraction for dilute solutions and separation of biomolecules. Emulsion liquid membrane and reverse micellar extraction have shown good potential for industrial application.

The additional methods, increasing the extraction of the substances are based on the principle of absorption (e.g., columns with silica gel, charcoal, and bentonite) and membrane permeabilization (e.g., high electric field pulses and high hydrostatic pressure) may be incorporated into the extraction process.

ACKNOWLEDGMENT

This work would not have been possible without the support of Prof. Dr. Dipl.-Ing. Dietrich Knorr and Mr. Mondaly for providing me with the latest information in the field of extraction technologies and giving suggestions and ideas for this entry.

BIBLIOGRAPHY

1. Aguilera, J.M.; Stanley, D.W. *Microstructural Principles of Food Processing and Engineering*, Aspen Food Science Text Series Book; Aspen Publishing: Gaithersburg, MD, 1999, 432.
2. Bart, H.J.; Stevens, G.W. Reactive solvent extraction. In: *Ion Exchange and Solvent Extraction. A Series of Advances*; Marcus Y., Segupta A.K., Eds.; Marcel Dekker, 2004; Vol. 17, 37–85.
3. Baggiani, C.; Anfossi, L.; Giovannoli, C. Solid phase extraction of food contaminants using molecular imprinted polymers. Anal. Chim. Acta **2007**, *591*, 29–39.
4. Cox, M.; Rydberg, J. *Introduction to Solvent Extraction*; Hatfield: Hertfordshire, UK, 2004; Vol. 1, 1–25.
5. Knorr, D. Process assessment of high pressure processing of foods: an overview. In *Processing Foods: Quality Optimisation and Process Assessment*; Oliveira, F.A.R., Oliveira, J.C., Eds.; CRC Press: Boca Raton, FL, 1999; 249–267.
6. Moret, S.; Conte, L.S. Polycyclic aromatic hydrocarbons in edible fats and oils: occurrence and analytical methods. J. Chromatogr. **2000**, *882*, 245–253.
7. Rastogi, N.K.; Raghavarao, K.S.M.S.; Balasubramaniam, V.M.; Niranjan, K.; Knorr, D. Opportunities and challenges in high pressure processing of foods. Crit. Rev. Food Sci. Nutr. **2007**, *47*, 69–112.
8. Srinivas, N.D.; Barhate, R.S.; Raghavarao, K.S.M.S. Aqueous two-phase extraction in combination with ultrafiltration for downstream processing of Ipomoea peroxidase. J. Food Eng. **2002**, *54*, 1–6.
9. Turner, C.H. Food and agricultural samples. Am. Chem. Soc., ACS Symposium Series. **2006**, *926*, 189.

Metabolomics and Genetically Modified Organisms (GMOs)

Howard V. Davies
Department of Science Co-ordination, Scottish Crop Research Institute, Invergowrie, U.K.

Louise VT Shepherd
Plant Products and Food Quality Programme, Scottish Crop Research Institute, Invergowrie, U.K.

Abstract

Concerns have been raised about the potential for unintended effects in genetically modified crops that could impact on their safety. Despite the fact that the risk assessment approaches used to date have proved to be effective, some argue that the use of contemporary, unbiased, and broad scale analytical approaches such as metabolomics should be applied to GM crops to further reduces any uncertainty. This review addresses the potential for using metabolomics to address such issues.

INTRODUCTION

In 2008, the global hectarage of biotech crops reached 125 million hectares (from 114 million hectares in 2007; ISAAA; http://www.isaaa.org) with the number of countries planting biotech crops reaching an historical milestone of 25 countries, and accumulatively, the second billionth acre (800 millionth hectare) of a biotech crop was planted—only 3 years after the first one-billionth acre of a biotech crop was planted in 2005. In 2008, developing countries outnumbered industrial countries growing biotech crops by 15–10, a trend expected to continue with 40 or more countries expected to adopt biotech crops by 2015, the final year of the second decade of commercialization. The evidence indicates that this is the fastest adopted crop technology in recent history. The outlook for the period till 2015 is for continued growth in the global hectarage of biotech crops (up to 2 million hectares) with at least 20 million farmers growing them. It is estimated that the global net economic benefits to biotech crop farmers in 2007 alone was $10 billion ($6 billion for developing countries and $4 billion for industrial countries). The accumulated benefits during the period 1996–2007 were $44 billion with $22 billion each for developing and industrial countries.

Owing to concerns primarily over the safety of GM crops to human health and the environment, uptake of the technology and GM crops has been extremely limited within the European Union (EU), despite the establishment of the European Food Safety Authority (EFSA) as an independent risk assessment body and the implementation of new EU Directives and Regulations dealing with food and feed safety (Regulation 1829/2003) and detection/traceability (Regulation 1830/2003) and releases into the environment (Directive 2001/18). There has been a history of a moratorium of GM imports and cultivation in Europe that has led to international trade disputes investigated by the WTO. While in recent years specific GM products for food and feed have been approved within the EU, it has been 10 years since a crop was approved for cultivation. However, scientific, commercial, and political pressures are mounting within the EU to facilitate the decision-making processes and to consider the value that these crops can and will bring to a sustainable European economy. These include: 1) mounting pressure on the supply chain to provide affordable animal feed; 2) the need for the European agriculture and downstream industries to compete more effectively in the global marketplace; 3) concerns over food security and food prices brought about by the development of the biofuel market and influenced by the impact of climate change; and 4) the need to reduce the environmental impact of agriculture.

GMOS—"OMICS" AND THE RISK ASSESSMENT PROCESS

Central to the risk assessment process for GM crops is the comparative analysis of the genetically modified organism (GMO) with its non-GM counterparts. The details of the approaches to be taken within the EU are provided in the EFSA guidance document of the Scientific Panel on Genetically Modified Organisms for the risk assessment of genetically modified plants and derived food and feed.[1] These are basically in line with approaches advocated by the OECD and Codex Alimentarius. A comparative analysis of key metabolites, both nutrients and antinutrients, is an important requirement and is aimed at ensuring that there are no significant unintended effects caused by the genetic modification process, which impinges on safety of the product to humans, animals, or the environment.

Traditionally, the analysis has involved a targeted quantification of nutrients and antinutrients using "tried and tested" analytical approaches, e.g., proximate analysis. The OECD is instrumental at helping to define, for given crop species, the metabolites that should be analyzed.[1]

Despite the fact that there are no indications of any safety issues concerning commercial GM crops currently on the market, questions continue to be raised over the adequacy of targeted analysis in helping to define the possibility for "unintended" effects in GM crops. This has opened the debate on the potential for using "omics" approaches such as transcriptomics, proteomics, and metabolomics to provide a broader, less biased analytical base to search for these effects. Since plants (depending on species) contain ca. 30,000–60,000 genes, with an estimated 100,000–120,000 proteins (perhaps more) and several thousand metabolites, the potential clearly exists to analyze many more parameters in GM crops than is current practice.[2] Whether or not our capacity to interrogate metabolomics information within a biological context to improve on current risk assessment practices remains open to question, at least as a routine approach.

METABOLOMICS—DEFINITIONS, TECHNOLOGIES, AND LIMITATIONS

There are several terms that are often used interchangeably when referring to broad scale, "unbiased" metabolite analysis. To be more specific, the term *metabolite profiling* applies to the analysis of a specific group of metabolites (e.g., amino acids) whereas *metabolomics* per se can be defined as a comprehensive analysis of the metabolome without particular bias to specific groups of metabolites. A *metabolite fingerprinting* approach typically involves the analysis of samples directly or with very little extraction and without lengthy chromatography. This is often used for an initial rapid screen to facilitate grouping of data based on inherent biological characteristics.

No single technology is fully comprehensive with regard to metabolite analysis and more detailed description of the various approaches used can be found in the literature.[3,4] Most laboratories use nuclear magnetic resonance spectroscopy (NMR), gas chromatography–mass spectrometry (GC-MS) and liquid chromatography–mass spectrometry (LC-MS) as the cornerstone of their analyses but a range of other sophisticated instrumentation and instrument combinations, e.g., LC-NMR, FT-IR, FIE-MS, FT-ICR/MS, and MALDI/TOF-MS can be deployed.[4]

To realize the potential of metabolomics for GM research we also need to be aware of the limitations. These include: 1) an incomplete of knowledge of secondary metabolic pathways in plants; 2) difficulties in identifying the majority of plant metabolites; and 3) the need for standard operating procedures and strategies to transfer information from many sources to build international metabolomics databases that can enhance comparative analyses. Despite the ever-expanding range of technological options, significant challenges remain related to the acquisition, interpretation, and statistical treatments of complex data sets. However, the development of metabolic pathway databases such as KEGG (http://www.genome.ad.jp/kegg/) and MapMan/PageMan[5] will certainly help as will the accumulation of metabolite databases as exemplified by MoToDB for tomato (>20,000 compounds).[6]

A number of initiatives are looking toward establishing standards for dealing with information-rich metabolomics data sets.[7] In 2005, the Metabolomics Society established the Metabolomics Standards Initiative (http://msi-workgroups.sourceforge.net) to address reporting standards related to biological context, chemical analysis performed, and data processing tasked to establish reporting standards. This includes minimal information requirements and exchange formats with defined semantics to enable efficient data sharing and meaningful data mining.

EXAMPLE STUDIES—METABOLOMICS AND GMOS

In the United Kingdom, the Food Standards Agency (FSA) commissioned a major research program to assess the potential use of transcriptomic, proteomic, and metabolomic approaches in the safety assessment of GM food crops.[8] With regard to metabolomics the program revealed that the methods developed were successful at detecting unintended changes resulting from transgene insertion into plants but that the vast majority of these changes were small (ca. twofold or less). Many of the differences observed between the GM crop and their non-GM counterparts may have been due to somaclonal variation resulting from the in vitro manipulation of plants rather than the presence of an inserted transgene per se. However, it is unlikely that these changes would have been detected by targeted metabolite analysis. It was also obvious from the FSA funded program that differences in the metabolomes between plants grown in different environments, and even different cultivars of the same species grown in the same environment, were often greater than the effects of the transgene itself. The issue of what drives variation in the metabolome has become an increasingly important topic in terms of benchmarking the extent of any "unintended" changes in metabolite pools of GMOs with "natural" variation in the metabolome of the species in question. Interestingly, a National Research Council (United States) taskforce estimated the probability of unintended effects resulting from a range of plant genetic modification processes and concluded that mutagenesis (a widely used approach for generating new commercial varieties) is the most genetically disruptive, producing a wide range of effects[9] (Fig. 1). Weight has been added to this argument by recent work on rice.[10]

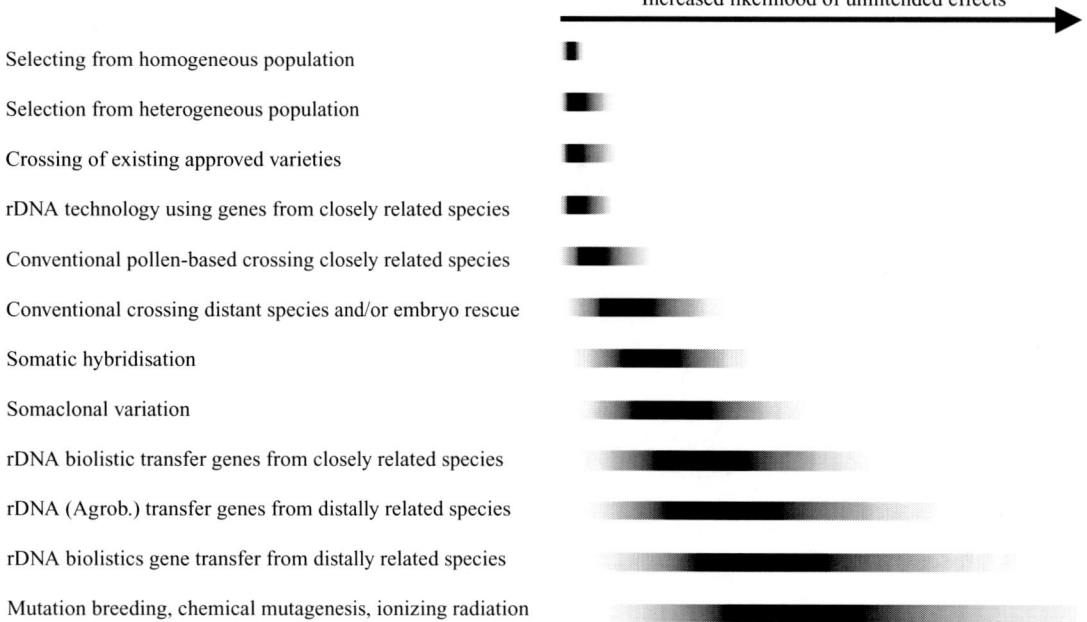

Fig. 1 Relative likelihood of unintended effects associated with various methods of plant genetic modification. The gray tails show the range of potential unintended changes; the black bars indicate the relative degree of genetic disruption for each method. Selection from a heterogeneous population is considered the least likely to express unintended effects, and the range of those that do appear is quite limited. By contrast, induced mutagenesis is the most genetically disruptive, and produces a wide range of effects. **Source:** Reproduced with permission of The National Academies Press.

Recent work with wheat[11] has revealed a stronger influence of site and year of production than of genotype, but some separation between the transgenic and parental lines was observed owing to increased levels of sugars in some transgenic lines. Differences in free amino acids were also apparent. The results again demonstrate that the environment affects the metabolome and that differences between the control and the specific transgenic lines used were generally within the same range as the differences observed between the control lines grown on different sites and in different years.

A comprehensive range of other examples of the use of metabolomics in GM analyses will appear in the journal *Food Control*.[12]

CONCLUSION

The growing body of literature on metabolomics and GMOs relates to a range of audiences from the more academic (basic research on plant metabolism and functional genomics) through to risk assessors, risk managers, and legislators with a keen eye on food safety issues. Metabolomic technologies, data handling, and comparative database building are clearly evolving in parallel to provide extremely powerful tools with which to dissect the processes that regulate the composition of food crops. Arguably, with regard to GMOs, these technologies are more likely to find a role in supplementing, rather than replacing, traditional targeted analysis. They may, e.g., prove useful in the analysis and risk assessment of second-generation food crops where composition has been intentionally modified through perturbations in endogenous metabolism. For first generation, input trait GM crops, current approaches of safety assessment have not proved wanting. However, if metabolomics technology is to be used more widely in risk assessment then some minimal standards need to be in place alongside appropriate databases for comparative purposes.

ACKNOWLEDGMENT

Howard Davies and Louise Shepherd acknowledge financial support from the Scottish Government Rural and Environment Research and Analysis Directorate.

REFERENCES

1. EFSA. Guidance document of the Scientific Panel on genetically modified organisms for the risk assessment of genetically modified plants and derived food and feed. EFSA J. **2004**, *99*, 1–94, http://www.efsa.europa.eu/en/science/gmo/gmo_guidance/660.html.

2. ILSI. Nutritional and safety assessments of foods and feeds nutritionally improved through biotechnology. Compr. Rev. Food Sci. Food Saf. **2004**, *3*, 36–104.
3. Hall, R.D. Plant metabolomics: from holistic hope, to hype, to hot topic. New Phytol. **2006**, *169*, 453–468.
4. Enot, D.P.; Beckmann, M.; Draper, J. Detecting a difference—assessing generalisability when modelling metabolome fingerprint data in longer term studies of genetically modified plants. Metabolomics **2007**, *3* (3), 335–347.
5. Sreenivasulu, N.; Usadel, B.; Winter, A.; Barley grain maturation and germination: metabolic pathway and regulatory network commonalities and differences highlighted by New MapMan/PageMan Profiling Tools. Plant Physiol. **2008**, *146*, 1738–1758.
6. Hoekenga, O.A. Using metabolomics to estimate unintended effects in transgenic crop plants: problems, promises, and opportunities. J. Biomol. Tech. **2008**, *19*, 159–166.
7. Hardy, N.W.; Taylor, C.F. A roadmap for the establishment of standard data exchange structures for metabolomics. Metabolomics **2007**, *3*, 243–248.
8. *Safety Assessment of Novel Foods*, Report of G02Research Programme Review, 2005, http://www.foodstandards.gov.uk/science/research/researchinfo/foodcomponentsresearch/novelfoodsresearch/g02programme.
9. Committee on Identifying and Assessing Unintended Effects of Genetically Engineered Foods on Human Health; National Research Council of the National Academies (U.S.). *Safety of Genetically Engineered Foods: Approaches to Assessing Unintended Health* (Report in Brief); The National Academies Press: Washington, D.C., 2004, http://www.nap.edu.
10. Batista, R.; Saibo, N.; Lourenço, T.; Oliveira, M.M. Microarray analyses reveal that plant mutagenesis may induce more transcriptomic changes than transgene insertion. Proc. Natl. Acad. Sci. U. S. A. **2008**, *105*, 3640–3645.
11. Baker, J.M.; Hawkins, N.D.; Ward, J.L.; Lovegrove, A.; Napier, J.A.; Shewry, P.R.; Beale, M.H. A metabolomic study of substantial equivalence of field-grown genetically modified wheat. Plant Biotechnol. J. **2006**, *4*, 381–392.
12. Davies, H.V. A role for "omics" technologies in food safety assessment? Food Control *in press*.

Microbial Molecular Biology

Harry J. Flint
Rowett Research Institute, Aberdeen, U.K.

Abstract

The powerful array of techniques that constitute molecular biology arose largely from the study of microorganisms. The ability to construct genomic DNA libraries—and cDNA libraries derived from expressed mRNA—in bacterial and fungal hosts remains a key approach for isolating genes. Dramatic technical developments, however, have now brought about a further revolution. In particular the polymerase chain reaction (PCR) allows precise amplification of DNA sequences, whereas developments in DNA sequencing, microarray, and proteomics technologies are making it more efficient to deal with whole microbial genomes rather than to search for individual genes. Molecular biology now pervades all areas of microbiology and is producing major advances in our understanding of the diversity and dynamics of the microbial ecosystems found in the animal gut, on plant surfaces, and in soils. Genes are being uncovered that define the interactions between microbes and animal hosts, notably the mechanisms involved in pathogenesis, survival, and the mutualistic relationships that allow herbivorous animals to gain energy from plant material. Molecular information underpins the quest to treat and prevent infectious diseases in animals, track and suppress microbes harbored by animals that cause disease in humans, optimize animal production, and minimize pollution.

MICROBIAL DIVERSITY

Ribosomal RNA

It has been difficult for microbiologists to know whether they can culture the full range of microorganisms present in a given habitat, but molecular approaches are now revealing the extent of previously unknown diversity. Most "culture-independent" approaches involve the sequencing of ribosomal genes that are amplified directly by PCR from environmental samples. Ribosomal genes (particularly those coding for the small subunit rRNA—16S in prokaryotes, 18S in eukaryotes) are suitable because they occur in all living organisms and contain highly conserved sequences (which are useful for such things as designing "universal" eubacterial or archaeal primers for PCR amplification) as well as regions that vary between strains and species. In soils, less than 1% of microbial rRNA sequence diversity appears to be represented by cultured species.[1] Analyses performed on the microbiota of the rumen and the pig and horse large intestine (Fig. 1) reveal enormous diversity; only 17% of eubacterial sequences recovered from the pig, and only 11% from the horse, correspond to known species.[2,3] The rapidly expanding sequence databases allow the design of probes and primers, specific to particular groupings, that are suitable for enumeration by dot blot hybridization, whole-cell fluorescent in situ hybridization (FISH), or real-time PCR.[4] Microarrays are also being developed in which panels of specific oligonucleotide probes can be used to describe the composition of microbial ecosystems.

Molecular profiling approaches such as denaturing gradient gel electrophoresis (DGGE) and terminal restriction fragment length polymorphism (T-RFLP), again usually based on amplified ribosomal sequences, are widely used to follow shifts in the composition of microbial communities. These produce bands characteristic of different DNA% G + C contents or sequences using primers that target broad phylogenetic groupings. These methods are contributing to our understanding of such topics as the impact of host variation and of diet upon the gut microflora and the impact of management practices upon soil microbial communities.

Gene Tracking

Polymerase chain reaction methods can be used to detect a variety of specific genes in environmental samples without prior cultivation and isolation of microorganisms. They have been applied particularly to virulence determinants (e.g., toxin genes) providing information on pathogen contamination in the food chain. PCR tracking of antibiotic resistance genes in the environment and in human and animal gut bacteria has implications for the debate over the impact of antibiotic use in animal husbandry, which centers on resistance to the antibiotics used in clinical and veterinary medicine.

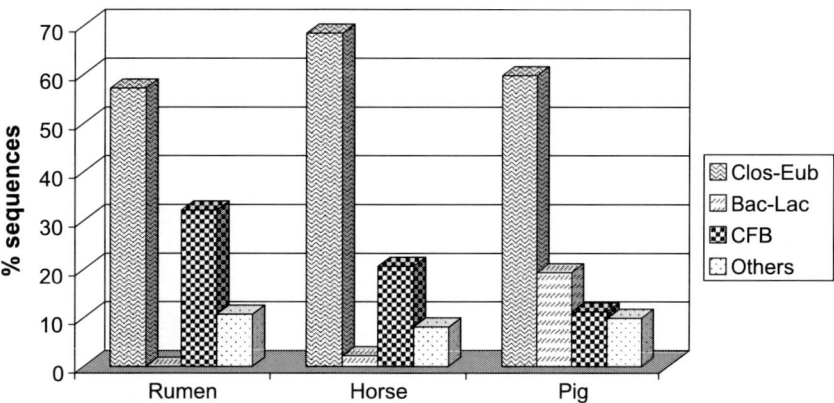

Fig. 1 Eubacterial diversity in gut samples as determined by amplification and sequencing of 16SrRNA genes. Data are from bovine rumen, horse large intestine, and pig intestine. Clos-Eub = *Clostridium*/*Eubacterium*/*Ruminococcus* relatives; Bac-Lac = *Bacillus*/*Lactobacillus*/*Streptococcus*; CFB = *Cytophaga*/*Flavobacterium*/*Bacteroides*. While independent of culture bias, it should be noted that PCR bias and rRNA gene copy number influence the apparent proportions of different types obtained by this approach. **Source:** From Appl. Environ. Microbiol.[4]

Strain Typing

A plethora of DNA-based methods are now available that enable more precise strain identification among the better known cultivable microorganisms, especially pathogens. Many rapid typing techniques rely on PCR amplification of randomly primed sequences, repeated sequences, or on ribotyping based on ribosomal RNA sequences. Alternatively, the whole genome can be profiled after restriction enzyme cleavage into large fragments that are separated by pulsed field gel electrophoresis (PFGE).

GENOMICS

Complete Genomes

Complete microbial genomes range in size from around 1 to 8 Mb for bacteria, and from 10 to 150 Mb for eukaryotes such as yeasts and protozoa. There has been an explosion in genome sequence information that now extends to most human and many animal bacterial pathogens (Table 1).[5] One important outcome has been to identify large genetic regions, or pathogenicity islands, that make particular strains infectious. The first projects are now underway to sequence genomes of mutualistic and commensal microorganisms from the animal gut.

Microarrays and Gene Expression

New technologies allow large numbers of sequences, e.g., representing all genes from a given species, to be arrayed on glass slides (microarrays) or on membranes. Genes showing differential expression can then be identified by comparative DNA hybridization. Allied to genome sequencing and rapidly developing proteomic methods for two-dimensional separation and identification of polypeptides, this provides unprecedented power for studies on microbial gene regulation.[6]

Metagenomics

Another powerful new approach, metagenomics, involves creating gene libraries from the DNA recovered from a mixed microbial community, followed by screening for specific functions. This allows the recovery of valuable and important genes from microorganisms that may never have been cultivated.[7]

GENE TRANSFER AND GENETIC ANALYSIS

Natural Gene Transfer Mechanisms

Microorganisms, especially bacteria, frequently carry genetic elements that are capable of transfer between cells independently of the main chromosome. Such elements, which include extrachromosomal plasmids and chromosomally located conjugative transposons, possess genes that promote their own transfer by cell–cell contact (conjugation), or that allow them to be mobilized by other elements. In addition many bacteria are able to take up DNA from their environment resulting in natural genetic transformation, or may acquire genes via bacteriophage virus-mediated transduction. Such genetic exchanges play a major role in microbial evolution. Traits including antibiotic resistance, heavy metal resistance, virulence factors, adhesion

Table 1 Some examples of fully sequenced bacterial genomes.

Species, strain	Chromosome (plasmid)		
	Size—base pairs	Predicted protein-coding genes	DNA % (G + C)
Streptomyces coelicolor A3(2)	8,667,507	7,825	71.1
Bacteroides thetaiotaomicron VPI5482	6,260,361 (33,038)[b]	4,779 (38)	42.8 (47.2)
Escherichia coli K12	4,639,221	4,288	50.8
Clostridium tetani E88[a]	2,799,250 (74,082)[b]	2,372 (61)	28.6 (24.5)
Bifidobacterium longum NCC2705	2,256,646	1,730	60.0
Campylobacter jejuni NCTC11168[a]	1,641,481	1,654	30.6
Mycoplasma pulmonis[a]	963,879	782	26.6

[a]Pathogenic strain.
[b]Plasmid.

properties, substrate utilization pathways, and the production of antimicrobials show evidence of natural horizontal transfer.

Genetic Analysis

Mobile genetic elements provide the basis for many molecular biology procedures, in particular, as vectors. In addition, transposons that insert randomly in the genome are used for insertional mutagenesis to identify microbial gene function. In many bacteria and fungi it is relatively straightforward to introduce and over express foreign genes through conjugal transfer or transformation, and to perform targeted gene knockouts. Refinements of these techniques allow the identification of genes that are switched on in particular environments, e.g., in mammalian host tissues (IVET, in vivo expression technology).[8] However, the lack of convenient gene transfer systems remains an obstacle to research in many less-studied species, including most anaerobes. Transformation can be induced artificially in bacteria, e.g., by electroporation, but endogenous nucleases normally present in wild-type strains tend to destroy the incoming DNA.

BIOTECHNOLOGY

Modified (mutant or genetically manipulated) strains of bacteria and fungi are used in a contained manner in the production of enzymes and other products (e.g., amino acids) used as animal feed additives or feed pretreatments. Modified microbial strains have also been considered for other applications that would require their release into the environment, e.g., as silage additives, probiotics, or for bioremediation. These latter possibilities are clearly subject to detailed risk assessments and to public acceptance. Recombinant vaccines and antibody engineering are increasingly important in the treatment and prevention of infectious diseases in animals caused by a wide range of viruses, bacteria, fungi, and protozoa. Again, genomic and proteomic approaches are helping to define new antigens, as well as possible targets for the development of new antimicrobial agents. Microbial genes continue to provide valuable proteins and enzymes for research, industry, and medicine.[9]

CONCLUSIONS

Molecular techniques are revolutionizing our understanding of the diversity and functioning of microbial ecosystems associated with the mammalian gut, soils, and wastes. These techniques provide new tools for tracking and identifying individual genes and species. Complete genome sequences are available for many human and animal pathogens and should soon become available for more nonpathogenic bacteria and eukaryotic microorganisms. The potential benefits for disease prevention, animal production, and environmental management are enormous, although many will take some time to be realized.

ACKNOWLEDGMENT

The author wishes to acknowledge the support of the Scottish Executive Environment and Rural Affairs Department.

REFERENCES

1. Amman, R.I.; Ludwig, W.; Schleifer, K.H. Phylogenetic identification and in situ detection of individual microbial cells without cultivation. Microbiol. Rev. **1995**, *59* (1), 143–169.
2. Leser, T.D.; Amenuvor, J.Z.; Jensen, T.K.; Lindecrona, R.H.; Boye, M.; Moller, K. Culture independent analysis of gut bacteria: The pig gastrointestinal tract revisited. Appl. Environ. Microbiol. **2002**, *68* (2), 673–690.
3. Daly, K.; Stewart, C.S.; Flint, H.J.; Shirazi-Beechey, S.P. Bacterial diversity within the equine large intestine as revealed by molecular analysis of cloned 16S rRNA genes. FEMS Microbiol. Ecol. **2001**, *38* (2–3), 141–151.

4. Tajima, K.; Aminov, R.I.; Nagamine, T.; Matsui, H.; Nakamura, M.; Benno, Y. Diet-dependent shifts in the bacterial population of the rumen revealed with real time PCR. Appl. Environ. Microbiol. **2001**, *67* (6), 2766–2774.

5. Bruggeman, H.; Baumer, S.; Fricke, W.F.; Wiezer, A.; Liesegang, H.; Decker, I.; Herzberg, C.; Martinez-Arias, R.; Merki, R.; Henne, A.; Gottshalk, G. The genome sequence of *Clostridium tetani*, the causative agent of tetanus disease. Proc. Natl. Acad. Sci. **2003**, *100* (3), 1316–1321.

6. *Functional Microbial Genomics*; Wren, B., Dorrell, N., Eds.; Methods in Microbiology; Academic Press: London, UK, **2002**, Vol. 33.

7. Rondon, M.R.; August, P.R.; Betterman, A.D.; Brady, S.F.; Grossman, T.H.; Liles, M.R.; Loiacono, K.A.; Lynch, B.A.; MacNeil, I.A.; Minor, C.; Tiong, C.L.; Gilman, M.; Osborne, M.S.; Clardy, J.; Handelsman, J.; Goodman, R.M. Cloning the soil metagenome: A strategy for accessing the genetic and functional diversity of uncultured microorganisms. Appl. Environ. Microbiol. **2000**, *66* (6), 2541–2547.

8. Handfield, M.; Levesque, R.C. Strategies for isolation of in vivo expressed genes from bacteria. FEMS Microbiol. Rev. **1999**, *23* (1), 69–91.

9. Vielle, C.; Zeikus, G.J. Hyperthermophilic enzymes: Sources, uses and molecular mechanisms for thermostability. Microbiol. Mol. Biol. Rev. **2001**, *65* (1), 1–43.

Microbial Polysaccharides

Graciela Font de Valdez
María Inés Torino
Fernanda Mozzi
Lactobacillus Reference Center, National Council of Scientific and Technical Research
(CERELA-CONICET), San Miguel de Tucumán, Argentina

Abstract
Microbial exopolysaccharides (EPS) have been extensively used in the food industry primarily because of their strong impact on the rheological properties of food. There exists a wide diversity of EPS, regarding their composition, molecular mass, yield, and functionalities; EPS not only serve as texturizers or gelling agents; but also as emulsifiers and stabilizers. Some microbial EPS produced by non-food-grade bacteria, yeast, or fungi must be recovered and added into foods as food additives. Currently, considerable interest has been shown on EPS produced in situ by the food-grade lactic acid bacteria in different fermented products.

INTRODUCTION

Exopolysaccharides (EPS) are long-chain polymers composed of repeating units of monosaccharides or monosaccharide derivatives. Certain plants, algae, fungi, and bacteria are able to synthesize EPS. Microbial EPS can be either capsular polysaccharides if they are associated with the cell surface or slime-EPS if they are secreted into the extracellular environment. These EPS can be divided into homopolysaccharides (HoPS), which are composed of a single type of sugar monomer, and heteropolysaccharides (HePS) that contain several types of monosaccharides. A wide diversity of microbial polysaccharides exists with respect to their monomer composition, structure, yield, and functionality. Microbial EPS synthesis is independent of the climate, season, and environment contamination and can be performed and optimized under controlled culture conditions; however, the high cost of production and the non-food-grade status of the majority of the EPS-producing microorganisms constitute the two main disadvantages. Nevertheless, these polymers have been extensively used for multiple industrial applications. This entry will specially deal with microbial polysaccharides for food use.

EPS FROM FOOD-GRADE MICROORGANISMS

Polysaccharides Produced by Lactic Acid Bacteria

Lactic acid bacteria (LAB) are industrially important microorganisms involved in the elaboration of diverse fermented foods worldwide. These food-grade bacteria are not only used for producing lactic acid and thus for preventing food spoilage and starting fermentation, but also for their contribution to other relevant characteristics of the final product such as aroma, flavor, texture, and nutrition.[1] Certain LAB produce EPS, either HoPS or HePS.[2,3] Homopolysaccharides from LAB can be divided into four groups: 1) α-D-glucans (dextrans, mutans, alternan, and reuterans); 2) β-D-glucans; 3) β-D-fructans (levans and inulin-type fructans); and 4) polygalactans. They are usually high (up to 10^7 Da) molecular mass (MM) EPS with different degree of branching, linking sides, and chain length. Most HoPS are synthesized by extracellular glycansucrases using sucrose as the glycosyl donor.

Heteropolysaccharides usually contain D-galactose, D-glucose, and L-rhamnose (at different ratios) and in few cases, fucose, ribose, acetylated amino sugars, uronic acid, and other non-carbohydrate compounds.[2] The majority of the industrially important LAB species synthesize HePS exhibiting a great diversity regarding composition, structure, MM, yield, and functionality.[3] The HePS of LAB, produced in small amounts (10–600 mg/L) through a complex, intracellular, energy-demanding process, display high thickening power at low concentrations in the food matrix.[2]

EPS in fermented milks

Exopolysaccharides from LAB play a major role in the dairy industry when they are produced in situ, contributing to the texture, mouth feel, taste perception, and stability of the product, especially fermented milks such as yogurt. Exopolysaccharide-producing LAB also affect the rheology of stirred fermented milks conferring a smooth and creamy texture. The amount of polysaccharide is dependent on the strain, culture conditions, and microbial interactions although no clear correlation between culture viscosity and EPS concentration has been observed.

The viscosity of stirred fermented milk is strongly influenced by the primary structure (stiffness), molecular weight, and radius of gyration of the EPS molecules.[1] Another example of fermented milk with EPS-producing LAB is the Finnish fermented milk "viili" prepared with a ropy strain of *Lactococcus lactis* subsp. *cremoris*, which produces the EPS "viilian" responsible for the thick and creamy characteristics of the product.[4] In "kefir," *Lactobacillus kefiranofaciens* is responsible for the synthesis of kefiran, a thick EPS matrix in which a microbial community of LAB, acetic acid bacteria, and yeasts is embedded.[4]

EPS in cheese making

Milkfat plays a key role in cheese making by reducing complete casein coalescence and increasing whey retention. The development of low-fat cheeses to meet consumer demands for healthier foods requires changes in technology or use of novel strategies to solve texture and body defects in the products. In this regard, the ability of EPS to bind water and increase cheese moisture with minimal adverse effects on whey ultra filtration processes was studied.[4] The EPS-producing LAB have been successfully used (as adjunct culture or included into the starter culture) to improve the texture, melting, and sensorial characteristics of low-fat Cheddar and Mozzarella cheeses.

EPS in sourdough fermentation

During sourdough fermentation, LAB produce diverse metabolites which have positive effects on the texture, staling, and shelf life of bread. The in situ production of EPS by LAB has also been documented.[4] These polysaccharides (HoPS mainly) may act as prebiotics, thus enhancing sourdough quality and positively influencing the lactic microbiota in the gut.

Exopolysaccharides from LAB have not yet been exploited industrially as food additives, mainly due to the low production yield compared to xanthan; however, selected LAB strains have been used for EPS production, e.g., *Leuconostoc mesenteroides* which has been used for the production of dextran, a glucan containing 50% α-1,6 osidic bonds; dextran is widely used in the pharmaceutical and chemical industries.

EPS FROM NON-FOOD-GRADE BACTERIA: THEIR USE AS FOOD ADDITIVES

Xanthan, gellan, and curdlan, in that order, were the first EPS produced by microbial fermentation approved for food use by the U.S. Food and Drug Administration (FDA). Out of these, xanthan and gellan are economically the most important. Curdlan, pullulan, and scleroglucan have fewer commercial uses, while other microbial EPS such as bacterial cellulose, succinoglucan, and levan are still promising. Hydrocolloids are selected for food application according to hydration, stability during food formulation and processing, and the desired characteristics of the final product.

Xanthan

Xanthan is a high molecular weight (5×10^6 Da) anionic HePS produced by the pathogenic plant bacterium, *Xanthomonas campestris*.[5] It has a pentasaccharide-repeating unit with a cellulosic backbone (two D-glucose units linked through the 1 and 4 positions) and a trisaccharide side-chain (two mannose and one glucuronic acid) attached on every second glucose residue. This unusual structure confers special physical properties. Xanthan is used in the food industry as an excellent thickener and stabilizer for suspensions, emulsions, or solid particles in water-based recipes. It is stable under a wide range of temperatures and pH, making it suitable for using in dressings and sauces, baked goods, beverages, dairy desserts, cream and processed cheeses, fruit preparations, and ice cream. It contributes to the suspension of solids (i.e., herbs and spices) and is a good stabilizing agent for oil/water emulsions. Combined with other EPS, xanthan acts as stabilizer in frozen foods to prevent the wheying off of proteins and formation of ice crystals. In fruit- and dairy-based beverages, xanthan prevents sediment formation during storage. In bakery goods, it is used for improving the volume in bread and cakes, retaining moisture, and suspending fruit portions. Xanthan provides the dough "stickiness" in gluten-free baking for products intended for celiac patients. It is also used as a thickening agent and stabilizer in the dairy industry, and provides a "fat feel" in low- or non-fat dairy products. Xanthan solutions are compatible with many food ingredients, salts, and chemicals. The most important features of xanthan are the viscosity increase at low EPS concentrations (usually 0.5% or less) and its pseudoplasticity. Xanthan is accepted as a safe food additive (designated as E415) in the United States, Canada, and Europe.

Gellan

Gellan is a linear, anionic HePS (molecular weight $>7 \times 10^4$ Da) produced by *Auromonas* (*Pseudomonas*) *elodea* (now *Sphingomonas paucimobilis*).[6] It is composed of a tetrasaccharide-repeating unit containing (1-3)-β-D-glucose, (1-4)-β-D-glucuronic acid, (1-4)-β-D-glucose, and (1-4)-α-L-rhamnose as the backbone with acyl substituents of L-glycerate and acetate. Its physical properties such as gel strength, texture, and melting and setting points are controlled by the addition of ions, acyl content, and pH. The most important traits of gellan are heat stability and acid resistance, elasticity and hardness, and good compatibility with other compounds. Gellan forms thermoreversible gels after heating and cooling to the gelling temperature. It is not only used as a broad-spectrum thickener or gelling agent, but also as a stabilizing and suspending agent in jams, fruit syrups, and sweet goods. Gellan (termed E418) is used for food purposes in the United States and the European Union.

Curdlan

Curdlan is an unbranched linear β-(1,3)-D-glucan of high molecular weight (10^5 Da) synthesized by strains of *Agrobacterium* sp. and *Alcaligenes faecalis*. It is insoluble in cold water, but aqueous suspensions plasticize and dissolve before producing reversible (elastic) gels at 55°C.[7] Aqueous dispersion forms two types of heat-induced gels depending on temperature. The low-set gel (obtained at 55–60°C) is thermoreversible, whereas the high-set gel (obtained at temperatures > 80°C) is irreversible. Salts tend to prevent curdlan from gelling. Curdlan gels are stable over a wide pH range and under severe food processing conditions. This EPS is used as a stabilizer and thickener, or as a texturizer and gelling agent to improve texture and water-holding capacity in meat, poultry, and seafood products. Also, it may be used to formulate reduced-calorie products. Curdlan gels are bland in taste, color, and odor and are thus, applicable for dairy foods. Curdlan (E424) is widely used in Japan, Korea, and Taiwan.

Pullulan

Pullulan is a linear α-(1,4)-; α-(1,6)-glucan (molecular weight 2×10^5 Da) produced from hydrolyzed starch by *Aureobasidium pullulans*. This EPS is water-soluble, odorless, and flavorless, and demonstrates low viscosity compared to other polymers.[8] It is used as a thickener, glazing agent, and starch replacement in low-calorie food formulations, jams and jellies, confectioneries, and some meat and fruit products. The major commercial interest in pullulan is its ability to produce strong, non-toxic, non-digestible films that are appropriate for food packaging. As a food additive, it is known as E1204.

Scleroglucan

Scleroglucan is a non-ionic moderately high molecular weight (5.4×10^5 Da) HoPS consisting of a backbone of β-(1,3)-D-glucose units with glucose side chains linked β-(1,6). It is produced by *Sclerotium, Schizophyllum,* and *Rhizobium* strains.[9] Scleroglucan offers a better stability and thickening performance than xanthan gum at high temperatures. Its remarkable rheological properties and stability over a wide range of pH 3–10, salinities, and temperature make it suitable for several applications.

Bacterial Cellulose

Bacterial cellulose is an unbranched HoPS of β-(1,4)-D-glucose residues synthesized by *Acetobacter xylinum* (now *Gluconacetobacter xylinus*) from glucose corn syrup. In the food industry, it is combined with coagents (sucrose and carboxymethylcellulose) to promote greater dispersion of product. It serves as a thickener, stabilizer, and binder in a variety of food products (sauces and salad dressings, ice cream, marshmallows, toppings, and liquid diet products), including low or non-fat applications. Bacterial cellulose has been accepted as "generally recognized as safe" by the FDA.[10]

Succinoglycan

Succinoglycan is a branched, acidic HePS with an octasaccharide-repeating unit, composed of glucose and galactose joined in β-glycosidic linkages; it may contain pyruvate, succinate, or acetate groups. Succinoglycan is synthesized by *Sinorhizobium* sp., *Agrobacterium* sp., and *Alcaligenes* sp. as either a high molecular weight ($>10^5$ Da) or low molecular weight ($<5 \times 10^3$ Da) EPS.[11] Succinoglycan displays pseudoplastic temperature-dependent rheological behavior in aqueous solution. The degree of polymerization and nature of the constituents affects its functional properties. It is allowed for food use only in Japan.

Levan

Levan is produced by both food-grade (LAB) and non-food-grade bacteria (*Bacillus polymyxa, Erwinia herbicola*). Levan from LAB has a great potential in the food industry since it may act as a prebiotic and cholesterol-lowering agent; it is proposed for developing novel functional foods.[12] In Japan, levan from non-food-grade bacteria is isolated and used as a thickener and stabilizer additive in milk products. Although there is a promising future of levan in food, its use is limited due to insufficient information on its properties and lack of feasible processes for large-scale production.

CONCLUSIONS

A wide diversity of functional microbial EPS has been described over the last decades, not only regarding the rheological characteristics that EPS impart in foods but also on potential health-promoting properties. Long-chain, high-molecular weight EPS that dissolve or disperse in water to give thickening, gelling, emulsifying, and/or suspending effects are indispensable tools in food formulation. Microbial EPS compete with plant and algae polysaccharides as well as with synthetic products due to their broad functional properties, which are determined by subtle structural characteristics. Better insight of the physiology of EPS-producing microorganisms as well as a detailed knowledge about polymer functionality (for in situ production or use as food additives) in the food matrix will allow developing novel functional cultures or foods with specific desirable properties.

ACKNOWLEDGMENTS

The authors acknowledge the financial support of CONICET, ANPCyT (PICTR20801), and CIUNT from Argentina.

REFERENCES

1. Hugenholtz, J.; Smid, E. Nutraceutical production with food-grade microorganisms. Curr. Opin. Biotechnol. **2002**, *13*, 497–507.
2. De Vuyst, L.; De Vin, F.; Vaningelgem, F.; Degeest, B. Recent developments in the biosynthesis and application of heteropolysaccharides from lactic acid bacteria. Int. Dairy J. **2001**, *11*, 687–708.
3. Mozzi, F.; Vaningelgem, F.; Hébert, E.M.; Van der Meulen, R.; Foulquié Moreno, M.R.; Font de Valdez, G.; De Vuyst, L. Diversity of heteropolysaccharide-producing lactic acid bacterium strains and their biopolymers. Appl. Environ. Microbiol. **2006**, *72*, 4431–4435.
4. Rúas-Madiedo, P.; Abraham, A.; Mozzi, F.; González de los Reyes Gavilán, C. Functionality of exopolysaccharides produced by lactic acid bacteria. In *Molecular Aspects of Lactic Acid Bacteria for Traditional and New Applications*; Mayo, B., López, P., Pérez-Martínez, G. Eds.; Research Signpost Editorial: Kerala, India, 2008, 137–166.
5. Sharma, B.R.; Naresh, L.; Dhuldhoya, N.C.; Merchant, S.U.; Merchant, U.C. Xanthan gum—a boon to food industry. Food Promotions Chronicle **2006**, *1*, 27–30.
6. Banik, R.M.; Kanari, B.; Upadhyay, S.N. Exopolysaccharide of the gellan family: prospects and potential. World J. Microbiol. Biotechnol. **2000**, *16*, 407–414.
7. Yotsuzuka, F. Curdlan. In *Handbook of Dietary Fiber*; Cho, S.S., Dreher, M.L. Eds.; CRC Press: Boca Raton, FL, 2001, 737–755.
8. Lochke, A.H.; Rale, V.B. Trends in microbial production of pullulan and its novel applications in the food industry. In *Food Biotechnology: Microorganisms*; Hui, Y.H., Khachatourians, G.G. Eds.; Wiley-IEEE: New Jersey, 1995, 589–604.
9. Survase, S.A.; Saudagar, P.S.; Singhal, R.S. Production of scleroglucan from *Sclerotium rolfsii* MTCC 2156. Biores. Technol. **2006**, *97*, 989–993.
10. Vega García, M.; Bontoux, L. *Food Applications of the New Polysaccharides Technology, 1995–2004*, The Institute for Prospective Technological Studies (IPTS) Report 20; Joint Research Centre European Commission: ESC-EEC-EAEC Brussels—Luxembourg, 1997.
11. Simsek, S.; Mert, B.; Campanella, O.H.; Reuhs, B. Chemical and rheological properties of bacterial succinoglycan with distinct structural characteristics. Carbohydr. Polymers. **2009**, *76*, 320–324.
12. Rhee, S.K.; Song, K.B.; Kim, Ch.H.; Park, B.S.; Jang, E.K.; Jang, K.H. Levan. In *Polysaccharides and Polyamides in the Food Industry. Properties, Production and Patents*; Steinbüchel, A., Rhee, S.K. Eds.; Wiley VCH Verlag GmbH & Co: Weinheim, 2005, 323–349.

Microbial Small Heat-Shock Proteins

Sang Yup Lee
Department of Chemical and Biomolecular Engineering, Korea Advanced Institute of Science and Technology (KAIST), Daejeon, South Korea

Mee-Jung Han
Department of Chemical and Biomolecular Engineering, Dongyang University, Gyeongbuk, South Korea

Abstract

Small heat-shock proteins (sHsps) are widely distributed in almost all organisms in nature from prokaryotes to eukaryotes. sHsps prevent uncontrolled protein aggregation in a multiplex refolding network, resulting in a general protection of the cellular proteome under extremely stressful conditions. sHsps are involved in important functions, such as thermotolerance, protein disaggregation, and proteolytic inhibition, and have various applications in the diverse field of biotechnology.

INTRODUCTION

Living cells require a sophisticated balance between folding of proteins and their proteolytic degradation, and therefore, molecular chaperones play a critical role in this quality control of the proteome by interacting with, stabilizing, and remodeling a wide range of proteins including non-native ones. (Chaperones are proteins that assist in the non-covalent folding and unfolding and the assembly and disassembly of other macromolecular structures.) Remarkable effort has been made in the elucidation of the two major chaperone families: 1) Hsp70 (DnaK); and 2) Hsp60 (GroEL). In addition, the correct folding of proteins requires the assistance of other chaperones. In *Escherichia coli*, for instance, cochaperones, e.g., DnaJ, GrpE, and GroES, disaggregating chaperone, e.g., ClpB, and holding chaperones, e.g., IbpAB play important roles.[1] Particularly, small heat-shock proteins (sHsps) have been attracting much research interest in recent years. The available information on sHsps has been constantly updated, and this information has been integrated into a model that postulates how sHsps cooperate in the framework of a complex multichaperone network.[2]

The sHsps are molecular chaperones of relatively small molecular mass (about 12–43 kDa) that are found in almost all organisms.[2,3] The most intensively studied sHsps are IbpA and IbpB of *E. coli*, the names of which originate from *i*nclusion *b*ody (IB)-associated *p*roteins A and B. Importantly, the sHsps have several properties that distinguish them from other heat-shock proteins.[3] They function as ATP-independent chaperones, and require flexible assembly and reassembly of oligomeric complex structures for their activation although the precise mechanism of sHsps remains controversial to date, and exhibit a wide range of substrate-binding capacities. As our knowledge on the molecular details of sHsps structure and function has been accumulated, a number of applications have recently emerged. Thus, this entry focuses on the use of sHsp in various biotechnological applications together with the challenges to be addressed in the individual areas.

IN VIVO APPLICATIONS

Increased Stress Tolerance

sHsps can protect cells from a variety of stresses and consequently increase cell survival rate. For example, *E. coli* cells, overproducing IbpA or IbpB, were found to be more resistant to heat and oxidative stresses.[3,4] and accumulated less aggregated proteins after exposure to high temperature as compared to normal *E. coli* strains. By contrast, deleting the *ibpAB* genes increased protein aggregation and negatively affected cell viability in vivo under extreme heat-shock conditions. The growth impairment during overproduction of recombinant proteins or in combination with mutations of *dnaK* was found to be more intense in the *ibpAB* mutant. Thus, sHsps could be utilized for the improvement of cellular tolerance under various stress conditions.

Enhanced Protein Production

Recombinant DNA technology enables us to obtain large amounts of desired proteins and simplify downstream purification processes in bacterial cells; however, overproduction of recombinant proteins in host cells causes the formation of insoluble protein aggregates known as IBs. One of the most widely tried methods to improve the yields of soluble recombinant proteins is coexpression of a combination of chaperones implicated in protein folding, such as DnaK, GroEL, or GroES. Similarly, sHsps could be used for controlling the quality and quantity of recombinant proteins. Han et al.[5] reported that knocking out the *ibpAB* genes in *E. coli* K-12 resulted in enhanced secretory production of alkaline phosphatase by secretory protein translocation (Sec)-dependent pathway; however, in the case of a twin-arginine translocation (Tat) pathway, the production of EGFP (enhanced variant of the *Aequorea victoria* green fluorescent protein) in the *ibpAB* mutant was not higher than that in its parent strain, but fluorescence intensity was much higher in the *ibpAB* mutant. These results suggest that the capacity of protein secretion by the Tat pathway is somewhat limited compared with the Sec-dependent system because only folded proteins can be secreted by the former. Also, the deletion of the *ibpAB* genes might cause degradation of non-secreted recombinant proteins that accumulate in the cytoplasm.[5] Thus, the *ibpAB* mutation not only allows efficient secretion of recombinant proteins by the Tat pathway without their accumulation in the cytoplasm, but also enhances Sec-dependent secretory production of recombinant proteins.

Alternatively, a strategy of producing proteins as IBs has some advantages, including simplicity of gene manipulation, high productivity, feasibility of toxic protein production, and simplicity of purification, if an effective refolding protocol is established. Overexpression of IbpA and/or IbpB could enhance the production of various recombinant proteins as IBs by protecting them from proteolytic degradation.[5] This strategy is useful for the production of toxic proteins in *E. coli*. The beneficial effect of IbpAB on the production of recombinant proteins is great when protease activities are high, and thus it becomes less at moderate temperatures, e.g., 30°C or lower, because of the decreased activities of cellular proteases.[6] Therefore, manipulation of *ibpAB* gene expression might prove to be an efficient way of fine-tuning the production of recombinant proteins. Again, some experimental optimization studies need to be performed to achieve maximum desired effects by examining temperature, strain species, or other environmental conditions since these parameters affect the *ibpAB* expression and subsequent effects.

IN VITRO APPLICATIONS

Chip or Array-Based Approaches

One of the most challenging works for chip or array technologies is the stable immobilization of proteins on surfaces in such a way that their 3-D structure, functionality, and binding sites are retained. A frequent problem encountered in these assays is that the proteins or peptides have a limited stability and shelf life. Especially, it arises when labile enzymes are used for diagnostic detection.

The sHsps can be applied for improving protein and peptide systems including chip-based and diagnostic immunological assays. Ehrnsperger et al.[7] reported a method utilizing murine Hsp25 to stabilize enzymatic activities and antigenicities of proteins used in immunogenic detection. Hsp25 is very effective in the binding and stabilization of a marker protein for myocardial damage, troponin T, and several peptides used in the diagnosis of viral infections such as hepatitis C, even after extensive incubation periods. Exploiting the ability of sHsps to keep proteins in their active conformation can overcome the biotechnological problems encountered in protein-based diagnostics of human diseases, and is applicable to a wide range of assays involving unstable proteins, including the generation of vaccines. Laksanalamai et al.[8] also reported that sHsp purified from hyperthermophilic archaeon *Pyrococcus fuiosus* (Pfu-sHsp) stabilized Taq polymerase and enzymes at high temperatures during the polymerase chain reaction (PCR). Pfu-sHsp confers thermotolerance on cellular cultures and proteins in cellular extracts during prolonged incubations at high temperatures, demonstrating that it is effective for combating enzymatic aggregation and intracellular precipitation during heat stress. Thus, sHsps enhance the utilities and stabilities of enzymes in various applications, such as Taq polymerase in PCR or high temperature stresses.

Additionally, the capability of using immobilized sHsp in achieving the protection and renaturation of enzymes might allow a wider range of applications including biosensor preparation. There is a good potential for using immobilized Hsps in combination with other chaperones to prevent the denaturation of entrapped and immobilized enzymes, which is a problem commonly encountered in biosensor development. Yang et al.[9] reported that a combination of immobilized Hsp70 and a mixture of α-crystallin and reticulocyte lysate enabled the recovery from luciferase of up to 10% of the original firefly luciferase activity after a long storage period at room temperature. This finding shows that immobilized chaperones show some potential for the stabilization and reactivation of enzymes that are trapped in thin aqueous films for applications in biosensors and reactors; however, it should be noted that optimal immobilization procedures without the creation of steric hindrance on the surface are required. A specific and oriented protein immobilization protocol might be appropriately selected depending on the specific proteins. Thus, the advantageous use of sHsps might best be achieved for systems that utilize properly immobilized sHsps, and free or entrapped substrate proteins or enzymes.

Gel- or Column-Based Protein Analysis and Purification

A proteolytic degradation is one of the most critical problems during separation and purification of proteins. The loss or modification of proteins in proteomic studies by gel- or non-gel-based approaches can be a substantial problem when the entire protein content of a sample must be analyzed because it reduces the number of proteins reproducibly observed and also results in erroneous conclusions caused by the artifacts. Recently, Han et al.[10] demonstrated that the sHsps were able to protect proteins from proteolytic degradation by trypsin or proteinase K under denaturation conditions, and that this strategy could be universally employed for the proteomic analysis of various biological samples, such as the extracts from bacteria, plant cells and human cells, and biofluids. Addition of sHsps, including E. coli IbpA and IbpB or Saccharomyces cerevisiae Hsp26 during 2-D electrophoresis enabled the detection of up to 50% more protein spots compared with the currently available protease inhibitors. The use of sHsps enabled the visualization of many more protein spots that were previously undetectable, thus opening up the possibility of identifying new protein targets.

Additionally, the sHsps might be applied to column-based proteomic studies for stabilizing and preventing the degradation of proteins or peptides. A promising column-based method is a catalytic column refolding. Column refolding has several advantages, including a rapid and effective refolding, and low cost because this strategy inhibits the illegitimate interactions leading to deleterious protein aggregation by the use of chaperones and foldase immobilized on an agarose gel. For example, an oxidative refolding chromatography column that has three components, the GroEL minichaperone, DsbA, and peptidyl-prolyl isomerase immobilized on an agarose gel was developed.[11] For further improvements of the refolding column system, the use of sHsps can circumvent the problem of aggregation during sample application by allowing the denatured protein to penetrate the column during refolding processes. It has been proposed that bovine-derived α-crystallin helps in the refolding of target proteins denatured owing to heat stress by preventing aggregation during the process of dialysis.[12] It demonstrated that the α-crystallin assists in refolding processes during the removal of the denaturing agent and prevents the irreversible loss owing to aggregation that occurs upon dialysis. Thus, the sHsps are effective in facilitating the refolding and stabilization of proteins or increasing solubility both in vitro and in vivo. They can be employed for various biotechnological applications, including column-based methods, if functional oligomeric sHsps can be properly immobilized.

CONCLUSION

Recent advances in our understanding on sHsps from bacteria, archaea, plants, and animals are allowing us to use the sHsps in various biotechnological applications. The sHsps have great potential to be used in protein engineering, diagnostic or detection devices, bioseparation or purification, proteomics, and other processes. It is expected that the sHsps will find more applications based on their general capabilities of protecting a wide range of proteins.

ACKNOWLEDGMENT

This work was supported by the Korean Systems Biology Research Grant (M10309020000-03B5002-00000) of the Ministry of Education, Science and Technology through the KOSEF. Further supports by LG Chem Chair Professorship and Microsoft are appreciated.

REFERENCES

1. Baneyx, F.; Mujacic, M. Recombinant protein folding and misfolding in *Escherichia coli*. Nat. Biotechnol. **2004**, *22* (11), 1399–1408.
2. Narberhaus, F. Alpha-crystallin-type heat shock proteins: socializing minichaperones in the context of a multichaperone network. Microbiol. Mol. Biol. Rev. **2002**, *66* (1), 64–93.
3. Han, M.J.; Yun, H.; Lee, S.Y. Microbial small heat shock proteins and their use in biotechnology. Biotechnol. Adv. **2008**, *26* (6), 591–609.
4. Kitagawa, M.; Miyakawa, M.; Matsumura, Y.; Tsuchido T. *Escherichia coli* small heat shock proteins, IbpA and IbpB, protect enzymes from inactivation by heat and oxidants. Eur. J. Biochem. **2002**, *269* (12), 2907–2917.
5. Han, M.J.; Park, S.J.; Park, T.J.; Lee, S.Y. Roles and applications of small heat shock proteins in the production of recombinant proteins in *Escherichia coli*. Biotechnol. Bioeng. **2004**, *88* (4), 426–436.
6. Kuczyńska-Wiśnik, D.; Zurawa-Janicka, D.; Narkiewicz, J.; Kwiatkowska, J.; Lipińska, B.; Laskowska, E. *Escherichia coli* small heat shock proteins IbpA/B enhance activity of enzymes sequestered in inclusion odies. Acta Biochim. Pol. **2004**, *51* (4), 925–931.
7. Ehrnsperger, M.; Hergersberg, C.; Wienhues, U.; Nichtl, A.; Buchner, J. Stabilization of proteins and peptides in diagnostic immunological assays by the molecular chaperone Hsp25. Anal. Biochem. **1998**, *259* (2), 218–225.
8. Laksanalamai, P.; Pavlov, A.R.; Slesarev, A.I.; Robb, F.T. Stabilization of Taq DNA polymerase at high temperature by protein folding pathways from a hyperthermophilic archaeon, *Pyrococcus furiosus*. Biotechnol. Bioeng. **2006**, *93* (1), 1–5.

9. Yang, Y.; Zeng, J.; Gao, C.; Krull, U.J. Stabilization and reactivation of trapped enzyme by immobilized heat shock protein and molecular chaperones. Biosens. Bioelectron. **2003**, *18* (2–3), 311–317.

10. Han, M.J.; Lee, J.W.; Lee, S.Y. Enhanced proteome profiling by inhibiting proteolysis with small heat shock proteins. J. Proteome Res. **2005**, *4* (6), 2429–2434.

11. Altamirano, M.M.; García, C.; Possani, L.D.; Fersht, A.R. Oxidative refolding chromatography: folding of the scorpion toxin Cn5. Nat. Biotechnol. **1999**, *17* (2), 187–191.

12. Horwitz, J. Alpha-crystallin can function as a molecular chaperone. Proc. Natl. Acad. Sci. U.S.A. **1992**, *89* (21), 10449–10453.

Microfluids for Assisted Reproduction

Henry Zeringue
Department of Bioengineering, University of Pittsburgh, Pittsburgh, Pennsylvania, U.S.A.

Abstract

For decades, Assisted Reproduction Technologies have focused on the health and development of in vitro embryos with relatively passive techniques (e.g., embryo culture and in vitro fertilization). Although high viability rates of these techniques have become routine, the increased use of more active techniques (e.g., intracytoplasmic sperm injection, preimplantation biopsy, cytoplasmic transfer) require another level of viability improvements. The use of techniques able to control the environment on the size and time scales of individual embryos would be beneficial for improved viability. Here we review developments in microtechnologies that have shown improvements in viability or efficiency for assisted reproduction. We present microfluidic devices for in vitro fertilization, cumulus removal, embryo culture and zona pellucida removal. Microfluidic technologies are able to control a number of environmental conditions (e.g., chemical concentrations, temperature, fluidic stress, and pressure) at micrometer length scales with sub-second precision. These abilities, with the experimental demonstrations described, should reinforce the case for continued research to develop microtechnologies for assisted reproduction.

INTRODUCTION

For centuries, scientists have explored the underlying biology of reproduction. The field assisted reproduction technologies (ART) began with the first successful transplant and development of a mammalian embryo in the early 1890s by Heape. The ability of ART to allow development of ova and embryos in an observable environment has permitted considerable progress toward the understanding of embryo development. Advances beyond simple embryo culture have allowed for more sophisticated techniques like intracytoplasmic sperm injection, chimera formation, and embryo biopsy. These techniques have allowed much exploration of biology and genetics as well as maintenance and preservation of genetic diversity among domesticated species. Still, these techniques impart undesirable artifacts on normal embryo development owing to innate properties of the artificial environment and current handling methods, including rapid environmental changes. These techniques have brought us a long way, but emerging technologies are demonstrating quite useful platforms for future ART.[1] This entry explores recent advances in microfluidic technology that suggests it may provide more gentle manipulation techniques and possibly a more in vivo-like environment.

MICROFLUIDICS

When exploring the utility of microfluidics within ART (uART), there are a number of phenomena of interest. For an in-depth review of these issues, see Walker et al.[2] The primary phenomenon utilized is laminar flow, which produces predictable and stable flow profiles. In laminar flow, as Fig. 1A shows, two streams can flow side by side with their only mixing occurring through diffusion at the interface (lightly shaded triangle). This allows precise temporal and spatial control of objects within any given stream. Manipulation of the inlet pressures can realize spatial (Fig. 1B) control of chemical delivery down to the micrometer scale. Appropriate channel geometries can also achieve precise temporal (Fig. 1C) control deliver to less than a second. Microfluidic manipulation is able to achieve both physical and chemical manipulation of an embryo. These manipulation capabilities allow for movement, orientation, holding, pressure control, and chemical influence on the size scale of a single embryo.

Microfluidic Methods

Recent advances in ART employing microfluidic designs to improve current in vitro manipulation techniques are explored. Most of this work has come out of a collaboration between researchers at the University of Illinois and University of Wisconsin.[1] They report the first attempt to manipulate an embryo and its local environment using microfabricated devices. Microfluidic channels were originally constructed using etched silicon channel structures sealed with glass. There were a number of issues with these first devices. The engineers did not like the obligatory design constraints due to the crystal structure of silicon nor the difficulty in controlling these first devices. The reproductive physiologists did not like the inability to visualize

Microfluids for Assisted Reproduction

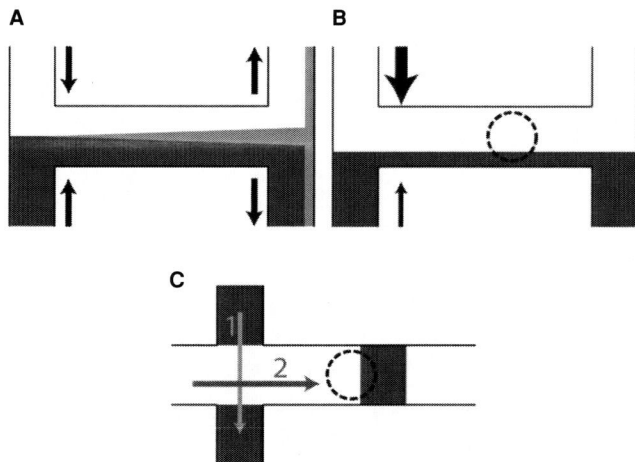

Fig. 1 Schematic of microfluidic laminar flow. (A) In the laminar regime, two fluid streams can flow side by side (clear and dark shade) with mixing occurring only through diffusion at the interface (light shade). The amount of mixing depends on the residency time in the chamber and the diffusivity of the soluble components. (B) By varying the relative inlet pressures, the proportion of each stream within the channel can be defined, allowing spatial definition of components on the size scale of micrometers. (C) Using "+" junctions, *plugs* of distinct fluid can be formed in a channel and delivered in a temporally discrete manner downstream. The size of the plug and velocity of the fluid help determine the duration of exposure to the plug.

through the device, or the long-term biocompatibility of some of the materials used. Duffy et al.[3] report a technique of rapid prototyping that allowed simple fabrication of complex microfluidic devices in poly(dimethylsiloxane) (PDMS). This general technique (Fig. 2) affords optical transparency, biocompatibility, and easy external connectivity for fluidic control. Murine embryos at various stages within a typical microfluidic channel are shown in Fig. 3.

MICROFLUIDICS FOR ART

Over the past decade, researchers have utilized improved microfabrication and microfluidic techniques to develop devices for assisted reproduction for improved efficiencies and/or yields. ART addressed to date including devices for in vitro fertilization (IFV), cumulus removal (CR), preimplantation embryo culture, and zona pellucida (ZP) removal.

In Vitro Fertilization

IVF of mammalian oocytes in microchannels was demonstrated by Clark et al.[4] using a microfluidic channel as the "reaction chamber" for porcine ovum fertilization. This study showed higher fertilization rates than traditional microdrop fertilization. Another collaboration at University of Michigan between Shuichi Takayama and Gary Smith has also explored the use of microfluidics for sperm

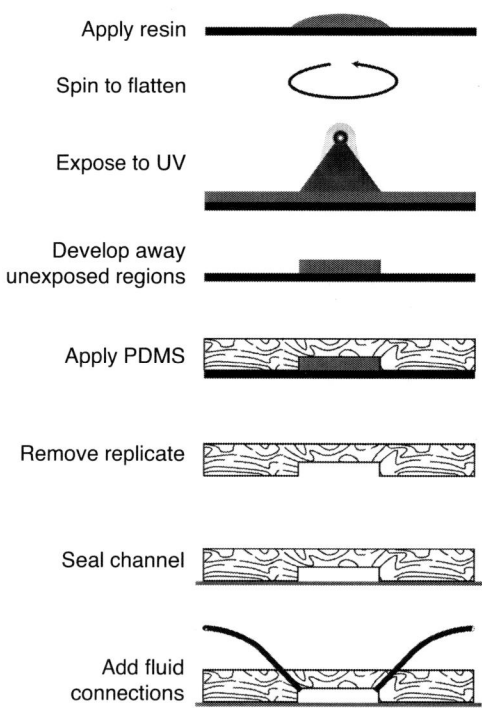

Fig. 2 Rapid prototyping. An epoxy resin (e.g., SU-8 from Microchem Corp., Massachusetts, United States) is applied to a clean silicon wafer. Spinning the wafer defines the thickness of the epoxy film. Exposure to UV in defined regions will solidify the resin in those regions, allowing unexposed regions to be developed away. This leaves the 3-D structure as the mold master. PDMS is prepared and poured onto the mold master. Once the PDMS cures, it can be pealed from the master. The resulting microfluidic channel impressions can be sealed with glass or another sheet of PDMS. Fluid control connections are made by placement of tubing near the ends of fluid channels.

manipulation. They further found that fertilization taking place in a microfluidic chamber allows fewer sperm to be used to achieve similar fertilization rates of oocytes.[5]

Additionally, they developed a passive microfluidic device to separate motile from non-motile sperm.[6] Using

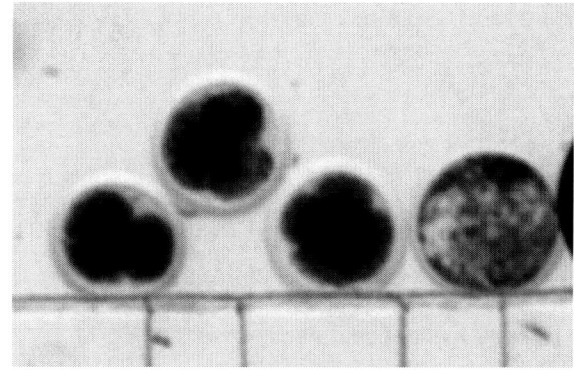

Fig. 3 Embryos in a microchannel. This figure shows multiple stages of bovine embryo development in a microchannel. From left to right, there are 2-cell, 16-cell, 8-cell, and blastocyst stages. To be sure, each of these has been developing the microchannel environment, but they were fertilized at different times.

previously developed techniques to separate soluble particles with high diffusivity from those with low diffusivity, a stream containing sperm is briefly brought into contact with a non-sperm stream, some of the motile sperm will diffuse across the boundary and end up in the previously empty stream.

Cumulus Removal

CR is performed in IVF techniques either before or after fertilization, depending on the animal model being used. Zeringue et al.[7] demonstrated a microfluidic technique for mechanical removal of the cumulus cell mass from fertilized zygotes (i.e., microfluidic cumulus removal, μFCR). Foregoing the need for extra enzymatic breakdown of the cumulus mass, this technique used mild pressure oscillations in a small "removal port" on the side of a microfluidic channel. Each pressure oscillation would withdraw a small patch of cumulus then perform a slight asymmetric rotation of the embryo. This technique allows for the stepwise removal of cumulus from the entire embryo.

Comparison of this technique with traditional vortex CR showed μFCR enhanced early bovine embryo development when compared to vortexing. Significant increases in 8-cell embryos at day 2 (35% vs. 20%) as well as blastocyst (57% vs. 33%) and expanded blastocyst (37% vs. 19%) at day 8 were seen after using μFCR, as opposed to vortexing.[8] More in-depth analysis of development directly following CR demonstrated an inverse relationship between development and duration of vortexing. A BrdU protocol to observe RNA transcription showed a spike in transcription directly following vortex but was not seen in μFCR. By contrast, a slow and steady increase in transcription was seen in the μFCR embryos. Further investigation by Reeder et al.[9] verified the increased activity corresponded with an increase in stress proteins, specifically HSP70.1. It is also worth mentioning that Reeder compared the microfluidic technique to both vortex and hand-stripping removal techniques, each yielding statistically similar results. These results implicate traditional CR techniques in decreased developmental efficiencies of in vitro produced embryos.

Embryo Culture

The first work demonstrating the utility of microfluidics to improve ART conditions was in vitro mammalian embryo culture in a microfluidic environment.[10] When cultured in microchannels vs. traditional microdrops for 96 hours, the microchannels produces significantly more viable blastocyst-stage embryos (87.5% vs. 63.3%) and significantly fewer degenerate embryos (12.5% vs. 36.7%). By using a mouse cross that has low yields of in vitro embryo development (ICR × B6SJL/F1), they were able to demonstrate not only that microfluidic devices can be used to culture preimplantation embryos but also that the microfluidic environment could be more beneficial than microdrop culture. Looking further at the data from this study, not only did the endpoint (96 hours of culture) demonstrate better results for the microchannels but the rate of development was faster in microchannels, something also seen in in vivo development compared to microdrop culture, suggesting better dynamics for the development of embryos in microchannels. More recently, Melin et al.[11] confirmed that robust mouse crosses (C57BL6 × DBA/2/F1) can also be cultured to the blastocyst stage in microfluidic chambers (81.8%) as well as microdrops (83.3%).

ZP Removal

Chemical manipulation of an embryo is explored as a final demonstration of microfluidic utility. The formation of chimeric animals is utilized in disease and genetic research areas. Formation of a chimera requires the removal of the ZP from one or more embryos with cells from multiple sources cultured together as a single embryo. Zeringue et al.[12] demonstrate the ability of microfluidics to remove the ZP with a brief (3–5 seconds) presentation of acidic Tyrode's medium of pH 2.5–2.8 (Fig. 4). Fluid manipulation at a

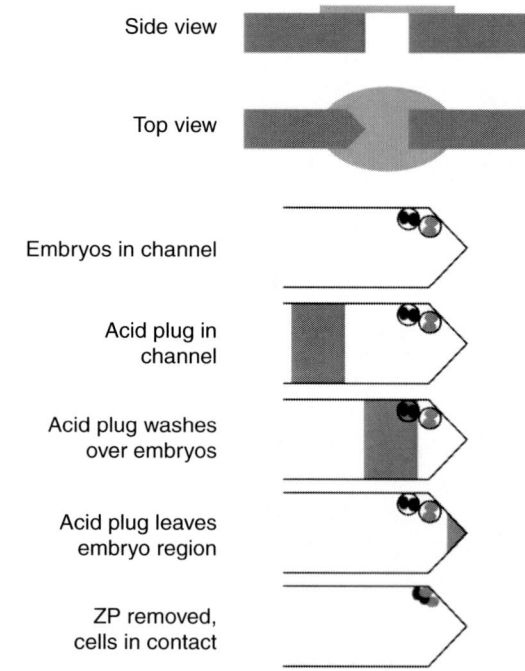

Fig. 4 Schematic of microfluidic ZP removal. The side view shows the large main channel for embryo holding with the thin connection channel for continual fluid flow. The top view shows the "arrow" channel design to maintain intimate contact of cells throughout the removal and subsequent culture steps. The acid plug flows down the main channel and over the embryos. Approximately 25 seconds after the acid has flown past the embryos, the zona breaks down and is carried with the fluid, while the embryonic cells are retained in the main channel. The channel design ensures the cells from multiple embryos will stay in intimate contact and be allowed to develop as a single embryo, as shown.

"+" junction allows presentation of a distinct plug of fluid (in this case, acidic medium) to wash downstream. In the case presented, acidic medium is presented to the embryos, after which fluid flow continues. ZP removal occurs approximately 25 seconds after acidic media is removed from the local embryo environment. This acid presentation of 3–5 seconds is significantly less than the typical time of 30–120-seconds exposure to acidic medium using microdrop techniques. Once the ZP is removed, the specific channel geometry encourages cells from distinct embryos to maintain intimate contact, as is necessary for chimera formation.

DISCUSSION

The examples presented above represent the groundwork for the development of a platform for ART to leverage emerging technologies for improved yields of in vitro processing and possibly new avenues for additional processing and technique development. There are certainly common themes running through the devices presented, regardless of the collaboration from which they came. All utilize the rapid prototyping of PDMS based on the ease of design and fabrication as well as the utility of the materials for biological work. There are also common designs, such as the inlet funnel well, allowing for easy introduction and retrieval of embryos. This design was presented in Zeringue's work[7], [12] and subsequently utilized by Suh et al.[5] The devices also all utilize similar pressure control connections and fluid handling techniques.

More importantly, they demonstrate striking examples of improved efficiencies for common techniques by replacing a single method with the microfluidic equivalent. The preimplantation embryo development in microfluidic channels shows up to a 25% improvement over traditional culture drop techniques. This is especially relevant for low-yield techniques like certain livestock breeds, embryo biopsy, and chimera formation. The IVF work shows improvements in both sperm concentration requirements as well as overall sperm usage. This is certainly relevant when stud semen sources can be costly and supplies limited. The CR results demonstrate another improvement of almost 25% blastocyst yield when microfluidic techniques are used in place of traditional means. Reeder's work reveals genetic transcription changes yielding possible insights to the causes of current in vitro yields compared to in vivo development. The ZP report demonstrates some of the precise temporal and spatial control microfluidics has over the embryo's local environment. The ability to robustly remove the ZP with 3–5-seconds exposure to acid should yield higher embryonic cell viability than typical 30–120-seconds exposure. Another benefit of this technique is that the embryonic cells are brought into intimate contact directly after ZP removal, eliminating the extra manipulation step of transferring the embryonic cells to a separate location to maintain contact during development.

CONCLUSIONS

It should be noted that each of the techniques described above was performed as a replacement for a single step within the ART process (e.g., microfluidic CR with traditional microdrop in vitro maturation, fertilization, and embryo culture). Significant improvements were seen even as a replacement for these individual steps of the process. Since they are built on a single fabrication platform, there is a promise of integration of multiple steps into a single device, possibly even a single holding chamber for each embryo with the "processes" coming to the embryo.

Looking forward, there are numerous areas of study for microfluidics in ART. The straightforward areas are improvements in the already described devices as well as integration of multiple steps into single devices. Subsequent work would be to develop additional devices to improve current techniques like delivery of cryoprotectant for gamete preservation as well as biopsy of single embryos. If this work is pursued successfully, more ambitious researchers might find new techniques for embryo manipulation not possible with current handling techniques. Wherever the research proceeds, it is sure that microfluidics demonstrates great promise for the future of ART.

REFERENCES

1. Beebe, D.J.; Wheeler, M.; Zeringue, H.C.; Walters, E.; Raty, S. Microfluidic technology for assisted reproduction. Theriogenology **2002**, *57* (1), 125–135.
2. Walker, G.M.; Zeringue, H.C.; Beebe, D.J. Microenvironment design considerations for cellular scale studies. Lab Chip **2004**, *4* (2), 91–97.
3. Duffy, D.; McDoanld, J.C.; Schueller, O.J.A.; Whitesides, G.M. Rapid prototyping of microfluidic systems in poly (dimethyl siloxane). Angew. Chem. Int. Ed. Engl. **1998**, *37*, 550–575.
4. Clark, S.G.; Beebe, D.J.; Wheeler, M.B. In vitro fertilization of porcine oocytes in polydimethylsiloxane (PDMS)-glass microchannels. Biol. Reprod. **2002**, *66* (Suppl. 1), 528.
5. Suh, R.S.; Zhu, X.; Phadke, N.; Ohl, D.A.; Takayama, S.; Smith, G.D. IVF within microfluidic channels requires lower total number and lower concentrations of sperm. Hum. Reprod. **2005**, *21* (2), 477–483.
6. Cho, B.S.; Schuster, T.G.; Zhu, X.Y.; Chang, D.; Smith, G.D.; Takayama, S. Passively driven integrated microfluidic system for separation of motile sperm. Anal. Chem. **2003**, *75* (7), 1671–1675.

7. Zeringue, H.C.; Beebe, D.J.; Wheeler, M.B. Removal of cumulus from mammalian zygotes using microfluidic techniques. Biomed. Microdevices **2001**, *3* (3), 219–224.

8. Zeringue, H.C.; Rutledge, J.J.; Beebe, D.J. Early mammalian embryo development depends on cumulus removal technique. Lab Chip **2004**, *5* (1), 86–90.

9. Reeder, A.; Schutzkus, V.; Wiebelhause-Finger, J.; Khatib, H.; Monson, R.L.; Wheeler, M.B.; Beebe, D.; Rutledge, J. Assessment of HSP70.1 transcription levels in bovine embryos after cumulus removal by different techniques. Reprod. Fertil. Dev. **2006**, *18* (2), 111.

10. Raty, S.; Davis, J.A.; Beebe, D.J.; Rodriguez-Zas, S.L.; Wheeler, M.B. Culture in microchannels enhance in vitro embryonic development of preimplantation mouse embryos. Theriogenology **2001**, *55* (1), 241.

11. Melin, J.; Lee, A.; Foygel, K.; Leong, D.E.; Quake, S.R.; Yao, M.W.M. In vitro embryo culture in defined, sub-microliter volumes. Dev. Dyn. **2009**, *238* (4), 950–955.

12. Zeringue, H.C.; Wheeler, M.B.; Beebe, D.J. A microfluidic method for removal of the zona pellucida from mammalian embryos. Lab Chip **2004**, *5* (1), 108–110.

Miso Fermentation

Naveen Chikthimmah
Jaison Karavally
Department of Food Science, Pennsylvania State University, University Park, Pennsylvania, U.S.A.

Abstract
Miso is a fermented soybean food product that is popularly consumed in Japan and Asia. It is made by the microbial fermentation of a soybean substrate that may contain one or more of three major cereal adjuncts—rice, wheat, or barley. Miso production is a three-step process, which includes making a koji culture, preparing a soybean substrate, and conducting the fermentation/ripening of the substrate. The entry discusses the scientific principles involved in miso production and outlines problems encountered during miso fermentation. Biotechnological approaches in starter culture selection and coculture usage to overcome fermentation problems are also discussed.

INTRODUCTION

Miso is a salted and fermented soybean food product made with or without the addition of rice, barley, or wheat. Miso is widely used in the culinary traditions of Japan, China, and East Asia and is often consumed at breakfast along with rice. Miso is also a popular soup base and is used in sauces, marinades, and dressings. It has been valued in tradition for its health-promoting properties and recent observational and interventional studies have investigated the potential health benefits of miso in preventing chronic diseases such as cancer, gastric carcinogenesis, and cardiovascular disease. In the United States, miso is commercially available in the form of a paste in large supermarkets, international food groceries, and health food stores.

In the following entry, the production operations of miso are described. Particular emphasis is placed on the chemistry of miso fermentation and the ripening processes. Microbiological problems and challenges associated with miso production are described and biotechnological methods to overcome fermentation problems are elaborated.

MISO: PRODUCT DESCRIPTION AND COMPOSITION

Miso is a soy-based fermented food product. It is made by the microbial fermentation of a soybean substrate that may in addition contain one or more of the three major cereal adjuncts—rice, wheat, or barley. The fungus, *Aspergillus oryzae*, and the yeast *Zygosaccharomyces rouxii* (formerly *Saccharomyces rouxii*), are the primary microorganisms involved in miso fermentation.[1] The major components of miso are water (43%), protein (11.7%), lipids (6.0%, mainly unsaturated), carbohydrates (26.5% by difference), and ash (12.8%). The high ash content is indicative of a high mineral content that mainly includes sodium (3.7%), potassium (0.2%), and phosphorous (0.16%).[2]

Miso can be broadly classified into three different categories on the basis of the cereal adjunct added to the soybean substrate. Kome miso is made with a rice adjunct, Mugi miso contains rye or barley as ingredients, and Mame miso is made only using soybeans without any added cereal adjuncts. Miso can be further classified on the basis of taste (sweet and salt taste profiles) and color.

MISO: MANUFACTURING PROCESS

Depending on the commercial context, two major methods, the traditional method and the modern method, are adopted in the miso manufacturing process. The traditional method is generally done in small batches at home and at cottage industry levels. The modern method is utilized by commercial miso producers who utilize advances in technology, microbiology, and engineering to manufacture and package miso and sell it in a large distribution network.

The production of miso is accomplished in three major steps. The first step is the preparation of the starter culture (koji) for the soybean substrate fermentation process. In the preparation of koji, a single-milled cereal grain (rice, wheat, or barley) or a combination of cereal grains is soaked in potable water overnight or until the moisture content of the grain reaches approximately 35%. The excess water is drained and the cereal grain is cooked, usually under pressure for 20–30 min, to gelatinize the cereal grain. The cooked cereal grain is spread and cooled in trays to 35°C and then inoculated with the mold, *A. oryzae*. Inoculation is done by dusting the mold spores at an inoculation rate of approximately 0.1%. The inoculated cereal grain is left in a ventilated room (aerobic conditions, room temperature) for 40–50 hours until the cooked cereal grain is covered by a moldy mycelial mat known as koji. Koji must be harvested while it is white and before it turns olive green, which is an indication of mold sporulation. Proper sanitation of the koji room and aeration are important factors in the production of the koji. Bulk production of koji at a depth of 30 cm or more may result in incomplete or stuck koji due to insufficient aeration in the bottom layers of the cooked cereal grain. Koji serves as the starter culture for the soybean fermentation step. Koji is also an important source of enzymes for the breakdown of the soybean substrate components during miso fermentation and ripening.[3]

The second step in miso production is the preparation of the soybean substrate and inoculation of the substrate with koji. To prepare the soybean substrate, whole soybeans are washed, soaked in water overnight, and steamed until they become tender. Proper hydration and adequate cooking of the soybeans are essential for preparing a substrate that will undergo protein hydrolysis during fermentation and ripening. This cooking step also helps to inactivate antinutrients in soybeans such as trypsin inhibitors, enhance softening of the soybean, and eliminate beany odors in the finished miso. Cooked and cooled soybean (60% w/w) is then mixed with koji (30% w/w), potable water, and sodium chloride (final salt concentration between 5 and 13%) to form the soybean substrate. Sometimes, for cultural or economic reasons, rice, wheat, or barley are also included into the soybean substrate. At this stage of production, commercial manufacturers commonly add a costarter inoculum of an osmophilic yeast such as Z. rouxii. The mashed substrate is blended and the mixture is placed in shallow wooden, concrete, or stainless steel fermentation tanks. It is not uncommon to leave the soybean substrate in an unblended form wherein the substrate is formed into little mounds or nuggets.

The primary microorganism involved in the miso fermentation process is A. oryzae. Based on its ability to grow in high salt concentrations and to produce flavor compounds specific for miso, Z. rouxii is often introduced as a costarter culture in industrial miso production.[1] Among bacterial species encountered in traditional miso fermentations conducted in open, genera from the lactic acid bacterial group are common isolates.

The third step in miso production is the soybean fermentation and ripening process to yield finished miso. The substrate in the fermentation tank is maintained at 28°C for about 7 days. During this time period, the growth of *Aspergillus* results in an increase in substrate temperature (to ~33°C). The growth of *Aspergillus* also results in the substrate turning anaerobic and inhibiting the growth of the mold. Stirring the substrate and/or transferring the substrate between tanks allow for sufficient aeration during fermentation. The fermentation step is followed by a prolonged ripening process period during which the characteristic miso flavors and colors develop in the fermented substrate. The duration of ripening can range from a few days to 1 year or more.

When the miso is fully ripened, it is blended, pressed, and packaged. Many commercial manufacturers often pasteurize the ripened miso before packaging. Since miso is prone to oxidation resulting in browning and off-flavor development, packaging materials with high oxygen barrier properties are preferred choices for packaging miso.

MISO: PROCESS AND FLAVOR CHEMISTRY

During fermentation and the early part of the ripening process (2–5 months), the hydrolysis of the substrate polysaccharides and proteins results in an increase in soluble sugar and soluble nitrogen concentrations, respectively. Glycinin, a major protein in soybeans, contains ~20% glutamic acid. Protein hydrolysis during fermentation and ripening liberates the glutamic acid as volatile sodium glutamate, which is a major component of miso flavor and aroma.[4] Soybean fatty acids released by the action of Koji lipases turn into fatty acid ethyl esters and contribute to the development of miso aroma.[5] During fermentation, osmophilic yeasts produce ethyl, butyl, and amyl alcohols. Esters produced by the reaction of the alcohols with organic acids also contribute to the miso aroma.[4] The mechanism of termination of polysaccharide and protein hydrolysis during the early and middle stages (up to 5 months) of ripening is currently unknown and is hypothesized to be because of the inactivation of the koji enzymes.[6]

Minimally ripened miso (10 days) is light brown in color and has a beany flavor that is not typical of miso. The characteristic sensory profile of miso with dark brown color and intense miso flavor and aroma does not develop until about 11 months of ripening. Hence, long ripening times (12–24 months) are an essential part of the miso manufacturing process. During miso ripening, the accumulation of Maillard reaction products with molecular weights in the range of 350–5000 suggests that the peptides that have undergone the Maillard reaction contribute to the typical flavor and sensory characteristics of the ripened miso. In this context, two important volatile compounds with low aroma thresholds, 4-hydroxy-2(or 5)-ethyl-5 (or 2)-methyl-3(2H)-furanone (HEMF) and 4-hydroxy-2, 5-dimethyl-3(2GH)-furanone (HDMF) have been shown to contribute to the characteristics flavor of miso.[7] Research indicates that HEMF and HDMF are products of both Maillard reactions as well as yeast secondary metabolites.[8] In miso ripened for long periods of time, lipid oxidation also contributes to the complex miso flavors and aroma.[9]

MISO: MICROBIOLOGICAL ISSUES AND METHODS IN BIOTECHNOLOGY TO OVERCOME FERMENTATION PROBLEMS

Since miso fermentation and ripening are traditionally carried out in an open tank, contamination by undesirable yeasts and bacteria often lead to failed fermentations or deterioration in the quality of miso. Spoilage yeasts and spores of *Bacillus* are leading causes of problems in miso fermentations. Osmophilic yeast spoilage is associated with filamentous white material on the surface of the fermentation tank accompanied by undesirable odor. *Bacillus* may also inhibit the growth of *A. oryzae* leading to stuck fermentations. The presence of a high concentration of salt, i.e., more than 10% (w/w), may also trigger the transformation of *Bacillus* cells into spores that can survive the fermentation and ripening processes[10] and result in eventual spoilage of packaged miso.

The use of starter cultures by commercial manufacturers has largely overcome problems associated with spoilage yeasts and bacteria. Starter culture usage results in a predictable fermentation with reduced fermentation times, improved antagonism against pathogens, and effective maintenance of the quality and consistency of miso. A previous study demonstrated that use of a pure culture of *Z. rouxii* NRRL Y-2547 as the primary microorganism in the fermentation resulted in high quality miso within a short period of about 75 days.[3] Utilization of starter lactic acid bacteria along with the *A. oryzae* in miso fermentation has also been investigated as a method to overcome problems associated with undesirable microorganisms that are commonly encountered during the production of low-salt miso. Kato et al.[11] demonstrated that a coculture of a nisin-producing strain of *Lactococcus lactis* along with *A. oryzae* had an inhibitory effect on contaminating *Bacillus* strains during miso fermentation without affecting the growth of *A. oryzae*. In another study, *Lactococcus* sp. GM005 isolated from miso was found to produce a bacteriocin with strong antibacterial activity.[12]

CONCLUSION

The entry has discussed the scientific principles involved in miso production with specific emphasis on the microbiology and chemistry of the manufacturing process. The discussion included mention of microbiological problems encountered during miso fermentation and describes biotechnological approaches in starter culture selection and coculture usage to overcome fermentation problems. While current research on miso is largely in the health domain, many aspects of the microbiology and chemistry of the miso fermentation and ripening processes remain less understood.

The growth in Japanese cuisine in the United States has triggered interest in miso and fermented soybean products and has also led to the product becoming commonly available. The demand for miso has also prompted large Japanese miso producers to set up production facilities in the United States to cater to the increasing local demands. Increased consumer interest has also resulted in home-based hobby miso producers. Research advances in miso starter cultures and fermentation/ripening process optimizations have the potential to spawn entrepreneurial initiatives to produce miso and other soy-based fermented food products in the United States.

REFERENCES

1. Sujaya, I.N.; Tamura, Y.; Tanaka, T.; Yamaki, T.; Ikeda, T.; Kikushima, N.; Yata, H.; Yokota, A.; Asano, K.; Tomita, F. Development of internal transcribed spacer regions amplification restriction fragment length polymorphism – method and its application in monitoring the population of *Zygosaccharomyces rouxii* M2 in Miso fermentation. J. Biosci. Bioeng. **2003**, *96* (5), 438–447.
2. United States Department of Agriculture, Agricultural Research Service. USDA National Nutrient Database for Standard Reference, Release 20. **2007**. http://www.ars.usda.gov/ba/bhnrc/ndl (accessed January 2009).
3. Hesseltine, C.W.; Shibasaki, K. Miso. III. Pure culture fermentation with *Saccharomyces rouxii*. Appl. Environ. Microbiol. **1961**, *9* (6), 515–518.
4. Beuchat, L.R. Traditional fermented food products. In *Food and Beverage Mycology*, 2nd Ed.; Springer: New York, **1987**; 269–302.
5. Yamabe, S.; Kaneko, K.; Inoue, H.; Takita, T. Maturation of fermented rice-koji Miso can be monitored by an increase in fatty acid ethyl ester. Biosci. Biotechnol. Biochem. **2004**, *68* (1), 250–252.
6. Ogasawara, M.; Yamada, Y.; Egi, M. Taste enhancer from the long-term ripening of Miso (soybean paste). J. Food Chem. **2006**, *99*, 736–741.
7. Slaughter, C. The naturally occurring furanones: formation and function from pheromone to food. Biol. Rev. **1999**, *74*, 259–276.
8. Sugawara, E.; Sakurai, Y. Effect of media constituents on the formation by halophilic yeast of the 2 (or 5)-ethyl-5 (or 2)-methyl-4-hydroxy-3 (2H)-furanone aroma component specific to Miso. Biosci. Biotechnol. Biochem. **1999**, *63* (4), 749–752.
9. Steinkraus, K.H. Industrialization of Japanese Miso fermentation. In *Industrialization of Indigenous Fermented Foods, Revised and Expanded*, 2nd Ed.; Food Science and Technology; CRC Press: New York, 2004; Vol. 136, 105–127.
10. Steinkraus, K.H. Indigenous amino acid/peptide sauces and pastes. In *Handbook of Indigenous Fermented Foods Revised and Expanded*, 2nd Ed.; Food Science and Technology; CRC Press: New York, 1995; Vol. 73, 509–654.
11. Kato, T.; Inuzuka, L.; Kondo, M.; Matsuda, T. Growth of Nisin-producing Lactococci in cooked rice supplemented with soybean extract and its application to inhibition of *Bacillus subtilis* in rice Miso. Biosci. Biotechnol. Biochem. **2001**, *65* (2), 330–337.
12. Onda, T.; Yanagida, F.; Tsuji, M.; Shinohara, T.; Yokotsuka, K. Production and purification of a bacteriocin peptide produced by *Lactococcus* sp. strain GM005, isolated from Miso-paste. Int. J. Food Microbiol. **2003**, *87*, 153–159.

Nanoscale Biology: Engineering Applications

Kaustubh Bhalerao
Goutam Nistala
Department of Agricultural and Biological Engineering, University of Illinois, Urbana, Illinois, U.S.A.

Abstract

Nanotechnology is the development of technology and applications based on the understanding of nanoscale processes. Biological nanotechnology (bionanotechnology/nanobiotechnology) is a significant subset of nanotechnology. It refers to tools and techniques used in precisely manipulating biological macromolecules to control the systems and processes they participate in, or create entirely new devices and systems whose functionality depends upon the biological properties of the macromolecules. Biological nanotechnology borrows techniques from molecular biology bioconjugation and synthetic chemistry. It is an extension to biotechnology with similar motivation: to improve healthcare, increase agricultural and industrial productivity, enable effective environmental stewardship, and of late, its applicability in biosafety and biosecurity concerns. This entry is organized into four sections covering biological components, most common tools and techniques, a few design principles unique to biological engineering and finally some examples that illustrate well-engineered applications.

INTRODUCTION

One nanometer (nm) is one billionth of a meter. By conventional definition, the "nano" prefix is used for size scales between 1 and 100 nm. Advances in our ability to visualize and understand nanoscale biomolecular phenomena and objects such as DNA and proteins drive the paradigm today in which we describe nature itself.

Nature provides a veritable array of nanoscale biological macromolecules such as nucleic acids, proteins, lipids, and other smaller molecules. Components from this vast library can be connected together in various novel configurations to realize systems and devices not known to exist currently in nature. The tools to manipulate these nanoscale building blocks of life must also be correspondingly nanoscaled. The prefabricated functionality of biological macromolecules provides both the object and the tool for biological nanotechnology. Biological nanotechnology not only provides solutions to problems in health, industry, agriculture, and the environment today, but also extends to various domains such as energy production and electronics that will enable applications of tomorrow. Engineering with biological macromolecules involves organizing their ready-made functionality into higher order devices and systems.[1] This notion of collocated functionality uses tools, techniques, and design philosophies borrowed from a variety of engineering and scientific disciplines. This document is, therefore, organized into four major sections, Components, Engineering Techniques, Design Philosophy, and Case Studies and attempts to provide a bird's-eye view of biological nanotechnology as it stands today.

NANOSCALE BIOLOGICAL COMPONENTS

Biology has been traditionally motivated by the discovery of fundamental processes that describe living processes in their native operational diseased states. The native function of biological macromolecules inspires applications for that molecule outside its native environment. (e.g., using an enzyme from fungi in a bioreactor process). To engineer a system with a biomolecule, it is critical to know the effect of the environment on its behavior (e.g., the kinetics of the enzyme under non-physiological pH levels).

Protein

Proteins form the building blocks of life. All living processes involve some participation from proteins. From subcellular structural molecules such as actin to enzymes that break down complex polymers to protein complexes that produce new proteins from DNA, nature provides a vast library of diverse functional proteins. Estimates in the human genome alone for the number of proteins vary from as few as around 26,000 to as many as 150,000. It is no coincidence that proteins are a major resource for components that provide prefabricated functionality to act directly on a biological system in highly diverse capacities.

Enzymes are an important group of nanocomponents that are employed in numerous scientific, industrial, and even day-to-day household environments. Enzymes are protein catalysts that have very specific actions on their substrates. These activities are governed thermodynamically by the immediate environment surrounding the enzyme and are amenable to some amount of engineering. The two commonly used strategies in modifying enzyme kinetics are to change the structure of the enzyme slightly to alter its function, or to modify the environment to effect changes in its thermodynamic characteristics.

Nucleic Acids

Ribonucleic (RNA) and deoxyribonucleic acids (DNA) are a linear repository of the three-dimensional structure of the proteins and their kinetics. DNA, in particular, is an extremely stable molecule. All the information on the DNA is encoded in a sequence of four bases, adenine, thymine, cytosine, and guanine. A single DNA strand is a polymeric chain of these four bases linked to each other on a sugar (ribose) and phosphate backbone. In the double-stranded structure of DNA, adenine is always found paired with thymine and cytosine with guanine by means of two and three hydrogen bonds, respectively. The exquisitely specific adenine–thymine and cytosine–guanine affinities create a redundancy in sequence information on the DNA strand that allows effective reproduction of not only the molecule itself but also the organism as a whole.

DNA-based nanosystems can be broadly defined into two categories: technologies that exploit the physical properties of DNA and those that harness its biological properties. For example, DNA is a negatively charged molecule on account of its phosphate backbone;[2] DNA strands have been used as a template for electrochemical deposition of metallic nanoparticles to serve as nanowire conductors. This exploits the negative charge of the DNA.

At room temperature, double-stranded DNA is usually in the double helical form, stabilized by hydrogen bonds and cations in solution. This is known as the hybridized state. As the temperature is increased, the hydrogen bonds break and the DNA can be dehybridized or "melted." If the DNA sequence is known, it is possible to predict the melting temperature of the double strand. A more complex engineered system is the DNA walker that makes use of the biochemical specificities among the base pairs to impart a walking motion to an engineered DNA strand relative to another DNA strand.[3]

Lipids

Lipid layers form the major structural component of cell membranes. They perform the important task of maintaining a potential difference between the inside and the outside of the cell by controlling the flow of various charged species (mostly ions) across the membrane. Phospholipids, in particular, have an interesting structure. They have a polar head, which is hydrophilic, and a non-polar tail, which makes it hydrophobic. The amphiphilic nature of the molecule allows it to self-assemble in different configurations in polar and non-polar solvents. One such structure is the liposome, a spontaneously self-assembled spherical structure where all the polar molecules face one direction either inward or outward forming a sphere.[4] Droplets of oil, for example, can be contained in liposomes, where the non-polar tails face inward. Such structures have their polar tails facing outward, allowing the liposomes to easily solubilize in water and act as a carrier device for their non-polar cargo.

ENGINEERING TECHNIQUES AND SYSTEM INTEGRATION

Bioconjugation

Bioconjugation provides the nuts and bolts to combine different nanoscale components to realize devices and systems manifesting the aforementioned built systems. Bioconjugation techniques rely on chemical bonding as well as biological self-assembly due to hydrogen bonding as seen in DNA base complimentarity, antigen–antibody affinity, protein–small molecule interactions, etc. Bioconjugation is used to impart a biochemical specificity to artificial systems such as sensors and biomolecular purification systems. This imparts a direct biological relevance to the artificial systems. The process in such cases is often referred to as surface functionalization. Biosensor surfaces are functionalized with specific antibodies that can bind to tiny amounts of their respective antigens allowing the biosensor to recognize the presence of the antigen through a suitable transducer. Such systems are becoming an important tool in disease, pathogen, and environmental contaminant diagnostics. The non-covalent bond formed between biotin and avidin is a popular bioconjugation technique. Biotinylating (conjugating biotin onto a protein) alters the affinity of the protein to avidin. This affinity is exploited in affinity chromatography, which has been used to separate proteins from a mixture of other proteins and chemicals.[5]

Bioconjugation can also be used to alter the physical or biological properties of the target biomolecule. For example, pegylation, wrapping up a protein in PolyEthylene Glycol (PEG) reduces the access of various degradation enzymes to the protein, modifying the kinetics with which it degrades in an environment where the enzymes are present. It also aids in increasing the solubility of the molecule through the same mechanism as liposomes, by providing an external hydrophilic sheath for the molecule.

The class of linkers commonly used to attach organic molecules to inorganic/metallic substrates is known as organometallic linkers.[6] Examples of such linkers include thiol bonding (the dative bonding between gold and sulfur),

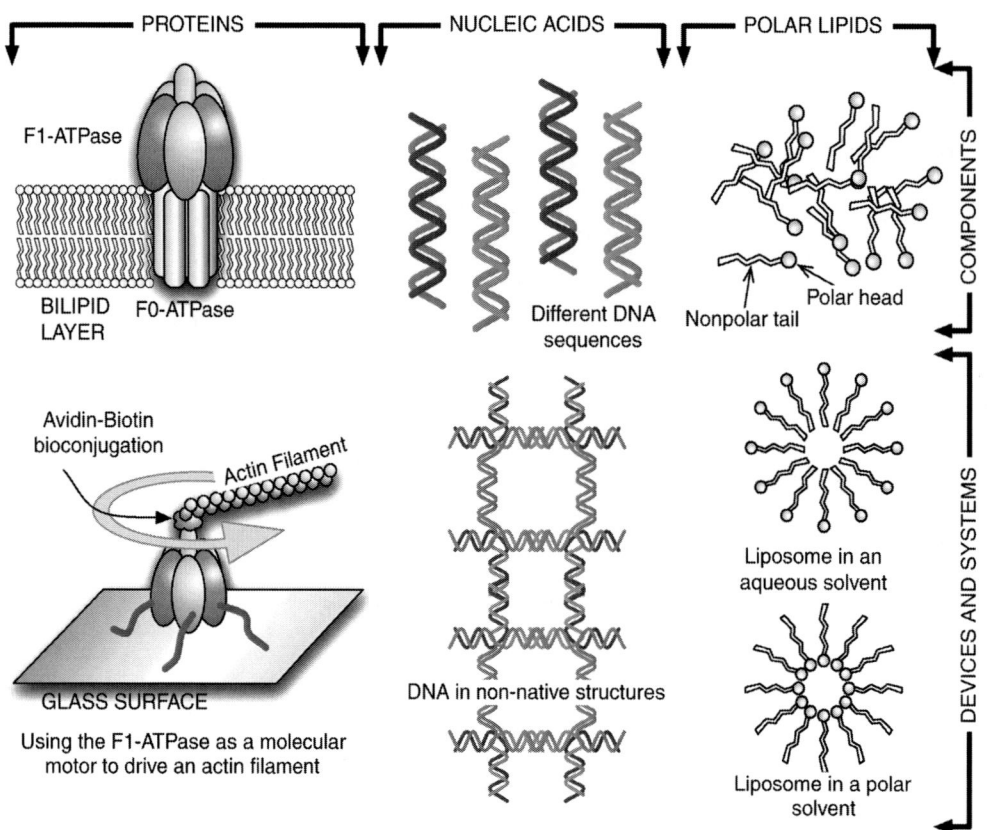

Fig. 1 Nanodevices made with biomolecular components. The F1 domain of the ATPase can be extracted and attached to a glass surface,[7] double helical structures of DNA can be reconstructed into non-native woven structures[3] and phospholipids can be assembled into liposomes under different solvent conditions.

silanization (covalently attaching molecules to oxide layers on the surface of silicon or titanium through a silane group), or phosphonic acid linkages (phosphorous–oxygen linkages). These linkers are usually hetero-bifunctional, i.e., they have two different functions (binding to organic moieties and binding to inorganic surfaces) on the same molecule. Fig. 1 shows a sample of different biomolecular components and the kinds of devices they can be assembled into.

Molecular Biotechnology

Where bioconjugation provides the tools to organize biological function into man-made constructs and systems, tools from molecular biotechnology allow us to isolate, manipulate, reproduce, and integrate biological components within biological systems. Molecular biotechnology provides the primary interface for dealing with nanoscale biology. Recombinant DNA technology in the last few decades has been the cornerstone of biotechnology. The various unit processes used in recombinant genetic engineering offer highly specific techniques to manipulate DNA using various enzymes. The DNA-based nanotechnologies mentioned here have been made possible using one or more of the following techniques:[8]

- *Isolation of specific genes*: Certain enzymes known as restriction endonucleases recognize specific nucleotide sequences in DNA and cut the strands of DNA along specific patterns in the neighborhood of the recognized sequences. A large number of different restriction enzymes have been identified that produce cuts in DNA along various nucleotide sequences.
- *Separation of fragments*: When DNA is placed in a strong electric field, it migrates towards the anode owing to its negative charge. When a mixture of fragmented DNA is placed in a viscous gel and subjected to an electric field, different fragments migrate at different rates. The larger fragments have more negative charges and experience a proportionally greater motive force, allowing spatial separation of the different fragments based on their size. This is known as gel electrophoresis.
- *Determination of sequence*: The nucleotide order of a DNA can be determined by traditional chemical sequencing or the enzymatic/chain termination sequencing methods that make use of gels, and are limited in their capabilities. New technologies for DNA sequencing make use of novel techniques like DNA capture beads and chemiluminescent enzymatic reactions.

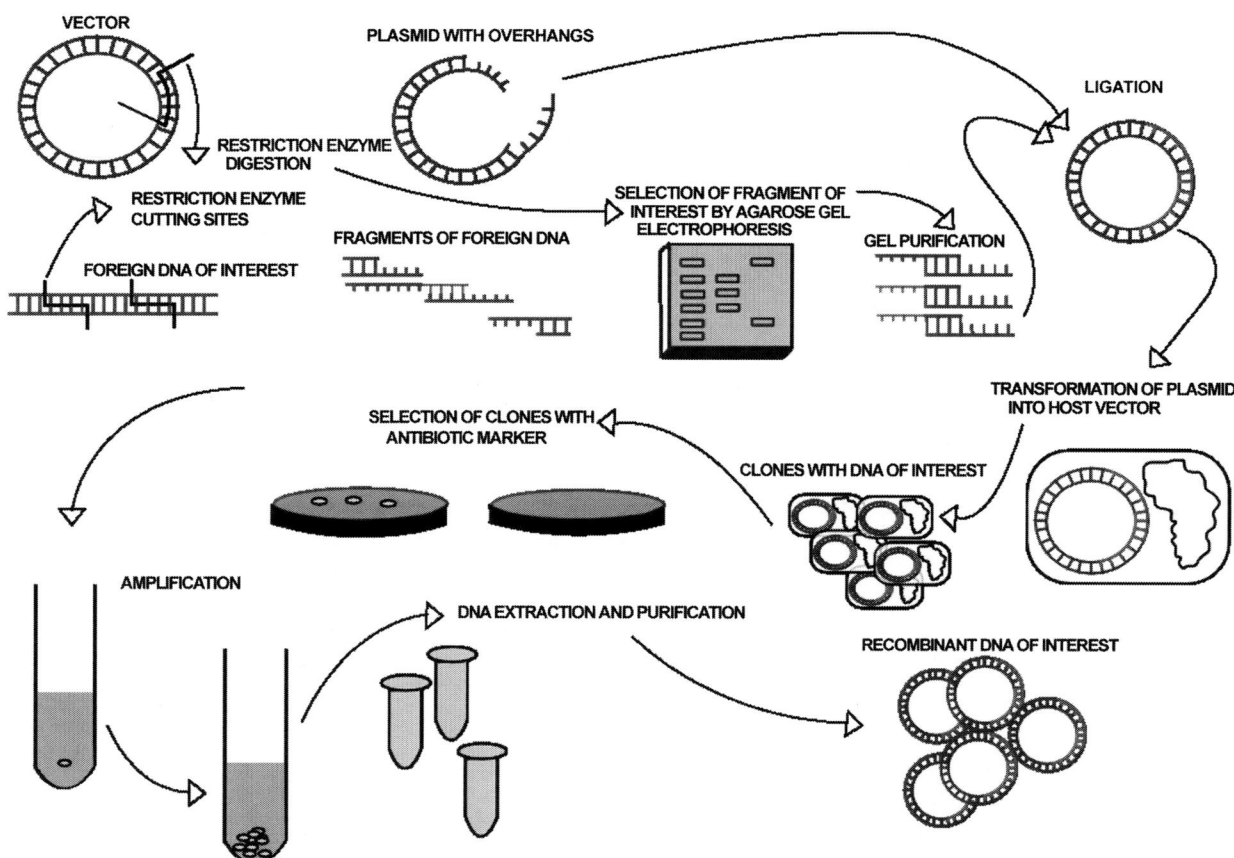

Fig. 2 Production of recombined DNA. A visual representation of extracting DNA from an organism, ligating it into a vector and reproducing it in bacteria.

- *Amplification*: The polymerase chain reaction is a process to amplify specific DNA sequences. Amplification of DNA involves melting it by heating it up, using "primers" to provide a starting location for DNA replicating enzymes known as DNA polymerases and allowing the polymerases to create two double strands where there was only one. This process is repeated multiple times to create a large number of copies of the desired DNA. The natural process of reproducing DNA is replicated in a thermal cycler and thermostable polymerase enzymes that can tolerate up to 95°C temperatures seen by the DNA in the process. PCR replicates the process of DNA replication in a thermostable polymerase enzyme to amplify any nucleotide sequence with modified ends to create huge number of copies of the oligonucleotide.
- *Creation of a new synthetic molecule*: The amplified gene is inserted into the carrier/vector such as a virus or a plasmid by using a ligase, another enzyme used in repairing broken DNA strands. This vector DNA can then be introduced/transfected into the cell by various means that include heat shock, electroporation, and chemical reagents. The gene of interest is now incorporated into the host. It can be reproduced along with the host itself and if everything works correctly, the function of the gene is imported into the host.

A visual representation of the process of DNA recombination is shown in Fig. 2. The tools of molecular biology have now been in existence for a few decades and have been used extensively to extract interesting genes from organisms and insert them into new ones. They allow an efficient way of reproducing fully functional biomolecular components, which can then be extracted and coupled with other devices.

Directed evolution

The theory of evolution depends on two related biological phenomena—random mutations occur in genes leading to a modification in their function, and the survival of the organism depends on its fitness to survive in its environment. Directed evolution is a powerful technology that exploits evolutionary phenomena to select for new proteins. For example, to optimize a known enzyme to operate more efficiently, the survival of the organism is tied to the action of the enzyme in a harsh environment. A colony of organisms is placed in an environment where efficient function of the enzyme is critical to the organism surviving. Only those organisms that have enzyme mutations that make them resilient to the environment adapt and survive.

After allowing the organisms to reproduce in the environment and subjecting them to progressively harsher environments selects for the organisms carrying the most efficient gene responsible for the survival of the organism.

Phage display is a variation on the directed evolution theme. Here, a library of bacteriophages with slightly different coat protein peptides is exposed to a substrate. Those phages with a coat protein peptide that bears affinity to the substrate are retained on the substrate, the others are washed away. The adhered phages are then eluted and allowed to transfect *Escherichia coli*. They are allowed to reproduce in the bacteria, are extracted, and the process is repeated. The procedure used to wash away the loosely bound/unbound phages is made progressively more stringent. Eventually, this procedure selects for phages that display a specific coat protein that binds to the desired substrate with a high degree of specificity.

Technologies from the Physico-Chemical Sciences

There are numerous other technologies from the physical sciences that have been fruitfully applied to manipulating biomolecules. An important tool in the development of bionanotechnology has been the atomic force microscope (AFM), which allows atomic level resolution in both visualizing and manipulating biomolecules. The AFM along with its technological variation, dip pen nanolithography (DPN),[9] has made possible true molecule-by-molecule assembly of materials from the bottom up. Other tools that allow single molecule level interrogation of biomolecular processes include the surface plasmon resonance (SPR) spectroscopy and nanopore-based devices to separate in sequence linear macromolecules like DNA.

DESIGN PHILOSOPHY

Every engineering discipline has its own philosophy for design. Biological molecules significantly constrain the space for design solutions because of their stringent environmental needs. Biomolecules have evolved to function in their native environment, under physiological pH, temperature, and other environmental conditions. The relationship between functionality and performance drives the design process: either the environment drives the choice of the biological component, or the biological component dictates the environment it operates in. Both routes require different knowledge bases, making design a little more challenging. An example of choosing a suitable biomolecule for a given environment is using the thermostable DNA polymerase from hydrothermal vent bacteria that can withstand the 95°C temperatures in a PCR thermal cycler rather from organisms that live at room temperatures. Directed evolution also provides a route to obtaining biomolecules that work at non-native environment.

The identification of suitable molecules based on the environment requires more familiarity with biology and ecology than does engineering the operational environment. As a result, this design approach depends on the intuition and experience of the biologist. The barrier to design has to do with the way biological information is organized in the bioinformatics framework. Traditionally, biomolecules have been organized based on their structural and their evolutionary relationships. While useful in determining ancestral relationships between organisms, the utility of these databases is limiting in biological nanodevice design, where the choice of the components is driven by function rather than structure.

Efforts like the Gene Ontology Consortium have made great strides to organize biomolecular components based on their function, the biological process, or the cellular compartment. Fields such as functional genomics and metabolic engineering complement this in terms of reaction kinetics under different environments. What is needed, however, is a resource to simulate and predict the behavior of multicomponent biological devices in a chosen environment.[10] Until then, the design of biological nanodevices will remain an iterative experimental exercise.

CASE STUDIES

Applications in bionanotechnology are inherently expensive to develop, requiring personnel with diverse and highly specialized knowledge bases. This cost has meant that nanotechnology, in general, has been explored in applications where solutions to problems are intractable by any other means. Until now, this has meant applications in complex and life-threatening diseases such as cancer, cardiovascular, and immune system diseases. However, as more information becomes available, the costs of finding nanobiotechnology solutions are decreasing rapidly, extending the scope of applications in a broader context.

Device Applications—Biosensors

In the area of food and agriculture, the problem of sensing is perhaps the most appropriate to develop biological nanotechnology applications. Biosensors are defined as devices in which the sensing element consists of some components of biological origin. Typically, the device is used for sensing a biological analyte, e.g., an antibody-based pathogen detector. However, biosensors may be designed to sense an analyte that is not necessarily biological in nature, e.g., chemicals in the environment. In any case, however, the biological component of the sensor is chosen to provide biological relevance to the measurement.

A biosensor has three functional layers. The topmost recognition layer, as the name implies, recognizes the analyte from the sample. The recognition layer affects an energy change on the transducer layer, which can then in a sequence of events encode this information onto an electrical signal. The final representation layer converts the information into a human interpretable indication.

Consider the example of a field effect transistor (FET)[11] based biosensor to detect a signature protein produced by a pathogen. An antibody is bioconjugated onto the gate of the FET. The specific affinity of the antibody to the protein antigen provides the biological recognition to the measurement. Since proteins are charged molecules, the charge on the FET gate changes, which then alters the current flowing through the device. This is the transducer. The numerical value of the current can then be interpreted to reflect the presence or absence of the antigen, and indirectly the pathogen.

The FET-based biosensor is an integrated hybrid device that is only partly biological in nature. Whole cell biosensors, on the other hand, are complete organisms that have the recognition, transduction, and the representation process entirely contained within the organism itself.[12] A whole-cell biosensor, also referred to as a bioreporter, naturally responds to an analyte through changes in gene expression. This response is usually not detectable in a straightforward manner. The transducer is a genetically engineered circuit that produces a visible (or otherwise detectable) signal such as a fluorescent protein. The presence of the fluorescence is an alias for the analyte of interest. While the development cost can be substantial, the reproduction cost for whole-cell biosensors can be vanishingly small. Whole-cell biosensors are being developed for use in spatially restricted environments, as rapid detection schemes before time-consuming chemical analyses, for portability and miniaturization purposes and largely due to their high sensitivity. Luminescent *E. coli*-based bioreporter strains have been reported that can detect concentrations of mercury as less as 0.1 fM.[12] However, most whole cell biosensors that are developed currently are in academic research and are yet to be optimized for commercial deployment.

SUMMARY

Biological nanotechnology is a fledgeling new science with promising solutions for many problems today. It is also an enabling field like any other branch of engineering, which will lead to unprecedented technologies, only a few of which we can foresee today. The pedagogy around biological nanotechnology will be created for a long time still. However, biological nanotechnology is its own discipline, which builds upon engineering and the life sciences, with its own class of components, tools and measurement devices, design philosophy, and information frameworks. The applications of biological nanotechnology in food and agricultural engineering are still in the research stages, with many exciting possibilities in the realm of biosensing. As our understanding of nanoscale phenomena grows, so will our ability to use it effectively in the context of improving health, agricultural and industrial productivity, and the environment.

REFERENCES

1. Merkle, R.C. Biotechnology as a route to nanotechnology. Trends Biotechnol. **1999**, *17*, 271–274.
2. Randall, M.S.; Woolley, A.T. DNA-templated nanowire fabrication. Biomed. Microdevices. **2004**, *6* (2), 105–111.
3. Seeman, N.C. DNA enables nanoscale control of the structure of matter. Q. Rev. Biophys. **2005**, *38* (4), 363–371.
4. Goodsell, D.S. *Bionanotechnology: Lessons from Nature*, Wiley-Liss Inc. Hoboken, NJ **2004**.
5. Hermanson, G.T. *Bioconjugate Techniques*, Academic PressSan Diego, CA **1996**.
6. Schwartz, J.; Bernasek, S.L. Organometallic chemistry at the interface with materials science. Catal. Today. **2001**, *66*, 3–12.
7. Soong, R.K.; Bachand, G.D.; Neves, H.P.; Olkhovets, A.G.; Craighead, H.G.; Montemagno, C.D. Powering an inorganic nanodevice with a biomolecular motor. Science. **2000**, *290* (5496), 1555–1558.
8. Glick, B.R.; Pasternak, J.J. *Molecular Biotechnology: Principles and Applications of Recombinant DNA*. 3rd, ASM Press **2003**.
9. Ginger, D.S.; Zhang, H.; Mirkin, C.A. The evolution of dip-pen nanolithography. Angew. Chem. Int. Ed. Engl. **2004**, *43* (1), 30–45.
10. Bhalerao, K.D.; Eteshola, E.; Keener, M.; Lee, S.C. Nanodevice design through the functional abstraction of biological macromolecules. Appl. Phys. Lett. **2005**, *87* (3).
11. Fang, Y.; Offenhaeusser, A. Admet biosensors: Up-to-date issues and strategies. Med. Sci. Monit. **2004**, *10* (12), MT127–MT132.
12. Harms, H.; Wells, M.C.; van der Meer, J.R. Whole-cell living biosensors—Are they ready for environmental application?. Appl. Microbiol. Biotechnol. **2006**, *70* (3), 273–280.

BIBLIOGRAPHY

1. Lee, S.C.; Reugsegger, M.; Barnes, P.D.; Smith, B.R.; Ferrari, M. Therapeutic nanodevices. *The Nanotechnology Handbook*; Springer-Verlag: Heidelberg, Germany, **2004**; 279–322.
2. Sudimack, J.; Lee, R.J. Targeted drug delivery via the folate receptor. Adv. Drug Delivery Rev. **2000**, *41* (26), 147–162.
3. Lee, S.C.; Reugsegger, M.; Ferrari, M. Biomolecules and nanodevices. *The Encyclopedia of Nanoscience and Nanotechnology*; American Scientific Publishers: Stevenson Ranch, CA, **2004**; Vol. 1, 309–327.

Nematodes: Biological Control

Simon Gowen
Department of Agriculture, University of Reading, Reading, U.K.

Abstract

Nematodes are a difficult group of pests to manage because generally they are inhabitants of soil and roots and are not easily influenced by soil treatments or cultural practices. The interest in the exploitation of natural enemies of nematodes has increased in recent years because of the demise of soil fumigants and nematicides through restrictions and withdrawals of registration of some products.

INTRODUCTION

Many pathogens and predators of nematodes are known but few have the necessary characteristics of specificity, mobility, or speed of colonization to have a significant influence on a pest population. Attempts at their commercial exploitation as field treatments have not been successful largely because of their inconsistency. Understanding the subtleties associated with the deployment of biocontrol agents will require considerable research effort. Additionally, the recommended rates of application and the formulation on suitable carriers and nutrient sources poses a problem in practicability and in the interpretation of the biological processes involved.

Contemporary research has shown that natural control does exist and that in certain crop/nematode/pathogen situations nematode populations will decline as they are attacked by components of the soil microflora. Soils where this occurs are known as suppressive, but well-documented examples of naturally occurring suppressiveness to particular nematode pests are uncommon.

During the life of a crop the population densities of many of the serious nematode pests can increase by 1000-fold. Economic damage may result from initial population densities of one nematode per gram of soil. To be effective therefore a biocontrol agent must have an impact on the numbers of nematodes that would invade a host and not simply eliminate the surplus individuals that may never locate or invade a root. This being so, those pathogens and predators that are relatively unspecific (trapping, ingesting, or parasitizing all types of free-living nematodes in soil) may be considered less promising than those that parasitize specific pests.

Significant progress has been made in the recognition and deployment of such microorganisms parasitic on some of the species of sedentary nematodes such as the root-knot nematodes, *Meloidogyne* spp., and some of the cyst nematodes, *Heterodera* spp., and *Globodera* spp.

Root-knot and cyst nematodes produce eggs either in clusters on roots or contained within or attached to the cuticle of the female nematode. Biocontrol agents that prevent these nematodes from reproducing may have more impact from an epidemiological point of view than those that kill the free-living individuals in the soil.

BIOCONTROL AGENTS SPECIFIC TO CERTAIN NEMATODE PESTS

Verticillium chlamydosporium is a facultative, soil-dwelling fungus that parasitizes eggs in egg masses exposed on the root surface. Under the right conditions, such fungi will have a significant effect on nematode populations. The efficacy of *V. chlamydosporium* is partly dependent on its root colonizing ability; this can vary according to the plant host. Skill is required in selecting crops that support and/or increase the root colonization by the fungus but are also less favored hosts of root-knot nematodes. *V. chlamydosporium* may be less effective when it is deployed with plants that are highly susceptible and large galls are produced in response to the nematode infection. In such cases, many egg masses may not be exposed on the root surface and so escape infection.

Paecilomyces lilacinus is another fungus commonly found infecting the eggs of sedentary nematodes such as the root-knot and the cyst nematodes, and, like *V. chlamydosporium* being relatively easy to produce on defined growth media, has good potential for commercial development.

Pasteuria penetrans, an obligate bacterial parasite of root-knot nematodes begins its life cycle on free-living juveniles in the soil. Spores attach to the juveniles as they move in search of host roots. Parasitic development begins after the nematode enters a root and continues in synchrony with that of its host. The nematode eventually is overcome by its parasite; it fails to produce eggs; and its body, filled with the spores of the bacterium, eventually ruptures releasing spores into the soil.

The efficiency of *P. penetrans* as a biocontrol agent of root-knot nematodes depends on the concentrations of spores in the soil, the chances of contact with the juvenile stage, and the specificity of the particular *P. penetrans* population. Commercial success will depend therefore on finding techniques for mass-producing the bacterium and on developing populations with a broad spectrum of pathogenicity.

Other *Pasteuria* species parasitic on some sedentary *(Heterodera)* and migratory *(Pratylenchus)* species have been described.

Biological control agents such as *V. chlamydosporium*, *P. lilacinus*, and *P. penetrans* could provide an adequate replacement for nematicides in some cropping systems but the lack of immediate effects, such as are provided by nematicide or fumigant treatments, is a disadvantage. Protection is normally needed in the early stages of plant growth such as in nursery beds. In this situation, integration with other practices such as nematicides, rotation, solarization, and mulches is necessary.

There are several reports of the successful deployment of these biocontrol agents. Small field plots treated once with *P. penetrans* spores (produced by an in vivo system) caused a decline in numbers of root-knot nematodes and increases in yield over a series of crop cycles using root-knot nematode susceptible crops. In other locations, where treatments with *P. penetrans* were combined with *V. chlamydosporium*, organic manures and grass mulches showed similar declines in nematode populations. These two organisms acted against root-knot nematodes in a complementary fashion. As part of this strategy, root systems containing spore-filled cadavers were deliberately left to disintegrate in the soil after each crop. No field treatments were effective after only one crop indicating that some crop loss must be expected during the development of suppressiveness. *P. penetrans* was also effective when used in combination with a nematicide in permanent beds within a plastic polytunnel. Better control of root-knot nematodes was achieved if the biocontrol agent was combined with other control strategies. With such treatments, beneficial effects may develop over one crop cycle.

The chlamydospores of *V. chlamydosporium* do not have the persistence of the spores of *P. penetrans*, which can remain viable for many years.

NONSPECIFIC BIOCONTROL AGENTS

There is a long history of interest in the fungi that trap nematodes in soil such as species of *Arthrobotrys*. These are commonly found in all soils but despite much research effort the problems of the unreliability of soil applications have not been solved and none have become established as successful commercial products.

There are several rhizosphere colonists that have potential for alleviating nematode damage. The precise mechanisms are not clear. Some produce toxins but others may affect root exudation and thus indirectly the attractiveness of roots to nematodes. Experiments have shown that strains of *Pseudomonas fluorescens* can reduce root invasion by different plant parasites but as with the trapping fungi, poor consistency hinders successful development of these microorganisms as commercial products.

FUTURE PROSPECTS

Recently, the nematicidal (and insecticidal) effects of the toxins produced by the bacteria associated with entomopathogenic nematodes (*Photorhabdus* spp., *Xenorhabdus* spp., and *Pseudomonas oryzihabitans*) have been demonstrated.

Success in the commercial development of biocontrol agents does appear promising with those microorganisms that can be formulated as a standard product with proven reliability; others may have a future as single treatment introductions in the more intensively managed protected cropping systems but commercialization may be difficult.

Research is still needed to develop reliable methods of production, formulation, and application. The challenge is to provide a sufficient duration of protection. Such treatments will need to be part of a package of control measures.

BIBLIOGRAPHY

Aalten, P.M.; Vitour, D.; Blanvillain, D.; Gowen, R.S.; Sutra, L. Effect of Rhizosphere fluorescent *Pseudomonas* strains on plant parasitic nematodes *Radopholus similis* and *meloidogyne* spp. Letters Appl. Microbiol. **1998**, *27*, 357–361.

Bourne, M.; Kerry, B.R.; De Leij, F.A.A.M. The importance of the host plant on the interaction between root-knot nematodes (*Meloidogyne* spp.) and the nematophagous fungus, *Verticillium chlamydosporium* Goddard. Biocon. SciTech. **1996**, *6*, 539–548.

Crump, D.J. A method for assessing the natural control of cyst nematode populations. Nematologica **1987**, *33*, 232–243.

Gowen, S.R.; Bala, G.; Madulu, J.; Mwageni, W.; Trivino, C.T. In *Field Evaluation of Pasteuria penetrans for the Management of Root-Knot Nematodes*, The 1998 Brighton Conference on Pests and Diseases, **1998**; 3, 755–760.

Kerry, B.R.; Jaffee, B.A. Fungi as Biocontrol Agents for Plant Parasitic Nematodes. In *The Mycota IV Environmental and Microbial Relationships*; Wicklow, D.T., Soderstrom, B.E., Eds.; Springer-Verlag: Berlin, Heidelberg, **1997**; 204–218.

Samaliev, H.Y.; Andreoglou, F.I.; Elawad, S.A.; Hague, N.G.M.; Gowen, S.R. The Nematicidal Effects of the Bacteria *Pseudomonas oryzihabitans* and *Xenorhabdus nematophilus* on the Root-Knot Nematode, *Meloidogyne javanica*. Nematology, **2000**, *2*(5), 507–514(8).

Stirling, G.R. Biological Control of Plant Parasitic Nematodes. CAB International: Wallingford, U.K., **1991**; 282.

Tzortzakakis, E.A.; Channer, A.G. de R.; Gowen, S.R.; Ahmed, R. Studies on the potential use of *Pasteuria penetrans* as a biocontrol agent of root-knot nematodes (*Meloidogyne* spp.). Plant Pathol. **1997**, *46*, 44–55.

Tzortzakakis, E.A.; Gowen, S.R. Evaluation of *Pasteuria penetrans* alone and in combination with oxamyl, plant resistance and solarization for control of *Meloidogyne* spp. on vegetables grown in greenhouses in Crete. Crop Prot. **1994**, *13*, 455–462.

Weibelzahl-Fulton, E.; Dickson, D.W.; Whitty, E.B. Suppression of *Meloidogyne incognita* and *M. javanica* by *Pasteuria penetrans* in field soil. J. Nematol. **1996**, *28*, 43–49.

Nicotiana benthamiana: Tobamoviral Vectors Redirect Carotenogenesis

Monto H. Kumagai
XtTremeSignPost, Inc., Davis, California, U.S.A.

Jennifer Lee Busto
Department of Plant and Environmental Protection Sciences, University of Hawaii at Manoa, Honolulu, Hawaii, U.S.A.

Abstract

Transcriptional profiling data for plants transfected with engineered RNA viral vectors is limited. We have used heterologous cDNA potato microarrays developed by The Institute for Genomic Research (TIGR) to study the potential of plant viral vector technology for metabolic engineering in Nicotiana benthamiana (Nb). Using carotenoid biosynthesis as a model pathway, we have shown that gene expression data can be rapidly filtered to gain further understanding of the role of phytoene in the regulation of this pathway. Unexpectedly, overexpressing a single enzyme using tobamoviral vectors can upregulate gene expression at various points in the carotenoid pathway, causing an accumulation of transcripts both upstream and downstream of targeted genes.

INTRODUCTION

Plant viral vector technology has a demonstrated application in the production of pharmaceuticals in transfected plants.[1] In addition, it is purported to be a valuable tool for gene discovery[2] and in the metabolic engineering of existing plant pathways.[3] In transfected animal systems, transcriptional studies are frequently employed to assess the efficacy or toxicity of viral vectors intended for gene therapy applications. However, there is limited information that relates to how engineered RNA viral vectors impact host gene expression in transfected crops.

Microarray studies examine the expression profiles of large subsets of genes in a given tissue under specific physiologic and environmental conditions. In plant systems, expression analyses have been used to discover novel floral fragrance-related genes;[4] genes involved in strawberry flavor;[5] and genes involved in regulation of plant defense responses.[6] Other microarray studies have focused on the diagnostic aspects of plant viral pathogens, seeking to design new techniques for identification purposes.[7] We have found that transcriptional profiling analyses can also be employed to determine how engineered tobamoviral vectors redirect plant pathways, particularly for those whose regulation is not well understood.

TRANSCRIPTIONAL STUDIES IN CAROTENOID BIOSYNTHESIS

One such pathway that has been intensively investigated for nutritional, commercial, and biomedical applications is carotenoid biosynthesis. Historically, the metabolic engineering of enhanced carotenoid content for dietary benefits has involved the protracted development of transgenic plants.[8] Transgenic manipulation of carotenoid biosynthesis pathways has been accomplished by creating plants that silence or overexpress genes encoding rate-limiting or rate-determining steps, using single or multiple genes. Ducreux et al. used microarray studies to study the effects of a bacterial phytoene synthase (*crtB*) overexpression on transcripts in potato tubers. Microarray analyses showed no increase in expression of the major carotenoid biosynthetic genes in transgenic tubers,[8] and that phytoene desaturase is not rate-limiting. Recently, Davuluri et al. reported improved levels of both carotenoids and flavonoids in transgenic tomatoes through an RNAi-mediated suppression of *DET1*, an endogenous, photomorphogenesis regulatory gene. Transcriptional analyses using quantitative real-time (QRT)-PCR showed simultaneous elevation of flux through two independent pathways.[9]

We present here a system that we have developed that rapidly analyzes changes in the levels of carotenogenic transcripts in *Nicotiana benthamiana* (Nb) using *tobamoviral* vectors. Engineered hybrid vectors of *tobacco mosaic virus* (*TMV*) carrying genes in sense or antisense orientations were introduced to plants to cause an overexpression or a cytoplasmic inhibition of endogenous genes encoding leaf carotenogenic enzymes. In the experiments described below, viral vector and transcriptional profiling technologies were combined to filter gene expression data and to gain further understanding of the role of phytoene in the regulation of this pathway. Transcriptional profiling data

VIRAL VECTOR AND TRANSFECTION DESIGN

Plant leaves were transfected with tobamoviral vectors carrying a *crtB* (phytoene synthase) cDNA derived from *Erwinia herbicola* (Fig. 1), or with viral vectors carrying an antisense construct of a gene encoding phytoene desaturase (*pds*) derived from ripening tomato. Both transfections result in an accumulation of phytoene, a colorless compound that has been purported to act as a potential signal in the regulation of the pathway.

The transfection design for the *crtB* transcriptional profiling studies is shown in Fig. 2. *Nb* plants were grown from seed and kept under lights at 25°C. Plants were rub-inoculated at the 6–8 leaf stage with in vitro transcripts (Ambion mMessage mMachine) of viral vector TTU51 CTP *CrtB*-RZ that carries a *crtB* (phytoene synthase) cDNA derived from *Erwinia herbicola* and a ribozyme that eliminates the need for linearization prior to in vitro transcription reactions. Control plants were transfected with TTOSA1 APE pBAD expressing green fluorescent protein (GFP). In knock-down experiments, plants were transfected with viral vector TTO1 PDS-[10] that contains a partial tomato phytoene desaturase cDNA derived from ripening tomato in the antisense orientation that is under the control of the TMV-U1 coat protein subgenomic promoter.

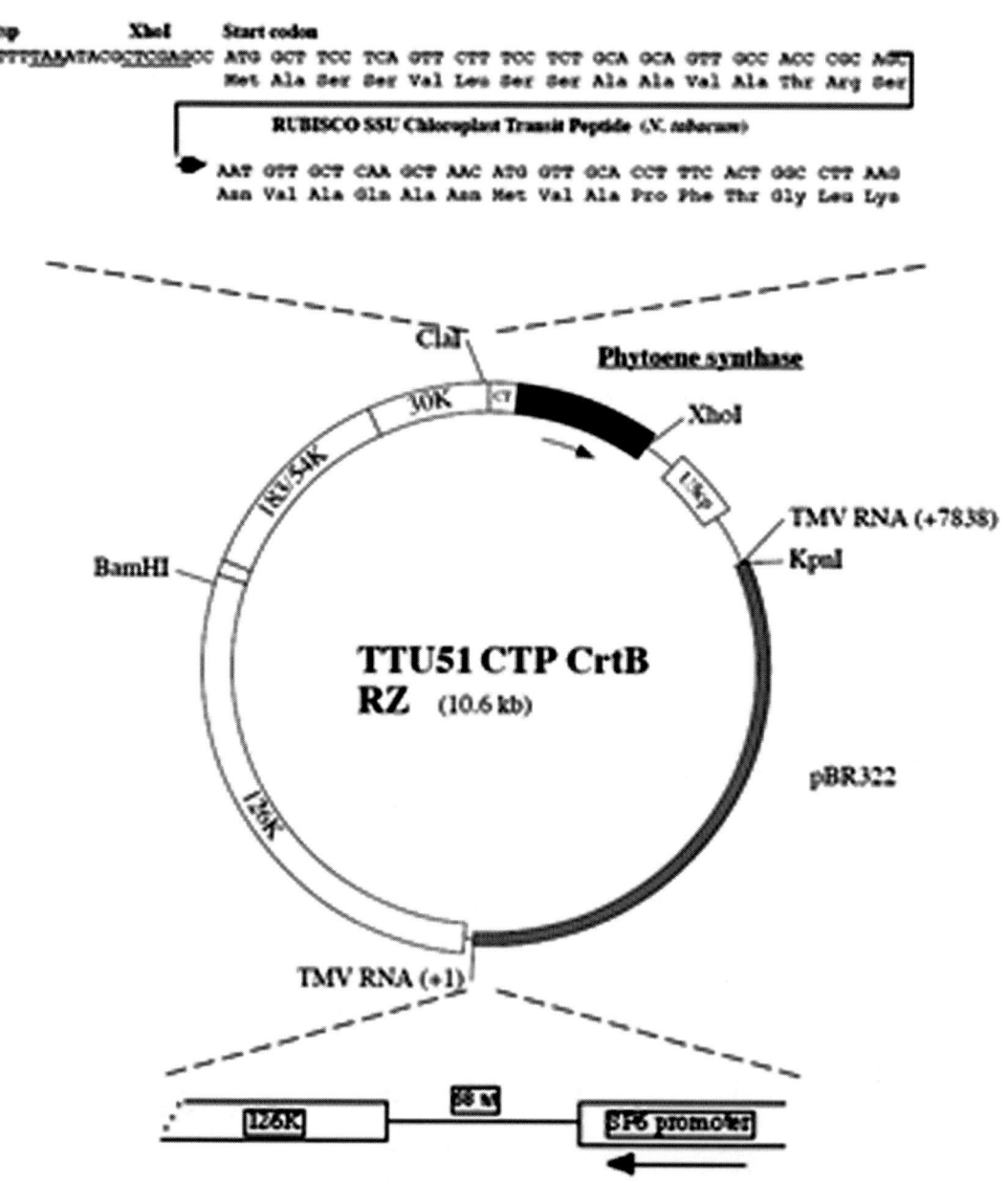

Fig. 1 Plasmid map of hybrid tobamoviral vector, TTU51 CTP *CrtB*-RZ.

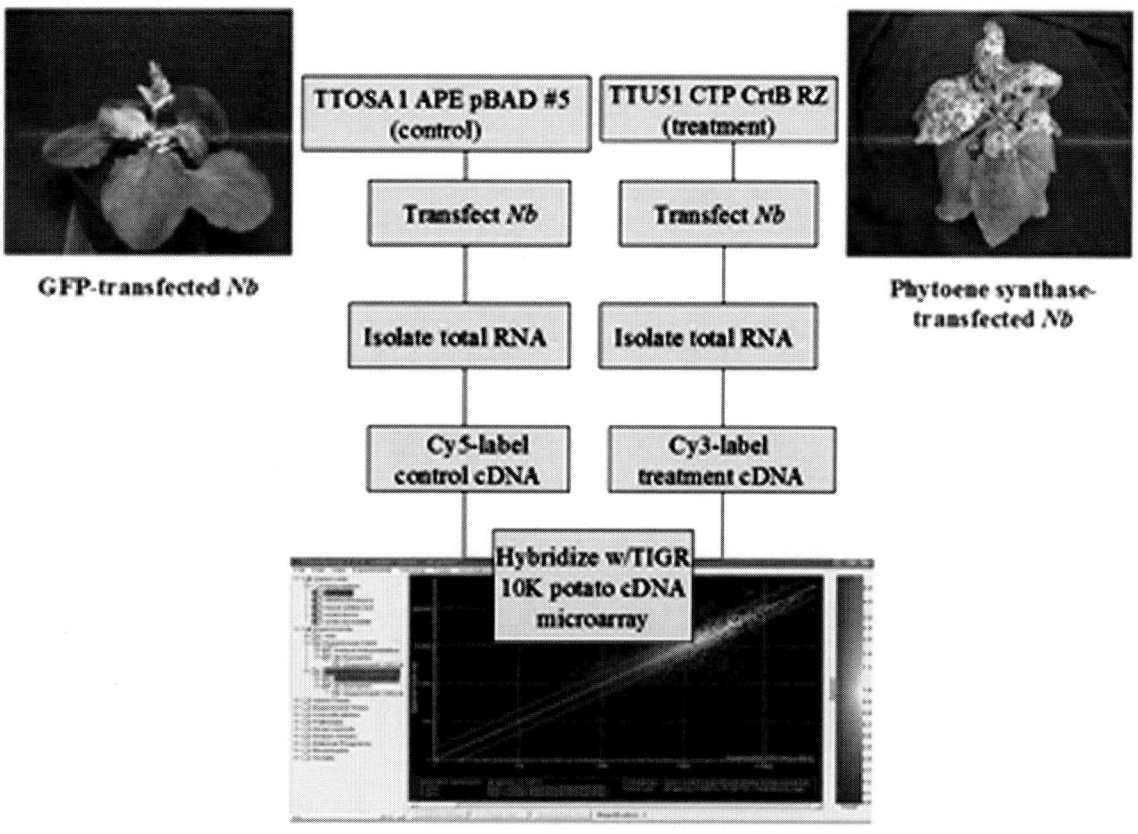

Fig. 2 Plant transfection design for transcriptional profiling of plants overexpressing phytoene synthase.

METHODS

Microarray analysis was performed using leaf RNA that was harvested at 10 dpi from transfected plants, using GFP-transfected plants as controls. Hybridization and post-hybridization washes were performed using recommended conditions found in protocols developed by TIGR. A detailed description of TIGR potato microarray optimized methods, the array design, and the cDNA description file can be found at http://www.tigr.org/tdb/potato/microarray_comp.shtml. Gene lists were generated for twofold upregulated genes using GeneSpring (Agilent, formerly Silicon Genetics) software, and QRT-PCR assays were employed to validate microarray data.[11] Efficiencies were determined for target and reference genes, and Pfaffl's method of calculating fold change ratios was used.

OVEREXPRESSION OF PHYTOENE SYNTHASE

Overexpression of phytoene synthase (*psy*) with the TTU51 CTP *CrtB*-RZ viral vector causes an increase in endogenous transcript accumulation at 10 dpi for genes encoding enzymes that are upstream of *psy* (IPP isomerase) and downstream of *psy* (*pds* and β-carotene hydroxylase).

Microarray data (Fig. 3) shows elevation of enzymes leading to the formation of ζ-carotene and β-carotene, two yellow-colored pigments in the pathway. The approximate 5-fold increase ($p < 0.03$) in expression of *pds* that catalyzes the formation of ζ-carotene and the 12-fold increase ($p < 0.05$) in expression of β-carotene hydroxylase that catalyzes the formation of β-carotene may contribute to the orange-yellow phenotype seen in leaves, stems, flowers, and roots. The use of a heterologous insert derived from *Erwinia herbicola* enables a clear distinction in the interpretation of microarray and QRT-PCR data for impact on endogenous *psy* transcripts, due to low homology of the bacterial sequence to *Nb* and potato. QRT-PCR data confirmed microarray data that the levels of endogenous *psy* *crtB*-transfected plants are also increased at 10 dpi. Pfaffl calculations that take into account efficiencies of amplification for both target and reference genes revealed a twofold accumulation of endogenous *psy* compared to the GFP-transfected controls. In *CrtB*-treated plants, efficiency of amplification of the *psy* target was 95% (correlation coefficient 1.00). Efficiency of amplification of the rubisco reference gene was 98.6% (correlation coefficient 0.999).

All carotenoids are derived from isopentenyl diphosphate (IPP). Isomerization of IPP to dimethylallyl diphosphate (DMAPP) is catalyzed by IPP isomerase, which

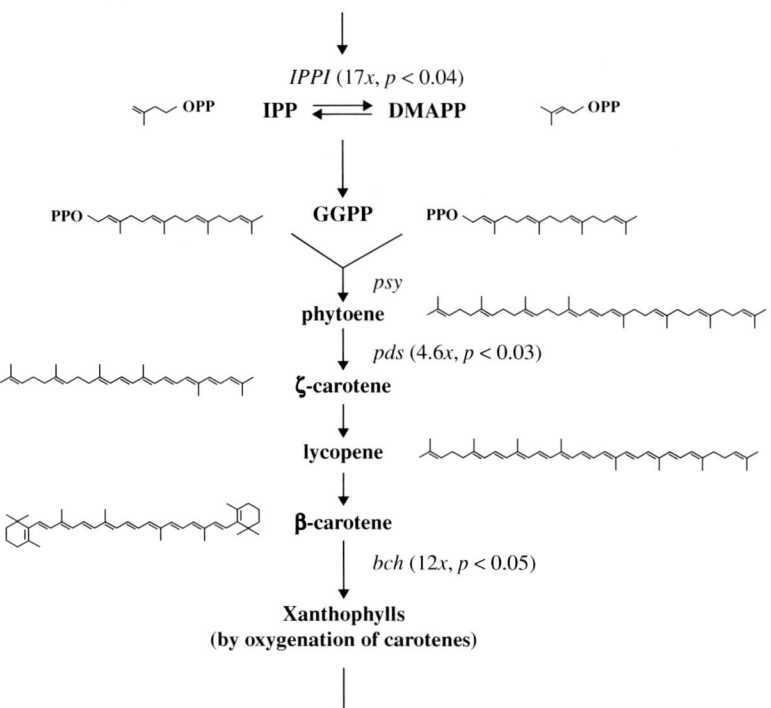

Fig. 3 crtB-transfection of Nb causes an increase in transcript levels of genes encoding enzymes in the plant leaf carotenogenesis pathway (*IPPI*, isopentenyl diphosphate isomerase, 17-fold increase; *pds* phytoene desaturase, 4.6-fold increase; *bch*, beta-carotene hydroxylase, 12-fold increase).

shows a 17-fold increased level of transcript ($p < 0.04$) in the *crtB*-transfected plants compared to the GFP-transfected controls. It is reasonable to suggest that the increase in IPP isomerase upstream of *psy* leads to a greater concentration of DMAPP, the activated substrate for the formation of geranylgeranylpyrophosphate (GGPP). Two molecules of GGPP condense to form 15-*cis* phytoene in a reaction catalyzed by *psy*. Therefore, overexpression of *crtB* appears to cause an increase in phytoene because of increased transcript levels of IPP isomerase, which then drives the pathway to proceed at elevated levels. It appears that phytoene accumulation is a by-product rather than the signal for upregulation of this gene in the pathway.

KNOCK DOWN OF PHYTOENE DESATURASE (*pds*)

At ten days post-inoculation, plants transfected with a partial *pds* antisense construct developed photobleaching in newly formed leaves (Fig. 4) due to cytoplasmic knock down of phytoene desaturase mRNA. *Nb* plants expressing an antisense *pds* accumulate phytoene at a level 51 times that of a non-transfected control.[10] Microarray analysis revealed high levels of virally derived *pds* transcript, suggesting that at this time point the virus is escaping RNA silencing. A differential QRT-PCR analysis (Fig. 4) discerned that although the viral vector *pds* transcript is high, there is a 78-fold decrease in the endogenous *pds* transcript accumulation compared to wild type *N. benthamiana*. Other genes in the pathway did not show significant changes in mRNA levels with the exception of a 5-fold decrease in transcript levels of a putative 9-cis-epoxycarotenoid dioxygenase (NCED), a key regulatory enzyme in abscisic acid biosynthesis. Additional experiments using norflurazon-treated plants that also accumulate phytoene, showed a 5-fold reduction in levels of *pds* transcripts by QRT-PCR (data not shown).

CONCLUSION

Overall, these data suggest that combining viral vector with transcriptional profiling technologies is a useful strategy to

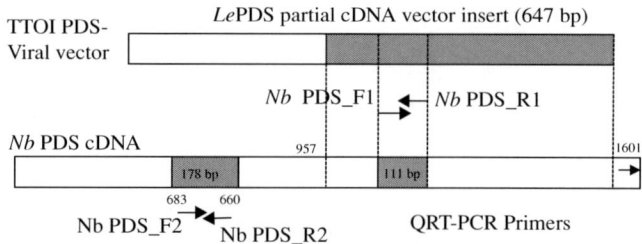

Fig. 4 TaqMan Primer Design. Regions of *pds* cDNA depicting amplification using primer sets that lie within the viral vector insert (QRT primer Set#1, 111 bp amplicon) and outside the viral vector insert (QRT primer Set#2, 178 bp amplicon) to distinguish between viral-derived and endogenous transcripts. Photo insert shows photobleaching effects of *PDS*-(as)-transfection causes of *Nb* leaves (not drawn to scale).

rapidly filter gene expression data and to gain information about plant pathways. Validation of targets using QRT-PCR indicates that heterologous potato cDNA microarrays for the hybridization of related solanaceous species such as *N. benthamiana* are effective tools in providing gene expression data.[12] Manipulation of the carotenoid pathway using *tobamoviral* vectors overexpressing a single enzyme clearly influences accumulation of transcripts both upstream and downstream of targeted enzymes. However, there was no consistent pattern of carotenogenic gene transcript accumulation among the transfected plants or chemical treatments in which phytoene accumulated. In contrast to published reports, this investigation does not support a direct role for phytoene in the regulation of the pathway. However, the data support a potential use of the TTU51 CTP *crtB*-RZ viral vector for metabolic engineering of leaf carotenogenesis in *N. benthamiana* and other solanaceous plants for targeted increase in pathway intermediates.

ACKNOWLEDGMENTS

The authors wish to thank Large Scale Biology Corporation (LSBC); Drs. Stan Kwang and Jon Miyake, at BioRad, for help with primer design, and technical assistance; and the USDA Special Grants Program for Tropical and Subtropical Agriculture, Agreement # 2002-34135-12791. The authors also wish to thank the faculty and staff at the University of Hawaii at Manoa Biotechnology Core Facility for use of the Affymetrix scanner and BioRad iCycler; the faculty and staff at the Hawaii Institute of Marine Biology (HIMB) Core Facility (Coconut Island) for the use of the VersArray ChipReader, and the faculty and staff, especially Dr. Pakieli Kaufusi, at the Research Retrovirology Laboratory (RRL) at Leahi Hospital, Oahu, for expert assistance as well as the use of the Beacon Designer 2.1 primer design software.

REFERENCES

1. Busto, J.L.; Kumagai, M.H. Tobamoviral vectors: developing a production system for pharmaceuticals in transfected plants. In *Encyclopedia of Plant and Crop Science*; Heldman, D.R., Ed.; Marcel Dekker: New York, 2004; 1229–1232.
2. Baulcombe, D.C. Fast forward genetics based on virus-induced gene silencing. Curr. Opin. Plant Biol. **1999**, *2* (2), 109–113.
3. Kumagai, M.H.; Keller, Y.; Bouvier, F.; Clary, D.; Camara, B. Functional integration of non-native carotenoids into chloroplasts by viral-derived expression of capsanthin-capsorubin synthase in *Nicotiana benthamiana*. Plant J. **1998**, *14* (3), 305–315.
4. Guterman, I.; Shalit, M.; Menda, N.; Piestun, D.; Dafny-Yelin, M.; Shalev, G.; Bar, E.; Davydov, O.; Ovadis, M.; Emanuel, M.; Wang, J.; Adam, Z.; Pichersky, E.; Lewinsohn, E.; Zamir, D.; Vainstein, A.; Weiss, D. Rose scent: genomics approach to discovering novel floral fragrance related genes. Plant Cell **2002**, *14* (10), 2325–2338.
5. Aharoni, A.; Giri, A.P.; Verstappen, F.W.; Bertea, C.M.; Sevenier, R.; Sun, Z.; Jongsma, M.A.; Schwab, W.; Bouwmeester, H.J. Gain and loss of fruit flavor compounds produced by wild and cultivated strawberry species. Plant Cell **2004**, *16* (11), 3110–3131.
6. Senthil, G.; Liu, H.; Puram, V.G.; Clark, A.; Stromberg, A.; Goodin, M.M. Specific and common changes in *Nicotiana benthamiana* gene expression in response to infection by enveloped viruses. J. Gen. Virol. **2005**, *86* (Pt 9), 2615–2625.
7. Bystricka, D.; Lenz, O.; Mraz, I.; Piherova, L.; Kmoch, S.; Sip, M. Oligonucleotide-based microarray: a new improvement in microarray detection of plant viruses. J. Virol. Methods **2005**, *128* (1–2), 176–182.
8. Ducreux, L.J.; Morris, W.L.; Hedley, P.E.; Shepherd, T.; Davies, H.V.; Millam, S.; Taylor, M.A. Metabolic engineering of high carotenoid potato tubers containing enhanced levels of beta-carotene and lutein. J. Exp. Bot. **2005**, *56* (409), 81–89.
9. Davuluri, G.R.; van Tuinen, A.; Fraser, P.D.; Manfredonia, A.; Newman, R.; Burgess, D.; Brummel, D.A.; King, S.R.; Palys, J.; Uhlig, J.; Bramley, P.M.; Pennings, H.M.; Bowler, C. Fruit-specific RNAi-mediated suppression of DET1 enhances carotenoid and flavonoid content in tomatoes. Nat. Biotechnol. **2005**, *23* (7), 890–895.
10. Kumagai, M.H.; Donson, J.; della-Cioppa, G.; Harvey, D.; Hanley, K.; Grill, L.K. Cytoplasmic inhibition of carotenoid biosynthesis with virus-derived RNA. Proc. Natl. Acad. Sci. U. S. A. **1995**, *92* (5), 1679–1683.
11. Busto, J.L. *Transcriptional Changes in Nicotiana benthamiana Induced by Tobamoviral Transfection*; Doctoral dissertation, University of Hawaii at Manoa: Honolulu, 2005.
12. Rensink, W.A.; Lee, Y.; Liu, J.; Iobst, S.; Ouyang, S.; Buell, C.R. Comparative analyses of six solanaceous transcriptomes reveal a high degree of sequence conservation and species-specific transcripts. BMC Genomics **2005**, *6*, 124.

Nutraceutical Compounds: Feruloyl Esterases as Biosynthetic Tools

Lisbeth Olsson

Gianni Panagiotou
Center for Microbial Biotechnology, BioCentrum, Technical University of Denmark, Kongens Lyngby, Denmark

Paul Christakopoulos
Evangelos Topakas
BIOtechMASS Unit, Biotechnology Laboratory, School of Chemical Engineering, National Technical University of Athens, Athens, Greece

Roberto Olivares
Department of Chemical and Biological Engineering, Chalmers University of Technology, Gothenburg, Sweden

Abstract

The development of new functional food ingredients is restricted by the limited availability of sufficiently characterized natural extracts; however, it is possible to speed up the discovery process for ingredients with biological activity by developing biocatalytic methods to make large numbers of derivatives of the naturally occurring compounds. The overall aim of combinatorial chemistry and biocatalysis is to make new molecules faster, cheaper, and in numbers large enough for high-throughput screening. In this entry, we discuss the applications of ferulic acid esterases (FAEs) on the development of nutraceutical and pharmaceutical compounds since the prospect of broad applications of FAEs has fueled much interest in these enzymes. Secondly, we provide a list of databases commonly used in protein informatics and discuss aspects of data integration important for capturing all data relevant to functional analysis. With the above tools, using the amino acid sequence for each FAE, it was possible to observe defined motifs that confer specific enzymatic activity.

INTRODUCTION

Biomass represents an unparallel reservoir of chemical structures with demonstrated pharma- or nutraceutical activities and a large source of unexplored chemical entities for promising lead compounds. There has recently been considerable interest in the potential exploitation of agroindustrial waste materials, such as those produced by the milling, brewing, and sugar industries, for the extraction of fine chemicals. Enzymatic synthesis can be used to assemble all possible combinations of this set of chemical building blocks to generate libraries of compounds that will be subsequently screened for useful functionality. Through the use of biocatalysts, otherwise cumbersome synthetic manipulations of complex molecules can be performed in an environmentally benign manner. In addition, many enzymes accept unnatural substrates, and protein engineering can further alter their stability, broaden their substrate specificity, and increase/modify their specific activity. The application of enzymes in synthesis thus represents a remarkable opportunity for the development of chemical and pharmaceutical processes for industry (Fig. 1). In this review, we provide an overview of one of the most promising enzymes of the biosynthetic toolbox, the FAEs [E.C. 3.1.1.73], and we give a short summary of advanced computational methods that are needed for integration, mining, comparative analysis, and functional interpretation of high-throughput proteomic data. It is very challenging to use a comparative approach between the known FAEs for identifying amino acids that are decisive for activity and specificity. In nature, FAEs have large diversification of its catalytic functions and relatively few FAEs have been characterized.

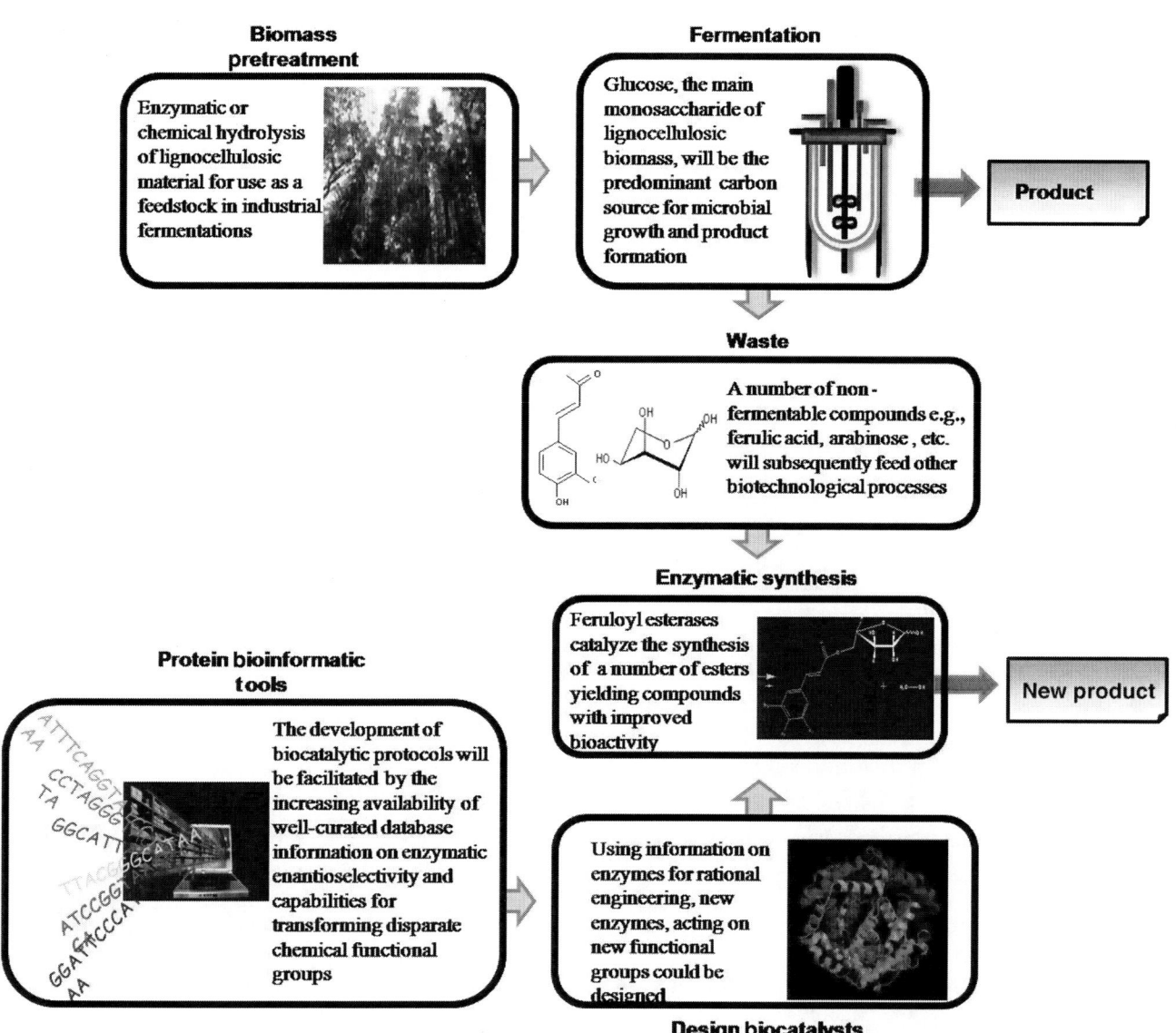

Fig. 1 The figure schematically depicts different sectors of industrial biotechnology and emphasizes the connectivity between biosynthetic and biocatalytic processes for the conversion of raw materials to biotechnological products.

BIOTECHNOLOGICAL APPLICATIONS OF FAEs IN ORGANIC SYNTHESIS

More than 30 FAEs have been purified and characterized from several microorganisms, including fungi and bacteria, showing significant variations in physical characteristics such as molecular weight, isoelectric points, and optimum hydrolytic reaction conditions.[1] Ferulic acid, released from plant cell wall by the action of FAEs, is an effective natural antioxidant with potential applications in the pharmaceutical and food industries. Studies on FAEs' action have shown improved conversion of lignocellulosic biomass to fermentable sugars; however, these enzymes have the ability not only to deconstruct plant biomass, but also to synthesize novel bioactive components (Fig. 2).

Various hydroxycinnamic acids (ferulic, p-coumaric, caffeic, and sinapic) have widespread industrial potential by virtue of their antioxidant properties. Apart from their use as model substrates for FAEs' characterization, methyl esters of hydroxycinnamic acids and their free acids have recently been used as acyl donor in transesterification or direct esterification reactions, respectively, with aliphatic alcohols and carbohydrates. Novel FAEs, purified from various filamentous fungi such as *Fusarium oxysporum*, *Sporotrichum thermophile*, and *Aspergillus niger*, have been used in the synthesis of alkyl ferulates (e.g., 1-propyl ferulate, 1-butyl ferulate, 1-pentyl ferulate, and 1-glyceryl ferulate) and feruloylated -arabino-oligosacharides (with up to six -arabinose units), showing regioselectivity for the primary hydroxyl group of the non-reducing -arabinofuranose ring.

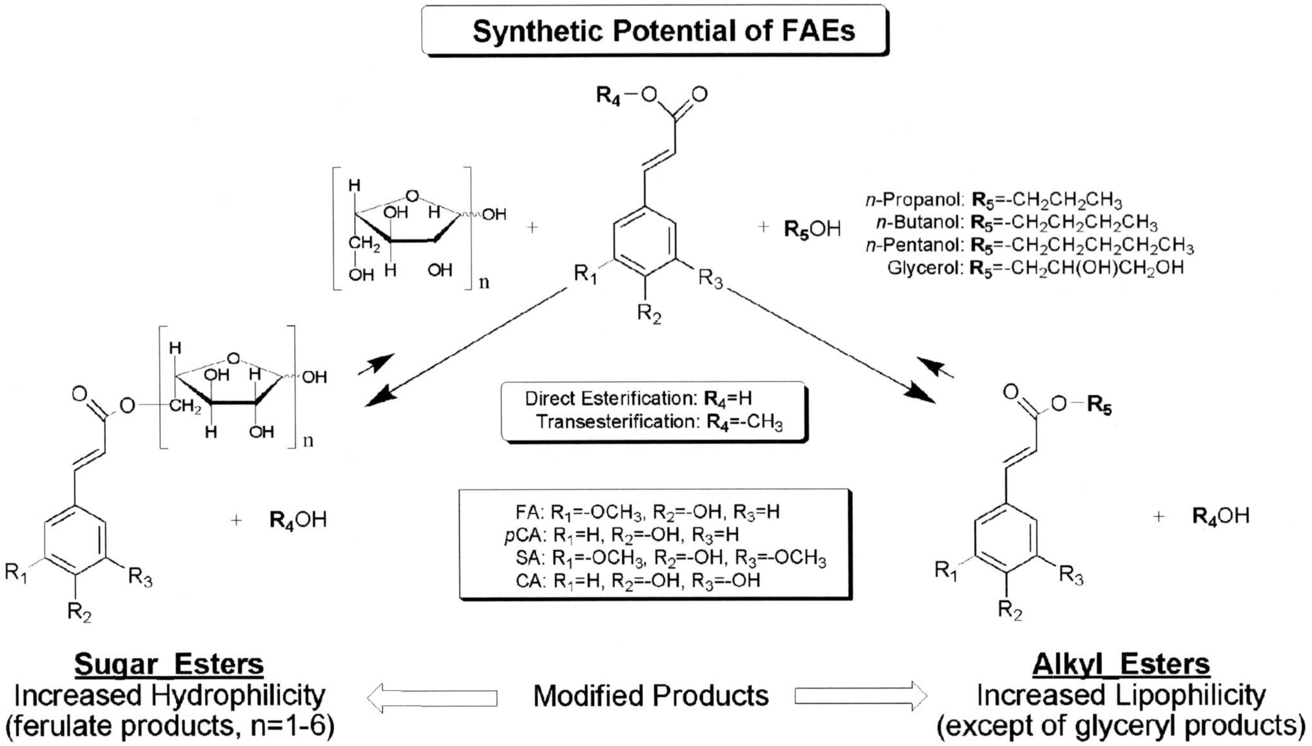

Fig. 2 Modification of hydroxycinnamic acids by FAEs in non-conventional reaction media. Abbreviations: FA, ferulic acid; pCA, p-coumaric acid; SA, sinapic acid; CA, caffeic acid.

The direct esterification of natural phenolic acids, including the above-mentioned hydroxycinnamic acids, with aliphatic alcohols catalyzed by various lipases in organic media has been reported, albeit with low reaction rate and yield. Several authors have demonstrated that electron-donating substituents conjugated to the carboxylic groups in hydroxylated derivatives of cinnamic acids, such as ferulic, p-coumaric, sinapic, and caffeic acid, inhibits lipases strongly.[2] Esterification can be carried out by lipases only if the aromatic ring of the substrate is not para-hydroxylated and the lateral chain is saturated. In contrast, the enzymatic esterification of hydroxycinnamoyl substrates can be obtained efficiently using FAEs as biocatalysts. Enzymatic esterification of phenolic acids with sugars offers an alternative to the poor selectivity of chemical synthesis.

The first synthetic reaction catalyzed by FAE was achieved using a water-in-oil microemulsion system for the synthesis of 1-pentyl ferulate. Detergentless microemulsions[3] have been employed for the synthesis of various alkyl and sugar esters, while in non- or low-aqueous organic solvents various transesterification reactions resulted in the production of various feruloylated glycosides and polyols. Detergentless microemulsions are organic–water mixtures formed in ternary systems consisting of a hydrocarbon, a short-chained alcohol and water, and they represent thermodynamically stable dispersions of aqueous microdroplets in the hydrocarbon solvent. An important advantage of these mixtures as reaction systems is that they allow the problem of separation of reaction products and enzyme reuse to be solved easily, while the solubility of relatively polar phenolic acids is high owing to the presence of large amount of polar alcohol. In this reaction medium, transesterification reactions using activated esters of ferulic acid as acyl donors have been proven to be faster and more efficient compared to the direct esterifications.

A thermophilic FAE from *S. thermophile* showed no chiral selectivity, feruloylating both L- and D-enantiomers of arabinose. The synthesized D-enantiomer has been shown to act as an inhibitor against the growth of *Mycobacterium bovis* BCG.[4] Mycobacterial infections, and most notably tuberculosis, have long been a cause of morbidity and mortality worldwide, and significant effort has been made in the quest for new antimycobacterial agents. With this example, FAEs have been proven to be a novel chemoenzymatic tool for the synthesis of new drugs.

PROTEIN INFORMATICS

This review considers the available and ready-to-use bioinformatics tools that can help end-users to interpret, validate, and generate biological information from their experimental data and gives an example of their utility in the identification of sequence motifs in FAEs. Bioinformatics tools for proteins, also called protein informatics tools, today span a large panel of very diverse applications

Table 1 Protein informatics tools using sequence information only.

Algorithm	Tool	Link
Similarity searching and finding of associated patterns and profiles to identify proteins domains, families, and functional sites. Discriminatory power for functional motifs	PROSITE MOTIF SCAN INTERPRO SCAN BLAST	http://www.expasy.ch/prosite/ http://myhits.isb-sib.ch/cgi-bin/motif_scan http://www.ebi.ac.uk/InterProScan/ http://www.ncbi.nlm.nih.gov/blast/Blast.cgi
Inductive machine learning algorithms. The biological patterns to train the algorithms are extracted from databases.	FindMod AutoMotif TargetP TMpred TopPred	http://www.expasy.org/tools/findmod/ http://ams2.bioinfo.pl/ http://www.cbs.dtu.dk/services/TargetP/ http://www.ch.embnet.org/software/TMPRED_form.html http://bioweb.pasteur.fr/seqanal/interfaces/toppred.html

ranging from simple tools for comparison of protein amino acid compositions to sophisticated software for large-scale protein structure determination.

The computational prediction of protein function is a central undertaking in computational biology. We are being deluged with genomic information from genomic and, more recently, metagenomic processes. When assessing function prediction programs, we should bear in mind that the definition of biological function is highly contextual. Different aspects of biological function of the same protein may be viewed as taking place in different scales of time and space. Considering the massive volume and diversity of experimental biological data available to characterize a protein, the algorithms developed for data-mining to identify biological patterns in such data sets are at the same level of diversity. The selection of the appropriate tool for data-mining depends of the characteristics and extension of data and the type of analysis. It is possible to find tools for protein identification and sequence similarity-searching, pattern and profile-searching, topology structure, and modification prediction. The nature of the algorithms used in sequence analysis can classify the application of protein tools into two main groups. The first group is applied to search biological patterns comparing and scoring according to internal databases, sequence, and experimental information. The accuracy of the prediction depends on the extension of the database. The second group consists in use of inductive machine-learning algorithms, such as SVM (support vector machines) and artificial neural networks, to predict protein profiles and classification. In this case, the predictions are correlated to the amount and quality of data used for training the tool. Table 1 contains a variety of computational resources available for the study of amino acid and nucleotide sequences.

IDENTIFYING PROPERTY-BASED SEQUENCE MOTIFS IN FAEs

One approach for motif-finding is based on the comparison of the amino acid sequence to the sequences available in internal libraries composing the databases (Table 1). These libraries contain the biologically significant patterns and profiles for each motif, and a score is calculated by matching the query sequence to the internal sequence to properly assign the correct motif and the corresponding activity description. Tools as PROSITE, MOTIF SCAN, INTERPROSCAN, among others, are databases with free access containing documentation entries describing protein domains, families, and functional sites.

With the aforementioned tools using the amino acid sequence for each FAE, it was possible to observe defined motifs that confer specific enzymatic activity (Table 2). The common active sites correspond to the enzymatic capability to hydrolyze ester bonds. The active site of the different *Aspergillus*, *Talaromyces*, *Neurospora*, *Penicillium*, and *Pyromyces* strains can be classified into several groups. Relatively few sequences for FAEs are available to date, and consequently FAEs fall into only one of the 16 carbohydrate esterase (CE) families (CAZy; http://www.cazy.org/fam/CE1.html), indicating that the full diversification found can still not be reflected in a well-defined sequence-based classification. Furthermore, *Penicillum* together with *Cellvibrio*, presents another important motif, the carbohydrate-binding module (CBM). This is a characteristic module found among endoglucanases, cellobiohydrolases, and xylanases. In case of the bacterial FAEs of *Clostridium*, it was possible to identify the *Clostridium* cellulosome motif, specific for the extracellular degradation of cellulose molecules, and a glycosidase motif to assist in biomass degradation.

Archaea, *Eukarya* and *Bacteria* proteins can be modified posttranslationally by the attachment of different classes of biochemical groups to the nascent polypeptide and/or by proteolytic cleavage. Around 20 different posttranslational modifications (PTM) are well known and they are commonly present in across many organisms.[5] PTM's are responsible for proper protein function, folding, turnover, subcellular localization and oligomerization. The most common protein PTM include ubiquitylation, phosphorylation, glycosylation, lipidation, methylation, and

Table 2 Active sites and functional classification of the 16 amino acid sequences coding for FAEs.

Organism and protein	Active site	N-glycosylation sites	Classification[a]	Motif finder tool
Aspergillus niger Y09330 (faeA)	Lipase, serine active site	1	Type A	PROSITE
Aspergillus awamori AB032760 (AwfaeA)		1	No info	MOTIF SCAN
Aspergillus tubigensis Y09331 (faeA)		1	No info	INTERPROSCAN
Aspergillus niger AJ309807 (faeB)	Tannase and FAE	12	Type C	PROSITE
Aspergillus nidulans EAA63948		5	Type B	MOTIF SCAN
Talaromyces stipitatus AJ505939 (faeC)		10	Type C	INTERPROSCAN
Butyrivibrio fibrisolvens U44893 (cinA)	Esterase/lipase/ thioesterase active site serine	NONE	No info	MOTIF SCAN
Neurospora crassa AJ293029 (fae-1)		1	Type B	
Piromyces equi AF164516 (estA)		NONE	Type D	
Penicillium funiculosum AJ312296 (faeA)		NONE	Type D	
Penicillium funiculosum AJ291496 (faeA)		2	Type B	
Penicillium funiculosum AJ312296 (faeA)	CBM1 (carbohydrate-binding type-1) domain profile	NONE	Type D	PROSITE
Penicillium funiculosum AJ291496 (faeA)		2	Type B	MOTIF SCAN
Clostridium thermocellum M22624 (XynZ)	*Clostridium* cellulosome enzymes repeated	10	No info	PROSITE
Clostridium thermocellum X83269 (XynY)		14	No info	MOTIF SCAN
Clostridium thermocellum EAM45408	Glycosyl hydrolases family 10 active site	14	No info	
Cellvibrio japonicus X58956 (xynD)	CBM2a (carbohydrate-binding type-2) domain signature CBM6 (carbohydrate-binding type-6) domain profile	10	Type D	PROSITE MOTIF SCAN
Butyrivibrio fibrisolvens U64802 (cinB)	Prolyl endopeptidase family serine active site	1	No info	PROSITE INTERPROSCAN

[a]Based on substrate utilization data and supported by primary sequence identity four subclasses have been characterized and termed—A, B,C, and D.
Source: From Appl. Microbiol. Biotechnol.[6]

acetylation. Among these PTM, the most widely studied is N-glycosylation. It confers solubility and biological activity to the protein, the relevant features of biotechnological application of FAEs. NetNGlyc[7] can be used and it is a sequence-based analysis tool that can predict the specific sites in the sequence of the protein. Its algorithm is based on artificial neural networks to discriminate the potential of the consensus sequence Asn-Xaa-Ser/Thr (where Xaa is not Pro) as a target of a N-glycosylation site. Furthermore, it is possible to find tools to predict phosphorylation, C-mannosylation, prenylation, and SUMO protein attachment. Although, NetNGLyc was elaborated to predict glycosylation in human, this tool has been used in other organisms resulting in accurate predictions. Summarizing from Table 2, the number of sequon sites appears to be organism-specific and NetNGLyc predicts that *Clostridium thermocellum* contains the highest number of potential sites. However, the conclusion on the number of predicted potential sites is indicative as glycosylation is a rare phenomena in bacteria and more biological evidence is necessary before any conclusions can be drawn.

CONCLUSIONS

As part of our responsibilities to the environment, clean technology is required to treat agroindustrial residues to reduce the mass of material available and, if possible, to facilitate the extraction of fine chemicals from them to provide a higher value-added product for further exploitation. Use of enzymes permits the potential of multiple products to be extracted from waste residues, while any residual material should still be suitable for animal feed (one end-use of waste residues already in practice). FAEs will have a profound role in exploiting agricultural waste for high value-added products. Candidate FAEs can be identified in many fungal and bacterial genomes using the known, identified sequences as search queries, making the choice of the resulting esterases, for aqueous or organic synthesis exploitation, an intriguing task. In the future, the aim is to use the technology platform for protein engineering and systems biology to develop novel enzymes not only as tools for the extraction of useful high value-added products, such as phenolic acids (for use as natural antioxidants), xylooligosaccharides (for use in functional foods as prebiotics), and processed cell wall polysaccharides (for use as novel food materials) but also for the synthesis of novel biological active derivatives with use in the medical and food industries.

ACKNOWLEDGMENT

G. Panagiotou would like to thank the Danish Research Council for Technology and Production for a postdoctoral grant.

REFERENCES

1. Topakas, E.; Vafiadi, C.; Christakopoulos, P. Microbial production, characterization and applications of feruloyl esterases. Process Biochem. **2007**, *42*, 497–509.
2. Figueroa-Espinoza, M.C.; Villeneuve, P. Phenolic acids enzymatic lipophilization. J. Agric. Food Chem. **2005**, *53*, 2779–2787.
3. Khmelnitsky, Y.L.; Hilhorst, R.; Veeger, C. Detergentless microemulsions as media for enzymatic reactions. Eur. J. Biochem. **1988**, *176*, 265–271.
4. Vafiadi, C.; Topakas, E.; Alderwick, L.J.; Besra, G.S.; Christakopoulos, P. Chemoenzymatic synthesis of feruloyl D-arabinose as a potential anti-mycobacterial agent. Biotechnol. Lett. **2007**, *29*, 1771–1774.
5. Jerry Eichler and Michael W. W. Adams, "Posttranslational Protein Modification in Archaea". Microbiology and Molecular Biology Reviews, September **2005**, p. 393–425, Vol. 69, No. 3
6. Crepin, V.F.; Faulds, C.B.; Connerton, I.F. Functional classification of the microbial feruloyl esterases. Appl. Microbiol. Biotechnol. **2004**, *63*, 647–652.
7. Blom, N.; Sicheritz-Pontén, T.; Gupta, R.; Gammeltoft, S.; Brunak, S. Prediction of post-translational glycosylation and phosphorylation of proteins from the amino acid sequence. Proteomics **2004**, *4* (6), 1633–1649.

Nutraceuticals and Functional Foods

Nicholas Kalaitzandonakes
Department of Agricultural Economics, University of Missouri, Columbia, Missouri, U.S.A.

James D. Kaufman
Lucy Zakharova
University of Missouri, Columbia, Missouri, U.S.A.

Abstract

Functional foods are foods that provide health benefits beyond basic nutrition. Functional foods cater to both ill and healthy, and target prevention and treatment of illness, as well as maintenance of one's well-being. Products in the market range from complete portfolios of prepared meals that battle cardiovascular and type II diabetes, to fortified drinks and cereals that provide supplemental minerals, vitamins, and fiber.

Nutraceuticals are isolates that provide concentrated nutrients in the form of pills, tablets, liquids, or powders for direct consumption or for use as ingredients in functional foods. Nutraceuticals include micro- and macronutrient isolates, herbs and botanicals, and isolated reagents (e.g., hormones).

DRIVERS OF INNOVATION

Traditional remedies utilizing plants and animals to cure disease and improve health and well-being have existed for centuries and have been passed down through generations. In the mid-1980s, clinical studies began to formalize the relationship between nutrition and healthfulness, attracting significant attention from both media and consumers. First among these studies was the 1984 National Cancer Institute certification that high-fiber cereals foster cardiovascular health. Since that time, the body of literature connecting nutrition to specific health benefits and disease treatment has increased substantially, and other foods (e.g., calcium, omega-3 fatty acids, whole oats, and soy protein) have gained similar certification by the U.S. Food and Drug Administration (FDA).

Scientific discovery in the development of new products capable of delivering specific health benefits has also advanced rapidly. Cholesterol reduction, cardiovascular disease, and osteoporosis have become the most frequent targets for functional foods and nutraceuticals, followed by child development, moodiness, low energy, high blood pressure, diabetes, gastrointestinal (GI) disorders, menopause, and lactose intolerance.

Consumer interest has bolstered new product development of nutraceuticals and functional foods.[1–3] An aging and wealthier population in Japan, the European Union (E.U.), and the United States has become increasingly interested in the prevention of disease and health maintenance, representing a receptive audience to health claims. Similarly, escalating health care costs and increasing interest in self-medication have shifted attention to health maintenance and disease prevention instead of treatment. Table 1 provides some examples of nutraceuticals and functional foods introduced in recent years.

MARKET SIZE AND GROWTH

In 2001, the global market for functional foods was estimated at over $47.6 billion. Well-developed food markets accounted for the majority of the global consumption of functional foods. The U.S. functional foods market alone was valued at $18.25 billion, representing 3.7% of the $495 billion spent for food at home by U.S. consumers. The European Union and Japan followed with $15.4 billion and $11.8 billion in functional food sales, respectively.[4] Consumer spending on functional foods is expected to grow at a fast rate although not evenly across all food categories. In recent years, functional beverages have experienced the fastest growth (Table 2).

Not all functional foods have been successful in the marketplace. Some products have seen diminished consumer demand due to unexpected negative side effects (e.g., the fat substitute Olestra). Others have been quietly removed from supermarket shelves due to lack of sufficient consumer demand (e.g., McNeil's Benecol in the United States; Kellogg's Ensemble; and Cambell Soup Company's Intelligent Cuisine).

The global market for nutraceuticals reached $50.6 billion in 2001. Vitamin/mineral supplements accounted for

Table 1 Nutraceuticals and functional foods.

Category	Product name	Product form	Claim
Health foods (Foods with approved health claims)			
	Tofu	Food	Cardiovascular health
	Oatmeal	Food	Cancer prevention
Nutraceuticals (Food ingredients with specific claims of benefit)			
	Glucosamine	Food ingredient	Joint health
	Soluble fiber	Food ingredient	Cancer prevention
	Ginseng	Herb/ingredient	Physical performance
	Soy protein	Food ingredient	Cardiovascular health
Functional foods (Foods or drinks with specific claims of benefit)			
	Yakult Honsha's Yakult	Probiotic active cultures	Improved digestion
	McNeil's Benecol	Margarinelike spread	Lowers LDL cholesterol
	Proctor & Gamble/Olean's Olestra	Fat substitute	Weight loss
	Tropicana's Tropicana calcium	Calcium fortified juice	Skeletal health
	Kellogg's Ensemble	High-fiber cereal	Cholesterol reduction
	Campbell Soup Company's Intelligent Cuisine	Line of prepared meals	Nutritionally balanced
Drug delivery foods (Food as clinical drug delivery mechanism)			
	(Not yet commercialized)	Potato, banana, tomato	Vaccine for hepatitis B, etc.

40% ($20.6 billion) of those sales, followed by herbs and botanicals ($19.6 billion) and sports/specialty supplements (10.4 billion).[7] Vitamins are especially important in the U.S. market, whereas herbs are relatively more important in the European and Asian markets, which have a long tradition of herbal use and remedies.

PRODUCT INNOVATION AND INDUSTRIAL DEVELOPMENT

For continuing growth in the nutraceuticals and functional foods markets, sustained technological innovation will continue to be critical. The scope of innovation will likely range from product functionality, to taste characterization and manufacturing. This will require the pooling of knowledge, skills, and infrastructure from industries that have not worked closely together in the past. For example, nutraceutical startups and biotechnology firms may contribute discoveries in new products and attributes. Food companies may provide product formulation, manufacturing, and distribution capabilities. Pharmaceutical companies may contribute discovery and knowledge in amplification of efficacy, as well as experience in the validation of safety (e.g., clinical trials).

Major food companies making significant investments in functional foods are already pursuing integration of skills and assets. Many of these companies have moved to acquire diverse skills through acquisitions or strategic alliances. For instance, PepsiCo recently acquired Gatorade and Sobe functional beverage brands; Heinz acquired a share in Hain Foods, a processor of soy and natural foods; and Quaker and Novartis entered into a joint venture to form Altus Food Company, which is dedicated to the development of functional foods.

To be successful, this emergent industry will have to learn how to develop products that deliver functionality in ways that appeal to changing consumer tastes and lifestyles. It will also have to learn how to protect and recoup its investments (e.g., R & D) through patents, brands and trademarks, and other marketing strategies. Finally, this new industry will have to learn how to position functionality claims within an uncertain and fragmented regulatory environment.

Table 2 U.S. retail sales of functional foods by category ($billion).

Category	1998[a]	2000[b]
Beverages	4.9	8.2
Breads and grains	6.1	4.8
Packaged/prepared foods	1.1	1.6
Dairy	1.7	1.1
Meat, fish, and poultry	0	0
Snack foods	0.8	1.4
Condiments	0.1	0.2
Total	**14.8**	**17.2**

[a]From Nutraceuticals & functional foods IVI.[5]
[b]Form Functional foods V.[6]

INSTITUTIONAL INNOVATION

It will be up to governments to establish relevant institutional environments that create appropriate incentives for research and development in nutraceuticals and functional foods while they protect public health and consumer interests. Meaningful institutional innovation will therefore be necessary in the coming years. Current regulation lacks clear standards on health claims (efficacy), product quality, and safety that can impede market development. For instance, the slow market development experienced in herbs/botanicals (a flat 1.2% rate of growth in 2000) has been attributed to the high incidence of low-quality (and occasionally unsafe) products introduced during the boom market of the late 1990s.[7] Herbs contaminated with heavy metals, heart problems attributed to Ephedra, and safety concerns with sports supplements (e.g., Androstenedione, or Andro) attracted significant negative public attention and reduced consumer interest.

Development of a proper regulatory framework has been complicated by the dualistic food and drug nature of nutraceuticals and functional foods. Complexities are likely to increase in the future as active ingredients become more powerful, requiring some foods to be regulated as pharmaceuticals. For instance, pharmaceutical foods (such as edible vaccines) are already under development. These bioengineered foods are being developed to express specific pharmaceutical compounds, providing a convenient drug manufacturing and delivery mechanism. Accordingly, an institutional environment capable of efficiently regulating both foods as drugs and drugs as foods will be necessary.[8]

Currently, there are few regulatory standards, and those that exist fail to transcend national borders. The U.S. system is characterized by a dichotomy of regulatory stringency, wherein some products can be commercialized with minimal regulatory oversight, while others are strictly regulated. Less-regulated products are allowed by the Dietary Supplement Health and Education Act (DSHEA), which shifted the burden of proving an ingredient unsafe from the manufacturer to the FDA. Under DSHEA guidelines, a dietary supplement may include a vitamin, mineral, herb/botanical, amino acid, metabolite, or an extract. Laws allow nutritional support statements such as those relating to classical nutrient-deficiency diseases, structure/function, or well-being, provided they include the following disclaimers: "This statement has not been evaluated by the Food and Drug Administration" or "This product is not intended to diagnose, treat, cure, or prevent any disease."

In order to make health claims linking a food or dietary supplement to a disease or health-related condition, pre-market approval by the FDA is required. The FDA approval process is dependent on scientific consensus identifying a specific functional component responsible for physiological action. The approved health claim must include the appropriate dietary context (e.g., low in saturated fat and cholesterol), which can then be used with any similar product, not just those produced by the petitioner.

The main obstacle for the European system is the lack of standard protocols across national borders. For example, France and Germany differ from the United States and many other European countries in that French and German herbal drugs are distributed via pharmacies, allowing for doctor mediation and reimbursement by health insurance.

In contrast to the U.S. and European models, the Japanese nutraceuticals and functional foods markets are highly regulated and supported. Since 1991, the Japanese government has actively promoted the development of functional foods through the Foods for Specialized Health Use (FOSHU) system. Foods in this product category can secure regulatory approval to make specific health claims, and are educationally supported by the program. FOSHU is an umbrella label for all functional foods, and ensures safety and efficacy; however, FOSHU is not mandatory and the manufacturers of some functional products opt to circumvent it, suggesting that subpar products may still be commercialized. This Japanese system has been popularly proposed as a potential template for the development of international standards. The development of such international standards, however, has progressed slowly (until now).

CONCLUSION

It is clear that both scientific discoveries and consumer interest are creating opportunities for an expanding menu of foods and supplements that claim health benefits beyond nutrition. In globalized markets where products are traded across national borders at an increasing pace, a proper regulatory framework for how such claims are brought forward will be increasingly important. Hence, consumers, scientists, entrepreneurs, and regulators will all shape the rate and direction of product innovation in the nascent markets of nutraceuticals and functional foods for many years to come.

REFERENCES

1. Heasman, M. Mellentin, J. *The Functional Foods Revolution: Healthy People, Healthy Profits?* Earthscan Publications, Ltd.: London, **2001**.
2. *Nutraceuticals: Developing, Claiming, and Marketing Medical Foods*; Defelice, S., Ed.; Marcel Dekker, **1998**.
3. Gilbert, L. Marketing functional foods: How to reach your target audience. Agbioforum **2001**, *3* (1). http://www.agbioforum.org.
4. Sloan, E. The top 10 functional food trends: The next generation. Food Technol. **2002**, *56* (4), 32–57.
5. Nutrition business journal. Nutraceuticals & functional foods IV. Nutr. Bus. J. **2000**, *5* (5), 1–6.
6. Nutrition business journal. Functional foods V. Nutr. Bus. J. **2001**, *6* (10), 1–8.
7. Gruenwald, J.; Herzberg, F. *The Global Nutraceuticals Market*; World Markets Research Centre, **2002**.
8. Leland, S. *Agriceuticals: Designing New Food Concepts. Rabobank International*; Food & Agribusiness Research: New York, **2001**.

Oil Crops: Genetically Modified

Denis J. Murphy
School of Applied Sciences, University of Glamorgan, Cardiff, U.K.

Abstract
For over five thousand years, oil crops have been sources of many useful nonfood products ranging from lubricants to lamp fuels. Since the late 19th century, such nonfood uses of plant oils have declined due to the twin pressures of relatively cheap petroleum-based alternatives and the need to feed the ever-increasing human population of our planet. By the late 20th century, less than 15% of plant-derived oils were used for industrial purposes. The vast bulk of vegetable oils are currently traded as commodities destined for human consumption in such products as margarines, cooking oils, and processed foods. More recently, however, interest has revived in the possible exploitation of oil crops for a wide spectrum of nonedible products including cosmetics, biodegradable plastics, and even high-value pharmaceuticals.

MARKETS AND CROPS

Global vegetable oil markets are dominated by the "big four" crops—soybean, oil palm, rapeseed, and sunflower—which together make up over 86% of the total global traded production of almost 90 million tonnes. The major aim of oil crop engineering has been to alter the fatty acid profiles of the major oil crops. As shown in Table 1, these crops have relatively narrow fatty acid profiles dominated by C16 and C18 groups that are not optimal for many industrial uses. On the other hand, there are many examples of minor crops and noncrop plants that accumulate very high levels of a diverse range of novel fatty acids with chain lengths from C8 to C24, and with useful chemical functionalities such as hydroxyl, epoxy, and acetylenic groups. Such oils can be used for the manufacture of products such as adhesives, paints, detergents, lubricants, and nylons, to name but a few. In Table 2, some examples of oil-bearing seeds that already produce some of these novel and potentially useful fatty acids are shown. Over the past decade, many of these plants have been used as sources of genes encoding fatty acid biosynthetic enzymes for transfer into mainstream oil crops in the hope that those crops would then accumulate the novel oils on a scale of up to millions of tonnes per year. This concept has been termed "designer oil crops."[1]

Rapeseed was successfully transformed as early as 1984, followed by sunflower and soybean in the 1990s. More recently, marker transgenes have been inserted into oil palm[2] and lipid biosynthetic genes will doubtless soon follow, although it will probably be another decade before alteration of the oil composition of this important tree crop has been achieved. Most of the genes related to fatty acid biosynthesis were isolated during the 1990s, either from the various exotic oilseeds or from model plants such as *Arabidopsis thaliana*, which as well as being an important research plant is also an oilseed that accumulates storage lipid as almost half of its seed weight. Therefore, in principle, it should now be possible to engineer transgenic oil crops to produce the desired range of novel fatty acids for a variety of industrial applications.

PROBLEMS AND PROSPECTS

Unfortunately, it appears that the accumulation of high levels of a single desired fatty acid in the storage oil, although fairly common in nature, is not readily achievable by simply inserting a few lipid biosynthetic genes into a given transgenic plant. There is increasing evidence that fatty acid modifications may behave as quantitative traits that are controlled to a greater or lesser extent by numerous genes.[3] It may be the case that in order to achieve levels of 90% of a particular fatty acid in a crop plant, the insertion of at least four and as many as ten transgenes may be required. This will add considerably to the cost and timescale of such manipulations.

Another problem in oil crops such as rapeseed is that exotic fatty acids designed for sequestration in the storage oils may sometimes also accumulate in membrane lipids, with possible deleterious consequences. Plants that naturally accumulate exotic fatty acids such as lauric acid (C12), which is a powerful membrane-destabilizing detergent, have evolved mechanisms to "channel" these deleterious fatty acids into the storage lipid pool. This mechanism, which probably involves specific phospholipases and acyltransferases, is found in lauric-accumulating seeds such as *Cuphea* spp. and *Umbelluria californica*, but does not appear to be particularly active in rapeseed.[4] A further potential technical challenge to the engineering of designer oil crops is the finding that some of the gene promoters most

Table 1 Percentage fatty acid composition of the "big four" oil crops

Fatty acid[a]	Soybean	Oil palm[b]	Rapeseed	Sunflower
16:0	11	45	5	6
18:0	4	5	1	5
18:1	22	38	61	20
18:2	53	11	22	69
18:3	8	0.2	10	0.1

[a]Fatty acids are denoted by their carbon chain length followed by the number of double bonds.
[b]Mesocarp.

Table 2 Accumulation of novel fatty acids by some oil-producing plants

Fatty acid[a]	Amount[b]	Plant species	Uses
8:0	94%	Cuphea avigera	Fuel, food
10:0	95%	Cuphea koehneana	Detergents, food
12:0	94%	Litsea stocksii	Detergents, food
14:0	92%	Knema globularia	Soaps, cosmetics
16:0	92%	Myrica cerifera	Food, soaps
18:0	65%	Garcinia cornea	Food, confectionery
20:0	33%	Nephelium lappaceum	Lubricants
22:0	48%	Brassica tournefortii	Lubricants
24:0	19%	Adenanthera pavonina	Lubricants
$18:1_{\Delta 6}$	76%	Coriandrum sativum	Nylons, detergents
$18:1_{\Delta 9}$	78%	Olea europaea	Food, lubricants
$22:1_{\Delta 13}$	58%	Crambe abyssinica	Plasticizers, nylons
$18:2_{\Delta 9,12}$	75%	Helianthus annuus	Food, coatings
$\alpha 18:3_{\Delta 9,12,15}$	60%	Linum usitatissimum	Paints, varnishes
$\gamma 18:3_{\Delta 6,9,12}$	25%	Borago officinalis	Therapeutic products
18:1–hydroxy	90%	Ricinus communis	Plasticizers, cosmetics
18:2–epoxy	60%	Crepis palestina	Resins, coatings
18:2–triple	70%	Crepis alpina	Coatings, lubricants
18:3–oxo	78%	Oiticica	Paints, inks
18:3–conj	70%	Tung	Enamels, varnishes
20:1/22;1wax	95%	Simmondsia chinensis	Cosmetics, lubricants

[a]Fatty acids are denoted by their carbon chain length followed by the number of double bonds or the nature of other functionalities.
[b]Percentage of total fatty acids; data are taken from Murphy.[10]

commonly used to achieve the seed-specific expression of transgenes may also direct gene expression in other tissues, including roots.[5–6] While it is possible that in many cases the accumulation of exotic fatty acids in nonseed tissues may not be problematic, this may not always be true, and it underlines the need for the thorough metabolic profiling of all transgenic varieties before their general release.

The consequence of these and other complexities of plant molecular genetics and metabolism is that, despite many impressive achievements in isolating oil-related genes and producing transgenic plants with modified seed oil compositions, it has not been possible yet to achieve the kind of high levels (80–90%) of novel fatty acids that will make possible their widespread commercial exploitation. In the case of the high-lauric transgenic rapeseed (canola) crop, difficulties in its commercialization are also due to the existence of a competing source of lauric oil, namely palm kernel oil from the Far East. Palm oil is both cheaper to produce than rapeseed oil and in much more plentiful supply. The lauric-oil variety of rapeseed was improved from 40% to 60% lauric content by the insertion of several additional transgenes,[3] but it still remains far from being a commercial success and is no longer under development as a crop variety in the United States.

The availability of many genes involved in fatty acid modification and the good progress in transforming the main oil crop species will doubtless encourage further efforts to resolve the challenge of low levels of novel fatty acid production. But even if such efforts are successful, the commercial success of transgenic oil crops will remain problematic. It will be necessary to identify and develop robust markets for transgenic oil products; simply substituting for low-cost petroleum-derived products is unlikely to be economical for many decades. The additional costs of identity preservation will probably preclude the use of such transgenic oils as large-scale, low-value commodities in competition with conventional plant oils, even for industrial applications. In summary, transgenic oil crops producing novel fatty acids may have promise for the long-term future, but their commercial prospects over the next few years remain decidedly uncertain.

An attractive alternative to novel fatty acid production in oil crops is to engineer them to accumulate biopolymers instead. Virtually all of our conventional plastics are made from nonrenewable petroleum-derived products such as adipic acid and vinyl chloride. Some soil bacteria such as *Ralstonia eutrophus* are able to accumulate up to 80% of their mass in the form of nontoxic biodegradable polymers called polyhydroxyalkanoates (PHAs). The PHAs are made up of β-hydroxyalkanoate subunits that are synthesized from acetyl-CoA via a relatively short pathway involving as few as three enzymes for the most common PHA, polyhydroxybutyrate.[7] The cost of PHAs could be considerably reduced if they were produced on an agricultural scale in transgenic crops. This prospect has led several companies, including Monsanto and Metabolix, to attempt to develop transgenic rapeseed plants containing the bacterial genes responsible for PHA biosynthesis. Provided the PHAs accumulate in the plastids, and not in the cytosol, it is possible to obtain modest yields of the polymer from either leaves or seeds.[8] A major and as yet unresolved technical hurdle is how to extract biopolymers from plant tissues in an efficient and cost-effective manner. Another complexity is that polyhydroxybutyrate, which is the most widespread PHA, is a rather brittle plastic and is not suitable for most applications. The best-performing plastics are copolymers of polyhydroxybutyrate with other PHAs, such as polyhydroxyvalerate. Although the production of such copolymers in transgenic plants is considerably more difficult than the production of single-subunit polymers, progress has recently been made in this area.[9] While there are several companies and academic labs attempting to make commercially extractable PHAs in plants (including one in oil palm), it seems unlikely that these environmentally friendly products will be commercially available for quite a few years to come.

Apart from these scientific and technical challenges, engineered oil crops also face considerable challenges regarding their management and economics. The major managerial problem concerns the need to segregate a transgenic crop variety producing a novel product from nontransgenic commodity crops and from other transgenic varieties of the same species that accumulate different products. This is a formidable task, given the intricacy of the supply chain from breeder to grower to crusher to processor and so on, all the way to the retailer and, ultimately, to the consumer. The difficulties in ensuring strict segregation of otherwise indistinguishable transgenic crops have consistently been underestimated by many in the industry. Several well-publicized failures in the segregation of transgenic rapeseed and maize crops in recent years (e.g., the STARLink affair, although note that this did not involve genes modified for oil composition)[10] have thrown this issue into much sharper focus. The contamination of a batch of seeds containing, for example, a hydroxy oil designed for industrial use with another batch of seeds containing a high-oleic oil for edible consumption (or vice versa) would result in a mixture that would be useless for both purposes. Efficient segregation is likely to be both difficult to control and expensive to implement. This may limit the cultivation of transgenic crops producing novel oils to geographically remote areas and/or to relatively high-value niche markets, where the additional costs of identity preservation can be met by the added value of the product.

CONCLUSION

The conclusion that engineered oil crops may be best suited to relatively low-volume, high-value markets allows the possible expansion of the target oilseed species beyond the "big four" oil crops to include minor oil crops like safflower or linseed, or even noncrop species like *Cuphea*. Use of such oil crops to produce novel products would have the advantage that segregation from sexually compatible food crop varieties would be fairly straightforward. Another innovative development of transgenic oil crops is the use of the oil as a carrier for recombinant high-value proteins such as pharmaceutical peptides and industrial enzymes.[11] Therefore, the prospects for oil crop biotechnology are now significantly different from the 1990s vision of large-scale "designer oil crops," but the prospects remain positive, albeit rather more long-term than was originally envisaged. For this new vision to be realized, more investment in research must be coupled with a better appreciation of the economic, managerial, and public acceptability challenges that will confront the new crops.

REFERENCES

1. Murphy, D.J. *Designer Oil Crops*; VCH Press: Weinheim, Germany, **1994**.
2. Parveez, G.K.A.; Masri, M.M.; Zainal, A.; Majid, N.A.; Yunus, A.M.M.; Fadilah, H.H.; Parid, O.; Cheah, S.C. Transgenic oil palm: Production and projection. Biochem. Soc. Trans. **2000**, *28*, 969–971.
3. Voelker, T.A.; Hayes, T.R.; Cranmer, A.M.; Turner, J.C.; Davies, H.M. Genetic engineering of a quantitative trait: Metabolic and genetic parameters influencing the accumulation of laurate in rapeseed. Plant J. **1996**, *9*, 229–241.
4. Wiberg, E.; Banas, A.; Stymne, S. Fatty acid distribution and lipid metabolism in developing seeds of laurate-producing rape (*Brassica napus* L.). Planta **1997**, *203*, 341–348.
5. Baumlein, H.; Boerjan, W.; Nagy, I.; Bassuner, R.; van Montagu, M.; Inze, D.; Wobus, U. A novel seed protein from *Vicia faba* is developmentally regulated in transgenic tobacco and *Arabidopsis* plants. Mol. Gen. Genet. **1991**, *225*, 459–467.
6. Murphy, D.J.; Hernandez-Pinzon, I.; Patel, K. Roles of lipid bodies and lipid-body proteins in seeds and other tissues. J. Plant Physiol. **2001**, *158*, 471–478.
7. Steinbüchel, A.; Fuchtenbusch, B. Bacteria and other biological systems for polyester production. Trends Biotechnol. **1998**, *16*, 419–427.
8. Snell, K.D.; Peoples, O.P. Polyhydroxyalkanoate polymers and their production in transgenic plants. Metab. Eng. **2002**, *4*, 29–40.
9. Slater, S.; Mitsky, T.A.; Houmiel, K.L.; Hao, M.; Reiser, S.E.; Taylor, N.B.; Tran, M.; Valentin, H.E.; Rodriguez, D.J.; Stone, D.A.; Padgette, S.R.; Kishore, G.; Gruys, K.J. Metabolic engineering of *Arabidopsis* and *Brassica* for poly(3-hydroxybutyrate-co-3-hydroxyvalerate) copolymer production. Nat. Biotechnol. **1999**, *10*, 960–961.
10. Murphy, D.J. Biotechnology, Its Impact and Future Prospects. In *Molecular to Global Photosynthesis*; Archer, M.A., Barber, J., Eds.; Imperial College Press: London, **2004**, 649–741.
11. van Rooijen, G.J.H.; Moloney, M.M. Plant seed oil bodies as carriers for foreign proteins. Bio/Technology **1995**, *13*, 72–77.

Oocytes and Embryos: Vitrification

José Luiz Rodrigues
*Laboratory of Embryology and Reproductive Biotechnology,
Federal University of Rio Grande do Sul, Porto Alegre, Brazil*

Abstract
Efficient cryopreservation of mammalian oocytes and embryos still a challenge in assisted reproductive technologies. The aim of this entry is to highlight the general principles and steps of mammalian oocyte and embryo freezing by vitrification.

INTRODUCTION

Cryobiology is the study of the effects of subfreezing temperatures on biological systems and stands at the interface between physics and biology. The technology to cryopreserve gametes and embryos, coupled with embryo transfer, is used for preservation of genetic variants in laboratory animals, for breeding and reproduction of farm animals, and for the treatment of infertility in humans. Furthermore, cryopreservation is a potential means for conservation of endangered species.

Vitrification is the process of cryopreservation that allows exposure of mammalian oocytes and embryos to the liquid nitrogen temperature ($-196°C$) without ice formation. It is a "glass like" state in which cells are suspended in high concentrations of cryoprotectants in liquid nitrogen. The aim of this entry is to describe vitrification procedures for mammalian oocytes and embryos.

BACKGROUND

The German chemist Gustav Tammann in 1898[1] wrote the first report about the possibility of exposing different solutions to very low temperatures without ice formation, and in 1937 Luyet[2] proposed preserving living cells using this cryoprocedure.

Since 1972, when successful mouse embryo cryopreservation was first reported, numerous studies have been performed using mammalian gametes and embryos as experimental models.[3] Most of these efforts have been directed toward improving survival of frozen-thawed mammalian oocytes and embryos and simplifying the method of cryopreservation. This has resulted in the development of several methods that are quick and relatively simple to perform.

PRINCIPLES OF CRYOPRESERVING CELLS

Cryobiological dogma holds that avoidance of large and numerous intracellular ice crystals is necessary but not sufficient for cells to survive freezing. The methods developed for cell freezing can be classified into three groups: 1) equilibrium; 2) quasi-equilibrium; and 3) non-equilibrium.[4] Equilibrium freezing occurs when embryos or other cells are cooled slowly at $\sim 1°C/min$ to about $-70°C$, to promote cell dehydration and allow equilibrium between the water chemical potential inside and outside of the cell. To achieve high survival rates, it is necessary to thaw the embryos slowly at $20°C/min$. This was the approach employed in 1972[3] to obtain the first live born mammalian after the embryo was exposed to liquid nitrogen.

Quasi-equilibrium freezing was described later,[5] when it was shown that embryo survival rate depended on the warming rate. Embryonic cells could survive slow cooling ($0.3-0.6°C/min$) to temperatures between $-30°C$ and $-40°C$, followed by direct transfer to liquid nitrogen temperature, if they were warmed rapidly at $\geq 500°C/min$. The freezable water inside the cells at the time of transition either vitrified or formed small innocuous ice crystals.

Non-equilibrium freezing is achieved when the embryos are cooled very rapidly without equilibrating intracellular and extracellular water chemical potentials. Non-equilibrium methods can be described as one-step freezing, two-step freezing, and vitrification.

STEPS OF STANDARD CRYOPRESERVATION

First, cells need to be exposed to a cryoprotectant solution to allow cell dehydration and penetration of the cryoprotectant through the cell membrane. The most frequently used cryoprotectants are dimethyl sulfoxide (DMSO), glycerol, ethyleneglycol, propyleneglycol, and various polysaccharides like sucrose. After that, cells are cooled at different rates, depending on the freezing method, and stored in liquid nitrogen. Finally, the cells are thawed and the cryoprotectant solution is removed, allowing the cell to return to normal metabolism.

VITRIFICATION

Vitrification is a physical process whereby a liquid solution is transformed into a solid, "glass like" state, in which the cells are suspended in high concentrations of cryoprotectants at very low temperature, usually in liquid nitrogen. The benefits of vitrification are the rapidity, simplicity, and low cost of the procedure. The first living cells cryopreserved by means of the vitrification were frog spermatozoa.[6] In 1985, the first successful vitrification of mouse embryos was published.[7] At that time the problem that needed to be overcome was the high cryoprotectant concentration, which induced osmotic and/or toxic effects. Arav and coworkers[8] developed a new method for vitrifying mammalian oocytes and embryos called "minimal drop size" (MDS). This technique allows cooling and warming oocytes and embryos very rapidly (1750°C/min) in small drops (0.06 μL) containing a low concentration of cryoprotectant(s). This technical approach was the basis for most vitrification methods developed. Different carrier vessels were used to minimize the volume and to expose the sample to liquid nitrogen very quickly; these included the open pulled straw (OPS), electron microscope grids, and cryoloops. Devices like the VitMaster™[9] were developed to increase cooling and warming rates using vacuum to decrease the liquid nitrogen temperature slightly below the boiling point (from −196°C to −200°C) to form a so-called "liquid nitrogen slush." The aim of this procedure is to minimize boiling of liquid nitrogen around the submerged sample to increase cooling rates.

Recently vitrification of human oocytes using a very efficient method called "cryotop" resulted in a breakthrough with a 91% survival rate, 50% developing to the blastocyst stage, and 10 babies born after 29 embryo transfers.[10]

To overcome sanitary control problems, owing to the possibility of cross-contamination through liquid nitrogen, safety devices for loading the embryos were also developed including the VitSet™ technique (Minitüb, Germany) and the CryoTip™ (Irvin Scientific, California, the United States) method.

During the last few years, vitrification has been used successfully to cryopreserve oocytes and embryos from a variety laboratory, domestic, and wild animal species as well as human beings.[11]

RESEARCH

The general approach to circumvent cell cryointolerance is to modify cryopreservation procedures, for example, by varying types and concentrations of cryoprotectants. Modifying the composition of oocytes and embryos is another recently described strategy to improve survival rates after vitrification.[12] Reduction of the cytoplasmic lipid content and induced changes in the ratio of cholesterol to phospholipids in cell membrane are examples of this new approach.

CONCLUSION

Improving long-term preservation by stopping biological time can be achieved by cryoprocedures that allow cells to be exposed to liquid nitrogen temperature, and after an indefinite time, possibility for centuries, to return to physiological conditions without loss of viability. Vitrification techniques are very useful tools to cryopreserve mammalian cells successfully, including oocytes and embryos, and offers new possibilities for assisted reproduction and the embryo transfer industry.

REFERENCES

1. Tammann, G. Der Glaszustand. Leipzig, Ed.; **1933**; 123 p.
2. Luyet, B.J. The vitrification of colloids and protoplasm. Biodynamica **1937**, *1* (39), 1–14.
3. Whittingham, D.G.; Leibo, S.P.; Mazur, P.; Survival of mouse embryos frozen to -196°C and -269°C. Science **1972**, *178*, 411–414.
4. Mazur, P. Equilibrium, quasi-equilibrium and nonequilibrium freezing of mammalian embryos. Cell Biophysics **1990**, *17*, 53–92.
5. Willadsen, S. Factors affecting the survival of sheep embryos during deep-freezing and thawing. In *The Freezing of Mammalian Embryos*, Ciba Foundation Symposium 52, 1977; 175–201.
6. Luyet, B.J.; Hodapp, R. Revival of frog's spermatozoa vitrified in liquid air. Proc. Soc. Exp. Biol. N. Y. **1938**, *39*, 433–434.
7. Rall, W.F.; Fahy, G.M. Ice free crypreservation of mouse embryos at -196°C by vitrification. Nature **1985**, *313*, 573–575.
8. Rubinsky, B.; Arav, A.; DeVries, A.L. Cryopreservation of oocytes using directional cooling and antifreeze glycoproteins. Cryo-Letters **1991**, *12*, 93–106.
9. Arav, A.; Yavin, S.; Zeron, Y.; Natan, Y.; Gracitua, H. New trend in gamete's cryopreservation. Mol. Cell Endocrinol. **2002**, *187*, 77–81
10. Kuwayama, M.; Vajta, G.; Kato, O.; Leibo, S. Highly efficient vitrification method for cryopreservation of human oocytes. Reprod. Biomed. Online **2005**, *11*, 300–308.
11. Vajta, G.; Nagy, Z.P. Are programmable freezers still needed in the embryo laboratory? Review on vitrification. Reprod. Biomed. Online **2006**, *12*, 779–796.
12. Seidel Jr., G.E. Modifying oocytes and embryos to improve their cryopreservation. Theriogenology **2006**, *65*, 228–235.

Organic Farming

Brenda Frick
Department of Plant Sciences, University of Saskatchewan, Saskatoon, Saskatchewan, Canada

Abstract
Organic farming strives to produce healthful food while maintaining or improving the health of the agro-ecosystem. Organic farmers emphasize a systems approach that manages, respects, and encourages natural, biological processes. Pests are managed through good husbandry practices such as crop rotation, residue management, cultivar selection, crop competition, soil fertility management, and, where necessary, through judicious use of biological and mechanical controls. Standards for organic certification focus on the production process, rather than the product. They emphasize natural processes rather than synthetic products in crop and livestock production, a program for soil building, and diverse crop rotations. Organically produced crops are kept separate from others during their journey from producer to consumer. An audit trail tracks the history of a given product, and helps to assure quality and consumer confidence.

ORGANIC FARMING STANDARDS

Certification standards for organic systems are complex. Minimum standards are set by regional certification bodies and international organizations such as the Organic Crop Improvement Association. National regulations for Canada were developed through the Canadian Organic Advisory Board. Standards are being developed in the United States through the U.S. Department of Agriculture National Organic Program Proposed Rule. A number of European countries and the European Union have developed or are developing standards. Different countries vary in the stringency of their regulations but, in general, the standards prohibit synthetically processed fertilizers, pesticides, growth regulators, antibiotics, and genetically modified or engineered organisms. They require a management plan that includes strategies for crop rotation, soil management, monitoring and problem solving for crop protection, and detailed record keeping. Farms must meet the regulations for a minimum period of time, usually three years, to qualify for certified organic status. The International Federation of Organic Agriculture Movements (IFOAM) offers a basic standard and accreditation for certification bodies that is highly regarded by international traders.

The word "organic" is protected by law in some countries. Equivalent terms in other countries include "biological" and "ecological." The term "organic" is more common in English-speaking countries; the latter terms are more common in mainland Europe.

PEST MANAGEMENT PRACTICES

Organic production is a systems approach to farming. Producers strive to understand the ecological relationships that influence the abundance of the various species in their systems, and to avoid outbreaks of those species that harm the crop. Many of the methods used to favor the crop in the ecological community can be summarized by good crop husbandry—appropriate timing, depth, and rate of seeding; management of soil fertility; selection of locally adapted and competitive crops and crop cultivars. Mechanical and biological pest controls are used as necessary. Off-farm inputs are considered only as a last resort.

Crop rotation is one of the strongest tools that the producer uses. A diverse rotation increases microbial and mycorrhizal activity in the soil and improves crop vigour. Populations of crop-specific pests, such as many diseases and insects, are severely reduced by years when a given crop is not grown. Crop rotations alter the timing and competitive relationships of crops and reduce the build-up of weed communities adapted to any given management practice. Crop and cultivar selection is also very important. Matching crops to fields improves their competitive relations with weeds, and helps them to resist both insect and pathogen attacks (Table 1).

Weeds

In organic systems, weeds are considered to be a part of the ecological system. They are often beneficial in moderating the soil environment, providing habitat and food for micro- and macrofauna, moving soil nutrients to the surface, and indicating soil or management problems. Of course, large weed populations are often detrimental. Weeds can harbor pests, reduce crop yield and quality, and interfere with harvest. When weeds are abundant and considered problematic despite prevention and crop rotation, an organic producer has several options.

Table 1 Influence of insects, diseases, and weeds on quality losses in crops.

Loss in size and thousand-kernel-weight (grain)
Loss in size of tubers and roots
Increase of moisture content
Changes in chemical composition of the kernels, tubers, and roots
Discoloration of kernels and tubers
Formation of mycotoxins by plant pathogens
Loss in germination of the seeds and tubers
Reduced processing properties (e.g., baking and malting properties)
Reduced feeding properties (nutrition value)

A strongly competitive crop is an excellent defense against weeds. Competitive crops, such as fall rye and sweet-clover, and perennial crops, such as alfalfa, are particularly effective at reducing the weed community. Less competitive crops such as flax and lentil are best saved for less weedy fields. Crop competition can be increased by appropriate cultivar selection and by crop management techniques such as heavy seeding, narrow row spacing, good seed-bed management, etc.

Tillage is commonly used to reduce weed populations. Tillage can be used after harvest, before seeding, and in fallow years. Concerns over the negative effects of tillage on soil quality, especially on erosion potential, have reduced the frequency of fall and fallow tillages, though these may be cautiously used for perennial weed control. Delayed seeding after spring tillage remains an important tool. Early tillage stimulates the germination of volunteer crop and weed seeds. These are destroyed with a second tillage at or before seeding. This strategy has been especially effective at reducing the abundance of early emerging species such as wild oats and winter annual weeds such as stinkweed.

Harrowing after seeding or even after crop emergence can also be effective. This strategy is most effective for weeds that emerge from shallow depths, such as green foxtail, in crops that have large, deeply placed seeds. For row crops, interrow cultivation can be effective. The combination of early harrowing across the rows and interrow cultivation can offer good weed control. Other mechanical weed control techniques include flaming, burning, and mowing. Chaff collection at harvest can remove significant numbers of weed seeds. It is most effective for volunteer crop seeds, and weed seeds such as lamb's quarters, that are largely retained on the plant at a height above the stubble.

Biocontrol of weeds includes the use of livestock, weed-eating insects, and weed-suppressing diseases. Livestock can be used for grazing or to consume mowed weeds. Livestock can also be used to consume chaff or seed screenings. Biocontrol insects generally target perennial weed species. For instance, the black-dot spurge beetle has been released for control of leafy spurge. Few fungal biocontrol agents are available. Examples include DeVine® for stranglervine, Collego® for northern joint vetch, and Biomal® for round-leaved mallow. Beneficial organisms can be encouraged by practices such as reducing tillage, maintaining shelter belts, and growing crops, such as legumes and some cereals, that encourage mycorrhizae. These can also be effective for weed control.

Insects and Other Invertebrates

A majority of insects found in crop fields are beneficial or are of no economic importance. Diversity of habitat and wildlife encourages positive interactions among species, and reduces the potential for outbreaks of insect pests.

Organically grown crops may be less attractive to insects than crops grown with abundant synthetic fertilizers. Synthetic fertilizers may result in the accumulation of excess nitrate in plant tissue; composted manure and green manures release nitrogen more slowly, reducing the potential for this accumulation. Excess nitrate makes plants more attractive to insect pests and can increase the reproductive rate of some insects, thus increasing the severity of an insect outbreak.

Crop rotations can be effective against insect pests with limited dispersal capabilities, such as the corn rootworm. For these types of insects, reducing crop residues and volunteer crop are also important.

For widely dispersed insects that are attracted to specific crops, large-scale cropping diversity is important. Monocultures favor these insects. Increasing the presence of nonattractive plant species reduces the incidence of attack. Practices such as strip cropping and intercropping reduce the attraction of insect pests by diluting the aroma of the crop or confusing the insect's search image. For instance, underseeding canola with yellow sweet-clover may reduce the incidence of flea beetles. Even weed populations may function this way. For instance, weeds in alfalfa may increase the habitat for parasitic wasps that control alfalfa caterpillar. Barrier strips of an unattractive crop around the outside of the susceptible crop may prevent insects from crossing into the susceptible crop. For instance, a border strip of peas may reduce the movement of grasshoppers into wheat. Trap strips of an attractive crop may be sown. After insects accumulate there, it can be mowed or cultivated, thus destroying many of the insects. Small-sized fields also limit the problem of insect outbreaks.

The time of seeding can sometimes be altered to avoid insect pests. For instance, late seeding of canola can reduce the severity of flea beetles and early seeding of wheat can reduce the attack of wheat midge.

Cultivar selection may help prevent insect problems. Wheat cultivars differ substantially in their susceptibility to wheat midge, in large part due to their rate of development.

Biocontrol agents are available for some insect pests. Predators such as ladybugs and lacewings can be used to reduce aphid populations. *Bacillus thuringiensis* (Bt) strains have been developed for control of several insects, including caterpillars and beetles. Some of these control organisms can be purchased. Maintaining a varied habitat in and around the crop field can also help to harbor such natural organisms.

Pheromone traps may be used to lure insects away from crops. Sticky traps can also be used. These trapping methods are particularly appropriate to monitoring insect populations. They are less effective at large-scale insect removal. In high-value crops, such as potatoes or strawberries, insect "vacuums" can be used for larger insects such as Colorado potato beetles or lygus bugs.

A few products can be used under organic certification standards. Soaps and oils can be used to suffocate insects such as aphids. Diatomaceous earth can be used to discourage soft-bodied insect larvae, rusty grain beetle, and slugs and snails. Natural products such as pyrethrum and rotenone are also acceptable under most certification standards.

Diseases

As with insects, many fungal and bacterial species are beneficial or benign. Practices that increase biodiversity will likely increase species that compete with or prey on disease species, and thus reduce the outbreak of disease epidemics. The severity of disease can be limited by reducing the population of the pathogen or the susceptibility of the host, or by changing environmental conditions that favor infection.

Rotation is an important key to reducing disease by reducing the inoculum level of the pathogen. Crop rotation is effective when pathogens are obligate and host specific, with low dormancy and poor aerial spread. Leaf blights of cereals and ascochyta blight of lentil can be reduced by rotations that include different crops. Crop rotation alone is not sufficient to eliminate diseases in perennial crops; diseases such as common root rots and seedling blights with a wide host range; diseases that persist in soil, such as fusarium wilt of flax; diseases with rapid spread, such as powdery mildew of peas; or diseases that are widespread in the air or by insect vectors, such as cereal grain rusts and aster yellows.

Straw, residue, and weed management can be important. Straw is a primary inoculum for some disease species. Incorporation of residues speeds their decomposition and thus reduces the pathogen population. The risks to soil quality with excess tillage need to be balanced against the risk of leaving inoculum in the field. Other sources of infection include volunteers of the target crop, and weeds that are closely related to it. These sources of inoculum can reduce the effectiveness of a rotation away from the target crop.

Where diseases spread slowly from adjacent fields, field edges can be treated separately. Barrier strips or early swathing or mowing of the severely affected area may reduce the spread of disease into the crop.

Seed quality can impact disease potential. Reducing seed damage during harvest, storage, and seeding can reduce the susceptibility to disease, especially seedling blights. This is especially important for seeds such as flax, rye, and pulses. Disease-free seed reduces the spread of seedborne diseases into new areas.

Host susceptibility can be reduced by the selection of appropriate cultivars. Resistance or relative tolerance to a number of diseases varies greatly among cultivars. Differences among cultivars in disease susceptibility may reflect differences in their rate of growth; differences in their architecture and thus canopy humidity; or physical, biochemical, or genetic properties that restrict disease entry. Crop timing may also be important. For instance, earlier seeding may reduce the incidence of diseases such as barley yellow dwarf, powdery mildew of pea, and pasmo of flax. Late seeding of fall-seeded crops can reduce spread of disease by reducing their overlap with similar spring-seeded crops. Late seeding into warm soil can increase seedling vigor, and thus reduce crop susceptibility to seedling blights. Reduced seeding rates can reduce the spread of disease by reducing contact among plants, and by altering the environment in the canopy. Intercropping or strip cropping can be effective as well.

Some environmental manipulations can make disease frequency less severe. Selection of an appropriate field for a given crop can be important. Nutrient imbalance can make diseases such as take-all in cereals more severe. The incorporation of manure and green manures in rotations can reduce the severity of disease by encouraging microbes that are antagonistic to crop pathogens.

FUTURE CONCERNS

Currently, most organic standards prohibit the use of genetically modified organisms. There is concern among organic producers that the popularity of biotechnology in crop breeding will result in the abandonment of traditional breeding programs. If all or even most future genetic disease and insect resistance is incorporated into a genetically modified background, it will be unavailable to organic producers. This will greatly reduce their pest management options.

BIBLIOGRAPHY

Altieri, M.A. *Agroecology: The Scientific Basis of Alternative Agriculture*; Westview Press: Boulder, CO, **1987**.

Earthcare: Ecological Agriculture in Saskatchewan; Hanley, P., Ed.; Earthcare Group: Regina, SK , **1980**.

Macey, A. *Organic Field Crop Handbook*; Canadian Organic Growers, Inc.: Ottawa, Canada **1992**.

Radesovich, S.; Holt, J.; Ghersa, C. *Weed Ecology. Implications for Management* 2nd Ed.; John Wiley & Sons: Toronto, ON, **1997**.

Food and Agriculture Organization of the United Nations. Special, organic agriculture and sustainability, defining organic agriculture, sustainable development dimensions, environmental policy, planning & management, 1998. http://www.fao.org/WAICENT/faoinfo/sustdev/epdirect/EPRE0056.HTM (accessed April 20, 1999).

Organic Crop Improvement Association, international certification standards, 1996. htttp://www.gks.com/library/standards/ocia/ociain.html (accessed April 21, 1999).

The national standard of Canada for organic agriculture, Canadian Organic Advisory Board, 1999. http://www.coab.ca/standard.htm (accessed May 1, 1999).

USDA National Organic Program Proposed Rule, United States Department of Agriculture, 1999. http://www.ams.usda.gov/nop/rule.htm (accessed May 1, 1999).

Organogenesis: In Vitro Plant Regeneration

Janet R. Gorst
Centre for Amenity and Environmental Horticulture, Benson Micropropagation, Brisbane, Queensland, Australia

Abstract

Efficient regeneration of plants from cells and tissues through organogenesis is an important prerequisite for the successful application of biotechnology to crop improvement. In vitro regenerability is a highly variable genetic trait that can be introgressed into nonregenerating (recalcitrant) lines by conventional breeding. The availability of mutants with distinctive regenerative characteristics and the rapid developments in molecular biology have considerably advanced our understanding of the phenomenon of in vitro organogenesis in the recent past. While the last 10 years have seen remarkable progress in defining molecular mechanisms underlying plant processes, the specific area of organogenesis in vitro has not yielded many molecular secrets.

TOTIPOTENCY

The concept of totipotency is central to understanding in vitro regeneration. The term is used in the context of differentiation not being an irreversible process as a cell undergoes maturation, i.e., a living plant cell with overt functional and structural specialization still carries all the information necessary to divide and undergo a morphogenetic process in the form of either organogenesis [which can be either rhizogenesis (root formation), caulogenesis (shoot formation) or, occasionally, flower formation] or embryogenesis, or to develop directly into a specialized cell type (e.g., as seen in xylogenesis). It is clear, however, from observations of regeneration in even highly regenerative explants that not all living differentiated cells of an explant participate in the regeneration process. This may be due to: 1) an inability to achieve in vitro the necessary conditions for totipotent expression; 2) genetic (physical changes to chromosomes, e.g., loss of DNA or nucleotide substitution) or epigenetic (changes in DNA pexpression as a consequence of development, e.g., DNA methylation or the isolation of DNA into heterochromatin) blocks that interfere with the expression of totipotency; 3) the fact that not all cells are totipotent, i.e., although all cells may appear to be the same in a particular tissue, only some possess special characteristics that enable them to regenerate plants when isolated and cultured under inductive conditions.

The first step in the expression of totipotency, where it occurs, is for mature cells to reenter the cell cycle and resume cell division (a process known as dedifferentiation). The next step is redifferentiation, either through direct formation of organized structures (direct regeneration) or by the formation of an intervening callus stage from which organized structures may later be induced (indirect regeneration). An early appreciation of the mechanisms underlying regeneration of whole plants, or parts of plants, from cells came with the classic observations of Skoog and Miller[1] that the direction of differentiation could be influenced by the ratio of the exogenously supplied growth regulators auxin and cytokinin. They observed in tobacco stem pith cultures that a high ratio of auxin to cytokinin led to initiation of roots, whereas a low ratio led to development of shoots. Although there are many species for which this simple manipulation will not work, in principle, this is the basis for regeneration in plant tissue culture systems. The two groups of growth regulators play a pivotal role in unlocking and realizing totipotent expression by influencing both dedifferentiation and redifferentiation. Note, however, that other medium conditions such as nitrogen, carbon source, and pH are also extremely important.

The process whereby differentiated cells respond to inductive phenomena leading to organogenesis involves two major phases—competence and determination. These phases reflect the two-stage practice of exposing cultures first to an "induction" medium and then to a "regeneration" medium during the regeneration process,[2] although there are cultures for which both phases will occur on the same medium, particularly in the case of direct regeneration.

COMPETENCE

This is a transient state in which cells can be induced to follow an organogenic pathway[2,3] and mechanical wounding is the most effective biological trigger for shifting cells into the competent state. Competence can be thought of as having two distinct components, one for cell division and the other for organogenesis.

Competence for Cell Division

In order to sustain cell division following wounding, exogenous auxin (+/− cytokinin) is usually required. Progress in understanding the molecular basis of the action of auxins and cytokinins in the initiation and maintenance of cell proliferation has been slow, and the complex interaction between exogenously applied growth regulators—overlayed with the unknown of endogenous synthesis—makes it difficult to differentiate the individual roles of auxin and cytokinin.

Proteins known as cyclin-dependent kinases (Cdks) govern the onset of S-phase and mitosis in all eukaryotic cells and require an activating cyclin subunit, which leads to the formation of Cyclin/Cdk complexes. In particular, homologues of a 34 kDa protein kinase known as p34 (coded for by the Cdk CDC) have been found in all eukaryotes that have been investigated. Induction of the competence of transformed protoplasts to divide in the presence of auxin and, to a lesser degree, cytokinin, was shown to be accompanied by expression of the Arabidopsis CDC2a gene, even if there was no subsequent cell division,[4] and led to the proposal of a linkage between the expression of CDC2a and competence for proliferation. A role for the Arabidopsis gene SRD2 in conferring competence for cell division has been hypothesized,[5] and the expression of CDC2a in the srd2 mutant is being undertaken. At another level of complexity, the gene AINTEGUMENTA (ANT) has also been implicated in cell cycle progression, but in addition it influences organ growth.[6] Ectopic expression of the gene gives rise to transformed plants that show spontaneous callus formation and regeneration (of roots, leaves, or shoots) at wound or senescence sites. Seen as a gene that maintains meristematic competence in cells, ANT could give a molecular basis to the frequent observation in tissue culture that explants derived from immature tissue (such as from embryos) are much more likely to succeed in producing regenerable cultures than those obtained from mature tissue. In other words, competence in vitro may be correlated with continuing meristematic activity in vivo.

Competence for Organogenesis

A gene in Arabidopsis (IRE1) that acts very early in dedifferentiation confers the ability of cells to respond later to specific regeneration stimuli such as auxin and cytokinin.[7] The work of Ozawa et al.[5] with Arabidopsis mutants srd1, srd2, and srd3, which are defective in their ability to regenerate, identified three sequentially acquired states associated with organogenic competence. Initially there is IC (incompetent with respect to both cell proliferation and organogenesis), which requires the gene SRD2 in order to progress to CR (competent with respect to rhizogenesis). Finally, SRD3 is involved in the progress from CR to CSR (competent with respect to shoot and root organogenesis).

DETERMINATION

This is a process in which cells follow a specific developmental pathway. The distinction between determination and competence can be illustrated by the work of Christianson and Warnick.[2] They found that callus produced on Convolvulus explants was initially developmentally interchangeable, i.e., it was competent to follow two developmental pathways—root formation and shoot formation. Once induction of shoots began, the cells involved in shoot formation became determined, and transfer to a root-inducing medium did not affect the formation of shoots. In other words, as determination proceeds, cells become more and more committed, and the developmental potential becomes restricted unless there is a catastrophic event—such as wounding—that cuts across the determined state. The realizing of commitment is considered to be a third phase in the process of organogenesis.[2]

In the Skoog and Miller model,[1] caulogenesis is stimulated by exogenous cytokinin. The work of Ozawa et al.[5] indicated that the genes SRD1 and SRD2 play essential roles in the caulogenesis induced by culturing competent explants on a medium containing cytokinin. Three genes, CKI1 isolated from Arabidopsis,[8] ESR1 isolated from Arabidopsis,[9] and PkMADS1 isolated from Paulownia,[10] have been identified as regulators of shoot regeneration. CKI1 is thought to function as a cytokinin receptor in the process of cytokinin induction of shoot organogenesis. ESR1 expression is induced by cytokinins, but transcripts of the gene accumulate only after acquisition of organogenic competence. PkMADS1 is hypothesised to be a shoot meristem identity gene whose expression is necessary in activating the developmental pathway leading to direct regeneration of shoots. Cytokinin also induces another shoot meristem gene, PASTICCINO, but the outcome is an inhibitory one that prevents excessive cell proliferation,[11] thus eliminating abnormal shoot development. The expression of homeobox genes in plants can be induced by cytokinin, and two such genes—KNOTTED1 and STM (SHOOT MERISTEMLESS)—have been implicated in shoot formation.[12]

The most frequent type of regeneration occurring in cultured cells is root formation, and this is stimulated by auxin.[1] However, there is some complexity in the action of auxin, because although it stimulates root initiation, its continued presence in the culture medium can inhibit the outgrowth of roots. Experimental systems looking at the molecular basis of rhizogenesis are few and deal almost exclusively with lateral root formation and the development of adventitious roots on stem cuttings. Lund et al.,[13] working with a tobacco root mutant (rac) that fails to initiate adventitious roots in response to exogenous auxin, concluded that the RAC gene is involved in an auxin signal transduction pathway acting prior to the first organized divisions that lead to the formation of root meristems. During the determination phase of rhizogenesis the LRP1

(*LATERAL ROOT PRIMORDIUM1*) gene is expressed,[14] and during both determination and commitment certain S-adenosylmethionine synthetase-encoding genes (*SAMS*) are also up-regulated.[15] The *Agrobacterium rhizogenes* infection system is a potentially useful tool for characterizing events in rhizogenesis. Cells infected with the bacterium show an increased sensitivity to auxin, consistent with the ability of such cells to undergo root meristem neoformation and proliferation.

CONCLUSION

Regeneration through organogenesis represents an amazing developmental plasticity that sets plant cells apart from most animal cells. It is an extraordinarily complex phenomenon influenced by an array of internal and external factors. The molecular work to date suggests that there is certainly no single "totipotency" gene, and the existence of a conserved suite of genes that defines a group of cells as organogenic or recalcitrant in vitro is not very evident either.

REFERENCES

1. Skoog, F.; Miller, C.O. Chemical regulation of growth and organ formation in plant tissues cultivated in vitro. Symp. Soc. Exp. Biol. **1957**, *11*, 118–131.
2. Christianson, M.L.; Warnick, D.A. Competence and determination in the process of in vitro shoot organogenesis. Dev. Biol. **1983**, *95*, 288–293.
3. Sugiyama, M. Organogenesis in vitro. Curr. Opin. Plant Biol. **1999**, *2*, 61–64.
4. Hemerly, A.S.; Ferreira, P.; de Almeda, J.; Van Montagu, M.; Engler, G.; Inzé, D. *cdc2a* expression in *Arabidopsis* is linked with competence for cell division. Plant Cell **1993**, *5*, 1711–1723.
5. Ozawa, S.; Yasutani, I.; Fukuda, H.; Komamine, A.; Sugiyama, M. Organogenic responses in tissue culture of *srd* mutants of *Arabidopsis thaliana*. Development **1998**, *125*, 135–142.
6. Mizukami, Y.; Fischer, R.L. Plant organ size control: *AINTEGUMENTA* regulates growth and cell numbers during organogenesis. Proc. Natl. Acad. Sci. **2000**, *97*, 942–947.
7. Cary, A.C.; Uttamchandani, S.J.; Smets, R.; Van Onckelen, H.A.; Howell, S.H. *Arabidopsis* mutants with increased organ regeneration in tissue culture are more competent to respond to hormonal signals. Planta **2001**, *213*, 700–707.
8. Kakimoto, T. CKI1, a histidine kinase homolog implicated in cytokinin signal transduction. Science **1996**, *274*, 982–985.
9. Banno, H.; Ikeda, Y.; Niu, Q.-W.; Chua, N.-H. Overexpression of *Arabidopsis ESR1* induces initiation of shoot regeneration. Plant Cell **2001**, *13*, 2609–2618.
10. Prakash, A.P.; Kumar, P.P. *PkMADS1* is a novel MADS box gene regulating adventitious shoot induction and vegetative shoot development in *Paulownia kawakamii*. Plant J. **2002**, *29*, 141–151.
11. Faure, J.D.; Vittorioso, P.; Santoni, V.; Fraiaier, V.; Prinsen, E.; Barlier, I.; Van Onckelen, H.; Caboche, M.; Bellini, C. The *PASTICCINO* genes of *Arabidopsis thaliana* are involved in the control of cell division and differentiation. Development **1998**, *125*, 909–918.
12. Rupp, H.-M.; Frank, M.; Werner, T.; Strnad, M.; Schmülling, T. Increased steady-state mRNA levels of the *STM* and *KNAT1* homeobox genes in cytokinin-overproducing *Arabidopsis thaliana* indicate a role for cytokinins in the shoot apical meristem. Plant J. **1999**, *18*, 557–563.
13. Lund, S.T.; Smith, A.G.; Hackett, W.P. Differential gene expression in response to auxin treatment in the wild type and *rac*, an adventitious rooting-incompetent mutant of tobacco. Plant Physiol. **1997**, *114*, 1197–1206.
14. Ermel, F.E.; Vizoso, S.; Charpentier, J.-P.; Jay-Allemand, C.; Catesson, A.-M.; Couée. Mechanisms of primordium formation during adventitious root development from walnut cotyledon explants. Planta **2000**, *211*, 563–574.
15. Lindroth, A.M.; Saarikoski, P.; Flygh, G.; Clapham, D.; Grönroos, R.; Thelander, M.; Ronne, H.; von Arnold, S. Two S-adenosylmethionine synthetase-encoding genes differentially expressed during adventitious root development in *Pinus contorta*. Plant Mol. Biol. **2001**, *46*, 335–346.

Oxygenases in Food

Estrella Nuñez-Delicado
Department of Science and Food Technology, San Antonio Catholic University of Murcia, Guadalupe, Spain

José Antonio Gabaldón-Hernández
National Technology Center for Preservatives and Food, Molina de Segura, Spain

Abstract

Oxygenases are among the most widely distributed enzymes that catalyze biological oxidation reactions. Because many chemical and biochemical transformations involve oxidative processes, development of practical biocatalytic applications for oxygenases has long been an important goal in biotechnology with food applications. Nowadays, significant progress has been made in the development of oxygenase-based diagnostic tests and improved biosensors in the construction of bioreactors for biodegradation of pollutants.

INTRODUCTION

Enzymes are naturally occurring proteins that allow all the biochemical processes of life to happen. The enzymes carry out an essential contribution in cellular activities and have a great number of biotechnological applications in the food industry. When purified and used in food preparation, some of these enzymes offer benefits such as improved flavor, texture, and digestibility.

For thousands of years, humans have used naturally occurring bacteria, yeasts, and molds, and the enzymes they produce to make foods such as bread, cheese, beer, and wine. Today, enzymes are used for an increasing range of applications: 1) baked goods; 2) cheese-making; 3) starch processing; and 4) production of fruit juices and other drinks. Here, enzymes can improve texture, appearance, and nutritional value and may generate desirable flavors and aromas.

In food production, enzymes have a number of advantages: 1) they are welcomed as alternatives to traditional chemical-based technology and can replace synthetic chemicals in many processes. This can allow advances in the environmental performance of production processes through lower energy consumption and biodegradability; 2) they are more specific in their action than synthetic chemicals. Processes that use enzymes have fewer side reactions and waste by-products, giving higher quality products and reducing the likelihood of pollution; and 3) they allow some processes to be carried out which would otherwise be impossible.

Although the presence of natural enzymes in food products, like cheese and meat, are advantageous to obtain desirable textures and tastes, natural enzymes may provoke undesirable reactions such as staling by lipases or browning by phenol oxidases or peroxidases. Sulfur dioxide is a more efficient technological agent in browning inhibition owing to its ability to penetrate readily into plant tissues; however, once a certain amount of quinone is formed by enzymatic reaction, SO_2 is oxidized and becomes inefficient as an oxidase inhibitor. Another widespread process used in the food industry for preventing phenol oxidase activity is the inactivation of enzymes by blanching or thermal processing of the raw material, which causes undesirable changes during the processing and subsequent storage of the products. However, the use of this method is somewhat limited, as the application of high temperatures changes the native properties and chemical composition of foods. In certain cases, high thermostabilities of phenol oxidase, and especially peroxidase, also represent a factor limiting the thermal treatment of plant raw material. Nowadays, the search for new compounds displaying a high ability to inhibit oxidases is in progress.[1] Occasionally, natural enzymes in food are used as process indicators of milk pasteurization (alkaline phosphatase or catalase detection) or evidence of incomplete scald treatment (presence of peroxidase in vegetables). Enzymes are being systematically developed as economically viable and environment-friendly industrial biocatalysts along with the fast advancement and expansion of modern biotechnology.[2]

Knowledge of the enzyme uses and functions to carry out desirable changes in foods has driven the production of commercial enzymes on a large scale. The industrial enzyme field is projected to grow fast, at a rate close to double digits annually in the near future. The majority of commercial enzymes are hydrolases (including proteases, carbohydrases, and esterases), whereas oxidoreductases account for a miniscule share. This is in contrast to the frequent occurrence of oxidoreductases in nature.

Oxidoreductases are widely distributed among microorganisms, plants, and animals and comprise the largest class of enzymes that catalyze biological oxidation/reduction reactions. As so many chemical and biochemical transformations involve oxidation/reduction processes, the notion of developing practical biocatalytic applications of oxidoreductase enzymes has been an attractive goal since the very early years of biotechnology.[3]

Oxidoreductases can be classified according to their sequence or 3-D structure as oxidases, peroxidases, oxygenases/hydroxylases, or dehydrogenases/reductases. For their application, oxidoreductases can also be classified according to their signature catalysis and/or coenzyme-dependence. A few oxidoreductases are now used in textile, food, and other industrial markets and more candidates are being actively developed for future commercialization. Applications envisioned for these enzymes have included asymmetric oxyfunctionalization of steroids and other pharmaceuticals, synthesis and modification of polymers, oxidative degradation of pollutants, oxyfunctionalization of hydrocarbons, and the construction of biosensors for a variety of analytical and clinical applications. Progress has been made in the utilization of oxidoreductases in improved biosensors and in diagnostic tests, important from the standpoint of public health.

OXYGENASES

Oxygenases are among the most widely distributed enzymes. Oxygenases incorporate molecular oxygen directly into organic substrates, and do so with high efficiency and selectivity. Among the reaction types catalyzed by oxygenases are hydroxylations, epoxidations, sulfoxidations, and oxygenative dealkylations. These enzymes readily carry out oxyfunctionalization of inactivated substrates, such as simple hydrocarbons. Coupled with their high stereo- and regio-selectivity, these capabilities have engendered much interest in oxygenases for possible biotechnological applications.

Applications of Oxygenases

Industrial Applications

Carbohydrates can serve as bio-based, renewable, and inexpensive raw materials, precursors, or additives for many industry products. Historically, useful organic acids, such as lactate, have been produced from sugars by whole-cell fermentation. Cell-free oxidoreductases may also be used for organic acid production.

Commodity sugars, such as glucose and sucrose, can be modified to produce value-added sugars or other substances. For example, D-glucose can be converted to 2-keto-D-glucose by glucose 2-oxidase. In another example, lactose (the milk sugar) can be converted by a carbohydrate oxidase to lactobionic acid, a valuable food additive, acidulant, chelator, drug formulant, and polymer precursor. Glucose and other carbohydrates are rich in (pro)chiral sites, which can be functionalized by specific carbohydrate oxidoreductases. Oligo- and polymeric carbohydrates may also be oxidatively modified to gain new properties. Adding reactive groups onto sugars, via oxidoreductase catalysis, can lead to the synthesis of new materials, such as glycolipid surfactants.

Food Applications

Many oxidoreductase substrates, such as carbohydrates, unsaturated fatty acids, phenolics, and thiol-containing proteins are important components of various foods and beverages. Their modification by oxidoreductases may lead to new functionality, quality improvement, or cost reduction. Sometimes O_2 is detrimental to the quality or storage of foods because of unwanted oxidation. Oxidases may be used as oxygen scavengers for better food packaging.

Glucose oxidase is a highly specific enzyme that catalyzes the oxidation of β-glucose to glucono-1,5-lactone, which spontaneously hydrolyzes to gluconic acid using molecular oxygen and releasing hydrogen peroxide. Glucose oxidase finds applications in the removal of either glucose or oxygen from foods to improve their storage capability and has been commercialized for bread-making. Hydrogen peroxide produced by the enzyme acts as a good bactericide and can be later removed using a second enzyme, catalase, which converts hydrogen peroxide to oxygen and water. Also, egg white, as a spray-dried powder, is an important raw material for the food industry. The original liquid form contains about 4 g/dm^3 of glucose. A problem occurring during the heat treatment of eggs is browning caused by the Maillard reaction. This occurs as a result of small amounts of glucose in the egg whites reacting with amino acids. This can be problematic for dried egg whites if the product is traditionally pasteurized after drying in a hot room for an extended period of time. In this sense, it is general practice to remove glucose from the egg white before spray drying to avoid browning of the product from caramel formation. In addition, the elimination of glucose enhances the resistance of the product against the growth of microorganisms, improving the storage stability of the egg white powder.[4] Glucose oxidase can be used to remove oxygen from the top of bottled beverages before they are sealed and to prevent discoloration and flavor loss of foods and beverages. Glucose oxidase can be used in the wine industry to lower the alcohol content of wine through removal of some of the glucose. Adding the enzyme to dough can lead to various physicochemical changes including cross-linking of wheat albumin, globulin, and to lesser extent, glutenin. Consequently the dough demonstrates better viscoelastic/rheological characteristics, and the baked bread has improved crumb, larger volume, and other improved properties. Some oxidoreductases, such as glucose oxidase and catalase, have been shown able to improve the freshness preservation of shrimp and fish. It is suggested that the active oxygen species generated by the enzymes can act as bactericides.

Lipoxygenase (LOX) is one of the most widely studied enzymes in the plant and animal kingdoms. LOX catalyzes the bioxygenation of polyunsaturated fatty acids (PUFA) containing cis,cis-1,4-pentadiene units to form conjugated hydroperoxydienoic acids. LOX is a promising candidate for baking applications. Modifying the endogenous lipids/unsaturated fatty acids (and their emulsification property) and generating oxidative peroxide, the enzyme may provide dough-strengthening and bread-whitening effects; however, adding the enzyme to certain foods may cause off-flavor or loss of endogenous antioxidants and color change. The latter is due to hydroperoxide and radical formation by oxidation of lipids, which can destroy chlorophyll and carotenoids during frozen storage.[5] With regard to the enzyme-mediated oxidation, lipoxygenases seem to be the major enzymes involved in the degradation of carotenoids. The process is called "co-oxidation" because the enzyme does not act directly on carotenoid molecules and occurs when lipoxygenase oxidizes polyunsaturated fatty acids and the oxidation products of enzyme activity (peroxyl radicals) react with carotenoids. Besides the bleaching of carotenoids, enzyme-catalyzed oxidation reactions are also responsible for browning and the formation of off-flavor compounds such as aldehydes, epoxides, and carbonilic compounds. Lipid oxidation is related to chlorophylls and carotenoids degradation in vegetables such as spinach.[6] Other applications of LOX include the bleaching of noodles, whey products, rice bran, and wheat bran.[7] Similar to the case of bread, these enzymes modify flour components either directly or indirectly through production of active oxygen species. Several oxidoreductases such as glucose oxidase, lipoxygenase, peroxidase, catalase, and combinations thereof, are used for baking. Traditionally, bakers strengthen gluten by adding ascorbic acid and potassium bromide. Some oxidoreductases can be used to replace potassium bromide in dough systems by oxidation of free sulfydryl units in gluten proteins. Disulfide linkages are formed resulting in stronger, more elastic doughs with greater resistance.

Laccases are polyphenol oxidases (PPOs) that catalyze the oxidation of various substituted phenolic compounds by using molecular oxygen as the electron acceptor.[8] The ability of laccases to act on a wide range of substrates makes them highly useful biocatalysts for various biotechnological applications and may be applied to certain processes that enhance or modify the color appearance of a food or beverage. One interesting case involves the processing of ripe olives in which laccase replaces conventional lye solution and oxidatively polymerizes various phenolics (such as oleuropein) in olive, resulting in color darkening and debittering.

Browning, haze formation, and turbidity development during the processing or storage of clear fruit juice, beer, and wine can be major problems for the industry. It is believed that phenolic compounds are involved in this process. Conventionally, undesirable phenolics are adsorbed and removed by various fining agents (e.g., gelatin and bentonite) that usually have low specificity, may affect color or aroma, and can pose disposal problems. Laccase and other oxidases may be used to remove or modify problematic phenolic saccharides and improve the clarity, color appearance, flavor, aroma, taste, or stability of fruit juice or fermented alcohol beverages.

A laccase has recently been commercialized for preparing cork stoppers for wine bottles. The enzyme oxidatively reduces the characteristic cork taint/astringency, which is frequently imparted to the bottled wine. One recent example of applying oxidoreductase to dairy is the use of a carbohydrate oxidase to convert lactose during cheese-making. Lactose, found in the whey fraction, is currently discarded as a cheese-making by-product. Lactobionic acid, the product of the enzymatic oxidation, is a valuable chemical widely used as a food additive, acidulant, chelator, drug formulant, and polymer precursor. Converting the lactose in situ to lactobionic acid may have many benefits in terms of added value and improved quality. Lactobionic acid may be generated during the manufacturing of milk products such as cheese and provide the resulting product with desired organoleptic properties and a reduced content of lactose. An advantage related to preparation of dairy products with reduced lactose content and manufacturing of processed cheese products is reduced problems with browning. Additionally, in cheese production lactobionic acid may be used to develop acidity. Generally, acidity is developed by fermenting milk with lactic acid bacteria that metabolize lactose to produce lactic acid. Consequently, by addition of lactobionic acid it is possible to produce relevant dairy products using reduced amounts of lactic acid bacteria.

The main mode of entry of caffeine, theophylline, theobromine, and other natural methylxanthines into the human systems is through the consumption of coffee, tea, caffeinated cola drinks, cocoa-derived beverages, and chocolate. Even low doses of caffeine can affect the quality and quantity of sleep. Common withdrawal effects of caffeine in humans are headache, fatigue, apathy, and drowsiness. Owing to such symptoms, decaffeinated beverages are popular.

The enzymes involved in the degradation of caffeine are demethylases and oxidases. In a mixed culture consortium containing *Klebsiella* sp. and *Rhodococcus* sp., caffeine can be directly oxidized by caffeine oxidase at the C-8 position leading to the formation of 1,3,7-trimethyluric acid. The oxidative degradation of caffeine to trimethyluric acid (a single step) appears to be efficient for development of enzymatic degradation of caffeine. Studies on enzyme stability, cloning, and overexpression of this enzyme in suitable hosts may lead to development of technological processes for efficient caffeine degradation using enzymes.

Biosensors

Biosensors were first defined as analytical devices incorporating a biological material, a biologically derived material or a biomimic, intimately associated with or integrated within a physicochemical transducer or transducing microsystem. Biosensors usually yield a digital electronic signal, which is proportional to the concentration of a specific analyte or group of analytes. Biosensors have been applied to a wide variety of analytical problems in medicine, drug discovery, the environment, food, process industries, security, and defense.

A glucose biosensor based on glucose oxidase immobilized by glutaraldehyde cross-linking on an overoxidized polypyrrole platinum modified electrode has been described. Ferri et al.[9] described the direct electrochemistry of horseradish peroxidase immobilized within a polymeric film at a pyrolytic graphite electrode. These investigators successfully carried out flow and flow injection voltammetric measurements and showed that glucose oxidase or choline oxidase could simultaneously be entrapped in the polymer along with the peroxidase for determination of the analytes, glucose or choline.

Cholesterol is one of the most frequently measured analytes in clinical studies, as well as in analysis of food samples. The development of efficient rapid analytical methods for cholesterol estimation in food and clinical samples is important. HPLC and gas–liquid chromatography methods used for the determination of total cholesterol offer sensitivity and selectivity but are neither suitable for rapid analysis nor cost-effective. Currently, enzymatic procedures have practically replaced the chemical methods based on the classical Libermann–Burchard reaction, used traditionally for free and total cholesterol determinations. Most cholesterol biosensors are applied for clinical analysis,[10] but application of these sensors in the field for food sample analysis is very limited.

L-Glutamate is a commercially important biotechnological product recognized as a flavor enhancer in many kinds of food. In certain cases, the high level of L-glutamate in the final product is desirable and can be achieved through the use of microorganisms producing high levels of glutaminase; however, some manufacturers merely add L-glutamate supplements to the final product to enhance the taste. Such practices have led government agencies to invoke regulations to prevent the addition of excessive amounts of various flavors enhancers to traditional fermented products. Udomsopagit et al.[11] developed a flow injection analysis system with a modified electrode for determination of L-glutamate in food samples using L-glutamate oxidase. The enzyme catalyzes the oxidative deamination of L-glutamate to α-ketoglutarate with the generation of an equimolar amount of hydrogen peroxide. L-Glutamate can be determined by monitoring either a decrease in dissolved oxygen or an increase in hydrogen peroxide during the course of the reaction.

Biodegradation, Bioremediation, and Environmental Applications

Biodegradation of widely used pesticides, herbicides, and other agrochemicals is an important priority in technologically advanced societies, and peroxidase enzymes have considerable potential for such applications. Immobilized manganese peroxidase from *Lentinula edodes* was recently employed in a two-stage bioreactor for oxidation of chlorophenols.

In a different approach to oxidation of phenolics, a capillary membrane bioreactor has been developed and tested for the removal of phenolic compounds from synthetic and industrial effluents. This bioreactor was based on PPO immobilized on two morphologically different polymeric membranes, one of which was designed so as to facilitate high-efficiency removal of reaction products.

Fermentation of wood hydrolysates to desirable products, such as ethanol, is made difficult by the presence of inhibitory compounds in these hydrolysates. Joensson et al.[12] reported that treatment of the hydrolysate with lactase or with lignin peroxidase results in improved subsequent fermentability by *Saccharomyces cerevisiae* as measured by glucose and ethanol productivity.

CONCLUSIONS

The notion of developing practical applications of oxidase biocatalysis to address real-world problems has been an attractive objective since the very early years of biotechnology. Indeed, one can point to great progress over the past few years in the utilization of oxidases in food industry, polymer synthesis, and bioremediation and in the construction of biosensors. Further improving the cost-competitiveness of existing oxidase products and enhancing the innovation effort in applying oxidases to new fields are vital for the future growth of industrial oxidoreductase biocatalysts.

REFERENCES

1. Mchedlishvili, N.I.; Omiadze, N.T.; Gulua, L.K.; Sadunushvili, T.A.; Zamtaradze, R.K.; Abutidze, M.O.; Bendeliani, E.G.; Kvesitadze, G.I. Thermostabilities of plant phenol oxidase and peroxidase determining the technology of their use in the food industry. Appl. Biochem. Microbiol. **2005**, *41*, 145–149.
2. Xu, F. Applications of oxidoreductases: recent progress. Ind. Biotechnol. **2005**, *1*, 38–50.
3. May, S.W. Applications of oxidoreductases. Curr. Opin. Biotechnol. **1999**, *10*, 370–375.

4. Sisak, C.; Csanádi, Z.; Rónay, E.; Szajáni, B. Elimination of glucose in egg white using immobilized glucose oxidase. Enzyme Microb. Technol. **2006**, *39*, 1002–1007.
5. Zhuang, H.; Barth, M.M.; Hildebrand, D.F. Packaging influenced total chlorophyll, soluble protein, fatty acid composition and lipoxygenase activity in broccoli florets. J. Food Sci. **1994**, *59*, 1171–1174.
6. Lopez-Ayerra, B.; Murcia, M.A.; Garcia Carmona, F. Lipid peroxidation and chlorophyll levels in spinach during refrigerated storage and after industrial processing. Food Chem. **1998**, *61*, 113–118.
7. Baysal, T.; Demirdöven, A. Lipoxygenae in fruits and vegetables: a review. Enzyme Microb. Technol. **2007**, *40*, 491–496.
8. Sharma, P.; Goel, R.; Capalash, N. Bacterial laccases. World J. Microbiol. Biotechnol. **2007**, *23*, 823–832.
9. Ferri, T.; Poscia, A.; Santucci, R. Direct electrochemistry membrane entrapped horseradish peroxidase part II: amperometric detection of hydrogen peroxide. Bioelectrochem. Bioenerg. **1998**, *45*, 221–226.
10. Basu, A.K.; Chattopadhyay, P.; Roychoudhuri, U.; Chakraborty, R. Development of cholesterol biosensor based on immobilized cholesterol esterase and cholesterol oxidase on oxygen electrode for the determination of total cholesterol in food samples. Bioelectrochemistry **2007**, *70*, 375–379.
11. Udomsopagit, S.; Suphantharika, M.; Künnecke, W.; Bilitewski, U.; Bhumiratana, A. Determination of L-glutamate in various commercial soy sauce products using flow injection analysis with a modified electrode. World J. Microbiol. Biotechnol. **1998**, *14*, 543–549.
12. Joensson, L.J.; Palmqvist, E.; Nivebrant, N.O.; Hahn-Haegerdal, B. Detoxification of wood hydrolysates with laccase and peroxidase from the white-rot fungus *Trametes versicolor*. Appl. Biochem. Microbiol. **1998**, *49*, 691–697.

Pediococcus

Jeff Broadbent
Department of Nutrition and Food Sciences, Utah State University, Logan, Utah, U.S.A.

James L. Steele
Department of Food Science, University of Wisconsin-Madison, Madison, Wisconsin, U.S.A.

Abstract

Pediococcus spp. are members of the industrially important lactic acid bacteria. This entry provides an overview of *Pediococcus* ecology and metabolism, genetics and gene transfer systems, and their application in food biotechnology.

INTRODUCTION

The genus *Pediococcus* has undergone changes in recent years, and currently holds 11 species. Although most commonly associated with fermenting plant material, pediococci have been isolated from a variety of fermenting foods, including alcoholic beverages and ripening cheese. This entry provides a summary of their distinguishing characteristics, important metabolic properties, and current knowledge of gene transfer systems and genetic elements. Finally, it outlines the role of these organisms in food biopreservation, safety, and spoilage.

CHARACTERISTICS OF PEDIOCOCCI

Ecology

Pediococcus spp. are Gram-positive, facultatively anaerobic, non-motile, and non-sporing members of the industrially important lactic acid bacteria (LAB). The genus is currently composed of 11 species whose identity and distinguishing characteristics are listed in Table 1. Two species once classified as pediococci, *P. halophilus* and *P. urinaeequi*, have been reassigned as *Tetragenococcus halophilus* and *Aerococcus urinaeequi*, respectively. Vegetable matter is the most common natural habitat, and pediococci may be isolated from a variety of fermenting vegetable foods and, less frequently, fermenting alcoholic beverages such as beer and wine, ripening cheese, and modified-atmosphere-packaged meat products.[1] Though generally viewed as non-pathogenic, pediococci are intrinsically resistant to vancomycin and infrequently have caused bacteremia, septicemia, and other ailments in debilitated or immunocompromised patients.[1] In addition, a few strains can produce biogenic amines.[2]

Classification and Morphology

Phylogenetically, *Pediococcus* falls into a subcluster of the *Lactobacillus* group that also includes *Leuconostoc* and *Oenococcus*.[3] A distinct morphological characteristic of pediococci (spherical cells; 0.4–1.4 μm in diameter) involves the formation of tetrads via cell division in two perpendicular directions in a single plane (Fig. 1); however, tetrad formation is not always apparent and cells may commonly be observed as pairs or single cocci.[1] Chains longer than cell pairs are not seen.

Metabolism

Like other LAB, pediococci are acid-tolerant chemoorganotrophs with complex growth requirements that typically include one or more vitamins and amino acids. Manganese (Mn^{+2}) is also required as its addition can significantly accelerate acid development in fermentations with pediococci.[2] Members of the LAB group also lack the ability to synthesize porphyrin and possess a strictly fermentative metabolism with lactic acid as the major metabolic end product of carbohydrate metabolism. Energy is derived from fermentation of simple mono- and disaccharides, but some species or strains possess enzymes to break down more complex sugars (*P. dextrinicus*, e.g., is able to utilize starch). Under microaerophilic or anaerobic conditions where substrate is not limiting, glucose and other hexose sugars are fermented to lactic acid at near-theoretical yields via the Embden–Meyerhof pathway (2 mol lactic acid plus 2 ATP per mole glucose). About half of the species in this genus (Table 1) can ferment ribose and perhaps other pentose sugars using enzymes from the lower half of the 6-phosphogluconate pathway,[3] which yields 1 mol each of lactic and acetic acids plus 1 mol ATP per mole ribose. Glucose, and probably most other

Table 1 Characteristics of individual *Pediococcus* species.[a]

Characteristic	*P. acidilactici* ATCC 33314	*P. cellicola* LMG 22956	*P. claussenii* ATCC BAA-344	*P. damnosus* ATCC 29358	*P. dextrinicus* ATCC 33087	*P. ethanolidurans* LMG 23354	*P. inopinatus* ATCC 49902	*P. parvulus* ATCC 19371	*P. pentosaceus* ATCC 33316	*P. siamensis* NRIC 0675	*P. stilesii* LMG 23082
Type strain[b]											
Mol% G + C	42	38	40.5	38.5	40.5	39.5	39.5	41	38	42	38
Lactate type	DL	DL	L-(+)	DL	L-(+)	DL	DL	DL	DL	DL	DL
Growth at: 35°C	+	+	+	−	+	+	+	+	+/−	+	+
45°C	+	+	−	−	−	+	−	−	+/−	+	+
48°C	+	NR	−	−	−	NR	−	−	−	NR	−
pH 4.5	+	+	+	+	−	+	+	−	+	NR	+
pH 7.0	+	+	+	−	+	+	+/−	+/−	+	+	+
pH 9.0	−	−	−	−	−	−	−	−	−	−	+
4% NaCl	+	+	+	−	+	−	+/−	+	+	+	+
6.5% NaCl	+	−	−	−	−	−	−	+	+	−	+
6.5% Ethanol	NR	+	NR	+	−	+	−	−	NR	NR	NR
7.5% Ethanol	NR	+	NR	+	−	−	−	−	NR	NR	NR
Ability to ferment ribose	+	+	+	−	−	−	−	−	+	−	+

[a] Abbreviations: +, 90% or more strains positive; −, 90% or more strains negative; +/−, 11–89% of strains positive; NR, not reported in the literature.
[b] ATCC, American Type Culture Collection, Manassas, VA, United States; LMG, Culture Collection of the Laboratorium voor Microbiologie Gent, Universiteit Gent, Belgium; NRIC, Nodai Research Institute Culture Collection, Tokyo University of Agriculture, Japan.

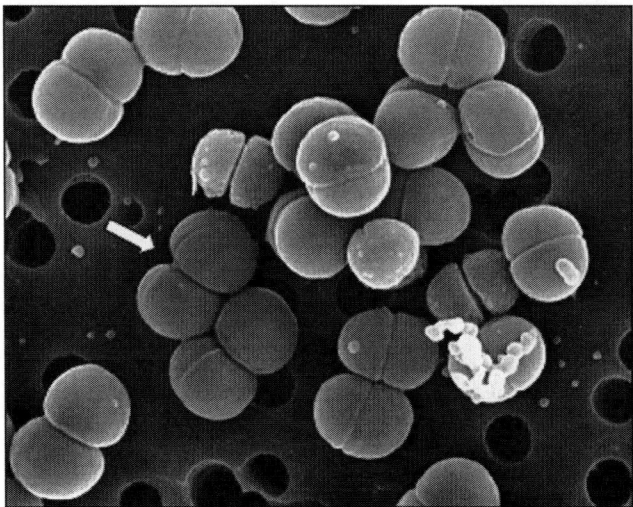

Fig. 1 Scanning electron micrograph of *Pediococcus pentosaceus* ATCC 25745. The arrow identifies one clearly visible tetrad.

mono- and disaccharides utilized, are taken up via sugar-specific phosphoenolpyruvate-dependent phosphotransferase systems (PTS); however, the *P. pentosaceus* genome sequence[3] provides evidence that in a few instances, sugar transport may occur via proton symport (e.g., arabinose) or facilitated diffusion (e.g., ribose).

The absence of enzymes for porphyrin synthesis translates into an inability to produce functional heme-containing proteins such as cytochromes or catalase. Despite this limitation, many pediococci are remarkably aerotolerant and one species, *P. pentosaceus*, may produce a "pseudo-catalase" that does not require heme, while another, *P. acidilactici*, may produce true catalase if heme is available in the medium.[4] Although detailed biochemical characterization of "pseudo-catalase" from pediococci has not been reported yet, two genes for non-heme Mn-catalases were identified in the genome sequence for *P. pentosaceus* ATCC 25745.[3] One of these genes is predicted to encode a protein with more than 90% identity to a well characterized Mn-catalase from *Lactobacillus plantarum*,[5] while the other shows strong homology (>50% identity) to several putative Mn-catalases in *Bacillus* sp.

GENETICS

Genetic information for most species is limited, but studies of two species that are of particular biotechnological interest, *P. pentosaceus* and *P. acidilactici*, have provided a sound knowledge base. Research to date has demonstrated gene transfer by transformation and conjugation, and characterized four major genetic elements: 1) the chromosome; 2) plasmid DNA; 3) transposable elements; and 4) bacteriophage.

The Chromosome

Because essential housekeeping, catabolic, and biosynthetic activities of the cell are encoded by the chromosome, detailed knowledge of its content and organization holds tremendous fundamental and applied value. The information that is available to date (which includes the genome sequence of *P. pentosaceus* ATCC 25745)[3] show that pediococci have a relatively small (~1.8–2.1 Mb pairs), single, and circular chromosome. DNA fingerprinting studies also reveal that there is considerable genomic diversity throughout the genus.[6]

The chromosome of *P. pentosaceus* ATCC 25745, a plant isolate, is 1,832,387 bp in length; it includes five rRNA operons, 55 tRNA genes, and has a G + C content of 37.4%.[3] The chromosome is predicted to contain 1,757 open-reading frames (ORFs) of which about 80% (a relatively high fraction) have a predicted function. Forty-three ORFs are forecasted to encode proteins found in no other organism. Public access to this sequence has opened the door for researchers worldwide to assemble a comprehensive view of the evolution, genetics, and physiology of this species, and examples of the insight already provided by this genome are distributed throughout this entry.

The tools of genomics can also identify strategies for industrial strain improvement. For example, Diep et al.[7] discovered a pediocin-like locus in the genome of ATCC 25745, even though bacteriocin production was not detected. Heterologous expression in *Lactobacillus sakei* of the *P. pentosaceus* genes confirmed that they did encode a novel broad-spectrum bacteriocin, named penocin A. Moreover, the absence of penocin A expression in ATCC 25745 was shown to be due to a frameshift mutation in the penocin A inducer gene (caused by a single nucleotide insertion), and restoration of the gene led to high level penocin production by ATCC 25745.[7] Thus, a ready means to significantly enhance the value of this culture in biopreservation and food safety was revealed through genomics.

Plasmid DNA

Plasmid DNA is a frequent component of the genome in several *Pediococcus* spp., including *P. acidilactici*, *P. damnosus*, *P. parvulus*, and *P. pentosaceus*. Some strains possess several resident plasmids, which may range in size from 1.8 kb to more than 190 kb.[4]

The presence of plasmids is fortuitous because it provides a ready source of extrachromosomal replicons to support the development of basic genetic tools for biotechnology (e.g., vectors for gene cloning and expression). Sequence analysis of several relatively small (3.3–19.5 kb) pediococcal plasmids suggest that most undergo rolling circle replication (RCR), but evidence for theta-type plasmid replication (which affords greater stability to large fragments of insert DNA) also exists.

Most of the plasmids identified in pediococci remain cryptic (encode no known functions beyond replication/segregation); however, phenotypic and genetic studies have identified plasmids that encode bacteriocin (i.e., pediocins) production, exopolysaccharide production, raffinose utilization, melibiose utilization, sucrose utilization, and antibiotic resistance.[1] In particular, the discovery that pediocin production is plasmid-coded in several *Pediococcus* sp. has fueled biotechnological interest in these organisms.

Insertion Sequences

Insertion sequences (IS) are small (1–2 kb) genetic elements that can move from one site to another in DNA and generally encode only transposase and *cis*-acting sequences required for mobility. Nucleotide sequence data for the *P. pentosaceus* genome and various *Pediococcus* sp. plasmids suggest that IS are not as abundant in this genus as in other LAB[3]; however, IS elements representing the *IS*30 and *IS*3 families have been identified in *P. pentosaceus* and an element from the *IS*30 family has been found in *P. damnosus*.

Bacteriophages

Although bacteriophages are known to cause fermentation failure in the dairy industry, little is known about their influence in vegetable or alcohol fermentations that involve pediococci. Descriptions of pediococcal bacteriophages are limited to two types of temperate bacteriophage in *P. acidilactici*[8] and two prophage gene clusters in the *P. pentosaceus* ATCC 25745 genome.[3] All belong to the *Siphoviridae* family.

Gene Transfer Systems

Access to efficient gene transfer mechanisms is critical for modern genetics research and industrial strain improvement. To date, two types of gene transfer, conjugation and transformation, have been described in both *P. acidilactici* and *P. pentosaceus*.[1] Conjugation is a natural form of gene transfer among bacteria that involves physical contact between viable donor and recipient cells. Transfer of plasmid DNA by conjugation between pediococci and several other LAB has been demonstrated.[1]

Transformation involves the uptake of naked DNA into cells, as development of an efficient transformation system enables use of recombinant DNA technology. At present, the only effective method for transformation of pediococci is electroporation, a process that involves exposure to a high voltage electric field. Electroporation of plasmid DNA into *P. acidilactici* and *P. pentosaceus* has been demonstrated by several groups.[1]

APPLICATIONS IN BIOTECHNOLOGY

Food Fermentation

Lactic acid fermentation is one of the oldest methods known for food biopreservation. Pediococci are important contributors to many natural lactic fermentations (see section "Ecology"), but addition of pure cultures accelerates acid development and provides greater control over spoilage and pathogenic microbes. In particular, *P. pentosaceus* and *P. acidilactici* have been used as acid-producing starter cultures for manufacture of cucumbers, sauerkraut, and olives, and for a variety of fermented moist, semi-dry, and dry sausages[4] These cultures have also been used in soy milk fermentations and for the production of silage, a fermented animal food.[1] Caldwell et al.[9] have demonstrated that *P. pentosaceus* and *P. acidilactici* genetically modified to rapidly ferment lactose, may have value as bacteriophage-resistant starter cultures in Mozzarella cheese fermentation.

Food Safety

One of the most active areas for the application of pediococci in biotechnology involves use of pediocin to enhance food safety. Pediocin-producing cultures have been used to inhibit *Listeria monocytogenes* and other pathogens in ready-to-eat meats, cheese, and other foods, and at least one pediocin-containing fermentate has been commercialized as a food preservative.[10]

Nearly all pediocins are class IIa bacteriocins (small heat-stable cationic peptides with antilisterial activity), but class I (lanthionine-containing peptides with relatively broad spectra of inhibition) and class III (large, heat-labile proteins) bacteriocins may also be produced by pediococci.[1] Class IIa bacteriocins have the following general structure: 1) a conserved N-terminus that includes a YGNGV motif; 2) two Cys residues involved in a disulfide linkage; 3) a region with a high density of cationic residues; and 4) a more divergent C-terminal region important for target cell specificity.[10] Recent research suggests that the membrane proteins IIC^{man} and IID^{man} of the mannose PTS system serve as the target/receptor molecule for several class IIa bacteriocins.[11] In this model, bacteriocin-producing cells synthesize an immunity protein that forms a strong complex with IIC/IID^{man} and the bacteriocin that shields these cells from the harmful effects of the bacteriocin. In the absence of immunity protein, bacteriocin alone binds to the target resulting in membrane permeabilization and, ultimately, cell death.[11]

Knowledge of the mechanism(s) by which pediocins inhibit *L. monocytogenes* and other cells is critical, as a major limitation to bacteriocin use in food involves the frequency at which undesirable bacteria develop resistance. A more detailed understanding of resistance and immunity should reveal new strategies to combat this limitation and extend the utility of pediocins and other bacteriocins as food preservatives.

Food Spoilage

Any discussion of pediococci in food would be incomplete without mention of the contribution of these bacteria to the spoilage of beer, wine, and other alcoholic beverages. Pediococci are known to cause acid off-flavor in beer and wine via fermentation of residual sugar, reduce the acidity of wine via the malolactic fermentation (which may or may not be desirable), produce an undesirable buttery odor in beer as the result of diacetyl production, and cause sliminess or thickening in wine due to polysaccharide production.[12] The *P. pentosaceu*s ATCC 25745 chromosome contains gene clusters associated with each of these capabilities, thereby providing new opportunities to understand the microbial physiology responsible for spoilage of these foods. This greater understanding may reveal new strategies to control spoilage of these products.

CONCLUSIONS

Pediococci have important roles in food biopreservation, safety, and spoilage, but basic knowledge of many important attributes remains limited. Genomic studies should help fill knowledge gaps by providing a comprehensive view of the enzymes and metabolic pathways related to key properties, and also help identify rational strategies for genetic improvements to industrial strains.

REFERENCES

1. Holzapfel, W.H.; Franz, C.M.A.P.; Ludwig, W.; Back, W.; Dicks, L.M.T. The genera *Pediococcus* and *Tetragenococcus*. In *Prokaryotes*; Dworkin, M., Falkow, S., Rosenberg, E., Schleifer, K.-H., Stackebrandt, E., Eds.; Springer: New York, 2006; Vol. 4, 229–266.
2. Raccach, M. Pediococci and biotechnology. Critical Reviews in Microbiology **1987**, *14* (4), 291–309.
3. Makarova, K.; Slesarev, A.; Wolf, Y.; Sorokin, A.; Mirkin, B.; Koonin, E.; Pavlov, A.; Pavlova, N.; Karamychev, V.; Polouchine, N.; Shakhova, V.; Grigoriev, I.; Lou, Y.; Rohksar, D.; Lucas, S.; Huang, K.; Goodstein, D.M.; Hawkins, T.; Plengvidhya, V.; Welker, D.; Hughes, J.; Goh, Y.; Benson, A.; Baldwin, K.; Lee, J.-H.; Díaz-Muñiz, I.; Dosti, B.; Smeianov, V.; Wechter, W.; Barabote, R.; Lorca, G.; Altermann, E.; Barrangou, R.; Ganesan, B.; Xie, Y.; Rawsthorne, H.; Tamir, D.; Parker, C.; Breidt, F.; Broadbent, J.; Hutkins, R.; O'Sullivan, D.; Steele, J.; Unlu, G.; Saier, M.; Klaenhammer, T.; Richardson, P.; Kozyavkin, S.; Weimer, B.; Mills, D. Comparative genomics of the lactic acid bacteria. Proceedings of the National Academy of Science USA **2006**, *103* (42), 15611–15616.
4. Ray, B. *Pediococcus* in fermented foods. In *Food Biotechnology: Microorganisms*; Hui, Y.H.; Khachatourians, G.G., Eds; VCH Publishers: New York, 1995; 745–795.
5. Barynin, V.V.; Whittaker, M.M.; Antonyuk, S.V.; Lamzin, V.S.; Harrison, P.M.; Artymiuk, P.J.; Whittaker, J.W. Crystal structure of manganese catalase from *Lactobacillus plantarum*. Structure **2001**, *9* (8), 725–738.
6. Simpson, P.J.; Stanton, C.; Fitzgerald, G.F.; Ross, R.P. Genomic diversity within the genus *Pediococcus* as revealed by randomly amplified polymorphic DNA PCR and pulsed-field gel electrophoresis. Applied and Environmental Microbiology **2002**, *68* (2), 765–71.
7. Diep, D.B.; Godager, L.; Brede, D.; Nes, I.F. Data mining and characterization of a novel pediocin-like bacteriocin system from the genome of *Pediococcus pentosaceus* ATCC 25745. Microbiology **2006**, *152* (pt 6), 1649–1659.
8. Caldwell, S.; McMahon, D.J.; Oberg, C.J.; Broadbent, J.R. Induction and characterization of *Pediococcus acidilactici* temperate bacteriophage. Systematic and Applied Microbiology **1999**, *22* (4), 514–519.
9. Caldwell, S.L.; McMahon, D.J.; Oberg, C.J.; Broadbent, J.R. Development and characterization of lactose-positive *Pediococcus* species for milk fermentation. Applied and Environmental Microbiology **1996**, *62* (3), 936–941.
10. Drider, D.; Fimland, G.; Héchard, Y.; McMullen, L.M.; Prévost, H. The continuing story of class IIa bacteriocins. Microbiology and Molecular Biology Reviews **2006**, *70* (2), 564–582.
11. Diep, D.B.; Skaugen, M.; Salehian, Z.; Holo, H.; Nes, I.F. Common mechanisms of target cell recognition and immunity for class II bacteriocins. Proceedings of the National Academy of Science USA **2007**, *104* (7), 2384–2389.
12. Lonvaud-Funel, A. Lactic acid bacteria in the quality improvement and depreciation of wine. Antonie Van Leeuwenhoek **1999**, *76* (1–4), 317–331.

Pest Management: Population Theory

Alan Andrew Berryman
Department of Entomology, Washington State University, Pullman, Washington, U.S.A.

Abstract
Integrated Pest Management (IPM) is concerned with the management of pest *populations*. The overall goal of IPM is to maintain pest populations at low, nondamaging, densities.

INTRODUCTION

Pest managers tend to be practical people. The reason is simple: their clients need immediate solutions to their pest problems, for without them their crops can be destroyed and their profits wiped out in a single growing season. Practical people tend to look for quick and effective solutions to their problems—the proverbial "silver bullet." Thus it was when DDT was discovered in 1939 and hailed as the miracle insecticide—the final answer to the insect problem.[1] Today there is the tendency to look to genetic engineering or biotechnology for the silver bullet.

Academics, on the other hand, are often theoretically minded. They are mainly interested in understanding how nature works rather than in the realities of crop production. Nature is often complex, and its understanding and explanation may require complex reasoning, complex jargon and, sometimes, complex mathematical models. It is not surprising that practical pest managers and theoretical academics often find it difficult to communicate.

WHY THEORY IS IMPORTANT TO PEST MANAGERS

Integrated Pest Management (IPM) students sometimes ask me why they should learn esoteric theory when their job is to solve practical problems. My usual answer is to draw on analogy: Engineers are also practical people but their education is burdened with mathematics, mechanics, and physics, much of it at the theoretical level. Engineers realize that they cannot build bridges and rockets without considering the laws of nature. How can one put a man on the moon without understanding the theories of gravity and planetary motion? Impossible! When engineers ignore or overlook theory, they sometimes reap the dire consequences. And so it was with the Tacoma Narrows Bridge, destroyed by harmonic motion during a violent windstorm. If the engineers had paid more attention to the theories of complex dynamics and harmonics, they might have avoided this disaster.

When DDT was first employed to control insects, some academics (those who understood the theory of evolution and took it seriously) predicted that insects would soon become resistant to the new pesticide. Although most pest managers now understand the theory of evolution and use it in their everyday operation (e.g., resistance management), they generally ignored or ridiculed the pessimistic academics when DDT first came on the scene. They had their silver bullet and it worked—for a time.

Theory provides us with an understanding of nature, and this understanding enables us to anticipate the consequences of our actions. Theory, therefore, is an essential part of any thoughtful and responsible human activity, including pest management.

WHY POPULATION THEORY?

IPM is concerned with the management of pest *populations*. It seems reasonable, therefore, to expect pest managers to understand what causes populations to change in time and space—what is generally known as *population dynamics*.[2] This is not to say that other theoretical knowledge is not important. We have already seen that an understanding of evolutionary theory is essential if we are to anticipate and manage pesticide resistance. I could also argue that a theoretical understanding of community ecology, food webs, and predator–prey dynamics is also necessary. What I am saying is that pest managers should, at the very least, have a good understanding of population dynamics theory.

Population theory concerns itself with the general explanation and understanding of changes in numbers, density, age-classes, and so on in populations of living organisms. However, because it is the density of pests per unit area of crop or cropland that determines the level of economic damage, pest managers are usually concerned more with changes in population density than with other population variables.

A theory can be viewed as a systematic statement of the principles and relationships underlying an observed natural phenomenon. Population dynamics theory involves two sets of principles and relationships, some pertaining to dynamic systems in general and others to populations of living organisms in particular.

The first general principle recognizes that changes in any dynamic system can be caused by either outside (*exogenous*) or internal (*endogenous*) effects. Exogenous effects cause changes in the system but are themselves unaffected by those changes. For example, weather can affect insect population dynamics, say, by killing insects, but is not influenced itself by insect numbers. In ecological parlay, exogenous effects are called *density-independent* because they act independently of pest density.

On the other hand, endogenous effects are caused by feedback loops within the dynamic system (notice that a dynamic system is thus defined as a group of variables linked together by feedback loops).[2] For example, suppose that a pest population increases for some reason or another. This provides more food for predators, causing their populations to increase. More predators eat more prey, and so the pest population is eventually suppressed. Notice that an original *increase* in pest density is followed by an eventual *decrease* in density through the action of predators. This is called a *negative* feedback loop because the initial change results in an opposing effect (increase in some quantity results in an eventual decrease in that quantity, or vice versa). Notice that the negative feedback action causes the density of the pest population to return towards its original level and, therefore, opposes changes in the characteristic state of the system. In other words, negative feedback loops tend to stabilize dynamic systems around what are called *equilibrium points* (Fig. 1A). Engineers would say that negative feedback is necessary to stabilize the dynamics of rockets and airplanes, or to keep them on track (the track being the equilibrium point). However, although negative feedback is necessary for stable dynamics, it is not sufficient. To be stable, negative feedback must act *rapidly*, otherwise the variables may oscillate around their equilibrium points, with the degree of oscillation being directly dependent on the length of the feedback time-delay (Fig. 1B). The speed of action of a negative feedback loop depends, to a large degree, on the number of variables involved in the loop because each variable needs time to change, quantitatively, in response to the other; e.g., predators require time to produce more offspring after being confronted with more prey.

Endogenous effects can also be created by *positive feedback*. However, unlike the stabilizing effects of negative feedback, positive feedback induces instability by accentuating or amplifying changes in the system. Positive feedback is the force behind population explosions, inflation spirals, arms races and, interestingly, the process of evolution. Positive feedback can also create unstable breakpoints or thresholds in dynamic systems and are important in the theory of insect outbreaks (Fig. 1C).[3]

In ecology, endogenous feedback loops are usually referred to as *density-dependent* because the feedback is induced by, and effects, population density. Negative feedback loops are then identified as *direct* density-dependent (or just plain density-dependent) and positive feedback

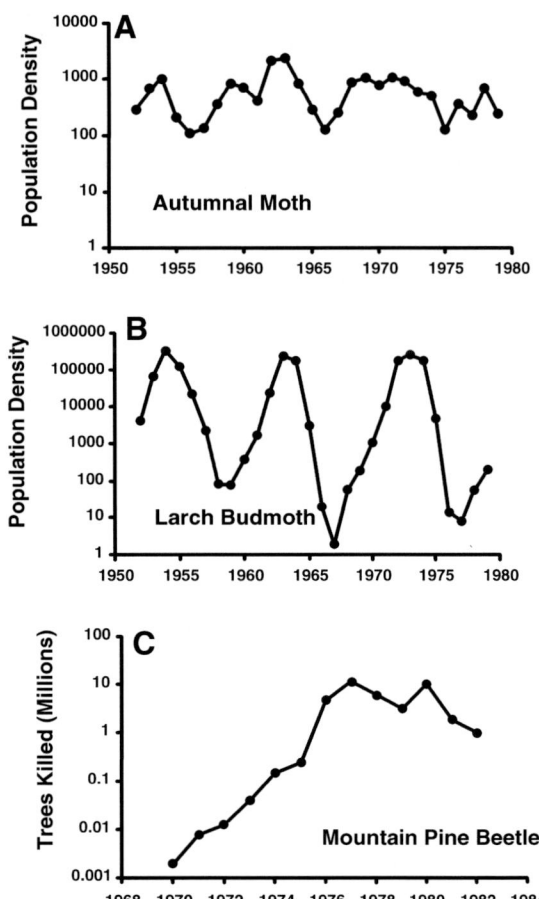

Fig. 1 Three major types of population dynamics: (A) Relatively stable population of the autumnal moth in the Swiss Alps. Notice that numbers fluctuate around a characteristic, or average, level of abundance. (B) Strongly oscillatory (cyclic) population of the larch budmoth in the Swiss Alps. Notice that densities oscillate by several orders of magnitude in a regular 9-year cycle. (C) Unstable outbreak of the mountain pine beetle in Glacier National Park. Notice that annual tree killing increased from a few thousand in 1970 to more than 10 million in 1977 as the beetle population erupted through the park.

inverse density-dependent. *Delayed* density-dependent refers to negative feedback loops that operate slowly and, thereby, create the conditions for oscillatory dynamics. Direct density-dependence usually results from competition between individuals for food, hiding places, or other resources.[2] It can also result from the feeding of general predators that switch to the most abundant prey species and/or aggregate in areas where prey are abundant. Delayed density-dependence often arises from feedback between a species and its food supply or natural enemies. Finally, inverse density-dependence usually results from cooperation between individuals during defense or hunting activities.

These general principles and relationships outlined above are important to the pest manager. The overall goal of IPM is to maintain pest populations at low, nondamaging, densities. In general systems parlay, this means creating a stable *and* low equilibrium. In order to do this we must either create our own negative feedback loop or take advantage of natural negative feedback loops. For example, the development of an economic action level, a pest density above which the grower initiates control actions, creates a negative feedback loop because the decision to control depends on pest density (direct density-dependent). Biological control, of course, is the use of natural feedback loops, as between the pest and its enemies, to control the pest population at subeconomic levels. A thorough understanding of this, however, may require one to delve into what is called predator-prey theory.[4,5] In contrast, the pest manager usually tries to avoid instabilities, such as pest population explosions (outbreaks) created by positive feedback. The theory of outbreaks[3] is yet another theoretical area important to the pest manager. Of course, there is much more I could say about the theory of population dynamics, but space is limited. Nevertheless, I hope to have convinced the reader that theoretical knowledge is important for the intelligent management of pest problems, and that population theory is perhaps of prime importance. Those who wish to dig deeper can consult the following references.

REFERENCES

1. Berryman, A.A. Population theory: an essential ingredient in pest prediction, management, and policy-making. Am. Entomol. **1991**, *37*, 138–142.
2. Berryman, A.A. *Principles of Population Dynamics and Their Application*; Stanley Thornes: Cheltenham, UK, 1999.
3. Barbosa, P.; Schultz, J.C. *Insect Outbreaks*; Academic Press: New York, 1987; 202–204.
4. Huffaker, C.B.; Gutierrez, A.P. *Ecological Entomology*, 2nd Ed.; John Wiley: New York, **1999**.
5. Hawkins, B.A.; Cornell, H.V. *Theoretical Approaches to Biological Control*; Cambridge University Press: Cambridge, UK, 1999.

Pesticides: History

Edward H. Smith
Cornell University, Ithaca, New York, U.S.A.

George Kennedy
Department of Entomology, North Carolina State University, Raleigh, North Carolina, U.S.A.

Abstract

Pests—insects, nematodes, plant pathogens, and weeds—destroy more than 40% of the world's food, forage, and fiber production. The struggle, pests versus people, grows ever more intense as population increases, arable land decreases, and human intervention disturbs biotic relationships on a global scale. Pesticides play a vital role in the struggle, but their use has not been without adverse ecological impacts and risk to the safety of those who apply them and those who consume treated products. This brief essay recounts the human experience with pesticides from the dawn of history to the dawn of the twenty-first century. It is the story of trial and error, old problems, and new lessons on nature's response to insult by ingenious synthetic molecules. It chronicles the intellectual probing and public debate of problems associated with the overreliance on chemical control that developed following World War II and led to adoption of integrated pest management (IPM) as an ecologically viable paradigm for crop protection. The success or failure of pest control programs may well hold the key to world order as the six billion peoples of the world compete for their place in the sun.

And he gave it for his opinion, that whoever could make two ears of corn or two blades of grass to grow upon a spot of ground where only one grew before, would deserve better of mankind, and do more essential service to his country, than the whole race of politicians put together.
—Jonathan Swift, 1726

THE INEVITABLE CONFLICT

Agriculture, which dates back a mere 15,000 years, requires the modification of natural systems. Agricultural practice imposes ecological simplicity on biota driven by natural selection toward diversity. Agriculture swims against the ecological tide and crops must be protected from the Darwinian struggle for existence. Intervention is required in many forms, including the use of pesticides. The conflict is inevitable.

THE ASCENDANCY OF PESTICIDES

The biblical records provide insight into the philosophy surrounding humankind's encounters with pests. According to Judeo-Christian beliefs, man was accorded dominion over the plants and animals. Departures from the laws of God were punished by plague. "I have smitten you with blasting and mildew your trees the palmer worm devoured them. Yet have ye not returned unto me," (Amos 4:9).

Development of the agricultural sciences progressed slowly. The emergence during the Renaissance of Natural Theology, which reconciled science and religion, followed by invention of the printing press, the microscope, and the Linnean system of biological nomenclature set in place the elements for rational thought and communication on natural history and on pest control.

The status of insect pest control in early nineteenth century Europe and North America is revealed by T.W. Harris's publication, *Report on the Insects of Massachusetts Injurious to Vegetation* (1841), prepared at the request of the state legislature. Harris, a Harvard librarian and meticulous scholar, drew upon European literature as well as American agricultural journals, which published pest control recommendations offered by their readers. His control measures included: hand picking; burning stubble and field refuse; smoke screens in orchards to drive moths away; running pigs in the orchards; poison baits; resistant varieties (wheat); favorable planting dates; dusting with ashes, quick lime, red pepper, sulfur, and tobacco; spraying with whitewash and glue; and encouraging woodpeckers in orchards. (How to do the latter was not specified.) While these measures were crude, they were based on reason and fragmentary knowledge of pest biology. They were free of the ridiculous nostrums proposed earlier out of ignorance, superstition, and fraud, and they represented early steps in cultural, biological, mechanical, and chemical control. A foundation had been laid drawing on knowledge from Europe and North America aided by state subsidy, a renowned educational institution, and an able scholar.

Recognition that abundant and reliable agricultural production was prerequisite to urbanization and industrial development helped to trigger the agricultural revolution in the United States, which began in the 1840s as canals and railroads linked the eastern population centers with expanding agricultural lands west of the Mississippi. The next great impetus to pest control in the United States came through congressional passage of the Morrell Act in 1862, establishing the Land Grant University System. This paved the way for professionals in the applied science of pest control, and was followed in 1887 by passage of the Hatch Act, establishing a coordinated system of State Experiment Stations devoted to the advancement of research. The Smith-Lever act of 1914 officially recognized the extension arm of the Land Grant University System and completed the American model that has proven to be one of the most innovative educational concepts of all time.

Annual reports on beneficial and injurious insects issued between 1856–1876 by early leaders such as Asa Fitch, (New York), B. D. Walsh (Illinois), and C. V. Riley (Missouri) became the backbone of applied entomology. These writers urged natural controls as the first line of defense and expressed their misgivings about the crude chemical controls of the time.

Pest control practices were strongly influenced by expanding commerce, which resulted in the introduction of exotic pests, and by the rapid, westward expansion of agriculture, which disrupted ecosystems and exposed crops to new pests. The Colorado potato beetle, *Leptinotarsa decemlineata* (Say), provides a prime example. It appeared as a devastating pest of potato in Iowa and Nebraska in 1861, having transferred from a native weed to an introduced relative, the potato. The beetle spread rapidly eastward, reaching the Atlantic coast in 1874, despite the use of traditional nonchemical means of control. In 1867, farmers in the west discovered that the Colorado potato beetle could be controlled with Paris Green, an arsenical. Paris Green was in general use by 1880 and became the first widely used pesticide in North America. Similar experiences followed with other major pests, such as the plum curculio *Conotrachelus nenuphar* (Herbst), boll weevil *Anthonomous grandis grandis* Boheman, gypsy moth *Lymantria dispar* (Linnaeus) and others.

During the first half of the nineteenth century, lime-sulfur and wettable sulfur gradually came into use for control of fungal pathogens, primarily of fruit trees and grapevines in Europe and the United States. In 1885, Pierre Milardet, professor of botany at Bordeaux, France, demonstrated control of downy mildew, *Plasmopora viticola*, on grapevines using a mixture of copper sulfate and lime, subsequently known as Bordeaux mixture. The success of Bordeaux mixture led to efforts to improve upon its effectiveness and to the expanded use worldwide of it and its variants.

Other components—petroleum oil, nicotine, pyrethrum, and organomercury fungicides for seed treatment—were soon added to the pesticide arsenal. By 1910, the arsenicals Paris Green, lead arsenate, and calcium arsenate were the most widely used pesticides. Herbicides were notably absent; they did not appear until the discovery of plant growth hormones paved the way for the synthesis of stable synthetic hormone analogues (2,4-D and 2,4,5-T) in the 1940s.

Farmers, their advisors in the fledgling Land Grant Universities, and an emerging chemical industry rallied behind pesticides, especially insecticides, for one pragmatic reason; they provided a degree of reliability in control programs that was absent with other available methods. World War I stimulated pesticide use for food production. It also stimulated the production of insecticides, such as dinitrophenols (DNOC) and paradichlorobenzene (PDB), as by-products of the manufacture of explosives from coal tar.

On the eve of World War II, insecticides were the backbone of insect control but their use was fraught with unease. A host of problems surfaced. Control was marginal. British markets rejected U.S. apples because of the high arsenical residues. There were concerns for the health of workers and consumers. The codling moth *Cydia pomonella* (Linneaus) acquired resistance to arsenicals; excessive pesticide treatments were phytotoxic to foliage causing reduced yields, and there were concerns about the build-up of residues in the soil. To many entomologists, it appeared that they were losing the fight. They clung to the early, idyllic hope for control by natural means but in the crunch of practical experience, they turned to pesticides because they worked, not well, but better than the alternatives.

DDT: DISCOVERY, DEVELOPMENT, AND IMPACT

The discovery and introduction of DDT, while purely a commercial enterprise, became immediately enmeshed in the intrigue and urgency of World War II. The Swiss chemist Paul Mueller, an employee of J.R. Geigy Co., discovered the insecticidal property of DDT in September 1939; this event coincided with the Nazi invasion of Poland. DDT found a vital military role in the control of insect-borne diseases. When the war ended in 1945, DDT, the shining chemical sword of World War II, found extensive peacetime use. It was distributed quickly for testing through the well-organized network of Agricultural Experiment Stations. Data poured in confirming the effectiveness of DDT against a wide spectrum of insect pests of agricultural and medical importance. In striking contrast to the prewar pessimism, DDT produced hope that at last the age-old insect scourges could be controlled and perhaps eradicated.

Such optimism had a profound effect on the crop protection sciences. In entomology and weed science especially, research shifted focus away from pest biology and on to pesticide technology. At this point, the birthright of pest control scientists as biologists became endangered. Insecticide use soared, based on the promise of DDT and the related chlorinated hydrocarbon insecticides that followed. New classes of insecticides, the organophosphates and the methylcarbamates, were discovered and exploited. The success of 2,4-D for control of broadleaf weeds stimulated the development and use of chemical weed control. Similarly, the discovery of the dithiocarbamate fungicides during the 1930s led to the development of an array of very effective fungicides and increased fungicide use. All this was catalyzed by a powerful coalition: the chemical industry with its high capitalization and integrated skills in synthesis, testing, and marketing; the agricultural community with considerable political clout; and the Land Grant Universities with their triple mission of teaching, research, and extension. Pesticide use and reliance on pesticides for crop production increased steadily.

REBUFF AND REASSESSMENT

The euphoria that accompanied the dominance of chemical control in the 1950s was short-lived. By the end of the decade, warnings about the adverse effects of pesticides were being expressed by environmentalists and some pest control specialists, but these were largely ignored. There was fear within the crop protection disciplines, especially entomology, that reliance on pesticides was placing agriculture on a "pesticide treadmill." There were problems with resurgence of targeted pest populations and outbreaks of secondary pest populations following destruction of their natural enemies, and with the development of pesticide resistance, all of which necessitated additional applications of pesticides. Similar concerns surfaced regarding control of medical and veterinary pests.

In 1962, Rachel Carson's book *Silent Spring* galvanized public attention on the problems spawned by pesticide use. She made her case with poetic beauty sounding the alarm that "we have put poisonous and biologically potent chemicals indiscriminately in the hands of persons largely or wholly ignorant of their potentials for harm." What had been a debate among scientists became a public debate. Drawing on lessons of the civil rights movement, the antipesticide forces headed by the Environmental Defense Fund turned to litigation in defense of the right of citizens to a clean environment. After long, contentious hearings, the Environmental Protection Agency banned DDT in 1972. This landmark decision placed the issue of pesticides in the forefront of the greatly energized environmental movement.

Increased public activism over environmental and food safety issues, which began during the 1960s and continues today, led to dramatic changes in pesticide regulation and to restrictions on pesticide use in both the United States and Europe. These actions dramatically strengthened the environmental and toxicological standards that pesticides must meet before they can be approved for use. In doing so, these changes provided strong impetus for the development of safer and more environmentally friendly pesticides.

The regulatory framework for pesticides continues to broaden as new knowledge is acquired and perceptions change. For instance, in 1996 the U.S. Congress passed the Food Quality Protection Act, which established more stringent safety standards aimed at protecting infants, children, and other sensitive subpopulations from risks associated with pesticide residues on food. Subsequently, several major food processors imposed their own more stringent tolerances for pesticide residues on the produce that they purchase. The process is expected to continue in response to the ebb and flow of new findings and public concern.

In the late 1950s and early 1960s, growing awareness of the problems associated with pesticide use and the specter of faltering pest control, viewed in the context of decreasing availability of arable land and dwindling supplies of fossil fuel to drive the technology of agribusiness, stimulated a reassessment of pest control. Earlier work in biological control in several countries provided points of departure but it was the intellectual probing in entomology at the University of California (Berkeley and Riverside) that ignited a great debate, which in time involved pest control specialists the world over. The topic of debate was the concept of integrated pest management (IPM). IPM emphasized that pest problems were under the influence of the total agroecosystem and that not all levels of pest abundance required treatment with pesticides. It also emphasized that pest management should be a multidisciplinary effort based on ecological principles and economic, social, and environmental considerations.

While the concept soon gained widespread acceptance, many factors impeded its implementation. The knowledge base was in most cases inadequate; the research and extension infrastructure required redirection; a corps of private consultants to supplement decision making by farmers had to be recruited and trained; and replacement of broad-spectrum pesticides was slow and costly. Federal and state governments were sold on the soundness of the concept and appropriated funds to overcome these constraints.

Four decades after the initiation of IPM, the steering mechanism for sound employment of pesticides, what is the score? The glass is half full. Great strides have been made on a worldwide scale. IPM has provided the framework to accommodate transition from singular reliance on broad-spectrum, long-residual pesticides to the use of highly selective, short-residual compounds as components of multifaceted crop protection programs, without an increase in losses to pests. Every phase of the University support network—teaching, research and extension—has been altered to reinforce the ecological foundations of IPM. The disappointing aspects are that adoption of programs has been slow, pesticides still predominate in many programs, overall use of pesticides has not declined, and successful interdisciplinary programs are few. Despite ongoing improvements in the characteristics of pesticides, the specter of pest resistance hangs like the sword of Damocles over the utility of pesticides.

Great challenges lie ahead as concepts of sustainable agriculture and the technology of genetic engineering meld with the ever-expanding scope of IPM. IPM has become a unifying catalyst, an intellectual quest that unites producers, plant protection disciplines, agribusiness, regulatory agencies, and the worldwide plant protection community concerned with the production of food and fiber for the six billion peoples of the world.

Genetic engineering technology will have broad application in control of pests of plants and animals. Using this technology, genes from one organism can be inserted into and expressed in totally different organisms. Genetic engineering is producing new kinds of insecticidal peptides and proteins, and is enabling plants, bacteria, and viruses to be used in novel ways to deliver toxins to targeted pests. The same technology has produced plants that are tolerant to broad spectrum, postemergence herbicides. The possibilities seem limitless. (For more comprehensive treatment of this subject see the entry by M.G. Paoletti in this encyclopedia.)

In a remarkably short span of three decades, the science of biotechnology became an applied technology and a new industry, involving new kinds of partnerships between university scientists and entrepreneurs. Overnight, genetically engineered crops were being planted on millions of acres in the United States.

The speed of scientific and technological advance in genetic engineering and the new partnership between universities and industry have given rise to a host of challenging issues: academic freedom in the context of university/industry partnerships; patenting of biological processes; monopolies; economic impact, particularly on developing countries; response of organisms to selective pressure (resistance). The most daunting questions focus upon risk assessment and regulatory procedures addressing the impact of organisms created outside the normal evolutionary pathways on the global biota.

Political debate on these issues has grown in intensity and rancor, first in Europe and then in the United States. While the time frame of debate and acceptance of genetic engineering is in doubt, it is clear that the tremendous pressures to meet food requirements for a world population of nine billion by 2050 are likely to force the incorporation of genetically engineered components into the arsenal of pest control.

The question germane to the present essay is what part will pesticides play in future IPM programs. They will be a vital component but in a modified role. Advances in toxicology, chemistry, biochemistry, physiology, molecular biology, and computer modeling are making possible the tailoring of pesticides to meet IPM requirements, which dictate low mammalian toxicity, high specificity conferring low environmental impact, and low residues on treated products. In the future, pesticides will be used with greater precision, made possible by improvements in pesticide application technology, pest and crop monitoring, weather prediction, and information processing, as well as by better understanding of population dynamics, microbial and weed ecology, and epidemiology.

It is important to note the influence of economics and elevated standards for pesticide potency and safety. These factors have dramatically increased the costs of pesticide discovery and development and have contributed to an internationalization and consolidation of the pesticide industry. They have also resulted in fewer new pesticides being introduced. While the agrichemical industry has not enjoyed a favorable public image in an era of environmental awareness, it should be remembered that it plays a vital role in the multifaceted IPM enterprise.

THE FUTURE

History should illuminate the future. We see pest control as a challenge woven into the economic, political, and social fabric of society. Major factors are shaping the new era of pest control.

Public Attitude

Growing environmental ills will further sensitize the public to problems arising from technology. This will find expression in stricter pesticide regulation and safer pesticides.

Global Commerce

Increased and increasingly rapid international movement of goods and people will intensify the introduction of exotic species and the spread of pesticide-resistant organisms.

Economics

The ever rising cost of developing new pharmaceuticals and pesticides will constrain research and product development, and the use of pesticides in developing countries.

Population Pressure

The environmental stress imposed by rapid growth of the human population will continue to exacerbate problems of agricultural production, including pest control. This is perhaps the most serious problem facing humankind, with no relief in sight.

Throughout the latter half of the twentieth century, pesticides contributed enormously to improvements in the quality and stability of the world's food supply and to the control of devastating insect-transmitted diseases of humans and livestock. Pesticides have also played a central role in fostering environmental awareness and public concern over food safety. The inevitable conflict between humans and pests will grow in intensity as the human population grows and arable land decreases. Pesticides, because of their ease and rapidity of use and the reliability with which they can rein in pest outbreaks, will continue to play an important role in IPM. Lessons having been learned, pesticides of the future will be safer and more environmentally friendly, and will be used more judiciously than in the past. Our crystal ball discerns no "silver bullet" of pest control, rather painstaking refinement of IPM, with further advances in established methods, including pesticides and biological control, and a melding of new technologies such as genetic engineering.

BIBLIOGRAPHY

Adler, E.F.; Wright, W.L.; Klingman, G.C. Development of the American Herbicide Industry. In *Pesticide Chemistry in the 20th Century*; Plimmer, J.R., Ed.; American Chemical Society: Washington, DC, **1977**; 39–55.

Brent, K.J. In *One Hundred Years of Fungicide Use, Fungicides for Crop Protection 100 Years of Progress,* Proceedings of The Bordeaus Mixture Centenary Meeting Smith, I.M., Ed.; British Crop Protection Council Publications: Croyden, UK, **1985** 11–22, **1985**; Monograph No. 31; 1.

Carson, R. *Silent Spring*; Houghton Mifflin: Boston, 1962.

Cassida, J.E.; Quistad, G.B. Golden age of insecticide research: past, present, or future. Annu. Rev. of Entomol. **1998**, *41*, 1–16.

Howard, L.O. *A History of Applied Entomology*; Smithsonian Institution: Washington, DC, **1930** .

Knight, S.C.; Anthony, V.M.; Brady, A.M.; Greenland, A.J.; Heany, S.P.; Murray, D.C.; Powell, K.A.; Shulz, M.A.; Spinks, C.A.; Worthington, P.A.; Youle, D. Rationale and perspectives on the development of fungicides. Annu. Rev. of Phytopathol. **1997**, *35*, 349–372.

Lever, B.G. *Crop Protection Chemicals*; Ellis Horwood: New York, **1990**.

Marco, G.J.; Hollingworth, R.M.; Plimmer, J.R. *Regulation of Agrochemicals: A Driving Force in Their Evolution*; American Chemical Society: Washington, DC, **1991**.

Perkins, J.H. *Insects, Experts, and the Insecticide Crisis: The Quest for New Pest Management Strategies*; Plenum: New York, **1982**.

Zimdahl, R. *Fundamentals of Weed Science,* 2nd Ed.; Academic Press: New York, **1999**.

Photosensitization and Food Safety

Zivile Luksiene
Institute of Materials Science and Applied Research, Vilnius University, Vilnius, Lithuania

Abstract
Photosensitization is a treatment involving the interaction of the two non-toxic factors, photoactive compound and visible light, which in the presence of oxygen, results in the selective destruction of the target cells. Different microorganisms, such as multidrug-resistant bacteria, yeasts, microfungi, and viruses are susceptible to this treatment. Therefore, a photosensitization phenomenon might open a new avenue for the development of non-thermal, effective, and ecologically friendly antimicrobial technology, which might be applied for food safety.

INTRODUCTION

Traditional thermal technologies are effective antimicrobial treatments, but usually they induce a lot of uncontrolled chemical reactions in the food matrix and later, reduce the nutrition quality of it. The emerging food safety technologies being effective decontamination tools make mostly different undesirable changes in the food matrix.[1]

A new approach to inactivate pathogenic and harmful microorganisms in cost-effective, non-thermal, and environmentally friendly way is highly needed. To this end, photosensitization might serve as a promising tool to decontaminate food and food-related surfaces.

Photosensitization is a treatment involving the administration of a photoactive compound that selectively accumulates in the target cells and the following illumination. The interaction of two non-toxic elements, photoactive compound, and visible light, in the presence of oxygen results in a plethora of cytotoxic reactions and consequently induces selective destruction of target microorganism.

PHOTOPHYSICAL AND PHOTOCHEMICAL PROCESSES

Absorption of light by a photosensitizer results in the excitation of molecule from the ground state (S_0) to the excited singlet state (S_1) (Fig. 1). The molecule can relax to the S_0 by fluorescence or by internal conversion, whereby the energy is lost as heat to the surroundings. The third way to relax is intersystem crossing, when the excitation transfers from the S_1 state to the lower excited triplet T_1 state with a longer lifetime (Fig. 1). Relaxation from the T_1 state results in either phosphorescence or induction of two types of photooxidative reactions. Type I pathway involves electron or hydrogen atom transfer, producing radical forms of the photosensitizer or the substrate. These intermediates may react with oxygen to form peroxides, superoxide ions, and hydroxyl radicals. Type II mechanism is mediated by an energy transfer process with ground state oxygen ($^3O_2$). The destruction of a cell is strictly localized due to a very short half-life of $^1O_2$ (nsec) and consequently short diffusion distance (20 nm).[2,3] As a consequence, triggered cytotoxic reactions induce the disruption of cell membrane, inactivation of enzymes, damage of DNA, and cell death.[4]

PHOTOSENSITIZERS

Photosensitizers, derived from vital stains, are known to be non-toxic in much higher concentrations than those required for effective pathogen killing. Hypericin, a natural pigment present in *Hypericum perforatum*, is described as "one of the most powerful photosensitizers in nature," and is devoid of toxic or genotoxic effects.[4]

Chlorophyll sodium salt exhibited high photosensitizing activity against food pathogens,[5,6] and can be used for the decontamination of fruits, vegetables, and food-related surfaces as it belongs to food additives in EU classification (E100-E199, it is E140[color]). Its concentration for food, cosmetics, and drugs is not limited as it is based on native product (regulated by directives 88/343 EEB, 97/60 EB ir 94/36 EB).

Some bacteria are known to produce significant amounts of endogenous porphyrins[7,8] (Fig. 2). Thus, it seems reasonable to exploit cell metabolism for the production of endogenous photosensitizers using a precursor 5-aminolevulinic acid (ALA) (Fig. 2). ALA is a naturally occurring metabolite produced during heme synthesis in eukaryotic as well as prokaryotic cells, which induces the production of endogenous photosensitizer protoporphyrin IX, uroporphyrin, and coproporphyrin. Hence, Gram

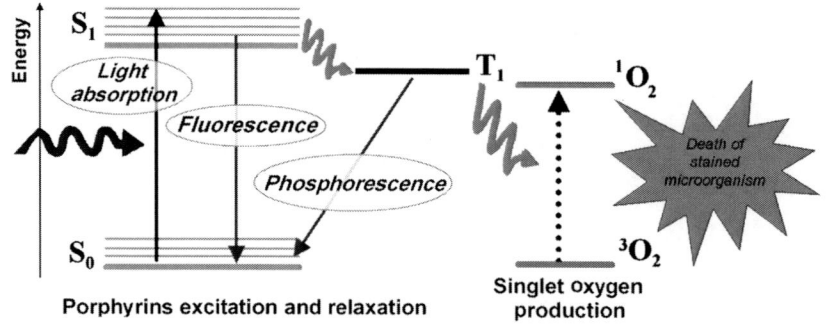

Fig. 1 Scheme of photosensitization.

(+) bacterium *Bacillus cereus* is producing endogenous porphyrins 10 times more than Gram (−) *Salmonella enterica*.[7,8] As the outer membrane of Gram (−) bacteria forms a physical and functional barrier for different compounds, one of the approaches for their inactivation can be the use of positively charged photosensitizers.[4]

Considering the application of ALA-based photosensitization to decontaminate food matrix, the question how does ALA interact with food matrix arises. Our previous experiments on the decontamination of wheat sprouts by ALA indicated that ALA stimulates the growth of wheat seedlings and roots without impairing the vigor of germination and the viability of seeds. Moreover, 5-ALA increases the rate of photosynthesis and the activities of antioxidant enzymes, which can be associated with enhanced cellular capacity to detoxify reactive oxygen species.[9] Moreover, ALA is an essential precursor of such tetrapyrole compounds as vitamin B12 and hemes, which serve as prosthetic groups of respiratory enzymes and chlorophyll in plants.[9] Suitable ALA concentrations have promotive effects on the growth rates and photosynthesis. Crop yields were enhanced by the application of ALA at the leaf stage for rice, barley, potato, garlic, and palm.[10]

To summarize, a desirable photosensitizer, which can be used for food microbial control, must be of high chemical purity, high killing efficiency with a lack of mutagenicity, or genotoxicity. It must be food additive or food component, low-cost, without strong color, taste, and flavor, and must work at very low concentrations with no negative effect on nutritional as well as organoleptic properties of the foods.

LIGHT SOURCES FOR PHOTOSENSITIZATION

The wavelength of light necessary for the induction of lethal photobiological reaction in a microorganism depends on the structure and electron absorption spectrum of the photosensitizer.[2] Common porphyrin and chlorine derivatives have a characteristic absorption band between 400 and 430 nm (Soret band).[3] Also, the wavelength determines the penetration depth of light into tissue, i.e., 400–500 nm light penetrates by about 300–400 μm (surface treatment), whereas 600–700 nm does by about 50–200% more (deeper treatment).

Historically, photosensitization has been performed using conventional gas discharge and incandescent lamps equipped color glass filters for narrowing the spectrum. Alternative light sources for activation of photosensitizers are light emitting diodes (LEDs) which feature numerous advantages over conventional sources of light, such as low driving voltage, robustness, shock and vibration resistance, the absence of mercury, compactness, lightweightness, and narrow-band emission. This makes them attractive for usage in photosensitizing luminaries that can be safe, portable, battery driven, free of thermal side effect, and low maintenance.[4]

MECHANISM OF MICROBIAL INACTIVATION

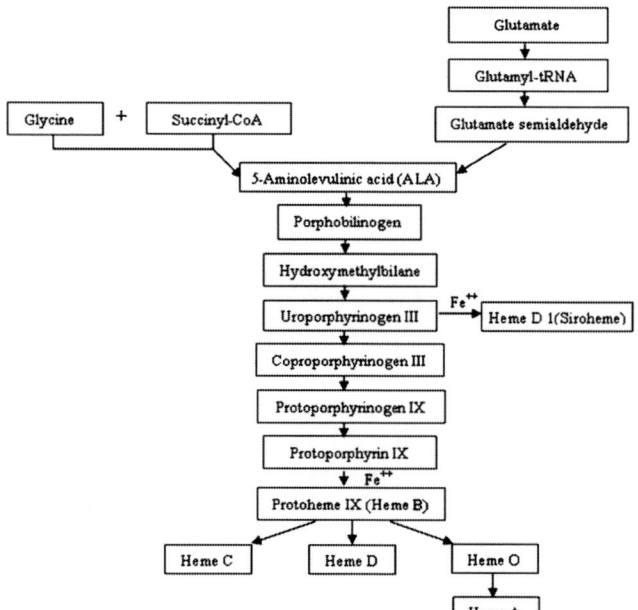

Fig. 2 Synthesis of endogenous porphyrins from exogenously applied ALA. **Source:** From Hamblin and Hasan (2004).

ALA-based photosensitization was used to effectively inactivate main food pathogens.[7,8] It is important to note that

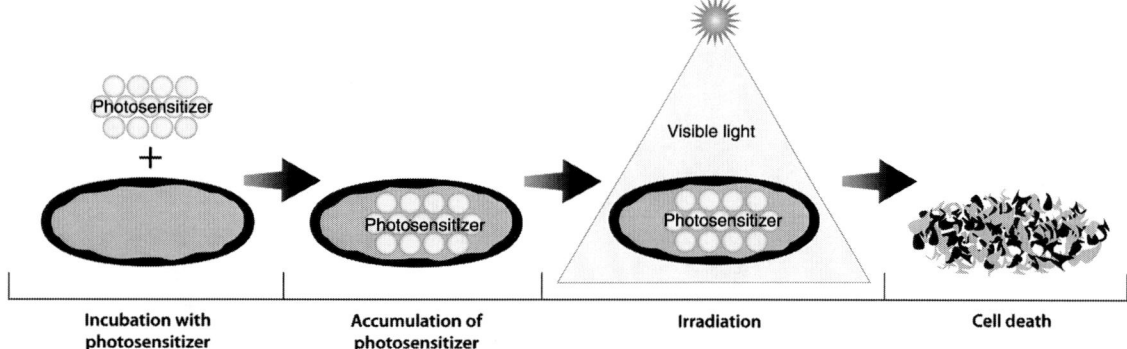

Fig. 3 Mechanism of destructive action of photosensitization in the cell: P, photosensitizer; P_1, excited state of photosensitizer after absorption of light; $^1O_2$, singlet reactive oxygen.

Bacillus spores being important target for food safety technologies can be inactivated by photosensitization. Moreover, *Listeria* biofilms which spread on different food-contact surfaces being resistant to hypochlorite can be destroyed by photosensitization as well.[5] In addition, plethora of yeasts, harmful and pathogenic micromycetes might be inactivated by photosensitization.[11]

Photosensitization is able to inactivate viruses. Positively charged photosensitizers cause nucleic acid damage, whereas anionic photosensitizers act against the viral envelope.[4]

There was significant difference in susceptibility to photosensitization between Gram (+) and Gram (−) bacteria. As a rule, neutral or anionic photoactive dyes might efficiently bind and after the illumination, inactivate Gram (+) bacteria. Gram (+) bacteria have a cytoplasmic membrane surrounded by relatively porous layer of peptidoglycan (15–80 nm) and lipoteichoic acid traversing this wall and allowing the photosensitizer to cross it.[4] Taking into account that the molecular weight of most photosensitizers does not exceed 1500–1800 kDa, the outer wall of Gram (+) bacteria is not permeability barrier. The cell envelope of Gram (−) bacteria consists of an inner cytoplasmic membrane and an outer membrane which are separated by the peptidoglycan-containing periplasm. It inhibits the penetration of several compounds, thus even hydrophilic 600–700 kDa molecules can diffuse through the porin channel.[4]

After irradiation by visible light, reactive oxygen species induce rapid disruption of the cell wall (Fig. 3). Reactive ROS interacts with unsaturated fatty acids, amino acid residues, and bases of nucleic acid. Breaks in both single- and double-stranded DNA have been detected in bacteria after photosensitization. *Deinococcus radiodurans*, having very efficient DNA repair mechanism, can be easily killed by photosensitization as well.[4]

A very attractive feature, peculiar to photosensitization as antimicrobial treatment, is the possibility of reactive oxygen species to destroy secreted virulence factors.

Photosensitization might help to overcome the problem of bacterial multidrug resistance. For instance, Gram (+) bacteria such as *Staphylococcus aureus*, *D. radiodurans*, or Gram (−) bacterium *Acinetobacter baumannii* are actually very sensitive to this treatment.[4]

DECONTAMINATION OF FOOD OR FOOD-RELATED SURFACES BY PHOTOSENSITIZATION

Photosensitization being environmentally friendly, effective, and free from chemical substances was used for the decontamination of germinated wheats (Fig. 4).[9]

Data obtained on freshly cut fruits and vegetables for ready to eat meals clearly show that the surface decontamination level from pathogens and naturally distributed mesophils after chlorophyll sodium salt-based photosensitization (3 log) is higher in comparison with hypochlorite, which in addition is instable, toxic, induces discoloration of products, and rapid regrowth of microorganisms. Decontamination of fruits and vegetables by photosensitization expanded their shelf life without any negative effects on their organoleptic properties.[6]

Photosensitization can serve as one of the possible alternatives to chemical decontamination of packaging. *B. cereus* as well as *Listeria monocytogenes* were inactivated by 3–4 log on the surface of food packaging material (polyolefine). In addition, inactivation of *Listeria* biofilms by 3.1 log suggests that this treatment has potential to combat biofilms. Moreover, population of Bacillus spores adhered on the surface of packaging material was reduced by 3.8 log after photosensitization.[5,8]

CONCLUSIONS

The field of antimicrobial struggle must be emphasized as one of permanent challenges. The discovery of novel, cost-effective, and human friendly non-thermal technology to inactivate harmful and pathogenic microorganisms seems an imperative. To this end, photosensitization is really effective technique. First, photosensitization has been shown to be effective against vegetative bacterial cells,

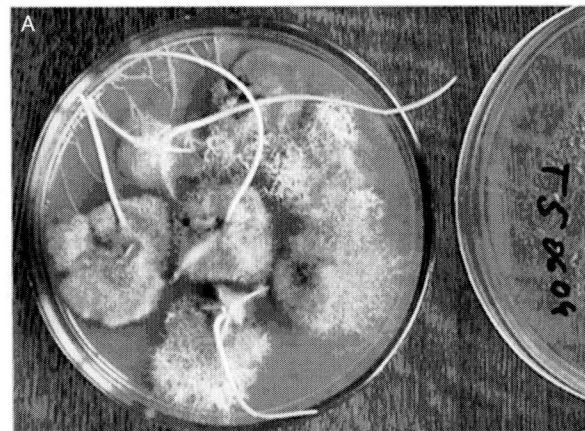
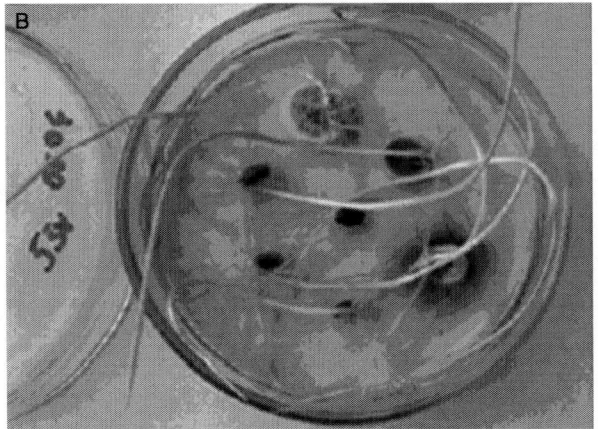

Fig. 4 Contamination of organic wheat sprouts by microfungi in control (A) and treated by photosensitization (B) samples.

their spores and biofilms, microfungi, yeasts, and viruses at non-thermal treatment conditions. Second, it induces loss of viral infectivity, is non-mutagenic or genotoxic. Third, no bacterial resistance to this treatment was detected.

In our opinion, this phenomenon opens a new and prospective avenue for the development of effective, human and environmentally friendly antimicrobial treatment. Its proper application for the treatment of food, packaging material, and processing equipments might be really effective to increase microbial food control.[12] Taking into account that the application of photosensitization treatment to decontaminate foods or food-related surfaces is still in its infancy, further deep research is needed on examination of shelf life and nutrition quality of treated foods.

REFERENCES

1. Manas, P.; Pagan, R. Microbial inactivation by new technologies of food preservation. J. Appl. Microbiol. **2005**, *98*, 1387–1399.
2. Durantini, E.N. Photodynamic inactivation of bacteria. Curr. Bioact. Compd. **2006**, *2*, 127–142.
3. Jori, G.; Brown, S.B. Photosensitized inactivation of microorganisms. Photochem. Photobiol. Sci. **2004**, *3*, 403–405.
4. Luksiene, Z. New approach to inactivate harmful and pathogenic microorganisms: photosensitization. Food Technol. Biotechnol. **2005**, *43*, 1–8.
5. Paskeviciute, E.; Buchovec, I.; Luksiene, Z. Control of food pathogens by chlorophyll-based photosensitization. J. Process. Energy Agric. **2009**, *13*, 181–183.
6. Paskeviciute, E.; Buchovec, I.; Luksiene, Z. Photosensitization as novel approach to decontaminate fruits and vegetables. J. Process. Energy Agric. **2009**, *13*, 50–53.
7. Buchovec, I.; Vaitonis, Z.; Luksiene, Z. Novel approach to control *Salmonella enterica* by modern biophotonic technology: photosensitization. J. Appl. Microbiol. **2009**, *106*, 748–754.
8. Luksiene, Z.; Buchovec, I.; Paskeviciute E. Inactivation of food pathogen *Bacillus cereus* by photosensitization *in vitro* and on the surface of packaging material. J. Appl. Microbiol. **2009**, *107*, 2037–2046.
9. Luksiene, Z.; Danilcenko, H.; Taraseviciene, Z.; Anusevicius, Z.; Maroziene, A.; Nivinskas H. New approach to the decontamination of germinated wheat from microfungi: effects of aminolevulinic acid. Int. J. Food Microbiol. **2007**, *116*, 153–158.
10. Hotta, Y.; Tanaka, T.; Tanaka, H.; Takeuchi, Y.; Konnai, M.; Al-Khateeb, H. Promotive effects of 5-aminolevulinic acid on the yield of several crops. Plant Growth Regul. **2002**, *2*, 109–114.
11. Luksiene, Z.; Peciulyte, D.; Jurkoniene, S.; Puras, R. Inactivation of possible fungal food contaminants by photosensitization. Food Technol. Biotechnol. **2004**, *43*; 335–341.
12. Luksiene Z.; Buchovec I. Method of decontamination of food and food related surfaces. Lithuanian patent, OB 2009/02, February 25, 2009.

Phytases

Xingen Lei
Department of Animal Science, Cornell University, Ithaca, New York, U.S.A.

Jesus M. Porres
Department of Physiology, University of Granada, Grenada, Spain

Abstract

Phytases have emerged as effective tools to improve phosphorus nutrition and to protect the environment from phosphorus pollution in animal production. Phytate is poorly available to simple-stomached animals such as swine, poultry, and the preruminant calves, due to the lack of phytases in their gastrointestinal tracts. As a result, a large portion of feed phosphorus is not utilized by them and ends up in manure, causing environmental pollution. Expensive and nonrenewable inorganic phosphorus needs to be added to diets for these species to meet their nutrient requirements for phosphorus if phytases are not provided.

INTRODUCTION

Phytases are meso-inositol hexaphosphate phosphohydrolases that catalyze the initiation of the stepwise phosphate splitting of phytic acid or phytate to lower inositol phosphate esters and inorganic phosphate (Fig. 1). These enzymes have emerged as effective tools to improve phosphorus nutrition and to protect the environment from phosphorus pollution in animal production. Although phosphorus is an essential nutrient to all species, 60–80% of phosphorus in feeds of plant origin is in the form of phytate (*myo*-inositol hexakisphosphate) that is poorly available to simple-stomached animals such as swine, poultry, and the preruminant calves, due to the lack of phytases in their gastrointestinal tracts. As a result, a large portion of feed phosphorus is not utilized by them and ends up in manure, causing environmental pollution. Meanwhile, expensive and nonrenewable inorganic phosphorus needs to be added to diets for these species to meet their nutrient requirements for phosphorus.

GENES, PROTEINS, AND PROPERTIES

A number of phytase genes and proteins have been identified from microorganisms and plants after the isolation of the very first phytase (PhyA) protein and DNA sequence from *Aspergillus niger*.[1] It remains unclear whether phytase is expressed in animal tissues. PhyA and most fungal phytases have molecular mass ranging from 80–120 kDa, with 10 or so *N*-glycosylation sites in the approximately 1.4-kb DNA sequences.[1] The average molecular masses of most bacterial phytases range from 40 to 55 kDa.[2] Plant phytases isolated from corn, wheat, lupine, oat, or barley have molecular sizes ranging from 47 to 76 kDa.[3]

Most identified phytases, but not all, belong to a group of histidine acid phosphatases (HAPs) that feature the conserved active site hepta-peptide motif RHGXRXP and the catalytically active dipeptide HD.[1] This group of phytases catalyzes phytic acid hydrolysis in a two-step mechanism via a nucleophilic attack from the histidine in the active site of the enzyme to the scissile phosphoester bond of phytic acid. In general, fungal phytases (E.C. 3.1.3.8) initiate the splitting of the phosphate group at the C_1 or C_3 carbon of the inositol ring, and are thus called 3-phytases, whereas plant phytases (E.C. 3.1.3.26) act preferentially at the C_6 carbon, and are named 6-phytase. However, phytases isolated from *Escherichia coli*, *Lupinus albus*, or *Peniophora lycii* are exceptions to this rule. Interestingly, a soybean phytase is a purple acid phosphatase with a dinuclear iron–iron or iron–zinc center in the active site. The phytase from *Bacillus subtilis* has a six-bladed folding scaffold, and calcium ion can affect its thermostability and catalysis. As a whole, phytases show a strong ability to cleave equatorial phosphate groups, but a limited ability to hydrolyze axial phosphate groups. The optimum pH for most known phytases is in the range of 4.5–6. Exceptions are phytases from mung bean, *Enterobacter* sp., or *B. subtilis* that have their pH optimum in the neutral to alkaline range. The temperature optimum of most plant and microbial phytases ranges from 45–65°C, higher than body temperatures of animals (37–40°C). Phytase activity unit is defined by the amount of inorganic phosphate released per minute from a selected substrate under certain pH and temperature. In the case of *A. niger* PhyA, the activity is determined in 0.05–0.2 M citrate or acetate buffer, pH 5.5, at 37°C; and one unit equals the amount of enzyme that releases 1 μmol of inorganic phosphorus per minute from sodium phytate. The biochemical properties of the currently available phytases for animal feeding are summarized in Table 1.

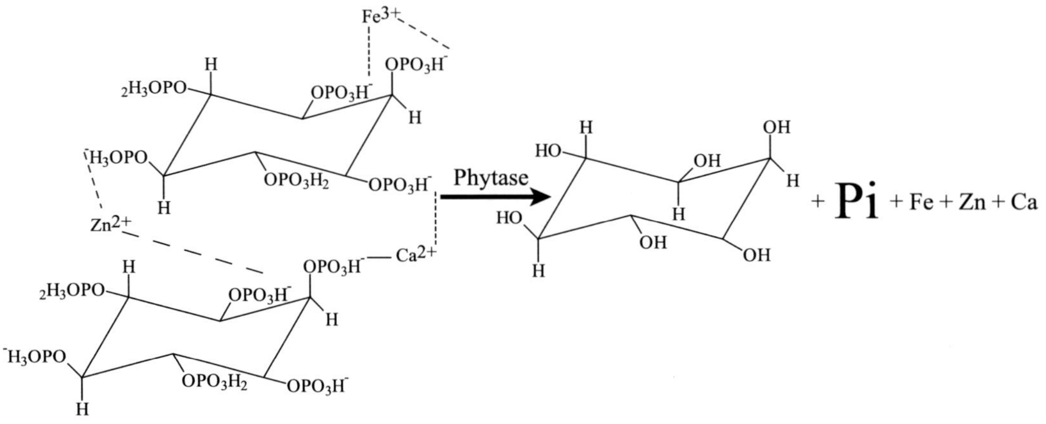

Fig. 1 Hydrolysis of phytate by phytase into inositol and phosphate. Phytate hydrolysis also releases chelated metals such as iron, zinc, and calcium.

SUBSTRATE OCCURRENCE

Phytases from *A. niger*, *Aspergillus terreus*, *E. coli* or *Bacillus* sp. seem to have a high specificity for phytic acid, whereas plant phytases and some fungal enzymes such as the one from *Aspergillus fumigatus* have a broader substrate specificity. Chemically, phytic acid refers to *myo*-inositol-1,2,3,4,5,6-hexakis dihydrogen phosphate, which contains approximately 30% phosphorus. Phytate and phytin refer to salts of phytic acid with individual or mixed metals such as calcium, sodium, potassium, magnesium, iron, zinc, copper, etc. In reality, all three of these compounds are indistinguishably called phytate. Total contents of phytate are 0.5–1.9% in cereals, 0.4–2.1% in legumes, 2.0–5.2% in oil seeds, and 0.4–7.5% in protein products.[4] Distribution of phytate varies with seeds. It is located in the germ of corn, in crystalloid-type globoids inside the protein bodies within the cotyledons of dicotyledoneous seeds (beans, soy, nuts, peanuts), or in the globoids of aleurone grains (protein bodies) present in the aleurone or bran layer of monocotyledoneous seeds (wheat, rice).[4] Phytate rapidly accumulates in seed ripening and serves mainly as storage of phosphorus, inositol, and minerals for the germinating seed. It may be involved in the control of inorganic phosphate levels in both developing seeds and seedlings, and it may also have antifungic and antioxidant roles.[1,4] Because of these functions of phytate and its universal abundance in all plants, caution should be given in developing low-phytate crops. However, phytate is an antinutrient factor in animal diets. With a poor availability of phosphorus, phytate also chelates divalent metals such as zinc and iron. There are twelve replaceable protons present in the phytic acid molecule: Six are dissociated in the strong acid range, one in the weak acid range, two with pK 6.8 to 7.6, and three with pK > 10.[5] At the neutral pH of small intestine, phytic acid is strongly negatively charged and is able to complex or bind to positively charged molecules. These complexes are rather insoluble, rendering the chelated metals unavailable for absorption.

NUTRITIONAL AND ENVIRONMENTAL BENEFITS

Numerous studies have demonstrated the effectiveness of microbial or plant phytase added to plant-based diets for swine, poultry, and fish in improving utilization of phytate-phosphorus and reducing phosphorus excretion by these animals.[6] The efficacy of different phytases varies, but the average amount of phytase needed to replace 1 g of inorganic phosphorus per kg of swine or poultry diet ranges from 500 to 1000 units.[7] With this efficacy, phytase can obviate inorganic phosphorus supplementation at least by half, saving the nonrenewable resource that may be exhausted in 80 years at the current extraction rate. More urgently, supplemental phytase reduces fecal phosphorus excretion by 30–50%, which can potentially eliminate 90,000 tons of phosphorus excreted to the environment by poultry and swine in the United States annually. In addition, phytase improves bioavailability of calcium, zinc, and iron, primarily by releasing these elements from binding to the phosphate groups of phytate. However, the effects of phytase on utilization of protein, amino acids, or energy are still controversial.

DIETARY DETERMINANTS OF EFFICACY

At least four dietary factors can modulate phytase efficacy. First, high levels of dietary calcium or calcium/phosphorus ratios reduce the effectiveness of phytase. In phytase-supplemented diets, the recommended calcium/phosphorus ratio is 1.2:1, not 2:1 as used in diets with adequate inorganic phosphorus added. Second, moderate to high levels of inorganic phosphorus may inhibit the full function of phytase. Third, supplemental organic acids such as citric acid or lactic acid enhance phytase efficacy. Those acids may reduce the pH of stomach digesta, thus providing a better environment for phytase to function, and/or to enhance the solubility of digesta phosphorus and modify the transit

Table 1 Biochemical properties of phytases currently available for use in animal diets

Origin	pH optimum	Temperature optimum (°C)	PI	M_r (kDa)[a]	Km (μM)[b]	Kcat (s^{-1})[b]	Kcat/Km ($s^{-1} M^{-1}$)[b]
A. niger PhyA[c,d,e,f]	2.5–3; 5–5.5	55–60	4.94	66–120 (50)	27	348	1.3×10^7
A. niger PhyB[g]	2.5	63	–	269 (65)	103	628	6.1×10^6
A. fumigatus PhyA[e,h,i,j,k]	4–6.5	58–70	7.04–7.3	60–76 (49)	30	46	1.5×10^6
P. lycii[l,m]	4–4.5	50–55	3.61–4.37	71–72 (44.6)	33	2200	6.6×10^7
E. coli AppA[h,i,n,o,p,w]	2.5–4.5	55–60	6.3; 6.5	42–55 (45–48)	130 (IP6) 15 (IP5)	6209 (IP6) 6926 (IP5)	4.8×10^7 (IP6) 5×10^8 (IP5)
Bacillus sp.[q,r,s,t,u]	6–9.5	55–65	5.0–5.1; 6.5–6.8	38–47	50	26.6	5.3×10^5
Wheat[v,w]	5.2	50	–	47–65	228–300	468	1.8×10^6

[a]The values in parentheses are calculated after deglycosylation of proteins or based on the deduced peptide sequence. The value shown for *A. niger phyB* is the molecular mass of the tetramer. The molecular mass of the monomer is shown within parentheses.

[b]Only phytic acid (IP6) is used as the substrate for all enzymes except for *E. coli* AppA. Assay conditions are as follows: *A. niger* PhyA: 58°C, pH = 5.0; *A. niger* PhyB: 63°C, pH = 2.5; *A. fumigatus* PhyA: 58°C, pH = 5.0; *P. lycii*: 58°C, pH = 5.0; *E. coli* AppA: 35–37°C, pH = 4.5; *Bacillus* sp: 37°C, pH = 7.0; Wheat: 55°C, pH = 5.15; 35°C, pH = 5.0.

[c]Ullah, A.H.; Sethumadhavan. K.; Mullaney, E.J.; Ziegelhoffer, T.; Austin-Phillips, S. Characterization of recombinant fungal phytase (phyA) expressed in tobacco leaves. Biochem. Biophys. Res. Commun. **1999**, *264*(1): 210–206.

[d]Han, Y.; Wilson, D.B.; Lei, X.G. Expression of an Aspergillus niger phytase gene (phyA) in Saccharomyces cerevisiae. Appl. Environ. Microbiol. **1999**, *65*(5): 1915–1918.

[e]Ullah, A.H.; Sethumadhavan, K.; Lei, X.G.; Mullaney, E.J. Biochemical characterization of cloned Aspergillus fumigates phytase (phyA). Biochem. Biophys. Res. Commun. **2000**, *275*(2): 279–285.

[f]Ullah, A.H.; Sethumadhavan, K.; Mullaney, E.J.; Ziegelhoffer, T.; Austin-Phillips, S. Cloned and expressed fungal phA gene in alfalfa produces a stable phytase. Biochem. Biophys. Res. Commun. **2002**, *290*(4): 1343–1348.

[g]Ullah; A.H.; Sethumadhavan, K. Differences in the active site environment of Aspergillus ficuum phytases. Biochem. Biophys. Res. Commun. **1998**, *243*(2): 458–462.

[h]Wyss, M.; Pasamontes, L.; Friedlin, A.; Rémy, R.; Tessier, M.; Kronenberger, A.; Middendorf, A.; Lehmann, M.; Schoebelen, L.; Röthlisberger, U.; Kusznir, E.; Wahl, G.; Müller, F.; Lahm, H.-W.; Vogel, K.; van Loon, A.P.G.M. Biophysical Characterization of Fungal Phytases (*myo*-Inositol Hexakisphosphate Phosphohydrolases): Molecular Size, Glycosylation Pattern, and Engineering of Proteolytic Resistance. Appl. Environ. Microbiol. **1999**, *65*(2): 359–366.

[i]Wyss, M.; Brugger, R.; Kronenberger, A.; Rémy, R.; Fimbel, R.; Oesterhelt, G.; Lehmann, M.; van Loon, A.P.G.M. Biochemical Characterization of Fungal Phytases (*myo*-Inositol Hexakisphosphate Phosphohydrolases): Catalytic Properties Appl. Environ. Microbiol. **1999**, *65*(2): 367–373.

[j]Mullaney, E.J.; Daly, C.B.; Sethumadhavan, K.; Rodriguez, E.; Lei, X.G.; Ullah, A.H. Phytase activity in Aspergillus fumigates isolates. Biochem. Biophys. Res. Commun. **2000**, *275*(3): 759–763.

[k]Rodriguez, E.; Mullaney, E.J.; Lei, X.G. Expression of the Aspergillus fumigates phytase gene in Pichia pastoris and characterization of the recombinant enzyme. Biochem. Biophys. Res. Commun. **2000**, *268*(2): 373–378.

[l]Lassen, S.F.; Breinholt, J.; Østergaard, P.R.; Brugger, R.; Bischoff, A.; Wyss, M.: Fuglsang, C.C. Expression, Gene Cloning, and Characterization of Five Novel Phytases from Four Basidiomycete Fungi: *Peniophora lycii*, *Agrocybe pediades*, a *Ceriporia* sp., and *Trametes pubescens*. Appl. Environ. Microbiol. **2001**, *67*(10): 4701–4707.

[m]Ullah; A.H.; Sethumadhavan K. PhyA gene product of Aspergillus ficuum and Peniophora lycii produces dissimilar phytases. Biochem. Biophys. Res. Commun. **2003**, *303*(2): 463–468.

[n]Greiner, R.; Konietzny, U.; Jany, K.D. Purification and characterization of two phytases from Escherichia coli. Arch. Biochem. Biophys. **1993**, *303*(1): 107–113.

[o]Golovan, S.; Wang, G.; Zhang, J.; Forsberg, C.W. Characterization and overproduction of the Escherichia coli appA encoded bifunctional enzyme that exhibits both phytase and acid phosphatase activities. Can. J. Microbiol. **2000**, *46*: 59–71.

[p]Rodriguez, E.; Wood, Z.A.; Karplus, P.A.; Lei, X.G. Site-Directed Mutagenesis Improves Catalytic Efficiency and Thermostability of *Escherichia coli* pH 2.5 Acid Phosphatase/Phytase Expressed in *Pichia pastoris*. Arch. Biochem. Biophys. **2000**, *382*(1): 105–112.

[q]Kerovuo, J.; Lauraeus, M.; Nurminen, P.; Kalkkinen, N.; Apajalahti, J. Isolation, Characterization, Molecular Gene Cloning, and Sequencing of a Novel Phytase from *Bacillus subtilis*. Appl. Environ. Microbiol. **1998**, *64*(6): 2079–2085.

[r]Kim, Y.-O.; Lee, J.-K.; Kim,H.-K.; Yu, J.-H.; Oh, T.-K. Cloning of the thermostable phytase gene (phy) from Bacillus sp. DS11 and its overexpression in Escherichia coli. FEMS Microbiol. Lett. **1998**, *162*(1): 185–191.

[s]Kim, Y.-O.; Kim, H.-K.; Bae, K.-S.; Yu, J.-H.; Oh, T.-K. Purification and properties of a thermostable phytase from *Bacillus* sp. DS11. Enz. Microb. Tech. **1998**, *22*(1): 2–7.

[t]Choi, Y.M.; Suh, H.J.; Kim, J.M. Purification and Properties of Extracellular Phytase from Bacillus sp. KHU-10. J. Prot. Chem. **2001**, *20*(4): 287–292.

[u]Tye, A.; Siu, F.; Leung, T.; Lim, B. Molecular clonging and the biochemical characterization of two novel phytases from B. subtilis 168 and B. licheniformis. Appl. Microbiol. Biotechnol. **2002**, *59*(2–3): 190–197.

[v]Peers, F.G. The phytase of wheat. Biochem J. **1953**, *53*: 102–110.

[w]Greiner, R.; Carlsson, N.-G.; Alminger, M.L.; Stereospecificity of *myo*-inositol hexakisphosphate dephosphorylation by a phytate-degrading enzyme of *Escherichia coli*. J. Biotech. **2000**, *84*(1): 53–62.

time of digesta in the small intestine. In addition, organic acids may release cations chelated by phytate, reducing the amount of insoluble phytate–cation complexes that are resistant to phytase action, thereby increasing the efficacy of endogenous or supplemented phytase. Last, inclusion of hydroxylated cholecalciferol compounds has been shown to improve dietary phosphorus and zinc utilization by chicks in an additive manner with phytase. Supplementing different phytases in combination has not shown any benefit over the singular additions. However, adding phytase with other hydrolytic enzymes seems to produce a synergism. Furthermore, there are several physical forms of phytase: powder, granule, and liquid. The chemical coating of phytase to improve heat stability may somewhat compromise its release in stomach.

STORAGE AND HANDLING

Phytase should be stored under dark, cool, and dry conditions. When this is done, the enzyme may maintain good stability for 3–4 months. Refrigeration or freezing may extend its shelf life, whereas high storage temperature certainly decreases its activity. Caution should be given in storing phytase mixed with vitamin and mineral premixes, as some of their components may have deleterious effects on phytase stability. There are few reports on immune responses of workers who have inhalation exposure to phytase. The hypersensitivity symptoms can be alleviated or avoided by implementing local exhaust systems and wearing protective clothing and masks with P2 filters.[8]

DEVELOPING IDEAL PHYTASES

A phytase would be considered ideal for feed application if it were catalytically effective, proteolysis-resistant, thermostable, and cheap. The catalytic efficiency and protease susceptibility of any given phytase decide its ability to release phytate-phosphorus in the digestive tract. The thermostability of phytase determines its feasibility in feed pelleting, and the overall cost to produce the enzyme ranks its final acceptance by industry. Although there are significant differences in these features among various naturally occurring phytases, no single wild-type enzyme possesses all of the desired properties. With advances of biotechnology, there are three ways to develop effective phytases with improved properties. First, site-directed mutagenesis, based on crystal structure of phytases, has been applied to improve pH profile, thermostability, and catalytic efficiency. Second, synthetic phytases such as the experimental consensus phytase have been generated based on homologous sequences of multiple phytases. Last, new phytases can be produced by directed evolution with efficient selections. A number of heterologous expression systems have been used for phytase production. The expression hosts include plants, bacteria, fungi, and yeast.[2][9] Recently, transgenic pigs overexpressing a bacterial phytase in salivary gland have been generated.[10] If approved by regulatory agencies, this approach may serve as a sustainable and economical delivery of phytase.

CONCLUSION

There is an increasing need for phytase to improve dietary phytate-phosphorus utilization by livestock, and thus reduce their phosphorus excretion to the environment worldwide, in particular in areas of intensive animal production. Although microbial phytase supplementation has been a widespread practice in swine and poultry feeding, and to a lesser extent, in fish feeding, continuous improvements in its property and reductions in its cost are warranted. Modern biotechnology has provided great potential to develop ideal phytases and effective deliveries for specific groups of animals.

REFERENCES

1. Mullaney, E.J.; Daly, C.B.; Ullah, A.B.J. Advances in phytase research. Adv. Appl. Microbiol. **2000**, *47*, 157–199.
2. Lei, X.G.; Stahl, C.H. Biotechnological development of effective phytases for mineral nutrition and environmental protection. Appl. Microbiol. Biotechnol. **2001**, *57*, 474–481.
3. Liu, B.; Rafiq, A.; Tzeng, Y.; Rob, A. The induction and characterization of phytase and beyond. Enzyme Microb. Technol. **1998**, *22*, 415–424.
4. Reddy, N.R.; Sathe, S.K.; Salukhe, D.K. Phytates in legumes and cereals. Adv. Food Res. **1982**, *28*, 1–92 (Academic Press, New York, NY).
5. Cheryan, M. Phytic acid interactions in food systems. CRC Crit. Rev. Food Sci. Nutr. **1980**, *13*, 297–336.
6. Lei, X.G.; Stahl, C.H. Nutritional benefits of phytase and dietary determinants of its efficacy. J. Appl. Anim. Res. **2000**, *17*, 97–112.
7. Kornegay, E.T. Chapter 18: Nutritional, Environmental, and Economic Considerations for Using Phytase in Pig and Poultry Diets. In *Nutrient Management of Food Animals to Enhance and Protect the Environment*; Kornegay, E.T., Ed.; CRC, Lewis Publishing: New York, NY, **1996**; 277–302.
8. Baur, X.; Melching-Kollmuss, S.; Koops, F.; Straburger, K.; Zober, A. IgE-mediated allergy to phytase—A new animal feed additive. Allergy **2002**, *57*, 943–945.
9. Pandey, A.; Szakacs, G.; Soccol, C.R.; Rodriguez-Leon, J.A.; Soccol, V.T. Production, purification and properties of microbial phytases. Bioresour. Technol. **2001**, *77*, 203–214.
10. Golovan, S.P.; Meidinger, R.; Ajakaiye, A.; Cottrill, M.; Wiederkehr, M.Z.; Barney, D.J.; Plante, C.; Pollard, J.W.; Fan, M.Z.; Hayes, M.A.; Laursen, J.; Hjorth, J.P.; Hackler, R.R.; Phillips, J.P.; Forsberg, C.W. Pigs expressing salivary phytase produce low-phosphorus manure. Nat. Biotechnol. **2001**, *19*, 741–745.

Phytoremediation

Adel Zayed
Paradigm Genetics, Inc., Research Triangle Park, North Carolina, U.S.A.

Abstract
Phytoremediation, the process by which various naturally occurring or genetically modified plants, including trees and grasses, are used to degrade, extract, detoxify, contain, and/or immobilize toxic pollutants from contaminated soil, water, and air, offers an attractive and cost-effective solution for the clean-up of contaminated sites. The potential economic benefits of using plants for remediation are impressive. Growing a crop on an acre of land can be accomplished at a cost ranging from two to four orders of magnitude less than the current engineering cost of excavation and reburial.

INTRODUCTION

Decades of unrestricted use, disposal, and release of industrial-, defense-, and energy production–related chemicals have considerably accelerated the pollution of our environment with radionuclides, toxic metals, and organic pollutants. Environmental exposure to such toxic chemicals is a serious health risk for humans and animals. Unfortunately, existing technologies for the cleanup of contaminated environments are very costly and/or limited in various ways. Cleanup of toxic metal contaminated soils with existing technologies costs up to $1,000,000 per acre without totally removing the pollutant from the environment. In the United States alone, the cost of cleaning up sites contaminated with toxic and radioactive metals is estimated to be $400 billion.

Plants provide a robust, solar-powered system that has little or no maintenance requirements. With their copious root systems, plants can scavenge large areas and volumes of soils, removing the toxic chemical. The rhizosphere soil (soil near plant roots) has microbial populations orders of magnitude greater than bulk soil (nonroot soil). Furthermore, plant-based systems are welcomed by the public due to their superior aesthetics and the societal and environmental benefits that their presence provides. There are several different ways that phytoremediation can be achieved (Table 1).

Research into phytoremediation may offer the remediation solution for at least 30,000 contaminated sites in the United States alone. Phytoremediation is applicable to a number of hazardous waste and other remedial scenarios, including remediation of metals, organics, and radionuclides from soils and water, which together offer a total U.S. market opportunity of up to $10 billion. Phytoremediation is also potentially applicable to all types of water treatments including industrial, agricultural, and municipal wastewaters, landfill leachate, stormwater, and drinking water with a potential market opportunity of up to $40.7 billion (Table 2). Similar markets exist overseas, which are smaller but which offer greater long-term potential for growth.

TRANSGENIC PHYTOREMEDIATION APPROACHES

The ideal plant species for phytoremediation is one that can accumulate, degrade, or detoxify large amounts of the toxic chemical, grow rapidly on contaminated sites and produce a large biomass, tolerate salinity and other toxic conditions, and provide a yield of economic value, e.g., fibers. Metal hyperaccumulator plants occur naturally on metal-rich soils and accumulate metals in their aboveground tissues to concentrations between one and three orders of magnitude higher than surrounding "normal" plants grown at the same site.[1,2] Unfortunately, most hyperaccumulator plants studied to date grow very slowly and accumulate little biomass. For phytoremediation to become a viable technology, dramatic improvements would be required in either hyperaccumulator biomass yield or nonaccumulator metal accumulation. The introduction of novel traits into high biomass or hyperaccumulator plants in a transgenic approach is a promising strategy for the development of effective phytoremediation technologies.

There are several conceivable strategies for the use of genetic engineering to improve phytoremediation. Strategies to improve metal phytoremediation involve the optimization of a number of processes, including trace element mobilization in the soil, uptake into the root, detoxification, and translocation within the plant.[3] Strategies for enhancing phytoremediation of organics are potentially more straightforward and involve the introduction and/or overexpression of genes encoding biodegradative enzymes in transgenic plants to enhance their biodegradative abilities.[4] In the last few years, progress has been

Table 1 Summary of phytoremediation processes and their definitions.

Process	Definition	Contaminated substrate	Target contaminant
Phytoextraction	Use of plants to take up contaminants from soil and water and accumulate them in aboveground plant tissues, which may then be harvested and removed from the site.	Soils Sediments Sludges Wastewater	Toxic trace elements Radionuclide
Phytostabilization	Use of plants to immobilize contaminants chemically and physically at the site, thereby preventing their movement to surrounding areas.	Soils Sediments Sludges	Heavy metals
Phytovolatilization	Use of plants and their associated microbes to remove contaminants, e.g., Se and Hg, from the environment in volatile forms.	Soils Sediments Sludges Industrial wastewater Agricultural wastewater	Metalloids (e.g., Se, Hg) Chlorinated solvents
Phytodetoxification	Use of plants to change the chemical species of the contaminant to a less toxic form.	Soils Sediments Sludges Industrial wastewater Agricultural wastewater	Chromium Organic compounds Explosives Pesticides
Rhizofiltration	Use of plant roots to absorb and adsorb pollutants from water and aqueous streams.	Surface water Industrial wastewater Agricultural wastewater Acid-mine drainage	Water soluble organics Heavy metals

made toward this goal and several types of genetic modifications of plants for phytoremediation have been reported. For example, bacterial genes were transferred to plants to enhance their potentials for 1) phytovolatilization (e.g., transfer of *merA* and *merB* genes from bacteria into three different plant species to increase their tolerance to and volatilization of the toxic mercuric ion after its transformation into the much less toxic elemental mercury); 2) phytoextraction (e.g., transfer of *gsh1* and *gsh2* genes from *Escherichia coli* to *Brassica juncea* plants to increase Cd accumulation in shoots); and 3) detoxification (e.g., transfer of bacterial genes for removal of nitroso groups from explosive compounds into plants to detoxify trinitrotoluene, TNT). Another approach was to express mammalian genes in plants, e.g., genes encoding cytochrome P450 2E1, the liver enzyme responsible for activating trichloroethylene (TCE), resulting in an increase in TCE oxidation by two orders of magnitude. A more straightforward approach is to overexpress existing plant genes that control key-limiting processes in order to overcome these limitations and accelerate the phytoremediation process (e.g., overexpression of AtAPS1 in *B. juncea* resulted in twofold increase in selenium accumulation and the overexpression of NtCBP4 in *Nicotiana tabacum* resulted in 2.5-fold increase in Ni tolerance and two-fold increase in Pb accumulation).[5] These examples provide dramatic evidence of the potential of genetically engineered plants for bringing the promise of phytoremediation to fruition.

Table 2 Summary of the size of potential U.S. markets for phytoremediation.

Market sector	Annual U.S. potential market
Metals from soils	$1.2–1.4 billion
Metals from groundwater	$1.2–1.4 billion
Organics from soils	$2.3–2.6 billion
Organics from groundwater	$2.3–2.6 billion
Radionuclides (all media)	$1.5–2.0 billion
Industrial wastewater	$7.0–10.0 billion
Municipal wastewater	$18–28 billion
Landfill leachate control	$0.6–1.2 billion
Agricultural runoff	$0.4–0.5 billion
Stormwater management	$0.2–1.0 billion
Treatment of drinking water	$0.6–1.0 billion
Total	$35.3–51.7 billion

Source: From *The 2000 Phytoremediation Industry*.[2]

THE PHYTOREMEDIATION GENOME/PROTEOME

The vast majority of plant genes affecting the remediation of toxic elements and organics have not yet been identified. Identification of these genes is vital to the success of the transgenic phytoremediation approaches. The recent completion of the *Arabidopsis* genome sequencing project has paved the way for the development of high throughput functional genomic approaches to accelerate the determination of gene function.[6] It is estimated that the *Arabidopsis* genome contains 80–90% of all the phytoremediation genes and gene families found in any macrophytes that could be used in phytoremediation applications.[4] The *Arabidopsis* genome is estimated to have 25,500 genes in 11,000 gene families.[4] Approximately 5% of the genome appears to encode membrane transport proteins responsible for the transport of metals and ions across plant plasma and organellar membranes.[7] These proteins are classified in 46 unique families containing approximately 880 members. In addition, several hundred putative transporters have not yet been assigned to families. Only a small number of these transporter proteins have their function fully characterized. The vast majority, however, have no known or presumed function, and genetic and physiological analyses will be needed to determine their functions and their relative contributions to toxic metal uptake and transport in plants.[4,7]

An analysis of the *Arabidopsis* genome using the amino acid sequences of proteins with known roles in remediation suggests that approximately 700 genes encode phytoremediation-related proteins (PRP). This portion of the *Arabidopsis* proteome has been termed the phytoremediation proteome and it consists of approximately 450 enzymes catalyzing redox reactions (e.g., cytochrome P-450s, oxygenases, and dehalogenases), approximately 250 transport proteins, and several metal-binding proteins, e.g., metallothioneins. The availability of the *Arabidopsis* genome and EST sequences and the identification of the phytoremediation proteome provide enormous opportunities to accelerate greatly our effort to clean up environmental pollution.[4]

PHYTOREMEDIATION IN CONSTRUCTED WETLANDS

Constructed wetlands represent a cost-effective phytoremediation solution for the cleanup of large volumes of wastewaters contaminated with low levels of toxic trace elements. They offer an efficient alternative to conventional water treatment systems because they are 1) relatively inexpensive to construct and operate, 2) easy to maintain, 3) effective and reliable wastewater treatment systems, 4) tolerant of fluctuating hydrologic and contaminant loading rates, and 5) providers of green space, wildlife habitat, and recreational and educational areas. In addition, ecosystems dominated by aquatic macrophytes are among the most productive in the world, largely as a result of ample light, water, nutrients, and the presence of plants that have developed morphological and biochemical adaptations enabling them to take advantage of these optimum conditions. Furthermore, these ecosystems are highly rich in microbial activities and therefore have high capacities to decompose organic matter and stabilize toxic trace elements. Constructed wetlands have been used for many years with great success to remove conventional pollutants (e.g., nitrate and phosphorus) from agricultural nutrient-laden runoff, drinking water, and domestic wastewater. Recently, there has been increasing interest in using constructed wetlands for the treatment of industrial wastewaters and acid-mine drainage containing heavy metals and toxic trace elements. A recent study demonstrated the successful use of wetlands in this area where a 36-ha constructed wetland removed 90% of the toxic selenium from 10 million liters/day of selenite-contaminated oil refinery effluent.[8] Constructed wetlands remove metals primarily through immobilization of the sulfide for Cu, Fe, Mn, Zn, and Cd. Sulfides of most metals are very stable under the anoxic waterlogged conditions of the wetlands.[9] In the case of metalloids such as Se, the soluble ions are taken up by plants and volatilized as a less toxic gas. Less is known about the use of constructed wetlands for the removal of toxic organics or pesticides although recent studies indicate that wetlands efficiently remove some chlorinated compounds present at low levels that are difficult to remove by other means.[9] A significant amount of research is currently underway to determine the best wetland plant species, including algae and vascular plants, that can be used to maximize pollutant removal by wetlands.[10]

FUTURE PROSPECTS

New uses of plants are emerging as a result of the recent advances in agricultural biotechnology. Phytoremediation is one of these new uses that has been extensively and effectively tested in the cleanup of real contamination sites and is becoming commercially feasible. The technology is currently being adopted and promoted by several government agencies (e.g., U.S. Environmental Protection Agency), and field demonstration projects are currently underway in many "superfund" sites across the country.[3] However, the existing knowledge about the rate and extent of degradation or extraction of pollutants by the phytoremediating crop is still limited, and specific data are needed on more plants, contaminants, and climate conditions. Additional progress in phytoremediation is likely to come from utilizing the recent genomic information to create new "superplants" overexpressing genes responsible for the removal, degradation, and/or sequestration of various contaminants. Further optimization of in-field performance of phytoremediation will require improvements in a number of agronomic practices, ranging from traditional crop management techniques to approaches more specific to phytoremediation, such as amendment of soil with chelators.

REFERENCES

1. Salt, D.E.; Smith, R.D.; Raskin, I. Phytoremediation. Annu. Rev. Plant Physiol. Plant Mol. Biol **1998**, *49*, 643–668.
2. Glass, D.J. *The 2000 Phytoremediation Industry*; D. Glass Associates, Inc.: Massachusetts, **1999**; 1–92.
3. Salt, D.E.; Blaylock, M.; Kumar, N.P.B.A.; Dushenkov, V.; Ensley, B.D.; Chet, I.; Raskin, I. Phytoremediation: A novel strategy for the removal of toxic metals from the environment using plants. Biotechnol **1995**, *13*, 468–474.
4. Cobbett, C.S.; Meagher, R.B. Arabidopsis and the Genetic Potential for the Phytoremediation of Toxic Elemental and Organic Pollutants. In *The Arabidopsis Book,* American Society of Plant Biologists; **2002**; 1–32. http://www.aspb.org/publications/arabidopsis/toc.cfm.
5. Krämer, U.; Chardonnens, A.N. The use of transgenic plants in the bioremediation of soils contaminated with trace elements. Appl. Microbiol. Biotechnol. **2001**, *55*, 661–672.
6. Boyes, C.D.; Zayed, A.M.; Ascenzi, R.; McCaskill, A.J.; Hoffman, N.E.; Davis, K.R.; Gorlach, J. Growth stage-based phenotypic analysis of *Arabidopsis*: A model for high throughput functional genomics in plants. Plant Cell **2001**, *13*, 1499–1510.
7. Maser, P.; Thomine, S.; Schroeder, J.; Ward, J.; Hirschi, K.; Sze, H.; Talke, I.; Amtmann, A.; Maathuis, F.; Sanders, D.; Harper, J.; Tchieu, J.; Gribskov, M.; Persans, M.; Salt, D.; Kim, S.; Guerinot, M. Phylogenetic relationships within cation transporter families of Arabidopsis. Plant Physiol. **2001**, *126*, 1646–1667.
8. Zayed, A.M.; Pilon-Smits, E.; deSouza, M.; Lin, Z.-L.; Terry, N. Remediation of Selenium-Polluted Soils and Waters by Phytovolatilization. In *Phytoremediation of Contaminated Soil and Water,* 1st Ed.; Terry, N., Banuelos, G., Eds.; Lewis Publishers: New York, **2000**; 61–83.
9. Horne, A.J. Phytoremediation by Constructed Wetlands. In *Phytoremediation of Contaminated Soil and Water*, 1st Ed.; Terry, N., Banuelos, G., Eds.; Lewis Publishers: New York, **2000**; 13–39.
10. Qian, J.-H.; Zayed, A.; Zhu, Y.-L.; Yu, M.; Terry, N. Phytoaccumulation of trace elements by wetland plants: III. Uptake and accumulation of ten trace elements by twelve plant species. J. Environ. Qual. **1999**, *28*, 1448–1455.

Plant DNA Virus Diseases

Robert L. Gilbertson
Maria R. Rojas
Department of Plant Pathology, University of California–Davis, Davis, California, U.S.A.

Abstract

The majority of plant virus diseases are caused by viruses with single-stranded RNA genomes. Three families of plant viruses have DNA genomes, and some DNA viruses cause economically important diseases, particularly in tropical and subtropical regions. Members of the family Caulimoviridae have a circular double-stranded (ds) DNA genome. The type member, *Cauliflower mosaic virus* (CaMV), is the source of the 35S promoter, which is widely used in plant biotechnology. Banana streak and rice tungro are important diseases caused by caulimoviruses. Members of the family Geminiviridae have a circular single-stranded (ss) DNA genome and distinctive twinned icosahedral virions. The whitefly-transmitted geminiviruses (genus *Begomovirus*) have emerged as one of the most economically important groups of plant viruses. Geminiviruses cause devastating diseases such as African cassava mosaic, bean golden mosaic, beet curly top, cotton leaf curl, maize streak, and tomato yellow leaf curl. Viruses in the family Circoviridae have a multipartite circular ssDNA genome, and members of the genus *Nanovirus* cause plant diseases. Banana bunchy top, the most important viral disease of banana, is caused by *Banana bunchy top virus*. Management of diseases caused by DNA viruses involves an integrated approach involving virus-free propagative material, resistant varieties, synchronized planting dates, insect vector management, sanitation, and host-free periods.

CAULIFLOWER MOSAIC: THE FIRST PLANT DISEASE SHOWN TO BE CAUSED BY A DNA VIRUS

In 1968, Shepherd and colleagues established that CaMV, the causal agent of cauliflower mosaic disease, has a genome composed of dsDNA.[1] Subsequently, a number of diseases were shown to be caused by dsDNA viruses related to CaMV. These viruses were placed in the family Caulimoviridae, which is the only recognized family of plant-infecting dsDNA viruses. The family Caulimoviridae includes two major genera: *Caulimovirus* and *Badnavirus* (Table 1). CaMV is the type species of the genus *Caulimovirus*. The genome of these viruses is composed of a single circular dsDNA (approximately 8.0 kilobase (kb) pair), which is encapsidated in isometric (spherical) particles approximately 50 nm in diameter.[1,2] Caulimoviruses cause mosaic-type diseases (mosaic, mottle, ringspots, and malformation in leaves and stunted plant growth) of dicot crop, ornamental, and weed plants; examples include cauliflower mosaic, carnation-etched ring, dahlia mosaic, and strawberry vein banding.[1,2] Caulimoviruses have narrow host ranges and are spread, plant-to-plant, by aphids or via propagative material, but not through seed.

BADNAVIRUSES: BACILLIFORM dsDNA VIRUSES CAUSING BANANA STREAK AND RICE TUNGRO DISEASES

In the late 1980s, a new type of DNA virus was identified with a circular dsDNA genome of approximately 7.5 kb, and bacilliform (bullet)-shaped virions (100–300 × 30 nm).[3,4] These viruses were placed into the genus *Badnavirus* (ba [bacilliform]-dna [DNA]-virus) in the family Caulimoviridae (Table 1).[4] Badnaviruses cause diseases including banana streak, rice tungro, and cacao swollen shoot (Table 2).[2,4] Banana streak is a mealybug-transmitted disease found in many banana growing regions. In addition to causing losses due to reduced size and production of banana fruit, the dsDNA genome of *Banana streak virus* (BSV) can become integrated into the banana genome, complicating the production of BSV-free propagative material. Rice tungro is a devastating disease of rice in certain areas of Asia, and its impact was particularly severe on high-yielding rice varieties introduced during the green revolution.[5] The disease is caused by a complex of two viruses: *Rice tungro bacilliform virus* (badnavirus) and *Rice tungro spherical virus* (ssRNA virus), and is spread by leafhoppers. The development of rice varieties resistant to the insect vector has been an effective management tool in certain areas

Table 1 Classification and characteristics of DNA viruses of plants.

Family	Genus	Genome	Virion shape and size	Insect vector and mode of transmission	Plants infected
Caulimoviridae	*Caulimovirus*	dsDNA monopartite	Spherical 50 nm in diameter	Aphids nonpersistent	Dicots, narrow host range
	Badnavirus	dsDNA monopartite	Bacilliform 130 × 30 nm	Mealybugs semipersistent	Monocots and dicots, individual viruses have narrow host range
	Rice tungro bacilliform virus	dsDNA monopartite	Bacilliform 130 × 30 nm	Leafhoppers semipersistent	Monocots, mostly rice, narrow host range
Geminiviridae	*Mastrevirus*	ssDNA monopartite	Germinate 18 × 30 nm	Leafhoppers persistent, nonpropagative	Monocots, including maize and wheat, narrow host range
	Curtovirus	ssDNA monopartite	Germinate 18 × 30 nm	Beet leafhopper persistent, nonpropagative	Dicots, wide host range
	Begomovirus	ssDNA most bipartite	Germinate 18 × 30 nm	Whiteflies (*Bemisia* spp.) nonpropagative?	Wide range of dicots, individual viruses have narrow host ranges
Circoviridae	*Nanovirus*	ssDNA multipartite	Spherical 18–20 nm in diameter	Aphids persistent, nonpropagative	Monocots, bananas, and *Musa* spp., narrow host range

(Table 2).[5] Cacao swollen shoot is a mealybug-transmitted disease of considerable importance in West Africa and Sri Lanka, where it reduces yield and quality of cacao by inducing swelling and necrosis of stems and roots.[2]

GEMINIVIRUSES: THE FIRST SINGLE-STRANDED DNA VIRUSES SHOWN TO CAUSE DISEASES IN PLANTS

Viruses in the family Geminiviridae have circular ssDNA genomes (3.0–5.5 kb) encapsidated in small (18 × 30 nm) twinned icosahedral virions.[3,6,7] The family name comes from the distinctive virion shape and from the latin word "geminus," meaning twin. Plant diseases caused by geminiviruses were recognized long before the nature of the causal agent was determined, and the first recorded observation of a plant disease—an aesthetically pleasing yellow vein symptom described in a Japanese poem in 752 A.D.—may have been caused by a geminivirus.[3,6] Today, the yellow variegation (mosaic) of the ornamental flowering maple (*Abutilon* spp.) is due to infection by the geminivirus *Abutilon mosaic virus* (Fig. 1A).

In the late 1970s, geminiviruses were identified and characterized and shown to be the causal agents of many important diseases, including African cassava mosaic, maize streak, and bean golden mosaic (Fig. 1).[2,6,7] Geminiviruses have been classified into four genera based on the insect vector, host range, and genome structure (Table 1).[2,3,6,7] and are now among the best characterized and most economically important plant viruses.

GEMINIVIRUS DISEASES HAVE EMERGED AS MAJOR THREATS TO CROP PRODUCTION

Geminiviruses cause many important diseases (Table 2, Fig. 1).[2,6,7] In southern Africa, India, and islands in the Indian Ocean, maize streak is the most damaging viral disease of maize, causing streaking of leaves, stunted growth, and reduced yields (Fig. 1C).[2,6] The development of moderately resistant varieties has helped reduce losses due to this leafhopper-vectored mastrevirus. Curly top disease is caused by three curtovirus species vectored by the beet leafhopper (*Circulifer tenellus*) (Table 1). Disease symptoms include twisted, crumpled, and yellowed leaves; stunted and distorted growth; and vascular discoloration (Fig. 1E).[6] The disease occurs in the western United States, certain Mediterranean countries and South America. In the early 1900s, curly top, nearly destroyed the sugar beet industry in the United States until resistant varieties were developed. The disease still causes losses in tomato and other crops in the U.S. State of California, and an annual spray program is used in an attempt to control the disease by reducing populations of the leafhopper vector.

Diseases caused by whitefly-transmitted geminiviruses (genus *Begomovirus*) (Table 1) have emerged as major constraints on vegetable and field crop production in tropical

Table 2 Plant diseases of major economic importance caused by DNA viruses.

Disease	Host of economic importance	Geographical distribution	Causal agent	Spread	Management
Rice tungro	Rice	South and Southeast Asia, China	Complex: *Rice tungro bacilliform virus* (caulimovirus) and *Rice tungro spherical virus* (ssRNA virus)	Leafhoppers, propagative material	Vector resistance, planting date, rice-free period
African cassava mosaic	Cassava	Africa	*African cassava mosaic virus* (geminivirus)	Whiteflies, propagative material	Virus-free planting material, sanitation and roguing, resistance
Bean golden mosaic	Common bean	South and Central America, Mexico, SE U.S.A.	*Bean golden mosaic virus* and *Bean golden yellow mosaic virus* (geminivirus)	Whiteflies	Host-free period, time of planting, resistance, vector management
Cotton leaf curl	Cotton	Pakistan, India	Complex: *Cotton leaf curl virus* (geminivirus) and nanovirus-like satellite DNA	Whiteflies	Host-free period, resistance
Tomato yellow leaf curl	Tomato	Asia, Africa, Caribbean basin, SE U.S.A.	Various species of *Tomato yellow leaf curl virus* and *Tomato leaf curl virus* (geminivirus)	Whiteflies, propagative materials (transplants)	Host-free period, time of planting, resistance, vector management
Banana bunchy top	Banana	Australia, Africa, Asia, South Pacific	*Banana bunchy top virus* (nanovirus)	Aphids, propagative material	Quarantine, virus-free planting material, roguing

and subtropical regions throughout the world (Table 2).[6,8] This is due to a worldwide increase in the population and distribution of whiteflies (*Bemisia* spp.) and monoculture of susceptible crops in areas with indigenous weed-infecting geminiviruses. Diseases caused by begomoviruses are characterized by mosaic/mottle, curling, crumpling, and/or yellowing of leaves; stunted and distorted growth; and a reduction in yield quantity and quality (Fig. 1).[2,3,7] African cassava mosaic is a devastating disease in many countries of Africa, causing significant yield losses (e.g., approximately 50% of total production) (Table 2, Fig. 1B).[6] In the late 1990s new, highly pathogenic forms of *African cassava mosaic virus* appeared, arising by genetic recombination. They have caused even greater losses. Bean golden mosaic (Fig. 1D) causes significant losses (as high as 100%) to common bean production in South and Central America and Mexico, and management options remain limited. Some bean varieties possess moderate resistance, but one of the most effective strategies has been a bean-free period of 2–3 months (Table 2).[2] Cotton leaf curl has devastated cotton production in some parts of Pakistan and India, and is caused by a complex of a begomovirus and a nanovirus-like satellite DNA (Table 2).

Tomato yellow leaf curl disease (TYLCD), caused by *Tomato yellow leaf curl virus* (TYLCV), was first recognized in Israel around 1940 and is the most damaging viral disease of tomato. The disease is characterized by stunted and erect growth; small, chlorotic leaves that roll upward; and flower abortion (Fig. 1F). In plants infected early in development, yield loss may reach 100%. In some regions, TYLCD limits the commercial cultivation of tomatoes. TYLCD is now found throughout the Middle East, Southeast Asia, India, many countries of Africa, southern Europe, the Caribbean Basin, and the southeastern United States (Table 2).[2]

In the early 1990s, TYLCV was inadvertently introduced into the Dominican Republic where it destroyed a flourishing tomato processing industry.[8,9] Using the tools of biotechnology, it was established that the virus in the Dominican Republic was identical to TYLCV from the eastern Mediterranean, and that tomato was the primary host.[9] A regional integrated management strategy was implemented that involved 1) a mandatory three-month whitefly host-free period; 2) planting of early maturing hybrid varieties (early season) and TYLCV-resistant varieties (late season); and 3) the selective use of new systemic

Fig. 1 Symptoms of diseases caused by various geminiviruses. A. Abutilon mosaic caused by *Abutilon mosaic virus*; B. African cassava mosaic caused by *African cassava mosaic virus*; C. Maize streak caused by *Maize streak virus*; D. Bean golden mosaic caused by *Bean golden yellow mosaic virus*; E. Curly top caused by *Beet mild curly top virus* (previously Worland strain of *Beet curly top virus*); F. Tomato yellow leaf curl caused by *Tomato yellow leaf curl virus*.

neonicotinoid insecticides (e.g., imidacloprid). The disease has been effectively managed in the Dominican Republic, and yields are higher than before the introduction of the virus.[9]

BANANA BUNCHY TOP DISEASE: AN ECONOMICALLY IMPORTANT DISEASE CAUSED BY A NANOVIRUS

Banana bunchy top disease (BBTD) is the most important viral disease of banana. First described in the South Pacific in the late 1800s, the disease spread to Australia, Africa, India, the Philippines, and Sri Lanka.[10] BBTD is characterized by the bunchy appearance of the upper portion of the plant, narrow and dwarfed leaves, and dark green streaks on leaves and stems. Infected plants become severely stunted and yield losses can reach 100%. The disease is transmitted by aphids and via propagative materials. Originally thought to be caused by a luteovirus, BBTD was subsequently shown to be caused by a spherical virus with a multipartite circular ssDNA genome. The virus was named *Banana bunchy top virus* (BBTV). Effective management of BBTD involves the use of quarantines, virus-free propagative material, and a strict regime of sanitation and eradication of diseased plants (Table 2). This approach has been highly effective in Australia, but less effective in other areas.[10] Other diseases caused by viruses similar to BBTV include faba bean necrotic yellows, coconut foliar decay, and subterranean clover stunt.[2,3] These viruses have been placed into the genus *Nanovirus* in the family Circoviridae (Table 1).[3]

CONCLUSION

Although DNA viruses represent a minority of the total viruses that infect plants, they cause many economically important plant diseases, particularly in tropical and subtropical regions. Management of these diseases has been difficult and successful cases have always involved an integrated regional approach that is based on a thorough understanding of viral biology and disease epidemiology.[5,9,10]

REFERENCES

1. Shepherd, R.J.; Lawson, R.H. Caulimoviruses. In *Handbook of Plant Virus Infections and Comparative Diagnosis*; Kurstak, E., Ed.; Elsevier/North Holland Biomedical Press: Amsterdam, **1981**; 847–878.
2. Agrios, G.N. *Plant Pathology,* 4th Ed.; Academic Press: San Diego, CA, **1997**; 1–635.
3. Hull, R. *Matthews' Plant Virology,* 4th Ed.; Academic Press: San Diego, CA, **2002**; 1–1001.
4. Lockhart, B.E.; Olszewski, N.E. Plant Pararetroviruses-Badnaviruses. In *Encyclopedia of Virology,* 2nd Ed.; Granoff, A., Webster, R.G., Eds.; Academic Press: San Diego, **1999**; 1296–3000.
5. Koganezawa, H. Present Status of Controlling Rice Tungro Virus. In *Plant Virus Disease Control*; Hadidi, A., Khetarpal, R.K., Koganezawa, H., Eds.; APS Press: St. Paul, MN, **1998**; 459–469.
6. Harrison, B.D. Advances in geminivirus research. Annu. Rev. Phytopathol. **1985**, *23*, 55–82.
7. Buck, K.W. Geminiviruses (*Geminiviridae*). In *Encyclopedia of Virology,* 2nd Ed., Granoff, A., Webster, R.G., Eds.; Academic Press: San Diego, **1999**; 597–606.
8. Polston, J.E.; Anderson, P. The emergence of whitefly-transmitted geminiviruses in tomato in the Western Hemisphere. Plant Dis. **1997**, *81*, 1358–1369.
9. Salati, R.; Nahkla, M.K.; Rojas, M.R.; Guzman, P.; Jaquez, J.; Maxwell, D.P.; Gilbertson, R.L. Tomato yellow leaf curl virus in the Dominican Republic: Characterization of an infectious clone, virus monitoring in whiteflies, and identification of reservoir hosts. Phytopathology **2002**, *92*, 487–496.
10. Dale, J.L.; Harding, R.M. Banana Bunchy Top Disease: Current and Future Strategies for Control. In *Plant Virus Disease Control*; Hadidi, A., Khetarpal, R.K., Koganezawa, H., Eds.; APS Press: St. Paul, MN, **1998**; 659–669.

Plant Genetic Resources: Effective Utilization

Hikmet Budak
Faculty of Engineering and Natural Science, Biological Science and Bioengineering Program, Sabanci University, Istanbul, Turkey

Abstract
Characterizing, understanding the genome organization better, and differentiating the identity of genotypes based on their morphology and genome characteristics are vital determinants in their commercialization, management of germplasm repositories, and genetic conservation. Morpho-agronomic characterization of plants is not always feasible or sometimes labor intensive. Employing chloroplast, mitochondrial, and nuclear genome diversity using molecular biology tools will enhance the effectiveness and efficiency of revealing identity differences between genotypes. Using organelle and nuclear genome diversity can also answer a broad range of genetic, evolutionary relationships, and ecological questions.

INTRODUCTION

Plant genetic resources are one of the most important aspects of agricultural society. The sustainability and productivity of ecological systems depend on their biodiversity. Hence, it is very crucial to investigate the biodiversity of plant genetic resources to meet food demand and food security. Plant genetic resources comprise landraces, modern and wild species, and their relatives. Due to (a) destruction of humid forests, (b) invasion by exotic plants, (c) urbanization, (d) changes in agricultural practices, and (e) long-term breeding efforts, loss of diversity in plant genetic resources is one of the biggest concerns today. For instance, genetic diversity currently used in cereals is much less than their wild relatives.[1] Consequently, a tremendous investment has been made in effective utilization of plant genetic resources research worldwide over the years.

The large-scale analysis of plant genome diversity with respect to plant morphology is one of the most important determinants for effective and efficient utilization of genetic resources. However, morpho-agronomic characterization and phenotypic profiling of most plants are labor intensive and time consuming. In addition, phenotypic profiling may not provide distinguishable and usable data. For instance, different ploidy levels in buffalograss [*Buchloe dactyloides* (Nutt.) Engelm.], a native North American C_4 turfgrass species with extremely favorable drought tolerance, cannot be distinguished morphologically.[2] Additionally, genetic information obtained from morphological traits is limited and environmental effects have a tremendous impact on expression of quantitative traits. Hence, increased information on degree and distribution of nuclear and organelle genome diversity is imperative to develop sampling strategies and base populations and to identify redundancies and genetic contamination in germplasm collections. The information on organelle and nuclear genome diversity are also important to fingerprint and quantify genetic drifts/shifts and manage gene pools, which are a source of alleles for sustainable genetic improvement of plant species.[2,3]

There are a number of molecular biology tools and analytical techniques to enhance the knowledge of nuclear and organelle genome diversity that help in defining the uniqueness of a species and their ranking and phylogenetic relatedness. Biochemical methods (based on protein and enzymes, isozyme or allozymes) and DNA-based techniques (novel global-scale technologies such as genome sequencing, gene expression analysis, and genetic engineering) have been applied in combination with traditional methods to evaluate biodiversity and differentiate the identity of genotypes.[4]

This entry discusses employing organelle [mitochondrial DNAs (*mt*DNA) and chloroplast DNA (*cp*DNA)] and nuclear genome diversity for effective and efficient use of plant genetic resources and molecular biology tools utilized for analyzing their diversity.

ORGANELLE GENOME DIVERSITY

Analysis of organelle and nuclear genomes and the information on the degree of diversity has been widely used at the interspecies and intergeneric level for investigating phylogenetic relationships, cultivar identification, the geographical distribution of the progenitor, domestication, and the evolution and natural history of plant species.[4,5] Due to their lower level of mutation rate, highly conserved organelle genomes are very well suited for fingerprinting genotypes and elucidating evolutionary relationships.[6] The distinct characteristics of plant mitochondrial and chloroplast genomes make it a powerful tool for population genetic analysis and tracing maternal lineages in most angiosperms. Organelle genome analysis is also very important to compare coding and non-coding regions, which may provide more information regarding rearrangements and mutations at the species level.[7]

Unlike the nuclear genome that is bi-parentally inherited, the chloroplast genome is uniparentally, mostly maternally, inherited.[8] This characteristic of the chloroplast genome is very important to identify specific parentage of hybrid species and thus can be traced to matriarchal lineages. The chloroplast is highly abundant in leaves and therefore isolation of large quantities in pure form is relatively easy,[9] and its sequence appears to be highly conserved in terms of size, structure, gene content, and order. The chloroplast genome size, ranging from 120 to 218 kb in higher plants,[10] is small enough to resolve the fragments but is large enough to get taxonomic information.[11] One of the most attractive features of chloroplast genome is that it is independent of polyploidy that is very important in the evolution of plants. However, allo- and autopolyploidy needs to be identified for accurate analysis because alloploidy involves hybridization of diverged taxa[10,11] that affects the inheritance of chloroplast genome. The majority of studies using sequence data from *cp*DNA have been focused on phylogenetic work at fairly high taxonomic levels (intergeneric and above). However, recently, primer pairs for *cp*DNA and sequencing of organelle genomes and the detailed analysis of homologies between plastid and nuclear genomes are of interest to elucidate evolutionary history of plant species.[7] For instance, the plastid *matK* gene has a high rate of substitution when compared to the other chloroplast genes and evolves about three times faster than the widely used *rbcL* and *atpB* genes.[7,12] Hence, the *matK* gene has been extensively used in plant species to elucidate plant evolution and address phylogenetic questions at various taxonomic levels.

The mitochondrial genome has not been used extensively to study genome diversity. There is less knowledge available on mitochondrial genomes, and these genomes are not considered to be as conserved as the chloroplast genome. Most plant mitochondrial genomes are too large to allow for entire genomes to be characterized.[11] In addition, the mitochondrial genome is less abundant in leaves, which makes nucleic acid extraction more difficult. Yet, the study of *mt*DNA sequences in the nucleus to identify and trace specific genes is a powerful tool for identifying new source of alleles and creating genetic diversity for genetic improvement of plant species. Mitochondrial DNA sequences can easily be used for identification of genetic shifts/drifts, for a better understanding of organelle genome structure and their differences for gene tracing, and for differentiating the identity of genetic materials.[7,13]

NUCLEAR GENOME DIVERSITY

Higher mutation rates of plant nuclear genomes[6] make nuclear genome diversity very helpful for revealing the identity of genotypes and ex situ conservation of plant genetic resources. Genetic fingerprints developed through nuclear genome diversity are used to establish the origin of the specific plants and the relatedness of one genotype to another. A rational classification, which is necessary for many collection of plant species that exist around the world, is possible by employing nuclear genome diversity. Isolation of large quantities of nucleic acids from the nuclear genome is easy. However, study of single copy genes is challenging because nuclear genomes are larger than organelle genomes and the presence of organelle DNA sequences in the nuclear genome is also sometimes a big problem. A wide range survey of angiosperms indicated a high frequency of functional, single-gene transfers from mitochondrial to nuclear genomes during evolution.[14] By analyzing nuclear genome diversity, geneticists and plant breeders can predict expected properties of the progeny by tracing the presence or absence of certain forms of genes. Knowledge of nuclear genome diversity will also help researchers to predict important phenotypic properties, such as tolerance to biotic and abiotic stresses.

OTHER USES OF NUCLEAR AND ORGANELLE GENOME DIVERSITY

Nuclear and organelle genome variation can be used to identify the geographical and ecological distribution of various plant species and their relatives. Certain plant genetic resources from one region can be separated by genetically distinct resources from other regions. This is very important to acquire and harbor germplasm with useful and rare genes for widening the genetic base, maintaining diversity, and understanding the dynamics and biological function of biodiversity in natural and agricultural ecosystems. Investigation of organelle and nuclear genome diversity can also provide taxonomic relationship, which will be necessary for conservation strategies. This information assists in developing in situ (maintaining genetic resources in native habitat where they occur) and ex situ (maintaining genetic resources outside the native habitat) conservation strategies for effective and efficient utilization of plant genetic resources. A suitable conservation and management strategy will help promote an effective and efficient use of plant genetic resources without wasting resources through the high cost of management, and identification of genotypes with enhanced agronomic traits, including maturity (flowering), grain yield, disease resistance, and stress resistance. Manipulation of these traits would certainly have tremendous impact on efficient use of plant genetic resources. Taxonomic relationships provide information on identification of genomic homologies among the plant species to devise an appropriate breeding approach and a better understanding of gene transfer from one species to another.

In molecular systematic studies, the discrepancy between the phylogeny based on nuclear, mitochondrial and chloroplast DNA data sets has been reported.[7,15] The most extensive differences are seen in comparisons between chloroplast and nuclear genome data and have been reported by a number of studies conducted using different plant species. The reason for this discrepancy is due to the likelihood that the nuclear genome and chloroplast genome had different evolutionary histories. It may also depend on introgressions or due to sorting of ancestral lineages.[15] This discrepancy is most serious at higher ploidy levels. Hence, solving these problems and having a better understanding is very important to accurately resolve problems of family and species relationships and taxonomic relatedness.

Analysis of nuclear and organelle genome diversity within and among plant species provides a suitable sampling strategy for germplasm acquisition. For instance, if most of the nuclear or organelle diversity is within populations when compared among populations, the sampling and conservation emphasis should be focused on collecting larger numbers of plants from few populations. On the other hand, emphasis should be placed on collecting a small number of plants from a large number of populations if any genome diversity is higher between populations.

GENETICS AND GENOMICS TOOLS USED FOR ORAGANELLE AND NUCLEAR GENOME DIVERSITY

Molecular biology tools have not only introduced new characters for the analysis of genome diversity, but they have provided characters that are not influenced by environment as is the case with morphological traits. PCR-based markers such as amplified fragment length polymorphism (AFLP), inter simple sequence repeat (ISSR), simple sequence repeat (SSR), sequence-related amplified polymorphism (SRAP), sequence tagged microsatellites (STMS), minisatellites, random amplified polymorphic DNA (RAPD), and single nucleotide polymorphism (SNP) have been extensively used to gain a better and enhanced knowledge on the nuclear genome variation and phenetic relationships among a broad range of plant species and subpopulations of single species[4] at the genomic level (Figs. 1 and 2).[2] PCR amplification of buffalograss genomic DNA from nine genotypes using two SRAP primer combinations is depicted in Fig. 2. Expressed sequence tags (ESTs) and putative genes controlling agronomic traits in genome sequence databases have been continually produced and are important resources to aid in exploiting wild relatives of plant species, which is an additional source of genes for domesticated plants.[16] The disadvantage of EST-SSR is that it is limited to the species where the sequence database is present. Transferability of molecular markers such as EST-SSR markers is very helpful for comparisons of genome structure among related species, high-resolution comparative maps and estimate genetic variation, which is indispensable to identify and manage unique plant genetic resources.

Universal primers that are available for amplification of specific organelle genome are being widely used in PCR-based RFLP analysis for either total DNA or extracted organelle DNA. Primers will work either directly, or with small alterations, across broad taxa. Organelle genome diversity can be assayed by direct sequencing for identification of new alleles and tracing genes during evolution. With genome sequence in hand, scientists are provided knowledge of all the genes within a plant species that will allow for the identification of unique germplasm and the ability to trace the ancestry of a specific gene in a gene pool.

Currently, the application of molecular biology tools has allowed scientists to remove species boundaries set by the traditional genetic improvement method of hybridization for a more effective utilization and improvement of plant genetic resources. Genetic engineering, embryo rescue,[17] somatic hybridization, and protoplast fusion techniques[18] are good examples of resolving problems of sexual incompatibility (begets the hybrid sterility) and lack of genetic recombination in distant wild relatives.[19] A gene from one organism can be transferred to any organism of choice for effective and efficient germplasm use. In vitro techniques through tissue culture has also proven to assist the conservation and management of plant genetic resources.

CONCLUSION AND PERSPECTIVES

Elucidation and an improved understanding of the degree and distributions of organelle and nuclear genome variation has enormous potential to benefit all phases of society, thus it provides improved, efficient, and effective genetics and breeding program. Understanding organelle and nuclear genome variation is ultimately needed to spawn a modern-day "Green Revolution" that is quite different from the Green Revolution of 40 years ago. The most exciting future prospects of employing a combination of both nuclear and organelle genome diversity consist of a comprehensive understanding of evolutionary relationship of plant genomes and their relationship with relatives. This will help in transferring new alleles from wild relatives to widely used plant species to effectively exploit wild germplasm resources. One should take care not to proclaim that organelle and nuclear genome diversity studies will feed the world, but that it will provide opportunities for improving plant genetic resources and will play a pivotal role in comparative studies in diverse fields such as ecology, molecular evolution, and comparative genetics. A challenge in the next decade will be to build integrated databases combining information on chloroplast, mitochondrial,

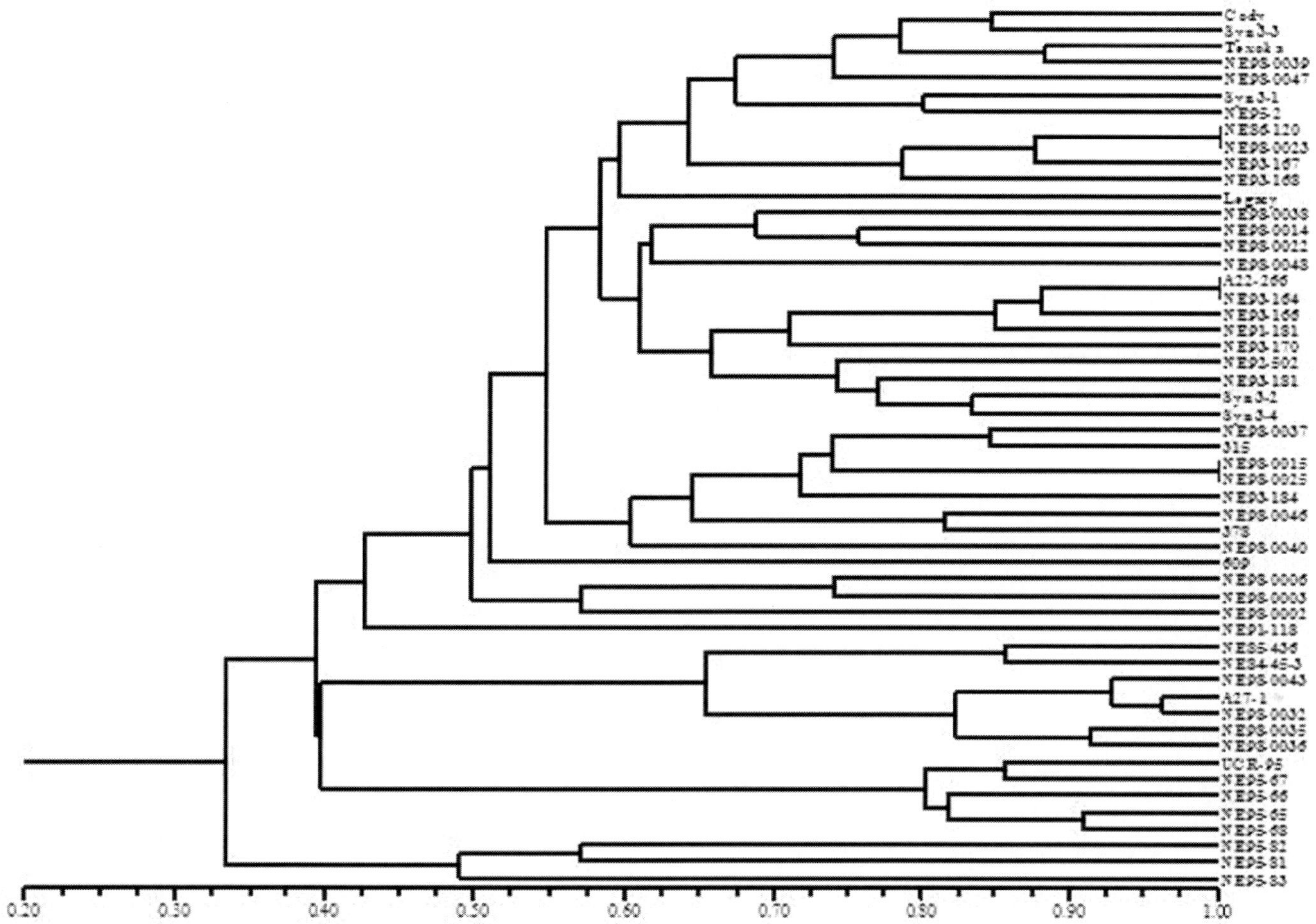

Fig. 1 An UPGMA dendrogram of genetic relationships among 53 buffalograss genotypes calculated based on genetic similarity by means of 34 SRAP primer combinations.[2]

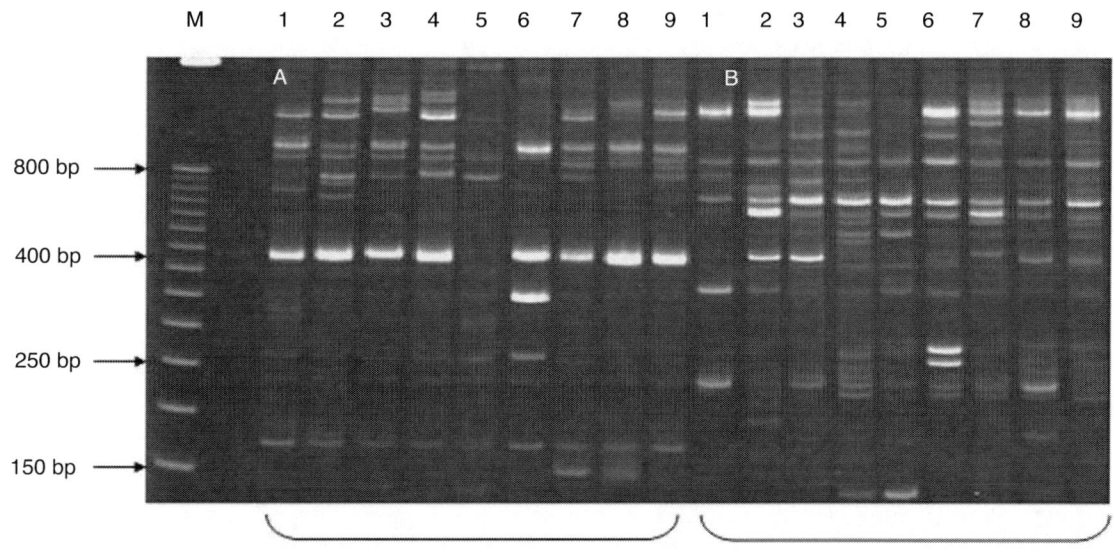

Fig. 2 PCR amplification of buffalograss genomic DNA from nine genotypes. Lanes: 1 = Cody, 2 = Texoka, 3 = NE 98-032, 4 = NE 85-436, 5 = NE 93-184, 6 = NE 95-2, 7 = NE 98-043, 8 = NE 98-0015, 9 = Syn3-1, and lane M contains a 50-bp size marker (Promega Corp., U.S.A.). Two SRAP primer combinations, (A) Em7 + Me6 and (B) Em7 + Me5, were assayed. The DNA samples were fractionated in 12% non-denaturing acrylamide gels stained with ethidium bromide.[2]

nuclear, and phenotypic data for effective and efficient use of plant genetic materials.

REFERENCES

1. Ellis, R.P.; Forsterö, B.P.; Robinson, D.; Handley, L.L.; Gordon, D.C.; Russell, J.R.; Powell, W. Wild barley: a source of genes for crop improvement in the 21st century? J. Exp. Bot. **2000**, *51* (342), 9–17.
2. Budak, H.; Shearman, R.C.; Parmaksiz, I.; Gaussoin, R.E.; Riordian, T.P.; Dweikat, I. Molecular characterization of buffalograss germplasm using sequence related amplified polymorphism markers. Theor. Appl. Genet. **2004**, *109* (2), 280–288.
3. Mort, M.E.; Soltis, D.E.; Soltis, P.S.; Francisco-Ortega, J.; Santos-Guerra, A. Phylogenetic relationships and evolution of Crassulaceae inferred from matK sequence data. Am. J. Bot. **2001**, *88* (1), 76–91.
4. Karp, A.; Edwards, K.J.; Bruford, M.; Funk, S.; Vosman, B.; Morgante, M.; Seberg, O.; Kremer, A.; Boursot, P.; Arctander, P.; Tautz, D.; Hewitt, G.M. Molecular technologies for biodiversity evaluation: opportunities and challenges. Nat. Biotechnol. **1997**, *15*, 625–628.
5. Karp, A.; Seberg, O.; Buiatti, M. Molecular techniques in the assessment of botanical diversity. Ann. Bot. **1996**, *78*, 143–149.
6. Wolfe, K.H.; Li, W.H.; Sharp, P.M. Rates of nucleotide substitution vary greatly among plant mitochondrial, chloroplast, and nuclear DNAs. Proc. Natl. Acad. Sci. U.S.A. **1987**, *84* (24), 9054–9058.
7. Budak, H.; Shearman, R.C.; Dweikat, I. Evolution of *Buchloë dactyloides* based on cloning and sequencing of *matK*, *rbcL*, and *cob* genes from plastid and mitochondrial genomes. Genome **2005**, *48* (3), 411–416.
8. Corriveau, J.L.; Coleman, A.W. Rapid screening method to detect potential biparental inheritance of plastid DNA and results for over 200 angiosperm species. Am. J. Bot. **1988**, *75*:1443–1458.
9. Palmer, J.D. Isolation and structural analysis of chloroplast DNA. Methods Enzymol. **1986**, *118*, 167–186.
10. Hilu, K.W. Chloroplast DNA in the systematics and evolution of the Poaceae. In *Grass Sytematics and Evolution*; International Grass Symposium; Washington, DC, 1987; 65–72.
11. Palmer, J.D.; Osorio, J.A.; Thomson, W.F. Chloroplast DNA evolution among legumes: loss of large inverted repeat occurred prior to other sequence rearrangements. Curr. Genet. **1987**, *11*, 275–286.
12. Hilu, K.W.; Borsch, T.; Müller, K.; Soltis, D.E.; Soltis, P.S.; Savolainen, V.; Chase, M.W.; Powell, M.P.; Alice, L.A.; Evans, R.; Sauquet, H.; Neinhuis, C.; Slotta, T.A.B.; Rohwer, J.G.; Campbell, C.S.; Chatrou, L.W. Angiosperm phylogeny based on *matK* sequence information. Am. J. Bot. **2003**, *90* (12), 1758–1776.
13. Wu, J.; Konstantin, V.K.; Strauss, S.H. Abundant mitochondrial genome diversity, population differentiation and convergent evolution in pines. Genetics **1998**, *150*, 1605–1614.
14. Adams, K.L.; Daley, D.O.; Qiu, Y.-L.; Whelan, J.; Palmer, J.D. Repeated, recent and diverse transfers of a mitochondrial gene to the nucleus in flowering plants. Nature **2000**, *408* (6810), 354–357.
15. Doyle, J.J.; Lucknow, M. The rest of the iceberg: legume diversity and evolution in a phylogenetic context. Plant Physiol. **2003**, *131* (3), 900–910.
16. Eujayl, I.; Sledge, M.K.; Wang, L.; May, G.D.; Chekhovskiy, K.; Zwonitzer, J.C.; Mian, M.A. *Medicago truncatula* EST-SSRs reveal cross-species genetic markers for *Medicago* spp. Theor. Appl. Genet. **2004**, *108* (3), 414–422.
17. Rao, N.K.; Reddy, L.J.; Bramel, P. Potential of wild species for genetic enhancement of semi-arid food crops. Genet. Resour. Crop Evol. **2003**, *50*, 707–721.
18. Zimmonch-Guzowska, E.; Lebecka, R.; Kryszczuk, A.; Maciejewska, U.; Szcerbakowa, A.; Wielgat, B. Resistance to phytophthora infestans in somatic hybrids of *Solanum nigrum* L. and diploid potato. Theor. Appl. Genet. **2003**, *107*, 43–48.
19. Rao, N.K. Plant genetic resources: advancing conservation and use through biotechnology. Afr. J. Biotechnol. **2004**, *3* (2), 136–145.

Plant Pathogens (Viruses): Biological Control

Hei-Ti Hsu
Floral and Nursery Plants Research, U.S. Department of Agriculture/Agricultural Research Service (USDA/ARS), Beltsville, Maryland, U.S.A.

Abstract
Prevention of virus and viroid infections in plants is based on biological means rather than chemical measures. In principle, there are no chemicals available for controlling plant diseases caused by viruses and viroids. The most feasible approaches for combating viruses and viroids are the elimination of source inoculum, prevention of secondary spread, cross protection, and use of crops bearing resistance traits.

ECONOMIC LOSS

Damage to crop plants due to virus and viroid infections is difficult to assess. The actual figures for global crop loss are not available. Plant disease losses are estimated at $60 billion annually. Losses due to virus and viroids have been considered second to those caused by fungi. Unlike diseases caused by fungi, bacteria, and nematodes, where control measures using chemical, biological, and integrated pest management approaches have been effective, diseases caused by viruses or viroids are far more difficult to manage.

Economic crop loss resulting from virus and viroid disease is due to the reduced growth and vigor of infected plants which, in turn, causes a reduction in yield. In some instances, a virus infection may kill a plant. Apart from yield reduction, the quality and market value of commercial end products may be affected. There are also costs of attempting to maintain crop health such as vector control, production of pathogen-free propagation materials, and quarantine and eradication programs. In addition, resources are being diverted to research, extension, and education as well as toward breeding for resistance to virus or viroid infection.

WORLD IMPACT

No single country is exempt from crop losses. Production of food, fiber, and horticultural crops are seriously affected worldwide by virus or viroid infection of plants.[1] This is even more so in developing countries that depend on one or a few major crops; for example, *Cassava mosaic virus* in cassava plants in Kenya, *Citrus tristeza virus* in citrus trees in Africa and South America, and *Cacao swollen shoot virus* in cacao trees in Ghana. Recently, *Papaya ringspot virus* (PRV) infection has affected every region where papaya plants are grown. The virus induces a lethal disease in papaya. The widespread aphid-transmitted PRV has changed the way papayas are grown in many parts of the world. Normally, papayas are produced annually for a number of years over the life of the papaya plant. For proper management of the disease due to PRV infection, papaya has now become an annual crop in which healthy seedlings are planted each year. Even so, productivity is still below the average yield obtained before PRV became a problem. Viroids infect a limited number of crops when compared with viruses. However, they can cause severe problems in specific crops, for instance, cadang-cadang disease of coconuts, potato spindle tuber disease, and chrysanthemum stunt disease.

CONTROL MEASURES

No direct chemical control means are available to combat virus infections in plants. Control of viral diseases is achieved primarily by sanitary practices that involve reducing sources of inoculum from outside, preventing spreading within the crop, and limiting the population of insects, mites, nematodes, and fungi that may serve as vectors for many plant viruses.[1] Virus disease testing programs are now common in many parts of the world where the economic importance of growing virus-free plants is recognized. Although seeds and seedlings certified as virus-free are more expensive than those that have not been tested for certain viruses, testing provides assurance of virus-free production materials. Early detection of virus in a field and removal of the infected plants minimizes spread of the virus.

Plants may be protected from development of severe disease symptoms by first introducing a mild strain of virus into a healthy plant. A plant systemically infected with a mild strain of virus is protected from infection by a severe strain of the same virus. This phenomenon in called "cross protection" and has been observed for many plant viruses.[2] It is also observed to occur between viroids or plant virus satellites. In practice, cross protection is of great interest since it has been utilized to protect plants against severe virus strains (*Citrus tristeza virus, Papaya ringspot virus, Zucchini yellow mosaic virus, Tomato mosaic virus*, etc.) in the field.

Another approach toward controlling plant virus diseases is to develop resistant or tolerant plants.[3] Historically, long-term manipulation of crop plants through breeding has produced many valuable commercial varieties resistant to plant viruses. Breeding plants resistant to vectors may also offer control of the virus they transmit. Conventional breeding of crossing and back crossing commercial varieties with plants bearing virus resistance traits takes years to develop. In order for a new variety to be commercially acceptable, undesirable traits from the resistant parent breeding line must be selected out. The process is labor intensive and time consuming. Advances in science have allowed new technology to precisely manipulate resistance genes at the molecular level.[4] Biotechnology represents the fastest growing area of biological research. The application of biotechnology in breeding for resistance to virus infection is a major area of research. Successful control of viral disease through resistance breeding will undoubtedly reduce the use of synthetic pesticides for vector control.[5]

Introducing virus resistance and vector resistance into a cultivar by gene transfer technology (genetic engineering) has been successful in combating plant viruses.[6] The technology has several major advantages over conventional cross breeding. It is a relatively fast procedure. Desirable genes can be introduced without disturbing the balanced genome of target plants. Furthermore, there is no restriction on the source of the transgenes allowing the use of genes from other plant species or even from outside the plant kingdom (Table 1).[1,7]

Several approaches for producing transgenic virus-resistant plants have been explored. Among these, plants expressing virus coat protein genes, parts of other viral genes, or virus satellite ribonucleic acids (RNAs) have been shown to offer the best control.[2,8,9] Plants expressing antisense viral RNAs, ribozymes, pathogen-related proteins, or virus-specific antibody genes may also confer resistance to virus infection. Control of virus vectors by introducing insect toxins such as trypsin inhibitor, lectin, and *Bacillus thuringiensis* (Bt) toxin genes into plants would undoubtedly contribute toward achieving the goal of controlling plant virus diseases.

PROSPECTS

Use of resistant cultivars is considered the best approach to combat virus infection in plants. Biotechnology, no doubt, will play a significant role in the economic growth of many countries. Molecular breeding, however, will not replace but complement the efforts of conventional cross breeding.

Table 1 Genes that contribute or may contribute toward control of virus diseases in plants.

Virus-derived gene sequences
    Coat proteins
    Replicases
    Movement proteins
    Polyprotein proteases
    Sense RNAs
    Antisense RNAs

Plant host-derived transgenes
    Pathogen-related proteins
    Anti-viral proteins
    Proteinase inhibitors
    Natural resistance genes
    Lectins

Other transgenes and sequences
    Satellite RNAs
    Virus-specific antibodies
    Interferon-induced mammalian oligoadenylate synthetase
    Insect toxins
    Anti-viral ribozymes (catalytic RNA)

Source: From *Plant Virus Disease Control*.[1]

Much attention has been given to engineering resistance to plant viruses. Recently, genetic engineering of crop plants has been closely scrutinized and criticized due to increasing public concerns regarding human health and environmental impact. Careful assessment of the benefits and potential risks involving the release of genetically modified plants into the environment and their consumption is necessary before these crops become widely accepted by the public.[10,11]

REFERENCES

1. Khetarpal, R.K., Koganezawa, H., Hadidi, A., Eds. *Plant Virus Disease Control*; APS Press: St. Paul, MN, **1998**, 1–684.
2. Beachy, R.N. Coat-protein-mediated resistance to tobacco mosaic virus: discovery mechanisms and exploitation. Phil. Trans. R. Soc. Lond. B. **1999**, *354*, 659–664.
3. Salomon, R. The evolutionary advantage of breeding for tolerance over resistance against viral plant disease. Israel J. Plant Sci. **1999**, *47*, 135–139.
4. Kawchuk, L.M.; Prufer, D. Molecular strategies for engineering resistance to potato viruses. Can. J. Plant Pathol. **1999**, *21*, 231–247.
5. Barker, I.; Henry, C.M.; Thomas, M.R.; Stratford, R. Potential Benefits of the Transgenic Control of Plant Viruses in the United Kingdom. In *Plant Virology Protocols: From Virus Isolation to Transgenic Resistance*; Foster, G.D., Taylor, S.C., Eds.; Humana Press, Inc.: Totowa, NJ, **1998**, *81*, 557–566.

6. Dempsey, D.A.; Silva, H.; Klessig, D.F. Engineering disease and pest resistance in plants. Trends Microbiol. **1998**, *6*, 54–61.
7. Gutierrez-Campos, R.; Torres-Acosta, J.A.; Saucedo-Arias, L.J.; Gomez-Lim, M.A. The use of cysteine proteinase inhibitors to engineer resistance against potyviruses in transgenic tobacco plants. Nat. Biotechnol. **1999**, *17*, 1223–1226.
8. Maiti, I.B.; Von Lanken, C.; Hong, Y.; Dey, N.; Hunt, A.G. Expression of multiple virus-derived resistance determinants in transgenic plants does not lead to additive resistance properties. J. Plant Biochem. Biotech. **1999**, *8*, 67–73.
9. Prins, M.; Goldbach, R. RNA-mediated virus resistance in transgenic plants. Arch. Virol. **1996**, *141*, 2259–2276.
10. Hammond, J.; Lecoq, H.; Raccah, B. Epidemiological risks from mixed virus infections and transgenic plants expressing viral genes. Adv. Virus Res. **1999**, *54*, 189–314.
11. Kaniewski, W.K.; Thomas, P.E. Field testing for virus resistance and agronomic performance in transgenic plants. Mol. Biotechnol. **1999**, *12*, 101–115.

Plant Pathogens: Pest Management

Kitty Cardwell
Chris J. Lomer
International Institute of Tropical Agriculture (IITA), Cotonou, Benin

Abstract
Plant diseases reduce global food and fiber production by an estimated 20% annually, but during severe epidemics, localized losses can reach 100%. This entry covers several control methods: chemical, cultural, biological, and biotechnological.

INTRODUCTION

Biotic plant diseases are caused by pathogenic microbial agents: fungi, bacteria, nematodes, viruses, and viroids. Plant diseases reduce global food and fiber production by an estimated 20% annually, but during severe epidemics, localized losses can reach 100%. From the 1930s to the 1960s, plant pathologists were confident that disease could not only be controlled, but perhaps even eliminated using chemicals and host plant resistance. Limitations to this approach emerged as host plant resistance was not durable if the genetic basis for it was too narrow, and chemicals were perceived as expensive, deleterious to the environment, or not durable if the target microbe could develop resistance. Thus, the concept of eradicating disease has given way to the concept of integrated disease management to hold damage to economically acceptable levels.[1] There are numerous control methods that may be used singly or in combination (Table 1).

HOST PLANT RESISTANCE

Host plant resistance to disease is a physiological response that limits the growth and development of a pathogen. The response to an invading pathogen may be highly specific so that the reaction is immunity (no disease development at all), or general, so that disease impact is reduced.[2,3]

Specific Host–Pathogen Interactions: Vertical Resistance

Many plant pathogens are specifically coevolved with their plant hosts; a specific gene within both the plant host and the pathogen will code for mutual recognition and response (gene for gene relationship). When a plant has resistance to a pathogen based on a single specific gene this is described as vertical resistance. In natural ecosystems, a plant population consists of resistant and susceptible individuals, while the pathogen population consists of individuals (races) that are either virulent or avirulent to the host.[4] Thus, some proportion of the plant population may be diseased, but the plant and pathogen populations are in equilibrium over time, allowing both to survive. In more intensive plant production systems, the aim is to minimize losses due to diseases. Through plant breeding, scientists strive to select only the resistant genotype and distribute this as a variety or hybrid. Unfortunately, such vertical resistance may be broken down when a pathogen gene locus mutates to virulence, or more frequently, when virulent strains already exist in the pathogen population. The virulent strain or mutant is specifically selected for as it is the only one that can infect and reproduce on the host. If the population of plants has been bred for uniformity in the resistance gene locus, then an epidemic can occur so that every plant becomes diseased.[4] There are a number of strategies to avoid "breakdown" of a single vertical resistance gene: 1) numerous resistance genes bred into a single variety; 2) several genes bred into isolines and the isolines mixed into multigenic varieties; and 3) geographic or temporal deployment of different resistant cultivars so that spread of a virulent race of pathogen will be checked.

General Resistance: Horizontal Resistance

Horizontal or general resistance to pathogens can occur in the absence of any specific genetic recognition/response interaction.[3] Disease may still occur, but the plant's nonspecific responses to external stresses reduce pathogen growth and development. The mechanism of response may be chemical or mechanical. A chemical response is a cascade of genetic signaling elicited by a pathogen (or other stresses) leading to the production and mobilization of pathogen suppressive metabolites in and around

Table 1 General methods of disease control.[1]

A. Avoidance of the pathogen
  (1) Choice of geographic area
  (2) Planting and harvesting date decisions
  (3) Use of disease-free planting stock
  (4) Modification of cultural practices
B. Exclusion of the pathogen
  (1) Treatment of seeds/planting material to kill resident pathogens
  (2) Meristem excision and tissue culture
  (3) Inspection, certification, and plant quarantine
  (4) Control of insect vectors
C. Management of the pathogen population
  (1) Biological control and microbial niche management
  (2) Crop rotation with nonhost plants or trap plants, and field fallowing
  (3) Intercropping with nonhosts
  (4) Removal of diseased plants and plant parts from the field
  (5) Understanding the role of weeds and weed hosts in and around the field
  (6) Sanitation and management of crop residues
  (7) Soil microbe management with chemical drenches, solarization, or mulching
  (8) Fungicide sprays
D. Protection of the plant
  (1) Application of chemical or biological control agents as prophylaxis against infection
  (2) Inoculation for induction of SAR
  (3) Control of insect vectors
  (4) Modification of plant nutrition
E. Development and deployment of host plant resistance
  (1) Selection and breeding for resistance (vertical, horizontal, and population resistance)
  (2) Genetically modifying plant with nonhost genes to induce a "nonhost" response
  (3) Genetically modify plant to contain viral coat protein to induce resistance to same virus
  (4) Use of external chemical or biological resistance inducers
  (5) Temporal and spatial mixing of different resistance genotypes
F. Therapy applied to diseased plant 1
  (1) Application of "curative" chemicals
  (2) Heat treatment (i.e., seed sanitation)
  (3) Surgery

the plant i.e., phytoalexins, phenolics, and chitinases. This is sometimes referred to as "systemically acquired resistance" (SAR). SAR can also be induced in response to chemicals, nonpathogenic microbes, and environmental stress.

Physical characteristics that give broad-based resistance such as waxy cuticle, quick growth habit, fewer stomata, etc., make it harder for the pathogen to gain entrance. This type of resistance, horizontal resistance, usually involves more than one gene and is considered to be more stable than vertical resistance.

CHEMICAL CONTROL

Fungicides

At the end of the twentieth century, in the Americas and Japan, fungicides comprise about 15% of the total pesticide market share, while in Europe, they are used more than other chemical pest control measures. In Africa, seed protectant and seed dressing fungicides are often used because of low cost. In China, antibiotic fungicides are produced regionally in small-scale factories. Modern fungicides are nonphytotoxic and have low toxicity for other organisms in the ecosystem.[5] The modes of action are specific to fungal metabolism such as inhibition of motility of swimming spore flagella, inhibition of respiration, thickening of cell walls, interference with cell division, or blockage of fungal lipid or protein synthesis. Nevertheless, anywhere that a fungicide is relied upon too heavily, and multiple sprays per cropping period are applied, it is possible to have problems of residual build-up. High levels of residue of any chemical can have unforeseeable deleterious effects on nontarget organisms.

Another unwanted effect of overreliance upon a chemical control is that loss of sensitivity to the pesticide may occur. Resistance to fungicides occurs through the development of and selection for strains that are insensitive to the mode of action of the chemical. Thus, useful, ecologically benign fungicides have to be managed carefully if they are to remain effective. The development of pathogen resistance or insensitivity may be delayed by avoiding repeated deployment of the fungicide as the sole control measure. Fungicides can be indispensable and sustainably effective components of integrated disease management systems when combined with other control measures.

Other Compounds

Antibiotics for control of fungal and bacterial diseases are rarely considered economical. Plant volatiles and oils with fungistatic effects have been reported as traditional or indigenous knowledge from Africa, Southeast Asia, and South America. The efficacy and consistency of effect have not been tested rigorously, but these local remedies may provide some options in small-scale agricultural systems. Chemical induction of systemically acquired resistance is an approach that has led to the development of commercial products.

CULTURAL CONTROL

All cropping systems involve different management practices, dependent on the agroeconomic scale of operation, which can be manipulated to influence disease levels. These may involve pathogen avoidance, for example, managed time of planting, and pathogen population management via interplanting, weeding, tillage, clearing plant litter, roguing, mulching, and the addition of organic matter to soil. For a more complete list see Table 1.[1,2,6]

BIOLOGICAL CONTROL

Although management of plant diseases is currently dominated by the use of resistant plant varieties, fungicides and cultural practices, other ecologically sustainable interventions are being developed.[7,8] Microorganisms inhabit soil, rhizosphere, or phyllosphere communities where interactions range from synergistic to antagonistic. Plant disease occurs on a susceptible host when a pathogen population finds little competition or challenge from the niche community and environmental factors favor the pathogen's growth and development. Conversely, disease development may be constrained by biological processes in the niche environment. We may be able to alter the balance in favor of the plant by the addition or augmentation of various beneficial microorganisms as biological or microbial pesticides.

  Hyperparasitic organisms are pathogenic to the pathogen (fungi on other fungi, fungi on nematodes, viruses on fungi, bacterial and viral pathogens of nematodes, etc.).
  Suppressive microorganisms function by crowding or direct antagonism (e.g., suppression of *Geaumannomyces*, a fungal pathogen that causes severe disease in wheat, by *Pseudomonas fluorescens*, a soil-inhabiting bacterium).
  Competitors—some strains or species of microbe— outcompete or displace another from the niche. A virulent plant pathogen can be displaced by an avirulent or "hypovirulent" strain of the same species. Temporal displacement can occur when one microbe occupies a niche before the other one arrives.

Mass production, formulation, and application techniques of microbial biological control agents have been developed, including various methods of liquid and solid-state culture for fungi and bacteria. Many fungi are known to store well in dry powder form. Carrier substances ranging from water, oil, kerosene, dust, and clay to agricultural wastes such as cereal hulls have been formulated for delivery of biological control agents. Globally, a serious constraint to development of this disease control technology is the economy of scale. Another constraint is the lack of uniformity in regulatory requirements and registration procedures for biological agents.[9]

BIOTECHNOLOGICAL CONTROL OPTIONS

Advances in biotechnology offer several control options.

  Cross-protection—when a plant cell is invaded by a virus, it is generally not susceptible to infection by another. Thus, an option for control of viral diseases is to genetically modify the plant so that each cell contains a piece of the viral coat protein, inducing permanent cross-protection.[10]
  Some pathogens pick up small pieces of satellite RNA or DNA, which can reduce their fitness and reduce disease. Vectors can be used to introduce these viral sequences into the host plant.
  Nonhost resistance—many plant pathogens can cause disease only on specific host plants. The relationship between a parasite and its host involves "recognition" of the host by the parasite. A gene from a nonhost plant placed in the genome of a host plant may render the modified plant a nonhost by changing some aspect of recognition.
  Finally, resistance genes transfer within and between species, and resistance gene amplification within a species are possible.

REFERENCES

1. Zadocks, J.C.; Schein, R.D. *Epidemiology and Plant Disease Management*; Oxford University Press: New York, **1979**; 427.
2. Chaube, H.S.; Singh, U.S. *Plant Disease Management: Principles and Practices*; CRC Press: Boca Raton, USA, **1991**; 319.

3. Vanderplank, E. *Disease Resistance in Plants*; Academic Press, Inc.: London, **1968**; 194.
4. Wolfe, M.S.; Caten, C.E. *Populations of Plant Pathogens, Their Dynamics and Genetics*; Blackwell Scientific Publications: Oxford, **1987**; 280.
5. Lyr, H. *Modern Selective Fungicides: Properties, Applications, Mechanisms of Action*; Gustav Fischer Verlag: Jena, New York, **1995**; 595.
6. Parker, C.A.; Rovira, A.D.; Moore, K.J.; Wong, P.T.W.; Kollmorgen, J.F. *Ecology and Management of Soilborne Plant Pathogens*; APS Press: St. Paul, MN, **1985**; 358.
7. Cook, R.J.; Baker, K.F. *The Nature and Practice of Biological Control of Plant Pathogens*; APS Press: St. Paul, MN, **1989**; 539.
8. Hornby, D. *Biological Control of Soil-borne Plant Pathogens*; CAB International: Wallingford, Oxon, UK, **1990**; 479.
9. Burges, H.D. *Formulation of Microbial Biopesticides, Beneficial Microorganisms, Nematodes and Seed Treatments*; Kluwer Academic Publishers: Dordrecht, **1998**; 412.
10. Kaniewski, W.K.; Thomas, P.E. *Molecular Biology 12*; 1999. http://www.biotech-info.net (accessed April 18, 2000).

Plant Sterols

Peter J.H. Jones
Vanu R. Ramprasath
Richardson Centre for Functional Foods and Nutraceuticals, University of Manitoba, Winnipeg, Manitoba, Canada

Abstract

Plant sterols (PS), structurally resembling cholesterol, are well known for their cholesterol-reducing properties. Plant sterols exist in several forms and the most abundant is sitosterol, followed by campesterol. Rich sources of PS include legumes and cereal grains. Consumption of 2 g/day of PS significantly reduces circulating cholesterol by suppressing intestinal absorption of cholesterol; hence, this level of consumption is recommended by the National Cholesterol Education Program (NCEP) in lowering LDL-cholesterol concentrations and is part of the standard therapy to reduce the risk for coronary heart disease. More recently, additional evidence has become available for the cholesterol-lowering efficacy of esterified or free PS incorporated in a wide variety of food formats, including low-fat or fat-free foods. Over 2400 subjects have participated in clinical trials involving PS supplementation up to 10 g/day, with no adverse effects. In addition to their cholesterol-lowering action, mounting evidence suggests that PS possess anticancer effects against various forms of cancer, including prostate, lung, stomach, ovary, and breast cancers. Plant sterols appear to act through multiple mechanisms, including inhibitions of carcinogen production, cancer cell growth, angiogenesis, invasion, and metastasis, as well as promoting apoptosis. This combined evidence strongly supports an anticarcinogenic action of PS and hence their dietary inclusion as an important strategy to prevent and treat cardiovascular diseases and cancer.

INTRODUCTION

Plant sterols (PS) are naturally occurring compounds that resemble cholesterol in both structure and biological function. Plant sterols exist in various forms in plants, including β-sitosterol, campesterol, stigmasterol, and cycloartenol. Among these, β-sitosterol is the most abundant PS followed by campesterol. Plant sterols possess a double bond at carbon-5 and when saturated by enzymatic hydrogenation form plant stanols, which are much less abundant in nature than sterols. Plant sterols are generally found in small quantities in various fruits, vegetables, nuts, seeds, cereals, legumes, vegetable oils, and other plant sources. These compounds cannot be synthesized by humans and, therefore, must be derived from the diet. Western populations consume about 150–350 mg/day of PS and 15–50 mg/day of stanols. Plant sterols are poorly absorbed in the intestine (0.4–3.5%), while phytostanol absorption (0.02–0.3%) is even lower.

Various food products enriched with PS are widely used as alternative therapeutic dietary options to control blood cholesterol levels and reduce the risk of cardiovascular disease. The FDA has allowed a health claim of reduction of cardiovascular disease risk, while the NCEP recommends 2 g/day of PS to reduce blood LDL-cholesterol (LDL-C) levels.

CHOLESTEROL REDUCTION BY PS

Cholesterol-lowering effects of PS have been demonstrated in both humans and animals. Plant sterols have been demonstrated to inhibit the uptake of both dietary and endogenously produced cholesterol from intestinal cells, which leads to reduction in serum total and LDL-C levels.[1–3] In randomized crossover trials, consumption of 1.7 g/day of PS[4] reduced LDL-C by 24.4%, whereas another study[5] showed that 1.8 g/day of PS reduced LDL-C by 10–15%. Although several theories have been proposed, the exact mechanisms of action of PS in reducing cholesterol have not been completely elucidated.

Plant sterols exert their effects at the intestinal level and reduce cholesterol absorption up to 40–60% by competing with cholesterol molecules for uptake into mixed micelles in the intestine. Plant sterols also play an important role in mediating intestinal membrane transport proteins, such as

ABCG5, ABCG8, and NPC1L1, which selectively pump PS from the enterocytes into the intestinal lumen, and hence, reduce circulating-cholesterol levels.[6] A positive dose-dependent effect is observed with the highest reduction in LDL-C levels when PS are consumed at 2.5 g/day. Furthermore, there is little additional effect of PS/stanols at doses above 2.5 g/day.

Beneficial Effects of Esterified PS

Esterification of free PS and stanols with fat-soluble compounds, such as fatty acids (FA), increases their fat solubility. Plant sterols esterified with fish oil FA simultaneously reduce plasma LDL-C and triacylglycerol concentrations. Supplementation of fish oil FA esters of PS effectively reduces levels of total and LDL-C, apolipoprotein-B, and plasminogen-activator inhibitor-I, an important marker for endothelial function. A water-soluble analog of phytostanols, disodium ascorbyl phytostanyl phosphates (DAPP), was developed through chemical modifications. Supplementation with DAPP also modulates lipid levels favorably and decreases body weight gain by reducing energy absorption at the intestinal level.

Factors Influencing the Effects of PS on Lipids

Factors that might influence the hypocholesterolemic effect of PS include their form, frequency of consumption, and dietary background. A meta-analysis by AbuMweis et al.[1] included 59 randomized clinical trials involving nearly 4500 subjects consuming PS/stanol-enriched food products. Results showed that the LDL-C lowering effect of PS/stanol was influenced by the baseline LDL level, food carrier, frequency, and time of intake.

Hypocholesterolemic effects of PS/phytostanols are influenced by the food carrier to which PS/stanols are incorporated. Plant sterols/stanols incorporated into fat spreads, mayonnaise, and salad dressing, or milk and yogurt, reduced LDL-C levels to a greater extent as compared to incorporation into other food products, such as chocolate, orange juice, cheese, non-fat beverages, meat, croissants and muffins, oil in bread, or cereal bars.[1]

Studies have demonstrated maximum efficacy of PS consumption when distributed in two or three doses daily, while other studies have recommended a single dose.[1,2] Therefore, until mechanisms and time of dose have been elucidated by which PS/stanols reduce LDL-C levels, PS should be consumed as three portions daily. The efficacy of PS/stanols in reducing cholesterol might vary with the time of consumption, typically coinciding with the diurnal rhythmic variation in cholesterol metabolism. Plant sterols were found to reduce the LDL-C more in subjects with higher baseline levels of LDL-C compared with borderline to optimal baseline levels.[1]

Differential Responses among Humans to PS Consumption

Although numerous studies have demonstrated substantial reductions in cholesterol absorption at intestinal levels after PS consumption, the responsiveness was found to be highly variable.[4] This variability was observed even when the compliance of the subjects was strictly controlled, owing to interindividual variations in intestinal sterol absorption and sterol metabolism.

Several intestinal membrane-transport proteins, including ABCG5, ABCG8, and NPC1L1, are involved in selectively pumping PS from the enterocytes into the intestinal lumen. Other intestinal membrane transporters have also been shown to be associated with cholesterol absorption, although their exact mechanisms are yet to be elucidated. It can be considered that polymorphisms associated with the genes coding these membrane transporters could explain variations in the responsiveness to PS consumption. Recently, Rudkowska et al.[7] demonstrated a rare single nucleotide polymorphism (SNP) in NPC1L1 of one non-responsive subject, although there were no common polymorphisms in ABCG5/8. Another recent study by Zhao et al.[8] observed that variations in ABCG5/8 and NPC1L1 polymorphisms have a profound effect on interindividual responsiveness of cholesterol concentrations. Results from these studies suggest that common polymorphisms of membrane-transporter genes together with metabolic biomarkers could be useful in predicting interindividual responsiveness to PS supplementation and devising individualized cholesterol-reducing strategies.

BENEFITS OF PS BEYOND CHOLESTEROL METABOLISM

Anticancer Effects of PS

Earlier evidence suggests that PS possess anticarcinogenic effects against lung, stomach, prostate, ovary, and breast cancers.[9] The effect of PS consumption on cancer development has been investigated in various human and animal models as well as in vitro. β-Sitosterol and campesterol have been shown to reduce the growth of various cancer cells including liver, prostate, and breast; however, there is a lack of effect of PS on the risk of colon cancer in vivo, which could be due to its inhibitory effect on cholesterol absorption in the small intestine, resulting in increased cholesterol flow to the colon, where it may induce and promote cancer development.

Anticancer Mechanism of Action

The exact mechanisms explaining anticancer effects of PS are not well understood; however, various proposed mechanisms have been reviewed by Awad and Fink.[10] Plant sterols inhibit carcinogen production, cancer cell growth and multiplication, invasion and metastasis, and induce apoptosis. Reactive oxygen species (ROS) produced by oxidatively stressed cells can damage DNA, and result in cancer. β-Sitosterol increased the activities of antioxidant enzymes in cultured macrophage cells under oxidative stress,[11] which suggests that PS can protect cells from damage by ROS and inhibit the production of carcinogens. Earlier studies on the effect of PS on cell growth and multiplication have shown an inverse relationship between PS and cancer development and progression. Plant sterols have been shown to induce apoptosis, an important mechanism in the inhibition of carcinogenesis. β-Sitosterol treatment increased apoptosis of prostate cancer and human leukemia cells in a dose-dependent manner. Angiogenesis plays a vital role in cancer cell growth and multiplication. Cells treated with PS undergo reduction in angiogenesis, invasiveness, and adhesiveness of breast cancer cells in vitro.[10,12]

SAFETY OF PS

Over 2400 subjects have participated in clinical trials involving supplementation of PS/stanols with dosages up to 10 g/day with no adverse events reported to date. The only group to caution against consumption of PS is those who have a highly rare autosomal recessive disorder termed phytosterolemia, also known as sitosterolemia, as consumption of PS may lead to significant increases of PS in the blood. Overall, aside from potential minor effects of PS consumption depressing circulating concentrations of fat-soluble vitamins, a sufficient body of evidence currently exists suggesting that intakes of PS at levels of 1–2 g/day do not affect any other biological parameters, while having the ability to significantly lower LDL-C levels. As such, PS consumption can be regarded as safe.

CONCLUSION

Phytosterols play a vital role in lipid metabolism by reducing the circulating serum cholesterol and also lowering cholesterol absorption in the intestine through several mechanisms. Beyond observed cholesterol-reducing effects, PS also possess a strong anticancer effect and alleviates various forms of this disease. Plant sterols appear to have no adverse effects and hence these data form a rationale for the use of PS as functional foods.

REFERENCES

1. AbuMweis, S.S.; Barake, R.; Jones, P. Plant sterols/stanols as cholesterol lowering agents: A meta-analysis of randomized controlled trials. Food Nutr. Res. **2008**, *52*. doi: 10.3402/fnr.v52i0.1811. Epub 2008 Aug 18.
2. AbuMweis, S.S.; Jones, P.J. Cholesterol-lowering effect of plant sterols. Curr. Atheroscler. Rep. **2008**, *10*, 467–472.
3. Jones, P.J.; AbuMweis, S.S. Phytosterols as functional food ingredients: linkages to cardiovascular disease and cancer. Curr. Opin. Clin. Nutr. Metab. Care **2009**, *12*, 47–51.
4. Jones, P.J.H.; Ntanios, F.Y.; Raeini-Sarjaz, M.; Vanstone, C.A. Cholesterol-lowering efficacy of a sitostanol-containing phytosterol mixture with a prudent diet in hyperlipidemic men. Am. J. Clin. Nutr. **1999**, *69*, 1144–1150.
5. Vanstone, C.A.; Raeini-Sarjaz, M.; Parsons, W.E.; Jones, P.J. Unesterified plant sterols and stanols lower LDL-cholesterol concentrations equivalently in hypercholesterolemic persons. Am. J. Clin. Nutr. **2002**, *76*, 1272–1278.
6. Lee, M.H.; Lu, K.; Hazard, S.; Yu, H.; Shulenin, S.; Hidaka, H.; Kojima, H.; Allikmets, R.; Sakuma, N.; Pegoraro, R.; Srivatsava, A.K.; Salen, G.; Dean, M.; Patel, S.B. Identification of a gene, ABCG5, important in the regulation of dietary cholesterol absorption. Nat. Genet. **2001**, *27*, 79–83.
7. Rudkowska, I.; AbuMweis, S.S.; Nicolle, C.; Jones, P.J. Association between non-responsiveness to plant sterol intervention and polymorphisms in cholesterol metabolism genes: a case-control study. Appl. Physiol. Nutr. Metab. **2008**, *33*, 728–734.
8. Zhao, H.L.; Houweling, A.H.; Vanstone, C.A.; Jew, S.; Trautwein, E.A.; Duchateau, G.S.; Jones, P.J. Genetic variation in ABC G5/G8 and NPC1L1 impact cholesterol response to plant sterols in hypercholesterolemic men. Lipids **2008**, *43*, 1155–1164.
9. Bradford, P.G.; Awad, A.B. Phytosterols as anticancer compounds. Mol. Nutr. Food. Res. **2007**, *51*, 161–170.
10. Awad, A.B.; Fink, C.S. Phytosterols as anticancer dietary components: evidence and mechanism of action. J. Nutr. **2000**, *130*, 2127–2130.
11. Vivancos, M.; Moreno, J.J. β-Sitosterol modulates antioxidant enzyme response in RAW 264.7 macrophages. Free Radic. Biol. Med. **2005**, *39*, 91–97.
12. Awad, A.B.; Fink, C.S.; Williams, H.; Kim, U. In vitro and in vivo (SCID mice) effects of phytosterols on the growth and dissemination of human prostate cancer PC-3 cells. Eur. J. Cancer Prev. **2001**, *10*, 507–513.

Plant Tissue Culture: Industrial Uses

Kirsten A. Hirneisen
Kalmia E. Kniel
Department of Animal and Food Sciences, University of Delaware, Newark, Delaware, U.S.A.

Abstract

Plant tissue culture has become an important industrial tool since the ability to culture plant cells in vitro in the early 1900s was discovered. Through the manipulation of growth media and hormones, whole plants can be regenerated from single plant cells. This micropropagation of plant species has played an important role in the reproduction of agricultural crops and also allows for the reproduction of pathogen-free plants. Plant tissue culture has also played an important role in secondary metabolite production of food ingredients. Large-scale cell culture has enabled the production of flavors, colorants, essential oils, sweeteners, and antioxidants in industrial bioreactors.

INTRODUCTION

Plant cells are totipotent, meaning whole plants can be regenerated from single non-sexual cells through the manipulation of growth media and hormones.

The concept of culturing plant cells in vitro was first proposed by Gottlieb Haberlandt in 1902, when he unsuccessfully tried to culture leaf mesophyll cells. Early plant tissue culture efforts used agricultural crops. In 1939, Roger Gautheret was able to culture tissues of carrots, and Philip White successfully cultured tomato roots. Both Gautheret and White were able to maintain their cultures for 6 years through subculturing on fresh media. These experiments demonstrated that tissue culture could be initiated and maintained over a long period of time.[1] In 1983, the first commercial production of plant cell and tissue culture-derived pigment, shikonin, was established by Mitsui Petrochemical Industries in Japan.[2] Plant cell and tissue cultures have many applications for the food and agricultural industry. This entry will focus the basics of plant cell culture and its use in micropropagation of agricultural crops. Secondary metabolite production by large-scale plant cell culture is also discussed due to the use of secondary metabolites as food ingredients.

PLANT CELL/TISSUE CULTURES

Plant cell and tissue cultures are obtained from explants that comprise specific tissues, especially with meristematic cells, and are placed in nutrient media under aseptic conditions.[3] Freshly cut pieces from surface-sterilized plants, called explants, are placed on solidified or liquid nutrient media with growth regulators. Explants develop from either differentiated tissues or from undifferentiated calli. A callus is an unorganized mass of cells that develops when cells are wounded; these cells can be continuously proliferated by controlling plant growth hormones.[1] Subculture of calli on an agar medium improves the friability of the callus that is a desirable trait for creating a liquid cell suspension.[3] Calli can be transferred to liquid media to form cell suspension cultures that can be grown in bioreactors similar to microbial fermentation and mammalian cell cultures. The cultivation of more differentiated tissues or root cultures requires the design of reactors that can accommodate the non-uniform morphology of the plant tissues.[3] Large-scale plant cell propagation has been achieved through the development of continuous cultures grown in bioreactors.

MICROPROPAGATION

Micropropagation is the clonal propagation of plants through tissue culture and allows for the production of plants in large numbers from a single individual plant.[3] The use of tissue culture for micropropagation was initiated in 1960 as the only commercially viable approach for orchid propagation, and as of today, many crop species have been micropropagated. Micropropagation is used for agricultural crops, including potatoes, bananas, pears, and walnuts.[3] Throughout the past two decades over 200 million plants have been produced via tissue culture worldwide. Fruit trees, vegetables, ornamental foliage, landscape plants, and other agricultural crops are the most widely produced.[4]

One advantage of using tissue culture for plant propagation is that virus-free plants can be produced. Most crops are infected by systemic diseases, and while pathogen infection does not always lead to plant death, it can reduce yield and quality of the crop as well as raise barriers to international trade.[3] During vegetative propagation, these pathogens are almost always transferred and viral diseases are in virtually all seed-propagated and vegetatively propagated crops; however, the exception is apical meristems, especially the most undifferentiated apical dome. Even the apical meristems of infected plants are usually free from virus infection. This phenomenon is believed to be due to the absence of the vascular system in meristems that transports viruses throughout the plant as well the high metabolic activity in this actively dividing area. Meristem tip culture of plants has allowed propagation of virus-free plants.[5]

Micropropagation of a species results in a monoculture of cloned cells; however, sporadic abnormalities can occur during culturing and lead to genetically varied plants. This is known as somaclonal variation and is a concern for commercial micropropagation where genetic instability can decrease regeneration rates of the plant tissue culture.[5] Sources of variation are due to the length of time in culture, growth regulators, stress response, epigenetic variation, and elimination of infectious agents. While somaclonal variation frequently tends to be deleterious, it can sporadically allow for useful variants to be selected.[3]

Another important use of plant tissue culture is their use in plant-germplasm banks, which are used to maintain an inventory of plant genetic resources. The most convenient form of storing germplasm is through seeds; however, over time seed stocks lose viability and the use of cryopreservation of tissue culture may be a better means of conserving plants.[6] In many species vegetative propagation is essential for maintenance of a desirable trait, and in such cases cryopreservation of stocks is an essential preservation strategy for food security. Plant tissue culture also plays a vital role in transgenic plant production and without tissue culture technologies there would be no further development of transgenic plants.[1] Recombinant protein production by large-scale transgenic plant cultures is gaining popularity and this is referred to as molecular farming. The tobacco suspension cell culture is the most popular system; however, pharmaceutical proteins have been produced in important agricultural crops such as soybeans, tomatoes, and rice cells.[1]

SECONDARY METABOLITES

Secondary metabolites are chemical compounds produced by plants as products of metabolism and have specialized functions such as an ecological role (e.g., interactions with other plants within a species as attractants or repellents of insects, nematodes, and microbes), function as plant hormones or growth regulators, or become structural components, such as lignin in the cell wall. Secondary metabolites may also be used by the plant for pollination/reproductive purposes, including color and smell; they are responsible for a wide variety of food ingredients. The most valuable food additives that can be obtained from plant tissue culture are colorants (anthocyanins and betalaines), flavors (saffron and vanilla), sweeteners (steviosides), and antibacterial food preservatives.[7] Secondary metabolites can be scaled-up and automated using bioreactors for commercial production. Many large-scale suspension cultures have been developed. These have several advantages over intact plants, including: large amounts of products can be obtained in a short period of time; maintenance of controlled conditions (temperature, humidity, and nutrients); ability to introduce precursors into cells; use of mutant cell lines for production of novel compounds; and economy.[3] An added benefit of using plant tissue culture for secondary metabolite production of food ingredients is that these products can be considered natural food ingredients.[7]

Many metabolites can be produced and accumulated in cell cultures; however, industrial-scale metabolite production from plant tissue culture is restricted to a few products. Mitsui Petrochemical Industries of Tokyo, Japan, has achieved success in commercially producing shikonin, which is utilized as both a dye and a pharmaceutical from *Lithospermum erythrorhizon* cell suspension cultures. Vanillin, the most abundant component of vanilla extract, has been produced in *Vanilla planifolia* tissue culture commercially.[8] Ginseng products from *Panax ginseng* derived from cell suspension culture have been used in wine, tonic drinks, and herbal liquors produced by the Nitto Denko, Co. in Japan since 1990.[9] Anthocyanins (water-soluble purple, blue, and red pigments with touted antioxidant properties) have been produced in *Aralia cordata* cell suspension culture in Japan and used as food additives in soft drinks and jams.[10] The signature aromas of cocoa and coffee have been produced by cell cultures of *Theobroma cacao* and *Coffea arabica*.[7] Tissue culture facilitates the production of the yellow food colorant, crocin, produced by the stigma of the flower saffron crocus (*Crocus sativus*). Extraction from plants would take 200,000 flowers grown for 12 months to produce 1 kg of saffron;[11] plant tissue culture allows increased production over a shorter period of time.

Cell suspension culture is the most convenient method for the mass propagation of cells and large-scale production of secondary metabolites. Secondary metabolites are a product of cell differentiation and research on the in vitro culture of differentiated plant cells including roots, shoots, and embryos has shown that these organ cultures produce similar patterns of secondary metabolites as the plant.[12] Disadvantages of plant cell cultures include the large size of plant cells, rigid cell wall, and large vacuole, which causes the cells to be sensitive to shear stresses that include mixing and aeration, making conventional microbial fermentation vessels unfavorable for mass cultivation of plant cells.[3] Also, plant cell growth is slow, and for rapid growth a high inoculum density is necessary.

Another disadvantage of plant cell culture is that secondary metabolites are generally not secreted by the plant cell but instead are maintained within the cytoplasm or cell vacuole. Improved methods for secondary metabolite release to the suspending medium from plant cells in large-scale culture is a current area of research and possible methods of elicitation of secondary metabolites include stress factors such as osmotic shock, addition of heavy metal ions, inorganic salts, micirbial homogenates, UV radiation, and infection with microorganisms.[12]

CONCLUSION

The food and agricultural industry has greatly benefited from plant cell and tissue culture technology. The generation of secondary metabolites along with micropropagation of plant species has greatly added to the variety of food ingredients. The utilization of these technologies is economical and sustainable. As future research develops, scale-up of these technologies will be more widely implemented in the agricultural and food industry.

REFERENCES

1. Cardoza, V. Tissue culture: the manipulation of plant development. In *Plant Biotechnology and Genetics: Principles, Techniques and Applications*; Stewart, C.N., Jr., Ed.; John Wiley & Sons: Hoboken, NJ, 2008; 113–134.
2. Curtin, M.E. Harvesting profitable products from plant tissue culture. Biotechnology **1983**, *1*, 649–657.
3. Razden, M.K. *Introduction to Plant Tissue Culture*, 2nd Ed.; Science Publishers, Inc.: Enfield, NH, 2003.
4. Rauther, G. Current status and future prospects of large-scale micropropagation in commercial plant production. Food Biotechnol. **1990**, *4*, 445.
5. Kartha, K.K; Gamborg, O.L. Elimination of cassava mosaic disease by meristem culture. Phytopathology **1975**, *65*, 826–828.
6. Lee, B.H. Plant biotechnology. In *Fundamentals of Food Biotechnology*; VCH Publishers, Inc.: New York, 1996; 355–370.
7. Smetanska, I. Production of secondary metabolites using plant cell cultures. Adv. Biochem. Eng. Biotechnol. **2008**, *111*, 187–228.
8. Dornenburg, H.; Knorr, D. Production of the phenolic flavor compound with cultured cells and tissues of *Vanilla planifolia* species. Food Biotechnol. **1996**, *10*, 75–92.
9. Fu, T.J.; Singh, G.; Curtis, W.R. Plant cell and tissue culture for food ingredient production. In *Plant Cell and Tissue Culture for the Production of Food Ingredients*; Fu, T.J., Singh, G., Curtis, W.R., Eds.; Kluwer Academic/Plenum Publishers: New York, 1999; 1–6.
10. Sakamoto, K.; Iida, K.; Sawamura, K.; Hajiro, K.; Asada, Y.; Yoshikawa, T.; Furuya, T. Anthocyanin production in cultured cells of *Aralia cordata* thumb. Plant Cell Tissue Organ Cult. **1994**, *36*, 21–26.
11. Chen, H.; Wang, X.; Zhao, B.; Yuan, X.; Wang, Y. Production of crocin using *Crocus sativus* callus by two-stage culture system. Biotechnol. Lett. **2003**, *25* (15), 1235–1238.
12. Verpoorte, R.; Contin, A.; Memelink, J. Biotechnology for the production of plant secondary metabolites. Phytochem. Rev. **2002**, *1*, 13–25.

Plant-Produced Recombinant Therapeutics

Lokesh Joshi
National Centre for Biomedical Engineering Science, National University of Ireland, Galway, Ireland

Miti M. Shah
C. Robert Flynn
Alyssa Panitch
Arizona State University, Tempe, Arizona, U.S.A.

Abstract

The technology of bioengineering plants to produce recombinant therapeutics has become commonplace. This entry briefly describes some representative proteins and the role of posttranslational modifications (PTMs) on these molecules.

INTRODUCTION

The technology of bioengineering plants to produce recombinant therapeutics has become commonplace. Transgenic plants are emerging as suitable alternatives to transgenic animals and bioreactor-based systems for recombinant protein production (e.g., bacterial, yeast, and insect cells). This trend is yielding a recombinant protein expression system that has increased economic viability and a greater capacity for producing lifesaving protein therapeutics and biopolymers. In addition, plant products are free of mammalian viruses, prions, or other adventitious organisms harmful to humans, making them safer production systems. A wide variety of recombinant protein molecules have been successfully produced in plants (Table 1). This section briefly describes some representative proteins and the role of PTMs on these molecules.

IMMUNOGLOBULINS

IgG and IgA are multifunctional glycoprotein immune molecules that bind to antigens, form immune complexes, and activate classical and alternative complement pathways, respectively. This activity leads to the destruction and clearance of the pathogen. Both IgG and IgA are glycosylated, and although glycosylation is not required for antigen binding, it is critical for activation of the complement pathway.[1] Deglycosylated or underglycosylated immunoglobulins (IgG and IgA) are unable to activate the effector mechanism and therefore fail in the clearance of the pathogen and other antigenic moieties. Plants are able to perform expression and assembly of immunoglobulin (Ig) heavy and light chains into functional antibodies.[2] Plants have been used for production of different forms of antibodies that include full-size IgG and IgA, chimeric IgG and IgA, single-chain Fv fragments (ScFv), Fab, and heavy-chain variable domains; the last three types do not require glycosylation. Among the antibodies that have been successfully produced in plants are those against surface antigen of *Streptococcus mutans*, the causative agent of dental caries;[3] herpes simplex virus;[4] carcinoembryonic antigen (CEA), a tumor-associated marker;[5] and single-chain Fv fragments against non-Hodgkins lymphoma.[6] The majority of plant-derived antibodies possessed high-mannose and hybrid-type, α1-3 fucose– and β1-2 xylose–containing structures that lack β1–4 galactose and the terminal sialic acid residues. Expression of mammalian β1-4 Galtransferase in plants resulted in synthesis of galactose-containing antibodies. About 30% of N-glycans of the antibodies were galactosylated.[7] Plant-derived antibodies do not show any alterations in affinity toward antigens or stability in vivo or in vitro.

VACCINES

Plants are a prime candidate for the production of vaccines because of the lower cost of production and the feasibility of producing "edible vaccines." This technology can be transferred to developing nations relatively easily compared to other expression systems. Some of the representative examples are provided here:

Rabies vaccine: The rabies virus glycoprotein G and nucleoprotein N are both very important in designing an appropriate vaccine. G protein is the major antigen responsible for induction of protective immunity, and N protein is responsible for induction of virus-specific T cells. A plant-derived oral vaccine against a fusion protein consisting of glycoprotein G, nucleoprotein N, and alfalfa mosaic virus coat protein was made, which produced significant immune response and rabies virus–specific antibodies in both mice and humans.[8]

Table 1 A brief list of representative biomolecules engineered in plants.

Biomolecule	Engineered plant
Human and animal vaccines	Tobacco, potato, corn, lettuce, tomato, alfalfa, *Arabidopsis*
Immunoglobulins	Tobacco, alfalfa, rice, wheat, soybean
Other therapeutic proteins	Tobacco, canola, rice, turnip, alfalfa, *Arabidopsis*
Biomaterial:	
Collagen	Tobacco
Spider silk	Tobacco, potato

See Daniell et al.[18] for a more complete list of plant-derived pharmaceuticals.

Human cytomegalovirus: Human cytomegalovirus (HCMV) is a member of the herpes virus family that is transmitted by blood and body secretions. In immuno compromised individuals, infection of HCMV can lead to damage of the central nervous system and death. Glycoprotein B, a transmembrane envelope protein, was used to produce oral vaccine against the viral infection. The affinity of recombinant glycoprotein B was examined in vitro by its ability to bind monoclonal antibodies. Plant-derived glycoprotein B showed a high affinity to antibodies.[9]

Gastroenteritis coronavirus: Swine-transmissible gastroenteritis virus (TGEV) is the causative agent of acute diarrhea of newborn piglets. Neutralizing antibodies against this virus is mainly directed toward a surface component, glycoprotein S. Glycoprotein S from TGEV, produced recombinantly in plants, can function as a vaccine against infection when injected intramuscularly in animals. Glycoprotein S has three glycosylation sites, which play an essential role in conformation of this protein. However, no further information on glycosylation of the recombinant glycoprotein S is available.[10]

Measles virus: Measles is a highly contagious viral disease. Severe infection may lead to pneumonia, encephalitis, and death. Hemagglutinin, a surface protein from measles virus, was expressed in plants. For recognition of hemagglutinin by B lymphocyte cells, an appropriate conformation of the protein is essential. Folding, stability, and protease susceptibility of this protein is dependent on four N-glycosylation sites. Plant-derived hemagglutinin was found to be stable and able to induce an immune response in animal model systems upon oral administration. This indicates that plants are able to achieve sufficient glycosylation required for stability and folding of the protein.[11]

OTHER PROTEIN THERAPEUTICS

Follicle-stimulating hormone: Follicle-stimulating hormone (FSH) is a pituitary heterodimeric glycoprotein hormone that requires N-glycosylation for subunit folding, assembly, targeting, and stability. FSH produced in plants possesses terminal mannose residues and is reported to be biologically active.[12]

Lactoferrin: Lactoferrin is a milk protein. In humans, it is known to contain two N-acetyllactosamine-type N-glycans that also contain fucose and sialic acid residues. Lactoferrin belongs to the family of transferrin with iron-binding properties. It has also been found to possess antibacterial, antifungal, antiviral, and antiinflammatory activities. Full-length lactoferrin with antimicrobial properties is produced in plants.[13]

Erythropoeitin: Erythropoeitin (EPO), an N-glycosylated protein that regulates the formation of erythrocytes in mammals, was recombinantly produced in plants. EPO produced in tobacco cells was glycosylated with N-linked oligosaccharides that did not possess terminal sialic acid residues. EPO produced in tobacco exhibited in vitro biological activities by inducing the differentiation and proliferation of erythroid cells. However, it did not show in vivo biological activities.[14] The authors speculated that this was attributable to the different glycosylation of EPO produced in tobacco cells as compared to authentic human EPO and was cleared from circulation by asialo-receptors.

Macrophage activating factor: Macrophage activating factor (MAF), also known as vitamin D–binding protein and Gc-globulin, is a multifunctional abundant serum protein. Threonine-420 of domain III of MAF is O-glycosylated. Upon infection with pathogen or tumor cells, activated T and B lymphocyte cell surface exoglycosidases remove sialic acid and galactose residues, leaving a GalNAc attached to the threonine.[15] This GalNAc is essential for MAF to bind to the C-type lectin on the cell surface of mononuclear phagocytes, which leads to their activation. Activated phagocytes remove pathogens and tumor cells from the body. Efforts are underway to express this serum glycoprotein in plants.

Acetylcholinesterase: Acetylcholinesterase (AChE) catalyzes the degradation of acetylcholine, a neurotransmitter. AChE is a potential therapeutic against organophosphate-based chemicals and warfare agents. Sialic acid residues on the N-glycans of AChE are crucial in determining the circulatory clearance rate. Production of AChE in HEK-293 cells led to hyposialylation of the protein, which was responsible for rapid clearance of recombinant AChE from the circulation. Recombinant human AChE was produced in tomato plants by Mor et al.[16] Kinetic studies of the recombinant AChE using inhibitors showed that the plant-produced AChE had an similar inhibition profile similar to the commercially available AChE purified from human erythrocytes. However, in vivo activity, serum half life, and glycosylation studies of plant-produced AChE have not yet been reported.

BIOMOLECULES

Collagen: Collagen is the major protein in the extracellular matrix and is thus the most prominent animal protein. There are more than 20 different types of collagens described so far. The triple-helical domains of collagens have the repeating amino acid sequence $(Gly-X-Y)_n$, where X and Y are frequently proline and hydroxyproline, respectively. Collagen molecules go through multiple PTMs involving at least eight enzymes and cofactors. Specific proline and lysine residues are hydroxylated, and some of the hydroxylysine residues are further modified by glycosylation with galactose-glucose disaccharide units. Hydroxylation of proline to hydroxyproline residues assists in trimer assembly of collagen polypeptide chains. Collagens and their derivative molecules (gelatin) have multiple applications but are currently only available from animal sources. There is a need for alternative sources to produce large quantities of pure and safe collagens. Human collagen α-1 (C1α1) has been expressed by itself and with the multisubunit recombinant prolyl-4-hydroxylase,[17] in transgenic tobacco, collagen trimers have been produced that are stable up to 37°C. These reports suggest that the production of human collagen is possible in plants. However, further research is required to understand and improve recombinant collagen synthesis and assembly in plants.

CONCLUSION

The discovery of and demand for medically important proteins are growing rapidly. It is becoming increasingly critical to ensure their availability in sufficient quantity for research, therapeutics, and diagnostic uses. Transgenic plants are emerging as suitable systems for the production of functional recombinant human proteins. Plants have the distinct advantages over other expression systems of product safety, economical production, and ease of scale-up. Plants are also able to perform most PTMs that are carried out by mammals, such as glycosylation, phosphorylation, and hydroxylation. However, the PTMs in plants are understudied and require more attention because it is important to control the addition of and evaluate the structural and functional roles of these modifications on recombinant products for regulatory purposes as well as for desired physiological activities, when used for therapeutic purposes in humans and animals.

REFERENCES

1. Miletic, V.D.; Frank, M.M. Complement-imunoglobulin interaction. Curr. Opin. Immunol. **Feb 1995**, *7* (1), 41–47.
2. Ma, J. Generation and assembly of secretory antibodies in plants. Science **1995**, *268*, 716–719.
3. Ma, J.; Hikmat, B.; Wykoff, K.; Vine, N.; Chargelegue, D.; Yu, L.; Hein, M.; Lehner, T. Characterization of a recombinant plant monoclonal antibody and preventive immunotherapy in humans. Nat. Med. **1998**, *4*, 601–606.
4. Zeitlin, L. A humanized monoclonal antibody produced in transgenic plants for immunoprotection of the vagina against genital herpes. Nat. Biotechnol. **1998**, *16*, 1361–1364.
5. Stoger, E.; Vaquero, C.; Torres, E.; Sack, M.; Nicholson, L.; Drossard, J.; Williams, S.; Keen, D.; Perrin, Y.; Christou, P.; Fischer, R. Cereal crops as viable production and storage systems for pharmaceutical scFv antibodies. Plant Mol. Biol. **2000**, *42*, 583–590.
6. McCormick, A.A.; Kumagai, M.H.; Hanley, K.; Turpen, T.H.; Hakim, I.; Grill, L.K.; Tuse, D.; Levy, S.; Levy, R. Rapid production of specific vaccines for lymphoma by expression of the tumor-derived single-chain Fv epitopes in tobacco plants. PNAS **1999**, *96*, 703–708.
7. Cabanes-Macheteau, M.; Fitchette.-Lain, A.-C.; Loutelier-Bourhis, C.; Lange, C.; Vine, N.D.; Ma, J.; Lerouge, P.; Faye, L. N-glycosyaltion of a mouse IgG expressed in transgenic tobacco plants. Glycobiology **1999**, *9*, 365–372.
8. Yusibov, V.; Hooper, D.C.; Spitsin, S.V.; Fleysh, N.; Kean, R.B.; Mikheeva, T.; Deka, D.; Karasev, A.; Cox, S.; Ransdall, J.; Koprowski, H. Expression in plants and immunogenicity of plant virus-based experimental rabies vaccine. Vaccine **2002**, *20*, 3155–3164.
9. Tackaberry, E.S.; Dudani, A.K.; Prior, F.; Tocchi, M.; Sardana, R.; Altosaar, I.; Ganz, P.R. Development of biopharmaceuticals in plant expression systems: Cloning, expression and immunological reactivity of human cytomegalovirus glycoprotein B (UL55) in seeds of transgenic tobacco. Vaccine **1999**, *17*, 3020–3029.
10. Gomez, N.; Carrillo, C.; Salinas, J.; Parra, F.; Borca, M.V.; Escribano, J.M. Expression of immunogenic glycoprotein S polypeptides from transmissible gastroenteritis Coronavirus in transgenic plants. Virology **1998**, *249*, 352–358.
11. Huang, Z.; Dry, I.; Webster, D.; Strungnell, R.; Wesselingh, S. Plant-derived measles virus hemagglutinin protein induces neutralizing antibodies in mice. Vaccine **2001**, *19*, 2163–2171.
12. Dirnberger, D.; Steinkellner, H.; Abdennebi, L.; Remy, J.-J.; Van de Wiel, D. Secretion of biologically active glycoforms of bovine follicle stimulating hormone in plants. Eur. J. Biochem. **2001**, *268*, 4570–4579.
13. Samyn-Petit, B.; Gruber, V.; Flahaut, C.; Wajda-Dubos, J.-P.; Farrer, S.; Pons, A.; Desmaizieres, G.; Slomianny, M.-C.; Theisen, M.; Delannoy, P. N-glycosylation potential of maize: The human lactoferrin used as a model. Glycoconjugate Journal **2001**, *18* (17), 519–527.

14. Matsumoto, S.; Ikura, K.; Ueada, M.; Sasaki, R. Characterization of a human glycoprotein (erythropoeitin) produced in cultured tobacco cells. Plant Mol. Biol. **1995**, *27*, 1163–1172.
15. Yamamoto, N.; Kumashiro, R. Conversion of vitamin D3 binding protein (group-specific component) to a macrophage activating factor by the stepwise action of beta-galactosidase of B cells and sialidase of T cells. J. Immunol. **Sept. 1, 1993**, *151* (5), 2794–2802.
16. Mor, T.S.; Sternfeld, M.; Soreq, H.; Arntzen, C.J.; Mason, H.S. Expression of recombinant human acetylcholinesterase in transgenic tomato plants. Biotechnol. Bioeng. **2001**, *75*, 259–266.
17. Merle, C.; Perret, S.; Lacour, T.; Jonval, V.; Hudaverdian, S.; Garrone, R.; Ruggiero, F.; Theisen, M. Hydroxylated human homotrimeric collagen I in *Agrobacterium tumefaciens*-mediated transient expression and in transgenic tobacco plant. FEBS Lett. **2002**, *515*, 114–118.
18. Daniell, H.; Streatfield, S.J.; Wykoff, K. Medical molecular farming: Production of antibodies, biopharmaceuticals and edible vaccines in plants. Trends Plant Sci. **2001**, *6*, 219–226.
19. Wamsley, A.M.; Arntzen, C.J. Plant cell factories and mucosal vaccines. Curr. Opin. Biotechnol. **Apr 2003**, *14* (2), 145–150.

Policy and Regulation: U.S. Regulations

Susan MacIntosh
MacIntosh & Associates, Inc., Saint Paul, Minnesota, U.S.A.

Abstract
The Federal government of the USA has developed a coordinated framework policy to use existing laws to regulate the products of modern biotechnology. Three agencies, the USDA-APHIS, FDA and the EPA, have roles and responsibilities to ensure the protection of public health and the environment. While the process is predictable, it is complex since new regulations, guidelines, and procedures are constantly being optimized as experience is gained with this new technology. This coordinated framework of USA regulatory agencies is described with additional comments about state registrations, labelling, combined traits and confidentiality.

INTRODUCTION

The products of modern biotechnology, known as genetically modified organisms (GMOs), are regulated in ways that differ by region and country. Legislation takes considerable time regardless of the country and has often lagged behind the rapid adoption rates of biotechnology products. Industrial countries were typically the first to establish domestic biosafety regimes; only recently, many of the developing countries have established their own national systems. The Cartagena Protocol for Biosafety (Cartagena Protocol on Biosafety is a supplemental agreement to the Convention on Biological Diversity.),[1] which was ratified by the required 50 countries in September 2003, requires many more countries to develop regulations for the import and environmental release of living modified organisms (LMOs), another legal term used for GMOs.

Some countries have developed completely new regulatory frameworks for biotechnology-derived products (e.g., European Union and Brazil), while others rely on existing laws and regulations (e.g., United States). In this entry, the system by which the U.S. Federal government regulates the modern products of biotechnology will be described. Special attention to the policy underpinning these laws and regulations will be highlighted.

COORDINATED FRAMEWORK OF BIOTECHNOLOGY

The products of modern biotechnology have been nothing short of a revolution for agriculture. Since the first commercial product in 1996, the global planting of GMOs has grown more than 50-fold to just under 90 M hectares. Between 2003 and 2004, the rate of increase in GMO plantings for the developing world (7.2 M hectares) exceeded the rate in the industrial countries (6.1 M hectares), reflecting the high value of these new traits in seeds.[2]

Even in the United States, where new technology is quickly adopted, the acceptance of GMOs has gone beyond expectations. For example, in the year of introduction, 1996, U.S. growers planted 1.7 M hectares. By 1999, the total was 20.5 M hectares, and by 2005, nearly 50 M hectares were planted in the United States.[2]

The President's Office of Science and Technology Policy (OSTP) established an interagency working group in 1984 to review the current legislation of conventional products that protect the environment, human, and animal health to determine if and/or how these laws could be utilized for the modern products of biotechnology. Two years later, OSTP, upon the advice of the working group, decided to take existing health and safety laws and statutes to produce a formal policy of the Coordinated Framework for Regulation of Biotechnology.[3] The experience of the existing agency with agriculture, pharmaceutical, and other commercial products provided a foundation, which was expected to evolve as experience was gained with these new biotechnology products. The regulation of GMOs, under this coordinated framework approach, utilizes risk assessments developed on a case-by-case basis, which are grounded in sound science.

Several laws were the basis for this coordinated framework policy to regulate GMOs: the Plant Protection Act (PPA),[4] the Federal Food, Drug, and Cosmetic Act (FFDCA),[5] the Federal Insecticide, Fungicide and Rodenticide Act (FIFRA),[6] and the Toxic Substances Control Act (TSCA).[7]

The U.S. Government agencies responsible for oversight of GMO products are the U.S. Department of Agriculture's Animal and Plant Health Inspection Service

(USDA-APHIS), the Department of Health and Human Services' Food and Drug Administration (FDA), and the U.S. Environmental Protection Agency (EPA). Depending on the unique traits of a specific product, it could be reviewed by one or more of these agencies.

Product-Specific Oversight

The responsibilities for product review are complementary and sometimes overlapping.[8] Only when all agencies have completed their review and have concurred with the risk assessment can a product be marketed and commercialized.

Non-Pesticide Traits

For an input trait that does not have pesticidal properties, such as a herbicide-tolerant crop, the USDA-APHIS would regulate all import, interstate movement, field releases, and ultimately grant non-regulated status to potential commercial products. The Biotechnology Regulatory Services (BRS) unit of the USDA-APHIS reviews all applications, governs compliance, and decides if non-regulated status is warranted. The completion of a biotechnology consultation with FDA is also needed prior to commercialization. Although this process remains voluntary, in practice all commercialized GMOs have successfully concluded the FDA consultation.

Pesticidal Traits

When a GMO expresses a pesticide (i.e., *Bacillus thuringiensis* insect control protein), then in addition to the reviews by both the USDA-APHIS and FDA, it is the EPA that takes the lead in reviewing the pesticidal protein and coordinates the overall agencies' review process. The EPA grants a product registration, usually with postmarketing conditions, which includes additional data requirements, such as annual sales figures and monitoring for insect susceptibility.

Plant-Made Pharmaceuticals/Plant-Made Industrial Products

Plant-made pharmaceuticals (PMPs) and plant-made industrial products (PMiP) are yet other categories of GMOs. The term PMP is limited to crops that express pharmaceutical products (i.e., antibodies for therapeutic needs or medical devices, lysozyme to fortify baby formula), while the term PMiP is more broadly utilized to define any non-food/feed, non-drug, industrial product expressed in a crop (i.e., specialty machine lubricants, biofuels, and modified starch for packaging materials). The USDA-APHIS regulates all field releases as well as the production of PMPs and PMiPs. To date, there is no process for gaining non-regulated status, since these crops cannot be utilized for food or feed uses, thus commercial production is carried out under permit. Likewise, the FDA will not allow for a consultation process. Aside from the production aspects, the actual approval for the pharmaceutical or veterinary product follows the usual process, either through the USDA or FDA depending on the product characteristics.

U.S. DEPARTMENT OF AGRICULTURE

The USDA was founded in 1862 when 58% of Americans were farmers. The agency has a very broad mandate to support agriculture and ranching, which includes topics such as protecting national forests and rangeland; guarding the safety of meat, poultry, and egg products; bringing housing, modern communications, and safe drinking water to rural America; leading antihunger efforts within the United States and abroad; ensuring open markets for agricultural products; and conducting a wide range of agricultural research. The role of APHIS is safeguarding the health and care of animals and plants while improving agricultural productivity.[9]

Permit for Import, Interstate Movement, and Environmental Release

The movement and environmental release of regulated articles, such as GMOs, is becoming more tightly regulated in the United States. New guidelines were developed in response to a request by OSTP[10] to update field trial requirements and to establish early voluntary food safety evaluation of new proteins produced in biotech crops. The USDA strengthened their regulatory oversight by establishing a new Compliance section within the BRS unit.

Permit Procedures

There are three main types of permits issued by USDA-APHIS that may be needed in order to perform a field trial with GMO. If seeds must be imported to the United States, then an import permit, with the proper labeling, is required. If plant materials must be moved across state lines, then an interstate movement is required. Should the plant materials be completely devitalized so that they are no longer viable and cannot reproduce, then no permit is needed for moving GMO materials. Finally, all environmental releases require a permit or through a more simplified system, a notification.

Compliance with all permit conditions is crucial during all aspects of the import, movement, and release of a regulated article. With the new compliance section of the BRS unit, a more rigorous process is now in place. Each permit application is ranked according to risk, which directly impacts the level of regulatory oversight in the field by the authorities.

Under the permitting system of the USDA-APHIS, states within the United States where testing will occur are consulted for their agreement with the field trial release conditions described by the USDA. Once granted, the state regulatory authorities conduct USDA-APHIS auditing responsibilities for the field release.

A Determination of Non-Regulated Status

Once all the data have been collected for food, feed, and environmental safety and no risks have been identified with the new GMO, a deregulation petition can be prepared. The applicant requests that the GMO should no longer be considered a regulated article since it is not materially different than the conventional crop except for the introduced trait and presents no plant pest risk. The procedure for gaining non-regulated status is generally well defined and predictable. More than 90 petitions have been approved and granted non-regulated status by the USDA-APHIS since 1992.[11]

U.S. FOOD AND DRUG ADMINISTRATION

The FDA dates back to June 1906 when the Food and Drugs Act created the Bureau of Chemistry within the USDA. This office eventually became the FDA, which is charged with regulating food ingredients, nutrition and dietary supplements, drugs, medical devices, biologics (e.g., vaccines and blood products), animal feed and drugs, cosmetics, and radiation-emitting devices (e.g., cell phones, lasers, etc.).[12]

The FDA has two different processes most applicable to GMOs.[13] New guidance for an early food safety evaluation[14] has recently been published to augment the biotechnology consultation that is completed prior to commercialization of a GMO product.

Early Food Safety Evaluation

According to the new guidance, an FDA review should occur prior to the stage in development at which a new protein might unintentionally enter the food supply due to pollen flow and/or commingling as field sizes increase. An early food safety evaluation file is only prepared for new, novel proteins, never before consumed in food or feed, but that will eventually be utilized for food/feed use. This could be a plant protein not normally consumed in a particular crop (e.g., rice protein expressed in soybean for the first time), or one that is found only in inedible plant parts, but now is expressed in edible portions of a plant, or a protein that is expressed at a much higher level that typical for that food crop.

The early food safety evaluation does not take the place of the biotechnology consultation with the FDA, which takes place prior to commercialization of a new plant variety. However, the data developed for the early food safety evaluation file can also be used during the later biotechnology consultation process. And the new guidance does not apply to plant-incorporated protectants (PIPs) that express a pesticidal ingredient, which are regulated by the EPA (discussed later in the entry).

Biotechnology Consultation

Historically, the food industry initiated premarket consultations with the FDA at very early stages of the development of a new technology even when there was no legal obligation to do so. This allowed both the agency and the food company to address any safety, labeling, or regulatory issue that new foods or food ingredients might present upon commercialization.[15] In 1996, the FDA established consultation procedures for biotechnology-based products under the FDA's 1992 Statement of Policy: Foods Derived from New Plant Varieties.[16] Several years later, in 2001, the FDA announced a draft rule to incorporate their experiences, refine the procedures for the biotechnology consultation, and make the process mandatory.[17] Although this process is voluntary until a final rule is issued, in practice all GMOs on the market today have successfully concluded the FDA biotechnology consultation. Up until November 2009, the FDA had favorably concluded 112 biotechnology consultations.

U.S. ENVIRONMENTAL PROTECTION AGENCY

The EPA is a relatively new agency, founded in 1970 to protect human health and the environment, and to provide cleaner air, water, and land. EPA develops and enforces environmental laws, of which FIFRA[6] and TSCA[7] are the cornerstones. They regulate all pesticides used in agriculture, commercial, and residential use, including pesticides expressed in plants (GMOs). It is important to note that only the new protein and the genetic material necessary to express the protein are regulated by the EPA, and not the GMO itself.

Experimental Use Permit

The EPA already has a well-established system for field trial permitting system that was developed for synthetic chemical pesticides that also applies to GMOs.[18] When the GMO contains a pesticidal trait and the area of the plot(s) exceeds 10 acres, further oversight by the EPA is required in the form of an application for an environmental use permit (EUP) in addition to the usual USDA-APHIS process described earlier. The intent of an EUP is to support research and development activities, in a highly controlled manner, especially in the development of regulatory data for gaining EPA registration.

Registration

A pesticide, including a PIP, cannot be marketed until the EPA decides that the pesticide poses no unreasonable risks of harm to human health or the environment.[19] A registration application contains data on human (and animal) health, gene flow, and outlines risks to non-target organisms and the environment. For GMOs containing an insecticide, a robust insect resistance management plan must also be described. The EPA must evaluate and balance the risks with the benefits of a specific PIP use and will only grant a license or registration when the benefits clearly outweigh the risks.

State Registration

The individual states of the United States are involved in two different processes related to GMOs, USDA permitting and EPA registration(s).[18] For both EUPs and registration actions, some states have additional requirements above and beyond those at the federal level. For some states, there is no additional action needed, for others a simple letter notifying the registration is sufficient and yet others, such as California, have an entire system that duplicates the EPA process and can only occur once the EPA registration is granted.

PRACTICAL ISSUES

Combined Traits

The applicant or registrant defines the GMO product for deregulation (USDA) or registration (EPA), which are typically one or more transformation events as long as they are phenotypically the same. Since the USDA grants non-regulated status to the event as well as all progeny of that event, there are no additional requirements if the event of interest is to be combined with another approved GM event, as long as traditional breeding practices are used. If, however, an already approved event is retransformed with another trait, then a new petition must be filed with the USDA-APHIS. The FDA follows the same policy.

The only exception to this policy is when two or more events with pesticidal properties are combined. Then the EPA requires an abbreviated data set, as compared to a full commercial registration, to evaluate any change in the risk assessment due to the presence of the different pesticidal elements. Protein expression, host specificity, efficacy of pest control, and a revised resistance management plan comprise the data package.

Transparency and Confidentiality

All agency decisions rely on sound science, consistency, fairness, and an open and transparent system to continue to support public trust in biotechnology. To that end, confidential business information should be minimized and must be fully justified. Nearly all interactions and data shared between an applicant and the agency are provided to the public either via the Federal Register,[20] the agency's Web site, or under the Freedom of Information Act.[21]

GMO Labeling

Foods produced through biotechnology are subject to the same rules and regulations that the FDA has established for all other foods.[22] FDA mandates labeling when there is a material difference in a food, such as the presence of known allergens or when a nutrient content claim is made (e.g., reduced fat and sugar). In 1992, the FDA set forth a policy that was confirmed in a draft guidance released in 2001, which states that there is no basis for concluding that foods derived from GMOs differ from other foods in any meaningful or uniform way. When significant changes have been made in a food, including GMOs, then labeling is required. In other cases, voluntary labeling is allowed as long as it is truthful, informative, and not misleading.

CONCLUSIONS

The federal government of the United States has developed a coordinated framework policy to use existing laws to regulate the products of modern biotechnology. Three agencies, the USDA-APHIS, FDA, and the EPA, have roles and responsibilities to ensure the protection of public health and the environment. Although the process is predictable, it is complex since new regulations, guidelines, and procedures are constantly being optimized as experience is gained with this new technology.

The regulation of GMOs under this coordinated framework approach utilizes risk assessments developed on a case-by-case basis, which are grounded in sound science. The role of USDA-APHIS is to ensure the health and care of animals and plants while improving agricultural productivity, and primarily focuses on any plant pest characteristics of the GMO. FDA reviews the food and nutritional safety of GMOs, while the EPA regulates all pesticides used in agriculture, commercial, and residential use, including pesticides expressed in plants. Only when all agencies have completed their review and have concurred with the risk assessment can a product be marketed and commercialized.

The responsibilities are complementary and often overlap since all three agencies may have oversight during the phases of research, development, and commercial launch. In recent years, the procedures for compliance of limited environmental releases of GMOs, such as field trials, have become more stringent. And the open transparency of agency processes and decisions has become ever more important to maintain public acceptance of agricultural biotechnology.

REFERENCES

1. Cartagena Protocol on Biosafety. http://bch.cbd.int/protocol/. Accessed November 5, 2009.
2. James, C. International Service for the Acquisition of Agri-biotech Applications (ISAAA), http://www.isaaa.org (2006).
3. Office of Science and Technology Policy's Coordinated Framework for Regulation of Biotechnology, June 26, 1986 (51 FR 23302).
4. Plant Protection Act; 7 U.S.C. 7701 et seq. (recently revised 2000).
5. Federal Food, Drug, and Cosmetic Act (FFDCA); 21 U.S.C. 301 et seq. (1938).
6. Federal Insecticide, Fungicide and Rodenticide Act (FIFRA); 7 U.S.C. s/s 135 et seq. (1972).
7. Toxic Substances Control Act (TSCA); 15 U.S.C. s/s 2601 et seq. (1976).
8. United States Regulatory Agencies Unified Biotechnology Website. http://usbiotechreg.nbii.gov/ (accessed November 5, 2009).
9. USDA-APHIS Website. http://www.aphis.usda.gov/biotechnology/brs_main.shtml (accessed November 5, 2009).
10. Fed Register, Aug 2, 2002 (67 FR 50578).
11. USDA-APHIS Status of Petitions. http://www.aphis.usda.gov/biotechnology/brs_main.shtml (accessed November 5, 2009).
12. FDA Website. http://www.fda.gov/ (accessed November 5, 2009).
13. FDA Plant Biotechnology Website. http://www.fda.gov/Food/Biotechnology/default.htm (accessed November 5, 2009).
14. Recommendations for the Early Food Safety Evaluation of New No-pesticidal Proteins Produced by New Plant Varieties Intended for Food Use, June 21, 2006 (71 FR 35688).
15. http://www.cfsan.fda.gov/~lrd/fr010118.html.
16. FDA Biotechnology Policy 1992 Policy (57 FR 22984).
17. Proposed rule, January 2001 (66 FR 4706).
18. EPA Registering Pesticides Website. http://www.epa.gov/pesticides/regulating/registering/ (accessed November 5, 2009).
19. EPA Plant Incorporated Protectants Website. http://www.epa.gov/pesticides/biopesticides/pips/index.htm.
20. EPA Federal Register Documents. http://www.epa.gov/fedrgstr/search.htm.
21. EPA Freedom of Information Act (FOIA). http://www.epa.gov/foia/.
22. FDA Food Website. http://www.cfsan.fda.gov/~dms/biolabgu.html.

Polymerase Chain Reaction (PCR)

Rolf D. Joerger
Department of Animal and Food Sciences, University of Delaware, Newark, Delaware, U.S.A.

Abstract
The polymerase chain reaction (PCR) is a technique for the synthesis of large quantities of specific DNA sequences. The amplified sequences are useful for genetic manipulations and for diagnostic purposes. In particular, PCR assays for detection of pathogens in food, food ingredients, animal feed, and agricultural commodities have been developed and some of them are used by farm and food processing operations. This entry explains different PCR techniques, such as quantitative and real-time PCR, and discusses their uses.

INTRODUCTION

The polymerase chain reaction (PCR) is a tool for exponential amplification of specific DNA sequences. Amplification is based on repeated cycles of synthesis of complementary DNA strands catalyzed by DNA polymerases. The complementary strands are produced by extension of primers that demark the beginning of the new strand. The enzymatic reaction is carried out in thermocyclers that are programmed to perform timed heating and cooling as required for the specific amplification of a DNA sequence.

The amplification process creates millions of identical copies of a DNA sequence that can be inserted into vectors for genetic manipulations or for becoming the basis of the systems for detection of the presence of diagnostic DNA sequences. Assays have been developed for the detection and sometimes also the quantitation of sequences specific for pathogenic, beneficial, genetically modified, or allergen-producing organisms. This entry describes the nature of these tests and the basic principles of PCR methods.

PCR BASICS

Biochemistry, Instrumentation, and Challenges

In the words of its inventors, the PCR "consists of repetitive cycles of denaturation, hybridization, and polymerase extension and is not a bit boring until the realization occurs that this procedure is catalyzing a doubling with each cycle in the amount of the fragment defined by the positions at the 5′ ends of the two primers on the template DNA."[1] PCR thus is a means of exponentially synthesizing (amplifying) a specific DNA sequence, often called an "amplicon." Biochemically, the amplification process is relatively simple, requiring template DNA, two primers, the four deoxynucleotides, a suitable buffer that includes Mg^{++}, and a heat-stabile DNA polymerase. The *Taq* polymerase, purified from the thermophilic bacterium *Thermus aquaticus* and named "molecule of the year" in 1989,[2] was the first such enzyme employed. *Taq* polymerase, like other DNA polymerases, does not recognize RNA templates, and therefore, RNA has to undergo reverse transcription (RT) into DNA prior to PCR. RT-PCR is achieved by reverse transcriptase added along with DNA polymerase or by *Tth* DNA polymerase alone since this enzyme exhibits RT and DNA polymerase activity.

PCR is performed in programmable thermocyclers that denature the template DNA, usually at 94–95°C, lower the temperature to allow specific binding of the primers to their complementary sequences, and finally increase the temperature to 72°C for optimal extension of the DNA strand by the polymerase. Denaturation and annealing times are generally in the order of seconds to 1 min, whereas extension times depend on length and sequence of the amplicon and the efficiency of the polymerase. The need for rapid temperature changes limits the volume of PCR reactions in most cases to 200 μL or less.

The schematic representation of the first three cycles of PCR in Fig. 1 illustrates how a single double-stranded DNA template gives rise to amplification products overwhelmingly consisting of a defined DNA species. This DNA appears in cycle 3 for the first time, and its number increases exponentially from then on. After 30 cycles, 1073741764 out of a total of 1073741824 synthesized DNA molecules are of the defined length. In practice, the yield of amplified molecules is lower since kinetic and enzyme-related factors reduce PCR efficiency especially at higher cycle numbers (Fig. 2).

The extraordinary synthesis power of PCR is responsible for its utility, but for its practical challenges as

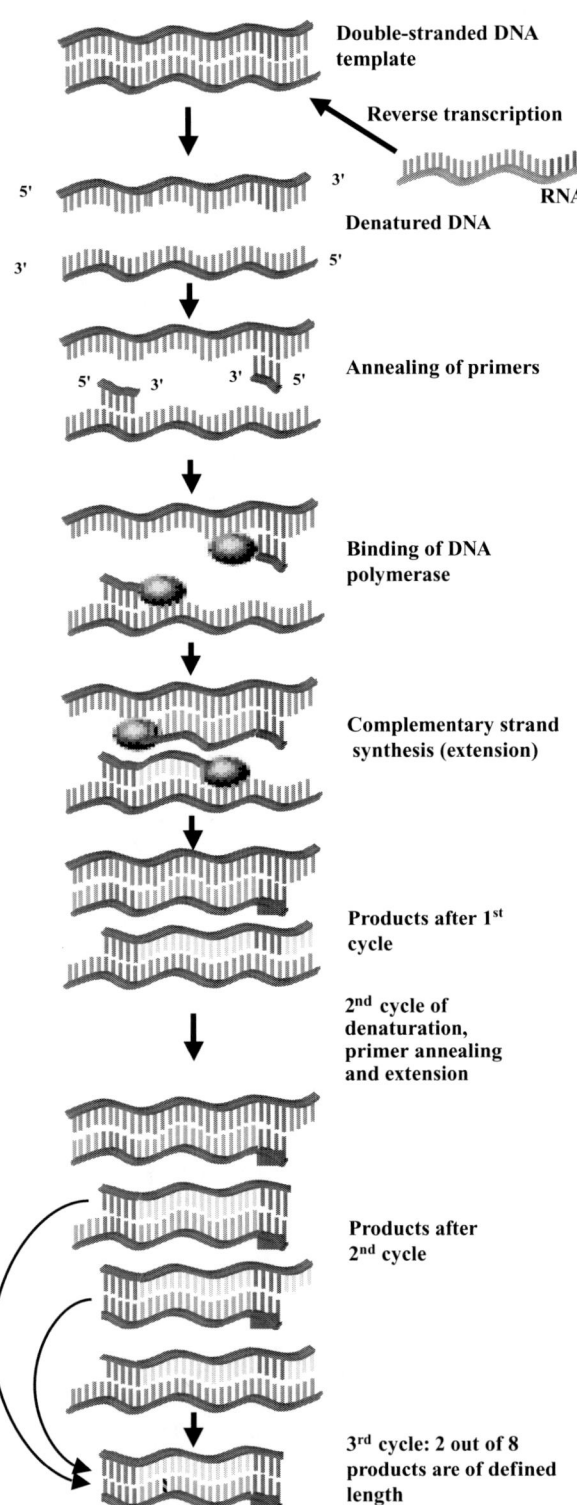

Fig. 1 Schematic representation of the steps and results of the first three cycles of PCR.

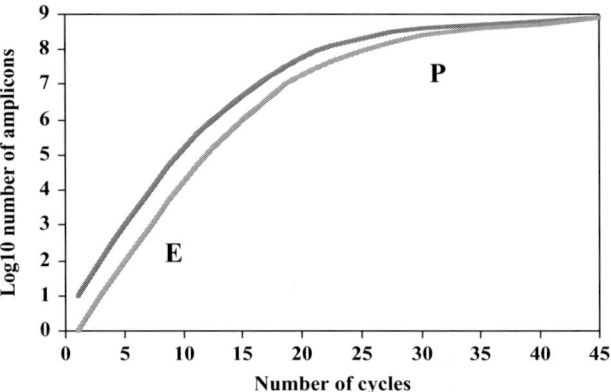

Fig. 2 Illustration of DNA amplification versus cycle number of beginning with 1 (light colored curve) and 10 (dark curve) original amplification targets. E and P denote exponential and plateau phases, respectively.

well. Non-specific binding of primers to template DNA occasionally causes exponential production of unwanted amplicons. Base pairing between primers followed by extension by DNA polymerase often leads to the production of short, highly amplifiable "primer-dimers" that compete with the intended amplicons for PCR reagents.

Techniques such as "hot start" and others are aimed at reducing non-specific primer binding and primer-dimer formation.[3] Aerosols generated when opening the PCR tubes or when handling their content can disperse amplification products into the lab environment where they become a source for contamination of reagents or reaction mixtures. Such contamination events often lead to the appearance of PCR products in reaction tubes that did not contain template DNA. The discovery that all PCR chemicals including *Taq* polymerase can be compressed in tablet form and stored for a year or more was a significant step toward reducing the incidence of contamination since handling of several reagents became unnecessary.

DETECTION OF PCR PRODUCTS

Gel Electrophoresis

For a long time, detection of PCR products was done almost exclusively by gel electrophoresis, staining with DNA-binding dyes such as ethidium bromide and observation under UV light. This approach revealed the quantity and quality of amplified DNA and the quality of the amplification products. Non-specifically amplified DNA and primer-dimers were usually distinguishable from specific product. Several disadvantages are associated with the gel electrophoresis technique: 1) the time required for electrophoresis and DNA staining runs counter to the notion of PCR as a rapid method; 2) opening of tubes after PCR and manipulation of amplified DNA, gels, and buffers are a common source of laboratory contamination with amplicons; and 3) gel electrophoresis is not suitable for determining quantity and quality of products during the exponential phase of PCR because of the limits in sensitivity and difficulties in retrieving samples during PCR. Analysis of amplified DNA is therefore done once the reaction is in the plateau phase where the amount of product is no longer proportional to the initial concentration of target (Fig. 2). A few strategies such as competitive PCR were developed to address this problem, but they were rather complex, cumbersome, and prone to errors.[4]

Quantitative and Real-Time Detection

The obvious need to eliminate contamination events in the laboratory and the desire for facile acquisition of quantitative data drove the invention of techniques for monitoring the progress of PCR without opening the reaction tubes (quantitative and real-time PCR). Two such methods are widely used. One method, illustrated in Fig. 3, is often referred to as the "TaqMan" method[5] named after the video game character, PacMan, that devours objects in its way. The technique relies on a third oligonucleotide in addition to a regular primer pair. This modified oligonucleotide is designed to bind to a sequence internal to the amplicon and carries a fluorescent moiety at the 5′ and a molecule that quenches fluorescence at the 3′ end. Modification of the 3′ end also prevents chain elongation. When DNA polymerase encounters the modified oligonucleotide during the extension phase, the enzyme cleaves it, releasing

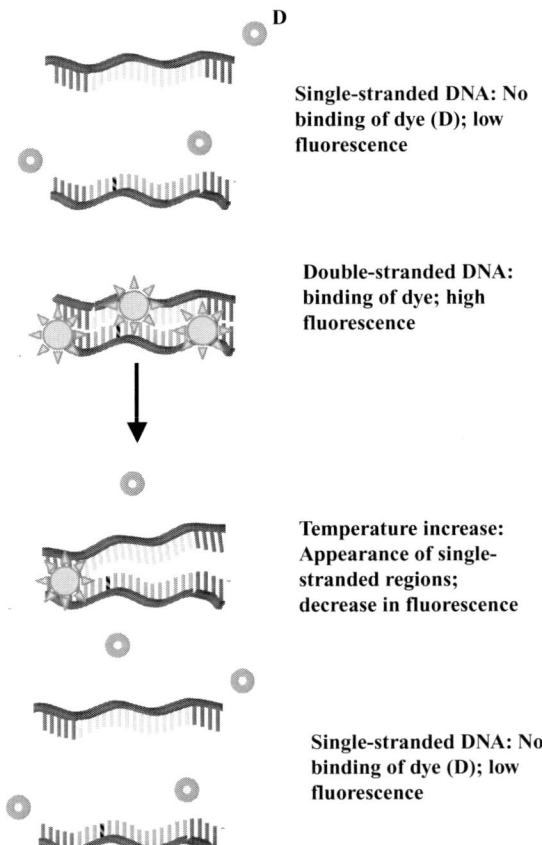

Fig. 4 Schematic representation of fluorescent dye-based quantitative PCR.

the nucleotide-bound fluorophor. The separation of fluorophor and quencher results in an increase in fluorescence. Total fluorescence in the PCR tube increases with each cycle in proportion to the amount of specific PCR product synthesized. Thermocyclers with built-in fluorometers monitor fluorescence continuously and thus, paint a "real-time" picture of PCR in progress.

The second technique does not require modification of primers or a third oligonucleotides, but utilizes a DNA binding dye, usually SYBR® Green, that fluoresces when bound to double-stranded, but not single stranded DNA[6] (Fig. 4). As the number of amplified DNA molecules increases, fluorescence increases each time after the elongation phase. Continuous monitoring of fluorescence in the PCR tubes provides a real-time view of the progress of PCR. Obviously, DNA-binding dyes do not discriminate between specifically and non-specifically amplified DNA and primer-dimers. Therefore, DNA melting curves are generated since melting of a specific double-stranded DNA species occurs at relatively narrow temperature range. Non-specific products almost always differ in their melting characteristics from those of the specific products, and short primer-dimers are easily recognized due to their low melting temperatures. The method even allows discrimination

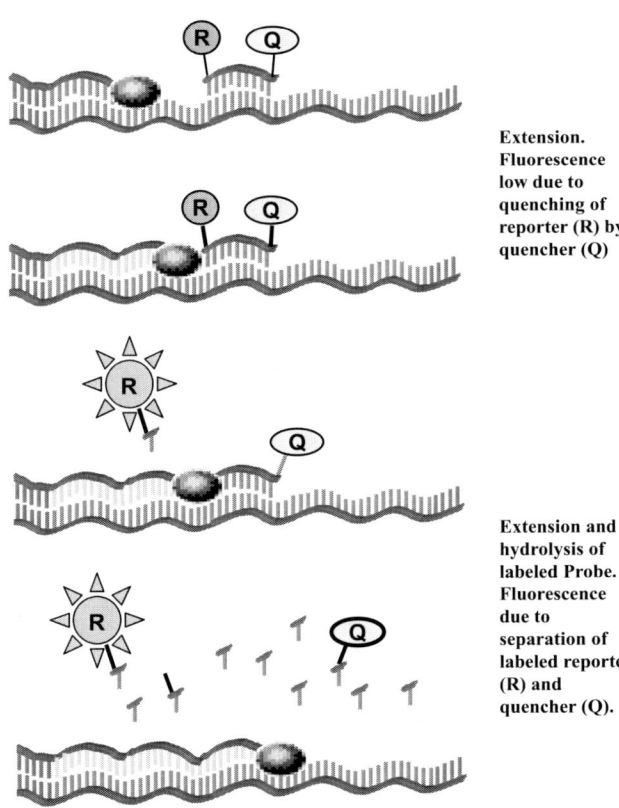

Fig. 3 Schematic representation of TaqMan PCR.

between two or more specific DNA products such as those generated during multiplex PCR.

APPLICATION OF PCR TO FOOD AND AGRICULTURAL BIOTECHNOLOGY

PCR as a Method for DNA Synthesis

The ease of amassing DNA molecules with PCR has revolutionized the way DNA is obtained for genetic studies and applications. In one way or another, PCR played a role in retrieving genetic material from *Adenovirus* to *Zymomonas* and from *Arabidopsis* to Zebrafish for cloning, sequencing, or probing purposes. For a number of reasons, it is often simpler and more useful to isolate fully processed mRNA from eukaryotes and convert it into an intron-free cDNA. Since most eukaryotic mRNAs have a 5′ polyA sequence, a polyT-primer can be employed, and if a primer-sized sequence toward the 3′ end of the mRNA is known, selective amplification of a specific mRNA sequence can be achieved. PCR is also instrumental in the synthesis of probe sequences for DNA microarrays, and is also harnessed to introduce mutations into gene sequences.[7]

PCR as a Detection Method

PCR was recognized early to ideally suit the detection of nucleic acid-containing targets present in low amounts, but only after the development of protocols and instrumentation that facilitate DNA or RNA extraction, PCR setup, and read-out did commercial PCR systems enter the market. Some of these assays are still performed only in specialized laboratories rather than on-site in food processing, agricultural, or biotechnology facilities. Competition from simpler immunoassay devices is fierce, but PCR assays have become entrenched when high sensitivity and specificity are required or when DNA is the only diagnostic molecule that survived processing or other treatments.

Allergen detection

PCR is, of course, unable to detect allergens, but it detects DNA from organisms known to produce allergens. Tests for the presence of DNA from almond, peanut, soy, fish, shellfish, and others are available, and some of these tests detect DNA in amounts corresponding to a potential presence of 1 or 2–200 ppm of allergens (the range that is of greatest concern to food manufacturers).[8]

GMO detection

Many countries require labeling of imported or domestic agricultural products, food, or ingredients derived from GMO, and often the labeling requirement is triggered when a threshold of 1% GMO-derived material is exceeded. In the United States, the label "organic" cannot be applied to GMO-derived food. PCR assays to verify GMO content are offered by several laboratories.[9] Generally, the assays test for DNA diagnostic of a common genetic modification such as the virus-derived 35 S promoter sequence or bacterial marker genes present in plants. More specific tests for the presence of Bt or herbicide resistance gene sequences are also available.

Detection of adulteration

The substitution of a part or whole of a declared product, for example, beef in hamburger, with another undeclared product, for example, pork, is illegal, but is being done nevertheless. PCR offers the potential to detect the presence and identity of even small amounts of undeclared ingredients.[10] Inclusion of banned ingredients, such as bone meal in feed for ruminants can be detected by PCR even after feed processing.

Detection of pathogens and other organisms

By far, the biggest market for diagnostic PCR is the detection of pathogens, and, less frequently beneficial bacteria, in food, animals and agricultural products.[11] Some viruses and parasites associated with food-borne illness are difficult or currently impossible to grow in the laboratory, and sensitive detection of, for example, Norwalk virus and *Cryptosporidium* sp. is frequently done by RT-PCR. A multitude of tests for bacteria causing food-borne illnesses has been developed and some of them are available commercially. Like other testing methods such as immuno- or hybridization-based assays, PCR shortens the time required for detection when compared to purely culture-based methods, but generally does not eliminate the need for culture steps entirely. These steps allow the target bacteria to multiply so that at least 10–50 colony-forming units per PCR assay volume are obtainable. Dilution by culture media reduces the concentration of PCR inhibitors such as the notorious plant-derived phenolics, but foremost, culturing addresses the problem of detecting DNA from dead cells that are no longer a health concern, but are common in food that has undergone processing. Multiplying cells produce detectable levels of DNA, whereas DNA from dead cells is degraded or remains at undetectable levels.

Another test to determine the presence of live or dead cells is based on the observation that, in contrast to more stabile DNA, mRNA is degraded rapidly. Detection of mRNA sequences by RT-PCR is, therefore, indicative of the presence of live cells. The presence of multiple copies of mRNA should, in theory, increase sensitivity, but frequently this is not the case because of the inefficiencies in mRNA isolation and RT. The inclusion of a step for capturing organisms in media or solutions by antibody-coated magnetic particles can improve detection sensitivity of RT- and DNA-based PCR assays since organisms are concentrated and separated from potential inhibitors.

PCR for typing of organisms and product certification

Although a method that is not based on amplification of DNA, pulsed-field electrophoresis of restriction enzyme-cleaved DNA, has gained official status as fingerprinting method for some bacterial pathogens for epidemiological purposes, simpler PCR-based methods are sometimes chosen to discriminate among isolates belonging to the same species. Electrophoretic patterns of random amplified polymorphic DNA (RAPD), produced with a single arbitrary primer whose sequence is not based on known sequences of the target DNA, allow facile typing of bacteria and eukaryotes such as cereal varieties. Difficulties in lab-to-lab reproducibility have limited the utility of RAPD, but a number of other methods less prone to variability have been developed.

CONCLUSION

Despite some of the technical and competitive challenges faced by PCR assays for diagnostics, these tests have become part of the repertoire of research and applied detection methods. In part this is due to the date of invention; it precedes isothermal and other amplification methods[12] by several years, but also because it has been possible to implement a number of powerful modifications (e.g., real-time PCR) that made it a more attractive tool.

REFERENCES

1. Mullis, F.; Faloona, F.; Scharf, S.; Saiki, R.; Horn, G.; Erlich, H. Specific enzymatic amplification of DNA in vitro: the polymerase chain reaction. Cold Spring Harb. Symp. Quant. Biol. **1986**, *51*, 263–273.
2. Guyer, R.L.; Koshland, D.E. The molecule of the year. Science **1989**, *246* (4937), 1543–1546.
3. Erlich, H.A.; Gelfand, D.; Sninsky, J.J. Recent advances in the polymerase chain reaction. Science **1991**, *252*, 1643–1650.
4. Wiesner, R.J.; Beinbrech, B.; Rüegg, J.C. Quantitative PCR. Nature **1993**, *366*, 416.
5. Heid, C.A.; Stevens, J.; Livak, K.J., Williams, P.M. Real-time quantitative PCR. Genome Res. **1996**, *6* (10), 986–994.
6. Morrison, T.B.; Weis, J.J.; Wittwer, C.T. Quantification of low-copy transcripts by continuous SYBR green I monitoring during amplification. Biotechniques **1998**, *24* (6), 954–962.
7. Williams, G.J.; Nelson, A.S.; Berry, A. Directed evolution of enzymes for biocatalysis and the life sciences. Cell. Mol. Life Sci. **2004**, *61*, (24), 3034–3046.
8. Poms, R.E.; Anklam, E. Polymerase chain reaction techniques for food allergen detection. J. AOAC Int. **2004**, *87* (6), 1391–1397.
9. Giesse, J. GMO testing. Food Technol. **2002**, *56* (11), 60–62.
10. Laube, I.; Spiegelberg, A.; Butschke, A.; Zagon, J.; Schauzu, M.; Kroh, L.; Broll, H. Methods for the detection of beef and pork in foods using real-time polymerase chain reaction. Int. J. Food Sci. Technol. **2003**, *38* (2), 111–118.
11. Hill, W.E. The polymerase chain reaction: applications for the detection of foodborne pathogens. Crit. Rev. Food Sci. Nutr. **1996**, *36* (1/2), 123–173.
12. Hagen-Mann, K.; Mann, W. RT-PCR and alternative methods to PCR for in vitro amplification of nucleic acids. Exp. Clin. Endocrinol. Diabetes **1995**, *103*, 150–155.

Poultry Genetics and Breeding

Samuel E. Aggrey
Department of Poultry Science, University of Georgia, Athens, Georgia, U.S.A.

Abstract

Poultry breeding has advanced tremendously over the past 50 years as a result of systematic developments of new technologies including pedigreeing, statistical and computational methodologies coupled with university trained geneticists. The marked improvement especially in growth has also resulted in increased skeletal and cardio-vascular problems. Advances in molecular biology and the release of the chicken sequence allow for the identification of trait loci, genes and their functions. Genetic markers in trait genes and/or quantitative trait nucleotides can aid selection for further improvements. Inclusion of molecular tools can assist to retool the future chicken to remain productive, resist pathogens, be free from metabolic diseases and meet welfare standards.

INTRODUCTION

This entry tracks poultry breeding from the early part of the 20th century, assesses advances made in poultry improvement programs, and reviews the subsequent health and welfare issues associated with such advances. With few companies breeding birds for different geographical and socio-cultural regions, genotype by environment interaction is important now and will be in the near future. The availability of the draft sequence of the chicken genome makes it possible for chicken genes to be identified and their functions studied. The prospects and limitations of incorporating molecular and genomic information in poultry breeding are also explored.

HISTORY OF BREEDING AND SELECTION

The documentation of poultry breeding in the first half of the 20th century is scarce, as poultry production occurred in small rural enterprises. In addition, knowledge of genetics was limited. However, the rediscovery of Mendelian principles and application of heredity, the invention of the trapnest in the 1930s that allowed for accurate individual egg production records, and the prevalence of pedigreeing in the 1940s all had a tremendous impact on the development of breeding for production. Until 1950, most commercial egg and meat stocks sold were pure breeds with only a few crossbreeds. From the 1950s onward, selection programs came under the control of university-trained geneticists. Towards the latter part of the 20th century, globalization of breeding and marketing reduced the number of independent breeding programs in commercial companies. At present, there are two major layer breeding companies (Hendrix Genetics and Hi-Line/Lohmann) and five major broiler breeding companies (Cobb-Vantress, Aviagen, Hubbard, Heritage, and Nutreco). The reduced number of companies poses a threat to the long-term health of the breeding industry. Existing lines can attain genetic uniformity and could become susceptible to new diseases that would damage genetically uniform strains.[1] The development of statistical tools from Henderson's methods[2] and the use of best linear unbiased prediction (BLUP) methodology in the 1990s, coupled with computer technology, have allowed for the efficient estimation of breeding values for selection. Genetic improvements in both meat-type and egg-type birds have been achieved without knowledge of the exact genes that control these traits. Advances in molecular genetics and the availability of the draft sequence of the chicken genome will facilitate research into identifying genes involved with traits of economic importance.

MEAT-TYPE BREEDING

The genetic changes in today's broiler are driven by the price the consumer is willing to pay. Since the 1950s, profitability has been dictated by the sale of whole birds. Fig. 1 shows broiler growth in the 20th century. In 1950, the average market age of broilers at 1.36 kg was 70 days. By 2000, a 2.27 kg broiler reached market age at 46 days. Growth rate in broilers increased five fold between 1950 and 2000.[3] Genetic gain in broiler growth was 58 g per year from 1957 to 1976,[4] 73 g per year from 1976 to 1991,[5] and 84 g per year from 1991 to 2001.[3] Improvements in feed conversion ratio (FCR) and carcass yield have contributed greatly to profitability. Havenstein, Ferket, Scheideler, and Larson[3] showed that carcass yield increased by sixfold between

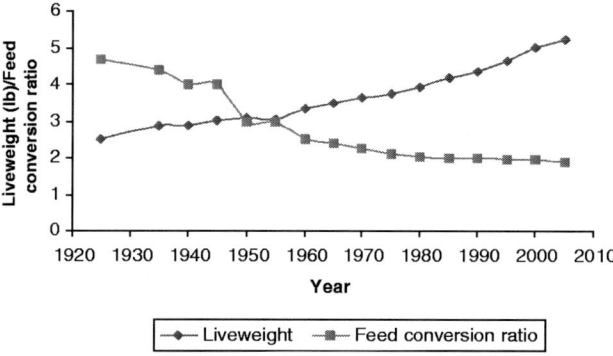

Fig. 1 Efficiency of broiler production. **Source:** From National Chicken Council.

1950 and 2000. The inclusion of individual FCR and yield characteristics as selection traits in elite lines has doubled genetic progress in the last ten years.

Poultry breeds (Plymouth Rock, New Hampshires, Rhode Island Reds, etc.) that were in use in the 1940s are no longer used by the commercial industry, as breeding companies now specialize in meat (broiler) and egg (layer) breeding. The current broiler lines originated from various genetic backgrounds. However, a major portion of the paternal grandparent lines originated from White Cornish lines, whereas maternal grandparents originated from White Plymouth Rock lines. Today's broiler is a hybrid from either a three- or four-way cross of closed pure-breeding strains. Pure breeding strains are a composite of various breeds synthesized by the breeding company. The breeding companies own these strains and subject them to intense selection regimens based on a range of broiler marketable products. The major traits of interest in broiler breeding are growth, breast meat yield, and FCR. Growth has consistently been a selection parameter because of its ease of measurement, high heritability, and impact on profitability. The premium on breast meat yield has justified the attention paid by the breeding industry to carcass yield traits. Feed cost constitutes about 70% of the total production cost. Therefore, improving FCR has a major impact on profitability for poultry meat producers. Great-grandparent strains are derived from the pure-breeding strains. This generation is mainly used to multiply pure strains needed to produce the grandparent strains. The grandparent strains are typically hybrids, the first generation of crossbreds, bred for complementation of traits from their respective parental lines and for their combination potential with other crossbreds for the final broiler products. The grandparent strains are distributed globally as parent stock. The parent stocks, are mated to produce commercial broilers.

Intensive selection for fast growth and breast meat yield has resulted in some undesirable consequences. The front half of today's broiler carries 58% of the body weight, compared to 40% in the 1950s. This has resulted in lameness, ruptured tendons, curved and twisted toes, and tibia dyschondroplasia. In turkey breeding, the male turkey cannot make cloacal contact with females during mating. As a result, artificial insemination has become the means of fertilization. Cardio-respiratory problems, carcass fatness, and nutrient absorption problems are also prevalent, and are directly related to the increased growth rate.

EGG-LAYER BREEDING

Modern commercial egg layers have been selected to produce either white or brown eggs. White egg-laying strains are derived from White Leghorns. A composite of selected strains (derived from White Leghorns, Rhode Island Reds, Rhode Island Whites, Light Sussex, and Barred Plymouth Rocks) are used to produce the brown egg laying strains. From the first quarter of the 20th century, egg production per hen has increased from 175 to 320 per calendar year. In layer selection, hen-housed production (HHP) is the most important trait, which is often based on the number of eggs laid when all birds were housed together until a fixed age. This means that HHP is confounded with sexual maturity and viability. Some breeders adjust egg records for sexual maturity and mortality. The act of adjusting records for sexual maturity and mortality is termed the hen day rate (HDR). In addition to HHP/HDR, breeders also select for body weight, sexual maturity, egg quality traits (i.e., shell strength, shell shape, shell color, albumen quality, and absence of blood or meat spots), egg weight, and FCR. Further processing of eggs has gained attention, and breeders have paid attention to percent solids and lipid content of the egg. Selection for high egg producing birds, coupled with housing layers in cages, has led to increases in aggressive behavior, cage fatigue, skeletal problems, and cannibalism.

Even though the commercial layer is usually a product of a four-way cross, breeding strategies may differ from one breeding company to another. Two inbred lines may be developed to produce the male parent, and another two inbred lines may be developed to produce the female parent. Other breeders use the reciprocal-recurrent selection procedure. Some breeding companies select from pure lines and use them to make a three-way or four-way cross depending on how the pure lines best combine for the final product. The BLUP procedure has become the standard methodology for estimating breeding values. However, in poultry breeding, it is debatable whether the extra computational cost merits the benefits of BLUP, as increases in efficiency from BLUP have been marginal.

MOLECULAR GENETICS AND POULTRY BREEDING

Genetics contributed to about 80%–85% of the improvements in poultry in the last century.[3] Most of the improvements in poultry genetics have come through traits with moderate to high heritability. However, for low heritability traits, such as disease resistance, improvements through conventional selection will be very slow. Also, direct measurement of carcass traits on live birds is impossible. Breeding values of sires for egg production can only be determined from daughter records. The selection programs that resulted in improvements in poultry for the past 50 years were conducted without knowledge of the genes involved in the selected traits. However, advances in molecular biology allow for identification of genes contributing to these traits. Linkage maps consisting of polymorphic genetic markers[6] are essential for identification of quantitative trait loci (QTL) that control and modify traits. QTL have been identified for growth, carcass traits and FCR,[7–9] Marek's disease,[10,11] feathering,[12] egg composition,[13] Salmonella,[14] and ascites.[15]

One major objective of QTL mapping is identification of specific genes responsible for traits. The description of the genetic architecture of a trait is not complete until sequence variations in gene(s), corresponding to the QTL that actually cause differences in the trait phenotype, quantitative trait nucleotides (QTN), are identified. The path to QTN identification is a multidisciplinary effort comprising functional and comparative genomics, bioinformatics, statistical and population genetics, molecular biology, physiology, nutrition, immunology, and biochemistry. The advent of expressed sequence tags and DNA microarray technology,[16] for example, allow for the simultaneous study of expression patterns of large number of genes. DNA microarrays have been useful in the study of global gene expression in the chicken during Marek's disease virus infection.[17] Progress will require the amalgamation of data from: (1) EST databases (GenBank, dbEST, http://www.chickest.udel.edu), (2) microarray and protein expression databases (http://www.ncbi.nih.gov.geo, http://www.ebi.ac.uk/arrayexpress), (3) QTL databases (http://acedb.asg.wur.nl, http://www.iastate.edu/chickmap), (4) whole-genome sequence, and (5) phenotypic data from a pedigreed population, together with the development of high-throughput genotyping methods.

The release of the draft chicken sequence in 2004 (http://genomeold.wustl.edu/projects/chicken),[18] should rapidly hasten the identification of genes and their functions in the chicken genome. The draft genome sequence is comprised of about 1 billion base pairs and is estimated to harbor about 20,000–23,000 genes. Fig. 2 shows the genomic geography of chromosome Z, illustrating the integration of physical, linkage, and sequence maps. Completion and integration of such emerging maps for each chromosome will rapidly lead to the identification of candidate genes in QTL regions. Marker assisted selection (MAS) should allow breeders to improve low heritability traits and assist in retooling the future chicken to remain productive, resist pathogens, be free from metabolic disease, and meet welfare standards. However, despite all the genetic and genomic tools at hand, there will be limitations. Among the limitations are non-random associations, undesirable genetic correlations, and economic factors. In poultry, it is expected that MAS will provide an additional 2%–4% gain over phenotypic selection.[19] Improvements in transgenic technologies and novel developments from emerging disciplines of bioinformatics, physiological genomics,

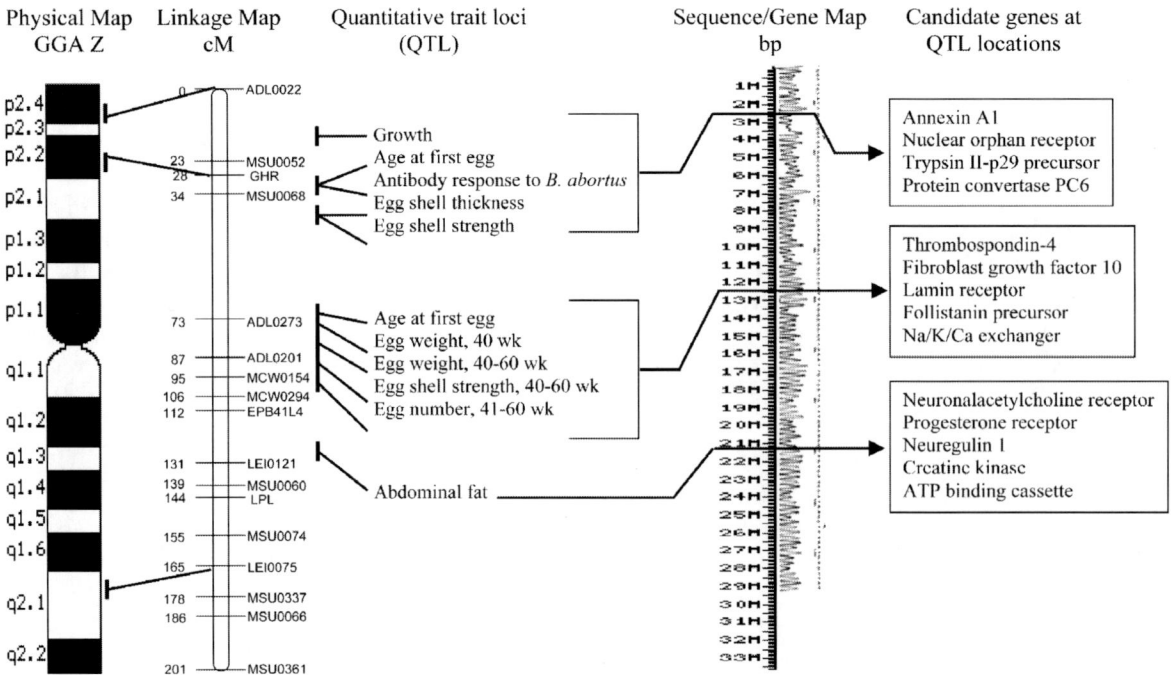

Fig. 2 Genomic geography of the chicken Z chromosome (GGA Z).

immunogenomics, and nutrigenomics will provide a vast array of additional tools for further improvement of poultry. Although the future of poultry breeding remains very bright, with the addition of new tools, the road to reaching the biological maximum remains a difficult one.

REFERENCES

1. Arthur, J.A.; Albers, G.A.A. Industrial perspective on problems and issues associated with poultry breeding. In *Poultry Genetics, Breeding and Biotechnology*; Muir, W.H., Aggrey, S.E., Eds.; CABI Publishing: Oxon, **2003**, 1–12.
2. Henderson, C.R. Estimation of variance and covariance components. Biometrics **1953**, *9*, 226–252.
3. Havenstein, G.B.; Ferket, P.R.; Qureshi, M.A. Growth, livability and feed conversion of 1957 versus 2001 broilers when fed representative 1957 and 2001 broiler diets. Poultry Sci. **2003**, *82*, 1500–1508.
4. Sherwood, D.H. Modern broiler feeds and strains: what two decades of improvement have done. Feedstuff **1977**, *49*, 70.
5. Havenstein, G.B.; Ferket, P.R.; Scheideler, S.E.; Larson, B.T. Growth, livability and feed conversion of 1991 versus 1957 broilers when fed typical 1957 and 2001 broiler diets. Poultry Sci. **1994**, *73*, 1795–1804.
6. Aggrey, S.E.; Okimoto, R. Genetic markers: prospects and applications in genetic analysis *Poultry Genetics, Breeding and Biotechnology*, Muir, W.H., Aggrey, S.E., Eds.; CABI Publishing: Oxon, **2003**, 419–438.
7. Van Kaam, J.B.C.H.M.; Groenen, M.A.M.; Bovenhius, H.; Veenendaal, A.; Verijken, A.L.J.; van Arendonk, J.A.M. Whole genome scan in chicken for quantitative trait loci affecting growth and feed efficiency. Poultry Sci. **1999**, *78*, 13–23.
8. Van Kaam, J.B.C.H.M.; Groenen, M.A.M.; Bovenhius, H.; Veenendaal, A.; Verijken, A.L.J.; van Arendonk, J.A.M. Whole genome scan in chicken for quantitative trait loci affecting carcass traits. Poultry Sci. **1999**, *78*, 1091–1099.
9. Ikeobi, C.O.N.; Wooliams, J.A.; Morrice, D.R.; Law, A.; Windsor, D.; Burt, D.W.; Hocking, P.M. Quantitative trait loci affecting fatness in the chicken. Animal Genetics **2002**, *33*, 421–425.
10. Vallejo, R.L.; Bacon, L.D.; Liu, H.L.; Witter, R.L.; Groenen, M.A.; Hillel, J.; Cheng, H.H. Genetic mapping of quantitative trait loci affecting susceptibility to Marek's disease virus inducing tumors in F_2 intercross chickens. Genetics **1998**, *148*, 349–360.
11. Yonash, N.; Bacon, L.D.; Witter, R.L.; Cheng, H.H. High resolution mapping and identification of new quantitative trait loci (QTL) affecting susceptibility to Marek's disease. Animal Genetics **1999**, *30*, 126–135.
12. Hamoen, F.F.A.; Van Kaam, J.B.C.H.M.; Groenen, M.A.M.; Vereijken, A.L.K.; Bovenhius, H. Detection of genes on the Z-chromosome affecting growth and feathering in broilers. Poultry Sci. **2001**, *80*, 527–534.
13. Honkatukia, M.; Tuiskula-Haavisto, M.; De Koning, D.J.; Virta, A.; Maki-Tanila, A.; Vilkki, J. A region of chicken chromosome 2 affects both egg white thinning and egg weight. Genetics, Selection, Evolution **2005**, *37*, 563–577.
14. Tilquin, P.; Barrow, P.A.; Marly, J.; Pitel, F.; Plisson-Petit, F.; Velge, P.; Vignal, A.; Baret, P.V.; Bumstead, N.; Beaumont, C. A genome scan for quantitative trait loci affecting the Salmonella carrier-state in chicken. Genetics, Selection, Evolution **2005**, *37*, 539–561.
15. Rabie, T. Pulmonary hypertension syndrome in chicken: peeking under QTL peaks Ph.D. Dissertation, 2004, Wageningen University, The Netherlands.
16. Cogburn, L.A.; Wang, X.; Carre, W.; Rejto, L.; Aggrey, S.E.; Duclos, M.J.; Simon, J.; Porter, T.E. Functional genomics in chickens: development of integrated systems microarray for transcriptional profiling and discovery of regulatory genes. Comparative and Functional Genomics **2004**, *5*, 161–253.
17. Liu, H.C.; Cheng, H.H.; Tirunagaru, V.G.; Soffer, L.; Burside, J. A strategy to identify positional candidate genes conferring Marek's disease resistance by integrating DNA microarrays and genetic mapping. Animal Genetics **2001**, *32*, 351–359.
18. International Chicken Genome Sequencing Consortium Sequence and comparative analysis of the chicken genome provide unique perspective on vertebrate evolution. Nature **2004**, *432*, 695–716.
19. Van der Beek, S.; Van Arendonk, J.A.M. Marker assisted selection in a poultry breeding program. Proceedings of the 5th World Congress on Genetics Applied to Livestock Production. University of Guelph, Guelph, Ont. **1994**, *21*, 237–240.

Probiotic Bacteria Preservation

Marcus Volkert
Antje Schulz
Katharina Schoessler
Dietrich Knorr
Department of Food Biotechnology and Food Process Engineering, Berlin University of Technology, Berlin, Germany

Abstract

Several preservation techniques can be used to produce probiotic powders on a large scale. However, these processes expose the cultures to extreme environmental stress. Preservation of probiotics aims to maintain the viability during processing and storage to ensure an adequate number of living cells in the final product. This entry presents the status quo in this field and discusses important developments in the areas of processing, protection mechanisms, and protective media design as well as stress responses.

INTRODUCTION

The importance of probiotic bacteria in nutrition has increased over the last years because of growing consumer interest in healthy food. Due to unfavorable conditions in most food matrices, probiotic organisms are separately grown and preserved for subsequent addition into the final product. This procedure aims for products with a long shelf life and a high number of living cells. The key mechanism in this respect is dehydration. It is either applied directly (drying) or indirectly (freezing). The water removal leads to an inactivation of the cell metabolism and provides stability during subsequent storage. This strategy is well known from nature, as a great variety of plants, animals, and microorganisms can survive harsh conditions in the so-called state of anhydrobiosis. Although these processes eventually lead to cell stabilization, they affect the microorganisms by subjecting them to numerous stresses, such as dehydration, osmotic pressure, oxidative stress, heat, or the formation of ice crystals. The latest research in this domain is therefore focused on the optimization of processes and the increase of survival rate and shelf life.

PRESERVATION PROCESSES

Non-preserved lactic acid bacteria (LAB) are fragile microorganisms that could not survive for long during storage. Various technologies have been proposed since the 1970s for preservation of bacterial viability. Essentially, the water activity reduction by freezing or drying is the aim of LAB preservation processes. Freeze-drying (FD) combines the two unit operations freezing and drying and is the most common technology applied to produce stable starter cultures with high viability and shelf life. However, due to high processing, transport, and storage costs associated with frozen cultures, alternative preservation technologies are in the focus of several research groups. Freezing exerts different injury effects. In particular, the freezing rate seems to be crucial in achieving high survival rates. However, the effect of freezing rate differs for different microbes. It was proposed that the freezing rate influences the size of ice crystals and the location of ice nucleation (intra- or intercellular), which ultimately determines the type of cellular damage of frozen cells.[1] Commonly liquid cultures are bulk frozen in the packaging or shock frozen as pellets in a liquid nitrogen freezing step. Due to the surface-to-volume ratio, the pellet shape is particularly suitable for rapid freezing and subsequent FD. Exploiting the benefits of small particles, Volkert et al.[2] developed the spray-freezing process for LAB cultures to produce a frozen snow-like product as a prestage for FD processes. The schematic flow diagram of the process is shown in Fig. 1. Survival rates of 90% after spray-freezing *Lactobacillus rhamnosus* in skim milk were achieved. The novel technology of high pressure low temperature treatment (HPLT) was investigated in the scope of the latter work. High pressure shift freezing offers high freezing rates and homogeneous nucleation irrespective of product size and geometry. However, the HPLT treatment involves pressure-induced enzyme inactivation and therefore limits the viability of the frozen bacteria.[2]

Very attractive from the economical point of view are continuous drying processes that produce a stable product with high viability in one process step to potentially replace the cost-intensive FD process. It has been estimated that

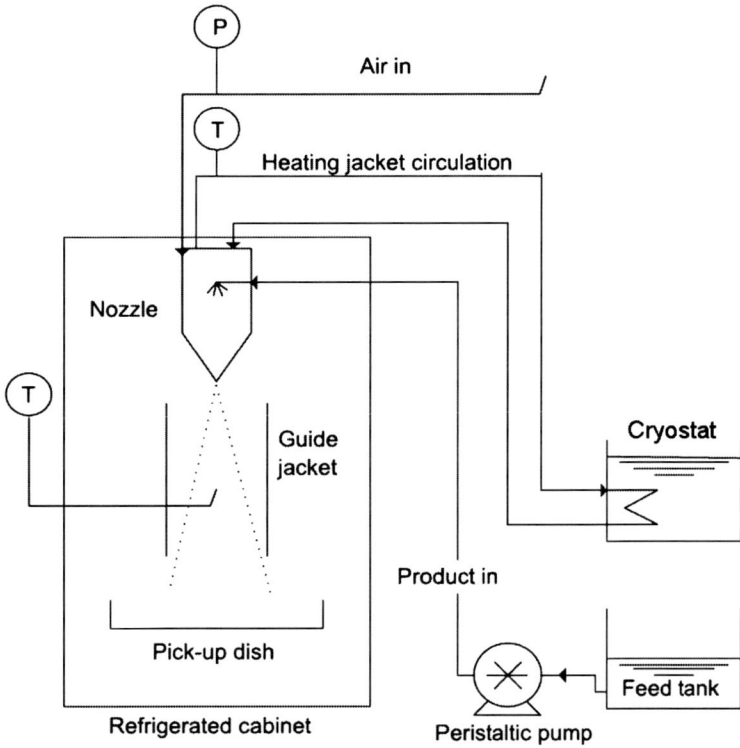

Fig. 1 Flow diagram of the spray-freezing process. **Source:** From J. Food Eng.[2]

the costs of spray-drying (SD) is 15–20 times lower per removed kilogram of water compared to FD. While FD is more suitable than SD for some cultures, researchers have found that for some strains the survival after FD is no higher compared to SD.[3] Studies have been undertaken on SD of yogurt cultures, cheese cultures, and bacteriocin-producing LAB. A critical parameter in SD of LAB is the drying temperature. Ananta et al.[4] investigated the impact of the outlet temperature on the survival of *L. rhamnosus* and found the survival of bacteria dried in skim milk at 100°C to 70°C in a range of 8% to 68%, respectively. The impact of different outlet temperatures on the survival rate in spray-dried powders is shown in Fig. 2.

PROTECTION MECHANISMS AND MEDIA

Although the presented preservation processes induce cell stabilization to some extend, they stress the microorganisms by subjecting them to dehydration, osmotic pressure, oxidative stress, heat, or ice crystal formation.

The protection of cell wall and cytoplasmic membrane from any alteration is of particular importance for the preservation of microbial vitality as these are the components protecting cells from external effects. The cytoplasmic membrane consists of a phospholipid bilayer stabilized by water molecules that prevent the phospholipids from agglomeration and keep the membrane in a fluid state. Dehydration may lead to a phase change from liquid-crystalline to gel, which can lead to separation of integral membrane proteins, loss of barrier function, and leakage of essential cell components due to non-uniform phase change during rehydration.[5]

Two different mechanisms are discussed that inhibit a change in membrane structure. According to the water replacement hypothesis described by Crowe et al.,[5] small hydrophilic compounds (e.g., disaccharides) are able to replace the water molecules between the polar head groups of the phospholipids during dehydration, preventing a membrane phase transition and its deleterious consequences (Fig. 3).

The second theory is based on the fact that the glass transition temperature of a substance increases with its molecular weight.[6] Accordingly disaccharides such as trehalose can facilitate the transition to the glassy state when added to a cell suspension prior to drying. The glassy state is characterized by extremely high viscosities inhibiting molecular movement and slowing down harmful reactions to a minimum. Consequently, cells in the glassy state are suggested to be well protected from cell membrane alteration.

The functionality of proteins and enzymes is closely linked to their three-dimensional structure that is stabilized by a water sheath surrounding the molecules. Dehydration may lead to a loss of this water sheath followed by protein or enzyme denaturation. Some substances such as amino acids, quaternary amines and sugars may act as so-called compatible solutes, molecules that serve for the osmotic adaptation of cells and can prevent macromolecules from

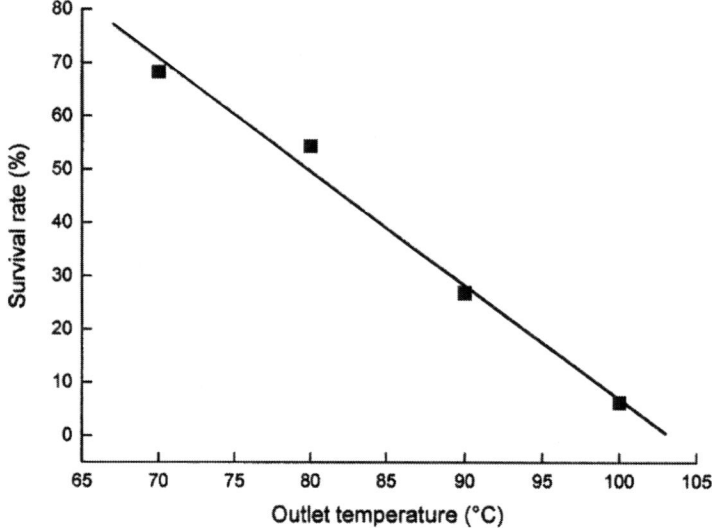

Fig. 2 Effect of outlet temperature during spray-drying on the survival of *Lactobacillus rhamnosus* GG in skim-milk powder prepared from 20% (w/v) reconstituted skim milk. Initial cell count of the feed solution was 109 cfu/mL.[4]

dehydration damage. Enzymes and proteins show a higher affinity toward water than toward these solutes. Consequently, the compatible solutes are arranged outside the hydration sheath, resulting in its preservation during drying. Hence, the three-dimensional structure and functionality of enzymes and proteins is maintained.[7]

The fact that anhydrobiotic organisms in nature show high concentrations of sugars (e.g., trehalose) has lead to a majority of research work in the area of protective media concentrating on the protective qualities of saccharides. Detailed research has been performed in the area of protective properties of sugars such as mannose, fructose, lactose, sorbitol, or glucose during LAB growth, freeze-drying, and storage and protective effects were found in most cases.[8]

The most important complex protection medium in the area of LAB preservation is reconstituted skim milk (RSM), which has evolved into a standard medium. Schössler[9] identified whey proteins as the essential RSM constituent assuring an inactivation of less than one log cycle after SD and four weeks of storage of *L. rhamnosus* (Fig. 4). These results show the research potential still existent in the area of protection mechanisms and media development and the numerous possibilities of research and improvement in this scientific domain.

STRESS RESPONSE

Apart from adding protective agents the natural ability of LAB to withstand adverse conditions can be exploited in the preservation process. Ananta et al.[10] demonstrated that bacteria respond to changes in their immediate

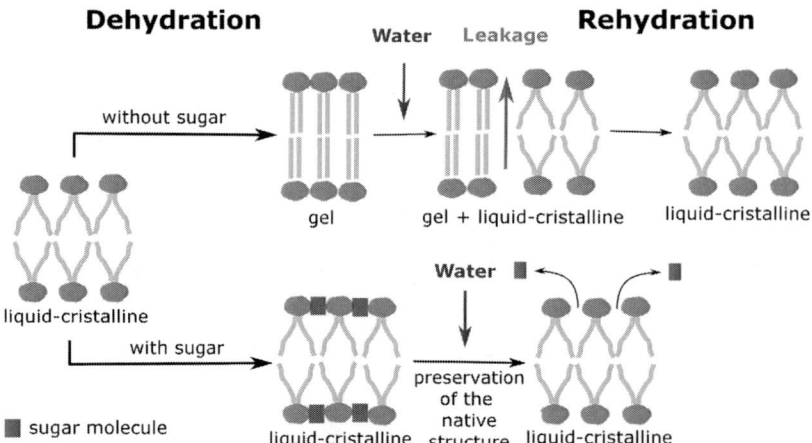

Fig. 3 Influence of sugars on membrane integrity during dehydration and rehydration of phospholipid bilayers.

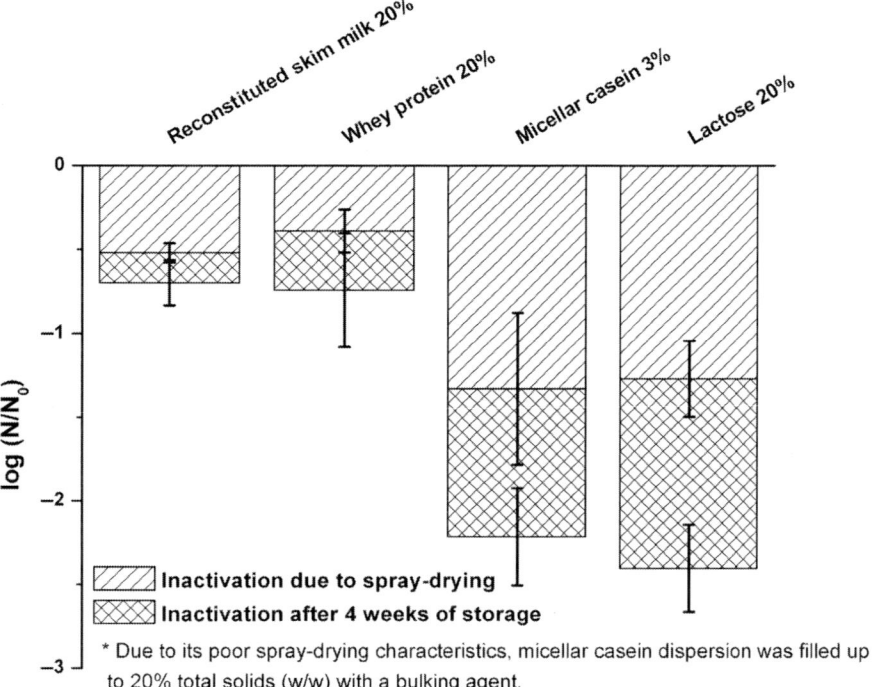

Fig. 4 Inactivation of *Lactobacillus rhamnosus* after spray-drying and 4 weeks of storage (25°C, 11% relative humidity) in the presence of skim milk and isolated RSM constituents. **Source:** From *Thesis*.[9]

surroundings by a metabolic reprogramming which leads to enhanced resistance.

Two different classes of such adaptations are known. The first one describes a specific system induced by a sublethal dose of a chemical or physical stress (e.g., sodium chloride, heat shock) that increases the survival rate after a crucial dose of the same agent.[11] The second class of adaptation comprises more general systems that prepare cells to survive very different environmental stresses, without the need for cultures to have had prior exposure to the very same stress. This indicates that cells exposed to one stress may develop resistance not only to that stress, but also to other unrelated stresses. This mechanism is known as cross protection.[11,12]

In the case of the first adaptation type, mild heat treatments can lead to adaptation of the cell membrane by increasing the saturation and the length of fatty acids in order to maintain optimal fluidity of the membrane and activity of intrinsic proteins.[13]

Viability of the heat-adapted *L. paracasei* NFBC 338 in RSM was enhanced 18-fold during SD at outlet temperatures of 95–105°C.[11] A pretreatment with heat or salt led to improved heat tolerance of LAB during SD. *L. paracasei* NFBC 338 pretreated with 0.3 M NaCl was significantly more resistant to SD at outlet temperatures between 95°C and 100°C than non-adapted control cells (33.46 ± 2.3% versus 8.27 ± 4.42% survival, respectively).[11] Although not as efficient as the homologous stress, the levels of cross protection were in the order heat > salt > hydrogen peroxide > bile.[14]

Furthermore, pressure pretreatment has also been shown to increase heat resistance of LAB. As shown in 5, it has been demonstrated that a pretreatment of *L. rhamnosus* GG at elevated pressure of 100 MPa for 5–10 min prior to exposure to heat at 60°C led to increased heat-resistance.[15]

CONCLUSIONS

With respect to increasing sellers' competition and consumer demands, the development of new preservation processes for LAB increasingly gains attractiveness in the food industry. Spray-drying seems an interesting technology to potentially replace the expensive FD process. However, the coherencies between different parameters that affect probiotic bacteria during preservation and storage are only partially understood. Much research in this field focuses on selected process steps, not looking at the whole production cycle of LAB, including fermentation, pretreatments, preservation, and storage. Very important in the development of new processes is the profound understanding of the mechanisms of fortification as well as inactivation during the whole process. A broad range of factors has to be taken into account, including process parameters, media composition and potentially fortifying pretreatments.

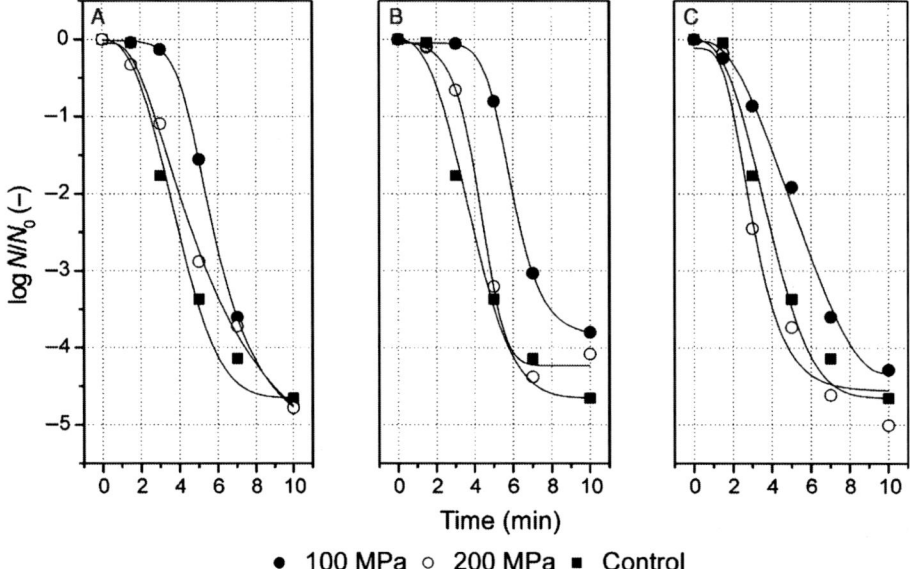

Fig. 5 Heat inactivation curves at 60°C of LGG, which were previously pretreated for 10 min at 100 MPa (●) and 200 MPa (○) in comparison with pressure-untreated population (■). The denotation of the three figures with A, B, and C corresponds to the temperature levels (37, 43, and 50°C, respectively) applied during pressure pretreatment. In case of pressure-treated cells N_0 was cell count after pressure treatment.

REFERENCES

1. Karlsson, J.O.M.; Toner, M. Long-term storage of tissues by cryopreservation: critical issues. Biomaterials **1996**, *17* (3), 243–256.
2. Volkert, M.; Ananta, E.; Luscher, C.M.; Knorr, D. Effect of air freezing, spray freezing, and pressure shift freezing on membrane integrity and viability of *Lactobacillus rhamnosus* GG. J. Food Eng. **2008**, *87*, 532–540.
3. Gardiner, G.E.; O'Sullivan, E.; Kelly, J.; Auty, M.A.E.; Fitzgerald, G.F.; Collins, J.K.; Ross, R.P.; Stanton, C. Comparative survival rates of human-derived probiotic *Lactobacillus paracasei* and *L. salivarius* strains during heat treatment and spray drying. Appl. Environ. Microbiol. **2000**, *66* (6), 2605–2612.
4. Ananta, E.; Volkert, D.; Knorr, D. Cellular injuries and storage stability of spray dried probiotic bacterium *Lactobacillus rhamnosus* GG. Int. Dairy J. **2005**, *15*, 399–409.
5. Crowe, J.H.; Carpenter, J.F.; Crowe, L.M. The role of vitrification in anhydrobiosis. Annu. Rev. Physiol. **1998**, *60*, 73–103.
6. Roos, Y.H. Glass-transition related physicochemical changes in foods. *Food Technol.* **1995**, *49*, 97–102.
7. Schwarz, T. Kompatible Solute: Ein genereller Mechanismus zum Schutz vor Umweltstress? In *Workshop Bio Processing*; Jülich, Germany, 2001.
8. Carvalho, A.S.; Silva, J.; Ho, P.; Teixeira, P.; Malcata, F.X.; Gibbs, P. Effect of various sugars added to growth and drying media upon thermotolerance and survival throughout storage of freeze-dried *Lactobacillus delbrückii* ssp. *bulgaricus*. Biotechnol. Prog. **2004**, *20*, 248–254.
9. Schössler, K. Differenzierung der Schutzwirkung einzelner Milchbestandteile bei der Sprühtrocknung von Probiotika am Beispiel von *Lactobacillus rhamnosus*. Thesis; Department of Food Biotechnology and Food Process Engineering, Technische Universität Berlin: Berlin, Germany; 2008.
10. Ananta, E.; Knorr, D. Evidence on the role of protein biosynthesis in the induction of heat tolerance of *Lactobacillus rhamnosus* GG by pressure pre-treatment. Int. J. Food Microbiol. **2004**, *96*, 307–313.
11. Desmond, C.; Stanton, C.; Fitzgerald, G.F.; Collins, K.; Ross, R.P. Environmental adaptation of probiotic lactobacilli towards improvement of performance during spray-drying. Int. Dairy J. **2001**, *11*, 801–808.
12. Pichereau, V.; Hartke, A.; Auffray, Y. Starvation and osmotic stress induced multiresistances: influence of extracellular compounds. Int. J. Food Microbiol. **2000**, *55*, 19–25.
13. Russell, N.; Fukanaga, N. A comparison of thermal adaptation of membrane lipids in psychrophilic and thermophilic bacteria. FEMS Microbiol. Rev. **1990**, *75*, 171–182.
14. Stanton, C.; Coakley, M.; Murphy, J.J.; Fitzgerald, G.F.; Devery, R.; Ross, R.P. Development of dairy-based functional foods. Sci. Aliments **2002**, *22*, 439–447.
15. Ananta, E.; Knorr, D. Pressure-induced thermotolerance of *Lactobacillus rhamnosus* GG. Food Res. Intern. **2003**, *36*, 991–997.

Probiotics

Dallas Hoover
Department of Animal and Food Sciences, University of Delaware, Newark, Delaware, U.S.A.

Abstract

The term "probiotics" means *for life*. It is applied to microorganisms specifically prepared and ingested by humans or animals to promote beneficial health effects primarily associated with the gastrointestinal tract. Popularity of food and beverages supplemented with probiotic cultures continues to rise as consumers seek a healthier diet and lifestyle. The majority of probiotics used as food additives are combinations of lactobacilli and bifidobacteria. These acidulating intestinal bacteria enhance the maintenance of a healthy gut ecology and immune system. This entry overviews the mechanisms of probiotic use.

INTRODUCTION

Consumption of microorganisms for health benefit is nothing new. Certainly, implied wellness from the consumption of fermented foods and beverages has been a part of the histories of people worldwide. Once the association and activities of microorganisms were established with cultured foods, the contributing microorganisms held health interest as well.

Although there were earlier reports of fermented milks with implied health benefits, in the early years of the 20th Century, Elie Metchnikoff, a Nobel Prize laureate biologist, was the first to put the subject on a scientific foundation. Metchnikoff spoke and published on his theories of sound health and longevity from the ingestion of lactobacilli (notably *Lactobacillus bulgaricus*) and other bacteria present in such foods as yogurt, kefir, and sour milk. Metchnikoff's colleague at the Institut Pasteur, Henri Tissier, showed that bifidobacteria were helpful in treating infant diarrhea. Such work fueled an increased demand for sour-milk products. Metchnikoff developed and perpetuated the theory that not only does the intestinal microbiota control the outcome of infection by enteric pathogens but it also regulates the natural chronic toxemia, which plays a major role in aging and mortality.

Today, the use of beneficial organisms as health additives in foods and beverages continues to grow. Dairy products remain the most popular food group to receive probiotic cultures as additives[1] with yogurt serving as the lead example.[2] First coined in 1960, the term "probiotics" (meaning *for life*) is no longer an unknown word to consumers. Probiotics refers to viable microorganisms that confer some kind of health benefit to the host when administered in appropriate amounts.[3] Prebiotics are compounds, usually carbohydrates such as oligofructose or inulin, that impart a physiological benefit to the host by selectively stimulating growth or activity of added and resident probiotic organisms present in the gastrointestinal tract. When an additive or product contains both probiotic and prebiotic elements, the term "synbiotic" is often used to recognize the implied synergy of probiotics and prebiotics.

While it is recognized that probiotic varieties are common to the intestinal tract, the premise is that supplementation of probiotic cultures in the diet elevates the total population of probiotic varieties in the gut to realize positive effects beyond conditions of non-supplementation. Probiotics are now used for both humans and animals. Their appeal is partly attributable to its use as a potential means to reduce our dependency on antibiotic use. In addition, use of probiotics carries a strong association with safety and designation as a "natural" therapy and are often referred to as a functional food.[4]

GUT ECOLOGY

The intestinal tract is the target site for delivery of probiotic cultures. Therefore, most of commercial strains of probiotic cultures were originally isolated from the gastrointestinal tract (usually from fecal or sewage samples) of humans or warm-blooded animals. Approximately 10^{14} microorganisms populate the human gastrointestinal tract weighing about two to three pounds.[5] This is over 10 times the total number of human cells in the body. Although estimates vary considerably, the human gut is home to at least 450 different species of microorganisms. Most of these organisms are located in the lower portion of the small intestine and the colon. The stomach and the upper intestine possess gastric acid, bile salts, and a highly propulsive motility to keep the concentrations and diversity of the microbiota low. Along the length of the small intestine the microbiota gradually increases. With healthy conditions the population of bacteria in the upper intestine is generally less than 10^5 organisms/mL of contents. The middle of the small intestine is a transitional zone between the sparse populations of the upper intestine and the luxuriant levels found in the large intestine. The ileum contains approximately 10^7 bacterial cells/mL. Once past the ileocaecal valve, the intestinal population of the microbiota increases dramatically. The total concentration of bacteria in the large intestine approaches the theoretical limit that can fit into a given volume, approximately 10^{11}–10^{12} organisms/mL.

Over time, the composition of the gut microbiota in adults is remarkably consistent; however, e.g., it has been determined that levels of bifidobacteria decline after weaning from mother's milk, and in the elderly, these levels diminish further with a corresponding increase in *Clostridium perfringens*. Bifidobacteria have demonstrated inhibitory effects toward many pathogenic organisms, both in vitro and in vivo. Examples include other clostridia besides *C. perfringens* (e.g., *C. difficile*), *Salmonella*, *Shigella*, *Campylobacter jejuni*, *Staphylococcus aureus*, *Bacillus cereus*, and the pathogenic yeast, *Candida albicans*.[5] The majority of studies have shown that administration of probiotic cultures will lower fecal pH and increase counts of the added probiotic, but with cessation of feeding, the gut microbiota will reestablish itself to previous levels and revert back to earlier conditions, suggesting that continued administration of probiotic cultures is necessary to maintain elevated levels of these bacteria in the gut.

PROBIOTIC TYPES

The two major types of probiotic cultures for commercial food use are lactobacilli and bifidobacteria; they are often used in combination. Members of both *Lactobacillus* and *Bifidobacterium* are found in the human intestinal tract, although gut lactobacilli primarily reside in the ileum of the small intestine and the bifidobacteria are most prevalent in the large intestine, especially in the area of the caecum. Both genera are non-sporeforming, non-motile, Gram-positive rods and acidulate from catabolism of carbohydrates although their biochemistry is distinctly different. Metabolism is fermentative, obligately saccharoclastic. Lactobacilli are grouped as either homofermentative or heterofermentative types. The homofermentative varieties, such as *Lactobacillus acidophilus*, produce only lactic acid from fermentation of glucose under optimal conditions, while heterofermentative types such as *Lactobacillus brevis* produce approximately only half lactic acid with acetate, ethanol, carbon dioxide, formate, and succinate as additional metabolic by-products. Bifidobacteria also produce lactic acid from glucose, but acetic acid is the predominate metabolite produced in a 3:2 ratio to lactate from the fructose-6-phosphate shunt.[6] Lactobacilli are termed aerotolerant anaerobes while bifidobacteria are generally described as obligate anaerobes although sensitivity to oxygen appears to vary among different strains and species.

Both the lactobacilli and bifidobacteria have complex nutritional requirements, which include amino acids, peptides, various vitamins, nucleic acid derivatives, salts, fatty acids or fatty acid esters and, of course, fermentable carbohydrates; nutritional requirements commonly vary among different strains and species.[6] Strains of both genera are usually lactose-positive and grow well in bovine milk. The favorite microbiological medium for growth of both in the laboratory is MRS broth or agar (i.e., deMan, Rogosa, and Sharpe medium).

The nomenclature frequently changes in microbial systematics. Also, commercial names now work their way into the scientific terms as witnessed from commercials with improvised species names touting immunity and regularity on the product label. The list of characterized species of lactobacilli and bifidobacteria used and studied as probiotic cultures is a long one. For *Lactobacillus*, considered probiotic species include *acidophilus*, *brevis*, *delbrueckii* (subsp. *bulgaricus*), *fermentum*, *gasseri*, *johnsonii*, *paracasei*, *plantarum*, *reuteri*, *rhamnosus*, and *salivarius*; for *Bifidobacterium*, considered probiotic species are *adolescentis*, *animalis* (subsp. *lactis*), *bifidum*, *breve*, *infantis*, and *longum*.

Other noted microbial varieties that have been used as probiotic cultures include strains and species of *Streptococcus* (*S. thermophilus* and *S. salivarius*), *Enterococcus* (*E. faecium*), *Pediococcus* (*P. acidilactici*), *Escherichia* (*E. coli*), sporeforming bacteria such as *Bacillus* (*B. coagulans* and *B. clausii*[7]), and the yeast, *Saccharomyces cerevisiae*.[8] Sometimes the genus/species name *Saccharomyces boulardii* is used for probiotic strains of *S. cerevisiae*, but the species name *boulardii* is not a valid descriptor.

While lactic acid bacteria (LAB) carry an identity as GRAS substances, there have been occasional instances of pathogenicity caused by LAB, e.g., in abdominal abscesses. Although rare, it does show that LAB can function as opportunistic pathogens or secondary invaders under some conditions. While bifidobacteria are not true LAB, they do contain a recognized pathogenic species, *Bifidobacterium dentium*, a cause of dental caries. Before a strain is fed to people in high numbers, clinical feeding trials and other characterizations are important to assure consumer safety. Anecdotal reports on feeding of probiotic cultures to people often report loose stools at least upon initial consumption.

Human safety is an obvious requirement for commercial probiotic strains, but other features are necessary if any health benefits are to be realized. While no firm agreement has been reached as to what level of probiotics should be fed to humans, usually a count in the range of 10^6–10^7 CFU/mL or g (or higher) of product is desired. In human efficacy studies, doses of 10^{10}–10^{11} CFU/day are typical; others have suggested a total of 10^9–10^{10} viable cells must reach the intestinal tract each day to be effective.[4] Therefore, a probiotic culture needs to maintain viability to a high degree in the product vehicle until consumed to survive transit in appreciable numbers to the ileum. Consequently, probiotic cultures should harbor heightened resistances to stresses in the product environment as well as resistances to stomach acidity and bile salts of the duodenum to reach their target sites in effective numbers and demonstrate measurable effect. Intestinal lactobacilli and bifidobacteria do well in their natural gut environment, but can suffer decline during storage in the product, while the yogurt-fermenting cultures (*S. thermophilus* and *L. bulgaricus*) survive well in refrigerated, acidified yogurt, but fare poorly and cannot replicate in the gastrointestinal tract.

IMPLIED BENEFITS AND MECHANISMS

Proscribed benefits from use of probiotic supplements or probiotics added to foods have included maintenance and restabilization of a healthy intestinal microbiota, enhanced immune response, improved digestion and utilization of lactose, and synthesis of vitamins and nutrients. Prevention of diseases and treatment of symptoms have also been suggested; specific reasons for probiotic administration include: 1) constipation; 2) irritable bowel syndrome and other diarrheal and inflammatory bowel diseases; 3) allergy development; 4) protection of liver function; 5) deconjugation of bile salts; 6) reduction of toxic metabolites and detrimental enzymes related to the aging process; 7) prevention of vaginal infections; 8) improvement of blood lipid profiles; 9) eradication of *Helicobacter pylori* infection of the stomach; 10) reduction of blood pressure; 11) body weight modulation; 12) important service in infant health and nutrition; 13) degradation of nitrosamines; 14) aid in absorption of calcium; 15) antitumorigenic activity; and 16) recolonization following antibiotic treatment, chemotherapy, or radiation treatment.[3,5,6,8,9] As noted earlier, probiotics have received considerable attention in the past few years as a potential replacement for antibiotic therapy given the issues of antibiotic-resistant pathogens and the perceived natural safety of probiotic use.

In the case of acidulating probiotic cultures, such as the lactobacilli and bifidobacteria, the administration of these additional bacteria tend to lower fecal pH, better stabilizing the gut against newly arrived or asymptomatic resident enteric pathogens, such as clostridia, coliforms, and enterococci. Added probiotic bacteria can compete for colonization of any existing intestinal sites as well as competition for consumption of key nutrients. In addition to organic acids, probiotic cultures can produce other antimicrobial substances such as bacteriocins and volatile fatty acids. Other attributed mechanisms for improving gut health are the reduction of gut transit time, stimulation of mucin production and immunomodulation.

PRODUCTS

Probiotic cultures are available as dietary supplements. Capsules are popular; these products can feature microencapsulation of cells to enhance storage viability. Usually refrigerated storage is recommended; these supplements are commonly found in such venues as health food stores. In the United States, yogurt is the favorite food for delivery of probiotics. As a fermented or cultured food, yogurt already has an established identity of 'natural goodness' associated with the presence of the yogurt cultures; however, probiotic cultures are used as food additives, not as starter cultures, since contribution of probiotic cultures to the fermentation profile is often judged to be minimal. With each passing year the number and kinds of probiotic-containing foods and beverages grows. Examples include fermented and unfermented milks and dairy beverages (e.g., smoothies), fruit juices, cheeses, cereal products, ice cream, candies, sports and nutrition meal replacement products, and infant formula.

IN CLOSING

Personal desire for health and well-being will continue to strongly influence purchasing choices for consumers into the 21st Century. The old perception that all microorganisms are germs to be avoided at all costs is no longer entrenched. Today's consumers have a better understanding of nutrition and food microbiology and avidly seek probiotic-containing foods and beverages to incorporate into their diets. With continued global interest in this product niche, it can be expected that standardization of probiotic usage will develop while research studies on probiotic mechanisms continue.[8,10,11] There is still much to be learned about optimization of health benefits conferred by probiotic microbiota.

REFERENCES

1. Tamine, A., Ed. 2005. *Probiotic Dairy Products*; Blackwell Publishing: Oxford, UK; Ames, IA.
2. Sarkar, S. Effect of probiotics on biotechnological characteristics of yogurt: A review. Br. Food J. **2008**, *110*, (6–7), 717–740.
3. Tannock, G.W., Ed. 2005. *Probiotics and Prebiotics: Scientific Aspects*; Academic Press: Wymondham, Caister, UK.
4. Weese, J.S. A review of probiotics: Are they really "functional foods"? AAEP Proc. **2001**, *47*, 27–31.
5. Hoover, D.G. Bifidobacterium. In *Encyclopedia of Food Microbiology*; Robinson, R., Batt, C., Panel, P., Eds.; Academic Press, Inc.: London, UK, 1999; 210–217.
6. Sneath, P.H.A.; Mair, N.S.; Sharpe, M.E.; Holt, J.G. *Bergey's Manual of Systematic Bacteriology*; Willams & Wilkins: Baltimore, MD, 1986; Vol. 2, 1209–1219, 1418–1424.

7. Ricca, E.; Henriques, A.O.; Cutting, S.M., Eds. *Bacterial Sporeformers: Probiotics and Emerging Applications*. Horizon Bioscience, Wymondham, Norfolk, UK, 2004.
8. Sanders, M.E.; Gibson, G.; Gill, H.S.; Guarner, F. Probiotics: Their potential to impact human health. Council for Agricultural Science & Technology, Ames, IA, 2007, Issue Paper 36, October.
9. Drisko, J.A. Probiotics in health maintenance and disease prevention. Altern. Med. Rev. May; **2003**, *8* (2), 143–155.
10. FAO/WHO. Published jointly by the food and Agriculture Organization of the United Nations and World Health Organization, New York, NY, 2001.
11. Raloff, J. Nurturing our microbes. Sci. News **2008**, *173*, 138–140.

Processing and Preservative Aids: Nisin and Other Bacteriocins

Dallas Hoover
Haiqiang Chen
Department of Animal and Food Sciences, University of Delaware, Newark, Delaware, U.S.A.

Abstract

This entry reviews information on the application of bacteriocins for their usage as preservative agents in our food supply. Bacteriocins are small peptides produced by bacteria to kill competing bacteria. Nisin is the featured bacteriocin in the entry owing to the vast quantity of information that describes its use in foods. Nisin has been added to food as a preservative for over 50 years by many countries; for decades it was the only bacteriocin applied as a food additive. The entry is an overview of how bacteriocins from lactic acid bacteria function and how they are best utilized in food products. Limitations for applications in foods are also described.

INTRODUCTION

It is generally believed that most bacteria produce antimicrobial compounds such as antibiotics, toxins, and bacteriocins. Bacteria synthesize bacteriocins to function as lethal agents that provide them with a competitive advantage over other bacteria inhabiting the same ecological niche. Most bacteriocins for use, or potential use, as food preservatives are peptides; that is, proteinaceous substances composed of less than 60 amino acids. Therefore, a majority of bacteriocins are antagonistic to bacteria very similar to the types that produce them. Normally, bacterial cells producing a bacteriocin are immune to its action.[1]

NISIN AND OTHER BACTERIOCINS

In the last 30 years, the emphasis of study has been on bacteriocins produced by lactic acid bacteria, such as *Lactococcus*, *Pediococcus*, *Lactobacillus*, and *Carnobacterium*. An important reason for focusing on these bacteria is that they are traditionally used to produce fermented foods, such as cheese and pickles. These products have been consumed by humans for countless generations with no evident health problems, so consequently these bacteria and their bacteriocins have an established reputation of safety. Lactic acid bacteria also have the distinction of producing the most commercially successful and most studied bacteriocin to date, nisin.[2]

Nisin

Nisin is the premier example of a bacteriocin successfully developed for use as a food additive. It has a mass of approximately 3500 daltons, and is produced by some strains of *Lactococcus lactis* subspecies *lactis*. Nisin is effective against a wide number of Gram-positive bacteria. (As a rule, bacteriocins produced by Gram-negative bacteria only inhibit certain Gram-negative bacteria, and bacteriocins produced by Gram-positive bacteria will only be effective against certain Gram-positive bacteria.) Awareness of nisin arose in the early part of the 20th century among cheesemakers who observed that some cheese fermentations were slow, slow to the point that the cheese fermentation had to be terminated because an acceptable product could no longer be obtained from the cheesemilk. Later it was found that some of these slow fermentations were caused by a subpopulation of the bacteria added as cheese starter cultures for the purpose of acidulating the cheesemilk and forming the curd. These inhibitory bacteria were bacteriocin-producing cells of *L. lactis*; and these nisin-producing lactococci were killing other lactococci, which severely reduced the total number of bacteria of the cheese starter culture and slowed acid production.[2]

With further study, it was realized that this antagonistic substance killed more than just lactococci. Indeed, it killed more than other lactic acid bacteria; other Gram-positive bacteria, such as clostridia, staphylococci, bacilli and listeriae also proved susceptible. Nisin was first used as a preservative agent in 1951 and gained worldwide acceptance

Table 1 Preferred qualities of bacteriocins with intended use in foods.

Documented safety for human consumption
Reasonable cost
Effective at low levels against bacteria of concern
Relatively stable over length of product shelf life
Stable at temperatures commonly used in thermal processing
No negative effects on sensory qualities of product
No common medical uses

---

under the brand name Nisaplin™. In the United States, it was first approved by the Food and Drug Administration as an additive in pasteurized processed cheese spreads in 1988.

Nisin was once referred to as an antibiotic because of its bacteriocidal activity, but the application of this term to nisin and other bacteriocins from lactic acid bacteria is discouraged to avoid confusion with therapeutic antibiotics used to treat diseases in people and animals. It is illegal to add therapeutic antibiotics to foods as preservatives. Part of the distinction of nisin from therapeutic antibiotics is the fact that nisin occurs naturally in milk and has been ingested by humans for thousands of years in cultured dairy products with no known toxicological problems. Nisin is rapidly digested in the gastrointestinal tract, and does not induce cross-resistance to therapeutic antibiotics. Nisin is neither used therapeutically in human medicine, nor as an animal feed additive or for growth promotion. Table 1 summarizes attributes expected for bacteriocins intended for food use.

When nisin was first utilized by the food industry, there was a concern that nisin applications might hide unsanitary manufacturing practices, but this concern was unfounded because the narrow target spectrum makes this bacteriocin ineffective against Gram-negative rapidly-growing spoilage bacteria as well as fungi. Also, the effectiveness of nisin against sensitive Gram-positive bacteria is heavily dependent on the bacterial concentration in the food, and the sensitivity of susceptible bacteria varies considerably. As the population of the microbiota of the food increases, the antibacterial effectiveness of nisin decreases. It is effective in extending the shelf life of canned soups, but it has no efficacy when poor-quality starting materials are used and the load of microorganisms is high.

Nisin is effective against vegetative cells and spores of clostridia and bacilli. Small amounts of nisin increase the sensitivity of spores to heat and consequently the addition of nisin to low-acid canned foods allows reduction of temperatures and treatment times normally used in thermal processing of these products. For example, the thermal processing D values of spores can be reduced by 50–60% in the presence of nisin. Nisin appears to block the pre-emergent swelling of spores, and this sporostatic effect is possibly due to a reaction between electrophilic groups within the nisin molecule and nucleophilic groups such as sulfhydryls in molecules within the spore.[1,2]

The mechanism by which nisin antagonizes vegetative cells is by binding to the cytoplasmic membrane followed by the dissipation of the electric potential and pH gradient across the membrane through the efflux of ions. The adsorption process is highly pH-dependent. As a secondary effect, nisin also blocks peptidoglycan synthesis of the cell wall. Disruption of the cell membrane is a mechanism nisin shares with a variety of other bacteriocins from lactic acid bacteria.[1]

As a food preservative, nisin functions best in acidic foods and beverages since it is most soluble and most stable at acidic pH. For example, in solutions of dilute HCl at pH 2.5 nisin can be autoclaved (121°C/15 min) without loss of activity. Nisin is effective in cheeses; in pasteurized processed cheese and cheese spreads, it controls "blowing" caused by *Clostridium butyricum* and *Clostridium tyrobutyricum*. Nisin's activity is dependent on the water activity, pH, the levels of sodium chloride, and phosphate salts. The amount of nisin that can be added to foods is subject to legal limits. The addition of nisin allows for higher moisture contents and lower levels of salts in processed cheese spreads while maintaining product shelf life. In parts of the world where refrigeration capacity is limited, nisin is added to milk to increase its shelf life and safety. For example, nisin-amended milks can maintain satisfactory quality for 21 days at 37°C with heat treatment or the pasteurization process reduced by 80%. In addition to dairy products and low-acid canned foods, nisin has also been used to preserve pasteurized liquid eggs, pasteurized soups, low-acid sauces, high-moisture reduced-fat foods, beer, fruit juices, smoked fish, and lobster.

While nisin affords protection against temperature abuse in these perishable products and allows replacement of some chemical preservatives, its effectiveness is limited. In foods, nisin is slowly degraded and the rate of degradation is dependent on storage conditions and the nature of the food. Nisin does not perform as well in meat products as it does in dairy products, apparently because it binds to lipid components, such as the phospholipids of meat. Nisin usually works better in liquid than in solid foods, where uneven distribution of nisin is a common problem. Sometimes there is interference from food additives such as sodium metabisulfite and titanium dioxide. Nisin-sensitive bacteria can develop resistance to the bacteriocin and lastly, the issue of cost is an obviously important factor affecting whether or not a food manufacturer will use nisin.

Nisin has been added to foods as a purified preparation (Nisaplin™ contains as an active ingredient 2.5% nisin, the remaining components are essentially inert although the 77.5% sodium chloride may carry some preservative effect) or as a metabolic by-product from a starter culture used in the production of a fermented food. The bacteriocin has also been blended with other preservative compounds. For example, MicroGARD™ contains in addition to nisin other bacteriocins, organic acids, enzymes, and preservative botanical compounds. Nisin has been incorporated into food-packaging films, spray onto the surface of foods, and incorporated together with other preservatives into casings. When hot dogs are cooked in these casings, nisin can guard against any *L. monocytogenes* that may have contaminated the food between the thermal processing and packaging steps.

Other Bacteriocins

Most bacteriocins from lactic acid bacteria are cationic, hydrophobic or amphiphilic molecules composed of 20–60 amino acid residues. These bacteriocins are commonly classified into three groups that also include bacteriocins from other Gram-positive bacteria.[3]

In class I, lantibiotics (from *lan*thionine-containing an*tibiotic*) are small (<5 kDa) peptides containing the unusual amino acids, lanthionine, ∀-methyllanthionine, dehydroalanine, and dehydrobutyrine. Nisin and lacticin 3147 are lantibiotics. Class I is further subdivided into type A and type B lantibiotics according to chemical structures and antimicrobial activities. Type A lantibiotics are elongated peptides with a net positive charge that exert their activity through the formation of pores in bacterial membranes. Type B lantibiotics are smaller globular peptides, and have a negative or no net charge. Antimicrobial activity of type B lantibiotics is related to the inhibition of specific enzymes.[3]

Class II contains small (<10 kDa), heat-stable, and non-lanthionine containing peptides. This is the largest group of bacteriocins in this classification system, and these peptides are divided into three subgroups. Class IIa includes pediocin-like peptides such as pediocin PA-1 and enterocin A having a specific terminal sequence. This subgroup of bacteriocins is known for its effectiveness against *Listeria monocytogenes*. Class IIb contains bacteriocins requiring two different peptides for activity, and class IIc contains the remaining peptides of the class.[3]

The class III bacteriocins are not as well characterized as classes I and II. Class III houses large (>30 kDa) heat-labile proteins that are of little interest to food scientists due to their inherent instability. A fourth class consisting of complex bacteriocins that require carbohydrate or lipid moieties for activity has also been suggested;[3] however, bacteriocins in this class have yet to be properly characterized adequately at the biochemical level.

Hurdle Technology

Different combinations of food processing technologies and preservative factors incorporating bacteriocins have been examined to search for valuable preservative synergies or other benefits for the processed food. The effectiveness of bacteriocins has been measured with accompanying mild heat treatments, addition of chelating agents, such as EDTA, and numerous antimicrobial agents (e.g., acetate, sorbate, monolaurin, sucrose fatty acid esters, carbon dioxide, etc.), vacuum packaging, modified atmosphere packaging, the lactoperoxidase system, high hydrostatic pressure processing, and pulsed electric fields. Examples of bacteriocins used in cocktails include nisin, pediocin AcH, lacticin 481, lactacin F, leucocin F10, and curvaticin. Results have been variable.

CONCLUSION

Whether as a natural preservative to slow food spoilage and extend product shelf life, or as a means to improve food safety against such food pathogens as *L. monocytogenes*, bacteriocins continue to generate interest.

Bacterial cultures are isolated and screened in the search for new bacteriocins, and characterized bacteriocins are further studied for possible improvements in application. Such research activities can be expected to continue.

Should additional information on bacteriocins be desired, several publications over the years have reviewed bacteriocins from lactic acid bacteria and applications of bacteriocins in food production. Noted examples of reviews of bacteriocins are referenced.[1-6]

REFERENCES

1. Chen, H.; Hoover, D.G. Bacteriocins and their food applications. Compr. Rev. Food Sci. Food Saf. **2003**, *2* (3), 81–100.

2. Hurst, A.; Hoover, D.G. Nisin. In *Antimicrobials in Foods*; Davidson, P.M.; Branen, A.L., Eds.; Marcel Dekker: New York, 1993; 369–407.
3. Klaenhammer, T.R. Genetics of bacteriocins produced by lactic acid bacteria. FEMS Microbiol. Rev. **1993**, *12*, 39–85.
4. Cleveland, J.; Montville, T.J.; Nes, I.F.; Chikindas, M.L. Bacteriocins: safe, natural antimicrobials for food preservation. Int. J. Food Microbiol. **2001**, *71*, 1–20.
5. Jack, R.W.; Tagg, J.R.; Ray, B. Bacteriocins of Gram-positive bacteria. Microbiol. Revi. **1995**, *59*, 171–200.
6. Sahl, H.G.; Bierbaum, G. Lantibiotics: biosynthesis and biological activities of uniquely modified peptides from gram-positive bacteria. Annu. Rev. Microbiol. **1998**, *52*, 41–79.

Product Labeling: Policy and Regulation

Bert Popping
Molecular Biology and Immunology, Eurofins Scientific Group, Pocklington, U.K.

Abstract

The labeling of biotech crops was first implemented in Europe in 1997, based on judgement of nutritional differences or equivalence. Some years later, several inconsistent regulations and introduction of threshold levels followed, where labeling was based on the presence of transgenic DNA or protein. In 2003, the current labeling regulations for food and feed were implemented in Europe, requiring labeling at a threshold of 0.9%. Although in Europe, labeling is based on starting-point (e.g., seed or grain) analysis, in other countries like Japan, end-point (e.g., finished products) analysis was introduced. In general, countries with large biotech crop production have less stringent regulations.

INTRODUCTION

Food products containing or derived from biotech crops have been around for the past 10 years. The first product seen in European supermarket shelves was genetically modified (GM) tomato purée. This was sold with a price incentive at Tesco's and Sainsbury's, clearly labeled as "genetically modified." According to the supermarkets' statements, these products sold well.

However, when soups containing products derived from biotech crops were not labeled and sold in the same supermarkets, environmental pressure groups campaigned and consumer pressure increased, which ultimately led to the withdrawal of all products containing transgenic material from the supermarket shelves and the implementation of stringent regulations across Europe.

EUROPEAN REGULATIONS

In 1997, the European Parliament and Council adopted the "Novel Food Regulation" (258/97/EC)[1] which applies to any type of new food product, including those derived from or containing biotech crops, and requires evidence that it is safe for human consumption. It also requires the labeling of novel foods and novel food ingredients if they are not "substantially equivalent" to their conventional counterparts. Through labeling, the regulation aims to inform the consumer whether such products differ from their conventional counterparts in composition, nutritional value or effects, or intended use. The presence of a novel protein, such as the EPSPS (EPSPS is a single copy of the gene coding for glyphosate tolerance CP4 5 enolpyruvylshikimate-3-phosphate synthase (CP4 EPSPS) from *Agrobacterium* sp. strain CP4) or the DNA coding for it, does not automatically constitute any of the above cases and therefore does not trigger labeling under this particular regulation.

Increasing pressure from environmental activists groups and consumers led to stringent but inconsistent labeling guidelines that were imposed through regulations 1139/98/EC[2] and 1813/97/EC.[3] These required labeling biotech soybeans and maize on the basis of the presence of transgenic protein or transgenic DNA, which is covered in decisions 96/281/EC[4] [soybean (*Glycine max* L. cv A5403) line (40-3-2), commonly known as Roundup Ready® soya (Monsanto, Creve Coeur, Missouri, U.S.A)] and 97/98/EC[5] [maize (*Zea mays* L.) line (CG 00256-176), commonly known as Bt176 maize]. None of the other transgenic crops approved under the Novel Food Regulation had to be labeled solely on the basis of the presence of transgenic DNA or protein. By then, Europe came to situation where biotech crops approved under the Novel Food Regulation required labeling only if they were not substantially equivalent, but the two crop lines regulated under 1813/97/EC (Roundup Ready soya and Bt176 maize) required labeling if transgenic DNA or protein could be detected. Council Regulation 1139/98/EC repealed the previous regulation 1813/97/EC in article 3 and requires labeling of the above-mentioned soybean and maize lines using specified terminology.

Inconsistency remains as the new regulation does not apply to crops regulated under the Novel Food Regulation or to highly processed products such as flavorings or additives (e.g., lecithin). As this regulation does not provide any threshold above which labeling is mandatory, it has ultimately led to a race for the lowest sensitivities in analytical laboratories, to trade impediments, and to confusion among all parties involved.

THRESHOLD LABELING

Two years later, the European Commission (EC) realized the implications of this situation and acted upon it by issuing two new regulations, 49/2000/EC[6] and 50/2000/EC.[7] Regulation 49/2000/EC amended Council Regulation 1139/98/EC and introduced a 1% labeling threshold for adventitious contamination—products that contain an amount of transgenic material of up to 1%, which is unavoidable and not deliberately added, do not require labeling. Again, although it became clear in consultations with the EC that the intention was to regulate all approved transgenic crops, it referred back to 1139/98/EC that only regulated Roundup Ready soya and Bt176 maize. This regulation applies to food products and ingredients made of or derived from transgenic crops.

The other regulation, 50/2000/EC, makes the labeling of additives and flavorings mandatory. No threshold is provided, and labeling is required only if transgenic DNA or protein can be detected.

Again, a number of inconsistencies remain and 3 years later, the EC acted by issuing another set of regulations.

The new regulations, 1829/2003/EC[8] and 1830/2003/EC,[9] are no longer limited to Roundup Ready soya and Bt176 maize and cover all crops. Another milestone in the regulation is the requirement to label feed in addition to food.

Also, new labeling thresholds have been set, requirements for traceability laid down, and approvals granted for a period of 10 years. Labeling is now required even if DNA or protein is no longer detectable.

Labeling thresholds have been set at 0% for unapproved biotech crops (zero tolerance), 0.5% threshold for scientifically positively evaluated crops that have not received final approval from the EC, and 0.9% for approved biotech crops. The 0.5% for scientifically positively assessed crops is a temporary approval for a period of 3 years. It also requires each biotech crop to have a unique identifier. The format of the identifier is outlined in regulation 65/2004/EC.[10] The regulations have a few other novelties before 1829/2003/EC came into force, applications for new biotech crops could be made in any European Union member state. The assessment was made (and approved) in that country and binding for any other EU member state. The assessments were apparently made differently with varying stringency by member countries. To harmonize this, the newly established European Food Safety Authority (EFSA) was tasked with it. Every new application for biotech crops now has to go centrally to EFSA for assessment and approval. This is laid down in 1829/2003/EC "whereas"

(9) The new authorisation procedures for genetically modified food and feed should include the new principles introduced in Directive 2001/18/EC.[11] They should also make use of the new framework for risk assessment in matters of food safety set up by Regulation (EC) No 178/2002[12] of the European Parliament and of the Council of 28 January 2002 laying down the general principles and requirements of food law, establishing the European Food Safety Authority, and laying down procedures in matters of food safety (1). Thus, genetically modified food and feed should only be authorised for placing on the Community market after a scientific evaluation of the highest possible standard, to be undertaken under the responsibility of the European Food Safety Authority (Authority), of any risks which they present for human and animal health and, as the case may be, for the environment. This scientific evaluation should be followed by a risk management decision by the Community, under a regulatory procedure ensuring close cooperation between the Commission and the Member States.

In addition, it has become mandatory for companies or organizations wanting to have their biotech crops approved in Europe to provide reference material as well as a specific detection method. This is then tested and validated by the European Network of GMO Laboratories (ENGL), in coordination with the Community Reference Laboratory of the Joint Research Centre in Ispra, Italy. This is laid down in 1829/2003/EC "whereas"

(36) To facilitate controls on genetically modified food and feed, applicants for authorisation should propose appropriate methods for sampling, identification and detection, and deposit samples of the genetically modified food and feed with the Authority; methods of sampling and detection should be validated, where appropriate, by the Community reference laboratory.

The methods need to pass the validation test successfully for the progress of the approval process in Europe.

The traceability requirement in regulation 1830/2003/EC demands that each person in the food production chain dealing with GM organisms keep a record of who they were bought from and whom they were sold to (one-up-one-down). As long as they are live (i.e., reproducible) organisms, the unique identifier has to be transmitted with the organism. After that the fact that the product contains GM material (e.g., GM soybean, maize, or canola) has to be relayed to the buyer. Products for the final consumer that contain transgenic material have to be labeled according to the wording in the regulation 1829/2003/EC. Article 13 of the regulation reads as follows:

(a) where the food consists of more than one ingredient, the words 'genetically modified' or 'produced from genetically modified (name of the ingredient)' shall appear in the list of ingredients provided for in Article 6 of Directive 2000/13/EC in parentheses immediately following the ingredient concerned;

(b) where the ingredient is designated by the name of a category, the words 'contains genetically modified (name of organism)' or 'contains (name of ingredient) produced

from genetically modified (name of organism)' shall appear in the list of ingredients;

(c) where there is no list of ingredients, the words 'genetically modified' or 'produced from genetically modified (name of organism)' shall appear clearly on the labelling;

(d) the indications referred to in (a) and (b) may appear in a footnote to the list of ingredients. In this case they shall be printed in a font of at least the same size as the list of ingredients. Where there is no list of ingredients, they shall appear clearly on the labelling;

(e) where the food is offered for sale to the final consumer as non-pre-packaged food, or as pre-packaged food in small containers of which the largest surface has an area of less than 10 cm2, the information required must be permanently and visibly displayed either on the food display or immediately next to it, or on the packaging material, in a font sufficiently large for it to be easily identified and read.

This information has to be kept for 5 years.

Once approved by the scientific committee and the EC, the biotech crops can be freely traded and, depending on their status, grown in Europe.

DIFFERENT IMPLEMENTATIONS IN EU MEMBER STATES

In some countries, national implementation of European regulations led to the creation of another category: the so-called without GM ("ohne Gentechnik" in German) category. Products labeled thus need to comply with even more stringent requirements. Although the use of, say, enzymes for the production of cheese that originates from GM bacteria does not require the cheese to be labeled as a GM product, a "without GM" labeling is not possible either. Such a labeling, for example, would be an option for milk from cattle that have not been fed GM products or when other GM products (e.g., pharmaceuticals, regardless of the method of production) were not used on the cattle.

PERCEPTION OF BIOTECHNOLOGY

Globally, the EU has the most comprehensive regulations on biotech products. The new regulations, 1829/2003/EC and 1830/2003/EC, were meant to end the *de facto* moratorium of the EU member states, which until then lasted for 4 years. However, a small number of new products have been approved until now. According to the 2006 Eurobarometer poll, which had 25,000 representative citizens from all European member states, the acceptance of transgenic crops remains considerably low. Overall, a majority of Europeans think that GM food should not be encouraged.

They see it as being not useful, morally unacceptable, and a risk for society. A look at a section of the European public—the "decided" public (~50%)—reveals that 58% oppose and 42% support. Only in Spain, Portugal, Ireland, Italy, Malta, the Czech Republic, and Lithuania do the supporters outnumber opponents. There are mixed opinions on the reported acceptability of buying GM food. The most persuasive reasons relate to health, the reduction of pesticide residues and environmental impacts. Whether GM food is approved by the relevant authorities or is cheaper are not convincing reasons. However, it is invalid to claim that European public opinion is a constraint to technological innovation and contributes to the technological gap between the United States and Europe. With the exception of nuclear energy, Europeans are more or less as optimistic as people in the United States and Canada about information technology, biotechnology, and nanotechnology. One exception is GM food, on which Europeans and Canadians have rather similar views, while people in the United States see it as much more beneficial and less risky. Europe's position is strikingly different on nanotechnology. In comparison to people in the United States and Canada, Europeans see nanotechnology as more useful and have greater confidence in regulations.

European consumers' attitude toward biotech crops will need to be reevaluated as economic conditions change. Current attitudes, however, have led to supermarkets carrying few products derived from transgenic organisms.

LABELING REGULATIONS IN JAPAN, KOREA, SAUDI ARABIA, AND THAILAND

In Japan, the focus is on the final products reaching the consumer rather than on labeling the products at the beginning of the production chain. Highly refined oils need to be labeled as GM in Europe even if no (GM-)DNA or protein can be detected. In comparison, labeling of such products is not required in Japan.

The Japanese Ministry of Health Labour and Welfare (MHLW) introduced regulations in 2000 that need to be followed by anybody wishing to have biotech crops authorized. The assessment, based on aspects similar to those in Europe, takes into account potential toxicology of new proteins or metabolites, allergenic aspects, and environmental issues.

Once the approval is passed, there is a de facto 5% labeling threshold. For unapproved biotech products, as in Europe, the zero-tolerance threshold applies. As in Europe, producers and retailers aim to avoid having products in the market that are labeled as GM. But here, unlike in Europe, products sold without a GM label may still contain significant quantities of GM-derived material (e.g., GM soybean oil), as highly refined products do not need to be labeled. Typical products that may contain GM but do not have to be labeled in Japan include some baked goods, cheese, soy sauce, and a number of manufactured foods.

The Japanese regulations are comparatively pragmatic and have not had such a significant impact on imports, especially from the United States. Similar regulations can be found in Korea, Saudi Arabia, and Thailand.

CONCLUSIONS

In general it appears that large producers of transgenic crops have less stringent regulations. Most are based on substantial equivalence rather than on the detection of DNA or protein. This means that products need only be labeled if, for example, the nutritional content changes in comparison with the conventional products. The presence of transgenic DNA or protein alone would not trigger labeling. This is similar to the European Novel Food Regulation 258/97/EC that was later tightened by other successively implemented regulations.

As labeling regulations tend to be a moving target and threshold levels for the labeling of transgenic crops have been set for some but not all countries, the following Web sites (noncomprehensive) are recommended for information on up-to-date regulations and thresholds:

Euro-Lex: http://eur-lex.europa.eu/en/index.htm
GMO Compass: http://www.gmo-compass.org/eng/home/
AgBioForum: http://www.agbioforum.org/
AgBios: http://www.agbios.com/main.php
FDA CFSAN: http://www.cfsan.fda.gov/
Belgian Biosafety Server: http://www.biosafety.be/
BATS Biosicherheit: http://www.bats.ch/bats/en/genfood.php

It has to be mentioned that while all of the above-mentioned Web sites provide excellent information, some are maintained by industrial or environmental lobby groups and may contain comments or interpretations favoring pro or con biotech aspects.

REFERENCES

1. 258/97/EC: Council Regulation (EC) No 258/97 of the European Parliament and of the Council of January 27, 1997 concerning novel foods and novel food ingredients.
2. 1139/98/EC: Council Regulation (EC) No 1139/98 of the European Parliament and of the Council of May 26, 1998 concerning the compulsory indication of the labeling of certain foodstuffs produced from genetically modified organisms of particulars other than those provided for in Directive 79/112/EEC.
3. 1813/97/EC: Commission Regulation (EC) No 1813/97 of the European Parliament and of the Council of September 19, 1997 concerning the compulsory indication on the labeling of certain foodstuffs produced from genetically modified organisms of particulars other than those provided for in Directive 79/112/EEC.
4. 96/281/EC: Commission Decision (EC) No 96/281 of the European Parliament and of the Council of April 3, 1996 concerning the placing on the market of genetically modified soya beans (*Lycine max* L.) with increased tolerance to the herbicide glyphosate, pursuant to Council Directive 90/220/EEC (text with EEA relevance).
5. 97/98/EC: Commission Decision (EC) No 97/98/ of the European Parliament and of the Council of January 23, 1997 concerning the placing on the market of genetically modified maize (*Zea mays* L.) with the combined modification for insecticidal properties conferred by the Bt-endotoxin gene and increased tolerance to the herbicide glufosinate ammonium, pursuant to Council Directive 90/220/EEC (text with EEA relevance).
6. 49/2000/EC: Commission Regulation (EC) No 49/2000 of January 10, 2000 amending Council Regulation (EC) No 1139/98 concerning the compulsory indication on the labeling of certain foodstuffs produced from genetically modified organisms of particulars other than those provided for in Directive 79/112/EEC, *Official Journal of the European Communities*.
7. 50/2000/EC: Commission Regulation (EC) No 50/2000 of January 10, 2000 on the labeling of foodstuffs and food ingredients containing additives and flavorings that have been genetically modified or have been produced from genetically modified organisms.
8. 1829/2003/EC: Commission Regulation (EC) No 1829/2003 of the European Parliament and of the Council of September 22, 2003 on genetically modified food and feed.
9. 1830/2003/EC: Commission Regulation (EC) No 1830/2003 of the European Parliament and of the Council of September 22, 2003 concerning the traceability and labeling of genetically modified organisms and the traceability of food and feed products produced from genetically modified organisms and amending Directive 2001/18/EC.
10. 65/2004/EC: Commission Regulation (EC) No 65/2004 of the European Parliament and of the Council of January 14, 2004 establishing a system for the development and assignment of unique identifiers for genetically modified organisms.
11. 2001/18/EC: Commission Directive 2001/18/EC of the European Parliament and of the Council of March 12, 2001 on the deliberate release into the environment of genetically modified organisms and repealing Council Directive 90/220/EEC.
12. 178/2002/EC: EC, 2002c. Commission Regulation (EC) No 178/2002 of the European Parliament and of the Council of January 28, 2002 laying down the general principles and requirements of food law, establishing the European Food Safety Authority and laying down procedures in matters of food safety.

Propionibacteria

Leon J. Spicer
Department of Animal Science, Oklahoma State University, Stillwater, Oklahoma, U.S.A.

Abstract

The animal feed industry's desire to control rumen fermentation, diseases, and improve animal efficiency has led researchers to continue to develop natural feed supplements containing live microorganisms such as rumen bacteria and fungi. This entry will summarize the theory, evidence, and use of various strains of *Propionibacterium* that have been successfully identified and manipulated as direct-fed microbials to improve production efficiency in cattle. In the animal industries, a specific strain of *Propionibacterium* is fed to feedlot cattle to prevent acidosis. Other propionibacterial strains are used as a feed supplement in beef cattle to prevent "nitrate" toxicity, and most recently studies have evaluated propionibacteria as a natural dietary supplement to increase milk production. In conclusion, various strains of *Propionibacterium* have been successfully adapted as direct-fed microbials or probiotics to improve production efficiency in cattle. Supplementation with *Propionibacterium* P169 may hold potential as a natural feed alternative to hormones and antibiotics to enhance lactational performance.

INTRODUCTION

One major metabolic challenge for cows once they calve is the tremendous nutrient demand to support milk production at a time when dry matter intake is reduced. This situation, combined with other stressors during the periparturient period, can increase the risk of metabolic diseases and health disorders. A goal of ruminant microbiologists and nutritionists is to manipulate the ruminal microbial ecosystem to improve efficiency of converting feed to animal food products consumable by humans. The animal feed industry's desire to control rumen fermentation, diseases, and improve animal efficiency has led researchers to continue to develop natural feed supplements containing live microorganisms such as rumen bacteria and fungi. This entry will summarize the theory, evidence, and use of various strains of propionibacteria that have been successfully identified and manipulated as direct-fed microbials to improve production efficiency in cattle.

SPECIES OF PROPIONIBACTERIA

Propionibacteria are mainly divided into two groups according to their habitats, the classical and the cutaneous strains. In general, propionibacteria are Gram-positive, rod-shaped, non-sporeforming, non-motile, and facultatively anaerobic bacteria. The classical strains include those bacteria found in cheese and other dairy products. After a primary culture ferments lactose to lactate, propionibacteria serve as the secondary culture to ferment lactate to propionate, acetate, and CO_2 during cheese-ripening, giving Swiss cheese its characteristic eyes and flavor. Propionibacteria are also used as inoculants for grain and silage (e.g., Biotal Plus II, Lallemand Animal Nutrition, Milwaukee, Wisconsin, United States). The cutaneous strains include those bacteria that are found on human skin. One acne bacillus, originally described as a *Corynebacterium* is an example of this group and is referred to as the "acne group strain" or the cutaneous propionibacteria. Although they can grow in the temperature range between 5 and 40°C, growth is most rapid at 37°C. Propionibacteria have an optimum pH for growth between 6.5 and 7.0 although the optimum range for different species ranged from 4.6 to 8.5. The four recognized dairy-related species in *Propionibacterium* include *P. freudereichii*, *P. acidopropionici*, *P. theonii*, and *P. jensenii* and less favorable species include *P. acnes*, *P. avidum*, *P. granulosum*, and *P. lymphophilum*. Because of their ability to produce vitamin B_{12} and propionic acid, members of the genus *Propionibacterium* have also been used for industrial production of these products.

USE OF PROPIONIBACTERIA IN CATTLE

Propionibacteria are natural inhabitants of the rumen and their population varies according to species, sampling times, individual animals, and the ration provided to the animals. In cattle, Davidson[1] reported that the population of propionibacteria ranges from 10^3 to 10^4 cfu/mL of rumen fluid. In the cattle industry, propionibacteria have been used as a feed supplement to prevent the risk of acidosis in feedlot cattle receiving high-concentrate diets because propionibacteria have the ability to convert lactic acid and glucose to acetic acid and propionic acid.[2] Parrott[3] determined that certain selected strains of *Propionibacterium* (strain P63) increase the lag time before lactic acid accumulated and suppressed the rate at which H+ concentration increased in ruminal fluid in vitro. Swinney-Floyd[4] reported that during the first 9 days of a 21-day experiment for the evaluation of adaptation to high-concentrate diets, acidosis never occurred and lactic acid concentrations did not accumulate in ruminally cannulated heifers inoculated with mixed culture of *P. acidipropionici* and *P. freudereichii*. One of the commercially available products using *Propionibacterium* strain P63 for feedlot cattle to prevent acidosis is called Micro-Cell PB (Lallemand Animal Nutrition, Milwaukee, Wisconsin, United States). Propionibacteria are also used as feed supplements in beef cattle to prevent "nitrate" toxicity,[5] and most recently studies have evaluated propionibacteria as a natural dietary supplement to increase milk production.[6,7] One of the commercially available products using *Propionibacterium* strain P5 for beef cattle to prevent nitrate poisoning is called Bova-Pro (Lallemand Animal Nutrition, Milwaukee, Wisconsin, United States) and is available as a feed additive or gel paste to be fed several days before anticipated nitrate exposure. One of the commercially available products using *Propionibacterium* strain P169 for dairy cattle to improve energy balance is called Dairy ProP169 (Bio-Vet Inc., Blue Mounds, Wisconsin, United States) and is available as a feed additive.

PHYSIOLOGICAL RESPONSES TO FEEDING PROPIONIBACTERIA

Recently, *Propionibacterium* P169 has been used in dairy cattle to enhance metabolism by increased ruminate propionate production.[6–10] Propionate is a major glucogenic substrate, spares glucogenic amino acids in gluconeogenesis, and consequently reduces the maintenance cost of metabolizable protein. The efficiency of utilization for maintenance of propionate is greater than for acetate and butyrate, and thus emphasizes the importance of propionate-producing bacteria as a potential feed supplement during the transition period. *Propionibacterium* P169 may alter metabolism via increases in plasma leptin, glucose, and insulin in Holstein cows.[6,8,11] Moreover, P169 has been shown to increase levels of propionate and consequently increase milk production and milk components in Holstein cows, particularly, in cows fed rations containing yeast (see Table 1). A recent study in steers compared the effect of P169 in conjunction with yeast to yeast alone and found that P169 affects ruminal fermentation, but not nutrient intake, digestibility, or flow of microbial protein to the lower gut.[9] In the absence of dietary yeast, dairy cows exhibit an increase in energetic efficiency without altering milk production.[10]

Overall, the milk production and component response to P169 has been positive (Table 1). Although two studies reported that *Propionibacterium* P169 fed daily to multiparous (i.e., having two or more calves) dairy cows had no significant effect on milk production or milk fat and lactose percentage,[10,11] other studies revealed positive effects of P169 on milk production (see Table 1). After one week of lactation, cows fed *Propionibacterium* P169 had a greater percentage of milk protein and solids-not-fat (SNF) and plasma non-esterified fatty acids (NEFA) concentrations than did control cows. Body weight and plasma leptin concentrations tended to be greater in cows fed P169 while plasma glucose, insulin, and cholesterol concentrations were not affected by feeding P169,[11] providing the first evaluation of use of propionibacteria in dairy cattle which suggested that supplemental feeding of P169 may alter some aspects of metabolism during lactation. In a second study that evaluated a higher dose of P169 in the presence of yeast culture, 4% fat-corrected milk (FCM) production over the first 30 weeks of lactation was found to average nearly 10% greater in high-dose and low-dose P169 vs. control multiparous cows (see Table 1). In a third study, feeding P169 in combination with yeast culture increased actual milk yield, solids-corrected milk and 4% FCM production by 8.5–16.6% above controls in multiparous cows, but not primiparous (i.e., having only one calf) cows, during mid but not early lactation,[6] and also between weeks 13 and 25 of lactation, milk urea nitrogen levels and SNF percentage were greater in high-dose and low-dose P169 vs. control cows. Cows fed high-dose P169 had higher milk lactose percentage than low-dose P169-fed cows,[7] and milk fat percentage was significantly greater in the low-dose P169 vs. high-dose multiparous cows (Table 1), but there was no difference among treatment groups in primiparous cows. Although weekly body weight from weeks 1 to 25 were greater in high-dose P169 than low-dose P169 and control multiparous cows, body weight in primiparous cows did not differ among groups.[7] These results show that multiparous and primiparous cows respond differently to supplemental feeding of P169, and this occurrence was probably due to the fact that primiparous cows utilize nutrients for body growth more than multiparous cows.

Feeding P169 to lactating dairy cows has increased blood glucose and insulin in some studies. Aleman et al.[8] reported that glucose concentrations in plasma samples collected weekly for 25 weeks were 9% greater in primiparous cows fed low-dose P169 (6×10^{10} CFU) vs. control primiparous cows, but glucose concentrations in multiparous cows did not differ among treatments. Consistent with this finding, Francisco et al.[11] found no effect of feeding P169 (6×10^{10} CFU) on plasma glucose concentrations measured weekly during weeks 1–12 of lactation in multiparous dairy cows. Insulin concentrations were greater in primiparous

Table 1 Summary of studies evaluating the effect of *Propionibacterium* strain P169 on milk production and milk components in lactating dairy cows.[a]

Author/reference	Group ID[b]	No. of cows	Milk production response		Milk fat response (% change)	Milk protein response (% change)	Milk lactose response (% change)
			Lbs.	%			
Francisco et al.[11]		20[c]	1.5	2.0	6.0	1.3	0.4
Stein et al.[7]	HD		6.8	8.9	−5.9	3.7	7.3
	LD	38	8.1	10.6	0.3	3.7	2.9
Lehloenya et al.[6]	VC		7.3	8.6	−7.9	1.0	3.5
	VY	31	1.5	1.7	0.5	−1.4	4.8
Weiss et al.[10]		50[c]	−0.9	−1.9	−2.9	0.4	0.6
Average			4.0	5.0	−1.6	1.4	3.2

[a]Responses shown are for multiparous cows only.
[b]HD, high-dose P169; LD, low-dose P169; VC, comparison vs. no-yeast control; VY, comparison vs. yeast-fed control.
[c]No yeast was included in ration.
Source: From J. Dairy Sci.[7]

cows fed high dose P169 vs. controls and in multiparous cows fed high-dose P169 vs. low-dose P169 during weeks 1–25,[8] but insulin did not differ between low-dose P169 fed and control multiparous cows during early lactation[11] giving further support to the notion that primiparous and multiparous cows respond differently to supplemental feeding of P169. Collectively, these studies support the idea that feeding P169 may improve metabolism and suggest that supplemental feeding of P169 in conjunction with yeast culture may improve milk production if fed for a period longer than 12 weeks postpartum.

Several studies have been conducted to evaluate the effect of propionate on plasma hormones and metabolites. In these studies, the responses differed depending on the method of administration and route of infusion, basal diet fed, dose given, and the length of time that the propionate was administered. In general, feeding or infusion of propionate transiently increases both glucose and insulin concentrations in cattle and these increases are short-lived (i.e., <120 min). Therefore, studies using low-sampling frequency of blood (e.g., weekly) may not detect these transient increases in glucose or insulin and may account for variability in glucose and insulin responses reported among previous studies.[8] Milk lactose (Table 1) and milk glucose were both greater in P169-fed cows than control cows, and thus the lack of treatment effects on plasma glucose levels may be partly due to the fact that insulin concentrations increased faster in P169-fed cows.[8] These results might be attributed to greater glucose demand and greater gluconeogenesis in P169-fed cows producing greater amounts of milk vs. control cows. In general, blood glucose status may affect energy balance and hence mobilization of body fat reserves. Hence, improved glucose status may enhance function of organs that depend on glucose, such as reproductive organs like the hypothalamus, pituitary, and ovary. Thus, additional benefits of feeding P169 may be improved reproductive efficiency, but more studies are needed to confirm this idea.

CONCLUSION

In conclusion, various strains of propionibacteria have been successfully adapted as direct-fed microbials to improve production efficiency in cattle. Supplementation of P169 may hold potential as a natural feed alternative to hormones and antibiotics to enhance lactational performance.

ACKNOWLEDGMENT

This work was approved for publication by the Director, Oklahoma Agricultural Experiment Station, and supported partly under projects H-2510 and grants from the Kellogg Foundation and Agtech Products Inc.

REFERENCES

1. Davidson, C. A. The isolation, characterization, and utilization of *Propionibacterium* as a direct-fed for beef cattle. M.S. Thesis. Oklahoma State University, Stillwater, 1998.
2. Krehbiel, C.R.; Rust, S.R.; Zhang, G.; Gilliland, S.E. Bacterial direct-fed microbials in ruminant diets: performance response and mode of action. J. Anim. Sci. **2003**, *81* (E. Suppl. 2), E120–E132.
3. Parrott, T.D. Selection of *Propionibacterium* strains capable of utilizing lactic acid from in vitro models. M.S. Thesis, Oklahoma State University, Stillwater, 1997.
4. Swinney-Floyd, D.L. The impact of inoculation with *Propionibacterium* on ruminal acidosis in beef cattle. Ph.D. Thesis, Oklahoma State University, Stillwater, 1997.
5. Swartzlander, J.H. Selection, establishment and in vivo denitrification of *Propionibacterium* strain in beef cattle. M.S. Thesis, Oklahoma State University, Stillwater, 1994.
6. Lehloenya, K.V.; Stein, D.R.; Allen, D.T.; Selk, D.E.; Jones, D.A.; Aleman, M.M.; Rehberger, T.G.; Mertz, K.J.; Spicer, L.J. Effects of feeding yeast and propionibacteria to dairy cows on milk yield and components, and reproduction.

J. Anim. Physiol. Anim. Nutr. (Berl). **2008**, *92* (2), 190–202.

7. Stein, D.R.; Allen, D.T.; Perry, E.B.; Bruner, J.C.; Gates, K.W.; Rehberger, T.G.; Mertz, K.; Jones, D.; Spicer, L.J. Effects of feeding propionibacteria to dairy cows on milk yield, milk components and reproduction. J. Dairy Sci. **2006**, *89* (1), 111–125.

8. Aleman, M.M.; Stein, D.R.; Allen, D.T.; Perry, E.; Lehloenya, K.V.; Rehberger, T.G.; Mertz, K.J.; Jones, D.A.; Spicer, L.J. Effect of feeding two levels of propionibacteria to dairy cows on plasma hormones and metabolites. J. Dairy Res. **2007**, *74* (2), 146–153.

9. Lehloenya, K.V.; Krehbiel, C.R.; Mertz, K.J.; Rehberger, T.G.; Spicer, L.J. Effects of propionibacteria and yeast culture fed to steers on nutrient intake and site and extent of digestion. J. Dairy Sci. **2008**, *91* (2) 653–662.

10. Weiss, W.P.; Wyatt, D.J.; McKelvey, T.R. Effect of feeding propionibacteria on milk production by early lactation dairy cows. J. Dairy Sci. **2008**, *91* (2), 646–652.

11. Francisco, C.C.; Chamberlain, C.S.; Waldner, D.N.; Wettemann, R.P.; Spicer, L.J. Propionibacteria fed to dairy cows: effects on energy balance, plasma metabolites and hormones, and reproduction. J. Dairy Sci. **2002**, *85* (7), 1738–1751.

Protein Bioseparation: Plant and Animal Products

Susan L. Woodard
Zivko Nikolov
Biological and Agricultural Engineering, Texas A&M University, College Station, Texas, U.S.A.

Abstract

The successful commercialization of therapeutic proteins and industrial enzymes expressed in mammalian cell culture and microbial hosts has inspired plant and animal scientists to develop a variety of new protein production systems. Improvements in expression technology have resulted in the accumulation of high levels of recombinant proteins in tissues of numerous plant hosts, including corn and rice seed, tobacco chloroplast, and in entire organisms such as the aquatic plant lemna. Improvements in transgenic animal production have also enabled the production of recombinant proteins in milk, urine, and eggs. As advances continue to increase expression levels of recombinant proteins in transgenic systems, the development of efficient bioprocesses becomes essential for successful product commercialization. Since each transgenic host and tissue presents a unique environment for protein production, there are a number of factors that may influence the choice of downstream processing steps. This entry describes the commonly used strategies to recover and purify recombinant proteins from transgenic plant and animal hosts.

INTRODUCTION

Plants and animals have been bred for thousands of years to obtain species with more desirable traits. More recently, plants and animals have been engineered to contain valuable output traits as well. That is, they can be manipulated to produce proteins that add value by enabling the protein to be recovered from the host and sold as an industrial enzyme or therapeutic. Heterologous protein production in transgenic plant and animals provides for a low-cost production system that is easily scaled up or down depending on demand. The use of transgenic plant and animal expression systems is gaining momentum as advances are made in generating high expression levels within these hosts. A number of strategies exist for achieving stable and transient expression in both plant and animal hosts. Recent reviews detail technologies used for transgenic plant expression[1–3] and transgenic animal expression.[4] As advances continue to result in increased expression levels for recombinant proteins, efficient recovery and purification of these commodities will be important in realizing the low-cost potential offered by these expression systems. The following sections present the strategies commonly used to isolate and purify recombinant proteins from various plant and animal host systems.

TRANSGENIC PLANT HOSTS

Plants offer the greatest host diversity for recombinant protein expression. In addition to terrestrial plants such as corn and tobacco, additional options are provided by plant cell culture and production from hairy roots, aquatic plants, microalgae, and moss expression systems. Unless the product of interest is secreted, bioprocessing of transgenic plants starts with biomass harvesting and pre-processing, which may include a wash step, oil removal, size reduction, and fractionation, followed by extraction, extract clarification, product concentration, and purification. The initial harvest step will vary depending on host type and the location of the product in the plant.

Terrestrial

The processing steps chosen for tissue from terrestrial expression systems depend on the plant size, organelle (leaf or seed), and growing environment. Pre-processing of seed (Fig. 1, left) may be useful for reducing biomass and enriching for the protein of interest. Examples include rice dehulling and germ and endosperm fractionation of corn seed. For oilseed crops or seed containing high amounts of oil, removal using hexane is often used to reduce membrane or chromatography resin fouling. Pre-processing such as oil removal, dry milling, flaking, dry germination, and seed pulverizing are generally accepted practices that do not appear to be detrimental to the recombinant protein.[5] Dry ground seeds could enter the extraction process immediately or could be stored prior to extraction.

Tobacco or other leaf tissue may also require a separate grinding step before extraction by homogenization

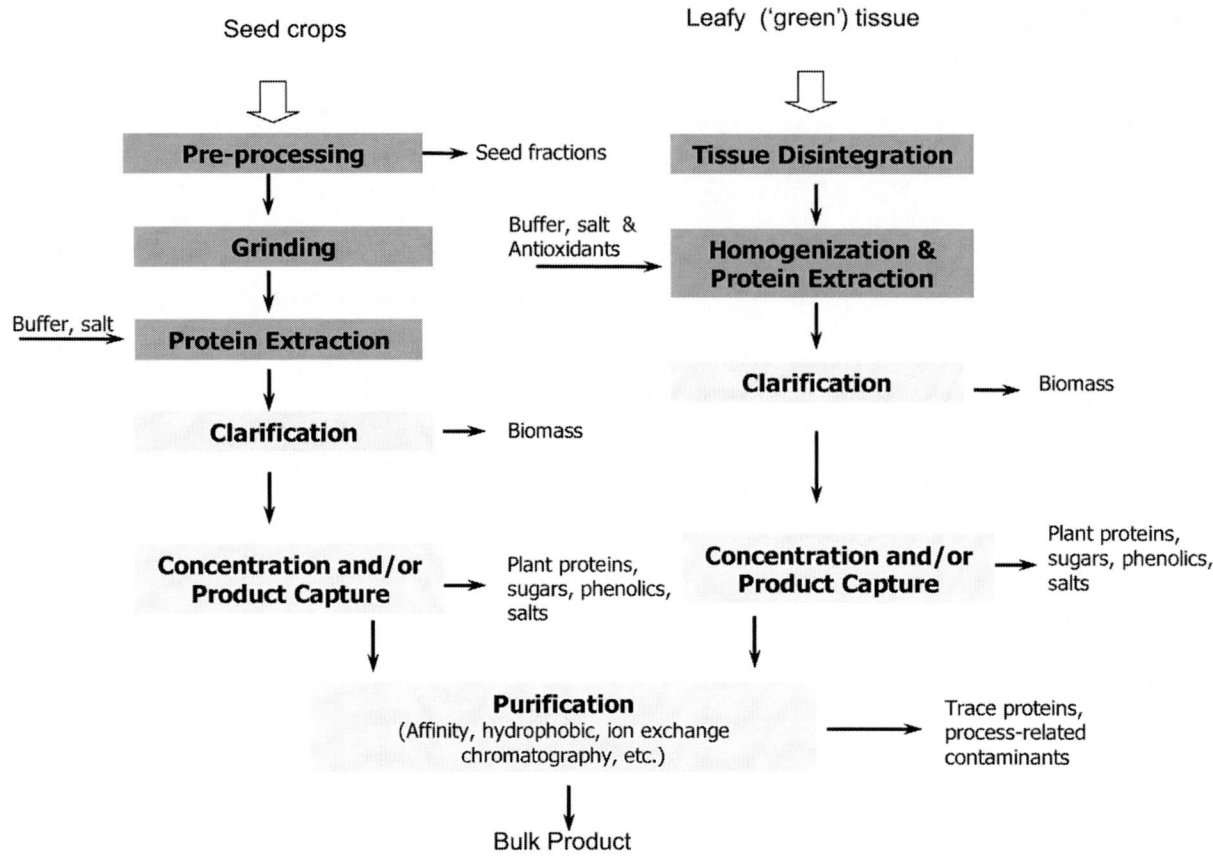

Fig. 1 Downstream processing diagram of recombinant proteins from transgenic terrestrial systems. Stages of downstream processing of seed crops are depicted on the left and those of leafy crops are shown on the right. **Source:** From Curr. Opin. Biotechnol.[10]

(Fig. 1, right). On a small scale, tobacco leaf can be ground frozen. More typically, green tissue is reduced in size by wet grinding. This size reduction can also be done as part of the extraction/homogenization process.

Aquatic

Aquatic expression systems such as hairy root cell culture, microalgae, moss, or lemna may secrete the protein, thus eliminating the need for biomass homogenization (see Fig. 2). Larger proteins (>40 kDa) expressed in these hosts require that plant tissue be harvested and homogenized to extract the protein as shown in Fig. 1 for leafy tissue.

BIOPROCESSING OF TRANSGENIC PLANTS

Particle Size Reduction

Seed is generally dry ground using a hammer mill or similar disintegration device to less than 255 μm to allow for maximum product recovery in the extraction phase.[6] Wet grinding and homogenization of seeds or seed fractions is possible but does not offer any advantages as recombinant proteins are more stable in a low moisture state than in aqueous slurries (possibility of proteolytic degradation, phenol oxidation, etc.).

Homogenization of the plant biomass in the presence of extraction buffer is used to further break cell walls and rupture individual cells for maximum release of the product. The homogenization step of green tissue or pre-ground seed fulfills a dual role—size reduction and continuous extraction or release of the intracellular components in a high-shear environment. Therefore, product protection from the shear and shear-induced temperatures has to be taken into consideration when running homogenizers.

Extraction

For the optimal extraction and recovery of the target product, both pH and ionic strength of the extraction buffer are critical.[7–9] Ionic strength is typically adjusted to a required level by using NaCl. The extraction conditions are product and plant tissue dependent and could be manipulated to maximize target protein extraction and minimize native protein solubilization. From all published studies so far, the consensus is that acidic (<pH 5) extraction conditions reduce native plant protein extractability. Important

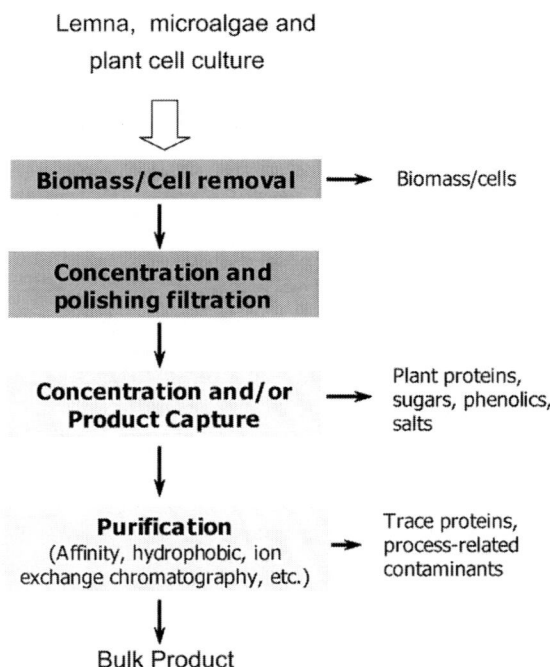

Fig. 2 Downstream processing diagram of aquatic culture systems where recombinant proteins are secreted in the growth medium. **Source:** From Curr. Opin. Biotechnol.[10]

additives to extraction buffers for green tissue (and plant cell culture) include metal chelators, antioxidants, protease inhibitors, and sometimes, small amounts of detergents. Whenever possible, temperature and pH could generally be employed to control proteolytic and phenol-oxidation activities in the extract. An extraction yield of greater than 80% in a single step is generally achievable assuming that particle size and extraction conditions have been previously optimized. Sometimes, a water or buffer wash of biomass solids after the solid–liquid separation is practiced to remove entrained fluid and increase the product yield.

Clarification (Solids Removal)

The removal of plant biomass solids after extraction utilizes traditional centrifugation and filtration.[10] Centrifugation works well for most types of plant biomass slurries whereas dead-end filtration (using plate and frame filter) is suitable only for ground seed solids that contain fiber and starch. Homogenized leafy tissue is quite compressible and dead-end filtration may require filter pre-coating and/or the addition of diatomaceous earth to the extract. Because transgenic biomass slurries typically contain between 10% and 30% by weight of solids, microfiltration would not be the most efficient way for clarification on a large scale. Clarified plant extracts after centrifugation contain 1–2% of solids that are easily removed by in-line dead-end polishing filters before chromatography. Mild heat pre-treatment or isoelectric and salt precipitation have been used for leafy tissue extract to remove native plant proteins like RuBisCo, particles, aggregates, and plant-derived pigments (chlorophyll).[11]

Purification of Plant-Expressed Proteins

Downstream purification for leafy tissue extracts often starts with a capture step, which has a primary goal of quick removal of active proteases, pigments, and carbohydrates, while also concentrating the product. If the target product is very dilute, or extract buffer and ionic strength are not suitable for direct capture chromatography, a UF/DF step may be included to concentrate the product and prepare the extract pH and ionic strength for maximal product adsorption. Seed extracts are more stable and process-forgiving as they typically contain native protease inhibitors co-extracted with the target protein and substantially less phenolics and other pigments. Downstream processing of leafy and other green tissue extracts seems to benefit most by employing membrane concentration in combination with a less-specific adsorption (ion exchange) step. UF/DF can also be used to remove phenolics and protect affinity resins. The potential downside for green tissue extracts and cell culture is the presence of residual proteolytic activity (if not previously completely removed or inhibited), which might then be co-concentrated with the product. After the initial capture step, purification strategies for the production of bulk and formulated products are similar to those practiced by biotech and pharmaceutical industries. The extent of downstream processing (number of stages) and final product purity is dependent on the intended use.

TRANSGENIC ANIMAL HOSTS

Milk and Other Fluids

To date, a wide range of proteins has been expressed in fluids of transgenic animals, including milk, egg, blood, urine, and semen. Since lactating animals produce large volumes of milk, the most common strategy is to use mammary gland-specific promoters to cause the expressed protein to be secreted into milk. Recovery of these proteins from milk poses some unique purification challenges due to the presence of a large amount of micelle-forming casein, fat globules, lipid, and lactose. The majority of published work in this area is from transgenic goats.[12]

Egg White

Recovery of recombinant proteins from egg white also poses unique challenges owing to the presence of the gel-forming glycoprotein, ovomucin. Depletion of ovomucin by diluting egg white with acidified water to precipitate it is a typical pre-treatment of egg white before purification.

BIOPROCESSING OF FLUIDS FROM TRANSGENIC ANIMALS

Clarification

Almost all processing schemes for transgenic milk products start with a centrifugation and/or microfiltration step to remove particulates, primarily fat globules and/or casein micelles. Clarified milk is then concentrated using ultrafiltration, precipitation, or adsorption. Downstream purification for milk-expressed proteins proceeds in a similar manner as that described for plant-produced products with the exception of the need to inactivate potential viruses at some point in the process for milk-produced proteins. Usually, a viral inactivation and removal step is required (Fig. 3) and integrated into the process somewhere before polishing chromatography steps. A final viral removal filtration (using nanofilters) is done on the purified product.

Purification of Animal-Expressed Proteins

As with transgenic plants, purification generally starts with a capture step. Affinity techniques used depend on the properties of the target protein and the properties of the potential impurities. Because milk contains whey, a combination of proteins with differing properties, a pass-through step may be needed to capture impurities prior to a target protein capture step. This strategy was described recently for the removal of the whey proteins α-lactalbumin, β-lactoglobulin, casein, and IgG using a Blue Sepharose column prior to an affinity capture step for α-fetoprotein.[12] The number of steps needed for purification of transgenic milk products will depend on the degree of purity required, generally dictated by the end-use.

Likewise, the goal of the first step in the purification of recombinant proteins from chicken egg white is to separate the protein of interest from the highly abundant egg white proteins such as ovalbumin, ovotransferrin, ovomucoid, and lysozyme. Even with ovomucin depletion, a gel can form from residual ovomucin and foul chromatography resins. It may be desirable, therefore, to use an initial precipitation, ion-exchange adsorption, or pass-through step to protect affinity resins used to capture and concentrate the protein product.

SUMMARY

One of the biggest advantages in producing recombinant proteins in transgenic plant and animal hosts is the lower cost of producing recombinant proteins in these hosts. The greater the expression levels attained in these hosts, the greater the impact the recovery and purification costs will have on the overall cost of the final protein product. Processes designed to minimize the co-extraction and/or co-purification of potential host impurities will result in more cost-effective purification strategies.

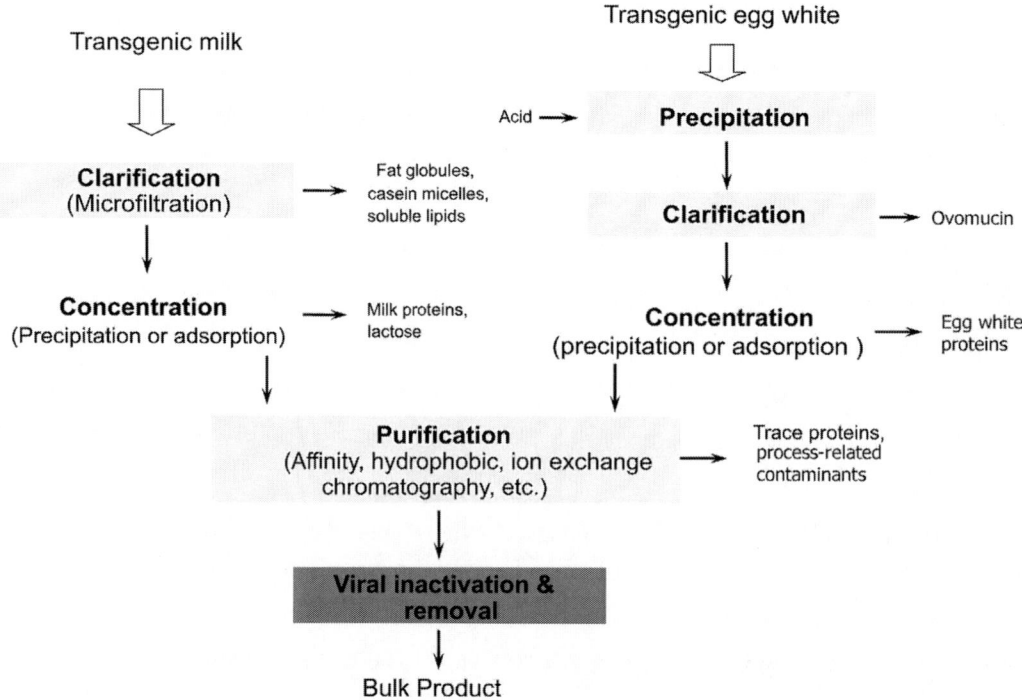

Fig. 3 Downstream processing diagram of recombinant proteins from transgenic animal systems. Stages of downstream processing of transgenic milk are shown on the left and those of eggs appear on the right. Note that viral inactivation could be executed at any stage in the process but is shown at the end for simplicity. **Source:** From Curr. Opin. Biotechnol.[10]

REFERENCES

1. Ma, K.-C.J.; Drake, P.M.W.; Christou, P. The production of recombinant pharmaceutical proteins in plants. Nature Rev. Genet. **2003**, *4* (10), 794–805.
2. Howard, J.A.; Hood, E.E. Bioindustrial and biopharmaceutical products produced in plants. In *Advances in Agronomy*; Sparks, D., Ed.; Academic Press: New York, **2005**; Vol. 85, 91–124.
3. Streatfield, S.J. Approaches to achieve high-level heterologous protein production in plants. Plant Biotechnol. J. **2007**, *5* (1), 2–15.
4. Houdebine, L.-M. Use of transgenic animals to improve human health and animal production. Reprod. Domest. Anim. **2005**, *40* (4), 269–281.
5. Menkhaus, T.J.; Bai, Y.; Zhang, C.; Nikolov, Z.L.; Glatz, C.E. Considerations for the recovery of recombinant proteins from plants. Biotechnol. Prog. **2004**, *20* (4), 1001–1014.
6. Bai, Y.; Nikolov, Z.L. Effect of processing on the recovery of recombinant beta-glucuronidase (rGUS) from transgenic canola. Biotechnol. Prog. **2001**, *17* (1), 168–174.
7. Azzoni, A.R.; Kusnadi, A.R.; Miranda, E.A.; Nikolov, Z.L. Recombinant aprotinin produced in transgenic corn seed: extraction and purification studies. Biotechnol. Bioeng. **2002**, *80* (3), 268–276.
8. Balasubramaniam, D.; Wilkinson, C.; Van Cott, K.; Zhang, C. Tobacco protein separation by aqueous two-phase extraction. J. Chromatogr. A **2003**, *989*, 119–129.
9. Wilken, L.R.; Nikolov, Z.L. Factors influencing recombinant human lysozyme extraction and cation exchange adsorption. Biotechnol. Prog. **2006**, *22* (3), 745–752.
10. Nikolov, Z.L.; Woodard, S.L. Downstream processing of recombinant proteins from transgenic feedstock. Curr. Opin. Biotechnol. **2004**, *15* (5), 479–486.
11. Pirie, N.W. *Leaf Protein and Its By-products in Human and Animal Nutrition* 2nd Ed.; University of Cambridge Press: Cambridge, UK, **1987**.
12. Parker, M.H.; Birck-Wilson, E.; Allard, G.; Masiello, N.; Day, M.; Murphy, K.P.; Paragas, V.; Silver, S.; Moody, M.D. Purification and characterization of a recombinant version of human alpha-fetoprotein expressed in the milk of transgenic goats. Protein Expr. Purif. **2004**, *38* (2), 177–183.

BIBLIOGRAPHY

1. Erickson, L.; Yu, W.-J.; Brandle, J. Eds. *Molecular Farming of Plants and Animals for Human and Veterinary Medicine*; Kluwer Academic Publishers: Dordrecht, The Netherlands, **2002**.

Protoplasts

Michael R. Davey
Department of Life Science, University of Nottingham, Nottingham, U.K.

P. Anthony

J.B. Power
School of Biosciences, Plant Science Division, University of Nottingham, Loughborough, U.K.

K.C. Lowe
School of Life and Environmental Sciences, University of Nottingham, Nottingham, U.K.

Abstract

Isolated plant protoplasts have several biotechnological applications, including their fusion to generate novel somatic hybrid and cybrid plants and the production of transgenic plants expressing specific characteristics. Normally, plant breeders rely upon sexual hybridization to combine useful genetic traits from different species or genera, but hybridization may be impeded by natural complex incompatibility barriers. The fusion of protoplasts isolated from somatic cells circumvents such barriers. Unique combinations of nuclear and organellar genomes generate novel germplasm; during protoplast fusion, there is no strict maternal inheritance of organelles, unlike in sexual hybridization.

FUSION OF ISOLATED PROTOPLASTS

Protoplast fusion can be induced chemically or electrically, or by a combination of these techniques, using small- or large-scale procedures.[1] Plasma membranes of protoplasts destabilize during fusion, establishing cytoplasmic continuity between tightly adhering protoplasts. Polyethylene glycol (PEG) is a common fusogen, sometimes in combination with high pH and Ca^{+2}.

Electrical fusion of protoplasts is more reproducible than chemical fusion and often gives greater fusion frequency. Versatile electrofusion instruments can be constructed.[2] Prior to electrofusion, isolated protoplasts are suspended in a medium of low conductivity [e.g., mannitol with $CaCl_2$] to stabilize plasma membranes. Exposure of protoplasts to an alternating current of 0.5–2.0 MHz at 100–400 V cm^{-1} induces protoplasts to align into "pearl chains" perpendicular to the electrodes; increasing the field strength induces close membrane-to-membrane contact. A subsequent direct current pulse of 10–200 micro-seconds and 500–2000 V cm^{-1} coalesces adhering plasma membranes at the poles of the protoplasts. Protoplast fusion normally occurs about 10 minutes after electrical treatment.

NOVEL HYBRIDS AND CYBRIDS GENERATED BY PROTOPLAST FUSION

Symmetric nuclear hybrid plants resulting from fusion are generally rare, with most being asymmetric hybrids. Plant breeders often require the introgression of a limited number of chromosomes, parts of chromosomes, or only organelles (chloroplasts and/or mitochondria) from one species into another. Consequently, effort has focused on generating asymmetric nuclear hybrids and cytoplasmic hybrids (cybrids). Cybrids are those with a nuclear genome of a given genus or species (recipient) with plastids from the other partner, or, more rarely, a mixed population of plastids from the recipient and donor. Cybrids harbor a mitochondrial genome partly or totally from the recipient or the donor, or novel mitochondria following recombination of recipient–donor mitochondrial DNA.

Asymmetric hybrid production is stimulated by treatment, before fusion, of protoplasts of one partner with X or gamma irradiation and exposure of the protoplasts of the other parent to a metabolic inhibitor (e.g., iodoacetamide). Irradiation stimulates partial genome transfer, the irradiated and nondividing but metabolically active protoplasts overcoming the inability of inhibitor-treated protoplasts to undergo mitosis. Frequently, somatic hybrid plants resemble the partner used as the source of the inhibitor-treated protoplasts.

GAMETOSOMATIC HYBRIDIZATION

Transfer of genetic information into cultivated plants can be achieved by generating addition or substitution lines via the sexual production of generating triploids. Infrequently, triploids arise during the fusion of diploid protoplasts, as in *Citrus*. They can also be generated by fusing diploid with haploid protoplasts. The latter can be isolated from haploid plants.[3]

PRODUCTS OF PROTOPLAST FUSION: SELECTION OF SOMATIC HYBRID TISSUES AND PLANTS

Fusion of protoplasts of the same genetic composition generates homokaryons, producing plants with increased ploidy. In contrast, heterokaryons from the fusion of protoplasts of different genetic composition have application in plant improvement, as they contain the nuclei of both parentals, initially in a mixed cytoplasm. The selection of heterokaryon-derived tissues and somatic hybrid plants remains a difficult aspect of somatic hybridization. In some combinations, heterosis (hybrid vigor) results in heterokaryon-derived tissues developing first in culture. Such tissues can be selected manually prior to plant regeneration.

Micromanipulation is useful when parental protoplasts are morphologically distinct. For example, fusion of suspension cell protoplasts with green leaf protoplasts generates heterokaryons initially with colorless plastids in one half of their cytoplasm and chloroplasts in the other half. Fluorochromes and flow cytometry can also be used in heterokaryon identification. Hormone autotrophism has also been exploited for hybrid selection.[4] Complementation systems have been devised to select somatic hybrids. The most simple involves fusion of protoplasts of nonallelic albino mutants to generate hybrid cells that complement to chlorophyll proficiency in the light.

Auxotrophic mutants have been exploited in selection, the nitrate reductase deficiency of the *nia-63* mutant of *Nicotiana tabacum* being complemented by the chlorate-resistant line (*cnx-68*). Protoplasts of the mutants fail to grow on nitrate-supplemented medium. However, somatic hybrid cells undergo complementation and can be selected on nitrate-containing medium. Antibiotic resistant cells are also useful in selection. Dominant antibiotic resistant genetic markers can be introduced by transforming parental cells prior to protoplast isolation.

In addition to antimetabolites being used to promote cybridization, such compounds that inhibit cell development have been exploited in hybrid selection. In the somatic hybridization of *Lactuca sativa* with the wild species *Lactuca virosa*, hybrids were selected by inactivation of *L. sativa* protoplasts with iodoacetamide, combined with the inability of *L. virosa* protoplasts to divide in culture.[5] Transformation to kanamycin resistance has also been combined with metabolic inhibition to select hybrid cells and plants.

CHARACTERIZATION OF SOMATIC HYBRID PLANTS

Characterization of plants generated by protoplast fusion necessitates morphological, cytological, and molecular analyses. Traits characteristic of both parents may be readily apparent in somatic hybrid plants, with leaves, flowers, and pigmentation being intermediate between those of both parents. In other cases, characteristics from one parent may be dominant.

Theoretically, plants regenerated by fusing two diploid somatic cell protoplasts should be tetraploid, which is the case in some combinations. However, plants with complete chromosome complements of both parents are generally rare. Hybrids often possess an asymmetric combination of parental chromosomes, following elimination of parts of genomes during culture.[6] The reasons for chromosome elimination are unclear. Some somatic hybrid plants are fertile; others may be sterile.[7] Cytological analyses of backcross progeny provide evidence of parental chromosomal behavior during mitosis and meiosis following introduction of somatic hybrids into breeding programs. Flow cytometry can estimate the ploidy of plants and provide a baseline for cytological analyses. DNA fingerprinting[8] permits detailed characterization of nuclear and organellar genomes of parental and somatic hybrid plants.

Organellar events are complex in somatic hybrids. Initially, heterokaryons contain a mixed population of organelles. Subsequently, plastids usually segregate, with those of one partner becoming dominant;[7] in some cases, the plastids of both partners persist in hybrids. Recombination of plastid DNA is rare. Analysis of ribulose biphosphate carboxylase–oxygenase, linked to the chloroplast DNA restriction enzyme profile, confirms the parental origin of chloroplasts in somatic hybrids. Recombination of mitochondrial DNA commonly generates "new" organelles, as evidenced by the restriction enzyme digestion patterns.[7] Importantly, DNA recombination in organelles increases genetic diversity arising from protoplast fusion.

Examples exist of the transfer of agronomical traits by protoplast fusion. Considerable effort has focused on the Solanaceae, such a potato and tobacco, together with the Brassicaceae. An excellent example is provided by studies in which protoplasts of *Solanum tuberosum* were fused with those of the wild potato, *Solanum bulbocastanum*,[9] generating hybrids having improved resistance to late blight. Importantly, this resistance was transferred to other breeding lines by back-crossing.

TRANSFORMATION OF ISOLATED PROTOPLASTS BY DNA UPTAKE

Induction of DNA Uptake

The induction of transient pores in the plasma membrane permits uptake of DNA by chemical and/or physical procedures.[10] Several agents induce DNA uptake into protoplasts, including salt solutions with Ca^{2+} at high pH and PEG. Electroporation, using short duration, high-voltage electrical pulses, often with PEG, is also used for DNA uptake into protoplasts. DNA has also been microinjected into protoplasts.[11] PEG and electroporation remain the most successful procedures for protoplast transformation, even though the frequency is low and, at best, only about 1 in 10^4 protoplasts develops into transformed tissues. Protoplasts can be transformed simultaneously with more than one gene, the genes being either on the same or on separate vectors.

Factors Influencing Protoplast Transformation

Parameters have been identified that influence protoplast transformation. The stage in the cell cycle is important, efficiency being higher when protoplasts are in the S or M phases. Consequently, it is beneficial to synchronize cells prior to or immediately following protoplast isolation. Heat shock or irradiation of protoplasts before DNA uptake also stimulates transformation, irradiation probably increasing recombination of genomic DNA with incoming DNA or initiating repair mechanisms that favor DNA integration. Complex integration patterns have been observed following DNA uptake into protoplasts.

MISCELLANEOUS STUDIES WITH PROTOPLASTS

Isolated protoplasts have also featured in physiological, ultrastructural, and genetic studies. Their development into colonies of single cell origin has been exploited to isolate clonal lines and plants, including those for increased secondary product synthesis. Exposure of isolated protoplasts to mutagenic agents or irradiation permits the induction and selection of mutants. Protoplasts take up macromolecules, which has been exploited in studies of endocytosis and virus infection and replication. The osmotic fragility of isolated protoplasts permits their controlled lysis for isolating cell components. In physiological studies, isolated vacuoles have been used to investigate sugar accumulation; protoplasts from barley aleurone cells contain protein storage vacuoles and a lysosome-like organelle, designated the secondary vacuole. Protoplasts are ideal for studying ion transport and regulation of the osmotic balance of cells.

Light-induced proton pumping has been investigated in guard cell protoplasts. Other studies have focussed on cell fusion and metabolism in microgravity, elicitor binding sites, binding of fungal phytotoxins to plasma membranes, and auxin accumulation and metabolism. Protoplasts have also provided unique material for studying cell wall synthesis and the role of microtubules during cell development. Protoplasts from totipotent cells, are useful for assessing the effects of pharmaceuticals, food additives, cosmetics, and agrochemicals on plant cells, whole plants, and their progeny over seed generations.[12]

REFERENCES

1. Blackhall, N.W.; Davey, M.R.; Power, J.B. Fusion and Selection of Somatic Hybrids. In *Plant Cell Culture, A Practical Approach*, 2nd Ed.; Dixon, R.A., Gonzales, R.A., Eds.; IRL Press at Oxford University Press: New York, **1994**; 41–48.
2. Jones, B. Lynch, P.T.; Handley, G.J.; Malaure, R.S.; Blackhall, N.W.; Hammatt, N.; Power, J.B.; Cocking, E.C.; Davey, M.R. Equipment for the large-scale electromanipulation of plant protoplasts. BioTechniques **1994**, *6*, 312–321.
3. Davey, M.R.; Blackhall, N.W.; Lowe, K.C.; Power, J.B. Gametosomatic Hybridisation In *Vitro Haploid Production in Higher Plants Volume 2: Application*; Mohan Jain, S., Sopory, S.K., Vielleux, R.E., Eds.; Kluwer Academic Publishers: Dordrecht, The Netherlands, **1996**; 309–320.
4. Carlson, P.S.; Smith, H.H.; Dearing, R.D. Parasexual somatic hybridization. Proc. Natl. Acad. Sci. U.S.A. **1972**, *69*, 2292–2294.
5. Matsumoto, E. Interspecific somatic hybridization between lettuce (*Lactuca sativa*) and wild-species *L. virosa*. Plant Cell Rep. **1991**, *9*, 531–534.
6. Oberwalder, B. Schilde-Rentschler, L. Loffelhardt-Ruoss, B. Ninnemann, H. Differences between hybrids of *Solanum tuberosum* L. and *Solanum circaeifolium* Bitt. obtained from symmetric and asymmetric fusion experiments. Potato Res. **2000**, *43*, 71–82.
7. Liu, J.H.; Dixelius, C.; Eriksson, I.; Glimelius, K. *Brassica napus* (+) *B. tournefortii*, a somatic hybrid containing traits of agronomic importance for rapeseed breeding. Plant Sci. **1995**, *109*, 75–86.
8. Oberwalder, B.; Ruoss, B.; Schilde-Rentschler, L.; Hemleben, V.; Ninnemann, H. Asymmetric fusion between wild and cultivated species of potato (*Solanum* spp.)-detection of asymmetric hybrids and genome elimination. Theor. Appl. Genet. **1997**, *94*, 1104–1112.
9. Helgeson, J.P.; Pohlman, J.D.; Austin, S.; Haberlach, G.T.; Wielgus, S.M.; Ronis, D.; Zambolim, L.; Tooley, P.; McGrath, J.M.; James, R.V.; Stevenson, W.R. Somatic hybrids between *Solanum bulbocastanum* and potato: A new source of resistance to late blight. Theor. Appl. Genet. **1998**, *96*, 738–742.
10. Davey, M.R.; Rech, E.L.; Mulligan, B.J. Direct DNA transfer to plant cells. Plant Mol. Biol. **1989**, *13*, 273–285.
11. Rakoczy-Trojanowska, M. Alternative methods of plant transformation—A short review. Cell. Mol. Biol. Lett. **2002**, *7*, 849–858.
12. Lowe, K.C.; Davey, M.R.; Power, J.B.; Clothier, R.H. Plants as toxicity screens. Pharm. News **1995**, *2*, 17–22.

Protoplasts: Culture and Regeneration

J.B. Power
School of Biosciences, Plant Science Division, University of Nottingham, Loughborough, U.K.

Michael R. Davey
Department of Life Science, University of Nottingham, Nottingham, U.K.

P. Anthony
School of Biosciences, Plant Science Division, University of Nottingham, Loughborough, U.K.

K.C. Lowe
School of Life and Environmental Sciences, University of Nottingham, Nottingham, U.K.

Abstract

Plant protoplasts represent the living contents of cells, each bounded by a plasma membrane and enclosed by a cell wall. The plasma membrane is involved in wall synthesis. Consequently, there is usually intimate contact between these two structures. However, when cells are stressed osmotically, their plasma membranes contract away from their surrounding walls. Subsequent removal of the walls enclosing plasmolyzed protoplasts enables the latter to be isolated as spherical, osmotically fragile, "naked" cells. The isolation of large populations of protoplasts from a range of plants is now routine. Such isolated protoplasts are ideal for studies of cell development, physiology, and cytogenetics. When cultured in the laboratory, isolated protoplasts undergo wall resynthesis and mitotic division. Protoplast-derived cells may express their totipotency, regenerating into fertile plants under the correct physiological and physical stimuli. This feature, unique to plant cells, is exploited in genetic manipulation through somatic hybridization and cybridization, both involving protoplast fusion, and transformation by direct uptake of foreign DNA.

ISOLATION OF PLANT PROTOPLASTS

The physiological status and age of tissues are crucial for the isolation of viable protoplasts. While leaves of glasshouse-grown plants are a convenient source of protoplasts, seasonal variation in illumination, temperature, and humidity may necessitate the use of environmental cabinets to ensure uniformity of material. Generally, cultured shoots and in vitro grown seedlings are more uniform as source material than pot-grown plants. Haploid pollen tetrads and mature pollen will release protoplasts; cell suspensions are convenient and frequently exploited as a source of protoplasts.[1]

Combinations of commercially available cellulase, hemicellulase and pectinase enzymes are used to release large populations of protoplasts. The cell wall composition of source tissues dictates the enzyme mixture required. Consequently, enzyme concentrations and conditions must be determined empirically for protoplast isolation from specific plant tissues.[2] The time of enzyme digestion, usually at 25–28°C, may be of short duration (e.g., 4–6 hours) or overnight (12–20 hours). Removal of the lower epidermis or dissection of leaves into thin strips facilitates tissue digestion and protoplast release. Preconditioning of donor plants or explants by exposure to reduced illumination or preculture of donor explants on suitable media may increase protoplast yield and viability.

Passage of enzyme-protoplast mixtures through sieves of suitable pore sizes following enzyme incubation removes undigested cells. Gentle centrifugation (e.g., 100 × g; 10 min) through a suitable osmoticum [e.g., 13% (w/v) mannitol] generally pellets the protoplasts, leaving fine debris in suspension. Mixing protoplasts with 21% (w/v) sucrose in a salts solution,[3] or with a solution of Percoll or Ficoll, followed by centrifugation, causes protoplasts of many species to float, facilitating their collection.

CULTURE OF ISOLATED PLANT PROTOPLASTS

Nutritional Requirements of Protoplasts: Culture Media

Protoplasts commence wall regeneration within hours of being introduced into culture. However, they require osmotic protection until they have regenerated a new primary wall of sufficient strength to counteract the turgor pressure exerted by the living cytoplasm/vacuoles.

Many media have been reported for protoplast culture. The nutrient-rich KM-type formulations[4] are beneficial for culture of protoplasts at low densities; other media are often based on the well-tested MS[5] and B5[6] formulations. Ammonium ions are detrimental to some protoplasts, particularly those of woody species. Sucrose is the most common carbon source, although glucose may act as both a carbon source and osmotic stabilizer. Maltose stimulates shoot regeneration from protoplast-derived cells, especially in cereals.[7] Most protoplasts require one or more auxins or cytokinins in the culture medium to sustain mitotic division.

Systems for Protoplast Culture

Incubation of isolated protoplasts in liquid medium in Petri dishes or in a shallow liquid layer overlaying semisolidified medium are simple methods of culture. The inclusion of a filter paper at the interface between the liquid and semisolid phases may stimulate cell colony formation. Isolated protoplasts can also be embedded in semisolidified media. Several gelling agents are available, with agarose often enhancing protoplast plating efficiencies (calculated as the percentage of the protoplasts originally plated that develop into cell colonies), compared with agar. Semisolid medium containing the protoplasts may be dispensed as layers or droplets (the latter about 100 μl in volume) in Petri dishes. Alginate is a useful gelling agent for heat-sensitive protoplasts. Following suspension of the protoplasts, the alginate-culture medium mixture is semisolidified by pouring the warm mixture in which the protoplasts are suspended over an agar layer containing Ca^{2+} or by gently dropping the molten medium into a solution of such ions.

Plating Density and Nurse Cells

The density at which protoplasts are plated in the culture medium is crucial for cell colony formation. Generally, the optimum plating density is $5 \times 10^2 – 1.0 \times 10^6$ ml^{-1}. At greater densities, protoplast-derived cells often fail to undergo sustained mitotic division because of rapid depletion of nutrients from the medium. Protoplasts also fail to grow when plated below a minimum density. Medium previously "conditioned" by supporting the culture of actively dividing cells for a limited period, will stimulate the growth of isolated protoplasts. "Nurse" cells are often employed to promote division of protoplasts in culture.[7] When using nurse cultures, the isolated protoplasts can be embedded in a semisolid layer, suspended in a thin layer of liquid medium, or spread in a liquid layer on a cellulose nitrate membrane, overlaying the semisolid medium containing the nurse cells.

INNOVATIVE APPROACHES TO PROTOPLAST CULTURE

Chemical Supplements for Culture Media: Surfactants and Antibiotics

Some antibiotics (e.g., cefotaxime) stimulate protoplast division, such compounds being thought to be metabolized to growth regulator-like molecule(s). Supplementation of medium with the nonionic, surfactant *Pluronic*® F-68, increases the plating efficiency of protoplasts in culture. Surfactants may increase the permeability of plasma membranes, stimulating uptake of nutrients from the culture medium into protoplasts and protoplast-derived cells.[8]

Manipulation of Respiratory Gases

Gassing of vessels with oxygen after introducing protoplasts into the culture medium increases the plating efficiency of protoplasts of jute and rice. A further novel approach for regulating the supply of respiratory gases to cultured protoplasts involves the use of inert, chemically stable perfluorocarbon (PFC) liquids.[9] Such compounds dissolve large volumes of respiratory gases and have been exploited in animal systems as oxygenation fluids.[10] PFC liquids, being about twice as dense as water, form a distinct layer beneath aqueous culture media. Consequently, protoplasts and protoplast-derived cells can be cultured at the interface between the lower PFC layer and the overlaying aqueous medium. Experiments revealed that protoplasts in such systems exhibited increased superoxide dismutase activity associated with oxygen detoxification. Other regulators of respiratory gases, notably chemically modified haemoglobin solutions, may also stimulate the growth of protoplasts in culture. Supplementation of culture media with commercial bovine haemoglobin solution (*Erythrogen*™) significantly increased the plating efficiency of rice protoplasts, compared to untreated controls.[11]

Physical Methods to Stimulate Protoplast Growth

Physical parameters have been shown to stimulate protoplast growth, including the use of cellulose nitrate filter membranes (0.2 μm pore size) at the liquid/semisolid medium interface, often in conjunction with nurse cells in the underlying semisolid medium. Insertion of glass rods vertically into semisolid agarose medium stimulated mitotic division of cassava leaf protoplasts in the overlaying liquid medium, the protoplasts probably receiving increased aeration where they aggregate in the liquid menisci around the glass rods. Electrical currents, both low and high voltage, also stimulate mitotic division and cell colony formation from protoplasts of many species.

PLANT REGENERATION FROM PROTOPLAST-DERIVED TISSUES

The induction and sustaining of plant regeneration in protoplast-derived tissues by different pathways of morphogenesis (organogenesis; somatic embryogenesis) is dependent, in part, upon the culture conditions, in particular, the composition of the culture medium, and the inherent totipotency of the donor species. Plant regeneration, via organogenesis, has been reported to occur for more than 70% of those species capable of regenerating plants from protoplast-derived tissues, most notably members of the Compositae, Cruciferae, Leguminosae, and Solanaceae. In contrast, plant regeneration from protoplast-derived tissues via somatic embryogenesis is restricted predominatly to members of the Curcurbitaceae, Gramineae, Leguminosae, Rutaceae, and Umbelliferae.[12] In a limited number of genera, protoplasts may develop directly into somatic embryos through early polar growth of their derived cells, as in *Medicago*, *Asparagus*, and *Persea*.

CONCLUSION

During the last 40 years, considerable progress has been made in regenerating plants from protoplast-derived tissues of an increasing number of genera and species, driven by a need for regeneration as a platform for many aspects of biotechnology. It is interesting to note that it is only recently that protoplast-to-plant systems have been developed for specific genera, an example being provided by *Sorghum*, in which the establishment of such a protoplast-to-plant system required nearly 20 years of on-going research.[13] An extensive literature is directed to plant regeneration from isolated protoplasts.[12] The expression of totipotency from protoplast-derived tissues remains an absolute requirement for the multifaceted applications of somatic cell technologies involving protoplasts, such as somatic hybridization, cybridization, and direct DNA uptake, in plant genetic improvement programs.

REFERENCES

1. Blackhall, N.W.; Davey, M.R.; Power, J.B. Isolation, Culture and Regeneration of Protoplasts. In *Plant Cell Culture, A Practical Approach,* 2nd Ed.; Dixon, R.A., Gonzales, R.A., Eds.; IRL Press at Oxford University Press: New York, **1994**; 27–39.
2. Enzymes for the Isolation of Plant Protoplasts. In *Biotechnology in Agriculture and Forestry. Plant Protoplasts and Genetic Engineering*; Ishii, S., Bajaj, Y.P.S., Ed.; Springer-Verlag: Berlin, **1997**; Vol. 8, 23–33.
3. Frearson, E.M.; Power, J.B.; Cocking, E.C. The isolation, culture and regeneration of *Petunia* leaf protoplasts. Dev. Biol. **1973**, *33*, 130–137.
4. Kao, K.N.; Michayluk, M.R. Nutritional requirements for growth of *Vicia hajastana* cells and protoplasts at a very low population density in liquid media. Planta **1975**, *126*, 105–110.
5. Murashige, T.; Skoog, F. A revised medium for rapid growth and bioassays with tobacco tissue cultures. Physiol. Plant. **1962**, *15*, 473–497.
6. Gamborg, O.L.; Miller, R.A.; Ojima, K. Nutrient requirements of suspension cultures of soybean root cells. Exp. Cell Res. **1968**, *50*, 151–158.
7. Jain, R.K.; Khehra, G.S.; Lee, S-H.; Blackhall, N.W.; Marchant, R.; Davey, M.R.; Power, J.B.; Cocking, E.C.; Gosal, S.S. An improved procedure for plant regeneration from indica and japonica rice protoplasts. Plant Cell Rep. **1995**, *14*, 515–519.
8. Lowe, K.C.; Anthony, P.; Davey, M.R.; Power, J.B. Beneficial effects of Pluronic F-68 and artificial oxygen carriers on the post-thaw recovery of cryopreserved plant cells. Art. Cells, Blood Subst. Immob. Biotech. **2001**, *23*, 221–239.
9. Lowe, K.C.; Davey, M.R.; Power, J.B. Perfluorochemicals: Their applications and benefits to cell culture. TIBTECH. **1998**, *16*, 272–277.
10. Lowe, K.C. Engineering blood: Synthetic substitutes from fluorinated compounds. Tissue Eng. **2003**, *9*, 389–399.
11. Al-Forkan, M.; Anthony, P.; Power, J.B.; Davey, M.R.; Lowe, K.C. Haemoglobin (*Erythrogen*™)—Enhanced microcallus formation from protoplasts of Indica rice (*Oryza sativa* L.). Art. Cells, Blood Subs., Immob. Biotech. **2001**, *29*, 399–404.
12. Xu, Z-H.; Xue, H-W. Plant Regeneration from Cultured Protoplasts. In *Morphogenesis in Plant Tissue Cultures*; Soh, W-Y., Bhojwani, S.S. Ed.; Kluwer Academic Publishers: Dordrecht, The Netherlands, **1999**; 37–70.
13. Sairam, R.V.; Seetharama, N.; Devi, P.S.; Verma, A.; Murthy, U.R.; Potrykus, I. Culture and regeneration of mesophyll—Derived protoplasts of sorghum [*Sorghum bicolour* (L.) Moench]. Plant Cell Rep. **1999**, *18*, 972–977.

Protozoa and Parasites: Food Safety

Kalmia E. Kniel
Department of Animal and Food Sciences, University of Delaware, Newark, Delaware, U.S.A.

Abstract
Zoonotic protozoa and helminth parasites may be transmitted to humans by food and water. While parasites were once considered a problem only related to areas of poverty and poor sanitation, present-day travel, a global food supply, and healthier eating habits have increased the occurrence of these organisms in foods in industrialized nations. Rapid diagnostic tests based on detection of unique nucleic acid sequences or proteins are widely used in clinical medicine and may be adapted for detection of parasites in contaminated food and water samples; however, one potential pitfall is interference of these tests by food ingredients.

INTRODUCTION

The World Health Organization (WHO) classifies parasites within the top six most harmful infective human diseases. Parasites can be contracted in a number of different ways, including the consumption of undercooked meats or contaminated raw produce, contaminated water, or through person-to-person transmission. Cases of parasitic infections, particularly protozoan infections, have increased over the past 20 years within the United States partly because of the better diagnostic capabilities along with the increase in imported fruits and vegetables.

Parasites are characteristically different from bacteria because they do not have the ability to multiply within foods. Instead, parasites use food as a vehicle to enter a human host and to replicate within the gut of the host. Parasites can either be consumed in an infective stage, such as oocysts of *Cryptosporidium parvum* unseen on a fresh vegetable, or become infective after being ingested, such as larvae of *Trichinella spiralis* undetected in a piece of undercooked meat. Parasites have a low infective dose that allows them to cause illness after ingestion of only a small amount of organisms, which in part may explain the spread of these microorganisms in food- and waterborne outbreaks.

Initial detection and subsequent confirmation of parasitic agents in food samples can be difficult and complicated since traditional enrichment strategies available for bacterial cells are not possible for detecting parasites. Rapid diagnostic methods are being used to determine environmental routes of transmission for many protozoan and helminth parasites, with water, soil, and food being of particular interest in the development of new applications. The possibly large numbers of transmissive stages and the environmental prevalence of parasites pose persistent threats to public and veterinary health.[1] Detection may require the concentration of large amounts of food or water, combined with molecular identification techniques, such as nucleic acid amplification. Typically, routine detection of parasites in a clinical laboratory is done by concentration of parasitic agents in stools, blood, or tissue, followed by immunofluorescent staining microscopy, nucleic acid amplification, and serodiagnostic techniques.

Rapid detection of parasites has gained greater interest in the last few years. While most foods are not routinely tested for parasites, rapid diagnostic tests may be useful during outbreak investigations. In general, rapid diagnostic methods for detecting pathogens in contaminated food products are based on detecting the genetic material (DNA) or specific surface antigens (Table 1). One advantage of nucleic acid detection is that DNA is not degraded by food processing methods that could destroy or alter surface proteins, making antigenic proteins undetectable by immunoassay. The largest need for these methods is to evaluate imported food products at food-entry ports. This capability would aid the U.S. Food and Drug Administration (FDA) and the United States. Department of Homeland Security in assessing the safety of imported foods, particularly in light of increased efforts to guard against bioterrorism. The following section describes information on current methods used predominantly in research laboratories for the detection of parasites in food.

RAPID DETECTION OF PROTOZOA

Protozoa are intricate single-celled eukaryotic organisms with complex life cycles, including one form which is encysted and called a cyst, oocyst, or spore. The most common food- and waterborne protozoa that infect humans are *Cryptosporidium* sp., *Cyclospora cayetanensis*, *Toxoplasma gondii*, and *Giardia* sp. Microsporidia may be considered emerging food- and waterborne pathogens with

Table 1 Overview of rapid testing methods for parasites.

Target	Signal	Time	Organism of interest
Nucleic acid-based	PCR fragment visualized on a gel or real-time PCR instrument	2–6 hr	Recover parasite from food product by blending, stomaching, or sonication; concentrate organism by centrifugation or IMS; lyse cells to release nucleic acid for PCR detection or apply directly to antigen detection assay
Protein-based	Antigen-antibody complex visualized by color indicator or fluorescence measured with an ELISA plate reader or microscopy, or using a dot blot apparatus	4–8 hr	

IMS, immunomagnetic separation.

several species known to cause human infection (*Nosema, Pleistophora, Enterocytozoon, Encephalitozoon, Brachiola, Vittaforma, Trachypleistophora,* and *Microsporidium*) and others known to cause infection in animals.[2]

The infective stage is a small, environmentally stable oocyst, cyst, or spore stage (ranging from 1.5 to 12 μm in size depending on the genus). Before the use of DNA-based methods in diagnostic laboratories, detection of protozoan oocysts was based on microscopy using immunofluorescence using specific antibodies, autofluorescence for *Cyclospora* oocysts, and histochemical stains. These techniques are still largely used in identification of oocysts from fecal specimens and may be combined with molecular techniques based on antigen and nucleic acid detection. The first step necessary for detection is to remove the oocysts from the food product through sonication, vortexing, washing, blending, or stomaching. Contemporary methods for analysis of food samples include use of one or more of the following techniques: 1) concentration by filtration, flotation, and/or centrifugation; and 2) recovery and identification with immunomagnetic separation (IMS) and polymerase chain reaction (PCR). Thus, this multi-step protocol combines isolation based on surface proteins with detection of a specific DNA sequence. Washing and homogenization of food samples may need to be optimized depending on the food product and manufacturer's protocol for the specific molecular test.

After overcoming the initial problems of low oocyst number and removal from the food product, PCR and/or immunofluorescence can be used to detect the protozoal pathogen of interest. The application of these rapid techniques is well illustrated for two specific human food- and waterborne parasites, the apicomplexan, *Cryptosporidium*, and the flagellate, *Giardia*.[1] Commercially available rapid immunoassays (EIA) used in clinical labs are based on detection of surface antigens and have a potential for use with blended food samples. A unique detection method uses an antibody directed to the internal symbiotic virus of *Cryptosporidium*.[3] The FDA Bacteriological Analytical Manual (BAM) includes techniques for these two organisms, along with *Cyclospora*, using filtration, IMS, and real-time PCR for quantification and detection from food samples.

Rapid diagnostics may be used for monitoring parasites relevant to animal health and the subsequent transmission to food products. For example, a number of tests are being developed to aid in the prevention of foodborne transmission of *T. gondii* in pork and other products. *Toxoplasma* is unique in which infection can be transmitted by ingestion of tissue cysts in undercooked meat or oocysts in water or on fresh produce. Current research is aimed at identifying the initial route of infection, either by oocyst or by tissue cyst (bradyzoite life stage), using genetic material specific to that life stage. Real-time PCR, which is used to quantify DNA, may also be used for the identification of *Toxoplasma* DNA in animal tissues.[4]

While Microsporidia have not been identified as the cause of a foodborne outbreak, they have been isolated from fresh produce, indicating their potential to cause illness through consumption of this commodity. In studies on produce, nested PCR, which includes two rounds of PCR and the further amplification of the first amplified sequence product, is several orders of magnitude more sensitive than standard single PCR.[5] Studies aimed at animal health may impact food safety. For example, animals have been screened for Microsporidia using a discriminatory PCR-RFLP (Restriction Fragment Length Polymorphism) assay.[12] This test uses restriction enzymes which recognize and excise specific DNA sequences and amplify that product of specific size. There is a potential that this type of test could be applied for surveillance and testing of food animals or foods at the pre- or postprocessing stage if necessary.

RAPID DETECTION OF NON-PROTOZOAN PARASITES

Non-protozoan parasites include myriad organisms, such as nematodes (roundworms) and trematodes (flukes). The majority of rapid tests available to detect these organisms is designed to evaluate animal health and is not specifically designed for food safety. Rapid diagnosis in animals using immunoassays or PCR may be performed for *Trichinella* and a variety of fish and animal nematodes or flukes, including *Fasciola* species. Combinations of PCR, RFLP, and dot blots (an immunoassay) have been assessed for detection of adult life stages, larvae, and eggs in animal tissues. A diverse number of assays have been applied to detection of this group of organisms including a DNA microarray for *Ascaris suum*.[6]

The well-known zoonotic parasite, *T. spiralis*, is no longer a problem in conventionally produced pork because of the success of the U.S. Trichinae Herd Certification Program, a preharvest pork safety program that uses documentation of swine management practices to minimize risk of exposure of swine to *Trichinella*. Before this program was in effect, an enzyme-linked immunosorbant assay (ELISA) was used extensively for testing at both pre- and postslaughter levels, and as a tool for determining or monitoring infection in herds.[7] In the early stages of the implementation of the Herd Certification Program, the ELISA test provided many advantages over visual macro- or microscopic inspection in terms of time and high-throughput analysis.

Unfortunately, the parasite still causes illness in humans from the consumption of infected undercooked wild boar, bear, and cougar. Testing for *Trichinella* species is based on serology (the presence of antibodies in the infected host), but the use of these tests to detect *Trichinella* infection in both animals and humans has not yet been standardized.[8] One example is a relatively new coagglutination test that is more useful compared to the traditional ELISA test in detecting low amounts of *Trichinella* in the urine of infected laboratory animals,[9] and could be useful in diagnosis of infection in humans and other animals. Another novel immunoassay, called a lateral card test, may be most useful in detecting infected carcass, blood, or tissue samples in parts of the world where *Trichinella* is widespread. Extensive studies performed in Romania during 1999–2000 showed this test to be highly specific, sensitive, rapid (3–12 min), and easy to use with no need for laboratory facilities.[10]

Fishborne trematode infections are a public health problem in parts of Asia where use of rapid methods, as opposed to conventional visual inspection, may help fish inspectors and public health workers to determine the safety of fish for human consumption. The largest obstacle in the development of antibody-based tests is identifying a strong antigen of the parasite in the correct life stage found in the infected fish.[11] For example, the eggs of flukes may contaminate aquaculture systems, and thereby infect other animals. While the adult fluke is highly immunogenic it is the metacercaria (larvae) which infect the fish and are less likely to react with an antibody assay, whether it be immunoblot assay or ELISA.

CONCLUSION

Undoubtedly, diagnostic laboratories need more precise rapid methods for the detection of all types of parasites. While many tests based on nucleic acid detection are available for use on clinical samples, these tests are not easily adapted to food samples. Current research is underway to develop improved methods of detection that will overcome the problems associated with concentration and detection owing to interference from food components. It is likely that this area of research will progress with the use of microarrays and biosensors, which have already shown potential in the water industry; however, methods of detection of these pathogens in foods will have to undergo extensive development and examination before reaching the breadth of use of rapid diagnostics for parasites now seen in medical fields.

REFERENCES

1. Caccio, S.M. Molecular techniques to detect and identify protozoan parasites in the environment. Acta Microbiol. Pol. **2003**, *52* (Suppl.), 23–34.
2. Mota, P.; Rauch, C.A.; Edberg, S.C. *Microsporidia* and *Cyclospora*: epidemiology and assessment of risk from the environment. Crit. Rev. Microbiol. **2000**, *26*, 69–90.
3. Kniel, K.E.; Jenkins, M.C. Detection of *Cryptosporidium parvum* oocysts on fresh vegetables and herbs using antibodies specific for a *C. parvum* viral antigen. J. Food Prot. **2005**, *68*, 1093–1096.
4. Jauregui, L.H.; Higgins, J.; Zarlenga, D.; Dubey, J.P.; Lunney, J.K. Development of a real-time PCR assay for detection of *Toxoplasma gondii* in pig and mouse tissues. J. Clin. Microbiol. **2001**, *39*, 2065–2071.
5. Bell, A.S.; Yokoyama, H.; Aoki, T.; Takahashi, M.; Maruyama, K. Single and nested polymerase chain reaction assays for the detection of *Microsporidium seriolae* (Microspora), the causative agent of 'Beko' disease in yellowtail *Seriola quinqueradiata*. Dis. Aquat. Organ. **1999**, *37* (2), 127–134.
6. Rhoads, M.L.; Fetterer, R.H.; Urban, J.F., Jr. Cuticular collagen synthesis by *Ascaris suum* during development from the third to fourth larval stage: identification of a potential chemotherapeutic agent with a novel mechanism of action. J. Parasitol. **2001**, *87*, 1144–1149.
7. Gamble, H.R. Detection of trichinellosis in pigs by artificial digestion and enzyme immunoassay. J. Food Prot. **1996**, *59*, 295–298.
8. Gamble, H.R.; Pozio, E.; Bruschi, F.; Nockler, K.; Kapel, C.M.; Gajadhar, A.A. International Commission on Trichinellosis: recommendations on the use of serological tests for the detection of *Trichinella* infection in animals and man. Parasite **2004**, *11* (1), 3–13.
9. Eissa, M.M.; El-Mansoury, S.T.; Allam, S.R. Co-agglutination (Co-A): a rapid test for the diagnosis of experimental trichinellosis. J. Egypt Soc. Parasitol. **2003**, *33* (2), 637–645.
10. Patrascu, I.; Gamble, H.R.; Sofronic-Milosavljevic, L.; Radulescu, R.; Andrei, A.; Ionescu, V.; Timoceanu, V.; Boireau, P.; Cuperlovic, K.; Djordjevic, M.; Murrell, K.D.; Noeckler, K.; Pozio, E. The lateral flow card test: an alternative method for the detection of *Trichinella* infection in swine. Parasite **2001**, *8* (2 Suppl.), S240–S242.
11. Rim, H. Field investigations on epidemiology and control of fish-borne parasites in Korea. Int. J. Food Sci. Technol. **1998**, *33*, 157–168.
12. Hogg, J.C.; Ironside, J.E.; Sharpe, R.G.; Hatcher, M.J.; Smith, J.E.; Dunn, A.M. Infection of *Gammarus duebeni* populations by two vertically transmitted microsporidia; parasite detection and discrimination by PCR-RFLP. Parasitology **2002**, *125* (Pt 1), 59–63.

Pulsed Electric Field Processing: Food Preservation

Huub L.M. Lelieveld
Unilever Research (Retired), Bilthoven, The Netherlands

Abstract

To have food available the year round for everybody, preservation is essential. Traditional preservation methods, such as drying, salting, acidifying, fermenting, and heat treatments, have been very successful for thousands of years and will probably continue to be of great value. These methods, however, also have disadvantages; to various degrees these processes affect color, flavor, and the concentration of nutrients. A technology invented in the 1960s, pulsed electric field treatment (PEF), offers the possibility to preserve food without these drawbacks. Because PEF can destroy vegetative bacteria, but not bacterial spores, it may replace traditional pasteurization, but not sterilization. PEF thus may be suitable for many products that are acidic or have a low water activity, and also for refrigerated low-acid products with a limited shelf life.

INTRODUCTION

Pulsed electric fields (PEF) may destroy microorganisms to the same effect as thermal pasteurization. This entry will discuss the inactivation of microorganisms by PEF and the influence of PEF on enzymes, color, flavor, nutrients, and structure. It will also discuss product safety aspects and a PEF process line.

INACTIVATION OF MICROORGANISMS AND ENZYMES

When vegetative microbial cells are exposed to electric fields with a strength of approximately 2 V/μm for a few microseconds, their membranes are perforated and the cells start to leak, leading to death.[1–3] Conditions for inactivation vary widely.[4] The higher the initial temperature, the more efficient the inactivation,[5] probably because of the increase in fluidity of the microbial membrane. The product to be treated with PEF should therefore be heated to the optimum for PEF inactivation. Bacterial spores are resistant to PEF, making PEF unsuitable for sterilization. Ascospores of molds and yeasts also show extreme resistance. Although ascospores are not a common source of microbial spoilage, it must be realized that ascospores of *Byssochlamys* and *Neosartorya* can be the only survivors relevant to high-acid products. PEF conditions aimed at destroying microorganisms have no significant effect on proteins, and thus on enzymes. Even under extreme conditions, there is no complete inactivation of enzymes.[6,7] PEF is unsuitable for products where enzymes can adversely affect product quality during the intended storage time. On the other hand, where enzyme preparations must be preserved or freed from vegetative microorganisms, PEF may be the ideal process.

INFLUENCE ON PRODUCT QUALITY

The change in color due to PEF treatment has been investigated for many products, including liquid whole egg, fruit and vegetable juices (e.g., orange, apple, and carrot), and yoghurt. Initially after treatment there is no change in color, and where the color does change with storage time, the changes are less than with thermally pasteurized products.[8] Generally, PEF does not influence flavor and taste. If any change after PEF treatment is observed, it can be attributed to the storage conditions (affecting products independent on the processing received) or to the growth of microorganisms (not all microorganisms may have been inactivated). Apple and orange juice retain their fresh taste for a long time if stored refrigerated (~4°C). Also the flavor and taste of milk products, egg products, and green pea soup are not adversely affected by PEF.[8]

Structure

PEF, like heating, affects cell membranes and thus may affect texture. The degree of change is a function of various process parameters such as field strength, number of pulses, length of pulses, and temperature. Most eukaryotic cells are very sensitive to PEF even under conditions that do not affect prokaryotic organisms.[9] PEF-induced cell damage may be desirable for vegetables that otherwise need boiling for some time to become digestible while destroying much of the nutrients in the process. Because heat causes coagulation of proteins, preservation by PEF may be of particular interest for eggs and liquid egg products and for meat- and fish-based products if the texture is not important issue.

Nutrients

Although little research has been done in this area, data available confirm that PEF under the conditions applied for microbial inactivation does not adversely affect nutrients, such as vitamin C, riboflavin, thiamine, cholecalciferol, and tocopherol.[8] The thermal processing of fruit juices reduces the bioavailability of flavonoids. Because PEF permeates cell membranes, the availability of intracellular nutrients such as lycopenes and flavonoids to the human body may be improved by replacing thermal pasteurization by PEF treatment. Also, while heat destroys desirable enzymes such as myrosinase, it is reasonable to assume that PEF leaves these enzymes intact.[8]

PRODUCT SAFETY

Microbiology

PEF is capable of inactivating vegetative microorganisms, including pathogens. Depending on the applied conditions, PEF may not eliminate all relevant microorganisms, but at least it should reduce the viable counts of pathogens sufficiently to render the product safe during the shelf life of the product under appropriate storage conditions. Confirming data have been produced for species of *Listeria*, *Salmonella*, *Escherichia* (including *E. coli* O157:H7), *Klebsiella*, *Pseudomonas*, *Staphylococcus*, and *Candida*.

Toxicology

Contrary to heat treatment, PEF, using conditions aimed at inactivation of microorganisms, does not or nominally denatures proteins. The protein profile of PEF-processed food products will therefore differ from that of heat-treated products. PEF-treated foods will not differ from the protein profile of the raw materials; however, heat treatment may be needed to destroy allergic or toxic proteins, or other undesirable heat-sensitive constituents, present in some crops. Therefore, in considering replacement of heat treatment by PEF, all potential effects of heat on product safety aspects must be taken into account. This does not apply if the product receives sufficient heat treatment (e.g., cooking) before consumption. Proton NMR fingerprinting has showed minor differences between the concentrations of a range of compounds in tomato purees treated with heat or PEF.[10] Therefore, depending on the application, it may be prudent to consider whether there may be any differences that may influence product safety. If so, it should be investigated whether these changes are substantial and, thus, are

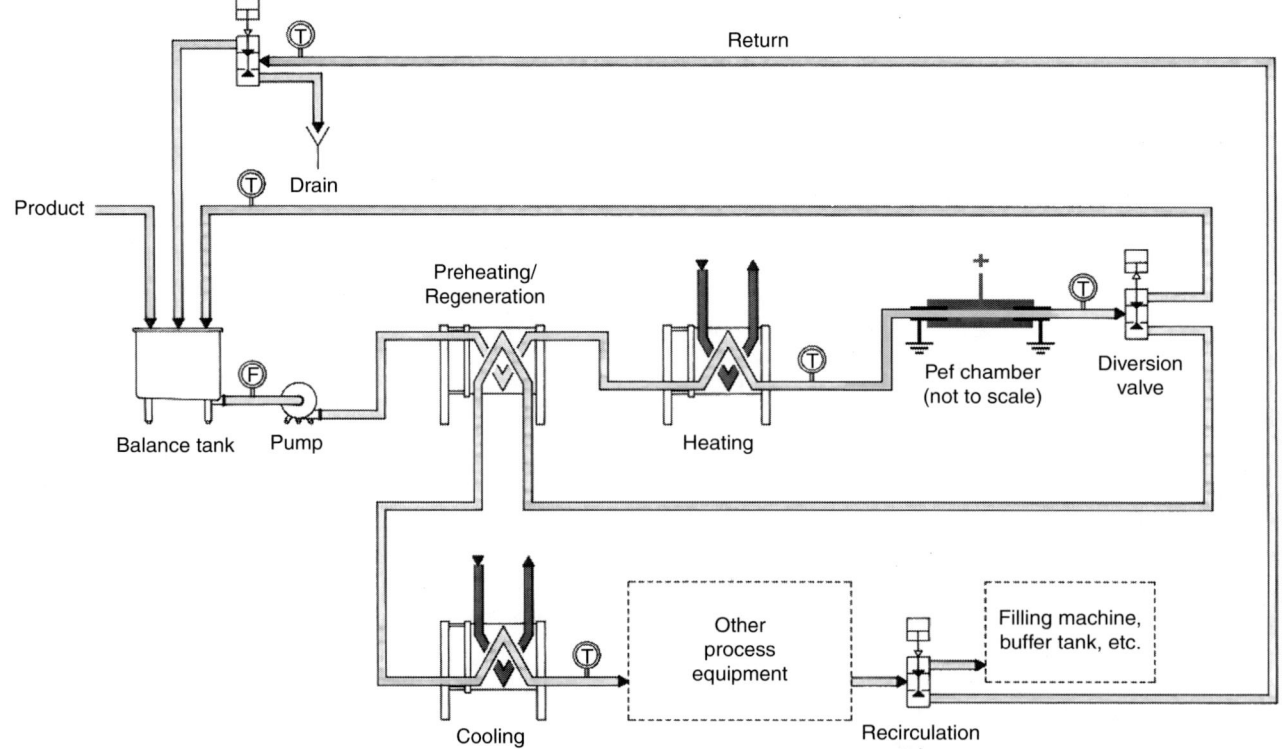

Fig. 1 Process line for liquid food where the holding section of the pasteurizer has been replaced by a PEF treatment chamber.

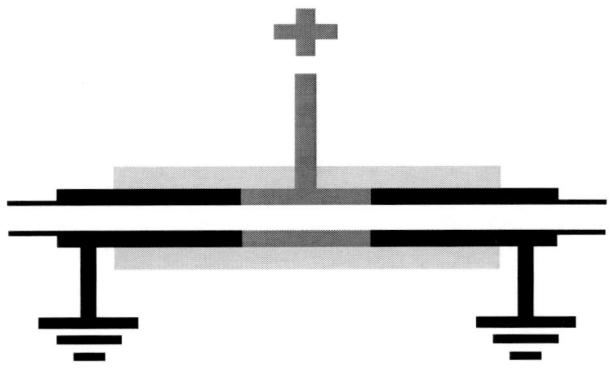

Fig. 2 PEF chamber, *black = steel, earthen, dark grey = high-voltage electrode, light grey = electric insulation, white (between the lines) = product*. Dimensions: Diameter is the same as that of the connecting pipeline and the total length maximum four times the diameter.

a reason for concern. Based mainly on theoretical considerations and calculations, taking into account field strength, current, time, polarity, and chemical composition of food products, FDA decided that there is no objection to PEF treatment of food for preservation purposes.[11]

PROCESSING ASPECTS

A process line for PEF treatment may be as shown in Fig. 1. In this example the product to be treated is preheated to the optimum temperature for the product, e.g., 40°C. It then passes the PEF chamber after which the product is cooled down to the desired end temperature for further processing or packaging. The holding section of the heat exchanger has been replaced by the PEF chamber. In reality, more changes are needed; in the case of a thermal treatment, the process line downstream from the holding section of the heat exchanger is pasteurized by the start-up liquid, usually water. In this manner it is ascertained that microorganisms that may have contaminated the process line from that point will not recontaminate the pasteurized product. To prevent contamination of the line downstream from the PEF chamber, start-up must be done with a vegetative-microbe-free solution with the same conductivity as the product to be treated. This may just be a thermally pasteurized aqueous solution of salt.[12] Alternatively such a solution may be freed from microorganisms by filtration.

Fig. 2 shows the cofield PEF chamber of Fig. 1 in detail. The chamber is based on the design of the Ohio State University.[13] The chamber consists of two separate parts so that the piping system can be connected to earth on both sides of the assembly, eliminating electrical shock hazards. For liquids with a relatively high conductivity, it may be best to use two of these assemblies in series, so that there are four chambers. The chambers can be connected electrically in such a way that the effect of ohmic heating during the PEF treatment and its influence on the product conductivity is used to improve the efficiency of the process. In a horizontal chamber, warm liquid product moves up and cool product moves downwards. Warm product is more conductive and thus heats faster; the cooler product has a lower conductivity and thus heats slower. The temperature difference increases and so does the movement of the product. The result is a non-uniform process that delivers insufficient treatments to different areas of the product and thus adversely affects product shelf life and safety. This effect disappears by placing the chamber in a vertical position.[8] Although possible, cooling between chambers should be avoided: it is clumsy and increases costs. Cooling becomes necessary if the energy dissipation is too high due to a too high conductivity. Such a product is not suitable for PEF.

Energy Requirements

An effective PEF treatment of an average liquid product, such as orange juice or milk, would require 150 J/mL of product. With correct design, ensuring that *all* products receive the intended treatment the energy requirement may even be lower.

CONCLUSION

For many products PEF may provide a serious alternative to thermal pasteurization. Compared to their heat-pasteurized equivalents, PEF-treated products generally possess a fresher taste and have retained the color and nutrient content of the raw product.

REFERENCES

1. Sale, A.J.H.; Hamilton, W.A. Effect of high electric fields on microorganisms: I. Killing of bacteria and yeast. Biochim. Biophys. Acta **1967**, *148*, 781–788.
2. Hamilton, W.A.; Sale, A.J.H. Effects of high electric fields on microorganisms: II. Killing of bacteria and yeasts. Biochim. Biophys. Acta **1967**, *148*, 789–800.
3. Sale, A.J.H.; Hamilton, W.A. Effect of high electric fields on microorganisms: III. Lysis of erythrocytes and protoplasts. Biochim. Biophys. Acta **1968**, *163*, 37–43.
4. Martín-Belloso, O.; Bendicho, S.; Elez-Martínez, P. Does high-intensity pulsed electric fields induce changes in enzymatic activity, protein confirmation, and vitamin and flavor stability? In *Novel Food Processing Technologies*; Barbosa-Cánovas, G.V., Tapia, M.S., Cano, M.P., Eds.; Marcel Dekker/CRC Press: New York, 2005; 87–105.
5. Wouters, P.C.; Dutreux, N.; Smelt, J.P.P.M.; Lelieveld, H.L.M. Effects of pulsed electric fields on inactivation kinetics of *Listeria innocua*. Appl. Environm. Microbiol. **1999**, *65*, 5364–5371.
6. Elez-Martínez, P.; Martín-Belloso, O.; Rodrigo, D.; Sampedro, F. Impact of pulsed electric fields on food enzymes and shelf life. In *Food Preservation by Pulsed Electric*

Fields – From Research to Application; Lelieveld, H.L.M., Notermans, S., De Haan, S.W.H., Eds.; Woodhead/CRC Press, 2007; 212–248.
7. Van Loey, A.; Verachtert, B.; Hendrickx, M. Effects of high electric field pulses on enzymes. Trends Food Sci. Technol. **2002**, *12*, 94–102.
8. Lelieveld, H.L.M. PEF – a food industry's view. In *Novel Food Processing Technologies*; Barbosa-Cánovas, G.V., Tapia, M.S., Cano, M.P., Eds.; Marcel Dekker/CRC Press: New York, 2005; 145–156.
9. Gudmundsson, M.; Hafsteinsson, H. Effect of electric field pulses on microstructure of muscle foods and roes. Trends Food Sci. Technol. **2001**, *12*, 122–128.
10. Lelieveld, H.L.M.; Wouters, P.C.; Léon, A.L. Pulsed electric field treatment of food and product safety assurance. In *Pulsed Electric Fields in Food Processing*; Barbosa-Cánovas, G.V.; Zhang, Q.H., Eds.; Technomic Publ. Co.: Lancaster, PA, 2001; 259–264.
11. Dunn, J. Pulsed electric field (CoolPure) cool pasteurization for milk, juices, and pumpable food. In *Tecnologías avanzadas en esterilización y seguridad de alimentos y otros productos*; Rodrigo, M., Martínez, A., Fiszman, S.M., Rodrigo, C., Mateu, A., Eds.; CSIC: Burjasot, Valencia, Spain, 1995; 293–296.
12. Smit, C.; De Haan, W. Hygienic design for pulsed electric field installations. In *Food Preservation by Pulsed Electric Fields – From Research to Application*; Lelieveld, H.L.M., Notermans, S., de Haan, S.W.H., Eds.; Woodhead/CRC Press; 108–117.
13. Yin, Y.; Zhang, Q.H.; Sastry, S.K. 1997. High voltage pulsed electric field treatment chambers for the preservation of liquid food products. United States Patent 5,690,978.

RNAs: Small Inhibitory

Ramanjulu Sunkar
Department of Biochemistry and Molecular Biology, Oklahoma State University, Stillwater, Oklahoma, U.S.A.

Abstract

Studies of endogenous small RNAs have uncovered their important role in regulating gene expression. In plants, unlike in animals, the endogenous small RNA component is highly complex and diverse, which can be attributed to the presence of multiple Dicer-like proteins (double-stranded RNA [dsRNA] binding proteins), RNA-dependent RNA polymerases (RdRps) and Argonaute proteins (AGOs), the proteins implicated in small-RNA biogenesis and function. Endogenous small RNAs can be divided into two major classes, microRNAs (miRNAs) and small-interfering RNAs (siRNAs). MicroRNAs (∼21 nt) arise from a long single-stranded RNA that can adopt a hairpin-like structure, whereas siRNAs (∼21–24 nt) are derived from long double-stranded RNAs. MicroRNAs, as well as *trans*-acting siRNAs (a subclass of endogenous siRNAs), can direct cleavage of the target mRNA and regulate gene expression at the posttranscriptional level, whereas 24-nt siRNAs can guide DNA and histone modifications and cause transcriptional gene silencing. Endogenous small RNAs play important roles in plant growth and development, but emerging evidence also implicates a role in biotic and abiotic stress resistance.

INTRODUCTION

Normal plant growth and development depend on the timely production of specific types of cells and organs at destined locations, which in turn depends on tightly controlled *spatial-* and *temporal*-gene expression programs. The involvement of proteins in gene regulatory processes is well known, and with the discovery of two classes of small regulatory RNAs, miRNAs and siRNAs, the importance of small RNAs in gene regulation at the transcriptional and posttranscriptional levels is now also widely recognized.[1,2] The underlying theme of small RNA-guided gene regulation is that it relies on specific interactions between small RNAs and mRNAs leading to posttranscriptional gene silencing or between small RNAs and DNA leading to transcriptional gene silencing. Here, the biogenesis of miRNAs and siRNAs and their involvement in plant development and other diverse physiological processes is discussed.

PROTEINS CRITICAL FOR SMALL-RNA BIOGENESIS AND FUNCTION

Genetic and biochemical dissection of small RNA-guided gene silencing pathways led to the identification of three key components—Dicers, AGOs, and RdRps—that are implicated in small-RNA biogenesis and function. Genome-wide sequencing projects have revealed that DCL (Dicer-like) and AGO proteins are evolutionarily conserved in a wide range of eukaryotic genomes.[3] The *Arabidopsis* genome encodes four DCL (DCL-1, DCL-2, DCL-3, and DCL-4) enzymes, which are essential for small-RNA biogenesis.[4] DCL-1 functions mainly in miRNA biogenesis, whereas DCL-2, DCL-3, and DCL-4 function in generating different types of siRNAs (Fig. 1). DCL-1 enzyme specifically acts on miRNA precursor transcripts that can adopt hairpins, liberating a duplex of 21-nt mature miRNAs and miRNA* (opposite strand of functional miRNA is referred to as miRNA*). DCL-2, DCL-3, and DCL-4 act on perfectly paired dsRNAs and release 21- to 24-nt siRNAs. RdRps are critical players in siRNA biogenesis, especially in converting single-stranded RNA (ssRNA) into dsRNA.[3] The *Arabidopsis* genome encodes six RdRps. Certain noncoding RNAs, such as *trans*-acting siRNA (ta-siRNA) precursor transcripts are subjected to miRNA-guided cleavage and the cleaved fragment serves as template for RdRp (RDR6) activity. Also, aberrant RNAs derived from transgenic plants overexpressing extra copies of a transgene frequently serve as templates for RdRp activity. MicroRNA and endogenous siRNA duplexes in plants are methylated by HEN-1, a methyl transferase and thereby protects these small RNA duplexes from the endogenous exonuclease activity.[4] AGO proteins are defined by a PAZ (Piwi, Argonaut, and Zwille) domain that binds to the 3′ end of the ssRNA and PIWI (P-element induced wimpy testis) domain with a structure similar to that of RNase H. AGO proteins are core components of RNA-induced silencing complex;

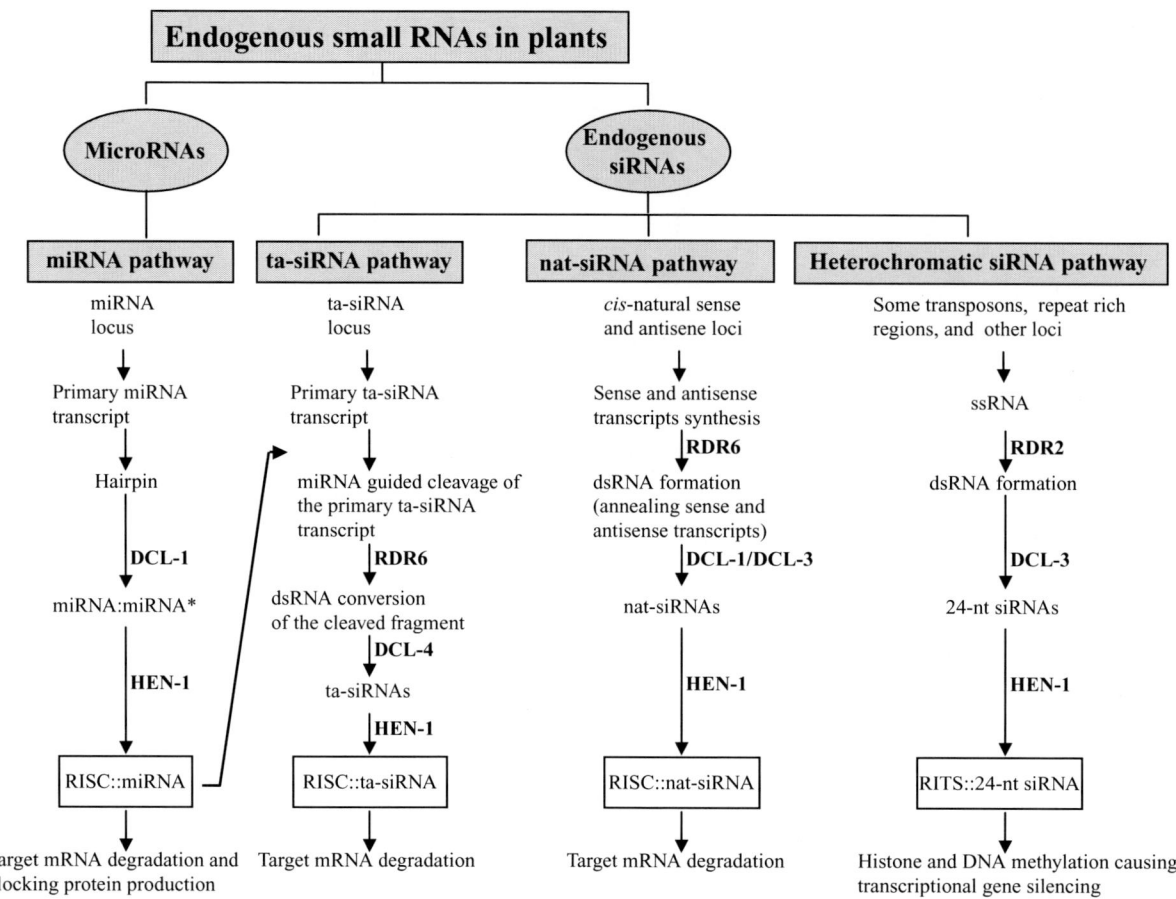

Fig. 1 Endogenous small RNA pathways in plants.

they possess the "slicer" activity and can direct the cleavage of mRNA at the site of mi/siRNA complementarity. The *Arabidopsis* genome encodes 10 AGO genes.[3] AGO1 functions in the miRNA pathway, and AGO4, AGO7, and AGO10 seem to be involved in the siRNA pathway.

MicroRNAs

MicroRNAs are an abundant class of single-stranded ~21-nt non-coding RNAs and are linked to almost all biological processes in multicellular eukaryotes and thus emerged as critical regulators of gene expression. The first miRNA, *lin-4*, was identified in *Caenorhabditis elegans*.[5] Later on, *lin-4* and *let-7* (a second miRNA also discovered in *C. elegans*) were found to be evolutionarily conserved from flies to humans, implying a more universal role for these genes in animals. This finding paved the way for the discovery of more miRNAs in diverse animals and in plants. MicroRNAs affect protein synthesis through base pairing with complementary mRNA sequences, resulting in cleavage of the target mRNA or prevention of protein translation. Whether the degradation of target mRNA or inhibition of protein synthesis occurs appears to depend on the degree of complementarity between the miRNA and mRNA. High complementarity, as observed in plants, leads to target mRNA degradation, and partial complementarity as observed in animals, results in attenuation of protein synthesis.[6]

miRNA HOMEOSTASIS IN PLANTS

Mutant plants defective in miRNA biogenesis (*dcl1*, *hen1*, and *hyl1*) and miRNA function (*ago1*) have striking developmental anomalies possibly caused by reduced miRNA accumulation and a concomitant increase in target mRNAs. Overexpression of several miRNAs also resulted in developmental abnormalities, possibly owing to depletion of miRNA targets.[1] These observations showed that a tightly controlled level of miRNA is critical for normal growth and development.

ROLES OF miRNAs IN PLANTS

The miRNA component in *Arabidopsis* is the most intensely studied, and currently, ~85 miRNA families are known (http://www.miRBase.com) in this model plant; however, only ~21 of these miRNA families are conserved between dicots and monocots. Many conserved miRNA families are targeting mRNAs encoding transcription factors implicated in diverse developmental aspects such as: auxin signaling; meristem boundary formation and organ separation; leaf development and polarity; seedling development; embryo development; phyllotaxy; lateral root formation; stomatal development; transition from juvenile to adult vegetative phase and from vegetative to flowering phase; floral organ identity; petal number; and reproduction.[1,2] Some other conserved miRNA families are targeting mRNAs encoding enzymes such as Cu/Zn-superoxide dismutases, laccases, ATP sulfurylases, and ubiquitin-conjugating enzyme or transporters such as phosphate and sulfate transporters and other proteins such as plantacyanin.[1,2] Furthermore, external environment affects the expression of a few miRNAs; the expression of miR398 is markedly decreased with oxidative stress; miR393 is induced in response to pathogen attack (*Pseudomonas syringae*); two other miRNAs, miR395 and miR399, are induced under conditions of low sulfate and low phosphate, respectively.[2] Thus, understanding miRNA-guided target gene regulation has implications for improving biotic and abiotic stress tolerance in crop plants.[2]

High-throughput sequencing of small-RNA libraries from *Arabidopsis*, rice, and other plants led to the discovery of many lineage-specific (monocot-specific or dicot-specific) and non-conserved *Arabidopsis*- or rice-specific miRNAs.[1,7] The lineage-specific miRNAs might have lineage-specific roles. Whether non-conserved *Arabidopsis*- or rice-specific miRNAs play important species-specific roles remains to be investigated.

ENDOGENOUS siRNAs

Endogenous siRNAs are derived from the processing of perfectly paired dsRNAs by the activity of the DCL family of enzymes (DCL-2, DCL-3, and DCL-4). In plants, siRNAs can be divided into two subclasses on the basis of their size and function: the 21-nt siRNAs (ta-siRNAs and natural antisense transcript-derived siRNAs [nat-siRNAs]) that cause target mRNA degradation (posttranscriptional gene silencing) and the 24-nt siRNAs (heterochromatic siRNAs) that can trigger methylation of homologous DNA leading to transcriptional gene silencing (Fig. 1).

trans-Acting siRNAs

Like miRNAs, ta-siRNAs are 21-nt long small RNAs and downregulate target mRNAs at the posttranscriptional level. The biogenesis of ta-siRNAs depends upon the miRNA pathway. ta-siRNA primary transcripts are initially targeted by miRNA for cleavage, and the RDR6 converts the cleaved fragment into dsRNA. Subsequently, DCL-4 recognizes and processes such dsRNAs into 21-nt ta-siRNAs. In *Arabidopsis*, TAS3 locus-derived ta-siRNAs target three auxin response factor (ARF) genes (ARF2, ARF3, and ARF4) and this targeting of ARF genes by Tas3-siRNAs is important for normal leaf development.[8]

nat-siRNAs

cis-Natural antisense transcripts (*cis*-NATs) are widespread in plant and animal genomes, and a significant number of such *cis*-NAT gene pairs seem to be involved in generation of siRNAs, so are referred to as nat-siRNAs (Fig. 1). The founding member of this type of siRNA was derived from a pair of *P5CDH* and *SRO5* overlapping genes and plays an important role in salt stress tolerance.[9] A second example is nat-siRNA-ATGB2, which is induced by the bacterial pathogen *P. syringae* and plays an important role in plant immunity.[2] It is tempting to speculate that many such nat-siRNAs may exist, which play regulatory roles important for plant adaptation to diverse stress conditions.

HETEROCHROMATIC siRNAs

Cytosine DNA methylation serves important roles in gene regulation and silencing of transposons or other repetitive elements. The bulk of DNA methylation in mammals is confined to the CG context. Similarly, DNA methylation in plants affects cytosine residues preceding guanines (CG), but plant cytosines are also modified at CNG (where N is any nucleotide) and asymmetrical sequence contexts (CHH) (where H could be A, C, or T but not G). These complex methylation patterns provide increased combinatorial power to the "DNA methylation code."[10] DNA methylation appears to be largely guided by 24-nt siRNAs in plants, also known as "heterochromatic siRNAs." Most of these siRNAs could be mapped to retroelements, transposons, repeat-rich regions, and heterochromatic regions but only rarely to the regions that code for genes. Genomic regions corresponding to siRNAs were also the regions with methylated cytosines in all contexts (CG, CHG, and CHH), a known mark of RNA-directed DNA methylation. Different pathways control the maintenance and establishment of cytosine methylation in plants.

CONCLUSIONS

Many miRNA families are conserved in higher plants, and some of their regulatory roles seem to be very ancient because they are found in primitive land plants such as moss. Conserved miRNAs are playing very important regulatory roles in plant growth and development and in plant stress responses. Recently, artificial miRNAs (amiRNAs) and artificial ta-siRNAs have been devised on the basis of the endogenous small RNA characteristics.[11,12] These amiRNAs or artificial ta-siRNAs seem to have notable specificity, specifically targeting a particular member of a gene family.[11,12] Thus, knowledge of small RNA-guided gene regulation not only broadens our basic understanding of posttranscriptional gene regulation but also has the potential for other biotechnological approaches such as silencing genes of interest in diverse plants.

REFERENCES

1. Jones-Rhoades, M.W.; Bartel, D.P.; Bartel, B. MicroRNAs and their regulatory roles in plants. Annu. Rev. Plant Biol. **2006**, *57*, 19–53.
2. Sunkar, R.; Chinnusamy, V.; Zhu, J.; Zhu, J.K. Small RNAs as big players in plant abiotic stress responses and nutrient deprivation. Trends Plant Sci. **2007**, *12* (7), 301–309.
3. Herr, A.J. Pathways through the small RNA world of plants. FEBS Lett. **2005**, *579* (26), 5879–5888.
4. Li, J.; Yang, Z.; Yu, B.; Liu, J.; Chen, X. Methylation protects miRNAs and siRNAs from a 3′-end uridylation activity in Arabidopsis. Curr Biol. **2005**, *15* (16), 1501–1507.
5. Lee, R.C.; Feinbaum, R.L.; Ambros, V. The *C. elegans* heterochronic gene lin-4 encodes small RNAs with antisense complementarity to lin-14. Cell **1993**, *75* (5), 843–854.
6. Bartel, D.P. MicroRNAs: genomics, biogenesis, mechanism, and function. Cell **2004**, *116* (2), 281–297.
7. Sunkar, R.; Zhou, X.; Zheng, Y.; Zhnag, W.; Zhu, J.K. Identification of novel and candidate miRNAs in rice by high throughput sequencing. BMC Plant Biol. **2008**, *8*, 25
8. Axtell, M.J.; Jan, C.; Rajagopalan, R.; Bartel, D.P. A two-hit trigger for siRNA biogenesis in plants. Cell **2006**, *127* (3), 565–577.
9. Borsani, O.; Zhu, J.; Verslues, P.E.; Sunkar, R.; Zhu, J.K. Endogenous siRNAs derived from a pair of natural *cis*-antisense transcripts regulate salt tolerance in Arabidopsis. Cell **2005**, *123* (7), 1279–1291.
10. Vaillant, I.; Paszkowski, J. Role of histone and DNA methylation in gene regulation. Curr. Opin. Plant Biol. **2007**, *10* (5), 528–533.
11. Ossowski, S., Schwab, R., Weigel, D. Gene silencing in plants using artificial microRNAs and other small RNAs. Plant J. **2008**, *53* (4), 674–590.
12. Gutiérrez-Nava, M.D.; Aukerman, M.J.; Sakai, H.; Tingey, S.V.; Williams, R.W. Artificial *trans*-Acting siRNAs confer consistent and effective gene silencing. Plant Physiol. **2008**, *147*, 543–551.

Secondary Metabolites: Plant Cell and Hairy Root Cultures

Iryna Smetanska
Department of Methods in Food Technology, Berlin University of Technology, Berlin, Germany

Abstract
Plant cell cultures represent a potential source of valuable metabolites, which can be used as food additives, nutraceuticals, and pharmaceuticals. The problems in obtaining phytochemicals from plants include environmental factors, uncontrollable variations in the crop quality, inability of authorities to prevent crop adulteration, losses in storage and handling. In many cases the chemical synthesis of these is either extremely difficult or economically infeasible. Moreover, the natural food additives are better accepted by consumers in contrast to those artificially produced. The synthesis of phytochemicals by the cell cultures in contrast to these in plants is not depending on environmental conditions and quality fluctuations.

APPLICATION OF PLANT CELL CULTURES

Plant cell culture has commercial applications as well as value in basic research into cell biology, genetics, and biochemistry.

The application of plant cell culture has three main aspects:

1. **Breeding and genetics:**
   - **micropropagation**—using meristem and shoot culture to produce large numbers of identical individuals;
   - **selection**—screening of cells, rather than plants for advantageous characters;
   - crossing distantly related species by **protoplast fusion** and regeneration of the novel hybrid;
   - production of dihaploid plants from **haploid cultures** to achieve homozygous lines;
   - **transformation**, followed by either short-term testing of genetic constructs or regeneration of transgenic plants;
   - **removal of viruses** by propagation from meristematic tissues;
2. **Model system** for study of plant cell genetics, physiology, biochemistry, and pathology;
3. **Production of secondary metabolites**—growth in liquid culture as a source of products.

APPLICATION FOR PRODUCTION OF SECONDARY METABOLITES

Cultured plant cells often produce different quantities with different profiles of secondary metabolites when compared with the intact plant and these quantitative and qualitative features may change with time.

As shown in Table 1, some metabolites in plant cell cultures can be accumulated with a higher titer compared with those in the parent plants, suggesting that the production of plant-specific metabolites by plant cell culture instead of whole plant cultivation possesses definite potential.

The most valuable food additives, which can be obtained from the plant cell cultures, are food colorants (anthocyanins and betalaines), flavors (saffron and vanillin), sweeteners (steviosides), pungent food additives (capsaicin), and anti-bacterial food preservatives (thiophene).

ADVANTAGES AND DISADVANTAGES OF PLANT CELL CULTURES

The advantages of plant cell cultures over the conventional production are as follows:

1. It is independent of geographical and seasonal variations and environmental factors—the synthesis of bioactive secondary metabolites runs in controlled environments and the negative biological influences that affect secondary metabolites production in nature are eliminated;
2. It offers a defined production system, which ensures the continuous supply of products, uniform quality, and yield;
3. It is possible to select cell lines with higher production of secondary metabolites;
4. It is possible to produce novel compounds that are not normally found in parent plant;

Table 1 Product yield from plant cell cultures compared with the parent plants.

		Yield (% dry weight)			
Product	Plant	Culture	Plant	Culture/Plant	Reference
Ajmalicine	*Catharanthus roseus*	1.0	0.3	3.3	[1]
Caffeic acid	*Vanilla planifolia*	0.02	0.05	4	[2]
Ginsenoside	*Panax ginseng*	27	4.5	6	[3]
Rosmarinic acid	*Coleus blumei*	27	3	9	[4]
Shikonin	*Lithospermum erythrorhizon*	20	1.5	13.5	[5]
Ubiquinone-10	*Nicotiana tabacum*	0.036	0.003	12	[6]

5. It allows the efficient downstream production;
6. Plant cell can perform stereo- and regio-specific biotransformations for the production of novel compounds from cheap precursors.

However, this technology is still being developed and despite the advantages, there are a variety of problems to be overcome before it can be adopted for the production of useful plant secondary metabolites.

OBSTACLES FOR THE CELL CULTURES

The problems with the plant cell cultures can be biological (slow growth rate, physiological heterogeneity, genetic instability, low metabolite content, and product secretion) and operational (wall adhesion, light requirement, mixing, shear sensitivity, and aseptic condition).

The large size of the plant cell contributes to its comparatively high doubling time, which thus prolongs the time required for a successful fermentation run. The vacuole is the major site of product accumulation, and since product secretion is uncommon, the high metabolite yields seen in microorganisms that secrete product cannot be expected.

STRATEGIES TO INCREASE SECONDARY METABOLITE PRODUCTION

The strategy for obtaining the secondary metabolites from the plant cell cultures can be represented as a multistage process (Fig. 1). Each link of it may be optimized, separately or in combination with other processes or treatments.

1. The initial step of this technology consists in the selection of the **parent plant** according to its molecular and biochemical characteristics, particularly to the high contents of the desired metabolites. In theory, any part obtained from any plant species can be employed to induce callus tissue; however, the successful production of callus depends upon plant species and their qualities. Dicotyledons are rather amenable for callus tissue induction as, compared to monocotyledons, the calluses of woody plants generally grow slowly. Stems, leaves, roots, flowers, seeds, and any other parts of plants are used, but younger and fresh explants are preferable as explant materials.
2. Afterwards the **selection of cell line** becomes important. It includes the establishment of **high-producing**

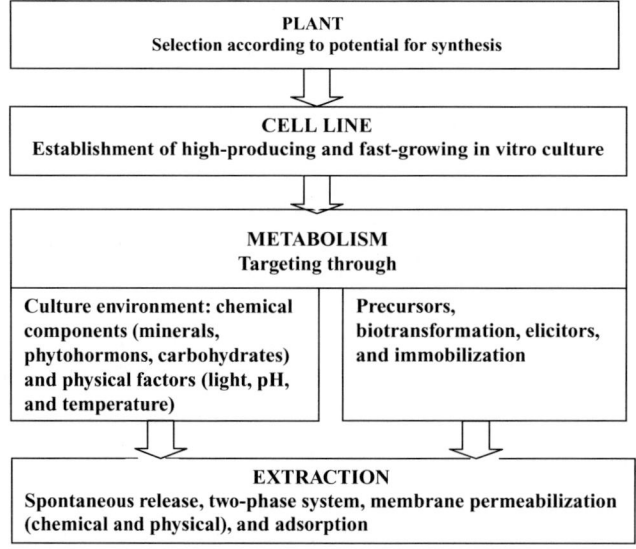

Fig. 1 Scheme of plant secondary metabolite production from the cell cultures.

and **fast-growing** in vitro cultures. It is possible to identify cell lines that can produce amounts of compounds equal or even higher as in the plant from which they derive. Moreover, increase of metabolite levels using **mutants** is possible and selection of suitable analogues for this purpose could be an important factor in order to produce a variety of products. Maximization of the production and accumulation of secondary metabolites by plant cultured cells requires production of new genotypes through **protoplast fusion** or **genetic engineering** but this presupposes the identification of the genes encoding key enzymes of secondary metabolic pathways and their expression once introduced in the plant cells; use of mutagens to increase the variability already existing in living cells. Furthermore, new molecules, which have not been found previously in plants, can be produced by cell cultures.

3. **Targeting metabolism**: A number of chemical and physical factors strongly affect the production of secondary metabolites. The expression of many secondary metabolite pathways is easily altered by external factors such as environmental conditions (chemical and physical) and special treatments (precursors and elicitors).

    a) Plant cell **culture medium** includes inorganic components, organics, and phytohormones. Changing of medium components (concentration, proportion, and form) is a very powerful way of enhancing the culture efficiency of plant cell cultures. Thus, high auxin level stimulates cell growth, but often negatively influences secondary metabolite production.

    Physical conditions such as light, temperature, medium pH have also been examined for their effect upon secondary metabolite accumulation in many types of cultures.

    b) **Special treatments** include feeding with precursors, application of elicitors, biotransformation, and immobilization.

    The concept of feeding with **precursors** is based upon the idea that supply with compounds, which are intermediate or at the beginning of biosynthetic route, gives a good chance of increasing the yield of the final product.

    Often the production of the desired metabolites is limited by the lack of particular precursors; **biotransformation** using an exogenous supply of biosynthetic precursors may improve the accumulation of compounds. Biotransformation is a process through which the functional groups of compounds are modified by cell cultures to chemically different product. Plant cells can transform natural or artificial compounds, introduced into the cultures, through a variety of reactions such as hydrogenation, dehydrogenation, isomerization, glycosylation, and hydroxylation.

    Plants and plant cells in vitro show physiological and morphological responses to microbial, physical, or chemical factors that are known as **elicitors**. Since the secondary metabolites protect plants from the environmental changes, the way to induce their synthesis is to apply unfavorable factors, i.e., simulate pathogen attack, herbivores, heavy metals, etc. Elicitation is a process of induced or enhanced synthesis of secondary metabolites by the plants to ensure their survival, persistence, and competitiveness. Biotic and abiotic elicitors are used to stimulate secondary metabolite product formation in plant cell cultures.

    Cell **immobilization** can result in much higher concentrations of the plant cells because of the certain grade of cell specialization while hundreds or thousands of them are immobilized in one aggregate. Most of the research in this area has utilized hydrocolloidal gels such as alginate and carrageenan that were used to entrap the plant cells into a gel matrix while allowing easy access of substrates.

4. Perhaps the most efficient bioprocessing concepts for the production of phytochemicals result in **spontaneous release into medium** where they can be more easily recovered. One of the most fruitful areas of research for the production of lower-value products may be the study of methods to induce product leakage from cells that normally accumulate the product. A study of the intracellular compartments in which synthesis of chemicals occurs may also be necessary, because the substances are transported to the vacuole for accumulation, then an alternative to be considered is prevention of vacuolar accumulation and consequently enhancement of substances released into the medium.

Plants often have sites of synthesis and storage of the secondary metabolites in separate cells or organs. Inhibition of metabolic enzymes as well as inhibition of membrane transport can be eliminated by the accumulation of synthesized products in a second phase introduced into the aqueous medium or **two-stage** system.

CONCLUSIONS AND OUTLOOK

The objectives are to develop techniques to the stage where it is possible to yield secondary products more cheaply from the plant cell culture than to either extract the whole plant grown under natural conditions or synthesize the product. Confronted with having to increase the amount of secondary metabolites in plant cell cultures, the need for biochemical and molecular research on the secondary metabolism of plants has been frequently emphasized. The research in this area could lead to the successful manipulation of secondary metabolism and significantly increase the amounts of the compounds. It should be possible to achieve the synthesis of a wide range of compounds such as alkaloids, flavonoids, terpenes, steroids, glycosides, etc. using plant cell culture technology.

ACKNOWLEDGMENT

This work would not have been possible without the continuous support of Prof. Dr. Dipl.-Ing. Dietrich Knorr providing me with the latest information on plant cell technology.

REFERENCES

1. Lee, C.W.T.; Shuler, M.L. The effect of inoculum density and conditioned medium on the production of ajmalcine and catharanthine from immobilized *Catharanthus roseus* cells. Biotechnol. Bioeng. **2000**, *67*, 61–71.
2. Knorr, D.; Caster, C.; Dörnenburg, H.; Dorn, R.; Gräf, S.; Havkin-Frenkel, D.; Podstolski, A.; Werrman, U. Biosynthesis and yield improvement of food ingredients from plant cell and tissue culture. Food Technol. **1993**, *47* (12), 57–63.
3. Matsubara, K.; Shigekazu, K.; Yoshioka, T.; Fujita, Y.; Yamada, Y. High density culture of *Coptis japonica* cells increases berberine production. J. Chem. Technol. Biotechnol. **1989**, *46*, 61–69.
4. Petersen, M.; Simmonds, M.S. Rosmarinic acid. Phytochemistry **2003**, *62* (2), 121–125.
5. Kim, D.J.; Chang, H.N. Enhanced shikonin production from *Lithospermum erythrorhizon* by *in situ* extraction and calcium alginate immobilization. Biotechnol. Bioeng. **1990**, *36* (5), 460–466.
6. Fujita, Y.; Tabata, M. Secondary metabolites from plant cells—pharmaceutical applications and progress in commercial production. In *Plant Tissue and Cell Culture*; Green, C.E., Somers, A. Hackett, W.P., Biesboer, D.D., Eds.; Liss: New York, 1987; 169–185.

BIBLIOGRAPHY

1. Alfermann, A.W.; Petersen, M. Natural products formation by plant cell biotechnology. Plant Cell Tissue Organ Cult. **1995**, *43*, 199–205.
2. Brodelius, P.; Pedersen, H. Increasing secondary metabolite production in plant-cell culture by redirecting transport. Trends Biotechnol. **1993**, *11* (1), 30–36.
3. Dicosmo, F.; Misawa, M. Plant cell and tissue culture: alternatives for metabolite production. Biotechnol. Adv. **1995**, *13*, 425–435.
4. Dornenburg, H.; Knorr, D. Challenges and opportunities for metabolite production from plant cell and tissue cultures. Food Technol. **1997**, *51*, 47–54.
5. Fontanel, A.; Tabata, M. Production of secondary metabolites by plant tissue and cell cultures. In *Present Aspects and Prospects*; Horisberger, M., Ed.; Nestlé Research News: Vevey, Switzerland, 1987; 93–103.
6. Knorr, D. Immobilization of microbial and cultured plant cells. In *Food Biotechnology*; Knorr, D., Ed.; Marcel Dekker: New York, 1987; 271–287.
7. Mulabagal, V.; Tsay, H. Plant cell cultures as a source for the production of biologically important secondary metabolites. Inter. J. Appl. Sci. Eng. **2004**, *2*, 29–48.
8. Namdeo, A.G. Plant cell elicitation for production of secondary metabolites. Pharmacognosy Rev. **2007**, *1*, 69, http://www.phcogrev.com.
9. Rao, R.S.; Ravishankar, G.A. Biotransformation of isoeugenol to vanilla flavour metabolites and capsaicin in freely suspended and immobilized cell cultures of *Capsicum frutescens*: study of the influence of b-cyclodextrin and fungal elicitor. Process Biochem. **1999**, *35*, 341–348.
10. Rao, R.S.; Ravishankar, G.A. Plant cell cultures: chemical factories of secondary metabolites. Biotechnol. Adv. **2002**, *20*, 101–153.
11. Raskin, I.; Ribnicky, D.M.; Komarnytsky, S.; Ilic, N.; Poulev, A.; Borisjuk, N.; Brinker, A.; Moreno, D.A.; Ripoll, C.; Yakoby, N.; O'Neal, J.M.; Cornwell, T.; Pastor, I.; Fridlender, B. Plants and human health in the twenty-first century. Trends Biotechnol. **2002**, *20*, 522–531.
12. Tabata, H. Production of paclitaxel and the related taxanes by cell suspension cultures of *Taxus* species. Curr. Drug Targets **2006**, *7* (4), 453–461.
13. Tepe, B.; Sokmen, A. Production and optimisation of rosmarinic acid by *Satureja hortensis* L. callus cultures. Nat. Prod. Res. **2007**, *21* (13), 1133–1144.
14. Vasconsuelo, A.; Giulietti, A.M.; Boland, R. Signal transduction events mediating chitosan stimulation of anthraquinone synthesis in *Rubia tinctorum*. Plant Sci. **2004**, *166*, 405–413.
15. Zhong, J.J. Biochemical engineering of the production of plant-specific secondary metabolites by cell suspension cultures. In *Advances in Biochemical Engineering/Biotechnology*; Scheper, T., Zhong, J.J., Eds; Springer-Verlag: Berlin, Heidelberg, 2001; Vol. 72, 26.

Seeds: Transgenes and Genetic Modification (GM)

Roberta Onori
Marzia De Giacomo
Department of Veterinary Public Health and Food Safety, GMOs and Mycotoxins Unit,
National Institute of Health (ISS), Rome, Italy

Abstract
Sampling and detection methods in the seed supply chain are needed to guarantee quality standards, verify product integrity, genetic purity in genetically modified (GM) varieties, guarantee coexistence of GM and non-GM farming system varieties in order to manage adventitious presence (AP) of approved and non-approved GM seeds, and monitoring the segregation strategies along the production and distribution chain. This entry focuses on management of AP and provides an overview of sampling and testing strategies of genetically modified organisms (GMOs) in seeds.

INTRODUCTION

In agriculture, new plant varieties are produced by different technologies from the traditional selective breeding. When conventional breeding and hybridization methods are not possible, the use of modern gene technologies is a useful tool to develop new varieties of plants with advantageous characteristics.

In order to provide a scientific regulatory structure to implement these techniques and to guarantee public health and the environmental safety, a specific GMO regulation has been introduced worldwide. The liberalization of crop cultivation allows the consumer to determine the market demand by allowing the free choice between GM and non-GM products and coexistence strategies for GM and non-GM seeds that have been implemented.[1]

With seed management, the legislative EU framework enforces the labeling of seed lots that contain any detectable trace of authorized GM seeds,[2] and any seed containing unauthorized GMOs are not placed on the market.

The implementation and maintenance of the regulations necessitate sampling and testing procedures that allow for accurate and reliable determination of GMO content in seed lots, where harmonized sampling and testing methods are required to facilitate worldwide trade.[3,4]

SEEDS SAMPLING METHODS

Seed sampling is the first significant part of seed quality control. Sampling procedures aim at obtaining a test sample that is consistent with the average quality of the entire seed lot, where sampling accuracy is fundamental. This can only be accomplished on the basis of a statistical inference that introduces an unavoidable degree of uncertainty.

Internationally, two organizations ensure uniformity in seed testing: the Association of Official Seed Analysts (AOSA) and the International Seed Testing Association (ISTA). The first operates in the United States and Canada and publishes rules only for the testing. The ISTA operates in 76 countries worldwide and has about 100 accredited member laboratories and sets rules for the sampling and testing. Both organizations provide seed quality certification. With ISTA certification, the samples have to be collected and prepared in accordance with ISTA Rules.[5] ISTA Rules have a broad application, for instance the EU Recommendation 787/2004 refers to ISTA rules for sampling of seed. In the ISTA Rules the prerequisite for accurate sampling is lot homogeneity. According to ISTA, the test on homogeneity could be performed based on visible signs. However, for GMOs there are no visible differences between GM and non-GM seeds, the homogeneity cannot be directly tested without additional costs.

The steps involved in sampling are shown in Fig. 1. According to ISTA Rules, to obtain a representative sample from a seed lot it is necessary to draw a minimum number of primary samples of approximately the same size based on lot and container size and selected at random or according to a systematic plan. The composite sample is obtained by combining the primary samples; the composite sample is then reduced in size to give the submitted sample by mixing and splitting. The laboratory-submitted sample has to be further subdivided to obtain working samples used in further tests. The methods used for the two sample reduction steps are according to ISTA Rules.

The statistical program Seedcalc (http://www.seedtest.org/en/home.html) can be used to subdivide the working

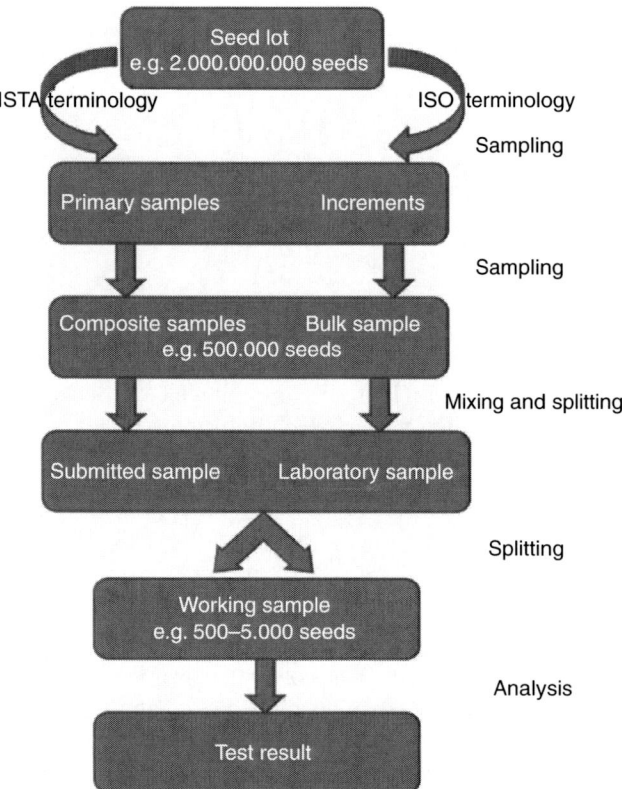

Fig. 1 Sampling plan schematic for GM seed detection in a seed lot; terminology conforming to ISTA and ISO (ISO 542:1990, ISO 13690:1999, ISO 6644:2002).

sample and to design testing plans for purity characteristics, including the estimation of GM seeds levels in conventional seed lots by sub-sampling strategies.[6]

The seed testing plans are based on the assumption of random sampling, several testing objectives, as well as budget and time resources with respect to consumer and producer risk.

Terminology has been based on standard acceptance sampling in the statistical quality control literature (General Guidelines on Sampling CAC/GL 50-200 11).

ANALYTICAL METHODS FOR GMO TESTING

Compliance with GM food labeling laws or seed purity assessments is dependent on the availability of test methods that can determine the presence or concentration of GMOs or both in agricultural products such as seed and grain. GMOs content can be determined by methods that detect either the novel protein or the inserted DNA. Numerous detection methods for GM DNA using the polymerase chain reaction (PCR) have been developed, and advances in molecular techniques for the detection and the quantification of biotech products have been recently reviewed.[7,8] Detection of the novel proteins produced by GM crops relies on the application of immunoassay technology. Commercial immunoassays are available for most of the GM crops on the market and have been used in a variety of large-scale applications.[9]

PCR-Based Methods

Generic or specific detection of new GMO sequences can be obtained, and PCR-based GMO tests can be grouped into at least four categories corresponding to their level of specificity (Fig. 2). Each category corresponds to the composition of the DNA fragment that is amplified in the PCR.[10]

Qualitative methods are commonly used for GMO detection and identification; for screening purposes, PCR methods are applied that can detect common genetic elements found in a range of GM plants, such as the 35S promoter and the NOS terminator present in most of the current authorized GMOs; event-specific PCR primers are used for GMOs identity determination.

Dedicated PCR approaches are also available for specific applications, such as nested PCR to improve the sensitivity and multiplex PCR to detect a range of target sequences. For the quantification of GMO content, real-time PCR is a more accurate and widely used quantitative PCR approach.

As the number of GM crops worldwide is constantly rising, there is an obvious need for methods that allow a one-step, simultaneous, broad sample screening of GM crops. High throughput methods such as microarrays based on DNA hybridization are the most recent tools for the detection of GMOs that have been developed and validated.

Protein-Based Detection Methods

For proteins detection in Agro–Biotechnology applications, enzyme-linked immunosorbent assay (ELISA), and lateral flow device (LFD) are the most commonly used test formats. LFDs are designed for qualitative "yes/no" testing. ELISA can be used as either a qualitative or a quantitative assay. The methods for GM plant detection that are currently commercially available have been developed specifically for insect-resistant Bt crops and for herbicide-tolerant GM plants. Some test methods have been described as cross-reactive because they cannot discriminate the difference between different GM events expressing the same protein.[9]

Protein-based methods are often considerably cheaper and faster than DNA-based GMO detection methods, and are consequently preferred for the practical screening of large unprocessed agricultural commodities in the field and during harvest, storage, and transportation; however, the limitation of this test should be taken into account.

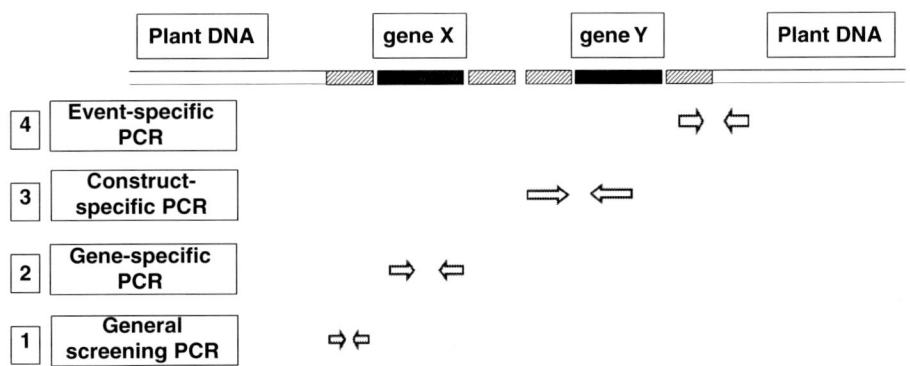

Fig. 2 A schematic representation of a typical gene construct and four types of PCR-based assays showing increasing specificity (from bottom to top).

HARMONIZING METHODS

Generally, the ISO (International Organization for Standardization) standards have been adopted as trade standards worldwide. ISO and its European counterpart (CEN—Comité Européen de Normalization) have worked to produce common standards for GMO detection, including a general document on performance criteria and laboratory organizational requirements. These standards currently serve as models in several other detection areas, starting with the modular approach and including the general requirements of PCR-based detection methods. These contribute to the global harmonization of molecular biology based detection methods (ISO 21569; ISO 21570; ISO 21572; ISO 21571).

The Codex Committee on Methods of Analysis and Sampling is presently working on criteria guidelines for analytical methods validation in the detection and identification of foods derived from biotechnology. In this approach the analytical method consists of steps forming independent modules. Each module should ideally be combinable, with each module in the next step providing cost advantages in the validation process.[11]

The strategy adopted by the ISTA is a performance-based approach; laboratories are free to choose their own methods, as long as these respect the ISTA minimum requirements that include proficiency testing, reliable performance data in term of accuracy, and repeatability.

UNKNOWN/UNAUTHORIZED GMOs

Differences in authorizations between jurisdictions, often referred to as "asynchronous authorizations," are increasingly likely to result in a low-level presence of unauthorized GMOs in exports. The problems associated with detecting unauthorized GMOs are inadequately supported by research investments, legislation, and enforcement. There are few ongoing systematic efforts to deal with unauthorized GMOs (http://www.coextra.eu).

The Codex Alimentarius ad hoc intergovernmental task force on foods derived from biotechnology is presently working to establish internationally harmonized guidelines to identify asynchronously authorized GMOs.

UNIT OF MEASUREMENT

Currently, different approaches for seed/grain testing are implemented. Sub-sampling approaches or individual testing of seeds by fast and cheap protein methods are used. Usually GM contaminations are determined in percentage GM seeds.

In the EU, event-specific real time PCR methods and the percentage DNA haploid genome copy measurement unit are recommended. The discrepancy between percentage GM seed number and percentage GM DNA copies creates a problem in the measurement and reporting of GM presence in heterozygous seeds and seeds with more complex genomes, because the real GM seeds content can be under- or overestimated.[12]

CONCLUSION

The increase of GMOs in world agriculture has been well documented. Traceability and detection of GMOs are a growing public concern in EU and other countries, including the United States.

In a recent full-scale experimental study (KeLDA—Kernel Lot Distribution Assessment), highly significant heterogeneities in 15 shiploads of soybean grains imported to Europe was documented, suggesting that sampling for GMO testing should be based on reliable distribution-free sampling strategies.

With regards to seed heterogeneity the KeLDA results cannot be applied directly as seeds are produced according to high quality standards following more stringent practices. However, different causes of heterogeneity for seed lots have been reported and heterogeneity increases with lot size.[12] The Theory of Sampling (TOS), developed by Pierre Gy covers all aspects for heterogeneity characterization and sampling error management. This approach could be a promising alternative to deal with heterogeneously distributed analytes.

The most important conclusions from this overview are that the reliability results for the detection of GM seed presence are a function of both analysis and sampling. Another factor is the fitness for purpose of a sampling plan that is dependent on the performance of the detection system and vice versa, particularly when a low level presence must be detected, namely the allowable threshold of GM material.

REFERENCES

1. http://www.agbioforum.org/v10n1/v10n1a06-gruere.htm (accessed May 2007).
2. http://eur-lex.europa.eu/LexUriServ/LexUriServ.do?uri=CELEX:32003R1830:EN:HTML (accessed Oct 2003).
3. http://www.daff.gov.au/__data/assets/pdf_file/0004/957469/for_printing_gm_sampling_and_testing.pdf (accessed December 2008).
4. http://ec.europa.eu/environment/biotechnology/pdf/seeds_study_2007.pdf (accessed December 2008).
5. http://www.seedtest.org/en/home.html (accessed 2003).
6. http://www.seeds.iastate.edu/publications/stat/Seed%20Purity%20Testing.pdf (accessed June 2001).
7. http://www.springerlink.com/content/p742733g55724783/ (accessed February 2008).
8. Marmiroli, N.; Maestri, E.; Gulli, M.; Malcevschi, A.; Peano, C.; Bordoni, R.; De Bellis, G. Methods for detection of GMOs in food and feed. Anal. Bioanal. Chem. **2008**, *392* (3), 369–384.
9. Grothaus, G.D.; Bandla, M.; Currier, T.; Giroux, R.; Jenkins, G.R.; Lipp, M.; Shan, G.; Stave, J.W.; Pantella, V. Immunoassay as an analytical tool in agricultural biotechnology. J. AOAC Int. **2007**, *85* (3), 780–786.
10. http://www.springerlink.com/content/fdvnvcyn343t96y7/fulltext.pdf?page=1 (accessed February 2003).
11. Holst-Jensen, A.; Berdal, K.G. The modular analytical procedure and validation approach and the units of measurement for genetically modified materials in foods and feeds. J. AOAC Int. **2004**, *87* (4), 927–936.
12. http://randd.defra.gov.uk/Default.aspx?Menu=Menu&Module=More&Location=None&Completed=0&ProjectID=12963 (final project) (accessed October 2007).

Soy Sauce Fermentation

Jean Guy LeBlanc
Lactobacillus Reference Center, National Council of Scientific and Technical Research (CERELA-CONICET), San Miguel de Tucumán, Argentina

Abstract
Soy sauce production is probably one of the earliest biotechnology industries since this valuable condiment has been mass-produced for over 500 years. Nowadays, commercial soy sauces can be classified as either fermented or non-fermented (chemical soy sauce). Fermented soy sauce is completely different in the constituents of aroma and flavor from chemical soy sauce. In this entry, the main processes involved during soy sauce fermentation will be discussed with special emphasis on microbial fermentation and health properties attributed to this important condiment.

INTRODUCTION

Soy sauce production is probably one of the earliest biotechnology industries since this important condiment has been shipped within Asia for over 500 years. The Chinese have been growing soybeans for food and animal feed for centuries; at the beginning salt was often added to preserve this valuable legume. These salted soybeans would then suffer changes in the holding barrels owing to microbial fermentation, resulting into a paste called miso. Humans have consumed miso for centuries because it is easier to digest than regular soybeans, probably owing to the removal of non-digestible sugars by the fermenting microorganisms. During the elaboration of miso, a black liquid accumulated at the bottom of the barrels, it was discovered afterward that this sauce could be used for cooking: soy sauce (shoyu) was invented. Nowadays, commercial soy sauces can be classified as either fermented or non-fermented (chemical soy sauce). Fermented soy sauce is completely different in the constituents of aroma and flavor from chemical soy sauce. In this entry, the main processes involved during soy sauce fermentation will be discussed with special emphasis on microbial fermentation and health properties attributed to this important condiment.

SOY SAUCE PRODUCTION

Authentic soy sauces are made from whole soybeans and result from fermentation with fungi, lactic acid bacteria, and other related microorganisms. Non-fermented soy sauces are obtained using the hydrolyzed method, where soybeans or soybean meal, and perhaps wheat, are degraded with hydrochloric acid and then neutralized with sodium hydroxide. These latter soy sauces do not have the natural color of authentic soy sauces and are typically colored with caramel coloring. The two products taste somewhat alike, but the fermentation process of the first technique produces a decidedly more aromatic and attractive product. Soy sauce production is a three-stage process that is illustrated in Fig. 1 and will be explained in the following sections.

Koji Preparation

Soybeans (or defatted soybeans) are moistened by soaking in water overnight at temperatures above 30°C, changing the water frequently to prevent possible growth of undesirable microorganisms. The moisturized beans are then steamed for several hours. In parallel, whole kernel wheat is toasted to: 1) transform β-starch to α-starch that can more easily be used by koji microorganisms (see below); 2) convert substances present in wheat to increase flavor compounds; and 3) produce Maillard reactions giving rise to the brownish color of the final product.[1] Toasted wheat is then crushed. In the first stage of soy sauce production, the moistened soybeans are mixed with toasted crushed wheat in a specific proportion that can vary depending on the desired product but normally in equal quantities. The molds, *Aspergillus oryzae* and *Aspergillus sojae*, are added to the mixture to make koji that is then left uncovered for two days. *A. oryzae* and *A. sojae* have been used for producing food-grade amylase and fermenting oriental foods such as saké, miso, and soy sauce for centuries.[2] The food and industrial importance of *A. oryzae* makes it essential to differentiate this species from the aflatoxigenic species *Aspergillus flavus* and *Aspergillus parasiticus* that can produce the potent carcinogens, aflatoxins, and pose a significant human health threat. The high relatedness between *A. flavus* and *A. oryzae* requires an underspecies classification to differentiate them, a task that can be accomplished by a limited number of techniques. Amplified fragment length polymorphism (AFLP) is a method for classifying strains that has been demonstrated to be useful in differentiating *A. flavus* from *A. oryzae* whereas the random amplified polymorphic DNA (RAPD) method could not do so.[2] Only

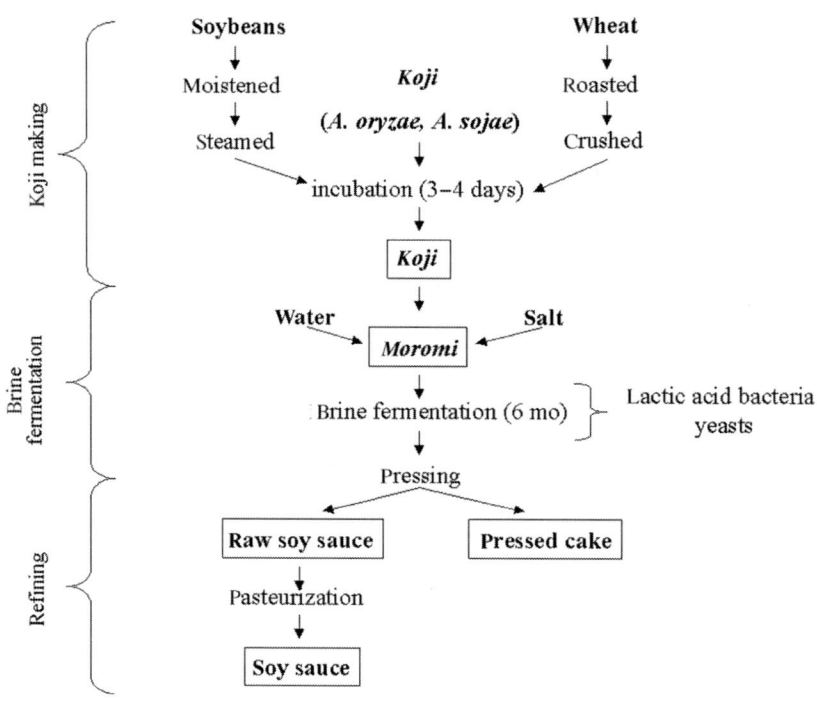

Fig. 1 Manufacturing process for typical Japanese types of soy sauce (koikuchi shoyu). Ingredients or additives are in boldface, processing steps in regular type, and resulting products are in boldface and boxed. **Source:** From *Soybeans: Chemistry, Technology, and Utilization*.[1]

properly characterized fungi strains should be used to prevent the occurrence of aflatoxin in the final product.

Brine Fermentation

The second step in the preparation of soy sauce is brine fermentation by osmophilic yeasts and lactic acid bacteria. In this step, a salt solution is added to the mature koji to obtain a mash called moromi. The final NaCl concentration in this brine should be 17–19% to prevent growth of putrefactive microorganisms and allow proper growth of beneficial halophilic and osmophilic yeasts and bacteria.[1] Moromi is then put in airtight containers where it is allowed to ferment for at least 6 months. At the beginning, the *Aspergillus* enzymes continue to hydrolyze the soybeans and wheat and, as a result, a surplus of different kinds of sugars and amino acids arise.[3,4] During brine fermentation these sugars and amino acids are consumed by salt-tolerant lactic acid bacteria, *Tetragenococcus halophila* (previously known as *Pediococcus halophilus* or *Pediococcus soyae*) and the yeasts, *Zygosaccharomyces rouxii* and *Candida versatilis*. In the first stage of this brine fermentation, *T. halophila* produces lactic acid to reduce the initial pH from about 7.0 to below 5.0. After the pH has dropped below 5.0, *T. halophila* is unable to grow and an alcoholic fermentation by *Z. rouxii* starts.[5] As a result, ethanol and many other flavor components, such as higher alcohols and 4-hydroxyfuranones, are produced.[5] At the last stage of the brine fermentation, when *Z. rouxii* is not active anymore, *C. versatilis* and other osmophilic yeasts, such as *Torulopsis versatilis* and *Torulopsis etchelsii*, start to grow, producing the specific aroma compounds of soy sauce such as phenolic components like 4-ethylguaiacol and 4-ethylphenol.[5] There are a number of reports stating that soy sauce yeasts can be inhibited by metabolites produced by the osmophilic lactic acid bacteria (*T. halophila*), so special care must be taken to prevent excessive lactic acid fermentation at the beginning of brine fermentation.[3] Recent reviews of flavor formation during the yeast brine fermentation have shown that research is needed to improve the flavor composition of the final soy sauce product.[5] Additionally, there have been a number of reports that have shown that microorganisms isolated after brine fermentation can be useful in numerous applications. For example, *Z. rouxii* could be employed in biodegradation processes since it has the ability to degrade xenobiotics into simpler molecules.[6] The yeast has also attracted the attention of industries for its utility as a biocatalyst in production-scale biotransformations (reviewed in Costello et al.).[7]

Final Processing

After fermentation, the resulting mash is then pressed to obtain the liquid soy sauce. Finally, the sauce is filtered, pasteurized, further clarified by sedimentation, and tightly bottled for distribution. The Japan Soy Sauce Association estimates annual production of fermented soy sauce and mixed soy sauce at about 9.5 million kiloliters, most of which is produced by China, Japan, and Korea. The statistics provided in 2006 by the Food and Agriculture

HEALTH PROPERTIES

It has been found that commercially available soy sauces, especially dark soy sauces, possess high antioxidant activity.[8,9] During human clinical trials, it was shown that dark soy sauce protects against lipid peroxidation, which is accompanied by vasodilatory and hemodynamic changes consistent with antioxidant effects on the endothelium or vascular smooth muscle.[8] The antioxidant activity of soy products has often been linked to its isoflavone content, although this seems unlikely owing to the low concentration of isoflavones in soy sauce. Recently it was reported that soy proteins, and not isoflavones, were responsible for lowering serum lipid levels and, thus, providing cardiovascular benefits.[10] Soy sauce is also considered a promising vehicle for iron fortification, since both naturally fermented and chemically hydrolyzed soy sauces could be fortified with different iron sources.[11] Recently it has been shown that shoyu polysaccharides may have antiallergic activities, making soy sauce a potentially promising seasoning for the treatment of allergic diseases through food.[12] Many beneficial attributes are now been associated with soy sauce; most of these properties are the direct result of the fermentation processes. Such health benefits are reasons for continuing research on the mechanisms that are involved in soy sauce fermentation.

CONCLUSION

In this entry, the main processes involved during soy sauce fermentation have been discussed with special emphasis on microbial fermentation and health properties attributed to this important condiment. Since soy sauce consumption is growing steadily all around the world, many researchers continue to study this complex dark liquid to hopefully discover additional benefits associated with the consumption of one of the most ancient yet effective form of biotechnology.

ACKNOWLEDGMENTS

Dr. Jean Guy LeBlanc would like to thank the Consejo Nacional de Investigaciones Científicas y Técnicas (CONICET), Agencia Nacional de Promoción Científica y Tecnológica (ANPCyT), and the Secretaría de Ciencia y Técnica de la Universidad Nacional de Tucumán for their financial support throughout the years.

REFERENCES

1. Liu, K. *Soybeans: Chemistry, Technology, and Utilization*; Chapman & Hall: New York, 1997.
2. Lee, C.L.; Liou, G.Y.; Yuan, G.F. Comparison of *Aspergillus flavus* and *Aspergillus oryzae* by amplified fragment length polymorphism. Bot. Bull. Acad. Sin. **2004**, *45*, 61–68.
3. Noda, F.; Hayashi, K.; Mizunuman, T. Antagonism between osmophilic lactic acid bacteria and yeasts in brine fermentation of soy sauce. Appl. Environ. Microbiol. **1980**, *40* (3), 452–457.
4. Yong, F.M.; Wood, B.J. Biochemical changes in experimental soy sauce Moromi. J. Food Technol. **1977**, *12*, 263–273.
5. Van Der Sluis, C.; Tramper, J.; Wijffels, R.H. Enhancing and accelerating flavour formation by salt tolerant yeasts in Japanese soy-sauce. Proc. Trends Food Sci. Technol. **2001**, *12*, 322–327.
6. Zadra, C.; Cardinali, G.; Corte, L.; Fatichenti, F.; Marucchini, C. Biodegradation of the fungicide iprodione by *Zygosaccharomyces rouxii* strain DBVPG 6399. J. Agric. Food Chem. **2006**, *54* (13), 4734–4739.
7. Costello, C.A.; Payson, R.A.; Menke, M.A.; Larson, J.L.; Brown, K.A.; Tanner, J.E.; Kaiser, R.E.; Hershberger, C.L.; Zmijewski, M.J. Purification, characterization, cDNA cloning and expression of a novel ketoreductase from *Zygosaccharomyces rouxii*. Eur. J. Biochem. **2000**, *267*, 5493–5501.
8. Lee, C.Y.J.; Isaac, H.B.; Wang, H.; Huang, S.H.; Long, L.H.; Jenner, A.M.; Kelly, R.P.; Halliwell, B. Cautions in the use of biomarkers of oxidative damage; the vascular and antioxidant effects of dark soy sauce in humans. Biochem. Biophys. Res. Com. **2006**, *344*, 906–911.
9. Wang, H.; Jenner, A.M.; Lee, C.Y.; Shui, G.; Tang, S.Y.; Whiteman, M.; Wenk, M.R.; Halliwell, B. The identification of antioxidants in dark soy sauce. Free Radic. Res. **2007**, *41* (4), 479–488.
10. McVeigh, B.L.; Dillingham, B.L.; Lampe, J.W.; Duncan, A.M. Effect of soy protein varying in isoflavone content on serum lipids in healthy young men. Am. J. Clin. Nutr. **2006**, *83*, 244–251.
11. Watanapaisantrakul, R.; Chavasit, V.; Kongkachuichai, R. Fortification of soy sauce using various iron sources: sensory acceptability and shelf stability. Food Nutr. Bull. **2006**, *27* (1), 19–25.
12. Kobayashi, M.; Matsushita, H.; Yoshida, K.; Tsukiyama, R.I.; Sugimura, T.; Yamamoto, K. In vitro and in vivo antiallergic activity of soy sauce. Int. J. Mol. Med. **2004**, *14*, 879–884.

Stem Cell and Germ Cell Technology

Matthew B. Wheeler
Department of Animal Sciences, Beckman Institute for Advanced Science and Technology, Institute for Genomic Biology, University of Illinois, Urbana, Illinois, U.S.A.

Samantha A. Malusky
Department of Animal Sciences, University of Illinois at Urbana-Champaign, Urbana, Illinois, U.S.A.

Abstract

In mammals, stem cells are defined as a unique cell population characterized by nearly unlimited self-renewal and capacity to differentiate via progenitor cells into terminally differentiated somatic cells. Stem cells may be of embryonic or adult origin. Adult stem cells are located in many specialized tissues, including the liver, skin, brain, fat, bone marrow, and muscle. As a result of stem cell activity, adult tissues are continuously renewed, even in the absence of injury, to ensure maintenance of cell type throughout the life of the animal. Pluripotency in mammals is restricted to the zygote, early embryonic cells, primordial germ cells, and the stem cells derived from embryonic carcinomas. Embryonic stem (ES) cells are derived from the inner cell mass of the blastocyst. In contrast to adult stem cells, ES cells are pluripotent, contribute to all three primary germ layers (endoderm, mesoderm, and ectoderm), indefinitely proliferate, and maintain an undifferentiated phenotype. Embryonic germ (EG) cells are derived from primordial germ cells, which are progenitor cells of the sperm and egg in the adult animal. EG cells reintroduced to the early embryo are capable, like ES cells, of colonizing fetal cell lineages and also possess the ability to differentiate in vitro to a variety of cell types. ES and EG cell lines from mammalian species other than the mouse have not been reported in published literature to successfully colonize the germ line of chimeric animals. The potential of stem-cell technology makes it a valuable and exciting science. Adult and embryonic stem cells may possess the ability to restore or replace tissue that has been damaged by disease or injury. Pluripotent, in vitro cell lines offer an opportunity to study the early stages of embryonic development not accessible in utero and are a powerful tool to facilitate genetic modification of animal genomes.

ADULT STEM CELLS

Adult stem cells are anchored permanent residents of a particular tissue that are involved in repair and maintenance. Adult bone marrow, brain, skeletal muscle, liver, pancreas, fat, skin, and gastrointestinal tract have all been shown to possess stem or progenitor cells.[1] It was originally believed that adult stem-cell function was restricted to cell lineages present in the organ from which they were derived. Recent studies suggest that these adult stem cells are multipotent and can transdifferentiate into different cell lineages.[2] Adult stem cells have a reduced differentiation potential compared to embryo-derived stem cells. Among all presently known adult stem or progenitor cells, cell populations from bone marrow have shown the highest potential with respect to multilineage differentiation. Potential adult stem cells are progenitors arising from a hierarchal pathway of cells dependent upon the tissue type and, in a healthy state, are part of the transit population. Injury can induce these potential stem cells to regenerate the entire lineage through clonal expansion, including the anchored stem cell, to maintain tissue integrity. Effective markers for anchored stem cells or potential stem cells do not exist outside of the hematopoietic lineage. The absence of a reliable identification system has made the study of these crucial cells limited.

EMBRYONIC STEM CELLS

Teratocarcinoma and Embryonic Carcinoma

The first pluripotent embryonic cells were isolated from teratocarcinomas, a spontaneous tumor of the testes in mice and humans. The tumors resembled a disorganized fetus consisting of a wide variety of tissues including hair, muscle, bone, and teeth. Developmental biologists discovered teratocarcinomas could be artificially induced by transferring mouse embryos to extrauterine sites and that they contained undifferentiated stem cells.[3] These embryonic carcinoma (EC) stem cells, once isolated, could grow in culture without losing the capacity to differentiate. When these cells were introduced into a blastocyst, they formed chimeras and could contribute to all somatic tissues. EC cells exhibit an unstable karyotype, which reduces their experimental value despite their capacity to differentiate into all three germ layers.

Embryonic Stem Cells

The early mammalian embryo is composed of cells that have the potential to contribute to all tissue types in the body, a property termed pluripotency. As the embryo develops to blastocyst stage, it forms an outer cell layer and an inner cluster of cells referred to as the inner cell mass (ICM). The outer cells become the trophectoderm and ultimately the placenta. The ICM cells create all tissues in the body, as well as nontrophoblast structures that support the embryo. Embryonic stem cells are derived in vitro from the ICM. ES cells were successfully developed from mouse blastocysts in 1981.[4,5] ES cells contribute to all three germ layers in the developing fetus, proving that they are pluripotent, but ES cells fail to contribute to the trophectoderm, revealing that they are not totipotent. ES cells, when removed from feeder layers, begin to differentiate into multilayered differentiated structures called embryoid bodies. In addition to blastocyst injection, the in vivo developmental potential of ES cells can be tested by injecting ES cells into severe combined immunodeficient (SCID) mice.[6] Benign teratomas form where the cells are injected and contain tumors representing all three germ layers.

Embryonic Stem Cell Criteria

Pluripotent cells do not all possess the same characteristics. Listed below are defining properties of mouse ES cells:

- Stable diploid karyotype
- Clonogenic property
- Ability to recover after freezing and thawing
- Ability to survive and proliferate in vivo indefinitely
- High telomerase activity
- Teratoma and embryoid body formation
- Chimera formation and germ line colonization
- Undifferentiated state

Embryonic Germ Cells

Pluripotent embryonic germ cells can be isolated from the genital ridge of the developing mammalian fetus. EG cells closely resemble ES cell lines in the morphology of colonies, response to induced differentiation, and ability to create chimeric offspring. ES and EG cells are not the same in all respects. Differences exist in the conditions required for their isolation, culture, lifespan in vitro, and differentiation capacity. An important difference is genetic modifications that occur in the deoxyribonucleic acid (DNA) of primordial germ (PG) cells that result in erasure of genomic imprints. The DNA modifications that occur can compromise the developmental potential of the EG cells.

Agricultural Applications for ES and EG Cells

The establishment of ES and/or EG cells from a wide variety of species will allow more flexibility in direct genetic manipulation of livestock as well as agricultural, gene-regulation, and developmental biology research. The use of ES cells in mouse developmental biology research is well documented. However, the production of a chimeric livestock species (swine and cattle) produced from ES cells has only recently been reported.[7,8] The use of ES or EG cells for the production of transgenic animals from DNA-transformed, individually derived and screened embryonic cell lines could allow large numbers of genetically identical animals to be established. There are many potential applications of stem cell–mediated transgenesis to develop new and improved strains of livestock. Practical applications of stem cell transgenic technology in livestock production include enhanced prolificacy and reproductive performance, increased feed utilization and growth rate, improved carcass composition, improved milk production and/or composition, and increased disease resistance (Fig. 1).

CONCLUSION

Stem cells have revolutionized many areas of biology, and with continued research more information will be learned regarding these unique cells. Comparisons of prospective applications between mammalian adult, ES, EG, and EC cells indicate varying levels of potential. Adult stem cell potential has not been as extensively investigated compared to ES cells due to the difficulty of identifying adult stem cells in a population of cells. EG cells have limited ability to recapitulate normal development due to genetic modifications affecting imprinting status of the cells. EC cells contain karyotypic abnormalities and have limited potential to transmit through the germ line of chimeric animals. ES cells do not possess any major limiting characteristics in comparison to the other cell types, which demonstrates their preferential use in scientific studies. ES cells offer a multitude of applications including access to a population of precursor cells difficult to identify in vivo, ability to identify novel genes during early embryonic development and differentiation processes, use as a standardized in vitro model to test embryotoxic effects of chemicals,[9] the study of targeted mutations of genes that may be lethal in vivo but can be studied in vitro, and less expensive teratogen testing that does not involve isolating embryos or sacrificing animals. The potential applications of stem-cell technology in livestock production are tremendous. The utility of this technology is limited only by our ability to identify

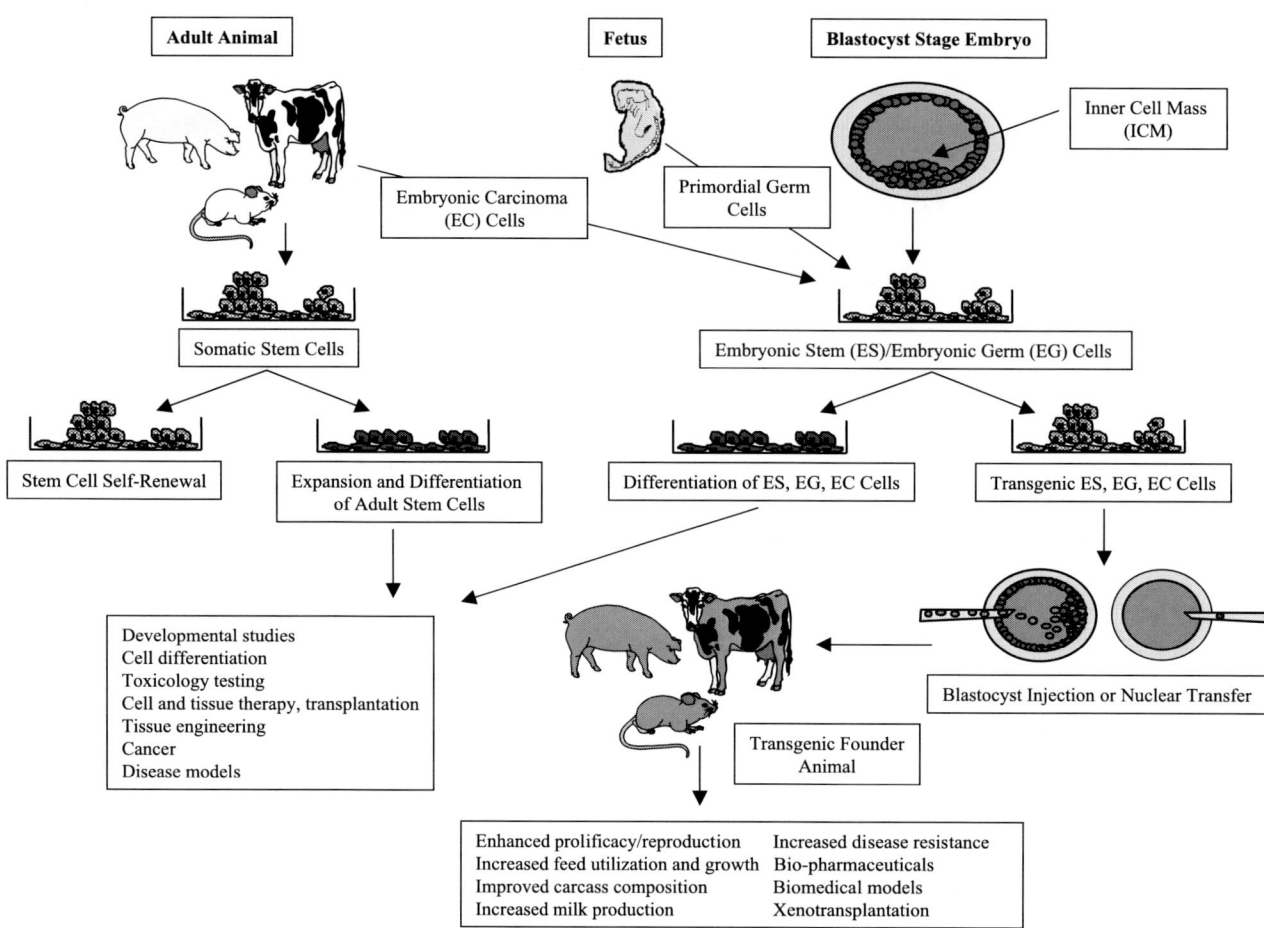

Fig. 1 Sources and applications of stem cells.

appropriate genes and gene functions to manipulate in our production of livestock species.

REFERENCES

1. Passier, R. Mummery, C. Origin and use of embryonic and adult stem cells in differentiation and tissue repair. Cardiovasc. Res. **2003**, *58*, 324–335.
2. Medvinsky, A. Smith, A. Fusion brings down barriers. Nature **2003**, *422*, 823–825.
3. Van der Heyden, M.A.; Defize, L.H. Twenty one years of P19 cells: What an embryonal carcinoma cell line taught us about cardiomyocyte differentiation. Cariovasc. Res. **2003**, *58*, 292–302.
4. Martin, G.R. Isolation of a pluripotent cell line from early mouse embryos cultured in medium conditioned by teratocarcinoma stem cells. Proc. Natl. Acad. Sci. U. S. A. **1981**, *787*, 634–638.
5. Evans, M.J.; Kaufman, M.H. Establishment in culture of pluripotent cells from mouse embryos. Nature **1981**, *292*, 154–156.
6. Foley, G.L.; Rund, L.A.; Wheeler, M.B. Factors affecting murine embryonic stem cell teratoma development. Biol. Reprod. **1994**, *50* Suppl 1, 291.
7. Wheeler, M.B.; Walters, E.M. Transgenic technology and applications in swine. Theriogenology **2001**, *56*, 1345–1370.
8. Cibelli, J.B.; Stice, S.L.; Golueke, P.J.; Kane, J.J.; Jerry, J. Blackwell, C.; Ponce deLeon, F.A.; Robl, J.M. Transgenic bovine chimeric offspring produced from somatic cell-derived stem-like cells. Nat. Biotechnol. **1998**, *16*, 642–646.
9. Spielmann, H.; Pohl, I.; Doring, B.; Liebsch, M.; Moldenhauer, F. The embryonic stem cell test, an in vitro embryotoxicity test using two permanent mouse cell lines: 3T3 fibroblasts and embryonic stem cells. Toxicol in Vitro **1997**, *10*, 119–127.

Stem Cell Research

Tetsuya S. Tanaka
Department of Animal Sciences, Institute for Genomic Biology, University of Illinois, Urbana, Illinois, U.S.A.

Abstract

A variety of stem cell types are identified in various organs and tissues as well as in developing embryos. The technological advances even shed light on the therapeutic use of stem cells. Now, the paramount importance is to generate a pure population of specialized cells from stem cells, which is a current challenge. A better understanding of gene expression at a cellular level will help improve manipulation of stem cell differentiation.

INTRODUCTION

The adult human body consists of about one hundred trillion cells, which are derived from a single egg fertilized by a sperm. Thus, the fertilized egg is defined as a totipotent cell and is the only totipotent cell existing naturally. As embryonic development proceeds most of the cells commit to becoming a limited variety of cell types. By contrast, there are groups of proliferating cells in the adult body, which are capable of replenishing specialized cells (i.e., differentiation) as well as generating cells that retain equal potential to produce differentiated cells (i.e., self-renewal). These groups of cells are defined as stem cells. Because of their characteristics, stem cells provide not only useful in vitro model systems to investigate underlying cell fate decision mechanisms but also artificial platforms to examine safety issues of drugs and foods on a cell type of interest in a less invasive manner. The outcome of these studies will be applicable to creating custom-made tissue replacements as a strategy to cure damaged tissues and organs. A survey of this emerging research field and limitation of its application to regenerative medicine is presented.

ADULT STEM CELLS

After the World War II, it became evident that whole-body exposure to radiation resulted in higher incidences of leukemia. By creating a marrow cell transplantation model with lethally irradiated mice, the marrow cells capable of self-renewal and differentiation to the cells of the hematopoietic lineage were first identified in 1964 (Table 1). Then, the term "stem cell" was applied to defining such a cell type in the bone marrow. Since then, a variety of stem cell types have been identified and isolated from various organs and tissues from a wide range of animals (Table 1).

Stem cells from adult tissues or organs, namely adult stem cells, provide differentiated cells continuously and repair minor damages in the organ to maintain homeostasis of the body. Generally, adult stem cells possess limited ability to proliferate and to generate differentiated cells in vitro. Thus, adult stem cells are multipotent. By contrast, a few studies have presented experimental evidence showing that some of the adult stem cells can differentiate into cell lineages other than the one that the stem cell is originated from, such as from neurons to blood cells, from blood cells to liver cells, and from skin cells to oocytes (Table 1). This so-called transdifferentiation event is well known to happen in pigmented epithelial cells in the iris of the newt, which regenerate lens cells when the lens is removed.[1] However, such a transdifferentiation event rarely occurs in other cell types of the body. Also, cells in the bone marrow spontaneously fuse with other cells and adopt the cellular properties of the fusion partners.[2] Therefore, the claims that some of the adult stem cells are capable of transdifferentiation need to be considered with caution. Reproducibility of these results has to be examined independently.

EMBRYO-DERIVED STEM CELLS

Developing embryos have been utilized as a source to derive stem cells (see Fig. 1A). These embryo-derived stem cells serve as a model system to understand the mechanism of embryonic development as well. Five different types of stem cell lines have been established, which are embryonic, trophoblast, and epiblast-derived stem cells, embryonic carcinoma cells, and embryonic germ cells (Table 1). Embryonic stem cells (ESCs, Fig. 1E) and embryonic germ cells (EGCs) self-renew indefinitely and maintain their normal number of chromosomes, while sustaining their ability to become almost all differentiated cells. Thus, ESCs and EGCs are pluripotent. Because of the pluripotentiality and relative ease in culture, ESCs are widely used to study the

Table 1 Stem cells in adults and embryos.

Source organs/tissues	References
Adults	
Bone marrow	J.E. Till et al. (1964);[16] M.F. Pittenger et al. (1999);[17] E. Lagasse et al. (2000)a;[18] S.H. Orkin and L.I. Zon (2008)[19]
Brain	M. Mayer-Proschel et al. (1997);[20] C.R. Bjornson et al. (1999)a;[21] S. Hitoshi et al. (2004);[22] R. Ravin et al. (2008)[23]
Retinab	V. Tropepe et al. (2000)[24] vs. S.A. Cicero et al. (2009)[25]
Fat	S. Gronthos et al. (2001)[26]
Heart	A. Moretti et al. (2006);[27] O. Bergmann et al. (2009)[28]
Intestine	N. Barker et al. (2007);[29] E. Sangiorgi and M.R. Capecchi (2008)[30]
Liver	T.S. Weiss et al. (2008)[31]
Muscle	D. Montarras et al. (2005)[32]
Ovaryb	J. Johnson et al. (2004)[33] vs. K. Eggan et al. (2006)[34]
Pancreasb	Y. Dor et al. (2004);[35] B.Z. Stanger et al. (2007)[36] vs. R.M. Seaberg et al. (2004);[37] X. Xu et al. (2008)[38]
Prostate	K.G. Leong et al. (2008)[39]
Skin	P.W. Dyce et al. (2006)a;[40] C. Blanpain and E. Fuchs (2009)[41]
Tooth	M. Miura et al. (2003)[42]
Testis	M. Kanatsu-Shinohara et al. (2004);[43] C. Chen et al. (2005);[44] S. Conrad et al. (2008)[45]
Amniotic fluid	P. De Coppi et al. (2007);[46] S. Kern et al. (2006)[47]
Cord blood	G. Kogler et al. (2004)[48]
Wharton's Jelly	M.L. Weiss et al. (2006)[49]
Embryos	
Inner cell mass	M.J. Evans and M.H. Kaufman (1981);[50] G.R. Martin (1981);[51] J.A. Thomson et al. (1998)[52]
Trophectoderm	S. Tanaka et al. (1998)[53]
Epiblast	I.G.M. Brons et al. (2007);[54] P.J. Tesar et al. (2007)[55]
Embryo-derived teratoma	L.C. Stevens (1970)[56]
Germ cells	Y. Matsui et al. (1992);[57] J.L. Resnick et al. (1992);[58] M.J. Shamblott et al. (1998)[59]

aEvidence of transdifferentiation.
bThe presence of stem cells in these organs is controversial.

cell fate decision and tissue regeneration mechanisms and to generate animals modified genetically.[3]

Since ESCs were derived from mouse embryos in 1981 (Fig. 1B–E), ESC lines have been established from embryos of a variety of species (Table 2). Derivation of ESC lines from non-human primate embryos led to the successful derivation of ESC lines from human embryos (Table 2). These studies have laid a path to the future of regenerative medicine and the stem cell-based therapy. However, it made every researcher of human ESCs develop his or her own ethical viewpoint more than ever to conduct research properly.

Differentiation of ESCs can be induced in vitro by culturing ESCs as clumps of cellular aggregates called embryoid bodies in combination with soluble factors that induce cell differentiation.[4] Typically, however, ESCs randomly differentiate into various cell types (see Fig. 1F). When ESCs are transplanted under the skin or the capsules of kidneys or testes of a host animal, ESCs grow into a benign tumor called a teratoma that contains various types of differentiated cells.[5] Therefore, generating teratomas from ESCs is a well-accepted method to validate their pluripotentiality. By contrast, pluripotentiality of mouse ESCs can be validated undeniably by introducing them into a host blastocyst, which will generate a fertile chimeric mouse having functional germ cells derived from ESCs.[5] However, chimeric animals are difficult to generate with blastocysts of large farm animals and non-human primates, and have never been generated with human blastocysts. Thus, although ESCs derived from these animal species are considered as pluripotent based on the in vitro differentiation and teratoma formation assays, it is a challenge to validate whether or not they possess the same differentiation potential as mouse ESCs do.

GENETIC STUDIES IN EMBRYONIC STEM CELLS

Because mouse ESCs can contribute to development of a normal chimeric mouse, ESCs have been used as a vehicle to deliver altered genetic information into the host. When the mutation is carried into sperm or eggs, the mice developed from the mutant sperm and eggs are expected to exhibit phenotypes associated with the mutation. By

Stem Cell Research

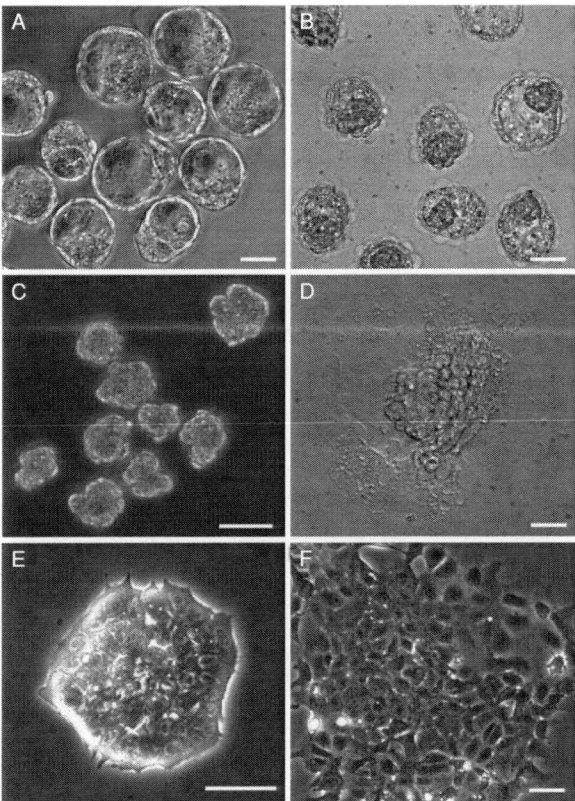

Fig. 1 Derivation and differentiation of mouse embryonic stem cells (ESCs). Bars, 50 μm. (A) Zona-free embryos at 3.5 days after fertilization called blastocysts. The outer layer consists of differentiated cells called trophectoderm (TE), which surrounds undifferentiated cells called inner cell mass (ICM). The TE will become part of the placenta and give rise to trophoblast stem cells, whereas the ICM will develop into the embryo proper and give rise to ESCs. (B) To isolate pure ICM cells, blastocysts are incubated with antibodies that recognize mouse cells, and an animal serum. The TE bound by the antibodies are lysed owing to the presence of an active cytotoxic protein component in the animal serum, the complement system. (C) Pure ICM cells are isolated mechanically from lysed blastocysts by pipetting and are ready to be cultured on the layer of embryonic fibroblasts to derive ESCs. (D) Alternatively, a blastocyst can be cultured directly on the embryonic fibroblasts, which undergoes a morphological change. A mass of the ICM is formed on top of the outgrown TE. When these ICM cells are isolated, trypsinized, and expanded, ESCs can be established. (E) A typical colony of self-renewing ESCs exhibits round and packed appearance. Each cell has smaller cytoplasmic space. (F) After a soluble factor, the leukemia inhibitory factor, that maintains self-renewal of ESCs is withdrawn from the culture, cell differentiation is induced randomly. Cytoplasmic space is noticeably increased.

using the inherent DNA repair mechanism, techniques to introduce gene specific mutation into ESCs have been established.[6] Furthermore, mutation can be introduced at random genomic loci with plasmids or retroviruses, which are designed to introduce a new splicing event that disturbs production of normal transcripts.[7] As a result, these approaches accelerated studying gene function in developing embryos as well as in ESCs.

Self-renewal of ESCs is maintained by soluble factors secreted from embryonic fibroblasts.[8,9] These extrinsic signals eventually lead to activation of DNA-binding proteins Oct3/4, Sox2, and Nanog, which further regulate expression of their target genes.[10,11] Interestingly, when transcriptional activity is analyzed at a single cell level, individual embryonic and adult stem cells exhibit heterogeneous transcriptional activity.[9,12] Therefore, a better understanding of gene expression in single stem cells will help improve therapeutic application of stem cells.

REGENERATIVE MEDICINE

Stem cell research provides tremendous opportunities to study potential application of stem cells to therapeutic use. However, stem cell-based therapy is still in its technical infancy. For example, stroma cells in the bone marrow and fat tissues, namely mesenchymal stem cells (MSCs), are capable of long-term proliferation and generation of differentiated cells in the bone, muscle, fat, and joint tissues in vitro. Furthermore, MSCs release anti-inflammatory cytokines allowing allogenic transplantation.[13] Because a large number of MSCs can be isolated from fat tissues after liposuction, MSCs provide a possibility to develop a strategy to heal wounded bones and cartilage efficiently in a less invasive manner. However, the same inflammatory signals from MSCs are often secreted from tumor cells, so that MSCs may have multiple roles in assisting or regulating cancer metastasis and angiogenesis of tumors.[13] Likewise, specialized cells differentiated from pluripotent embryo-derived stem cells have been transplanted into animal disease models.[4] However, the transplants also developed tumors, because they had undifferentiated stem cells remaining, which became teratomas. Thus, to improve quality of replacement tissues, better understanding of gene expression at a cellular level is a prerequisite, which will make cell differentiation uniform and eliminate expression of threatening signals before transplantation.

Therapeutic use of isologous cells can be achieved in collaboration with techniques to rejuvenate or reprogram differentiated cells into stem cell-like cells.[14] Because differentiated cells keep genetic information intact, the developmental program that the nuclei of differentiated cells have gone through can be reversed by transferring these nuclei into enucleated eggs. Mouse and non-human primate ESCs have been successfully derived from embryos developed from reconstituted eggs as such (Table 2). However, this approach is a challenge with human eggs both technically and ethically. To circumvent these issues, Takahashi and Yamanaka originally developed a simple, non-invasive method to reprogram differentiated cells into pluripotent stem cell-like cells.[15] Reprogramming is achieved in 0.1%

Table 2 Animal species from which embryonic stem cells have been derived.

Animal species	References
Fish	
Medaka	Y. Wakamatsu et al. (1994)[60]
Zebrafish	C. Ma et al. (2001)[61]
Rodents	
Hamster	T. Doetschman et al. (1988)[62]
Mouse	M.J. Evans and M.H. Kaufman (1981);[63] G.R. Martin (1981);[64] T. Wakayama et al. (2001)a[65]
Rabbit	K.H. Graves and R.W. Moreadith (1993)[66]
Rat	P.M. Iannaccone et al. (1994);[67] N. Ouhibi et al. (1995)[68]
Companion animals	
Cat	X. Yu et al. (2008)[69]
Dog	S. Hatoya et al. (2006);[70] M.R. Schneider et al. (2007)[71]
Farm animals	
Cattle	N.L. First et al. (1994);[72] S.L. Stice et al. (1996)a[73]
Chicken	B. Pain et al. (1996)[74]
Horse	S. Saito et al. (2002)[75]
Mink	M.A. Sukoyan et al. (1993)[76]
Pig	N.C. Talbot et al. (1993);[77] M.B. Wheeler (1994)[78]
Sheep	A.H. Handyside et al. (1987);[79] E. Notarianni et al. (1991)[80]
Primates	
Common marmoset	J.A. Thomson et al. (1996);[81] E. Sasaki et al. (2005)[82]
Monkey	J.A. Thomson et al. (1995);[83] H. Suemori et al. (2001);[84] J.A. Byrne et al. (2007)a[85]
Human	J.A. Thomson et al. (1998);[86] K. Hasegawa et al. (2006)[87]

aThe somatic cell nuclear transfer technique was applied to deriving embryonic stem cells from blastocysts developed from reconstituted eggs.

of differentiated cells introduced with a set of DNA-binding proteins (Oct3/4, Sox2, Klf4, and c-Myc), which play important roles in self-renewal and pluripotency of ESCs. Practically, cells in a hair follicle can be reprogrammed into such induced pluripotent stem cell-like cells. Thus, the current critical need is to induce uniform cell differentiation from stem cells.

CONCLUSION

Collectively, stem cell research requires better understanding of gene expression at a cellular level during differentiation of specialized cells from stem cells. Successful application of stem cells to clinical use depends on improved recapitulation and manipulation of developmental processes in vitro.

ACKNOWLEDGMENTS

I would like to express my special gratitude to Dr. Matthew B. Wheeler for the critical reading of this manuscript. I would like to express my apologies to colleagues whose original studies could not be cited owing to the strict page limit. Images shown in Fig. 1 are obtained under the generous support from the department, which includes the USDA Cooperative State Research, Education, and Extension Service, Hatch project number ILLU-538–323.

REFERENCES

1. Kosaka, M.; Kodama, R.; Eguchi, G. In vitro culture system for iris-pigmented epithelial cells for molecular analysis of transdifferentiation. Exp. Cell Res. **1998**, *245*, 245–251.
2. Rodic, N.; Rutenberg, M.S.; Terada, N. Cell fusion and reprogramming: resolving our transdifferences. Trends Mol. Med. **2004**, *10*, 93–96.
3. Wheeler, M.B. Agricultural applications for transgenic livestock. Trends Biotechnol. **2007**, *25*, 204–210.
4. Murry, C.E.; Keller, G. Differentiation of embryonic stem cells to clinically relevant populations: lessons from embryonic development. Cell **2008**, *132*, 661–680.
5. Solter, D. From teratocarcinomas to embryonic stem cells and beyond: a history of embryonic stem cell research. Nat. Rev. Genet. **2006**, *7*, 319–327.
6. Capecchi, M.R. Gene targeting in mice: functional analysis of the mammalian genome for the twenty-first century. Nat. Rev. Genet. **2005**, *6*, 507–512.
7. Skarnes, W.C.; von Melchner, H.; Wurst, W.; Hicks, G.; Nord, A.S.; Cox, T.; Young, S.G.; Ruiz, P.; Soriano,

P.; Tessier-Lavigne, M.; Conklin, B.R.; Stanford, W.L.; Rossant, J.; International Gene Trap Consortium. A public gene trap resource for mouse functional genomics. Nat. Genet. **2004**, *36*, 543–544.
8. Ohtsuka, S.; Dalton, S. Molecular and biological properties of pluripotent embryonic stem cells. Gene Ther. **2008**, *15*, 74–81.
9. Tanaka, T.S. Transcriptional heterogeneity in mouse embryonic stem cells. Reprod. Fertil. Dev. **2009**, *21*, 67–75.
10. Aiba, K.; Carter, M.G.; Matoba, R.; Ko, M.S. Genomic approaches to early embryogenesis and stem cell biology. Semin. Reprod. Med. **2006**, *24*, 330–339.
11. Jaenisch, R.; Young, R. Stem cells, the molecular circuitry of pluripotency and nuclear reprogramming. Cell **2008**, *132*, 567–582.
12. Graf, T.; Stadtfeld, M. Heterogeneity of embryonic and adult stem cells. Cell Stem Cell **2008**, *3*, 480–483.
13. Uccelli, A.; Moretta, L.; Pistoia, V. Mesenchymal stem cells in health and disease. Nat. Rev. Immunol. **2008**, *8*, 726–736.
14. Gurdon, J.B.; Melton, D.A. Nuclear reprogramming in cells. Science **2008**, *322*, 1811–1815.
15. Yamanaka, S. A fresh look at iPS cells. Cell **2009**, *137*, 13–17.
16. Till, J.E.; McCulloch, E.A.; Siminovitch, L. A stochastic model of stem cell proliferation, based on the growth of spleen colony-forming cells. Proc. Natl. Acad. Sci. U.S.A. **1964**, *51*(1): 29–36.
17. Pittenger, M.F.; Mackay, A.M.; Beck, S.C.; Jaiswal, R.K.; Douglas, R.; Mosca, J.D.; Moorman, M.A.; Simonetti, D.W.; Craig, S.; Marshak, D.R. Multilineage potential of adult human mesenchymal stem cells. Science. 1999, **284**(*5411*): 143–147.
18. Lagasse, E.; Connors, H.; Al-Dhalimy, M.; Reitsma, M.; Dohse, M.; Osborne, L.; Wang, X.; Finegold, M. Weissman, I.L.; Grompe, M. Purified hematopoietic stem cells can differentiate into hepatocytes in vivo. Nat. Med. **2000**, *6*(11): 1229–1234.
19. Orkin, S.H.; Zon, L.I. SnapShot: hematopoiesis. Cell. **2008**, *132*(4): 712.
20. Mayer-Proschel, M.; Kalyani, A.J.; Mujtaba, T.; Rao, M.S. Isolation of lineage-restricted neuronal precursors from multipotent neuroepithelial stem cells. Neuron **1997**, *19*(4):773–785.
21. Bjornson, C.R.; Rietze, R.L.; Reynolds, B.A.; Magil, M.C.; Vescovi, A.L. Turning brain into blood: a hemapoietic fate adopted by adult neural stem cells in vivo. Science. **1999**, *283*(501): 534–537.
22. Hitoshi, S.; Seaberg, R.M.; Koscik, C.; Alexson, T.; Kusunoki, S.; Kanazawa, I.; Tsuji, S.; van der Kooy, D. Primitive neural stem cells from the mammalian epiblast differeftiate to definitive neual stem cells under the control of Notch signaling. Genes and Dev. **2004**, *18*: 1806–1811.
23. Ravin, R.; Hoeppner, D.J.; Munno, D.M.; Carmel, L.; Sullivan, J.; Levitt, D.L.; Miller, J.L.; Athaide, C.; Panchision, D.M.; McKay, R.D. Potency and fate specification in CNS stem cell populations in vitro. Cell Stem Cell. **2008**, *3*(6): 670–680.
24. Tropepe, V.; Coles, B.L.; Chiasson, B.J.; Horsford, D.J.; Elia, A.J.; McInnes, R.R.; van der Kooy, D. Retinal stem cells in the adult mammalian eye. Science. **2000**, *287*(5460): 2032–2036.
25. Cicero, S.A.; Johnson, D.; Reyntjens, S.; Frase, S.; Connell, S.; Chow, L.M.L.; Baker, S.J.; Sorrentino, B.P.; Dyer, M.A. Cells previously identified as retinal stem cells are pigmented ciliary epithelial cells. Proc. Natl. Acad. Sci. U.S.A. **2009**, *106*(16): 6685–6690.
26. Gronthos, S.; Franklin, D.M.; Leddy, H.A.; Robey, P.G.; Storms, R.W.; Gimble, J.M. Surface protein characterization of human adipose tissue-derived stromal cells. J Cell Physiol **2001**, *189*(1): 54–63.
27. Moretti, A.; Caron, L.; Nakano, A.; Lam, J.T.; Bernshausen, A.; Chen, Y.; Qyang, Y.; Bu, L.; Sasaki, M.; Martin-Puig, S.; Sun, Y.; Evans, S.M.; Laugwitz, K.-L.; Chien, K.R. Multipotent Embryonic Isl1$^+$ Progenitor Cells Lead to Cardiac, Smooth Muscle, and Endothelial Cell Diversification. Cell. **2006**, *127*(6): 1151–1165.
28. Bergmann, O.; Bhardwaj, R.D.; Bernard, S.; Zdunek, S.; Barnabé-Heider, F.; Walsh, S.; Zupicich, J.; Alkass, K.; Buchholz, B.A.; Druid, H.; Jovinge, S.; Frisen, J. Evidence for Cardiomyocyte Renewal in Humans. Science. **2009**, *324*(*5923*): 98–102.
29. Barker, N.; van Es, J.H.; Kuipers, J.; Kujala, P.; van den Born, M.; Cozijnsen, M.; Haegebarth, A.; Korving, J.; Begthel, H.; Peters, P.J.; Clevers, H. Identification of stem cells in small intestine and colon by marker gene Lgr5. Nature. **2007**, *449*(7165): 1003–1007.
30. Sangiorgi, E.; Capecchi, M.R. Bmi1 is expressed in vivo in intestinal stem cells. Nat. Genet. **2008**, *40*(7): 915–920.
31. Weiss, T.S.; Lichtenauer, M.; Kirchner, S.; Stock, P.; Aurich, H.; Christ, B.; Brockhoff, G.; Kunz-Schughart, L.A.; Jauch, K.-W.; Schlitt, H.-J.; Thasler, W.E. Hepatic progenitor cells from adult human livers for cell transplantation. Gut. **2008**, *57*(11): 1129–1138.
32. Montarras, D.; Morgan, J.; Collins, C.; Relaix, F.; Zaffran, S.; Cumano, A.; Partridge, T.; Buckingham, M. Direct Isolation of Satellite Cells for Skeletal Muscle Regeneration. Science. **2005**, *309*(5743): 2064–2067.
33. Johnson, J.; Canning, J.; Kaneko, T.; Pru, J.K.; Tilly, J.L. Germline stem cells and follicular renewal in the postnatal mammalian ovary. Nature. **2004**, *428*(6979): 145–150.
34. Eggan, K.; Jurga, S.; Gosden, R.; Min, I.M.; Wagers, A.J. Ovulated oocytes in adult mice derive from non-circulating germ cells. Nature. **2006**, *441*(7097): 1109–1114.
35. Dor, Y.; Brown, J.; Martinez, O.I.; Melton, D.A. Adult pancreatic beta-cells are formed by self-duplication rather than stem-cell differentiation. Nature. **2004**, *429*(6987): 41–46.
36. Stanger, B.Z.; Tanaka, A.J.; Melton, D.A. Organ size is limited by the number of embryonic progenitor cells in the pancreas but not the liver. Nature. **2007**, *445*(7130): 886–891.
37. Seaberg, R.M.; Smukler, S.R.; Kieffer, T.J.; Enikolopov, G.; Asghar, Z.; Wheeler, M.B.; Korbutt, G.; van der Kooy, D. Clonal identification of multipotent precursors from adult mouse pancreas that generate neural and pancreatic lineages. Nat. Biotechnol. **2004**, *22*(9): 1115–1124.
38. Xu, X.; D'Hoker, J.; Stangé, G.; Bonné, S.; De Leu, N.; Xiao, X.; Van de Casteele, M.; Mellitzer, G.; Ling, Z.; Pipeleers, D.; Bouwens, L.; Scharfmann, R.; Gradwohl, G.; Heimberg, H. Beta cells can be generated from endogenous progenitors in injured adult mouse pancreas. Cell. **2008**, *132*(2): 197–207.

39. Leong, K.G.; Wang, B.E.; Johnson, L.; Gao, W.Q. Generations of a prostate from a single adult stem cell. Nature. **2008**, *456*(7223): 804–808.
40. Dyce, P.W.; Wen, L.; Li, J. In vitro gremlin potential of stem cells derived from fetal porcine skin. Nat. Cell. Biol. **2006**, *8*(4): 384–390.
41. Blanpain, C.; Fuchs, E. Epidermal homeostasis: a balancing act of stem cells in the skin. Nat. Rev. Mol. Cell. Biol. **2009**, *10*(3): 207–217.
42. Miura, M.; Gronthos, S.; Zhao, M.; Lu, B.; Fisher, L.W.; Robery, P.G.; Shi, S. SHED: Stem cells from human exfoliated deciduous teeth. Proc. Natl. Acad. Sci. U.S.A. **2003**, *100*(10): 5807–5812.
43. Kanatsu-Shinohara, M.; Inoue, K.; Lee, J.; Yoshimoto, M.; Ogonuki, N.; Miki, H.; Baba, S.; Kato, T.; Kazuki, Y.; Toyokuni, S.; Toyoshima, M.; Niwa, O.; Oshimura, M.; Heike, T.; Nakahata, T.; Ishino, F.; Ogura, A.; Shinohara, T. Generation of pluripotent stem cells from neonatal mouse testis. Cell. **2004**, *199*(7): 1001–1012.
44. Chen, C.; Ouyang, W.; GRigura, V.; Zhou, Q.; Carnes, K.; Lim, H.; Zhao, G.Q.; Arber, S.; Kurpios, N.; Murphy, T.L.; Cheng, A.M.; Hassell, J.A.; Chandrashekar, V.; Hofmann, M.C.; Hess, R.A.; Murphy, K.M. ERM is required for transcriptional control of the spermatogonial stem cell niche. Nature. **2005**, *436*(7053): 1030–1034.
45. Conrad, S.; Renninger, M.; Hennenlotter, J.; Wiesner, T.; Just, L.; Bonin, M.; Aicher, W.; Bühring, H.-J.; Mattheus, U.; Mack, A.; Wagner, H.-J.; Minger, S.; Matzkies, M.; Reppel, M.; Hescheler, J.; Sievert, K.-D.; Stenzl, A.; Skutella, T. Generation of pluripotent stem cells from adult human testis. Nature. **2008**, *456*(7220): 344–349.
46. De Coppi, P.; Bartsch, G. Jr.; Siddiqui, M.M.; Xu, T.; Santos, C.C.; Perin, L.; Mostoslavsky, G.; Serre, A.C.; Snyder, E.Y.; Yoo, J.J.; Furth, M.E.; Soker, S.; Atala, A. Isolation of amniotic stem cell lines with potential for therapy. Nat. Biotechnol. **2007**, *25*(1): 100–106.
47. Kern, S.; Eichler, H.; Stoeve, J.; Klüter, H.; Bieback, K. Comparative analysis of mesenchymal stem cells from bone marrow, umbilical cord blood, or adipose tissue. Stem Cells. **2006**, *24*(5): 1294–1301.
48. Kogler, G.; Sensken, S.; Wernet, P. Comparative generation and characterization of pluripotent unrestricted somatic stem cells with mesenchymal stem cells from human cord blood. Exp. Hemat. **2004**, *34*(11): 1589–1595.
49. Weiss, M.L.; Medicetty, S.; Bledsoe, A.R.; Rachakatla, R.S.; Choi, M.; Mercgav, S.; Luo, Y.; Rao, M.S.; Velagaleti, G.; Troyer, D. Human umbilical cord matrix stem cells: preliminary characterization and effect of transplantation in a rodent model of Parkinson's disease. Stem Cells. **2006**, *24*(3): 7810792.
50. Evans, M.J.; Kaufman, M.H. Establishment in culture of pluripotential cells from mouse embryos. Nature. **1981**, *292*(5819): 154–156.
51. Martin, G.R. Isolation of a pluripotent cell line from early mouse embryos cultured in medium conditioned by teratocarcinoma stem cells. Proc. Natl. Acad. Sci. U.S.A. **1981**, *78*(12): 7634–7638.
52. Thomson, J.A.; Itskovitz-Eldor, J.; Shapiro, S.S.; Waknitz, M.A.; Swiergel, J.J.; Marshall, V.S.; Jones, J.M. Embryonic stem cell lines derived from human blastocysts. Science. **1998**, *282*(5391): 1145–1147.
53. Tanaka, S.; Kunath, T.; Hadjantonakis, A.K.; Nagy, A.; Rossant, J. Promotion of trophoblast stem cell proliferation by FGF4. Science. **1998**, *282*(5396): 2072–2075.
54. Brons, I.G.; Smithers, L.E.; Trotter, M.W.; Rugg-Gunn, P.; Sun, B.; Chuva de Sousa Lopes, S.M.; Howlett, S.K.; Clarkson, A.; Ahrlund-Richter, L.; Pedersen, R.A.; Vallier, L. Derivation of pluripotent epiblast stem cells from mammalian embryos. Nature. **2007**, *448*(7150): 191–195.
55. Tesar, P.J.; Chenoweth, J.G.; Brook, F.A.; Davies, T.J.; Evans, E.P.; Mack, D.L.; Gardner, R.L.; McKay, R.D. New cell lines from mouse epiblast share defining features with human embryonic stem cells. Nature. **2007**, *448*(7150): 196–199.
56. Stevens, L.C. Environmental influence on experimental teratocarcinogenesis in testes of mice. J. Exp. Zool. **1970**, *174*(4): 407–414.
57. Matsui, Y.; Zsebo, K.; Hogan, B.L. Derivation of pluripotential embryonic stem cells from murine primordial germ cells in culture. Cell. **1992**, *70*(5): 841–847.
58. Resnick, J.L.; Bixler, L.S.; Cheng, L.; Donovan, P.J. Long-term proliferation of mouse primordial germ cells in culture. Nature. **1992**, *359*(6395): 550–551.
59. Shamblott, M.J.; Axelman, J.; Wang, S.; Bugg, E.M.; Littlefield, J.W.; Donovan, P.J.; Blumenthal, P.D.; Huggins, G.R.; Gearhart, J.D. Derivation of pluripotent stem cells from cultured human primordial germ cells. Proc. Natl. Acad. Sci. U.S.A. **1998**, *95*(23): 13726–13731.
60. Wakamatsu, Y.; Ozato, K.; Sasado, T. Establishment of a pluripotent cell line derived from a medaka (Oryzias latipes). Mol. Mar. Biol. Biotechnol. **1994**, *3*(4): 185–191.
61. Ma, C.; Fan, L.; Ganassin, R.; Bols, N.; Collodi, P. Production of zebrafish germ-line chimeras from embryo cell cultures. Proc. Natl. Acad. Sci. U.S.A. **2001**, *98*(5): 2461–2466.
62. Doetschman, T.; Williams, P.; Maeda, N. Establishment of hamster blastocyst-derived embryonic stem (ES) cells. Dev. Biol. **1988**, *127*(1): 224–227.
63. Evans, M.J.; Kaufman, M.H. Establishment in culture of pluripotential cells from mouse embryos. Nature. **1981**, *292*(5819): 154–156.
64. Martin, G.R. Isolation of a pluripotent cell line from early mouse embryos cultured in medium conditioned by teratocarcinoma stem cells. Proc. Natl. Acad. Sci. U.S.A. **1981**, *78*(12): 7634–7638.
65. Wakayama, T.; Yanagimachi, R. Effect of cytokinesis inhibitors, DMSO and the timing of oocyte activation on mouse cloning using cumulus cell nuclei. Reproduction. **2001**, *122*(1): 49–60.
66. Graves, K.H.; Moreadith, R.W. Derivation and characterization of putative pluripotential embryonic stem cells from preimplantation rabbit embryos. Mol. Reprod. Dev. **1993**, *36*(4): 424–433.
67. Iannaccone, P.M.; Taborn, G.U.; Garton, R.L.; Caplice, M.D.; Brenin, D.R. Pluripotent embryonic stem cells from the rat are capable of producing chimeras. Dev. Biol. **1994**, *163*(1): 288–292.
68. Ouhibi, N.; Sullivan, N.F.; English, J.; Colledge, W.H.; Evans, M.J.; Clarke, N.J. Initial culture behavior of rat blastocysts on selected feeder cell lines. Mol. Reprod. Dev. **1995**, *40*(3): 311–324.
69. Yu, X.; Jin, G.; Yin, X.; Cho, S.; Jeon, J.; Lee, S.; Kong, I. Isolation and characterization of embryonic stem-like cells

69. derived from in vivo-produced cat blastocysts. Mol. Reprod. Dev. **2008**, *75*(9): 1426–1432.
70. Hatoya, S.; Torii, R.; Kondo, Y.; Okuno, T.; Kobayashi, K.; Wijewardana, V.; Kawate, N.; Tamada, H.; Sawada, T.; Kumagai, D.; Sugiura, K.; Inaba, T. Isolation and characterization of embryonic stem-like cells from canine blastocysts. Mol. Reprod. Dev. **2006**, *73*(3): 298–305.
71. Schneider, M.R.; Adler, H.; Braun, J.; Kienzle, B.; Wolf, E.; Kolb, H.J. Canine embryo-derived stem cells—toward clinically relevant animal models for evaluating efficacy and safety of cell therapies. Stem Cells. **2007**, *25*(7): 1850–1851.
72. First, N.L.; Sims, M.M.; Park, S.P.; Kent-First, M.J. Systems for production of calves from cultured bovine embryonic cells. Reprod. Fertil. Dev. **1994**, *6*(5): 553–562.
73. Stice, S.L.; Strelchenko, N.S.; Keefer, C.L.; Matthews, L. Pluripotent bovine embryonic cell lines direct embryonic development following nuclear transfer. Biol. Reprod. **1996**, *54*(1): 100–110.
74. Pain, B.; Clark, M.E.; Shen, M.; Nakazawa, H.; Sakurai, M.; Samarut, J.; Etches, R.J. Long-term in vitro culture and characterization of avian embryonic stem cells with multiple morphogenetic potentialities. Development. **1996**, *122*(8): 2339–2348.
75. Saito, S.; Ugai, H.; Sawai, K.; Yamamoto, Y.; Minamihashi, A.; Kurosaka, K.; Kobayashi, Y.; Murata, T.; Obata, Y.; Yokoyama, K. Isolation of embryonic stem-like cells from equine blastocysts and their differentiation in vitro. FEBS Lett. **2002**, *531*(3): 389–396.
76. Sukoyan, M.A.; Vatolin, S.Y.; Golubitsa, A.N.; Zhelezova, A.I.; Semenova, L.A.; Serov, O.L. Embryonic stem cells derived from morulae, inner cell mass, and blastocysts of mink: comparisons of their pluripotencies. Mol. Reprod. Dev. **1993**, *36*(2): 148–158.
77. Talbot, N.C.; Rexroad, C.E. Jr.; Pursel, V.G.; Powell, A.M.; Nel, N.D. Culturing the epiblast cells of the pig blastocyst. In Vitro Cell Dev. Biol. Anim. **1993**, *29A*(7): 543–554.
78. Wheeler, M.B.; Development and validation of swine embryonic cells: a review. Reprod. Fertil. Dev. **1994**, *6*(5): 563–568.
79. Handyside, A.; Hooper, M.L.; Kaufman, M.H.; Wilmut, I. Towards the isolation of embryonal stem cell lines from the sheep. Roux's Arch. Dev. Biol. **1987**, *196*(3): 185–190.
80. Notarianni; E. Galli, C.; Laurie, S.; Moor, R.M.; Evans, M.J. Derivation of pluripotent, embryonic cell lines from the pig and sheep. J. Reprod. Fertil. Suppl. **1991**, *43*: 255–260.
81. Thomson, J.A.; Kalishman, J.; Golos, T.G.; Durning, M.; Harris, C.P.; Hearn, J.P. Pluripotent cell lines derived from common marmoset (Callithrix jacchus). Biol. Reprod. **1996**, *55*(2): 254–259.
82. Sasaki, E.; Hanazawa, K.; Kurita, R.; Akatsuka, A.; Yoshizaki, T.; Ishii, H.; Tanioka, Y.; Ohnishi, Y.; Suemizu, H.; Sugawara, A.; Tamaoki, N.; Izawa, K.; Nakazaki, Y.; Hamada, H.; Suemori, H.; Asano, S.; Nakatsuji, N.; Okano, H.; Tani, K. Establishment of novel embryonic stem cell lines derived from the common marmoset (Callithrix jacchus). Stem Cells. **2005**, *23*(9): 1304–1313.
83. Thomson, J.A.; Kalishman, J.; Golos, T.G.; Durning, M.; Harris, C.P.; Becker, R.A.; Hearn, J.P. Isolation of a primate embryonic stem cell line. Proc. Natl. Acad. Sci. U.S.A. **1995**, *92*(17): 7844–7848.
84. Suemori, H.; Tada, T.; Torii, R.; Hosoi, Y.; Kobayashi, K.; Imahie, H.; Kondo, Y.; Iritani, A.; Nakatsuji, N. Establishment of embryonic stem cell lines from cynomolgus monkey blastocysts produced by IVF or ISCI. Dev. Dyn. **2001**, *222*(2): 273–279.
85. Byrne, J.A.; Pedersen, D.A.; Clepper, L.L.; Nelson, M.; Sanger, W.G.; Gokhale, S.; Wolf, D.P.; Mitalipov, S.M. Producing primate embryonic stem cells by somatic cell nuclear transfer. Nature. **2007**, *450*(7169): 497–502.
86. Thomson, J.A.; Itskovitz-Eldor, J.; Shapiro, S.S.; Waknitz, M.A.; Swiergiel, J.J.; Marshall, V.S.; Jones, J.M. Embryonic stem cell lines derived from human blastocysts. Science. **1998**, *282*(5391): 1145–1147.
87. Hasegawa, K.; Fujioka, T.; Nakamura, Y.; Nakatsuji, N.; Suemori, H. A method for the selection of human embryonic stem cell sublines with high replating efficiency after single-cell dissociation. Stem Cells. **2006**, *24*(12): 2649–2660.

Superovulation in Mammals

George E. Seidel, Jr.
Department of Animal Reproduction and Biotechnology Laboratory, Colorado State University, Fort Collins, Colorado, U.S.A.

Abstract

Most mammals ovulate fewer than 100 oocytes over their lifetime, yet many species have 100,000 or more oocytes in their ovaries at puberty. Superovulation permits harvesting 3 to >10 times as many mature oocytes per session as are ovulated in normal reproductive cycles. This usually is accomplished using exogenous follicle-stimulating hormone (FSH) to stimulate FSH receptors on the somatic cells inside growing follicles. For most applications, the resulting oocytes or preimplantation embryos are removed from the donor's reproductive tract and used via various reproductive biotechnologies. Most such oocytes and embryos are normal, but a considerable number can be abnormal. Abnormal oocytes and embryos usually are screened out by failure to fertilize or develop normally, so that the offspring produced from superovulation are similar to those produced without superovulation.

INTRODUCTION

Superovulation refers to pharmacological treatment of animals so that more oocytes are ovulated than normal. Procedures for superovulation vary enormously, depending on the species, the reproductive condition of the female, and the objectives at hand. Superovulation almost always is coupled with procedures to remove oocytes or resulting embryos from the female reproductive tract, since uterine capacity is usually insufficient for normal pregnancy to term if excess embryos implant. Superovulation procedures may vary depending on whether oocytes, early embryos, late embryos, or fetuses are to be recovered, and whether the superovulated donor will also be the recipient of the embryos in the same reproductive cycle, as usually is the case with infertile women.

FOLLICULAR GROWTH AND DEVELOPMENT

The mammalian ovary contains many thousands of follicles in various stages of follicular development, from the primordial follicle, which contains a primary oocyte plus 6–8 follicular cells, to the preovulatory Graafian follicle, which contains the oocyte plus many thousands to millions of follicular cells, depending on the species. Regulation of follicular growth and development depends on many hormones and growth factors; for superovulation, one manipulates only the last few days to a week (depending on the species) of follicular growth before ovulation. I will use the cow as the main example, but similar principles apply to all species, although timing may vary considerably. The cow has been an especially useful model because changes in follicular growth on ovaries can be studied relatively non-invasively by transrectal ultrasonography.

In cattle, from well before puberty, during reproductive cycles, and continuing for months after initiation of pregnancy, there is a 6- to 12-day (usually 7- to 10-day) ovarian cycle of follicular growth and attrition. In cows with normal reproductive cycles, a group of follicles, termed a cohort, starts to grow after estrus.[1] This cohort has from 4 or 5 to 20 or more follicles (sum of both ovaries) that are approximately 4–5 mm in diameter, each containing tens of thousands of follicular cells plus an oocyte. Continued growth of these follicles depends on FSH secreted from the anterior pituitary.

RELEVANT ENDOCRINOLOGY

FSH secretion is regulated by at least four molecules, gonadotropin-releasing hormone (GnRH), activin, inhibin, and follistatin. The first two of these hormones increase FSH secretion, and the last two inhibit it, follistatin via inactivating activin. Because of FSH stimulation, the cohort of follicles continues to grow, with most follicles nearly doubling in diameter (equals an eightfold increase in volume) after 4–5 days. At this time in cows, one (occasionally two) follicle becomes dominant and no longer dependent on FSH for continued growth. This follicle then also secretes copious amounts of inhibin, which greatly decreases the secretion of FSH by the anterior pituitary. This "starves" the subordinate follicles in the cohort of FSH, on which they are still dependent, and they regress by a process termed

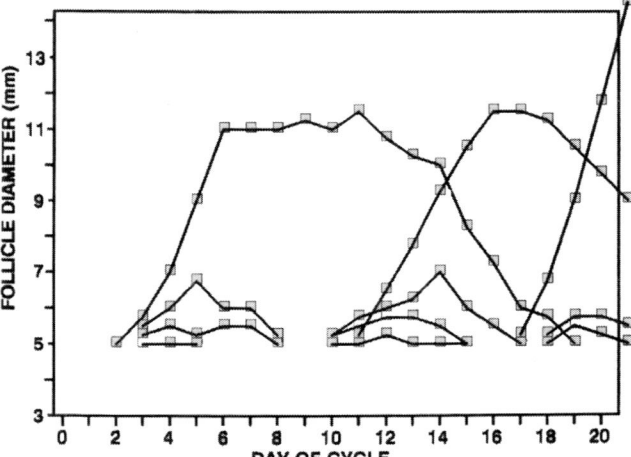

Fig. 1 Waves of bovine follicular growth. **Source:** From Theriogenology.[2]

atresia, in which the follicle cells and oocytes eventually die because of apoptosis.

Meanwhile, the dominant follicle, now dependent on luteinizing hormone (LH), continues to grow. However, the final spurt of follicular growth to >11 mm in diameter requires steady pulses of LH secretion, and this is prevented by the high concentrations of progesterone at early stages of the estrous cycle. Therefore, the dominant follicle also degenerates and stops secreting inhibin. Because of the lack of inhibin, the anterior pituitary now secretes FSH again, starting a new cohort of follicles, which goes through the same cycle again (Fig. 1). Escaping from just repeating this cycle of growth and degeneration of follicles requires removing progesterone from the system. This occurs naturally on approximately Days 16–17 of the reproductive cycle of the cow because of the lysis of the corpus luteum (the source of progesterone) by prostaglandin F-2-α secretion from the uterus. One can also lyse the corpus luteum pharmacologically with an injection of prostaglandin F-2-α earlier in the reproductive cycle. Normally, lysis of the corpus luteum occurs during the third follicular wave of the estrous cycle (occasionally the second or fourth waves), and the dominant follicle then does not become atretic, but grows and ovulates a normal oocyte (Fig. 1). These same events occur in litter-bearing species, but there are multiple dominant follicles and many subordinate ones.

EXOGENOUS FSH RESCUES SUBORDINATE FOLLICLES

The principle of superovulation (perhaps more appropriately termed superstimulation of follicular growth) is to inject exogenous FSH (or equine chorionic gonadotropin [eCG], which stimulates FSH receptors) to make most of the follicles in a cohort into dominant follicles. The dominant follicle will still secrete inhibin, and as the subordinate follicles continue development, they will also secrete inhibin, which really shuts down FSH secretion by the anterior pituitary. However, inhibin does not affect the FSH being injected, so most of the follicles in the cohort continue development. It is still necessary to decrease progesterone secretion so that final maturation and ovulation can occur. This is usually done with a prostaglandin F-2-α injection; a less reliable option is to time FSH injections to coincide with natural luteolysis.

Ideally, FSH injections are timed to start just before selection of the dominant follicle occurs and continue for about 4 days to the point at which the follicles no longer are dependent on FSH. A major problem is that the exact stage of follicular growth will not be known unless ovaries are examined daily by ultrasound, which usually is impractical. Empirically, it has been found that starting FSH injections between Days 9 and 12 of the bovine estrous cycle[3] hits the second follicular wave sufficiently precisely to get a reasonable superovulatory response most of the time. An alternative is to reset follicular waves by removing the dominant follicle, if present, by transvaginal aspiration, or to cause it to ovulate or at least luteinize prematurely. This is done by stimulating an LH surge with GnRH. A third option is to cause regression of the dominant follicle by starving it of FSH, which can be done by injecting an estrogen to repress FSH release. With these approaches to regulating the dominant follicle, one can time FSH injections accordingly. If combined with a progesterone-releasing device, this regimen can be used without regard to the stage of the estrous cycle.[4] Another option commonly used in women is to "downregulate" anterior pituitary function with a GnRH analog[5] so the anterior pituitary stops secreting LH and FSH, and then regulating follicular growth entirely with exogenous hormones.

SPECIES-SPECIFIC CONSIDERATIONS

For many species and situations, efficacious superovulation requires injection of a large dose of LH or human chorionic gonadotropin (hCG) to cause ovulation of the follicles stimulated to grow with the FSH or eCG.[6–8] For postpubertal ruminants and horses, the endogenous LH surge triggered by the high estradiol-17-β secretion of the preovulatory follicles (which causes pulsatile GnRH secretion) is usually sufficient to cause reliable ovulation. In women, hCG is routinely given to induce final maturation of the follicle and oocyte; this is particularly essential if the gonadotropin secreting cells of the anterior pituitary are downregulated. For sheep, often a combination of FSH plus a small amount of eCG is used, usually in conjunction with an intravaginal progestin-releasing device such as a sponge.[9]

For mice and hamsters, which have 4-day reproductive cycles, routine superovulation is induced with an injection of eCG followed by hCG 2 days later. Occasionally, researchers time these injections to fit the mouse reproductive cycle which can be monitored by vaginal changes. However, the more common approach is to superovulate mice between 3 and 4 weeks of age, while they are still prepubertal. This results in 2–3 times more oocytes and embryos than the same treatment at random times of the reproductive cycle after puberty. Superovulating other species just before puberty also works well, although if done way before puberty, oocyte quality seems to be suboptimal. There are numerous other species-specific considerations, e.g., pigs superovulate well immediately after their litters are weaned.

HORMONE DOSES AND TIMING

I have not presented specific information on hormone doses as these depend on many factors, including purity of hormones. Perhaps the most important principle is that of half-life. In the circulation, half-lives of LH and FSH are in the order of 1 and 4 hours, respectively, whereas half-lives of eCG and hCG are around a day. These half-lives depend on the species of origin in the case of LH and FSH and the species being superovulated. Because of these considerations, a single injection of eCG is sufficient to support follicular growth, whereas multiple injections of FSH are used routinely, usually at half-day intervals for 3 or 4 days, but for more days for horses and women. Less frequent FSH injections are possible, especially if steps are taken to delay uptake into the circulation, either by injecting into sites with less blood supply or adding a matrix to the injection bolus, or increasing half-life by covalently linking the alpha and beta subunits of FSH.

The long half-life of eCG has the disadvantage that considerable eCG remains after appropriate follicular growth has occurred. This can interfere with transport of gametes and embryos in the female reproductive tract owing to premature stimulation of progesterone production. One approach is to neutralize eCG with an antibody at the appropriate time, although this is not a common practice.

SUCCESS RATES

In most species, responses to superovulation are extremely variable.[10] Typically superovulation increases oocyte production 3- to 10-fold over the normal rate of ovulation, but there is no increase at all in some individuals, and occasionally 30 or more normal embryos are recovered from a superovulated cow. There is similar variation in other species. When too many follicles are stimulated to grow, the overresponse often results in sufficient premature progesterone production to interfere with normal function of the reproductive tract. Typically, with high progesterone, sperms are not transported to the site of fertilization in the ampullar region of the oviduct, so fertilization does not take place. Embryos from successful fertilization may also be transported from the oviduct to the uterus prematurely, resulting in a suboptimal environment. Even under optimal superovulatory conditions, many fewer sperm reach the site of fertilization than with normal reproduction.[11]

Thus, superovulation requires compromise. The optimum is to increase oocyte (and embryo) production as much as possible without overdoing it, the latter often resulting in complete failure of the process. While superovulatory treatments can be optimized for a population, this is practically impossible for an individual animal, hence the huge variation. Superovulation fails in nearly one-third of outbred animals. A special problem is obtaining reliable FSH preparations, since FSH usually is extracted from pituitaries of slaughtered animals, primarily pigs and sheep. It is not unusual to use porcine or ovine FSH to superovulate other species. One other characteristic of such extracted FSH is that the product is very impure, often less than 1% pure, and thus, contaminated with many other substances, including LH. This contamination usually is of little significance for routine superovulation, but can be a problem for research. Reliable suppliers of FSH discard batches of product that have inappropriate contamination. This entire problem is beginning to be circumvented by production of recombinant FSH by cell lines in vitro.

NORMALITY OF SUPEROVULATED OOCYTES

Superstimulation of follicular growth results in rescuing follicles at various stages of maturation, nearly all of which would have otherwise degenerated by atresia. Under the best of circumstances, fertilization rates for superovulated oocytes are lower than normal, and more resulting embryos develop abnormally than without superovulation, usually manifested by retarded development. This problem usually is exacerbated with prepubertal donors. However, the majority of embryos resulting from superovulation develop into normal animals after appropriate embryo transfer, although pregnancy rates are slightly lower than without superovulation.[3] Abnormal oocytes are screened out at the various steps by not being fertilized, being discarded as abnormal or retarded embryos, and failing to establish pregnancy after embryo transfer.

Many millions of cattle have been produced by transfer of embryos resulting from superovulation, over a million human babies from superstimulation, and over one million offspring from other species combined. When offspring produced via superovulation are studied intensively, a small increase in abnormalities is observed, most of them subtle. For the vast majority of applications, this is a minor issue.[12,13] Superovulation has a major advantage of requiring fewer donor animals for a given project.

REFERENCES

1. Merton, J.S.; de Roos, A.P.W.; Mullaart, E.; de Ruigh, L.; Kaal, L.; Vos, P.L.; Dieleman, S.J. Factors affecting quality and quantity in commercial application of embryo technologies in the cattle breeding industry. Theriogenology **2003**, *59*, 651–674.
2. Fortune, J.R.; Sirois, J.; Quirk, S.M. The growth and differentiation of ovarian follicles during the bovine reproductive cycle. Theriogenology **1988**, *29*, 95–109.
3. Elsden, R.P.; Nelson, L.D.; Seidel, G.E., Jr. Superovulating cows with follicle stimulating hormone and pregnant mare's serum gonadotropin. Theriogenology **1978**, *9*, 17–26.
4. Mapletoft, R.J.; Stewart, K.B.; Adams, G.P. Recent advances in superovulation in cattle. Reprod. Nutr. Dev. **2002**, *42*, 601–611.
5. Davis, O.K.; Rosenwaks, Z. Superovulation strategies for assisted reproductive technologies. Semin. Reprod. Med. **2001**, *19*, 207–212.
6. Cognié, Y.; Baril, G.; Poulin, N.; Mermillod, P. Current status of embryo technologies in sheep and goat. Theriogenology **2003**, *59*, 171–188.
7. Foote, R.H.; Carney, E.W. 2000. The rabbit as a model for reproductive and developmental toxicity studies. Reprod. Toxicol. **2000**, *14*, 477–493.
8. Youngs, C.R. Factors influencing the success of embryo transfer in the pig. Theriogenology **2001**, *56*, 1311–1320.
9. de Graaf, S.P.; Beilby, K.; O'Brien, J.K.; Osborn, D.; Downing, J.A.; Maxwell, W.M.; Evans, G. Embryo production from superovulated sheep inseminated with sex-sorted ram spermatozoa. Theriogenology **2007**, *67*, 550–555.
10. Hasler, J.F. The current status and future of commercial embryo transfer in cattle. Anim. Reprod. Sci. **2003**, *79*, 245–264.
11. Hawk, H.W. Gamete transport in the superovulated cow. Theriogenology **1988**, *29*, 125–142.
12. King, K.K.; Seidel, G.E., Jr.; Elsden, R.P. Bovine embryo transfer pregnancies. I. Abortion rates and characteristics of calves. J. Anim. Sci. **1985**, *61*, 747–757.
13. Elmazar, M.M.; Vogel, R.; Spielmann, H. Maternal factors influencing development of embryos from mice superovulated with gonadotropins. Reprod. Toxicol. **1989**, *3*, 135–138.

Temperate Climate Fruit Crop Pest Management (Plant Pathogens)

David F. Ritchie
Department of Plant Pathology, North Carolina State University, Raleigh, North Carolina, U.S.A.

Abstract

Temperate-climate fruit crops, such as pome fruits (e.g., apples and pears), stone fruits (e.g., peaches, nectarines, cherries, plums), brambles (e.g., blackberries, raspberries), blueberries, grapes, kiwifruit, and strawberries, generally are among the highest value crops grown ($1000 per hectare). However, establishment costs are also high, and except for strawberries, most fruit crops do not begin to bear until at least two years after establishment; consequently, growers need to do as much as possible to protect their investment. It is therefore important, firstly, to select a planting site with good air movement and with access to a full day of sunshine to reduce risks from late spring freezes and to promote optimal coloring and sugar content of fruit as ripening occurs. Secondly, the soil type and structure are also important for good plant growth and the reduction of soil-borne diseases. Thirdly, a certified pathogen-free plant material should be used.

Quality, including "eye appeal," is essential for fruits destined for fresh market. Pathogens, especially foliar pathogens and/or some viruses, can affect marketable yield, thus decreasing profits. Thus, for fruit to be competitive, disease-causing pathogens and other pests must be managed successfully. Disease management is compounded by the array of pathogens that can attack the fruit, foliage, branches, and roots. Occurrence and severity of these diseases are greatly influenced by moisture and temperature. They are primarily managed by a combination of cultural and chemical controls.

PATHOGENS

Representatives of all major pathogen groups, including fungi, bacteria, nematodes, viruses, and phytoplasmas (formally known as mycoplasmalike), cause diseases of fruit crops. Fruit diseases result in direct crop loss; however, viruses, phytoplasmas, other foliar and soil-borne diseases, and nematodes indirectly affect fruit yield and quality as well as the productive longevity of the plant. Many of the pathogens that infect fruit also infect leaves, branches, and stems; thus, the pathogens are able to complete their life cycles on the crop or on nearby reservoir and alternate hosts. Nematodes are parasites that may directly damage the plant by causing severe root damage (e.g., root knot), function as predisposition agents (e.g., lesion and ring nematodes) of the plant to other biological and environmental factors, and can serve as virus vectors (e.g., dagger nematodes). Some of the most common fruit diseases and their pathogens/parasites are listed in Table 1.

MANAGEMENT STRATEGIES

Successful disease management starts with the selection of varieties adapted to the geographical growing region. Equally important is the selection of an appropriate site for growing the fruit crop. Fruit crops and varieties planted in areas and soils to which they are not adapted do not grow well and are more prone to diseases. Once the varieties and a growing site are selected, disease management is built on four basic principles: the use of genetic resistant plants, if available, adapted to the region; the use of disease/pathogen-free plants or planting material, cultural and chemical control of pathogens, and vectors of pathogens; and the use of good sanitation practices throughout the year.

Biological control agents (BCAs) have been most successful where traditional chemical controls are lacking or cannot be used because of concerns for human health and safety reasons. One of the most successful uses of a BCA has been in the management of the bacterial disease crown gall. BCAs also have shown efficacy for management of fire blight, some foliar and fruit fungal diseases, and postharvest fruit diseases.

Soil-borne problems caused by nematodes and fungi pose management challenges with the loss of soil fumigants because of environmental and human safety concerns. Thus, management tactics focus on planting site selection for pathogen avoidance when possible; the use of soil

Table 1 Common diseases and pathogens of fruit crops and the major plant organ(s) infected.[a]

		Primary plant organ(s) infected				
Crop and disease	Pathogen or parasite	Blossom	Fruit	Leaves	Stems, branches, limbs, or trunk	Crown and/ or roots
Brambles						
Anthracnose	*Elsinoe veneta*	+	+	+	+	
Crown/cane gall, hairy root	*Agrobacterium* spp.				+	+
Cane blight and canker	*Botryosphaeria* sp.				+	
	Botrytis sp.				+	
	Leptosphaeria sp.				+	
Rusts	*Arthuriomyces* sp.			+	+	
	Phragmidium sp.			+	+	
	Pucciniastrum sp.			+	+	
Fruit rots	*Botrytis* sp.		+			
	Rhizopus spp.		+			
Root rot	*Phytopthtora* spp.					+
Root-lesion nematode	*Pratylenchus* spp.					+
Dagger nematode	*Xiphinema* spp.					+
Virus and phytoplasma diseases[b]		+	+	+	+	+
Blueberry						
Mummy berry	*Monilinia* sp.	+	+	+	+	
Fruit rot	*Phomopsis* spp.		+			
	Botrytis spp.		+			
	Colletotrichum spp.		+			
Stem cankers and leaf spots	*Botryosphaeria* spp.			+	+	
	Phompopsis spp.			+	+	
	Septoria spp.			+	+	
Root rot	*Phytophthora* spp.					+
Virus and phytoplasma diseases[b]		+	+	+	+	
Grapes						
Powdery mildew	*Uncinula nectator*	+	+	+	+	
Downy mildew	*Plasmopara viticola*	+	+	+	+	
Fruit rot	*Botrytis* sp.		+			
	Guignardia sp.		+	+		
	Colletotrichum spp.		+	+		
Crown gall	*Agrobacterium* sp.				+	+
Pierce's disease	*Xylella fastidiosa.*			+	+	
Root-knot nematode	*Meloidogyne* spp.					+
Dagger nematode	*Xiphinema* spp.					+
Virus and viruslike diseases[b]			+	+	+	+
Kiwifruit						
Fruit rot	*Botrytis cinerea*	+	+			
Bacterial blight and bleeding canker	*Pseudomonas syringae*				+	
	Pseudomonas viridiflava				+	
Crown gall	*Agrobacterium* sp.				+	+
Crown and root rot	*Phytophthora* spp.					+
	Armillaria sp.					+
Pome fruits						
Scab	*Venturia* spp.		+	+	+	
Rusts	*Gymnosporangium* spp.			+	+	

(Continued)

Table 1 Common diseases and pathogens of fruit crops and the major plant organ(s) infected (*Continued*)

		Primary plant organ(s) infected				
Crop and disease	Pathogen or parasite	Blossom	Fruit	Leaves	Stems, branches, limbs, or trunk	Crown and/ or roots
Powdery mildew	*Podosphaera* spp.		+	+	+	
Fire blight	*Erwinia amylovora*	+	+	+	+	
Fruit rots and cankers	*Botryosphaeria* spp.		+	+	+	
	Colletotrichum spp.		+	+	+	
Crown and root rot	*Phytophthora* spp.					+
	Armillaria spp.					+
Root-lesion nematode	*Pratylenchus* spp.					+
Dagger nematode	*Xiphinema* spp.					+
Virus and phytoplasma diseases[b]			+	+	+	+
Strawberry						
Fruit rot	*Botrytis* sp.	+	+			
	Colletotrichum spp.		+	+		
	Phytophthora sp.		+			
	Rhizopus sp.		+			
	Phomopsis sp.	+	+	+		
Powdery mildew	*Sphaerotheca* sp.			+		
Red stele	*Phytophthora* sp.					+
Angular leaf spot	*Xanthomonas* sp.		+	+	+	
Leaf, stem nematode	*Ditylenchus* sp.			+	+	
Root-knot nematode	*Meloidogyne* spp.					+
Root-lesion nematode	*Pratylenchus* spp.					+
Dagger nematode	*Xiphinema* spp.					+
Virus and phytoplasma diseases[b]		+	+	+	+	
Stone fruits						
Brown rot	*Monilinia* spp.	+	+	+	+	
Leaf curl	*Taphrina* sp.		+	+		
Powdery mildew	*Podosphaera* spp.		+	+		
	Sphaerotheca spp.		+	+		
Bacterial canker	*Pseudomonas syringae*	+	+	+		
Bacterial spot	*Xanthomonas arboricola* pv.		+	+	+	
Crown gall	*Agrobacterium* sp.				+	+
Root rot	*Armillaria* sp.					+
	Phytophthora spp.				+	+
Root-knot nematodes	*Meloidoyne* spp.					+
Ring nematode	*Criconemella* sp.					+
Lesion nematode	*Pratylenchus* sp.					+
Dagger nematode	*Xiphinema* sp.					+
Virus and phytoplasma diseases[b]			+	+	+	+

[a]Listing of diseases and pathogens is not meant to be exhaustive. More information may be obtained in other sources such as the Crop Diseases Compendium series published by the American Phytopathological Society, St. Paul, MN, USA.
[b]Fruit crops, most of which are vegetatively propagated, are prone to numerous viral, viral-like, and phytoplasma diseases.

amendments, groundcover, and secondary host management; host resistance; and biocontrol.

Although the use of fruit varieties having disease resistance is ideal, it is a difficult goal to achieve because of the characteristics of both the crop and the pathogens. Most fruit crops do not come into bearing for several years; they do so when a dormancy period is fulfilled. Thus, breeding programs for fruit crops are long-term endeavors. Additionally, many varieties are selected for "consumer appeal" and shipping-and-storage qualities rather than disease resistance. For complete reliance upon host resistance, normally a high level of resistance is needed because of market demands for blemish-free fruit. Also, fruit crops are affected by multiple pathogens; thus, resistance to one pathogen still may not negate the need to use fungicides to manage others. On the pathogen side, pathogen populations evolve or are selected; they can defeat host resistance. High levels of pathogen resistance often are associated with genotype-specific resistance (i.e., vertical resistance), which is conferred by single genes and is prone to nondurability. Biotechnology and the use of transgenic methods hold great promise for developing disease resistance in crops that have long cycles of development when traditional breeding methods are used. However, there may be potential biological (e.g., lack of single-gene durability) and social (e.g., lack of consumer acceptance) pitfalls regarding the use of transgenics.

Starting with certified disease-free plants is essential for successful production of fruit crops. This is essential for management of virus-caused diseases. Once the crop is established, sanitation plays a major role in successful disease management. Most fruit crops are perennial, and many pathogens survive from one bearing season to the next either on the host plant or within or near the crop site in reservoir and alternate hosts. Thus, sanitation practices that remove diseased fruit and plant parts from the crop area aid in inoculum and usually subsequent disease reduction.

Fungicides

Fungicides have traditionally and still play an important role in disease management of fruit crops because many diseases cannot be controlled adequately otherwise. The appearance of disease symptoms and signs shows that the pathogen has successfully infected the host; thus, monitoring for disease symptoms and signs per se is not adequate for managing many diseases. This is particularly true for diseases of fruit that develop rapidly, causing rots such as brown rot of stone fruits. In contrast, some diseases, such as powdery mildew, occur on the foliage or do so prior to infecting fruit. Other diseases have latent periods of days or weeks between infection and the occurrence of symptoms, thus allowing time for use of an eradicant fungicide before economic damage occurs. Development of fungicides having eradicative or curative properties in the last half of the 20th century further stimulated the development of forecast models. Although models may accurately predict the occurrence of disease, they have little practical value if effective interventions (e.g., eradicant fungicides) are not available to either prevent or eradicate the infection.

Forecast Models

Most disease-forecast models are based upon defining the relationship between particular weather conditions as they influence the infection process and crop phenology. The Mill's model, published in 1944, was designed to aid in timing sulfur dust applications for the control of apple scab. It was based on the concept that the occurrence and severity of apple scab was related to the time length of leaf wetness and the temperature during the wetting period. Based on the concept of this model, predictive disease models have been developed for many other diseases of fruit crops. These include fire blight of pome fruits, rusts and many of the fruit rot and blemishing diseases of apples, downy and powdery mildews and fruit rots of grapes, and leaf spots of cherries. These models have been used with varying levels of disease management and economic success. Some of the success or failure of predictive models is associated with the number of diseases occurring on a given fruit crop, the accuracy in measurement of environmental conditions and weather forecasts, the severity of conditions for infection and disease development, and the effectiveness of intervention tactics.

INFORMATION DELIVERY

Since the mid-1990s, the World Wide Web has established itself as a rapid source of information for aiding in making disease management decisions. Information, such as aid in diagnosis, review of management strategies, the latest on pesticides, access to scientific and production journals, and real-time weather forecasts, are available. Some of this information can be availed at no cost, while others are available on a subscription basis.

CONCLUSION

The management of fruit crop diseases will become more complex, but will increase in efficiency. Improved methods for pathogen detection and quantification combined with more accurate weather forecasts will aid in the prediction of infection and subsequent disease occurrences. Crop protection chemicals that are highly specific and have low toxicity to nontarget organisms and the environment will continue to be developed and are incorporated into spray programs. Synthetic chemicals and BCA that activate natural plant defense systems will receive increased investigation and applications. The use of molecular biology will aid in shortening the time required to breed disease-resistant varieties and possibly in the development of highly effective BCAs.

FURTHER READING

Briggs, A.R.; Grove, G.G. Role of the World Wide Web in extension plant pathology: Case studies in tree fruits and grapes. Plant Dis. **1998**, *82*, 452–464.

Compendium of Apple and Pear Diseases; Jones, A.L., Aldwinckle, H.S. Eds.; APS Press: St. Paul, MN, **1990**, 100.

Compendium of Stone Fruit Diseases; Ogawa, J.M., Zehr, E.I., Bird, G.W., Ritchie, D.F., Uriu, K., Uyemoto, J.K., Eds.; APS Press: St. Paul, MN, **1995**, 98.

Funt, R.C.; Ellis, M.A.; Madden, L.V. Economic analysis of protective and disease-forecast-based fungicide spray programs for control of apple scab and grape black rot in Ohio. Plant Dis. **1990**, *74*, 638–642.

Janisiewicz, W.J. Biocontrol of Postharvest Diseases of Temperate Fruits—Challenges and Opportunities. In *Plant-Microbe Interactions and Biological Control*; Boland, G.J., Kuykendall, D., Eds.; Marcel Dekker: New York, **1998**; 171–198.

Sikora, R.A. Management of the antagonistic ecosystem for the biological control of plant parasitic nematodes. Annu. Rev. Phytopathol. **1992**, *30*, 245–270.

Steiner, P.W.; Lightner, G. *MARYBLYT: A Predictive Program for Forecasting Fire Blight Disease in Apples and Pears. Version 4.0 of a Copyrighted Computer Program*; Univ. Maryland: College Park, **1992**.

Stockwell, V.O.; Johnson, K.B.; Loper, J.E. Establishment of bacterial antagonists of *Erwinia amylovora* on pear and apple blossoms as influenced by inoculum preparation. Phytopathology **1998**, *88*, 506–513.

Sutton, T.B. Changing options for the control of deciduous fruit tree diseases. Annu. Rev. Phytopathol. **1996**, *34*, 527–547.

UCPMG. Pest Management and Identification. Pests of Agricultural Crops, Floriculture, and Commercial Turf. In *UC Pest Management Guidelines*; **1999**, Davis, CA. http://www.ipm.ucdavis.edu/PMG/crops-agriculture.html (accessed April 5, 2002).

Transgenes: Plant Breeding

Duska Stojsin
Trail Development Group, Monsanto Company, St. Louis, Missouri, U.S.A.

Abstract
Transgenic crop breeding and evaluation approaches are based on the same general principles as those practiced by breeders working with non-transgenic crops. There are, however, specific considerations associated with evaluation of transgenic crops due to inheritance differences, evaluation criteria, and regulatory requirements.

Generally, there are two distinct breeding stages associated with the commercial development of a transgene. The first stage involves evaluating and selecting a transgenic event, and the second stage involves integrating the event into commercial germplasm.

STAGE 1: CONSTRUCT AND EVENT SELECTION

A construct contains an engineered DNA sequence that is inserted into the plant genome via a transformation method. Numerous transformants are usually generated for a construct, each resulting in unique insertion site(s). These independent transformation events need to be evaluated in order to identify those with superior performance. The construct and event evaluation process is based on five major criteria[1]: 1) transgene expression/efficacy; 2) molecular characterization of the insert; 3) segregation of the trait of interest; 4) agronomics of the developed lines; and 5) stability of the transgene expression. Screening criteria usually stay the same throughout the evaluation process, but stringency of testing generally increases for each criterion in more advanced generations. As with traditional breeding, a large population of independent transformation events and a more thorough evaluation would result in a greater chance of success.[2]

1. Expression of a transgene is defined by the level of protein or protein product generated or modified by the inserted DNA construct. Efficacy of a transgene is defined as the phenotype observed in plants that contain the gene of interest (e.g., herbicide tolerance and insect resistance). Factors that can affect the level of expression and efficacy of a given event are insertion site, transgene copy number, intactness of the transgene, zygosity of the gene of interest (GOI), level of inbreeding associated with a genotype, the genetic background, and environmental conditions.
2. Molecular characterization is based on a variety of assays to determine presence of GOI, copy number, insert number, insert complexity, presence of the vector backbone, and development of event-specific assays. These assays assist in identification of events that exhibit simple inheritance and are more likely to contribute to stable expression of the transgene. Molecular characterization also provides tools for determining event purity and identity, and is a required component of various regulatory submissions.
3. Segregation of the trait of interest is tested in order to identify transgenic events that follow a single-locus segregation pattern. A direct approach is to evaluate the segregation of the trait of interest; an indirect approach is to assess the selectable marker segregation (if one is associated with the transgene). Regardless of the approach, this criterion becomes more important once breeders can evaluate large number of plants, family segregation, and the self-pollinated progeny of a given event.
4. Agronomic characteristics may vary among events due to somaclonal variation, insertional effects, homozygosity of a transgene, level of inbreeding of the plant genome, and genetic background. In advanced generations, agronomic performance should be evaluated in several genetic backgrounds in replicated trials. For hybrid crops, agronomic trials should be conducted in both inbred and hybrid background.
5. Stability of transgene expression needs to be tested in different generations, environments, and in different genetic backgrounds. Event instability over generations is often caused by transgene inactivation due to multiple transgene copies, zygosity level, highly methylated insertion site, or level of stress.[3] Such events should be discarded even when gene silencing is reversible.

STAGE 2: COMMERCIAL PLANT BREEDING INVOLVING TRANSGENES

Transgenes used in commercial breeding programs are generally selected for the criteria discussed above before being transferred to commercial plant breeding programs. Generally, events with a single intact insert will be inherited as a single dominant gene and follow Mendelian segregation ratios. Commercial breeding strategies include: 1) backcrossing and 2) forward breeding.

1. Backcrossing is an efficient breeding method when the goal is to recover the genotype of an elite inbred line with the addition of a transgene. In each backcross generation, plants that contain the transgene are identified and crossed to the recurrent parent. Several backcross generations with selection for recurrent parent phenotype are generally used by commercial breeders to recover the genotype of the elite parent. During backcrossing, the transgene is kept in a hemizygous state; therefore, at the end of backcrossing it is necessary to self or sib pollinate plants containing the transgene to fix the transgene in a homozygous state. The number of backcross generations can be reduced by molecular-assisted backcrossing (MABC). The MABC method uses molecular markers to identify plants that are most similar to the recurrent parent in each backcross generation. With the use of MABC and appropriate population size, it is possible to identify plants that have recovered over 98% of the recurrent parent genome after only two or three backcross generations. By eliminating several generations of backcrossing, it is often possible to bring a commercial transgenic product to market one year earlier vs. conventional backcrossing. Backcrossing and MABC are routinely used to develop transgenic crop products.
2. Forward breeding is any breeding method that has the goal of developing a transgenic variety, inbred line, or hybrid that is genotypically different, and superior, to the parents used to develop the improved genotype. When forward breeding a transgenic crop, selection pressure for the efficacy of the transgene is usually applied during each generation of the breeding program. Additionally, it is usually advantageous to fix the transgene in a homozygous state during the breeding process as soon as possible to fully realize the benefit of the transgene or uncover potential agronomic problems caused by unfavorable transgene-by-genotype interactions. Forward breeding has been used extensively to develop Roundup Ready® varieties of soybean [*Glycine max* (L.) Merr.] and cotton (*Gossypium hirsutum* L.), as well as Bollgard™ cotton.

After integrating the transgene into commercial germplasm, the final product should be tested in multiple locations. Testing typically includes yield trials in trait neutral environments (e.g., insect-free environments for insect protected plants), as well as typical environments in the target market. If the new transgenic product has been derived from backcrossing, it is usually tested for performance equivalency by comparing it to the non-transgenic version in all environments.

QUALITY ASSURANCE AND QUALITY CONTROL

Different government agencies worldwide are regulating transgenic crops. In the United States, approval of biotechnology crops falls under the jurisdiction of three agencies: U.S. Department of Agriculture (USDA), Environmental Protection Agency (EPA), and Food and Drug Administration (FDA). One of the regulatory requirements is that commercial transgenic crops should not contain unintended transgenic events. Breeders working with transgenes must have appropriate quality assurance (QA) and quality control (QC) programs in place to avoid and/or detect errors that may result during the event evaluation and selection process.

- Quality assurance is the set of best practices used to ensure the overall quality of the final product from the standpoint of transgene purity.[4] The setup of a transgenic breeding nursery should be designed to minimize the possibility of contamination between non-transgenic and transgenic material or among different transgenic events. Nursery designs should use spatial and/or temporal isolation between non-transgenic and transgenic plants as well as among different transgenic events. Other important QA methods include utilizing a consistent and concise pedigree nomenclature that clearly identifies the germplasm and transgene being integrated, checking phenotypic segregation ratios for the gene(s) of interest in each generation, and utilizing differences in germplasm (e.g., flower color, maturity, and plant height) to maximize the identification of off-types due to outcrossing. Additional QA measures are often practiced in cross-pollinated crops due to the increased risk of outcrossing. Some of these include restricting movement of a pollinating crew from one nursery to another, or increasing the spatial isolation between different transgenic events.
- Quality control is the testing conducted to verify that the desired quality is actually achieved.[4] Quality control is done with the objective to test the seed purity and integrity for the presence of an intended event. In addition, it is important to test for the absence of unintended transgenes, regardless of whether they control

the same trait as the event of interest or different traits. Stringent QC standards ensure that the event purity and integrity is maintained in the evaluation, breeding, and seed increase portion of transgene development.

CONCLUSIONS

The product of transgenic crop breeding is a new cultivar containing a transgene at a locus that controls or modifies expression of a given trait. During the development and screening, events with single intact inserts, high and stable transgene expression should be selected. Efficacy and agronomic performance of the transgene needs to be demonstrated in different genetic backgrounds and in different environments. After incorporating the transgene into commercial germplasm, the final product should also be tested in multiple locations. Regulated handling of transgenic seed adds to the complexity of developing transgenic crops. However, social, economic, and environmental benefits associated with these products[5–7] make these efforts worthwhile.

REFERENCES

1. Stojšin, D.; Behr, C.F.; Heredia, O.; Stojšin, R. Evaluation and breeding of transgenic corn. Genetika **2000**, *32* (3), 419–430.
2. Conner, A.J.; Christey, M.C. Plant breeding and seed marketing options for the introduction of transgenic insect-resistant crops. Biocontrol Sci. Tech. **1994**, *4*, 463–473.
3. Zhong, G.Y. Transgene genetics and breeding: what do we know? Proc. 53rd Annu. Corn Sorghum Res. Conf. **1998**, *53*, 250–269.
4. Hall, M.; Moeghji, M.; Parker, G.; Peterman, C.; Yates, D. Strategies to maximize quality assurance (QA) when integrating transgenes in breeding programs. Illinois Corn Breeders School **2002**, *38*.
5. James, C.; *Global Status of Commercialized Biotech/GM Crops: 2005;* ISAAA briefs No. 34-2005; ISAAA: Ithaca, NY, **2005**.
6. Brookers, G.; Barfoot, P.G.M. Crops: the global economic and environmental impact—the first nine years 1996–2004. AgBioForum **2005**, *8* (2/3), 187–196.
7. Fernandez-Cornejo, J; Caswell, M. *The First Decade of Genetically Engineered Crops in the United States;* Economic Research Service/USDA Bulletin No. 11, **2006**.

Transgenic Animals

Vernon G. Pursel
Agricultural Research Service, U.S. Department of Agriculture (USDA-ARS), Beltsville, Maryland, U.S.A.

Abstract
This entry describes each of the four primary methods that have been used to transfer genes into farm animals and the methods used to subsequently evaluate the animals that have resulted from the transfer of genes to prove that the genes are present in the genome, are functioning, and are being transmitted to the next generation.

INTRODUCTION

The genetic composition of all living creatures is continually undergoing alteration by mutation, natural selection, and genetic drift. Beginning at the onset of plant and animal domestication, humans have further manipulated the genetic composition of plant and animals to enhance their health and usefulness to humans by selecting for specific phenotypic traits. Development of recombinant DNA technology has enabled scientists to isolate single genes, analyze and modify their nucleotide structures, make copies of these isolated genes, and insert copies of these genes into the genome of plants and animals. The procedure used to insert these isolated genes is called "gene transfer," an animal that contains the inserted gene or genes is called a "transgenic animal," and the transferred gene is called a "transgene."

The first intentional transfer of a transgene into the genome of an animal was achieved in 1980 in mice.[1] Gene transfer methodology was subsequently successfully applied to pigs, sheep, and rabbits.[2] Gene transfer has thus far been used most extensively for basic research on all aspects of biology and genetics, but it has numerous potential applications for genetic improvement of farm animals. Practical applications of gene transfer in livestock production include improved milk production and composition, increased growth rate, improved feed use and carcass composition, increased disease resistance, enhanced reproductive performance, and increased prolificacy. Gene transfer in farm animals has also been investigated extensively for potential to produce human pharmaceutical products, and alteration of cell or tissue characteristics for potential use in organ transplantation in humans.

GENE TRANSFER METHODS

Pronuclear Microinjection

The primary method used to produce transgenic farm mammals has been the direct microinjection of the transgene into the pronucleus of a zygote (recently fertilized ovum or egg). As in the mouse, pronuclei of rabbit, sheep, and goat zygotes can be readily seen using phase-contrast microscopy or differential interference contrast (DIC) microscopy. Lipid granules in the cytoplasm interfere with visualization of pronuclei of pig and cow zygotes. Centrifugation of pig and cow zygotes can be used to stratify the cytoplasm so that pronuclei are visible with use of DIC microscopy.[3]

To permit microinjection, ova are placed on a depression slide in a microdrop of media that is overlaid with silicone or paraffin oil to prevent evaporation. The microscope must be equipped with two micromanipulators, one for an egg-holding pipette and the other for an injection pipette. The holding pipette and injection pipette are each fitted with a tube leading to a syringe that permits either gentle suction or carefully controlled fluid injection. As an ovum is held with light suction by the holding pipette, the tip of the injection pipette is inserted through the zona pellucida and cytoplasm into the most visible pronucleus. Several hundred copies of the gene are expelled into the pronucleus. The person performing the injection carefully observes the pronucleus and withdraws the pipette when the pronuclear structure has visibly enlarged. After microinjection, cow embryos are usually cultured in vitro until they are morulae or blastocysts before non-surgical transfer into the uterus of a synchronous host cow. The injected zygotes of the other species are usually cultured only a few hours before they are surgically transferred directly into the oviduct of synchronous host females.

The mechanism by which a transgene integrates into a chromosome is unknown. Usually multiple copies of a transgene integrates in a head to tail array in a single site on a chromosome but multiple integrations can occur. If a transgene integrates in a zygote after DNA replication has occurred or at a subsequent stage of development, then the transgenic founder will be genetically mosaic since the transgene will only be present in a portion of the cells. Breeding studies with transgenic pigs and sheep indicate mosaicism is a definite problem. As a consequence, about

20% of transgenic founder animals fail to transmit the transgene to progeny and another 20–30 % transmit the transgene to less than 50 % of their progeny.

The efficiency is usually lower for integration of transgenes into farm animals than into mice. The percentage of gene-injected zygotes that develop into transgenic animals varied from 0.3 to 4.0% for pigs, 0.1 to 4.4% for sheep, 1.0 to 1.7% for goats, and 0.3 to 2.6% for cattle. A few transgenic chickens have been produced by microinjection of genes into the germinal disk of the recently fertilized egg.[4] After microinjection, the chick embryos were cultured in a host eggshell until hatching time.

Retroviral Insertion

Retroviruses can be modified by recombinant DNA techniques to make them replication-defective and replace part of the viral DNA with a desired transgene so they can then be used as a gene vector. Retroviral-mediated gene transfer was originally used to insert transgenes into mouse embryos[5] and blostodermal cells of chicken eggs.[6] In comparison to microinjection, retroviral infection offers advantages of: 1) integration of single copies of the gene; and 2) retroviral DNA integrates into a high percentage of embryos when exposed to high concentrations of viral stock by coculture with infected cells in vitro, or in the case of chickens, by microinjection into the blastodisk. The disadvantages are: 1) added work to produce a retrovirus carrying the transgene; 2) the gene being transferred must be smaller than 10 kb in size; 3) resulting transgenic animals are frequently highly mosaic, which necessitates extensive outbreeding to establish pure transgenic lines; and 4) unresolved problems with hypermethylation interfering with expression of the transgene.

More recently, retrovirus-mediated gene transfer has been used to produce transgenic cattle by insertion of retroviruses into the metaphase II oocytes to avoid mosaicism and ensure that a high percentage of the offspring are transgenic.[7]

Many laboratories involved in production of transgenic livestock have not embraced the use of retroviral insertion technology because of concerns about public perception and the potential consequences of recombination events between the viral vectors and endogenous retroviruses to generate new pathogenic agents.

Cellular Insertion

The third method of introducing genes into the germ line is a two-step process involving first the transfection of a transgene into embryonic stem (ES) cells, embryonic germ (EG) cells (also known as primordial germ cells), or fetal somatic cells during in vitro culture, and then incorporating the transgenic ES or EG cells into an inner cell mass of an embryo or inserting the transgenic cell's nucleus into an enucleated oocyte by nuclear transfer (NT). The advantage of this procedure is that a particular genotype can be selected during in vitro culture before introduction of the cells into the embryo or NT. In addition, this technique provides the ability for site-specific insertion of a transgene by homologous recombination. This approach with ES cells has been used extensively in the mouse but has not been effective in other mammalian animals because of extreme difficulty in isolating and maintaining ES or EG cells in the undifferentiated state during in vitro culture. Nuclear transfer (also known as animal cloning) is currently being extensively investigated in cattle, goats, and pigs. Consequently, cellular insertion by NT may become the method of choice for gene transfer in these species because relatively few recipient hosts are required to produce transgenic founder animals.

Sperm-Mediated Gene Transfer

The simplest but most controversial method of gene transfer involves merely mixing a transgene with spermatozoa and using them to fertilize oocytes, either in vitro or by artificial insemination. The use of sperm-mediated transfer in mice by Dr. Lavitrano[8] was initially discounted as unrepeatable by many investigators. During the past decade, research on this procedure has persisted and many investigators report successful gene transfers by this technique. Only a few studies have provided convincing evidence that the transgene was unaltered before or during the integration process and capable of expressing appropriately in the resulting transgenic animals.[9] In addition, recent investigations indicate that most of genes transferred by this method are not integrated into the host genome but remain extrachromosomal and replicate independent from the host genome.[10]

EVALUATION OF TRANSGENIC ANIMALS

Integration

The primary way that the presence of the transgene is confirmed is by removing a piece of tail tissue at birth, extracting the DNA from the tissue, and analysis of the DNA for the presence of the transgene by Southern blot hybridization, slot-blot hybridization, or polymerase chain reaction using a unique segment of the transgene as a probe. If performed correctly, the Southern blot analysis provides information on presence, intactness, copy number, and orientation of the transgene.

Expression

Mere presence of the transgene in the transgenic animal does not guarantee that the transgene will be expressed or expressed appropriately. Expression is usually evaluated by assay of appropriate tissues or fluids recovered from the transgenic animal for presence of the transgene transcription by Northern blot analysis of the mRNA, by presence of the protein using Western immunoblot analysis, or by some other assay that is appropriate for specific transgene.

Transmission

The primary aim of transgenesis is to establish a new genetic line of animals in which the trait is stably transmitted to succeeding generations. This can only be determined by mating the transgenic animal to a non-transgenic animal and subsequent evaluation of the progeny for presence and expression of the transgene.

CONCLUSION

Transgenic animals were initially produced by pronuclear microinjection. During the past two decades, a number of other gene transfer methods have been developed and include use of retroviral vectors, cellular insertion, and sperm-mediated transfer. Additional details regarding each of these procedures can be found in *Transgenic Animal Technology: A Laboratory Handbook*.[11]

REFERENCES

1. Gordon, J.W.; Scangos, G.A.; Plotkin, D.J.; Barbosa, J.A.; Ruddle, F.H. Genetic transformation of mouse embryos by microinjection of purified DNA. Proc. Natl. Acad. Sci. U.S.A. **1980**, *77* (12), 7380–7384.
2. Hammer, R.E.; Pursel, V.G.; Rexroad, C.E., Jr.; Wall, R..J.; Bolt, D.J.; Ebert, K.M.; Palmiter, R.D.; Brinster, R.L. Production of transgenic rabbits, sheep and pigs by microinjection. Nature **1985**, *315* (6021), 680–683.
3. Wall, R.J.; Pursel, V.G.; Hammer, R.E.; Brinster, R.L. Development of porcine ova that were centrifuged to permit visualization of pronuclei and nuclei. Biol. Reprod. **1985**, *32* (3), 645–651.
4. Love, J.; Gribbin, C.; Mather, C.; Sang, H. Transgenic birds by DNA microinjection. Biotechnology **1994**, *12* (1), 60–63.
5. Soriano, P.; Cone, R.D.; Mulligan, R.C.; Jaenisch, R. Tissue-specific and ectopic expression of genes introduced into transgenic mice by retroviruses. Science **1986**, *234* (4782), 1409–1413.
6. Salter, D.W.; Smith, E.J.; Hughes, S.H.; Wright, S.E.; Crittenden, L.B. Transgenic chickens: insertion of retroviral genes into the chicken germ line. Virology **1987**, *157* (1), 236–240.
7. Chan A.W.; Homan E.J.; Ballou L.U.; Burns, J.C.; Bremel, R.D. Transgenic cattle produced by reverse-transcribed gene transfer in oocytes. Proc. Natl. Acad. Sci. U. S. A. **1998**, *95* (24), 14028–14033.
8. Lavitrano, M.; Camaioni, A.; Fazio, V.M.; Dolci, S.; Farace, M.G.; Spadafora, C. Sperm cells as vectors for introducing foreign DNA into eggs: genetic transformation of mice. Cell **1989**, *57* (5), 717–723.
9. Wall, R.J. New gene transfer methods. Theriogenology **2002**, *57* (1), 189–201.
10. Smith, K.; Spadafora, C. Sperm-mediated gene transfer: application and implications. BioEssays **2005**, *27* (5), 551–562.
11. Pinkert, C.A. *Transgenic Animal Technology: A Laboratory Handbook*, 2nd Ed.; Academic Press: San Diego, 2002.

Transgenic Animals: Mitochondrial Genome Modification

Carl A. Pinkert
Department of Pathobiology, College of Veterinary Medicine, Auburn University, Auburn, Alabama, U.S.A.

Lawrence C. Smith
University of Montreal, Montreal, Quebec, Canada

Ian A. Trounce
University of Melbourne, Melbourne, Victoria, Australia

Abstract

In comparison to the techniques successfully employed for nuclear gene transgenesis in livestock over the past 20 years, the lack of comparable recombination in mitochondrial DNA (mtDNA) has, until recently, prevented its direct in vivo manipulation. The coordinated expression of single-copy nuclear gene products, together with the polyploid mtDNA gene products, is required for normal mitochondrial biogenesis and respiratory chain function. It is of great current interest to seek improved technologies for manipulating the mitochondrial genome, so that interactions of nuclear and mtDNA genotypes can be studied in experimental systems.

MITOCHONDRIAL GENETICS AND ANIMAL MODELING

Mammalian mitochondria contain between one and approximately ten copies of a closed, circular, supercoiled, double-stranded DNA that is bound to the inner mitochondrial membrane and is not associated with histones or a scaffolding protein matrix. The mtDNAs of all vertebrates are highly conserved and quite small (~16.5 kb in length) in comparison to the nuclear genome. Mammalian mitochondria have their own genetic systems, replete with a unique genetic code, genome structure, transcriptional and translational apparatus, and tRNAs. Perhaps, because of a postulated less-extensive mitochondrial DNA (mtDNA) repair system and because of the absence of protective histones, the mitochondrial genome is subject to an increased sensitivity to mutations due to metabolic (e.g., oxidative stress) and environmental (e.g., toxins, mutagens, and UV light) sources. Mitochondrial genes encode for 13 of the protein subunits that function in the mitochondrial oxidative phosphorylation system, along with two ribosomal RNAs (rRNAs) and 22 transfer RNAs (tRNAs). Accordingly, directed modification of mitochondrial genes and/or their function would provide a powerful tool in production agriculture.[1]

Cytoplasmic-based traits in domestic animals have included growth, reproduction, and lactation. In addition, mitochondrial restriction fragment-length polymorphisms (RFLPs) were identified and associated with specific lactational characteristics in a number of dairy cattle lineages. The matrilineal inheritance of mammalian mtDNA has also been used to advantage in studies exploring the timing and geography of domestication events, as recently demonstrated for horses, where multiple domestication events appear to have occurred in the Eurasian steppe.[2] In addition, metabolic and cellular abnormalities in humans were correlated to mutations arising exclusively within the mitochondrial genome. Indeed, various diseases have been associated with mtDNA point mutations, deletions, and duplications (e.g., diabetes mellitus, myocardiopathy, and retinitis pigmentosa) as well as age-associated changes in the functional integrity of mitochondria (as seen in Parkinson's, Alzheimer's, and Huntington's diseases). As such, for both agricultural and biomedical research efforts, the ability to manipulate the mitochondrial genome and to regulate the expression of mitochondrial genes would provide one possible mode of genetic manipulation and therapy.

The creation of heteroplasmic transmitochondrial animals has developed along three lines: 1) direct mitochondrial injection into oocytes or embryos; 2) embryonic stem (ES) cell-based technologies; and 3) in relation to karyoplast or cytoplast transfer (including consequences associated with nuclear transfer or cloning experimentation; Table 1). These techniques have illustrated model systems that will provide a greater understanding of mitochondrial dynamics, leading to the development of genetically engineered production animals, and therapeutic strategies for human metabolic diseases affected by aberrations in mitochondrial function.

As described in a number of recent reports,[3,4] nuclear-encoded genes and knock-out modeling have been informative in identifying novel models in mitochondrial

Table 1 Methods for creating mitochondrial modifications in animals.

Method	Heteroplasmy/homoplasmy detected	Germline transmission	Limitations
Mitochondrial injection into ova	Heteroplasmy	Yes	Low-level heteroplasmy
Karyoplast fusion (nuclear transfer)	Heteroplasmy	Yes	Varying efficiencies using PEG or electrofusion
Karyoplast or cytoplast transfer into ES cells and transfer	Yes	Yes	Availability of germline competent/efficient cell lines
Cytoplast/ooplasm transfer	Heteroplasmy	Yes	Varying efficiencies and low-level heteroplasmy
Sperm mediated	?	?	Rare event, aberrant recombination, or programmed destruction postfertilization

disease pathogenesis as well as critical pathways associated with mitochondrial function. With initial characterization of these nuclear gene-encoded models, our search for a greater understanding of mitochondrial interactions and function would eventually lead us to a desire to develop methodology for mitochondria and mitochondrial gene transfer. As a first step, efficient methods to introduce foreign or altered mtDNA or genomes into somatic or germ cells would be needed.

TRANSMITOCHONDRIAL ANIMALS

To make a transmitochondrial animal, the ability to manipulate normal and mutant mitochondria in vivo has been a critical and difficult first step. In vivo mitochondrial gene transfer remains a technological hurdle in the development of mitochondria-based genetic therapies and in the generation of experimental animal models for the study of mitochondrial dynamics and mitochondria-based traits. While gene transfer has been performed in a host of cell types and organisms, transfer of nuclear DNA has been the only demonstrable form of mammalian gene transfer, short of cell fusions, to date.

Rapid segregation of mtDNA genotypes could occur in mammals and was first demonstrated in Holstein cattle where pedigree records in the industry allowed detailed analysis of maternally related individual genotypes.[5] Segregation of mtDNA was investigated in maternal lineages of heteroplasmic mice created by cytoplast fusion[6,7] and by embryonic karyoplast transplantation.[8] Although mitochondrial segregation in somatic tissues is effective in some tissues and with increasing age, the preceding studies have shown that mtDNA heteroplasmy is maintained at stable levels throughout several generations. This would suggest that the mouse germline is not very effective in segregating mtDNA haplotypes. In cattle, however, highly heteroplasmic females will produce homoplasmic oocytes, whereas heteroplasmic bulls produce mostly heteroplasmic sperm, indicating that mtDNA segregation is very stringent in the female and practically absent in the male germline.[9] Together, these results suggest that mammalian species show variable patterns of mtDNA segregation.

In contrast to these techniques, our efforts to devise a direct mitochondria transfer technique offered certain advantages. Principally, the ability to use isolated mitochondria for the production of heteroplasmic mice would allow for investigations into the feasibility of genetic manipulation of mtDNA in vitro prior to mitochondria microinjection into zygotes.

CONCLUSIONS

Through the early 1990s, various early attempts to create transmitochondrial strains of mammalian species by introduction of foreign mitochondria into germ cells were largely unsuccessful. A number of constraints have been identified or postulated, from perturbations of biological pathways to mechanistic aspects of the specific protocols used. Since 1997, a number of laboratories have reported on methodologies used to create transmitochondrial animals. To date, methods for mitochondria isolation and interspecific transfer of mitochondria have been reported both in laboratory and domestic animal models.[3,10,11] Interestingly, early reports on development of cloned animals by nuclear transfer resulted in conflicting consequences when retrospective studies on mitochondrial transmission were reported.[12–16] Indeed, dependent upon the specific methodology employed for nuclear transfer and cytoplasm/ooplasm transfer to rescue low-quality embryos, additional models of heteroplasmy may or may not have been characterized as a consequence of mitochondrial dysfunction. As such, research independent of targeted mitochondrial genomic modifications may also help unlock mechanisms underlying the dynamics related to persistence of foreign mitochondria and maintenance of heteroplasmy in various cloning protocols.

REFERENCES

1. Pinkert, C.A. Genetic Engineering of Animals. In *Handbook of Biomedical Technology and Devices*; Moore, J.E., Jr., Zouridakis, G. Ed., CRC Press: Boca Raton, **2004**; 18-1–18-12.
2. Vila, C.; Leonard, J.A.; Gotherstrom, A.; Marklund, S.; Sandberg, K.; Liden, K.; Wayne, R.K.; Ellegren, H. Widespread origins of domestic horse lineages. Science **2001**, *291* (5503), 474–477.
3. Pinkert, C.A.; Trounce, I.A. Production of transmitochondrial mice. Methods **2002**, *26* (4), 348–357.
4. Wallace, D.C. Mouse models for mitochondrial disease. Am. J. Med. Genet. **2001**, *106* (1), 71–93.
5. Olivo, P.D.; Van de Walle, M.J.; Laipis, P.J.; Hauswirth, W.W. Nucleotide sequence evidence for rapid genotypic shifts in the bovine mitochondrial DNA D-loop. Nature **1983**, *306* (5941), 400–402.
6. Jenuth, J.P.; Peterson, A.C.; Fu, K.; Shoubridge, E.A. Random genetic drift in the female germline explains the rapid segregation of mammalian mitochondrial DNA. Nat. Genet. **1996**, *14* (2), 146–151.
7. Jenuth, J.P.; Peterson, A.C.; Shoubridge, E.A. Tissue-specific selection for different mtDNA genotypes in heteroplasmic mice. Nat. Genet. **1997**, *16* (1), 93–95.
8. Meirelles, F.V.; Smith, L.C. Mitochondrial genotype segregation in a mouse heteroplasmic lineage produced by embryonic karyoplast transplantation. Genetics **1997**, *145* (2), 445–451.
9. Smith, L.C.; Bordignon, V.; Garcia, J.M.; Meirelles, F.V. Mitochondrial genotype segregation and effects during mammalian development: Applications to biotechnology. Theriogenology **2000**, *53* (1), 35–46.
10. Meirelles, F.V.; Bordignon, V.; Watanabe, Y.; Watanabe, M.; Dayan, A.; Lobo, R.B.; Garcia, J.M.; Smith, L.C. Compete replacement of the mitochondrial genotype in a *Bos indicus* calf reconstructed by nuclear transfer to a *Bos taurus* oocyte. Genetics **2001**, *158* (1), 351–356.
11. McKenzie, M.; Trounce, I.A.; Cassar, C.A.; Pinkert, C.A. Production of homoplasmic xenomitochondrial mice. Proc. Natl. Acad. Sci. USA **2004**, *101* (6), 1685–1690.
12. Hiendleder, S.; Zakhartchenko, V.; Wenigerkind, H.; Reichenbach, H.D.; Bruggerhoff, K.; Prelle, K.; Brem, G.; Stojkovic, M.; Wolf, E. Heteroplasmy in bovine fetuses produced by intra- and inter-subspecific somatic cell nuclear transfer: Neutral segregation of nuclear donor mitochondrial DNA in various tissues and evidence for recipient cow mitochondria in fetal blood. Biol. Reprod. **2003**, *68* (1), 159–166.
13. Evans, M.J.; Gurer, C.; Loike, J.D.; Wilmut, I.; Schnieke, A.E.; Schon, E.A. Mitochondrial DNA genotypes in nuclear transfer-derived cloned sheep. Nat. Genet. **1999**, *23* (1), 90–93.
14. Hiendleder, S.; Schmutz, S.M.; Erhardt, G.; Green, R.D.; Plante, Y. Transmitochondrial differences and varying levels of heteroplasmy in nuclear transfer cloned cattle. Mol. Reprod. Dev. **1999**, *54* (1), 24–31.
15. Steinborn, R.; Schinogl, P.; Zakhartchenko, V.; Achmann, R.; Schernthaner, W.; Stojkovic, M.; Wolf, E.; Muller, M.; Brem, G. Mitochondrial DNA heteroplasmy in cloned cattle produced by fetal and adult cell cloning. Nat. Genet. **2000**, *25* (3), 255–257.
16. Takeda, K.; Takahashi, S.; Onishi, A.; Goto, Y.; Miyazawa, A.; Imai, H. Dominant distribution of mitochondrial DNA from recipient oocytes in bovine embryos and offspring after nuclear transfer. J. Reprod. Fertil. **1999**, *116* (2), 253–259.

Transgenic Animals: Secreted Products

Michael J. Martin
David A. Dunn
Carl A. Pinkert
Department of Pathobiology, College of Veterinary Medicine, Auburn University, Auburn, Alabama, U.S.A.

Abstract

Interest in modifying traits that determine productivity of domestic animals was greatly stimulated by early experiments in which body size and growth rates were dramatically affected in transgenic mice expressing growth hormone transgenes driven by a metallothionein (MT) enhancer/promoter. From that starting point, similar attempts followed to enhance growth in farm animals by introduction of various growth factors, modulators, and their receptors. It soon became apparent that transgene regulation was an exquisite balancing act, where precise regulation of transgenes was crucial to normal development. Yet, the overexpression of various transgene products illustrated that such animals could produce biologically important molecules as efficient mammalian bioreactors, with efficiencies far greater than conventional bacterial or cell culture systems. From early studies in the mid-1980s through the 1990s, one of the main targets of genetic engineering or gene pharming efforts has involved attempts to direct expression of transgenes encoding biologically active human proteins in farm animals. To date, expression of foreign genes encoding various protein products was successfully targeted to the mammary glands of goats, sheep, cattle, and swine, yet the jump from model to achieving regulatory approval has proven most challenging.

ADVANTAGES OF RECOMBINANT PROTEIN SYNTHESIS IN TRANSGENIC ANIMALS

Several different organisms have been harnessed to produce recombinant proteins. Bacteria, yeast, fungi, plants, and cultured mammalian cells can all be reprogrammed and, if properly managed, yield relatively large amounts of recombinant proteins. Problems begin to arise, however, when one examines the ability of these organisms to posttranslationally modify and even release recombinant proteins. Bacteria, for example, are often unable to package and secrete recombinant proteins. In these instances, the recombinant protein must be physically extracted from the bacteria, a process that can be difficult and costly. Whereas yeast can secrete recombinant proteins that are glycosylated, the enzymatic pathway(s) that they utilize to accomplish protein glycosylation differs from that employed in higher plants and animals. As a result, many of the recombinant proteins produced by yeast exhibit inadequate glycosylation. Posttranslational modification of recombinant proteins produced in fungi appears to be aberrant in many instances as well. Mammalian cell lines, in contrast, typically perform posttranslational modifications of recombinant proteins that are quite similar to those observed in indigenous proteins. Primary drawbacks to the synthesis of recombinant proteins in animal cell lines include cost and the logistical challenge associated with developing and managing cell cultures for large-scale protein production.

In contrast, transgenic animals, as Louis-Marie Houdebine describes,[1] share most of the properties of animal cells in culture, exhibit appropriate posttranslational modifications of recombinant proteins, and synthesize and secrete proteins extremely efficiently. Indeed, mammary gland epithelia typically have a cell density that is 100- to 1000-fold greater than that used in mammalian cell culture bioreactors. In one recent example, 35 transgenic goats that produced a human monoclonal antibody at a concentration of 8 g/L in their milk were equivalent to an 8500-liter batch cell culture running 200 days/year with a 1 g/L final production level.[2] Thus, from a production standpoint, the amount of antibody synthesized in 170,000-liter cell culture yield was equivalent to that generated in 21,000 liters of milk from transgenic goats. Assuming a process yield of 60%, both systems would generate 100 kg of purified monoclonal antibody, yet the transgenic bioreactor was significantly more efficient.

Another obvious incentive for the production of biopharmaceuticals in transgenic livestock is their potential economic value (Table 1). The cost of human proteins

Table 1 Molecular pharming projects: potential biomedical and commercial products from transgenic farm animals.

Products	Use	Commercializing firm(s)
α-1 antitrypsin	Hereditary emphysema/cystic fibrosis	(Bayer/PPL)
α-1 proteinase inhibitor	Hereditary emphysema/cystic fibrosis	(Bayer/PPL)
α-fetoprotein (rhAFP)	Myasthenia gravis, multiple sclerosis, and rheumatoid arthritis	(Merrimack/GTC)
Antithrombin III (rhATIII)	Emboli/thromboses	(GTC)
β-glucosidase	Glycogen storage disease	(Pharming)
Collagen	Rheumatoid arthritis	(Pharming)
CFTR	Ion transport/cystic fibrosis	(GTC)
Factor VIII	Hemophilia A	(ARC)
Factor IX	Blood coagulation/hemophilia	(GTC, PPL)
Fibrin, fibrinogen	Tissue sealant development	(ARC, PPL, Pharming)
Hemoglobin	Blood substitute development	(Baxter)
Lactalbumin	Food additive	(Univ. Illinois)
Lactoferrin	Immunomodulatory, antiinflammatory	(Pharming)
MSP-1 (Merozoite Surface Protein 1)	Malarial vaccine	(GTC)
Phytase (Enviropig™)	Bioremediation, pollution control	(Univ. Guelph)
Human antibodies	Biotherapeutics, biodefense	(Abgenix, Hematech, Medarex)
Human C1 inhibitor	Hereditary angioedema	(Pharming)
Human lysozyme	Antimicrobial, immune modulator	(UC-Davis)
Human protein C	Blood coagulation	(ARC, PPL)
Human serum albumin	Blood pressure, trauma/burn treatment	(Pharming; GTC)
Spider silk (Biosteel®)	Materials development	(Nexia)
tPA	Dissolve fibrin clots/heart attacks	(Genzyme)
Tissues/organs	Engineered for xenotransplantation	(Alexion, Bresagen, Novartis)
Monoclonal antibodies and immunoglobulin fusion proteins:		
5G1.1	Rheumatoid arthritis, nephritis	(Alexion/GTC)
Antegren™	Neurological disorders	(Elan/GTC)
CTLA4Ig	Rheumatoid arthritis	(Bristol Myers Squibb/GTC)
D2E7	Rheumatoid arthritis	(Abbott/GTC)
huN901	Small-cell lung cancer	(ImmunoGen/GTC)
MM-093	Myasthenia gravis, multiple sclerosis, and rheumatoid arthritis	(Merrimack/GTC)
PRO 542	HIV/AIDS	(Progenics/GTC)
Remicade®	Crohn's disease, rheumatoid arthritis	(Centocor/GTC)

Source: From *The History and Theory of Transgenic Animals*.[3]

obtained from donated plasma and used in replacement therapy has ranged from $4/g for serum albumin and $5000/g for antithrombin III to $150,000/g for human blood clotting factor VIII (FVIII).[4] Although the individual values of these seem dramatic, they pale in comparison to the projected worth of a number of recombinant structural products. Biomedical applications of Biosteel™ (Nexia Inc.), a recombinant form of dragline spider silk, produced in the milk of transgenic goats, is projected to represent $150 to $450 million in annual earnings (exclusive of military and other industrial applications).

EXPRESSION OF RECOMBINANT PROTEINS IN MILK

Since the introduction of the first exogenous genes into mice, more than 60 proteins have been produced in milk of transgenic animals. In order to target protein expression specifically to the mammary gland, a transgene typically consists of the desired protein gene fused to one of several available mammary-specific regulatory sequences.[3–7] These sequences have included: ovine BLG; murine, rat, and rabbit whey acidic protein (WAP); bovine α-s$_1$ casein; rat, rabbit, and goat β-casein; and guinea pig, ovine and caprine, and bovine α-lactalbumin. While expression of the target protein can be achieved using either a genomic DNA or cDNA coding sequence(s), the former normally yields higher levels of protein expression.

Therapeutic monoclonal antibodies produced in the mammary gland of a transgenic animal line present a potentially valuable technology. Transgenic monoclonal antibodies are produced by cloning genetic sequences for both heavy- and light-chain genes downstream of mammary gland-specific regulatory elements. Chimeric antibodies may also be produced by ligating antigen-binding region sequences from a (usually murine) monoclonal antibody to constant region sequences from a different species and/or isotype. The first transgenic mice harboring immunoglobulin genes were made in the mid-1980s.[8] Though the majority of effort and funding in this field is currently focused toward human therapeutics, veterinary use of monoclonal antibodies also shows significant promise as a developing application.

Whereas several therapeutic monoclonal antibodies have been approved for use by the U.S. Food and Drug Administration, none as yet has been approved where a transgenic animal was used as a production vehicle. Using antibody production technologies in transgenic bioreactor systems, these products target a wide range of clinical ailments and are mostly in the preclinical stage of development.

EXPRESSION OF RECOMBINANT PROTEINS IN MEDIA OTHER THAN MILK

Secretion of transgene-encoded proteins in the urine of transgenic animals was demonstrated using recombinant genes under the control of kidney-[9] or bladder-[10] specific regulatory sequences. Expression of transgenes in the kidney or bladder of transgenic animals and subsequent secretion in the urine may provide some advantages over the mammary gland as a bioreactor, as the purification of proteins from urine may be facilitated by lower lipid and protein levels in comparison to milk. Additionally, such animals can be used for production of recombinant proteins over the course of their entire life span.

RECOMBINANT PROTEIN PRODUCTION: HEALTH AND SAFETY ISSUES

In addition to being structurally and functionally analogous to the natural plasma-derived protein, purified recombinant proteins must be free of pathogenic organisms. Viral and bacterial contamination of human biopharmaceutical products produced in the blood or milk can be minimized by focusing prevention/eradication efforts on at least three levels of production: the transgenic animal donor, the medium in which the recombinant protein is produced, and the final product.[4] An initial key to minimizing the risk of contamination is to derive the transgenic donor animals from a source herd that is free from as many pathogens as possible. Maintenance of these animals in a closed facility, the implementation of strict monitoring procedures for various pathogens, and the use of animal husbandry practices that follow generally accepted practices (GAPs) and standard operating procedures should greatly reduce the entry of pathogens. Though quite costly, one can develop pathogen-free herds of transgenic livestock. Such a feat was recently achieved by introducing hysterotomy-derived transgenic piglets into an elaborate SPF barrier facility.[11] Diagnostic testing of this herd over the past 3 years in this facility had revealed the absence of 35 major and minor swine pathogens including PRRS, parvovirus, leptospira, parainfluenza, and *Streptococcus suis*.

CONCLUSION

While transgenic animal technology continues to open new and unexplored agricultural frontiers, molecular pharming efforts raise questions concerning regulatory and commercialization issues. Although significant advances have been made since the inception of various clinical trials, the resources required to move the projects forward and the attendant financial risks have led a number of companies to curtail product development. Various societal issues exist and will continue to influence the development of value-added animal products produced through transgenesis—until transgenic products and foodstuffs are proven safe for human use and are accepted by a wide cross section of society.

REFERENCES

1. Houdebine, L.-M. Production of pharmaceutical proteins from transgenic animals. J. Biotechnol. **1994**, *34* (3), 269–287.
2. Young, M.W.; Okita, W.B.; Brown, M.; Curling, J.M. Production of biopharmaceutical proteins in the milk of transgenic dairy animals. BioPharm. **1997**, *10* (6), 34–38.
3. Pinkert, C.A. The history and theory of transgenic animals. Lab. Anim. **1997**, *26* (8), 29–34.
4. Clark, A.J.; Simons, P.; Wilmut, I.; Lathe, R. Pharmaceuticals from transgenic livestock. Trends Biotechnol. **1987**, *5* (1), 20–24.
5. Palmiter, R.D.; Brinster, R.L.; Hammer, R.E.; Trumbauer, M.E.; Rosenfeld, M.G.; Birnberg, N.C.; Evans, R.M. Dramatic growth of mice that develop from eggs microinjected with metallothionein-growth hormone fusion genes. Nature **1982**, *300* (5893), 611–615.
6. Martin, M.J.; Pinkert, C.A. Production of Transgenic Swine by DNA Microinjection. In *Transgenic Animal Technology: A Laboratory Handbook,* 2nd Ed.; Pinkert, C.A., Ed.; Academic Press: San Diego, **2002**; 307–336.
7. Simons, J.P.; McClenaghan, M.; Clark, A.J. Alteration of the quality of milk by expression of sheep β-lactoglobulin in transgenic mice. Nature **1987**, *328* (6130), 530–532.
8. Storb, U.; Pinkert, C.; Arp, B.; Engler, P.; Gollahon, K.; Manz, J.; Brady, W.; Brinster, R.L. Transgenic mice with mu and kappa genes encoding antiphosphorylcholine antibodies. J. Exp. Med. **1986**, *64* (2), 627–641.
9. Zbikowska, H.M.; Soukhareva, N.; Behnam, R.; Chang, R.; Drews, R.; Lubon, H.; Hammond, D.; Soukharev, S. The use of the uromodulin promoter to target production of recombinant proteins into urine of transgenic animals. Transgenic Res. **2002**, *11* (4), 425–435.
10. Kerr, D.E.; Liang, F.; Bondioli, K.R.; Zhao, H.; Kreibich, G.; Wall, R.J.; Sun, T.T. The bladder as a bioreactor: Urothelium production and secretion of growth hormone into urine. Nat. Biotechnol. **1998**, *16* (1), 75–79.
11. Risdahl, J.; Edgerton, S.; Adams, C.; Martin, M.; Wiseman, B. Establishing a Designated Pathogen Free Swine Colony for Xenotransplantation, Proc. 17th International Pig Veterinary Society Congress (IPVS), Ames, IA, June 2–4, **2002**.

Transgenic Crops: Environmental Concerns

Alan Frank Raybould
Product Safety, Syngenta, Bracknell, U.K.

Abstract

Concern has been raised that cultivation of transgenic crops may harm biodiversity and other ecosystem services. Properties of transgenic crops that raise those concerns include toxicity to non-pest species, and increased weediness and invasiveness; adverse effects on soil quality and increased agricultural intensification through use of transgenic crops are also postulated. Environmental risk assessments indicated that cultivation of current commercially available transgenic crops had a low probability of increasing environmental harm compared with current agricultural practice; these predictions have been corroborated, and in some cases environmental benefits have accrued. The categories of environmental risk associated with transgenic plants are not unique and are also posed by conventionally bred crops and all methods of pest control. It follows that regulators may require that new non-transgenic production methods, such as perennial crops for biofuels, to be assessed similarly to transgenic crops.

INTRODUCTION

In most countries, seeds of transgenic crops must be approved by governmental regulatory authorities before they can be sold for unrestricted cultivation. Decisions to approve or forbid sale of seeds take into account the potential for their cultivation to cause environmental harm, along with potential harm to human and animal health of food and feed derived from crop, the benefits of the crop to farmers and the public, and any other factors deemed relevant. What constitutes environmental harm cannot be deduced scientifically; harm must be determined by public policy, which may be controversial owing to differing values people place on environmental attributes.

Once environmental harm has been defined, scientific methods can be used to make predictions about the likelihood and magnitude of that harm; in other words, a scientific risk assessment can be performed. The purpose of this entry is to describe some of the environmental concerns about cultivating transgenic crops, the properties of transgenic crops that raise those concerns, and methods to assess whether transgenic crops pose risks to the environment.

WHAT SHOULD WE BE CONCERNED ABOUT?

Cultivation of a transgenic crop will inevitably lead to environmental differences compared with alternative land uses; however, this does not mean that the cultivation of a transgenic crop will inevitably cause harm and is thereby inherently risky. Cultivation of a transgenic crop poses a risk only if it is likely to result in *harmful* change. The degree of risk is a combination of the likelihood and magnitude of a harmful change: risks are negligible if the likelihood of even a small harmful change is low; and risks are severe when there is a high probability of a very harmful change. Categorizing risk is more difficult when one factor—likelihood or harm—is low and the other is high; although in general, unless the likelihood is extremely low, risk perception tends to follow the magnitude of the potential harmful change.

Because not all change is harmful, environmental risk assessments are ineffective and inefficient if they seek to catalogue all detectable changes that result from cultivation of transgenic crops. Good risk assessment predicts changes in things we value—the protection goals—as indicated by public policy and the laws under which transgenic plants are regulated. Protection goals are usually general concepts such as water quality, biodiversity, or ecosystem services; risk assessments must derive specific objects for protection, called assessment endpoints, to focus on the risk assessment and make scientific analysis possible. Typical assessment endpoints in environmental risk assessments of many potentially harmful activities include the population sizes of particular species, and the concentration of certain chemicals in water or air. Failure to make the protection goals and assessment endpoints explicit in risk assessments for the cultivation of transgenic plants has led to protracted and inconsistent decision-making because the criteria for judging risk are unclear.[1]

Protection goals, and the assessment endpoints derived from them, may vary among jurisdictions. It is not possible, therefore, to describe a generic risk assessment for transgenic plants; however, it is possible to describe properties of transgenic plants that have raised environmental concerns, to suggest categories of assessment endpoints that may be affected, and to formulate hypotheses that could be tested to assess risk.

TOXICITY

Many transgenic crops are designed to be toxic to pest species that eat them. Current commercial transgenic insect-resistant crops produce proteins derived from the soil bacterium *Bacillus thuringiensis* (Bt), which have activity against certain Lepidoptera, certain Coleoptera, or both. Transgenic plants could cause environmental harm through toxicity to non-pest species, which may be exposed to toxins through crop tissues, such as pollen, or by eating pests that have fed on crop tissues.

A typical assessment endpoint that may be adversely affected by toxicity of transgenic plants is the population size of arthropods that provide biological control in crops.[2] Typically, two hypotheses are tested to assess toxicity: first, that the concentration of endogenous toxins, anti-nutrients, and nutrients in the transgenic crop are within the normal range of the non-transgenic crop; and second, that the transgenic proteins are not toxic to non-pest species at concentrations in the crop.[3]

The first hypothesis is tested by comparison of the composition of the transgenic crop and a genetically similar non-transgenic crop grown side-by-side in a range of environments. The transgenic and non-transgenic crops are tested for differences in concentrations of nutrients, anti-nutrients, and toxins, which will vary depending on the crop and the assessment endpoint; for example, if the abundance of biological control insects is the main assessment endpoint, tests could focus on insecticidal compounds such as gossypol in cotton, glucosinolates in oilseed rape, and glycoalkaloids in potatoes.[1]

The second hypothesis for assessing risk is usually tested by laboratory studies that test the effect of high concentrations of the transgenic proteins on non-pest species that are representative of those likely to be exposed to the transgenic crop.[3] A "high" concentration is usually judged to be about 10 times a conservative estimate of the concentration of the protein to which non-pest species will be exposed as a result of cultivation of the transgenic crop. Exposure estimates are made from measurements of several variables including the concentration of the protein in crop tissues, the dispersal of crop tissues within and outside fields, predicted concentrations of the protein in pest species that feed on the crop, the rate of degradation of the protein in soil, and estimates of gene flow from the crop to wild species.[3]

If the hypotheses are corroborated after rigorous testing, the environmental risks from toxicity may be judged negligible; if the hypotheses are falsified, further evaluation is necessary to assess whether the identified effects will cause harm under realistic conditions of cultivation. In other words, tests of the initial hypotheses determine whether there is significant risk under conservative ("worst-case") exposures; if there is a significant risk under those conditions, subsequent work tests hypotheses that these risks are mitigated under field conditions.[1]

Risk assessments based on laboratory measurements of the toxicity of insecticidal proteins and their predicted environmental concentrations indicate low risk to non-pest species from current commercially cultivated transgenic crops. The predictions of low risk have been corroborated in field studies, and in many studies the abundance of non-pest species in transgenic insecticidal crops has been found to be higher than in comparable conventionally managed non-transgenic crops.[2]

WEEDINESS AND INVASIVENESS

Crops can become weeds of agriculture and reduce yields or the quality of subsequent crops, and occasionally invade non-agricultural land, where they may displace other species or damage ecosystem services.[4] Hybrids between crops and wild species may also be weedy and invasive; a related concern is that gene flow from transgenic crops may harm genetic variation of crop relatives in their centers of origin (see *Transgenic Plants: Economic and Environmental Risks and Gene Flow*, p. 641). Long experience of cultivation means that the environmental risks associated with the weediness and invasiveness of common crops and their hybrids are well known and are generally considered acceptable; therefore, if the potential for weediness and invasiveness of a transgenic crop is not significantly different from that of a non-transgenic counterpart, its associated risks can also be considered acceptable.

The hypothesis that weediness and invasiveness of a transgenic crop are no greater than that of a non-transgenic counterpart is tested in multilocation field trials by comparisons of many traits that potentially control plant population dynamics; these tests concentrate on possible unintended effects of transformation. Separate experiments may be required to test that the intended effects of transformation do not increase weediness or invasiveness potential. For example, to test that an insect-resistant transgenic crop does not have increased invasiveness potential, non-transgenic crop seed could be sown outside fields and protected from insect attack with insecticide or netting, or left unprotected. Comparison of the survival and growth of plants between treatments would indicate whether insect resistance increased invasiveness potential (see *Transgenic Plants: Economic and Environmental Risks and Gene Flow*, p. 641.

Current commercially available traits, such as insect resistance, herbicide tolerance, and improved nutritional quality, are predicted not to increase the weediness or invasiveness of field crops such as soybeans, maize, and cotton. Traits that confer tolerance to salt or drought, or recently domesticated perennial crops for biofuels that have greater dispersal and competitive ability than field crops, may pose greater environmental risks from weediness and invasiveness.

EFFECTS ON SOIL

There are concerns that cultivation of transgenic plants may reduce soil quality because of adverse effects on soil biodiversity, and from delayed and long-term effects on soil microorganisms in particular. Soil quality is "the continued capacity of soil to function as a vital living system, within ecosystem and land use boundaries, sustain biological productivity, to promote the quality of air and water environments, and to maintain plant, animal, and human health,"[5] and hence reduction in soil quality could be extremely harmful. It is important, therefore, to have high confidence that transgenic crops are no more harmful to soil quality than are non-transgenic counterparts.

The effects of transgenic and non-transgenic plants on soil microbes have been compared in many studies.[6] In some cases, soils containing a transgenic crop and a non-transgenic crop differed in the genetic structure of microbial populations or in the rate of soil respiration; in many studies no difference was detectable. Differences in soil microbe populations were restricted to the regions adjacent to the roots and were transient; and differences between the structure or function of soils amended with transgenic and non-transgenic material were small compared with the effects of crop species, soil management, such as ploughing, and season. A crucial additional factor is that environmental harm, defined as a reduction in soil quality, is unlikely to result from changes the microbial community structure of soil, because the relationship between soil biodiversity and function is weak due to large functional redundancy in soil.[7]

INDIRECT EFFECTS

With the concerns described above, putative environmental harm results from direct contact between the assessment endpoint and the transgenic crop or its derivatives. Cultivation of the transgenic crop may also have effects on assessment endpoints without direct contact; such indirect effects may be meditated through management of the crop, or from knock-on effects of the control of pests. Application of pesticides may differ between transgenic and non-transgenic crops.[8] Insect-resistant plants may require fewer applications of insecticides, resulting in increased abundance of biological control organisms, or of secondary pests that the transgenic plant does not target and were previously controlled by insecticide. Insect-resistant plants may also decrease the abundance of pests that were previously difficult to control, which may indirectly decrease the abundance of specialist predators and parasitoids of those pests. A further indirect effect of insect-resistance is a potential human and animal health benefit from lower concentrations of fungal toxins in grain of transgenic maize protected against attack by European corn borer; control of corn borer damage leads to fewer sites for fungi to colonize.[9]

The pattern of use of herbicides may differ between herbicide-tolerant and sensitive crops, which may change the abundance and temporal distribution of weeds, possibly reducing the abundance of organisms that feed on weeds. Nevertheless, if increased production results from better weed control, less land may be needed for cultivation, hence mitigating the effects of intensification; under some scenarios of demand for agricultural products and the relationship between crop yield and species persistence, intensive agriculture may preserve more biodiversity than extensive farming.[10] Use of herbicide-tolerant crops may also alter other farming practices such as increasing use of no-till methods leading to improved soil carbon sequestration and water retention, and reduced use of fuel.

ARE CONCERNS ABOUT TRANSGENIC PLANTS UNIQUE?

Transgenic plants in current commercial use do not have properties that should raise unique environmental concerns: non-transgenic plants can be toxic, weedy and invasive, resistant to herbicides, and influence soil quality; and indirect effects result from any method of pest control. Nevertheless, the environmental risks of new varieties of non-transgenic crops are rarely assessed, whereas transgenic plants undergo extensive environmental risk assessments before their cultivation is permitted. Detailed comparisons of the genetic changes associated with plant transformation with those related to other methods of plant breeding[11] suggest that there is nothing intrinsic about transgenic plants that should trigger different risk assessments from non-transgenic crops with similar phenotypes. This has led to the suggestion that the environmental risks of all new methods of crop production, however developed, should be assessed similarly.[12]

CONCLUSION

Concerns have been raised that transgenic plants may cause environmental harm through toxicity to non-pest species, by being weeds of agriculture or non-agricultural habitats, by reducing soil quality, or through indirect adverse effects on biodiversity and other ecosystem services as a result of crop management and pest control. Environmental risk assessments indicated that cultivation of current commercially available transgenic crops had a low probability of increasing such adverse effects compared with currently accepted agricultural practice; these predictions have been corroborated.[2] Transgenic crops do not pose unique risks; therefore, risk assessments for new transgenic products should be based on their phenotype, not their method of production. A corollary is that regulators may evaluate new non-transgenic production methods, such as perennial crops for biofuels, for the risks currently only assessed for transgenic crops.

REFERENCES

1. Raybould, A. Ecological versus ecotoxicological methods for assessing the environmental risks of transgenic crops. Plant Sci. **2007**, *173* (6), 589–602.
2. Romeis, J.; Meissle, M.; Bigler, F. Transgenic crops expressing *Bacillus thuringiensis* toxins and biological control. Nature Biotechnol. **2006**, *24* (1), 63–71.
3. Raybould, A.; Stacey, D.; Vlachos, D.; Graser, G.; Li, X.; Joseph, R. Non-target organism risk assessment of MIR604 maize expressing mCry3 A for control of corn rootworm. J. Appl. Entomol. **2007**, *131* (6), 391–399.
4. Pimentel, D.; McNair, S.; Janecka, J.; Wightman, J.; Simmonds, C.; O'Connell, C.; Wong, E.; Russel, L.; Zern, J.; Aquino, T.; Tsomondo, T. Economic and environmental threats of alien plant, animal, and microbe invasions. Agric. Ecosyst. Environ. **2001**, *84* (1), 1–20.
5. Schloter, M.; Dilly, O.; Munch, J.C. Indicators of soil quality. Agric. Ecosyst. Environ. **2003**, *98* (1–3), 255–262.
6. Widmer, F. Assessing effects of transgenic crops on soil microbial communities. Adv. Biochem. Eng. Biotechnol. **2007**, *107* (1), 207–234.
7. Ritz, K. Underview: origins and consequences of belowground biodiversity. In *Biological Diversity and Function in Soils*; Bardgett, R.D., Usher, M.B., Hopkins D.W. Eds.; Cambridge University Press: Cambridge, UK, **2005**; 381–401.
8. Kleter, G.A.; Bhula, R.; Bodnaruk, K.; Carazo, E.; Felsot, A.S.; Harris, C.A.; Katayama, A.; Kuiper, H.A.; Racke, K.D.; Rubin, B.; Shevah, Y.; Stephenson, G.R.; Tanaka, K.; Unsworth, J.; Wauchope, R.D.; Wong, S.S. Altered pesticide use on transgenic crops and the associated general impact from an environmental perspective. Pest Manag. Sci. **2007**, *63*, 1107–1115.
9. Kershen, D.L. Health and food safety: the benefits of Bt-corn. Food Drug Law J. **2006**, *61* (2), 197–235.
10. Green, R.E.; Cornell, S.J.; Scharlemann, J.P.W.; Balmford, A. Farming and the fate of wild nature. Science **2005**, *307*, 550–555.
11. Filipecki, F.; Malepszy, S. Unintended consequences of plant transformation. J. Appl. Genet. **2006**, *47* (4), 277–286.
12. Butler, S.J.; Vickery, J.A.; Norris, K. Farmland biodiversity and the footprint of agriculture. Science **2007**, *315*, 381–384.

Transgenic Crops: Perennials

Abhaya Dandekar
Department of Plant Sciences, University of California–Davis, Davis, California, U.S.A.

Matthew Escobar
Department of Biological Sciences, California State University San Marcos, San Marcos, California, U.S.A.

Abstract

The biology and cultivation of perennial crops substantially differ from that of the annual species that are the subject of the vast majority of plant biotechnology research. These differences, from clonal propagation to self-incompatibility, present not only unique challenges, but also unique opportunities for the application of transgenic approaches. Despite poorly developed transformation methodologies for many perennial species, the past several years have seen the production of transgenic perennials displaying reduced juvenility, delayed ripening, suppressed self-incompatibility, and enhanced resistance to pathogens and pests. However, these impressive scientific advances in perennial crop biotechnology will likely remain in the laboratory and greenhouse until public acceptance issues and economic hurdles associated with deregulation can be resolved. For now, "indirect" biotechnology applications such as transgenic rootstocks and trap crops may provide the only realistic means to introduce biotech into the orchard.

INTRODUCTION

Though not globally accepted by consumers, transgenic crops are of major importance in world agriculture, with about 200 million acres of biotech canola, cotton, maize, and soybean cultivated in 2004. However, the commercial success of biotechnology in annual crop improvement has not been reproduced in perennial crops, as only a single transgenic perennial (Rainbow papaya) is currently grown commercially. Perennials are generally high-value, low-acreage crops, and present a unique set of cultural and management practices as compared to annuals. Of particular note are the prevalence of vegetative (esp. graft) propagation, a long juvenile (non-reproductive) growth period, the prevalence of self-incompatibility, and a high investment:plant ratio. This entry discusses how these characteristics represent both unique problems and unique opportunities for the application of biotechnology to perennial crop species, with an emphasis on the fruit and nut tree crops.

TRANSFORMATION OF PERENNIAL CROPS

Any commercially viable transformation system for perennial crops must maintain the genetic identity of the clonally propagated parent. For this reason, the transformation of cells derived from mature, rather than seed or seedling, tissues is required, as is the operation of the transgene in hemizygous plants, since breeding to homozygosity will result in the loss of the parental genotype. As is the case with annual crops, *Agrobacterium*-mediated transformation is the preferred method for the generation of transgenic perennials due to its technical simplicity, low cost, and comparatively simple patterns of DNA integration into the plant genome. Unfortunately, many perennial crop species, especially in the genera *Prunus* and *Citrus*, have proven highly recalcitrant to both *Agrobaceterium*-mediated transformation and alternatives such as microprojectile bombardment.[1] In other perennial crop species, such as walnut and avocado, transformation systems are established but utilize juvenile or juvenile-derived explant material. Thus, for currently intractable or juvenile-dependent transformation systems, new methods must be developed allowing efficient production of transgenic plants that possess the genetic and phenotypic characteristics of current cultivars. Toward this end, further work must be done to optimize *Agrobacterium* transformation (e.g., utilizing *virG*-overexpressing hypervirulent strains or strains that suppress plant defense responses,[2]) to test alternatives to antibiotic selection systems (e.g., mannose selection, D-amino acid selection,[1]) and to minimize time in culture, since this is correlated with the development of substantial genetic and epigenetic changes.[3]

PATHOGEN CONTROL

Because each mature perennial represents a large monetary investment, potentially lethal plant pathogens such as

Phytophthora cinnamomi, plum pox potyvirus, *Xylella fastidiosa*, and *Erwinia amylovora* represent a severe threat to growers. The two most common approaches to increase pathogen resistance in perennials have been the introduction of pathogen-specific resistance (R) genes cloned from other plant species or the constitutive expression of one or more genes encoding proteins with direct antipathogen properties. Pathogen resistance mediated by R genes is dependent upon the recognition of a specific pathogen elicitor by the R protein, causing activation of plant defense responses and the suppression of pathogen infection. This approach was recently utilized to generate resistance to apple scab through transformation of the commercial apple cultivar Gala with the resistance gene *HcrVf2* from Japanese crabapple.[4] Ectopic expression of genes encoding proteins with direct antipathogen activities, such as chitinase and stilbene synthase, has also proven effective in generating de novo pathogen resistance. For example, grapevines overexpressing a polygalacturaonse-inhibiting protein from pear displayed increased resistance to disease caused by both the fungal pathogen *Botrytis cinerea* and the bacterial pathogen *X. fastidiosa*.[5] The most notable example of the application of biotechnology to perennial crop improvement has been the Rainbow papaya, which is resistant to papaya ringspot virus (PRSV), a major pest of the Hawaiian papaya industry. Overexpression of the PRSV coat protein in Rainbow papaya causes the activation of post-transcriptional gene silencing, leading to degradation of the viral genome and near-complete resistance to disease symptoms.[6] As mentioned previously, Rainbow papaya is the only commercially grown transgenic perennial and currently accounts for approximately 50% of Hawaiian papaya production.

WEED CONTROL

Herbicide-resistant annual crops currently represent the largest share of the transgenic crop market, and herbicide resistance genes could be easily transferred to a variety of perennials. However, in many perennial crop systems weed control is not a significant problem. Canopy shading severely limits weed growth in most mature orchards, and in younger orchards wide plant spacing allows weed control by mowing, discing, or flaming. Thus, while herbicide resistance could be beneficial for forestry and the nursery trade, it would likely be unutilized by growers of perennial crops.

INSECT CONTROL

Biotech strategies to increase insect resistance in perennial crops have focused primarily on the *Bacillus thuringiensis* (Bt) insecticidal crystal proteins (ICPs). Bt has been exploited commercially for more than 40 years as a highly specific and rapidly biodegradable component of integrated pest management systems. Bt ICPs disrupt insect midgut membranes and are toxic in the parts-per-billion range

Table 1 Engineered insect resistance in perennial crops.

Crop/Transgene	Insects targeted
Apple/*CryIAb*, *CryIAc*	Lepidoptera, Coleoptera
Apple/Cowpea trypsin inhibitor	Lepidoptera, Coleoptera
Apple/chitinase, chitobiosidase	Homoptera
Cranberry/*CryIAa*	Lepidoptera
Grapefruit/GNA lectin	Homoptera
Grapevine/*CryIAc*	Lepidoptera
Grapevine/GNA lectin	Homoptera, Lepidoptera
Pear/unspecified *Cry*	Lepidoptera, Coleoptera?
Pear/*D5C1* lytic peptide	Homoptera
Persimmon/*CryIAc*	Lepidoptera
Sugarcane/unspecified *Cry*	Lepidoptera, Coleoptera?
Sugarcane/GNA lectin, WGA lectin	Homoptera, Coleoptera
Walnut/*CryIAc*	Lepidoptera

against a variety of Coleoptera and Lepidoptera, including important orchard pests, such as codling moth and navel orange worm. Codon-optimized ICP-encoding genes have been expressed in apple, cranberry, grapevine, persimmon, and walnut, conferring resistance to several insect pests (see Table 1). Several other proteins that interfere with insect digestion, such as lectins, chitinases, and proteinase inhibitors, have also been expressed in transgenic plants. Though they possess an acute toxicity several orders of magnitude lower than Bt ICPs, these proteins can target Homoptera (e.g., leafrollers and aphids), which are not effectively controlled by Bt ICPs.

INCREASING FRUIT QUALITY

The development of transgenic crops that provide direct benefits to the consumer (output traits) could be a critical step in improving worldwide public opinion of agricultural biotechnology. Several aspects of crop quality, including vitamin content, oil composition, and ripening rate have been modified in transgenic annuals, though none of these crops has yet been commercialized. A few reports have described modification of fruit quality in perennials, such as increasing β-carotene content in grapefruit through the introduction of novel carotenoid biosynthesis genes,[1] and increasing shelf life in apple by suppressing ethylene synthesis[7] (see Fig. 1). These approaches will likely grow in importance in both annuals and perennials as the agricultural biotechnology industry begins to shift from input to output trait development.

OTHER APPLICATIONS

Over the past 5 years, major progress has been made in the genetic manipulation of several important aspects of perennial biology, including tree architecture, juvenility, and self-incompatibility. Many of these advances have arisen from

Fig. 1 Transgenic apples with suppressed ethylene biosynthesis. (A) A three-year-old apple tree expressing an antisense *ACC oxidase* (*ACO*) transgene. ACO activity is reduced >100-fold in this transgenic line. (B, C) *ACO*-silenced (B) and wild-type (C) apples stored at room temperature for 1 months. (D, E) *ACO*-silenced (D) and wild-type (E) apples stored at room temperature for 3 months. **Source:** From *Effect of down-regulation of ethylene biosynthesis on fruit flavor complex in apple fruit.*[7]

the expression of the *rol* genes from *Agrobacterium rhizogenes* T-DNA. Decreased tree size, with potential cultural and harvest applications, has been reported in apple, walnut, and trifoliate orange plants expressing some combination of the *rol*A, *rol*B, and/or *rol*C genes. Likewise, increased rooting efficiency, which can be important for clonal rootstock propagation, has been reported in *rol*B-expressing apple and pear. Expression of the *rol*D gene can accelerate flowering in tobacco and *Arabidopsis*; however, the most impressive results to date in deceasing juvenility in perennials have been achieved with the *Arabidopsis LFY* and *AP1* floral meristem identity genes. *AP1*-expressing citrange plants generated from seedling explants displayed decreased juvenile growth traits (e.g., decreased thorniness, adult leaf shape) and flowered after 1 year, instead of the 6–7 years generally required for citrange to reach reproductive maturity (see Fig. 2).[8] Finally, an initial report described the suppression of the self-incompatibility system in apple through cosuppression (silencing) of a stylar S-RNase. As compared to non-transgenic controls, the S_3 RNase-silenced "Elstar" apples displayed an ~8-fold increase in fruit set when self-pollinated.[9]

CONCLUSIONS

Biotechnology research in perennials is still in its infancy, but encouraging recent progress has been made in the important areas of pathogen/pest protection, fruit quality, and self-incompatibility. Unfortunately, non-scientific barriers pose huge challenges that may impede the progress of these innovations into the orchard. Perhaps the most concerning aspect of the future of transgenic perennial plants is the economics of commercialization. It has been estimated that the current cost to develop and deregulate a single transgene allele is $1–5 million.[10] Because most perennial crops are clonally propagated, a single allele cannot be introgressed into additional crop cultivars by crossing (as is common practice in annuals), so modification of each existing cultivar will require development and deregulation of a new allele. This would represent a huge investment for relatively low-acreage perennial crops, especially considering that the end product may not be embraced by consumers and growers. A large proportion of the harvest of many perennial crops is exported, primarily to the European Union and Japan, both of which have imposed de facto moratoria on genetically modified foods. In addition, transgenic seed has generally been priced based upon its "value added" compared to conventional seed. For perennial crops, the "value added" for 3–4 decades of decreased orchard management or increased fruit quality may be hundreds of dollars per tree—a price that would probably not be acceptable to current growers.

One potential strategy to avoid several of the pitfalls described above is to utilize "indirect" biotechnology applications for crop improvement. One such approach is the transformation and improvement of rootstocks, rather than scion cultivars. Since rootstocks produce neither flowers nor

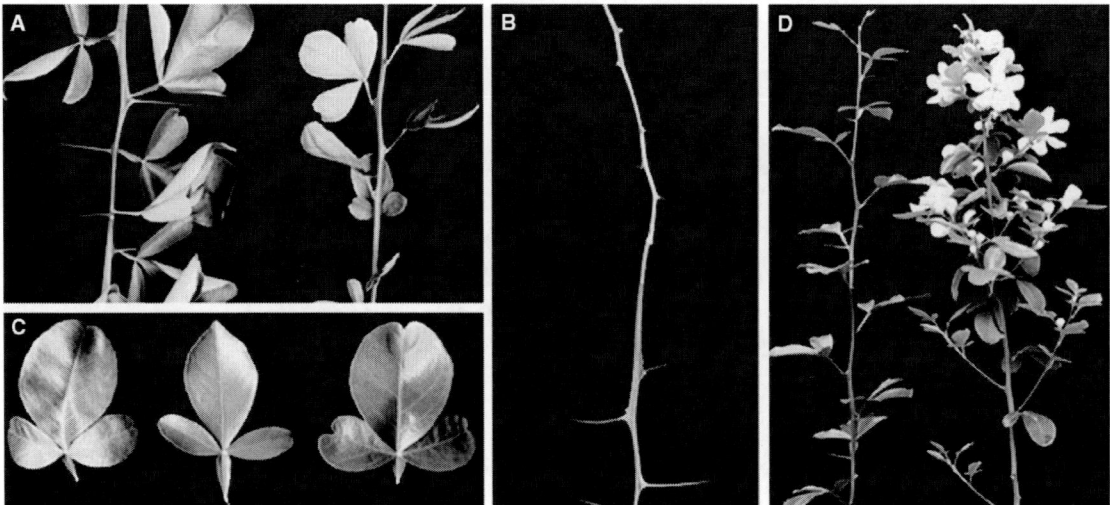

Fig. 2 Shortened juvenile phase in transgenic citrange expressing the floral meristem identity gene *AP1*. (A) Comparison of vegetative shoots from wild type (left) and *AP1*-expressing (right) plants. The reduction in thorns in the transgenic plant is a morphological indicator of reproductive maturity. (B) Rapid phase change from juvenile to adult tissue in *AP1*-expressing shoots, as indicated by the reduction in thorn size in new growth. (C) Leaves from *AP1*-expressing plants (right) are morphologically similar to leaves from mature citrange tissues (left), not juvenile leaves (center). (D) Flowering in a 13-month-old *AP1*-expressing plant (right) with comparison to a wild-type plant of the same age (left). **Source:** From *Constitutive expression of Arabidopsis LEAFY or APETALA1 genes in citrus reduces their generation time.*[8]

fruit, consumer acceptance issues may improve, and government deregulation should be simpler and less expensive. This approach has recently been utilized to improve walnut rootstocks through the development of RNA interference-mediated resistance to crown gall disease.[11] In the field of pest protection, the use of transgenic "trap crops" can substantially reduce insect damage without genetic alteration of the harvested crop species. For example, the codling moth's preference for oviposition on apple trees has recently been exploited by planting a small number of Bt ICP-expressing apples in non-transgenic walnut orchards, resulting in near-complete control of codling moth damage in the harvested walnuts.[12] Thus, until regulatory and consumer acceptance hurdles are overcome, indirect application strategies may be the most feasible means to apply biotechnology in the commercial orchard.

REFERENCES

1. Petri, C.; Burgos, L. Transformation of fruit trees. Useful breeding tool or continued future prospect? Transgenic Res. **2005**, *14*, 15–26.
2. Wroblewski, T.; Tomczak, A.; Michelmore, R. Optimization of *Agrobacterium*-mediated transient assays of gene expression in lettuce, tomato and *Arabidopsis*. Plant Biotechnol. J. **2005**, *3* (2), 259–273.
3. Labra, M.; Vannini, C.; Grassi, F.; Bracale, M.; Balsemin, M.; Basso, B.; Sala, F. Genomic stability in *Arabidopsis thaliana* transgenic plants obtained by floral dip. Theor. Appl. Genet. **2004**, *109* (7), 1512–1518.
4. Belfanti, E.; Silfverberg-Dilworth, E.; Tartarini, S.; Patocchi, A.; Barbieri, M.; Zhu, J.; Vinatzer, B.A.; Gianfranceschi, L.; Gessler, C.; Sansavini, S. The *HcrVf2* gene from a wildapple confers scab resistance to a transgenic cultivated variety. Proc. Natl. Acad. Sci. U. S. A. **2004**, *101* (3), 886–890.
5. Agüero, C.B.; Uratsu, S.L.; Greve, C.; Powell, A.L.T.; Labavitch, J.M.; Meredith, C.P.; Dandekar, A.M. Evaluation of tolerance to Pierce's disease and Botrytis in transgenic plants of *Vitus vinifera* L. expressing the pear PGIP gene. Mol. Plant Pathol. **2005**, *6* (1), 43–51.
6. Chiang, C.-H.; Wang, J.-J.; Jan, F.-J.; Yeh, S.-D.; Gonsalves, D. Comparative reactions of recombinant papaya ringspot viruses with chimeric coat protein (CP) genes and wild-type viruses on CP-transgenic papaya. J. Gen. Virol. **2001**, *82* (11), 2827–2836.
7. Dandekar, A.M.; Teo, G.; Defilippi, B.G.; Uratsu, S.L.; Passey, A.J.; Kader, A.A.; Stow, J.R.; Colgan, R.J.; James, D.J. Effect of down-regulation of ethylene biosynthesis on fruit flavor complex in apple fruit. Transgenic Res. **2004**, *13*, 373–384.
8. Peña, L.; Martín-Trillo, M.; Juárez, J.; Pina, J.A.; Navarro, L.; Martínez-Zapater, J.M. Constitutive expression of *Arabidopsis LEAFY* or *APETALA1* genes in citrus reduces their generation time. Nat. Biotechnol. **2001**, *19*, 263–267.
9. Broothaerts, W.; Keulemans, J.; Van Nerum, I. Self-fertile apple resulting from S-RNase gene silencing. Plant Cell Rep. **2004**, *22*, 497–501.
10. Redenbaugh, K.; McHughen, A. Regulatory challenges reduce opportunities for horticultural biotechnology. Calif. Agric. **2004**, *58* (2), 106–115.
11. Escobar, M.A.; Leslie, C.A.; McGranahan, G.H.; Dandekar, A.M. Silencing crown gall disease in walnut (*Juglans regia* L.). Plant Sci. **2002**, *163* (3), 591–597.
12. Driver, J.; Castillón, J.; Dandekar, A. Transgenic trap crops and rootstocks show potential. Calif. Agric. **2004**, *58* (2), 96–97.

Transgenic Crops: Regulatory Standards and Procedures of Research and Commercialization

Qifa Zhang
National Key Lab of Crop Genetic Improvement, Huazhong Agricultural University, Wuhan, China

Abstract

Although there have been widespread controversies regarding many safety aspects of transgenic plants as crops and food, there is an increasingly strong belief that transgenic plants can provide safe crops and food. This is because, in reality, transgenic technique does not pose any higher risk than traditional breeding methods to the safety of the crop products, and also because of the rigorous governmental regulations implemented by the countries that conduct transgenic research and grow GM crops. This entry presents a brief overview of the regulatory standards and procedures for the environmental release and commercialization of GM crops.

INTRODUCTION

The estimated global area of transgenic or genetically modified (GM) crops reached 52.6 million hectares in 2001.[1] The rapid growth of the area planted to GM crops in both developed and developing countries strongly indicates that GM crops are welcome by growers and consumers alike. Utilization of transgenic crops will be a vital alternative to provide the world with adequate food and other agricultural products.

REGULATORY STANDARDS AND PROCEDURES FOR TESTING AND CULTIVATING GM CROPS

The development of a transgenic crop consists of the following stages: laboratory research, confined field tests, environmental release, and commercialization. Governmental regulations in general address the following concerns in the process of commercialization of GM crops:

- Is the transgenic plant safe for the environment?
- Is the transgenic plant safe for agriculture?
- Is the transgenic plant product safe for use in foods, feeds, or other consumption?

Several strategies have been widely adopted in evaluating the safety of GM crops and foods. One of the strategies is based on the concept of substantial equivalence.[2] This approach acknowledges that the goal of the assessment is not to establish the absolute safety of GM crops and foods, but to evaluate their level of safety relative to their traditional counterparts, where such counterparts exist. Another important strategy commonly used in regulation is that the evaluation be done on a case-by-case basis.

For safeguarding the use of transgenic crops, science- and risk-based regulatory standards have now been established and implemented in many countries, and are being developed in many others. It should be noted that governmental regulation standards are also frequently bound to international agreements. Although there is substantial scientific commonality in the regulatory standards, the procedures followed in one country may be very different from those in another country. The cases of two countries, the United States and China, are given below as examples for demonstrating the common ground and differences in regulation of GM crops.

The United States was the first country to establish functional regulatory machinery for the biosafety of transgenic crops. A coordinated regulatory framework was set up in 1986 that assigned the responsibility to the Environmental Protection Agency (EPA), the U.S. Department of Agriculture (USDA), and the Food and Drug Administration (FDA), based on existing regulatory authorities of the U.S. government. A good starting point for information about regulation in the United States is the Web site "United States Regulatory Oversight in Biotechnology Responsible Agencies—Overview" (http://www.aphis.usda.gov/ppq/biotech/usregs.html). The basic premise for regulation of GM crops is that such crops shall not be fundamentally different from unmodified organisms or those produced by conventional methods. The key principle adopted by the regulatory framework is that it is the product, rather than the method of producing the product, that should be regulated.

The regulatory body of the USDA is the Animal and Plant Health Inspection Service (APHIS). This agency regulates the importing, transportation, and field testing of GM crops and determines the likelihood of a transgenic plant having negative agricultural or environmental effects (http://www.aphis.usda.gov/ppg/biotech). For common crops and traits, researchers need only notify the agency of their intention to transport or field test a transgenic plant. The researcher is responsible to ensure that the gene introduced meets certain technical criteria. For less common crops and for genes or traits that may pose greater risk, the researchers are required to file formal applications for permission to transport or plant the materials. Measures are required in field testing to prevent the spread of the transgene to the environment or into the food supply. Before commercializing a transgenic plant, the developer petitions APHIS for nonregulated status, which requires data on the introduced gene construct, plant biology, likely effects on the ecosystem, and field test reports, as well as data and information for any unfavorable effects. APHIS also has the authority to halt the sale of the GM crop if there is evidence that the GM crop is becoming a pest.

The FDA has authority to determine the safety of foods and food ingredients based on the concept of substantial equivalence, and relevant information can be found at the Web site (http://www.cfsan.fda.gov/~lrd/biotechm.html#label). The FDA adopts a voluntary consultation process with the GM crop developer, and reviews safety and nutritional data. If the introduced gene is from a known allergenic source, the GM food is required to be assessed for allergenicity. The FDA also has the authority to order the GM food's removal from the market, if there is evidence that it is unsafe.

The EPA regulates transgenic plants that contain plant-incorporated protectants, including plants engineered for insect or disease resistance that the agency refers to as plant pesticides (http:/www.epa.gov/pesticides/biopesticides/). This is implemented by granting an experimental use permit for testing, plant propagation registration, and full commercial registration, based on extensive reviews of data for the plant pesticide including its biochemical characteristics, toxicity, and environmental effects. The agency may require a resistance management plan in order to prevent or slow the development of resistance in the target pest. The EPA regulates herbicides, but not herbicide-tolerant plants. In the case of engineering a plant for herbicide tolerance, herbicides must be registered for a new use.

In addition to regulation at the federal level, many states in the United States also apply regulations on GM crops that require additional review and approval at the state level. Moreover, most research institutions have biosafety committees that monitor potentially hazardous transgenic research and ensure compliance with biosafety procedures.

In China, the biosafety of GM crops is regulated by the Ministry of Agriculture (MOA) (http://www.agri.gov.cn/ztzl/2001/0820/0820.htm). The MOA regulates the research, development, field tests, environmental release, and commercialization of GM crops. The MOA also regulates the import of GM crops and their products.

The process from laboratory research to commercialization of GM crops is divided into four stages in the Chinese regulation framework: laboratory research, confined field tests, environmental release, and product demonstration and commercialization. An MOA biosafety committee is responsible for reviewing the applications filed by researchers. In this system, notification of the MOA is required for transgenic research that may pose higher risk. For conducting field tests of a transgenic plant, a formal application must be filed with the MOA, which will be reviewed by the biosafety committee; a permit will subsequently be issued. In filing this application, the researcher is required to provide data about the gene construct, biology of the plant, test site, measures of confinement, and data to be collected. After a two-year field test, the transgenic plant can advance to environmental release when another application is filed with the MOA, submitting all data gained in the field test. Experimentation of environmental release in general will be conducted in another two consecutive years, after which another application should be filed with the MOA to advance the transgenic plant to product demonstration in large scale. At the end of the product demonstration, which usually requires two years to complete, the researcher can apply for a biosafety certificate for commercialization, which may require extensive data and comprehensive reviews. The transgenic plants can be used as parents for breeding purposes only when the biosafety certificate is issued, and the progenies from crosses with the transgenic parent must undergo the product demonstration procedure again and reapply for the biosafety certificate.

Information can also be found at Web sites for regulatory standards and procedures of the European Community (http://www.biosafety.be/Menu/BiosEur.html) and Japan (http://www.s.affrc.go.jp/docs/sentan/).

LABELING OF GM FOODS

Labeling of GM foods is a widely debated issue in many parts of the world. Although different countries' regulatory standards (described in previous sections) address all concerns of consumers and thus provide strong assurance for food safety, there are still diverse opinions about the necessity of labeling GM foods. Again, countries are divided over requirements and procedures for labeling the GM foods.

As described in the previous section, food safety in the United States is regulated by the FDA. The FDA requires that all labeling be truthful, informative and not misleading, and identifies no characteristics of the GM food would justify labeling it as a special class. The FDA requires labeling for GM foods in the same way as for other foods, including information about the composition, nutrition, and allergenicity concerns, but does not require special labeling of GM foods. Additional voluntary labeling is allowed, provided that it is truthful, informative, and not misleading.

The FDA has recently developed industry guidance on voluntary labeling.

In Japan, implementation of new regulations on GM food labeling was initiated in April 2001 (http://www.maff.go.jp/soshiki/syokuhin/hinshitu/organic/eng_yuki_top.htm). According to the authority, labeling is not for safety concerns but for consumers' right to choose or right to know. The labeling is implemented according to a positive list system, in which a committee is formed in the Ministry of Agriculture, Forestry, and Fishery (MAFF) to review the positive list of foods that should be labeled. In 2001, 15 soybean products and nine corn products were listed. The regulation requires that the food should be labeled if the DNA/protein of the transgene can be detected, the raw materials are among the top three constituents, and the GM materials are 5% by weight.

In China, labeling of transgenic organisms was enforced in March 2002 (http://www.agri.gov.cn/xxfb/2002/0107/dt0110-2.doc). Labeling is administered by the MOA, and regulation covers the transgenic organisms included in a list that is updated from time to time. The current list includes soybean, corn, rapeseed, cotton, and tomato. It regulates not only GM foods, but also seeds and other products from GMOs that are sold in the marketplace.

In Europe, a series of regulations applies concerning traceability and labeling of GMOs and of GMO food and feed products. (http://www.biosafety.be/Menu/BiosEur4.html). The regulation requires labeling of products consisting of or containing GMOs, foods and food ingredients (including food additives and flavorings produced from GMOs), feed materials, and compound feeding stuffs and feed additives produced from GMOs placed on the market in accordance with European Community legislation. Traceability requires that information about GMO products placed on the market be transmitted from one operator to the other in transaction, and that operators have in place systems and procedures to allow the identification of persons involved in the transactions for a period of five years.

CONCLUSION

Functional regulatory machinery has been established in many countries in the last decade. However, regulatory standards and procedures are still evolving, even in countries with a relatively long history of GM crop cultivation. With rapidly accumulating large-scale adoption and public recognition of the advantages of GM crops, it is believed that goals, standards, and procedures of regulation will gradually become globally harmonized and will promote the utilization and exploitation of the full benefits of GM crops, in order to better meet the demands of the ever-increasing world population.

REFERENCES

1. James, C. Global Review of Commercialized Transgenic Crops: 2001. In *International Service for the Acquisition of Agri-Biotech Application No. 24-2001*; **2001**.
2. Food and Agriculture Organization of the United Nations and World Health Organization Safety Aspects of Genetically Modified Foods of Plant Origin. In *Report of a Joint FAO/WHO Expert Consultation on Foods Derived from Biotechnology*; **2000**.

Transgenic Livestock: RNA Interference (RNAi)

Charles R. Long
Department of Veterinary Physiology and Pharmacology, Texas A&M University, College Station, Texas, U.S.A.

Abstract
Utilization of genetically engineered (GE) livestock species to enhance the production of economically important traits lags behind the advances made in GE crops. Nonetheless, the potential benefits of GE livestock are tremendous to the producer, the consumer, and also to the animal. Stably initiating the RNA interference pathway in animal cells can target both endogenous and exogenous genes and has paved the way for production of livestock with enhanced characteristics in animal health and production. Exploitation of these technologies could provide new opportunities to provide high quality nutrition to human populations where traditional animal agriculture is limited.

INTRODUCTION

The utilization of breeding and selection to alter the physical and biochemical characteristics of agriculturally important plants and animals has revolutionized production agriculture around the world. Livestock represent an important component of the world's food supply, and enhancing production through genetic selection has significantly improved the production traits in nearly every sector of the livestock industry. In livestock, innovations in production performance via GE lags behind the advances made in GE crops but, nonetheless, offer similar opportunities to increase the production capacity and efficiency of animal agriculture. To date, most applications of GE livestock have been to produce animals producing a protein that imparts a benefit either to the health of the animal or to make human pharmaceuticals.

The ability to specifically block gene function through knockout technology has been severely limited owing to the lack of robust embryonic stem cell techniques in livestock species. Knockout techniques in livestock have been developed using very inefficient techniques of homologous recombination in fetal fibroblasts and subsequent somatic cell nuclear transfer to produce the GE offspring. Although, these procedures are possible, they are highly inefficient and extremely expensive. Furthermore, knockout techniques can only be directed against endogenous genes and thus are ineffective against invading virus targets. However, by utilizing novel RNA interference based mechanisms, expression vectors can be inserted into the animal's genome that can suppress the expression of endogenous genes or viral genes infecting the host cell.

RNA interference (RNAi) is an evolutionarily conserved mechanism in eukaryotic cells where double-stranded RNA acts to silence the translation of mRNA(s) for which it has perfect or near perfect complementary sequence.[1,2] Endogenous genes known as microRNAs (miRNAs) fold into a double-stranded RNA when transcribed and processed. Previously thought to be genomic junk, miRNAs are now recognized as key regulators of cell function through their role in altering the expression of perhaps more than 30% of the entire genome. Double-stranded RNA capable of activating the RNAi pathway can be transfected into the cell from an exogenous source to elicit transient activation of RNAi. However, in the context of producing GE animals utilizing this phenomenon, the RNAi pathway can be activated by expression of double-stranded RNA molecules from an expression vector incorporated into the DNA of the animal. Transcription from this expression construct produces a short hairpin RNA (shRNA), which is then processed by the existing cellular machinery in a manner similar to the miRNAs. In all cases, the double-stranded RNA is ultimately processed by the enzyme dicer to form the functional unit of RNAi, the short interfering RNA (siRNA). The mechanism of how these non-coding RNAs produce such a profound cellular response has been reviewed elsewhere.[3] Recent studies have shown that the target of the siRNA can be either an endogenous gene product or the mRNA of viruses that have invaded the host cell. The ability to produce genetically modified (transgenic) animals that express shRNAs targeting viral genomes could be a novel approach to producing livestock and companion animals that are resistant to a whole range of viruses.

Although many important areas of livestock production can be addressed using transgenic technologies, here we will focus on the production of livestock with resistance to viral disease. Specifically, this entry will outline the utilization genetic engineering of livestock genomes to activate the RNAi pathway to prevent viral propagation in the host animal.

VIRUS DISEASE IN LIVESTOCK

Viruses pose a continuous and very serious threat to the health of livestock, companion animals, and humans. Viruses infect the cells of the host animal and utilize the cells' molecular mechanisms to replicate, producing more viruses that are then released to infect more and more cells of the host and often become transmitted to other animals.

Virus outbreaks in livestock are currently controlled by vaccination (when available), implementation of prudent livestock transportation and handling procedures, or ultimately treatment or eradication of infected animals. Vaccination protocols have been extremely effective against many virus diseases; however, there are numerous viral diseases affecting livestock that are not controlled by effective vaccines. Many viruses such as foot and mouth disease virus (FMDV) and equine infectious anemia virus (EIAV) continually evolve via random genetic mutation, which makes any vaccination to the plethora of virus serotypes and substrains extremely inefficient and rarely effective.

Thus, a novel method for controlling these viruses is needed at the molecular level, a method that is capable of blocking the production of new virus and allowing the host animal's immune system to eliminate remaining viruses. RNAi-based approaches to block the expression of viral genes have shown great promise in mammalian cells in vitro and are now ready for testing in GE animals.

INDUCTION OF RNAi IN ANIMAL CELLS

Transient Exposure

Activation of the cells RNAi pathway is initiated by the presence of double-stranded RNA. In non-mammalian systems, RNAi can be mediated by transient transfection of long double-stranded RNA molecules, which would then be processed by the enzyme dicer into the functional siRNAs. However, in most mammalian cells, the transfection of long double-stranded RNA activates the interferon response in most mammalian cells and leads to cell death. Transient transfection of double-stranded RNA less than 30 nucleotides in length and most often 19–23 are utilized to activate RNAi, without inducing cell death. Transient transfection of siRNAs in cultured mammalian and avian cells has been shown to be extremely effective at blocking the expression of viral genes, leading to decreased levels of virus production in transfected cells. Therefore, many different viruses are susceptible to RNAi-based therapeutics, which demonstrates the potential of these cells to block viral production.

Stable Expression

Genetic engineering strategies can be applied to modify the genome of mammalian cells and direct the expression of a variety of different gene products. Short "hairpin" RNA, which mimics the endogenous miRNA, can be effectively produced from an introduced DNA construct (transgene). Although several methods are available to stably integrate the transgene into the mammalian genome, retroviral vectors, and more specifically, lentiviral vectors have become the preferred method for delivering the shRNA expression constructs into the host cell genome. Lentiviruses modified to be replication incompetent and self-inactivating can integrate the transgene with high efficiency in both dividing and non-dividing cells, including early embryos and ova, which improves the efficiency of transgenic animal production.[4] Also, modification of these recombinant lentiviral vectors has made them a safe and effective delivery system for the small transgenes required to express the shRNA.[5]

Typically transcription of transgenic shRNA is initiated from a polymerase III type promoter, but more recently, methods for using more efficient polymerase II promoters have become available. By correct design of the promoter used to express the shRNA, tissue specific and possibly even inducible expression of the shRNA is now possible.[6] These innovations will help circumvent potential off-target effects and toxicity issues. Furthermore, because the shRNAs do not code for any protein products, immunological reactions in the animal or for potential human consumption are not a concern.

AGRICULTURE AND VETERINARY APPLICATIONS OF RNAi

Inhibiting viral gene translation, to block various aspects of virus function in the host cell has been demonstrated in a variety of in vitro studies for various classes of viruses.[7–9] Consequently, mammalian and avian cells have the capacity to effectively limit viral processes through the RNAi mechanism. Thus, the key components are present to effectively produce animals that have innate viral resistance via expression of shRNA targeting virus genes.

Preventing the virus from ever entering the host cell would also be an effective strategy to stem the spread of viral disease. Targeting endogenous genes via RNAi that are critical for viral binding and incorporation would produce animals in which viruses are unable to enter the host, thus, unable to replicate and would be quickly eliminated from the animal by the immune system. The first steps to this approach have now been accomplished for the human immunodeficiency virus in vitro and now require further testing in an in vivo animal model.[10]

A primary concern of any defense against viruses is the possibility of viral escape from the treatment owing to the innate and ongoing random mutation of the viral genome. This ability to form functional virus with a slightly altered genome is common among some of the most devastating viruses and leads to difficulties in producing effective vaccines and RNAi mediated defenses. However, novel approaches have been utilized to combine multiple shRNAs into a single lentiviral vector which when expressed from the host cell generates multiple unique targeting regions in the virus genome. By blocking virus replication simultaneously in a range of distinctive sites, the chance of the random virus mutation circumventing this combinatorial approach is greatly reduced. In fact, these strategies have been used to successfully block HIV replication when previous single shRNA expression studies allowed HIV escape.[11]

Non-viral disease is also an important consideration for genetic engineering strategies. Targeting of the endogenous genes that provide a health advantage both to the animal and potentially to humans is an obvious application. One such application is to initiate a downregulation of cellular proteins that are necessary for disease progression in the host. One such application is the downregulation of the prion protein, which is the causative agent in transmissible spongioform encephalopathies (TSE). Initial studies suggest that stable expression of shRNA targeting the prion mRNA can reduce prion protein levels to less than 10% of control in a goat model.[12] Therefore, the likelihood of an animal succumbing to a TSE or to transmit aberrant prions to other animals, including humans, is greatly diminished.

CONCLUSION

Future utilization of RNAi in veterinary medicine and animal agriculture could be realized through a variety of different methods. Products and animals are being developed to take advantage of both therapeutic and prophylactic applications of RNAi. Production of livestock with enhanced health and production traits by utilization of RNAi offers a tremendous opportunity to supply food to people in environments not currently suitable for conventional animal agriculture.

REFERENCES

1. Fire, A.; Xu, S.; Montgomery, M.K.; Kostas, S.A.; Driver, S.E.; Mello, C.C. Potent and specific genetic interference by double-stranded RNA in *Caenorhabditis elegans*. Nature **1998**, *391*, 806–811.
2. Hannon, G.J. RNA interference. Nature **2002**, *418*, 244–251.
3. Bartel, D.P. MicroRNAs: genomics, biogenesis, mechanism, and function. Cell **2004**, *116*, 281–297.
4. Hofmann, A.; Kessler, B.; Ewerling, S.; Weppert, M.; Vogg, B.; Ludwig, H.; Stojkovic, M.; Boelhauve, M.; Brem, G.; Wolf, E.; Pfeifer, A. Efficient transgenesis in farm animals by lentiviral vectors. EMBO Rep. **2003**, *4*, 1054–1060.
5. Pfeifer, A. Lentiviral transgenesis—a versatile tool for basic research and gene therapy. Curr. Gene Ther. **2006**, *6*, 535–542.
6. Szulc, J.; Aebischer, P. Conditional gene expression and knockdown using lentivirus vectors encoding shRNA. Methods Mol. Biol. **2008**, *434*, 291–309.
7. Chen, W.; Yan, W.; Du, Q.; Fei, L.; Liu, M.; Ni, Z.; Sheng, Z.; Zheng, Z. RNA interference targeting VP1 inhibits foot-and-mouth disease virus replication in BHK-21 cells and suckling mice. J. Virol. **2004**, *78*, 6900–6907.
8. Hu, W.Y.; Myers, C.P.; Kilzer, J.M.; Pfaff, S.L.; Bushman, F.D. Inhibition of retroviral pathogenesis by RNA interference. Curr. Biol. **2002**, *12*, 1301–1311.
9. McCaffrey, A.P.; Nakai, H.; Pandey, K.; Huang, Z.; Salazar, F.H.; Xu, H.; Wieland, S.F.; Marion, P.L.; Kay, M.A. Inhibition of hepatitis B virus in mice by RNA interference. Nat. Biotechnol. **2003**, *21*, 639–644.
10. Anderson, J.; Akkina, R. HIV-1 resistance conferred by siRNA cosuppression of CXCR4 and CCR5 coreceptors by a bispecific lentiviral vector. AIDS Res. Ther. **2005**, *2*, 1.
11. ter Brake, O.; Konstantinova, P.; Ceylan, M.; Berkhout, B. Silencing of HIV-1 with RNA interference: a multiple shRNA approach. Mol. Ther. **2006**, *14*, 883–892.
12. Golding, M.C.; Long, C.R.; Carmell, M.A.; Hannon, G.J.; Westhusin, M.E. Suppression of prion protein in livestock by RNA interference. PNAS **2006**, *103*, 5285–5290.

Transgenic Plant Risk: Coexistence and Economy

Gretchen Mosher
Charles Hurburgh
Department of Agricultural and Biosystems Engineering, Iowa State University, Ames, Iowa, U.S.A.

Abstract
Transgenic crops have caused much debate in global agriculture. To allow farmers the freedom of choice regarding transgenic products, coexistence policies have been developed and implemented in many countries. This entry will examine scientific findings related to coexistence of transgenic and non-transgenic crops, review economic implications of coexistence measures, and evaluate the importance of these concepts to the scientific community.

INTRODUCTION

Although the development of transgenic plant technologies were heralded as a major scientific breakthrough in plant development, they have not been universally accepted. To preserve the freedom of choice for agricultural producers, coexistence policies have been developed and implemented in many countries. This entry will review scientific considerations influencing coexistence practices and strategies, examine economic implications of these practices, and summarize their importance to the scientific community.

COEXISTENCE

The use of transgenic crops has increased globally each year since their commercial introduction in 1996, with 25 countries growing the crops in 2008.[1] Although accepted and grown by many developed and developing nations, acceptance in the European Union (EU) and some Asian nations has been limited. Because of this, European Union policymakers have had a sustained debate on practices and regulations for the coexistence of transgenic and non-transgenic crops.[2] In 2003, the Commission of European Communities (CEC) defined coexistence as:

> The ability of farmers to make a practical choice between conventional, organic, and GM crop production, in compliance with the legal obligations of European labeling and/or purity standards.[3]

Coexistence does not address environmental safety or risks to human health because full regulatory approval of the transgenic crop is assumed. The main focus of coexistence in the European Union is the potential economic impact of accidental mixture of transgenic crops with crops whose value depends heavily on solid verification of their non-transgenic status. Accidental and trace components of transgenic material found in non-transgenic products is defined as adventitious presence (AP), and the aim of coexistence measures is to keep AP levels below specified tolerance levels. Although absolute purity is often the goal for non-transgenic crops, a purity of 100% is generally unattainable, given the open air environment of most agricultural fields. In 2003, the European Union declared a tolerance level of 0.9% for AP of transgenic material in food and feed.

SOURCES OF ADVENTITIOUS PRESENCE

Several scenarios have been identified as possible sources of adventitious presence in non-transgenic crops. Three major sources of AP are listed in Table 1 along with suggested mitigation strategies and factors which may affect the probability of occurrence.

Adventitious presence resulting from cross pollination or gene flow can be more difficult to contain than that from impure seeds or unclean machinery. Although several factors influence pollen dispersion in plants, some species of plants harbor natural restrictions to gene transfer because of their pollination properties. Self-pollinating crops such as soybeans and cotton limit gene transfer and pollen exchange, but open pollination plants such as maize and canola have a greater chance of facilitating cross pollination under certain conditions.[5]

CONTAINMENT OF POLLEN FLOW

Research in cross pollination and gene transfer has concentrated on maize because of its wide usage, global economic importance, and open pollination properties. Based on the

Table 1 Sources of adventitious presence.

Source	Suggested mitigation strategies	Variability factors
Presence of transgenic seeds in non-transgenic lots	• Implement quality management system for seed production • Planting and flowering lag time • Isolation distances • Use of dedicated farm equipment • Use of non-transgenic buffer zones	• Size of field • Wind direction and strength • Plant pollination properties • Planting and flowering time lag
Cross-pollination of transgenic plants to non-transgenic plants	• Isolation distances • Buffer zones of non-transgenic plants • Flowering and planting lag time	• Size of fields • Proximity of fields • Wind direction and strength • Size and heaviness of pollen • Regional weather patterns
Comingling due to harvest machinery residue	• Thorough cleaning of machines • Harvesting non-transgenic plants before transgenic plants • Use of dedicated machinery	• Machine design • Handling practices

Data compiled from various sources.[4–6]

examination of many scenarios of coexistence with maize, several key factors have been identified as having a large influence on the level of pollen and genetic material that is transferred between plants. Isolation distance between fields, staggered planting and flowering times, wind patterns, field size and proximity, and the properties of buffer zones have all been shown to play a significant role in controlling cross pollination.[4,6,7]

Typically researchers manipulate variables in either a field-based environment or within a computer simulation to determine how each factor must be managed to meet specified tolerance levels. Overwhelmingly, data have shown that coexistence of transgenic and non-transgenic plants is possible at a 0.9% tolerance level. However, adjustments must be made to manage controllable factors such as isolation and buffer distances and planting and flowering lag times.[4,6,7]

Generally, findings indicate the greater the isolation or buffer distances, the less important staggered planting and flowering times are because of the pollen's limited ability to travel long distances. Smaller isolation and buffer distances require greater efforts to differentiate planting and flowering times of maize. Wind patterns and the size and proximity of the field to transgenic planted areas may add uncontrollable and often unpredictable effects.

ECONOMIC IMPLICATIONS

Although researchers have found that transgenic and non-transgenic crops can coexist in a variety of settings and scenarios, the second major question when determining coexistence policy involves the calculation of economic implications of these measures.

Although the specific economics vary between farms and regions, costs of segregation across all regions have three major elements. First are costs that involve cleaning of planting and harvesting equipment. These are typically the smallest of all segregation costs. A second component is the cost involved with dedicating equipment for use with only one handling channel (transgenic or non-transgenic), often the largest cost. Additional costs are incurred by non-transgenic growers when they elect not to plant transgenic varieties, which have been shown to reduce overall planting cost. With maize, additional costs result from the required isolation and forfeiture of crops grown with buffer zones and the loss of potential crops not grown in isolation areas.[4,8] Tolerance levels have a strong negative relationship with segregations costs, with lower tolerance levels causing higher segregation costs.[8] In some cases, segregation costs may be up to 35% higher than baseline costs of production.[9] An unresolved debate concerns which party should bear these increased costs.

ECONOMIC EFFECTS

Although economists have attempted to model innovation and welfare effects for the introduction, production, and coexistence measures of transgenic crops, the various models have multiple effects and are often accompanied by several uncertain assumptions, including the incentive for non-transgenic products, consumer demand for the differentiated product, and a willingness to pay for the incentives.

Several economic models that have reported positive welfare effects for the European Union have failed to account for differentiated consumer demand for non-transgenic products and the potential costs of segregation.[10] When these effects are included in the model or calculated as externalities, welfare tends to decrease.[10] However, if consumers do not perceive the non-transgenic product as more valuable, the basis for coexistence measures and segregation do not exist, and this could raise the welfare of transgenic cultivation dramatically.[2]

The continuing debate on who pays for segregation remains unresolved by economics. In many economic models, non-transgenic producers, as marketers of the presumed superior (albeit weakly superior) and presumptive recipients of a portion of any market incentive for the product, would assume the task of necessary segregation and identity preservation activities. However, current coexistence measures within the EU have shifted the responsibility for the segregation procedures to transgenic producers, as users of the technology in question. This mode of regulation places additional liability on innovators within the market, reducing social welfare and limiting further innovation within European agriculture.[10,11]

Strict regulations limiting innovation have been shown to decrease social welfare.[9] However, if the innovation of widespread introduction of transgenic crops is considered with indifferent consumer demand and less restrictive regulatory requirements, European farmers could be afforded substantial gains from transgenic plants. A more flexible system of regulation allowing for the heterogeneous environment in European agriculture would allow for greater gains in innovation and welfare.[2,10]

The magnitude of trade-offs between transgenic and non-transgenic is still under considerable debate, in part because of the complexity of measuring economic impact of the transgenic technology with uncertainties in consumer acceptance, regulatory limitations, and economic incentives for both transgenic and non-transgenic products. Although positive welfare gains are noted when portions of the model are assumed constant and the innovation is considered alone, negative welfare is noted when externalities such as regulatory requirements and tolerance levels are considered at current levels.

In democratic societies, scientific innovations often drive changes in the economics of large-scale systems, and the case of transgenic plants is no different. The scientific community's role will supply new data and refine current information to sharpen future policy decisions and improve existing products. The role of science in determining best practices for coexistence policies and procedures is imperative for the field of agriculture to be able to take strategic advantage of transgenic technology, allowing customer choice while maximizing economic welfare of both farmers and consumers.

CONCLUSION

To preserve the freedom of choice for all types of agriculture within the European Union, science-based coexistence policies must be developed. Although research has demonstrated the possibility of coexistence between transgenic and conventional plants, several factors must be managed so tolerance levels can be met. Although the introduction of transgenic crops in Europe has been modeled as an increase in social welfare, heavy regulation, uncertain consumer demand, and unknown incentives for differentiated products in the marketplace have led to negative impacts on welfare.

REFERENCES

1. James, C. *Global Status of Commercialized Biotech/GM Crops: 2008*, Brief 39; International Service for the Acquisition of Agri-Biotech Applications: Ithaca, New York, 2008.
2. Demont, M.; Devos, Y. Regulating coexistence of GM and non-GM crops without jeopardizing economic incentives. Trends Biotechnol. **2008**, *26* (7), 353–358.
3. Commission of the European Communities. *Commission Recommendation on Guidelines for the Development of National Strategies and Best Practices to Ensure the Co-Existence of Genetically Modified Crops with Conventional and Organic Farming*, Document No. C (2003) 2624; Commission of European Communities: Brussels, Belgium, 2003; 6–7.
4. Messean, A.; Angevin, F.; Gomez-Barbero, M.; Menrad, K.; Rodriguez-Cerezo, E. *New Case Studies on the Coexistence of GM and Non-GM Crops in European Agriculture*, EUR 22102 EN; European Join Commission Joint Research Centre: Seville, France, 2006; 21–54.
5. Council for Agricultural Science and Technology. *Implications of Gene Flow in the Scale-Up and Commercial Use of Biotechnology-Derived Crops: Economic and Policy Considerations*, Issue Paper Number 37; Council for Agricultural Science and Technology: Ames, IA, 2007; 3–11.
6. Weber, W.E.; Bringezu, T.; Broer, I.; Eder, J.; Holz, F. Coexistence between GM and non-GM maize crops—tested in 2004 at the field scale level (Erprobungsandbau 2004). J. Agron. Crop Sci. **2007**, *193*, 79–92.
7. Halsey, M.E.; Remund, K.M.; Davis, C.A.; Qualls, M.; Eppard, P.J.; Berberich, S.A. Isolation of maize from pollen-mediated gene flow by time and distance. Crop Sci. **2005**, *45* (6), 2172–2185.
8. Bullock, D.S.; Desquilbet, M. The economics of non-GMO segregation and identity preservation. Food Policy **2002**, *27* (1), 81–99.
9. Kalaitzandonakes, N.; Magnier, A. Biotech labeling standards and compliance costs in seed production. Choices Magazine **2004**, 2nd quarter, 1–5.
10. Moschini, G.C. Pharmaceutical and industrial traits in genetically modified crops: coexistence with conventional agriculture. Am. J. Agric. Econ. **2006**, *88* (5), 1184–1192.

11. Beckmann, V.; Soregaroli, C.; Wesseler, J. Coexistence rules and regulations in the European Union. Am. J. Agric. Econ. **2006**, *88* (5), 1193–1199.

BIBLIOGRAPHY

1. Bock, A.; Lheureux, K.; Libeau-Dulos, M.; Nisagard, H.; Rodriguez-Cerezo, E. *Scenarios for Coexistence of Genetically Modified, Conventional, and Organic Crops in European Agriculture*, Report EUR 20394 EN; European Commission Joint Research Centre: Seville, France, 2002, 52–64, 101–122.
2. Moschini, G.; Lapan, H. Labeling regulations and segregation of first generation GM products: innovation incentives and welfare effects. In *Regulating Agricultural Biotechnology: Economics and Policy*; Just, R.E., Alston, J., Zilberman, D., Eds.; Springer: New York, 2006; 263–282.

Transgenic Plants: Economic and Environmental Risks and Gene Flow

Alan Frank Raybould
Product Safety, Syngenta, Bracknell, U.K.

Abstract
The economic and environmental risks of gene flow from transgenic crops are described. The economic risks are the presence of transgenes above permitted thresholds in agricultural commodities, which may trigger fines and disruption of trade. The environmental risks include damage to crop production or non-agricultural habitats that could arise from increased weediness of wild relatives of crops. Methods to assess and manage the risks are described.

INTRODUCTION

Gene flow is the movement of genetic information among populations of the same or closely related species. In plants, pollen and seed are the main agents of gene flow. Many plants have traits that promote the dispersal of pollen, such as flowers that attract pollinators, or pollen grains able to travel long distances by wind. Other plants restrict pollen dispersal, for example by self-pollinating before their flowers open. Plants in general have traits that encourage seed dispersal: structures to help seeds stay airborne or catch the fur of animals, hard coatings to survive digestion by birds and mammals, and so on. Seeds of some agricultural weeds mimic crop seeds to aid dispersal by man. Some seeds can also disperse in time as well as space by remaining dormant in soil for many years before germinating. Plant genes can also move into unrelated organisms by direct uptake of DNA from the environment; such movement of genes by asexual reproduction is called horizontal gene transfer.

Transgenes in crops have the potential to move like any other gene; therefore, quantification of gene flow among non-transgenic crop varieties and establishment of the limits of gene flow between non-transgenic crops and other species helps to predict the movement of transgenes, and can be important when assessing the risks from cultivating transgenic crops. The purpose of this entry is to describe the main economic and environmental risks associated with gene flow from transgenic crops.

CROP-TO-CROP GENE FLOW

Transgenic crops are strictly regulated: usually approvals from government regulatory authorities are needed before seed of a transgenic crop can be sold for commercial cultivation and before the crop or its derivatives can be used for human food or animal feed. Regulations may also specify limits for transgenic material in certain commodities. Crop-to-crop gene flow via pollen and the mixing of seed during harvesting, transportation, or processing (comingling) can cause the unintended presence of small amounts of one type of seed in batches of another. Such "adventitious presence" poses serious economic risks to those who are liable when transgenic products are found above regulatory thresholds.

Because regulatory decisions are not made simultaneously worldwide, a transgenic crop can be approved for cultivation in some countries, while having no approvals for food or feed use elsewhere. In such cases, the legal threshold of unapproved transgenic material in commodities for import may be zero. The sensitivity of detection methods for transgenes means that extremely low amounts of adventitious presence can be detected; hence, if there are cultivation approvals for a transgenic commodity crop in one country, but no import approvals in some countries that import commodities from that country, strict segregation ("channeling") of that product may be needed to manage the risk to trade that could result from detection of unapproved transgenic material in commodities for import.[1]

Management of gene flow through spatial or temporal isolation, or by use of barrier crops, may also be needed to manage coexistence of transgenic and non-transgenic crops within countries that have granted a cultivation approval.[2] Some jurisdictions require labeling of agricultural products that exceed a certain content of approved transgenic material, creating the possibility of liability if a product loses value through being labeled as transgenic because of adventitious presence. If plants are used to produce industrial or pharmaceutical products, strict risk management will be required to prevent these products entering the food chain. Likely zero tolerance of industrial and pharmaceutical chemicals in food may mean that non-food crops must be used to produce these products.[3]

Crop-to-crop gene flow has also raised concerns about loss of genetic variation in traditional crops, particularly in their centers of origin or diversity. Loss of genetic resources is more likely from replacement of landraces by high-yielding crops, whether transgenic or not, than by crop-to-crop gene flow from transgenic crops.[4] Nevertheless, crop-to-crop gene flow may need to be controlled by strict risk management to comply with laws that seek to minimize the frequency of transgenes in local varieties for cultural or socioeconomic reasons.[4]

GENE FLOW FROM CROPS TO WILD RELATIVES

Gene flow between crops and wild relatives (wild plants sexually compatible with the crop) is common and has been implicated in the evolution of several agricultural weeds.[5] Despite gene flow from wild relatives to crops appearing to be more common than from crops to wild relatives in weed evolution, it is understandable why the possibility of problematic weeds arising from the movement of transgenes into wild relatives was one of the earliest concerns raised about the cultivation of transgenic crops;[6] however, what began as a genuine scientific problem, eventually developed into unsubstantiated scares about wild plants evolving into "superweeds," and the implication that there is such uncertainty about their properties that the cultivation of any transgenic plant amounts to a dangerous uncontrolled experiment.[7] Environmental risk assessments suggest that these fears are overstated.

In many countries there are no known wild relatives of the main commercially cultivated transgenic crops. For example, in the United States and Europe, there are no wild relatives of maize and soybean; hence, here the likelihood of more harmful weeds evolving as the result of gene flow from these crops is remote. Risk is minimal because exposure is minimal.

If there is uncertainty about whether there are wild relatives of a crop in a region, a simple series of sequential ("tiered") tests can be carried out:[8]

Tier I—test for hybrids between the crop and the putative wild relative using laboratory methods (hand pollination, embryo rescue, etc.). If no hybrids are detected, stop testing; if hybrids are detected, go to tier II.
Tier II—test for "spontaneous" hybrid production by placing the crop and the putative wild relative together in the laboratory or the field. If no hybrids are detected, stop testing; if hybrids are detected, go to tier III.
Tier III—make a systematic search for naturally produced hybrids.

If the possibility of gene flow from a transgenic crop cannot be ruled out with sufficient confidence, some regulators implement risk management and restrict cultivation of the transgenic crop to areas where wild relatives are absent (Fig. 1). For example, the U.S. Environmental Protection Agency (EPA) forbids cultivation of transgenic insect-resistant cotton in Florida, Hawaii, Puerto Rico, and the U.S. Virgin Island because cotton grows wild in these places;[9] and regulators in centers of origin and diversity may restrict cultivation of transgenic crops if gene flow to wild relatives raises concerns about crop genetic resources.[4]

Risk management may not be necessary if it can be shown that transgenic wild relatives are unlikely to cause environmental harm; that is, they pose low hazard (Fig. 1). Invasive alien plant species cause environmental harm in several ways:[10] by displacing native species, including wild relatives of crops that provide valuable genetic resources;[4] by causing physical changes, such reduced water supply, changed nutrient cycles, and increased incidence of fires; by reducing yields in arable agriculture and in semi-natural pasture; and through the costs of control. Similar changes could result if transgenic wild relatives become more abundant or more persistent than non-transgenic wild relatives; thus, an effective method to show low hazard of transgenic hybrids is to test the hypothesis of no change in plant traits that are likely to determine the population size of wild relatives.

Hazard may be assessed by testing two hypotheses: there are no unintended effects of transformation on characters likely to determine weediness, and the intended effects of the transgene will not increase the abundance or persistence of the wild relative. Although it is not possible to prove these hypotheses and hence be certain that there is no risk, if the hypotheses are corroborated under conditions most likely to falsify them, low hazard, and hence low risk of gene flow can be concluded with confidence.[11]

The hypothesis of no unintended effects of transformation can be rigorously tested by comparing transgenic and non-transgenic crops that differ only by the presence or absence of the transgene and DNA tightly linked to the transgene in controlled field trials. Many such trials are performed during the development of a transgenic crop. If the transgenic and non-transgenic lines do not differ significantly in characters such as growth rate, height, seed dispersal, and dormancy, the hypothesis of no unintended effects on weediness potential is corroborated. Testing cannot prove that unintended effects of transformation will not appear in the genetic background of the wild relative; however, testing for unintended effects in several crop genetic backgrounds increases confidence that no effects will arise in the wild relative.

Methods to test the hypothesis that intended effects of the transgene will not increase the abundance of wild relatives will vary greatly depending on the trait conferred. Concern about intended effects has focused on protection from

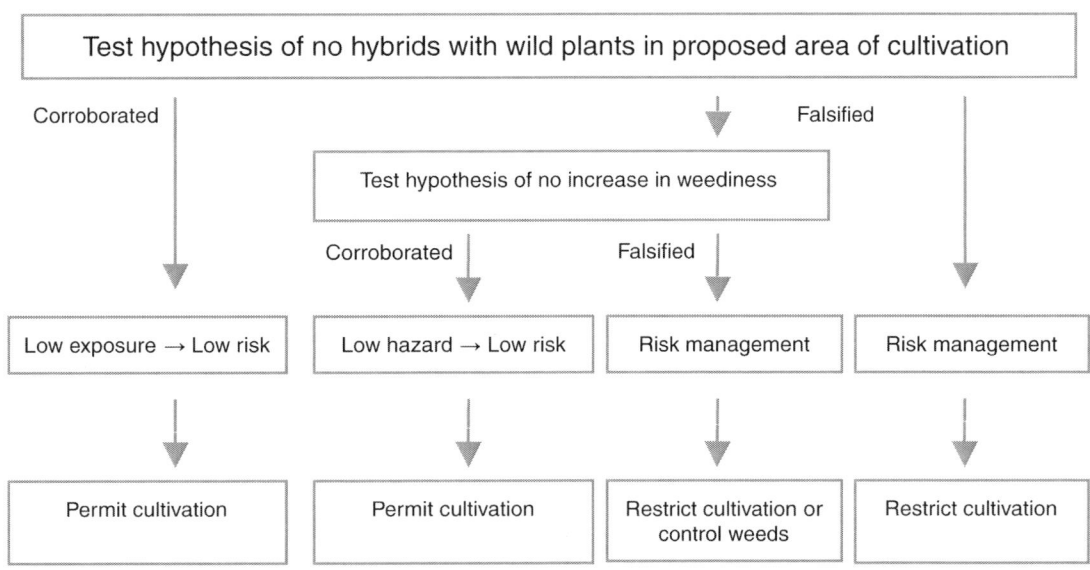

Fig. 1 A conceptual model for determining the environmental risks of hybrids between transgenic crops and wild plants.

environmental factors that may control the abundance wild relatives; for example, if insect damage controls population size, a transgene that confers insect resistance may lead to increased abundance. The hypothesis that this effect does not occur could be tested by observing whether populations of the wild relative increase when protected from insect damage by netting or insecticides. The potential effects of a disease-resistant gene can be predicted by testing whether the wild relative is already immune to the disease, and if so, whether the disease reduces growth or survival, and whether the wild relative is infected by the disease in the field.[8] Results of such experiments indicate that pest and disease resistance transgenes are unlikely to be hazardous in wild relatives.[8,11] Transgenes for drought and salt tolerance will require similar tests to assess their weediness hazard.

Risk management by restricting cultivation could be implemented if a weediness hazard is revealed. Alternatively, cultivation could be permitted but with measures to reduce risk (Fig. 1); for example, risk management of herbicide-tolerant crops could involve removal of wild relatives that flower concurrently in and around the crop. Periodic use of herbicides with a different mode of action from the one to which the crop is tolerant would control any wild relatives that acquired the herbicide-tolerance transgene.

HORIZONTAL GENE TRANSFER

Horizontal gene transfer (HGT) is the movement of genes between organisms without sexual reproduction. It is most common in microorganisms that can take up DNA from the environment and incorporate it into their genomes. There have been concerns that transgenes from crops could be incorporated into the genomes of bacteria in soil in which transgenic crops are cultivated, and that the transformed microorganisms could cause ecological problems.

The likelihood of harm from HGT of transgenes to soil microorganisms from transgenic crops is negligible. HGT depends on high sequence similarity between the DNA taken up and a gene in the microorganism; however, even when plant transgenes are derived from soil bacteria, sequence similarity is likely to be too low for HGT because genes from bacteria are highly modified before transfer to plants to ensure efficient protein expression. In addition, should HGT occur, protein expression from the transgene is improbable because the gene is optimized for expression in plants, and because plant gene promoters are unlikely to function in bacteria. Finally, it is not clear what selective advantage plant transgenes could confer to soil bacteria: most transgenes in current commercial transgenic crops are derived from plants or soil bacteria, and therefore there has been substantial prior exposure of soil bacteria to similar genes. Also, any effects on soil microflora are likely to be temporary and minor compared with variation associated with changing seasons, crop species, and soil management.[12]

CONCLUSION

Gene flow from transgenic crops poses economic and environmental risks. In some situations, the economic risks are high: crop-to-crop gene flow and mixing of seed can lead to adventitious presence of transgenes above permitted thresholds in agricultural commodities, which may trigger fines and disruption of trade. The environmental risks of gene flow from current commercial cultivation of transgenic crops are low. In many cases, there are no wild relatives in the regions where the crops are cultivated. If wild relatives are present, risk can be managed by forbidding the cultivation of transgenic crops. Methods to assess the hazards of transgene flow are available, and as confidence in these methods grows, cultivation of transgenic crops that pose low hazard may be increasingly permitted in areas where wild relatives occur.

REFERENCES

1. Newell-McGloughlin, M. A retrospective prospective perspective on agricultural biotechnology ten years on. J. Commer. Biotechnol. **2006**, *13* (1), 20–27.
2. Devos, Y.; Reheul, D.; De Schrijver, A. The co-existence between transgenic and non-transgenic maize in the European Union: a focus on pollen flow and cross-fertilization. Environ. Biosafety Res. **2005**, *4* (2), 71–87.
3. Moschini, G. Pharmaceutical and industrial traits in genetically modified crops: co-existence with conventional agriculture. Am. J. Agric. Econ. **2006**, *88*, 1184–1192.
4. Engels, J.M.M.; Ebert, A.W.; Thormann, I.; de Vicente, M.C. Centres of crop diversity and/or origin, genetically modified crops and implications for plant genetic resources conservation. Genetic resources and Crop Evolution **2006**, *53* (8), 1675–1688.
5. Ellstrand, N.C.; Prentice, H.C.; Hancock, J.F. Gene flow and introgression from domesticated plants into their wild relatives. Annu. Rev. Ecol. Syst. **1999**, *30* (1), 539–563.
6. Colwell, R.K; Norse, E.A.; Pimentel, D.; Sharples, F.E.; Simberloff, D. Genetic engineering in agriculture. Science **1985**, *229* (4709), 111–112.
7. Trewavas, A.J.; Leaver, C.J. Is opposition to GM crops science or politics? EMBO Rep. **2001**, *2* (6), 455–459.
8. Raybould, A.; Cooper, J.I. Tiered tests to assess the environmental risk of fitness changes in hybrids between transgenic crops and wild relatives: the example of virus resistant *Brassica napus*. Environ. Biosafety Res. **2005**, *4* (3) 127–140.
9. Mendelsohn, M.; Kough, J.; Vaitusis, Z; Matthews, K. Are Bt crops safe? Nat. Biotechnol. **2003**, *21* (9), 1003–1009.
10. Pimentel, D.; McNair, S.; Janecka, J.; Wightman, J.; Simmonds, C.; O'Connell, C.; Wong, E.; Russel, L.; Zern, J.; Aquino, T.; Tsomondo, T. Economic and environmental threats of alien plant, animal and microbe invasions. Agric. Ecosyst. Environ. **2001**, *84* (1), 1–20.
11. Raybould, A. Ecological versus ecotoxicological methods for assessing the environmental risks of transgenic crops. Plant Sci. **2007**, *173* (6), 589–602.
12. Widmer, F. Assessing effects of transgenic crops on soil microbial communities. Adv. Biochem. Eng. Biotechnol. **2007**, *107*, 207–234.

Transgenic Plants: Protein Quality Improvements

Jesse M. Jaynes
Department of Biological and Physical Sciences, Kennesaw State University, Kennesaw, Georgia, U.S.A.

Abstract
With respect to human and animal nutrition, most seeds do not provide a balanced source of protein because of deficiencies in one or more of the essential amino acids in the storage proteins. Consumption of proteins of unbalanced composition of amino acids can lead to a malnourished state which is most often found in people inhabiting developing countries where plants are the major source of protein intake. Thus the development of a more nutritionally balanced protein for introduction into plants takes on extreme importance.

INTRODUCTION

Past expansions of the world's food supply have relied primarily on plant breeding directed toward improving yields, increases in available cultivable lands, and augmentation of irrigation techniques. Because we are now encountering further constraints in all of these areas, future emphasis must include enhancing the "nutritional content" of the world's basic food and feed crops, especially those that are indigenous to developing nations. Such nutritional enhancement can result in lowering per capita intake of plant-based food crops, ultimately making more food available for expanding populations. These developments are made possible through advances in the fields of biochemistry and molecular biology, which has caused the "biotechnology revolution."

The composition of plant storage proteins, a major food reservoir for developing seeds, roots, and tubers, determines the nutritional value of plants and grains when they are used as foods and feed for humans and domestic animals. The amount of protein varies with genotype or cultivar, but in general, cereals contain 10% of the dry weight of the seed as protein, while in legumes, the protein content varies between 20% and 30% of the dry weight. Roots and tubers retain far less, generally around 2–3%. In many seeds, storage proteins account for 50% or more of the total protein and thus determine the protein quality of seeds. Each year, the total world cereal harvest amounts to some 1700 million tons of grain. This yields about 85 million tons of cereal storage proteins harvested each year and contributes about 55% of the total protein intake of humans. It has been difficult to produce significant increases in the level of protein and essential amino acids of crop plants utilizing classical plant breeding approaches. This is primarily because of the fact that the genetics of plant breeding is complex and that an increase in either trait may be offset by a loss in other agronomically important characters. In addition, it is probable that the storage proteins are very conserved in their structure and their essential amino acid composition would be little modified by these conventional techniques.

PAST WORK

Over the last two decades, much work has been performed in an attempt to improve the nutritional quality of plant storage proteins by transferring heterologous storage protein genes from other plants.[1–3] The development of genetic engineering and the various gene transfer systems have made this approach possible. Genes encoding storage proteins containing a more favorable amino acid balance, by and large, do not exist in the genomes of major crop plants. Furthermore, modification of native storage proteins has met with difficulty because of their instability, low level of expression, and limited host range. However, there has been some success in recent years in improving the content of single amino acids using this approach. For example, 2S methionine-rich Brazil nut albumin (18% methionine) has been used to enhance levels of seed protein methionine in canola. A chimeric gene regulated by a phaseolin promoter was fused to the 17-kDa Brazil nut albumin and expressed in transgenic canola plant seeds. The methionine-rich protein exhibited temporal regulation with significant accumulation of the protein late in development, thereby correlating with that of wild-type 11S-canola seed protein. There was a 33% increase in the methionine level, as well as a 4% increase in the total protein level.[4] In the case of Brazil nut 2S albumin, the highly allergenic nature of the protein, however, renders it unsuitable for use in food plants. A possible alternative to the chimeric gene approach would be to design de novo a more nutritionally balanced protein that retains certain characteristics of the natural storage proteins of plants, yet contains all of the essential amino acids at their proper ratio for the feeding of humans and animals.

The biosynthesis of amino acids from simpler precursors is a process vital to all forms of life as these amino acids are the building blocks of proteins. Organisms differ markedly with respect to their ability to synthesize amino acids. In fact, virtually all members of the animal kingdom are incapable of manufacturing some amino acids. There are 20 common amino acids that are utilized in the fabrication of proteins and essential amino acids are those protein building blocks that cannot be synthesized by the animal. It is generally agreed that humans require 8 of the 20 common amino acids in their diet: isoleucine, leucine, lysine, methionine, phenylalanine, threonine, tryptophan, and valine, to maintain good health.[5] Protein malnutrition can usually be ascribed to a diet that is deficient in one or more of the essential amino acids. Therefore a nutritionally adequate diet must include a minimum daily consumption of these amino acids.

When diets are high in carbohydrates and low in protein, over a protracted period, essential amino acid deficiencies result. The name given to this undernourished condition is "kwashiorkor" which is an African word meaning "deposed child" (deposed from the mother's breast by a newborn sibling). This debilitating and malnourished state, characterized by a bloated stomach and reddish-orange discolored hair, is more often found in children than in adults because of their greater need for essential amino acids during growth and development. In order for normal physical and mental maturation to occur, a daily source of essential amino acids is a requisite. Essential amino acid content, or protein quality, is as important a feature of the diet as total protein quantity or total calorie intake.

Some foods, such as milk, eggs, and meat, have very high nutritional values because they contain a disproportionately high level of essential amino acids. As mentioned previously, many plants are notoriously deficient in essential amino acids. The amino acid composition of most plants is insufficient to sustain proper human growth and development. To rely solely on plants as a source of food (as so many people in developing countries must do) requires large intakes and mixtures of plant material to obtain all of the essential amino acids required to sustain life. To satisfy the minimum daily requirement of essential amino acids of a human child, a very unbalanced amount of plant foodstuffs are required as compared with the amounts necessary to consume from egg and beef (Table 1). As we know from experience, obtaining such essential amino acids from animal products creates an increasing demand on basic food crops such as corn, soybeans, and wheat. At this time, increases in animal production to meet future food needs are not viable options, at least through traditional methods.

PRODUCTION OF NOVEL PROTEIN

De novo artificial plant storage proteins have been designed to accomplish this nutritional goal.[6,7] These proteins can be adjusted to accommodate any composition of essential amino acids desired for the consumption by animals

Table 1 Consumption necessary.

Food stuff	Requirement in grams/day[a]
Cassava	4400
Corn	1800
Plantain	6100
Potato	2100
Rice	3100
Sweet potato	5760
Wheat	2300
Beef	170
Egg	180

[a]The values are what are necessary to consume in grams/day to achieve minimum daily requirement for all essential amino acids for a 10-year-old child. This assumes that the protein, in each foodstuff, is 100% bioavailable, and we know it is not, so these numbers should be increased to an even higher level.

and humans, based on any parent crop. Moreover, unlike many storage proteins found naturally in plants—that are only "partially" bioavailable to those consuming them—the proteins produced as a result of these designs are near 100% bioavailable. In collaboration with Dr. Marceline Egnin and Dr. C.S. Prakash, of Tuskegee University, we have introduced one of these artificial plant storage protein genes into sweet potatoes. Several years of field trials have been completed and small animal feeding studies have been conducted. The results of this work have been most promising. The roots of this transgenic plant T5 contain a more balanced amino acid composition provided by the new gene, as well as substantially higher levels of overall protein content[8] (Fig. 1). Thus we have within our grasp the capability of producing indigenous, edible plant foodstuffs and feedstuffs for humans and domesticated animals that would be efficient, cost-effective, and provide complete

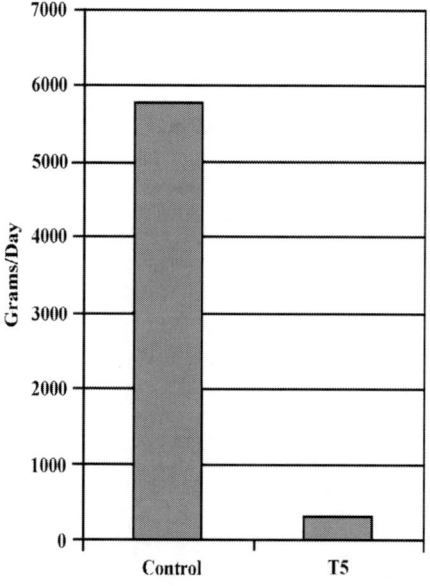

Fig. 1 Amount necessary to consume.

protein and essential amino acid sources; no supplementation with animal protein sources would be necessary. It is estimated that the total food or feed intake necessary to meet these daily needs could be reduced by more than 75% after this technology is implemented.

CONCLUSION

Improving the essential amino acid composition of basic food and feed crops, as well as increasing their overall protein content, can make a major contribution toward helping to meet the world's future food needs. That advancement combined with conference of disease and stress resistance could result in a better-fed world in the future.

REFERENCES

1. Agros, P.; Pederson, K.; Marks, D.; Larkins, B.A. A structural model for maize zein proteins. J. Biol. Chem. **1982**, *257*, 9984–9990.
2. Rahman, S.; Kreis, M.; Forde, B.G.; Shewrry, R.; Miflin, B.J. Hordein-gene expression during development of the barley (*Hordeum vulgare*) endosperm. Biochem. J. **1984**, *223*, 315–322.
3. Sharma, S.B.; Hancock, K.R.; Ealing, P.M.; White, D.W.R. Expression of a sulfur-rich maize seed storage protein in white clover (shape *Trifolium repens*) to improve forage quality. Mol. Breed. **1998**, *4*, 435–448.
4. Altenbach, S.B.; Chiung-Chi, K.; Staraci, L.C.; Pearson, K.W.; Wainwright, C.; Georgescu, A.; Townsend, J. Accumulation of a Brazil nut albumin in seeds of transgenic canola results in enhanced levels of seed protein methionine. Plant Mol. Biol. **1992**, *18*, 235–246.
5. FAO Nutr. Meet. Rep. Ser. **1985**, *724*.
6. Yang, M.S.; Espinoza, N.O.; Dodds, J.H.; Jaynes, J.M. Expression of a synthetic gene for improved protein quality in transformed potato plants. Plant Sci. **1989**, *64*, 99–111.
7. Kim, J.H.; Cetiner, S.; Jaynes, J.M. Enhancing the Nutritional Qualities of Crop Plants. In *Molecular Approaches to Improving Food Quality and Safety*; AVI Book: New York, **1992**; 1–36.
8. Prakash, C.S.; Egnin, M.; Jaynes, J. Increasing the protein content in sweet potato using a synthetic storage protein gene. Abst. Pap. Am. Chem. Soc. **2000**, *219*, 69-AGFD.

Transgenic Plants: Wax Esters from

Kathryn D. Lardizabal
Monsanto Company, Davis, California, U.S.A.

Abstract

Wax esters are long-chain linear esters of fatty acids and fatty alcohols. They are commonly found as a component of the cuticle that covers plant surfaces where they help provide protection against stresses such as desiccation, wetting, and pathogen attack. A small number of plants produce and store wax esters in the seed instead of the more commonly found triacylglycerols. These seed storage lipids provide energy for use during germination. Of these plants, jojoba (*Simmondsia chinensis*) is the only one that is commercially cultivated; however, the economics involved in the growth and harvesting of this crop have limited the use of jojoba oil as an industrial feedstock. With the advent of biotechnology, the opportunity arose to overcome this limitation and produce wax esters in a temperate crop, such as *Brassica napus*, which is grown on far greater acreage with more favorable economics. The genes involved in the production of wax esters in jojoba were cloned and introduced into *Brassica napus* in order to produce wax esters transgenically.

WAX BIOSYNTHESIS IN JOJOBA

Jojoba (*Simmondsia chinensis*) produces long-chain wax esters exclusively as a seed storage lipid rather than triacylglycerols.[1–3] A cross section of developing jojoba embryos reveals the presence of lipid bodies analogous to those present in all oilseeds.[4] Two proteins are responsible for this biosynthesis.[5,6] A fatty acyl-CoA reductase (FAR) carries out the four-electron reduction of an acyl-CoA to form a primary alcohol. The enzyme requires NADPH as the electron donor and prefers very-long-chain (C20–C24) fatty acyl-CoA substrates. This preference is reflected in the wax composition of jojoba, where greater than 93% of the fatty acids are C20–C24. Though the reaction proceeds through an aldehyde intermediate, free aldehyde is not released in the process. Next, a fatty alcohol acyltransferase (wax synthase, WS) transfers a second acyl-CoA molecule to the primary alcohol to form the wax ester (Fig. 1). Since the FAR prefers very-long-chain substrates, an efficient mechanism for fatty acid elongation system is essential.[7] The endoplasmic reticulum contains such a system, which involves four enzymatic activities; however, the first enzyme, beta-ketoacyl-CoA synthase (KCS), was found to be the rate-limiting and chain-length-determining step.[8]

IDENTIFICATION OF THE GENES RESPONSIBLE FOR WAX BIOSYNTHESIS

The enzyme activities involved in wax production in jojoba were described in the late 1970s; however, the membrane-associated nature of the proteins hindered their identification for another 20 years. Following cell fractionation of developing jojoba embryos, the activities are found in both the membrane and floating wax fractions. Purification of the proteins was achieved by detergent solubilization of the activities and chromatographic separation of the proteins followed by peptide sequencing and cloning of the corresponding genes.

The jojoba FAR was identified as a 1.7 kb gene that encoded a protein with a molecular mass of 56.2 kD and a pI of 8.76.[9] Hydropathy analysis suggested the presence of 1–2 transmembrane domains. Expression of the protein in *E. coli* resulted in the production of a small amount of primary alcohol in the cells. The jojoba FAR was expressed in *Brassica napus* in combination with the KCS from *Lunaria annua*, in order to maximize the substrate available to the enzyme. Low levels of alcohol were detected to about 6%. Further investigation, using nuclear magnetic resonance spectroscopy and GC/mass spectroscopy, showed that a majority of the alcohols were esterified to fatty acids, though the presence of unesterified alcohol indicated the incorporation into wax was incomplete.[9] This confirmed the ability of the jojoba FAR to produce alcohols transgenically in another plant and also identified the presence of an endogenous wax-ester-forming activity in *Brassica napus*.

The jojoba WS was also purified from membrane fractions.[10] This enzyme proved to be more versatile, demonstrating activity toward a wide range of acyl-CoA and alcohol (C8–C24; saturated; monounsaturated and polyunsaturated) substrates in in vitro assays. It was also more difficult to purify due to its inhibition by detergents and

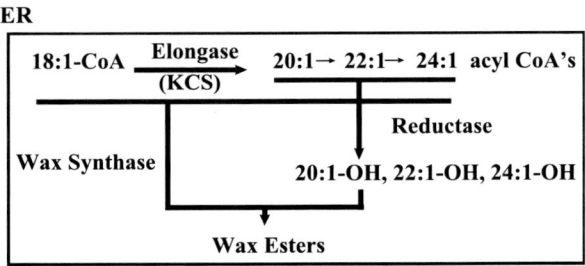

Fig. 1 Wax biosynthetic pathway in jojoba.

Table 1 Data from the first (R1) generation of plants transformed with cDNAs encoding wax biosynthetic enzymes.

R1 transformant	% Wax
8559-26	13.79
8559-13	9.91
8559-32	9.10
8559-28	7.85
8559-16	5.82
8559-34	5.05
8559-3	4.71
8559-8	4.36
8559-37	3.24
8559-1	3.24
Reston control	0.22

a requirement for phospholipid reconstitution in order to restore enzymatic function. The jojoba WS was identified as a 1.3 kb gene that encoded a protein with a molecular mass of 40.2 kD and a pI of 9.86. Hydropathy analysis suggested the presence of 7–9 transmembrane domains. The jojoba WS was cloned and expressed in the model oilseed plant *Arabidopsis thaliana* in combination with jojoba FAR and *Lunaria* KCS. All three genes were under the control of napin regulatory sequences that directed gene expression to the seed and coordinated the timing of expression to coincide with fatty acid deposition during embryo development. Intact wax levels were determined on seed pools using ^{13}C-NMR. The highest expressing plants stored approximately half of their lipid in the form of wax. Further analysis showed that up to 68% wax was produced in individual seeds.[10]

PRODUCTION OF WAX ESTERS IN *BRASSICA NAPUS*

Identification of the genes involved in wax production provided the tools necessary to produce wax esters transgenically in a commercial crop. The three-gene construct (jojoba WS: jojoba FAR: *Lunaria* KCS) was transformed into *Brassica napus* var. Reston, and a range of wax levels (0.2–14%) was detected in the transgenic R1 seed pools (Table 1). Single seed analysis of the highest wax-containing lines indicated that up to 33% wax ester and 68% very-long-chain-fatty-acids (VLCFA) were made. Selected lines were

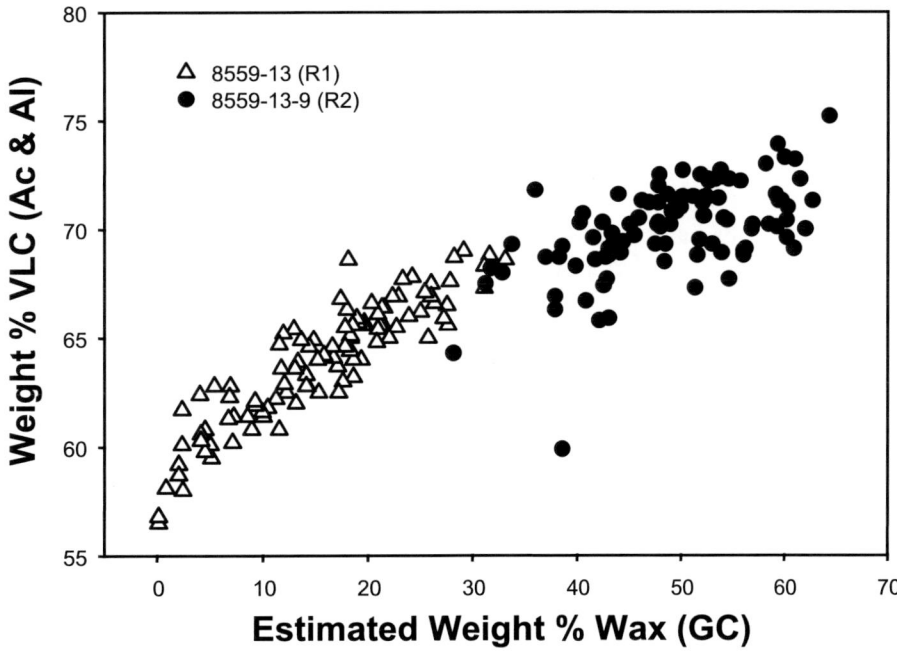

Fig. 2 The weight percent of very-long-chain (VLC) fatty acids and alcohols of individual half-seeds of transgenic *Brassica* determined by GC analysis is plotted against the estimated weight percent wax determined by GC analysis. Data from the first (R1) generation of seed (8559-13) and data from the second (R2) generation of seed (8559-13-9).

advanced to the next generation (R2) where wax ester content increased to 64% and VLCFA content increased to 75%. It is important to note that wax and VLCFA levels for a given population represent an average of the single seed content. In R2 seed, wax levels range from 28–64% with an average of 49%, and VLCFA levels range from 64–75% with an average of 70% (Fig. 2).

CONCLUSION

Overexpression of wax biosynthetic genes (WS, FAR, KCS) in the seeds of *Brassica napus* resulted in the storage of approximately 50% of the lipid as wax ester. In jojoba, these genes contribute to the high levels of wax esters (50–60% by weight of the seed) found there. The proteins encoded by these genes clearly are optimized for this production since wax esters represent the entire stored form of lipid in jojoba seeds. Since acyl-CoAs are the substrates for both wax ester biosynthesis and triacylglycerol biosynthesis, the achievement of a 50% diversion of substrate, in developing *Brassica* seeds, into a novel storage product (wax ester) indicates that the wax biosynthetic enzymes compete efficiently for these substrates. The stored wax ester also appears to be stable and nontoxic to *Brassica* embryos since the seeds can be propagated.

The capacity to produce wax esters resides in the genetic makeup of nearly all plants since they are found as a component of the cuticle. In *Brassica*, expression of the jojoba FAR in combination with the *Lunaria* KCS produced a limited amount of wax ester in the absence of jojoba WS. A similar observation was made in *Arabidopsis*. The source of this wax-ester-forming activity was presumed to come from one or more of the seven genes identified in *Arabidopsis* that share homology to jojoba WS, an indication that these genes are present in *Brassica* and, most likely, other plant species.

Many waxes found in nature are used in the lubricant, food, and cosmetic industries. In particular, jojoba oil is used in cosmetics because of its moisturizing ability, despite its high cost. By demonstrating the production of high levels of wax esters in transgenic *Brassica*, the possibility for commercial production of wax esters in a temperate crop has become reality.

REFERENCES

1. Post-Beittenmiller, D. Biochemistry and molecular biology of wax production in plants. Ann. Rev. Plant Physiol. Plant Mol. Biol. **1996**, *47*, 405–430.
2. Miwa, T.K. Jojoba oil wax esters and derived fatty acids and alcohols: Gas chromatographic analyses. J. Am. Oil Chem. Soc. **1971**, *48*, 259–264.
3. Kartha, A.R.; Singh, S.P. The in vivo "quantum" synthesis of reserve waxes in seeds of *Murraya koenigii*. Chem. Ind. **1969**, 1342–1343.
4. Muller, L.L.; Hensarling, T.P.; Jacks, T.J. Cellular ultrastructure of jojoba seed. J. Am. Oil Chem. Soc. **1975**, *52*, 164–165.
5. Pollard, M.R.; McKeon, T. Gupta, L.M.; Stumpf, P.K. Studies on biosynthesis of waxes by developing jojoba seed. II. The demonstration of wax biosynthesis by cell-free homogenates. Lipids **1979**, *14*, 651–662.
6. Wu, X.-Y.; Moreau, R.A.; Stumpf, P.K. Studies of biosynthesis of waxes by developing Jojoba seed: III. Biosynthesis of wax esters from acyl-CoA and long chain alcohols. Lipids **1981**, *6*, 897–902.
7. Lassner, M.W.; Lardizabal, K.; Metz, J.G. A jojoba β-ketoacyl-CoA synthase cDNA complements the canola fatty acid elongation mutation in transgenic plants. Plant Cell **1996**, *8*, 281–292.
8. Millar, A.A.; Kunst, L. Very-long-chain fatty acid biosynthesis in controlled through the expression and specificity of the condensing enzyme. Plant J. **1997**, *12*, 121–131.
9. Metz, J.G.; Pollard, M.R.; Anderson, L. Hayes, T.; Lassner, M.W. Purification of a jojoba embryo fatty acyl-coenzyme A reductase and expression of its cDNA in high erucic acid rapeseed. Plant Physiol. **2000**, *122*, 635–644.
10. Lardizabal, K.; Metz, J.G.; Sakamoto, T.; Hutton, W.C.; Pollard, M.R.; Lassner, M.W. Purification of a jojoba embryo wax synthase, cloning of its cDNA, and production of high levels of wax in seeds of transgenic Arabidopsis. Plant Physiol. **2000**, *122*, 645–655.

Tropical Agriculture: Pest Management

Charles Muangirwa
Tropical Pesticides Research Institute, Arusha, Tanzania

Abstract

Integrated pest management has a bright future in tropical agriculture on the basis of adoption of available options of cultural practices, as well as numerous pest management options that are likely to emerge from understanding the interaction of the tropical biological diversity. Pesticides are used for pest management in the tropics, more so on cash crops than on food crops. While the use of pesticides is on a decline elsewhere, it is likely to increase in the tropics even on food crops and threaten basic biological components of pest management, the environment, and safety to people. Generally, there has been an improvement from the use of persistent pesticides over wide areas to application of less persistent pesticides on specified sites based on pest ecology. There is a need to undertake training on safe use of pesticides at all levels, create awareness that pesticide use is only one among other pest management options, and establish poison information centers.

Overall, efforts should continue to promote the concept of integrated pest management as a system. The strategy should be to produce sustainable pest management options based on an understanding of tropical biological diversity and, at the same time, improve decision making for the adoption of existing and emerging options.

BASIS OF PEST MANAGEMENT IN TROPICAL AGRICULTURE

Pest management in tropical agriculture is basically influenced by: 1) high diversity of organisms (biological diversity), 2) diversity of physical features and ecosystems, 3) moderate seasonal weather changes, and 4) the human factor.

The high biological diversity of the tropics includes a wide range of interacting organisms from which have been identified pests that affect health, crops, and livestock, as well as cultural practices for pest management. Today, the interaction is known to include chemical, visual, and audible cues; and has strong implication on known and potential pests, natural enemies of pests, natural products for pest management, and pest/host/crop resistance.

Physical features of the tropics are also diverse, and include high-altitude temperate zones, rain forests, savannah woodlands, grasslands, and desert zones. Each zone has its own biological diversity either existing in isolation or overlapping with others. Moderate seasonal weather changes in the tropics affect populations indirectly, through (modification of) interactions between organisms. An understanding of the dynamics of such interactions would strengthen the contribution of pest management to sustainable tropical agriculture (Fig. 1).

Generally, humans influence pest management through their objectives, perceptions, problems, and adoption of options for attaining those objectives or solving those problems.[2] Most tropical countries are involved in subsistence agriculture, which is dominated by rudimentary cultural practices of pest management. Pest management on commercial farms is mainly based on temperate models.

As can be seen above, pest management for sustainable tropical agriculture should be regarded as a system of options that is dominated by tropical biological diversity, and should be processed starting and ending with the consideration of human factors.[3,4]

CURRENT PEST MANAGEMENT PROBLEMS IN TROPICAL AGRICULTURE

Key problems in tropical pest management include: 1) low application of pest management options that are based on tropical pest ecology, 2) lack of information on tropical pest ecology, 3) dependency on synthetic pesticides, 4) exotic pests, and 5) migrant pests.

Low Application of Pest Management Options That Are Based on Tropical Pest Ecology

Subsistence farmers in the tropics have applied rudimentary cultural practices in pest management on their own, perhaps in desperation. Even where there has been a scientific breakthrough, say in biological control, such options

> Generations of insect pests (i.e. rice stem borer brown plant hopper) overlap when rice is grown intensively in Southeast Asia, thus making it difficult to time insecticide applications. Besides, insecticides kill natural enemies of these pests and hence create futher room for increases in pest populations.
>
> The rice stem borer and brown plant hopper are better controlled by pesticides in temperate Japan and Korea where pest populations are synchronized and less dependent on natural enemies.

Fig. 1 Difference in the control of insect pests of rice (rice stem borer and brown plant hopper) in tropical Southeast Asia and temperate Japan and Korea. **Source:** From *A Synopsis of Integrated Pest Management in Developing Countries in the Tropics.*[1]

have been viewed as risky by commercial farmers, and too costly to be adopted by subsistence farmers (Fig. 2).

Lack of Information on Tropical Pest Ecology

The rudimentary cultural pest management practices used by subsistence farmers and the overall interaction of tropical biological diversity have not been researched to exploit their full potential for formulating sustainable pest management options. Information, expertise, and facilities are lacking from the taxonomy level to the biology, behavior, and ecology of most organisms in the tropics. In effect, this has left room for dependency on pesticides. On the other

> Lack of taxonomic expertise and facilities has often resulted in delay in intervention or misidentification, particularly of exotic pest; for example, coffee mealy bug in Kenya, floating aquatic fern in Lake Kariba (Zimbabwe), larger grain borer in Tanzania, and cassava mealy bug in Uganda.
>
> However; prompt and correct identification can result in successful control, as was the case with the New World scew worm. The worm entered Africa through Libya and was a threat to the continent's livestock and game.
>
> Ironically, in most cases, farmers have been the first to note and report new damaging organisms, sometimes in dispute with authorities.

Fig. 2 Need of prompt identification of pests. **Source:** Adapted from Castella et al.[4] and BioNET–International.[5]

> Control of tsetse flies, the vectors of trypanosomiasis in Africa, has evolved through various stages:
> i. Environment manipulation—cutting tress that constitute tsetse habitat and shooting vertebrate hosts.
> ii. Spraying residual insecticides on resting sites on trees or spraying aerosol insecticides to whole vegetation with aircraft.
> iii. Use of traps, specifically designed for tsetse flies, baited with tsetse attractants and pesticides.
>
> In the process it has been possible to avoid environment destruction resulting from cutting tress and killing animals. Also the use of pesticides has been reduced from spraying the whole emvironment to specific targets and even total avoidance of pesticides.
>
> Exposure of people to pesticides has also been reduced while participation of affected communities has enhanced. This has happened because of the growing understanding of biology, behavior, and ecology of tsetse files.

Fig. 3 Understanding of biology, behavior, and ecology of tsetse flies leads to enviromentally friendly pest management options. **Source:** From *Biocontrol in Coffee Pest Management in Tropical Entomology.*[6]

hand there has been some progress in designing and using novel pest management options (Fig. 3).

Dependency on Synthetic Pesticides

Dependency on synthetic pesticides of today's agriculture has negative effects on: 1) safety of people, 2) pest management, and 3) environment.

Pesticide safety

About 60–85% of the workforce in the tropics is engaged in agriculture. About 2.9 million cases of acute exposure to pesticides are reported annually from the tropics, out of which about 220,000 deaths occur.[7] Commonly, exposure occurs during mixing, loading, and applying of pesticides. Exposure during mixing and loading is greatest when pesticides that penetrate the skin are used. Exposure is mainly due to poor knowledge of pesticide safety at farm level among extension workers, and is complicated by lack of skills among health workers to attend pesticide safety related problems. Although some populations have been exposed to pesticides for long periods, there are no direct and concrete records of chronic symptoms. Generally, farm workers are more exposed to pesticides than peasants. Peasants are likely to be exposed to pesticides in activities related to cash crop production than food crop production.

The growing demand for food may contribute to an increase in the use of pesticides on food crops as well.

As there is no structured reporting system on pesticide poisoning, the available data is only indicative.

Effects of pesticides on pest management

The effect of pesticides on pest management has mainly been recorded in relation to natural enemies of pests. Decline in natural control mechanisms has resulted in an increase in status of present pests and occurrence of new pests.[6] Pesticide resistance has been associated with application of low doses that kill susceptible individuals of a pest population selectively, leaving resistant individuals to reproduce and increase to high proportions.

Pesticides and environment

Pesticide residues have been traced in fish and fish-eating birds;[8] human bodily fluids including milk; soil, and water. Most of the residues observed were traces of organochlorine pesticides, which accumulate in the fat of organisms and in the environment. Biological assays based on levels of acetylcholinesterase in blood have indicated exposure to organophosphates and carbamate pesticides as well. There is a growing concern that the breakdown of pesticides in the tropics results in lesser-known compounds that are more toxic than the original pesticides.[9]

Migrant Pests and Tropical Pest Management

Major migrant pests affecting tropical agriculture include birds (Queleas) and insects (locusts and armyworms). Migrant pests are a problem because they can travel in large numbers over long distances and denude vast areas of crops and other vegetation. Control of migrant pests is so far dependent on pesticides applied at virtually any cost, so as to prevent disasters. Environmental implications of the control measures are yet to be assessed. Studies on physiology, ecology, and behavior of queleas and locusts in relation to migration gives some hope on the use of nonpesticidal control measures of the pest.

Exotic Pests of Tropical Agriculture

Exotic pests that invade the tropics flourish to become key pests because of 1) lack of interaction with resident organisms—particularly natural enemies, and 2) favorable conditions throughout all seasons. Invasion of the tropics by exotic pests is often due to inadequate enforcement of quarantine requirements and general ignorance about quarantine. Routes of exotic pests include food aid, germ plasm movement, and traveling. Careful introduction of natural enemies from native localities controls exotic pests in their new homes.[9]

TRENDS IN TROPICAL PEST MANAGEMENT

Trends in pest management in tropical agriculture include three eras: 1) before synthetic pesticides, 2) synthetic pesticides, and 3) integrated pest management.

Before Synthetic Pesticides Era

Tropical pest management before the synthetic pesticides era was characterized by the use of cultural practices and the use of plants with pesticidal properties, all based on farmers' experiences on the tropical physical and biological diversity. Experts of the presynthetic pesticides era were involved in taxonomy, biology of pests, and advocacy of cultural practices.

Synthetic Pesticides Era

Synthetic pesticides were used globally for the first time in the mid-1940s. At that time cash crops were introduced to most tropical countries by colonialists. Synthetic pesticides were adopted almost immediately for the management of pests in cash crops. Farmers and peasants were taught about pesticides, but as part of crop agronomy and animal husbandry, and not on personal or environmental safety.

During this era, experts drifted from being naturalists to pesticide applicators. There was also an increase in training of natives at various levels, as a postcolonial activity.[10] However, training emphasized the use of pesticides more than a naturalistic approach to pest management, and on the control aspects of pesticide rather than on safety to users and environment. Research on pesticides was initially aimed at testing if a chemical killed a pest and would reduce pest population. Subsequently, such studies evolved to include ecology of pests and related organisms, pesticide application, and pesticide residues. Problems of pesticide poisoning, pest resistance, and emergence of new pests were observed some four decades ago. Over the years there was low use of pesticides on food crops, but the situation changed in response to demand for food to feed the increasing human population. It appears that peasants transferred their experiences from use of pesticides on cash crops to the control of pests on food crops without technical considerations.

Integrated Pest Management Era

The synthetic pesticides era is coming to an end or taking a different outlook in tropical agriculture because of

1. negative impact of pesticides on users' health and on the environment;
2. pesticides failing to control pests;
3. likelihood of applying pesticides on specified targets, based on pest ecology, and hence reduce environmental impact;

4. lessons learnt during the before-pesticides era that cultural practice and use of plants with pesticidal properties are effective pest management options;
5. emerging technologies based on cultural practices and studies on interactions between organisms revealing a wide range of pest management options;[1,5,7,11]
6. increasing interest in pest management by various parties. A number of governments have declared pest management as an official policy, various institutions (governmental, NGOs, and private, etc.) are involved in various aspects of pest management including pest ecology and biology research, safe use of pesticides, and adoption of the use of alternative pesticides. Multinational pesticide companies have also started taking interest in pest management, particularly in areas of biotechnology. 7) Increase in action by governments to undertake pesticides registration and control activities as a move to minimize negative effects of synthetic pesticides.

DISCUSSION

By observing the integrated pest management era in the tropics through the problem-free cultural practices (i.e., item 4 above) and the emerging knowledge on the interaction of organisms (5 above) it is now clear that reliance on pesticides as a single option for achieving desired health, crop, or livestock protection is unnecessary. In effect, interaction between organisms in the tropics represents a wide range of "suppressing" factors including natural enemies of pests, natural products, cultural practices, crop, and host and pest resistance, etc. We can add pesticides to this list, but only to play a complimentary role. The various options for achieving the desired protection are presently limited, but would increase depending on the increase in knowledge of interactions between organisms in the tropics. Socioeconomic considerations and training are increasingly becoming important components of pest management. Considering the multiplicity and interaction between the various options in the tropics, it is now increasingly acceptable that pest management be implemented as an integrated system rather than a single act of reducing pest population or decreasing damage. Generally, the systems approach to pest management in tropical agriculture would be sustainable and be optimized if implemented through two complementary strategies: 1) Reduction of constraints to adoption of sustainable pest management options at various levels, i.e., to improve decision making, and 2) research aimed at producing sustainable pest management options, based on understanding of tropical pest ecology.

REFERENCES

1. N.R.I. *A Synopsis of Integrated Pest Management in Developing Countries in the Tropics*; Natural Resources Institute: Chatham, U.K., **1999**; 20.
2. Norton, G.A. A decision-analysis approach to integrated pest control. Crop Prot. **1982**, *1* (2), 196–199.
3. Norton, G.A.; Adamson, D.; Aitken, L.G.; Bilston, L.J.; Foster, J.; Fronk, B. Facilitating IPM: role of participatory workshops. Int. J. Pest Manage. **1999**, *45* (2), 85–90.
4. Castella, J.C.; Jourdain, D.; Trebuil, G.; Napompeth, B.; Jourdain, G.A.; Napompheth, B.A. Systems approach to understanding obstacles to effective implementation of IPM in Thailand: key issues for the cotton industry. Agric. Ecosyst. Environ. **1999**, *72* (1), 16–34.
5. BioNet–International. Taxonomy in the Biological Control of Pests, Diseases, and Weeds. In *BioNET–International: The Business Plan*; **1999**; 3–65. http://www.bionet.intl.org.
6. Nyambo, B.; Murphy, S.T.; Barker, P.; Walker, J. In *Biocontrol in Coffee Pest Management in Tropical Entomology*, Proceedings of the 3rd International Conference on Tropical Entomology, Nairobi, Kenya, Oct. 30–Nov. 4, **1994**; ICIPE Science Press, **1998**; 155–168.
7. Hargrove, J.W. In *Trypanosomiasis Management Using Bait: Some Implications of Tsetse Behavior and Ecology*, Proceedings of the 3rd International Conference of Tropical Entomology, Nairobi, Kenya, Oct. 30–Nov. 4, **1994**; ICIPE Science Press, **1998**; 155–168.
8. Ijani, A.S.M.; Katondo, J.M.; Malulu, J.M. In *Effects of Orgonochlorine Pesticides in Birds in United Republic of Tanzania, Environmental Behavior of Crop Protection Chemicals*, Proceedings of IAEA/FAO Conference, Viena, Italy, July 1–5, **1996**; IAEA/FAO: Rome, Italy, **1996**; 460–461.
9. Lehtinen, S. Pesticides. In *African Newsletter on Occupational Health and Safety*; **1999**; 9 (1), 23.
10. Paasivirta, J.; Palm, H.; Paukku, R.; Ak'habuhaya, J.; Lodenius, M. Chlorinated insectides residues in Tanzania environment Tanzadrin. Chemosphere **1988**, *17* (10), 2055–2062.
11. Hilder, V.A.; Gatehouse, A.M.R. Biological control in developing countries: towards its wider application in sustainable pest management. Med. Fac. Landb. Rijksaniv. Gent. **1990**, *55* (2a), 216–223.

Utility Patents and Plant Innovation

Jay P. Kesan
College of Law, University of Illinois, Urbana, Illinois, U.S.A.

Abstract
This entry provides a systematic summary of utility patent protection in the United States, including the requirements for obtaining utility patent protection for plant innovation; the rights associated with an utility patent; the test for patent infringement; the defenses against a claim of infringement; the remedies available once infringement is proven; and finally, a quick review of some the emerging issues in this arena.

INTRODUCTION

Historically, plant innovations have not fit into traditional intellectual property regimes. Plant nursery operators of the early twentieth century complained of others copying plants, but there was no remedy.[1] Early proposals for intellectual property protection included trademark and unfair competition laws.[2] Congress responded by passing the Townsend-Purnell Plant Patent Act (PPA) of 1930, which prohibits unauthorized asexual propagation.[1] Four decades later, in 1970, Congress passed the Plant Variety Protection Act (PVPA), which protects against copying through sexual reproduction.[1] Although plant innovations are eligible for protection under both the PPA and PVPA, the question remained for several years whether plant innovations where also eligible subject matter for traditional utility patents. In the landmark cases *Diamond v. Chakrabarty* and *J.E.M. Ag Supply v. Pioneer Hi-Bred*, the Supreme Court extended utility patent protection to biotechnology and agricultural biotechnology innovations, which led to a patent gold rush. The creator of a plant innovation can now secure intellectual property rights through the PPA, PVPA, trade secret protection, or a traditional utility patent. Such a patent is crucial to plant innovation because it protects research and development costs from downstream appropriation by third parties who did not contribute to development costs (Fig. 1).[3]

ELIGIBLE SUBJECT MATTER

The *J.E.M. Ag Supply v. Pioneer Hi-Bred* case was a landmark decision for agricultural utility patents. In this case, Pioneer Hi-Bred, the world's leading corn seed producer, sued J.E.M. Ag Supply (J.E.M.) for infringement of utility patents relating to hybrid and inbred corn lines.[4] J.E.M. defended the lawsuit by claiming that Congress intended the PPA and PVPA to be the only means of plant intellectual property protection.[4] The Supreme Court rejected this argument. They relied on the famous *Diamond v. Chakrabarty* decision, which holds that biotechnology, specifically a human-made microorganism, is an eligible subject matter for patent protection[4] Based on this precedence, Justice Clarence Thomas noted that the language of the relevant statutory provision was "extremely broad," covering almost anything under the sun made by man.[4] Justice Thomas also noted that there was nothing in either the PPA or the PVPA that precluded the use of utility patents in the plant context.[4] Thus, an overlapping intellectual property regime now exists for plants, including the PPA, PVPA, and utility patent protection.

OTHER PATENT REQUIREMENTS

Although the Supreme Court has firmly held that plant innovations are eligible subject matter for patent protection, establishing eligible subject matter is simply the first step in obtaining a patent. There are four additional requirements.

The first requirement is utility. The Patent and Trademark Office (PTO) requires that a patent applicant demonstrate "specific, credible, and substantial" utility.[5] This requirement leads to the denial of patent applications that fail to identify a specific purpose or that are too far-fetched to be credible, such as a perpetual-motion machine.[5]

The second requirement is the presence of novelty and the absence of a statutory bar. Novelty requires that the invention does not already exist in the public domain. In the United States patent system, the first person to invent a device has the right to patent it. In other countries, patents are often awarded to the first person to file the invention with the patent office. In *Kridl v. McCormick*, the Court of Appeals for the Federal Circuit invalidated a patent that used antisense recombinant DNA technology to make virus-resistant plants because another inventor had already invented this technology and reduced it to practice only a

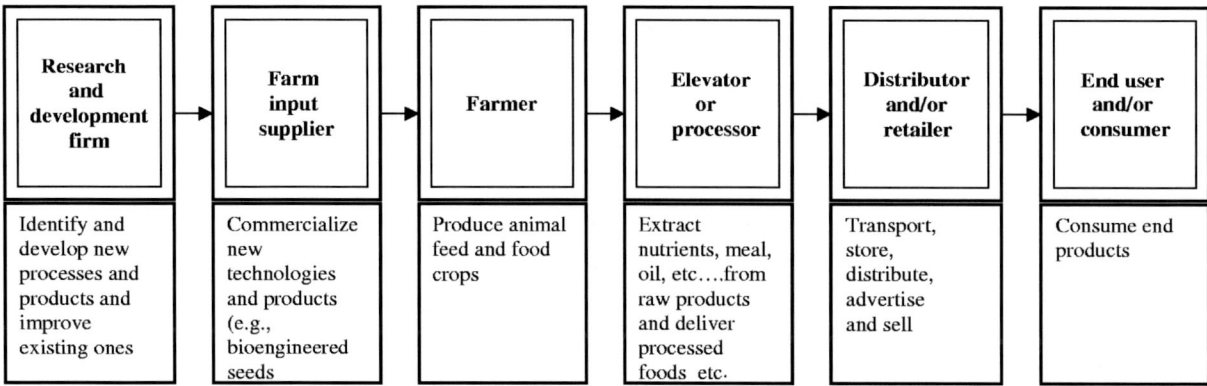

Fig. 1 Value chain in the modern food system. The value chain begins with research and development. The products of research and development are commercialized into farm inputs, which are then passed on to the farmer. Farmers convert these inputs into raw products that are distributed to wholesalers. Wholesalers pass these products on to the end user or consumer.

few years earlier.[6] The statutory bar denies patentability to an inventor who has published, sold, or used the invention in the public domain for more than one year. This requirement gives an inventor the incentive to promptly file the patent while designating the invention for public use, publication, or sale.

The third requirement for obtaining a patent is nonobviousness. This requirement imposes a burden on the inventor to demonstrate that his or her invention signifies an inventive step over pre-existing technology. In other words, the invention cannot be an obvious variation of a previously patented device. In *Graham v. John Deere Company*, the Supreme Court confronted this issue when the patent holder for a shank plow tried to obtain a patent for a simple modification of the device.[7] The Court held that this minor improvement was an obvious variation and upheld the denial of the patent because it did not constitute an inventive step.[7,9]

The fourth and final requirement is describing and enabling the invention. The purposes of this requirement are to prove to the world that the patent applicant, in fact, invented the item and to enable people to make and use the invention.[5] The patent applicant must also describe the "best mode" for implementing the invention.[5] The best mode requirement prevents the patentee from holding back what is necessary to capture all of the benefits of the invention. In *Plant Genetic Systems v. DeKalb Genetics Corporation*, the patentee held a patent for a device that genetically engineered corn, but the court invalidated the patent because the description of the device had not enabled others to correctly use the device to insert heterologous DNA into a plant cell without extensive experimentation.[8]

OBTAINING A PATENT AND PATENT RIGHTS

In the United States, filing an application with the PTO is the first step to obtaining a patent. The application must contain all of the required elements described in the preceding section. The application must also include drawings and illustrations of the invention if they are necessary to understanding the invention.[3] In the biotechnology field, the patent applicant must describe the structure of the DNA to be used.[3] The patent application must consist of at least one claim. Claims are statements that define the patentee's right to exclude future users.[3]

The patent application process takes an average of 2.77 years.[5] Once a patent is granted, the patent holder has the exclusive right to use, make, or sell the claimed invention.[3] The patent holder can also license the use of the patent to a third party. These patent rights last for the term of the patent, which is 20 years from the date of filing. Note, however, that a patent confers rights only in the jurisdiction in which the patent is issued. For example, a patent that is issued in the United States grants no protection for an invention in Japan. Outside of the jurisdiction where a patent is issued, the invention may be freely used, patent-free, by others. To obtain protection in multiple jurisdictions, an applicant must file separate patent applications in each jurisdiction in which he or she desires protection. While no single patent protects an invention in all jurisdictions around the world, international agreements, such as the Patent Cooperation Treaty (PCT), the Paris Convention, and the Trade-Related Aspects of Intellectual Property Rights (TRIPS) Agreement, have been drafted and adopted to facilitate the application process for patentees who want to obtain individual patents in multiple jurisdictions.[3]

INFRINGEMENT, DEFENSES, AND REMEDIES

Patentees hold a variety of remedies against those that infringe upon their patents. The first step in determining an actual infringement is to interpret the claims of the patent. This process includes referencing the language of the claims of the patent, diagrams and technical descriptions of the device, and public records of the filing process. After the court has determined the actual scope of the claims, infringement can be found through: (1) literal infringement or (2) the doctrine of equivalents. Literal infringement occurs when someone copies all elements of one claim of a patent. Therefore, if a claim consists of components A, B, and C, literal infringement occurs only if someone else copied A, B, and C into their device.[3] Infringement also occurs under the doctrine of equivalents when the second device is technically different but substantially the same. For example, if the original claim consisted of A, B, and C, and someone else simply substituted D for C, when D is substantially the same as C, infringement exists under the doctrine of equivalents.[3]

Historically, infringement defenses are almost nonexistent in patent law. The most prevalent defense is for patent misuse or experimental use. Patent misuse occurs when the patent holder exceeds the scope of the patent. Courts uphold this defense to infringement in limited circumstances, e.g., when the patentee goes beyond the scope of the patent grant. The experimental use defense is a very narrow, judicially created exception to infringement for using a patent for experimental purposes. This defense is usually not successful either. In contrast, the PVPA contains a broad research exemption for bona fide experimentation.

The range of remedies for patent infringement includes injunctions, damages, and attorneys' fees. Injunctions are much more common in patent law than other areas of law because of the limited duration of a patent life and the devastating effects of infringement. If damages are awarded, they come in the form of lost profits or reasonable royalties. The imposition of reasonable royalties requires the court to award a royalty to the patent holder for use of the device by the infringing defendant. In order to recover lost profits, the patent holder must be able to project and prove the profits that he or she would have received if the infringing product was not on the market. The United States is one of the few countries to award damages for lost profits, which is a much more attractive form of damages because of the potential for recovery.

EMERGING ISSUES

The *J.E.M. Ag Supply v. Pioneer Hi-Bred* decision and the rise of agricultural biotechnology have brought several emerging utility patent issues to the surface. Some seed-producing companies have sought to protect their intellectual property by including licensing agreements on the seed bags. This is known as seed-wrap licensing. These agreements usually prohibit farmers from saving seeds for longer than one year. This requirement may be seen to impose hardship on farmers who cannot afford to purchase expensive seeds every year.[9] The motive behind seed-wrap licensing has led to the creation of GURTs (Genetic Use Restriction Technologies), which protect the intellectual property of the seed by essentially destroying the seed after one year. Again, this technology presents problems for smaller farmers that cannot afford the costs of seeds every year and for seeds sold in developing countries, where the GURTs could lead to starvation by essentially wiping out farms.[10] This concern has led to protests of GURTs across the world.[10] Multilateral international agreements, such as the Cartagena Protocol on Biosafety, however, permit countries to remain GURTs free by blocking the import of genetically modified seeds. The enforceability of these licenses is another emerging issue. Bona fide seed purchasers are normally entitled to the full rights of a purchased item, including the right to save and sell the device under the first-sale doctrine of intellectual property law.[11] This doctrine states that the patent creator's interests are exhausted upon sale.[11] Nevertheless, patent holders have limited purchaser rights through the types of licenses mentioned above. Unlike inanimate objects, a genetically-modified plant is a self-replicating invention that produces copies of itself through the natural growth process.[2] U.S. Courts have upheld these licenses for the most part, even though they may be seen as conflicting with the first-sale doctrine.[11]

Another emerging issue is whether patent infringement can occur through pollen drift.[12] Canadian courts have confronted this issue and have held that accidental infringement is still infringement; but, there has been no definite resolution in the United States.[13] Another unanswered question is the continued relevance of the PVPA or trade secret protection. PVPA protection is more easily obtained because the utility patent novelty and nonobviousness requirements are not prerequisites for obtaining PVPA protection. But the PVPA provides less protection than utility patents because it contains a seed-saving exemption and research exemption. Trade secret protection also provides less security because there are exceptions for independently creating the device or legally creating it through reverse engineering. Therefore, it is doubtful that any agricultural conglomerates will opt for sole intellectual property protection under the PVPA or trade secret law. Finally, there are emerging international issues to consider. The World Trade Organization's (WTO) TRIPS provides for a *sui generis* approach to the protection of new plant varieties.[14] This means that WTO-member countries are entitled to implement their own system of protection or use patent law or a combination of both.[14] This has implications for developing countries that maintain a deep suspicion of intellectual property protection. They view intellectual property as another form of colonialism, where researchers frequently misappropriate plants and knowledge from indigenous people.[15] This biopiracy has led to a North–South split over the appropriate scope and use of patent protection for plants. India, for example, has addressed the biopiracy of its traditional knowledge by implementing a Plant Variety Protection and Farmers' Rights Act that recognizes the contributions of traditional farmers and provides protection for native crops.[16]

REFERENCES

1. Janis, M.; Kesan, J.; Weed-Free, I.P. The Supreme Court, intellectual property interfaces, and the problem of plants, Illinois Public Law and Legal Theory Research Papers Series, 2001; 5–13.
2. Janis, M.; Kesan, J. U.S. plant variety protection: sound and fury...? Hous. L. Rev. **2002**, *39*, 731.
3. Kesan, J. Intellectual property protection and agricultural biotechnology, a multidisciplinary perspective. Am. Behav. Sci. **2000**, *44* (3), 466–468.
4. *J.E.M. Ag Supply v. Pioneer Hi-Bred Int'l, Inc.*, 534 U.S. 124, 2001.
5. Lemley, M.; Menell, P.; Merges, R. *Intellectual Property in the New Technological Age*, 3rd Ed.; Aspen Publishers: New York, **2003**, see pages 139–183, 196–201.
6. *Kridl v. McCormick*, 105 F.3d 1446 (Fed. Cir. 1997).
7. *Graham v. John Deere Company*. 383 U.S. 1, 1966.
8. *Plant Genetic Systems v. DeKalb Genetics Corporation*. 175 F. Supp. 2d 246 (2nd Cir. 2001).
9. Roberts, M. National Aglaw Center Research Article, *J.E.M. Ag Supply, Inc. v. Pioneer Hi-Bred International, Inc.*: its meaning and significance for the agricultural community. S. Ill. U. L.J. **2003**, *28*, 116–117.
10. Ohlgart, S. Note, the terminator gene: intellectual property rights vs. the farmers common law right to save seeds. Drake J. Agric. L. **2002**, *7*, 473–490.
11. Janis, M. Supplemental forms of intellectual property protection for plants. Minn. J.L. Sci. Technol. **2004**, *6*, 327–329.
12. Smyth, S.; Khachatourians, G.G.; Phillips, P.W.B. Liabilities and economics of transgenic corps. Nat. Biotechnol. **2002**, *20*, 537–541.
13. *Monsanto Canada Inc. v. Shmeiser*, S.C.C. 34 (Can. 2004), 2002.
14. Tansey, G. *A Discussion Paper: Trade, Intellectual Property, Food and Biodiversity: Key Issues and Options for the 1999 Review of Article 27.3(b) of the TRIPS Agreement*; Quaker Peace and Service: London, **2000**; 8–9.
15. McManis, C.R. Re-engineering patent law: the challenge of new technologies. Wash U. J.L. Pol'y **2000**, *2*, 1–22.
16. Ruiz, M.; Lapeña, I.; Clark, S.E. The protection of traditional knowledge in Peru: a comparative perspective. Wash. U. Global Stud. L. Rev. **2004**, *3*, 755–797.

Vaccine Production: Plants as Biofactories

Luca Santi
Hugh S. Mason
Biodesign Institute, Arizona State University, Tempe, Arizona, U.S.A.

Abstract
The use of molecular biology to genetically modify plants has led to promising applications in biotechnology. In this entry, we analyze the advantages and challenges of plant-derived vaccines. We discuss the current state of the art on methods for heterologous protein expression in plants as well as the immunological activity of the plant-produced antigens used as subunit vaccines against various infectious diseases in animal and human studies.

INTRODUCTION

Old and new or re-emerging infectious agents, together with the worrisome increase in resistance to existing therapies, pose an enormous threat that will become even greater in the future. Vaccines are one of the great successes of modern medicine. However, the development and implementation of new vaccines remains prohibitively expensive for economically depressed countries (where such measures are needed most often) or when the pandemic significance of some of the infections is considered. In the last 15 years, a growing number of research groups worldwide have studied different plant expression systems and antigen delivery strategies for selected subunit vaccines, some showing very promising potential.[1]

ADVANTAGES AND CHALLENGES OF PLANT EXPRESSION SYSTEMS FOR VACCINE DEVELOPMENT

There are two main systems for recombinant protein expression in plants: stable genetic transformation and transient expression. Stable transformation causes integration of foreign DNA into the nuclear or chloroplast chromosomal DNA. These lines can be propagated vegetatively or by seeds and thus readily scaled up for protein production. Transient expression uses non-integrated gene constructs that may or may not replicate in the host. Replicating vectors use a plant virus that carries the vaccine gene, which is replicated and expressed during the infection of the plant host.

Plants systems are far less likely to harbor microbes pathogenic to humans than are mammalian cells or whole animal systems, and plants, moreover, offer the opportunity of large scalability. Both field and greenhouse production, using stable or transient transgenic plants, do not require large investments in hardware and culture media, thus making scale-up more economical than fermentation culture. In particular, the technology developed for stable genetic transformation of plants has provided the opportunity to use large-scale agriculture for the production of recombinant proteins.

Stable nuclear transformation is often achieved using *Agrobacterium tumefaciens*; the "biolistic" method (microprojectile bombardment) is frequently used for plant hosts that are inefficiently transformed by *Agrobacterium*.[2] Stable transformation of the chloroplast genome can yield high levels of recombinant protein. Another advantage of chloroplast transformation is maternal inheritance, which limits the potential for transgene escape by dissemination of pollen. A possible problem with this strategy is that some eukaryotic cellular processing events (e.g., glycosylation) may not be obtained in chloroplasts.[3] Stably transformed edible plants are considered to be an attractive possibility for the delivery of vaccine antigens orally, thus obviating the costly purification process required of injectable vaccines. Two of the biggest challenges for orally delivered vaccines reside in the digestive system: 1) the antigens must be stable in the acidic and proteolytic environment in order to resist degradation; and 2) antigens must possess characteristics that permit their efficient transport across the epithelium for presentation to the underlying lymphoid tissues.

The highest yielding transient systems rely on autonomously replicating viruses for trans-gene amplification and expression. In the last 20 years, there has been a great deal of molecular engineering on various plant viruses to create new efficient viral vectors. Advantages of transient viral vectors include very high levels of recombinant protein production in a very short time; the possibility to produce a large number of different constructs that can be quickly tested, and, in the case of fully functional vector systems, the opportunity for mechanical inoculation for large scale infections.[4] Most of the transient expression systems are

Table 1 Clinical studies using plant-derived vaccines.

Antigen	Plant system	Plant material administered	Antigen delivered	Dose	Number of volunteers	Year	Reference
LT-B	Potato	100 g FW	750 μg	3	14	1998	[8]
LT-B	Corn	2.1 g DGM	1000 μg	3	13	2004	[11]
NVCP	Potato	150 g FW	750 μg	2/3	24	2001	[10]
HBsAg	Lettuce	200 g FW	1 μg	2	5	1999	[9]
HBsAg	Potato	100 g FW	850 μg	2/3	42	2005	[12]

LT-B, B subunit of the heat-labile enterotoxin of enterotoxigenic *Escherichia coli*; NVCP, Norwalk virus capsid protein; HBsAg, hepatitis B surface antigen; FW, fresh weight; DGM, defatted corn grain meal.

based on non-food non-feed plants like tobacco, thus requiring purification steps prior to administration. Processing should comply with good manufacturing practice (GMP) standards, which would increase cost of manufacturing but, on the other hand, would guarantee homogeneity and quality control procedures typical of standard pharmaceuticals products.

ANIMAL AND HUMAN STUDIES

Several vaccine candidates have been expressed in plants and tested in animals, showing immune responses after oral delivery. Most importantly, the number of reports describing protection after viral, bacterial, or toxin exposure is rapidly growing.

In a notable recent study[5] to evaluate a vaccine candidate for anthrax, a zoonosis shared by animals and humans, the protective antigen (PA) from *Bacillus anthracis* was expressed in tobacco plants by inserting the coding sequence into the chloroplast genome. Mature leaves from selected transgenic lines expressed very high levels of the recombinant protein, and mice injected subcutaneously with the partially purified PA showed 100% survival when challenged with a lethal dose of the anthrax toxin.

Alfalfa stably transformed with the VP6 coding sequence from human group A rotavirus was used to conduct an animal study orally, administering 10 mg of total protein from transgenic alfalfa (24 μg of VP6) together with a mucosal adjuvant. The treated mice developed potent systemic and mucosal immune responses that provided passive protection against infection with simian rotavirus SA-11 in neonatal mice born to immunized dams.[6]

The evaluation of the plague antigens F1, V, and their recombinant fusion F1-V provide a recent example of protection against a bacterial infectious disease. The three proteins were rapidly produced in very high levels using a tobacco mosaic virus-based system in *Nicotiana benthamiana* plants. Each purified antigen, administered subcutaneously to guinea pigs, provoked systemic immune responses that protected the animals against a lethal aerosol dose of the virulent bacteria.[7]

Five human studies using edible plant vaccines have been performed[8–12] (Table 1). Three vaccines were directed against agents of gastroenteritis (enterotoxic *E. coli* and Norwalk virus), while the other two vaccines targeted the hepatitis B virus.

Cholera and enterotoxigenic *E. coli* (ETEC) cause diarrhea by secretion of cholera toxin (CT) and labile toxin (LT), respectively. The labile toxin B subunit (LT-B) expressed in transgenic potatoes produced toxin-protective gut antibody responses after ingestion by mice. The LT-B potatoes became the first human test for plant vaccines[8] in which subjects ate 100 g of raw LT-B potato slices containing 750 μg LT-B at each of three weekly doses. The production of toxin-neutralizing serum antibodies in 10 of 11 subjects proved that ingestion of transgenic plant material could be used effectively to deliver a vaccine in humans. Oral administration of transgenic corn seeds expressing LT-B showed similar results in a recent study involving 13 human volunteers.[11] In this case, the antigen was extremely concentrated and the administration of just 2.1 g of corn was sufficient to deliver 1 mg of recombinant LT-B. After three weekly doses, seven of the nine subjects (78%) developed serum IgG anti-LT and four of nine (44%) developed stool IgA.

Most cases of viral gastroenteritis are caused by rotavirus and Norwalk-like viruses, which are potential targets for oral subunit vaccines owing to the ability of their capsid proteins to assemble virus-like particles (VLP). These recombinant VLP can mimic the structure and cell binding of the authentic virus particles, are acid-stable, and stimulate serum responses in humans. The Norwalk virus capsid protein (NVCP) expressed in plants formed VLP that were orally immunogenic in mice, and NVCP potatoes were used in a second clinical trial.[10] The volunteers ate 150 g of raw potato containing up to 750 μg NVCP. Although the average antibody levels were less impressive than in the LT-B potato study, 19 of the 20 subjects showed significant immune responses.

The currently used vaccine for hepatitis B is recombinant viral surface antigen (HBsAg) purified from transgenic yeast cultures. HBsAg was described in the first report of a plant-derived (tobacco) vaccine, and later showed immunogenicity by potato ingestion in mice. HBsAg expressed in transgenic lettuce and delivered to humans by ingestion of 200 g (containing ∼1 μg antigen) caused production of serum antibodies at protective levels in 2 of

3 volunteers.[9] Since the dose was quite low, it is likely that the HBsAg had assembled into VLP structure, which enhanced sampling by M cells of the gut associated lymphoid tissue. In a recent broader and more comprehensive clinical trial, HBsAg transgenic potatoes were ingested by volunteers who had been vaccinated earlier with the commercial injectable HBsAg in order to test the boosting effect of the ingested antigen. Two groups of volunteers ate three and two doses of 100 g of raw potatoes respectively, each dose containing 850 μg HBsAg. The three-dose regimen stimulated an increase of antibody titers in 62.5% of the cases, while the two-dose regimen affected 52.9%.[12] This study demonstrated the safety and immunogenicity of the HBsAg plant-derived vaccine in humans, and confirmed that antigen from a non-enteric pathogen could be immunogenic via oral delivery.

CONCLUSIONS

Improvements in antigen design, expression levels, antigenic and immunogenic properties, manufacturing, processing, and delivery strategies indicate a strong potential for plant vaccine technology. Different systems for production of recombinant proteins in plants and different strategies of vaccine administration are available, spanning the range from oral delivery of minimally processed plant material all the way to injectable delivery of highly purified antigens. An evaluation a priori of the best strategy is not always possible, and will usually need to be assessed on a case-to-case basis considering the recipient subjects, the specific infectious disease, and the particular protective antigen used. In any case production systems must maintain rigorous containment to prevent contamination of food supplies using, for example, dedicated greenhouses and postharvest processing facilities for pharmaceutical production. In addition, pollen-mediated gene flow should be limited in open field production settings, e.g., with male-sterile lines, chloroplast expression, or virus-vectored transient expression.

ACKNOWLEDGMENTS

The authors thank Charles Arntzen, Zhong Huang, and Michelle Kilcoyne for research, critical reading of the manuscript, and helpful discussions.

REFERENCES

1. Mason, H.S.; Chikwamba, R.; Santi, L.; Mahoney, R.T.; Arntzen, C.J. Transgenic plants for mucosal vaccines. In *Mucosal Immunity*, 3rd Ed.; Mestecky, J., Bienenstock, J., Lamm, M.E., Mayer, L., McGhee, J.R., Strober, W., Eds.; Elsevier: San Diego, CA, 2004; 1053–1060.
2. Hansen, G.; Wright, M.S. Recent advances in the transformation of plants. Trends Plant Sci. **1999**, *4* (6), 226–231.
3. Bock, R.; Khan, M.S. Taming plastids for a green future. Trends Biotechnol. **2004**, *22* (6), 311–318.
4. Canizares, M.C.; Nicholson, L.; Lomonossoff, G.P. Use of viral vectors for vaccine production in plants. Immunol. Cell Biol. **2005**, *83* (3), 263–270.
5. Koya, V.; Moayeri, M.; Leppla, S.H.; Daniell, H. Plant-based vaccine: mice immunized with chloroplast-derived anthrax protective antigen survive anthrax lethal toxin challenge. Infect. Immun. **2005**, *73* (12), 8266–8274.
6. Dong, J.-L.; Liang, B.-G.; Jin, Y.-S., Zhang, W.-J.; Wang, T. Oral immunization with pBsVP6-transgenic alfalfa protects mice against rotavirus infection. Virology **2005**, *339* (2), 153–163.
7. Santi, L.; Giritch, A.; Roy, C.J.; Marillonnet, S.; Klimyuk, V.; Gleba, Y.; Webb, R.; Arntzen, C.J.; Mason, H.S. Protection conferred by recombinant *Yersinia pestis* antigens produced by a rapid and highly scalable plant expression system. PNAS **2006**, *103* (4), 861–866.
8. Tacket, C.O.; Mason, H.S.; Losonsky, G.; Clements, J.D. Immunogenicity in humans of a recombinant bacterial antigen delivered in a transgenic potato. Nat. Med. **1998**, *4* (5), 607–609.
9. Kapusta, J.; Modelska, A.; Figlerowicz, M.; Pniewski, T.; Letellier, M.; Lisowa, O.; Yusibov, V.; Koprowski, H.; Plucienniczak, A.; Legocki, A.B. A plant-derived edible vaccine against hepatitis B virus. FASEB J. **1999**, *13* (13),1796–1799.
10. Tacket, C.O.; Mason, H.S.; Losonsky, G.; Estes, M.K.; Levine, M.M.; Arntzen, C.J. Human immune responses to a novel Norwalk virus vaccine delivered in transgenic potatoes. J. Infect. Dis. **2000**, *182* (1), 302–305.
11. Tacket, C.O.; Pasetti, M.F.; Edelman, R.; Howard, J.A.; Streatfield, S. Immunogenicity of recombinant LT-B delivered orally to humans in transgenic corn. Vaccine **2004**, *22* (31/32), 4385–4389.
12. Thanavala, Y.; Mahoney, M.; Pal, S.; Scott, A.; Richter, L.; Natarajan, N.; Goodwin, P.; Arntzen, C.J.; Mason, H.S. Immunogenicity in humans of an edible vaccine for hepatitis B. PNAS **2005**, *102* (9), 3378–3382.

BIBLIOGRAPHY

1. Busto, J.L.; Kumagai, M.H. Tobamoviral vectors: developing a production system for pharmaceuticals in transfected plants. In *Encyclopedia of Plant and Crop Science*; New York: Marcel Dekker, 2004; 1229–1232.
2. Fujiyama, K. Pharmaceuticals in plants. In *Encyclopedia of Plant and Crop Science*; New York: Marcel Dekker, 2004; 1–3.
3. Daniell, H. Medical molecular pharming: therapeutic recombinant antibodies, biopharmaceuticals and edible vaccines in transgenic plants engineered via the chloroplast genome. In *Encyclopedia of Plant and Crop Science*; New York: Marcel Dekker, 2004; 705–710.
4. Joshi, L.; Shah, M.M.; Flynn, C.R.; Panitch, A. Plant-produced recombinant therapeutics. In *Encyclopedia of Plant and Crop Science*; New York: Marcel Dekker, 2004; 969–972.
5. Fischer, R.; Emans, N.J.; Twyman, R.M.; Schillberg, S. Molecular farming in plants: technology platforms. In *Encyclopedia of Plant and Crop Science*; New York: Marcel Dekker, 2004; 753–756.

Vaccines and Anti-Infective Agents: Delivery via Lactic Acid Bacteria

Tri Duong
Department of Poultry Science, Texas A&M University, College Station, Texas, U.S.A.

Abstract
Lactic Acid Bacteria (LAB) are microorganisms with important food, agricultural, and industrial uses. LAB are important components of the gut microbiota and have long been recognized for their health-promoting properties. Advances in recombinant genetic techniques have made it possible to engineer these organisms for the expression of heterologous proteins. Thus, much interest and knowledge have been gained in the use of LAB for the delivery of bioactive molecules including vaccines, anti-infective agents, and biotherapeutic compounds. This entry will focus on recent advances in the development of LAB for the delivery of vaccines and anti-infective agents.

INTRODUCTION

Lactic Acid Bacteria (LAB) are Gram-positive, acid-tolerant, non-sporulating bacteria that include species of *Lactobacillus*, *Lactococcus*, *Pediococcus*, *Leuconostoc*, and *Streptococcus*. LAB are naturally found in a wide range of environments including dairy products, meats, and vegetable and plant materials, and are associated with animal and human mucosal surfaces. LAB are important in the production and preservation of foods; LAB serve as starter cultures for fermented dairy products, meats, and vegetables, and are important for the production of coffee, chocolate, and wine. In addition to their role in food production, LAB are important in the production of renewable, low-carbon-footprint energy, and related materials. L-lactic acid is an important building block for the synthesis of biodegradable polymers and is the substrate for the production of other compounds such as acrylic acid, acetic acid, propylene glycol, and ethanol.

LAB have long been considered beneficial to human health, and their history of use and consumption has resulted in their status as "generally recognized as safe" (GRAS) organisms from the U.S. Food and Drug Administration (FDA). The lactobacilli in particular occupy important niches in the gastrointestinal (GI) tracts of humans and animals and are increasingly recognized as modulators of human and animal health. The use of LAB in probiotic products is rapidly expanding due to increasing public awareness of the role of nutrition and microbiota in maintaining health. Because of their acid tolerance, record of safety, and importance in human health, considerable interest has developed for the use of LAB as vectors for the delivery of vaccines and biotherapeutic compounds to mucosal surfaces.

In March of 2001, the European Commission funded a three-year project known as the LABDEL project (for LAB delivery) to develop and test prototype products for oral delivery of vaccines and other therapeutic agents using LAB.[1] As a result of the efforts of the LABDEL consortium and independent research groups, many advances have been made in the use of recombinant LAB in human health, including advances in vaccinations, modulation of type I allergies, and treatment of inflammatory bowel diseases. Additionally, the first trial of a recombinant LAB used in the delivery of a biotherapeutic agent in humans has been published.[2] Also, a number of comprehensive reviews of advances in this area have been published.[1,3–5] This entry will focus on a few of important publications of recent advances in the areas of mucosal delivery of vaccines and anti-infective agents providing the reader with broad overview of the topic.

LAB AS VACCINES

The development of bacteria as live vaccines and vaccine vehicles has focused primarily on the use of attenuated strains of pathogenic bacteria, including *Salmonella*, *Bortedella*, and *Listeria*.[6] While many properties related to their pathogenicity make them attractive candidates, the potential for reversion of attenuated strains to virulence is a significant safety concern. Also, it is unclear whether these attenuated strains would retain their ability to colonize and replicate effectively. LAB have been consumed by humans for centuries in fermented foods. Thus, these organisms can be orally administered, are well-tolerated by recipients, and can be easily and economically provided to large populations. Due to their GRAS status and the availability of genetic tools for recombinant expression of proteins, LAB are an attractive alternative to the use of attenuated pathogenic bacteria.

Table 1 Examples of lactic acid bacteria as vaccine vectors.

Organism	Antigen	Model
Lactobacillus acidophilus	Bacillus anthracis PA	Mouse
Lactobacillus casei	B. anthracis PA	Mouse
	Clostridium tetani TTFC	Mouse
	Coronovirus glycoprotein S	Mouse
	HPV-16 E7	Mouse
	Salmonella enterica FliC	Mouse
	Streptococcus pneumoniae PsaA and PspA	Mouse
	SARS CoV spike protein	Mouse
Lactobacillus helveticus	S. pneumoniae PsaA	Mouse
Lactobacillus plantarum	Helicobacter pylori UreB	Mouse
	S. pneumoniae PsaA	Mouse
Lactococcus lactis	C. tetani TTFC	Mouse
	Erysipelothrix rhusiopathiae SpaA	Mouse
	Helicobacter pylori UreB	Mouse
	HIV Env	Mouse
	HPV-16 E7	Mouse
	Plasmodium falciparum MSA2	Rabbit
	Rotavirus VP7	Mouse
	S. pneumoniae PsaA	Mouse

There is a growing body of work studying the potential of using live LAB in vaccines with several examples in which LAB have elicited antigen-specific immune responses (Table 1).[3,4] A model antigen commonly used to evaluate the potential of LAB as vaccines is tetanus toxin fragment C (TTFC). Orally administered recombinant *Lactococcus lactis*, *Lactobacillus plantarum*, *Lactobacillus casei*, and *Lactobacillus fermentum* expressing TTFC have been able to elicit potent immune responses in mice.[3]

The ability of LAB to colonize and persist in the gastrointestinal tract is strain-dependent and may be important in the effectiveness of LAB-based vaccines. In a comparison of *L. plantarum*, a persisting LAB, and *L. lactis*, a non-persisting LAB, Grangette et al.[7] found *L. plantarum* to be more effective at eliciting an immune response suggesting that the capacity of the bacterial vector to persist in the gastrointestinal tract impacted its immunogenicity.

A probiotic functionality that is important to the potential use of LAB vectors in vaccines is the ability to stimulate or modulate the immune system. LAB are thought to strengthen non-specific defenses against infection, increase IgA production, increase phagocytic activity of white blood cells, and regulate the balance of T-helper 1 (Th1) and T-helper 2 (Th2) cells. It is proposed that probiotic cultures are able to influence the immune system possibly by stimulating the expression of cytokines as shown in both cell culture and BALB/c mice.

Adjuvant-like effects on intestinal and systemic immunity have also been demonstrated using LAB. Individuals consuming fermented milk containing *Lactobacillus acidophilus* and *Bifidobacterium* species showed a four-fold higher specific serum IgA titer to *Salmonella* Typhi Ty21a than individuals not receiving fermented milk. Also, children receiving *Lactobacillus* GG during acute rotavirus infection showed enhanced IgA-specific antibody-secreting cell response as compared to those not receiving *Lactobacillus* GG.[4]

Of particular interest is the ability of *Lactobacillus* species to alter cytokine profiles and induce maturation of dendritic cells (DCs). DCs are antigen-presenting cells that comprise a critical component of the immune system and can be found at mucosal surfaces of the gastrointestinal tract.[4] Christensen et al.[8] evaluated the ability of six different *Lactobacillus* strains to activate murine DCs. All six strains up-regulated surface MHC class II and B7–2 (CD86), indicating DC maturation. *Lactobacillus* was also found to modulate expression of IL-6, IL-10, IL-12, and TNF-α in a species- and dose-dependent manner, suggesting that species of *Lactobacillus* may differentially determine whether DCs favor a Th1 or Th2 immune response. Mohamadzadeh et al.[9] showed that *Lactobacillus*-exposed human DCs secret the proinflammatory cytokines, IL-12 and IL-18, enhanced proliferation of CD4$^+$ and CD8$^+$ T cells, and thus skewed them toward a Th1 response. The ability of lactobacilli to promote DCs to regulate T cell responses toward Th1 pathways may be an important adjuvant property.

One promising strategy is the targeting of immunogenic antigens to DCs using probiotic microorganisms. Using a phage-display peptide library, Curiel et al.[10] identified 12-mer DC-specific binding peptides. One such peptide, when fused to human hepatitis 3 non-structural protein (NS3), showed significantly improved immunogenicity as compared to a NS3 control fusion protein or NS3 alone.

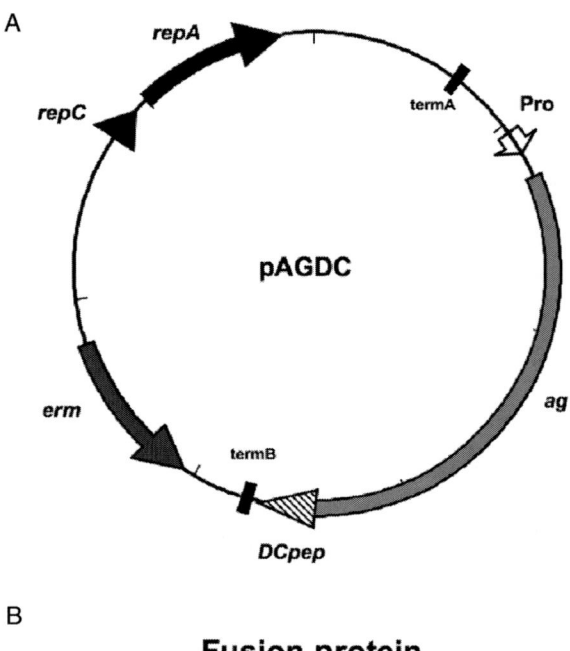

Fig. 1 Expression of DC-targeted antigens in lactobacilli. Twelve-mer DC-binding peptides were identified using a phage-display peptide library. DNA-encoding antigens can be transcriptionally fused to the DC-binding peptide sequence for expression in lactobacilli. (A) Schematic of an antigen:DC-binding peptide expression construct. Replication determinants shown as black arrows; antibiotic-resistance markers shown as dark gray arrows; promoter shown as white arrow; terminators shown as black boxes; antigen sequence shown in light gray; DC-peptide sequence shown as a hatched triangle. (B) Schematic of fusion protein.

It has been proposed that DNA sequences encoding DC-peptides can be fused genetically with antigen-coding sequences and expressed in *Lactobacillus* strains (Fig. 1).[4] These engineered bacteria can be administered orally and colonize the gastrointestinal tract where they would express and release immunogenic fusion proteins into the intestinal lumen for processing by DCs. Recently, Mohamadzadeh et al.[11] demonstrated that *L. acidophilus* expressing DC-targeted *Bacillus anthracis* protective antigens, protected mice from lethal challenge with anthrax.

DELIVERY OF ANTI-INFECTIVE MOLECULES

The potential of recombinant LAB in the expression and delivery of anti-infective molecules for the prevention of infection by microbial pathogens at mucosal surfaces has been evaluated. The goal of such a "bioshield" strategy is the inactivation and removal of microbial pathogens before they reach a host cell-receptor site.[3] Much of the research in this area has focused on the prevention of HIV infection at the vaginal mucosa.

Cell-binding and fusion processes serve as important targets for the inhibition of HIV infection. Binding of gp120 to CD4 receptors on immune cells is the first step in viral entry and can be blocked by cyanovirin-N (CV-N). Using a vaginal isolate of *Lactobacillus jensenii*, Liu et al.[12] described a strain capable of colonizing the vagina and producing high levels of CV-N when administered intravaginally to mice. Additionally, the *Lactobacillus*-produced CV-N dramatically reduced the infectivity of HIV in vitro. In a separate study, expression of two gp120-binding domains of CD-4 (2D CD4) using *L. jensenii* has also been evaluated as a potential bioshield strategy.[13] While these studies represent a major step toward the development of vaginal delivery of microbiocides to prevent the transmission of HIV, protection studies using these strains have not been published.

BIOLOGICAL CONTAINMENT OF RECOMBINANT STRAINS

The use of genetically modified organisms raises concerns about their release and propagation in the environment and potential transfer of transgenes, such as virulence determinants, from a live vaccine strain or antibiotic-resistance markers to other microorganisms. Effective biological containment systems should be simple, robust, and prevent the escape of both transgenes and recombinant organisms into the environment. A promising control strategy is the complementation of auxotrophy by supplementation with an essential metabolite. One example, based on a thymidine ($thyA^-$) auxotrophic mutant of *L. lactis*, combines active and passive approaches.[14] Thymine starvation results in activation of the SOS-repair system and DNA fragmentation acting as a suicide system. By replacing the *thyA* gene with a transgene of interest, any reversion of the *thyA* mutation by homologous recombination acquired by horizontal gene transfer would restore a wild-type phenotype and result in loss of the transgene. The effectiveness of this strategy has been evaluated and applied to the delivery of human IL-10 by live *L. lactis*.[14] The constructed strain produced potentially therapeutic levels of IL-10 but died rapidly in the environment in the absence of thymidine or thymine. The effectiveness of such biological containment systems has the potential to help overcome regulatory hurdles in the application of this technology.

CONCLUDING REMARKS

The development of LAB for the delivery of vaccines and biotherapeutic compounds has advanced dramatically over the last decade, culminating in the first clinical trial of recombinant LAB in humans. Vaccine delivery by LAB has been demonstrated to be effective in animal models; however, only a few protection studies have been published. Thus, much more work is required to determine their efficacy prior to additional human trials. Bioshield strategies also show much potential, and in addition to vaccines, will be an important strategy for prevention of infectious diseases. Clearly, these and other applications of LAB delivery show much potential. It is hoped that the next decade will see such products brought to the marketplace.

ACKNOWLEDGMENTS

The author would like to thank Dr. Michael Konkel for critical review of this manuscript.

REFERENCES

1. Hanniffy, S.; Wiedermann, U.; Repa, A.; Mercenier, A.; Daniel, C.; Fioramonti, J.; Tiaskolova, H.; Israelsen, H.; Madsen, S.; Vrang, A.; Hols, P.; Delcour, J.; Bron, P.; Kleerebezem, M.; Wells, J. Potential and opportunities for use of recombinant lactic acid bacteria in human health. Adv. Appl. Microbiol. **2004**, *56*, 1–64.
2. Braat, H.; Rottiers, P.; Hommes, D.W.; Huyghebaert, N.; Remaut, E.; Remon, J.P.; van Deventer, S.J.; Neirynck, S.; Peppelenbosch, M.P.; Steidler, L. A phase I trial with transgenic bacteria expressing interleukin-10 in Crohn's disease. Clin. Gastroenterol. Hepatol. **2006**, *4* (6), 754–759.
3. Wells, J.M.; Mercenier, A. Mucosal delivery of therapeutic and prophylactic molecules using lactic acid bacteria. Nat. Rev. Microbiol. **2008**, *6* (5), 349–362.
4. Mohamadzadeh, M.; Duong, T.; Hoover, T.; Klaenhammer, T.R. Targeting mucosal dendritic cells with microbial antigens from probiotic lactic acid bacteria. Expert Rev. Vaccines **2008**, *7* (2), 163–174.
5. Mercenier, A.; Pavan, S.; Pot, B. Probiotics as biotherapeutic agents: present knowledge and future prospects. Curr. Pharm. Des. **2003**, *9* (2), 175–191.
6. Detmer, A.; Glenting, J. Live bacterial vaccines—a review and identification of potential hazards. Microb. Cell Fact. **2006**, *5*, 23.
7. Grangette, C.; Muller-Alouf, H.; Geoffroy, M.; Goudercourt, D.; Turneer, M.; Mercenier, A. Protection against tetanus toxin after intragastric administration of two recombinant lactic acid bacteria: impact of strain viability and in vivo persistence. Vaccine **2002**, *20* (27–28), 3304–3309.
8. Christensen, H.R.; Frokiaer, H.; Pestka, J.J. Lactobacilli differentially modulate expression of cytokines and maturation surface markers in murine dendritic cells. J. Immunol. **2002**, *168* (1), 171–178.
9. Mohamadzadeh, M.; Olson, S.; Kalina, W.V.; Ruthel, G.; Demmin, G.L.; Warfield, K.L.; Bavari, S.; Klaenhammer, T.R. Lactobacilli activate human dendritic cells that skew T cells toward T helper 1 polarization. Proc. Natl. Acad. Sci. U.S.A. **2005**, *102* (8), 2880–2885.
10. Curiel, T.J.; Morris, C.; Brumlik, M.; Landry, S.J.; Finstad, K.; Nelson, A.; Joshi, V.; Hawkins, C.; Alarez, X.; Lackner, A.; Mohamadzadeh, M. Peptides identified through phage display direct immunogenic antigen to dendritic cells. J. Immunol. **2004**, *172* (12), 7425–7431.
11. Mohamadzadeh, M.; Duong, T.; Sandwick, S.J.; Hoover, T.; Klaenhammer, T.R. Dendritic cell targeting of *Bacillus anthracis* protective antigen expressed by *Lactobacillus acidophilus* protects mice from lethal challenge. Proc. Natl. Acad. Sci. U.S.A. **2009**, *106* (11), 4331–4336.
12. Liu, X.; Lagenaur, L.A.; Simpson, D.A.; Essenmacher, K.P.; Frazier-Parker, C.L.; Liu, Y.; Tsai, D.; Rao, S.S.; Hamer, D.H.; Parks, T.P.; Lee, P.P.; Xu, Q. Engineered vaginal lactobacillus strain for mucosal delivery of the human immunodeficiency virus inhibitor cyanovirin-N. Antimicrob. Agents Chemother. **2006**, *50* (10), 3250–3259.
13. Liu, X.; Lagenaur, L.A.; Lee, P.P.; Xu, Q. Engineering of a human vaginal Lactobacillus strain for surface expression of two-domain CD4 molecules. Appl. Environ. Microbiol. **2008**, *74* (15), 4626–4635.
14. Steidler, L.; Neirynck, S.; Huyghebaert, N.; Snoeck, V.; Vermeire, A.; Goddeeris, B.; Cox, E.; Remon, J.P.; Remaut, E. Biological containment of genetically modified *Lactococcus lactis* for intestinal delivery of human interleukin 10. Nat. Biotechnol. **2003**, *21* (7), 785–789.

Vegetables: Fermentation Applications

Tony Savard
Food Research and Development Center, Agriculture and Agri-Food Canada, St-Hyacinthe, Quebec, Canada

Abstract

Fermentation is one of the oldest methods for preserving food. Contrary to other industries, vegetable fermentation is mostly uncontrolled. Nevertheless, it represents one of the most important forms of nutrition around the world ensuring safety and well-being. Recent studies have shown that starter cultures could be developed, salt level modified, and safety or health aspect emphasized in the functioning of the process. Knowledge in this field needs to be improved and links with human health are to be correlated.

INTRODUCTION

Lactic acid fermentation is a cheap and effective means of preserving perishable vegetables. It prevents growth of most foodborne pathogens and obviates the need for pasteurization and the addition of preservatives. Numerous studies have shown the health benefits of fermented food. The Food and Agricultural Organization of the United Nations (FAO) recognizes that fermentation technologies play an important role in ensuring the food security of millions of people around the world. Fermentation improves nutrition by the detoxification of raw materials, by the removal of antinutritional factor, and by the enrichment of vitamins. This entry will review the basic principles of fermentation and provide a summary of scientific knowledge in the field, including the latest research on some vegetables.

ORIGIN OF FERMENTATION

The development of fermentation technologies is lost in the mists of time. The first fermentation was probably applied to alcoholic drinks 7000 years ago in Babylon. Bread, milk, and meat fermentation were developed subsequently, and the first written trace of vegetable fermentation has been reported under the Chou dynasty (1126–1256 B.C.) in China.[1]

Most of the actual fermentation technologies have been handed down from generation to generation, and it is only in the latter part of the nineteenth century that fermentation has acquired a scientific basis with Louis Pasteur, who was the first to relate living cells (yeasts) with fermentation in 1854. Intensive research on vegetable fermentation did not begin before the middle of the twentieth century, and knowledge is still limited in comparison to the dairy and alcoholic fermentation.

SOCIAL AND ECONOMIC IMPORTANCE

Fermented foods are popular throughout the world and in some regions make a significant contribution to the diet and well-being of millions of individuals. The most popular are sauerkraut/kimchi, pickles, olives, ogi, idli, gari, fufu, miso, natto, tempeh, and soy sauce. In Asia, fermented products supply proteins, minerals, and other nutrients that contribute to a nutritional fortification to otherwise starchy, bland diets. Soy sauce has a worldwide distribution and is part of the fundamental diet from Indonesia to Japan. Fermented cassava (gari and fufu) is the major component of the diet for more than 800 million people in Africa and often constitutes more than 50% of the daily diet. Kimchi is part of the daily Korean diet and is eaten as a side dish or incorporated in soup and main meals. Lately, intensive research has been done on the fermentation process and its nutritional and health aspects. Contrary to fermented cabbage (sauerkraut), where the process has been highly standardized and industrialized to facilitate a worldwide distribution, kimchi is usually eaten locally thereby dispensing with the need for pasteurization or chemical stabilization.

BASIC PRINCIPLE

The lactic acid fermentation of vegetable occurs spontaneously with the commensal microflora when salt and anaerobiosis are used. Vegetables harbor a high sugar content and neutral pH favorable to microbial growth. Although processing varies depending on the product, three to four stages are commonly identified.

Pretreatment

This phase is characterized by the preparation of vegetable manipulations like sorting and grading, cleaning, cutting, and shredding or punching, all of which facilitate the release of nutrient and liquid from vegetal vacuoles. It may also include a blanching process or a lye-treatment, as in olives, to ensure the removal of oleuropein that is responsible for the bitter taste.

Creation of the Fermentation Environment

Following pretreatment, salt is generally used to enhance the creation of the fermentation media based on plasmolysis action. The use of salt also helps to inhibit pectinase associated with the softening of vegetables and contributes to the elimination of non-fermentative and spore-forming bacteria. Dissolved oxygen will disappear from the media as a result of the respiration of indigenous microorganisms, vegetal cells, and the production of carbon dioxide, creating an anaerobic environment favorable to the development of lactic acid bacteria (LAB). Temperature is a crucial factor in determining the microbial ecology and sequence of the fermentation.

Microbial Fermentation

Although LAB are the least prevalent group of microorganisms at the surface of vegetables (<0.1% of the 10^5–10^7 microorganisms/g), anaerobiosis, salt, and temperature favor their growth to become dominant in the initial days, resulting in fast acidification of the media. The catabolism of sugar (mostly glucose, fructose, and sucrose) by LAB generates lactic acid (homofermentative pathway) or a mixture of lactic and acetic acid, ethanol, and CO_2 (heterofermentative pathway). Minor compounds like hydrogen peroxide, esters, diacetyl, and bacteriocins are also produced and contribute to organoleptic qualities or in the elimination of other bacteria. Mannitol is also produced by the conversion of fructose.

Microbial fermentation is characterized by a sequential and mixed process. The main actors are *Leuconostoc mesenteroides*, *Pediococcus acidilactici*, and *Lactobacillus plantarum*. *Leuconostoc*, above all, initiate the fermentation process with a gaseous phase generated by the heterolactic fermentation associated with the production of carbon dioxide. The reason why *Leuconostoc* dominates the early fermentation has been associated to its shorter lag and generation times as opposed to other LAB. Subsequently, acidification continues with homofermentative bacteria, mainly *Lactobacillus* and *Pediococcus*, to reach a pH between 3.5 and 4.0. A final concentration of 2–3% of lactic and acetic acid can be produced with an appropriate lactic/acetic ratio of three. At this level, *Leuconostoc* is no longer alive and *Lactobacillus* is metabolically inactive.

Postfermentation

Although the microbial fermentation is over at this stage, fermented vegetables undergo a maturing phase where complex reactions like esterification and modification of chemical compounds will contribute to their organoleptic qualities. Depending on the vegetables, all sugars may or may not have been depleted. If there are still fermentable sugars, a postfermentation may occur with the growth of acido-tolerant yeasts like *Saccharomyces*, *Hansenula*, and *Kluyveromyces*. Secondary fermentation by yeasts can result in gaseous spoilage such as bloater damage in cucumbers and in the depletion of organic acids corresponding to an increase in pH.[2] This problem could explain why most of the commercial sauerkraut is pasteurized or has stabilizers.

CONTROL OF FERMENTATION: OLD AND NEW TRENDS

Traditional fermentations of vegetables have been done with most of the parameters being uncontrolled. Examination of the process has identified multiple criteria that affect fermentation and the resulting products. Salt, temperature and microorganisms are the most important criteria.

Salt and Substitutes

Sodium chloride (NaCl) is present in every kind of vegetable fermentations. Dry-salting or brining usually initiates fermentation. In sauerkraut, the cabbage is dry-salted at a level of 2–2.5% by weight allowing it to self-brine through its own moisture. In kimchi, cabbage is also dry-salted but to a level of approximately 10% followed by washing and draining because of the lower water content of the cabbage. Olives and cucumbers are soaked in 5% brine, and the brine strength is gradually increased up to 15%. Low levels of salt promote heterofermentative bacteria whereas higher levels favor homofermentative bacteria.[3] *Leuconostoc* is active in sauerkraut and kimchi but not prominent in olives and cucumbers partially, at least, because of the higher salt concentration in the latter product.

With increasing concerns about high sodium content in many foods, reduced salt levels have been tried, mostly with sauerkraut. The product had good flavor even though it differed in acid content, texture, and shelf life. The speed of fermentation is also decreased by lower salt concentration. Cucumber fermentation without NaCl resulted in loss of firmness and severely bloated tissue.[4] Replacement of NaCl by potassium, calcium chloride, or a mixture of the two with a low level of NaCl reduced fermentation speed and resulted in sweeter products. It was therefore suggested to accelerate the fermentation by the addition of starter cultures. Additional research is necessary because salt is important not only for plasmolysis and taste but also for restricting the activities of undesirable Gram-negative bacteria.

Temperature of Fermentation

Temperature is one of the most important factors in controlling the fermentation process and the quality of the final product. The temperature range for vegetable fermentation is 10–35°C, but the usual optimal temperature is between 10 and 20°C. Low temperature favors *Leuconostoc* while higher temperature favors *Lactobacillus* and *Pediococcus*.[5] In sauerkraut, where a lactic/acetic ratio of three is considered appropriate for attaining good quality, the optimal temperature is around 18–19°C to control the microbial sequence between *Leuconostoc* and *Lactobacillus*. At a lower temperature, *Leuconostoc* is predominant conferring a lower ratio and a higher pH while fermentation at 32–37°C favors the homofermentative bacteria *Lactobacillus* and *Pediococcus* producing a higher ratio and a lower pH.[6] Cucumber fermentation is done as rapidly as possible to reduce purging costs and preferably at 26–29°C to favor homofermentative bacteria to reduce the risk of bloating associated with overproduction of CO_2 by *Leuconostoc*. At the other extreme, kimchi fermentation is done at 10–15°C to favor *Leuconostoc* over *Lactobacillus* to achieve a less acidic product with a pH of no less than 4.

Microorganisms

Fresh vegetables harbor numerous and various types of microorganisms, mostly aerobic bacteria and yeasts. LAB, the principal actors of fermentation, represent only 0.15–1.5%.[7] Vegetable fermentation has to provide conditions resulting in a succession of different LAB becoming the predominant bacteria.

Spontaneous fermentation

The major LAB are located in three genera, *Leuconostoc*, *Pediococcus*, and *Lactobacillus*. These bacteria play a sequential role and were first studied by Pederson[8] in a spontaneous sauerkraut fermentation. It is known that the dominant species and the sequence of their dominance can be influenced by temperature and salt levels, carbonation, or pH control with buffering solutions, but not all the LAB involved in different vegetable fermentations are known. As an example, Cho et al.[9] have recently identified a *Weissella* species linked to kimchi fermentation.

Starter cultures

The introduction of juice from a previous kraut fermentation (backslopping) can be viewed as supplying a starter culture, but this technique is dependent on the age of the juice as juice changes with the bacterial sequence. Pederson was the first to incorporate a pure culture of *Lactobacillus* in cabbage.[8] The introduction of pure and mixed cultures to produce sauerkraut of consistent quality gave poor results because vegetable fermentation involves a succession of bacteria. It is therefore not surprising that commercial starter cultures are rare, even today. The list of attributes that a starter cultures should have is long and includes the acceleration of the fermentation process resulting in better control of the initial microflora, the promotion or avoidance of carbon dioxide production, the total depletion of fermentable sugars with pH control and buffering solutions, the prolonging of shelf life, the change in flavor of the products, the production of different ratios of organic acids and of more or less acidic products, the production of bacteriocins or the expression of resistance to these antimicrobials, the resistance to bacteriophage, the reduction of the synthesis of biogenic amines, and to the control of secondary fermentation.

NEW TRENDS IN VEGETABLE FERMENTATION

New trends concern safety, nutrition, and health. Although fermented foods are generally recognized as safe, some starter cultures have been selected to control foodborne pathogens. LAB, in particular, show efficacy in controlling these pathogens and in the decontamination of raw materials or minimally processed vegetables.[10]

The selection of probiotic bacteria for incorporation in fermented vegetable products for their health benefits is still in its infancy. It is not certain if such probiotics are needed since fermented vegetables are themselves considered healthy by increasing nutritional availability and diet value[11] as well as having some protective functionality against different diseases.[12] The validation of these safety, nutritional, and health claims by rigorous research will be a task for the future.

REFERENCES

1. Yokotsuka, T. Fermented protein foods in the Orient, with emphasis on Shoyu and Miso in Japan. In *Microbiology of Fermented Foods*; Wood, B.J.B., Ed.; Elsevier Applied Science: London, 1985; 197–247.
2. Etchells, J.L.; Bell, T.A.; Fleming, H.P.; Kelling, R.E.; Thompson, R.L. Suggested procedure for the controlled fermentation of commercially brined pickling cucumbers—the use of starter cultures and reduction of carbon dioxide accumulation. Pickle Pak. Sci. **1973**, *3*, 4–14.
3. Bamforth, C.W. Vegetable fermentations. In *Food, Fermentation and Micro-Organisms*. Blackwell Sciences, Ed.; Blackwell Publishing: Oxford, 2005; 193–197.
4. Fleming, H.P.; Kyung, K.H.; Breidt, F. Vegetable fermentations. In *Biotechnology*, 2nd Ed.; Rehm, H.-J., Reed, G. Eds.; VCH Publishers, Inc.: New York, 1995; 629–661.
5. Fleming, H.P.; McFeeters, R.F.; Daeschel, M.A. The lactobacilli, pediococci, and leuconostocs: vegetable products. In

Bacterial Starter Cultures for Foods; Gilliland, S.E., Ed.; CRC Press: Boca Raton, FL, 1985; 97–118.
6. Pederson, C.S.; Albury, M.N. Factors affecting the bacterial flora in fermenting vegetables. Food Res. **1953**, *18*, 290–300.
7. Buckenhuskes, H.J. Fermented vegetables. In *Food Microbiology—Fundamentals and Frontiers*; Doyle, M.P., Beuchat, L.R., Montville, T.J., Eds.; ASM Press: Washington, DC, 1997; 595–609.
8. Pederson, C.S. The effect of pure culture inoculation on the quality and chemical composition of sauerkraut. N.Y. State Agric. Exp. Stn. Geneva Tech. Bull. **1930**, *169*.
9. Cho, J.; Lee, D.; Yang, C.; Jeon, J.; Kim, J.; Han, H. Microbial population dynamics of kimchi, a fermented cabbage product. FEMS Microbiol. Lett. **2006**, *257*, 262–267.
10. Breidt, F. Safety of minimally processed, acidified, and fermented vegetable products. In *Microbiology of Fruits and Vegetables*; Sapers, G.M., Gorny, J.R., Yousef, A.E., Eds.; CRC Press: Boca Raton, FL, 2005; 313–335.
11. Svanberg, U.; Lorri, W. Fermentation and nutrient availability. Food Control **1997**, *8*, 319–327.
12. Lee, C.H. Lactic acid fermented foods and their benefits in Asia. Food Control **1997**, *8* (5/6), 259–269.

Vertebrates: Biological Control

Peter Kerr
Vertebrate BioControl Cooperative Research Center, Wildlife and Ecology, Commonwealth Scientific and Industrial Research Organisation (CSIRO), Canberra, Australian Capital Territory, Australia

Abstract

Conceptually, biological control of vertebrate pests is a very attractive, cost-effective alternative to other methods of control such as poisoning, shooting, trapping, and exclusion. In support of this, theoretical models and practical experience show that pathogens and parasites can regulate host populations either by increasing mortality rates or by decreasing the reproduction rate. Unfortunately, most potential biological control agents identified in laboratory or small-scale field trials have been ineffective. This can be for several reasons depending on the epidemiology and distribution of the parasite and the population biology of the pest species. The organism may already be present and adapted to the pest species, it may become established at too low a prevalence to reduce the rate of increase of the host population, or it may fail to persist and transmit. Even following successful establishment of a biological control agent in a pest population, subsequent adaptation of host and parasite or specific aspects of the epidemiology of the pathogen may significantly reduce the impact of biological control over quite short time periods.

INTRODUCTION

Biological control can be defined as the use of one species to reduce the population of another species, in this case vertebrate pests. This definition could include the use of vertebrate predators to control other vertebrates, e.g., cats to control rodents. However, biological control is more commonly considered as the use of microparasites: viruses, protozoa or bacteria; or macroparasites: helminths and arthropods. The introduction of the viral disease myxomatosis to control European rabbits (*Oryctolagus cuniculus*), initially in Australia and later in Europe, is the best example of the success and limitations of biological control of vertebrates. In fact, the European rabbit is the only vertebrate pest for which biological control has had a major impact.

MYXOMATOSIS AND EUROPEAN RABBITS

The wild European rabbit was introduced into Australia in 1859 and over the next 60 years spread across the continent, establishing itself as the major vertebrate pest species and a highly significant cause of agricultural and ecological depredation. In 1995, it was estimated that effective rabbit control could be worth 3% of agricultural production in Australia. The ecological costs of rabbits to Australia have never been quantified. Myxoma virus[1] is a poxvirus of the South American forest rabbit (*Sylvilagus brasiliensis*). In this species, it causes a benign cutaneous fibroma at the site of inoculation. Mosquitoes or other biting arthropods have their mouthparts contaminated with virus when probing through this fibroma and then transmit the virus at subsequent feedings. In European rabbits, myxoma virus causes the lethal disease myxomatosis and its potential as a biological control agent for rabbits was recognized as early as 1919. Extensive testing demonstrated that myxoma virus could only infect lagomorphs. However, successful use of the virus in European rabbits initially proved elusive. In trials in Australia and Europe the virus spread locally but then died out. In retrospect, the reasons for this are obvious. The virus is rapidly lethal, there were no reservoir hosts and rabbits that survive infection are neither persistent carriers of the virus nor susceptible to reinfection. Thus the virus becomes extinct when there are too few susceptible animals to support the epidemic and no insect vectors to transmit virus into high density populations of susceptible rabbits.

In the summer of 1950–1951 a strain of myxoma virus succeeded in establishing in the Australian rabbit population. These rabbits had never been exposed to myxoma virus so all were susceptible and mosquito vectors were plentiful due to unusually wet seasons. During the next two years the virus spread to all the rabbit-infested areas of Australia. An estimated 99.8% of infected rabbits were killed and the rabbit population decreased by 95% across most of southern Australia. In the wool industry alone, it was estimated that myxomatosis allowed the production of 32,000,000 kg of additional wool in 1953. However, this high lethality was not maintained. Moderately attenuated strains of virus that killed 70–90% of infected rabbits were more efficiently transmitted than either highly virulent or highly attenuated strains and became predominate in the field. These attenuated strains of virus allowed survival of rabbits with a degree of innate resistance, which led to a rapid selection for resistance to myxomatosis.

Myxomatosis epidemics in the Australian rabbit population now provide an additional level of mortality above that due to other diseases and predation. It is difficult to quantify the current impact of myxomatosis. However, large-scale field immunization trials in Australia, and trials in Britain removing the flea vector of myxoma virus, indicated that the virus still has a substantial effect on rabbit populations. Some releases of virulent myxoma virus are still made by property holders and pest control agencies in Australia. These releases may have some local effect on reducing rabbit numbers, with the virus acting as a biocide, but are epidemiologically insignificant. Of more impact has been the deliberate establishment of two exotic arthropod vectors of myxomatosis in the Australian wild rabbit population. The European rabbit flea *Spilopsyllus cuniculi* acts as a vector in the temperate zones of Australia while the Spanish rabbit flea *Xenopsylla cunicularis* provides a vector in the arid and semi arid regions where mosquitoes may be absent for long periods. These fleas have significantly altered the epidemiology of myxomatosis by removing the reliance on mosquitoes for transmission of myxoma virus.

RABBIT HEMORRHAGIC DISEASE VIRUS

Rabbit hemorrhagic disease virus (RHDV), a calicivirus that appears to be specific for European rabbits, emerged in China in 1984. It has subsequently spread to most areas of the world where European rabbits are farmed or occur in the wild. The virus causes rapid death in rabbits more than 4–6 weeks old and is spread by contact and by insect vectors. RHDV was investigated as a potential biological control for rabbits in Australia. In October 1995, RHDV escaped from a quarantine site off the mainland of Australia and spread rapidly (more than 300 km in 2 weeks) through high density rabbit populations in South Australia. During the next 12 months it spread throughout the rabbit populations of southern Australia.[2] RHDV has had a major impact on the rabbit population of Australia particularly in arid and semi-arid areas. In some areas population densities were reduced by 95% allowing regeneration of native trees and shrubs. So far, emergence of attenuated strains of RHDV or development of resistance in rabbits has not been documented in Australia and RHDV has not significantly altered the epidemiology of myxomatosis. However, some resistance to RHDV may be present in wild rabbits in Europe. There are also naturally attenuated strains of the virus present in Europe. These probably predate the RHDV epidemics and at least in Britain appear to have limited the spread of the virulent virus by immunizing the rabbit population. Recent serological evidence indicates that a related virus may be present in Australia and New Zealand. However, the epidemiological consequences of this are not yet known. The myxoma virus model suggests that some form of resistance to and attenuation of RHDV will eventually emerge in Australia but the time scale and selection pressures for these may be distinctly different from those operating for myxomatosis.

BIOLOGICAL CONTROL OF CATS USING FELINE PANLEUCOPENIA VIRUS

On the sub-Antarctic Marion Island, a population of feral cats (*Felis catus*), estimated at more than 3400 cats in 1977, was descended from five cats released in 1948–1949. These cats killed more than 450,000 seabirds in 1975. Feline panleucopenia virus is a parvovirus that is endemic in cat populations around the world but was not present in the cats on Marion Island. A virulent isolate of this virus was used to inoculate 93 trapped cats in 1977. Over the next five years the population decreased to an estimated 615 cats. But by 1982 the rate of decrease had slowed from 29% per year to 8%. Shooting and trapping to eradicate the cats followed the population knockdown.[3] The release of panleucopenia virus to control cats on other islands has been ineffective. Surveys of feral cats in mainland locations have usually demonstrated that this virus is endemic and its release as a biological control is unlikely to have a significant impact.

BIOLOGICAL CONTROL OF RODENTS

Rodents are the major vertebrate cause of crop loss and spoilage worldwide. They are also significant vectors of human and animal diseases and when introduced onto islands can cause severe ecological damage. The difficulties of controlling rodents by conventional means, the development of resistance to rodenticides that has occurred, and the dangers posed to other species by many rodenticides make biological control a very attractive option. This has been attempted with bacteria, viruses, protozoa, cestodes, and nematodes but without significant success.[4]

GENETIC MODIFICATION OF BIOLOGICAL CONTROL AGENTS

Biological control agents that reduce reproduction rates have the potential to be just as effective at population reduction as lethal agents. A novel approach to population control is to genetically engineer viruses or other microorganisms to deliver contraceptive vaccines to vertebrate pest species. In laboratory trials, mice inoculated with recombinant mousepox virus[5] or, recombinant murine cytomegalovirus expressing a murine oocyte antigen developed long-term infertility. Parallel studies are being undertaken with recombinant myxoma virus expressing a rabbit oocyte antigen. Whether it will be possible for such recombinant viruses to spread in the field remains to be determined. However, using molecular genetics, the potential for biological control of vertebrate pests may finally be realized.

REFERENCES

1. Fenner, F.; Ross, J. Myxomatosis. In *The European Rabbit. The History and Biology of a Successful Colonizer*; Thompson, H.V., King, C.M., Eds.; Oxford University Press: Oxford, **1994**; 205–240.
2. Mutze, G.; Cooke, B.; Alexander, P. The initial impact of rabbit hemorrhagic disease on european rabbit populations in South Australia. J. Wildl. Dis. **1998**, *34*, 221–227.
3. Van Rensburg, P.J. J.; Skinner, J. D.; Van Aarde, R. J. Effects of feline panleucopaenia on the population characteristics of feral cats on Marion Island. J. Appl. Ecol. **1987**, *24*, 63–73.
4. Singleton, G. R. In *The Prospects and Associated Challenges for the Biological Control of Rodents*, Proceedings of the 16th Vertebrate Pest Conference, Halverson, W. S., Crabb, A.C., Eds.; University of California: Davis, CA, **1994.**
5. Jackson, R.J.; Maguire, D.J.; Hinds, L.A.; Ramshaw, I.A. Infertility in mice induced by a recombinant ectromelia virus expressing mouse zona pellucida glycoprotein 3. Biol. Reprod. **1998**, *58*, 152–159.

Vinegar

Wendu Tesfaye
Department of Food Technology, Polytechnical University of Madrid, Madrid, Spain

A.M. Troncosco
Department of Nutrition and Bromatology, University of Seville, Seville, Spain

Abstract

Vinegar is neither merely acidulated water nor a spontaneously fermented product. Its production follows a well-defined acetification method. Different factors determine vinegar quality; among the most important factors are the substrate used and the production method employed. Vinegar is more than just a condiment; it can serve as basis for a health drink or as a cosmetic agent.

INTRODUCTION

Vinegar as a spontaneously fermented natural product has existed among some of our oldest civilizations, such as ancient Babylon (around 5000 B.C.), Egypt (3000 B.C.), China (1200 B.C.), biblical times (Old and New Testaments), and Greece (around 400 B.C.). Vinegar has been used in food preservation and flavoring, as an energizing drink, as a medicine, and as a solvent. Thanks to technological development and scientifically acquired knowledge, the use of vinegar as a condiment is highly specialized. The aim of this entry is to describe the composition and characteristics of vinegar as well as describe the factors affecting production, including information on acetic acid bacteria, the many uses of vinegar, and the future of the product.

VINEGAR

Etymologically, the word "vinegar" is derived from two Latin words, *vinum* that means wine and *acre* meaning sour. Sourness is common to all vinegars produced from different raw materials and diverse production techniques. Acetic acid (CH_3COOH) is the major component of vinegar; however, acetic acid is a sign of spoilage from an oenological point of view. Acetic acid is the result of two fermentations. First, sugar is converted to ethanol by yeasts (alcoholic fermentation) and then acetic acid bacteria convert the ethanol to acetic acid (acetic acid fermentation or acetification). Acetification is a two-step process in which ethanol is converted to acetaldehyde and then the acetaldehyde is converted to acetic acid.

There are two primary methods of vinegar production. The first method is the surface culture system or traditional method. As the name implies, acetic acid bacteria accomplish the acetification process on the surface of the liquid or at the liquid–air interface. With this method, high-quality vinegars, such as traditional balsamic vinegar from Modena[1] and sherry wine vinegars,[2] are produced. This method mainly employs wood casks where both acetification and aging occur simultaneously. The second method is known as the submerged culture system where the acetic acid bacteria are suspended in the fermenting liquid. To maintain viability of the acetic bacteria, the acetifying liquid must be saturated with oxygen or air on a continuous basis. In this way the acetification surface area is increased. Consequently, this method has the highest yield of acetic acid in the shortest amount of time. Most vinegars for commercial use are produced in this way. These vinegars are the cheapest to produce (Fig. 1).

Raw Material

Based on the Joint FAO/WHO Food Standards Programme of 1987,[3] vinegar is defined as "a liquid fit for human consumption, produced from a suitable raw material of agricultural origin, containing starch, sugars, or starch and sugars by the process of double fermentation, alcoholic and acetous, and contains a specified amount of acetic acid." The raw material is normally wine grapes for wine vinegars, apples for apple cider vinegar, cereals for malt vinegars, honey for honey vinegars, and rice for rice vinegars. Most vinegars are produced from wine.

Acetic Acid Bacteria

Acetic acid bacteria are obligately aerobic, Gram-negative rods that can be ellipsoidal-shaped. Without them, vinegars suitable for human consumption would not be possible (Fig. 2). Currently the family *Acetobacteraceae* is classified as containing 10 genera and 45 species;[4] however, only two genera, *Acetobacter* and *Gluconobacter* are involved in commercial vinegar production. The main difference between these two genera is that all *Acetobacter*

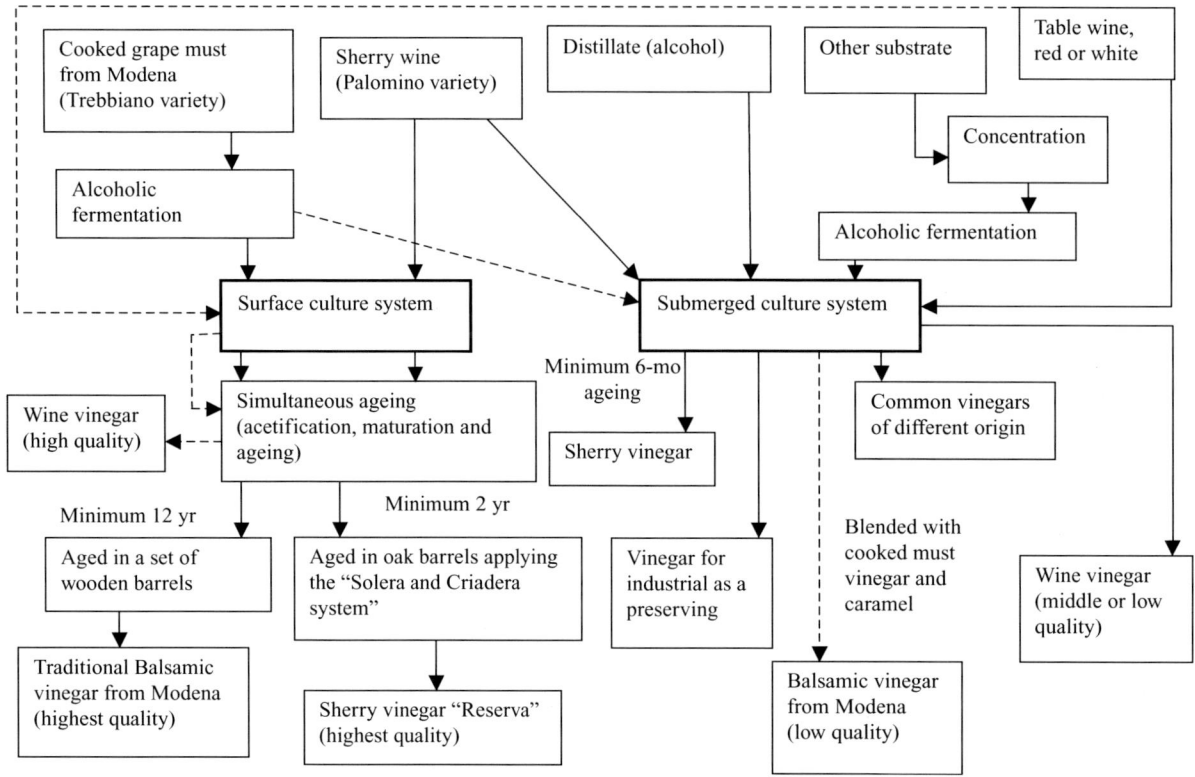

Fig. 1 Vinegar production methods and the resulting vinegars.

species over-oxidize ethanol to acetic acid and finally to CO_2 and H_2O, while species of *Gluconobacter* are unable to oxidize acetic acid to CO_2 and H_2O. *Gluconobacter* is capable of oxidizing glucose to gluconic acid, and as the result balsamic vinegars from Modena contain higher amounts of gluconic acid in comparison to other vinegars. Recent research has been focused on the selection of a pure strain that is resistant to higher temperatures, and higher substrate and product concentrations, along with selection of improved phenotypic traits relevant for use as a vinegar starter culture.

FACTORS IMPLICATED IN VINEGAR ELABORATION

In order for the acetic acid bacteria to accomplish the acetification process, the following factors should be strictly monitored: 1) continuous oxygen or air supply should be guaranteed; 2) the proportion of unloaded finished product is controlled, where a portion of the total volume of a submerged culture is withdrawn with the finished product on a periodic basis and the remaining volume, which acts as an inoculum for the next cycle, is replenished with fresh substrate that is added in small portions to obtain the final working volume; 3) the ideal working temperature (30°C) should be maintained since the process itself is exothermic; and 4) the total working volume must be taken into account.[5] Other factors such as pH, alcohol, and acetic acid concentrations of the acetification medium are factors that mainly affect the process at the beginning; however, the bacteria become less sensitive to these factors as repeated cycles are realized. For submerged culture acetification systems, the size of air bubbles that are formed when air is introduced into the acetifying medium has to be as small as possible for maximum oxygen transfer efficiency.[6]

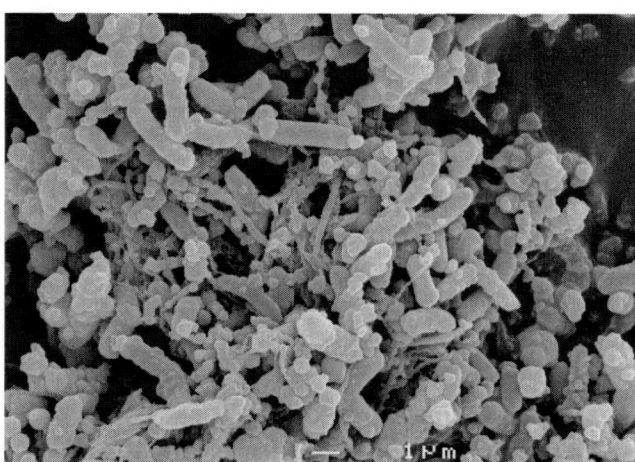

Fig. 2 Acetic acid bacteria during the acetification process (acetic degree 9.7% and alcohol residual 0.5%). **Source:** Photographed by Francisco Barja.

COMPOSITION OF VINEGAR

The composition of a vinegar is mainly dependent on what it was made from. Some of the components are formed during the acetification process while others occur during aging. The chemical composition is a good indicator in predicting the quality, the production method employed, and the classification of the vinegar. The aromatic profile is mainly indicative of the production method. Vinegars produced using the traditional method are more aromatic than those made using submerged culture system. For example, sherry wine vinegars are easily differentiated from other vinegars by the higher concentration of ethyl acetate; whereas higher concentration of gluconic acid is a better indicator of traditional balsamic vinegars from Modena.

There are different chemical components in different types of vinegars. The flavor of wine vinegars is mainly influenced by compounds known as polyphenols. Polyphenols are common secondary metabolites of plants. Their content in plants varies greatly among different species, cultivars, levels of maturity, the time of season, and region. These compounds play an important role in the color and flavor of the vinegar, as well as the defense mechanisms and bioactive elements of the plant. Polyphenols are classified according to their structure as phenolic acid derivates, flavonoids, or tannins.

The presence of these compounds in vinegars is directly related to the original raw plant material and the wooden barrels used in acetification and aging.[7] Vinegar polyphenols reported by different authors are summarized in Tables 1 and 2.

Volatile components of vinegar are also play an important role to ameliorate the pungent sensation of the acetic acid. These compounds are from the original substrate used (e.g., odors wine, apple, or honey), as well as odors formed from the acetification process (e.g., odors produced from the metabolic activity of *Acetobacter*). Some compounds are condensation products (e.g., ethyl acetate formed from acetic acid and ethanol); the remaining volatile products are extracted from the aging material employed (e.g., odors of vanilla, caramel, and wood). Those vinegars produced by the surface culture system have the most aromatic and volatile components.[8]

Vitamins are minor components of vinegar. Vitamins found in apple cider vinegar include vitamins E, A, B_1, B_2, B_6, and C.[9] Amino acids are also among the minor components of vinegar that are derived from the lysed yeast and acetic acid bacteria during the process. L-leucine, arginine, and L-proline are the most common amino acids found in vinegar.[10]

The ash of vinegar is composed of minerals. The common minerals present in vinegar include sodium, potassium, calcium, magnesium, copper, iron, manganese and zinc.[11]

USES OF VINEGAR

Babylonians used vinegar made from the fruits of date palms as a food and food preservative agent in 5000 B.C.. It is mentioned in the Bible as a food and medicine in both the Old and New Testaments. Hippocrates, the father of modern medicine, used cider vinegar with honey to treat cough and colds (about 400 B.C.). From anecdotes of Hannibal and Cleopatra, vinegar was considered a strong acid with powerful dissolving power. Hannibal crossed the Alps with elephants to invade Italy in 218 B.C. It is said he poured vinegar over hot rocks to disintegrate them and allow his troops to march through. Cleopatra reportedly dissolved a priceless pearl in vinegar and then drank the resulting

Table 1 Non-flavonoid compounds of vinegars.

Main group (Non Flavonoids)	Common polyphenols to vinegars		Source
Phenolic acids	$R_1=R_2=H$ $R_1=OH, R_2=H$ $R_1=OCH_3, R_2=H$ $R_1=R_2=OH$ $R_1=R_2=OCH_3$	p-hydroxybenzoic acid Protocatechic acid Vanillic acid Gallic acid Syringic acid	White grape, Red grape, Fruit vinegars
Cinnamic acid	$R_1=R_2=H$ $R_1=OH, R_2=H$ $R_1=OCH_3, R_2=H$ $R_1=R_2=OCH_3$	p-Cumaric acid Caffeic acid Ferulic acid Sinapic acid	White grape, Red grape, Fruit vinegars
Coumarins	$R_1=H, R_2=OH, R_3=H$ $R_1=OCH_3, R_2=OH, R_3=H$	Umbelliferone Escopoletin	Apple vinegar, barrel during aging

Table 2 Flavonoid compounds of vinegar.

Structure	Substituents	Compound	Source
Anthocyanidin	$R_1 = R_2 = OCH_3$	Malvidin acid	Red grape, Fruit vinegars
Catechin and Proanthocyanidins	R=H R=OH	(+)-Catechin (+)-Catechin gallate	Red grape, White grape
	R=H R=OH	(−)-Epicatechin (+)-Epicatechin gallate	
Flavonol	$R_1 = OH, R_2 = H$	Quercetin	White grape

solution. By doing this she proved that she could provide a feast that would cost a fortune.

Nowadays, vinegar is included as a recipe and food-flavoring agent in human gastronomy. The quality of the vinegar is acknowledged by its place of origin in different vinegar-producing areas. The health benefits of vinegar have recently been considered in addition to its use as a seasoning agent. The antioxidant activities of phenolic components and other health benefits of vinegars are current research topics. There is sufficient evidence that supports the health benefits from consuming vinegar.[12] Roman legionnaires drank wine vinegar-based drinks for endurance on the march. In colonial times, a vinegar-based drink called "shrub" was very popular. This old favorite is becoming popular again. Vinegar bars are now springing up in Japan, where consumers eagerly down such concoctions as soy and vinegar shakes and fruit-flavored vinegars over ice.[13]

FUTURE DIRECTIONS

Recent and future research tendencies are focused on:

- Molecular approaches to evaluate the diversity of acetic acid bacteria independent of their capacity to grow in culture media.
- Modeling gas–liquid and liquid–gas transfers in vinegar production.
- Production and evaluation of healthy vinegar-based drinks.
- Isolation and selection of thermotolerant acetic acid bacteria from different sources.
- Development of acetic acid bacteria starter cultures.
- Accelerated aging of vinegar.
- Continuous production of vinegar and gas recirculation systems to increase vinegar production yields.

CONCLUSION

Vinegar was once considered an undesirable by-product of alcoholic beverages, and its price was consequently undervalued. Vinegar actually plays an important role in gastronomy and its quality has improved greatly as a result of technological development. Active research is ongoing to characterize and optimize different aspects of vinegar production. Presently, the quality of vinegar is protected in part by defining places of origin in different vinegar-producing areas. The potential to use vinegar as a cosmetic ingredient is one research topic being pursued to expand on the utility of vinegar.

REFERENCES

1. Maria, P.; Davide, B.; Francesca, M. Extraction and identification by GC-MS of phenolic acids in traditional balsamic vinegar from Modena. J. Food Comp. Anal. **2006**, *19*, 49–54.
2. Palacios, V.; Valcarcel, M.; Caro, I.; Perez, L. Chemical and biochemical transformations during the industrial process of

sherry vinegar aging. *J. Agric. Food Chem.* **2002**, *50* (15), 4221–4225.

3. Joint FAO/WHO Food Standards Programme, Commission of Codex Alimentarius. Codex standard for vinegar (Regional European standard, Codex Stan 162–1987). In *Codex Standards for Sugars, Cocoa Products and Chocolate and Miscellaneous*, 3rd Ed.; Joint FAO/WHO Food Standards Programme: Rome, Italy, 1995; Vol. 11, 116–121.
4. Cleenwerck, I.; De Vos, P. Polyphasic taxonomy of acetic acid bacteria: an overview of the currently applied methodology. Int. J. Food Microbiol. **2008**, *125*, 2–14.
5. Tesfaye, W.; García Parrilla, M.C.; Troncoso, A.M. Set up and optimization of a laboratory scale fermenter for the production of wine vinegar. J. Inst. Brew. **2000**, *106* (4), 215–219.
6. Tesfaye, W.; Morales, M.L.; García Parrilla, M.C.; Troncoso, A.M. Optimising wine vinegar production: fermentation and aging. Appl. Biotechnol. Food Sci. Pol. **2003**, *1* (2), 109–114.
7. Natera, R.; Castro, R.; Garcia-Moreno, M.D.; Hernandez, M.J.; Garcia-Barroso, C. Chemometric studies of vinegars from different raw materials and processes of production. J. Agric. Food Chem. **2003**, *51*, 3345–3351.
8. Callejón, R.M.; Tesfaye, W.; Torija, M.J.; Mas, A.; Troncoso, A.M.; Morales, M.L. Volatile compounds in red wine vinegars obtained by submerged and surface acetification in different woods. Food Chem. **2008**, *113* (4), 1252–1259.
9. Moallem, S.A.; Barahoyee, A. Evaluation of acute and chronic anti-nociceptive and anti-Inflammatory effects of apple cider vinegar. Iranian J. Pharm. Res. **2004**, *2*, 57.
10. Valero, E.; Berlanga, T.M.; Roldan, P.M.; Jimenez, C.; Garcia, I.; Mauricio, J.C. Free amino acids and volatile compounds in vinegars obtained from different types of substrate. J. Sci. Food Agric. **2005**, *85* (4), 603–608.
11. Diaz Marquina, A.; Zapata Revilla, M.A.; Lopez Tallada, M. Preliminary study of wine, sherry and cider vinegars. Alimentaria **2003**, *340*, 113–117.
12. Takahiro, S.; Shigeru, M.; Sachiko, T.; Yueqin, T.; Toru, S.; Kenji, K. Antioxidant activity of vinegar produced from distilled residues of the Japanese liquor Shochu. J. Agric. Food Chem. **2008**, *56* (10), 3785–3790.
13. Sachiko, H.; Atsushi, S.; Akira, T.; Yoshioki, S.; Yuji, N.; Keitaro, H. A red wine vinegar beverage can inhibit the renin–angiotensin system: experimental evidence *in vivo*. Biol. Pharm. Bull. **2005**, *28* (7), 1208–1210.

Viruses: Food Safety

Lee-Ann Jaykus
Department of Food, Bioprocessing, and Nutrition Sciences, North Carolina State University, Raleigh, North Carolina, U.S.A.

Doris H. D'Souza
Department of Food Science and Technology, University of Tennessee, Knoxville, Tennessee, U.S.A.

Abstract

The early and rapid detection of human enteric viruses in food is critical in the prevention and control of foodborne disease outbreaks. This short entry will address the methodologies and challenges in the concentration and detection of the epidemiologically significant human enteric viruses, Noroviruses and hepatitis A virus, in food commodities.

INTRODUCTION

From an epidemiological standpoint, the most significant foodborne viruses are the noroviruses (NoVs), because of their frequent association with outbreaks, and hepatitis A virus (HAV) because it causes a more severe disease, comparatively. Initial work on the detection of these foodborne viruses was directed at clinical specimens (stool or acute/convalescent sera) using techniques of electron microscopy, immune electron microscopy, or enzyme immunoassay (EIA). In the 1990s, the first molecular amplification methods were reported and have become the methods of choice for detecting enteric viruses in the clinical realm.[1,2] For a variety of reasons, the detection of viruses in foods differs significantly from traditional methods used for the detection of bacterial agents of foodborne disease. Similar to many bacterial pathogens, viruses are typically present at low levels in contaminated foods; however, unlike most bacterial pathogens, the foodborne viruses are much smaller, have RNA rather than DNA genomes, and do not replicate in foods. Furthermore, the naturally occurring (wild-type) foodborne viruses of epidemiological significance cannot be cultivated in vitro, lacking both animal and mammalian cell culture hosts. Consequently, traditional food microbiological techniques such as cultural enrichment and selective plating are not possible. When considering the detection of viruses in contaminated foods, one must take into account the need to separate and concentrate the viruses from the food matrix before detection, as well as the need for a sensitive and specific method to amplify the virus or its nucleic acid. Invariably, we rely on nucleic acid amplification strategies to provide such detection sensitivity. The major steps for the detection of viruses in foods consist of the following: 1) virus concentration and purification; 2) nucleic acid extraction; 3) detection; and 4) confirmation. These individual steps will be described below.

VIRUS CONCENTRATION AND PURIFICATION

The presence of low numbers of viruses in contaminated food and their low infectious doses make virus concentration and purification from the food matrix critical preceding detection. Other challenges to sample preparation include the need to process large sample volumes to assure adequate sample representation and the fact that residual food components can compromise downstream detection methods.[3–5] Accordingly, virus concentration methods are designed to achieve reduction in sample volume with recovery of virus and elimination of matrix-associated interfering substances.[3–5] Many sample manipulations depend on the behavior of viruses to act as proteins in solutions, to co-sediment by simple centrifugation when adsorbed to larger particles, and to remain infectious at extremes of pH and/or in the presence of organic solvents.

Viruses can be separated from the food matrix through the use of filtration, centrifugation, adsorption, elution, solvent extraction, precipitation, and/or organic flocculation. Polyethylene glycol (PEG) is the most common virus precipitant, with acid precipitation a viable alternative. Both of these methods capitalize on the fact that viruses behave as proteins and "fall out" of solution with the removal of water (PEG) or upon exposure to pH values approximating the virus' isoelectric point (acid precipitation).[3–5] Organic flocculation utilizes flocculating agents commonly used in water treatment; these interact with organic matter in the food to cause virus adsorption to a gelatinous "floc," which can be separated by centrifugation. Polar components of foods such as lipids can be removed using organic solvents such as chloroform, tri-chloro tri-fluoroethane (Freon®) and more environmentally friendly solvents such as Vertrel® (DuPont, New Britain, Connecticut, United States). Alternative commercial virus purification agents[4–6] as well as the cationic detergent cetyltrimethylammonium bromide (CTAB) are known to eliminate

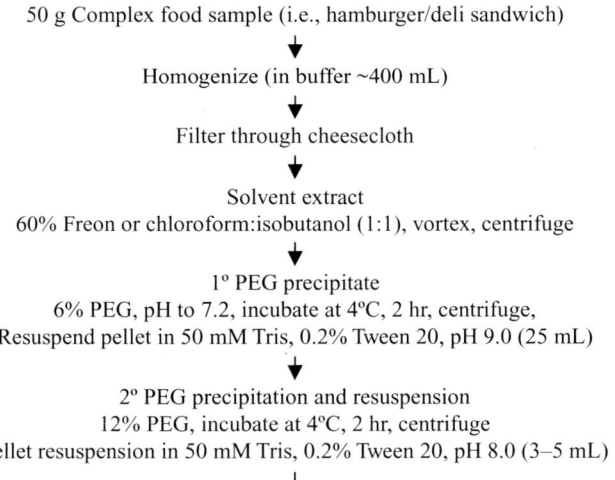

Fig. 1 Concentration of enteric viruses from complex foods. **Source:** From *Foodborne Diseases Handbook, Viruses, Parasites, Pathogens and HACCP*.[4]

polysaccharides. Sephadex®, cellulose, or Chelex® can remove salts and proteins. Ultrafiltration reduces sample volumes while simultaneously purifying the sample and can be applied as a final concentration step. Some investigators have incorporated an immunocapture step to concentrate viruses from foods.[7] Fig. 1 is an illustration of a prototype virus concentration method as applied to foods.[4] It should be noted that combined steps are frequently applied, and both the choice and sequence of virus extraction steps depends on the food product being processed.

After virus extraction, sample volume reductions from 10- to 1000-fold can be obtained and theoretically, a 25-g food sample can be reduced to 25 μL–2.5 mL in volume. The recovery of infectious virus ranges from as low as 1–2% to as high as 90%;[3–5] however, recovery efficiency is both virus and matrix-specific. For example, HAV recovery tends to be low in comparison to relatively high recovery of human enteroviruses (e.g., poliovirus);[3–5] simple sample matrices such as lettuce tend to be easier to work with than more complex matrices such as sandwiches or ready-to-eat (RTE) salads. Presently, there is no universal extraction method that can be applied to all foods, particularly when viral loads are low.

NUCLEIC ACID EXTRACTION

Since nucleic acid amplification efficiency is dependent upon both the purity and the quantity of target molecules obtained from the sample, one must extract the nucleic acid from the sample concentrate before molecular amplification. This step provides additional sample concentration and removal of residual matrix-associated inhibitory compounds. Methods based on the chaotropic agent, guanidinium thiocyanate (GuSCN) are most often used, since this agent is effective at deproteinizing and purifying nucleic acids while protecting RNA against degradation by native RNases.[4,5] Combinations of multiple extraction methods can also be used to purify nucleic acids. For example, rapid and simple methods have been reported that utilize a combination of GuSCN and silica particles, or alternatively, GuSCN followed by RNA binding to glass powder.[4,5] In both cases, RNA is released by a subsequent elution step. The metal-chelating agent Chelex-100, or alternatively, Sephadex G200 column chromatography or magnetic poly (dT) beads, have also been used for further purification of previously extracted RNA. A few investigators have attempted to bypass upstream sample preparation by applying direct nucleic acid extraction to contaminated foods without prior virus concentration and purification steps. Although easy and well suited to compositionally simple food matrices, residual amplification inhibitors frequently remain.[8] In all cases, a major issue with RNA extraction is the destruction of virion integrity, which leads to the inability to directly correlate infectivity to reverse transcription polymerase chain reaction (RT-PCR) detection limits.

DETECTION

Clinical enteric virus detection methods such as electron microscopy and EIA are not usually applied to the detection of viruses in foods because their detection limits exceed the levels expected in naturally contaminated products. Early detection methods applied to food concentrates were based on mammalian cell culture techniques, but these are cumbersome, expensive, and not available for the epidemiologically relevant foodborne viruses (i.e., NoVs and HAV). Consequently, we rely on nucleic acid amplification methods for detection. Because most of the enteric viruses have an RNA genome, the most commonly used detection method is RT-PCR. More recently, nucleic acid sequence-based amplification (NASBA), a novel RNA amplification method, also has been applied to the detection of viruses in foods.[9] Regardless of the method chosen, the sensitivity and the specificity of molecular amplification as applied to the detection of viral contamination in foods depends upon a number of factors, including the efficiency of the upstream processing methods, the purity and yield of RNA, and the choice of primers. Since even the best virus concentration and extraction methods frequently leave behind residual matrix-associated material that can result in significant nonspecific amplification, primers of low degeneracy and high melting temperature should be used. A single primer pair that will detect all the epidemiologically relevant viruses within a single genus is ideal. For HAV this is possible, and broadly reactive HAV primers that amplify sequences within the viral VP1/2A junction are usually used; however, the genetic diversity of the NoVs makes broadly reactive primer design a challenge for this virus genus. Early NoV primers were designed to amplify conserved regions in the RNA-dependent RNA polymerase gene of the virus and tended to be strain-specific.[4,5] With the increased availability of sequence data from a variety of NoV strains, more broadly reactive primers have been designed, some of which target the viral capsid region.[10] These primers tend to be somewhat degenerate, so they must be used with caution when attempting to detect viral contamination of foods, particularly if anticipated contamination levels are low.

CONFIRMATION

In most cases, the products of molecular amplification are visualized by agarose gel electrophoresis; however, since non-specific products of amplification are a consistent concern, confirmation of amplicon sequence is highly recommended. The most widely applied confirmatory tool is Southern hybridization using specific oligoprobes internal to the amplicon that can be enzyme-labeled for colorimetric, luminescent, or fluorescent detection. Liquid hybridization using an electrochemiluminescent format has also been applied for amplicon confirmation.[9] For the genetically diverse NoVs, an oligonucleotide array dot-blot format has been reported, which facilitates amplicon confirmation and strain genotyping in a single assay.[10] Other confirmation methods include "nested" reactions, restriction endonuclease digestion of RT-PCR products and direct sequencing of the amplicon. Nested reactions use a second pair of primers internal to the first amplicon sequence to perform a second round of amplification, where the product of the first amplification acts as template. Although nested reactions can improve assay sensitivity, they are prone to cross-contamination.[4,5]

Recently, real-time molecular amplification methods have come to the forefront. The term "real time" refers to the simultaneous detection and confirmation of amplicon identity as the amplification progresses.[4,5] Real-time RT-PCR uses compounds that fluoresce when associated with double-stranded DNA and exposed to suitable wavelength of light, or fluorescently labeled probes that hybridize to internal regions of the amplicon. Prototype real-time RT-PCR methods have been developed for the detection of HAV and the NoVs[11] and are currently an active area of research.

CONCLUSIONS

An important component in the prevention of foodborne viral outbreaks is early recognition. To facilitate this, improved methods with increased availability to detect viral contamination are essential. In this regard, progress has been made, but challenges remain that must be addressed in the future. For additional information, the interested reader is directed to more comprehensive reviews.[3-5,12]

ACKNOWLEDGMENTS

The use of trade names in this entry does not imply endorsement by the North Carolina Agricultural Research Service nor criticism of similar ones not mentioned.

REFERENCES

1. Sair A.I.; D'Souza, D.H.; Moe, C.L.; Jaykus, L.A. Improved detection of human enteric viruses in foods by RT-PCR. J. Virol. Methods **2002**, *100*, 57–69.
2. Sair, A.I.; D'Souza, D.H.; Jaykus, L. Human enteric viruses as causes of foodborne disease. Comp. Rev. Food Sci. Saf. **2002**, *1*, 73–89.
3. D'Souza, D.H.; Jean, J.; Jaykus, L.A. Methods for detection of viral and parasitic pathogens in food. In *Handbook of Food Science, Technology and Engineering*; Hui, Y.H., Castell-Perez, E., Cunha, L.M., Guerrero-Legarreta, I., Liang, H.H., Lo, Y.M., Marshall, D.L., Nip, W.K., Shahisi, F., Sherkat, F., Winger, R.J.; Yam, K.L., Eds.; CRC Press: Boca Raton, FL, **2006**, 188–1 to 188–23.
4. Jaykus, L. Detection of human enteric viruses in foods. In *Foodborne Diseases Handbook, Viruses, Parasites, Pathogens and HACCP*, 2nd Ed.; Sattar, S. Ed.; Marcel Dekker: New York, 2000; Vol. 2, 137–163.
5. Jaykus, L.A.; Hemard, M.T.; Sobsey, M.D. Human enteric pathogenic viruses. In *Environmental Indicators and Shellfish Safety*; Hackney, C.R., Pierson, M.D., Eds.; Chapman and Hall: New York, 1994; 92–153.
6. Leggitt, P.R.; Jaykus, L.A. Detection methods for human enteric viruses in representative foods. J. Food Prot. **2000**, *63*, 1738–1744.
7. Bidawid, S.; Farber, J.M.; Sattar, S.A. Rapid concentration and detection of hepatitis A virus from lettuce and strawberries. J. Virol. Methods **2000**, *88*, 175–185.
8. Schwab, K.J.; Neill, F.H.; Le Guyader, F.; Estes, M.K.; Atmar, R.L. Development of a reverse transcription-PCR-DNA enzyme immunoassay for detection of "Norwalk-like" viruses and hepatitis A virus in stool and shellfish. Appl. Environ. Microbiol. **2001**, *67*, 742–749.
9. Jean, J.; D'Souza, D.H.; Jaykus, L. Multiplex nucleic acid sequence-based amplification (NASBA) for the simultaneous detection of enteric viruses in ready-to-eat food. Appl. Environ. Microbiol. **2004**, *70*, 6603–6610.
10. Vinje, J.; Koopmans, M.P. Simultaneous detection and genotyping of "Norwalk-like viruses" by oligonucleotide array in a reverse line blot hybridization format. J. Clin. Microbiol. **2000**, *38*, 2595–2601.
11. Loisy, F.; Atmar, R.L.; Guillon, P.; Le Cann, P.; Pommepuy, M.; Le Guyader, F.S. Real-time RT-PCR for norovirus screening in shellfish. J. Virol. Methods **2005**, *123* (1), 1–7.
12. D'Souza, D.H. Detection of human enteric viruses in foods. J. Assoc. Food Drug Off. **2003**, *67* (4), 26–47.

Volatile Flavor Generation: Genetic Methods

Tim Coolbear
Marie-Laure Delabre
Fonterra Research Centre, Fonterra Cooperative Group Limited, Palmerston North, New Zealand

Abstract
Genetic tools present opportunities to alter the types and relative levels of the wide range of volatile flavors that are produced enzymatically (and chemically) in food. This entry presents a brief overview of the genetic technologies applicable to volatile flavour profile manipulation and notes regulatory, labeling and consumer perception issues.

INTRODUCTION

Flavor is perceived through orthonasal aroma, retronasal odor, taste, and trigeminal perception. Of these, volatile compounds impact on aroma and odor—the taste sensations of sweet, sour, salt, bitter, and umami are generally not given by volatile compounds. While the term "flavor," in the context of this entry on volatile flavors, is limited to olfactory rather than gustatory perception, the many interactions among the many volatile and non-volatile compounds mean that flavor combinations are not perceived as a simple summing of the individual components. In addition, there is wide variation in the relative rates of release of flavor compounds from different foods. The generation of flavors themselves is, therefore, only a part of the whole picture; it is how the consumer perceives flavors that is important, and this is dependent not only on what is present in a food but also on the matrix in which the flavor compounds are contained (and how that matrix is presented). As will become clear, consumer perception of flavors is also relevant to the origin of those flavors, and this has a strong impact on the application of genetic methods for the generation of volatile flavor compounds.

Volatile flavors intrinsic to food are essentially derived from biological and chemical modification of proteins, lipids, sugars, and organic acids; however, it is the biological conversion of these substrates to flavor compounds that is, by far, the dominating factor, and it is the biochemical understanding of the enzyme pathways involved that underpins the ability to manipulate volatile flavor production using biotechnological methods.[1,3] Such methods include both genetic and non-genetic techniques, and the former do not necessarily require genetic engineering. However, while genetic techniques can be applied to living organisms to influence the generation of volatile flavors in terms of the breadth of compounds produced and their relative and absolute levels, the science hurdles that this approach presents are possibly lower than the hurdles of consumer perception and acceptance.

Genetic technology has expanded tremendously over the past two decades, and improvements to the techniques and development of bioinformatic tools have contributed to make genetic methods affordable, wide-ranging, and safe. In this entry, we will cover: 1) selective breeding—crossing individuals that each have separate desired traits to produce new varieties having a mixture of these traits, and random mutagenesis of an organism to select for a specific activity; 2) genomics—the study of an organism's genomic sequence to gain understanding of existing metabolic pathways to generate flavors; and 3) genetic engineering—the manipulation of an organism's genes to direct pathways in a required direction. In doing so, we will comment on how these techniques impact on the use of growing and non-growing cells in flavor bioconversions, and also on the use of these cells as factories for the production of enzymes for use in flavor generation. We will also comment briefly on regulatory, labeling, and consumer perception issues.

GENETIC METHODS

Selective Breeding and Random Mutagenesis

Selective breeding is not just a common practice with animals (whether livestock or companion animals) or plants (whether cropping or ornamentals), it is also the basis of the growth and development of the respective industries. In all cases, selective breeding is for enhancement or diminishment of a trait or set of traits to improve the organism for economic reasons (e.g., production volume and conversion efficiency) or for aesthetic value (e.g., shape, fragrance, and vibrancy). As a consequence, the genetic diversity gained is different from that would have been found in nature and artificially protected from the adverse pressures that nature would impose. The robustness of many breeds developed in all spheres is consequently limited, although obviously increased robustness, e.g., against disease or drought, is also the end point in some selective breeding. How often such breeding has to be undertaken to reverse the diminishment of aroma that has been an unintended consequence of selective breeding for shape, size, color, and disease resistance is a moot point.

While selective breeding has been done for thousands of years without scientific understanding, a scientific basis for improving and enabling the specific targeting of required traits is now emerging. In the fruit industry, selective breeding has been applied to improve fruit aroma, which is de facto an effect on the amounts and relative proportions of volatile flavor compounds. Until relatively recently, plants had to be taken to maturity to determine the ultimate outcome of the breeding. Genetic technologies can now be used in some cases to determine which progeny possess the required genetic information and which do not. This can be used to selected out unusable progeny and so reduce the number of plants that have to be taken through to maturity.

In the microbial world, the speed at which selective breeding can be accomplished is accelerated manyfold by the speed of growth of the organism and by the technique of random mutagenesis. In this technique, mutations in a microbial genome are introduced in a random manner by exposing the organism to ultraviolet light or chemical mutagens. Thus, many different mutations are induced in a single population of cells out of which the required mutated trait can be rapidly selected by challenging the population with, for instance, a metabolic hurdle that only a particular mutation can overcome. Alternatively, high-throughput screening can be used to screen for mutants with the desired traits where an appropriate assay for these traits exists.

Genomics

The advances in genomic tools are quite remarkable and reflect the rate at which capabilities can be developed given the right impetus and appropriate levels of funding. New high-throughput sequencing technologies have considerably lowered the cost of DNA sequencing, and the sequence of thousands of complete microbial and plant-annotated genomes and billions of nucleotide sequences are publicly available. This expedites the elucidation of the pathways of flavor compound formation and the regulatory mechanisms that govern their expression. Comparison of the genomic sequences of related organisms is a powerful tool to enable the understanding of the pathways and could be a very useful tool for describing and predicting the performance of mixed cultures in microbial flavor fermentations.

By screening in silico genomic sequences, new enzymes can be identified and these enzymes may have specificities, kinetic profiles, and stabilities different from those already known, providing opportunities to further manipulate the end points of flavor pathways. In addition, the protein sequences derived from the genomic information can be subjected to computational models to provide structural information. This process has the potential to enable selection of mutations required to modify the protein sequence of an enzyme to fit an activity profile required for a particular flavor pathway.

Genetic Engineering

Genetic engineering is based on recombinant DNA technologies using molecular cloning to alter directly the structure of genes by deletions or insertions. The techniques can be applied to various end points, such as enzyme engineering, molecular evolution, microbial fermentation, and transgenic plant generation.

Although there are, or have been, considerable technical difficulties to overcome in all genetic engineering, at least some microbial systems present relatively lower hurdles than other systems. Furthermore, the production of flavor compounds presents a complexity of biochemical and chemical reactions, compounded by a complexity of interrelated metabolic pathways. Overall, the greatest amount of knowledge that has been gained with respect to the complexity of pathways that produce volatile flavor compounds has pertained to cheese manufacture and ripening.[2,4,5] The organisms that have been studied are mainly starter and non-starter lactic acid bacteria; the substrates that they convert are proteins (caseins—the whey proteins are essentially removed in the formation of the curds), lipid (milkfat), and carbohydrate (lactose). Therefore, this system provides an example that is broadly indicative of the way in which genetic engineering technologies can be applied to volatile flavor pathways; however, it should be pointed out that the opportunities for actual application of such genetic engineering, at least as is reflected by literature in the public domain, are limited. This situation is essentially due to public concerns over genetic manipulation technologies. These concerns are reflected in legislation that constrains food applications of the technology. Because of this situation, commercial opportunities for genetically engineered products are limited and food companies may well be expected to monitor consumer attitudes, legislation, and scientific progress that is rate-limited by available research funding.

In cheese-ripening, proteolysis ultimately gives rise to amino acids that are the precursors of many volatile flavor compounds; there are various enzymes that initiate these subsequent conversions. Enhancement of transaminase levels in starter bacteria, through overproduction of the organism's own enzymes, knockout mutants, or insertion of a gene encoding for a complementary enzyme from a different bacterium, showed that transaminases are key enzymes, provided they are not rate-limited by the levels of α-keto acids.

This caveat illustrates an important point for genetic engineering of flavor pathways; it is not just the levels of the various enzymes that determine the success of the engineering, it is also the availability of substrates (and the removal of inhibitory products). With interspecies constructs, if the host organism does not have the biosynthetic capacity to provide the substrates for introduced enzymes, then these will have to be provided exogenously, which has ramifications for process design, process costs, and product labeling.

The potential to more specifically target the production of certain volatile flavor compounds using genetic engineering can be illustrated by looking at the pathways for the volatile sulfur compounds important in cheese-flavor profiles. Transaminase overproduction may lead to a range of compounds owing to the breadth of specificity of the enzymes, whereas overproduction of methionine lyase, which has a narrow substrate range, would specifically favor production of volatile sulfur compounds such as dimethyldisulfide and dimethyltrisulfide.

It has been shown that decarboxylase is pivotal for the conversion of branched-chain amino acids to aldehydes important in cheese flavor. In dairy starter lactococci, the functional gene encoding decarboxylase activity has been lost as a consequence of evolution in the dairy environment. In contrast, non-dairy lactococci have been shown to have decarboxylase activity. Comparisons between these two groups have indicated that the gene in dairy starter strains is truncated and inactive. This example illustrates the opportunities that gene technologies present to determine the root cause of a block in a flavor pathway and the necessary genetic information that has to be inserted to overcome the block and open up new opportunities for flavor compound production.

Recently, new techniques have been developed that allow cloning from a non-culturable organism. This protocol opens the door to exploitation of a previously inaccessible pool of genetic diversity that could provide unique opportunities in flavor compound production.

IMPACT OF REGULATORY ISSUES, LABELING, AND CONSUMER PERCEPTION

Neither selected breeding nor random mutagenesis is considered to result in a genetically engineered organism because there has been no cloning or manipulation of mutated genes; however, as commented above, biochemical pathways dominate flavor production, and genomics and genetic engineering can be used, respectively, to identify key pathway enzymes and then modify their levels and activities in a specifically targeted manner. Further, since enzymes are usually present in small amounts in an organism, genetic methods such as gene cloning and heterologous expression allow considerably larger quantities of pure enzyme to be obtained for fermentative production of compounds without the occurrence of contaminant activities. Such approaches present considerable advantages over selected breeding.

The use of genetic engineering has provoked significant controversies in many areas. Some groups see the generation and use of genetically modified organisms and their products as fundamentally intolerable meddling with biological states or processes that have been created or have naturally evolved over long periods of time (depending on beliefs). Others are concerned about the limitations of modern science to comprehend fully all of the potential ramifications of genetic manipulation. The safety of genetically modified organisms and derived compounds is subject to extensive questioning and debate.

Acceptance of any new technology is generally based on the perceived benefits. It is clear that for genetic engineering to be accepted in foods, the benefits of the products produced through the technologies must be made clear for favorable consumer response. A lack of benefits or confusion as to what the benefits are will stymie acceptance.

It is within the complex contexts of public knowledge, perception, beliefs, and scientific advancement that legislation for the use, release, and labeling of genetically modified processes and products is formulated. Different countries take different approaches, definitions differ of what is and is not a genetically modified organism, and the techniques of genetic engineering and modification are rapidly developing and changing. All these factors will make progress on the introduction of genetic technologies into the flavor arena, slow indeed. Whether gene technologies will move from fear to dependence in a manner reflecting the introduction of electricity or computers may be known in a few decades!

BIBLIOGRAPHY

1. Berger, R.G.; Krings, U.; Zorn, H. Biotechnological flavour generation. In *Food Flavour Technology*; Taylor, A.J., Ed.; Sheffield Academic Press: Sheffield, UK, 2002; 60–104.
2. Pastink, M.I.; Sieuwerts, S.; de Bok, F.A.M.; Janssen, P.W.M.; Teusink, B.; van Hylckama Vlieg, J.E.T.; Hugenholtz, J. Genomics and high-throughput screening approaches for optimal flavour production in dairy fermentation. Int. Dairy J. **2008**, *18*, 781–789.
3. Smit, G. Engineering of important flavour compounds. In *Food Fermentation*; Nout, R.M.J., de Vos, W.M., Zwietering, M.H., Eds.; Wageningen Academic Publishers: Wageningen, The Netherlands, 2005; 103–113.
4. Weimer, B.C. Genomics and cheese flavor. In *Improving the Flavour of Cheese*; Weimer, B.C., Ed.; Woodhead Publishing Limited: Cambridge, UK, 2007; 219–235.
5. Yvon, M. Key enzymes for flavour formation by lactic acid bacteria. Aust. J. Dairy Technol. **2006**, *61*, 88–96.

Weed Science

Patrick J. Tranel
University of Illinois, Urbana, Illinois, U.S.A.

Federico Trucco
Department of Crop Sciences, University of Illinois at Urbana-Champaign, Urbana, Illinois, U.S.A.

Abstract
Weed science is a very practical discipline that has the general goal of improving weed management. To achieve this goal, however, weed scientists historically have relied on many basic scientific disciplines, such as plant physiology, biochemistry, chemistry, and genetics. In more recent years, the tools afforded by molecular biology techniques also have been brought to weed science. In fact, among the first and most widely adopted outcomes of biotechnology were weed science products, namely herbicide-resistant crops. Molecular biology tools are also being widely used in weed science to investigate fundamental questions regarding the biology, ecology, and evolution of weeds. The application of molecular biology tools to weed science is described and illustrated with several examples.

RECENT EVENTS IN THE HISTORY OF WEED SCIENCE

Although humans have contended with weeds for millennia, rapid growth of weed science as a discipline did not occur until the middle of the 20th century, after the discovery of the first synthetic herbicides. Throughout much of the latter half of the 1900s, weed science was focused on herbicide discovery and herbicide physiology. Weed scientists were tremendously successful, increasing the efficacy and range of new herbicidal chemistries and thereby simplifying and improving weed management. As a result, soil conservation practices expanded and yields improved. Herbicides also aided basic biological research. For example, much of what we now know about the shikimic acid pathway (which is disrupted by the herbicide glyphosate) and photosynthetic electron transport (disrupted by triazine and other herbicides) was discovered by weed scientists investigating herbicide phytotoxicity.

Herbicide-Resistant Weeds

A repercussion of the wide adoption of herbicides for weed management was the evolution of herbicide-resistant weed populations. This phenomenon continues to be an area of intense weed science research, and is greatly aided by molecular biology. Illustrative of this point are investigations of resistance to herbicides that inhibit the enzyme, acetolactate synthase (ALS).

The ALS enzyme is necessary for the production of certain amino acids, so inhibition of this enzyme leads to plant death. Numerous herbicides, most of which belong to either the sulfonylurea or imidazolinone chemical group, target this enzyme and have been widely used since the early 1980s. Although these herbicides have been very effective, populations of more than 70 weed species have evolved resistance to these herbicides.[1] In most cases in which it has been investigated, resistance was determined to be due to an altered target site. More specifically, mutations in the gene encoding ALS result in the production of herbicide-insensitive versions of the ALS enzyme. Multiple mutations have now been identified from resistant weed populations and are being catalogued.[2] Identification of these mutations and the corresponding patterns of resistance to the various ALS-inhibiting herbicides has greatly improved our understanding of how these herbicides interact with their target site.

Herbicide-Resistant Crops

In recent years, application of molecular biology research methods to herbicide physiology and herbicide resistance in weeds has furthered our understanding in these areas. Taking this idea a step further, our understanding of herbicide phytotoxicity at the molecular level has enabled the most recent revolution in weed science: the development and commercialization of herbicide-resistant crops.

Herbicide-resistant crops are perhaps the most consequential outcome of molecular biology applied to weed science. Among these, glyphosate resistance technology stands as the best example. Glyphosate's ability to control a broad spectrum of weeds, its low toxicity to humans and other nontarget organisms, and its limited environmental persistence have made it a good candidate for efforts at engineering herbicide-resistant crops.

Glyphosate targets the enzyme 5-enolpyruvylshikimate-3-phosphate synthase (EPSPS). First attempts to engineer glyphosate resistance involved overexpression of the gene encoding EPSPS and expression of mutated forms of the gene encoding glyphosate-insensitive variants of the enzyme.[3] Although marginally successful, such attempts did not result in commercially acceptable levels of glyphosate resistance. The screening of microorganisms resulted in identification of a second type of EPSPS enzyme (Class II EPSPS) that was highly resistant to glyphosate. A Class II EPSPS gene was cloned from *Agrobacterium* spp. strain CP4 and expressed in crop plants. Success in using the CP4 EPSPS gene led to the commercialization of glyphosate-resistant soybean in the United States in 1995, followed by canola and cotton in 1997 and corn in 1998.[4] Since 1996, the worldwide adoption of glyphosate-resistant crops has been immense (Fig. 1).

Transgenic approaches also have been used to obtain other commercial forms of herbicide-resistant crops. Additionally, mutagenesis has been used with success in non-transgenic attempts to obtain herbicide resistance. In this approach, resistant variants are selected from a population of the crop of interest. To improve the chance for success, genetic variation in the population is increased by using one of several techniques to induce mutations in the plants.

Regardless of the specific technique used, the development of herbicide-resistant crops has increased options for weed management, and several such crops are now commercially available (Table 1). Although often criticized by environmental groups, such crops—if used wisely—offer great promise for reducing detrimental impacts of weeds and for developing more sustainable weed management systems.

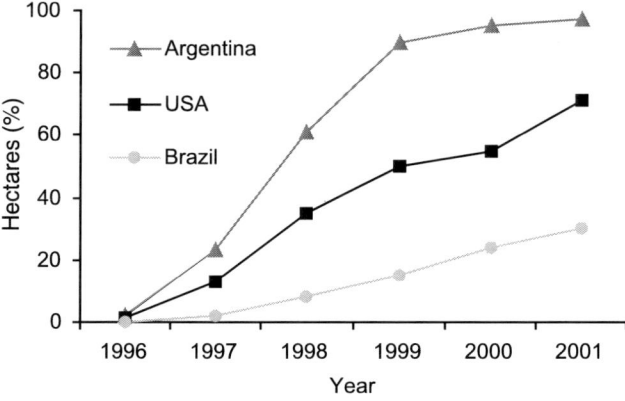

Fig. 1 Glyphosate-resistant soybean hectares as a percentage of total soybean hectares. The United States, Argentina, and Brazil produce 90% of the world's soybean exports. Note the high adoption in Brazil, despite the fact that glyphosate-resistant soybean is not legal in that country. **Source:** From http://soystats.com, accessed March 2003.

Table 1 Partial list of herbicide-resistant crops obtained using molecular biology tools.

Herbicide	Resistance mechanism	Crops
Bromoxynil	Herbicide detoxification	Cotton
Glufosinate	Herbicide detoxification	Corn, canola
Glyphosate	Altered herbicide target site	Cotton, corn, canola, soybean
Imidazolinones/ sulfonylureas	Altered herbicide target site	Corn, canola, sunflower, wheat
Sethoxydim	Altered herbicide target site	Corn

MOLECULAR BIOLOGY APPLIED TO THE BIOLOGY, ECOLOGY, AND GENETICS OF WEEDS

Weed scientists have often wondered what the underlying mechanisms are that contribute to the success of weeds. Tools of molecular biology offer novel ways of approaching this question.

Tools to Study Genetic Diversity

It is often hypothesized that large genetic diversity in weed species allows them to survive and adapt to changing agroecosystems. In the past, phenotypic markers were used to measure such diversity. However, because such markers may be influenced by environmental conditions, they do not always reflect genetic diversity. Molecular markers are a significant improvement for the study of genetic diversity.[5]

The first widely used molecular marker system used isozymes. Different versions of a particular enzyme can be separated by electrophoresis, due to different migration rates, and thus can provide a scoreable genetic marker. The discovery of restriction endonucleases (enzymes that cleave DNA) led to development of DNA markers based on restriction fragment length polymorphisms (RFLPs). The RFLP system detects genetic diversity based on the presence or absence of specific nucleotide sequences that are recognized by the restriction endonucleases.

Development of polymerase chain reaction (PCR) technology led to the improvement of DNA-based molecular markers into finer-scale genetic assessment techniques, such as amplified fragment length polymorphism (AFLP) and random amplified polymorphic DNA (RAPD). In general, AFLPs are based on the same principles as RFLPs, but with enhanced resolution provided by PCR amplification. RAPD markers rely on random amplification of DNA by means of short, arbitrary primers. Other variants of these PCR-based techniques have been used to target extremely diverse regions of genomes constituted by tandemly repeated sequences. Genetic diversity can be examined at the finest possible level by comparing nucleotide sequence information (i.e., by direct DNA sequencing).

Application of Genetic Diversity Tools: Studying Gene Flow

Studies that explore the mechanisms by which genetic diversity is acquired gained new force with the use of molecular biology tools. In particular, molecular biology has been invaluable in research examining gene flow among populations of the same species and between different species. Interspecific gene flow has been studied between some crops and their weedy relatives and among weed species themselves.[6] In most cases, gene flow has been studied in the context of herbicide resistance transmission. Hybrid intermediates between herbicide-susceptible and herbicide-resistant populations have been detected using a variety of molecular techniques. Such studies have made it possible to understand the significance of gene flow in the evolution of weed species. Furthermore, such studies are instrumental in predicting the risk that a herbicide-resistant crop may outcross with a weedy relative.

Gene Expression Profiling

Molecular biology can greatly facilitate understanding of how weeds respond to their environment, whether it be growing in competition with crops, surviving herbicide treatment, or initiating flowering after an early frost. In particular, a variety of techniques that can be loosely grouped in a category called gene expression profiling are ideally suited for such studies. The aim of gene expression profiling is to determine what genes are "turned on" or "turned off" in response to a particular treatment or environment. Results from such studies could provide insight at the molecular level as to how a weed responds to a particular treatment.

Modern gene expression profiling techniques (e.g., DNA microarrays) allow monitoring of thousands of genes simultaneously.[7] Such so-called "genomics" analyses have been adopted recently with model organisms—including plants such as Arabidopsis and several crop species—but have seen scant use in weed science research. As this technology becomes more commonplace, it likely will be applied to several aspects of weed science. Basic questions regarding biological phenomena important to weediness, such as induction of flowering, seed dormancy, and control of vegetative reproduction could be readily addressed with genomic techniques.

CONCLUSION

As illustrated by the foregoing examples, molecular biology research techniques provide many opportunities for advancements in weed science. In the future, molecular approaches will continue to provide new tools for managing weeds and will provide new information about the weeds themselves.

REFERENCES

1. http://www.weedscience.org/in.asp (accessed March 2003).
2. http://www.weedscience.org/mutations/MutDisplay.aspx (accessed March 2003).
3. Bradshaw, L.D.; Padgette, S.R.; Kimball, S.L.; Wells, B.H. Perspectives on glyphosate resistance. Weed Technol. **1997**, *11* (1), 189–198.
4. http://www.monsanto.com (accessed March 2001).
5. Jasieniuk, M.; Maxwell, B.D. Plant diversity: New insights from molecular biology and genomics technologies. Weed Sci. **2001**, *49* (2), 257–265.
6. Darmency, H. Movement of Resistance Genes Among Plants. In *Molecular Genetics and Evolution of Pesticide Resistance;* Brown, T.M., Ed.; American Chemical Society: Washington, DC, **1996**; 209–220.
7. Shu-Hsing, W.; Ramonell, K.; Gollub, J.; Somerville, S. Plant gene expression profiling with DNA microarrays. Plant Physiol. Biochem. **2001**, *39* (11), 917–926.

Wine Fermentation

Christopher Curtin
Australian Wine Research Institute, Adelaide, South Australia, Australia

Paul Chambers
Australian Wine Research Institute, Adelaide, South Australia, Australia

Sakkie Pretorius
Department of Plant Science, University of the Free State, Bloomfontein, South Africa

Abstract

Winemaking, as the oldest form and use of biotechnology, is imbued with deep tradition; however, globalization of the wine industry has necessitated modernization of this ancient craft, and this has largely been achieved by the application of science and technology. An ancient cultural practice has thus been brought into the modern era. Biotechnology has contributed to this modernization by providing winemakers with, among other things, knowledge of how to improve the reliability of fermentations, novel fermentation starter cultures, and processing enzymes. Recent advances in the biological sciences have the potential to dramatically accelerate this modernization, enabling biotechnologists to generate tailored starter cultures for production of wines targeted to specific markets. This entry explores the potential of 21st century biotechnology to accelerate development of new tools for winemakers at a time when the need for innovation driven by global competition is high.

INTRODUCTION

Louis Pasteur once said that "*A bottle of wine contains more philosophy than all the books in the world.*" In the era of wine industry globalization, the more traditional winemaking philosophies have been and will continue to be challenged, as scientific innovations are embraced by consumer-oriented producers. Increased efficiency and enhanced wine quality beckons when the complex system of wine fermentation is simplified through biotechnological improvements such as fit-for-purpose starter cultures. Genetically modified (GM) starter cultures are yet to gain widespread acceptance, but have paved the way for new generation biological disciplines known as *systems biology* and *synthetic biology* to revolutionize industrial starter culture development. Systems biology promises a "complete" understanding of cells in all their complexity, while synthetic biology provides the fundamental building blocks to reprogram cellular functions. In the future, this approach will enable strain development with unmatched precision and speed, and facilitate the tailoring of yeast metabolism to meet the ever-increasing demands and preferences of winemakers and consumers.

WINE FERMENTATION: A BLEND OF ART AND INNOVATION

Transformation of grapes into wine involves the same basic processes today that have been applied by humans for the past 7000 years (Fig. 1). The business of wine production, on the contrary, has changed profoundly, and the driver for these changes is encapsulated in one strategic goal of the globalized wine industry—*meeting the consumer challenge*. Increasingly discerning consumers have high expectations regarding wine quality at all price points, a pressure that has seen winemaking morph from a cottage industry to large-scale production where scientific innovation underpins the winemaker's art. As a consequence, 21st century winemakers have, at their disposal, an array of traditional and technological tools with which to shape a wine. Choosing "what," "when," and "how" to apply these is dictated by winemaking philosophy and market requirements.

Wine fermentation is a complex system with two highly variable biological factors: 1) grape juices that are prone to seasonal and regional influences; and 2) microbiota composition. The fact that endlessly diverse wines can be produced from the same variety of grape in different regions or vintages is part of its allure. The challenge for modern winemakers is to balance the desire for diversity and regional expression with the need for reliable fermentation and a predictable final product that consumers will trust and enjoy. To this end, modern wine fermentations are typically initiated by the addition of commercially available, fit-for-purpose starter cultures. This is in contrast to traditional wine fermentations that occur "spontaneously" through the action of indigenous yeast and bacteria arising from the grapes and/or winery. Such fermentations

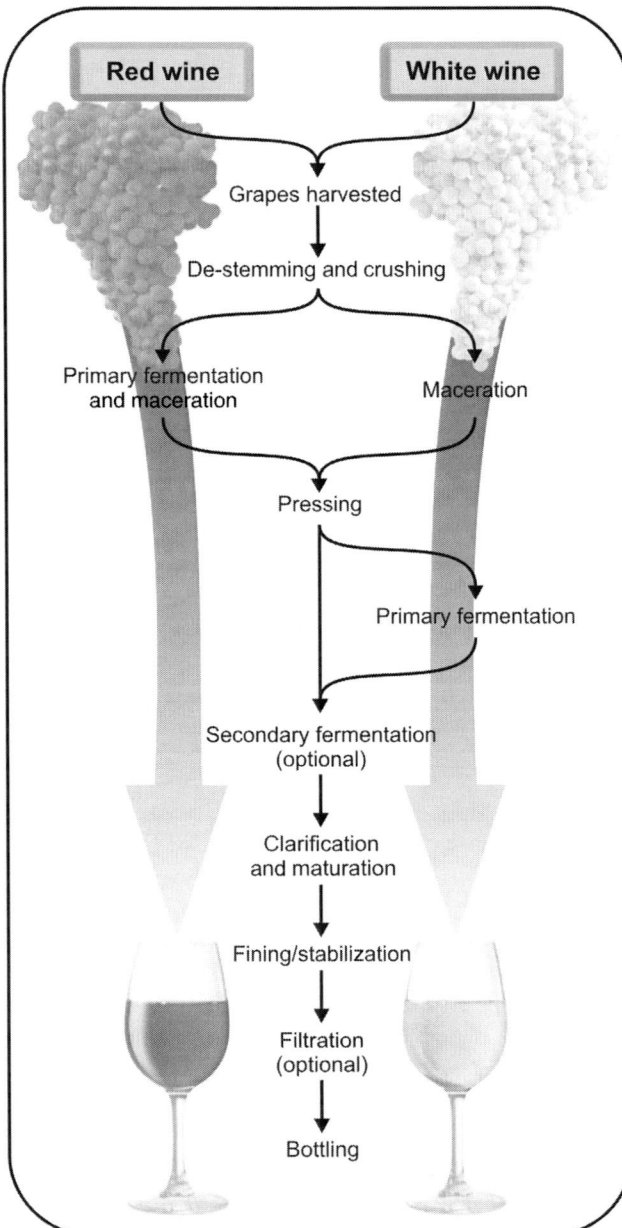

Fig. 1 Processes involved in the production of wine. Processing of red and white wine follows the same basic principles with the key difference being that red wine primary (or alcoholic) fermentations are conducted "on skins." Secondary fermentation to convert malic acid into lactic acid (malolactic fermentation) may be conducted in both red and white wines depending upon wine style. Some processing steps such as filtration before bottling may not be practiced in wineries where traditional approaches are embraced.
Source: From *Tailoring wine yeast for the new millennium: novel approaches to the ancient art of winemaking.*[1]

involve yeast from a range of genera including *Candida*, *Hanseniaspora* (*Kloeckera*), *Kluyveromyces*, *Pichia*, *Saccharomyces*, and *Torulaspora*[1] and tend to progress at a slower rate than inoculated fermentations. Inoculation of wine fermentations with selected single strains of *Saccharomyces cerevisiae* was first performed in 1890, and various strains have been available as commercial starter cultures in "Active Dry Wine Yeast" form since 1965.

In addition to alcoholic fermentation, production of many wines involves a secondary, malolactic fermentation in which the malic acid is converted to lactic acid. Indigenous grape or winery associated lactic acid bacteria from the *Lactobacillus*, *Leuconostoc*, *Oenococcus*, and *Pediococcus* genera may be involved in spontaneous malolactic fermentation; however, starter culture technology for this process has centered on the species, *Oenococcus oeni*, which is favored owing to its robust performance in wine and the desirable aroma and flavor attributes these bacteria can impart.[2]

Despite technological improvements that bring a level of control to winemaking, the outcomes of alcoholic and malolactic fermentations continue to be difficult to predict, both in terms of production efficiency and product quality. Winemakers require greater control over wine fermentation to meet the consumer challenge of the 21st century (see Table 1 for list of targets for starter culture improvement) and this represents a flow-on challenge to biotechnologists—develop innovative starter cultures that will enable production of wine according to market specifications.

DEVELOPMENT OF FIT-FOR-PURPOSE STARTER CULTURES

Biotechnology, in its broadest sense, has developed hand-in-hand with applications of the yeast species, *S. cerevisiae*. This amazingly versatile organism enables us to produce a vast range of alcoholic beverages, raise dough, degrade waste materials into biofuels, study disease, and produce pharmaceuticals. Unsurprisingly, individual strains of this species are more or less appropriate for each of these purposes—just as an Olympic sprinter is more suited to run 100 m than an Olympic swimmer, an *S. cerevisiae* strain suited to baking is not going to perform well in wine fermentation. The concept of applying fit-for-purpose starter cultures is therefore rather obvious, and there is evidence of this process occurring empirically over time during human domestication of *S. cerevisiae*.[3] The ability of researchers in the postgenomic era to evaluate and catalog natural genetic variation among wine yeasts means that "natural" polymorphisms can be readily linked to target properties. This catalog enhances traditional approaches to starter culture development by enabling researchers to preselect highly divergent strains for breeding or to target specific genes and pathways for screening of intentionally mutated populations.

The advent of modern biotechnology in the 20th century and subsequent utilization of *S. cerevisiae* as one of

Table 1 Targets for improvement of wine yeast and bacterial starter cultures.

Driven by consumer preferences	Driven by production efficiency
Sensory attributes	*Processing efficiency*
Production of acetate esters[Y]	Improved nitrogen assimilation efficiency[Y]
Liberation of grape-derived flavor precursors[YB]	Fructose utilization[Y]
Decreased sulphide production[YB]	Minimal foam formation[Y]
Decreased acetic acid production[YB]	Sulphite tolerance[YB]
Enhanced mouthfeel[YB]	Efficient malic acid conversion[YB]
Enhanced wine color[YB]	Microorganism compatibility[YB]
Wholesomeness	*Fermentation performance*
Decreased ethanol production[Y]	Rapid initiation of alcoholic fermentation[Y]
Decreased sulphite formation[Y]	Rapid initiation of malolactic fermentation[B]
Minimal biogenic amine formation[YB]	Ethanol tolerance[YB]
Minimal ethyl carbamate potential[Y]	Osmotolerance[YB]
Production of antioxidants[YB]	Low temperature optimum[YB]

[Y] Applicable to yeast starter cultures.
[B] Applicable to bacterial starter cultures.
Source: From *Tailoring wine yeast for the new millennium: novel approaches to the ancient art of winemaking.*

the key model eukaryotic organisms for studies in molecular biology have meant that starter culture developers now have access to knowledge and readily available techniques for strain improvement. GM approaches to starter culture development are appealing in terms of the potential for novelty, compressed timeframes for development, and specificity of alterations—preserving other desired attributes. Two GM wine yeast strains are commercially available,[4,5] exhibiting highly novel traits such as the ability to degrade malic acid—eliminating the need for a secondary malolactic fermentation. Both have GRAS (generally recognized as safe) status in the United States, although the use of GM strains in commercial wine production is effectively banned in many countries. The advent of protocols for integration of novel genes into chromosomal DNA of yeast strains, followed by the removal of all exogenous DNA (i.e., plasmid material containing selectable antibiotic resistance markers), has moved this technology closer to acceptance. Indeed, self-cloned yeast strains that contain only DNA from *S. cerevisiae* have been developed for brewing applications.[6] Interestingly, it is acceptable to most people to inject a GM pharmaceutical into the body (e.g., hepatitis B vaccine or insulin), yet the thought of consuming a food or beverage where GM technology has been used appears to be cause for concern.

Other barriers to uptake of GM wine yeast strains include the perceived risk of environmental dispersal, marketing and trade restrictions, safety, and cultural and traditional beliefs. These factors are comprehensively discussed by Vivier and Pretorius,[7] while a framework for assessing risk of environmental release has been proposed recently.[8] The concerns that surround GM technology should not be discounted, but dealt with through comprehensive research and communication of outcomes.

THE APPLICATION OF ENZYMES IN WINE PRODUCTION

As an adjunct to the properties of starter cultures, approved enzymes can be added directly to facilitate the processing of grape juice and have potential as processing aids throughout the winemaking process. Commonly used classes of enzyme include proteases, polyphenol oxidases, and pectinases.[9] Other enzymes, such as β-glucosidases, could be used to release bound aroma-compound precursors.

Impurities, such as cinnayml-esterase in pectinase preparations, have the potential to cause loss of wine quality; therefore, recombinant production should be considered. Furthermore, novel enzymes of interest to the wine industry, such as the *Escherichia coli* tryptophanase enzyme that is capable of releasing compounds imparting "passion-fruit" aroma,[10] are not stable at low pH. Directed evolution of recombinant enzymes[11] may facilitate development of a broad suite of pH stable enzymes for use in grape juice and wine.

NEW GENERATION BIOLOGICAL TECHNOLOGIES: WINE AND SYSTEMS BIOLOGY

Science revolves around forming a simple, answerable hypothesis and experimentally proving or disproving it. Typically, a large system is broken down into smaller components to facilitate the development of an answerable hypothesis. In recent years, it has become feasible to expand the scope of biological science through whole-system studies, built on platforms that enable researchers to observe what is occurring inside the cell at different layers of organization. Rather than studying one gene, it is possible to

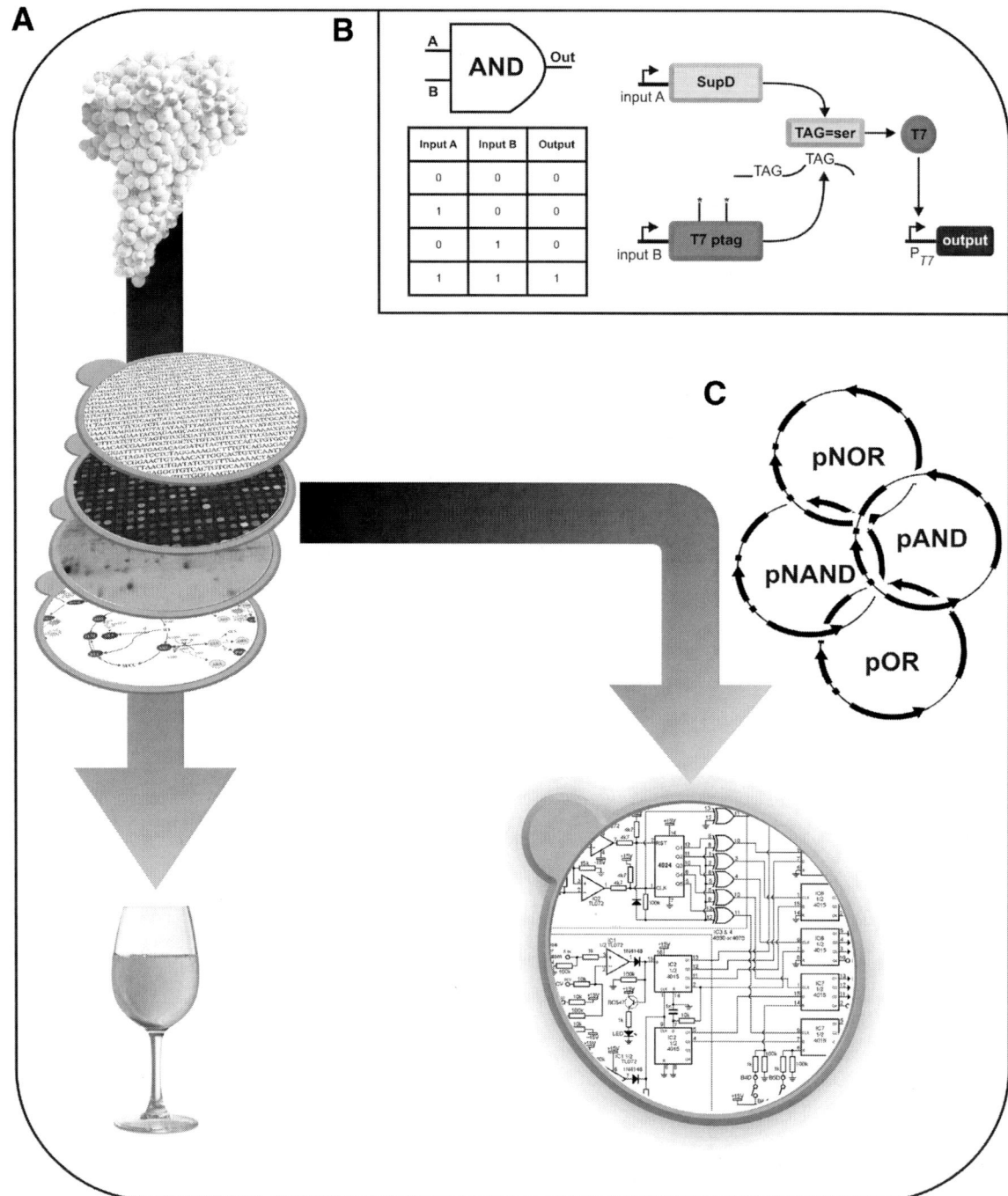

Fig. 2 New generation biological technologies: systems biology and synthetic biology. (A) The wine fermentation system from "grape to glass." Technologies facilitating analysis of the yeast genome, transcriptome, proteome, and metabolome are being integrated into "whole system" studies within the wine fermentation context. In this way, it will be possible to mathematically model the impact of alterations to the genome upon other levels of cellular organization and ultimately upon wine quality. (B) Development of modular regulatory networks for use in synthetic biology. A modular "AND" gate was constructed for use in *Escherichia coli*, whereby activation of two input promoters is required to drive "output." **Source:** From J. R. Soc. Interface.[13] Functional T7 polymerase is translated only when the SupD tRNA is expressed, through its suppression of early stop codons in the T7ptag sequence. The T7 polymerase then activates expression of the "output" gene. (C) Implementation of synthetic biology to reprogram yeast metabolism. Modular, reusable logic gates constructed in "core" plasmids will, in the future, be used to fabricate novel molecular circuits inside the yeast cell. Programming of logic gates with promoters responsive to the physiological changes that occur in yeast during fermentation will enable fine-tuning of both the timing and the magnitude of "output" gene expression. This represents a quantum leap from standard recombinant DNA approaches where the "output" gene is driven by a constitutive (always on) promoter. Guided by systems biology modeling, synthetic biology has the capacity to deliver "tailored" wine fermentation starter cultures.

study the entire DNA nucleotide sequence of an organism containing all genes (the *genome*); rather than investigating how one transcript is regulated, all transcripts (the *transcriptome*) can be monitored simultaneously. Similarly, the entire protein complement (the *proteome*) and the entire complement of small chemicals (excluding nucleic acids and proteins) and metabolites (the *metabolome*) present in the cell can be analyzed as complete sets. The mathematical, statistical, and computing methods that support the unraveling and interpretation of genomic, transcriptomic, proteomic, and metabolomic data sets are referred to as bioinformatics.

Without question, the so-called "*omics*" platforms underpin the future of biotechnology, particularly for applications such as industrial yeast strain development. Comparative genomics is already beginning to transform our understanding of what makes a wine-yeast tick.[12] Bioinformatic integration of data from multiple "omics" platforms takes us into an even more holistic way of understanding biological systems such as wine fermentation, in a cutting edge scientific discipline known as *systems biology*. So how can systems biology approaches change the way biotechnology contributes to the wine industry?

Current approaches to strain development involve iterative empirical studies of cause and effect, usually focusing on one gene at a time. This gene will typically be deleted and overexpressed to see what phenotype arises. Then to avoid GM acceptance issues, natural polymorphisms that provide the same functionality are sought and, if necessary, the target phenotype is combined with other desirable attributes through breeding programs. Systems biology has the potential to enable scientists to achieve the same goal more efficiently by asking the questions in a different order—identify the chemical or physiological changes required to manifest the desired phenotype, and then use mathematical modeling to predict which combinations of genes require manipulation. To efficiently generate such a tailored strain, the application of cutting edge GM technology will be required. In particular, methodologies to integrate modular regulatory networks into cells[13] will enable the fine-tuning of biochemical pathways to deliver the required phenotypes. Metabolic engineering in this vein is part of another modern biotechnology discipline—*synthetic biology*.

The combination of systems biology and synthetic biology has the potential to revolutionize industrial yeast strain development (Fig. 2). For the wine industry this will mean availability of fit-for-purpose starter cultures that facilitate production of wine according to market requirements. In addition, shorter lead times for strain development will enable rapid exploitation of emerging markets with specific preferences.

CONCLUSION

Meeting the consumer challenge requires innovation along the entire wine production and marketing chain, working backwards from an understanding of what consumers prefer. It is crucial that the end product poured into a glass meets the consumer expectation of quality, particularly in emerging markets,[14] because this is the only path to sustainable consumer acceptance. Biotechnology has contributed innovation to wine fermentation through the development of fit-for-purpose starter cultures; the future will see this contribution become more prominent. The combination of systems biology and synthetic biology will make it possible to design tailored starter cultures, enabling winemakers to target specific consumer markets.

ACKNOWLEDGMENTS

The Australian Wine Research Institute, a member of the Wine Innovation Cluster in Adelaide, is supported by Australia's grapegrowers and winemakers through their investment body, the Grape and Wine Research and Development Corporation, with matching funds from the Australian Government.

REFERENCES

1. Pretorius, I.S. Tailoring wine yeast for the new millennium: novel approaches to the ancient art of winemaking. Yeast **2000**, *16*, 675–729.
2. Bartowsky, E.J.; Pretorius, I.S. Microbial formation and modification of flavour and off-flavour compounds in wine. In *Biology of Microorganisms on Grapes, in Must and Wine*; Unden, H., König, G., Fröhlich, J., Eds.; Springer: Heidelberg, 2008; 211–233.
3. Liti, G.; Carter, D.M.; Moses, A.M.; Warringer, J.; Parts, L.; James, S.A.; Davey, R.P.; Roberts, I.N.; Burt, A.; Koufopanou, V.; Tsai, I.J.; Bergman, C.M.; Bensasson, D.; O'Kelley, M.J.; van Oudenaarden, A.; Barton, D.B.; Bailes, E.; Nguyen, A.N.; Jones, M.; Quail, M.A.; Goodhead, I.; Sims, S.; Smith, F.; Blomberg, A.; Durbin, R.; Louis, E.J. Population genomics of domestic and wild yeasts. Nature **2009**, *458*, 337–341.
4. Coulon, J.; Husnik, J.I.; Inglis, D.L.; van der Merwe, G.K.; Lonvaud, A.; Erasmus, D.J.; van Vuuren, H.J.J. Metabolic engineering of *Saccharomyces cerevisiae* to minimize the production of ethyl carbamate in wine. Am. J. Enol. Vitic. **2006**, *57*, 113–124.
5. Husnik, J.I.; Volschenk, H.; Bauer, J.; Colavizza, D.; Luo, Z.L.; Van Vuuren, H.J.J. Metabolic engineering of malolactic wine yeast. Metab. Eng. **2006**, *8*, 315–323.
6. Wang, Z.Y.; He, X.P.; Zhang, B.R. Over-expression of *GSH1* gene and disruption of *PEP4* gene in self-cloning industrial brewer's yeast. Int. J. Food Microbiol. **2007**, *119*, 192–199.
7. Vivier, M.A.; Pretorius, I.S. Genetically tailored grapevines for the wine industry. Trends Biotechnol. **2002**, *20*, 472–478.

8. Schoeman, H.; Wolfaardt, G.M.; van Rensburg, P.; Pretorius, I.S. Establishing a risk assessment process for potential release of genetically modified wine yeast into the environment. Can. J. Microbiol. **2009**, *55*, 8, 990–1002.

9. Ugliano, M. Enzymes in winemaking. In *Wine Chemistry and Biochemistry*; Victoria Moreno-Arribas, M., Carmen Polo, M., Eds.; Springer: New York, 2009; 103–126.

10. Swiegers, J.H.; Capone, D.L.; Pardon, K.H.; Elsey, G.M.; Sefton, M.A.; Francis, I.L.; Pretorius, I.S. Engineering volatile thiol release in *Saccharomyces cerevisiae* for improved wine aroma. Yeast **2007**, *24*, 561–574.

11. Eijsink, V.G.H.; Gaseidnes, S.; Borchert, T.V.; Van Den Burg, B. Directed evolution of enzyme stability. Biomol. Eng. **2005**, *22*, 21–30.

12. Borneman, A.R.; Forgan, A.H.; Pretorius, I.S.; Chambers, P.J. Comparative genome analysis of a *Saccharomyces cerevisiae* wine strain. FEMS Yeast Res. **2008**, *8*, 1185–1195.

13. Boyle, P.M.; Silver, P.A. Harnessing nature's toolbox: regulatory elements for synthetic biology. J. R. Soc. Interface, **2009**, 6, Suppl 4, S535–S546.

14. Osidacz, P.; Francis, L. What the Chinese want. Wine Business Monthly **2009**, *February*, 58–61.

Xanthan Gum: Bioreactors in Production

Y. Martin Lo
Pavan Kumar Soma
Department of Nutrition and Food Science, University of Maryland, College Park, Maryland, U.S.A.

Abstract

Production of xanthan gum at the industrial scale remains cost-ineffective and environmentally unfriendly. These problems are primarily due to the high viscosity of xanthan polymers produced in the bioreactor that inherently hinders aeration efficacy critical to the growth of *Xanthomonas campestris* and synthesis of xanthan gum. A wide spectrum of bioreactors currently designed for xanthan gum production are evaluated, compared, and contrasted in this entry. Emphases are placed on the aeration and agitation capacity of each of the systems, while issues pertinent to downstream recovery, product quality, and scale-up operations are discussed.

INTRODUCTION

Widely used as a thickening or stabilizing agent in the food, pharmaceutical, and oil-recovery industries, xanthan gum (Fig. 1) is known for its distinctive rheological properties, namely high viscosity at low shear, shear-thinning, stability over a broad range of temperature and pH, and high resistance to shear degradation in aqueous solutions.[1,2] The gum is produced in large, agitated fermentation units and sent to a holding tank, followed by alcohol precipitation using up two to three volumes of isopropanol to each volume of the broth (a method in compliance with FDA food grade requirements) before it is dried, milled, and packaged. Production of xanthan gum by fermentation of *Xanthomonas campestris*, an aerobic strain, depends on many parameters and variables, including medium composition, temperature, pH, and oxygen transfer.[3–5] Since it is not possible to provide a microbial culture with all the oxygen it needs for the complete respiration of the glucose or any other carbon source in one addition, a microbial culture must be supplied with oxygen during growth at a rate sufficient to satisfy the organisms' demand. The high-broth viscosity causes agitation and aeration in the conventional stirred-tank reactor (STR) to be extremely difficult and consequently limits its productivity to <25 g/L (2.5%, w/v).[6]

Considerable efforts have been made to improve oxygen transfer during xanthan fermentation, including new STR impellers, novel reactor designs, and alternative aeration strategies such as water-in-oil emulsions and liquid-phase conversion of hydrogen peroxide. Several attempts have shown increased oxygen transfer at xanthan concentrations <2% (w/v); however, the performance of these systems is still greatly hindered at xanthan concentrations >3% (w/v). Moreover, the specific rates and yields of xanthan production as well as the quality of xanthan polymers have been found to be highly dependent on the dissolved oxygen tension (DOT), and more specifically the volumetric oxygen transfer rate (OTR) in the fermentor. The development of a novel, centrifugal, packed-bed reactor (CPBR) followed by ultrafiltration (UF) to reach high xanthan concentration presents an integrated approach that could significantly reduce downstream recovery and purification costs.[7,8] Nevertheless, the feasibility of the system has only been demonstrated at bench-top and pilot scales, leaving considerable discrepancies for industrial implementation. This entry explores the current state-of-the-art as well as promising advancements of bioreactors for xanthan gum production in hope to stimulate development of bioreactor systems capable of effectively handling such a complicated fermentation process at the industrial scale.

XANTHAN GUM FERMENTATION

It is known that under growth-limiting conditions *X. campestris* produces its gum at the cell surface, secreting it into the surrounding medium, and thus encapsulating the cell. During batch fermentation, cells undergo a physiological change from trophophase to idiophase. After entering idiophase, the cells stop growing, but xanthan gum continues to accumulate.[9] Four steps of reactions are considered to be responsible for the synthesis of xanthan gum during fermentation.[10] The first step is the uptake of the substrate (carbohydrate), which may occur by active transport, group translocation, or facilitated diffusion, by phosphorylation. The second step in the biosynthesis is the intermediary metabolism, involving the formation of precursors for assembly of the xanthan-repeating unit. The third step occurs at the cell periplasmic membrane. The sugar nucleotides are transferred to isoprenoid alcohol phosphate (a lipid acceptor) to form a pentasaccharide-P-P-lipid. The pyruvic acid acetal residues in xanthan

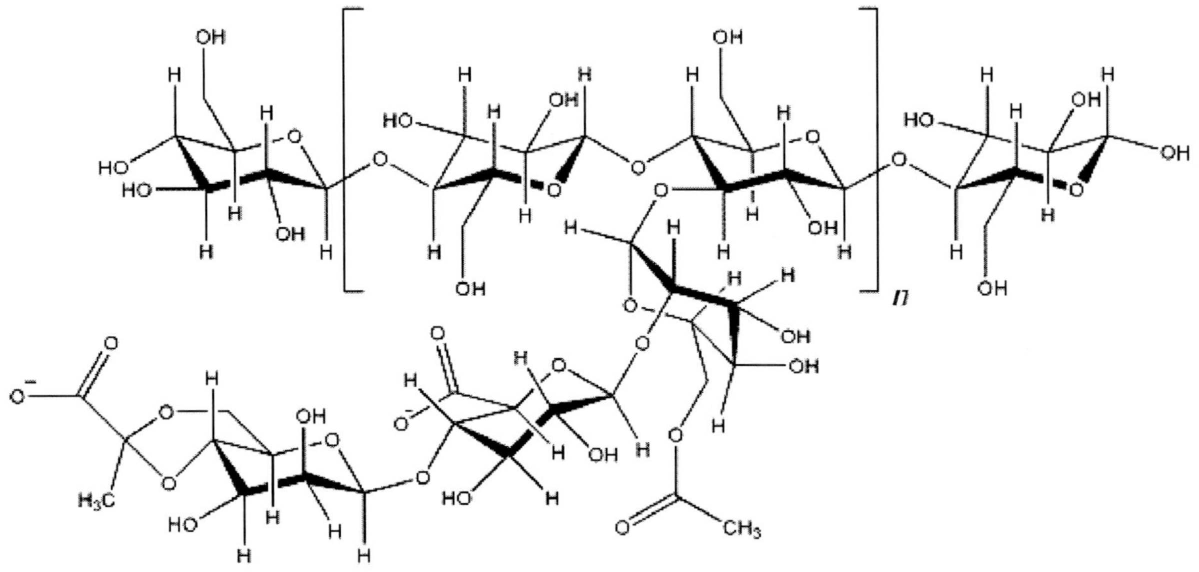

Fig. 1 Conformation of xanthan gum polymers.

are formed by transfer from phosphoenolpyruvate (PEP) to the pentasaccharide-P-P-lipid. The addition of acetyl residues also occurs at the lipid intermediate stage and the donor is acetyl-CoA. The fourth step is polymerization that occurs by transfer of the pentasaccharide unit from the lipid carrier to the growing xanthan.[11]

At the early stage of fermentation, the inoculated culture showed exponential growth. The culture then reached the stationary phase. Xanthan gum concentration increased with time until all glucose was consumed. Dissolved oxygen tension (DOT) monitored in the medium decreased as the fermentation progressed, dropping gradually from 100% initially to approximately 20%. As the broth became gel-like and viscous, the DOT dropped to approximately 10%. This phenomenon mainly depended on the oxygen consumption for both cell growth and xanthan production. The OTR in the medium was reduced owing to increased broth viscosity resulting from increased xanthan concentration. On the other hand, it is also recognized that the aeration conditions necessary for the optimum production of a fermentation product may differ from that favoring biomass accumulation, making optimization of xanthan fermentation extremely challenging. Efforts on creating mutant strains for xanthan production have been reported;[12] however, it is beyond the scope of this entry and their industrial applications remain unclear without convincing stability of the strains and product quality.

STIRRED-TANK REACTORS

There have been many attempts to increase xanthan concentration and to lower the energy costs by using new types of bioreactors or agitation designs with improved aeration and oxygen transfer for viscous xanthan fermentation. It is generally agreed that a mechanically agitated fermentor is required to achieve good heat and mass transfer at high viscosities. Efforts to improve xanthan fermentation have mainly been focused on the type of impellers used as well as the oxygen transfer performance by each type of the impellers, such as Rushton turbine and twin-impellers. However, the performance of stirred-tank bioreactors with various impellers is limited because of the formation of stagnant zones as xanthan concentration exceeds 20 g/L, which subsequently leads to a high power consumption and torque oscillation. Moreover, the high degree of aeration and agitation required in fermentation frequently gives rise to the undesirable formation of foams. It is necessary to break down foam before it causes any process difficulties, and this may be achieved by the use of mechanical foam breakers or chemical antifoams. Antifoams were found to decrease the oxygen transfer rate in solution due to their surfactant nature. For fermentations that are limited by the availability of oxygen, the minimal level of antifoam necessary to control foaming should be used.

Since the limitations due to aeration and agitation difficulties were observed in agitated fermentors, other options were also explored to improve xanthan fermentation. The characteristics and performance of various bioreactor designs and systems reported in the literature regarding xanthan gum fermentation, including stirred-tank bioreactors, foam fermentors, bubble column and air-lift bioreactors, plunging jet bioreactors, and water-in-oil emulsion systems are summarized in Table 1 with corresponding references.

REACTORS WITH LOW POWER INPUTS

Unconventional bioreactors, such as foam fermentors, bubble column or air-lift bioreactors, and plunging jet

Table 1 Comparison of different bioreactor systems for xanthan gum fermentation.

System type	Maximum OTR[a] (g/L hr)	$k_L a$[b] (sec)	Xanthan concentration (g/L)	Fermentation time (hr)	Productivity (g/g cell/hr)	Remarks	Reference
Stirred-tank bioreactor							
Double disc turbine impeller	0.6	0.03	22–30	80	0.1–0.2	High OTR at low xanthan concentration; Low torque fluctuation; Low power drop; Stagnant zone in wall region	[13–17]
45° pitch downward-pumping 6-bladed agitator with disc turbine	–	0.05–0.1	20	56	–	Better OTR than disc turbine; Stagnant zone still existed	[18]
Intermig impeller	0.7	0.02–0.2	50	57	0.107	Reach high xanthan concentration in short fermentation time; High power input	[16,19,20]
Intermig impeller with internal draft tube	0.36	0.01–0.1	20	135	0.1	Ineffective at low viscosity; Advantage at high viscosity; Reached low xanthan concentration	[21]
Twin impeller	–	–	23–27	50	0.2	Large high shear region; High-specific production rate; High power consumption	[22]
Lightnin A-315 impeller	–	–	35	–	–	Severe torque oscillation at high xanthan concentration	[23]
Scaba 6SRGT agitator	–	–	35	–	–	Improved bulk mixing; Instability in torque	[24]
Helical ribbon screw impeller	–	0.01	30	–	–	No dead zones throughout the length of the impeller; Limited capacity to disperse the bubble	[25]
Foam fermentor	–	0.01–0.1	20	120	–	High-specific growth rate; Final yield only 70%; Need surfactant	[26]
Bubble column	–	0.01	20	120	–	Homogeneous mixing; High-specific growth rate; Hold only small volume; Very difficult to scale-up	[27,28]
Air-lift bioreactor							
External circulation loop	–	0.01	25	36	0.05	Low power input; Poor OTR at high viscosity; Low production rate and yield; Hard to scale-up	[28,29]
External circulation loop with short draft tubes covered by perforated plates	–	0.02	–	–	–	Better OTR than with external circulation loop only; Still needs improvement	[30]
Plunging jet bioreactor	0.5	0.02	20	100	–	Very low power input; High-specific growth rate; Poor mixing near wall; Hard to scale-up	[31]
Water-in-oil emulsion							
Isoparaffin	–	–	50	–	0.075–0.1	High liquid–liquid interfacial area; Low power requirement; Agglomeration at high xanthan concentration; Contamination by emulsifier	[32]
n-hexadecane	–	–	65	100	0.13	Fine water-in-oil dispersion; Hard to recover product	[33]
Cell immobilization							
Entrapment using porous support particles	–	–	55	120	–	Can reach high xanthan concentration; Only 25% total xanthan was produced at 55 g/L; Oxygen transport limitation; Hard to recover product	[34]
Centrifugal, packed-bed reactor (CPBR)	0.4	0.02–0.05	25	50	0.3	High xanthan productivity; ~85% xanthan yield from glucose; High cell density (~7 g/L)	[35]

OTR: oxygen transfer rate, i.e., the change in oxygen concentration over a period of time, in mmoles O_2 dm^3/hr.
$k_L a$: the volumetric mass transfer coefficient, normally per hr. The volumetric mass transfer coefficient is a measure of the aeration capacity of a fermentor under the test conditions.

bioreactors, are not applicable in the industry because of the difficulties in process scale-up (Table 1). The water-in-oil emulsion, on the other hand, is able to provide high liquid–liquid interfacial area that consequently benefits the oxygen transfer at high xanthan concentrations. Emulsion fermentations for xanthan production as described in various patents have proven to provide better aeration than the conventional STR. The water-in-oil emulsion system exhibits the most improved oxygen transfer at high xanthan concentrations (e.g., 3%, w/v). However, the contaminations caused by the use of emulsifier, in addition to the difficulties in emulsion breaking, need to be overcome before it becomes a viable option.

Enhanced oxygen transfer and conditions for improved mixing in a centrifugal film bioreactor, particularly of very viscous liquid media, have been demonstrated. The use of centrifugal bioreactors, developed for bioseparation, has been the subject of studies for applications in enzyme and fermentation technology. Therefore, the centrifugal bioreactor could become a potential candidate for the oxygen-limited xanthan fermentation to reach higher productivity and xanthan concentration than existing systems. Other bioreactors with improved oxygen transfer efficiency include liquid-impelled loop reactor, deep jet bioreactor, multicompartment bioreactor, and down-flow liquid jet loop reactor.

IMMOBILIZED CELL BIOREACTORS

A transport-controlled bioreactor using immobilized cell technology for simultaneous production and concentration of xanthan gum was reported to reach an upper limit of ~5% (w/v) (Table 1). Such immobilized cell bioreactors were not considered feasible for long-term continuous xanthan production, mainly due to the oxygen limitation caused by the support material, until the development of a CPBR.[7] The mixing and aeration problems in xanthan fermentation were overcome by continuously pumping and circulating the medium broth through the rotating fibrous matrix to ensure intimate contact of gas and liquid with the cells and to separate the xanthan polymer from the cells. Almost all of the cells were immobilized on the fiber surfaces, resulting in a cell-free fermentation broth with ~85% xanthan yield from glucose. Consistent xanthan production rate and gum quality were obtained.

The improved xanthan productivity achieved is primarily attributed to the innovative aeration strategies—liquid-continuous (LC) and gas-continuous (GC) operation modes plus a recirculation loop—besides the high cell density (~7 g/L) achieved in the matrix via natural adhesion for cell immobilization. Metabolic flux analysis (MFA), a tool capable of assessing the invisible intracellular metabolic pathways involved in xanthan synthesis, revealed that CPBR is a preferable system to STR in terms of xanthan yield, productivity, and the utilization of glucose. Despite the high xanthan productivity and glucose uptake achieved in CPBR-GC, CPBR-LC with 5.0% initial glucose concentration was found to be the most energy-efficient operation.

CONCLUSION

Capable of producing xanthan gum at a productivity of twice that for the present industrial process, CPBR utilizes cell immobilization in combination with centrifugal force to separate xanthan polymers from the support matrix. Integration of CPBR with UF to increase xanthan concentration to ~15% (w/v) or higher is considered the most feasible option to ensure the cost effectiveness of the process while reducing environmental pollutions. Further research is needed to overcome obstacles in scaling up the process for industrial production.

REFERENCES

1. Margaritis, A.; Pace, G.W. Microbial polysaccharides. In *Comprehensive Biotechnology*; Moo-Young, M., Ed.; Pergamon: New York, 1985; Vol. 3, 1005–1043.
2. Podolsak, A.K.; Tiu, C.; Saeki, T.; Usui, H. Rheological properties and some applications for rhamsan and xanthan gum solutions. Polym. Int. **1996**, *40*, 155–167.
3. Lo, Y.M.; Yang, S.T.; Min, D.B. Effects of yeast extract and glucose on xanthan production and cell growth in batch culture of *Xanthomonas campestris*. Appl. Microbiol. Biotechnol. **1997**, *47*, 689–694.
4. Garcia-Ochoa, F.; Castro, E.G.; Santos, V.E. Oxygen transfer and uptake rates during xanthan gum production. Enzyme Microb. Technol. **2000**, *27* (9), 680–690.
5. Leela, J.K.; Sharma, G. Studies on xanthan production from *Xanthomonas campestris*. Bioprocess Eng. **2000**, *23* (6), 687–689.
6. Flores, F.; Torres, L.G.; Galindo, E. Effect of the dissolved oxygen tension during cultivation of *Xanthomonas campestris* on the production and quality of xanthan gum. J. Biotechnol. **1994**, *34*, 165–173.
7. Yang, S.T.; Lo, Y.M.; Min, D.B. Xanthan gum fermentation by *Xanthomonas campestris* immobilized in a novel centrifugal fibrous-bed bioreactor. Biotechnol. Prog. **1996**, *12*, 630–637.
8. Lo, Y.M.; Yang, S.T.; Min, D.B. Kinetic and feasibility studies of ultrafiltration of viscous xanthan gum fermentation broth. J. Memb. Sci. **1996**, *117*, 237–249.
9. Behrens, U.; Klima, M.; Fiedler, S. Growth and accumulation of polysaccharide by *Xanthomonas campestris*. Z. Allg. Mikrobiol. **1980**, *20* (3), 209–213.
10. Garcia-Ochoa, F.; Santos, V.E.; Alcon, A. Xanthan gum production—an unstructured kinetic model. Enzyme Microb. Technol. **1995**, *17* (3), 206–217.
11. Hsu, C.H.; Lo, Y.M. Characterization of xanthan gum biosynthesis in a centrifugal, packed-bed reactor using metabolic flux analysis. Process Biochem. **2003**, *38*, 1617–1625.

12. Schroter, K.; Flaschel, E.; Puhler, A.; Becker, A. *Xanthomonas campestris* pv. *Campestris* secretes the endogluconases ENGXCA and ENGXCB: construction of an endogluconase-deficient mutant for industrial xanthan production. Appl. Microbiol. Biotechnol. **2001**, *55*, 727–733.
13. Moraine, R.A.; Rogovin, P. Xanthan biopolymer production at increased concentration by pH control. Biotechnol. Bioeng. **1971**, *13*, 381–391.
14. Moraine, R.A.; Rogovin, P. Kinetics of the xanthan fermentation. Biotechnol. Bioeng. **1973**, *15*, 225–237.
15. Funahashi, H.; Maehara, M.; Taguchi, H.; Yoshida, T. Effects of agitation by flat-bladed turbine impeller on microbial production of xanthan gum. J. Chem. Eng. Jpn. **1987**, *20* (1), 16–22.
16. Peters, H.-U.; Suh, I.-S.; Schumpe, A.; Deckwer, W.-D. Modeling of batch wise xanthan production. Can. J. Chem. Eng. **1992**, *70*, 742–750.
17. Flores, F.; Torres, L.G.; Galindo, E. Effect of the dissolved oxygen tension during cultivation of *X. campestris* on the production and quality of xanthan gum. J. Biotechnol. **1994**, *34*, 165–173.
18. Nienow, A.W. Mixing studies on high viscosity fermentation processes xanthan gums. World Biotech. Rep. **1984**, *1*, 293–304.
19. Himmelsbach, W. Studies on mass and heat transfer and on homogenization in stirred fermentors using xanthan as an example. Chem. Ing. Tech. **1985**, *57* (6), 548–549.
20. Herbst, H.; Schumpe, A.; Deckwer, W.-D. Xanthan production in stirred tank fermentors: oxygen transfer and scale-up. Chem. Eng. Technol. **1992**, *15*, 425–434.
21. Herbst, H.; Peters, H.-U.; Suh, I.-S.; Schumpe, A.; Deckwer, W.-D. Oxygen transfer during xanthan fermentation. In *AIChE Annual Meeting*, New York, Paper no. 175a, 1987.
22. Nakajima, S.; Funahashi, H.; Yoshida, T. Xanthan gum production in a fermentor with twin impellers. J. Ferment. Bioeng. **1990**, *45*, 165–175.
23. Galindo, E.; Nienow, A.W. Mixing of highly viscous simulated xanthan fermentation broths with the Lightin A-315 impeller. Biotechnol. Prog. **1992**, *8*, 233–239.
24. Galindo, E.; Nienow, A.W. Performance of the scaba 6SRGT agitator in mixing of simulated xanthan gum broths. Chem. Eng. Technol. **1993**, *16*, 102–108.
25. Tecante, A.; Choplin, L. Gas–liquid mass transfer in non-Newtonian fluids in a tank stirred with a helical ribbon screw impeller. Can. J. Chem. Eng. **1993**, *71*, 859–865.
26. Misra, T.K.; Barnett, S.M. Evaluation of a novel foam fermentor in the production of xanthan gum. In *Biotechnology Processes: Scale-up and Mixing*; Ho, C.S.; Oldshue, J.Y., Eds.; American Institute of Chemical Engineering, New York, 1987, 227–237.
27. Pons, A.; Dussap, C.G.; Gros, J.B. Modeling *Xanthomonas campestris* batch fermentation in a bubble column. Biotechnol. Bioeng. **1989**, *33*, 394–405.
28. Suh, I.S.; Schumpe, A.; Deckwer, W.-D. Xanthan production in bubble column and air-lift reactors. Biotechnol. Bioeng. **1991**, *39*, 85–94.
29. Kessler, W.R.; Popovic, M.K.; Robinson, C.W. Xanthan production in an external-circulation-loop airlift fermentor. Can. J. Chem. Eng. **1993**, *71*, 101–106.
30. Kawase, Y.; Tsujimura, M. Enhancement of oxygen transfer in highly viscous non-Newtonian fermentation broths. Biotechnol. Bioeng. **1994**, *44*, 1115–1121.
31. Zaidi, A.; Ghosh, P.; Schumpe, A.; Deckwer, W.-D. Xanthan production in a plunging jet reactor. Appl. Microbiol. Biotechnol. **1991**, *35*, 330–333.
32. Schumpe, A.; Diedrichs, S.; Hesselink, P.G.M.; Nene, S.; Deckwer, W.-D. Xanthan production in emulsions. Proceedings of 2nd International Symposium on, Biochemical Engineering, Stuttgart, Germany, 1991; 196–199.
33. Ju, L.-K.; Zhao, S. Xanthan fermentations in water/oil dispersions. Biotechnol. Tech. **1993**, *7* (7), 463–468.
34. Robinson, D.K.; Wang, D.I.C. A transport controlled bioreactor for the simultaneous production and concentration of xanthan gum. Biotechnol. Prog. **1988**, *4*, 231–241.
35. Yang, S.T.; Lo, Y.M.; Min, D.B. Xanthan gum fermentation by *Xanthomonas campestris* immobilized in a novel centrifugal fibrous-bed bioreactor. Biotechnol. Prog. **1996**, *12*, 630–637.

Xenogenic Tissue Use in Clinical Medicine

Nicholas Zavazava
Internal Medicine, University of Iowa, Iowa City, Iowa, U.S.A.

Abstract

The ideal treatment of degenerative human diseases is to replace the damaged tissue with new functional cellular transplants. For example, type I diabetes leads to the destruction of insulin-producing cells by an autoimmune process. One way to deal with this diabetes is to replace the damaged cells with new ones from a non-related donor. This is, however, hampered by the lack of adequate donors and in some cases by the recurrence of disease post-transplantation. Therefore, xenotransplantation, i.e., the transplantation of animal tissues in humans, has been viewed as an alternative, which could capitalize the many farm animals available. Such a strategy is likely to avoid the recurrence autoimmune processes. Here, I will discuss the pros and cons of xenotransplantation and elaborate on new molecular and cellular options to engineer and design animals that may be more suitable to organ transplantation in humans. The problems of immunological rejection will be presented and the potential role of stem cells in minimizing rejection discussed.

INTRODUCTION

Analysis of studies on the use of farm animals to improve human health date back to the pioneering work by Edward Jenner who worked with cowpox in the eighteenth century and paved the way for the development of vaccines against smallpox and other plagues. The horse has been used for the developments of anti-venom sera for decades now, in addition to the use of farm animals, in general, for testing biodegradable implants. In more recent years, bovine and porcine insulin have been used for the treatment of diabetes in humans. More significantly, the porcine has been a source of cardiac valves that are successfully used to treat cardiac valve failure in humans.[1] The use of farm animals to improve human health is much more broad. For example, animals are now used for the production of pharmaceuticals, as disease models, and in some cases to solve environmental problems.[2] Thus, there has been an enormous gain in the use of animal models and tissues to improve human health.

An exciting area in the last few years has been the production of transgenics and the advancement of reproductive technologies that have led to nuclear transfer, a technology that many anticipate could become a powerful tool in the use of stem cells. The first milestone in this technology was the microinjection of DNA into one pro-nucleus of a zygote.[3] In more recent years, the cloning of Dolly has seen this area of research propel forward as nuclear transfer was able to generate healthy sheep capable of reproduction.[4] The technology remains poorly efficient with success rate a little over 1%.[2] More that 10 species have been successfully generated by somatic cell nuclear transfer showing that the technology can be applied in large animals. However, nuclear transfer in humans is widely considered very difficult and has yet to be demonstrated.

Embryonic stem cells (ESC) have now been widely utilized in many different models. The advantage of these cells is their pluripotency, i.e., ability to differentiate into all three germ layers: mesoderm, endoderm, and ectoderm. ESC derived from large farm animals allow study of their applicability in deriving tissues that can be applied in the treatment of human disease. For example, Dr. Wheeler of the University of Illinois, Champaign, Illinois, has established a huge cell bank of porcine embryonic stem cell lines that can be used to study human disease.[5] These cells can be used to generate models for disease in a large animal model. One of the problems that, however, need to be overcome in the porcine is the lack of available antibodies specific for porcine antigens and cell markers. Although a few antibodies raised against other species have been shown to cross-react with swine antigens, these antibodies are not sufficient to allow comprehensive studies on cell trafficking, reaction to antigen, and elucidation of mechanisms. Therefore, greater efforts need to be made to exploit these cells in disease models. An obvious advantage in developing porcine models is that the physiology of the porcine is very similar to that in humans, making it ideal for use in experimentation of human disease. Generally, the porcine is not difficult or expensive to maintain, making it an ideal source of tissues that could be used to treat human disease.

TISSUE ENGINEERING FOR HUMAN THERAPIES

Cardiac Valves

For decades now, porcine cardiac valves have been successfully used in humans to treat defective cardiac valves. These valves have the advantage of being biological and, therefore, not marred with problems such as the clicking noise made by metal valves. Further, they do not lead to cell lysis of hematopoietic cells as has been reported with mechanical valves. A concern has been the possibility of immunological reaction to this xenogenic tissue. Luckily the valves are fixed in formaldehyde that destroys any viable tissue capable of triggering an anti-porcine reaction to the tissue. My laboratory reported in 1995 that we had evidence that these xenogeneic valves are re-endothelialized after implantation.[1] Not only did we find an endothelial sheet of cells on the valves expressing human MHC antigens, but the data seemed to suggest that degeneration of valves could be prevented by re-endothelialization of porcine valves with human cells prior to implantation. This tissue engineering could be useful in preserving the integrity of the valves and also in strengthening the fixed valves.

To determine the immunogenicty of cardiac valves, endothelial cells were recovered from valves by enzyme digestion and used in mixed lymphocyte reactions. Endothelial cells induced a very powerful immunological response 4 to 5 times stronger than that induced by peripheral blood lymphocytes. It appears that in the allogeneic setting, cardiac valves would provoke an immunological reaction that might accelerate rejection and eventual degeneration. Therefore, the use of porcine valves that have been re-endothelialized with recipient andothelial cells is an approach that could be used to prolong the half-life of the valves post-transplantation.

Porcine Pancreatic Islets

The strong homology of porcine insulin to that of humans has opened the opportunity to utilize isolated porcine islets for use in humans. A number of groups around the world have been using porcine islets in vivo to treat human disease. Although here too rejection has been of concern, different groups have come up with different solutions. For example, Valdes-Gonzalez et al. have come up with a solution to immunoprotect neonatal porcine islets from rejection.[6] In their series, they found that the majority of their patients successfully reduced intake of insulin or even partially became insulin independent. In this series, the cells were protected from rejection by mixing them with Sertoli cells that have been shown to be immune privileged by the expression of FasL (CD95L).[7] In another study, patients were found to be more reluctant of such porcine transplants when informed about the possible transmission of animal viruses and the disadvantages of possible cancer induction by the higher amount of immunosuppression that may be required.[8] Indeed zooneses has been of great concern in such animal farm-derived transplants.[9,10] A number of studies seem to suggest that there is limited danger, however, in this day and age of bird flu and a possible pandemic, that message is hard to sell.

Groth et al.[11] transplanted fetal porcine islet-like cell clusters into diabetic patients. All patients received standard immunosuppression, antithymocyte globulin or 15-deoxyspergualin. The cells were injected intraportally or under the kidney capsule. In four patients, there was clear evidence of insulin and glucagons production and lack of infiltration of the transplants. However, all patients developed anti-porcine antibodies, the majority of which were directed against the Gal epitope.[12] These data generated enthusiasm that xeno-transplantation is a modality in the treatment of diabetes. More recently though, concerns of zooneses have slowed progress, although there is an overwhelming body of data now that the problem is much smaller than anticipated.[13] Clearly the advantages of successful islet transplantation are obvious: better control of glucose levels that ultimately control the onset of secondary complications of diabetes and freedom from insulin injections. In neural studies, the FDA has allowed the transplantation of porcine neural tissue for the treatment of Parkinson's disease, Huntington's disease, and spinal cord injuries. These treatment forms have been ongoing without apparent problems with zooneses.

Gene-Modified Pigs and Embryonic Stem Cells

The naturally occurring antibodies in the xenogeneic setting are anti-Gal antibodies. Pigs transgenic for human complement inhibitors[14] were hailed as the approach for preventing pre-formed antibodies from damaging xenogenic tissues. In more recent years, animals have been cloned that are α-1,3 galactosyltransferase gene knockouts.[15] There was a marked prolongation of graft survival when renal grafts of these knockout swine were transplanted in baboons. Although hyperacute rejection can be avoided, there is still concern that these cells may be rejected by cellular processes. The management of cellular rejection, however, now appears to be well controlled and better manageable than hyperacute rejection. Interestingly, fetal islets appear to express Gal, but not tissue from adult animals. Gal-expression is found on the non-endocrine tissue that accompanies islet transplants. A newer approach is to differentiate porcine embryonic stem cells into islets that are free of any other accompanying non-endocrine tissues. The advantage of embryonic stem cells is that they proliferate very well, providing a huge source of cells that can be used for the treatment of patients. Another approach would be to perform mixed transplants. Human islets would be mixed with porcine islets and transplanted. An allogeneic reaction to human tissue would leave the xenogenic tissue intact.

CONCLUSION

The rapid progress in the development of stem-cell-based technologies has greatly benefited the development of new strategies for therapies of untreatable diseases such as diabetes. In the future, improved differentiation of embryonic stem cells may solve the conundrum of lack of tissue availability. Further, there is evidence that embryonic stem-cell-derived tissues may be less immunogenic than adult tissues. This may make it even more attractive to pursue embryonic stem cells as a possible rescue therapy in the future.

REFERENCES

1. Zavazava, N.; Simon, A.; Sievers, H.; Bernhard, A.; Mueller-Ruchholtz, W. Porcine valves are reendothelialized by human recipient endothelium in vivo. J. Thorac. Cardiovasc. Surg. **1995**, *109*, 702–706.
2. Kues, W.A.; Niemann, H. The contribution of farm animals to human health. Trends Biotechnol. **2004**, *22*, 286–294.
3. Hammer, R.E.; Pursel, V.G.; Rexroad, C.E., Jr.; Wall, R.J.; Bolt, D.J.; Ebert, K.M.; Palmiter, R.D.; Brinster, R.L. Production of transgenic rabbits, sheep and pigs by microinjection. Nature **1985**, *315*, 680–683.
4. Wilmut, I.; Schnieke, A.E.; McWhir, J.; Kind, A.J.; Campbell, K.H. Viable offspring derived from fetal and adult mammalian cells. Nature **1997**, *385*, 810–813.
5. Wheeler, M. Development and validation of swine embryonic stem cells: a review. Reprod. Fertil. Dev. **1994**, 6, 563–568.
6. Valdes-Gonzalez, R.A.; Dorantes, L.M.; Garibay, G.N.; Bracho-Blanchert, E.; Mendez, A.J.; Dávila-Pérez, R.; Elliott, E.B.; Terán, L.; White, D.J. Xenotransplantation of porcine neonatal islets of Langerhans and Sertoli cells: a 4-year study. Eur. J. Endocrinol. **2005**, *153*, 419–427.
7. Bellgrau, D.; Gold, D.; Selawry, H.; Moore, J.; Franzusoff, A.; Duke, R. A role for CD95 ligand in preventing graft rejection. Nature **1995**, *377*, 630–632.
8. Deschamps, J.Y.; Roux, F.A.; Gouin, E.; Sai, P. Reluctance of French patients with type 1 diabetes to undergo pig pancreatic islet xenotransplantation. Xenotransplantation **2005**, *12*, 175–180.
9. Patience, C; Takeuchi, Y.; Weiss, R.A. Infection of human cells by an endogenous retrovirus of pigs. Nat. Med. **1997**, *3*, 282–286.
10. Patience, C.; Takeuchi, Y.; Weiss, R.A. Zoonosis in xenotransplantation. Curr. Opin. Immunol. **1998**, *10*, 539–542.
11. Groth, C.G., Korsgren, O., Tibell, A.; Tollemar, J.; Möller, E.; Bolinder, J.; Ostman, J.; Hellerström, C.; Andersson, A. Transplantation of porcine fetal pancreas to diabetic patients. Lancet **1994**, *344*, 1402–1404.
12. Satake, M.; Kawagishi, N.; Kumagai-Braesch, M.; Samuelsson, B.E.; Rydberg, L.; Tibell, A.; Andersson, A.; Korsgren, O.; Groth, C.G.; Möller, E. Specificity of human xenoantibodies formed in response to fetal porcine isletlike cell clusters. Transplant. Proc. **1994**, *26*, 1122.
13. Korsgren, O.; Buhler, L.H.; Groth, C.G. Toward clinical trials of islet xenotransplantation. Xenotransplantation **2003**, *10*, 289–292.
14. Cozzi, E.; White, D. The generation of transgenic pigs as potential organ donors for humans. Nat. Med. **1995**, *1*, 964–966.
15. Yamada, K.; Yazawa, K.; Shimizu, A.; Iwanaga, T.; Hisashi, Y.; Nuhn, M.; O'Malley, P.; Nobori, S.; Vagefi, P.A.; Patience, C.; Fishman, J.; Cooper, D.K.; Hawley, R.J.; Greenstein, J.; Schuurman, H.J.; Awwad, M.; Sykes, M.; Sachs, D.H. Marked prolongation of porcine renal xenograft survival in baboons through the use of alpha1,3-galactosyltransferase gene-knockout donors and the cotransplantation of vascularized thymic tissue. Nat. Med. **2005**, *11*, 32–34.

Xenotransplantation

Jeffrey L. Platt
Transplantation Biology Program, University of Michigan, Ann Arbor, Michigan, U.S.A.

Abstract
Transplantation of living cells, tissues and organs from individuals of one species into individuals of anther is called "xenotransplantation." Xenotransplantation could overcome the severe shortage of human cells, tissues and organs for the treatment of disease. Application of xenotransplantation has been sought for more than one hundred years but success has been impaired by the immune response of the recipient against the graft. Renewed interest in xenotransplantation has been spurred by the advent of genetic engineering and other technologies that might lower or eliminate the immune and other biological barriers.

INTRODUCTION

Xenotransplantation refers to the transplantation of *living* cells, tissues, or organs from individuals of one species into individuals of another species. Xenotransplantation has long been envisioned as a way of treating diseases because animals are available in large numbers. Availability of human sources of organs and tissues for transplants into human recipients (allotransplants) is quite limited. In fact, the number of transplants that can be performed using human sources of cells, tissues, and organs is as low as 5% of the number that would be performed if sources were unlimited.[1]

The first serious efforts at xenotransplantation were made in the early years of the 20th century. With the development of surgical techniques that could enable organ transplants to be performed (the vascular anastomosis), experimental surgeons sought a ready source of organs. Humans, living and recently deceased, were not viewed as potential sources, and therefore experimental surgeons turned to animals for this purpose. The first animals used for xenotransplants were pigs and sheep (not shown) (Table 1).

The early attempts at xenotransplantation failed, as did early attempts at human-to-human transplantation (allotransplantation). The reason for failure of xenografts and allografts was not understood until the 1940s, when the immune response of the recipient against the graft was found to be the principal biological barrier to transplantation. This immunological barrier for allografts was overcome with the advent of immunosuppressive drugs in the late 1950s, and the era of clinical allotransplantation began. However, as human donors were quite limited, efforts were made once again to use animals, such as chimpanzees and baboons, as a source of transplants. When the recipient was treated with immunosuppressive agents, primate-to-human xenografts functioned for weeks to months, but ultimately failed; whereas, human-to-human transplants would sometimes function indefinitely. This experience suggested that while xenotransplantation was a potential solution to organ failure, it would be impaired by an immunological barrier more severe than the barrier hindering human-to-human transplantation. Recent years have brought a better understanding of the immunological barriers to xenotransplantation, and the possibility of applying genetic engineering to address that barrier.[2]

THE RATIONALE FOR XENOTRANSPLANTATION

The main rationale for xenotransplantation is that animals could potentially provide a limitless source of organs and tissues for transplantation. However, xenotransplantation might be pursued perhaps in preference to allotransplantation for other reasons. Where organ failure is caused by a viral infection such as hepatitis, xenotransplantation might be preferred because the transplant would resist reinfection by the virus that caused organ failure.[3] Xenotransplantation might also be preferred as a way of delivering genes possibly introduced by genetic engineering into lines of animals. For example, an animal source might be genetically engineered to express a gene encoding a coagulation

Table 1 Some clinical attempts at xenotransplantation.

Year	Donor	Organ	Maximum survival
1906	Pig	Kidney	2 days
1964	Chimpanzee	Kidney	9 mo
1964	Baboon	Kidney	60 days
1984	Baboon	Kidney	20 days
1992	Baboon	Liver	70 days

Table 2 Technologies for organ replacement.

Technology	Potential applications[a]
Allotransplantation	Heart, lung, liver, kidney, pancreas
Implantable devices	Heart, pancreas
Cell transplantation	Heart, liver, pancreas
Stem cells	Heart, liver, pancreas
Tissue engineering	Liver
Organogenesis	Kidney, lung
Transplantation	Heart, lung, liver, kidney, pancreas

[a]Adapted from Ogle and Platt.[6]

protein or vital enzyme at a high level or under regulated conditions.

An important consideration today is how to weigh xenotransplantation against other potential approaches to treating organ failure (Table 2).[4,5] While some new technologies, such as stem cells, tissue engineering, and cardiac assist devices, have received attention, they have also received less scrutiny than xenotransplantation because they are so recent in their development. Most likely, these technologies will be applied in ways that fill therapeutic niches, such as repairing local defects or injury of tissues. Devices may eventually be used to replace the heart, but application of these technologies for other organs is more remote. On the other hand, cell transplantation, stem cell transplantation, and tissue engineering seem less promising for replacement of the function of structurally complex organs such as the kidney, lungs, and heart. For replacement of these organs, organogenesis (the de novo formation of organs) or xenotransplantation may be necessary. Xenotransplantation may also find application in conjunction with organogenesis. For example, one might envision "growing" human organs, perhaps derived from stem cells as a xenograft in an animal host, and then transplanting the organs to human patients.[5]

SOURCE OF XENOGRAFTS

Many species have been used as sources of tissues and organs for xenotransplantation. Xenografts from sources phylogenetically closer to the recipient would be expected to provoke less intense immunity and to be more physiologically compatible with the recipient. Consistent with that idea, experimental cardiac xenografts from monkeys to baboons have survived greater than a year and renal xenografts from chimpanzees to humans have survived and functioned up to 9 months (Table 1). However, some biological barriers to xenotransplantation derive from expression of one or very few genes in the donor or recipient. Expression of these genes does not necessarily reflect genetic difference.

Table 3 Phylogeny of expression of Galα1-3Gal.[a]

Species	Galα1-3Gal	Anti-Galα1-3Gal antibodies
Mouse	+	−
Pig	+	−
New World monkey	+	−
Baboon	−	+
Human	−	+

[a]Adapted from Galili.[7]

The genetic barrier of greatest current interest and importance is the expression of α1,3-galactosyltransfersase, a glycosyltransferase that catalyzes synthesis of Galα1-3Gal. (Table 3). Galα1-3Gal is a saccharide expressed by lower mammals and New World monkeys, but not by humans and Old World monkeys. Humans and Old World monkeys have "natural" antibodies specific for this saccharide, and these antibodies trigger severe reactions against organs containing Galα1-3Gal. Indeed, many of the efforts in genetic engineering and immunosuppression for xenotransplantation are directed respectively at eradicating expression of the sugar or suppressing immunity directed against it.

Today most efforts in xenotransplantation focus on the pig as a potential source of tissues and organs. Pigs are favored as sources of xenografts because they are available in large numbers (it is estimated that more than one million pigs would be needed for transplants each year on a worldwide basis) and can be bred and genetically manipulated, as described below. Still another reason for favoring the pig is that the organs are large enough to replace the organs of full sized humans and some strains of pigs, such as the mini-pig, may at maturity approximate human size. Finally, the microorganisms harbored by pigs and potentially infectious for humans are well known, and can be detected using established methods.[8] In contrast, some viruses of non-human primates are poorly known and potentially lethal for humans.

Although the experience is limited, the best present evidence would suggest that the heart, lung, and kidneys of the pig would function sufficiently well to sustain the life of a human. Whether a swine liver would function in this way is a matter of controversy because of the metabolic complexity of that organ and because of the possibility that complex cascades, such as complementing coagulation, could be incompatible between pig and human.

HURDLES TO XENOTRANSPLANTATION

The main difficulty with using the pig as a source of xenografts is the expression of Galα1-3Gal, and the severe immune reaction it provokes.[7] Another problem with the pig is that the proteins of that species are so different from the proteins of humans that they also provoke strong immune reactions, thus increasing the challenge of developing an effective approach to immunosuppression. Still another hurdle to using pigs as a source of xenografts in humans is the possibility that infectious agents might transfer from the graft to the recipient. As indicated above, most infectious agents potentially transmitted from pigs to humans are known, and the means exist to eliminate those agents from potential sources of xenografts. One exception may be the porcine endogenous retrovirus (PERV). PERV is found in the genome of all or nearly all pigs and has been transmitted from porcine cells to the cells of humans and non-human primates under a variety of experimental circumstances. Although the subject of much investigation, to date no evidence has emerged that can prove that PERV can be transmitted to human cells in vivo as a result of xenotransplantation.[9] This subject will be an active area of investigation until a decisive answer is achieved.

GENETIC ENGINEERING AND XENOTRANSPLANTATION

The ability to genetically engineer and breed animals has stimulated recent excitement about xenotransplantation. This excitement stems in part from the idea that genetic engineering might make the pig more compatible with humans, decreasing or eliminating the need for immunosuppression. This hope has been achieved only in part.

Transgenic techniques have been used *to introduce into pigs* human genes for complement regulatory proteins and carbohydrate modifying enzymes. For example, the human decay-accelerating factor and CD59 genes, which inhibit complement reactions, have been introduced into lines of pigs in an effort to prevent complement-mediated tissue injury. These efforts have prevented the severest form of complement-mediated rejection, hyperacute rejection.[10] Genes that would modify synthesis of Galα1-3Gal, such as H-transferase, have also been expressed, but expression of that sugar has not been fully eradicated by this approach.

Gene targeting by homologous recombination has been used to eliminate α1,3-galactosyltransferase in pigs. Gene targeting in pigs was recently made feasible by advances in reproductive cloning. Organs from pigs deficient in α1,3-galactosyltransferase, have survived for months in non-human primates, although the primates were treated with severe regimens of immunosuppression.[11] While these advances have stimulated excitement, preliminary work in several laboratories suggests that targeting this gene will not fully eradicate immune reactions against pig cells.[12]

Genetic engineering may be applied in other ways in xenotransplantation. One potential application is to eliminate endogenous viruses, such as in PERV, should these viruses be found to cause human pathology. Another potential application of genetic engineering is to eliminate or replace porcine genes that pose a physiologic hurdle to the success of transplantation or toxicity to the recipient. For example, porcine coagulation factors may be incompatible with human proteins that control coagulation, and hence genes for human coagulation factors might replace the genes for these factors in pigs. Still another application might be to introduce genes of therapeutic benefit for the recipient of a transplant. For example, a xenotransplant might express genes encoding an enzyme, a growth factor or antitumor agent of therapeutic importance.

CONCLUDING REMARKS

At the time of this writing, xenotransplantation is not applied routinely for any medical purpose. However, clinical trials using porcine cells for the treatment of Parkinson's disease and various metabolic diseases have been conducted and new trials of various types are planned. Because the immune reactions injure transplanted cells far less than they injure transplanted organs, xenogeneic cellular transplants may provide a first step toward full clinical application. It is hoped that new insights gained from the immune response to cellular xenografts and application of gene targeting in swine will eventually allow the transplantation of intact porcine organs into humans. Such an advance would certainly revolutionize the treatment of organ failure.

REFERENCES

1. Evans, R.W. Coming to terms with reality: why xenotransplantation is a necessity. In *Xenotransplantation*; Platt, J.L., Ed.; ASM Press: Washington, DC, 2001; 29–51.
2. Cascalho, M.; Platt, J.L. Xenotransplantation and other means of organ replacement. Nat. Rev. Immunol. **2001**, *1*, 154–160.
3. Mueller, Y.M.; Davenport, C.; Ildstad, S.T. Xenotransplantation: application of disease resistance. Clin. Exp. Pharmacol. Physiol. **1999**, *26* (12), 1009–1012.
4. Polak, J.M.; Hench, L.L.; Kemp, P., Eds. *Future Strategies for Tissue and Organ Replacement*; Imperial College Press: London, 2002.
5. Cascalho, M.; Platt, J. New technologies for organ replacement and augmentation. Mayo Clin. Proc. **2005**, *80* (3), 370–378.
6. Ogle, B.M.; Platt, J.L. Approaches to the replacement of the function of failing organs. Curr. Opin. Organ Transplant. **2002**, *7* (1), 28–34.
7. Galili, U. Interaction of the natural anti-Gal antibody with a-galactosyl epitopes: a major obstacle for xenotransplantation in humans. Immunol. Today **1993**, *14*, 480–482.
8. National Research Council of the National Academies. *Animal Biotechnology: Science-Based Concerns*; The National Academies Press: Washington, DC, 2002.

9. Paradis, K.; Langford, G.; Long, Z.; Heneine, W.; Sandstrom, P.; Switzer, W.M.; Chapman, L.E.; Lockey, C.; Onions, D.; Otto, E. Search for cross-species transmission of porcine endogenous retrovirus in patients treated with living pig tissue. Science **1999**, *285*, 1236–1241.
10. McCurry, K.R.; Kooyman, D.L.; Alvarado, C.G.; Cotterell, A.H.; Martin, M.J.; Logan, J.S.; Platt, J.L. Human complement regulatory proteins protect swine-to-primate cardiac xenografts from humoral injury. Nat. Med. **1995**, *1*, 423–427.
11. Yamada, K.; Yazawa, K.; Shimizu, A.; Iwanaga, T.; Hisashi, Y.; Nuhn, M.; O'Malley, P.; Nobori, S.; Vagefi, P.A.; Patience, C.; Fishman, J.; Cooper, D.K.; Hawley, R.J.; Greenstein, J.; Schuurman, H.J.; Awwad, M.; Sykes, M.; Sachs, D.H. Marked prolongation of porcine renal xenograft survival in baboons through the use of alpha1,3-galactosyltransferase gene-knockout donors and the cotransplantation of vascularized thymic tissue. Nat. Med. **2005**, *11* (1), 32–34.
12. Zhong, R.; Platt, J.L. Current status of animal-to-human transplantation. Expert Opin. Biol. Ther. **2005**, *5* (11) 1415–1420.

Yogurt Microbiology

Ashraf N. Hassan
Sanjeev Anand
Department of Dairy Science, South Dakota State University, Brookings, South Dakota, U.S.A.

Abstract

Yogurt is a fermented food that is produced by culturing milk with *Lactobacillus delbrueckii* subsp. *bulgaricus* and *Streptococcus thermophilus*. Other lactic acid bacteria may be included to yield product variations. Some yogurt cultures produce exopolysaccharides (EPS), which improve texture of the final product. The distinctive flavor of yogurt is due to lactic acid and a variety of volatile organic compounds. The low pH in yogurt creates a selective medium for yeasts. Consumption of yogurt is known to provide several health benefits. If probiotic cultures are added, the end products may preferably contain live bacteria, because of their potential beneficial effects to human health. Health hazards from yogurt consumption would only arise if a high number of pathogens contaminated milk after pasteurization, which is highly unlikely in large modern plants.

INTRODUCTION

Fermented milks have been consumed throughout the world. Milk is fermented to develop aroma and flavor characteristics that result directly or indirectly from fermenting organisms that extend shelf life and improve digestibility. The word "yogurt" is derived from the Turkish word *yoğurt*. The earliest yogurts were spontaneously fermented by wild bacteria living on animal skin. The origin of yogurt is debated but likely it has been independently discovered in different regions. Most yogurts contain not less than 3.25% milkfat and not less than 8.25% non-fat milk solids, and have a titratable acidity of not less than 0.9% expressed as lactic acid. Yogurt is characterized by a smooth and viscous gel with a delicate walnutty flavor. Codex regulations define yogurt as the product obtained by fermenting milk with cultures that include *Lactobacillus delbrueckii* subsp. *bulgaricus* and *Streptococcus thermophilus*. Sweeteners, dye, fruit, and stabilizer may also be added. The types of commercially available yogurt include plain, flavored, liquid, carbonated, low-lactose, and low-calorie. There are two main styles of flavored yogurt in the United States, sundae style and Swiss style. In the sundae style, milk is cultured in a cup, and fruit is found in the bottom. The Swiss style is a prestirred yogurt in which milk is cultured in bulk and filled into cups after fermentation and stirring. This chapter briefly explains the manufacturing steps, population dynamics, probiotic microorganisms, production of texture and flavor compounds, and growth of spoilage and pathogenic microorganisms in yogurt.

YOGURT MANUFACTURE

In yogurt-making, milk is standardized to a high level of solids (around 15%) to achieve the desired body characteristics of the finished product. Milk is usually fortified with milk powder, buttermilk powder, casein, whey powder, whey protein concentrate, or concentrated milk. Standardized milk is then homogenized and heat-treated at 175–185°F with a 30-min holding time. This severe heat treatment denatures whey proteins, which is important in obtaining a desirable body and texture and destroying undesirable microorganisms. Heat treatment also lowers the redox potential, which encourages growth of the microaerophilic starter culture. Milk is then inoculated with the starter culture and incubated at 42°C until a pH value of 4.6 is attained. This is followed by a gradual cooling to 10°C in 5–6 hours. The final product temperature should be reached in 24 hours.

MICROBIAL POPULATION DYNAMICS IN YOGURT

Yogurt cultures always contain a co-culture of *L. delbrueckii* subsp. *bulgaricus* and *S. thermophilus*. Other permitted species include *Lactobacillus helveticus*, *Lactobacillus jugurti*, *Lactobacillus delbrueckii* subsp. *lactis*, *Lactobacillus acidophilus*, and *Bifidobacterium* spp. Yogurt cultures provide a classic example of synergism. Both cultures achieve a much higher final number in a mixed-strain culture than in single culture. During the early hours of fermentation, the streptococci rapidly outgrow the lactobacilli and can dominate by 3:1, even if they started out as 1:1. The reduction in redox potential and production of lactic acid stimulate growth of *L. delbrueckii* subsp. *bulgaricus*. The lactobacilli are also stimulated by the formic acid and carbon dioxide produced by *S. thermophilus*. Amino acids and small peptides produced by *Lactobacillus* are important stimulants for *Streptococcus*. Usually there is no increase in counts of *L. delbrueckii* subsp. *bulgaricus* during the first hour of fermentation. It starts to grow rapidly when the pH reaches 5.5. In most cases, the *Streptococcus* population is greater than the population of *Lactobacillus* throughout fermentation; however, the specific growth rate of *Lactobacillus* is greater than that of *Streptococcus* as the pH drops below 5.0. Fresh yogurt contains about 10^9 organisms per gram, but this number declines upon storage. Determination of the exact ratio of rods and cocci in yogurt is difficult because of cell aggregation. Chains of streptococci and pairs of rods grow on solid media as a single colony. The optimal growth temperature for *L. delbrueckii* subsp. *bulgaricus* and *S. thermophilus* is 44 and 39°C, respectively; however, during yogurt fermentation, the optimal temperature does not remain constant but follows a time-dependent profile. The temperature profile that gives the best acidification rate and optimizes the production of flavor compounds has yet to be elucidated.

Yogurt cultures are facultative anaerobes. *L. delbrueckii* subsp. *bulgaricus* is one of the least oxygen-tolerant lactic acid bacteria. It produces enough hydrogen peroxide to inhibit its own growth as well as other co-cultured microorganisms; however, *S. thermophilus* does not produce sufficient hydrogen peroxide to activate the lactoperoxidase system in milk that makes them less sensitive to oxygen. Unlike *L. delbrueckii* subsp. *bulgaricus*, *S. thermophilus* can grow and coagulate milk in the presence of up to 30% dissolved oxygen. Modern molecular techniques, such as terminal restriction fragment length polymorphism (T-RFLP) are currently being used to analyze the small subunit ribosomal gene to examine the complex bacterial dynamics of this relative fast fermentation process.[1] The complete gene sequences of both cultures are already available. The sequences reveal substantial number of pseudogenes, incomplete metabolic pathways, and relatively few regulatory functions with signs of ongoing specialization.

PRODUCTION OF EXOPOLYSACCHARIDES

Some strains of yogurt cultures produce exopolysaccharides (EPS). These polymers can remain attached to the producing cells as capsules or be secreted into the growth medium.[2] Some strains produce both types of EPS, while the others produce only unattached EPS. Strains producing only the capsular type have not been characterized yet.[2] Some types of unattached EPS produce ropiness in fermented milk. Capsule formation does not cause ropiness, nor does the production of unattached EPS ensure ropiness.[2] Exopolysaccharides increase viscosity, smoothness, and water-holding capacity of yogurt; however, they do not necessarily decrease syneresis (whey separation). Strains producing the ropy type of EPS can be used in making stirred (low viscosity) yogurt. In set yogurt, strains producing the capsular type or low levels of ropiness are recommended for consumers who do not like the appearance and slippery mouthfeel of set yogurt made with highly ropy cultures.

FLAVOR COMPOUNDS

In addition to acidification and texture development, yogurt cultures develop the typical yogurt flavor that is a combination of lactic acid and various carbonyl compounds. A great number of volatile compounds, such as aldehydes, ketones, and alcohols, identified in yogurt originate from the milk. Lactic acid imparts the acidic and refreshing taste. Acetaldehyde is the major flavor compound produced by yogurt cultures (5–21 mg/L). It is responsible for the green apple flavor characteristic of yogurt. Several pathways have been proposed for the production of acetaldehyde by lactic acid bacteria. It can be produced directly from pyruvate, deoxyribose-5-phosphate, acetate or the amino acid, threonine.[3] Acetaldehyde can also be produced indirectly from pyruvate through the intermediate compound, acetyl coenzyme A. Both *L. delbrueckii* subsp. *bulgaricus* and *S. thermophilus* produce acetaldehyde; however, production of this flavor compound is strain-dependent. In addition to acetaldehyde, diacetyl is an important flavor compound. Also, volatile fatty acids play a specific role in flavor development. The mixed cultures of *L. delbrueckii* subsp. *bulgaricus* and *S. thermophilus* contain higher levels of acetaldehyde and diacetyl than individual cultures. The acetaldehyde:diacetyl ratio of 7–10:1 is important for the development of the desirable organoleptic properties of yogurt.[4] Acetoin and ethanol are also found in yogurt. Levels of carbonyl compounds decrease during cold storage that may be due to hydrolysis to other compounds by the yogurt cultures. In addition to lactic acid, citric acid, uric acid, and formic acid have been found in yogurt.[5] Unlike carbonyl compounds, levels of organic acids do not change during cold storage unless the product is contaminated by yeasts.

PROBIOTIC YOGURT AND HEALTH BENEFITS

Yogurt cultures confer several health benefits. Although, the number of yogurt bacteria that survive the acidic environment of gastric tract is yet debatable, the yogurt strains are generally considered to be sufficient to impart some probiotic benefit.[6] Both *L. delbrueckii* subsp. *bulgaricus* and *S. thermophilus* have high lactase activities. It is well established that yogurt consumption improves lactose digestion in human. It can also cure acute diarrheal disorders, stimulate the immune system, and decrease allergic symptoms. Other probiotic cultures such as *L. acidophilus* and *Bifidobacterium* sp. have been incorporated into yogurt formulations. *Bifidobacterium* sp. of human origin are preferred and include *Bifidobacterium breve*, *Bifidobacterium longum*, *Bifidobacterium infantis*, and *Bifidobacterium bifidum*. It has been recently suggested that the required minimum level of probiotic bacteria in foods is 1×10^7 colony-forming units (CFU)/g to elicit health benefits. Incorporation of bifidobacteria can lead to counts $>10^6$ CFU/g the time of purchase; however, high inoculum levels are used because bifidobacteria are slow acid producers. Since viability decreases during storage, some products may contain less than the stated desired level of bacteria. Cultured milk containing bifidobacteria shows several advantages such as mild taste, limited postacidification, formation of the L (+) type of lactic acid (which is physiologically desirable), and the therapeutic benefits prescribed to bifidobacteria. Yogurts containing bifidobacteria have the characteristic vinegar taint due to acetic acid production that varies from one strain to another. This acid, in addition to other microbial inhibitors produced by bifidobacteria, may inhibit yogurt cultures and prevent postacidification.

GROWTH OF YEASTS

The low pH in yogurt creates a selective medium for yeasts. In addition, the ability of yeasts to metabolize organic acids and galactose produced during the fermentation enables them to compete in the acidic environment. Assimilation of lactic acid by yeasts leads to an increase in yogurt pH and the viability of yogurt cultures. Yeasts also stimulate yogurt cultures by excretion of growth factors and metabolites. *Kluyveromyces marxianus* ferments lactose and utilizes citric and succinic acids, while *Yarrowia lipolytica* and *Debaromyces hansenii* have strong proteolytic and lipolytic activities.[7] The increase in pH during yogurt storage due to yeast growth ranges from 0.05 to 0.09.[7]

DEFECTS

Microbially induced defects occur in yogurt. Repeated transfers of mixed culture can lead to an overabundance of rods that lead to excessive acid production. The termination pH of fermentation and the rate of cooling should be adjusted to prevent continuation of growth of the lactobacilli; however, too rapid cooling can cause shrinkage of the curd and induce syneresis. Bitterness may also be caused by some strains of *Lactobacillus* or sporeforming bacteria. Yeasts cause undesirable changes in yogurt such as gas production, formation of visible colonies (white color of *K. marxianus* and pink of *Rhodotorula mucilaginosa*), and ethanol and citric acid formation.[7,8]

SURVIVAL OF PATHOGENIC BACTERIA

The presence of pathogens in high-acid foods indicates their potential to survive adverse conditions such as low pH, low water activity, and cold storage. Although the chance of pathogens to resist the severe heat treatment given to yogurt during pasteurization is very low, postmanufacture contamination is always a possibility. The survival of *Escherichia coli* O157:H7 in yogurt at pH values lower than 4.1 and stored at refrigeration temperatures is unlikely; however, yogurt can still be a vehicle for this bacterium if contaminated with high numbers and consumed soon after manufacture. Counts of *Staphylococcus aureus* do not necessarily decrease during yogurt manufacture but would significantly decrease during storage.[9] The inhibitory effect of yogurt on *S. aureus* is due to the presence of organic acids, hydrogen peroxide, and bacteriocins produced by lactic cultures, and not the low pH. *Listeria monocytogenes* can survive the yogurt fermentation, but the pathogen will significantly decrease during cold storage.[10] A health hazard would arise only if a high number of cells of *L. monocytogenes* contaminated the milk after pasteurization, which is unlikely in large modern plants. The same would be true for other pathogens too, making yogurt a safe and wholesome product.

CONCLUSION

Yogurt is popular fermented milk produced by the combination of *L. delbrueckii* subsp. *bulgaricus* and *S. thermophilus* with or without the incorporation of probiotic bacteria. It is consumed in many parts of the world for its acidic (and perhaps sweetened) taste and desired health benefits. The low pH, production of microbial inhibitors by lactic cultures, and cold storage prevent growth of most pathogens and make it a safe and wholesome product.

REFERENCES

1. Rademaker, J.L.W.; Hoolwerf, J.D.; Wagendorp, A.A.; te Giffel, M.C. Assessment of microbial population dynamics during yogurt and hard cheese fermentation and ripening by DNA population fingerprinting. Int. Dairy J. **2006**, *16*, 457–466.
2. Hassan, A.N. Possibilities and challenges of exopolysaccharide-producing lactic cultures in dairy foods. J. Dairy Sci. **2008**, *91*, 1282–1298.
3. Chaves, A.C.S.D.; Fernandez, M.; Lerayer, A.L.S.; Mierau, I.; Kleerebezem, M.; Hugenholtz, J. Metabolic engineering

of acetaldehyde production by *Streptococcus thermophilus*. Appl. Environ. Microbiol. **2002**, *68*, 5656–5662.
4. Beshkova, D.; Simova, E.; Frengova, G.; Simov, Z. Production of flavor compounds by yogurt starter cultures. J. Ind. Microbiol. Biotechnol. **1998**, *20*, 180–186.
5. Ekinci, F.Y.; Gurel, M. Effect of using propionic acid bacteria as an adjunct culture in yogurt production. J. Dairy Sci. **2008**, *91*, 892–899.
6. Elli, M.; Callegari, S.; Ferrari, S.; Bessi, E.; Cattivelli, D.; Soldi, S.; Morelli, L.; Feuillerat, N.G.; Antoine, J.-M. Survival of yogurt bacteria in human gut. Appl. Environ. Microbiol. **2006**, *72*, 5113–5117.
7. Lourens-Hattingh, A.; Viljoen, B.C. Survival of dairy-associated yeasts in yogurt and yogurt-related products. Food Microbiol. **2002**, *19*, 597–604.
8. Canganella, F.; Ovidi, M.; Paganini, S.; Vettraino, A.M.; Bevilacqua, L.; Trovatelli, L.D. Survival of undesirable microorganisms in fruit yogurt during storage at different temperatures. Food Microbiol. **1998**, *15*, 71–77.
9. Pazakova, J.; Turek, P.; Laciakova, A. Survival of *Staphylococcus aureus* during the fermentation and storage of yogurt. J. Appl. Microbiol. **1997**, *82*, 659–662.
10. Massa, S.; Trovatelli, L.D.; Canganella, F. Survival of *Listeria monocytogenes* in yogurt during storage at 4°C. Lett. Appl. Microbiol. **1991**, *13*, 112–114.

Zebu Cattle: Timed Artificial Insemination

Luiz Nasser
Department of Animal Reproduction, University of Sao Paolo, Sao Paulo, Brazil

Abstract
Nowadays, most of the world's large cattle herds are found in tropical regions. Because of their adaptation, *Bos indicus* (Zebu) cattle predominate in this environment. Biotechnologies of reproduction such as artificial insemination can be used as tools to spread and induce innovations in genetics into its herd.

INTRODUCTION

Because of climate and management conditions, nature has built the physiology and endocrinology system of Zebu cattle in such a way that they are able to survive and reproduce in much adverse conditions. One has to consider the differences between *Bos indicus* and *Bos taurus* cattle before implementing the reproductive biotechnologies. Strategies to improve their application in large scale, as are usually managed in tropical areas, include the keys to improve the production efficiency in these areas.

ESTROUS BEHAVIOR AND REPRODUCTIVE PHYSIOLOGY IN *B. INDICUS* CATTLE

Both *B. taurus* and *B. indicus* females present a homosexual behavior during estrus. However, *B. indicus* cattle express a temperament that can interfere in its detection, with animals presenting "silent" or "missed" heats.[1] Study utilizing radiotelemetry has also reported that Zebu animals express more mounting activity during the night period, with 30.7% of these beginning and ending during the night itself (06:00 P.M.–6:00 A.M.).[2]

Seasonality can also affect cyclicity in *B. indicus* cattle. There was a decrease in preovulatory LH-surge and their luteal cells in vitro were less responsive to LH during the winter.[3] An increased occurrence of anestrous and anovulatory estrus in Brahman females during the winter has also been reported.[3]

The characteristics of the estrus cycle related to follicular dynamics of *B. taurus* are similar to those of *B. indicus* cattle. The pattern of follicular growth, number of waves per cycle, and dominance are equivalent in both types of females,[2,4] except that *B. indicus* cattle seemed to have smaller maximum diameter of the dominant follicle (10–12 mm) and consequently, a tendency emerged to form a smaller CL (17–21 mm)[5] presenting a lower serum progesterone concentrations than *B. taurus*.[6]

The heifers of Nelore, a Brazilian Zebu breed, are able to ovulate follicles with a diameter between 7 and 8.4 mm[7] after getting treatment with exogenous LH, whereas Holstein cows acquire this same capability when follicles reach at least 10 mm.[8] It was also reported that at the time of deviation, the diameter of the dominant follicle was smaller in Nelore (6.0–6.3 mm)[7] than in Holstein (8.5 mm) cattle.[8]

During follicular emergence, it has been shown that the ovaries from *B. indicus* heifers bear more follicles <5 mm in diameter than ovaries from *B. taurus*.[6] There is evidence as well which states that ovarian insulin and insulin like growth factor (IGF) system differs between both types of cattle,[5] with *B. indicus* cows having a lower FSH concentration than *B. Taurus* cows. The differences in the FSH and IGF-I system may also explain why *B. indicus* donors are so highly sensitive to the FSH dosages currently used for superstimulation of *B. taurus* cattle.[9]

MANIPULATION OF FOLLICULAR DEVELOPMENT FOR SYNCHRONIZATION OF WAVE EMERGENCE AND OVULATION

Follicular-wave development occurs early in the postpartum period in both beef and dairy cattle. Gonadotropin secretion patterns in the postpartum period have been shown to differ between beef and dairy cattle, and this difference has appeared to increase over time after parturition. Although there is an increase in FSH followed by the emergence of a follicular wave soon after parturition, a decrease in LH-pulse frequency is also seen, leading to an inadequate exposure to LH, disrupting follicle growth, LH surge, and ovulation.[10] This situation is more evident in *B. indicus* than in *B. taurus* cattle, especially in suckled beef cows kept on pasture conditions, which could be explained by the low LH circulating concentration encountered in cows with poor body condition and poor pasture quality.[10]

Prostaglandins F2α (PGF) has been the most commonly used treatment for synchronization of estrus in cattle (Odde

1990). Although it is the cheapest method of estrous synchronization, the necessity to observe the signs of behavioral estrus and a functional CL on early postpartum make this method unfeasible to be used in large scale *B. indicus* beef herds.

Oestradiol and progestogen/progesterone treatments have been increasingly used over the past years in estrous synchronization programs in both types of cattle. Oestradiol treatments suppressed antral follicle growth[5] and depending on its half-life, different types of oestradiol esters will induce a more or a less predictable emergence of a new follicular wave. Currently, there are two types of devices in the market, the norgestomet ear implants and the vaginal P4-releasing devices. The ear implant is left in place for 9 days and an i.m. injection of 5 mg of Oestradiol Valerate (EV) and 3 mg norgestomed is given at the time of implanting. It has also been demonstrated in Zebu beef heifers that pregnancy results after EV synchronization protocols were variable and they could be related with the occurrence of anovulatory estrus, owing to a prolonged suppression of follicular growth failing to induce an LH-surge. However, satisfactory results were obtained in *B. indicus* adult cattle.[5] The most commonly used treatment with intravaginal P4-releasing device in *B. indicus* cattle is in association with a small dose (2 mg) of Estradiol Benzoate (EB) i.m. at the time of device insertion (Day 0), PGF treatment at device removal (Days 7 or 8), and 1 mg EB i.m. 24 hours later. Females are fixed timed artificially inseminated (FTAI) between 52 and 56 hours after device removal.[5] The factors that most affected pregnancy rates were body condition score and cyclicity of the cows.[5] More recently, the addition of eCG at the time of progestin/progesterone device removal has been proven to increase pregnancy rates in these categories of Zebu beef cows.[11] The use of 400 IU of eCG at the time of progestin/progesterone device removal has been largely used on synchronization protocols for FTAI improving reproductive efficiency in large beef herds.[11] It has been shown that the CL formed after an eCG-based protocol produce more progesterone then a regular CL and that the LH bioactivity present on the eCG preparation could also increase the final growth rate of the preovulatory follicle.[11]

In one experiment performed in Brazil using suckled Nelore cows,[12] a protocol using FTAI plus clean-up bulls was compared to traditional AI scheme and clean-up bulls or to a regular bulls exposure during a 90 days breeding season. Results showed that FTAI and bulls had not only 8% more pregnancy rate than bulls, but also hastened the conception by about 22 days. Conversely, the application of a traditional AI scheme of estrus detection and AI for 45 days and clean-up bulls for more 45 days was the least efficient program, a reflection of the difficult of estrus detection in suckled *B. indicus* cows.

Using the best program tested in the previous experiment, a large practical trial was done testing the efficiency of the use of FTAI for *B. indicus* cows in a large cow–calf operation in central Brazil. In this farm, a FTAI program was implemented in 5579 suckled Nelore cows. They were early in the postpartum period (i.e., 35 to 45 days postpartum), and the protocol used was with Crestar ear implant plus EV (5 mg) and 400 IU eCG at the time of implant removal, and cows were FTAI 54 hours later. Ten days later, they were exposed to clean-up bulls for the remainder of the breeding season. Overall pregnancy rate after FTAI was 50 (2817/5579) and the final pregnancy rate after two cycles with clean-up bulls was 80.7% (4390/5579). An anticipation of the calving season was also seen in this trial allowing a calving interval of 12 months under pasture conditions.

Research is now concentrating on developing a protocol to reach heifers and first calf cows, age categories that have given the worst results. One other objective is to decrease the amount of time that they have spent in a corral to be treated; this will decrease labor and animal stress.

CONCLUSION

While reproductive efficiency is crucial for producer's survival, the world's economic situation requires efficient management practices in order to increase the profitability of large beef cattle operation in tropical and subtropical areas. The use of a FTAI is a tool that could allow a production of one calf per cow per year with 50% of that by artificial insemination, improving herd genetic gain thus, increasing farm profitability.

It is in the hands of reproductive technicians to apply the most suitable hormone protocol, according to animal category, body condition score, and hormones available in the local market.

REFERENCES

1. Galina, C.S.; Arthur, G.H. Review on cattle reproduction in the tropics. Part. 4. Oestrus cycles. Anim. Breed. Abstr. **1990**, *58*, 697–707.
2. Barros, C.M.; Figueiredo, R.A.; Pinheiro, O.L. Estro, ovulação e dinâmica folicular em Zebuínos. Revista Brasileira de Reprodução Animal **1995**, *19* (1–2), 9–12.
3. Randel, R.D. Seasonal effects on female reproductive functions in the bovine (Indian breeds). Theriogenology, **1984**, *21*, 170–185.
4. Ginther, O.J.; Kastelic, J.P.; Knopf, L. Composition and characteristics of follicular waves during the bovine estrous cycle. J. Anim. Reprod. Sci. **1989**, *20* (3), 187–200.
5. Bo, G.A.; Baruselli, P.S.; Martinez, M.F. Pattern and manipulation of follicular development in *Bos indicus* cattle. Anim. Reprod. Sci. **2003**, *78*, 307–326.
6. Segerson, E.C.; Hansen, T.R.; Libby, D.W.; Randel, R.D.; Getz, W.R. Ovarian and uterine morphology and function in Angus and Brahman cows. J. Anim. Sci. **1984**, *59*, 1026–1046.

7. Gimenes, L.U.; Carvalho, N.A.T.; Sá Filho, M.F.; Ayres, H.; Torres-Júnior, J.; Souza, A.H.; Bó, G.A.; Barros, C.M.; Sartorelli, E.S.; Baruselli, P.S.; Mapletoft, R.J. Capacidade ovulatória em novilhas *Bos indicus* [Abstract]. Acta Scientiae Veterinariae. **2005**, *33*, 209.
8. Sartori, R.; Fricke, P.M.; Ferreira, J.C.P.; Ginther, O.J.; Wiltbank, M.C. Follicular deviation and acquisition of ovulatory capacity in bovine follicles. Biol. Reprod., **2001**, *65*, 1403–1409.
9. Barros, C.M.; Nogueira, M.F.G. Embryo transfer in *Bos indicus* cattle. Theriogenology **2001**, *56*, 1483–1496.
10. Wiltbank, M.C.; Gumen, A.; Sartori, R. Physiological classification of anovulatory conditions in cattle. Theriogenology **2002**, *57*, 21–52.
11. Baruselli, P.S.; Reis, E.L.; Marques, M.O.; Nasser, L.F.; Bo, G.A. The use of hormonal treatments to improve reproductive performance of anestrous beef cattle in tropical climates. Anim. Reprod. Sci. **2004**, *82–83*, 479–486.
12. Madureira, E.H., Baruselli, P.S., Pimentel, J.R.V., Almeida, A.B. A IATF possui custo benefício favorável. Acta Scientiae Veterinariae. **2005**, *33* (Suppl. 1), 141–143.

Zygote Intrafallopian Transfer (ZIFT)

Albert L. Smith
Fertility Lab Consulting, Deming, New Mexico, U.S.A.

Abstract

The transfer of zygotes into the fallopian tubes (ZIFT) is used as a method of assessing the fertilization potential of human gametes in a less hostile environment. ZIFT was also popular for a time because human zygotes could be cryopreserved successfully at the pronuclear stage, thus minimizing the amount of time taken to culture human embryos in vitro. Due to improvements in oocyte quality, in vitro fertilization, and in vitro embryo culture methods, ZIFT has largely been replaced in the field of human assisted reproduction.

INTRODUCTION

Zygote intrafallopian transfer, or ZIFT, was a procedure that was popular in the field of human assisted reproduction for a number of years during the late 1980s to the early 1990s. ZIFT was the marriage of in vitro fertilization (IVF) and gamete intrafallopian transfer (GIFT). Unlike the GIFT procedure, ZIFT assured that fertilization had occurred and was used as a method to minimize the in vitro culture time of human embryos. In the ZIFT procedure ova were collected one day, fertilized overnight in vitro, and the zygotes were replaced into the oviducts the following day. This ensured that fertilization had occurred. It was also logically concluded that the oviducts were more suitable environments for early embryonic development than in vitro culture conditions. It was also assumed that transferring zygotes back into the oviducts would also decrease the asynchrony which develops between development of in vitro fertilized embryos and in vivo fertilized embryos.

Although ZIFT was utilized successfully for a number of years in human assisted reproduction, it has largely been replaced by newer techniques. The origins of the ZIFT procedure, the use of ZIFT, advantages and disadvantages for the use of ZIFT, and the ultimate replacement of the ZIFT procedure by newer in vitro culture methodologies will be discussed.

Human assisted reproduction began with the birth of Mary Louise Brown over 20 years ago in England, although in vitro fertilization (IVF), with the birth of live offspring, had been done in a wide range of species prior to that point. Unlike the lab species, which had been studied extensively, little was known (and still is known) about human IVF. In lab species it was common to inseminate the ova from one female and transfer the fertilized embryos to other recipient females. IVF and in vitro culture conditions resulted in embryos, which were slower in developing compared to in utero derived embryos. Therefore, recipient females were synchronized to be about a day behind the donor females, so that the embryos and uteri of the recipient females would be synchronized. Optimal culture conditions for these species were well documented in the literature, as are cryopreservation techniques for preserving embryos in liquid nitrogen. In humans, virtually nothing was known about stimulation protocols for retrieval of eggs, in vitro fertilization and in vitro culture conditions of embryos, embryo cryopreservation, or embryo transfer itself. So, the development of the ZIFT procedure was based more on serendipity and dogma rather than scientific research.

Gamete intrafallopian transfer, the forerunner of the ZIFT procedure, was initially developed in response to a specific case where a couple going through an in vitro fertilization program was leaving the country immediately after the IVF procedure was to be performed. The stimulation cycle ran longer than anticipated and the IVF team was faced with developing a procedure that could be done in one day, allowing the couple to leave the next day. It was determined that since the patient had patent oviducts (fallopian tubes) the ova collected from the procedure could be mixed with the husbands' sperm and returned to the fallopian tubes for fertilization and early development to occur, since this is normally the site of fertilization and early development. Four ova were collected, mixed with the husband's sperm, and replaced into the oviduct via the fimbriated end. The patient became pregnant with twins, so thus the GIFT procedure was born.[1]

The GIFT procedure also found favor among human IVF labs, which had difficulty achieving pregnancies through IVF. It was also reasoned, quite correctly, that the human oviducts were a less hostile environment than the in vitro culture conditions of that time.

One problem with the GIFT procedure was that it was not possible to determine whether or not fertilization actually occurred, unless a pregnancy occurred. The ZIFT procedure was developed in response to this problem. By fertilizing oocytes in vitro, and transferring the zygotes, it was possible to determine whether or not fertilization had occurred.

By transferring zygotes into the oviducts the next day, in vitro culture was minimized, since in vitro cultured embryos developed more slowly than embryos developing in utero in the early stages of IVF. Cryopreservation of pronuclear embryos had been reported,[2] so it was possible to cryopreserve excess zygotes for future frozen/thaw cycles. This success combined with the greatly shortened culture time for each patient resulted in reduction of the time spent in the lab, making ZIFT popular with embryologists as well. Of course, the ZIFT procedure could only be used on patients with patent oviduct(s). By coupling the GIFT procedure with the IVF procedure, it was now possible to determine whether fertilization had actually occurred.

Because of the large cumulus mass associated with human oocytes, it was necessary to strip the remaining cumulus cells from the oocytes to in order to see the two pronuclei. In most cases this was done manually by using a glass pipette pulled down to a diameter of approximately 100–120 microns. The oocyte was aspirated in and out of the pipette until the cumulus cells were knocked off and the pronuclei visualized. In the early days of assisted human infertility, this step was certainly the most nerve-wracking for embryologists, since a pipette with too small a diameter could rupture the oocyte, thus destroying it. Other alternatives include using a 30 gauge needles attached to 1 cc tuberculin syringes. The cumulus could be trimmed off by using the needles like a knife and a fork to trim the cumulus cells off enough to visualize the pronuclei.

After all the ova were examined, the zygotes to be transferred were selected. Zygotes with the largest pronuclei, and those where the pronuclei were closest together were chosen, since these were assumed to be the most fully developed ones out of all the zygotes. Zygotes to be transferred were rinsed in a drop of fresh medium, and placed in a holding drop of fresh medium until the time of transfer. The remaining zygotes were either frozen at that point or cultured a few hours further before freezing. At that time in human assisted reproduction, pregnancies had been reported with cryopreservation of zygotes and 4-cell embryos, but not 2-cell embryos.[2] Therefore, freezing zygotes was preferable to freezing 4-cell embryos after another 24 hours or so in culture. ZIFT was advantageous to the embryologist because culture time was reduced to a minimum and excess embryos could be frozen at that point, thus minimizing the time from oocyte aspiration until transfer.

The transfer catheter was a long flexible piece of tubing, small enough to fit into the oviduct but large enough to hold the embryos safely. The catheter was generally attached to a 1 cc tuberculin syringe, although other types of syringes were sometimes used. The embryos were subsequently drawn into the catheter in 20–30 μL of sterile medium. The transfer medium used was dependent on the preference/superstition of the embryologist/physician.[3] To reduce the possibility of a pH change while going from a 5% CO_2 atmosphere to room air, Dulbecco's modified PBS, a HEPES-based medium, or a medium containing a lower concentration of sodium bicarbonate was used. Usually, 4–6 zygotes were transferred at a time, sometimes into one oviduct, sometimes two zygotes were placed into each oviduct. The zygotes were drawn into the catheter immediately before the transfer to minimize the time during which the zygotes were potentially exposed to the ambient conditions. The transfer itself was done either via through the fimbriated end of the oviduct or with the aid of an ultrasonically guided catheter through the cervix, uterus, fundus, and into the oviducts. In the early days of ZIFT, a laparoscope and fiber optics were used to guide the end of the catheter approximately 2 cm into the fimbria. The end of the fimbria was held with a pair of pick-ups as the end of the catheter was guided into the end of the oviduct. The contents of the catheter were gently expelled into the oviduct. In extreme cases, a small incision was made, allowing the physician to place a hand through to guide the catheter into the end of the fimbria. This procedure was actually a minor surgical procedure, which entailed the use of an operating room and sterile conditions. This procedure added considerably to the added expense of the ZIFT procedure.[4] over other assisted reproductive technology procedures.

The second procedure for transferring zygotes involved the use of an ultrasound machine and specially designed catheters. The tip of the catheter was coated with a compound, which made the tip of the pipette visible while using ultrasound guidance to direct the catheter through the cervix, the body of the uterus, and through the fundus into the oviduct. The tip of the catheter was specially coated to be visible with ultrasound, and could thus be guided into the oviduct non-surgically. After the insertion of the embryos into the oviduct, the catheter was rinsed. The rinse medium was examined to determine that all zygotes had been introduced into the oviduct, and that none had gotten stuck in the catheter. Several weeks after the procedure, a pregnancy test was performed.

Although ZIFT was first reported in the late 1980s and represented about 7% of the total assisted reproduction cases performed in the United States during the early 1990s,[5] the overall results with the ZIFT procedure were not statistically significantly different than conventional IVF or GIFT. By the late 1990s, the programs reporting ZIFT procedures dwindled down to a handful. Today, less than 1% of the assisted reproductive technology procedures done in the United States are ZIFT procedures.[6] In other countries, ZIFT remains popular for patients with a history of repeated infertility failures.[7,8] In these cases ICSI is performed on the oocytes and the microinjected oocytes are transferred back into the oviducts.

While one would expect a high pregnancy rate when replacing four (or more) embryos into the oviduct, this was simply not the case The average pregnancy rate when replacing four embryos, reported in 1995 was approximately 27%, with an implantation rate of 10–15% per embryo.[9] Because ZIFT involved extra procedures, i.e., the oviductal transfer, it was more expensive than conventional IVF, especially when the surgical transfer procedure was utilized. Thus the ZIFT procedure quickly lost popularity with patients and insurance companies that were paying for the procedure.[4] ZIFT pregnancy rates were only minimally higher than conventional IVF (30% for ZIFT vs. 27% for conventional IVF). Because patent oviducts were required for the ZIFT procedure, not all patients qualified for the ZIFT procedure because approximately 20%[9] of patients undergoing ART procedures had a tubal disorder, which prevented the procedure from being used on those patients.

Another, and quite unexpected, problem with the ZIFT procedure was the incidence of ectopic pregnancies.[10] In large animals, recipient females are not hyperstimulated with gonatrophins to induce the production of a large number of ova. Recipient females are normally cycling, and not exposed to the hyperstimulation treatments. The hyperstimulation of the ovaries in women undergoing infertility treatment causes a 100-fold increase in the normal levels of circulating estrogen. This abnormally high level of estrogen may stimulate the muscle cells of the uterus and oviduct, resulting in retrograde movement embryos out of the oviduct and even the uterus into the abdominal cavity. Even in conventional IVF, where embryos are placed into the uterus, ectopic pregnancies occur because of a retrograde movement.

Another factor which helped contribute to the demise of the ZIFT procedure was the simple fact that the human uterus is not the hostile environment it is in many other species, because zygotes, 2-cell embryos, and 4-cell embryos, can be transferred directly into the uterus, implant, and produce live offspring. Published reports indicate that transfer of zygotes, 2-cell, 4-cell human embryos results in about the same implantation rates. However, implantation rates are higher with the transfer of 8-cell embryos, morulae, or blastocysts.[10] The trend in the last few years has thus been to culture for longer periods of time in vitro and do transfers at 3–5 days post oocyte retrieval, rather than 1–2 days post oocyte retrieval.

Although the initial logic behind doing ZIFT was to return embryos to an in situ environment as quickly as possible, improvements in culturing human embryos have led the infertility field toward extended culture, rather than shorter culture. However, long-term culture of human embryos does still have one major limitation: it works on only a small subset of the population undergoing infertility treatment.[10] Patients who stimulate well, resulting in the retrieval of 20 or more oocytes, benefit the most from long-term culture and transfer of blastocysts because human embryos do not culture as well or are as viable as other species. In vitro, 20 oocytes, after fertilization and long-term culture may yield only 3–4 blastocysts to transfer.

It is estimated that approximately 50% of human embryos carry a chromosomal defect of some kind. Long-term culture may simply be weeding out those abnormal embryos. In effect, long-term culture is a selective tool, which does not improve the overall implantation rates, but simply provides a better pool of embryos from which to select.

CONCLUSIONS

What is the future of ZIFT given the trend towards long-term culture? Not very promising, although there are reports that selection of embryos at the pronuclear stage may yield benefits. Determining which embryos had the highest potential for development at the pronuclear stage would certainly be advantageous, so some form of the ZIFT procedure may reappear in the field of human infertility at some time in the future.

REFERENCES

1. Asch, R.H. GIFT and associated techniques. In *Gamete Physiology*; Asch, R.H., Balmaceda, J.P., Johnson, I., Eds.; Serono Symposium 75 publisher; Norwell, MA US, 1990.
2. Cohen, J.; DeVane, G.W.; Elsner, C.W.; Kort, H.I.; Massey, J.B.; Norbury, S.E. Cryopreserved zygotes andembryos and endocrinologic factors in the replacement cycle. Fertil. Steril. **1988**, *50* (1), 61–67.
3. Braude, P.R. Fertilization of human oocytes and culture of human pre-implantation embryos. In *Mammalian Development a Practical Approach*; Monk, M., Ed.; IRL Press: Washington, 1987; 281–305.
4. Griffin, M.; Panak, W.F. The economic cost of infertility-related services: an examination of the Massachusetts infertility mandate. Fertil. Steril. **1998**, *70* (1), 22–29.
5. Toner, J.P. Progress we can be proud of: U.S. trends in assisted reproduction over thepast 20 years. Fertil. Steril. **2002**, *78* (5), 943–950.
6. 2007 Assisted Reproductive Technology Success Rates. National Summary and Fertility Clinic Report, http://apps.nccd.cdc.gov/ART/NSR.aspx?SelectedYear=2007 (accessed December 29, 2009) 75.
7. Fahri, J.; Weissman, A.; Nahum, H.; Levran, D. Zygote intrafallopian transfer inpatients with tubal factor infertility after repeated failure ofimplantation with in vitro fertilization-embryo transfer. Fertil. Steril. **2000**, *74* (2), 390–393.
8. Alleyassin, A.; Khademi, A.; Aghahosseini, M.; Safdarian, L.; Saeidabadi, H.S.; Hedayati, H. Comparison of unilateral and bilateral transfer of injected oocytes intofallopian tubes: a prospective, randomized trial. Fertil. Steril. **2006**, *85* (1), 96–100.
9. Assisted reproductivetechnology in the United States and Canada: 1995 results generated from the American Society for Reproductive Medicine/Society for Assisted Reproductive Technology Registry. Fertil. Steril. **1998**, *69* (3), 389–398.
10. Smith, A.L. Blastocyst culture in human IVF: the final destination or a stop along theway. Proceedings of the Annual Conference of the International Embryo Transfer Society, Foz do Iguassu, Parana, Brazil, January 12–15, 2002. Theriogenology *57* (1), 97–107.

Index

α-acetolactate decarboxylase, 221
α-amylase, 4, 178, 179, 181, 221, 223, 225
α-cyclodextrins, 187, 188
 chemical structure and shape of, 188
α-d-glucans, 418
α-d-glucopyranosyl α-d-glucopyranoside, 181
α-glucanases, 4
α-glucanotransferase, 221
α-glucosidase, 80, 179, 181
α-glucuronidase, 80
α-ketoglutarate, 192
α-linolenic acid, 374
α-pinene, 368
α1,3-galactosyltransferase gene, 164, 704
β-1-3-glucan, 381
β-amylase, 4, 179, 180, 181, 223, 224, 225
β-carotene, 154–155, 445
β-casein, 331
β-conglycinin, 13
β-cyclodextrins, 187, 188, 189
β-d-fructans, 418
β-d-glucan, 418
 with 1–6 branching, 77, 80
β-galactosidase, 192, 213, 221
β-glucanases, 1, 71, 72
β-glycosidase enzyme, 389, 691
β-lactoglobulin, 30, 330
β-mercaptoethanol, 125
β-myrcene, 368
β-pinene, 368
β-sitosterol, 517, 518
β-thalassemia, 294
β-thujene, 368
β-phospho-galactosidase (β-PGal) enzyme, 192
γ-linolenic acid, 374
γ-terpinene, 368
γ-cyclodextrins, 187, 188
γ-decalactone, 48
δ-aminolevulinic acid, 85
κ-casein, 228
1,4-α-d-glucan 4-α-d-(1, 4-α-d-glucano)-transferase, 223
1,4-α-d-glucan glucanohydrolase, 223
1,4-α-d-glucan maltohydrolase, 223
2-D protein gel electrophoresis, 62
2.5-Furandicarboxylic acid, 84
2,2(azino-di-[3-ethylbenzthiazoline sulphonate] (ABTS), 213
2,3-butanediol, 192
2,3-butanedione, 74
2,3-pentanedione, 74
2,4-D, 109
2,4,4-trinitrotoluene, 109
2-ethyl-3,5-dimethyl- and 2,3-diethyl-5-methylpyrazine, 172
2-methyl-3-(methyldithio)furan, 172
$2n$ gametes, 160
2S albumins, 13
3-hydroxybutyrolactone, 84
3-hydroxypropionaldehyde (3-HPA), 86
3-hydroxypropionic acid, 84
3-methylbutanol, 172
3-hydroxy-3-methylglutaryl-coenzymeA reductase (HMG-CoA reductase), 77
4-carene, 368
4-O-α-d-glucopyranosyl-d-glucose, 178
4-hydroxy-2(or 5)-ethyl-5 (or 2)-methyl-3(2H)-furanone (HEMF), 432
4-hydroxy-2,5-dimethyl-3(2GH)-furanone (HDMF), 432
5-enolpyruvylshikimate-3-phosphate synthase (EPSPS), 687
5-aminolevulinic acid (ALA), 487–488
6-pentyl-α-pyrone, 48
7S globulins of soybean, 13
11S globulins of peanut (Ara h 3), 13
18 S rDNA, 149
49/2000/EC Regulation, 554
50/2000/EC Regulation, 554
1829/2003/EC Regulation, 554, 555
1830/2003/EC Regulation, 554–555

A

ABB Environmental Service, 111
ABCG5, 517
ABCG8, 517
Abelson virus, 80
Absidia, 156
Absidia atrospora IF09471, 157
Absidia coerulea, 157
Absidia coerulea IF04011, 157
Absidia coerulea IF04012, 157
Absidia glauca, 157
Absidia glauca IF04002, 157
Absidia glauca IF04003, 157
Absidia glauca IF04004, 157
Absidia glauca var. *pardoxa IF04007*, 157
Absidia glauca var. *paradoxa IF04431*, 157
Abutilon mosaic, 502
Abutilon mosaic virus, 500, 502
Acetaldehyde, 195, 201, 708
Acetate, 708
Acetic acid (CH3COOH), 44, 138, 675
Acetic acid bacteria, 387, 675–676
 during acetification process, 676
Acetification, 675
Acetobacter, 141, 676
Acetobacter xylinum, 44, 420
Acetobacteraceae, 676
Acetohydroxy acids, 74
Acetoin, 192, 708
Acetolactate decarboxylase, 4
Acetolactate synthase (ALS), 686
Acetyl-coenzyme A acyltransferase 1 (ACAA1), 317
Acetylcholinesterase (AChE), 523–524
Acid-fermented leavened dough sourdough, 364–365
Acid-leavened breads
 enjera, 365
 idli and dosa, 365–366
 kisra, 365–366
Acne group strain, 557
Act c 1 of kiwi, 13
Actinobacteria, 39
Actinomucor taiwanensis, 45
Actinomycetes, 39, 107, 108
Active Dry Wine Yeast, 690
Activin, 604
Acyl-CoAs, 652
Acyl-coenzyme A oxidase 1, palmitoyl (ACOX1), 317
Acylglycerolases, 370
Acylhydrolases, 370
Adai, 135
Adenosine deaminase, 294
Adenovirus type 12, 80
Adult stem cells, 23, 594, 597
Adulteration, detection of, 534
Adventitious presence (AP), 639, 640
Aegilops species, 185
Aequorea Victoria, 423
Aerococcus urinaeequi, 474
Aflatoxins, 591
African cassava mosaic virus, 501, 502
Agaricus bisporus, 76, 79, 157

Agaricus blazei, 79
Agarose gel electrophoresis, 682
Agbagba (AAB), 68
Ageniaspis citricola, 101
Agriculture and veterinary applications, of RNAi, 637–638
Agro-industrial residues, 48
Agrobacterium radiobacter, 109
Agrobacterium rhizogenes, 468
Agrobacterium sp., 420, 687
Agrobacterium tumefaciens, 14, 30, 293, 294, 301, 661
Agrobacterium-mediated transformation, 629
Agrocybe aegerita, 79
AINTEGUMENTA (*ANT*), 467
Akkadian Empire, 205
Albatrellus ovinus, 78
Alcalase, 331
Alcaligenes faecalis, 420
Alcaligenes sp., 420
Alcohol production, 221
Alcoholic and lactic acid fermentations, 7
Alcoholic Beverage Labeling Act 1998, 260
Alcoholic beverages. *See* Starter cultures
 brewing, enzymes used in, 2–3
 cider-making, enzymes used in, 4
 exogenous enzymes uses, 1
 wine-making, enzymes used in, 4
Alcoholic fermentation
 inoculation after, 388
 inoculation during, 388
Ale yeasts, 134
Ales, 9
Alfalfa, 51, 463
Algae, 109
Algal pond, basic design of, 79
Alkali catalysis, conventional feedstocks and, 96
Alkaline phosphatase, 213
Allergen detection, 534
Allergens transfer, 118
Allergic asthma, 383
Allergies, food and peanut risk reduction, 11–16
Alpha-parvalbumins allergens, 12
Altus Food Company, 455
Amano Enzyme Inc., 373
Amaranthin, 209, 211
Amaranthus rudis, 327
American bison, 343
Amino acids, 85, 125, 192, 708
Aminopeptidases, 237
Ammonia, 22
Amplicon, 531
Amplified fragment length polymorphism (AFLP), 307, 506, 591–592, 687
Amycolatopsis (*Streptomyces*) *mediterranei*, 41

Amycolatopsis (*Streptomyces*) *orientalis*, 41
Amylase, 71, 223–225. *See also* Enzymes and lactic acid-producing organisms, 356–357
Amyloglucosidases, 179, 221, 225
Amylolytic lactic acid bacteria
 direct lactic acid fermentation by, 357
Amylopectin, 178, 223
Amylose, 178
Anabaena CH3, 109
Anabaena doliolum, 109
Anaerobic phases, 172
Analytical breeding, 159–160
Anarshe, 135
Angelman syndrome (AS), 277–278
Angus, 316
Animal and Plant Health Inspection Service (APHIS), 282, 283, 303, 633, 634
Animal agriculture
 DNA microarrays, functional genomics through, 316–318
 systems biology approach in, 318, 319
Animal biotechnology, 286
 clones, 287–288
Animal cell culture
 biopharmaceutical products from, 23
 bioreactors types, 21–23
 cell source and characteristics, 18, 19
 cellular assays for high throughput drug screening, 24
 clinical applications, cell and tissue types examples for, 23
 culture media, 20
 culturing conditions, 19–21
 dissolved oxygen (DO), 21
 industrial applications, 23–24
 pH, 20–21
 solid substrate, 19–20
 temperature, 21
Animal cloning. *See* Nuclear transfer (NT)
Animal food allergens, 12–13
Animal medicines
 antibiotics for, 39–41
 delivery systems, 26–27
 diagnostic detection and techniques using, 25–26
 discovery and development of, 25
 vaccine development and applications, 26
Animal models, 217
Animal-expressed proteins, purification of, 564
Animals
 genetic engineering in, 294–295
 plants and genetically modified organisms. *See* Genetically modified organisms (GMOs)
 RNA interference (RNAi), 33–37
Anthocyanins, 520

Anthraquinones, 209
Anti IgE vaccines, 13
Antibiotic-resistant genes, 118
Antibiotics, 85
 legislature, European approach, 40
 nature and use of, 39–40, 41
 toxicological models of assessment, maturing perspectives, 41
 used in food-producing animals, 39
Antidiabetic effect, and mushrooms, 80
Anti-Gal antibodies, 701
Antigen, 213
Antigen-presenting cells (APCs), 12
Antihistamines, 13
Anti-infective molecules, delivery of, 666
Antimicrobial effects, and mushrooms, 80
Antimicrobial packaging
 bacteriocins, 43
 bio-based polymers, 45
 cellulose, 44
 chitosan, 45
 edible biopolymers, 44–45
 films, 45
 lysozyme, 44
 organic acids, 44
 polyhydroxyalkanoates (PHA), 45
 polylactic acid (PLA), 45
 starch, 44–45
Antioxidant activity, and mushrooms, 80
Antioxidant enzyme, cellular signaling for induction by chemopreventive phytopharmaceuticals, 153–154
Antioxidant response elements (ARE), 153–154
Antisera, 212
Antitumor, 39
Antlerogenic periosteum (AP), 169
Aphelinidae, 336
APHIS Biotechnology Regulatory Services (APHIS-BRS), 282
Api g 1 of celery, 13
Api g 4 of celery, 13
Apical meristems, 520
Apolipoprotein A-I (APOA1), 317
Apple, 13, 14, 51
Apple pomace, 48
Aquatic expression systems, 562
Aqueous two-phase extraction (ATPE), 408
Ara h 1, 14
Ara h 2, 14
Ara h 3, 14
Ara h 3/4, 14
Ara h 4, 14
Ara h 5 of peanut, 13
Ara h 6, 14
Ara h 7, 14
Ara h 8 of peanut, 13
Arabidopsis, 308, 383, 579, 580, 581, 652, 688
The Arabidopsis Information Resource (TAIR), 308

Arabidopsis thaliana, 205, 383, 384, 457, 651
   gametophytic mutations and genes affecting pollen development in, 385
Arachidonic acid, 374
Aralia cordata, 520
Arber, W., 292
Archer Daniels Midland, 373–374
Argentina
   GM crops in, 301
Arginine, 677
Argonaute proteins (AGOs), 579
Arizona Genomics Institute (AGI), 307
Arizona Genomics Institute Computational Laboratory (AGCoL), 308
Armenian Red sheep, 343
Armillariella mellea, 79
Aroma compounds
   applications of, 48
   microbial production, 47–48
   plant cell cultures, 48–49
   solid-state fermentation, 48
Aromas, 47
Aromatic polyols, 84
Aromatics, 82
Arroz fermentado, 135
Arteriosclerosis process, 77
Artificial insemination (AI), 394, 403. *See also* Insemination
   semen collection, 56–57
   semen extension and preservation, 57
Arthobacter ilicis, 109
Arthrobotrys, 441
Arthropod host-plant resistant crops
   crops percentage and pest resistance degree, 50, 51
   economic and social benefits, 51
   plant resistance effect, on pest populations, 51
Artificial chromosomes, in plants, 53–54
Artificial vagina uses, of semen collection, 57
Ascorbic acid, 76, 225
Ashbya gossypii, 157
Asian lady beetle, 336
Asilomar, 293
Asilomar conference, 295
Asn, 13
Asparagus, 51
Aspartame, 85
Aspartic acid, 84
Aspartyl proteinases, 228
Aspergilli, 138
Aspergillus, 138, 139, 141, 156, 365, 451
Aspergillus awamori, 356, 357, 452
Aspergillus clavatus, 157
Aspergillus flavus, 140, 157, 591
Aspergillus nidulans, 157, 452
Aspergillus niger, 3, 4, 44, 140, 157, 180, 196, 449, 452

Aspergillus oryzae, 4, 9, 135, 141, 157, 178, 431, 432, 433, 591
Aspergillus parasiticus, 591
Aspergillus sojae, 591
Aspergillus terreus, 157
Aspergillus terricola, 157
Aspergillus tubigensis, 452
Aspergillus usamii, 140, 157
Assisted reproductive technologies (ART), 347, 394
Association of Official Seed Analysts (AOSA), 587
Asynchronous authorizations, 589
Atherosclerosis, 294
Atmospheric plasma, 174
Atomic force microscope (AFM), 438
ATP-driven translocase system (PTS), 192
AtpB genes, 505
Atrazine, 109
Auger de-excitation process, 175
Augmentation implementation strategy, in genetic improvement, 88
Aureobasidium pullulans, 420
Aureobasidium sp., 8
Auricularia auricula, 76
Auricularia auricula-judae, 80
Auricularia auricula-judas, 78
Auromonas (*Pseudomonas*) *elodea*, 419
Authorized GMOs, 302–303
AutoMotif, 451
Auxin, 466, 467
Avena sativa, 184, 185
Aventis Crop Science Pty. Ltd., 293
Aventis Starlink corn, 303
Aviagen, 536
Avian pathogenic *Escherichia coli* (APEC), 25
Aya-bisbaya, 135

B

B chromosome, 54
Baby hamster kidney (BHK21), 18, 19
Babylonians, 677
BAC library, 307
Bacillus acidipullulyticus, 4
Bacillus anthracis, 60, 662, 666
Bacillus cereus, 139, 177, 488
Bacillus licheniformis, 178
Bacillus macerans, 224
Bacillus polymyxa, 420
Bacillus popilliae, 103
Bacillus spp., 60, 62, 198, 358
Bacillus stearothermophilus, 178, 223
Bacillus subtilis, 1, 3, 4, 60, 62, 80, 196, 491
Bacillus thuringiensis (Bt), 103, 285, 293, 337, 464, 527, 626, 630
Bacillus thuringiensis insecticidal crystal proteins (Bt ICPs), 630
Back-slopping, 6, 195, 366, 670

Backcrossing, 614
Bacteria, 107, 108. *See also* Gram-positive bacteria
Bacterial artificial chromosomes (BACs), 307
Bacterial cellulose, 420
Bacteriocins, 43
   in fermented dairy products, 198
Bacteriophages, 477
   mechanisms of, 64–65
   pathogen control, 64
   and their control, 202–203
Bacteroides thetaiotaomicron, 416
Badnaviruses, 499–500
"Bad" cholesterol, 77
Baechu kimchi, 348, 349, 351
Bakers' yeast
   basic characteristics and production, 133–134
   commercial formats, 134
   flavours, 134
Baking, 221
   lipases in, 372, 373
Balanced salt solutions (BSS), 20
Balao balao, 135
Balsamic vinegars 675, 676
Banana bunchy top disease (BBTD), 501, 502–503
Banana streak diseases, 499–500
Banana streak virus (BSV), 499
Bananas, 161. *See also* Somatic cell genetics
   genetic transformation
   protoplast isolation and culture, 69–70
   somaclonal variation and mutation induction, 68–69
   somatic embryogenesis and organogenesis, 69
Banku, 135
Bantu beer. *See* Kaffir beer
Barley, 13, 51, 224
Barley *mlo* gene, 377
Basta®, 293, 326
Batch cell culture process, 22
Bayer Crop Science, 303
Bean, 51
Bean golden mosaic disease, 501, 502
Bean golden yellow mosaic virus, 502
Beckwith–Wiedemann syndrome (BWS), 277–278
Bedouins of North Africa, 138
Beer and beverages, 224
Beer fermentation
   brewhouse processes, 72–73
   fermentation, 73
   malting, 71–72
   maturation and finishing, 73–74
Beer product, 8
Beet mild curly top virus, 502
Begomovirus, 500
Benzene, 84, 189

Benzene, Toluene, Ethylene, and Xylene (BTEX), 109
Benzoic acid, 44
Best linear unbiased predictions (BLUP), 217, 537
Bet v 1, 13
Bet v 2, 13
Beta vulgaris, 184
Beta vulgaris ssp. *maritima*, 184
Beta-ketoacyl-CoA synthase (KCS), 650
Beta-parvalbumin allergens, 12–13
Betapol, 372
Beverages and beer, 224
Bhatura, 135
Bicellular and tricellular pollen, 383
Bifidobacteria, 546, 709
Bifidobacterium bifidum, 709
Bifidobacterium breve, 709
Bifidobacterium dentium, 546
Bifidobacterium infantis, 709
Bifidobacterium longum, 139, 416, 709
Bifidobacterium sp., 196, 197, 707, 709
Bioactive compounds, from mushrooms, 76–81
Bioactive peptides production, 270
  fermented dairy products, 197–198
Bioamines, 9
Bioavailability, 270
Bio-based chemicals
  renewable carbon for biorefinery production. *See* Renewable carbon for biorefinery production
Bio-based polymers
  polyhydroxyalkanoates (PHA), 45
  polylactic acid (PLA), 45
Biocatalysts Ltd., 373
Biochemical changes effects and general practices
  during cocoa fermentation, 171–172
Bioconjugation, 435–436
Biocontrol agents
  genetic improvement history, 87
  transgenic technologies, 88–89
  two main implementation strategies, 87–88
Biocontrol, of weeds, 463
Bioconversions. *See* Ethanol
Biodiesel (enzymatic production)
  alternative feedstocks and enzymatic alcoholysis, 96
  conventional feedstocks and alkali catalysis, 96
  lipases used directly with lipid feedstock, 96–97
  microorganisms for biodiesel production, 96
Biofungicides, 104–105
Biogenesis Technologies, 111
Bioherbicides, 105
Bioinformatics, 318, 693
Bioinsecticides, 103–104

Biolistic method, 661
Biological basis, of sperm sexing, 390
Biological control. *See also* Plant pathogens
  definition of, 335, 672
  evidence for improved efficacy, 99–100
  future prospects, 101
  general principles, 99
  of insects and mites, 335
  natural enemies, types of, 335–337
  new associations, 99
  of plant diseases, 514
  in practice, 337–338
  recent cases employing new associations, 100–101
  strategies for, 335
  successes and failures, 99–101
Biological control agents (BCAs), 608
  genetic modification of, 673–674
Biological nanotechnology, 434
Biomal®, 463
Biomass, 83
Bio-oils, 83
Biopesticides
  biofungicides, 104–105
  bioherbicides, 105
  bioinsecticides, 103–104
  future directions, 105
Biopharmaceutical market, growth rate, 18
Biopim, 111
Bioreactors types, in animal cell culture, 21–23
Biorefinery production, renewable carbon for. *See* Renewable carbon, for biorefinery production
Bioremediation
  constructed strains, 109–110
  future developments, 111
  pesticides. *See* Pesticides System, 111
  technology, 110–111
Bioreporter, 439
BioSeed®, 23
Biosensors, 438–439
  in agriculture and food products, 114–115
  to measure pungency, 114
  immobilization, 113–114
  membranes, 113
  metabolites, 114
  pathogen detection, 114–115
  pesticide residues, 114
  principal components of, 113
  sampling, 113
  technology, 112–114
Bioshield strategy, 666
Biotechnological aroma compounds. *See* Aroma compounds
Biotechnological control options, of pathogen, 514

Biotechnology, 416
  allergens transfer, 118
  antibiotic-resistant genes, 118
  applications and prospects, 116–118
  consultation, 528
  consumer choice, 122
  diagnostic detection and techniques using, 25–26
  dilemmas and concerns, 118
  discovery and development, of medicines, 25
  ethical aspects, 120–121
  food fermentation, 477
  food safety, 477–478
  food spoilage, 478
  framework of, 526–527
  gene escape and genetic pollution, 118
  of genomic resources, 306
  herbicide and insecticide resistance, 118
  moral objections, 122
  non-pesticide traits, 527
  pesticidal traits, 527
  plant-made industrial products (PMiP), 527
  plant-made pharmaceuticals (PMPs), 527
  product-specific oversight, 527
  public concerns, 120
  revolution, 647
  risk, 121–122
  social, moral, and ethical issues, 118
  usefulness, 121
Biotechnology Regulatory Services (BRS), 527
Biotin, 76
Biotinylating, 435
Biotransformation process, 47, 585
BioVECTOR, 104
Black-dot spurge beetle, 463
Blackberry, 160
Blakeslea trispora, 157
BLAST, 451
Blastomeres, 167
Blueberry, 160
Boar sperm cryopreservation, 353, 354
Board of Appeals and Interferences of U.S. Patent and Trademark Office, 340
Bogobe, 135
Bollgard™, 614
Boltzmann factor, 174
Bom-kimchi, 348
Bordeaux mixture, 483
Borlaugh, N., 296
Bortedella, 664
Bos gaurus gaurus, 343
Bos indicus cattle, 711, 712
  estrous behavior and reproductive physiology in, 711
Bos javanicus, 343
Bos taurus, 165, 711
Bos taurus indicus, 342, 343, 344

Bos taurus taurus, 342, 343, 344
Botrytis cinerea, 4, 105, 157, 630
Bova-Pro, 558
Bovine 50K Illumina™ iSelect chip, 314
Bovine, 700
Bovine embryos, in vitro culture of
   assessing culture effectiveness, 126, 127
   culture systems, 123
   environment, 125–127
   medium, 123–124
   supplements, 124–125
Bovine milk, 227
Bovine serum albumin (BSA), 124, 212, 213
Bovine spongiform encephalopathy (BSE) gene, 164, 265
Boza, 139
Brachiola, 573
Braconidae, 100
Brassica, 652
Brassica napus, 184, 185, 650, 651, 652
   production of wax esters in, 651–652
Brassica napus L., 328
Brassica rapa, 186
Brazil, 91
   GM crops in, 301
Brazil nut (Ber e 1 and Ber e 12), 13
BrdU protocol, 427
Bread and bakery products, 224
Breeding. *See* Cloning
Brevibacterium linens, 144, 145, 201
Brewers, 73
Brewers' yeast, 134–135
Brewing, 221
   α-glucanases, 4
   β-glucanases, 1
   acetolactate decarboxylase, 4
   endoxylanases, 1, 3
   enzymes used in, 2–3
   oxidoreductases, 4
   proteinases, 3-4
Brine fermentation, 592
Broccoli, 270
Bromoxynil, 327
Bromoxynil-resistant cotton, in United States, 327
Brown, Mary Louise, 714
Brown, Patrick, 403
Brownian movement, 64
Bt toxins, 31
Bt10, 303, 304
Bt176 maize, 553, 554
BTEX, 111
Bulgaria, 139
Bulk cultures, 202
Burghol, 138
Burkholderia cepacia, 96–97
Burukutu, 135
Butyrivibrio fibrisolvens, 452
Byssochlamys, 575

C

c-Myc, 599, 602
C-NMR, 651
Cacao swollen shoot virus, 500, 509
Caenorhabilitis elegans, 13, 33, 34, 580
Caffeine, 471
Calcium, 180, 269–270
Calf chymosin, 196, 227
Calli, 519
Calpain, 330
Calpastatin genes, 312
Caltech, 292
CAMBIA intellectual property resource, 308
Camembert, 144
Campesterol, 517
Campylobacter jejuni, 215, 416
Canada, 11, 51, 118
   GM crops in, 301
Cancer, 41, 294
Candida, 138, 365, 690
Candida albicans, 80
Candida antarctica, 97
Candida cylindracea, 96, 232
Candida guilliermondii, 365
Candida humilis, 365
Candida krusei, 138, 366
Candida milleri, 365
Candida mycoderma, 139
Candida rugosa, 96
Candida spp., 9
Candida stellata, 8
Candida tropicalis, 139
Candida versatilis, 592
Canidae, 321
Canola, 328
Capra hircus, 165, 343, 344
Capra pyrenaica, 343
Capra pyrenaica hispanica, 344
Captive breeding and wildlife conservation, 320, 322
   adaptive and detrimental variation, 322–323, 324
   genome-enabled species, 320–321
   introgressive hybridization, 321
Carbohydrate-binding module (CBM), 451
Carbohydrates, 84
Carbon/nitrogen (C/N) ratio, 48
Carboxymethylcellulose, 44
Carboxypeptidases, 237
Cardiotoxicity test, 24
Cardiovascular and cholesterol-lowering effects, 77
Cargill Bioactives US, LLC, 373
Carnitine palmitoyltransferase 1A (CPT1A), 317
Carotenoid biosynthesis, 443–444
Carotenoids, 152
Carrots, 28
Carson, Rachel, 484
Cartagena Protocol for Biosafety, 526, 659
CartiCel®, 23
Casein micelle, 330
Caseins, 13, 227
Cassava, 51, 160
Cassava bagasse, 48
Cassava mosaic virus, 509
Catalase, 221
Cathepsin B, 330
Cathepsin L, 330
Cationic detergent cetyltrimethylammonium bromide (CTAB), 680
Cattle
   muscle and adipose tissue development, insights into, 316–317
   semen from, 57, 58
   sex-sorted sperm, utilization of, 391
Cattle embryo transfer
   cloning embryo, 131–132
   cryopreservation of embryos, 130
   in vitro fertilization, 130–131
   procedures, 129–130
   reasons to, 129
   sexing embryos, 130
Cauliflower mosaic virus (CaMV), 499
Cauliflower Mosaic Virus 35S promoter (CaMV 35S), 14
Caulimovirus, 499, 500
Caviar, 251
CC, cat clone, 165
CD4+ T cells, 12
CD8+ T cells, 12
CD59 genes, 705
CDC2a gene, 467
CDNA microarray, 307, 317, 403
Celera genomics, 312
Cell culture, animal. *See* Animal cell culture
Cell transplantation, 704
Cell-based assays, animal cell culture application
   used in high throughput drug screening, 24
Cellophane, 85
Cellular insertion, 617
Cellulases, 221
Cellulose, 44, 83, 92
Cellulose binding domain, 117
Cellulose esters, 85
Cellulosic crops, 84
Cellvibrio, 451
Cellvibrio japonicus, 452
CEN (Comité Européen de Normalization), 589
Center for Science in the Public Interest, 260
Center for Veterinary Medicine (CVM), 286
Centromere, 53

Ceratocystis fimbriata, 48
Ceratocystis ips, 157
Cereal alpha-amylase/trypsin inhibitors, 13
Cereal-based foods. *See* Starter cultures
Cereal-based grain products, 138
  fermented alcoholic beverages. *See* Fermented alcoholic beverages
  fermented foods. *See* Fermented foods
Cerrena unicolor, 78
Certified Reference Material (CRM), 302
Chaetomium thermophilum var. *coprophilum*, 181
Chaff collection, 463
Chambers Science and Technology Dictionary, 116
Chaperones, 422
Chardonnay, 389
Cheddar cheese, 145, 419
Cheese flavor, enhancement and production of, 232
Cheese making, 234–235
Cheese production, 228
  biochemical properties of technological interest, 144
  culture-independent fungal community analyses, molecular tools for, 149
  culture-independent methods limitations, 149
  dairy fungi, functional genomics of, 149
  ecology in, 144
  functionality and role in, 145
  genera and species, methods for identifying, 147–148
  nonstarter culture bacteria, 143–145
  non-starter LAB, 143
  propionic acid bacteria (PAB), 143
  smear bacteria, 143–144
  strain discrimination and tracking, 148–149
  taxonomy, physiology, and habitat, 143–144
  yeasts and molds, 147
Chelex-100, 681
Chemical production examples, in biorefinery
  biocatalysis, 85–86
  conventional processing, 85
Chemical protecting agents (CPAs), 352
Chemoprevention with dietary phytopharmaceuticals, 151–152
  cellular signaling for antioxidant enzyme induction by, 153–154
  fruits and vegetables promise, 151
  functional foods development containing, 154–155
  intracellular signal network as novel target of, 152–153
  nutrigenomic perspectives, 154–155
Chenopodium rubrum, 209
Chicha, 135, 141
Chicken pepsin, 228

Chickens, semen from, 57, 58
Chickpea, 51
Chilton, M.D., 293
Chimera formation, 343
China, 141
  GM crops in, 301
Chinese hamster ovary (CHO), 18, 19
Chinese medicine, 13
Chip. *See* Microarray
Chip/array-based approaches, 423
Chitin, 156
Chitin deacetylase (CDA), 156–157
Chitinase synthase, 630
Chitosan, 45
  chitin deacetylase (CDA), 156–157
  microorganisms, produced by, 156
Chlamydomonas humicola, 109
Chloramphenicol, 40, 41
Chlorella minesstissima, 109
Chlorella vulgaris, 109
Chloroform, 680
Chlorophenol, 109
Chocolate flavor quality. *See* Cocoa fermentation
Cholera, 662
Cholera toxin (CT), 662
Cholesterol, 77, 472
Chorleywood Bread Process (CBP), 364
Chromolaena odorata, 100
Chromosomes, 131
ChTOG, 384
Chymosin, 196, 221, 227, 229, 234
  cheese making, 234–235
  milk protein system, 227
  and other milk coagulants, 227
  preparation, 234
  rennet coagulation of milk, 228
  rennet role, in cheese ripening, 228–229
  rennets and rennet substitutes, 227–228
  substitutes, 235
Cider-making, enzymes used in, 4
Cinnayml-esterase, 691
Circulifer tenellus, 500
cis-Natural antisense transcripts (cis-NATs), 581
cis-β-ocimene, 368
Citrate metabolism, 192–193
Citric acid, 44, 85
Citrobacter, 365
Citrus leafminer, 101
Citrus tristeza virus, 509
CKI1, 467
Cladosporium cucumerinum, 157
Cladosporium ladosporioides, 157
Clarification, 563, 564
Clarified rice wine. *See* Saké
Classic Maya, 205
Classical biological control, 335
Clavulanic acid, 41
Clemson University Genomics Institute (CUGI), 307

Cleopatra, 677
Clinical propionic acid bacteria (PAB), 143
Cloning
  analytical breeding, 159–160
  crossbreeding, 159
  embryo, 131–132
  endangered species cloning, 165
  future prospects of biotechnology in improving, 161–162
  multi- and unipotent stem cells, 168–169
  nuclear transfer, in farm animals, 163–164
  nuclear transfer, in pets, 165
  pluripotent stem cells, 167–168
  rabbit cloning, 164–165
  somaclonal variation, 160–161
  therapeutic SCNT for human therapy, 165
  totipotent stem cells, 167
  and transgenics, 245
Clostridium acetobutylicum, 62
Clostridium botulinum, 43, 60
Clostridium butyricum, 550
Clostridium perfringens, 62
Clostridium spp., 198
Clostridium tetani, 416
Clostridium thermocellum, 452
Clostridium tyrobutyricum, 198, 550
CO_2, 22, 48, 195, 387
Coal, 82, 84
Cobalt, 180
Cobb-Vantress, 536
Cocoa butter, 233, 372
Cocoa butter equivalents (CBE), 372
Cocoa fermentation
  drying effects, 173
  enzymatic changes effects, 172–173
  general practices and biochemical changes effects during, 171–172
Code of Federal Regulations (CFR), 47
Codex and WTO Agreements, 265
Codex Alimentarius, 31, 410
Codex Alimentarius Commission (CAC), 263, 265, 288
Coffea arabica, 520
Coffee husk, 48
Cold plasmas
  applications of, 176
  for food processing, 174
  interaction with surfaces and use in diagnostics, 175–176
  nitrogen-based atmospheric plasma, 174–175
  production of, 174, 175, 176
  surface decontamination, 176–177
Cold Spring Harbor Laboratory, 292
Coleoptera, 336, 337
Collaboration in Animal Health and Food Safety Epidemiology (CAHFSE), 25–26
Collagen, 44, 524

Collego®, 463
Colletotrichum gloeosporioides f.sp. *aeschynemene*, 105
Colletotrichum graminicola, 377
Colloidal calcium phosphate (CCP), 227
Colony-forming units (CFU), 709
Colorado State University, 333
Column refolding, 424
Commercial oocyte technologies, 332
    dead mares, collection and transfer of oocytes from, 333
    gamete intrafallopian transfer (GIFT), 334
    intracytoplasmic sperm injection (ICSI), 334
    oocyte collections, 332–333
    oocyte transfer, 333
Commission of European Communities (CEC), 639
Community Reference Laboratory (JRC/CRL), 302
Comparative mapping, 312
Competent, 60
Competitive crops, 463
Complementary DNA (cDNA), 36, 307
Complete microbial genomes, 415
Comté, 144
Conjugated linoleic acid (CLA), 269, 374
Conjugation technique
    DNA transfer in Gram-positive bacteria, 60
Conservation, of genomic resources, 306
Consultative Group on International Agricultural Research (CGIAR), 308, 341
Consumer attitudes, changes in, 299
Consumer satisfaction, 402
Control measures, of plant pathogens, 509–510
Conversion for renewable carbon, in biorefinery production, 83
Convolvulus, 467
Coordinated Framework for Regulation of Biotechnology, 281
Co-oxidation, 471
Coprinellus micaceus, 79
Coprinus comatus, 79
Coprinus domesticus, 79
Copyrights, 339–340
Cor a of hazel nut, 13
Cordyceps sinensis, 79, 80
Corn, 92
Corn kernels, 84
Corn sweeteners
    enzymatic starch hydrolysis, 178–180
    future prospects, 181
    glucose–fructose isomerization, 180
    liquefaction, 178
    low-intensity and alternative sweeteners, 181
    saccharification process, 178, 180
    use of enzyme, 178
Corn zein, 44
Corolase, 331
Coronary heart disease (CHD) and fish oil consumption, 269
Corpus luteum (CL), 240
Corynebacterium, 139
Coryneforms, 143–144
Cotton, 31, 51
Cotton leaf curl disease, 501
Council Regulation EEC No. 2377/90, 40
Country of origin labeling (COOL), 260–261
CPBR, 698
CR1aa, 123, 124
Crabtree effect, 134
Cremoris, 419
Creutzfeldt–Jakob disease, 294
Criollo, 171
Crop rotation, 462, 463, 464
Crop-to-crop gene flow, 643–644, 645
Crops
    documented ferality, 184–185
    domestication and ferality, 183, 184
    feral de-domestication, 183–186
    feral plants, 183
    to wild relatives gene flow, 644–645
    transgenic preventing/mitigating ferality evolution, 185–186
    volunteers, first step to ferality, 184
Cross protection, 509
Crossbreeding, 159
Crude peanut extract (CPE), 14–15
Cry9C protein, 118
Cryobiology, 352, 460
Cryopreservation. *See also* Embryo manipulation
    cryobiology, 352
    of manipulated embryos, 397
    of oocytes and sperm, 352
    rodents, 352–353
    swine, 353–354
Cryopreserving cells, 460
CryoTip™, 461
Cryotop, 461
Cryphonectria parasitica, 235
Cryptococcus sp., 8
Cryptosporidium sp., 572
Cucumber fermentation, 669
Cultivar selection, 463
Cultivation, of transgenic crops, 625–626
Cultural control, of pathogen, 514
Culture independent approaches, 149
Culture media, 20
Cumulus removal (CR), 428
Cuphea spp, 457
Cupin superfamily, 13
Curdlan, 420
Curly top, 502
Curtovirus, 500
Cutaneous propionibacteria, 557
Cyanobacteria, 109
Cyanovirin-N (CV-N), 666
Cyclin-dependent kinases (Cdks), 467
Cyclodextrin glycosyltransferase (CGTase), 187, 223, 224
Cyclodextrins, 187–190, 224
Cyclohexanes, 84
Cyclospora cayetanensis, 572
Cydia pomonella, 483
Cys, 13
Cyst, 572
Cysteine, 125
Cystic fibrosis, 294, 295
Cytokinin, 466, 467
Cytokins, 77
Cytoplasmic hybrids (cybrids), 566
Cytoplasmic male sterility (CMS), 54
Cytotoxic macrophages, 77

D

Daedaleopsis confragosa, 78
Dairy, 221
    in 21st century, 203
    bacteriocins in, 198
    bacteriophage and their control, 202–203
    bioactive peptides production, 197–198
    engineered phage resistance, 203
    fermented food products, 195–198
    functions and applications, 201–202
    high pressure treatment for, 329–330
    industrial production, 202
    Lactobacillus plantarum in, 361
    lipases in, 372
    microorganisms and their properties, 200–201
    probiotics concept, 196–197
    production steps for, 202
    recombinant chymosin, 196
    starter cultures, 195–196, 200–203
Dairy fungi, functional genomics of, 149
Dairy lactococci
    citrate metabolism, 192–193
    extracellular polysaccharides (EPS), 193
    functionality of, 191–193
    gene cloning and expression, 193
    genomics and proteomics, 193–194
    lactose metabolism, 192
    nisin and other bacteriocins, 193
    phage and phage resistance, 193
    proteolytic system, 192
Dairy ProP169, 558
Dairy propionic acid bacteria (PAB), 143
Dairy proteins, 239
DANISCO, 373
Danish Integrated Antimicrobial Resistance Monitoring and Research Programme (DANMAP), 25–26
Dau c 1 of carrot, 13
Daunomycin, 41

DCL (Dicer-like), 579, 581
DDT, 479, 483–484
Dead mares, collection and transfer of oocytes from, 333
Debaromyces hansenii, 140, 147, 149, 365, 709
Debaryomyces hansenii var. *hansenii*, 366
Debaryomyces occidentalis, 140
Debaryomyces sp., 138
Decarboxylase, 685
Dedifferentiation, 466
Deinococcus radiodurans, 489
Delage, Yves, 163
Delayed density-dependent, 480
Denaturing gradient gel electrophoresis (DGGE) analysis, 400, 414
Dendritic cells (DCs), 77, 665
Dendropolyporus umbelatus, 78, 80
Denmark, 6
Density-dependent, 480
Density-independent, 480
Deoxyribose-5-phosphate, 708
Department of Health and Human Services (DHHS), 281
Dermagraft®, 23
Desert Bighorn, 343
Designer fats and oils, lipases in, 372
Designer oil crops, 457
DET1, 443
Detergentless microemulsions, 450
DeVine®, 105, 463
Dextranicum, 139
Dextrose equivalent (DE), 178, 224
Diacetyl, 4, 74, 192, 195, 201, 389, 708
Diacids, 84
Diacylglycerol acetyltransferase (DGAT), 312
Diacylglycerols (DAGs), 233, 373
Diadzain, 209, 211
Diamond v. Chakrabarty, 657
Diatraea saccharalis, 100
Dicarboxylic acid, decarboxylation of, 387
Dicers, 34, 579
Dielectric breakdown model, 209
Diesel, Rudolf, 95
Dietary importance, of kimchi, 348–349
Dietary ingredients, 259
Dietary Supplement Health and Education Act (DSHEA), 257, 259, 456
Diethylaminoethyl (DEAE), 19
Digital pressure/massage of semen collection, 56
Diglyceride oils, 233
Dimethyl sulfoxide (DMSO), 69, 85, 396
Dimethylallyl diphosphate (DMAPP), 445, 446
Dinitrophenols (DNOC), 483
Dip pen nanolithography (DPN), 438
Diphenolic acid, 85

Diptera, 336, 337
Direct density-dependence, 480
Direct lactic acid fermentation, 355
  advantages of, 355–356
  amylase and lactic acid-producing organisms, co-cultures of, 356–357
  by amylolytic lactic acid bacteria, 357
  different processes of, 356
  by lactic acid-producing fungi, 357
  multistepped conventional fermentation, 356
  by simultaneous saccharification and fermentation, 357–358
Direct Vat Set (DVS), 197
Direct-to-vat cultures, 202
Directed evolution, 437–438
Disease-forecast models, 611
Disodium ascorbyl phytostanyl phosphates (DAPP), 517
Dissolved oxygen (DO), 21
Dissolved oxygen tension (DOT), 695, 696
Distiller's dried grain (DDG), 93
Distiller's dried grain with solubles (DDGS), 93
DNA (deoxyribonucleic acid), 26, 130
  array technology, 403–404
  chip. See Microarray
  ligases, 292, 293
  polymerases, 437
  test for livestock species, 313
DNA Database of Japan (DDBJ), 307
DNA microarrays, functional genomics through
  cattle muscle and adipose tissue development, 316–317
  metabolic disease from liver transcriptomics in dairy cattle, 317–318
DNA-based microarray, 403
DNA-based technologies, 403
Docosahexaenoic acids, 374
Doctrine of substantial equivalence, 259
Dokla, 135
Dolly, 131, 163
Domestic cattle, 343
Domestic dog, genome of, 321
Domestic goats, 344
Domestic sheep, 245
Domestication syndrome, 183
Dominant follicle, 240
Donor cell reprogrammability assays, 167, 168
Dosa, dosai, 135
Double-stranded RNAs (dsRNA), 13, 33, 34
Dough development, 363, 364
Dough fermentation, 363
DREB, 205
Drosophila heatshock, 89
Drosophila melanogaster, 88
Drosophila sp., 34

Drought and drought resistance breeding for, 205–206
Drug absorption test, 24
Drying effects, in cocoa fermentation, 173
DSM, 373
Durable resistance breeding, 375
  breakdown of divide, 377
  durably resistant plants, development of, 378–379
  genetic basis of, 377–378
  plant disease resistance, bilateral view of, 375–377
Durably resistant plants, development of, 378–379

E

E418, 420
EC DG Agri, 304
EC Joint Research Centre, 302
EC-funded Co-Extra program, 304
Economic and environmental risks, of gene flow. See Transgenic plants
Economically important trait loci (ETL), 312
EcoRI, 292
Ectaga Garcia, 100
Ectomycorrhizal fungi, 109
Edible biopolymers
  cellulose, 44
  chitosan, 45
  starch, 44–45
Edible oils, refining of
  lipases in, 374
EDTA, 125
EEC No. 2377/90, 40
EF-hand proteins, 12
Efficacy of transgene, 613
Egg-layer breeding, 537
Egnin, Marceline, 648
Egypt, 139
Eicosapentaenoic acids, 374
Electric field stress, on plant systems
  pulsed electric fields (PEFs), 209–210, 211
  secondary metabolites, 208
  stress responses in nature, 208–209
Electrical fusion, of protoplasts, 566
Electrochemical transduction methods, for biosensor, 113
Electron microscopy, 680, 681
Electrophile response elements (EpRE), 153–154
Electroporation technique
  DNA transfer in Gram-positive bacteria, 60–61
Electrospinning, 19
Elicitation, 585
Elicitors, 585
Embden–Meyerhof pathway, 7
Embryo biopsy, 394–395

Embryo cryopreservation, 395
  approaches for, 395
  interrupted slow freezing (ISF), 395–396
  rodents, 353
  swine, 354
  vitrification, 396–397
Embryo culture, 428
Embryo-derived stem cells, 597–598
Embryo manipulation, 394–395
  cryopreservation of, 397
  embryo cryopreservation, 395–397
Embryo transfer, 243, 244. *See also* Cattle embryo transfer; Multiple ovulation/embryo transfer (MOET)
  cloning and transgenics, 245
  embryo recovery, 244
  embryos evaluation, 244
  freezing, 244–245
  industry, 243
  meaning of, 342
  micromanipulation, 245
  research, 245
  superovulation, 243–244
  in vitro fertilization, 245
Embryogenic cell suspensions (ECSs), 69
Embryogenic cells, 70
Embryonic germ (EG) cells, 595
  agricultural applications, 595
Embryonic stem cells (ESCs), 167, 594–595, 597, 598, 700
  agricultural applications, 595
  and gene-modified pigs, 701
  genetic studies, 598–599
Embryotoxicity test, 24
Emmental, 144
Empoasca sp., 50
Emulsion liquid membrane, 408
Encarsia formosa, 336
Encephalitis virus, 80
Encephalitozoon, 573
Endangered species cloning, 165
Endo/exo-feral evolution, 184, 185
Endogenous effects, 480
Endogenous proteolytic enzymes, 253–254
Endogenous siRNAs, 581
Endomyces lactis, 141
Endonucleases, 292
Endopeptidases, 3, 237
Endosperm balance number (EBN), 160
Endoxylanases, 1, 3
Enhanced protein production, 422–423
Enjera, 136, 365
Enteric viruses concentration, from complex foods, 681
Enterobacter, 365
Enterococcus, 143
Enterococcus faecalis, 60, 61, 144, 145, 349
Enterococcus faecium, 144, 145
Enterocytozoon, 573

Enterohemorrhagic *Escherichia coli* O157:H7, 44
Enterotoxigenic *Escherichia coli* (ETEC), 662
Environmental benefits and risks of HRCs, 328
Environmental Defense Fund, 484
Environmental Protection Agency (EPA), 281, 283–284, 295, 484, 527, 528–529, 614, 633, 634, 644
  experimental use permit, 528
  registration, 529
  state registration, 529
Environmental use permit (EUP), 528
Enzymatic alcoholysis, alternative feedstocks and, 96
Enzyme immunoassay (EIA), 680, 681
Enzyme modified cheeses (EMCs), 232
Enzyme-linked immunosorbant assay (ELISA), 574, 588
  in crude peanut extracts (CPE), 14–15
  direct and indirect competitive techniques, 214
  enzymes uses in, 212–213
  for foodborne bacterial pathogens detection and identification, 214–215
  fundamental techniques, 214
  haptens and their conjugation to proteins, 212
  microorganisms and toxins in foods, 212–215
  microtiter plates used in, 213–215
  for toxins detection in foods, 214
Enzymes
  amylases, 223–225
  applications in food industry, 224–225
  applications of, 221
  beverages and beer, 224
  bread and bakery products, 224
  in cider-making, 4
  cyclodextrins, 224
  in ELISA assays, 212–213
  food chemical analysis, 224
  future applications and innovation, 225
  history and use in food processing and production, 220
  industrially-relevant enzymes produced using gene technology, 221
  infant formula, 224
  occurrence, biological significance, and commercial production, 223–224
  for sweeteners production from corn starch. *See* Corn sweeteners
  syrups, 224
  used in brewing, 1–4
  uses, 224
  in wine-making, 4
EpiCel®, 23
Epicoccum nigrum, 157
Epidinicarsis lopezi, 336
Epinephrine, 13

Equilibrium freezing, 460
Equilibrium points, 480
Equine chorionic gonadotropin (eCG), 242, 605, 606
Equine embryos, 244, 245
Equine infectious anemia virus (EIAV), 637
Equus assinus, 343
Equus caballus, 343
Equus przewalski, 343
Equus zebra, 343
Eragrostis tef, 365
Erinacins, 80
Eritadenine, 77
Erwinia amylovora, 630
Erwinia herbicola, 420, 444, 445
Erysiphe graminis f. sp. *hordei*, 377
Erythrogen™, 570
Erythropoeitin (EPO), 523
Escherichia coli, 37, 60, 65, 81, 86, 96, 114, 175, 196, 292, 293, 307, 416, 422, 491, 650
Escherichia coli O157:H7, 215, 709
Escherichia coli tryptophanase, 691
Essential amino acids, 125
EST clones, 307
EST databases, 538
Esterases, 3, 373
Estimated breeding value (EBV), 315
Estradiol Benzoate (EB), 712
Estradiol benzoate, 241
Estrus synchronization and ovulation, 240
  using estradiol and progesterone, 241–242
  using GnRH and prostaglandin, 241
  ovarian follicular dynamics during estrous cycle, 240
  using progesterone, 241
  using prostaglandin, 240–241
Ethanol, 48, 85–86, 139, 708
  current process, 92–94
  fuel, 91–92
  introduction, 91
  need, 92
Ethers, 85
Ethyl acetate, 48, 139, 389
Ethyl hexanoate, 389
Ethyl lactate, 389
Ethyl methanesulphonate (EMS), 68
Ethyl octanoate, 389
Ethyl-2-methylbutanoate, 172
Ethylene glycol (EG), 245, 396
EtOH, 85
Euhrychiopsis lecontei, 100
Eulopidae, 100
Eurasian Watermilfoil, 100
Eurobarometer, 120, 298
Europe, 120, 121
  GM crops in, 301
  GMO regulations, 302
European approach, legislation, 40

European Bioinformatics Institute (EMBL-EBI), 307
European Commission, 664
European Community
  GMO food and feed in, 302–303
  GMO regulations in, 301–302
European consumer attitudes, 298–299
European countries, 11
European Food Safety Authority (EFSA), 40, 265, 302, 410, 554
European Network of GMO Laboratories (ENGL), 302, 554
European rabbit, 672–673
European regulations, 553
European Union (EU), 30, 121, 410, 639
Exocellular polysaccharides (EPS), 196
Exogenous effects, 480
Exogenous enzymes, 72
  uses in alcoholic beverages. See Alcoholic beverages
Exogenous FSH rescues subordinate follicles, 605
  species-specific
Exopeptidases, 237
Exopolysaccharides (EPS), 135, 418, 707, 708
Exotic pests, of tropical agriculture, 655
Experimental use permits (EUPs), 284
Expressed clones, 306, 307
Expressed sequence tags (ESTs), 307, 403, 506
Expression of transgene, 613
Expression QTL (eQTL) mapping, 314
Expression vectors
  DNA transfer, in Gram-positive bacteria, 61
Extracellular matrix (ECM), 19
Extracellular polysaccharides (EPS), 193

F

Facultatively heterofermentative lactobacilli (FHL), 143, 144, 145
Fair Packaging and Labeling Act (FPLA), of 1966, 257, 258
Fall rye, 463
Farm Bill 2002, 260
  and 2008, 257–258
Farmer, risks and benefits for
  of HRCs, 326–327
Fat-corrected milk (FCM), 558
Fat-free skimmed milk (FFSM), 213
Fatty acid binding protein 4 (FABP4), 317
Fatty acyl-CoA reductase (FAR), 650
Fatty alcohol acyltransferase, 650
Fatuoids, 185
Fed-batch cell culture process, 22
Federal Food, Drug and Cosmetic Act (FFDCA) 1938, 257, 258
Federal Insecticide, Fungicide, and Rodenticide Act (FIFRA), 283

Federal Plant Protection Act, 282
Feed conversion ratio (FCR), 536, 537
Feline panleucopenia virus, biological control of cats using, 673
Felis cattus, 165, 673
Felis silvestrus ornate, 343
Felis sylvestris lybica, 343
Feral de-domestication, of crops. See Crops
Fermentation. See also Beer fermentation; Vegetable fermentation
  definition of, 251
  environment creation, 669
Fermentation-produced chymosins, 228
Fermented alcoholic beverages
  boza, 139
  chicha, 141
  kaffir, 140–141
  pito, 139–140
  sake, 141
  takju, 139
Fermented dairy products. See Dairy
Fermented fish roe products, 251–253
  biochemical changes, 253–254
  health benefits, 255
  salted dried fish roe, 252
  salted fermented fish roe, 252–253
  salted fish roe, 253
  trace element and cholesterol contents, fermentation impact on, 254–255
  utilization, of fish roe, 251
Fermented foods
  bioactive properties in. See Lemon essential oils
  kenkey (kenky), 138
  kishk (fugush), 138
  medida, 139
  ogi, 139
  tarhana (trahana), 138–139
  togwa, 139
Ferulic acid, 48, 449
Ferulic acid esterases (FAEs), 448
  in organic synthesis, 449–450
  identifying property-based sequence motifs, 451–452
  protein informatics, 450–451
Fibrous bed bioreactor (FBB), 22–23
Field effect transistor (FET), 439
FindMod, 451
Fine mapping, 312
Finger print contig (FPC), 307
Fish eggs, 254
Fish oil consumption and coronary heart disease (CHD), 269
Fish roe, 252–253, 255
Fit-for-purpose starter cultures, development of, 690–691
Five-a-Day for Better Health, 151
Fixed timed AI (FTAI), 242
Fixed timed artificially inseminated (FTAI), 712
Flaked barley, 1

Flamulina velutipes, 79
Flavonoid compounds, of vinegars, 678
Flavonoids, 189
Flavours, 47
  and fragrances, lipases in, 372–373
Flax, 463
Fleming, Alexander, 39
Flesh-eating bacteria, 60
Florida, 101
Florida Department of Agriculture and Consumer Services, 89
Florida panther, 323
Flow cytometer, 403
Flow sorting system, 390
Fluorescent in situ hybridization (FISH), 62
Fluorescent pseudomonads, 104
Foley catheters, 244
Folic acid, 225
Follicle-stimulating hormone (FSH), 523, 604, 605, 606
Follicular growth and development, 604
Follicular wave development, 711–712
Follistatin, 604
Fomesfomentarius, 78
Fomitopsispinicola, 78
Food and Agriculture Organization (FAO), 116, 306, 341, 668
  World Health Organization (FAO/WHO) International Conference, 263, 264
Food and Drug Administration (FDA), 47, 257, 258, 259, 281, 284, 285, 303, 456, 527, 528, 529, 614, 633, 634, 635
  biotechnology consultation, 528
  coexistence, 286
  food safety evaluation, 528
Food Allergen Labeling and Consumer Protection Act (FALPA), 257, 259
Food allergy. See also Peanut allergens
  genes, strategy to eliminating
  animal food allergens, 12–13
  food proteins identified as allergens, 12–13
  genetic modification to alleviate risks of, 13–14
  mechanisms of, 12
  peanut allergy, 11, 14
  plant food allergens, 13
  treatment of, 13
Food applications, of lipases, 372–374
  baking, 372, 373
  dairy products, 372
  designer fats and oils, 372
  edible oils, refining of, 374
  flavors and fragrances, 372–373
  infant formulas, fats for, 372
  nutraceuticals, 374
  surfactants, 373–374
Food biosafety
  genetically modified organisms (GMOs), 30–31

Food, Drug, and Cosmetic Act (FDCA), 284
Food export, 263
Food fermentation, 477
Food labeling, 257
   of alcohol beverages, 260
   country of origin labeling (COOL), 260–261
   dietary supplements, 259
   of food allergens, 259
   of genetically modified food products, 259–260
   of meat products, 260
   organic, eco-labeling, fair trade, and sustainability, 261
   of transesterified fats, 260
Food legislation, 264
   Codex and WTO Agreements, 265
   Codex Alimentarius Commission (CAC), 265
   historical background, 264–265
   International Health Regulations, 266–267
   International Organization for Standardization, 266
   WTO and dispute settlement, 265
Food preservation. *See* Pulsed electric field (PEF) processing
Food processing, 13
Food Quality Protection Act, 484
Food regulations, 262
   purpose, 262–263
Food safety, 263, 477–478. *See also* Protozoa and parasites
   evaluation, 528
   of GMO food and feed, 302
Food Safety and Inspection Service (FSIS), 260
Food spoilage, 478
Food Standards Agency (FSA), 411
Food-grade bacteria, 418
Food-grade lipases, 373
Foods for Specialized Health Use (FOSHU), 456
Foot and mouth disease virus (FMDV), 637
Forastero, 171
Ford, Henry, 83
Forward breeding, 614
FOSHU (foods for specified health issues), 269
Foxtail millet, 184, 185
Fragrances, 47
France, 11
"Frankenfoods," 302
Fraudulent marketing, 258
Free aldehyde, 650
Free amino nitrogen concentration (FAN), 74
Free fatty acids (FFA), 232
Freedonia Group, Inc., 47
Freeze-dried sperm, 346

Freeze-drying (FD), 540
Freon. *See* Tri-chloro tri-fluoroethane
Fresh semen, 57, 58
Frozen dough development and stability, 364
Frozen-leavened dough
   dough development and stability, 364
   frozen-dough yeast, 364
Frozen semen, 57, 58
Fructose, 85, 180
Fruit and vegetables, high pressure treatment for, 330
Fruity flavours, 48
FSH (follicle-stimulating hormone), 217
FTAI program, 712
Fuel ethanol. *See* Ethanol
Fugush. *See* Kishk
Full-length complementary DNA (FLcDNA) clones, 307
Fumaric acid, 84
Functional foods, 269
Functional genomics, 311, 314–315, 402
   definition of, 316
   through DNA microarrays, 316–318
Functional proteomics, 402
Fungi, 107, 108, 109, 387
Fungicides, 513–514, 611
Fusarium, 138, 139
Fusarium oxysporium, 374, 449
Fusarium oxysporum f. sp. *cubense*, 68
Fusarium wilt, 68

G

G1.2, 123, 124
G2.2, 123, 124
GalNAc, 523
Galα1-3Gal, 705
   phylogeny of expression of, 704
Gamete cryopreservation, 352
Gamete intrafallopian transfer (GIFT), 334, 714
Gametes and embryos, 272
Gametosomatic hybridization, 566–567
Ganoderic acid, 77
Ganoderma applanatum, 79
Ganoderma lucidum, 77, 79, 80
Ganosporeric acid A, 77
Gas chromatography–mass spectrometry (GC-MS), 411
Gas-continuous (GC) operation modes, 698
Gasoline, 84
Gastroenteritis coronavirus, 523
Gastrointestinal (GI) tracts, 664
   human, 12
Gautheret, Roger, 519
Gc-globulin. *See* Macrophage activating factor (MAF)
GC/mass spectroscopy, 650
GDF8 gene, 312
Geaumannomyces, 514

Gel electrophoresis, 532–533
Gel- or column-based protein analysis and purification, 424
Gelatine, 44
Gellan, 419–420
GEM1 protein, 384
Gemini pollen1 (*gem1*) mutation, 384
Geminiviruses, 500–502
Gene banking, of genomic resources, 307–309
Gene chip. *See* Microarray
Gene cloning and expression
   lactococci for, 193
Gene escape and genetic pollution, 118
Gene expression patterns, 276–278
Gene expression profiling, 688
Gene flow, 643
Gene-for-gene (GFG) resistance, 375, 377
Gene-modified pigs and embryonic stem cells, 701
Gene of interest (GOI), 613
Gene ontology (GO) analysis, 318
Gene Ontology Consortium, 438
Gene silencing mechanism, 33
   by RNA interference (RNAi), 34–35
Gene therapy. *See* Human gene therapy
Gene tracking, 414
Gene transfer, 477, 616
   cellular insertion, 617
   genetic analysis, 416
   natural gene transfer mechanisms, 415–416
   pronuclear microinjection, 616–617
   retroviral insertion, 617
   sperm-mediated gene transfer, 617
Genencor Division, 373
General Food Law, 302
General resistance, to pathogens, 512–513
Generally accepted practices (GAPs), 624
Generally recognized as safe (GRAS), 47, 193, 285, 664
Generative cells, 381, 382
GeneSpring software, 445
Genetic analysis, 416
Genetic conservation. *See* Genomic resources
Genetic engineering, 292
   animal biotechnology, 286–288
   in animals, 294–295
   background, 292
   and biotechnology, 279
   current U.S. regulations, 281–282
   Environmental Protection Agency (EPA), 283–284
   Food and Drug Administration (FDA), 284–286
   human gene therapy, 295
   international perspective, 288–290
   in microorganisms, 292–293
   in plants, 293–294
   of volatile flavor generation, 684–685

Genetic engineering (*Continued*)
United States Department of Agriculture (USDA), 282–283
and xenotransplantation, 705
Genetic improvement, of arthropod natural enemies. *See* Biocontrol agents
Genetic knowledge, of microbial starters, 400–401
Genetic manipulation. *See* genetic engineering
Genetic modification (GM), 247
background, 247
livestock feeding studies, 249
transgenic crops for food and feed, safety aspects of, 249
transgenic crops impacts on livestock feed quality and cost, 247–249
Genetic parameters estimation
for multiple ovulation/embryo transfer (MOET), 216–218
Genetical genomics approach, 314
Genetically modified (GM) approaches, 701
Genetically modified (GM) crops, 293–294, 295–296
cultivation of, 301
Genetically modified (GM) starter cultures, 689
Genetically modified animals, 294
Genetically modified food
labeling of, 259, 529, 634–635
United States consumer awareness and attitudes, 297–299
Genetically modified organisms (GMOs), 28, 238, 247, 301, 410, 526, 528, 529, 635. *See also* Biotechnology
authorized GMOs, 302–303
control steps to, 31
detection, 534
European Community, regulations in, 301–302
example studies, 411–412
food biosafety, 30–31
food safety of, 302
gene transfer methods, 29–30
GM plant material, mixtures of, 303
intellectual property of, 31
plant material mixtures, 303
risk assessment process, 410–411
side effects of released, 31
unauthorized GMOs, 303–304
unintended release, 31
unknown GMOs, 304
used as feed/food, 30
Genetically modified pigs, 294
Genistein, 209, 211
Genome-enabled species, 320–321
Genome sequencing, of Gram-positive bacteria, 62
Genomic estimated breeding value (GEBV), 218, 315

Genomic resources, 306–307
conservation and biotechnology, 306
gene banking of, 307–309
Genomic shadowing. *See* Phylogenetic shadowing
Genomics, 688. *See also* Animal agriculture
complete microbial genomes, 415
functional genomics, 311, 314–315
implications of, 315
metagenomics, 415
microarrays and gene expression, 415
and proteomics, 193–194
research, 310
structural genomics, 311–314
of volatile flavor generation, 684
Gentio-oligosaccharides, 180, 181
Geo-Microbial Technologies Inc., 111
Geotrichum candidum, 9, 147, 149
Geranylgeranylpyrophosphate (GGPP), 446
Germplasm, access to, 341
Germplasm cryopreservation efficiency, by species and germplasm, 353
Ghana, 138, 139
Giant palm bran, 48
Giardia sp., 572
Gibberella fujikuroi var. *intermedia* (*ATCC 42052*), 157
Gliocladium catenulatum, 157
Gln, 13
Global harmonization, 262
achievements in, of food safety regulations, 264
importance of, 263
progress in, of food legislation, 264–267
Globodera spp., 440
GloFish®, 287, 295
Glucan 1,4-α-glucosidase, 223
Glucaric acid, 84
Glucoamylase, 4, 92, 179, 223, 224
Gluconacetobacter xylinus, 420
Gluconic acid, 85
Gluconobacter, 676
Glucose, 20, 22, 48, 85, 125, 180, 224
Glucose–fructose isomerization, 180
Glucose isomerase, 221
Glucose oxidase, 221, 470
Glucose syrups, 180
Glucosidases, 178
Glucosyl hydrolases acting on starch, 179
Glufosinate, 293, 326, 327
Glutamate cysteine ligase, 153, 154
Glutamate dehydrogenase (GDH), 144
Glutamic acid, 84
Glutamine, 20, 22
Glutathione, 125
Glutathione peroxidise, 153, 154
Glutathione *S*-transferase, 153, 154
Glutathione synthetase, 153, 154
Gluteraldehyde, 213

Glycerol, 84, 396
Glycine max (L.) Merr., 614
Glycinin, 432
Glycogen synthase kinase, 153
Glycoproteins types, 23
Glym3 of soybean, 13
GlymBd, 14
Glyphosate, 293, 326, 327
Glyphosate-resistant cotton, in United States, 327
Glyphosate-resistant crops, 326
GnRH analog, 332
GO Consortium, 318
Goats, semen from, 57, 58
Golden rice, 117
Gonadotropin-releasing hormone (GnRH), 241, 604
Gongronella butleri, 157
Gongronella butleri IF08080, 157
Gongronella butleri IF08081, 157
Gongronella butleri USDB 0201, 157
"Good" cholesterol, 77
Good manufacturing practice (GMP) standards, 662
Gossypium hirsutum L., 614
Graham v. John Deere Company, 657
Graingenes, 308
Gram-positive bacteria
conjugation technique in, 60
electroporation technique in, 60–61
expression vectors in, 61
fluorescent in situ hybridization (FISH), 62
genetic manipulation and analysis of, 61–62
genome sequencing of, 62
integration vectors in, 61
mechanisms for DNA transfer into, 60–61
microarray, 62
mutagenesis procedures, 61–62
natural transformation technique in, 60
proteomics, 62
protoplast transformation technique in, 61
quantitative PCR and real-time PCR, 62
shuttle vectors in, 61
transduction technique in, 60
vehicles for DNA transfer into, 61
Gramene, 308
"Grand Nain" (AAA), 68, 69
Grape, 51
GRAS (generally recognized as safe), 691
Green beer, 74
Green Revolution, 205–206
Green tea, 269
Green technology, 117
Green vegetable bug, 100–101
Grifola frondosa, 80, 78
Group A streptococci, 60
Growth factors types, 23

Gruyère, 144
Guanidinium thiocyanate (GuSCN), 681
Guehomyces pullulans, 366
Guinness, 9
GURTs (Genetic Use Restriction Technologies), 659
Gwent Electronic Materials, 114
Gypsy moth, 337

H

H-transferase, 705
Haberlandt, Gottlieb, 519
Hafnia, 365
Hain Foods, 455
Hale, William, 83
Hannibal, 677
Hansen, Christian, 6
Hanseniaspora, 690
Hanseniaspora uvarum, 140
Hanseniospora sp., 8
Hansenula, 669
Hansenula anomala, 139
Hansenula sp., 8
Haploid gene expression, 383–384
Haploids, 160
Hapten–protein conjugate, 213
Harmonia axyridis, 336
Harmonizing methods, 589
Harris, T.W., 482
Harrowing, 463
Harvard mouse. *See* Oncomouse®
Hatch Act, 483
Hayfever, 383
HcrVf2 gene, 630
Health claim, 259
Heifers, 403
HeLa cells, 196
Helianthus annuus, 184, 185
Hematopoietic stem cells (HSCs), 168
Heme oxygenase-1, 153, 154
Hemicellulase, 221
Hemiptera, 336
Hen day rate (HDR), 537
Hen-housed production (HHP), 537
Hendrix Genetics, 536
HepaMate™, 23
Hepatitis 3 non-structural protein (NS3), 665
Hepatitis, 77
Hepatitis A virus (HAV), 680
Hepatitis B surface antigen (HBsAg), 662
Hepatitis B virus, 662
Hepatoprotective effects, and mushrooms, 77, 80
Herbicide and insecticide resistance, 118
Herbicide-resistant crops (HRCs), 326, 686–687
  environmental benefits and risks, 328
  farmer, risks and benefits for, 326–327
  weed management, current impact on, 326
Herbicide-resistant weeds, 686
Herbicide-tolerant crops, 627. *See also* Herbicide-resistant crops (HRCs)
Heredity laws, 28
Hericenones, 80
Hericium erinaceum, 78, 80
Heritage, 536
Heterochromatic siRNAs, 581
Heterodera spp., 440
Heteropolysaccharides (HePS), 418
Heterorhabditis, 337
Heterozygotes, 294
Hi-Line/Lohmann, 536
High fructose corn syrup (HFCS), 180
High hydrostatic pressure (HHP)
  oocytes, 274–275
  preimplantation stage embryos, 275
  spermatozoa, 274
  as sublethal stress, 273–274
High pressure (HP) processing and enzymatic reactions, in food, 329
  dairy products, pressure on enzymes in, 329–330
  fruit and vegetables, high pressure on enzymes in, 330
  proteolysis, by high-pressure processing, 330–331
  seafood, high pressure on enzymes in, 330
High pressure low temperature treatment (HPLT), 540
High-density lipoprotein (HDL), 77
Hippocrates, 677
His, 13
Histamine, 9
Histidine acid phosphatases (HAPs), 491
Hite, Bert, 329
Hoechst 33342, 403
Hollow-fiber bioreactor (HFB), 22
Holstein skeletal muscle, 317
Homopolysaccharides (HoPS), 418
Hopper, 136
Hops (*Humulus lupulus*), 73
Hordein, 3
Horizontal gene transfer (HGT), 645
Horizontal resistance, 375, 376
  to pathogens, 512–513
Hormesis, 41
Hormodendrum, 365
Hormones types, 23
Horse embryos, 245
Horseradish peroxidase (HRP), 213
Horses
  semen from, 57, 58
  sex-sorted sperm, utilization of, 391
Host cell membrane, 64
Host plant resistance, 512
  horizontal/general resistance, 512–513
  specific host–pathogen interactions, 512
HP treatment, principle of, 329
Hsp25, 423
Hubbard, 536
Human chorionic gonadotropin (hCG), 332, 605
Human cytomegalovirus (HCMV), 523
Human embryo kidney (HEK-293), 18, 19
Human embryonic stem (hES) cells, 163
Human gene therapy, 295
Human infertility, ICSI in, 346–347
Human interleukin-10, 193
Human retina-derived (PER C6) cells, 18, 19
Human therapies, tissue engineering for
  cardiac valves, 701
  gene-modified pigs and embryonic stem cells, 701
  porcine pancreatic islets, 701
Humicola grisea, 157
Humicola insolens, 3
Humulus lupulus, 73
Hurdle technology, 551
Hwang, Woo Suk, 165
Hyaluronan, 124
Hydrocolloids, 419
Hydrogen peroxide, 470
Hydrostatic pressure, 274
Hydroxycinnamic acids, 449
Hydroxymethylfurfural (HMF), 85
Hymenoptera, 336
Hyperacute rejection (HAR), 164
Hypercholesterolemia, 77, 294
Hypericum perforatum, 487
Hypsizigus marmoreus, 79
Hysteroscopic insemination technique, 392

I

In vitro applications, of sHsps
  chip/array-based approaches, 423
  gel- or column-based protein analysis and purification, 424
In vitro culture (IVC), 394
In vitro fertilization (IVF), 130–131, 245, 391, 394, 427–428, 714
In vitro plant regeneration. *See* Organogenesis
In vitro production (IVP) systems, 276
In vitro-produced (IVP) bovine embryos. *See* Bovine embryos, in vitro culture of
In vivo applications, of sHsps
  enhanced protein production, 422–423
  increased stress tolerance, 422
IbpA, 422, 423
IbpAB genes, 422, 423
IbpB, 422, 423
IBs, 423

Idli, 136
  and dosa, processing steps for, 365–366
IgA antibodies, 12, 522
IgD antibodies, 12
IgE antibodies, 11, 12
IgG antibodies, 12, 522
IgM antibodies, 12
Ile-Pro-Pro, 198
Illumina™ 18K iSelect chip, 314
Illumina™ iSelect pig DNA chip, 60K, 314
Illumina™ OvineSNP50, 314
Imidazolinones, 327
Immobilized cell bioreactors, 698
Immortalized cell lines, 18
Immune electron microscopy, 680
Immunomodulating and antitumor effects, of mushrooms, 77
Immunotherapy, 13
Indigo, 85
Indirect effects, of transgenic crops, 627
Indoles, 152
Induced pluripotent stem (iPS) cells, 167, 168
Industrial cell lines, 18, 19
Infant formulas, fats for
  lipases in, 232, 372
Inhibin, 604
Injera. See Enjera
Inner cell mass (ICM), 344, 595, 599
Inoculation implementation strategy, in genetic improvement, 87
Inoculation, of MLF
  alcoholic fermentation, inoculation after, 388
  alcoholic fermentation, inoculation during, 388
  yeast and bacteria, simultaneous inoculation of, 388
Inoculative augmentation, 335
Inonotus obliquus, 79
Insect control, 630
Insect-resistant plants, 627
Insemination. See also Artificial insemination (AI)
  from cattle, 58
  from chickens, 58–59
  dose and regimen, 57
  from goats, 58
  from horses, 58
  from sheep, 58
  from swine, 58
  from turkeys, 58–59
  technique, 57–59
Insertion sequences (IS), 477
The Institute for Genomic Research (TIGR), 308, 444, 445
Institute for Reference Materials and Measurements (IRMM), 302

Integrated pest management (IPM), 479, 484, 485
  era, 655–656
Integration vectors
  DNA transfer in Gram-positive bacteria, 61
Intellectual property, and plant science, 339
  copyrights, 339–340
  of genetically modified organisms (GMOs), 31
  geographical indications, 340
  germplasm, access to, 341
  meaning of, 339
  patenting life, 340
  patents, 339
  Plant Breeders' Rights, 340
  plant patent, 340
  trade secrets, 340
  trademarks, 340
  traditional knowledge, 341
Inter simple sequence repeat (ISSR), 506
Interesterification reactions, 371
Interfering RNA (siRNA), 636
Interferon-gamma, 193
Interferons, 77
International embryo transfer industry, 243
International Embryo Transfer Society, 344
International Federation of Organic Agriculture Movements (IFOAM), 462
International Food Safety Authorities Network (INFOSAN), 266
International Health Regulations (IHR), 266–267
International Institute of Tropical Agriculture, 160
International Organization for Standardization (ISO), 266, 589
International Sanitary Conference, 266
International Seed Testing Association (ISTA), 587, 589
International Treaty of Plant Genetic Resources (ITPGR), 341
International Union for the Protection of New Varieties of Plants (UPOV), 339, 340
International Union of Biochemistry and Molecular Biology (IUBMB), 223, 231
INTERPRO SCAN, 451
Interrow cultivation, 463
Interrupted slow freezing (ISF), 395–396
Interspecies embryo transfer, 342–344
Intertaxon embryo transfer. See Interspecies embryo transfer
Intracellular pH (pHi), 127
Intracytoplasmic sperm injection (ICSI), 334, 345, 394
  basics, 345
  early attempts of, 346

  in human infertility, 346–347
  ultimate success of, 346
Intravaginal P4-releasing devices, 712
Introgressive hybridization, 321
Intron hairpin RNA (ihpRNA) vector, 14
Inulinases, 221
Inundative augmentation, 335
Investigational new animal drug exemption (INAD) regulations, 287
Invitrogen, 125
Iosthiocyanates, 152
IRE1, 467
Irpex lacteus, 78
Islet sheet, 23
ISO (International Organization for Standardization), 266, 589
ISO 22000, 266
Isoamyl acetate, 139
Isoamylase, 179
Isoflavones, 152
Isologous cells, 599
Isomalto-oligosaccharides, 180, 181
Isopentenyl diphosphate (IPP), 445–446
Iso-α-acids, 73
Issatchenkia orientalis, 139
IT Corporation, 111
Itaconic acid, 84

J

Jalebi, 136
Japan, 141
  product labeling in, 555–556
Japanese Black muscle tissue, 316
JD Special (AAA), 68
J.E.M. Ag Supply v. Pioneer Hi-Bred, 657, 659
Jenner, Edward, 700
Jeotgal, 252
Joint FAO/WHO Expert Committee on Food Additives (JECFA), 264
Joint FAO/WHO Expert Meetings on Microbiological Risk Assessment (JEMRA), 264
Joint FAO/WHO Food Standards Programme, 675
Joint FAO/WHO Meetings on Pesticide Residues (JMPR), 264
Jojoba FAR, 651
Jojoba oil, 652
Jojoba, wax biosynthesis in, 650, 651
Jojoba WS, 650

K

Kaffir beer, 9, 140–141
  production of, 140
Kanamycin, 293
Kanga kopiro, 136
Kanji, 136
KAO Corporation, 374

Karashi mentaiko, 252
Karasumi, 254
Karyotyping method, 148
Kefir, 419
KEGG, 411
Kelch-like ECH-associated protein 1 (Keap1), 154
Kenkey (Kenky) in Ghana, 138
Keratinocyte progenitors (KPCs), 169
Keratinocyte stem cells (KSCs), 169
Ketosis, in dairy cattle, 317–318
Keyhole limpet hemocyanin (KLH), 212
Khanomjeen, 136
Kichudok, takju, 136
Kimchi fermentation, 348, 668, 669
   acidity and storage, of kimchi, 350–351
   dietary importance of, 348–349
   distribution, of kimchi, 351
   factors affecting, 350
   microorganisms associated with, 349, 350
   physicochemical change during, 349, 350
   preparation process, 349
   raw materials, 350
   salt concentration and temperature, 350
   varieties of, 348
Kimjang-kimchi, 348
Kinetics of reactions catalyzed, by lipases, 371
Kinetochore, 53
Kishk, 138
Kisra, 136, 365–366
Klebsiella aerogenes, 4
Klebsiella sp., 365, 471
Klf4, 559, 602
Kloeckera apiculata, 8
Kloeckera. See Hanseniaspora
Kluveromyces, 138, 669, 690
Kluyveromyces africanus, 140
Kluyveromyces lactis, 147, 196
Kluyveromyces marxianus, 149, 709
Kluyveromyces marxianus, 48, 147, 709
KNOTTED1, 467
Kocuria varians, 139
Koji, 141, 431, 432
   preparation, 591–592
Kome miso, 431
Korea
   product labeling in, 555–556
Kossel, Albrecht, 208
Kraft pulp sludge, 358
Kridl v. McCormick, 657
KSOMaa, 123, 124
Kulcha, 136
Kwashiorkor, 648
Kwuna-zaki, 136
Kyoto Accord friendly, 91
Kyoto Protocol, 95

L

L-ascorbic acid, 85
L. coryniformis, 139
L-Glutamate, 472
L-lactic acid, 85, 368, 387, 664
L-leucine, 677
L-lysine, 85
L-malic acid, 368, 387
L-proline, 677
LABDEL project, 664
Labeling of food. See Food labeling
Labile toxin (LT), 662
Labile toxin B subunit (LT-B), 662
Laccases, 471
Lactate, 22, 44
Lactation, 29
Lactic acid, 44, 85–86, 138, 708
Lactic acid bacteria (LAB), 8, 133, 143, 147, 195, 200, 360, 361, 363, 365, 387, 389, 474, 540, 546, 664, 669
   anti-infective molecules, delivery of, 666
   in cheese making, 419
   as dairy starter cultures, 201
   in fermented milks, 418–419
   recombinant strains, biological containment of, 666
   in sourdough fermentation, 419
   as vaccines, 664–666
Lactic acid fermentation, 668
Lactic acid–producing fungi
   direct lactic acid fermentation by, 357
Lacticin 3147, 196,198
Lactobacillales, 360
Lactobacilli, 546
Lactobacillus, 118, 135, 200, 201, 365, 387, 664, 665, 666, 668, 670, 690, 708, 709
Lactobacillus acidophilus, 196, 197, 201, 358, 665, 666, 707, 709
Lactobacillus amylophilus, 357
Lactobacillus amylovorus, 357, 366
Lactobacillus brevis, 138, 139, 349, 360, 546
Lactobacillus bulgaricus, 139, 196, 545
Lactobacillus casei, 138, 144, 145, 358, 665
Lactobacillus curvatus, 144, 361, 399
Lactobacillus delbreuckii, 198, 358, 366
Lactobacillus delbrueckii subsp. *bulgaricus*, 201, 707, 709
Lactobacillus delbrueckii subsp. *lactis*, 707, 708
Lactobacillus fermentum, 138, 139, 366, 357, 665
Lactobacillus GG, 665
Lactobacillus helveticus, 198, 201, 665, 707
Lactobacillus jensenii, 666
Lactobacillus jugurti, 707
Lactobacillus kefiranofaciens, 419
Lactobacillus manihotivorans, 357
Lactobacillus paracasei, 144, 145, 543
Lactobacillus plantarum, 138, 139, 140, 144, 349, 357, 368, 476, 665, 669
Lactobacillus plantarum 299v, 361, 362
Lactobacillus plantarum BCC 9546, 400
Lactobacillus plantarum G11, 400
Lactobacillus plantarum, in foods, 360
   in dairy and meat products, 361
   diversity and metabolism, 360
   and health benefits, 361–362
   in plant-based foods and beverages, 361
Lactobacillus plantarum NCIMB8826, 361
Lactobacillus plantarum WCFS1, 360, 361
Lactobacillus reuteri, 366
Lactobacillus rhamnosus, 144, 145
Lactobacillus rhamnosus, 144, 542, 543
Lactobacillus sakei, 361, 399, 400–401, 476
Lactobacillus sanfranciscensis, 139
Lactobacillus spp., 60, 62, 349, 366, 388, 389
Lactobionic acid, 471
Lactococci, 685. See also Dairy lactococci
Lactococcus, 200, 664
Lactococcus garvieae, 191
Lactococcus lactis, 43, 60, 61, 62, 139, 191, 192, 193, 196, 198, 357, 361, 419, 433, 549, 665, 666
Lactococcus lactis subsp. *cremoris* 200, 201
Lactococcus lactis subsp. *Lactis*, 200, 201
Lactococcus lactis subsp. *lactis* biovar. *diacetylous*, 201
Lactococcus piscium, 191
Lactococcus plantarum, 191
Lactococcus raffinolactis, 191
Lactococcus spp., 62
Lactoferrin, 523
Lactose metabolism, 192
Laetiporus sulphurous, 78
Lagering, 73–74
Lama pacos, 343
Land Grant University System, 483
Langmuir probes technique, 176
Lantana, 100
Lantana camara, 100
Lantana montevidensis, 100
Lao-chao, 136
Large offspring syndrome (LOS), 131, 276
Lateral flow device (LFD), 588
Lauter tun, cross-section of, 72
Lautering. See Wort separation
Lawrence Livermore National Laboratory, 390
LD_{50}, 68
Leaching. See Solid–liquid extraction
Leavened breads, 363
   acid-fermented leavened dough, 364–365
   acid-leavened breads, 365–366

Leavened breads (*Continued*)
  dough development, 363, 364
  frozen-leavened dough, 364
  leavening, 363
  wheat gluten, 363
  yeast-leavened doughs, 363–364
Lecitase®, 374
Legislation (European approach), 40
Legislative period, 264
Lemon essential oils, 367–368
Lentil, 463
Lentinan, 80
Lentinula edodes, 77, 79, 80, 472
Lepidoptera, 336, 337
Leprosy, 39, 41
Leptinotarsa decemlineata, 483
Lettuce, 51
Leuconostoc, 118, 141, 143, 144, 200, 365, 387, 474, 664, 669, 670, 670
Leuconostoc lactis, 144, 201
Leuconostoc mesenteroides, 138, 144, 181, 349, 361, 366, 419, 669
Leuconosotoc mesenteroides subsp. *Cremoris*, 201
Leuconostoc oenos, 9
Leuconostoc paramesenteroides, 139
Leuconosotoc lactis, 201
Levan, 420
Levels of concern, MPRL, 40
Levoglucosan, 85
Levoglucosenone, 85
Levuana iridescens, 100
Levulinic acid, 84, 85
Lignin, 84, 85
Lime sulfur, 483
Limit dextrinase, 223, 224
Limonene, 368
Linkage maps, 312
Lipases, 96, 221, 231, 370
  applications in food industry, 232–233
  biological significance, 231
  cheese flavor, enhancement and production of, 232
  cocoa butter substitutes, 233
  diglyceride oils, 233
  effects on food quality, 231–232
  food applications of, 372–374
  future innovations and applications, 233
  infant formulae, 232
  kinetics of reactions catalyzed by, 371
  medium-chain triglycerides (MCTs), 232–233
  in non-aqueous media, 371–372
  occurrence and commercial production, 231
  omega-3 concentrates, 232–233
  over-expression and modification, 370
  reactions catalyzed by, 371
  three-dimensional structures of, 370
  trans fat alternatives, 233
  with lipid feedstock, 96–97

Lipid oxidation, 254
Lipids, 435
Lipolysis, 231
Lipoprotein lipase (LPL), 232
Lipoxygenase (LOX), 471
Lipozyme RM-IM, 97
Lipozyme TL-IM, 97
Liquid chromatography–mass spectrometry (LC-MS), 411
Liquid hybridization, 682
Liquid–liquid extraction, 407
Liquid membrane extraction, 408–409
Liquid nitrogen slush, 461
Liquid-continuous (LC) operation modes, 698
Listeria monocytogenes, 43, 45, 60, 62, 66, 80, 144, 198, 215, 273, 477, 709
Listeria P100 bacteriophage, 66
Listeria spp., 61, 62
Lithospermum erythrorhizon, 520
Livarot cheese, 149
Livestock production. *See* Genomics:research
Living modified organisms (LMOs), 526
Llambazi, lakubilisa, 136
LLrice601, 303, 304
LNT model (linear non-threshold model), 41
LOAEL (lowest observed adverse effect level), 40
Locus-RFLP method, 148
Loders Croklaan, 372
Loessner, 66
Long-chain omega-3 acids, 233
Lovastatin, 77
Low power inputs, reactors with, 696, 698
Low-density lipoprotein (LDL), 77
LRP1 (*LATERAL ROOT PRIMORDIUM1*) gene, 467–468
LT model (linear threshold model), 41
"Luddite", 295
Lunaria annua, 650
Lunaria KCS, 651, 652
Lupinus albus, 491
Luteinizing hormone (LH), 605
Lyc e 3, 14
Lygus lineolaris, 101
Lymantria dispar, 337
Lymphocytes, 77
Lyngbya gracilis, 109
Lysozyme, 44
Lytic phage, 65

M

Macaca mulatta, 343
Macmillan Dictionary of Biotechnology, 116
Macromolecules, 84
Macrophage activating factor (MAF), 523

Mae1, 9
Magnesium, 180
Mahewu, mogou, 136
Maillard reaction, 470
Maize, 13, 31, 51
Maize B chromosome, 54
Maize genetics and genomics database, 308
Maize ogi. *See* Ogi
Maize streak, 502
Maize streak virus, 502
Major histocompatibility complex (MHC), 12, 323
Mal d 1, 13, 14
Malate decarboxylase (*mleA*), 387
Malaysia, 68
Malbranchea sulfurea, 138
Male gametogenesis, 381
  haploid gene expression, 383–384
  microgametogenesis, 381–383
  microsporogenesis, 381
  mutants and development, 384–385
  pollen development, 381–383
  pollen wall–vital interface, 383
Malic acid, 84, 387
Mally, 87
Malolactic enzyme, 387
Malolactic fermentation (MLF), 387
  and bacterial starter cultures, 9
  commercial MLF, advantages of, 388
  factors influencing, 388–389
  inoculation, time of, 388
  risks associated with, 387–388
  starter cultures, 387–388
  wine aroma, impact on, 389
Malolactic yeast ML01, 9
Malting and brewing process, enzyme action in, 2-3
Malto-oligosyl trehalose synthase, 181
Malto-oligosyl trehalose trehalohydrolase, 181
Maltogen amylase, 221
Maltose, 4, 180
Mame miso, 431
Mammalian cells, 18
Mammalian species
  for genome sequencing, 322
Mammalian sperm sexing, 390
  biological basis of, 390
  flow sorting system, 390
  sex-sorted sperm, utilization of, 391–392
Mammary cells, 131
Manganese, 180
Mannitol, 669
Mantou, 136
MapMan/PageMan, 411
Marasmius androsaceus, 79
Marek's disease, 314
Margarine fat, 96
Marker-assisted selection (MAS), 312, 538
Mastrevirus, 500
Material transfer agreement (MTA), 341

Maternal to embryonic transition (MET), 277
MatK gene, 505
Mawe, 136
Maximum residue limit (MRL), 40, 264
MBMOC, 123, 124
Me, 136
Measles, 523
Measles virus, 523
Meat fermentation, 399
    genetic knowledge, of microbial starters, 400–401
    identification and characterization, of microorganisms, 399–400
    investigation of microbial groups/specific microorganisms, 400
    manufacture and quality control, 399, 400
Meat products
    biotechnology, 402
    labeling, 260
    Lactobacillus plantarum in, 361
Meat-type breeding, 536–537
Mechanically agitated columns, 407
Medida, 139
Medium-chain fatty acids (MCFAs), 232–233
Medium-chain triglycerides (MCTs), 232–233
Melengestrol acetate (MGA), 241
Meloidogyne spp., 440
Mendel, G., 28
MerA genes, 496
MerB genes, 496
Meristems, 70
Mesenchymal stem cell, 23
Mesenchymal stem cells (MSCs), 169, 599
Mesenteroides, 139
Messenger RNA (mRNA), 33, 276
Metabolic disease from liver transcriptomics in dairy cattle, insights into
    ketosis, 317–318
    prepartum energy nutrition, 317
Metabolic flux analysis (MFA), 698
Metabolism and toxicity test, 24
Metabolite extraction, from plant tissues, 407
    aqueous two-phase extraction (ATPE), 408
    liquid membrane extraction, 408–409
    reverse micelle extraction, 409
    solvent extraction, 407–408
    supercritical fluid extraction (SFE), 408
Metabolites
    biosensors in agriculture and food products, 114
    fingerprinting, 411
    profiling, 411
    secondary, 208
Metabolix, 459
Metabolomics, 410
    definitions, 411
    example studies, 411–412
    limitations, 411
    technologies, 411
Metabolomics Standards Initiative, 411
Metagenomics, 415
Metallothioneins, 41
Metarhizium anisopliae, 103
Metaseiulus occidentalis, 88, 89
Metchnikoff, Elie, 103, 545
Methane, 82
Methane thiol, 144
Methicillin-resistant *Staphylococcus aureus* (MRSA), 41
Methionine lyase, 685
Methylcellulose, 44
Methyltetrahydrofuran, 85
Metschnikowia sp., 8
Mevinolin, 77
Micelles, 234
Michaelis-Menten rate expressions, 371
Microalgae, 109
Microarray
    and gene expression, 415
    and protein expression databases, 538
    for Gram-positive bacteria, 62
    screening of cDNAs, 307
Microbial aroma compounds, 47–48
Microbial diversity
    gene tracking, 414
    ribosomal RNA (rRNA), 414, 415
    strain typing, 415
Microbial exopolysaccharides (EPS), 418
    bacterial cellulose, 420
    in cheese making, 419
    curdlan, 420
    in fermented milks, 418–419
    gellan, 419–420
    levan, 420
    pullulan, 420
    scleroglucan, 420
    in sourdough fermentation, 419
    succinoglycan, 420
    xanthan, 419
Microbial fermentation, 669
Microbial inactivation, mechanism of, 488–489
Microbial molecular biology, 414
    gene transfer and genetic analysis, 415–416
    genomics, 415
    microbial diversity, 414–415
    modified microbial strains, 416
Microbial population dynamics, in yogurt, 707–708
Microbial proteolytic enzymes, 254
Micro-cell PB, 558
Micrococcaceae, 399
Micrococci, 144
Micrococcus lenuteus, 80
Microdrops, culturing in, 126
Microfluidic cumulus removal (μFCR), 428
Microfluidics, for assisted reproduction technologies (ART), 426
    cumulus removal (CR), 428
    embryo culture, 428
    methods, 426–427
    in vitro fertilization (IFV), 427–428
    ZP removal, 428–429
Microgametogenesis, 381–383
MicroGARD™, 551
Micromanipulation, 245
Microorganisms, 670
    and toxins in foods, 212–215
    for biodiesel production, 96
    genetic engineering in, 292–293
Microprojectile bombardment, 661
Micropropagation, 519–520
MicroRNAs (miRNAs), 34, 580, 636
    homeostasis, in plants, 582
    role in plants, 580–581
Microsatellites, 403
Microsporidia, 573
Microsporidium, 573
Microsporogenesis, 381
Microtiter plates, in ELISA assays, 213
    for foodborne bacterial pathogens detection and identification, 214–215
    for toxins detection in foods, 214
Migrant pests and tropical pest management, 655
Milardet, Pierre, 483
Milk coagulation, 234
Milk fat globule membrane (MFGM), 232
Milk protein system, 227
Milk-clotting activity, and mushrooms, 80–81
Milkfat, 419
Millet, 51
Mill's model, 611
Minchin, 136
Minimal drop size (MDS), 461
Minimum inhibitory concentration (MIC), 368
Minimum Required Performance Limit (MPRL), 40
Minisatellites, 506
Ministry of Agriculture (MOA), 634
Ministry of Agriculture, Forestry, and Fishery (MAFF), 635
Ministry of Health Labour and Welfare (MHLW), 555
MiRNA*, 579
Misbranding, 258
Miso, 135, 591
Miso fermentation, 431
    manufacturing process, 431–432
    overcome fermentation problems, 432–433

Miso fermentation (*Continued*)
  process and flavor chemistry, 432
  product description and composition, 431
Mitochondrial DNA (mt-DNA), 148, 322
Mitochondrial genetics and animal modeling, 619–620
Mitochondrial genome modification, of transgenic animals, 619
  mitochondrial genetics and animal modeling, 619–620
  transmitochondrial animals, 620
Mitogen-activated protein (MAP) kinases, 153
Mitsubishi rayon process, 85
Mitsui Petrochemical Industries, 520
ML01, 9
MleA, 9
Mobile genetic elements, 416
Moche IV–V Transformation, 205
Modena, 675, 676
Modified microbial strains, 416
Mold fermentation, 135–137
Molecular assisted backcrossing (MABC), 614
Molecular biotechnology, 436–437
Monilia candida, 141
Monocarboxylic acid, 387
Monochlorobenzoate, 111
Monoclonal antibodies (mAbs), 23, 212
Monocytes, 77
Monoglycerols, 373
Monosodium glutamate, 85
Monoterpenes, 389
Monsanto, 459
Monsanto Company, 293, 301
Moo-kimchi, 348
Morinda citrifolia, 209
Moromi, 592
Morrell Act, 483
MOTIF SCAN, 451
MoToDB, 411
Mozzarella cheeses, 419
MTBE (methyl tertiary butyl ether), 91
Mucor hiemalis, 157
Mucor javanicus, 96
Mucor miehei, 235
Mucor pusillus, 235
Mucor rouxii, 156, 157
Mueller, Paul, 483
Mugi miso, 431
Muller-Thurgau, Hermann, 6
Mullet roe, 252
Multi- and unipotent stem cells, 168–169
Multiparous cows, 558, 559
Multiple ovulation/embryo transfer (MOET)
  genetic parameters estimation, 216–218
  limitations, 216
  synergistic model for, 217
Multi-stage extractors, 408

MultiStem®, 23
Multiwells, 21
Munich Information Center for Protein Sequences (MIPS), 308
Murine lymphoid (NS0, Sp2/0), 18, 19
Mus caroli/Mus musculus, 343
Musa acuminata, 68
Musa balbisiana, 68
Musa species, 161
Muscidifurax raptor, 336
Mushrooms
  antidiabetic effect, 80
  antimicrobial effects, 80
  antioxidant activity, 80
  bioactive compounds from, 76
  cardiovascular and cholesterol-lowering effects, 77
  hepatoprotective effects, 77, 80
  immunomodulating and antitumor effects, 77
  milk-clotting activity, 80–81
  nerve tonic activity, 80
Mustard (Bra j 1, Sin a 1), 13
Mustela lutreola, 343
Mutagenesis procedures
  in Gram-positive bacteria, 61–62
Mutiara (AAB) with Fusarium resistance, 68
Mutwiza, 136
Mycobacterium leprae, 39
Mycobacterium tuberculosis, 39, 80
Mycoderma vini, 141
Mycoherbicides, 105
Mycoplasma pulmonis, 416
Mycosphaerella spp., 68
Myosin genes, 317
Myostatin gene, 37, 312
Myriophyllum spicatum, 100
Myrothecium verrucaria, 157
Myxoma virus, 672
Myxomatosis, 672–673

N

Nacional, 171
NAD(P)H:quinone oxidoreductase 1, 153, 154
NADPH, 650
Nan, 136
Nanog, 599
Nanoscale biology, 434
  case studies, 438–439
  components, 434–435
  design philosophy, 438
  engineering techniques and system integration, 435–438
Nanotechnology, 434
Nanovirus, 500
NARMS (National Antimicrobial Resistance Monitoring System), 25–26
Nasha, 136

nat-siRNAs, 581
Nathans, D., 292
National Center for Biotechnology Information (NCBI), 307, 320
National Cholesterol Education Program (NCEP), 516
National Environmental Policy Act (NEPA), 281
National Human Genome Research Institute (NHGRI), 312
National Institutes of Health, 293
National Research Council (NRC), 287, 411
National Science Foundation (NSF), 309
Natural aroma, 47
Natural enemies, 335
  parasitoids, 336
  pathogens, 336–337
  predators, 335–336
Natural gas, 82, 84
Natural gene transfer mechanisms, 415–416
Natural killer cells, 77
Natural transformation technique
  DNA transfer in Gram-positive bacteria, 60
Nematodes, 440
  biocontrol agents, 440–441
  nonspecific biocontrol agents, 441
Neosartorya, 575
Nerve growth factors (NGF), 80
Nested reactions, 682
NetNGlyc, 452
Neural stem cells (NSC), 168
Neurospora, 451
Neurospora crassa, 452
Neurotoxicity test, 24
Neutrase, 331
Neutrophils, 77
New Zealand, 164
Nezara viridula, 100–101
Niacin, 76, 225
NICE (NIsin-Controlled gene Expression) system, 193
Nicotiana benthamiana (Nb), 443, 662
  carotenoid biosynthesis, 443–444
  methods, 445
  phytoene desaturase (pds), knock down of, 446
  phytoene synthase (*psy*), overexpression of, 445–446
  viral vector and transfection design, 444
Nicotine, 85, 139
Nisaplin™, 549
Nisin, 43, 193, 198, 549–551. *See also* Processing and preservative aids
Nitrogen, 157
Nitrogen-based atmospheric plasma, 174–175
"No-dose no-illness" approach, 41

NOAEL (no observed adverse effect level), 40
Non-aqueous media, lipases in, 371–372
Non-carcinogens, 41
Non-equilibrium freezing, 460
Non-essential amino acids, 125
Non-flavonoid compounds, of vinegars, 677
Non-GM plants, 296
Non-homologous end-joining mechanism, 53
Nonidet-40, 213
Non-IgE-mediated food hypersensitivity reactions, 12
Non-inhibiting concentration (NIC), 368
Non-pesticide traits, 527
Non-protozoan parasites, rapid detection of, 573–574
Non-specific lipid transfer proteins (nsLTPs), 13
Nonstarter culture bacteria. *See* Cheese production
Non-starter LAB (NSLAB), 143, 144, 361
Non-viral disease, 637
Norgestomet ear implants, 712
Norisoprenoids, 389
Noroviruses (NoVs), 680
North America, 91, 185
Norwalk virus, 662
Norwalk virus capsid protein (NVCP), 662
Nosema, 573
Novaria, 68
Novel Food Regulation, 553
Novel protein, production of, 648–649
Novozymes, 97, 373, 374
NoVs, 682
NPC1L1, 517
Nrf2, 154
Nuclear genome diversity, 505–506
  genetics and genomics tools used for, 506, 507
Nuclear magnetic resonance spectroscopy (NMR), 411, 650
Nuclear transfer (NT) cloning technology, 617. *See also* Cloning
Nuclear transfer ES (ntES) cell lines, 165
Nucleic acid extraction, 681
Nucleic acid sequence-based amplification (NASBA), 681
Nucleic acids, 435
Nuruk, 140
Nutraceuticals and functional foods, 454
  drivers of innovation, 454
    institutional innovation, 455–456
    lipases in, 374
    market size and growth, 454–455
    product innovation and industrial development, 455
Nutreco, 536
Nutrigenomic perspectives, 154–155
Nutrigenomics, science of, 270

Nutrition Facts panel, 258
Nutrition Labeling and Education Act (NLEA), 258, 257, 259

O

Oats, 1, 51, 184, 185
Obesity, 269–270
Occupational Safety and Health Administration (OSHA), 281
Oct3/4, 599, 602
Odorants, 47
Oenococcus, 474, 690
Oenococcus oeni, 9, 368, 387, 388, 690
Oestradiol, 712
Oestradiol valerate (EV), 712
Office of Science and Technology Policy (OSTP), 526
Ogi, 136, 139
"Ohne Gentechnik," 555
Oil crops, 457
  markets and crops, 457
  problems and prospects, 457–459
Oilseed rape, 184, 185
Olefins, 82
Olestra®, 374
Oligopeptide uptake system (OppA), 192
Oligosaccharides, 85
Omega-3 acids, 233
Omega-3 concentrates, 232–233
"omics" platforms, 693
Oncomouse®, 295
One-dimensional (1D) electrophoresis vs. two-dimensional (2D) electrophoresis, 404–405
O-nitrophenyl-β-d-galactopyranoside (ONPG), 213
Oocyst, 572
Oocyte collections, 332–333
Oocyte cryopreservation
  rodents, 353
  swine, 354
Oocytes, 163, 165, 274–275
  and embryos, 460
  background, 460
  cryopreserving cells, 460
  research, 461
  standard cryopreservation, 460–461
  transfer, 333
  vitrification, 461
Oospora, 139
Open pulled straw (OPS), 461
Open-reading frames (ORFs), 193, 476
Opines, 293
Optical transduction methods, for biosensor, 113
Oral tolerance, 12
Organ replacement, technologies for, 704
Organelle genome diversity, 504–505, 506
  genetics and genomics tools used for, 506, 507

Organic acid, 44, 111
Organic farming, 462
  pest management practices, 462–464
  standards, 462
Organization for Economic Cooperation and Development (OECD), 288, 410, 411
Organogenesis, 466
  cell division, competence for, 467
  competence for, 467
  competence, for organogenesis, 467
  determination, 467–468
  totipotency, 466
Organometallic linkers, 435
Organosulfur compounds, 152
Orthophenylene diamine (OPD), 213
Oryctolagus cuniculus, 165, 672
Oryza sativa, 184, 366
Oscillatoria animalis, 109
Oscillatoria annae, 109
Oscillatoria pseudogeminata, 109
Osmoprotectants, 382
Ova, 130–131
Ovalbumin (OVA), 212
Ovarian follicular dynamics during the estrous cycle, 240
Ovis aries, 343
Ovis musimon, 343
Ovis orientalis gmelini, 343
Ovis orientalis musimon, 343
Ovsynch protocol, 241
Oxidoreductases, 4, 470, 471
Oxydia trychiata, 100
Oxygen mass transfer, 22
Oxygen transfer rate (OTR), 695
Oxygen uptake rate (OUR), 21
Oxygenases, in food, 469
  biodegradation, bioremediation, and environmental applications, 472
  biosensors, 472
  food applications, 470–472
  industrial applications, 470
Oyi-kimchi, 348
Ozone, 176

P

P-chlorophenol, 189
P-cymene, 368
P-hydroxibenzaldehyde, 49
P-hydroxybenzyl methyl ether, 49
P-nitrophenylphosphate (PNPP), 213
P100 phage, 66
P34, 467
P5CDH gene, 581
PacMan, 533
Paecilomyces farinosus, 337
Paecilomyces lilacinus, 440, 441
Palm wine, 9
Palmitic acid, 232
Panax ginseng, 520

Papain, 3, 238
Papaya ringspot virus (PRSV), 509, 630
Papillae, 56
Paradichlorobenzene (PDB), 483
Parasitoids, 87, 336
Para-κ-casein, 228, 234
Pareuchaetes aurata aurata, 100
Pareuchaetes pseudoinsulata, 100
Paris Convention, 658
Paris Green, 483
Participatory plant breeding, 206
Parvalbumins, 12
Pasteur Institute, 292
Pasteur, Louis, 6, 133, 668, 689
Pasteuria penetrans, 440–441
PASTICCINO, 467
Patent, 339
 and patent rights, 658
 requirements, 657–658
Patent and Trademark Office (PTO), 657
Patent Cooperation Treaty (PCT), 658
Patent infringement, 659
Patenting life, 340
Pathogen control, 629–630
Pathogen detection
 biosensors in agriculture and food products, 114–115
Pathogen-specific resistance (R) genes, 630
Pathogenic bacteria, survival of, 709
Pathogens, 336–337, 608, 609–610
Pathogens and superantigens detection, 24
PAZ (Piwi, Argonaut, and Zwille), 579
PCBs, 109
PCR-based methods, 588, 589
PDK28 vector, 14
Pea, 51
Peach (Pru p 1), 13
Peanut, 13, 51
Peanut allergens genes, strategy to eliminating. *See also* Food allergy
 Ara h 2 protein reduction level, immunological analyses of transgenic seeds to determine, 14–15
 fertile transgenic peanut production, 14
 genomic DNA isolation encoding the major peanut allergens, 14
 kanamycin-resistant plantlets regeneration, 14
 molecular analyses to identify transgenic peanut lines, 14
 and reducing increasing risks, 14–16
 RNAi application to silence the major peanut allergen Ara h 2, 14–16
 RNAi transformation vector construction, 14
Pectin methylesterase (PME), 330
Pectinase, 4, 221, 691
Pedigree data, 217

Pediocin, 43, 477
Pediococci, characteristics of, 474
 classification and morphology, 474
 ecology, 474
 metabolism, 474–476
Pediococcus, 143, 387, 388, 389, 474, 664, 669, 670, 690
 biotechnology, applications in, 477–478
 genetics, 476–477
 pediococci, characteristics of, 474–476
Pediococcus acidilactici, 43, 144, 476, 477, 669
Pediococcus damnosus, 476
Pediococcus halophilus, 474, 592
Pediococcus parvulus, 476
Pediococcus pentosaceous, 43, 139, 144, 349, 360, 401, 476, 477, 478
Pediococcus soyae, 592
Pediococcus urinaeequi, 474
Penicillia, 39, 138
Penicillin chrysogenumare, 157
Penicillium, 138, 365, 451
Penicillium camemberti, 147, 201
Penicillium citrinum, 140
Penicillium culosum, 140
Penicillium digitatum, 157
Penicillium funiculosum, 1, 452
Penicillium notatum, 39
Penicillium roqueforti, 147, 195, 201
Penicillium sp., 139
Peniophora lycii, 491
Penocin A, 476
Pentosan, 1, 3
Pentosanase, 221
PepsiCo, 455
Pepsin, 198, 235, 331
Peptide immunotherapy, 13
Peptides, 270
Percentage of transferable embryos (PTE), 217, 218
Perennial crops. *See also* Transgenic crops
 engineered insect resistance in, 630
 transformation of, 629
Perfluorocarbon (PFC), 570
Perfusion cell culture process, 22
Peristenus digoneutis, 101
Permeating cryoprotective agents (CPAs), 396
Peroxisome proliferator-activated receptor γ (*PPARG*), 317
Pest management, 479, 653. *See also* Plant pathogens
 before synthetic pesticides era, 654
 basis of, 653, 654
 diseases, 464
 exotic pests of tropical agriculture, 655
 insects and other invertebrates, 463–464
 integrated pest management era, 655–656
 lack of information, on tropical pest ecology, 654

 migrant pests and tropical pest management, 655
 pest management on tropical pest ecology, low application of, 653–654
 pest managers, 479
 population theory, 479–481
 practices, 462
 problems, 653–655
 synthetic pesticides, dependency on, 654–655
 synthetic pesticides era, 655
 terminology hierarachy in, 104
 trends in, 655–656
 weeds, 462–463
Pest managers, 479
Pesticidal traits, 527
Pesticides, 482
 and environment, 655
 actinomycetes, 107, 108
 ascendancy of, 482–483
 bacteria, 107, 108
 cyanobacteria, 109
 DDT, 483–484
 degradation, microbial potential of, 107–109
 fungi, 107, 108, 109
 future, 485–486
 inevitable conflict, 482
 pest management, effects on, 655
 pollution, 107
 rebuff and reassessment, 484–485
 residues on foodstuffs, 114
 safety of, 654–655
Petri dish, 23
Petroleum, 84
Pfaffl, 445
Pfam protein family database, 12
pH, 20–21, 48, 74, 127
Phage resistance
 of lactic acid bacteria (LAB), 203
Phage resistance, 193
Phages. *See* Bacteriophages
PHANNIBAL vector, 14
Phase-contrast microscopy, 616
 differential interference contrast (DIC) microscopy, 616
Phaseolus mungo, 366
Phellinus linteus, 79, 80
Phellinus robustus, 79
Phenazine ethosulfate (PES), 397
Phenol, 84, 111, 189
Phenotypic data, 538
Phenotypic trait (*P*), 216, 217
Phenylactaldehyde, 172
Pheromone traps, 464
Pholiota adipose, 79
Pholiota nameko, 79
Phormidium faveolarum, 109
Phormidium laminosum, 109
Phormidium tenue, 109
Phosphates, 20

Phosphatidylinositol-3-kinase, 153
Phosphoenolpyruvate (PEP), 696
Phospholipases, 373
Phosphonic acid linkages, 436
Phosphotungstic acid (PTA)-soluble nitrogen, 229
Photomultiplier tube (PMT), 390
Photosensitation and food safety, 487
   decontamination of food/food-related surfaces by, 489, 490
   light sources for, 488
   microbial inactivation, mechanism of, 488–489
   photophysical and photochemical processes, 487
   photosensitizers, 487–488
Photosensitizers, 487–488
Phycomyces blakesleeanus, 157
Phyllocnistis citrella, 101
Phylloquinone, 225
Phylloxera vittifolae, 50
Phylogenetic shadowing, 320, 321
Phytases (PhyA), 221, 491
   dietary determinants, of efficacy, 493
   genes, proteins, and properties, 491, 492
   ideal phytases, developing, 493–494
   nutritional and environmental benefits, 493
   storage and handling, 493
   substrate occurrence, 492–493
Phytochemicals, 151, 209
Phytodetoxification, 496
Phytoene desaturase (*pds*), 443, 444
   knock down of, 446
Phytoene synthase (*crtB*), 443, 444
Phytoene synthase (*psy*), 445, 446
   overexpression of, 445–446
Phytoextraction, 496
Phytopharmaceuticals, 151
   as guardians of our health, 151–152
   fight with diseases, 152
Phytophthora cinnamomi, 630
Phytoremediation, 495
   in constructed wetlands, 497
   future prospects, 497
   genome/proteome, 496–497
   transgenic phytoremediation approaches, 495–496
Phytoremediation-related proteins (PRP), 497
Phytostabilization, 496
Phytosterolemia, 518
Phytovolatilization, 496
Pichia, 690
Pichia anomala, 140
Pichia fabianii, 140
Pichia membranefaciens, 8
Piedmontese × Hereford (PH) cattle, 317
Pig and rodent embryo cryopreservation, differences between, 354

Ping Pong Bi Bi mechanism, 371
Pinolenic acid, 374
Piptoporus betulinus, 78, 80
Piromyces equi, 452
Pito, 139–140
   production of, 140
PIWI (P-element induced wimpy testis), 579
PkMADS1, 467
Plague antigens, 662
Plant and animal products. See Protein biosepearation
Plant array database, 308
Plant-based foods and beverages
   Lactobacillus plantarum in, 361
Plant Breeders' Rights (PBR), 340
Plant breeding, transgenes used in, 613
   agronomic characteristics, 613
   backcrossing, 614
   commercial plant breeding, involving transgenes, 614
   construct and event selection, 613
   forward breeding, 615
   molecular characterization, 613
   quality assurance (QA), 614–615
   quality control (QC), 614–615
   segregation of trait, 613
   stability of transgene expression, 613
   transgene expression/efficacy, 613
Plant cell and hairy root cultures, 583
   advantages and disadvantages, 583–584
   application of, 583
   for aroma compounds, 48–49
   obstacles for, 584
   secondary metabolites production584, 583, 584–585
Plant disease resistance, bilateral view of, 375–377
Plant DNA virus diseases, 499
   badnaviruses, 499–500
   banana bunchy top disease (BBTD), 502–503
   banana streak diseases, 499–500
   cauliflower mosaic disease, 499
   geminiviruses, 500–502
   rice tungro diseases, 499–500
Plant-expressed proteins, purification of, 563
Plant food allergens, 13
Plant genetic resources, 504
   nuclear genome diversity, 505–506
   organelle genome diversity, 504–505, 506
Plant Genetic Systems v. DeKalb Genetics Corporation, 657
Plant genome database, 308
Plant genomics and genefinder genomic resources, laboratory for, 308
Plant-incorporated protectants (PIPs), 528, 529
Plant-made industrial products (PMiP), 527

Plant-made pharmaceuticals (PMPs), 527
Plant patent, 340
Plant pathogens, 509, 512
   biological control, 514
   biotechnological control options, 514
   chemical control, 513–514
   control measures, 509–510
   cultural control, 514
   economic loss, 509
   host plant resistance, 512–513
   prospects, 510
   world impact, 509
Plant pest, 282
Plant Pest Act, 282
Plant-produced recombinant therapeutics
   acetylcholinesterase (AChE), 523–524
   biomolecules, 524
   erythropoietin (EPO), 523
   follicle-stimulating hormone (FSH), 523
   immunoglobulins, 522
   lactoferrin, 523
   macrophage activating factor (MAF), 523
   vaccines, 522–523
Plant resistance
   to manage arthropod pest populations, 50–51
Plant sterols (PS), 516
   anticancer effects of, 518
   anticancer mechanism, of action, 518–519
   beneficial effects of, 517
   cholesterol metabolism, 517–518
   cholesterol reduction by, 516–517
   differential responses among humans to, 517
   effects on lipids, 517
   safety, 518
Plant storage proteins, 647, 648
Plant tissue culture, 519
   micropropagation, 519–520
   plant cell/tissue cultures, 519
   secondary metabolites, 520–521
Plant variety protection (PVP), 340
Plant Variety Protection and Farmers' Rights Act, 659
Plant Variety Protection Act (PVPA), 657
Plant viral vector technology, 443
Plantain, 161
Plants
   artificial chromosomes in, 53–54
   biofactories for vaccine production, 661–663
   genetic engineering in, 293–294
   genetically modified organisms (GMOs). See Genetically modified organisms (GMOs)
   protein expression in, 661
Plasma non-esterified fatty acids (NEFA), 558
Plasmid DNA, 476–477

Plasmopora viticola, 484
Pleistophora, 573
Pleurotus, 77
Pleurotus cornucopiae, 79
Pleurotus ostreatus, 76, 79, 80
Ploidy manipulations, 159, 160, 162
Plum pox potyvirus, 630
Pluripotent stem cells, 167–168
Pluronic F68, 22
Pluronic®F-68, 570
PMSP3535, 61
Pnigalio minio, 101
Policy and regulation. *See* Product labeling; U.S. regulations
Pollen development, 381–383
Pollen flow, containment of, 639–640
Pollen mitosis I (PMI), 381–382
Pollen mitosis II (PMII), 382
Pollen wall–vital interface, 383
Pollen-expressed genes, 384
Poly(dimethylsiloxane) (PDMS), 426
Poly(ethylene terephthalate) (PET), 19
 astrocytes on, 20
 colon cancer HT29 cells in, 20
 embryonic stem cell on, 20
Poly(lactic-co-glycolic acid) (PLGA), 19
 breast cancer cells in, 20
Poly(urethane), 19
 hepatocyte in, 20
Polycaprolactone (PCL), 19
Polyclonal antibodies (pAbs), 212
Polycyclic aromatic hydrocarbons (PAH), 107
Polyethylene (PE), 44
Polyethylene glycol (PEG), 435, 566, 568, 680
Polygalactans, 418
Polyhydroxyalkanoates (PHAs), 45, 458–459
Polyhydroxybutyrate, 459
Polyhydroxyvalerate, 459
Polylactic acid (PLA), 45
Polylysine, 212
Polymerase chain reaction (PCR), 14, 16, 26, 36, 130, 414, 437, 531, 573, 687
 adulteration, detection of, 534
 allergen detection, 534
 biochemistry, 531–532
 challenges, 531–532
 as detection method, 534–535
 detection of products, 532–534
 for DNA synthesis, 534
 gel electrophoresis, 532–533
 GMO detection, 534
 instrumentation, 531–532
 organisms and product certification, 535
 pathogens and other organisms, detection of, 534–535
 quantitative and real-time detection, 533–534

Polymeric materials, 19
Polymers, 156
Polyphenol oxidases (PPOs), 330, 471, 691
Polyphenols, 152, 270, 677
Polypropylene (PP), 44
Polysaccharide-based biopolymers, 44
Polysaccharides, 44, 80
Polyunsaturated fatty acids (PUFA), 232, 471
Polyunsaturated omega-3 fatty acids, 374
Polyvinyl pyrrolidine (PVP), 346
Polyvinylpolypyrrolidone (PVPP), 74
Population dynamics, 479
Population theory, 479–481
Porcine, 700
Porcine cardiac valves, 701
Porcine endogenous retrovirus (PERV), 705
Porcine pancreatic islets, 701
Porcine pepsin, 228
Poria cocos, 79, 80
Porphyrin, 474, 476, 487
Postfermentation, of vegetables, 669
Posttranscriptional gene silencing (PTGS) process, 33
Posttranslational modifications (PTM), 451, 452
Potato, 51, 160
Potato leaf roll virus (PLRV), 161
Potato virus Y (PVY), 161
Potato-derived hepatitis B vaccine, 117
Poultry genetics and breeding, 536, 537–539
 egg-layer breeding, 537
 meat-type breeding, 536–537
 selection, 536
Pozol, 136
Prakash, C.S., 648
Prebiotics, 270
Precautionary Principle, 121
Predators, 87, 335–336
Preimplantation stage embryos, 275
Pre-miRNAs, 34
Prepartum energy nutrition, in dairy cattle, 317
Preprochymosin, 234
Pri-miRNA, 34
Primary cells, 18
Primary vegetables, 348, 349
Primiparous cows, 558, 559
Probiotic bacteria preservation, 540
 preservation processes, 540–541
 protection mechanisms and media, 541–542, 543, 544
 stress response, 542–543
Probiotic yogurt and health benefits, 708–709
Probiotics, 201, 545
 benefits and mechanisms, 547
 definition of, 361
 gut ecology, 545–546

 in closing, 547
 in fermented dairy products, 196–197
 products, 547
 types, 546–547
Processing and preservative aids, 549
 hurdle technology, 551
 nisin, 549–551
 other bacteriocins, 551
Prochymal™, 23
Prodigene corn, 303
Product labeling, 553
 EU member states, implementations in, 555
 European regulations, 553
 in Japan, 555–556
 in Korea, 555–556
 in Saudi Arabia, 555–556
 in Thailand, 555–556
 threshold labeling, 554–555
Progesterone, 242
Progestogen/progesterone treatments, 712
Prolamin superfamily, 13
Prolyl endopeptidase, 3
Pronuclear microinjection, 616–617
Propionate, 558
Propionibacteria, 557
 physiological responses to feeding, 558–559
 species of, 557
 use of, 557–558
Propionibacterium, 143
Propionibacterium acidipropionici, 144, 557
Propionibacterium acnes, 557
Propionibacterium avidum, 557
Propionibacterium freudenreichii, 144
Propionibacterium freudenreichii subsp. *shermani*, 201
Propionibacterium freudereichii, 557
Propionibacterium granulosum, 557
Propionibacterium jensenii, 144, 557
Propionibacterium lymphophilum, 557
Propionibacterium strain P169, 558, 559
Propionibacterium strain P5, 558
Propionibacterium strain P63, 558
Propionibacterium theonii, 144, 557
Propionic acid, 44
Propionic acid bacteria (PAB), 143, 144, 145
Propylene glycol (PG), 396
PROSITE, 451
Prostaglandins F2α (PGF), 240, 605, 711–712
Proteases, 23, 71, 221, 237, 691
 biological significance, 238
 future innovations and applications, 239
 in new functional protein and peptide ingredients/products manufacture, 238–237

occurrence and commercial production, 237–238
as processing aids, 238
Protein, 434–435
Protein arrays, 404
Protein bioseparation, 561
   animal-expressed proteins, purification of, 564
   aquatic expression systems, 562
   bioprocessing, of transgenic animals, 564
   bioprocessing, of transgenic plants, 562–563
   clarification (solids removal), 563, 564
   egg white, 563
   extraction, 562–563
   milk and other fluids, 563
   particle size reduction, 562
   plant-expressed proteins, purification of, 563
   terrestrial expression systems, 561–562
   transgenic animal hosts, 563
   transgenic plant hosts, 561–562
Protein degradation
   endogenous proteolytic enzymes, 253–254
   microbial proteolytic enzymes, 254
Protein expression, in plants, 661
Protein expression systems, 26
Protein informatics, 450–451
Protein kinase B/Akt, 153
Protein kinase C, 153
Protein quality improvements, from transgenic plants, 647
   novel protein, production of, 648–649
   past work, 647–648
Proteinases, 3-4
Proteoglycans, 77
Proteolysis, 228, 229, 684
   by HP processing, 330–331
Proteomic approach
   of Gram-positive bacteria, 62
Proteomics, 192, 402, 404
   DNA array technology, 403–404
   DNA-based technologies, 403
   emergent technologies, 402–403
   and genomics, 193–194
   one-dimensional (1D) electrophoresis vs. two-dimensional (2D) electrophoresis, 404–405
   protein arrays, 404
   sexing sperm, 403
Proteus, 365
Protoplasts, 566, 568, 569
   culture media, 569–570
   culture of, 569–570
   DNA transfer, in Gram-positive bacteria, 61
   DNA uptake, induction of, 567–568
   factors influencing, 568
   fusion of, 566
   gametosomatic hybridization, 566–567
   innovative approaches to culture, 570
   isolation and culture, 569, 69–70
   novel hybrids and cybrids generated by, 566
   plating density and nurse cells, 570
   protoplast-derived tissues, 570–571
   respiratory gases, manipulation of, 570
   somatic hybrid plants, characterization of, 567
   somatic hybrid tissues and plants, selection of, 567
   stimulate protoplast growth, methods to, 570
   surfactants and antibiotics, 570
   systems for, 570
   transformation of, 567–568
Protozoa and parasites, 572
   non-protozoan parasites, rapid detection of, 573–574
   rapid detection of, 572–573
ProViva, 136
Proximate analysis, 411
PrtM, 192
PrtP, 192
Pru av 1 of cherry, 13
PS/stanols, 517, 518
Pseudo-catalase, 476
Pseudomonas cepacia, 96, 97
Pseudomonas fluorescens, 96, 441, 514
Pseudomonas syringae, 581
Public outcry subsided, 295
Puccinia triticina, 377
Puda, pudla, 136
Pullaria, 365
Pullulan, 420
Pullulanase, 179, 221, 223, 224
Pulsed electric field (PEF) processing, 209, 210, 575
   energy requirements, 577
   inducing stress, 209–210, 211
   microbiology, 576
   microorganisms and enzymes, inactivation of, 575
   processing aspects, 577
   product quality, influence on, 575–576
   product safety, 576–577
   toxicology, 576–577
Pulsed-field gel electrophoresis (PFGE), 148, 415
Pure Food and Drugs Act of 1906, 257, 258
Purines, 85
Puta, 136
PVPA protection, 659
Pyr c 4 of pear, 13
Pyrococcus fuiosus (Pfu-sHsp), 423
Pyromyces, 451
Pyruvate concentration, 74
Pyruvate dehydrogenase kinase 4 (PDK4), 317

Q

Q-bot, 309
QTL databases, 538
QTL mapping, 314
Quantitative and real-time detection, 533–534
Quantitative PCR (qPCR), 149
Quantitative PCR and real-time PCR, 62
Quantitative real-time PCR (Q-PCR), 302, 303, 443, 445, 446
Quantitative resistance loci (QRL), 377, 378
Quantitative trait loci (QTL), 205, 310, 312, 315, 538
Quantitative trait nucleotides (QTN), 538
Quasi-equilibrium freezing, 460
Quinones, 84

R

Rabbit cloning, 164–165
Rabbit hemorrhagic disease virus (RHDV), 673
Rabdi, 136
Rabies vaccine, 522
RAC gene, 467
Radish, 184, 185
Rainbow papaya, 630
Ralstonia eutrophus, 458
Random amplified polymorphic DNA (RAPD), 506, 687
Random amplified polymorphic DNA—polymerase chain reaction (RAPD-PCR) method, 148, 149
Randomized clinical trial (RCT), 270
Rapeseed seeds, 31
Raphanus raphanistrum, 185
Raphanus sativus, 184, 185
Raspberry, 51
Raw barley, 1, 2
Rayleigh, Lord, 174
Rayons, 85
RbcL gene, 505
Rcg1, 377
Reactions catalyzed, by lipases, 371
Reactive oxygen species (ROS), 125, 153, 518
Ready-to-eat (RTE) foods, 45
Real-time molecular amplification methods, 682
Real-time RT-PCR, 682
Recombinant bovine somatotropin (rBST), 286
Recombinant strains, biological containment of, 666
Redifferentiation, 466
Regenerative medicine, 599–600
"Regulated articles," 282
Regulation 178/2002, 302
Regulation 178/2002/EC, 40

Regulation 1929/2003, 303
Regulation EC No. 470/2009, 40
Remediation Technologies Inc., 111
Renaissance of Natural Theology, 482
RENCO, 373
Renewable carbon, for biorefinery production
    case for, 83–84
    chemical production examples, in biorefinery, 85–86
    conversion, 83
    economic considerations, 84
    separation, 83
    supply, 82
    technology for bio-based products, in biorefinery, 84–85
    use of, 82–83
Renewable Fuels Standard, 84
Rennets, 196
    in cheese ripening, 228–229
    coagulation of milk, 228
    and rennet substitutes, 227–228
Rennin. *See* Chymosin
Report on the Insects of Massachusetts Injurious to Vegetation, 482
Reprogramming plant metabolism, 270
Resins, 73
Restricted maximum likelihood (REML), 217
Restriction fragment length polymorphism (RFLP), 148, 307, 687
Retroviral insertion, 617
Retrovirus-mediated gene transfer, 617
Reverse micelle extraction, 409
Reverse transcription-polymerase chain reaction (RT-PCR), 531, 681
Rhizobium, 420
Rhizofiltration, 496
Rhizomucor pusillus, 138
Rhizopus, 138, 139, 140, 156
Rhizopus arrhizus, 357
Rhizopus azygosporus, 45
Rhizopus niceus, 4
Rhizopus niveus, 96
Rhizopus oryzae, 44, 48, 140, 157, 357
Rhizopus sp., 356, 357
Rhizopus stolonifer, 157
Rhodococcus sp., 471
Rhodotorula, 365
Rhodotorula mucilaginosa, 709
Rhodotorula sp., 8, 139
Riboflavin (B_2), 76, 225
Ribosomal DNA sequencing method, 148
Ribosomal RNA (rRNA), 33, 414, 415
Rice (Ory s 12), 13, 51, 184, 185
Rice tungro bacilliform virus, 499, 500
Rice tungro disease, 501, 499–500
Rice tungro spherical virus (ssRNA virus), 499
Rifampicin, 41

Riley, C. V., 483
RNA interference (RNAi), 13, 14, 33, 636
    agriculture and veterinary applications, 637–638
    animal biotechnology, applications in, 36–37
    antiviral therapies, 36–37
    disease prevention, 37
    gene silencing mechanism, 34–35
    in animal cells, 637
    other uses, 37
    production traits, 37
    siRNA as a tool for functional genetics, 35–36
    stable expression, 637
    transient exposure, 637
    virus disease in, 637
RNA-dependent RNA polymerases (RdRps), 579
RNA-induced gene activation (RNAa), 34
RNA-induced silencing complex (RISC), 13, 34
RNAi application to silence major peanut allergen Ara h 2, 14–16
Roast barley, 1
Robl, James, 164
Rodents
    biological control of, 673
    embryo cryopreservation, 353
    oocyte cryopreservation, 353
    and pig embryo cryopreservation, 354
    spermatozoa, 352–353
*Rol*A genes, 631
*Rol*B genes, 631
*Rol*C genes, 631
Rolling circle replication (RCR), 476
Root formation, 467
Roslin Institute in Glasgow, 163
Rotating biological contactor (RBC), 111
Rotation, 464
Roundup Ready® soya, 553, 554, 614
Roundup®, 293, 326
Russet Burbank, 159
Rye, 13, 51
Rye B chromosome, 54
Rye bread, 365, 368
Rye, crop, 184, 185

S

S-adenosylmethionine synthetase-encoding genes (SAMS), 468
S. exiguous, 366
S. halepense, 185
Saccharification process, 178, 180
Saccharomyces, 1, 138, 349, 669, 690
Saccharomyces bayanus, 8
Saccharomyces carlsbergensis, 6, 7
Saccharomyces cerevisiae, 6, 7, 9, 73, 80, 133, 134, 138, 139, 140, 141, 196, 293, 363, 364, 365, 366, 387, 472, 690, 691

Saccharomyces chevalieri, 364
Saccharomyces diastaticus, 6
Saccharomyces fibuligera, 140
Saccharomyces paradoxus, 7
Saccharomyces pastorianus, 9, 134
Saccharomyces rouxii, 431
Saccharomyces sake, 141
Saccharomyces spp., 366
Saccharomyces uvarum, 8, 73, 139, 141, 365
Saccharomyces yeasts, 91
St. Augustine grass, 51
Saké, 135, 137
    steps in, 141
Salmonella, 44, 215, 664
Salmonella Typhi Ty21a, 665
Salted dried fish roe, 252
Salted fermented fish roe, 252–253
Salted fish roe, 253
San Francisco-style (French bread) sourdough, 365
Sanitary and Phytosanitary Measures (SPS) Agreement, 265
Sarcocystis neurona, 25
Saudi Arabia
    product labeling in, 555–556
Scalps, 69
Scendesmus spp., 109
Schell, J, 293
Schizophyllan, 80
Schizophyllum, 420
Schizophyllum commune, 78, 80
Schizosaccharomyces pombe, 9
Scleroglucan, 420
Sclerotinia sclerotiorum, 157
Sclerotium, 420
SDS-PAGE, 15
Seafood, high pressure treatment for, 330
Seasoning ingredients, 348
Secale cereale, 184, 185
Secondary biota, 143
Secondary fermentation, 669
Secondary metabolites, 208, 211, 520–521, 584, 583, 584–585
Secondary vegetables, 348, 349
Secreted products, of transgenic animals, 619
    advantages of recombinant protein synthesis, 622–623
    expression of recombinant proteins, in milk, 623–624
    recombinant protein production, 624
    recombinant proteins, in media, 624
Seed-wrap licensing, 659
Seedcalc, 587–588
Seeds, 587
    for GMO testing, 588–589
    harmonizing methods, 589
    PCR-based methods, 588, 589
    protein-based detection methods, 588

sampling methods, 587–588
unit of measurement, 589
unknown/unauthorized GMOS, 589
Selective breeding and random mutagenesis, of volatile flavor generation, 683–684
Semen collection
artificial vagina, 57
digital pressure, 56
Semen extension and preservation, 57
Sensitive crops, 627
Separation technology, biorefinery, 83
Sephadex G200, 681
Sequence drafts, 312, 314
Sequence tagged microsatellites (STMS), 506
Sequence-characterized amplified region (SCAR), 149
Sequence-related amplified polymorphism (SRAP), 506
Serial nuclear transfer, 131
Sertoli cells, 169
Serum, 20
Sesame (Ses i 1), 13
Setaria italica, 184, 185
Sethoxydim, 327
Severe combined immunodeficient (SCID), 295, 595
Sexing embryos, 130
Sexing sperm, 403
Sex-sorted sperm, utilization of, 391–392
cattle, 391
horses, 391–392
other mammals, 392
sheep, 391
swine, 391
SH 3436 (AAAA), 68
Shear stress, 22
Sheep
semen from, 57, 58
sex-sorted sperm, utilization of, 391
Sherry wine vinegar, 675
Shoot formation, 467
Short chain alcohols, 408
Short hairpin RNA (shRNA), 34, 36, 636, 637
Short tandem repeats (STRs), 321, 322
Shrub, 678
Shuttle vectors
DNA transfer in Gram-positive bacteria, 61
Sigatoka leaf spots, 68
Silanization, 436
Silicon Genetics, 445
Silk worms, 28
Simmondsia chinensis, 650
Simple sequence repeat (SSR), 307, 506
Single nucleotide polymorphisms (SNPs), 154, 218, 310, 314, 506
Single-stage extractors, 408

Single-strand conformation polymorphism (SSCP) analysis, 149
Single-stranded DNA(ssDNA) molecule, 149
Sinorhizobium sp., 420
Sitosterolemia, 518
Small heat-shock proteins (sHsps), 422
in vitro applications, 423–424
in vivo applications, 422–423
Small interfering RNAs (siRNAs), 13, 33, 34, 37
as tool for functional genetics, 35–36
Small plasmid (pSC101), 292, 293
Small RNAs, 579
biogenesis and function, 579–580
endogenous siRNAs, 581
heterochromatic siRNAs, 581
microRNAs (miRNAs), 580
miRNA homeostasis, in plants, 582
miRNAs in plants, roles of, 580–581
nat-siRNAs, 581
trans-acting siRNAs, 581
Smear bacteria, 143–144, 145
Smith, Gary, 427
Smith, H.O., 292
Smith–Lever Act, 483
SNP chip, 314
Snuppy, 165
Sodium bicarbonate, 20, 21
Sodium chloride (NaCl), 669
SOFaa, 123, 124
Sola dosis facit venenum, 41
Solanines, 29
Solanum bulbocastanum, 567
Solanum tuberosum, 567
Solid–liquid extraction, 407–408
Solid substrate
animal cell culture *in vitro*, 19–20
Solid-state fermentation (SSF), 48
Solids-not-fat (SNF), 558
Solvent extraction, 407–408
Somaclonal variation, 160–161, 520
Somatic cell count (SCS), 218
Somatic cell genetics. See Bananas
Somatic cell nuclear transfer (SCNT), 131, 163, 164
for human therapy, 165
Somatic hybrid tissues and plants
characterization, 567
selection, 567
Sorbic acid, 44
Sorbitol, 84
Sorghum, 51, 184, 185
Sorghum bicolor, 184
Sourdough, 364–365
South American alcoholic maize beverage, 141
Southern green stink bug, 100–101
Southern hybridization, 14, 16, 682

Sox2, 599, 602
Soy oil, 84
Soy protein isolate, 44
Soy sauce, 668
brine fermentation, 592
fermentation, 591
final processing, 592–593
health properties, 593
koji preparation, 591–592
production, 591–593
Soybean (glycinin), 13, 31, 561
Soybean Gly m Bd 30 K, 13
Spanish ibex, 344
Sparassis crispa, 78
Specialty Enzymes and Biochemicals, Inc., 373
Specific host-pathogen interactions, 512
Sperm sexing, biological basis of, 390.
See also Mammalian sperm sexing
Sperm-mediated gene transfer, 617
Spermatozoa, 274
rodents, 352–353
swine, 353–354
Sphingomonas paucimobilis, 419
Spilopsyllus cuniculi, 673
Spinner flasks, 21
Spirits product, 8
Spirulina maxima, 109
SPME-GC technique (solid phase microextraction–GC mass spectrometry), 368
Spoilage yeasts, 8
Sponge and dough process, of leavened breadmaking, 364
Spontaneous fermentation, 670
Spore, 572
Sporobolomyces sp., 8
Sporopollenin, 383
Sporotrichum thermophile, 449, 450
Spray-drying (SD), 541
Srd1, 467
Srd2, 467
Srd3, 467
SRO5 gene, 581
SSCP method, 148
Stable genetic transformation, 661
Stacked genes, meaning of, 303
Stacked GM crops, 303
Staling effect, 224–225
Standard cryopreservation, steps of, 460–461
Staphylococcaceae, 399
Staphylococci, 144
Staphylococcus aureus, 60, 62, 80, 114, 192, 709
Staphylococcus carnosus, 401
Staphylococcus spp., 62, 198
Staphylococcus xylosus, 401
Starch, 44–45, 178, 221
Starlink, 304
StarLink corn, 118

Starter cultures, 670. *See also* Dairy
  in alcoholic beverages, 6
  alcoholic fermentation and yeast starter cultures, 7–8
  bacteria, yeasts, and molds used as, 9
  bakers' yeast, 133–134
  brewers' yeast, 134–135
  for cereal foods, 133
  concepts and definition, 6
  endogenous microflora, 133
  fermented dairy products, 195–196
  flow chart for industrial production of, 10
  food industry, importance in 6–7
  history of, 6
  lactic acid bacteria, 135
  malolactic fermentation (MLF) and bacterial starter cultures, 9
  molds, 135–137
  molecular biology application in development of, 9–10
  production of, 7–9, 10
  technology of, 9
  traditional and modern methods comparison, 7
Starter LAB (SLAB), 143
Static extraction columns, 407
Stearoyl-CoA desaturase (SCD), 317
Steers, 403
Steinernema, 337
Stem cell and germ cell technology, 594
  adult stem cells, 594
  agricultural applications for ES and EG cells, 595, 596
  of different developmental potency, 167–169
  embryonic germ (EG) cells, 595
  embryonic stem (ES) cells, 594–595
  teratocarcinoma and embryonic carcinoma, 594–595
Stem cell research, 597
  adult stem cells, 597
  embryo-derived stem cells, 597–598
  genetic studies, 598–599
  regenerative medicine, 599–600
Stem cell transplantation, 704
Sterol regulatory element binding factor 1 (SREBF1), 317
Sticky traps, 464
Stilbene synthase, 630
Stimulate protoplast growth, methods to, 570
Stirred-tank reactor (STR), 695, 696, 697
STM (*SHOOT MERISTEMLESS*), 467
Straight dough bulk fermentation, 363–364
Strain typing, 415
Straw, 464
Strawberry, 160
Strawberry aroma, 49
Streptococcus, 118, 200, 365, 664, 708

Streptococcus mutans, 60, 62, 522
Streptococcus pneumonia, 60
Streptococcus sp., 62, 139
Streptococcus thermophilus, 138–139, 196, 201, 707, 708, 709
Streptomyces, 39, 40
Streptomyces clavuligerus, 41
Streptomyces coelicolor, 416
Streptomyces venezuelae, 41
Streptomycin, 40
Stress, electric field
  on plant systems, 208–211
Structural genomics, 311–314
Sublethal hydrostatic pressure stress treatment, 273
Sublethal stress, 272
  HHP as, 273–274
  HHP treatment in ART, 274
Submerged culture system, 675
Sub-Saharan Africa, 139
Succinic acid, 84
Succinoglycan, 420
Suckled Nelore cows, 712
Sucromalt, 180, 181
Sucrose, 180
Sugar beet, 51, 184, 185
Sugar cane, 51, 160
Sugarcane bagasse, 48
Sui generis system, 340
Sulfonylureas, 327
Sulfur dioxide, 469
Sundae style yogurt, 707
Sudan, 139
Sunflower, 51, 184, 185
Supercritical fluid extraction (SFE), 408
Superovulation, in mammals, 129, 243–244, 604
  endocrinology, 604–605
  exogenous FSH rescues subordinate follicles, 605
  follicular growth and development, 604
  hormone doses and timing, 606
  species-specific considerations, 605–606
  success rates, 606
  superovulated oocytes, normality of, 606
Supplement Facts panel, 259
Supply issues for
  renewable carbon, in biorefinery production, 82
Supported liquid membrane, 408
Suppressive, 440
Surface plasmon resonance (SPR), 438
Surfactant liquid membrane. *See* Emulsion liquid membrane
Surfactants, lipases in, 373–374
Surlyn®, 44
SVM (support vector machines), 451
Swaminathan, M.S., 296
Swann Committee, in 1969, 39
Sweet-clover, 51, 463
Sweet potato, 51, 160

Sweeteners production from corn starch enzymes uses for, 178–181
Sweetzyme IT, 180
Swine
  embryo cryopreservation, 354
  oocyte cryopreservation, 354
  semen from, 57, 58
  sex-sorted sperm, utilization of, 391
  spermatozoa, 353–354
Swine-transmissible gastroenteritis virus (TGEV), 523
Swiss Alps, 480
Swiss style yogurt, 707
SYBR® Green, 533
Sylvilagus brasiliensis, 672
Synchronization of estrus and ovulation
  using progesterone, 241
  using progesterone and estradiol, 241–242
  using prostaglandin, 240–241
  using prostaglandin and GnRH, 241
Synergistic model for MOET technology, 217
Syngenta, 303
Synthetic alleles, 379
Synthetic biology, 689, 692, 693
Synthetic chemical pesticides, 335
Synthetic pesticides, dependency on
  pest management, effects on, 655
  pesticide safety, 654–655
  pesticides and environment, 655
Synthetic pesticides era, 655
Syrups, 224
Systemically acquired resistance (SAR), 513
Systems biology, 689, 692, 693
  in animal agriculture, 318, 319
  bioinformatics, 318
  general considerations, 318
  and wine, 690–693

T

T-flasks, 21
T-helper 1 (Th1) cells, 665
T-helper 2 (Th2) cells, 665
T_1 coliphage, 64
Tachinidae, 100
Tai-Chiao No 1 (AAA), 68
Takayama, Shuichi, 427
Takju, 140
Talaromyces, 451
Talaromyces stipitatus, 452
Taleggio. *See* Camembert
Tammann, Gustav, 460
Tapetal cells, 381
Taq polymerase, 531
TaqMan, 533
TargetP, 451
Tarhana (trahana), 138
Tarnished plant bug, 101

TC1-299 (AAA), 68
Technical Barrier to Trade (TBT) Agreement, 265
Telomeres, 53
Temperate climate fruit crop pest management, 608
  disease-forecast models, 611
  fungicides, 611
  information delivery, 611
  management strategies, 608, 610–611
  pathogens, 608, 609–610
Temperature
  in animal cell culture, 21
Teratocarcinoma and embryonic carcinoma, 594–595
Teratoma, 598
Terminal restriction fragment length polymorphism (T-RFLP), 414, 708
Terpenoids, 152, 189
Terpinolene, 368
Terrestrial expression systems, 561–562
Tetanus toxin fragmentC (TTFC), 665
Tetracycline, 293
Tetracycline-/zinc-activated promoters, 37
Tetragenococcus halophila, 592
Tetragenococcus halophilus, 474
Tetramethylbenzidine (TMB), 213
Tetramethylpyrazine, 172
Tetraploidy, 69
Texas pumas, 323
TGF-B, 12
Thailand, 555–556
  product labeling in, 555–556
Theobroma cacao, 171, 173, 520
Theory of evolution, 437–438
Therapeutic SCNT for human therapy, 165
Thermoanaerobacter, 224
Thermococcus, 224
Thermomyces lanuginose, 97
Thermomyces lanuginosus, 3, 138
Thermus aquaticus, 531
Thiamine (B_1), 76, 225
Thiol bonding, 435
Thioredoxin reductase, 153, 154
Thomas, Clarence, 657
Threonine, 708
Threshold labeling, 554–555
Thyroglobulin, 312
Thyroid hormone responsive SPOT14 (THRSP), 317
Tibet, 185
Tillage, 463
Time temperature indicators (TTI), 225
Tissue engineering, 704
TMpred, 451
Tn916, 61
Tn917, 61
TNF-α, 665
Tn*phoZ*, 61–62
Tobacco mosaic virus (TMV), 443
Togwa, 136, 139

Toluene, 84
Tomato, 14
Tomato yellow leaf curl, 502
Tomato yellow leaf curl disease (TYLCD), 501
Tomato yellow leaf curl virus (TYLCV), 501, 502
TopPred, 451
Torrefied barley, 1
Torulaspora, 690
Torulaspora delbrueckii, 8
Torulopsis, 349
Torulopsis candida, 139
Torulopsis etchelsii, 592
Torulopsis glabrata, 366
Torulopsis versatilis, 592
Total polyphenolics, 210, 211
Totipotent cell, 597
Totipotent stem cells, 167
Townsend-Purnell Plant Patent Act (PPA), 657
Toxic radioactive materials, 111
Toxic Substances Control Act (TSCA), 283, 284
Toxicity, of transgenic crops, 626
Toxoplasma, 573
Toxoplasma gondii, 572
TPH, 111
Trachypleistophora, 573
Trade Related Aspects of Intellectual Property Rights (TRIPS) Agreement, 339, 340, 658
Trade secret protection, 659
Trade secrets, 340
Trademarks, 340
Tragelaphus euryceros, 343
Trahanas, tarhanas, kishk, 136
Trametes ochracea, 78
Trametes versicolor, 78, 80
Trans-acting siRNAs, 581
Transgenic animals. *See also* Mitochondrial genome modification, of transgenic animals; Secreted products, of transgenic animals
Transaminase, 685
Transcriptome, 316
TransCyteTM, 23
Transduction technique
  DNA transfer in Gram-positive bacteria, 60
Transesterified fats, labeling of, 260
Transfer RNA (tRNA), 33
Transformed cell lines. *See* Immortalized cell lines
Transgene, 616
Transgene efficacy. *See* Efficacy of a transgene
Transgene expression. *See* Expression of a transgene
Transgenesis process, 132
Transgenic animal hosts

egg white, 563
milk and other fluids, 563
Transgenic animals, 616
  animal-expressed proteins, purification of, 564
  cellular insertion, 617
  clarification, 564
  defined, 616
  evaluation of, 617–618
  expression, 617–618
  gene transfer methods, 616–617
  integration, 617
  mitochondrial genome modification, 619–620
  pronuclear microinjection, 616–617
  retroviral insertion, 617
  secreted products, 622–624
  sperm-mediated gene transfer, 617
  transmission, 618
Transgenic crops, 625
  cultivation, 625–626
  effects on soil, 627
  for food and feed, safety aspects of, 249
  fruit quality, increasing, 630, 631
  GM foods, labeling of, 634–635
  impacts on livestock feed quality and cost, 247–249
  indirect effects, 627
  insect control, 630
  other applications, 630–631, 632
  pathogen control, 629–630
  perennial crops, transformation of, 629
  product (output) traits, 248–249
  production (input) traits, 247
  public safeguards, 249
  testing and cultivating GM crops, 633–634
  toxicity, 626
  transgenic crop safety assessment, 249
  unique environmental concerns, 627
  weed control, 630
  weediness and invasiveness, 626–627
Transgenic livestock. *See* RNA interference (RNAi)
Transgenic mouse, construction of, 295
Transgenic phytoremediation approaches, 495–496
Transgenic plants, 561, 639, 643
  adventitious presence (AP), 639, 640
  aquatic expression systems, 562
  bioprocessing, 562–563
  clarification (solids removal), 563
  coexistence, 639
  construction of, 294
  crop-to-crop gene flow, 643–644, 645
  crops to wild relatives gene flow, 644–645
  economic effects, 640–641
  economic implications, 640
  extraction, 562–563
  horizontal gene transfer (HGT), 645

Transgenic plants (*Continued*)
  particle size reduction, 562
  plant-expressed proteins, purification of, 563
  pollen flow, containment of, 639–640
  terrestrial expression systems, 561–562
Transgenic plant-based vaccines, 26
Transient expression, 661–662
Transmissible spongioform encephalopathies (TSE), 638
Transmitochondrial animals, 620
Transoposons, 61
Trehalose, 80, 180, 181
Tremella aurantia, 80
Tremella fuciformis, 80
Tremella mesenterica Ritz., 78
Tremellafuciformis, 78
Triacylglycerol acylhydrolase, 96
Triacylglycerol hydrolases, 370
Triacylglycerols (TAGs), 232
Triazines, 327
Trichapatum laricinum, 78
Trichinae Herd Certification Program, 574
Trichinella, 574
Trichinella spiralis, 574
Tri-chloro tri-fluoroethane, 680
Trichloroacetic acid (TCA), 111
Trichloroethylene (TCE), 111, 496
Trichoderma harzianum, 48, 105
Trichoderma viride, 1, 105, 157
Trichogramma, 336
Trichopoda giacomellii, 101
Trichopoda pennipes, 101
Trichosporon, 138
Trichothecium roseum, 157
Triffid (Siam) weed, 100
Trinitario, 171
TRIPS, 659
Trissolcus basalis, 101
Triticale, 28
Triticum aestivum, 184, 185
Triton X-100, 213
Tropical pest ecology
  lack of information on, 654
  low application of pest management on, 653–654
Tropomyosins, 12
Trypanosoma congolense, 314
Trypsin, 198
Tryptophanase, 80
Tsukemono, 136
TTU51 CTP *CrtB*-RZ, 444, 445
Tuberculosis (TB), 39, 41
Turkey, 139
Turkeys, semen from, 57, 58
Tween 20, 213
Twin-arginine translocation (Tat) pathway, 423
Type I allergic reactions, 12
Type I food allergy, 11

Type IV-/cell-mediated hypersensitivity reactions, 12
Typhoid (*Salmonella enterica* serovar Typhimurium), 41

U

UDP-glucuronyltransferase, 153
Uji, 136
UK cropnet, 308
Ultrasonography, 245
Umbelluria californica, 457
Unauthorized GMOs, 303–304
Unconventional bioreactors, 696
Unilever, 372
Unique environmental concerns, of transgenic crops, 627
United States, 47, 50, 51, 91, 101, 105, 120
  and European regulatory attitude, 302
  attitude change, 299
  consumers awareness and attitudes, 297
  European consumer attitudes, 298–299
  GM crops in, 301
  in situ groundwater biodegradation in, 111
United States Regulatory Oversight in Biotechnology Responsible Agencies—Overview, 633
University of Florida Biosafety Committee, 89
University of Ghent, 301
University of Washington, 301
Unknown GMOs, 304
Uruguay Round of Multilateral Negotiations, 265
U.S. Department of Agriculture (USDA), 257, 260, 282–283, 295, 303, 527–528, 614, 633
U.S. Department of Agriculture Animal and Plant Health Inspection Service (USDA–APHIS), 89, 282, 285, 286, 303, 526–527, 528
U.S. Department of Energy, 96
U.S. Department of Fish and Wildlife, 89
U.S. Food and Drug Administration (FDA), 295, 298, 402, 419, 664
U.S. regulations, 526
  biotechnology consultation, 528
  combined traits, 529
  environmental protection agency (EPA), 528–529
  experimental use permit, 528
  food and drug administration (FDA), 528
  food safety evaluation, 528
  framework of biotechnology, 526–527
  GMO labeling, 529
  import, interstate movement, and environmental release, permit for, 527
  non-pesticide traits, 527
  non-regulated status, determination of, 528

  permit procedures, 527–528
  pesticidal traits, 527
  plant-made industrial products (PMiP), 527
  plant-made pharmaceuticals (PMPs), 527
  practical issues, 529
  product-specific oversight, 527
  registration, 529
  state registration, 529
  transparency and confidentiality, 529
  U.S. Department of Agriculture (USDA), 527–528
U.S. Supreme Court, 340
USDA Beltsville Agricultural Research Center, 390
Utility patents and plant innovation, 657
  eligible subject matter, 657
  emerging issues, 659
  infringement, defenses, and remedies, 658–659
  patent and patent rights, 658
  patent requirements, 657–658

V

Vaccine development and applications
  in animal medicines, 26
  lactic acid bacteria (LAB) as, 664–666
Val-Pro-Pro, 198
Van Montagu, M., 293
Vancomycin, 41
Vanilla planifolia, 49, 520
Vanillic acid, 49
Vanillin, 48, 85
Vedalia beetle, 336
Vegetable fermentation, 668
  basic principle, 668–669
  control, 669–670
  fermentation environment, creation of, 669
  microbial fermentation, 669
  microorganisms, 670
  new trends in, 670
  origin, 668
  postfermentation, 669
  pretreatment, 668–669
  salt and substitutes, 669
  social and economic importance, 668
  spontaneous fermentation, 670
  starter cultures, 670
  temperature of, 670
Vegetative cells, 381, 382
Verenium, 373
Vertebrates, biological control of, 672
  biological control agents, genetic modification of, 673–674
  feline panleucopenia virus, biological control of cats using, 673
  myxomatosis and European rabbits, 672–673

rabbit hemorrhagic disease virus, 673
rodents, biological control of, 673
Vertical resistances, 375–377
Verticillium chlamydosporium, 440, 441
Vertrel, 680
Very low-density lipoprotein (VLDL), 77
Very-long chain-fatty-acids (VLCFA), 651, 652
Vesicular stomatis virus, 80
Vibrios, 215
Vicinal diketone (VDK), 73, 74
"viili," 419
Vinegar, 675
  acetic acid bacteria, 675–676
  composition of, 677
  definition of, 675
  factors implicated in, 676
  flavonoid compounds of, 678
  future directions, 678
  non-flavonoid compounds of, 677
  production methods, 675, 676
  raw material, 675
  uses of, 677–678
  vinegar elaboration, factors implicated in, 676
Viral vaccines, 23
Viral vector and transfection design, 444
Virus concentration and purification, 680–681
Virus detection, 680, 681–682
Virus disease, in livestock, 637
Virus-like particles (VLP), 662
Viruses, 680
  confirmation, 682
  detection of, 680, 681–682
  nucleic acid extraction, 681
  virus concentration and purification, 680–681
Vitamin B1, 225
Vitamin B2, 225
Vitamin B3, 225
Vitamin B9, 225
Vitamin C, 225
Vitamin D–binding protein. *See* Macrophage activating factor (MAF)
Vitamin K1, 225
Vitamins, 85, 677
VitMaster™, 461
Vitrification, 460, 461
Vitrification protocol, 396–397
VitSet™ technique, 461
Vittaforma, 573
Volatile flavor generation, 683
  genetic engineering, 684–685
  genetic methods, 683–685
  genomics, 684
  regulatory issues, labeling, and consumer perception, impact of, 685
  selective breeding and random mutagenesis, 683–684

Volunteer weeds
  first step to ferality, 184
Volvariela volvacea, 79
VP6 coding sequence, 662

W

"Without GM," 555
Wagyu × Hereford (WH) cattle, 317
Walnut (Jug r 1), 13
Walnut (Jug r 2), 13
Walsh, B. D., 483
Wax biosynthesis
  in jojoba, 650, 651
  identification of genes responsible for, 650–651
Wax esters, 254, 650
  Brassica Napus, production of wax esters in, 651–652
  identification of genes responsible, for wax biosynthesis, 650–651
  wax biosynthesis, in jojoba, 650
Wax synthase (WS), 650
Weed control, 630
Weed management, current impact on, 326
Weed science, 686
  gene expression profiling, 688
  gene flow, studying, 688
  genetic diversity tools, 687–688
  herbicide-resistant crops, 686–687
  herbicide-resistant weeds, 686
  history of, 686–687
  molecular biology, 686–688
Weediness and invasiveness, of transgenic crops, 626–627
Weeds, 462–463
Weight management, 269–270
Weissella confuse, 139
Weissella koreanis, 349
Weissella species, 670
Western blot analysis, 14, 15, 16, 26
Westhusin, Mark, 165
Wettable sulfur, 483
Wheat, 13, 51, 184, 185
Wheat gluten, 44, 363
Wheat *Lr34* gene, 377
Wheeler, 700
Whey protein, 330
  in milk, 227
  isolates, 44
White, Philip, 519
Whole genome clones, 306–307
Whole-genome sequence database, 538
Wild germplasm, 341
Wild panthers, 323
Wild/endangered species, genome of, 321
Wilmut, Ian, 163
Wine and fruit juice, 221
Wine and systems biology, 691–693
Wine aroma, impact on, 389

Wine fermentation, 689
  art and innovation, blend of, 689–690, 691
  biological factors, 689–691
  fit-for-purpose starter cultures, development of, 688–689
  wine and systems biology, 691–693
  wine production, application of enzymes in, 691
Wine-making, 387
  enzymes used in, 4
Wine product, 8
Wine production
  application of enzymes in, 691
  processes involved in, 690
Wood Buffalo National Park, 321
World Health Organization (WHO), 572
World Intellectual Property Organization (WIPO), 339
World Trade Organization's (WTO), 31, 659
World Wide Web, 611
Wort separation, 72
WTO Agreements and Codex, 265
WTO and dispute settlement, 265

X

Xanthan, 85, 419
  fermentation, 695
Xanthan gum, 695
  bioreactor systems for, 697
  conformation of, 696
  fermentation, 695–696
  immobilized cell bioreactors, 698
  low power inputs, reactors with, 696, 698
  stirred-tank reactors, 696, 697
Xanthomonas campestris, 419, 695
Xenogenic tissue use, in clinical medicine, 700
  cardiac valves, 701
  gene-modified pigs and embryonic stem cells, 701
  human therapies, tissue engineering for, 701
  porcine pancreatic islets, 701
Xenografts, source of, 704
Xenopsylla cunicularis, 673
Xenopus XMAP215, 384
Xenorhabdus, 104
Xenotransplantation, 700
  and genetic engineering, 705
  clinical attempts at, 703
  hurdles to, 704–705
  meaning of, 703
  rationale for, 703–704
  xenografts, source of, 704
Xianyou Bt63, 303
Xylanase, 3, 221
Xylella fastidiosa, 630
Xylene, 84

Xylitol, 85
Xylitol/arabinitol, 84
Xylose, 180

Y

Y chromosome, 130
Yang, Xiangzhong Jerry, 164
Yarrowia lipolytica, 48, 147, 149, 709
Yeast, 363, 387, 389, 669
    and bacteria, simultaneous inoculation of, 388
    growth of, 709
    starter cultures, 7–8
Yeast-fermented beverages, 361
Yeast-leavened doughs
    Chorleywood Bread Process (CBP), 364
    sponge and dough, 364
    straight dough bulk fermentation, 363–364
Yellowstone National Park, 321
Yogurt microbiology, 419, 707
    defects, 709
    exopolysaccharides, production of, 708
    flavor compounds, 708
    manufacture, 707
    microbial population dynamics in, 707–708
    origin of, 707
    pathogenic bacteria, survival of, 709
    probiotic yogurt and health benefits, 708–709
    yeasts, growth of, 709
    yogurt manufacture, 707
Yosa®, 136

Z

Z. rouxii, 432
Zebu cattle, 711
    B. indicus cattle, estrous behavior and reproductive physiology in, 711
    synchronization of wave emergence and ovulation, follicular development for, 711–712
Zona pellucida (ZP), 395, 397, 428–429
Zygomycetes, 156
Zygomycetes mycelia, 45
Zygosaccharomyces rouxii, 139, 431, 592
Zygote intrafallopian transfer (ZIFT), 714–716
Zygotes, 715, 716